The GALE
ENCYCLOPEDIA
of SCIENCE

THIRD EDITION

The GALE ENCYCLOPEDIA of SCIENCE

THIRD EDITION

VOLUME 3
Factor – Kuru

K. Lee Lerner and
Brenda Wilmoth Lerner,
Editors

GALE®

THOMSON
GALE

Detroit • New York • San Diego • San Francisco • Cleveland • New Haven, Conn. • Waterville, Maine • London • Munich

Gale Encyclopedia of Science, Third Edition

K. Lee Lerner and Brenda Wilmoth Lerner, Editors

Project Editor
Kimberley A. McGrath

Editorial
Deirdre S. Blanchfield, Chris Jeryan, Jacqueline Longe, Mark Springer

Editorial Support Services
Andrea Lopeman

Indexing Services
Synapse

Permissions
Shalice Shah-Caldwell

Imaging and Multimedia
Leitha Etheridge-Sims, Lezlie Light, Dave Oblender, Christine O'Brien, Robyn V. Young

Product Design
Michelle DiMercurio

Manufacturing
Wendy Blurton, Evi Seoud

LIBRARY OF CONGRESS CATALOGING-IN-PUBLICATION DATA

Gale encyclopedia of science / K. Lee Lerner & Brenda Wilmoth Lerner, editors.— 3rd ed.
   p. cm.
   Includes index.
   ISBN 0-7876-7554-7 (set) — ISBN 0-7876-7555-5 (v. 1) — ISBN
   0-7876-7556-3 (v. 2) — ISBN 0-7876-7557-1 (v. 3) — ISBN 0-7876-7558-X
   (v. 4) — ISBN 0-7876-7559-8 (v. 5) — ISBN 0-7876-7560-1 (v. 6)
   1. Science—Encyclopedias. I. Lerner, K. Lee. II. Lerner, Brenda Wilmoth.

Q121.G37 2004
503—dc22
                                                            2003015731

This title is also available as an e-book.
ISBN: 0-7876-7776-0 (set)
Contact your Gale sales representative for ordering information.

Printed in Canada
10 9 8 7 6 5 4 3 2 1

CONTENTS

TOPIC LIST

A

Aardvark
Abacus
Abrasives
Abscess
Absolute zero
Abyssal plain
Acceleration
Accelerators
Accretion disk
Accuracy
Acetic acid
Acetone
Acetylcholine
Acetylsalicylic acid
Acid rain
Acids and bases
Acne
Acorn worm
Acoustics
Actinides
Action potential
Activated complex
Active galactic nuclei
Acupressure
Acupuncture
ADA (adenosine deaminase)
　deficiency
Adaptation
Addiction
Addison's disease
Addition
Adenosine diphosphate
Adenosine triphosphate
Adhesive

Adrenals
Aerobic
Aerodynamics
Aerosols
Africa
Age of the universe
Agent Orange
Aging and death
Agouti
Agricultural machines
Agrochemicals
Agronomy
AIDS
AIDS therapies and vaccines
Air masses and fronts
Air pollution
Aircraft
Airship
Albatrosses
Albedo
Albinism
Alchemy
Alcohol
Alcoholism
Aldehydes
Algae
Algebra
Algorithm
Alkali metals
Alkaline earth metals
Alkaloid
Alkyl group
Alleles
Allergy
Allotrope
Alloy

Alluvial systems
Alpha particle
Alternative energy sources
Alternative medicine
Altruism
Aluminum
Aluminum hydroxide
Alzheimer disease
Amaranth family (Amaranthaceae)
Amaryllis family (Amaryllidaceae)
American Standard Code for
　Information Interchange
Ames test
Amicable numbers
Amides
Amino acid
Ammonia
Ammonification
Amnesia
Amniocentesis
Amoeba
Amphetamines
Amphibians
Amplifier
Amputation
Anabolism
Anaerobic
Analemma
Analgesia
Analog signals and digital signals
Analytic geometry
Anaphylaxis
Anatomy
Anatomy, comparative
Anchovy
Anemia

C

Calibration
Caliper
Calorie
Calorimetry
Camels
Canal
Cancel
Cancer
Canines
Cantilever
Capacitance
Capacitor
Capillaries
Capillary action
Caprimulgids
Captive breeding and reintroduction
Capuchins
Capybaras
Carbohydrate
Carbon
Carbon cycle
Carbon dioxide
Carbon monoxide
Carbon tetrachloride
Carbonyl group
Carboxyl group
Carboxylic acids
Carcinogen
Cardiac cycle
Cardinal number
Cardinals and grosbeaks
Caribou
Carnivore
Carnivorous plants
Carp
Carpal tunnel syndrome
Carrier (genetics)
Carrot family (Apiaceae)
Carrying capacity
Cartesian coordinate plane
Cartilaginous fish
Cartography
Cashew family (Anacardiaceae)
Cassini Spacecraft
Catabolism
Catalyst and catalysis
Catastrophism

Catfish
Catheters
Cathode
Cathode ray tube
Cation
Cats
Cattails
Cattle family (Bovidae)
Cauterization
Cave
Cave fish
Celestial coordinates
Celestial mechanics
Celestial sphere: The apparent motions of the Sun, Moon, planets, and stars
Cell
Cell death
Cell division
Cell, electrochemical
Cell membrane transport
Cell staining
Cellular respiration
Cellular telephone
Cellulose
Centipedes
Centrifuge
Ceramics
Cerenkov effect
Cetaceans
Chachalacas
Chameleons
Chaos
Charge-coupled device
Chelate
Chemical bond
Chemical evolution
Chemical oxygen demand
Chemical reactions
Chemical warfare
Chemistry
Chemoreception
Chestnut
Chi-square test
Chickenpox
Childhood diseases
Chimaeras
Chimpanzees

Chinchilla
Chipmunks
Chitons
Chlordane
Chlorinated hydrocarbons
Chlorination
Chlorine
Chlorofluorocarbons (CFCs)
Chloroform
Chlorophyll
Chloroplast
Cholera
Cholesterol
Chordates
Chorionic villus sampling (CVS)
Chromatin
Chromatography
Chromosomal abnormalities
Chromosome
Chromosome mapping
Cicadas
Cigarette smoke
Circle
Circulatory system
Circumscribed and inscribed
Cirrhosis
Citric acid
Citrus trees
Civets
Climax (ecological)
Clingfish
Clone and cloning
Closed curves
Closure property
Clouds
Club mosses
Coal
Coast and beach
Coatis
Coca
Cocaine
Cockatoos
Cockroaches
Codeine
Codfishes
Codons
Coefficient
Coelacanth

D

Density
Dentistry
Deoxyribonucleic acid (DNA)
Deposit
Depression
Depth perception
Derivative
Desalination
Desert
Desertification
Determinants
Deuterium
Developmental processes
Dew point
Diabetes mellitus
Diagnosis
Dialysis
Diamond
Diatoms
Dielectric materials
Diesel engine
Diethylstilbestrol (DES)
Diffraction
Diffraction grating
Diffusion
Digestive system
Digital Recording
Digitalis
Dik-diks
Dinosaur
Diode
Dioxin
Diphtheria
Dipole
Direct variation
Disease
Dissociation
Distance
Distillation
Distributive property
Disturbance, ecological
Diurnal cycles
Division
DNA fingerprinting
DNA replication
DNA synthesis
DNA technology
DNA vaccine

Dobsonflies
Dogwood tree
Domain
Donkeys
Dopamine
Doppler effect
Dories
Dormouse
Double-blind study
Double helix
Down syndrome
Dragonflies
Drift net
Drongos
Drosophila melanogaster
Drought
Ducks
Duckweed
Duikers
Dune
Duplication of the cube
Dust devil
DVD
Dwarf antelopes
Dyes and pigments
Dysentery
Dyslexia
Dysplasia
Dystrophinopathies

E

e (number)
Eagles
Ear
Earth
Earth science
Earth's interior
Earth's magnetic field
Earth's rotation
Earthquake
Earwigs
Eating disorders
Ebola virus
Ebony
Echiuroid worms

Echolocation
Eclipses
Ecological economics
Ecological integrity
Ecological monitoring
Ecological productivity
Ecological pyramids
Ecology
Ecosystem
Ecotone
Ecotourism
Edema
Eel grass
El Niño and La Niña
Eland
Elapid snakes
Elasticity
Electric arc
Electric charge
Electric circuit
Electric conductor
Electric current
Electric motor
Electric vehicles
Electrical conductivity
Electrical power supply
Electrical resistance
Electricity
Electrocardiogram (ECG)
Electroencephalogram (EEG)
Electrolysis
Electrolyte
Electromagnetic field
Electromagnetic induction
Electromagnetic spectrum
Electromagnetism
Electromotive force
Electron
Electron cloud
Electronics
Electrophoresis
Electrostatic devices
Element, chemical
Element, families of
Element, transuranium
Elements, formation of
Elephant
Elephant shrews

Elephant snout fish

Elephantiasis

Elevator

Ellipse

Elm

Embiids

Embolism

Embryo and embryonic development

Embryo transfer

Embryology

Emission

Emphysema

Emulsion

Encephalitis

Endangered species

Endemic

Endocrine system

Endoprocta

Endoscopy

Endothermic

Energy

Energy budgets

Energy efficiency

Energy transfer

Engineering

Engraving and etching

Enterobacteria

Entropy

Environmental ethics

Environmental impact statement

Enzymatic engineering

Enzyme

Epidemic

Epidemiology

Epilepsy

Episomes

Epstein-Barr virus

Equation, chemical

Equilibrium, chemical

Equinox

Erosion

Error

Escherichia coli

Ester

Esterification

Ethanol

Ether

Ethnoarchaeology

Ethnobotany

Ethyl group

Ethylene glycol

Ethylenediaminetetra-acetic acid

Etiology

Eubacteria

Eugenics

Eukaryotae

Europe

Eutrophication

Evaporation

Evapotranspiration

Even and odd

Event horizon

Evolution

Evolution, convergent

Evolution, divergent

Evolution, evidence of

Evolution, parallel

Evolutionary change, rate of

Evolutionary mechanisms

Excavation methods

Exclusion principle, Pauli

Excretory system

Exercise

Exocrine glands

Explosives

Exponent

Extinction

Extrasolar planets

Eye

F

Factor

Factorial

Falcons

Faraday effect

Fat

Fatty acids

Fault

Fauna

Fax machine

Feather stars

Fermentation

Ferns

Ferrets

Fertilization

Fertilizers

Fetal alcohol syndrome

Feynman diagrams

Fiber optics

Fibonacci sequence

Field

Figurative numbers

Filtration

Finches

Firs

Fish

Flagella

Flame analysis

Flamingos

Flatfish

Flatworms

Flax

Fleas

Flies

Flightless birds

Flooding

Flora

Flower

Fluid dynamics

Fluid mechanics

Fluorescence

Fluorescence in situ hybridization (FISH)

Fluorescent light

Fluoridation

Flying fish

Focused Ion Beam (FIB)

Fog

Fold

Food chain/web

Food irradiation

Food poisoning

Food preservation

Food pyramid

Foot and mouth disease

Force

Forensic science

Forestry

Forests

Formula, chemical

Halley's comet
Hallucinogens
Halogenated hydrocarbons
Halogens
Halosaurs
Hamsters
Hand tools
Hantavirus infections
Hard water
Harmonics
Hartebeests
Hawks
Hazardous wastes
Hazel
Hearing
Heart
Heart diseases
Heart, embryonic development and
   changes at birth
Heart-lung machine
Heat
Heat capacity
Heat index
Heat transfer
Heath family (Ericaceae)
Hedgehogs
Heisenberg uncertainty principle
Heliocentric theory
Hematology
Hemophilia
Hemorrhagic fevers and diseases
Hemp
Henna
Hepatitis
Herb
Herbal medicine
Herbicides
Herbivore
Hermaphrodite
Hernia
Herons
Herpetology
Herrings
Hertzsprung-Russell diagram
Heterotroph
Hibernation
Himalayas, geology of
Hippopotamuses

Histamine
Historical geology
Hoatzin
Hodgkin's disease
Holly family (Aquifoliaceae)
Hologram and holography
Homeostasis
Honeycreepers
Honeyeaters
Hoopoe
Horizon
Hormones
Hornbills
Horse chestnut
Horsehair worms
Horses
Horseshoe crabs
Horsetails
Horticulture
Hot spot
Hovercraft
Hubble Space Telescope
Human artificial chromosomes
Human chorionic gonadotropin
Human cloning
Human ecology
Human evolution
Human Genome Project
Humidity
Hummingbirds
Humus
Huntington disease
Hybrid
Hydra
Hydrocarbon
Hydrocephalus
Hydrochlorofluorocarbons
Hydrofoil
Hydrogen
Hydrogen chloride
Hydrogen peroxide
Hydrogenation
Hydrologic cycle
Hydrology
Hydrolysis
Hydroponics
Hydrosphere
Hydrothermal vents

Hydrozoa
Hyena
Hyperbola
Hypertension
Hypothermia
Hyraxes

I

Ibises
Ice
Ice age refuges
Ice ages
Icebergs
Iceman
Identity element
Identity property
Igneous rocks
Iguanas
Imaginary number
Immune system
Immunology
Impact crater
Imprinting
In vitro fertilization (IVF)
In vitro and in vivo
Incandescent light
Incineration
Indicator, acid-base
Indicator species
Individual
Indoor air quality
Industrial minerals
Industrial Revolution
Inequality
Inertial guidance
Infection
Infertility
Infinity
Inflammation
Inflection point
Influenza
Infrared astronomy
Inherited disorders
Insecticides
Insectivore

Lorises
Luminescence
Lungfish
Lycophytes
Lyme disease
Lymphatic system
Lyrebirds

M

Macaques
Mach number
Machine tools
Machine vision
Machines, simple
Mackerel
Magic square
Magma
Magnesium
Magnesium sulfate
Magnetic levitation
Magnetic recording/audiocassette
Magnetic resonance imaging (MRI)
Magnetism
Magnetosphere
Magnolia
Mahogany
Maidenhair fern
Malaria
Malnutrition
Mammals
Manakins
Mangrove tree
Mania
Manic depression
Map
Maples
Marfan syndrome
Marijuana
Marlins
Marmosets and tamarins
Marmots
Mars
Mars Pathfinder
Marsupial cats
Marsupial rats and mice

Marsupials
Marten, sable, and fisher
Maser
Mass
Mass extinction
Mass number
Mass production
Mass spectrometry
Mass transportation
Mass wasting
Mathematics
Matrix
Matter
Maunder minimum
Maxima and minima
Mayflies
Mean
Median
Medical genetics
Meiosis
Membrane
Memory
Mendelian genetics
Meningitis
Menopause
Menstrual cycle
Mercurous chloride
Mercury (element)
Mercury (planet)
Mesoscopic systems
Mesozoa
Metabolic disorders
Metabolism
Metal
Metal fatigue
Metal production
Metallurgy
Metamorphic grade
Metamorphic rock
Metamorphism
Metamorphosis
Meteorology
Meteors and meteorites
Methyl group
Metric system
Mice
Michelson-Morley experiment
Microbial genetics

Microclimate
Microorganisms
Microscope
Microscopy
Microtechnology
Microwave communication
Migraine headache
Migration
Mildew
Milkweeds
Milky Way
Miller-Urey Experiment
Millipedes
Mimicry
Mineralogy
Minerals
Mining
Mink
Minnows
Minor planets
Mint family
Mir Space Station
Mirrors
Miscibility
Mistletoe
Mites
Mitosis
Mixture, chemical
Möbius strip
Mockingbirds and thrashers
Mode
Modular arithmetic
Mohs' scale
Mold
Mole
Mole-rats
Molecular biology
Molecular formula
Molecular geometry
Molecular weight
Molecule
Moles
Mollusks
Momentum
Monarch flycatchers
Mongooses
Monitor lizards
Monkeys

Organic farming
Organism
Organogenesis
Organs and organ systems
Origin of life
Orioles
Ornithology
Orthopedics
Oryx
Oscillating reactions
Oscillations
Oscilloscope
Osmosis
Osmosis (cellular)
Ossification
Osteoporosis
Otter shrews
Otters
Outcrop
Ovarian cycle and hormonal regulation
Ovenbirds
Oviparous
Ovoviviparous
Owls
Oxalic acid
Oxidation-reduction reaction
Oxidation state
Oxygen
Oystercatchers
Ozone
Ozone layer depletion

P

Pacemaker
Pain
Paleobotany
Paleoclimate
Paleoecology
Paleomagnetism
Paleontology
Paleopathology
Palindrome
Palms
Palynology

Pandas
Pangolins
Papaya
Paper
Parabola
Parallax
Parallel
Parallelogram
Parasites
Parity
Parkinson disease
Parrots
Parthenogenesis
Particle detectors
Partridges
Pascal's triangle
Passion flower
Paternity and parentage testing
Pathogens
Pathology
PCR
Peafowl
Peanut worms
Peccaries
Pedigree analysis
Pelicans
Penguins
Peninsula
Pentyl group
Peony
Pepper
Peptide linkage
Percent
Perception
Perch
Peregrine falcon
Perfect numbers
Periodic functions
Periodic table
Permafrost
Perpendicular
Pesticides
Pests
Petrels and shearwaters
Petroglyphs and pictographs
Petroleum
pH
Phalangers

Pharmacogenetics
Pheasants
Phenyl group
Phenylketonuria
Pheromones
Phlox
Phobias
Phonograph
Phoronids
Phosphoric acid
Phosphorus
Phosphorus cycle
Phosphorus removal
Photic zone
Photochemistry
Photocopying
Photoelectric cell
Photoelectric effect
Photography
Photography, electronic
Photon
Photosynthesis
Phototropism
Photovoltaic cell
Phylogeny
Physical therapy
Physics
Physiology
Physiology, comparative
Phytoplankton
Pi
Pigeons and doves
Pigs
Pike
Piltdown hoax
Pinecone fish
Pines
Pipefish
Placebo
Planck's constant
Plane
Plane family
Planet
Planet X
Planetary atmospheres
Planetary geology
Planetary nebulae
Planetary ring systems

Plankton
Plant
Plant breeding
Plant diseases
Plant pigment
Plasma
Plastic surgery
Plastics
Plate tectonics
Platonic solids
Platypus
Plovers
Pluto
Pneumonia
Podiatry
Point
Point source
Poisons and toxins
Polar coordinates
Polar ice caps
Poliomyelitis
Pollen analysis
Pollination
Pollution
Pollution control
Polybrominated biphenyls (PBBs)
Polychlorinated biphenyls (PCBs)
Polycyclic aromatic hydrocarbons
Polygons
Polyhedron
Polymer
Polynomials
Poppies
Population growth and control
(human)
Population, human
Porcupines
Positive number
Positron emission tomography
(PET)
Postulate
Potassium aluminum sulfate
Potassium hydrogen tartrate
Potassium nitrate
Potato
Pottery analysis
Prairie
Prairie chicken

Prairie dog
Prairie falcon
Praying mantis
Precession of the equinoxes
Precious metals
Precipitation
Predator
Prenatal surgery
Prescribed burn
Pressure
Prey
Primates
Prime numbers
Primroses
Printing
Prions
Prism
Probability theory
Proboscis monkey
Projective geometry
Prokaryote
Pronghorn
Proof
Propyl group
Prosimians
Prosthetics
Proteas
Protected area
Proteins
Proteomics
Protista
Proton
Protozoa
Psychiatry
Psychoanalysis
Psychology
Psychometry
Psychosis
Psychosurgery
Puberty
Puffbirds
Puffer fish
Pulsar
Punctuated equilibrium
Pyramid
Pythagorean theorem
Pythons

Q

Quadrilateral
Quail
Qualitative analysis
Quantitative analysis
Quantum computing
Quantum electrodynamics (QED)
Quantum mechanics
Quantum number
Quarks
Quasar
Quetzal
Quinine

R

Rabies
Raccoons
Radar
Radial keratotomy
Radiation
Radiation detectors
Radiation exposure
Radical (atomic)
Radical (math)
Radio
Radio astronomy
Radio waves
Radioactive dating
Radioactive decay
Radioactive fallout
Radioactive pollution
Radioactive tracers
Radioactive waste
Radioisotopes in medicine
Radiology
Radon
Rails
Rainbows
Rainforest
Random
Rangeland
Raptors
Rare gases
Rare genotype advantage

Rate

Ratio

Rational number

Rationalization

Rats

Rayleigh scattering

Rays

Real numbers

Reciprocal

Recombinant DNA

Rectangle

Recycling

Red giant star

Red tide

Redshift

Reflections

Reflex

Refrigerated trucks and railway cars

Rehabilitation

Reinforcement, positive and negative

Relation

Relativity, general

Relativity, special

Remote sensing

Reproductive system

Reproductive toxicant

Reptiles

Resins

Resonance

Resources, natural

Respiration

Respiration, cellular

Respirator

Respiratory diseases

Respiratory system

Restoration ecology

Retrograde motion

Retrovirus

Reye's syndrome

Rh factor

Rhesus monkeys

Rheumatic fever

Rhinoceros

Rhizome

Rhubarb

Ribbon worms

Ribonuclease

Ribonucleic acid (RNA)

Ribosomes

Rice

Ricin

Rickettsia

Rivers

RNA function

RNA splicing

Robins

Robotics

Rockets and missiles

Rocks

Rodents

Rollers

Root system

Rose family (Rosaceae)

Rotation

Roundworms

Rumination

Rushes

Rusts and smuts

S

Saiga antelope

Salamanders

Salmon

Salmonella

Salt

Saltwater

Sample

Sand

Sand dollars

Sandfish

Sandpipers

Sapodilla tree

Sardines

Sarin gas

Satellite

Saturn

Savanna

Savant

Sawfish

Saxifrage family

Scalar

Scale insects

Scanners, digital

Scarlet fever

Scavenger

Schizophrenia

Scientific method

Scorpion flies

Scorpionfish

Screamers

Screwpines

Sculpins

Sea anemones

Sea cucumbers

Sea horses

Sea level

Sea lily

Sea lions

Sea moths

Sea spiders

Sea squirts and salps

Sea urchins

Seals

Seamounts

Seasonal winds

Seasons

Secondary pollutants

Secretary bird

Sedges

Sediment and sedimentation

Sedimentary environment

Sedimentary rock

Seed ferns

Seeds

Segmented worms

Seismograph

Selection

Sequences

Sequencing

Sequoia

Servomechanisms

Sesame

Set theory

SETI

Severe acute respiratory syndrome (SARS)

Sewage treatment

Sewing machine

Sex change

Sextant

Stroke
Stromatolite
Sturgeons
Subatomic particles
Submarine
Subsidence
Subsurface detection
Subtraction
Succession
Suckers
Sudden infant death syndrome
   (SIDS)
Sugar beet
Sugarcane
Sulfur
Sulfur cycle
Sulfur dioxide
Sulfuric acid
Sun
Sunbirds
Sunspots
Superclusters
Superconductor
Supernova
Surface tension
Surgery
Surveying instruments
Survival of the fittest
Sustainable development
Swallows and martins
Swamp cypress family
   (Taxodiaceae)
Swamp eels
Swans
Sweet gale family (Myricaceae)
Sweet potato
Swifts
Swordfish
Symbiosis
Symbol, chemical
Symbolic logic
Symmetry
Synapse
Syndrome
Synthesis, chemical
Synthesizer, music
Synthesizer, voice
Systems of equations

T

T cells
Tanagers
Taphonomy
Tapirs
Tarpons
Tarsiers
Tartaric acid
Tasmanian devil
Taste
Taxonomy
Tay-Sachs disease
Tea plant
Tectonics
Telegraph
Telemetry
Telephone
Telescope
Television
Temperature
Temperature regulation
Tenrecs
Teratogen
Term
Termites
Terns
Terracing
Territoriality
Tetanus
Tetrahedron
Textiles
Thalidomide
Theorem
Thermal expansion
Thermochemistry
Thermocouple
Thermodynamics
Thermometer
Thermostat
Thistle
Thoracic surgery
Thrips
Thrombosis
Thrushes
Thunderstorm
Tides

Time
Tinamous
Tissue
Tit family
Titanium
Toadfish
Toads
Tomato family
Tongue worms
Tonsillitis
Topology
Tornado
Torque
Torus
Total solar irradiance
Toucans
Touch
Towers of Hanoi
Toxic shock syndrome
Toxicology
Trace elements
Tragopans
Trains and railroads
Tranquilizers
Transcendental numbers
Transducer
Transformer
Transgenics
Transistor
Transitive
Translations
Transpiration
Transplant, surgical
Trapezoid
Tree
Tree shrews
Trichinosis
Triggerfish
Triglycerides
Trigonometry
Tritium
Trogons
Trophic levels
Tropic birds
Tropical cyclone
Tropical diseases
Trout-perch
True bugs

True eels
True flies
Trumpetfish
Tsunami
Tuatara lizard
Tuber
Tuberculosis
Tumbleweed
Tumor
Tuna
Tundra
Tunneling
Turacos
Turbine
Turbulence
Turkeys
Turner syndrome
Turtles
Typhoid fever
Typhus
Tyrannosaurus rex
Tyrant flycatchers

U

Ulcers
Ultracentrifuge
Ultrasonics
Ultraviolet astronomy
Unconformity
Underwater exploration
Ungulates
Uniformitarianism
Units and standards
Uplift
Upwelling
Uranium
Uranus
Urea
Urology

V

Vaccine

Vacuum
Vacuum tube
Valence
Van Allen belts
Van der Waals forces
Vapor pressure
Variable
Variable stars
Variance
Varicella zoster virus
Variola virus
Vegetables
Veins
Velocity
Venus
Verbena family (Verbenaceae)
Vertebrates
Video recording
Violet family (Violaceae)
Vipers
Viral genetics
Vireos
Virtual particles
Virtual reality
Virus
Viscosity
Vision
Vision disorders
Vitamin
Viviparity
Vivisection
Volatility
Volcano
Voles
Volume
Voyager spacecraft
Vulcanization
Vultures
VX agent

W

Wagtails and pipits
Walkingsticks
Walnut family
Walruses

Warblers
Wasps
Waste management
Waste, toxic
Water
Water bears
Water conservation
Water lilies
Water microbiology
Water pollution
Water treatment
Waterbuck
Watershed
Waterwheel
Wave motion
Waxbills
Waxwings
Weasels
Weather
Weather forecasting
Weather mapping
Weather modification
Weathering
Weaver finches
Weevils
Welding
West Nile virus
Wetlands
Wheat
Whisk fern
White dwarf
White-eyes
Whooping cough
Wild type
Wildfire
Wildlife
Wildlife trade (illegal)
Willow family (Salicaceae)
Wind
Wind chill
Wind shear
Wintergreen
Wolverine
Wombats
Wood
Woodpeckers
Woolly mammoth
Work

Wren-warblers
Wrens
Wrynecks

X

X-ray astronomy
X-ray crystallography
X rays
Xenogamy

Y

Y2K
Yak
Yam
Yeast
Yellow fever
Yew
Yttrium

Z

Zebras
Zero
Zodiacal light
Zoonoses
Zooplankton

ORGANIZATION OF THE ENCYCLOPEDIA

The *Gale Encyclopedia of Science, Third Edition* has been designed with ease of use and ready reference in mind.

- Entries are alphabetically arranged across six volumes, in a single sequence, rather than by scientific field

- Length of entries varies from short definitions of one or two paragraphs, to longer, more detailed entries on more complex subjects.

- Longer entries are arranged so that an overview of the subject appears first, followed by a detailed discussion conveniently arranged under subheadings.

- A list of key terms is provided where appropriate to define unfamiliar terms or concepts.

- Bold-faced terms direct the reader to related articles.

- Longer entries conclude with a "Resources" section, which points readers to other helpful materials (including books, periodicals, and Web sites).

- The author's name appears at the end of longer entries. His or her affiliation can be found in the "Contributors" section at the front of each volume.

- "See also" references appear at the end of entries to point readers to related entries.

- Cross references placed throughout the encyclopedia direct readers to where information on subjects without their own entries can be found.

- A comprehensive, two-level General Index guides readers to all topics, illustrations, tables, and persons mentioned in the book.

AVAILABLE IN ELECTRONIC FORMATS

Licensing. *The Gale Encyclopedia of Science, Third Edition* is available for licensing. The complete database is provided in a fielded format and is deliverable on such media as disk or CD-ROM. For more information, contact Gale's Business Development Group at 1-800-877-GALE, or visit our website at www.gale.com/bizdev.

ADVISORY BOARD

A number of experts in the scientific and libary communities provided invaluable assistance in the formulation of this encyclopedia. Our advisory board performed a myriad of duties, from defining the scope of coverage to reviewing individual entries for accuracy and accessibility, and in many cases, writing entries. We would therefore like to express our appreciation to them:

CONTRIBUTORS

Nasrine Adibe
Professor Emeritus
Department of Education
Long Island University
Westbury, New York

Mary D. Albanese
Department of English
University of Alaska
Juneau, Alaska

Margaret Alic
Science Writer
Eastsound, Washington

James L. Anderson
Soil Science Department
University of Minnesota
St. Paul, Minnesota

Monica Anderson
Science Writer
Hoffman Estates, Illinois

Susan Andrew
Teaching Assistant
University of Maryland
Washington, DC

John Appel
Director
Fundación Museo de Ciencia y
   Tecnología
Popayán, Colombia

David Ball
Assistant Professor
Department of Chemistry
Cleveland State University
Cleveland, Ohio

Dana M. Barry
Editor and Technical Writer
Center for Advanced Materials
   Processing
Clarkston University
Potsdam, New York

Puja Batra
Department of Zoology
Michigan State University
East Lansing, Michigan

Donald Beaty
Professor Emeritus
College of San Mateo
San Mateo, California

Eugene C. Beckham
Department of Mathematics and
   Science
Northwood Institute
Midland, Michigan

Martin Beech
Research Associate
Department of Astronomy
University of Western Ontario
London, Ontario, Canada

**Julie Berwald, Ph.D. (Ocean
   Sciences)**
Austin, Texas

Massimo D. Bezoari
Associate Professor
Department of Chemistry
Huntingdon College
Montgomery, Alabama

John M. Bishop III
Translator
New York, New York

T. Parker Bishop
Professor
Middle Grades and Secondary
   Education
Georgia Southern University
Statesboro, Georgia

Carolyn Black
Professor
Incarnate Word College
San Antonio, Texas

Larry Blaser
Science Writer
Lebanon, Tennessee

Jean F. Blashfield
Science Writer
Walworth, Wisconsin

Richard L. Branham Jr.
Director
Centro Rigional de
   Investigaciones Científicas y
   Tecnológicas
Mendoza, Argentina

Patricia Braus
Editor
American Demographics
Rochester, New York

David L. Brock
Biology Instructor
St. Louis, Missouri

Leona B. Bronstein
Chemistry Teacher (retired)
East Lansing High School
Okemos, Michigan

Contributors

Brandon R. Brown
Graduate Research Assistant
Oregon State University
Corvallis, Oregon

Lenonard C. Bruno
Senior Science Specialist
Library of Congress
Chevy Chase, Maryland

Janet Buchanan, Ph.D.
Microbiologist
Independent Scholar
Toronto, Ontario, Canada.

Scott Christian Cahall
Researcher
World Precision Instruments, Inc.
Bradenton, Florida

G. Lynn Carlson
Senior Lecturer
School of Science and
    Technology
University of Wisconsin—
    Parkside
Kenosha, Wisconsin

James J. Carroll
Center for Quantum Mechanics
The University of Texas at Dallas
Dallas, Texas

Steven B. Carroll
Assistant Professor
Division of Biology
Northeast Missouri State
    University
Kirksville, Missouri

Rosalyn Carson-DeWitt
Physician and Medical Writer
Durham, North Carolina

Yvonne Carts-Powell
Editor
Laser Focus World
Belmont, Massachustts

Chris Cavette
Technical Writer
Fremont, California

Lata Cherath
Science Writer
Franklin Park, New York

Kenneth B. Chiacchia
Medical Editor
University of Pittsburgh Medical
    Center
Pittsburgh, Pennsylvania

M. L. Cohen
Science Writer
Chicago, Illinois

Robert Cohen
Reporter
KPFA Radio News
Berkeley, California

Sally Cole-Misch
Assistant Director
International Joint Commission
Detroit, Michigan

George W. Collins II
Professor Emeritus
Case Western Reserve
Chesterland, Ohio

Jeffrey R. Corney
Science Writer
Thermopolis, Wyoming

Tom Crawford
Assistant Director
Division of Publication and
    Development
University of Pittsburgh Medical
    Center
Pittsburgh, Pennsylvania

Pamela Crowe
Medical and Science Writer
Oxon, England

Clinton Crowley
On-site Geologist
Selman and Associates
Fort Worth, Texas

Edward Cruetz
Physicist
Rancho Santa Fe, California

Frederick Culp
Chairman
Department of Physics
Tennessee Technical
Cookeville, Tennessee

Neil Cumberlidge
Professor
Department of Biology
Northern Michigan University
Marquette, Michigan

Mary Ann Cunningham
Environmental Writer
St. Paul, Minnesota

Les C. Cwynar
Associate Professor
Department of Biology
University of New Brunswick
Fredericton, New Brunswick

Paul Cypher
Provisional Interpreter
Lake Erie Metropark
Trenton, Michigan

Stanley J. Czyzak
Professor Emeritus
Ohio State University
Columbus, Ohio

Rosi Dagit
Conservation Biologist
Topanga-Las Virgenes Resource
    Conservation District
Topanga, California

David Dalby
President
Bruce Tool Company, Inc.
Taylors, South Carolina

Lou D'Amore
Chemistry Teacher
Father Redmund High School
Toronto, Ontario, Canada

Douglas Darnowski
Postdoctoral Fellow
Department of Plant Biology
Cornell University
Ithaca, New York

Sreela Datta
Associate Writer
Aztec Publications
Northville, Michigan

Sarah K. Dean
Science Writer
Philadelphia, Pennsylvania

Sarah de Forest
Research Assistant
Theoretical Physical Chemistry
    Lab
University of Pittsburgh
Pittsburgh, Pennsylvania

Louise Dickerson
Medical and Science Writer
Greenbelt, Maryland

Marie Doorey
Editorial Assistant
Illinois Masonic Medical Center
Chicago, Illinois

Herndon G. Dowling
Professor Emeritus
Department of Biology
New York University
New York, New York

Marion Dresner
Natural Resources Educator
Berkeley, California

John Henry Dreyfuss
Science Writer
Brooklyn, New York

Roy Dubisch
Professor Emeritus
Department of Mathematics
New York University
New York, New York

Russel Dubisch
Department of Physics
Sienna College
Loudonville, New York

Carolyn Duckworth
Science Writer
Missoula, Montana

Laurie Duncan, Ph.D.
    (Geology)
Geologist
Austin, Texas

Peter A. Ensminger
Research Associate
Cornell University
Syracuse, New York

Bernice Essenfeld
Biology Writer
Warren, New Jersey

Mary Eubanks
Instructor of Biology
The North Carolina School of
    Science and Mathematics
Durham, North Carolina

Kathryn M. C. Evans
Science Writer
Madison, Wisconsin

William G. Fastie
Department of Astronomy and
    Physics
Bloomberg Center
Baltimore, Maryland

Barbara Finkelstein
Science Writer
Riverdale, New York

Mary Finley
Supervisor of Science Curriculum
    (retired)
Pittsburgh Secondary Schools
Clairton, Pennsylvania

Gaston Fischer
Institut de Géologie
Université de Neuchâtel
Peseux, Switzerland

Sara G. B. Fishman
Professor
Quinsigamond Community
    College
Worcester, Massachusetts

David Fontes
Senior Instructor
Lloyd Center for Environmental
    Studies
Westport, Maryland

Barry Wayne Fox
Extension Specialist,
    Marine/Aquatic Education
Virginia State University
Petersburg, Virginia

Ed Fox
Charlotte Latin School
Charlotte, North Carolina

Kenneth L. Frazier
Science Teacher (retired)
North Olmstead High School
North Olmstead, Ohio

Bill Freedman
Professor
Department of Biology and
    School for Resource and
    Environmental Studies
Dalhousie University
Halifax, Nova Scotia

T. A. Freeman
Consulting Archaeologist
Quail Valley, California

Elaine Friebele
Science Writer
Cheverly, Maryland

Randall Frost
Documentation Engineering
Pleasanton, California

Agnes Galambosi, M.S.
Climatologist
Eotvos Lorand University
Budapest, Hungary

Robert Gardner
Science Education Consultant
North Eastham, Massachusetts

Gretchen M. Gillis
Senior Geologist
Maxus Exploration
Dallas, Texas

Larry Gilman, Ph.D. (Electrical
    Engineering)
Engineer
Sharon, Vermont

Kathryn Glynn
Audiologist
Portland, Oregon

David Goings, Ph.D. (Geology)
Geologist
Las Vegas, Nevada

Natalie Goldstein
Educational Environmental
    Writing
Phoenicia, New York

David Gorish
TARDEC
U.S. Army
Warren, Michigan

Louis Gotlib
South Granville High School
Durham, North Carolina

Hans G. Graetzer
Professor
Department of Physics
South Dakota State University
Brookings, South Dakota

Jim Guinn
Assistant Professor
Department of Physics
Berea College
Berea, Kentucky

Steve Gutterman
Psychology Research Assistant
University of Michigan
Ann Arbor, Michigan

Johanna Haaxma-Jurek
Educator
Nataki Tabibah Schoolhouse of
 Detroit
Detroit, Michigan

Monica H. Halka
Research Associate
Department of Physics and
 Astronomy
University of Tennessee
Knoxville, Tennessee

Brooke Hall, Ph.D.
Professor
Department of Biology
California State University at
 Sacramento
Sacramento, California

Jeffrey C. Hall
Astronomer
Lowell Observatory
Flagstaff, Arizona

C. S. Hammen
Professor Emeritus
Department of Zoology
University of Rhode Island

Lawrence Hammar, Ph.D.
Senior Research Fellow
Institute of Medical Research
Papua, New Guinea

William Haneberg, Ph.D.
 (Geology)
Geologist
Portland, Oregon

Beth Hanson
Editor
The Amicus Journal
Brooklyn, New York

Clay Harris
Associate Professor
Department of Geography and
 Geology
Middle Tennessee State
 University
Murfreesboro, Tennessee

Clinton W. Hatchett
Director Science and Space
 Theater
Pensacola Junior College
Pensacola, Florida

Catherine Hinga Haustein
Associate Professor
Department of Chemistry
Central College
Pella, Iowa

Dean Allen Haycock
Science Writer
Salem, New York

Paul A. Heckert
Professor
Department of Chemistry and
 Physics
Western Carolina University
Cullowhee, North Carolina

Darrel B. Hoff
Department of Physics
Luther College
Calmar, Iowa

Dennis Holley
Science Educator
Shelton, Nebraska

Leonard Darr Holmes
Department of Physical Science
Pembroke State University
Pembroke, North Carolina

Rita Hoots
Instructor of Biology, Anatomy,
 Chemistry
Yuba College
Woodland, California

Selma Hughes
Department of Psychology and
 Special Education
East Texas State University
Mesquite, Texas

Mara W. Cohen Ioannides
Science Writer
Springfield, Missouri

Zafer Iqbal
Allied Signal Inc.
Morristown, New Jersey

Sophie Jakowska
Pathobiologist, Environmental
 Educator
Santo Domingo, Dominican
 Republic

Richard A. Jeryan
Senior Technical Specialist
Ford Motor Company
Dearborn, Michigan

Stephen R. Johnson
Biology Writer
Richmond, Virginia

Kathleen A. Jones
School of Medicine
Southern Illinois University
Carbondale, Illinois

Harold M. Kaplan
Professor
School of Medicine
Southern Illinois University
Carbondale, Illinois

Anthony Kelly
Science Writer
Pittsburgh, Pennsylvania

Amy Kenyon-Campbell
Ecology, Evolution and
    Organismal Biology Program
University of Michigan
Ann Arbor, Michigan

Judson Knight
Science Writer
Knight Agency
Atlanta, Georgia

Eileen M. Korenic
Institute of Optics
University of Rochester
Rochester, New York

Jennifer Kramer
Science Writer
Kearny, New Jersey

Pang-Jen Kung
Los Alamos National Laboratory
Los Alamos, New Mexico

Marc Kusinitz
Assistant Director Media
    Relations
John Hopkins Medical Institution
Towsen, Maryland

Arthur M. Last
Head
Department of Chemistry
University College of the Fraser
    Valley
Abbotsford, British Columbia

Nathan Lavenda
Zoologist
Skokie, Illinios

Jennifer LeBlanc
Environmental Consultant
London, Ontario, Canada

Nicole LeBrasseur, Ph.D.
Associate News Editor
Journal of Cell Biology
New York, New York

Benedict A. Leerburger
Science Writer
Scarsdale, New York

Betsy A. Leonard
Education Facilitator

Reuben H. Fleet Space Theater
    and Science Center
San Diego, California

Adrienne Wilmoth Lerner
Graduate School of Arts &
    Science
Vanderbilt University
Nashville, Tennessee

Lee Wilmoth Lerner
Science Writer
NASA
Kennedy Space Center, Florida

Scott Lewis
Science Writer
Chicago, Illinois

Frank Lewotsky
Aerospace Engineer (retired)
Nipomo, California

Karen Lewotsky
Director of Water Programs
Oregon Environmental Council
Portland, Oregon

Kristin Lewotsky
Editor
Laser Focus World
Nashua, New Hamphire

Stephen K. Lewotsky
Architect
Grants Pass, Oregon

Agnieszka Lichanska, Ph.D.
Department of Microbiology &
    Parasitology
University of Queensland
Brisbane, Australia

Sarah Lee Lippincott
Professor Emeritus
Swarthmore College
Swarthmore, Pennsylvania

Jill Liske, M.Ed.
Wilmington, North Carolina

David Lunney
Research Scientist
Centre de Spectrométrie
    Nucléaire et de Spectrométrie
    de Masse
Orsay, France

Steven MacKenzie
Ecologist
Spring Lake, Michigan

J. R. Maddocks
Consulting Scientist
DeSoto, Texas

Gail B. C. Marsella
Technical Writer
Allentown, Pennsylvania

Karen Marshall
Research Associate
Council of State Governments
    and Centers for Environment
    and Safety
Lexington, Kentucky

Liz Marshall
Science Writer
Columbus, Ohio

James Marti
Research Scientist
Department of Mechanical
    Engineering
University of Minnesota
Minneapolis, Minnesota

Elaine L. Martin
Science Writer
Pensacola, Florida

Lilyan Mastrolla
Professor Emeritus
San Juan Unified School
Sacramento, California

Iain A. McIntyre
Manager
Electro-optic Department
Energy Compression Research
    Corporation
Vista, California

Jennifer L. McGrath
Chemistry Teacher
Northwood High School
Nappanee, Indiana

Margaret Meyers, M.D.
Physician, Medical Writer
Fairhope, Alabama

G. H. Miller
Director
Studies on Smoking
Edinboro, Pennsylvania

J. Gordon Miller
Botanist
Corvallis, Oregon

Kelli Miller
Science Writer
NewScience
Atlanta, Georgia

Christine Miner Minderovic
Nuclear Medicine Technologist
Franklin Medical Consulters
Ann Arbor, Michigan

David Mintzer
Professor Emeritus
Department of Mechanical
   Engineering
Northwestern University
Evanston, Illinois

Christine Molinari
Science Editor
University of Chicago Press
Chicago, Illinois

Frank Mooney
Professor Emeritus
Fingerlake Community College
Canandaigua, New York

Partick Moore
Department of English
University of Arkansas at Little
   Rock
Little Rock, Arkansas

Robbin Moran
Department of Systematic Botany
Institute of Biological Sciences
University of Aarhus
Risskou, Denmark

J. Paul Moulton
Department of Mathematics
Episcopal Academy
Glenside, Pennsylvania

Otto H. Muller
Geology Department

Alfred University
Alfred, New York

Angie Mullig
Publication and Development
University of Pittsburgh Medical
   Center
Trafford, Pennsylvania

David R. Murray
Senior Associate
Sydney University
Sydney, New South Wales,
   Australia

Sutharchana Murugan
Scientist
Three Boehringer Mannheim
   Corp.
Indianapolis, Indiana

Muthena Naseri
Moorpark College
Moorpark, California

David Newton
Science Writer and Educator
Ashland, Oregon

F. C. Nicholson
Science Writer
Lynn, Massachusetts

James O'Connell
Department of Physical Sciences
Frederick Community College
Gaithersburg, Maryland

Dúnal P. O'Mathúna
Associate Professor
Mount Carmel College of
   Nursing
Columbus, Ohio

Marjorie Pannell
Managing Editor, Scientific
   Publications
Field Museum of Natural History
Chicago, Illinois

Gordon A. Parker
Lecturer
Department of Natural Sciences
University of Michigan-Dearborn
Dearborn, Michigan

David Petechuk
Science Writer
Ben Avon, Pennsylvania

Borut Peterlin, M.D.
Consultant Clinical Geneticist,
   Neurologist, Head Division of
   Medical Genetics
Department of Obstetrics and
   Gynecology
University Medical Centre
   Ljubljana
Ljubljana, Slovenia

John R. Phillips
Department of Chemistry
Purdue University, Calumet
Hammond, Indiana

Kay Marie Porterfield
Science Writer
Englewood, Colorado

Paul Poskozim
Chair
Department of Chemistry, Earth
   Science and Physics
Northeastern Illinois University
Chicago, Illinois

Andrew Poss
Senior Research Chemist
Allied Signal Inc.
Buffalo, New York

Satyam Priyadarshy
Department of Chemistry
University of Pittsburgh
Pittsburgh, Pennsylvania

Patricia V. Racenis
Science Writer
Livonia, Michigan

Cynthia Twohy Ragni
Atmospheric Scientist
National Center for Atmospheric
   Research
Westminster, Colorado

Jordan P. Richman
Science Writer
Phoenix, Arizona

Kitty Richman
Science Writer
Phoenix, Arizona

Vita Richman
Science Writer
Phoenix, Arizona

Michael G. Roepel
Researcher
Department of Chemistry
University of Pittsburgh
Pittsburgh, Pennsylvania

Perry Romanowski
Science Writer
Chicago, Illinois

Nancy Ross-Flanigan
Science Writer
Belleville, Michigan

Belinda Rowland
Science Writer
Voorheesville, New York

Gordon Rutter
Royal Botanic Gardens
Edinburgh, Great Britain

Elena V. Ryzhov
Polytechnic Institute
Troy, New York

David Sahnow
Associate Research Scientist
John Hopkins University
Baltimore, Maryland

Peter Salmansohn
Educational Consultant
New York State Parks
Cold Spring, New York

Peter K. Schoch
Instructor
Department of Physics and
    Computer Science
Sussex County Community
    College
Augusta, New Jersey

Patricia G. Schroeder
Instructor
Science, Healthcare, and Math
    Division
Johnson County Community
    College
Overland Park, Kansas

Randy Schueller
Science Writer
Chicago, Illinois

Kathleen Scogna
Science Writer
Baltimore, Maryland

William Shapbell Jr.
Launch and Flight Systems
    Manager
Kennedy Space Center
KSC, Florida

Kenneth Shepherd
Science Writer
Wyandotte, Michigan

Anwar Yuna Shiekh
International Centre for
    Theoretical Physics
Trieste, Italy

Raul A. Simon
Chile Departmento de Física
Universidad de Tarapacá
Arica, Chile

Michael G. Slaughter
Science Specialist
Ingham ISD
East Lansing, Michigan

Billy W. Sloope
Professor Emeritus
Department of Physics
Virginia Commonwealth
    University
Richmond, Virginia

Douglas Smith
Science Writer
Milton, Massachusetts

Lesley L. Smith
Department of Physics and
    Astronomy
University of Kansas
Lawrence, Kansas

Kathryn D. Snavely
Policy Analyst, Air Quality Issues
U.S. General Accounting Office
Raleigh, North Carolina

Charles H. Southwick
Professor
Environmental, Population, and
    Organismic Biology
University of Colorado at Boulder
Boulder, Colorado

John Spizzirri
Science Writer
Chicago, Illinois

Frieda A. Stahl
Professor Emeritus
Department of Physics
California State University, Los
    Angeles
Los Angeles, California

Robert L. Stearns
Department of Physics
Vassar College
Poughkeepsie, New York

Ilana Steinhorn
Science Writer
Boalsburg, Pennsylvania

David Stone
Conservation Advisory Services
Gai Soleil
Chemin Des Clyettes
Le Muids, Switzerland

Eric R. Swanson
Associate Professor
Department of Earth and Physical
    Sciences
University of Texas
San Antonio, Texas

Cheryl Taylor
Science Educator
Kailua, Hawaii

Nicholas C. Thomas
Department of Physical Sciences
Auburn University at
    Montgomery
Montgomery, Alabama

W. A. Thomasson
Science and Medical Writer
Oak Park, Illinois

Marie L. Thompson
Science Writer
Ben Avon, Pennsylvania

Laurie Toupin
Science Writer
Pepperell, Massachusetts

Melvin Tracy
Science Educator
Appleton, Wisconsin

Karen Trentelman
Research Associate
Archaeometric Laboratory
University of Toronto
Toronto, Ontario, Canada

Robert K. Tyson
Senior Scientist
W. J. Schafer Assoc.
Jupiter, Florida

James Van Allen
Professor Emeritus
Department of Physics and
   Astronomy
University of Iowa
Iowa City, Iowa

Julia M. Van Denack
Biology Instructor
Silver Lake College
Manitowoc, Wisconsin

Kurt Vandervoort
Department of Chemistry and
   Physics
West Carolina University
Cullowhee, North Carolina

Chester Vander Zee
Naturalist, Science Educator
Volga, South Dakota

Rashmi Venkateswaran
Undergraduate Lab Coordinator
Department of Chemistry
University of Ottawa
Ottawa, Ontario, Canada

R. A. Virkar
Chair
Department of Biological
   Sciences
Kean College
Iselin, New Jersey

Kurt C. Wagner
Instructor
South Carolina Governor's
   School for Science and
   Technology
Hartsville, South Carolina

Cynthia Washam
Science Writer
Jensen Beach, Florida

Terry Watkins
Science Writer
Indianapolis, Indiana

Joseph D. Wassersug
Physician
Boca Raton, Florida

Tom Watson
Environmental Writer
Seattle, Washington

Jeffrey Weld
Instructor, Science Department
   Chair
Pella High School

Pella, Iowa

Frederick R. West
Astronomer
Hanover, Pennsylvania

Glenn Whiteside
Science Writer
Wichita, Kansas

John C. Whitmer
Professor
Department of Chemistry
Western Washington University
Bellingham, Washington

Donald H. Williams
Department of Chemistry
Hope College
Holland, Michigan

Robert L. Wolke
Professor Emeritus
Department of Chemistry
University of Pittsburgh
Pittsburgh, Pennsylvania

Xiaomei Zhu, Ph.D.
Postdoctoral research associate
Immunology Department
Chicago Children's Memorial
   Hospital, Northwestern
   University Medical School
Chicago, Illinois

Jim Zurasky
Optical Physicist
Nichols Research Corporation
Huntsville, Alabama

Factor

In **mathematics**, to factor a number or algebraic expression is to find parts whose product is the original number or expression. For instance, 12 can be factored into the product 6×2, or 3×4. The expression $(x^2 - 4)$ can be factored into the product $(x + 2)(x - 2)$. Factor is also the name given to the parts. We say that 2 and 6 are factors of 12, and $(x-2)$ is a factor of $(x^2 - 4)$. Thus we refer to the factors of a product and the product of factors.

The fundamental **theorem** of **arithmetic** states that every positive integer can be expressed as the product of prime factors in essentially a single way. A prime number is a number whose only factors are itself and 1 (the first few **prime numbers** are 1, 2, 3, 5, 7, 11, 13). **Integers** that are not prime are called composite. The number 99 is composite because it can be factored into the product 9×11. It can be factored further by noting that 9 is the product 3×3. Thus, 99 can be factored into the product $3 \times 3 \times 11$, all of which are prime. By saying "in essentially one way," it is meant that although the factors of 99 could be arranged into $3 \times 11 \times 3$ or $11 \times 3 \times 3$, there is no factoring of 99 that includes any primes other than 3 used twice and 11.

Factoring large numbers was once mainly of interest to mathematicians, but today factoring is the basis of the security codes used by computers in military codes and in protecting financial transactions. High-powered computers can factor numbers with 50 digits, so these codes must be based on numbers with a hundred or more digits to keep the data secure.

In **algebra**, it is often useful to factor polynomial expressions (expressions of the type $9x^3 + 3x^2$ or $x^4 - 27xy + 32$). For example $x^2 + 4x + 4$ is a polynomial that can be factored into $(x + 2)(x + 2)$. That this is true can be verified by multiplying the factors together. The **degree** of a polynomial is equal to the largest **exponent** that appears in it. Every polynomial of degree n has at most n polynomial factors (though some may contain **complex numbers**). For example, the third degree polynomial $x^3 + 6x^2 + 11x + 6$ can be factored into $(x + 3) (x^2 + 3x + 2)$, and the second factor can be factored again into $(x + 2)(x + 1)$, so that the original polynomial has three factors. This is a form of (or corollary to) the fundamental theorem of algebra.

In general, factoring can be rather difficult. There are some special cases and helpful hints, though, that often make the job easier. For instance, a common factor in each **term** is immediately factorable; certain common situations occur often and one learns to recognize them, such as $x^3 + 3x^2 + xy = x(x^2 + 3x + y)$. The difference of two squares is a good example: $a^2 - b^2 = (a + b)(a - b)$. Another common pattern consists of perfect squares of binomial expressions, such as $(x + b)^2$. Any squared binomial has the form $x^2 + 2bx + b^2$. The important things to note are: (1) the **coefficient** of x^2 is always one (2) the coefficient of x in the middle term is always twice the **square root** of the last term. Thus $x^2 + 10x + 25 = (x+5)^2$, $x^2 - 6x + 9 = (x-3)^2$, and so on.

Many practical problems of interest involve polynomial equations. A polynomial equation of the form $ax^2 + bx + c = 0$ can be solved if the polynomial can be factored. For instance, the equation $x^2 + x - 2 = 0$ can be written $(x + 2)(x - 1) = 0$, by factoring the polynomial. Whenever the product of two numbers or expressions is **zero**, one or the other must be zero. Thus either $x + 2 = 0$ or $x - 1 = 0$, meaning that $x = -2$ and $x = 1$ are solutions of the equation.

Resources

Books

Bittinger, Marvin L, and Davic Ellenbogen. *Intermediate Algebra: Concepts and Applications.* 6th ed. Reading, MA: Addison-Wesley Publishing, 2001.

Davison, David M., Marsha Landau, Leah McCracken, Linda Immergut, and Brita and Jean Burr Smith. *Arithmetic and Algebra Again.* New York: McGraw Hill, 1994.

Larson, Ron. *Precalculus.* 5th ed. New York: Houghton Mifflin College, 2000.

McKeague, Charles P. *Intermediate Algebra.* 5th ed. Fort Worth: Saunders College Publishing, 1995.

J.R. Maddocks

Factorial

The number n! is the product $1 \times 2 \times 3 \times 4 \times ... \times n$, that is, the product of all the **natural numbers** from 1 up to n, including n itself where 1 is a natural number. It is called either "n factorial" or "factorial n." Thus 5! is the number $1 \times 2 \times 3 \times 4 \times 5$, or 120.

Older books sometimes used the symbol In for n factorial, but the numeral followed by an exclamation **point** is currently the standard symbol.

Factorials show up in many formulas of **statistics**, probability, **combinatorics**, **calculus**, **algebra**, and elsewhere. For example, the formula for the number of permutations of n things, taken n at a **time**, is simply n!. If a singer chooses eight songs for his or her concert, these songs can be presented in 8!, or 40,320 different orders. Similarly the number of combinations of n things r at a time is n! divided by the product r!(n - r)!. Thus the number of different bridge hands that can be dealt is 52! divided by 13!39!. This happens to be a *very* large number.

When used in conjunction with other operations, as in the formula for combinations, the factorial **function** takes precedence over **addition**, **subtraction**, negation, **multiplication**, and **division** unless parentheses are used to indicate otherwise. Thus in the expression r!(n - r)!, the subtraction is done first because of the parentheses; then r! and (r - n)! are computed; then the results are multiplied.

As n! has been defined, 0! makes no sense. However, in many formulas, such as the one above, 0! can occur. If one uses this formula to compute the number of combinations of 6 things 6 at a time, the formula gives 6! divided by 6!0!. To make formulas like this work, mathematicians have decided to give 0! the value 1. When this is done, one gets 6!/6!, or 1, which is, of course, exactly the number of ways in which one can choose all six things.

As one substitutes increasingly large values for n, the value of n! increases very fast. Ten factorial is more than three million, and 70! is beyond the capacity of even those calculators which can represent numbers in scientific notation.

This is not necessarily a disadvantage. In the series representation of sine x, which is $x/1! - x^3/3! + x^5/5! -...$, the denominators get large so fast that very few terms of the series are needed to compute a good decimal **approximation** for a particular value of sine x.

Fahrenheit *see* **Temperature**

Falcons

Falcons are **birds of prey** in the family Falconidae. There are 39 **species** of true falcons, all in the genus *Falco*. Like other species in the order Falconiformes (which also includes **hawks**, **eagles**, osprey, and **vultures**), falcons have strong raptorial (or grasping) talons, a hooked beak, extremely acute **vision**, and a fierce demeanor. Falcons can be distinguished from other **raptors** by the small toothlike serrations (called tomial teeth) on their mandibles and by their specific coloration. They also have distinctive **behavior** patterns, such as killing their **prey** by a neck-breaking bite, head-bobbing, defecating below the **perch** or nest, and an often swift and direct flight pattern.

Falcons can be found on all continents except **Antarctica**. Some species have a very widespread distribution. In particular, the **peregrine falcon** (*F. peregrinus*) is virtually cosmopolitan, having a number of subspecies, some of them specific to particular oceanic islands. Other falcons are much more restricted in their distribution: for example, the Mauritius kestrel (*F. puctatus*) only breeds on the **island** of Mauritius in the Indian Ocean. At one time, fewer than ten individuals of this **endangered species** remained, although populations have since increased as a result of strict protection and a program of captive breeding and release.

Species of falcons exploit a very wide variety of **habitat** types, ranging from the high arctic **tundra** to boreal and temperate forest, **prairie** and **savanna**, and tropical **forests** of all types. Falcons catch their own food. Most species of falcons catch their prey in flight, although kestrels generally seize their food on the ground, often after hovering above. As a group, falcons eat a great range of foods; however, particular species are relatively specific in their feeding, limiting themselves to prey within certain size ranges. The American kestrel (*F. sparverius*), for example, eats mostly **insects**, earthworms, small **mammals**, and small **birds**, depending on their seasonal availability.

A peregrine falcon taking flight. *Photograph by Alan & Sandy Carey. The National Audubon Society Collection/Photo Researchers. Reproduced by permission.*

Heavier, more powerful falcons, such as the peregrine, will eat larger species of birds, including **ducks**, seabirds, **grouse**, pigeons, and shorebirds.

The nests of many falcons are rather crudely made, often a mere scrape on a cliff ledge or on the ground. Some species, however, nest in natural cavities or old woodpecker holes in trees, as is the case with the American kestrel. Most kestrels will also use nest boxes provided by humans. Peregrines, which sometimes breed in cities, will nest on ledges on tall buildings, a type of artificial cliff.

The **courtship** displays of falcons can be impressive, in some cases involving spectacular aerial displays and acrobatics. Those of the peregrine are most famous. To impress a female (properly called a falcon), the male bird (called a tiercel) will swoop down from great heights at speeds as high as 217 MPH (350 km/h) and will execute rolls and other maneuvers, including midair exchanges of

food with its intended mate. Although this species undertakes long-distance seasonal migrations, the birds return to the same nesting locale and, if possible, will mate with the same partner each year. Incubation of falcon eggs does not begin until the entire clutch is laid, so all young birds in a nest are about the same size. This is different from many other birds of prey, which incubate as soon as the first egg is laid, resulting in a great size range of young birds in the nest. In falcons, the female (which is always larger than the male) does most of the incubating, while the tiercel forages widely for food.

The most northerly species is the gyrfalcon (*F. rusticolus*), a large white species that breeds throughout the Arctic of **North America** and Eurasia. This bird usually has its nest, or aerie, high on a cliff. The nest site is typically reused for many years, and can often be discerned from miles away by the colorful orange and white

An American kestrel (*Falco sparverius*) at the Arizona Sonora Desert Museum, Arizona. Not much larger than a blue jay, the kestrel is the smallest of the North American falcons. *Potograph by Robert J. Huffman. Field Mark Publications. Reproduced by permission.*

streakings of guano and rock **lichens** growing in a fertilized zone extending several meters beneath the nest. Depending on the nearby habitat, gyrfalcons may feed on ptarmigan, seabirds, or small migratory birds such as buntings and shorebirds.

Other familiar falcons of North America include the **prairie falcon** (*F. mexicanus*), which ranges widely in open habitats of the southwestern region, and the merlin or pigeon hawk (*F. columbarius*), which breeds in boreal and subarctic habitats and winters in the southern part of the **continent** and Central America.

Interaction of falcons with humans

Falcons fascinate many people, largely because of their fierce, predatory behavior. As a result, sightings of falcons are considered to be exceptional events for bird watchers and many other people. Some species of falcons, such as kestrels, are also beneficial to humans because they eat large numbers of **mice**, **grasshoppers**, and locusts that are potential agricultural **pests**.

However, as recently as the middle of this century, some species of falcons were themselves regarded as major pests—dangerous predators of game birds. As a result, falcons, especially peregrines, were killed in large numbers by professional gamekeepers and hunters. Fortunately, this practice ended, and falcons are now rarely hunted by humans. However, young falcons are still taken from wild nests, often illegally, for use in falconry.

Falconry is a sport with a three-thousand-year history, in which falcons are free-flown to catch and kill game birds, such as grouse, ptarmigan, **pheasants**, and ducks. Falcons are rather wild birds, however, and they must be well trained or they may not return to the falconer's hand. Because of their power, speed, and fierce and independent demeanor, the most highly prized species in falconry are the largest, most robust falcons, especially the gyrfalcon and the peregrine.

Some birds trained in falconry are not only used for sport. Falcons are also used in some places to drive birds such as **gulls** away from airports, to help prevent potentially catastrophic collisions with **aircraft**.

Some species of falcons have suffered considerable damage from the widespread usage of certain types of **insecticides**. Most harmful has been the use of persistent bioaccumulating chlorinated-hydrocarbon insecticides, such as DDT and dieldrin. These and other related chemicals (such as polychlorinated biphenyls, or PCBs) have caused the collapse of populations of peregrines and other species of birds. For example, populations of the most widespread subspecies of the peregrine falcon in North America (*F. peregrinus anatum*) were widely destroyed by these toxic exposures, and the northern subspecies (*F. p. tundrius*) suffered large declines. However, because of restrictions in the use of these chemicals since the 1970s, they now have less of an effect on falcons and other birds. In fact, some breeding and migratory populations of peregrine falcons in North America have significantly increased since the late 1970s. This recovery has been aided by large captive breeding programs in the United States and Canada aimed at releasing these birds into formerly occupied or underpopulated habitats.

Still, the populations of many species of falcons is greatly reduced, and some species are threatened or endangered. Protecting these species would best be accomplished by ensuring that extensive tracts of appropriate natural habitat always remain available for falcons and other **wildlife**. However, in more acute cases, expensive management of the habitat and populations of falcons is necessary to protect these fascinating birds.

Current status of North American falcons

• Aplomado falcon (*Falco femoralis*). Endangered (sub)Species. Has been reintroduced in Texas. Decline in population is thought to have been due to agricultural expansion and to eggshell thinning resulting from the use of **pesticides**. Now considered a Southwestern stray.

• Collared forest falcon (*Micrastur semitorquatus*). Southwestern stray.

• Peregrine falcon (*Falco peregrinus*). Pesticides and PCB poisoning caused widespread reproductive failure from the 1940s to 1970s, causing species to disappear from many of the former nesting grounds. It has since been reintroduced in many areas, and appears to be doing well locally.

• Prairie falcon (*Falco mexicanus*). Species has experienced some eggshell thinning and mercury poisoning (mainly built up from feeding on the seed-eating Horned lark). Has declined in some areas (including Utah, western Canada, and agricultural regions of California), but the current population appears stable.

• American kestrel (*Falco sparverius*). Decline in population in the northeast in recent years, but otherwise the population appears stable. Nest boxes have helped maintain populations in some areas.

• Gyrfalcon (*Falco rusticolus*). Rare. Has declined in parts of Arctic **Europe**, but appears stable in North America. Illegal poaching for falconry may be a problem in some areas, but fortunately most nest sites are out of range of human disturbance.

• Merlin (*Falco columbarius*). There were earlier indications that this bird was experiencing adverse effects from the use of pesticides in eastern Canada, and from mercury buildup in western Canada. Numbers now appear to be increasing in the northern prairies, and to be remaining stale elsewhere.

• Crested caracara (*Polyborus plancus*). Has declined due to loss of habitat due to agricultural expansion and hunting. There has been some evidence of an increase in population in Texas. The population on Guadalupe Island, Mexico, became extinct in 1900.

Resources

Books

Cade, T.J. *The Falcons of the World.* Ithaca, NY: Cornell University Press, 1982.

Ehrlich, Paul R., David S. Dobkin, and Darryl Wheye. *The Birder's Handbook.* New York: Simon & Schuster Inc., 1988.

Freedman, B. *Environmental Ecology.* 2nd ed. San Diego: Academic Press, 1994.

Peterson, Roger Tory. *North American Birds.* Houghton Miflin Interactive (CD-ROM), Somerville, MA: Houghton Miflin, 1995.

Randall Frost

Faraday effect

The Faraday effect is manifest when a changing magnetic field induces an electric field. Hence the effect is also known as "induction." It is most simply exemplified by a loop of wire and a bar magnet. If one moves the magnet through the loop of wire, the changing magnetic field within the loop gives rise to an electrical current in the wire. The current is larger for stronger magnets, and it can also be augmented by moving the magnet more quickly. In other words, the size of the electric field created depends directly on the **rate** at which the magnetic field changes. In principal, by moving a very strong magnet quickly enough, the induced current could illuminate a common **light** bulb. To really understand the effect, note that the bulb would only be lit as long as the magnet was moving. As soon as a magnetic field quits changing, the Faraday effect disappears.

Many useful devices exploit the Faraday effect. Most notably, an electric **generator** relies on it to derive **electricity** from mechanical **motion**. A generator uses the **energy** of burning **fossil fuels**, for example (or falling **water** in the case of a hydroelectric plant), to rotate a loop of wire between two magnets. Since the loop spins, it perceives that the magnetic field is changing and, via the Faraday effect, yields electricity which can then be sent out to traffic lights, **radio** alarm clocks, hair dryers, et cetera.

Michael Faraday discovered the effect in 1831 at the Royal Institution Laboratories in London. When he powered up an electromagnet, a nearby coil of wire (in no way physically connected to the magnet) registered a sizable but brief current. While the electromagnet remained on, no further current could be detected in the nearby coil. However, when he turned his magnet off he again observed a short-lived burst of electrical activity in the otherwise dormant coil. He reasoned that by turning the electromagnet on and off, he had created abrupt changes in the magnetic field inside the coil and that these changes had, in turn, created the fleeting **electric current**. For Faraday, this discovery carved both prestige in the **physics** community and, moreover, a place in scientific history. Neither of these was a small feat because his fellow physicists considered his educational background to be inferior. He lacked any formal learning of **mathematics**, and his training in **chemistry** was (in the eyes of his colleagues) a questionable preparation for his career as a physicist.

Farm machinery *see* **Agricultural machines**

Fat

A fat is a solid triester of **glycerol**. It is formed when a **molecule** of glycerol, an **alcohol** with three hydroxyl groups, reacts with three molecules of **fatty acids**. A fatty acid is a long-chain aliphatic carboxylic acid. The more correct name for a fat is a triglyceride.

The three fatty acid fragments in a fat may be all the same (a simple triglyceride) or they may be different from each other (a mixed triglyceride). The fat known as glyceryl tripalmitate, for example, is formed when a molecule of glycerol reacts with three molecules of palmitic acid. Glyceryl palmitate distearate, on the other hand, is produced in the reaction between one molecule of glycerol, one molecule of palmitic acid and two molecules of **stearic acid**.

Fats and oils are closely related to each other in that both are triesters of glycerol. The two families differ from each other, however, in that fats are solid and oils are liquid. The difference in physical state between the two families reflects differences in the fatty acids of which they are made. Fats contain a larger fraction of saturated fatty acid fragments and have, therefore, higher melting points. Oils contain a larger fraction of unsaturated fatty acid fragments and have, as a result, lower melting points.

As an example, beef tallow contains about 56% saturated fatty acid fragments and about 44% unsaturated fatty acid fragments. In comparison, corn oil contains about 13% saturated fatty acid fragments and 87% unsaturated fatty acid fragments.

Both fats and oils belong to the family of biochemicals known as the lipids. The common characteristics that all lipids share with each other is that they tend to be insoluble in **water**, but soluble in organic solvents such as **ether**, alcohol, **benzene**, and **carbon tetrachloride**.

Fats are an important constituent of **animal** bodies where they have four main functions. First, they are a source of **energy** for **metabolism**. Although carbohydrates are often regarded as the primary source of energy in an **organism**, fats actually provide more than twice as much energy per calories as do carbohydrates.

Fats also provide insulation for the body, protecting against excessive heat losses to the environment. Third, fats act as a protective cushion around bones and organs. Finally, fats store certain vitamins, such as vitamins A, D, E, and K, which are not soluble in water but are soluble in fats and oils.

Animal bodies are able to synthesize the fats they need from the foods that make up their diets. Among humans, 25-50% of the typical diet may consist of fats and oils. In general, a healthful diet is thought to be one that contains a smaller, rather than larger, proportion of fats.

The main use of fats commercially is in the production of soaps and other cleaning products. When a fat is boiled in water in the presence of a base such as **sodium hydroxide**, the fat breaks down into glycerol and fatty acids. The **sodium salt** of fatty acids formed in this process is the product known as **soap**. The process of making soap from a fatty material is known as saponification.

See also Lipid.

Fatty acids

A fatty acid is a combination of a chain of **carbon** and **hydrogen atoms**, known as a **hydrocarbon**, and a particular acid group (-COOH). Three fatty-acid molecules combined with a **glycerol** form a triglyceride **fat** or oil.

While several varieties of fatty acid occur in nature, all belong in one of two categories—saturated or unsaturated. In a saturated fatty-acid **molecule**, all the carbon atoms in the chain are attached to two hydrogen atoms, the maximum amount. All the bonds between the carbon atoms in the chain are single **electron** bonds. An example of fat made of saturated fatty acids is butter.

Unsaturated fatty-acid molecules have one or more carbon atoms with only a single hydrogen atom attached. In these chains, one or more bonds between the carbon atoms are double. A molecule with one double bond is called monounsaturated, and two or more double bonds is called polyunsaturated. An example of unsaturated fat is vegetable oil.

Generally, fats consisting of saturated fatty acids are solid, and those made up of unsaturated molecules are liquid. An unsaturated fatty acid may be converted into saturated through a process called **hydrogenation**. While most modern diets are aimed at the reduction of fatty acids (fats), it is important to recognize that several of them, such as oleic, butyric, and palmitic acid, are important parts of the human diet. Another, linoleic acid, is absolutely essential to human life. It is an important part of a vital chemical reaction in the body, and is obtained solely through ingestion. It is found in corn, **soybean**, and peanut oils.

Recently, concern about the amount of trans fatty acids present in food has caused debate. Trans fatty acids are formed during the process of partial hydrogenation of unsaturated fatty acids (like vegetable oil) into margarine and vegetable shortening. Some research suggests that levels of trans fatty acids can alter the amount of **cholesterol** found in **blood**, which can be a significant risk to people suffering from high cholesterol levels and **heart disease**. In addition to being found in margarine, trans fatty acids are also found naturally in small quantities in beef, pork, lamb, and milk. There is conflicting evidence, however, of the dangers of trans fatty acids in daily diets. Generally, it is recommended to limit the total daily amount of fat eaten, rather than focusing solely on trans fatty acid consumption.

See also Carboxylic acids.

Fault

A fault is a geologic term describing a fracture at which two bodies of rock have been displaced relative to each other. **Bedrock** faults are those in which bodies of rock meet; small, local movements may occur on bedrock faults. Much larger movements or displacements occur along Faults where plates of Earth's crust abut each other. Faults may be inches (centimeters) to hundreds of miles (kilometers) in length, and movements or displacements have the same range in length. Major fault systems are typically found where plates meet; for example, the San Andreas Fault in California, is really a fault system including many smaller faults that branch off of the main trace of the San Andreas as well as faults that **parallel** the main fault. It may be more accurate to call these systems "fault zones" or "fault belts" that contain known and unknown faults. The Northridge **earthquake** in the Los Angeles, California, area in January 1994, occurred along a thrust fault that had not previously been known but is within the San Andreas zone. A fault zone may be hundreds of feet (meters) wide and each has a unique character; some include countless faults and others have very few.

Plate tectonics

To understand faults, it is helpful to understand **plate tectonics**. Earth's crust is not a solid skin. Instead, it is made up of huge blocks of rock that fit together to form the entire surface of the **planet**, including the continents or land masses and the floors of the oceans. Scientists believe the crust is composed of about 12 of these plates. Each plate is relatively rigid, and, where the plates meet, they can spread apart, grind against each other, or ride one over the other in a process called subduction. Spreading plates most commonly occur in the oceans in the phenomenon known as sea-floor spreading; when plates spread within land masses, they create huge valleys called rifts. The process of plates grinding together causes near-surface earthquakes, and the collision and subduction of plates causes the most intense earthquakes much deeper in the crust.

The engine driving the movement of the plates originates deep in the **earth**. The mantle, a zone underlying the crust, is very dense rock that is almost liquid. Deeper still is Earth's core, which is molten rock. Because it is fluid, the core moves constantly. The mantle responds to this, as well as to centrifugal **force** caused by the **rotation** of Earth on its axis and to the force of gravity. The slower motions of the mantle pulse through the thin crust, causing earthquakes, volcanic activity, and the movement of tectonic plates. Together, the pulses caused by the **heat** engine inside Earth result in over a million earthquakes per year that can be detected by instruments. Only one third of these can be felt by humans, most of which are very small and do not cause any damage. About 100–200 earthquakes per year cause some damage, and one or two per year are catastrophic.

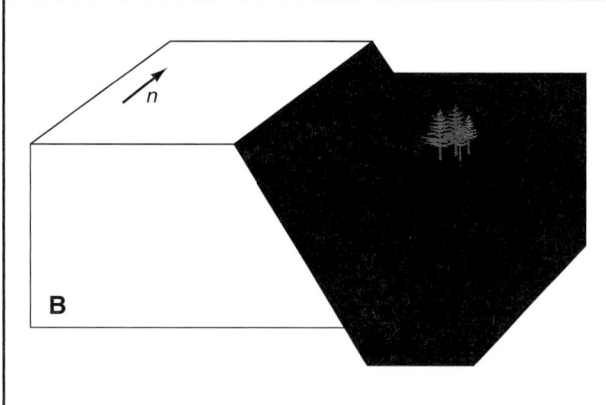

Figure 1. Normal fault striking north. The solid square represents the slip vector showing the motion of block A relative to block B. *Illustration by Hans & Cassidy. Courtesy of Gale Group.*

History of our understanding of faults

In the history of the study of faults, Robert Mallet, an Irish engineer, was the first to believe that simple mechanics of the earth's crust cause earthquakes. Until 1859, when he proposed his theory, earthquakes were believed to be caused by huge explosions deep within the earth, and the origin of these explosions was never questioned. Mallet knew that **iron**, which appears indestructible, ruptures under extreme stress, and Mallet theorized that earthquakes are caused "either by the sudden flexure and constraint of the elastic materials forming a portion of the earth's crust, or by their giving way and become fractured." Mallet was not supported, primarily because he was not a scholar and lived in Ireland where earthquakes seldom occur. In 1891, however, Professor Bunjiro Koto, a Japanese specialist in seismology, or the study of earthquakes, endorsed Mallet's theory. After the Mino-Iwari earthquake, which occurred along a remarkably clear fault line crossing the **island** of Honshu, he said the shaking earth caused quakes and not the other way around. Harry Fielding Reid, an American scientist, was the first to relate the stresses along faults to tectonic plate boundaries after the 1906 Great San Francisco Earthquake.

Types of faults

Faults themselves do not cause earthquakes; instead, they are the lines at which plates meet. When the plates press together (compress) or pull apart (are in tension), earthquakes occur. The fault line is essentially a stress concentration. If a rubber band is cut partially through then pulled, the rubber band is most likely to break at the cut (the stress concentration). Similarly, the "break" (stress release or earthquake) occurs along a fault when

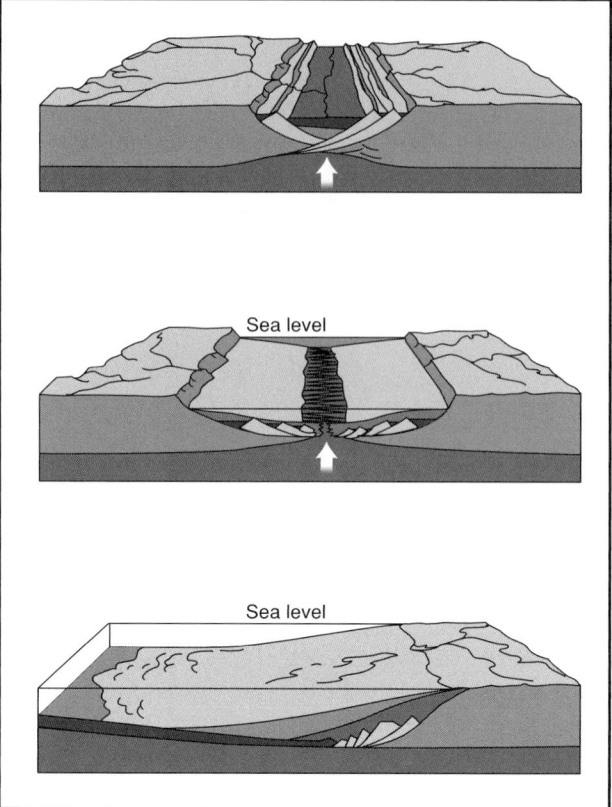

Figure 2. Formation of oceanic basin. *Illustration by Hans & Cassidy. Courtesy of Gale Group.*

the plates or rock bodies that meet at the fault press together or pull apart.

Movement along a fault can be vertical (up and down, changing the surface elevation), horizontal (flat at the surface but with one side moving relative to the other), or a combination of motions that inclines at any **angle**. The angle of inclination of the fault **plane** measured from the horizontal is called the dip of the fault plane. This movement occurs along a fault surface or fault plane. Any relative vertical **motion** will produce a hanging wall and a footwall. The hanging wall is the block that rests upon the fault plane, and the footwall is the block upon which you would stand if you were to walk on the fault plane.

Dip-slip faults are those in which the primary motion is parallel to the dip of the fault plane. A normal fault is a dip-slip fault produced by tension that stretches or thins Earth's crust. At a normal fault, the hanging wall moves downward relative to the footwall. Two normal faults are often separated by blocks of rock or land created by the thinning of the crust. When such a block drops down relative to two normal faults dipping toward each other, the block is called a graben. The huge troughs or

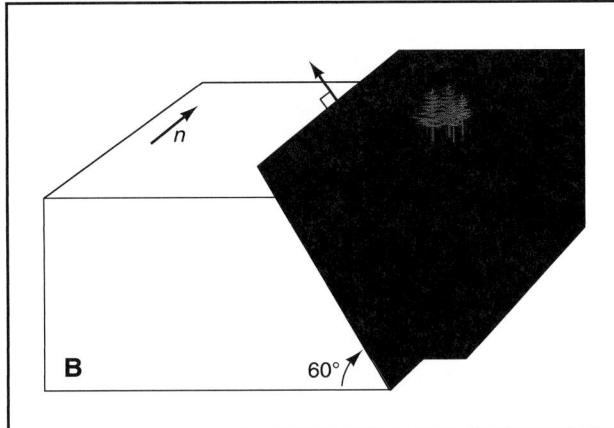

Figure 3. Thrust fault striking north. The solid square represents the slip vector showing the motion of block A relative to block B. *Illustration by Hans & Cassidy. Courtesy of Gale Group.*

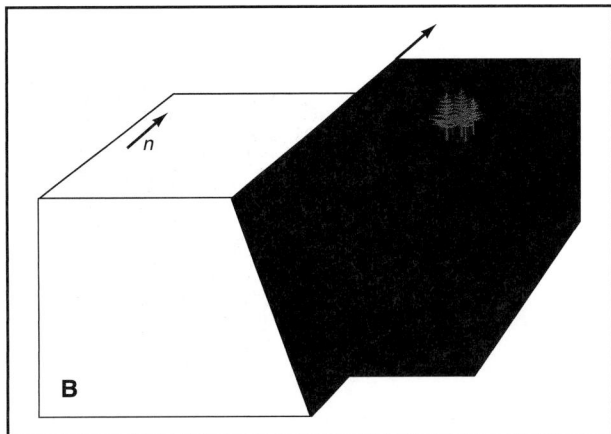

Figure 4. Strike-slip fault. *Illustration by Hans & Cassidy. Courtesy of Gale Group.*

rift valleys created as plates move apart from each other are grabens. The Rhine Valley of Germany is a graben. An extreme example is the Atlantic Ocean; over 250 million years ago, **North America** and **Africa** were a single mass of land that slowly split apart and moved away from each other (a process called divergence), creating a huge graben that became the Atlantic Ocean **basin.** Two normal faults dipping away from each other can create an uplifted block between them that is called a horst. Horsts look like raised plateaus instead of sunken valleys. If the block between normal faults tilts from one side to the other, it is called a tilted fault block.

A reverse fault is another type of dip-slip fault caused by compression of two plates or masses in the horizontal direction that shortens or contracts the earth's surface. When two crustal masses butt into each other at a reverse fault, the easiest path of movement is upward. The hanging wall moves up relative to the footwall. When the dip is less than (flatter than) 45°, the fault is termed a thrust fault, which looks much like a ramp. When the angle of dip is much less than 45° and the total movement or displacement is large, the thrust fault is called an overthrust fault. In terms of plate movement, the footwall is slipping underneath the hanging wall in a process called subduction.

Strike-slip faults are caused by shear (side-by-side) stress, resulting in a horizontal direction, parallel to the nearly vertical fault plane. Strike-slip faults are common in the sea floor and create the extensive offsets mapped along the mid-oceanic ridges. The San Andreas Fault is perhaps the best-known strike-slip fault, and, because much of its length crosses land, its offsets are easily observed. Strike-slip faults have many other names including lateral, transcurrent, and wrench faults. Strike-slip

faults located along mid-oceanic ridges are called transform faults. As the sea floor spreads, new crust is formed by **magma** (molten rock) that flows up through the break in the crust. This new crust moves away from the ridge, and the plane between the new crust and the older ridge is the transform fault.

Relative fault movement is difficult to measure because no point on the earth's surface, including **sea level** is fixed or absolute. Geologists usually measure displacement by relative movement of markers that include **veins** or dikes in the rock. **Sedimentary rock** layers are especially helpful in measuring relative **uplift** over **time.** Faults also produce rotational movements in which the blocks rotate relative to each other; some sedimentary **strata** have been rotated completely upside down by fault movements. These beds can also be warped, bent, or folded as the comparatively soft rock tries to resist compressional forces and **friction** caused by slippage along the fault. Geologists look for many other kinds of evidence of fault activity such as slickensides, which are polished or scratched fault-plane walls, or fault gouge, which is clayey, fine-grained crushed rock caused by compression. Coarse-grained fault gouge is called fault breccia.

Mountain-building by small movements along faults

Compression of land masses along faults has built some of the great mountain ranges of the world. Mountain-building fault movements are extremely slow, but, over a long time, they can cause displacements of thousands of feet (meters). Examples of mountain ranges that have been raised by cumulative lifting along faults are the Wasatch Range in Utah, the uplifting of layer upon layer of sedimentary **rocks** that form the eastern front of

The San Andreas fault extends almost the full length of California. The landscape consists of pressure ridges formed by hundreds of fault movements. *JLM Visuals. Reproduced by permission.*

the Rocky Mountains in Wyoming and Montana, the large thrust faults that formed the Ridge and Valley Province of the Appalachian Mountains in Virginia and Tennessee, and the Himalayas (including Mount Everest and several of the other tallest mountains in the world) that are continuing to be pushed upward as the tectonic plate bearing the Indian Subcontinent collides with the Eurasian plate. Tension along smaller faults has created the mountain ranges that bracket the Great Basin of Nevada and Utah. These mountains may have been formed by the hanging walls of the many local faults that slid downward by thousands of feet (meters) until they became valley floors.

Earthquake generation by large, sudden movements along faults

The majority of fault motion are slow and creeping movements, unlikely to be felt by humans at ground surface. Some movements occur as rapid spasms that happen in a few seconds and can cause ground displacements of inches or feet (centimeters or meters). These movements are resisted by friction along the two faces of the fault plane until the tensional, compressional or shear stress exceeds the frictional force. Earthquakes are caused by these sudden jumps or spasms. Severe shaking can result, and ground rupture can create fault scarps.

Famous or infamous faults

The San Andreas Fault

The San Andreas Fault may well be the best known fault in the world. It marks a major fracture in the Earth's crust, passing from Southern through Northern California for a length of about 650 mi (1,050 km) and then traversing under part of the northern Pacific Ocean. The San Andreas does mark a plate boundary between the Northern Pacific and North American plates, and, because this transform fault extends to the surface in a heavily populated area, movement along the fault causes major earthquakes. The forces that cause these movements are the same ones responsible for **continental drift**. The Great San Francisco Earthquake of 1906 occurred along the main San Andreas, and the Loma Prieta earthquake of 1989 was caused by movement on a branch of the San Andreas. The motion of the Northern Pacific plate as it grinds past the North American plate causes strike-slip fault movements. The plate is moving at an average of about 0.4-in (1 cm) per year, but its speed accelerated during the 1900s to between 1.6–2.4 in (4–6 cm) per year as it pushes Los Angeles northward toward San Francisco. Much more rapid jumps occur during earthquakes; in 1906, movements as great as 21 ft (6.4 m) were measured in some locations along the San Andreas Fault.

The San Andreas Fault is infamous for another reason. The major cities of California including Los Angeles, Oakland, San Jose, and San Francisco, home to millions of people, straddle this fault zone. Such development in this and other parts of the world puts many at risk of the devastation of major fault movements. Sudden fault movements fill the headlines for weeks, but, over the course of **geologic time**, they are relatively rare so the chances to study them and their effects are limited. Similarly, our knowledge and ability to predict fault motions and to evacuate citizens suffers. An estimated 100 million Americans live on or near an active earthquake fault.

The New Madrid Fault is more properly called a seismic zone because it is a large fracture zone within a tectonic plate. It is a failed rift zone; had it developed like the East African Rift Valley, it would have eventually split the North American **continent** into two parts. The zone crosses the mid-section of the United States, passing through Missouri, Arkansas, Tennessee, and Kentucky in the center of the North American Plate. The zone is about 190 mi (300 km) long and 45 mi (70 km)

KEY TERMS

Continental drift—A theory that explained the relative positions and shapes of the continents, and other geologic phenomena, by lateral movement of the continents. This was the precursor to plate tectonic theory.

Core—The molten center of the earth.

Crust—The outermost layer of the earth, situated over the mantle and divided into continental and oceanic crust.

Dip—The angle of inclination (measured from the horizontal) of faults and fractures in rock.

Footwall—The block of rock situated beneath the fault plane.

Graben—A block of land that has dropped down between the two sides of a fault to form a deep valley.

Hanging wall—The block of rock that overlies the fault plane.

Horst—A block of land that has been pushed up between the two sides of a fault to form a raised plain or plateau.

Mantle—The middle layer of the earth that wraps around the core and is covered by the crust. The mantle consists of semi-solid, partially melted rock.

Normal fault—A fault in which tension is the primary force and the footwall moves up relative to the hanging wall.

Plate tectonics—The theory, now widely accepted, that the crust of the earth consists of about twelve massive plates that are in motion due to heat and motion within the earth.

Reverse fault—A fault resulting from compressional forces and the hanging wall moves up relative the footwall.

Seismic gap—A length of a fault, known to be historically active, that has not experienced an earthquake recently and may be storing strains that will be released as earthquake energy.

Strike-slip fault—A fault at which two plates or rock masses meet and move lateral or horizontally along the fault line and parallel to the compression.

Subduction—In plate tectonics, the movement of one plate down into the mantle where the rock melts and becomes magma source material for new rock.

Thrust fault—A low-angle reverse fault in which the dip of the fault plane is 45° or less and displacement is primarily horizontal.

wide, and it lies very deep below the surface. The zone is covered by alluvial material (**soil** and rock carried and deposited by **water**) from the Mississippi, Ohio, and Missouri **rivers**; because this alluvial material is soft and unstable, movement within the fracture zone transmits easily to the surface and is felt over a broad area.

On December 16, 1811, and January 23 and February 7, 1812, three earthquakes estimated to have measured greater than magnitude 8.0 on the Richter scale had their epicenters near the town of New Madrid, Missouri, then part of the American Frontier. An area of 3,000–5,000 sq mi (7,800–13,000 sq km) was scarred by landslides, fissures, heaved-up land, leveled **forests**, and lakes, swamps, and rivers that were destroyed, rerouted, or created. These earthquakes were felt as far away as the East Coast, north into Canada, and south to New Orleans.

On January 16, 1995, the city of Kobe, Japan was struck by a magnitude 7.2 earthquake that killed more than 4,000 people and left almost 275,000 homeless. Like the California cities along the San Andreas, Kobe is a port city, so the earthquake also caused tremendous losses to the economy of the region. Also like Oakland and San Francisco, California, Kobe is located next to a deep bay. Osaka Bay is encircled by a host of faults and fault zones with complicated relationships. The Nojima Fault on Awaji Island appears to have been the fault that hosted the Hyogogen-Nambu Earthquake of 1995. The North American Plate, Pacific Plate, Eurasian Plate, and Philippine Sea Plate all impact each other near the islands united as Japan. Thick, relatively young deposits of alluvial soil overly the faults that pass under Osaka Bay; these amplified the earth's movements along the fault in this highly populated area.

Earthquakes caused by human activities

Although the most devastating earthquakes occur in nature, humans have been able to learn more about faults and earthquake mechanisms since we have had the power to produce earthquakes ourselves. **Nuclear weapons** testing in the **desert** near Los Alamos, New Mexico, was the first known human activity to produce measurable earthquakes that were found to propagate along existing faults.

Our ability to build major **dams** that retain huge quantities of water has also generated earthquakes by so-called "hydrofracturing," in which the weight of the water stresses fractures in the underlying rock. Pumping of oil and **natural gas** from deep wells and the disposal of liquid wastes through injection wells have also produced small motions along faults and fractures.

Advances in fault studies

Our understanding of how faults move has improved greatly with modern technology and mapping. **Laser** survey equipment and **satellite** photogrammetry (measurements made with highly accurate photographs) have helped measure minute movements on faults that may indicate significant patterns and imminent earthquakes. Seismic gaps have been identified along plate boundaries. Through detailed mapping of tiny earthquakes, zones where strains in the earth have been relieved are identified; similarly, seismic gap areas without those strain-relieving motions are studied as the most likely zones of origin of coming earthquakes.

Resources

Books

Erickson, Jon. "Quakes, Eruptions, and Other Geologic Cataclysms." *The Changing Earth Series.* New York: Facts on File, 1994.

Halacy, D. S., Jr. *Earthquakes: A Natural History.* Indianapolis, IN: The Bobbs-Merrill Company, Inc., 1974.

Keller, Edward. *Environmental Geology.* Upper Saddle River, NJ: Prentice-Hall, Inc., 2000.

Japanese Geotechnical Society. *Soils and Foundations: Special Issue on Geotechnical Aspects of the January 17, 1995, Hyogoken-Nambu Earthquake.* Tokyo: Japanese Geotechnical Society, January 1996.

Verney, Peter. *The Earthquake Handbook.* New York: Paddington Press Ltd., 1970.

Walker, Bryce and the Editors of Time-Life Books. *Planet Earth: Earthquake.* Alexandria, VA: Time-Life Books, 1982.

Gillian S. Holmes

Fauna

Fauna is a generic term for the list of **animal** species occurring in a particular, large region. Fauna can refer to a prehistoric collection of animals, as might be inferred from the fossil record, or to a modern assemblage of **species** living in a region. The botanical analogue is known as **flora**. More locally, a faunation refers to the communities of individuals of the various animal species and occurring in a particular place. Because many zoologists are specialized in the animals they study, faunas are often considered on the basis of systematic groups, as is the case of bird species (avifauna) or **reptiles** and **amphibians** (herpetofauna).

A faunal region is a zoogeographic designation of large zones containing distinct assemblages of species that are more-or-less spatially isolated from other provinces by physical barriers to **migration**, such as a large body of **water**, a mountain range, or extensive **desert**. Faunal provinces are less distinct sub-units of faunal regions. These various designations are typically separated by zones of rapid transition in species types.

In the Americas, for example, there are two major faunal regions, with a zone of rapid transition occurring in Central America. The South American zoofauna includes many species and even families that do not occur naturally in **North America**, and vice versa. The South and North American faunal regions are divided by the narrow Isthmus of Panama, which has been submerged by oceanic waters at various times in the geological past, or has otherwise presented a significant barrier to the migration of many species of animals. However, during periods in the past when animals were able to pass through this barrier, significant mixtures of the two faunas occurred. Lingering evidence of relatively recent episodes of prehistoric faunal blending include the presence of the opossum (*Didelphis virginiana*) and California condor (*Gymnogyps californianus*) in North America, and white-tailed **deer** (*Odocoileus virginianus*) and cougar (*Felis concolor*) in **South America**.

Another famous faunal transition is known as Wallace's Line, after the nineteenth century naturalist who first identified it, A. R. Wallace (he was also with Charles Darwin, the co-publisher of the theory of **evolution** by natural **selection**). Wallace's Line runs through the deepwater oceanic straits that separate Java, Borneo, and the Philippines and Southeast **Asia** more generally to the north, from Sulawesi, New Guinea, and **Australia** to the south. The most extraordinary faunistic difference across Wallace's Line is the prominence of marsupial animals in the south, but there are also other important dissimilarities.

One of the most famous faunal assemblages in the fossil record is that of the Burgess Shale of southeastern British Columbia. This remarkable fauna includes 15-20 extinct phyla of metazoan animals that existed during an evolutionary **radiation** in the early Cambrian era, about 570 million years ago. Most of the phyla of the Cambrian marine fauna are now extinct, but all of these lost animals represented innovative and fantastic experiments in the form and function of the invertebrate body plan (and also undoubtedly, in invertebrate **physiology**, **behavior**, and **ecology**, although these cannot be inferred from the fossil record).

Faunal dating *see* **Dating techniques**

Fax machine

The facsimile, or fax, machine is both a transmitting and receiving device that "reads" text, maps, photographs, fingerprints, and graphics and communicates via **telephone** line. Since 1980s, fax machines have undergone rapid development and refinement and are now indispensable communication aids for news services, businesses, government agencies, and individuals.

The fax was invented by Alexander Bain of Scotland in 1842. His crude device, along with scanning systems invented by Frederick Bakewell in 1848, evolved into several modern versions. In 1869 a Frenchman, Ludovic d'Arlincourt, synchronized transmitters and receivers with tuning forks and thus aided further developments. In 1924, faxes were first used to transmit wire photos from Cleveland to New York, a boom to the newspaper industry. Two years later, RCA inaugurated a trans-Atlantic **radio** photo service for businesses.

The use of faxes, and fax technology itself, remained comparatively limited until the mid-1980s. By that time, models either required an electrolytic or photosensitive **paper**, which changed **color** when current passed through it; or thermal paper, a material coated with colorless dye, which became visible upon contact with a toner. Updated models from the 1990s employ plain paper (which, unlike thermal paper, avoids curling) and are preferred for their superior reproduction. Another improvement is the invention of a scrambler, an encoder that allows the sender to secure secrecy for documents, particularly those deriving from highly sensitive government projects or secret industrial or business dealings.

Some fax machines are incorporated into telephone units; others stand alone; and still others are part of personal computers. These last models contain a fax board, an electronic circuit that allows the computer to receive messages. In the most common models, the user inserts the material to be transmitted into a slot, then makes a telephone connection with another facsimile machine. When the number is dialed, the two machines make electronic connection. A rotating drum advances the original before an optical scanner. The scanner reads the original document either in horizontal rows or vertical columns and converts the printed image into a pattern of several

Fax machine. *Photograph by David Young-Wolff. PhotoEdit. Reproduced by permission.*

million tiny electronic signals, or pixels, per page. The facsimile machine can adjust the number of pixels so that the sender can control the sharpness and quality of the transmission. Within seconds, the encoded pattern is converted into **electric current** by a **photoelectric cell**, then travel via **telegraph** or telephone wires to the receiving fax, which is synchronized to accept the signal and produce an exact replica of the original by reverse process.

Feather stars

Feather stars, or comatulids, are echinoderms that belong to the class Crinoidea (phylum Echinodermata) which they share with the sea lilies. Unlike the latter group, however, feather stars are not obliged to remain in one place; instead they can swim or even crawl over short distances before attaching themselves to some support. Swimming movements are achieved by waving the arms up and down in a slow, controlled manner. Feather stars are widely distributed throughout tropical and warm-temperate waters, with the main center of their distribution being focused on the Indo-Pacific region. An estimated 550 **species** are known.

A feather star's body consists of a basal plate known as the calyx which **bears** a number of specialized cirri that are used for holding onto **rocks** or other objects when the **animal** is feeding or at rest. The mouth is situated on the upper surface of the calyx. Also arising from this small body are the jointed arms which are usually quite short but may measure more than 11.8 in (30 cm) in some species. Cold-water species tend to have much shorter arms than tropical feather stars. Each arm bears a large number of short pinnules (featherlike appendages).

By far the most striking part of a feather star's **anatomy** is their delicate, ostrich-plumelike arms that are usually highly colored. Some species can have more than 200 arms. Feather stars are suspension feeders and, when feeding, unfurl their arms and extend the many pinnules into the **water** current. Feather stars usually carefully choose their feeding site; typically they select a site on a high vantage point at an **angle** to the current. As the water flows between the pinnules, additional tiny tentacles (known as podia) that are covered with mucus trap the many tiny food particles. These are then transferred to special grooves that run the inner length of the pinnules, where they are transferred down and inward to the mouth region. From here the food passes directly through the esophagus which opens into an intestine. Most feeding takes place at night, when the majority of reef fishes are resting, in order to avoid the grazing effects of these predators. Although feather stars react almost immediately to being touched and hurriedly **fold** away their arms, most species can also shed their arms if attacked. The arms will regenerate in **time**. As daylight approaches, however, and the risk of predation increases, they move away and hide among the crevices of the reef face.

Feather stars are either male or female and **fertilization** is usually external. Some species retain their eggs on their arms but this is not the usual pattern of **behavior**. When the larvae hatch they pass through a series of development stages as free-swimming animals known as vitellaria. Eventually these settle and undergo a transformation which initially restricts them to a sessile (attached, not free-moving) state, as in sea lilies. Eventually, however, they develop the familiar arms of the adult feather **star** and are able to move around.

See also Sea lily.

Feldspar *see* **Minerals**

Fermentation

In its broadest sense, fermentation refers to any process by which large organic molecules are broken down to simpler molecules as the result of the action of **microorganisms**. The most familiar type of fermentation is the conversion of sugars and starches to **alcohol** by enzymes in **yeast**. To distinguish this reaction from other kinds of fermentation, the process is sometimes known as *alcoholic* or *ethanolic fermentation.*

History

Ethanolic fermentation was one of the first **chemical reactions** observed by humans. In nature, various types of food "go bad" as a result of bacterial action. Early in history, humans discovered that this kind of change could result in the formation of products that were enjoyable to consume. The "spoilage" (fermentation) of fruit juices, for example, resulted in the formation of primitive forms of wine.

The mechanism by which fermentation occurs was the subject of extensive debate in the early 1800s. It was a key issue among those arguing over the concept of *vitalism*, the notion that living organisms are in some way inherently different from non-living objects. One aspect in this debate centered on the role of so-called "ferments" in the conversion of sugars and starches to alcohol. Vitalists argued that ferments (what we now know as *enzymes*) are inextricably linked to a living **cell**. Destroy a cell and ferments can no longer cause fermentation, they said.

A crucial experiment on this issue was carried out in 1896 by the German chemist Eduard Buchner. Buchner ground up a group of cells with **sand** until they were totally destroyed. He then extracted the liquid that remained and added it to a sugar **solution**. His assumption was that fermentation could no longer occur since the cells that had held the ferments were dead, so they no longer carried the "life-force" needed to bring about fermentation. He was amazed to discover that the cell-free liquid did indeed cause fermentation. It was obvious that the ferments themselves, distinct from any living **organism**, could cause fermentation.

Theory

The chemical reaction that occurs in fermentation can be described quite easily. Starch is converted to simple sugars such as sucrose and glucose. Those sugars are then converted to alcohol (ethyl alcohol) and **carbon dioxide**. This description does not adequately convey the complexity of the fermentation process itself. During the 1930s, two German biochemists, G. Embden and O. Meyerhof, worked out the sequence of reactions by which glucose ferments.

In a sequence of twelve reactions, glucose is converted to ethyl alcohol and **carbon** dioxide. A number of enzymes are needed to carry out this sequence of reactions, the most important of which is zymase, found in yeast cells. These enzymes are sensitive to environmental conditions in which they live. When the **concentration** of alcohol reaches about 14%, they are inactivated. For this reason, no fermentation product (such as wine) can have an alcoholic concentration of more than about 14%.

Uses

The alcoholic beverages that can be produced by fermentation vary widely, depending primarily on two factors—the **plant** that is fermented and the enzymes used for fermentation. Human societies use, of course, the materials that are available to them. Thus, various peoples have used **grapes**, berries, corn, **rice**, **wheat**, honey, potatoes, **barley**, hops, **cactus** juice, cassava roots, and other plant materials for fermentation. The products of such reactions are various forms of beer, wine or distilled liquors, which may be given specific names depending on the source from which they come. In Japan, for example, rice wine is known as sake. Wine prepared from honey is known as mead. Beer is the fermentation product of barley, hops, and/or malt sugar.

Early in human history, people used naturally occurring yeast for fermentation. The products of such reactions depended on whatever enzymes might occur in "wild" yeast. Today, wine-makers are able to select from a variety of specially cultured yeast that control the precise direction that fermentation will take.

Ethyl alcohol is not the only useful product of fermentation. The carbon dioxide generated during fermentation is also an important component of many baked goods. When the batter for bread is mixed, for example, a small amount of sugar and yeast is added. During the rising period, sugar is fermented by enzymes in the yeast, with the formation of carbon dioxide gas. The carbon dioxide gives the batter bulkiness and texture that would be lacking without the fermentation process.

Fermentation has a number of commercial applications beyond those described thus far. Many occur in the food preparation and processing industry. A variety of **bacteria** are used in the production of olives, cucumber pickles, and sauerkraut from the raw olives, cucumbers, and cabbage, respectively. The selection of exactly the right bacteria and the right conditions (for example, acidity and **salt** concentration) is an art in producing food products with exactly the desired flavors. An interesting line of research in the food sciences is aimed at the production of edible food products by the fermentation of **petroleum**.

In some cases, **antibiotics** and other drugs can be prepared by fermentation if no other commercially efficient method is available. For example, the important drug cortisone can be prepared by the fermentation of a plant steroid known as diosgenin. The enzymes used in the reaction are provided by the mold *Rhizopus nigricans.*

KEY TERMS

Vitalism—The concept that compounds found within living organisms are somehow inherently differ from those found in non-living objects.

Wastewater—Water that carries away the waste products of personal, municipal, and industrial operations.

One of the most successful commercial applications of fermentation has been the production of ethyl alcohol for use in gasohol. Gasohol is a mixture of about 90% gasoline and 10% alcohol. The alcohol needed for this product can be obtained from the fermentation of agricultural and municipal wastes. The use of gasohol provides a promising method for using renewable resources (plant material) to extend the availability of a nonrenewable resource (gasoline).

Another application of the fermentation process is in the treatment of wastewater. In the activated sludge process, **aerobic** bacteria are used to ferment organic material in wastewater. Solid wastes are converted to carbon dioxide, **water**, and mineral salts.

See also Ethanol; Enzyme.

Resources

Books

Baum, Stuart J., and Charles W. J. Scaife. *Chemistry: A Life Science Approach.* New York: Macmillan Publishing Company, Inc., 1975, Chapter 28.

Brady, James E., and John R. Holum. *Fundamentals of Chemistry.* 2nd edition. New York: John Wiley & Sons, 1984, p. 828A.

Loudon, G. Mark. *Organic Chemistry.* Oxford: Oxford University Press, 2002.

David E. Newton

Fermions *see* **Subatomic particles**

Fermium *see* **Element, transuranium**

Ferns

Ferns are plants in the Filicinophyta phylum, also called the Pteridophyta phylum. They are intermediate in complexity between the more primitive (i.e., evolutionarily ancient) bryophytes (mosses, liverworts, and hornworts) and the more advanced (or recent) seed plants. Like bryophytes, ferns reproduce sexually by making spores rather than **seeds**. Most ferns produce spores on the underside or margin of their leaves. Like seed plants, ferns have stems with a vascular system for efficient transport of **water** and food. Ferns also have leaves, known technically as megaphylls, with a complex system of branched **veins**. There are about 11,000 **species** of ferns, most of them indigenous to tropical and subtropical regions.

General characteristics

A fern **plant** generally consists of one or more fronds attached to a **rhizome**. A frond is simply the **leaf** of the fern. A rhizome is a specialized, root-like stem. In most temperate-zone species of ferns, the rhizome is subterranean and has true roots attached to it. Fronds are generally connected to the rhizome by a stalk, known technically as the stipe. The structures of the frond, rhizome, and stipe are important characteristics for species identification.

The sizes of ferns and their fronds vary considerably among the different species. **Tree** ferns of the Cyatheaceae family are the largest ferns. They are tropical plants which can grow 60 ft (18 m) or more in height and have fronds 15 ft (5 m) or more in length. In contrast, species in the genus *Azolla*, a group of free-floating aquatic ferns, have very simple fronds which are less than 0.2 in (0.5 cm) in diameter.

The fern frond develops from a leaf bud referred to as a crozier. The crozier is coiled up in most species, with the frond apex at the middle of the coil. This pattern of coiled leaf arrangement in a bud is called circinate vernation. Circinate vernation is found in a few other seed plants, but not in any other free-sporing plants. During growth of a bud with circinate vernation, the cells on one side of the coil grow more rapidly than those on the other, so the frond slowly uncoils as it develops into a full-grown leaf.

The **horsetails** (phylum Sphenophyta) and **club mosses** (phylum Lycodophyta) are known colloquially as fern allies. The fern allies also reproduce sexually by making spores and have stems with vascular systems. However, there are two principal differences between ferns and fern allies. First, unlike the ferns, the leaves of fern allies, known technically as microphylls, are small, scale-like structures with a single mid-vein. Second, fern allies make their spores at the bases of their leaves or on specialized branches. There are about 1,500 species of fern allies in the world.

The reproductive cells of ferns are microscopic spores which are often clustered together in the brown spots visible on the fronds' undersides. Since fern spores are microscopic, fern reproduction was not well understood until the mid-1800s. This led some people to attribute mystical powers to the ferns. According to folklore, ferns made invisible seeds and a person who held these would also become invisible. Even Shakespeare drew upon this folklore and wrote in *Henry IV*; "we have the receipt of fern seed; we walk invisible." Nowadays, anyone with a simple **microscope** can tease apart the brown spots on the underside of a fern frond and see the tiny spores.

Natural history

There are about 11,000 species of ferns in the world. Ferns are found throughout the world, from the tropics to the subarctic region. The greatest species diversity is in the tropical and subtropical region from southern Mexico to northern **South America**.

In temperate **North America**, most ferns are terrestrial plants and grow in woodlands. However, in the tropics, many ferns grow as epiphytes. Epiphytes are plants which rely upon other plants, such as trees, for physical support, while obtaining their **nutrition** from organic debris and rain water that falls through the forest canopy.

Ferns can be found in very different habitats throughout the world. Some species are free-floating aquatic plants, some species grow in moist woodlands, and a few species grow in arid or semiarid regions. Most species require some rainfall because their sperm cells must swim through a fluid to reach the egg cells.

Interestingly, sperm cells of the resurrection fern (*Polypodium polypodioides*) swim through a fluid exuded by the fern itself to reach the female's egg. This species is widely distributed in semi-arid and arid regions of the world, such as central **Australia**, central Mexico, and central **Africa**. The resurrection fern is sold in some garden shops as a brown, dried-out, and curled-up ball of fronds. When this dried-out fern is soaked in water, it rapidly expands and becomes green in a day or so, attesting to the remarkable desiccation tolerance of this species.

At the other extreme are water ferns of the genus *Azolla*, which grow free-floating on fresh water. *Azolla* is particularly interesting because it has special pockets in its leaves which apparently have evolved to accommodate symbiotic cyanobacteria of the genus *Anabaena*. These cyanobacteria transform atmospheric **nitrogen** (N_2) to **ammonia** (NH_3), a chemical form useful to plants. This process is called **nitrogen fixation**. Many Asian farmers encourage the growth of *Azolla* and its associated *Anabaena* in their **rice** paddies to increase the amount of nitrogen available to their rice plants.

Many species of ferns can act as alternate hosts for species of rust **fungi** that are pathogenic to **firs**, economically important timber trees. The rust fungi are a large and diverse group of fungi, which have very complex life cycles, often with four or five different reproductive stages. In the species of rust fungi that attack ferns, part of the life cycle must be completed on the fern, and part on the fir tree. These parasitic fungi can usually be eradicated by simply eliminating one of the hosts, in this case, either the fern or the fir tree.

Life cycle

Like all plants, the life cycle of ferns is characterized as having an alternation of a gametophyte phase and a sporophyte phase. A typical fern sporophyte is the large, familiar plant seen in nature. Its cells have the unreduced number of chromosomes, usually two sets. Most fern gametophytes are not seen in nature. A typical gametophyte is about 0.4 in (1 cm) in diameter, multicellular, flat, heart-shaped, and green. Its cells have the reduced number of chromosomes, usually one set.

Interestingly, the gametophyte and sporophyte are about equally dominant in the life cycle of ferns. In contrast, the gametophyte is dominant in the more evolutionarily primitive bryophytes (mosses, liverworts, and hornworts), whereas the sporophyte is dominant in the more evolutionarily advanced seed plants.

Gametophyte

The gametophyte phase of the fern life cycle begins with a **spore**. A fern spore is a haploid reproductive **cell**, which unlike the seeds of higher plants, does not contain an embryo. Fern spores are often dispersed by the **wind**. Upon **germination**, a spore gives rise to a green, thread-like **tissue**, called a protonema. The protonema develops into a prothallus, a small, green, multicellular tissue that is rarely seen in nature. The prothallus has numerous subterranean rhizoids to anchor it to the substrate and absorb **nutrients**.

Light and other environmental factors control the development of fern gametophytes. In many species, gametophytes kept in darkness do not develop beyond the thread-like protonemal stage. However, illumination with blue or ultraviolet **radiation** causes the protonema to develop into a heart-shaped prothallus. This is an example of photomorphogenesis, the control of development by light.

Male and female reproductive structures form on the prothallus, and these are referred to as antheridia and archegonia, respectively. Each antheridium produces many flagellated sperm cells which swim toward the archegonia. The sperm cells of some ferns have up to

Uluhe ferns (*Dicranopteris emarginata*) in Hawaii. *JLM Visuals. Reproduced with permission.*

several hundred **flagella** each. Each archegonium produces a single egg which is fertilized by a sperm cell.

Sporophyte

Fusion of the egg and sperm nuclei during **fertilization** leads to the formation of a zygote, with the unreduced number of chromosomes, usually two sets. The zygote develops into a sporophyte, the most familiar stage of the fern life cycle. As the sporophyte grows, the prothallus to which it is attached eventually decays. Most fern sporophytes in temperate North America are green and terrestrial.

As the sporophyte continues to grow, it eventually develops numerous structures with spores inside, referred to as sporangia. The sporangia form on the underside of fronds or on specialized fertile fronds, depending on the species. In many species, the sporangia develop in clusters referred to as sori (singular, sorus). The size, shape, and position of the sori are frequently used in species identification. As development proceeds, the sporangium dries out, releasing the many spores inside for dispersal into the environment.

Most ferns are homosporous, in that all their spores are identical and all spores develop into a gametophyte with antheridia and archegonia. However, some water ferns are heterosporous. In these species, separate male and female spores develop on the sporophyte. The smaller and more numerous male spores germinate and develop into male gametophytes with antheridia. The female spores germinate and develop into female gametophytes with archegonia.

Polyploidy

In many species of ferns, the sporophyte phase is diploid (two sets of chromosomes) and the gametophyte phase is haploid (one set of chromosomes). However, many other ferns are considered polyploid, in that their sporophyte contains three or more sets of chromosomes. In polyploid ferns, the gametophyte and sporophyte phases are said to have the "reduced" and the "unreduced" number of chromosomes, respectively.

Apospory and apogamy are special types of **asexual reproduction** which have important roles in the generation and proliferation of polyploidy. In apospory, the gametophyte develops directly from special cells on the sporophyte, so that the gametophyte and sporophyte both have the unreduced number of chromosomes. The sperm and egg cells produced by such a gametophyte have twice the original number of chromosomes. In apogamy, the sporophyte develops directly from special cells on the gametophyte, so that the sporophyte and gametophyte have the same reduced number of chromosomes. Apogamy typically occurs in gametophytes which themselves have arisen by apospory.

Evolution

Most botanists believe that the ferns and fern allies are descendants of the Rhyniopsida, an extinct group of free-sporing plants which originated in the Silurian period (about 430 million years ago) and went extinct in the mid-Devonian period (about 370 million years ago). The Rhyniopsida were primitive vascular plants which were photosynthetic, had branched stems, and produced sporangia at their stem tips, but had no leaves or roots.

The Cladoxylales is a group of plants known colloquially as the "pre-ferns." They also evolved from the Rhyniopsida, but went extinct in the lower Carboniferous period (about 340 million years ago). Some botanists previously considered these as ancestors of the ferns, because they had leaves somewhat similar to fern fronds. However, most botanists now believe the evolutionary line which led to the Cladoxylales went extinct, and that the modern ferns evolved from a separate lineage of the Rhyniopsida.

As a group, the ferns were the first plants to have megaphylls. A megaphyll is a leaf with a complex system of branched veins. Many botanists believe that the ferns evolved megaphylls by developing a flattened and webbed version of the simple, three-dimensional branching system of the Rhyniopsida. The **evolution** of the megaphyll was a major event in plant evolution, and nearly all ecologically dominant plants in the modern world have megaphylls.

Modern ferns

There are two evolutionarily distinct groups of modern ferns, the leptosporangiates and the eusporangiates.

In the leptosporangiates, the sporangium develops from one cell and is usually only one cell thick. In the eusporangiates, the sporangium develops from several cells and is usually several cells thick. Most botanists believe that the leptosporangiate and eusporangiate ferns separated evolutionarily in the lower Carboniferous (about 340 million years ago) or earlier. Modern leptosporangiate ferns are often placed into the Filicales class, and eusporangiate ferns into the Marattiales or Ophioglossales classes.

While there is general agreement about the natural division between the leptosporangiate and eusporangiate ferns, there is considerable uncertainty about other relationships among the modern ferns. Thus, there have been many proposed classification schemes. The widespread occurrence of polyploidy (see above) and hybridization (see below) in ferns has complicated the determination of evolutionary relationships.

Hybridization

Many species of ferns form hybrids in nature and hybridization is believed to have had a major role in fern evolution. A **hybrid** species is the offspring of a sexual union between two different species. Most hybrids cannot engage in **sexual reproduction** because they lack homologous (corresponding) chromosomes, which typically pair off during production of sperm and egg cells. However, since many fern species can engage in apogamy and apospory, fern hybrids can often reproduce and proliferate.

A hybrid species is often identified by the number of chromosomes in its cells and by the presence of aborted spores. **Chromosome** number is often used to infer evolutionary relationships of hybrid ferns. The ferns, as a group, tend to have very high chromosome numbers due to the widespread occurrence of polyploidy. One fern species, *Ophioglossum reticulatum*, has 631 chromosomes, the largest number of any **organism**.

Psilotum and *Tmesipteris*

Lastly, the evolutionary status of two additional genera of modern plants must be considered: *Psilotum* and *Tmesipteris*. These free-sporing tropical and subtropical plants have very simple morphologies. In particular, species in the genus *Psilotum* superficially resemble plants of the Rhyniopsida in that their sporophytes consist of three-dimensional branched stems, with tiny scalelike appendages believed to be leaf derivatives. Moreover, like the Rhyniopsida, *Psilotum* does not have true roots. Thus, some botanists have suggested that *Psilotum* is a direct descendant of the Rhyniopsida. Others reject this hypothesis and point to the lack of a fossil record con-

KEY TERMS

. . . .

Apogamy—Development of a sporophyte directly from the gametophyte without fusion of sex cells.

Apospory—Development of a gametophyte directly from the sporophyte without sex cell production.

Epiphyte—A plant which relies upon another plant, such as a tree, for physical support, but does not harm the host plant.

Flagellum—Thread-like appendage of certain cells, such as sperm cells, which controls their locomotion.

Megaphyll—Leaf with a complex system of branched veins, typical of ferns and seed plants.

Microphyll—Scale-like leaf with a single midvein, typical of fern allies.

Prothallus—Gametophyte phase of ferns and fern allies, rarely seen in nature.

Rhizome—This is a modified stem that grows horizontally in the soil and from which roots and upward-growing shoots develop at the stem nodes.

Sorus—Group of many sporangia, which often appear as a brown spot on the margin or underside of a fern frond.

Sporangium—Structure that produces spores.

Spore—Small reproductive cell that develops into a gametophyte.

Symbiosis—A biological relationship between two or more organisms that is mutually beneficial. The relationship is obligate, meaning that the partners cannot successfully live apart in nature.

necting these two groups. They suggest that *Psilotum* and *Tmesipteris* may have evolved by evolutionary simplification of an ancestor of the modern fern genus, *Stromatopteris*. Clearly, further research is needed to resolve the relationships of these fascinating, fern-like plants.

Importance to humans

In general, ferns are of minor economic importance to humans. However, ferns are popular horticultural plants and many species are grown in ornamental gardens or indoors.

Most people can recognize ferns as understory or groundcover plants in woodland habitats. However, several hundred million years ago ferns and fern allies were the dominant terrestrial plants. Thus, the fossils of these plants have contributed greatly to the formation of our fossil fuels—coal, oil and **natural gas**.

Various non-western cultures have used the starch-rich rhizome and stems of some fern species as a food. Westerners generally eschew ferns as a food. However, those who frequent restaurants known for their *haut cuisine* will occasionally find croziers or fiddleheads (unfurled fern leaves, see above) of the ostrich fern (*Matteuccia struthiopteris*) served in salads, around a bowl of ice cream, or as a steamed vegetable.

Herbalists have advocated some fern species for treatment of **ulcers**, rheumatism, intestinal infections, and various other ailments. Although many modern pharmaceuticals are derived from chemicals produced by plants, there is little scientific evidence that ferns are useful as treatments for these or other ailments.

See also Maidenhair fern; Seed ferns.

Resources

Books

Cobb, B. *A Field Guide to Ferns and Their Related Families: Northeastern and Central North America.* New York: Houghton Mifflin Company, 1975.

Jones, D. *Encyclopedia of Ferns.* vol. 1. Portland, OR: Timber Press, 1987.

Margulis, L., and K.V. Schwartz. *Five Kingdoms.* New York: W. H. Freeman and Company, 1988.

McHugh, A. *The Cultivation of Ferns.* North Pomfret, VT: Trafalgar Press, 1992.

Peter A. Ensminger

Ferrets

Ferrets are small carnivores belonging to the weasel family (Mustelidae). The name is most commonly given to the fitch, or European polecat (*Mustela putorius*), which has been domesticated and used for hunting **rodents** and as a pet for hundreds of years. Like most **weasels**, ferrets are long and slender, and are determined hunters. Their **color** varies from yellowish to all black, and they are about 2 ft (60 cm) long, including the tail. Like all weasels, the male is considerably larger than the female.

In **North America**, a close relative of the European polecat inhabits **prairie habitat**. The black-footed ferret (*Mustela nigripes*) looks like its European relative, but all four feet are black under a yellowish or tan body. It also has a black mask on its small, triangular head and a black tip on its tail, whereas the tail of the polecat is entirely dark.

A black-footed ferret. *JLM Visuals. Reproduced by permission.*

The black-footed ferret is probably descended from the steppe polecat (*Mustela eversmanni*) of Russia, which looks almost identical and leads a similar life in open grassland. However, the steppe polecat is still numerous, and even welcomed by people because it eats rodent **pests**. The black-footed ferret has developed specialized eating habits that place its future survival in jeopardy.

The black-footed ferret lives in close association with prairie dogs (*Cynomys* species). It depends on these rodents as a source of food and also uses their burrows for shelter. Over many decades, prairie dogs throughout their range have been killed by poisoned grain spread by farmers determined to keep these rodents from tearing up their land when they dig their burrows. As the prairie dogs have disappeared, so has the black-footed ferret.

Ferrets are solitary animals, except during the spring mating season. One to six kits are born after a gestation period of about six weeks. Males do not help females raise the young. Young ferrets begin to venture out of their burrow in July. One of their first trips out might be to a nearby burrow where their mother has placed a dead **prairie dog** for them to eat. She will continue to provide them with meat until they are able to hunt on their own in the autumn.

As prairie dogs were eradicated from most of the American prairie, young ferrets in search of territory after separating from their mother often could not find a new prairie-dog colony to inhabit. Or if they could find a new colony, it was too isolated from other ferrets for breeding to occur. Populations of black-footed ferrets dwindled, and by the 1960s it was clear that there were very few of these animals left in the wild. In fact, the **species** was considered extinct.

An experiment in captive breeding

Fortunately, in 1981 a tiny population of nine black-footed ferrets was found in the Absaroka Mountains of Wyoming. These animals were protected, and the number quickly grew to more than 100. Unfortunately, an introduced **disease** (canine distemper) then killed all but 18 animals. The few survivors were taken into captivity by the U.S. Fish and Wildlife Service to establish a captive breeding program, in a last-ditch effort to save the black-footed ferret from **extinction**. Part of the attempt to save the species involved studies of the steppe polecat, to learn about the **biology** and **behavior** of this closely related **animal**. This information was then applied to breeding the black-footed ferret in captivity.

By 1987, several captive ferrets had succeeded in raising young. The captive population increased during the next several years, and by 1996 there were more than 500 black-footed ferrets in captivity. Since 1991, several hundred black-footed ferrets have been returned to the wild, mostly in Wyoming, Utah, and Colorado. It will take many years and good stewardship by landowners to determine whether the reintroduction of black-footed ferrets into the American prairie will succeed. The black-footed ferret is considered an **endangered species** by all **conservation** organizations, including the U.S. Fish and Wildlife Service.

See also Weasels.

Resources

Books

Burton, John A., and Bruce Pearson. *The Collins Guide to the Rare Mammals of the World.* Lexington, MA: Stephen Greene Press, 1987.

Casey, Denise. *Black-footed Ferret.* New York: Dodd, Mead & Company, 1985.

Fox, J.G., ed. *Biology and Diseases of the Ferret.* Lippincott, Williams & Wilkins, 1998.

Knight, Lindsay. *The Sierra Club Book of Small Mammals.* San Francisco: Sierra Club Books, 1993.

Nowak, Ronald M., ed. *Walker's Mammals of the World.* 5th ed. Baltimore: Johns Hopkins University Press, 1991.

Schreiber, A., et al. *Weasels, Civets, Mongooses and Their Relatives: An Action Plan for the Conservation of Mustelids and Viverrids.* Gland, Switzerland: IUCN-The World Conservation Union, 1989.

Jean F. Blashfield

Ferromagnetism *see* **Magnetism**

Fertilization

In animals, fertilization is the fusion of a sperm **cell** with an egg cell. The penetration of the egg cell by the chromosome-containing part of the sperm cell causes a reaction, which prevents additional sperm cells from entering the egg. The egg and sperm each contribute half of the new organism's genetic material. A fertilized egg cell is known as a zygote. The zygote undergoes continuous **cell division**, which eventually produces a new multicellular **organism**.

Fertilization in humans occurs in oviducts (fallopian tubes) of the female reproductive tract and takes place within hours following sexual intercourse. Only one of the approximately 300 million sperm released into a female's vagina during intercourse can fertilize the single female egg cell. The successful sperm cell must enter the uterus and swim up the fallopian tube to meet the egg cell, where it passes through the thick coating surrounding the egg. This coating, consisting of sugars and **proteins**, is known as the zona pellucida. The tip of the head of the sperm cell contains enzymes which break through the zona pellucida and aid the penetration of the sperm into the egg. Once the head of the sperm is inside the egg, the tail of the sperm falls off, and the perimeter of the egg thickens to prevent another sperm from entering.

The sperm and the egg each contain only half the normal number of chromosomes, a condition known as haploid. When the genetic material of the two cells fuses, the fertilization is complete.

In humans, a number of variables affect whether or not fertilization occurs following intercourse. One factor is a woman's ovulatory cycle. Human eggs can only be fertilized a few days after ovulation, which usually occurs only once every 28 days.

In other **species**, fertilization occurs either internally (as above) or externally, depending on the species involved. Fertilization outside of the animal's body occurs in aquatic animals such as **sea urchins**, **fish**, and **frogs**. In sea urchins, several billion sperm are released into the **water** and swim towards eggs released in the same area. Fertilization occurs within seconds of sperm release in the vicinity of the eggs. Sea urchins have been used greatly in research on fertilization.

Artificial insemination, in humans or animals, occurs when sperm is removed from the male and injected into the vagina or uterus of the female. In the latter case, the sperm must first be washed to remove the semen. This is a common treatment for human **infertility**. The development of **gamete** intra-fallopian transfer (GIFT) technology has further improved the treatment of infertility. In this procedure, sperm and eggs are placed together in the woman's fallopian tube and fertilization progresses naturally.

Fertilization occurring outside of the body is *in vitro* (in a dish or test tube) fertilization, IVF. Eggs are removed surgically from the female's reproductive tract and fertilized with sperm. At the 4-cell (day 2) stage, the embryos are returned to the female's fallopian tube or uterus where development continues. Mammalian IVF has been performed successfully on animals since the 1950s, and the first human **birth** following IVF occurred in Great Britain in 1978. This procedure has since become a routine treatment for infertility. If the sperm is too weak to penetrate the egg, or if the male has a very low sperm count, an **individual** sperm can be injected directly into the egg. Both eggs and sperm can be frozen for later use in IVF. A mechanical "sperm sorter" that

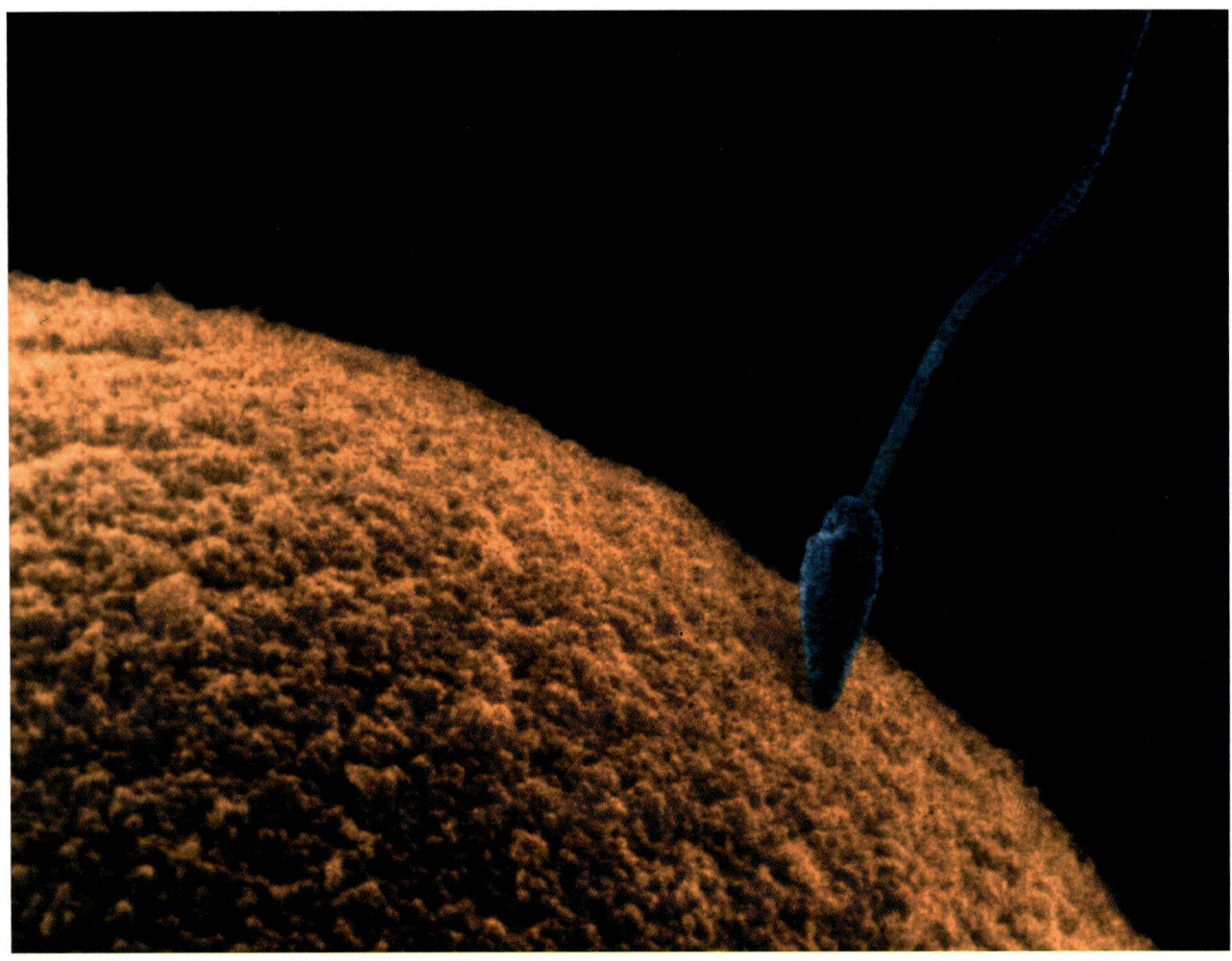

Magnified image of a human sperm penetrating a female egg (ovum). *D.W. Fawcett/Photo Researchers, Inc. Reproduced by permission.*

separates sperm according to the amount of DNA each contains, can allow couples to choose the sex of their child. This is because sperm containing an X **chromosome**, which results in a female embryo, contains more DNA than sperm with a Y chromosome, which would yield a male embryo.

See also Reproductive system; Sexual reproduction.

Fertilizers

A fertilizer is any substance applied to land to increase **plant** growth and produce higher crop yield. Fertilizers may be made from organic material, such as **animal** manure or compost, or it may be chemically manufactured. Manufactured fertilizers contain varying amounts of inorganic **nitrogen**, **phosphorus**, and potassium, all of which are **nutrients** that plants need to grow.

Since the 1950s crop production worldwide has increased dramatically because of the use of fertilizers. In combination with the use of **pesticides** and improved varieties of **crops**, fertilizers have greatly increased the quality and yield of such important foods as corn, **rice**, and **wheat**, as well as fiber crops such as **cotton**. However excessive and improper use of fertilizers have also damaged the environment and affected the health of humans.

It has been estimated that as much as 25% of applied agricultural fertilizer in the United States is carried away as runoff. Fertilizer runoff has contaminated **groundwater** and polluted bodies of **water** in agricultural areas. **Ammonia**, released from the decay of nitrogen fertilizer, causes minor irritation to the **respiratory system**. High concentrations of nitrate in drinking water are more serious, especially for infants, because of **interference** with the ability of the **blood** to transport **oxygen**. High and unsafe nitrate concentrations in drinking water have been reported in all countries that practice intense agriculture, including the United States. The accumulation of nitrogen and phosphorus from chemical fertilizers in waterways has also contributed to the **eutrophication** of lakes and ponds.

Few people would advocate the complete elimination of the use of chemical fertilizers. However, most environmentalists and scientists urge that more efficient ways be found of using fertilizers. For example, some farmers apply up to 40% more fertilizer than they need for specific crops. Frugal applications, occurring at small rates and on an as-needed basis for specific crops, helps to reduce the waste of fertilizer and the **pollution** of runoff and the environment. The use of organic fertilizers, including animal waste, crop residues, grass clippings, and composted food waste, is also encouraged as an alternative to the use of chemical fertilizers.

See also Composting.

Fetal alcohol syndrome

Fetal alcohol **syndrome** (FAS) represents a preventable pattern of clinical abnormalities that develop during embryogenesis (the developmental stages shortly after conception) due to exposure to **alcohol** during pregnancy. FAS is currently the leading cause of **birth defects** and developmental delay, with as many as 12,000 babies born affected in the United States each year. Although the prevalence of FAS is not known for certain, it is estimated that there are between 0.5 to three cases of FAS per 1,000 liveborns in most populations. Alcohol is a **teratogen** in that exposure to the fetus during pregnancy can result in physical malformations of the face and head, growth deficiency and mental retardation. Exposure to excessive amounts of alcohol can even cause embryonic death. It is often difficult to quantify the amount of alcohol that is associated with developmental and physical abnormalities and even subtle amounts might cause varying degrees of developmental delay that are not immediately recognized. For this reason, abstinence from alcohol during pregnancy it is often recommended

Alcohol as a teratogen

Infants, young children, and young adults who were exposed to alcohol during pregnancy often have lower than average **birth** weight and height. Cardinal clinical manifestations include physical abnormalities such as

hypotonia (low muscle tone), smaller than normal skull, irregularities of the face including small **eye** sockets, mid-face hypoplasia (arrested development of the nose, or "flat-face" syndrome), and a very thin upper lip with either an elongated or absent lip indentation. Neurological or central **nervous system** disorders such as hyperactivity, **learning** and intellectual deficits, temper tantrums, short attention and **memory** span, perceptual problems, impulsive **behavior**, seizures, and abnormal electroencephalogram (EEG, or **brain** wave patterns) become apparent after the infant stage. Usually, the severity of the physical manifestations correlates with the severity of the intellectual deficits. Children exposed to alcohol during pregnancy may lack the typical physical features that characterize FAS, but manifest behavioral and neurological defects known as alcohol-related birth defects (ARBD).

Even for FAS-affected children with almost normal intelligence, learning problems become evident by the second grade. By third and fourth grade, affected children experience increasing difficulty with **arithmetic**, organizational skills, and abstract thinking. By the time they reach middle or junior high school, children with FAS display a delayed level of independence and self-control leading to persistent social adjustment problems. Impaired judgment and decision-making abilities often results in an inability to sustain independent living later in life.

The affects of FAS range from severe to mild and correlate to the amount and frequency of alcohol consumed by the pregnant woman and the stage of pregnancy in which drinking takes place. Also, drinking in the first three months of pregnancy may have more serious consequences than drinking the same quantities later in the pregnancy. The recurrence risk in the case of a woman who has had one child with FAS is approximately 25% higher than the general population, increasing as she continues to reproduce. The most severe cases seem to be children of long-term, chronic alcoholic mothers.

A historical and research perspective of FAS

In 1899, the first observation connecting children of alcoholic mothers to the associated risks was shown in a study comparing these children to children of non-alcoholic relatives. However, alcohol consumption during pregnancy was not considered to be a risk to the fetus until it was formally concluded as a risk factor in 1973. During the late 1960s, federally funded studies investigating causes of mental retardation and neurological abnormalities did not include alcohol as a possible teratogen. In fact, intravenous alcohol drips were used to help prevent premature birth. However, by the 1970s, concerns began to grow regarding the adverse effects of toxic substances and diet during pregnancy. Cigarette smoking was known to produce babies of low birth-weight and diminished size and **malnutrition** in pregnant women seriously impaired fetal development. When the effects of prenatal exposure to alcohol were first discovered, studies were launched internationally to determine long-term effects. It is now considered that alcohol consumption during pregnancy causes neurological and behavioral problems that affect the quality of life for the child.

In 1974, a United States study compared the offspring of 23 alcoholic mothers to 46 non-drinking mothers with participants that were defined using the same general characteristics such as geographic region, socioeconomic group, age, race, and marital status. By the age of seven years, children of alcoholic mothers earned lower scores on math, reading, and spelling tests, and lower IQ scores (an average of 81 versus 95). Although 9% of the children born to non-drinking mothers tested 71 or lower 44% of children of alcoholic mothers fell into this range. Similar percentages of reduced weight, height, and head circumference were also observed. A Russian study in 1974 demonstrated that siblings born after their mothers became alcoholics had serious disabilities compared to children born before the mother became an alcoholic. Fourteen of the 23 children in this category were considered mentally retarded. A 1982 Berlin study reported for the first time that FAS caused hyperactivity, distractibility and **speech**, eating, and sleeping disorders. In a study that began in 1974 and followed subjects until the age of 11 years, children of "low risk" mothers who simply drank "socially" (most not even consuming one drink per day after becoming pregnant) found deficits in attention, intelligence, memory, reaction time, learning ability, and behavior were often evident. On average, these problems were more severe in children of women who drank through their entire pregnancy than those who stopped drinking. A 1988 study confirmed earlier findings that the younger child of an alcoholic mother is more likely to be adversely affected than the older child. In 1990, a Swedish study found that as many as 10% of all mildly retarded school-age children in that country suffered from FAS.

Until recently, most studies regarding FAS have been with children. In 1991, a major report done in the United States on FAS among adolescents and adults aged 12–40 years with an average chronological age of 17 years revealed that physical abnormalities of the face and head as well as size and weight deficiencies were less obvious than in early childhood. However, intellectual variation ranged from severely retarded to normal. The average level of intelligence was borderline or mildly retarded, with academic abilities ranging between the second and fourth grade level. Adaptive living skills av-

eraged that of a seven-year-old, with daily living skills rating higher than social skills.

Since the 1990s, studies that involve the specific effects of alcohol on brain cells have been undertaken. In order to understand the specific mechanisms that lead the developmental abnormalities, studies in 2002 demonstrated that in **rodents**, the time of greatest susceptibility to the effects of alcohol coincides with the growth-spurt period. This is a postnatal period in rodents but extends from sixth months of gestation to several years after birth in humans. It is during this time that alcohol can trigger massive programmed brain **cell death** and appears to be the period in which alcohol can have the greatest damaging effects on brain development.

Diagnosis and prevention

Accurate **diagnosis** of FAS is extremely important because affected children require special education to enable them to integrate more easily into society. Mild FAS often goes unnoticed or mimics symptomatology caused by other birth defects. It is important, therefore, that children with abnormalities, especially in cases where the mother consumes alcohol during pregnancy, be fully evaluated by a professional knowledgeable about birth defects. Evidence of the characteristic facial abnormalities, growth retardation, and neurodevelopmental abnormalities are critical for diagnosing FAS. Neuroimaging techniques, such as CT or MRI scans provide a visual representation of the affected areas of the brain and studies using these techniques support observations that alcohol has specific rather than global effects on brain development.

Genetic differences in an individual's ability to metabolize alcohol may contribute to the variability in clinical manifestations. For example, in comparing the effects on the offspring of a woman who ingests moderate amounts of alcohol to the offspring of another woman who drinks the same amount can be **variable**.

Alcohol is a legal psychoactive drug with a high potential for abuse and **addiction**. Because it crosses the placenta (and enters the **blood** stream of the unborn baby), the level of blood alcohol in the baby is directly related to that of the mother, and occurs within just a few short minutes of ingestion. Despite warnings about alcohol consumption by pregnant women placed on the labels of alcoholic beverages initiated during the early 1980s, more than 70,000 children in the ensuing 10 years were born with FAS in the United States. The Centers for Disease Control and Prevention estimates that in the United States, more than 130,000 pregnant women per year consume alcohol at levels known to considerably increase the risk of having a infant with FAS or FAS-related disorder.

See also Childhood diseases; Embryo and embryonic development.

Resources

Books

Steissguth, Ann P., Fred L. Bookstein, Paul D. Sampson, and Helen M. Barr. *The Enduring Effects of Prenatal Alcohol Exposure on Child Development.* Ann Arbor: The University of Michigan Press, 1993.

Stratton, K., C. Howe, and F. Battaglia. *Fetal Alcohol Syndrome: Diagnosis, Epidemiology, Prevention, and Treatment.* Washington, DC: National Academy Press, 1996.

Periodicals

Ebrahim, S.H., S.T. Diekman, L. Floyd, and P. Decoufle. "Comparison of Binge Drinking Among Pregnant and Nonpregnant Women, United States, 1991–1995." *Am J Obstet Gynecol* 180(1 pt. 1):1–7, 1999.

Armstrong, Elizabeth M. "Diagnosing Moral Disorder: the Discovery and Evolution of Fetal Alcohol Syndrome." *Social Science & Medicine* 47, no. 12 (Dec 15, 1998): 2025.

Johnson, Jeannette L., and Michelle Leff. "Children of Substance Abusers: Overview of Research Findings." *Pediatrics* 103, no. 5. (May 1999): 1085.

Other

Fetal Alcohol Syndrome by Anuppa Caleekal B.A., M.Sc. (Health Science and Technology Gallery). 2002 [cited January 15, 2003]. <http:// www.digitalism.org/hst/fetal.html.>.

National Drug Strategy Fetal AS: A National Expert, Advisory Committee on Alcohol, Colleen O'Leary, December 8, 2000 [cited January, 10, 2003]. <http://www.health.gov.au/pubhlth/publicat/document/fetalcsyn.pdf.>.

Marie L. Thompson
Bryan R. Cobb

Fetus *see* **Embryo and embryonic development**

Feynman diagrams

American physicist Richard Feynman's (1918–1988), work and writings were fundamental to the development of quantum electrodynamic theory (QED theory). With regard to QED theory, Feynman is perhaps best remembered for his invention of what are now known as Feynman diagrams, to portray the complex interactions of atomic particles. Moreover, Feynman diagrams allow visual representation and calculation of the ways in which particles can interact through the exchange of virtual photons and thereby provide a tangible picture of processes outside the human capacity for observation. Because Feynman diagrams allow physicists to depict subatomic processes and develop theories regarding particle interactions, the diagrams have become an indispensable and widely used tool in particle **physics**.

Feynman diagrams derive from QED theory. **Quantum electrodynamics (QED)**, is a fundamental scientific theory that is also known as the quantum theory of **light**. QED describes the quantum properties (properties that are conserved and that occur in discrete amounts called quanta) and mechanics associated with the interaction of electromagnetic **radiation** (of which visible light is but one part of an **electromagnetic spectrum**) with **matter**. The practical value of QED rests upon its ability, as set of equations, to allow calculations related to the absorption and **emission** of light by **atoms** and thereby allow scientists to make very accurate predictions regarding the result of the interactions between photons and charged atomic particles (e.g., electrons).

Feynman diagrams are form of shorthand representations that outline the calculations necessary to depict electromagnetic and weak interaction particle processes. QED, as quantum field theory, asserts that the electromagnetic **force** results from the quantum behavior of the **photon**, the fundamental particle responsible for the transmission electromagnetic radiation. According to QED theory, particle vacuums actually consist of electron-positron fields and electron-positron pairs (positrons are the positively charged **antiparticle** to electrons) are created when photons interact with these fields. QED accounts for the subsequent interactions of these electrons, positrons, and photons. Photons, unlike the particles of everyday experience, are **virtual particles** constantly exchanged between charged particles. As virtual particles, photons cannot be observed because they would violate the laws regarding the conservation of **energy** and **momentum**. QED theory therefore specifies that the electromagnetic force results from the constant exchange of virtual photons between charged particles that cause the charged particles to constantly change their **velocity** (speed and/or direction of travel) as they absorb or emit virtual photons. Accordingly, only in their veiled or hidden state do photons act as mediators of force between particles and only under special circumstances do photons become observable as light.

In Feynman time-ordered diagrams, **time** is represented on the x axis and a depicted process begins on the left side of the diagram and ends on the right side of the diagram. All of the lines comprising the diagram represent particular particles. Photons, for example, are represented by wavy lines. Electrons are denoted by straight lines with arrows oriented to the right. Positrons are depicted by a straight line with the arrow oriented to the left. Vertical y axis displacement in Feynman diagrams represents particle **motion**. The representation regarding

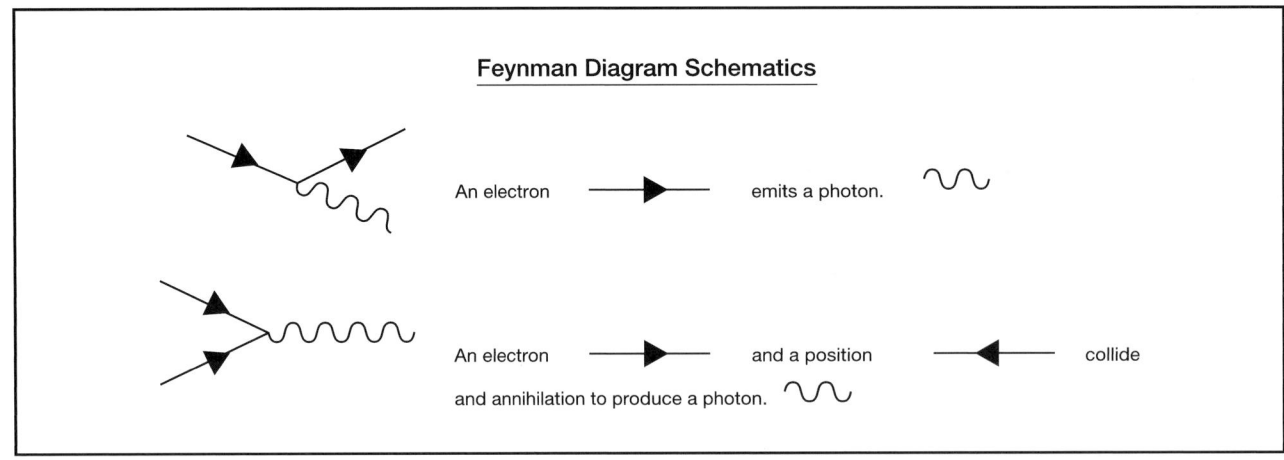

The Feynman diagram schematics. *Courtesy of K. Lee Lerner and the Gale Group.*

motion is highly schematic and does not usually reflect the velocity of a particle.

Feynman diagrams depict electromagnetic interactions as intersections (vertices) of three lines and are able to describe the six possible reactions of the three fundamental QED particles (i.e., the **electron**, positron, and the photon). Accordingly, the diagram can depict the emission and absorption of photons by either electrons or positrons. In addition, is possible to depict the photon production of an electron-positron pair. Lastly, the diagrams can depict the collisions of electrons with positrons that results in their mutual annihilation and the production of a photon.

Feynman diagrams presuppose that energy and momentum are conserved during every interaction and hence at every vertex. Although lines on the diagram can represent virtual particles, all line entering or leaving a Feynman diagram must represent real particles with observable values of energy, momentum and **mass** (which may be **zero**). By definition, virtual particles acting as intermediate particles on the diagrams do not have observable values for energy, momentum or mass.

Although QED theory allows for an infinite number of processes (i.e., an infinite number of interactions) the theory also dictates that interactions of increasing numbers of particles becomes increasingly rare as the number of interacting particles increases. Correspondingly, although Feynman diagrams can accommodate or depict any number of particles, the mathematical complexities increase as the diagrams become more complex.

Feynman developed a set of rules for constructing his diagrams (appropriately named Feynman rules) that allow QED theorists to make very accurate calculations that closely match experimental findings. The Feynman Rules are very simple, but as greater **accuracy** is de-

manded the diagrams become more complex. Feynman diagrams derive from the Feynman path **integral** formulation of **quantum mechanics**. Using asymptotic expansions of the integrals that describe the interactions, physicists are able to calculate the interactions of particles with great (but not unlimited) accuracy. The mathematical formulae associated with the diagrams are added to arrive at what is termed a Feynman amplitude, a value that is subsequently used to calculate various properties and processes (e.g., decay rates).

The development of QED theory and the use of Feynman diagrams allowed scientists to predict how subatomic processes create and destroy particles. Over the last half-century of research in particle physics, QED has become, arguably, the best-tested theory in science history. Most atomic interactions are electromagnetic in nature and, no matter how accurate the equipment yet devised, the predictions made by modern QED theory hold true. Some tests of QED, predictions of the mass of some **subatomic particles**, for example, offer results accurate to six significant figures or more. Although some predictions can be made using one Feynman diagram and a few calculations, others may take hundreds of Feynman diagrams and require the use of high speed computers to complete the necessary calculations.

Resources

Books

Feynman, R. P. *QED: The Strange Theory of Light and Matter.* Princeton. NJ: Princeton University Press, 1985.

Gribbin, John and Mary Gribbin. *Q is for Quantum.* Touchstone Books, 2000.

Johnson, G. W., and M. L. Lapidus. *The Feynman Integral and Feynman's Operational Calculus.* Oxford, England: Oxford University Press, 2000.

Mattuck, R. D. *A Guide to Feynman Diagrams in the Many-Body Problem,* 2nd ed. New York: Dover, 1992.

Periodicals

Feynman, R. P. "Space-Time Approaches to Quantum Electrodynamics." *Phys. Rev.* 76 (1949): 769-789.

Other

Egglescliffe School Physics Department. "Feynman Diagrams. Elementary Particle Physics [cited January 2003]. <http://www.egglescliffe.org.uk/physics/particles/parts/parts1.html>.

K. Lee Lerner

Fiber optics

Optical fiber is a very thin strand of **glass** or plastic capable of transmitting **light** from one point to another. Optical fiber can also be called an optical waveguide, since it is a device that guides light.

Optical fibers consist of a light-carrying core and a cladding surrounding the core. There are generally three types of construction: glass core/cladding, glass core with plastic cladding, or all-plastic fiber. Optical fibers typically have an additional outside coating which surrounds and protects the fiber (see Figure 1).

Commonly available glass fiber diameters range from 8 micron core/125 micron cladding to 100 micron core/140 microns cladding, whereas plastic fibers range from 240 micron core/250 micron cladding to 980 micron core/1,000 micron cladding. The human hair, by comparison, is roughly 100 microns in diameter.

The principles behind fiber optics

Fiber **optics** work on the principle of total internal reflection. Light reaching the boundary between two materials is reflected such that it never leaves the first material. In the case of fiber optics, light is reflected from the optical fiber core-cladding interface in such a way that it propagates down the core of the fiber. This can be explained by a brief discussion of Snell's law of refraction and law of reflection, and a physical quantity known as index of bottom material. According to Snell's law, the light will be bent from its original path to a larger **angle** in the second material. As the incoming, or incident angle increases, so does the refracted angle. For the properly chosen materials, the incident angle can be increased to the point that the ray is refracted at 90 degrees and never escapes the first medium. The equation can be solved to give the incoming, or incident, angle which will result in a refracted angle of 90 degrees.

$$q_2 = \sin^{-1}(n_2 / n_1)$$

This is known as the critical angle (see Figure 2).

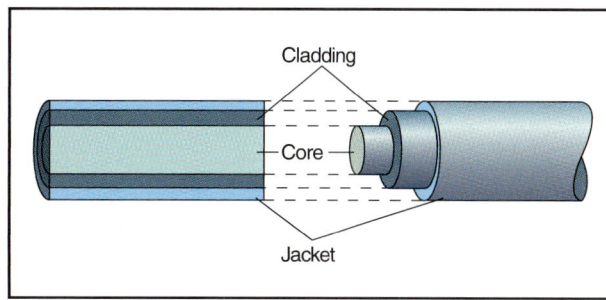

Figure 1. A cable cross section. *Illustration by Hans & Cassidy. Courtesy of Gale Group.*

Light hitting the boundary or interface at angles greater than or equal to this value would never pass into the second material, but would rather undergo total internal reflection.

Now change the model slightly so that the higher index material is sandwiched between two lower index layers (see Figure 3).

Light enters the higher index material, hits the upper interface and is reflected downward, then hits the second interface and is reflected back upward, and so on. Like a marble bouncing off **rails**, light will make its way down the waveguide. This picture essentially corresponds to an optical fiber in cross-section. Light introduced to the fiber at the critical angle will reflect off the interface, and propagate down the fiber.

The second law of **thermodynamics** cannot be disregarded, however. Light will not travel down the fiber indefinitely. The strength of the signal will be reduced, or attenuated. Some light will be absorbed by impurities in the fiber, or scattered out of the core. Modern fibers are made of very pure material so that these effects are minimized, but they cannot be entirely eliminated. Some light will be diverted by microbends and other imperfections in the glass. Recall the law of reflection. If a microbend is encountered by light traveling through the fiber, the light may hit the interface at an angle smaller than the critical angle. If this happens, the light will be reflected out of the core and not continue propagating (see Figure 4).

Fabrication of optical fibers

Optical fibers are fabricated in a multi-step process: preform fabrication, fiber drawing, fiber coating, and spooling. A preform is a giant-sized version of the final fiber, with central core and cladding refractive indices equal to those of desired product. Preform diameters are typically 0.4-1 in (1-2.5 cm). They are produced by one of several variations on chemical vapor

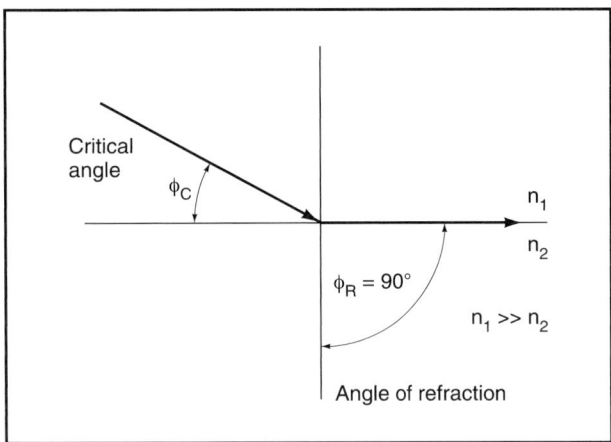

Figure 2. *Illustration by Hans & Cassidy. Courtesy of Gale Group.*

deposition, in which chemicals (primarily silica, with other exotic compounds) are vaporized and allowed to deposit on a tube or rod. The porous form produced is heated to release trapped gases and **water** vapor that might otherwise compromise the performance of the final fiber.

In the drawing stage, the end of the preform is lowered into a furnace heated to roughly 3,632°F (2,000°C). The tip softens until it is drawn down by gravity, shrinking in diameter. Sensors constantly monitor the fiber diameter and concentricity to assure optimal results. An acrylic coating is applied to protect the fiber from damage and preserve its strength. Finally, it is wound onto a takeup spool.

Fiber classifications

Optical fiber falls into three basic classifications: step-index multimode, graded-index multimode, and sin-

gle **mode**. A mode is essentially a path that light can follow down the fiber. Step-index fiber has a core with one index of refraction, and a cladding with a second index.

A graded-index fiber has a varying core index of refraction, and a constant cladding index (see Figure 6).

In general, the beam diameters of light sources for optical fibers are larger than the diameter of the fiber itself. Each fiber has a cone of light that it can propagate, known as the cone of acceptance of the fiber. It is driven by the critical angle of the fiber, which in turn varies according to the refractive index of the material. Light outside the cone of acceptance will not undergo total internal reflection and will not travel down the fiber.

Now, if light in the cone of acceptance is entering the fiber at a variety of angles greater than or equal to the critical angle, then it will travel a number of different paths down the fiber. These paths are called modes, and a fiber that can support multiple paths is classified as multimode. Notice that the light hitting at the smallest possible angle travels a longer path than the light at the largest angle, since the light at the largest angle is closest to a straight line. For step-index multimode fiber in which light travels the same speed everywhere, the rays running the longest path will take longer to get to the destination than the light running the shortest path. Thus a sharp pulse, or packet of light, will be spread out into a broad packet as it travels through the fiber. This is known as modal dispersion and can be a disadvantage in many applications. This type of fiber is used for in-house phone lines and data links.

Graded-index fiber offers one method for minimizing dispersion. The index of refraction of the core of graded index fiber increases toward the center. Remember, the refractive index of a material controls the speed of light traveling through it. Light propagating in the

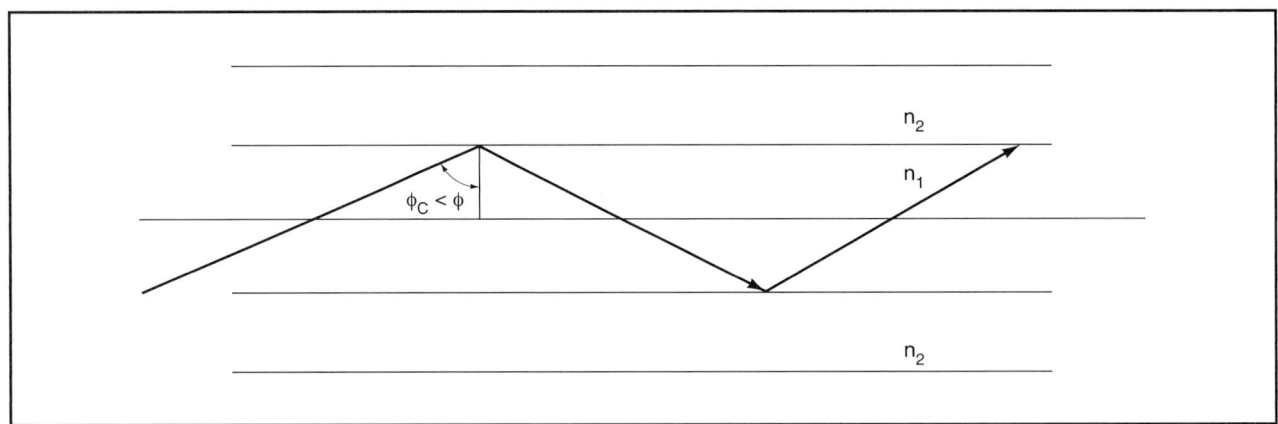

Figure 3. *Illustration by Hans & Cassidy. Courtesy of Gale Group.*

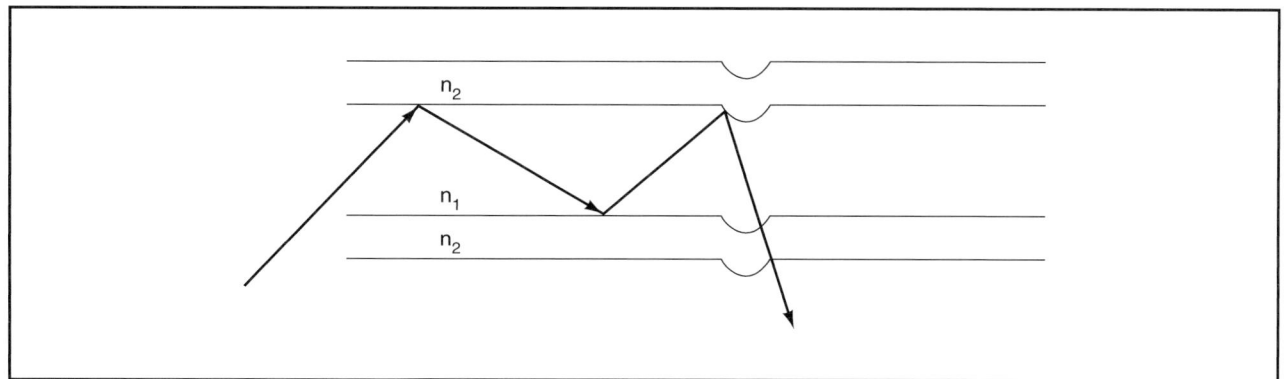

Figure 4. *Illustration by Hans & Cassidy. Courtesy of Gale Group.*

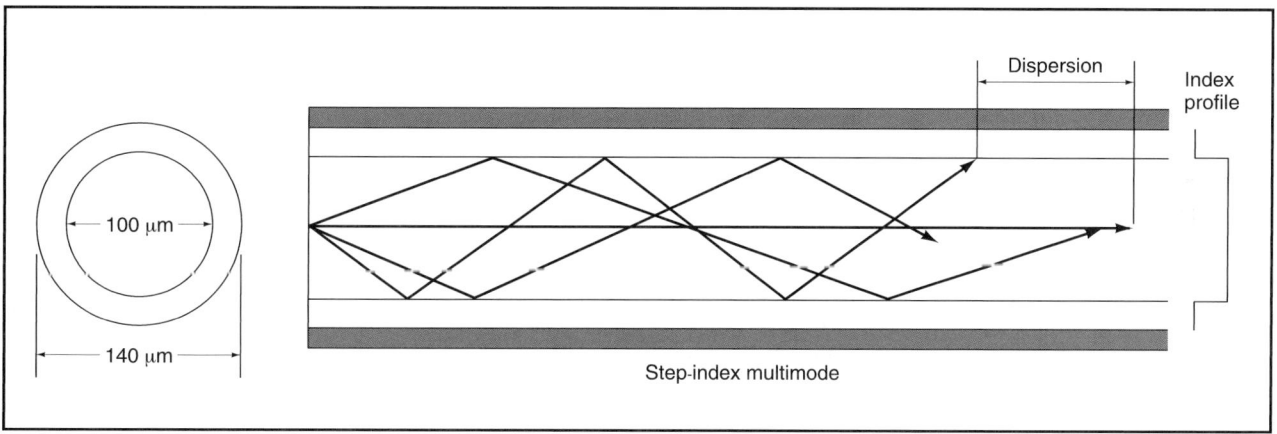

Figure 5. *Illustration by Hans & Cassidy. Courtesy of Gale Group.*

center of the fiber thus goes more slowly than light on the edges. This reduces the pulse spread caused by differing path lengths. While not a perfect transmission, the transmitted pulse is dramatically improved over the step-index multimode fiber output. Graded-index fiber requires very specialized fabrication and is more expensive than step-index multimode. It is commonly used for mid-length communications (see Figure 6).

The best way to avoid modal dispersion, however, is to restrict transmission to only one mode. Single mode fiber is very narrow, with core diameters typically 8 microns, allowing light to propagate in only one mode (see Figure 7). The cone of acceptance is dramatically decreased, however, which makes light injection difficult. Splicing fiber together is more challenging, as well. Single-mode fiber is more costly than step-index multimode but less so than graded-index multimode. Single-mode fiber is used for long **distance** communication such as transoceanic **telephone** lines.

Plastic fiber is available in all three types. It is less expensive and lightweight but experiences more signal

attenuation. It is practical for very short distance applications such as in automobiles.

Fiber optic communications

Why is the propagation of pulses of light through optical fibers important? Voice, video, and data signals can be encoded into light pulses and sent across an optical fiber. Each time someone makes a phone call, a stream of pulses passes through an optical fiber, carrying the information to the person on the other end of the phone line.

A fiber optic communication system generally consists of five elements: the encoder or modulator, the transmitter, the fiber, the detector, and the demodulator (see Figure 8).

Electrical input is first coded into a signal by the modulator, using signal processing techniques. The transmitter converts this electrical signal to an optical signal and launches it into the fiber. The signal experiences attenuation as it travels through the fiber, but it is amplified periodically by repeaters. At the destination, the detector receives the signal, converting it back to an

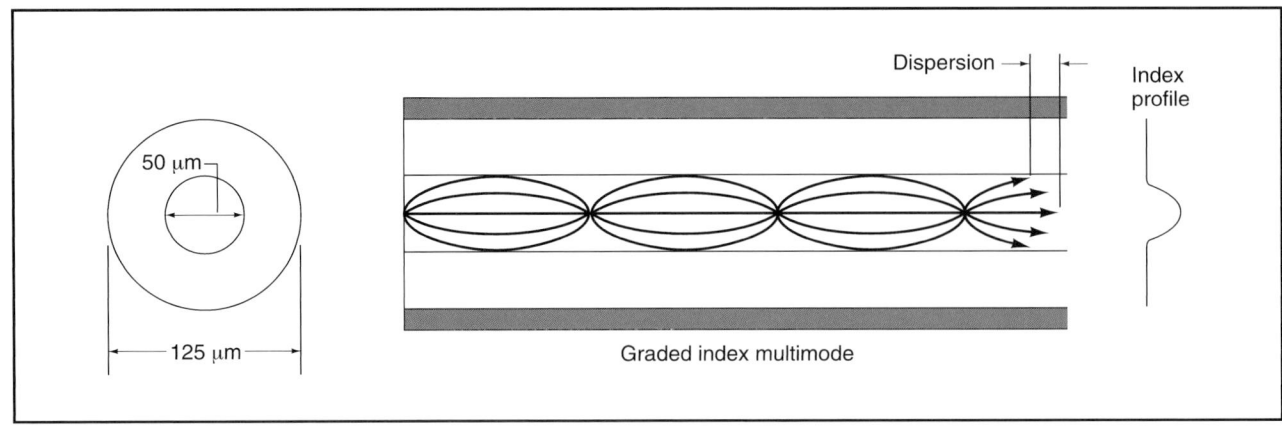

Figure 6. *Illustration by Hans & Cassidy. Courtesy of Gale Group.*

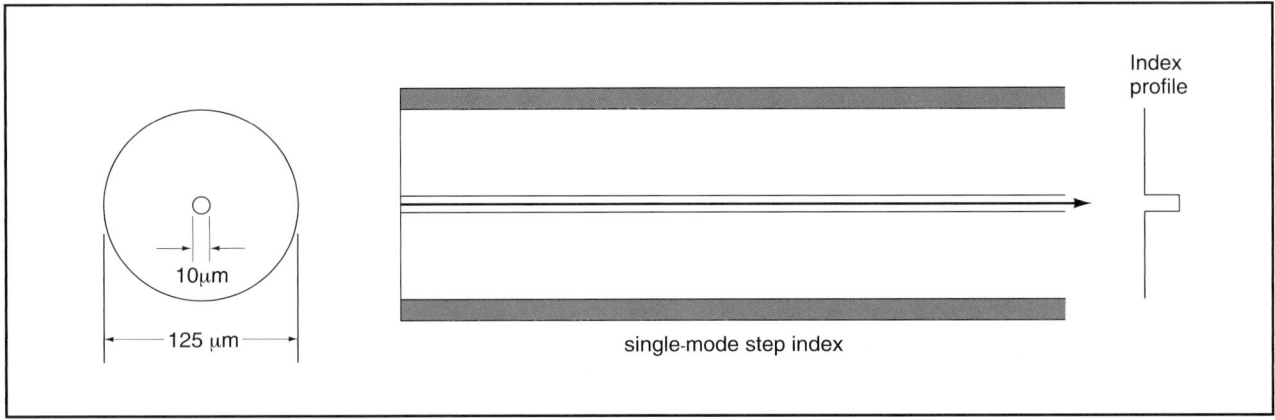

Figure 7. *Illustration by Hans & Cassidy. Courtesy of Gale Group.*

electrical signal. It is sent to the demodulator, which decodes it to obtain the original signal. Finally, the output is sent to the computer or to the handset of your telephone, where electrical signals cause the speaker to vibrate, sending audio waves to your **ear**.

Advantages of fiber optic cable

Communication via optical fiber has a number of advantages over **copper** wire. Wires carrying electrical current are prone to crosstalk, or signal mixing between adjacent wires. In addition, copper wiring can generate sparks, or can overload and grow hot, causing a fire hazard. Because of the electromagnetic properties of current carrying wires, signals being carried by the wire can be decoded undetectably, compromising communications security. Optical fiber carries light, no **electricity**, and so is not subject to any of these problems.

The biggest single advantage that optical fiber offers over copper wire is that of capacity, or bandwidth. With the rising popularity of the Internet, the demand for

bandwidth has grown exponentially. Using a technique called wavelength **division** multiplexing (WDM), optical networks can carry thousands of times as much data as copper-based networks.

Most copper networks incorporate a technique known as time division multiplexing (TDM), in which the system interleaves multiple conversations, sending bits of each down the line serially. For example, the system transmits a few milliseconds of one conversation, then a few milliseconds of the next, a few milliseconds of a the next, then returns to transmit more of the first conversation, and so on. For many years, network designers increased **carrying capacity** by developing **electronics** to transmit shorter, more closely spaced data pulses.

Electronics can operate so quickly, however, and eventually copper wire hit a maximum carrying capacity. To increase bandwidth, network operators had to either lay more copper cable in already packed underground conduits, or seek another method. Enter fiber optics.

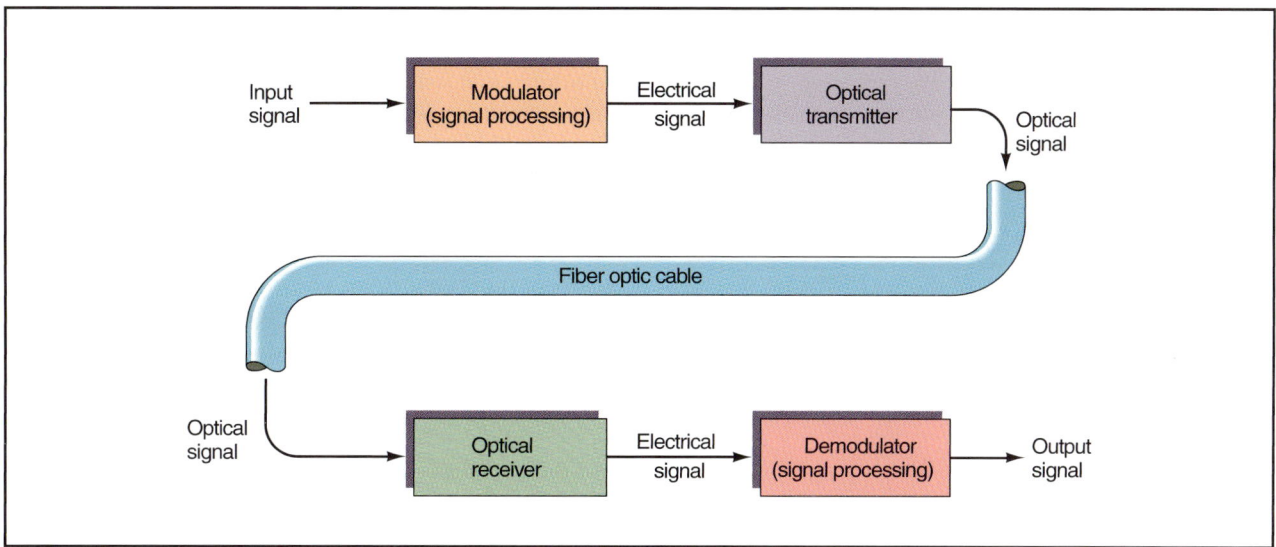

Figure 8. *Illustration by Hans & Cassidy. Courtesy of Gale Group.*

The electrons in copper wire can only carry one stream of time-division multiplexed data at a time. Optical fiber, on the other hand, can transmit light at many wavelengths simultaneously, without **interference** between the different optical signals. Fiber optic networks can thus carry multiple data streams over the same strand of optical fiber, in a technique known as wavelength division multiplexing. A good analogy is a that of a ten-lane expressway compared to a one-lane county road.

Wavelength division multiplexing is an incredibly powerful technique for increasing network capacity. Transmitting data over two wavelengths of light instantly doubles the capacity of the network without any additional optical fiber being added. Transmitting over sixteen wavelengths of light increases the capacity by sixteen times. Commercially deployed WDM systems feature 64 wavelengths, or channels, spaced less than 1 nanometer (nm) apart spectrally. Researchers have built WDM networks that operate over hundreds of channels, sending the equivalent of the amount of data in the Library of Congress across the network in a single second.

Attenuation, dispersion, and optimal communications wavelengths

As mentioned previously, signals carried by optical fiber eventually lose strength, though the loss of attenuation is nowhere near as high as that for copper wire. Singlemode fiber does not incur as much attenuation as multimode fiber. Indeed, signals in high quality fiber can be sent for more than 18.6 mi (30 km) before losing strength. This loss of signal strength is compensated for by installing periodic repeaters on the fiber that receive, amplify, and retransmit the signal. Attenuation is mini-

mized at 1,550 nm, the primary operating wavelength for telecommunications.

Signals in optical fiber also undergo dispersion. One mechanism for this is the modal dispersion already discussed. A second type of dispersion is material dispersion, where different wavelengths of light travel through the fiber at slightly different speeds. Sources used for fiber optics are centered about a primary wavelength, but even with lasers, there is some small amount of variation. At wavelengths around 800 nm, the longer wavelengths travel down the fiber more quickly than the shorter ones. At wavelengths around 1,500 nm, the shorter wavelengths are faster. The **zero** crossing occurs around 1,310 nm: shorter wavelengths travel at about the same speed as the longer ones, resulting in zero material dispersion. A pulse at 1,310 nm sent through an optical fiber would arrive at its destination looking very much like it did initially. Thus, 1,310 nm is an important wavelength for communications.

A third kind of dispersion is wavelength dispersion, occurring primarily in single-mode fiber. A significant amount of the light launched into the fiber is leaked into the cladding. This amount is wavelength dependent and also influences the speed of propagation. High **volume** communications lines have carefully timed spacings between individual signals. Signal speed variation could wreak havoc with data transmission. Imagine your telephone call mixing with someone else's! Fortunately, wavelength dispersion can be minimized by careful design of refractive index.

Based on dispersion and attenuation considerations, then, the optimal wavelengths for fiber-optic communi-

KEY TERMS

Attenuation—Loss of energy in a signal as it passes through the transmission medium.

Cladding—The outer layer of an optical fiber.

Core—The inner portion of an optical fiber.

Dispersion: modal, material, wavelength—Spreading of a signal pulse in an optical fiber.

Index of refraction—The ratio of speed of light in a vacuum to speed of light in a given material.

Modulation—Variation, a method of varying a signal such that information is coded in.

Refraction—The bending of light that occurs when traveling from one medium to another, such as air to glass or air to water.

Total internal reflection—When light reaches an interface between two materials and is reflected back into the first material.

cations are 1,300 and 1,550 nm. Despite the dispersion advantages of operating at 1,310 nm, most modern fiber optic networks operate around 1,550 nm. This wavelength band is particularly important to the WDM networks that dominate the major cross-country fiber optic links because the erbium-doped fiber amplifiers (EDFAs) incorporated in the repeaters provide signal amplification only across a range of wavelengths around 1,550 nm. Thus, most modern fiber optic networks operate around the so-called EDFA window. These signals are in the infrared region of the **spectrum**, that is, these wavelengths are not visible. **Diode** lasers are excellent sources at these wavelengths.

Telecommunications companies have developed singlemode optical fiber that addresses the problem of dispersion. Dispersion-shifted fiber is designed so that the region of maximum dispersion falls outside of the so-called telecommunications window. Although dispersion-shifted fiber is sufficient for basic transmission, in the case of WDM systems with tightly spaced channels, the fiber triggers nonlinear effects between channels that degrades signal integrity. In response, fiber manufacturers have developed non-zero dispersion-shifted fiber that eliminates this problem.

Other applications

Optical fiber has a variety of other applications. Fiber-optic stress and strain sensors are in common use on structures, bridges, and in monitoring industrial

processes. Researchers have developed fiber-optic lasers that are tunable throughout the visible and fiber-optic amplifiers that will further increase capacity in the communications network. Fiber-optic endoscopes allow doctors to perform non-invasive internal examinations, and fiber-optic chemical sensors allow researchers to monitor **pollution** levels remotely.

Fiber-optic technology is continually improving and growing more and more an invisible part of our daily lives. In 1854, when John Tyndall demonstrated light guided in a curved path by a parabolic stream of water, he could never have guessed at the ramifications of his discovery. By the same token, we can only guess what applications will be found for optical fiber in the future.

Resources

Books

Sterling, D. *Technician's Guide to Fiber Optics (AMP)*. Albany, NY: Delmar Publishers Inc., 1987.

Kristin Lewotsky

Fibonacci sequence

The Fibonacci sequence is a series of numbers in which each succeeding number (after the second) is the sum of the previous two. The most famous Fibonacci sequence is 1, 1, 2, 3, 5, 8, 13, 21, 34, 55, 89.... This sequence expresses many naturally occurring relationships in the **plant** world.

History

The Fibonacci sequence was invented by the Italian Leonardo Pisano Bigollo (1180-1250), who is known in mathematical history by several names: Leonardo of Pisa (Pisano means "from Pisa") and Fibonacci (which means "son of Bonacci").

Fibonacci, the son of an Italian businessman from the city of Pisa, grew up in a trading colony in North **Africa** during the Middle Ages. Italians were some of the western world's most proficient traders and merchants during the Middle Ages, and they needed **arithmetic** to keep track of their commercial transactions. Mathematical calculations were made using the Roman numeral system (I, II, III, IV, V, VI, etc.), but that system made it hard to do the addition, **subtraction**, **multiplication**, and **division** that merchants needed to keep track of their transactions.

While growing up in North Africa, Fibonacci learned the more efficient Hindu-Arabic system of arithmetical

TABLE 1							
	Newborns (can't reproduce)		One-month-olds (can't reproduce)		Mature Pairs (can reproduce)		Total Pairs
Month 1	1	+	0	+	0	=	1
Month 2	0	+	1	+	0	=	1
Month 3	1	+	0	+	1	=	2
Month 4	1	+	1	+	1	=	3
Month 5	2	+	1	+	2	=	5
Month 6	3	+	2	+	3	=	8
Month 7	5	+	3	+	5	=	13
Month 8	8	+	5	+	8	=	21
Month 9	13	+	8	+	13	=	34
Month 10	21	+	13	+	21	=	55

Each number in the table represents a pair of rabbits. Each pair of rabbits can only give birth after its first month of life. Beginning in the third month, the number in the "Mature pairs" column represents the number of pairs that can bear rabbits. The numbers in the "Total Pairs" column represent the Fibonacci sequence.

notation (1, 2, 3, 4...) from an Arab teacher. In 1202, he published his knowledge in a famous book called the *Liber Abaci* (which means the "book of the abacus," even though it had nothing to do with the **abacus**). The *Liber Abaci* showed how superior the Hindu-Arabic arithmetic system was to the Roman numeral system, and it showed how the Hindu-Arabic system of arithmetic could be applied to benefit Italian merchants.

The Fibonacci sequence was the outcome of a mathematical problem about rabbit breeding that was posed in the *Liber Abaci*. The problem was this: Beginning with a single pair of rabbits (one male and one female), how many pairs of rabbits will be born in a year, assuming that every month each male and female rabbit gives **birth** to a new pair of rabbits, and the new pair of rabbits itself starts giving birth to additional pairs of rabbits after the first month of their birth?

Table 1 illustrates one way of looking at Fibonacci's solution to this problem.

Other Fibonacci sequences

Although the most famous Fibonacci sequence is 1, 1, 2, 3, 5, 8, 13, 21, 34, 55..., a Fibonacci sequence may be *any* series of numbers in which each succeeding number (after the second) is the sum of the previous two. That means that the specific numbers in a Fibonacci series depend upon the initial numbers. Thus, if a series be-

gins with 3, then the subsequent series would be as follows: 3, 3, 6, 9, 15, 24, 39, 63, 102, and so on.

A Fibonacci series can also be based on something other than an integer (a whole number). For example, the series 0.1, 0.1, 0.2, 0.3, 0.5, 0.8, 1.3, 2.1, 3.4, 5.5, and so on, is also a Fibonacci sequence.

The Fibonacci sequence in nature

The Fibonacci sequence appears in unexpected places such as in the growth of plants, especially in the number of petals on flowers, in the arrangement of leaves on a plant stem, and in the number of rows of **seeds** in a sunflower.

For example, although there are thousands of kinds of flowers, there are relatively few consistent sets of numbers of petals on flowers. Some flowers have 3 petals; others have 5 petals; still others have 8 petals; and others have 13, 21, 34, 55, or 89 petals. There are exceptions and variations in these patterns, but they are comparatively few. All of these numbers observed in the **flower** petals—3, 5, 8, 13, 21, 34, 55, 89—appear in the Fibonacci series.

Similarly, the configurations of seeds in a giant sunflower and the configuration of rigid, spiny scales in pine cones also conform with the Fibonacci series. The corkscrew spirals of seeds that radiate outward from the center of a sunflower are most often 34 and 55 rows of seeds in opposite directions, or 55 and 89 rows of seeds

KEY TERMS

. . . .

Phyllotaxis—The arrangement of the leaves of a plant on a stem or axis.

Radially—Diverging outward from a center, as spokes do from a wagon wheel or as light does from the sun.

in opposite directions, or even 89 and 144 rows of seeds in opposite directions. The number of rows of the scales in the spirals that radiate upwards in opposite directions from the base in a pine cone are almost always the lower numbers in the Fibonacci sequence—3, 5, and 8.

Why are Fibonacci numbers in plant growth so common? One clue appears in Fibonacci's original ideas about the **rate** of increase in rabbit populations. Given his **time** frame and growth cycle, Fibonacci's sequence represented the most efficient rate of breeding that the rabbits could have if other conditions were ideal. The same conditions may also apply to the propagation of seeds or petals in flowers. That is, these phenomena may be an expression of nature's efficiency. As each row of seeds in a sunflower or each row of scales in a pine cone grows radially away from the center, it tries to grow the maximum number of seeds (or scales) in the smallest **space**. The Fibonacci sequence may simply express the most efficient packing of the seeds (or scales) in the space available.

See also Integers; Numeration systems.

Resources

Books

Gies, Joseph, and Frances Gies. *Leonardo of Pisa and the New Mathematics of the Middle Ages.* New York: Thomas Y. Crowell Co., 1969.
Swetz, Frank J. *Capitalism & Arithmetic: The New Math of the 15th Century.* LaSalle, Illinois: Open Court Press, 1987.

Periodicals

Stewart, Ian. "Mathematical Recreations: Daisy, Daisy, Give Me Your Answer, Do." *Scientific American* 272.1 (January 1995): 96-99.
Stewart, Ian. "Mathematical Recreations: Fibonacci Forgeries." *Scientific American* 272.5 (May 1995): 102-105.

Patrick Moore

Field

A *field* is the name given to a pair of numbers and a set of operations which together satisfy several specific

laws. A familiar example of a field is the set of rational numbers and the operations addition and **multiplication**. An example of a set of numbers that is not a field is the set of **integers**. It is an "integral domain." It is not a field because it lacks multiplicative inverses. Without multiplicative inverses, **division** may be impossible.

The elements of a field obey the following laws:

1. Closure laws: a + b and ab are unique elements in the field.

2. Commutative laws: a + b = b + a and ab = ba.

3. Associative laws: a + (b + c) = (a + b) + c and a(bc) = (ab)c.

4. Identity laws: there exist elements 0 and 1 such that a + 0 = a and a $\times$ 1 = a.

5. Inverse laws: for every a there exists an element -a such that a + (-a) = 0, and for every a $\neq$ 0 there exists an element a^{-1} such that a $\times$ a^{-1} = 1.

6. Distributive law: a(b + c) = ab + ac.

Rational numbers (which are numbers that can be expressed as the **ratio** a/b of an integer a and a natural number b) obey all these laws. They obey closure because the rules for adding and multiplying fractions, a/b + c/d = (ad + cb)/bd and (a/b)(c/d) = (ac)/(bd), convert these operations into adding and multiplying integers which are closed. They are commutative and associative because integers are commutative and associative. The ratio 0/1 is an additive identity, and the ratio 1/1 is a multiplicative identity. The ratios a/b and -a/b are additive inverses, and a/b and b/a (a, b $\neq$ 0) are multiplicative inverses. The rules for adding and multiplying fractions, together with the distributive law for integers, make the distributive law hold for rational numbers as well. Because the rational numbers obey all the laws, they form a field.

The rational numbers constitute the most widely used field, but there are others. The set of **real numbers** is a field. The set of **complex numbers** (numbers of the form a + bi, where a and b are real numbers, and i^2 = -1) is also a field.

Although all the fields named above have an infinite number of elements in them, a set with only a finite number of elements can, under the right circumstances, be a field. For example, the set constitutes a field when addition and multiplication are defined by these tables:

+	0	1		x	0	1
0	0	1		0	0	0
1	1	0		1	0	1

With such a small number of elements, one can check that all the laws are obeyed by simply running down all the possibilities. For instance, the **symmetry** of

KEY TERMS

· ·

Field—A set of numbers and operations exemplified by the rational numbers and the operations of addition, subtraction, multiplication, and division.

Integral domain—A set of numbers and operations exemplified by the integers and the operations addition, subtraction, and multiplication.

the tables show that the commutative laws are obeyed. Verifying associativity and distributivity is a little tedious, but it can be done. The identity laws can be verified by looking at the tables. Where things become interesting is in finding inverses, since the addition table has no **negative** elements in it, and the multiplication table, no fractions. Two additive inverses have to add up to 0. According to the addition table 1 + 1 is 0; so 1, curiously, is its own additive inverse. The multiplication table is less remarkable. **Zero** never has a multiplicative inverse, and even in ordinary **arithmetic**, 1 is its own multiplicative inverse, as it is here.

This example is not as outlandish as one might think. If one replaces 0 with "even" and 1 with "odd," the resulting tables are the familiar **parity** tables for catching mistakes in arithmetic.

One interesting situation arises where an algebraic number such as $\sqrt{2}$ is used. (An algebraic number is one which is the root of a polynomial equation.) If one creates the set of numbers of the form $a + b\sqrt{2}$, where a and b are rational, this set constitutes a field. Every sum, product, difference, or quotient (except, of course, $(a + b\sqrt{2})/0$) can be expressed as a number in that form. In fact, when one learns to rationalize the denominator in an expression such as $1/(1 - \sqrt{2})$ that is what is going on. The set of such elements therefore form another field which is called an "algebraic extension" of the original field.

J. Paul Moulton

Resources

Books

Birkhoff, Garrett, and Saunders MacLane. *A Survey of Modern Algebra.* New York: Macmillan Co., 1947.

McCoy, Neal H. *Rings and Ideals.* Washington, DC: The Mathematical Association of America, 1948.

Singh, Jagjit, *Great Ideas of Modern Mathematics.* New York: Dover Publications, 1959.

Stein, Sherman K. *Mathematics, the Man-Made Universe.* San Francisco: W. H. Freeman, 1969.

Fig *see* **Mulberry family (Moraceae)**

Figurative numbers

Figurative numbers are numbers which can be represented by dots arranged in various geometric patterns. For example, triangular numbers are represented by the patterns shown in Figure 1.

The numbers they represent are 1, 3, 6, 10, and so on.

Figurative numbers were first studied by the mathematician Pythagoras in the sixth century B.C. and by the Pythagoreans, who were his followers. These numbers were studied, as were many kinds of numbers, for the sake of their supposed mystical properties rather than for their practical value. The study of figurative numbers continues to be a source of interest to both amateur and professional mathematicians.

Figurative numbers also include the square numbers which can be represented by square arrays of dots, as shown in Figure 2.

The first few square numbers are 1, 4, 9, 16, 25, etc.

There are pentagonal numbers based on pentagonal arrays. Figure 3 shows the fourth pentagonal number, 22.

Other pentagonal numbers are 1, 5, 12, 22, 35, and so on.

There is, of course, no limit to the number of polygonal arrays into which dots may be fitted. There are hexagonal numbers, heptagonal numbers, octagonal numbers, and, in general, n-gonal numbers, where n can be any number greater than 2.

One reason that figurative numbers have the appeal they do is that they can be studied both algebraically and geometrically. Properties that might be hard to discover by algebraic techniques alone are often revealed by simply looking at the figures, and this is what we shall do, first with triangular numbers.

If we denote by T_n the n-th triangle number, Figure 1 shows us that $T_1 = 1$ $T_2 = T_1 + 2$ $T_3 = T_2 + 3$ $T_4 = T_3 + 4$.

$$T_n = T_{n-1} + n$$

These formulas are *recursive*. To compute T_{10} one must compute T_9. To compute T_9 one has to compute T_8, and so on. For small values of n this is not hard to do: $T_{10} = 10 + (9 + (8 + (7 + (6 + (5 + (4 + (3 + (2 + 1))))))))) = 55$.

For larger values of n, or for general values, a formula that gives T_n directly would be useful. Here the use of the figures themselves comes into play: From Figure 4 one can see that $2T_3 = (3)(4)$; so $T_3 = 12/2$ or 6.

The same trick can be applied to any triangular number: $2T_n = n(n + 1)$ $T_n = n(n+ 1)/2$ When n = 10, $T_{10} = (10)(11)/2$, or 55, as before.

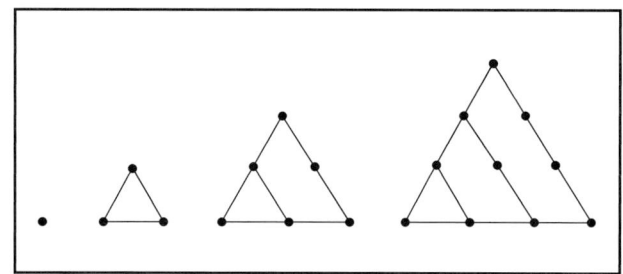

Figure 1. *Illustration by Hans & Cassidy. Courtesy of Gale Group.*

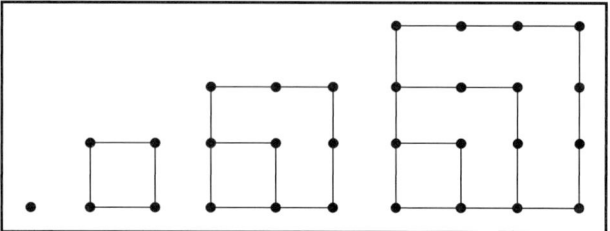

Figure 2. *Illustration by Hans & Cassidy. Courtesy of Gale Group.*

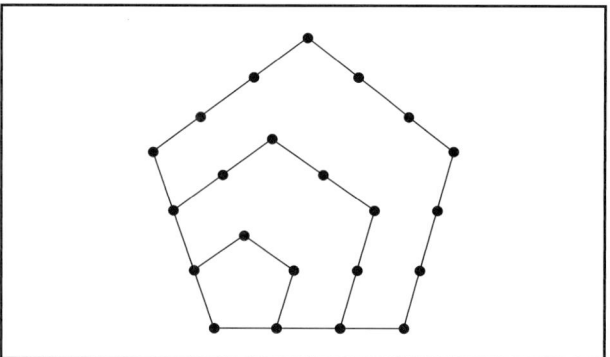

Figure 3. *Illustration by Hans & Cassidy. Courtesy of Gale Group.*

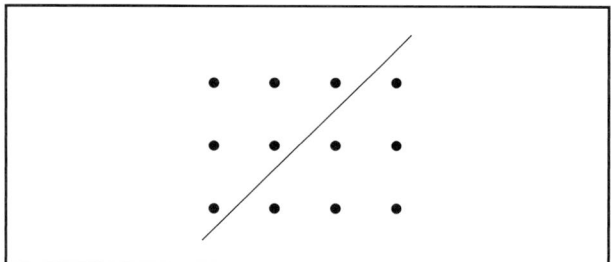

Figure 4. *Illustration by Hans & Cassidy. Courtesy of Gale Group.*

In the case of square numbers, the general formula for the n-th square number is easy to derive, It is simply n^2. In fact, the very name given to n^2 reflects the fact that it is a square number. The recursive formulas for the var-

ious square numbers are a little less obvious. To derive them, one can turn again to the figures themselves. Since each square can be obtained from its predecessor by adding a row of dots across the top and a column of dots down the side, including one dot in the corner, one gets the recursive pattern $S_1 = 1$ $S_2 = S_1 + 3$ $S_3 = S_2 + 5$ $S_4 = S_3 + 7...$

$$S_n = S_{n-1} + 2n - 1$$

or, alternatively

$$S_{n+1} = S_n + 2n + 1$$

Thus $S_8 = 15 + (13 + (11 + (9 + (7 + (5 + (3 + 1))))))$ or 64.

Because humans are so fond of arranging things in rows and columns, including themselves, square numbers in use are not hard to find. Tic-tac-toe has S_3 squares; chess and checkers have S_8. S_{19} has been found to be the ideal number of points, says a text on the oriental game of "go," on which to play the game.

One of the less obvious places where square numbers-or more correctly, the recursive formulas for generating them-show up is in one of the algorithms for computing square roots. This is the **algorithm** based on the formula $(a + b)^2 = a^2 + 2ab + b^2$. When this formula is used, b is not 1. In fact, at each stage of the computing process the size of b is systematically decreased. The *process* however parallels that of finding S_{n+1} recursively from S_n. This becomes apparent when n and 1 are substituted for a and b in the formula: $(n + 1)^2 = n^2 + 2n + 1$. This translates into $S_{n+1} = S_n + 2n + 1$, which is the formula given earlier.

Formulas for pentagonal numbers are trickier to discover, both the general formula and the recursive formulas. But again the availability of geometric figures helps. By examining the array in Figure 5, one can come up with the following recursive formulas, where P_n represents the n-th pentagonal number: $P_1 = 1$ $P_2 = P_1 + 4$ $P_3 = P_2 + 7$ $P_4 = P_3 + 10$. One can guess the formula for P_n: $P_n = P_{n-1} + 3n - 2$ and this is correct. At each stage in going from P_{n-1} to P_n one adds three sides of n dots each to the existing pentagon, but two of those dots are common to two sides and should be discounted.

To compute P_7 recursively we have $19 + (16 + (13 + (10 + (7 + (4 + 1)))))$, which adds up to 70.

To find a general formula, we can pull another trick. We can cut up the array in Figure 5 along the dotted lines. When we do this we have $P_5 = T_5 + 2T_4$ or more generally $P_n = T_n + 2T_{n-1}$.

If we substitute algebraic expressions for the triangular numbers and simplify the result we come up with $P_n = (3n^2 - n)/2$.

The fact that pentagonal numbers can be cut into triangular numbers makes one wonder if other polygonal numbers can, too. Square numbers can, and Figure 6 shows an example.

It yields $T_5 + T_4$, or in general $S_n = T_n + T_{n-1}$. Arranging these formulas for triangular dissections suggests a pattern: $T_n = T_n$ $S_n = T_n + T_{n-1}$ $P_n = T_n + 2T_{n-1}$ $H_n = T_n + 3T_{n-1}$ where H_n is the n-th hexagonal number. If we check this for H_4 in Figure 7, we find it to be correct.

In general, if N_k represents the k-th polygonal number for a polygon of N sides (an N-gon) $N_k = T_k + (N - 3)T_{k-1}$.

The Pythagoreans were also concerned with "oblong" arrays (here "oblong" has a more limited meaning than usual), having one column more than its rows.

In this case the concern was not with the total number of dots but with the **ratio** of dots per row to dots per column. In the smallest array, this ratio is 2:1. In the next array it is 3:2; and the third, 4:3 (Figure 8). If the arrays were further enlarged, the ratios would change in a regular way: 5:4, 6:5, and so on.

These ratios were related, by Pythagoreans, to music. If, on a stringed instrument, two notes were played whose frequencies were in the ratio of 2:1, those notes would be an octave apart and would sound harmonious when played together. (Actually the Pythagoreans went by the lengths of the strings, but if the string lengths were in the ratio 1:2, the frequencies would be in the ratio 2:1.) If the ratio of the frequencies were 3:2, the notes would be a perfect fifth apart and would also sound harmonious (in fact, violinists use the harmony of perfect fifths to tune the strings of the violin, which sound a perfect fifth apart). Notes in the ratio 4:3 were also considered harmonious. Other ratios were thought to result in discordant notes.

The Pythagoreans went well beyond this in developing their musical theories, but what is particularly interesting is that they based a really fundamental musical idea on an array of dots. Of course, for reasons having little to do with figurative numbers, they got it right.

Figurative numbers are not confined to those associated with **plane** figures. One can have pyramidal numbers based on figures made up of layers of dots, for example a **tetrahedron** made up of layers of triangular arrays. Such a tetrahedron would have T_n dots in the first layer, T_{n-1} dots in the next, and so on. If there were four such layers, the tetrahedral number it represents would be 1 + 3 + 6 + 10, or 20.

A general formula applicable to a **pyramid** whose base is an N-gon with N_k points in it is $(k + 1)(2N_k + k)/6$. In the example above N = 3 and k = 4. Using these values in the formula gives 5(20 + 4)/6 = 20.

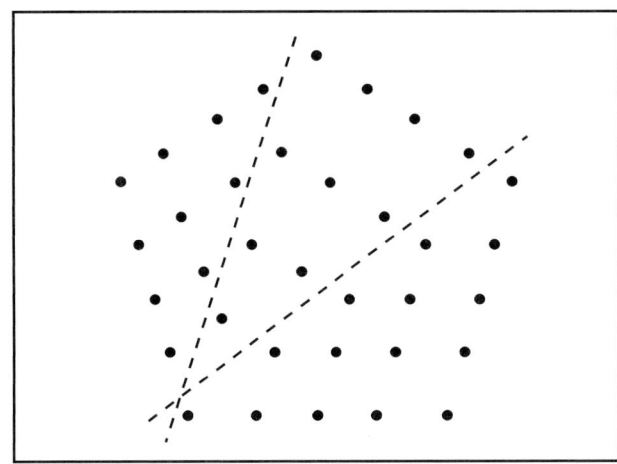

Figure 5. *Illustration by Hans & Cassidy. Courtesy of Gale Group.*

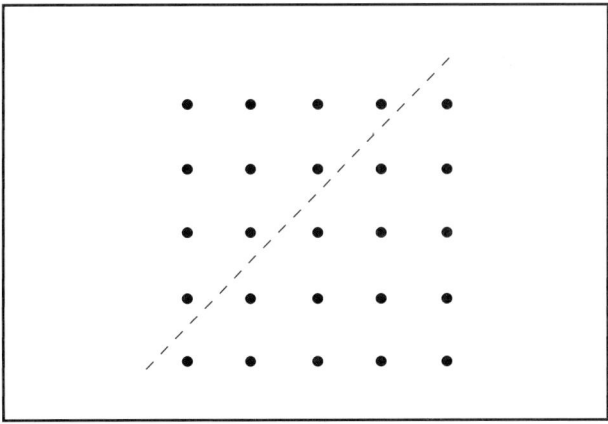

Figure 6. *Illustration by Hans & Cassidy. Courtesy of Gale Group.*

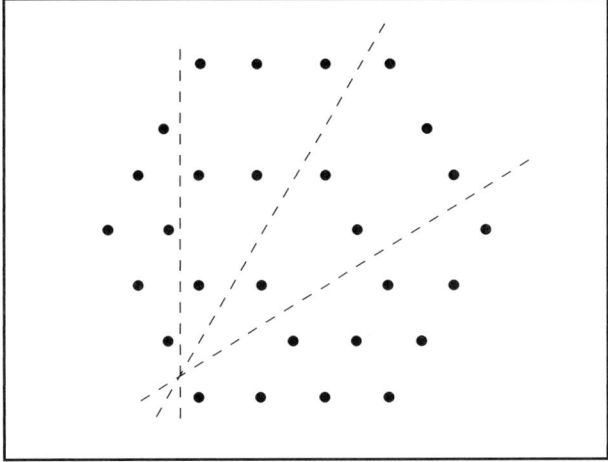

Figure 7. *Illustration by Hans & Cassidy. Courtesy of Gale Group.*

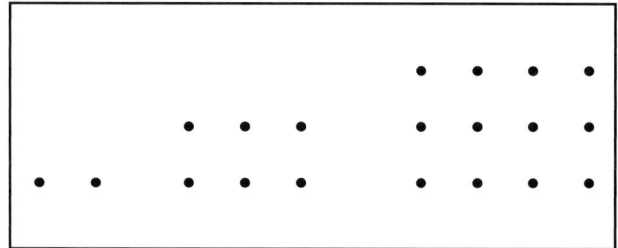

Figure 8. *Illustration by Hans & Cassidy. Courtesy of Gale Group.*

If the base is a square with 5 points on a side, N = 4 and k = 5. Then the total number of points is 6(50 + 5)/6, or 55. To arrange cannon balls in a pyramidal stack with a square array of 25 balls for a base, one would need 55 balls.

See also Imaginary number; Natural numbers; Transcendental numbers.

Resources

Books

Gullberg, Jan, and Peter Hilton. *Mathematics: From the Birth of Numbers.* W.W. Norton & Company, 1997.
Rosen, Kenneth. *Elementary Number Theory and Its Applications.* 4th ed. Boston: Addison-Wesley, 2000.

J. Paul Moulton

Filtration

Filtration is the process by which solid materials are removed from a fluid mixture, either a gas or liquid mixture. One of the most familiar kinds of filtration is that which students of **chemistry** encounter in their early laboratory experiences. In these experiences, suspensions of a solid precipitate in **water** are passed through filter **paper** supported in a **glass** funnel. The filter paper traps solid particles in the mixture, while a clear **solution** passes through the filter, down the funnel, and into a receiving container.

Filtration is carried out for one of two general purposes: in order to capture the solid material suspended in the fluid or in order to clarify the fluid in which the solid is suspended. The general principle is the same in either case although the specific filtration system employed may differ depending on which of these objectives is intended.

In the world outside of chemistry laboratories, a very great variety of filtration systems are available. These systems can be categorized according to the fluids on which they operate (gaseous or liquid) and the driving

force that moves fluids through them (gravity, **vacuum**, or **pressure**). They can also be sub-divided depending on the type of material used as a filter.

Liquid filtration

Liquid filtration occurs when a suspension of a solid in a liquid passes through a filter. That process takes place when the liquid is pulled through the filter by gravitational force (as in the laboratory example mentioned above) or is forced through the filter by some applied pressure or by a pressure differential supplied by the existence of a vacuum.

One of the most familiar gravity filters in the industrial world is that used for the purification of water. A water filtration system generally makes use of a thick layer of granular materials, such as **sand**, gravel, and charcoal. Such a filter may be many feet thick and is known, therefore, as a deep-bed filter. When impure water passes through such a filter, suspended solids are removed, allowing relatively pure water to be collected at the bottom of the filter. In commercial water purification plants, the deep-bed filter may be modified so as to remove other impurities. For example, dissolved gases that add unpleasant odors and taste to the water may be removed if activated **carbon** (finely divided charcoal) is included in the filter. The gases responsible for offensive odor and taste are absorbed on particles of charcoal, leaving an effluent that is nearly odorless and tasteless.

The filtration of smaller volumes of solution than those normally encountered in a water filtration **plant** is often accomplished by means of positive pressure systems. A positive pressure system is one in which the fluid to be filtered is forced through a filtering medium by an external pressure. A number of variations on this concept are commercially available. For example, in one type of apparatus, the fluid to be filtered is introduced under pressure at one end of a horizontal tank and then forced through a series of vertical plates covered with thin filtering cloths. As the fluid passes through these filters, solids are removed and collect on the surface of the cloths. The material that builds up on the filters is known as a cake, and the filters themselves are sometimes called cake filters.

In another type of pressure filter a series of filter plates is arranged one above the other in a cylindrical tank. Liquid is pumped into the tank under pressure, which forces it downward through the filters. Again, solids suspended in the liquid collect on the filters while the clear liquid passes out of the tank through a drain pipe in the center of the unit.

A variety of vacuum filters have also been designed. In a vacuum filter, the liquid to be separated is poured onto a filtering medium and a vacuum is created below the medium. **Atmospheric pressure** above the filter then forces the liquid through the medium with suspended solids collecting on the filter and the clear liquid passing through.

Probably the most common variation of the vacuum filter is the continuous rotary vacuum filter. In this device, a drum with a perforated surface rotates on a horizontal axis. A cloth covering the drum acts as the filter. The lower part of the drum is submerged in the liquid to be separated and a vacuum is maintained within the drum. As the drum rotates, it passes through the liquid and atmospheric pressure forces liquid into its interior. Solids suspended in the liquid are removed by the filter and collect as a cake on the outside of the drum. Because the cake can constantly be removed by a stream of water, the drum can continue to rotate and filter the suspension in the pan below it.

Clarifying filters

The filters described thus far are used most commonly to collect a solid material suspended in a liquid. Clarifying filters, on the other hand, are designed to collect a liquid that is as free from solid impurities as possible. The most important feature of a clarifying filter, then, is the filter itself. It must be constructed in such a way as to remove the very smallest particles suspended in the liquid. A number of systems have been developed to achieve this objective. Some rely on the use of wires or fibers very closely spaced together. Others make use of finely powdered materials, such as diatomaceous **earth**.

Gas filtration

Examples of gas filtration are common in everyday life. For example, every time a vacuum cleaner runs, it passes a stream of dust-filled air through a filtering bag inside the machine. Solid particles are trapped within the bag, while clean air passes out through the machine.

The removal of solid particles from air and other gases is a common problem in society. Air conditioning and heating systems today not only change the **temperature** of a room, but also remove dust, pollen, and other particles that may cause respiratory problems for humans.

The cleansing of waste gases is also a significant problem for many industrial operations. Effluent gases from coal- and oil-burning power plants, for example, usually contain solid particles that cause **air pollution** and **acid rain**. One way to remove these particles is to pass them through a filtering system that physically collects the particles leaving a clean (or cleaner) effluent gas.

See also Bioremediation; Hydrologic cycle; Sustainable development; Water pollution.

KEY TERMS

Cake filter—A type of filter on which solid materials removed from a suspension collect in a quantity that can be physically removed.

Clarification—The process by which unwanted solid materials are removed from a suspension in order to produce a very clear liquid.

Diatomaceous earth—A finely divided rock-like material obtained from the decay of tiny marine organisms known as diatoms.

Fluid—A gas or liquid.

Precipitate—A solid material that is formed by some physical or chemical process within a fluid.

Suspension—A temporary mixture of a solid in a gas or liquid from which the solid will eventually settle out.

Resources

Books

Filters and Filtration Handbook. 3rd ed. Tulsa, OK: Penn Well Books, 1992.

Orr, Clyde, ed. *Filtration Principles and Practices.* Ann Arbor, MI: Books on Demand, 1992.

Trefil, James. *Encyclopedia of Science and Technology.* The Reference Works, Inc., 2001.

David E. Newton

Finches

Finches are **species** of arboreal, perching **birds** that make up the large, widespread family, the Fringillidae. There are three subfamilies in this group, the largest being the Carduelinae or carduacline finches, a geographically widespread group that contains about 122 species. The subfamily Fringillinae or fringillid finches consists of three species breeding in woodlands of Eurasia, while the Drepanidinae or Hawaiian **honeycreepers** (which are sometimes treated as a separate family, the Drepanididae) are 23 species of native tropical **forests** on the Hawaiian Islands.

Species of finches occur in North and **South America**, **Africa**, **Europe**, and **Asia**. In addition, a few species have been introduced beyond their natural range, to Australasia.

Species of finches can occur in a wide range of habitats, including **desert**, steppe, **tundra**, savannas, woodlands, and closed forests. Finches that breed in highly seasonal, northern habitats are migratory, spending their non-breeding season in relatively southern places. A few other northern species wander extensively in search of places with abundant food, and breed there. Other species of more southerly finches tend to be residents in their **habitat**.

It should be noted that in its common usage, the word "finch" is a taxonomically ambiguous term. Various types of seed-eating birds with conical bills are commonly referred to as finches, including species in families other than the Fringillidae. For example, the zebra finch (*Taeniopygia guttata*) of Asia is in the waxbill family, Estrildidae, and the snow finch (*Montifringilla nivalis*) is in the weaver-finch family, Ploceidae. The "typical" finches, however, are species in the family Fringillidae, and these are the birds that are discussed in this entry.

Biology of cardueline finches

The cardueline or typical finches are smallish birds, with a strong, conical beak, well designed for extracting and eating **seeds** from cones and **fruits**. These finches also have a crop, an oesophageal pouch used to store and soften ingested seeds, and a muscular gizzard for crushing their major food of seeds. They eat buds, soft fruits, and some **insects**.

Most species of finches are sexually dimorphic, with male birds having relatively bright and colorful plumage, and the female being more drab, and cryptic. This coloration helps the female to blend into her surroundings while incubating her eggs, a chore that is not shared by the brightly colored male.

Species of finches occur in Europe, Asia, Africa, and the Americas. However, the greatest diversity of species occurs in Eurasia and Africa.

Cardueline finches typically occur in flocks during their non-breeding season. Many of the northern species of finches are highly irruptive in their abundance, sometimes occurring in unpredictably large populations, especially during their non-breeding season. These events of great abundance are associated with a local profusion of food, for example, at times when **conifer** trees are producing large quantities of cones and seeds. Crossbills, siskins, and redpolls are especially notable in this respect.

The flight of cardueline finches is strong, and often occurs in an up-and-down, undulating pattern. This flight pattern may be exaggerated during nuptial and territorial displays.

Male cardueline finches defend a territory during their breeding season, mostly by singing, often while the bird is in flight. The songs of most species are loud and melodious. The nest is cup-shaped, and may be placed in a **tree**, shrub, on the ground, and sometimes in a cavity in piles of **rocks**. The clutch size is larger in northern species and populations, and can be as few as three and as large as six. The female incubates the bluish-tinged eggs, but she is fed by the male during her seclusion. Both sexes share in the rearing of the young birds.

Cardueline finches in North America

Fifteen species in the Fringillidae breed regularly in **North America**, all of them cardueline finches. The most prominent of these are discussed below.

The pine grosbeak (*Pinicola enucleator*) breeds in conifer-dominated and mixedwood forests across northern North America, and as far south as California and New Mexico. The pine grosbeak is a relatively large, robin-sized finch. Males are a pinkish red **color**, with black wings, while females are a more cryptic grayish olive. This species has a holarctic distribution, also occurring widely in Europe and Asia, ranging from Scandinavia to Japan.

The purple finch (*Carpodacus purpureus*) breeds in a wide range of coniferous and mixedwood forests, and also in open but treed habitats, such as orchards and regenerating cutovers. The plumage of males is a bright purple-red, especially around the head, while females are a streaked olive in coloration. The purple finch breeds widely across the central regions of North America. Cassin's finch (*C. cassinii*) is a closely related, similar-looking species of open coniferous forests of the western **mountains**. The house finch (*C. mexicanus*) is also a western species, with males being rather rosy in their plumage. In recent decades, the house finch has greatly expanded its range into eastern North America. This process was initiated by introductions of this species to Long Island in 1940, by cagebird dealers who were hoping to establish a local population of house finches to supply the pet trade.

The rosy finch (*Leucostricte arctoa*) breeds in western North America, from the Aleutian Islands of Alaska, south through British Columbia and Alberta to Oregon and Montana. This species breeds in upland, rocky tundras, and then descends in the winter to lowlands with a more moderate climate. The rosy finch will frequent bird feeders in the wintertime.

The crossbills (*Loxia* spp.) are interesting finches, with unique mandibles that cross at their rather elongated tips. This unusual bill is very effective at prying apart the scales of conifer cones, particularly those of species of **pines**, to extract the nutritious seeds that are contained inside. The red crossbill (*L. curvirostra*) ranges very widely, breeding in coniferous forests in North America,

A large ground finch (*Geospiza magnirostris*) on Santa Cruz Island in the Galapagos Islands. *Photograph by Tim Davis. Photo Researchers, Inc. Reproduced by permission.*

across Europe and Asia, in North Africa, even in montane forests in the Philippines. The white-winged crossbill (*L. leucoptera*) occurs in more-open coniferous and mixedwood forests, and it also breeds widely across North America and in Eurasia. Males of both species of crossbills are red colored, with black wings, while females are a dark olive-gray.

The crossbills are also interesting in that they are irruptive breeders and will attempt to nest at almost any time of year, as long as there is a good, local supply of conifer seeds. Northern populations will even breed in the wintertime. Crossbills are great wanderers and can show up unpredictably in large numbers in years when their food is locally abundant, and then disappear for several years, breeding elsewhere until the local crop of pine cones increases again.

The common redpoll (*Carduelis flammea*) breeds in coniferous boreal forests and high-shrub tundras of northern Canada, and in similar habitats in northern Europe and Asia. The hoary redpoll (*C. hornemanni*) breeds further to the north in more sparsely vegetated tundras, also in both North America and Eurasia. The hoary redpoll breeds as far north on land as is possible in

North America, at the very tip of Ellesmere Island. The pine siskin (*C. tristis*) breeds further to the south in coniferous and mixedwood forests, as far south as the highlands of Guatemala in Central America.

The American goldfinch (*Carduelis tristis*), sometimes known as the wild canary, is a familiar species that breeds widely in North America. The usual habitat is in disturbed situations, such as regenerating burns and cutovers, weedy fields, and shrubby gardens. Male American goldfinches are brightly colored with a yellow-and-black pattern, while females are a paler, olive-yellow. This species is rather partial to the seeds of thistles, which are herbaceous plants that tend to fruit late in the growing season. As a result, goldfinches breed rather late in the summer, compared with almost all other birds within its range. The lesser goldfinch (*C. psaltria*) occurs in the southwestern United States and Mexico.

The evening grosbeak (*Hesperiphona vespertinus*) breeds in conifer-dominated forests of southern Canada and the western United States. This yellow-and-black bird wanders widely in the wintertime in search of food, and it can sometimes occur in the southern states during this season.

Cardueline finches elsewhere

The common canary or wild serin (*Serinus canaria*) is a famous songster, native to the Azores, Madeira, and Canary Islands off northwestern Africa. Wild birds have a gray-olive, streaked back, and a yellowish face and breast. However, this species has been domesticated, is available in a wide range of colors, and is commonly kept as a singing cagebird.

The European goldfinch (*Carduelis carduelis*) is a common and widespread species in Europe, western Asia, and northern Africa. This red-faced bird is much-loved in Europe, and as a result several attempts were made by homesick European immigrants to introduce the species to North America. These introductions failed, which is probably just as well, because this species might have caused ecological damages by competing with native species of finches. The European greenfinch (*Carduelis chloris*) is another closely related species that is also widespread in Europe. Both of these finches are commonly kept as cagebirds.

The hawfinch (*Coccothraustes coccothraustes*) is a widespread Eurasian species. The hawfinch has a large beak, used for crushing large, hard fruits of trees.

Fringillinae finches

The subfamily Fringillinae is comprised of only three species of finches that breed widely in Europe and Asia. These are superficially similar to the cardueline finches, but they do not have a crop, and they feed their young mostly insects, rather than regurgitated seeds and other **plant** matter.

The common chaffinch (*Fringilla coelebs*) has a wide breeding range across northern Eurasia. The male chaffinch has a black head and back, an orange-buff breast, and a white belly, while the coloration of the female is less bright.

The brambling (*Fringilla montifringilla*) is also a widespread breeder across northern Eurasia. The brambling also occurs during its migrations in western Alaska, particularly on some of the Aleutian islands, where flocks of this species may be observed. The male brambling has a blue head, pinkish brown face and breast, a greenish rump, and black-and-white wings, while the female is a relatively drab, olive-gray.

The blue chaffinch (*F. teydea*) only occurs in conifer forests on the Canary Islands. The male is a rather uniformly slate-blue, and the female is a darker gray.

Finches and humans

Because they are attractive and often abundant birds, are easy to feed, and usually sing well, species of

KEY TERMS

Holarctic—This is a biogeographic term, used in reference to species that occur in suitable habitat throughout the northern regions of North America, Europe, and Asia.

Irruption—A periodic, sporadic, or rare occurrence of a great abundance of a species. Some species of migratory finches are irruptive, especially in their winter populations.

finches have long been kept as cagebirds. The most famous of the pet finches is, of course, the canary, but goldfinches and other species are also commonly kept, particularly in Europe. The canary is available in a wide variety of plumages, postures, and song types, all of which have been selectively bred from wild-type birds to achieve some aesthetic goal, which as often as not is focused on the development of birds that are "different" and unusual. The most commonly kept variety of canary is colored bright yellow, and has a richly cheerful, melodic song.

Species of cardueline finches are among the more common species of birds that visit seed-bearing feeders. This is particularly true during the wintertime, when natural seeds can be difficult to find, because they are hidden under accumulated snow. Most of the finches of North America will visit feeders, but their abundance can vary tremendously from week to week and from year to year, depending on the regional availability of wild foods, and also on the **weather**.

Bird feeding has a rather substantial economic impact. Each year, millions of dollars are spent by North American homeowners to purchase and provision backyard feeders. This money is rather well-spent, in view of the aesthetic pleasures of luring finches and other wild birds close to the home, while also helping these attractive, native species of **wildlife** to survive their toughest time of the year.

A few species of finches are considered to be agricultural **pests**. The bullfinch (*Pyrrhula pyrrhula*) of Eurasia can be especially important, because it eats the buds of fruit trees, and can cause considerable damages in orchards in some places within its range.

Some finches have become rare and endangered because of changes that humans have caused to their habitats.

See also Waxbills; Weaver finches.

Resources

Books

Ehrlich, P., D. Dobkin, and D. Wheye. *The Birders Handbook.* New York: Simon and Schuster, 1989.

Farrand, J., ed. *The Audubon Society Master Guide to Birding.* New York: A.A. Knopf, 1983.

Forshaw, Joseph. *Encyclopedia of Birds.* New York: Academic Press, 1998.

Trollope, J. *The Care and Breeding of Seed-eating Birds.* Dorset, U.K: Blandford Press, 1983.

Bill Freedman

Firs

The true firs are about 40 **species** of **conifer** trees in the genus *Abies*, occurring in cool-temperate, boreal, and montane **forests** of the northern hemisphere. Firs are members of the pine family (Pinaceae).

Firs are characterized by flattened needles, usually having two white lines running the length of the **leaf**. Firs do not have a petiole joining the needles to the twigs, and after the foliage is shed large scars are left on the twigs. The cones of firs are held upright, and they shed their scales soon after the winged **seeds** have been dispersed, leaving a spike-like axis on the twig. Fir trees generally have a dense, spire-like crown. The **bark** of most species is rather smooth on younger trees, becoming somewhat scaly on older trees. Many species develop resin-containing blisters on the surface of their bark. Firs are not a prime species for sawing into lumber, but they are excellent as a source of pulpwood for the manufacturing of **paper**, and are also cultivated as Christmas trees and as ornamentals.

Douglas firs (*Pseudotsuga* spp.) are a closely related group of six species that occur in western **North America** and eastern **Asia**. Douglas firs are distinguished from true firs by their small, raised leaf scar, a petiole joining the leaf to the twig, and the distinctive, three-pointed bracts (scale-like leaves) that occur immediately below and close to the scales of their oval-shaped, hanging cones.

Firs of North America

Nine species of true firs grow naturally in North America. The most widespread species is balsam fir (*Abies balsamea*), a prominent **tree** in boreal and north-temperate forests of eastern Canada and the northeastern United States. On moist sites with a moderate climate, this species grows as tall as 65 ft (20 m). In some places, balsam firs occur above the timber-line in a depressed growth-form known as krummholtz. Balsam fir is highly intolerant of fire, and it tends to be a relatively short-lived tree. Balsam fir is the major food species of the **spruce** budworm (*Choristoneura fumiferana*), a moth that periodically causes extensive forest damage in northeastern North America. Fraser fir (*A. fraseri*) is closely related to balsam fir, but occurs in montane forests of the southern Appalachians.

The other seven species of true firs in North America all occur in western forests. The subalpine fir (*Abies lasiocarpa*) grows in montane forests from southern Alaska to northern Texas, sometimes occurring past the timberline in a krummholtz growth form. Grand fir (*A. grandis*), Pacific silver fir (*A. amabilis*), and white fir (*A. concolor*) are species of moist, western rain forests, growing on sites of moderate altitude, and achieving heights of as much as 164 ft (50 m). Species with relatively restricted distributions in the western United States are bristlecone fir (*A. bracteata*), noble fir (*A. procera*), and California red fir (*A. magnifica*).

The Douglas fir (*Pseudotsuga menziesii*) is a common, fast-growing, and valuable timber species in western North America, where it can grow as tall as 262 ft (80 m) and attain a diameter of more than 6.5 ft (2 m). Some taxonomists divide the species into two races, the coastal Douglas fir (*P. m. menziesii*), which grow in humid western forests, and the Rocky Mountain or interior Douglas fir (*P. m. glauca*), which grows in drier forests further to the east. The big-cone Douglas fir (*Pseudotsuga macrocarpa*) is a locally occurring species in extreme southern California and northern Baja.

Economic uses of firs

The most important use of true firs is for the production of pulp for the manufacturing of paper. All of the abundant firs are used in this way in North America, especially balsam fir and white fir.

True firs are used to manufacture a rough lumber, suitable for framing buildings, making crates, manufacturing plywood, and other purposes that do not require a fine finish. The Douglas fir is an important species for the manufacturing of a higher-grade lumber.

Canada balsam is a viscid, yellowish turpentine that is secreted by balsam fir, and can be collected from the resinous blisters on the stems of these trees. Canada balsam is now a minor economic product, but it used to be important as a clear-drying, mounting fixative for **microscope** slides, and as a cement for optical lenses. Oregon balsam, collected from Douglas fir, was similarly used.

Some species of firs are grown as ornamental trees around homes and in public parks. White fir, grand fir, and Douglas fir are native species commonly used in this way. The European white fir (*Abies alba*) and Himalayan

KEY TERMS

Boreal—This refers to the conifer-dominated forest that occurs in the sub-Arctic, and gives way to tundra at more northern latitudes.

Krummholtz—A stunted, depressed growth form that some conifers develop above the tree-line on mountains, in the arctic, and sometimes along windy, oceanic coasts. Krummholtz trees are extremely slow-growing, and can be quite old, even though they are small in diameter and less than 3 ft (1 m) tall.

Montane—A conifer-dominated forest occurring below the alpine tundra on high mountains. Montane forest resembles boreal forest, but is affected by climate changes associated with altitude, rather than latitude.

silver fir (*A. spectabilis*) are also sometimes cultivated as ornamentals in North America.

Firs are highly desirable for use as Christmas trees, and in some areas they are grown on plantations established for this purpose. They can be pruned to develop a thick canopy with a pleasing shape, and firs retain their foliage for a rather long time, even inside dry homes during the winter.

See also Pines.

Resources

Books

Brockman, C.F. *Trees of North America.* New York: Golden Press, 1968.

Fowells, H.A. *Silvics of Forest Trees of the United States.* Washington, DC: U. S. Department of Agriculture, 1965.

Hosie, R.C. *Native Trees of Canada.* Ottawa: Canada Communications Group, 1985.

Judd, Walter S., Christopher Campbell, Elizabeth A. Kellogg, Michael J. Donoghue, and Peter Stevens. *Plant Systematics: A Phylogenetic Approach.* 2nd ed. with CD-ROM. Suderland, MD: Sinauer, 2002.

Petrides, G.A. *A Field Guide to the Trees and Shrubs of North America.* New York: Houghton Mifflin, 1986.

Van Gelderen, D.M., and J.R.P. Van Hoey Smith. *Conifers.* Eugene, OR: Timber Press, 1989.

Bill Freedman

Fish

More than three quarters of Earth's surface is covered by **salt water**; in addition, large areas are inundated with **freshwater** in the form of lakes, **rivers**, canals, swamps, and marshes. It is therefore not surprising that animals and plants have undergone a wide **radiation** in such habitats. One of the most successful groups of animals that have evolved to fill all of these habitats is the fish. Today it is possible to find different sorts of fish at all depths of the oceans and lakes-from the shoreline to the base of the deepest **ocean** trenches.

There are two types of fish on **Earth**: those that have a skeleton comprised of cartilage and those with a bony skeleton. The former include the **sharks**, dogfish, **skates**, and **rays**. The remainder, and by far the most abundant in terms of numbers and **species**, are known as the bony fishes. More than 25,000 species have been described. The majority of these are streamlined to reduce water resistance, with specialized fins that provide propulsion. Fins are basically of two types: vertical, or unpaired fins, and paired fins. The former include a dorsal fin in the midline of the back, an anal fin along the underside and a caudal fin at the rear end of the fish. The paired fins are known as pectoral and pelvic fins; they correspond to the limbs of terrestrial **vertebrates**.

In the majority of species, there is no neck, and all external appendages, with the exception of the fins, have been reduced. The body is covered with tiny, smooth scales that offer no resistance to the water. The form, size, and number of fins varies considerably according to the individual's **habitat** and requirements. In fast-swimming species such as **tuna** or **mackerel**, the dorsal and anal fins form sharp thin keels that offer little resistance to water flow. Departures from this body shape, however, are very common. Puffer or porcupine fish, for example, have short, round bodies with greatly reduced fins that are more effective in brief, sculling movements than rapid movement. Yet other species such as eels have lost almost all traces of external fins and swim instead by rhythmic movements of their muscular bodies.

In exploiting the aquatic and marine habitats, fish have evolved a number of unique features. One of these is the manner in which they breathe. The respiratory surface of fish forms special gills which are highly convoluted and well supplied with **blood**. Water is passed over the gills as the body moves through the water. As it does, the highly dissolved **oxygen** in the water meets the respiratory surface, diffuses across the **membrane** and into the blood where it is taken up by hemoglobin pigment in the blood cells.

Another important **adaptation** which has meant that fish have been able to thrive in the rich waters of the seas and rivers has been the development of the swim bladder-a special **organ** which has arisen from an outgrowth of the alimentary **canal**. This gas-filled chamber fulfills several functions, but one of the most important is in providing buoyancy, a feature that enables **bony fish** to remain at the same level in the water column without expending any **energy**. Sharks and rays do not possess a swim bladder.

In conquering the water environment, fish have developed a wide range of behavioral specializations that include feeding adaptations, **courtship**, and breeding behaviors, and defensive and attacking postures. Many of these are assisted or augmented through special sensory organs, most of which have evolved independently in many of these species. Altogether, they combine to provide the fishes at all stages of their lives with a wide range of specialized adaptations that enable them to live and reproduce so successfully on Earth.

See also Cartilaginous fish.

Fission *see* **Nuclear fission**

Flagella

Flagella are long, thread-like appendages which provide some live single cells with the ability to move, motility. **Bacteria** which have flagella are either rod or spiral-shaped and are known as bacilli and spirochetes, respectively. Cocci, or round bacteria, are almost all nonmotile. **Animal** sperm cells also have flagella. However, prokaryotic cells (such as bacteria) have flagella made up of the protein flagellin. Whereas, eukaryotic cells (such as sperm) which have a nucleus have flagella composed of tubulin **proteins**.

Bacteria can have a single flagellum or multiple flagella in a number of patterns. A single flagellum is lo-

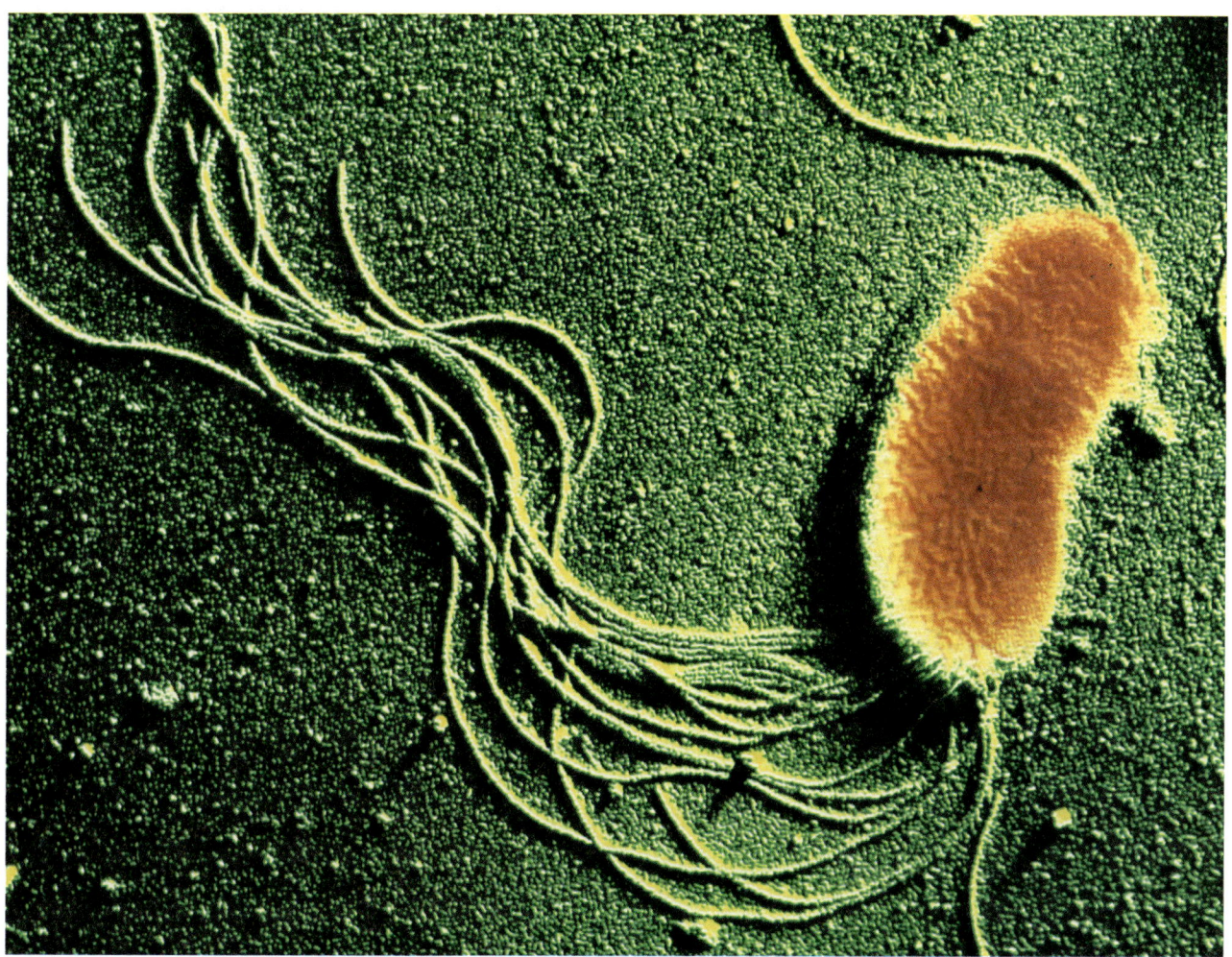

False-color transmission electron micrograph (TEM) of the aerobic soil bacterium *Pseudomonas fluorescens*, flagellae curving behind the orange cell, magnification at x10,000. *Photograph by Dr. Tony Brain. Photo Researchers, Inc. Reproduced by permission.*

cated at the tail end of a bacterium similar to the position occupied by a propeller on a submarine. Monotrichous is the term used to describe a single flagellum with this polar orientation. Bacteria with multiple flagella may have one at each end (amphitrichous), several at one end (lophotrichous), or several all around their perimeter (peritrichous).

Flagellar movement is chemically driven. Environmental **nutrients** attract motile bacteria; while other substances repulse them. This reactive motility is called chemotaxis. The sensation of chemical gradients translates into a **proton** flow into the bacterial **cell**. These protons power a pump which rotates the flagellar base (or hook) at about 150 revolutions per second. When the hook rotates counter-clockwise, a bacterium will move toward a chemical attractant. If the hook rotates clockwise, then a bacterium falls, or tumbles, aimlessly until it senses a more favorable position.

Eukaryotic flagella are very different from bacterial flagella. The tubulins of eukaryotic flagella are arranged in a microtubule array of nine doublets surrounding two singlets along the length of the flagella, sort of like straws standing up in a cylindrical straw-container. These "straws" slide along each other to generate movement and are connected by protein spokes to regulate their interaction. This sliding **motion** generates the flagellar beat which begins at the base (next to the cell) and is propagated away from the cell (distally) in a standing wave. This beat occurs in a single **plane** with a whip-like movement.

Flagellar movement can be visualized using specialized microscopic techniques. Flagella are usually 12-30 nm in diameter and much longer than the cell which they move. Because they are so thin, they can not be seen with normal **light microscopy**. Instead scientists use staining techniques or phase-contrast microscopy to visualize them. Phase-contrast microscopy accentuates differences in how light bends as it passes through the specimen observed. Motile bacteria will appear either oval, oblong or **spiral**; whereas sperm look triangular with rounded corners. Cells such as the photosynthetic protist *Chlamydomonas reinhardtii*, which have two flagella at one end, will appear as though they are doing the breaststroke.

A number of environmental factors greatly influence the stability of the flagellar structure. In both prokaryotes and eukaryotes, an acidic **pH** will cause flagella to fall off. In addition, very cold temperatures will lead to disassembly of the flagellar proteins. However, flagella will reassemble with a change to an environment with a neutral pH or normal **temperature**.

See also Eukaryotae.

Flame analysis

Allowing analysis of the **light** (photons) from excited **atoms**, flame analysis is a form of atomic **emission spectroscopy** (AES).

German chemist Robert Bunsen's (1811–1999) invention of the Bunsen burner—a tool now commonly used in modern **chemistry** laboratories—also spurred the development of flame analysis. Working with Gustav Kirchhoff (1824–1887), Bunsen helped to establish the principles and techniques of spectroscopy. Bunsen's techniques also enabled his discovery of the elements cesium and rubidium.

Bunsen's fundamental observation that flamed elements emit light only at specific wavelengths, and that every element produced a characteristic spectra, paved the way for the subsequent development of quantum theory by German physicist Maxwell Planck (1858–1947), Danish physicist Niels Bohr (1885–1962), and others. Using techniques pioneered by Bunsen, scientists have since been able to determine the chemical composition of a variety of substances ranging from bioorganic debris to the composition of the stars.

Analysis of emission spectra

Bunsen examined the spectra; the colors of light emitted when a substance was subjected to intense flame. When air is admitted at the base of a **Bunsen burner** it mixes with **hydrocarbon** gas to produce a very hot flame at approximately 3,272°F (1,800°C). This **temperature** is sufficient to cause the emission of light from certain elements. Often termed "spectral fingerprints," the **color** of the flame and its spectral distribution of component colors is unique for each element.

To examine the spectra of elements, Bunsen used a simple apparatus that consisted of a **prism**, slits, and a magnifying **glass** or photosensitive film. Bunsen determined that the spectral patterns of elements that emitted light when subjected to flame analysis differed because each pattern represented limited portions of the total possible **spectrum**.

Flame analysis or atomic emission spectroscopy is based on the physical and chemical principle that atoms—after being heated by flame—return to their normal **energy** state by giving off the excess energy in the form of photons of light. The mathematically related frequencies and wavelengths of the photons emitted are characteristic for each element and this is the physical basis of the uniqueness of spectral fingerprints.

Qualitative testing

Flame analysis is a qualitative test, not a quantitative test. A qualitative chemical analysis is designed to identify the components of a substance or mixture. Quantitative tests measure the amounts or proportions of the components in a reaction or substance. The unknown **sample** subjected to flame analysis is either sprayed into the flame or placed on a thin wire that is then introduced into the flame.

Highly volatile elements (chlorides) produce intense colors. The yellow color of **sodium**, for example, can be so intense that it overwhelms other colors. To prevent this obscuration, the wire to be coated with the unknown sample is usually dipped in hydrochloric acid and subjected to flame to remove the volatile impurities and sodium.

Standard or Bunsen burner based flame tests do not work on all elements. Those that produce a measurable spectrum when subjected to flame include, but are not limited to: **lithium**, sodium, potassium, rubidium, cesium, **magnesium**, **calcium**, strontium, **barium**, zinc, and cadmium. Other elements may need hotter flames to produce measurable spectra.

Analysts use special techniques to properly interpret the results of flame analysis. The colors produced by a potassium flame (pale violet) can usually be observed only with the assistance of glass that can filter out interfering colors. Some colors are similar enough that a line spectrum must be examined to make a complete and accurate identification of the unknown substance, or the presence of an identifiable substance in the unknown.

Flame analysis can also be used to determine the presence of **metal** elements in **water** by measuring the spectrum produced by the metals exposed to flame. The water is first vaporized to allow observation of the emissions of the subsequently vaporized residual metals.

Flame tests are useful means of determining the composition of substances. The colors produced by flame tests are compared to known standards to identify or confirm the presence of certain elements in the sample.

See also Atomic spectroscopy; Electromagnetic field; Electromagnetism; Forensic science; Geochemical analysis; Spectral classification of stars; Spectral lines; Spectroscope; Spectroscopy.

Resources

Books

American Water Works Association. *Water Quality and Treatment.* 5th ed. Denver: American Water Works Association, 1999.

Daintith, John, and D. Gjertsen, eds. *A Dictionary of Scientists.* New York: Oxford University Press, 1999.

Hancock, P. L., and B. J. Skinner, eds. *The Oxford Companion to the Earth.* New York: Oxford University Press, 2000.

Keller, E. A. *Introduction to Environmental Geology.* 2nd ed. Upper Saddle River: Prentice Hall, 2002.

Klaassen, Curtis D. *Casarett and Doull's Toxicology.* 6th ed. Columbus: McGraw-Hill, Inc., 2001.

Klein, C. *The Manual of Mineral Science.* 22nd ed. New York: John Wiley & Sons, Inc., 2002.

Lide, D. R., ed. *CRC Handbook of Chemistry and Physics.* Boca Raton: CRC Press, 2001.

Other

Helmenstine, A. M. "Qualitative analysis—Flame Tests." About.com. <http://chemistry.about.com/library/weekly/aa110401a.htm> [cited October 20, 2002].

K. Lee Lerner

Flamingos

Flamingos are five **species** of large, colorful, very unusual-looking wading **birds** that encompass the family Phoenicopteridae. The flamingo lineage is ancient, with fossils of these birds being known from the early Tertiary. These birds occur in tropical and temperate regions of **Africa**, Madagascar, India, southern **Europe**, Caribbean coasts, highlands of the Andes in **South America**, and on the Galapagos Islands. The usual **habitat** of flamingos is shallow lakes, lagoons, and estuaries with fresh, alkaline, **brackish**, or fully saline **water**.

Flamingos range in height from 36-50 in (91-127 cm). Flamingos have a very long neck and long legs, with webbed toes on their feet. Their bill is quite unique,

Lesser flamingos (*Phoenicopterus minor*) in Transvaal, South Africa. *Photograph by Nigel Dennis. National Audubon Society Collection/Photo Researchers, Inc. Reproduced by permission.*

being bent downwards in the middle, with the relatively smaller, lid-like, lower mandible being rigid and the trough-like upper mandible being mobile. (In terms of mobility, the reverse is true of virtually almost all other jawed vertebrates.)

The unusual structure of the bill is adaptive to the feeding habits of flamingos. These birds feed while standing and bending their neck downwards to hold their head upside-down in shallow water or while swimming in somewhat deeper water. The flamingo uses its large, muscular tongue to pump water and mud into and out of the mouth. As this is done, their food of small **invertebrates** or **algae** is strained from the fluid using sieve-like structures known as lamellae, located on the inside edges of the upper mandible. Depending on the species of flamingo, the water column may be filtered for **zooplankton** and algae, or the sediment may be processed for invertebrates and **seeds**.

Flamingos have long, strong wings, and a short tail. Depending on the species, the coloration may be a solid pink or white, except for the primary flight feathers, which are black. The sexes are similar in shape and **color**, although males tend to be somewhat larger.

Flamingos fly with their neck extended forward and their legs and feet extended backward. They commonly fly in groups, with the flock organized into lines or a V-shaped pattern. During flight and at other times, the groups of flamingos organize themselves with the aid of their raucous, goose-like honkings. Flamingos **sleep** while standing on one leg, the other leg folded up and stowed under the body, and the head laid over the back.

Flamingos court using highly ritualized displays, which resemble stiff renditions of preening and stretching movements. These displays are sometimes undertaken in social groups that can contain hundreds of birds displaying together in unison, often marching stiffly in compressed, erect troops. Both sexes display, but the males are more enthusiastic about this activity.

Flamingos nest communally in very shallow water or on recently dried, muddy **lake** beds, sometimes in colonies exceeding a million pairs of birds. The nest of flamingos is placed on the top of a cone-shaped structure made of mud scooped up from shallow water using the bill. Parents will vigorously defend their nests, and sites are spaced, conveniently enough, about two neck-lengths apart. Both sexes incubate the one or two eggs. The

1620

young can walk rather soon after they hatch, but they do not leave the nest until they are 5-8 days old. The young are tended by both parents.

Flamingos that breed in temperate climates, that is, at high latitude or high altitude, migrate to more tropical conditions during their non-breeding season.

Species of flamingos

The largest species is the common or greater flamingo (*Phoenicopterus ruber*). This is an extremely widespread species, with populations breeding in subtropical or tropical climates in the West Indies, northern South America, southwestern France, East and South Africa, India, and in the vicinity of the Caspian and Black Seas and Kazakhstan. Flamingos do not breed in **North America**, but on rare occasions individuals of this deep-pink colored species can be observed in south Florida after severe windstorms.

The greater flamingo is also commonly kept in theme parks and zoos, and these may also escape into the wild. Chemicals occurring in their food appear to be important in the synthesis of the pink pigments of flamingos. The color of these birds becomes washed-out and whitish in captivity, where a fully natural diet is unavailable.

The Chilean flamingo (*Phoenicopterus chilensis*) is a smaller species, occurring from central Peru through the Andes to Tierra del Fuego.

The lesser flamingo (*Phoeniconaias minor*) breeds on alkaline lakes in East and South Africa, Madagascar, and northwestern India. This species breeds in **saltwater** lagoons and brackish lakes, and colonies can achieve numbers as large as one million pairs.

The Andean flamingo (*Phoenicoparrus andinus*) occurs above 8,200 ft (2,500 m) in the Andean highlands from Peru to Chile and northwestern Argentina. James's flamingo (*P. jamesi*) is a smaller species that only occurs above 11,500 ft (3,500 m) in about the same range.

See also Cranes; Ibises; Storks.

Resources

Books

Bird Families of the World. Oxford: Oxford University Press, 1998.

Brooke, M. and T. Birkhead. *The Cambridge Encyclopedia of Ornithology.* Cambridge, U: Cambridge University Press, 1991.

Bill Freedman

Flatfish

Flatfish are a group of mostly **saltwater**, carnivorous, bottom-dwelling **fish** in which both eyes are located the same side of the head. The under side of a flatfish is white while the upper side with the two eyes may be brightly colored. Many of these fish can change **color** to match their surroundings, making them hard to detect. When flatfish hatch, the eyes are located normally on each side of the head. However, when a young flatfish reaches a length of about 0.8 in (2 cm), one **eye** moves close to the other eye, and the mouth is twisted. Many **species** of flatfish, such as halibut, sole, and turbot, are popular food fish and are commercially valuable.

The flatfish family Pleuronectidae includes mainly right-sided species (i.e. both eyes are found on the right side of the head), although there are some left-sided species. The largest flatfish is the Atlantic halibut (*Hippoglossus hippoglossus*), which is found on the European and North American sides of the North Atlantic in Arctic and sub-Arctic waters. The halibut is especially prolific north of Scotland and in the northern North Sea. This species may reach a length of about 7 ft (2.1 m) and a weight of 720 lb (325 kg). It is brown, dark green or blackish on the eyed side.

The Pacific halibut (*Hippoglossus stenolepis*) is somewhat smaller and slimmer than its Atlantic relative and is found on both sides of the Pacific Ocean. It is greenish brown and may reach a weight of about 440 lb (200 kg).

Some species of flatfish are considerably smaller. For example, the common or winter flounder (*Pseudopleuronectes americanus*), found in shallow coastal waters of the Atlantic Ocean from Georgia to Labrador, reaches about 1 ft (30 cm) long. The American plaice or rough dab (*Hippoglossoides platessoides*) reaches a length of 2 ft (60 cm) and a weight of 4 lb (1.8 kg). This reddish or brownish fish is found in the Atlantic Ocean from Massachusetts to the cold waters of **Europe**. The larger European plaice (*Pleuronectes platessa*) reaches 3 ft (90 cm) in length and weighs about 20 lb (9 kg).

A windowpane flounder (*Scophthalmus aquosus*) on the sea floor in the Gulf of Maine. *Photograph by Andrew J. Martinez. National Audubon Society Collection/Photo Researchers, Inc. Reproduced by permission.*

The more than 100 species of sole (family Soleidae) have a thin body with a downward curved mouth. Of all the flatfish, soles demonstrate the most efficient **adaptation** to a bottom-living environment. They possess small, paired fins, and the dorsal and anal fins are considerably extended. Unlike the flatfish in the family Pleuronectidae, soles prefer more southern waters, and some are found in the tropics. Soles are found in the Mediterranean Sea and in the Atlantic Ocean extending northward to the North Sea. The most well-known species in this family is the European or Dover sole (*Solea solea*). It may reach a weight of 3 lb (1.4 kg) and a length of 20 in (50 cm).

The lefteye flounders are classified in the family Bothidae. One species in this family, the summer flounder (*Paralichthys dentatus*), is found in the Atlantic Ocean from Maine to Florida. The southern flounder (*P. lethstigma*) is found in the Gulf of Mexico. The turbot (*Scophthalmus maximus*), another member of the family Bothidae, has a thick, diamond-shaped body, and may weigh more than 44 lb (20 kg). It is found in the Mediterranean and in the European side of the Atlantic Ocean to the southern part of the North Sea.

Resources

Books

The Great Book of the Sea: A Complete Guide to Marine Life. Philadelphia: Running Press, 1993.

Migdalski, E.C., and G.S. Fichter. *The Fresh and Salt Water Fishes of the World.* New York: Greenwich House, 1994.

Whiteman, Kate. *World Encyclopedia of Fish & Shellfish.* New York: Lorenz Books, 2000.

Nathan Lavenda

Flatworms

Flatworms are small, multicelled animals with elongated bodies that have clearly defined anterior (front) and posterior (rear) ends. These worms are bilaterally symmetrical, meaning that their two sides reflect each other. They usually have a recognizable head, which houses gravity and light-receptive organs, and **eye** spots. They lack circulatory and respiratory systems and have only one opening that serves both as their anus and mouth. Most flatworm **species** live in fresh and marine waters, although some live on land.

Their soft, flattened bodies are composed of three layers-the ectoderm, endoderm, and mesoderm. They may be covered by a protective cuticle or by microscopic hairs, called cilia. Their internal organs are comprised of a **nervous system**, usually **hermaphrodite** sexual organs, and an **excretory system**.

As members of the phylum Platyhelminthes, flatworms belong to four classes: Turbellaria, Monogenea, Trematoda, and Cestoidea. Within these four classes, there are hundreds of families and some 10,000 species, including animals with common names like free-living flatworms, parasitic flatworms, tapeworms, and flukes.

Class Turbellaria

Containing the most primitive flatworms, the class Turbellaria consists of nine orders and a total of about 3,000 species, most of which are free-living. While some species live in moist, dark areas on land, most live at the bottom of marine **water**. These flatworms are found in all seas. While the aquatic species seldom grow more than 0.4-0.8 in (1-2 cm) long, some land varieties can reach lengths of 19.7 in (50 cm). The aquatic species have relatively flat, leaf-shaped bodies and are usually gray, brown, or black, although some species have a green tint. The turbellaria's head possesses one or more pair of eyes and tentacles. These flatworms are covered by microscopic hairs (cilia) that they beat continuously, creating turbulence in the water-an activity that gave them their name. Their cilia are important in their locomotion; they also crawl along the ground gripping it with sticky secretions from their **glands**.

Turbellarians are hermaphrodites, possessing the complex reproductive apparatus of both male and female. The fertilized eggs usually produce a small worm, although sometimes larvae are produced. The majority of turbellarians are carnivores.

Planarians

Probably the most familiar Turbellarians are the planarians, soft-bodied, aquatic, flattened worms that appear to have crossed eyes and **ear** lobes. In fact, the crossed eyes are eye spots with which the worm can detect **light**. The lobes to each side are sensory and also are equipped with glands to secrete an **adhesive** substance used in capturing **prey**.

The single opening on the ventral (bottom) surface of the worm serves as both mouth and anus. Internally the worm has a complex, branching gut that courses nearly the full length of the body. Since the worm has no **circulatory system**, the elongated gut brings food to nearly all areas of the worm's body. Planaria have no skeletal or respiratory systems.

These animals possess great powers of regeneration. If a planaria is cut in half, the front half will grow a new tail section and the rear half will generate a new head. If cut into thirds, the middle third will regrow a head and tail and the other two sections will regenerate as described.

Class Monogenea

Species in the class Monogenea are **parasites**, completing their life cycles within the body of a single living host, such as **fish**, **frogs**, **turtles**, and **squid**. Most of the 400 species in this class are ectoparasites, meaning that they cling to the outside of their host, for example, to the gills, fins, or skin of fish. Their bodies are covered by a protective cuticle and have adhesive **suckers** at each end. They eat by sucking **blood** through their mouths, which open beside their suckers.

Class Trematoda

Commonly known as flukes, there are over 6,000 species of flatworm in this class. Descended from the parasitic flatworm, flukes grow slightly larger, to about 0.8-1.2 in (2-3 cm) long. A fluke must live in two or more hosts during its lifetime because its developmental needs are different than its adult needs. The first host is usually a mollusk and the final host-which houses the fluke during its mature, sexual stage-is invariably a vertebrate, sometimes a human. In general, flukes lay tens of thousands of eggs to increase their offspring's chances of survival.

Three families in this class contain blood flukes, those that live in the bloodstream of their hosts. Blood flukes, called schistosomes, are of particular importance to humans, since an estimated 200 million people are affected by them. Second only to **malaria** among human parasites, they usually do not kill their victims immediately; rather, they make their hosts uncomfortable for years until a secondary illness kills them.

As larvae, some species inhabit **snails** but, upon destroying their hosts' livers, leave and swim freely for several days. They are then absorbed through the skin of

GALE ENCYCLOPEDIA OF SCIENCE 3

KEY TERMS
· ·

Bilateral symmetry—The flatworm is divisible into two identical halves along a single line down the center of its body, called the sagittal plane.

Ectoderm—The skin covering the flatworm on the outside of its body.

Ectoparasites—Parasites that cling to the outside of their host, rather than their host's intestines. Common points of attachment are the gills, fins, or skin of fish.

Endoderm—The tissue within the flatworm that lines its digestive cavity.

Free-living species—Nonparasitic; feeding outside of a host.

Hermaphrodite—Having the sex organs of both male and female.

Mesoderm—A spongy connective tissue where various organs are embedded.

Schistosomes—Blood flukes that infect an estimated 200 million people.

---

a second host, such as a human, and live in **veins** near the stomach. There they mature and can live for 20 years or more. Unlike other species in the phylum, blood flukes have clearly defined genders.

Class Cestoidea

Tapeworms are the dominant member of the class Cestoidea. They are ribbon-like, segmented creatures living in the intestines of their vertebrate hosts. There are a dozen orders in this class, most living in fish but two that use humans as hosts. Tapeworms cling to the intestinal wall of their hosts with suckers, hooks, or other adhesive devices. Having no mouth or gut, they receive their nourishment through their skin. Further, they have no type of sensory organs. White or yellowish in **color**, species in this class vary from 0.04 in (1 mm) long to over 99 ft (30 m).

The broad fish tapeworm (*Diphyllobothrium latum*), a large tapeworm present in humans, can illustrate the typical life of a tapeworm. As an adult, it attaches itself to the intestinal wall of the human host. Its body, composed of roughly 3,500 sections, probably measures 33-66 ft (10-20 m) long. At this point, it lays about one million eggs each day. Each egg, encased in a thick capsule so that it will not be digested by the host, leaves the host through its feces. When the egg capsule reaches water, an embryo develops and hatches. The larva swims until it is eaten by its first host, a minute crustacean called a copepod. The larva feeds on the copepod and grows. When a fish eats the copepod, the young tapeworm attaches itself to the fish's gut. The tapeworm continues to grow and develop until the fish is eaten by a suitable mammal, such as a human. Only at this point can the tapeworm mature and reproduce.

Resources

Books

George, David, and Jennifer George. *Marine Life: An Illustrated Encyclopedia of Invertebrates in the Sea.* New York: John Wiley and Sons, 1979.

Grzimek, H.C. Bernard, Dr., ed. *Grzimek's Animal Life Encyclopedia.* New York: Van Nostrand Reinhold Company, 1993.

The Illustrated Encyclopedia of Wildlife. London: Grey Castle Press, 1991.

The New Larousse Encyclopedia of Animal Life. New York: Bonanza Books, 1987.

Pearse, John, et. al. *Living Invertebrates.* Palo Alto, CA: Blackwell Scientific Publications and Pacific Grove, The Boxwood Press, 1987.

Kathryn Snavely

Flax

The flax **plant**, genus *Linum*, family Linaceae, is the source of two important commodities. Linen is a historic, economically important cloth made from the fiber of flax. Linseed oil is obtained from the pressed **seeds** of the plant. There are about 200 **species** of Linum. The species that is cultivated most extensively is *L. usitatissimum,* an annual plant grown for its fiber and seed. Varieties of *L. usitatissimum* grown as a fiber crop have been selected to have stems that are tall, which ensures long fibers. Varieties grown for seed are shorter, with extensive branching, and thus bearing more flowers and yielding more seed.

Flax plants have gray-green, lanceolate (long and tapered), alternate leaves. Their height ranges from 1-4 ft (0.3-1.2 m). Many cultivated varieties of flax have blue flowers, although some have white, yellow, pink, or red flowers. The flowers are self-pollinating and symmetrical, with five sepals, five petals, five stamens, and a pistil with five styles. The fruit is a capsule with five carpels, each containing two brown, yellow, or mottled, shiny seeds. Flax **crops** are grown in rotation with other crops to avoid fungal **pathogens** that cause diseases in flax plants.

Linum angustifolium is a wild, perennial flax, is thought to be a "parent" of cultivated flax. There is evidence that this species was used by prehistoric peoples in Switzerland about 10,000 years ago. The ancient

Flax (*Linum usitatissimum*). *JLM Visuals. Reproduced with permission.*

Egyptians wrapped their mummies in linen. Today, Russia is the largest producer of flax. Flax grown in the United States (mainly for seed) is raised in the northern plains states.

Fiber flax

Flax plants grown for fiber require well-weeded well-drained **soil**, and a cool, humid environment. The plant is harvested when the stems begin to turn brown. Any delay in harvesting results in deterioration of the fiber, causing it to lose its luster and softness. The plants are often harvested by hand, uprooting the plant to preserve the length of the fiber. Flax is also harvested mechanically, but fiber length is sacrificed to some degree. Good fiber is 12-20 in (20-30 cm) long. The seed pods (bolls) are separated from the uprooted plants, either mechanically or by hand, a process called rippling.

The uprooted plants, now called straw, are then retted. This is a process by which **bacteria** and **fungi** are

KEY TERMS

Hackling—A combing procedure to straighten flax fibers.

Linen—Thread or cloth made from the long fibers of the flax plant.

Linseed oil—Oil obtained from the seeds of the flax plant.

Retting—A process whereby flax plants are moistened, allowing the stem covering to rot, and to break down the substances that hold the fiber together.

Rippling—A manual method of removing seed pods from the flax stalks, by drawing the stalks through combs.

Scutching—A process in which the flax fiber is beaten, separating it from the stem.

Shives—Small pieces of flax stem obtained after putting the retted straw through a machine called a flax brake, which crushes the woody part of the stem without damaging the fiber.

allowed to rot the semi-woody stalk tissues, and break down the gummy substance (pectin) that binds the fibers together. If the straw is not retted enough, removal of the semi-woody stalk is difficult, but if the straw is over-retted, the fiber is weakened. In pool or dam retting, the straw is placed in a tank of warm **water**, while in dew retting it is spread out in a field, allowing the straw to become dampened by dew or rain. Stream retting is a method where the flax bundles are put into flowing streams, and this produces the best linen fiber. Straw can also be retted chemically. The various retting processes are used to create various shades and strengths of fiber.

After retting, the straw is dried and put through a machine called a flax brake, which crushes the stems into small, broken pieces called shives. The shives are removed from the fiber by a process called scutching, done either mechanically or by hand. The fibers are then straightened out by hackling or combings, sorted according to length, and baled. The long fibers, called line fiber, are used to make fine fabrics, threads used for bookbinding and shoe making, and twine. The short, damaged or tangled fibers, called tow, are used for products such as rope, and rougher linen yarns.

The finest, strongest linen is made from flax immersed in hot water, and spun while wet; dry spinning produces a rougher, uneven yarn. Linen yarn is very strong, but inflexible. Flax fiber is basically pure **cellulose**, and is not very porous, making it difficult to dye unless the cloth is bleached first. The manufacturing of linen is very labor intensive and its price reflects this fact. France and Belgium have the reputation of producing the highest quality linens.

Seed flax

Seed flax grows best in a warm climate, but hot temperatures and **drought** can reduce the crop yield and oil content. The soil should be fertile and well weeded. To obtain the seed, the flax plants are allowed to over-ripen, which destroys the plant's value for its fiber as linen. Flax seed contains about 40% oil, and the seeds are crushed and pressed to remove this product. Linseed oil, which hardens by oxidation, is used to manufacture paints, varnishes, patent leather, linoleum, and oilcloth.

The remaining seed and hull wastes after pressing are used for **livestock** feed. Fiber can also be obtained from seed flax plants. This fiber is made into special papers.

See also Natural fibers.

Resources

Books

Lewington, Anna. *Plants for People.* New York: Oxford University Press, 1990.

Periodicals

Akin, D.E. "Enzyme-Retting of Flax and Characterization of Processed Fibers." *Journal Of Biotechnology* 89, no. 2-3 (2001): 193-203.

"Nontraditionally Retted Flax for Dry Cotton Blend Spinning." *Textiler Research Journal* 71, no. 5 (2001): 375-380.

Christine Miner Minderovic

Fleas

Fleas are about one thousand **species** of small **insects** in the order Siphonaptera, including several hundred species in **North America**. Adult fleas are external **parasites** (that is, ectoparasites) of **mammals** or **birds**, living on skin or in fur or feathers, and feeding on the **blood** of their hosts. Some fleas are serious parasites of birds or mammals, and may cause important damage to domestic animals, and sometimes great discomfort to humans. Some species of fleas are vectors of serious diseases of humans or other animals. Infestations of fleas often must be dealt with using topical applications of an insecticide.

Biology of fleas

Fleas have a laterally compressed body, a tough, smooth cuticle with many backward-projecting bristles, and relatively long legs. The mouth parts of fleas include stylets that are used to pierce the skin of the host **animal**, so that a blood meal can be obtained by sucking.

Fleas have a life cycle characterized by four developmental stages: egg, larva, pupa, and adult. The eggs are usually laid close to the body of the host in a place where the host commonly occurs, for example, on the ground, in a bird or mammal nest, or in carpets or soft furniture in homes. Larval fleas have chewing mouth parts and feed on organic debris and the feces of adult fleas, while adults require meals of bird or mammal blood.

Fleas commonly spend a great deal of time off their hosts, for example, in vegetation or on the ground. They can generally survive for a long time without feeding, while waiting for a suitable opportunity to parasitize a host animal. Fleas are wingless, but they walk well and actively travel over the body of their hosts, and between hosts as well.

Fleas are well known for their jumping ability, with their hind legs providing the propulsive mechanism. As a defensive measure, a flea can propel itself many times its body length through the air. The human flea (*Pulex irritans*), for example, can jump as high as 7.9 in (20 cm) and as far as 15 in (38 cm), compared with a body length of only a few millimeters.

Species of fleas

There are numerous species of fleas, occurring in various parts of the world. Although most species of fleas are specific to particular host animals for breeding, they are less specific in their feeding and may use various species of birds or mammals for this purpose.

The cat flea (*Ctenocephalides felis*) can be quite abundant during the hotter months of summer, as can the dog flea (*C. canis*), when their populations may build up in homes with pet animals. These fleas will also avidly feed on humans, biting especially commonly around the feet and ankles.

The human flea (*Pulex irritans*) is an important pest of worldwide distribution that can be quite common in human habitations, especially in tropical and sub-tropical countries. The oriental rat flea (*Xenopsylla cheopis*) is an especially important species, because it is the vector by which the deadly **bubonic plague** can be spread to humans.

Fleas and diseases

Many of the species of fleas that infest domestic mammals and birds will also utilize humans as a host, al-

---

though people are not the generally preferred host of these blood-sucking parasites.

The most deadly **disease** that can be spread to humans by fleas is bubonic plague or black death, caused by the bacterium *Pasteurella pestis*, and spread to people by various species of fleas, but particularly by the plague or oriental rat flea (*Xenopsylla cheopis*). Bubonic plague is an extremely serious disease, because it can occur in epidemics that afflict large numbers of people, and can result in high mortality rates. During the European Black Death of medieval times, millions of people died of this disease. There have been similarly serious outbreaks in other places where **rats**, plague fleas, and humans were all abundant. Bubonic plague is mostly a disease of **rodents**, which serve as a longer-term reservoir for this disease. However, plague can be transmitted to humans when they serve as an alternate host to rodent fleas during times when rodent populations are large. Plague is mostly spread to humans when infested flea feces are inadvertently scratched into the skin, but transmission can also occur more directly while the fleas are feeding, or when a host accidentally ingests an infected flea.

Another disease that can be spread to humans by fleas is known as **endemic** or murine flea-borne **typhus**. This disease is caused by a microorganism known as *Rickettsia*, and is passed to humans by various species of fleas and **lice**, but especially by the oriental rat flea. Fleas are also the vector of a deadly disease that afflicts rabbits, known as myxomatosis.

Fleas may also serve as alternate hosts of several tapeworms that can infect humans. These include *Dipylidium caninum*, which is most commonly a parasite of dogs, but can be passed to humans by the dog flea (*Ctenocephalides canis*). Similarly, the tapeworm *Hymenolepis diminuta* can be passed to people by the rat flea (*Xenopsylla cheopis*).

Resources

Books

Borror, D.J., C.J. Triplehorn, and N. Johnson. *An Introduction to the Study of Insects.* New York: Saunders, 1989.

Carde, Ring, and Vincent H. Resh, eds. *Encyclopedia of Insects.* San Diego, CA: Academic Press, 2003.

Davies, R.G. *Outlines of Entomology.* New York: Chapman and Hall, 1988.

Bill Freedman

Flies

Flies belong to the order Diptera, a group that also includes **mosquitoes**, gnats, and midges. Flies make up the fourth largest order of **insects**, with about 100,000 **species** recognized. Dipterans are amongst the most advanced insects in terms of morphology and biological adaptations. Their versatility and extreme range of anatomical and behavioral adaptations have enabled them to thrive in almost every corner of the globe—in soils, plants, and around **water** bodies. A large number of species have developed special relations with other animals as well as plants: many are free-living, feeding off a wide range of plants, while others are **parasites** and scavengers. A number are of economic importance in controlling pest species, while others serve as vectors for a range of human, **animal**, and **plant diseases**.

Dipterans are characterized by a single pair of functional wings positioned high on the thorax, behind which rest a pair of knoblike vestigial wings known as halteres. The head is free-moving and attached by a slender neck to the thorax. Two large compound eyes are prominent features on the head, as are a pair of segmented antennae. Also attached to the thorax are three pairs of legs, each ending in a pair of claws. In most species, these are short and powerful, even capable of grasping and carrying **prey** their own size. Crane flies (Tipulidae) are exceptional in having extremely long, delicate legs, an **adaptation** which together with their slender bodies has evolved in species that frequent damp habitats, frequently around streams and lakes.

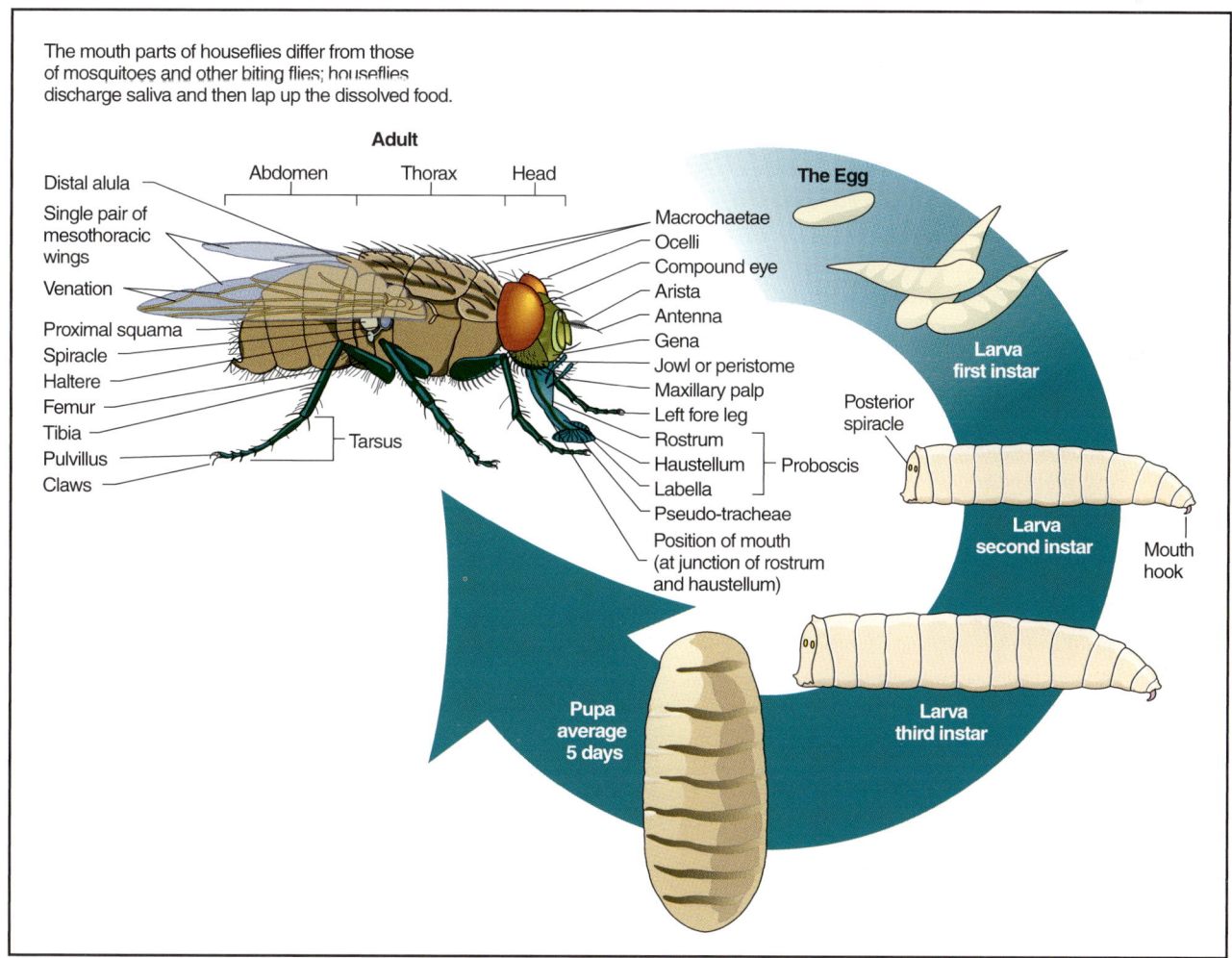

The life cycle of a common housefly. *Illustration by Hans & Cassidy. Courtesy of Gale Group.*

Many species are a dull dung **color** which assists equally well as camouflage for avoiding the ever-watchful eyes of predators, as well as for ambushing prey. Hover flies (Syrphidae) though are among the most colorful species, many of which are boldly colored in similar patterns to **bees** and **wasps** that carry a venomous sting. While hoverflies carry no such defenses, the act of mimicking the garish black and yellow colors of these other insects guarantees them a greater security from predators than many other species might enjoy. Most hover flies feed on **nectar** and pollen.

One part of the **anatomy** which exhibits considerable variation among flies is the structure of the mouth, a feature which dictates the way of life of many species. The proboscis in blood-feeding and other predatory flies, for example, is in some species a hollow piercing needle, while in others it resembles a broad, dagger-shaped weapon. Only the females of the bloodsucking species practice this habit and the mouthparts of the males are therefore quite different. In other species the proboscis is quite short and equipped with an absorbent type of soft pad through which liquids and small solid objects can pass. The mouthparts of some other species are nonfunctional and no food is taken during the adult stage, which is usually of short duration and intended only for dispersion and reproduction.

One of the most widely known groups of flies is the family Culcidae, which consists of mosquitoes and gnats. Most are slender built with long thin legs and narrow wings. All members of this family have well-developed toughened mouthparts for piercing **plant** or animal cells and long slender probosces for sucking up fluids. The probosces are used in some species for sucking up nectar and others for sucking up **blood** from animal prey. Mosquitoes and gnats are lightweight flies so that their animal victims rarely feel them when they alight to feed. Once settled, they make a tiny incision in the skin and inject a small amount of anticoagulant to the wound; this prevents the blood from clotting in and around the wound and guarantees the free flow of blood through the proboscis.

Other predatory and often blood-feeding flies include the much larger and more robust horse flies (Tabanidae) and the highly specialized robber flies (Asilidae) that lie in ambush for a passing insect; once the robberfly has captured its prey, the prey is killed immediately with an injection of poison. The body fluids from the hapless prey are then sucked up through the versatile proboscis.

Reproduction in diptera proceeds in a series of well-defined stages, from egg to adult. Eggs are laid on or near a food source (plant or animal material). Some species such as mosquitoes lay their eggs in standing water bodies. When they hatch, the larvae live suspended in the water in a horizontal position, just below the surface, feeding on tiny food particles. The pupae may also develop and remain in water up to the stage when they finally hatch into adults. Hover fly larvae, in contrast, may lay their eggs on plants infested with **aphids**. When the larvae hatch they devour the harmful aphids and are therefore important for many gardeners and horticulturalists. Some species of parasitic flies bypass the egg-laying phase and lay larvae directly into their hosts. Most larvae, or maggots, have short or reduced legs and the head is also quite indistinct. The only other remarkable feature about most larvae is the range of siphonlike appendages near the hind end, which assist with **respiration**. As they grow, the larvae shed their skin, a feature which may be repeated four or five times before it finally pupates. During the pupal phase, the larvae undergoes a complete transformation in a process known as **metamorphosis**. After several weeks in a cocoon that the larvae spins about itself, an adult fly emerges to begin the life cycle all over again.

Many species of flies are an economic concern to humans. Bot flies (Muscidae) and horse flies can be serious **pests** among **livestock** and other wild animals, such as **deer**. Although males feed exclusively on nectar and plant juices, females are blood-suckers and lay their eggs on the hair of cattle, **horses**, and other species. As the animals groom themselves, the eggs are taken into the mouth from where they move through the digestive tract to the stomach. Here they hook onto the stomach lining and feed until they are ready to pupate, at which time they release their grip and pass further along the intestine where they are deposited with the animal's feces. Other species when they hatch chew their way directly through the skin where they cause a small swelling. Here they will remain, feeding off flesh, until such time as they pupate and leave their host. Such species are not only a painful irritation for the host animal, but also seriously reduce the beast's condition and overall value of its hide.

Other fly species are important from a medical point of view as many are vectors of diseases such as **malaria**, **yellow fever**, typhoid, sand-fly fever, and others. Considerable amounts of money continue to be spent in the tropics in an attempt to eradicate malaria, which is caused by a tiny protozoan, but which is spread by female mosquitoes of the genus *Anopheles*, which require blood **proteins** for the development of their eggs.

Not all species of flies are harmful. Many species fulfil an important role in pollinating fruit **crops**, as well as a great many wild flowers, while even the distasteful actions of many scavenging flies serve an important role as they remove and recycle carrion and other animal and plant wastes that might otherwise pose a serious health hazard. Increasing attention is now being given to the possibility of using some flies, such as predatory species, to control a number of destructive flies and their larvae, such as those which attack livestock and plants.

See also True flies.

Flightless birds

Ratites are flightless **birds** that lack the keel (high ridge) on the breastbone to which the flight muscles of flying birds are attached. Instead, the entire breastbone looks rather like a turtle's shell. It has also been described as a raft, which gives this group of flightless birds its name, Ratitae (*Ratis* means raft in Latin). Ratites have heavy, solid bones, while flying birds have lightweight, hollow ones. Several ratites, such as ostriches, rheas, emus, and cassowaries, are the largest living birds. The kiwis of New Zealand, however, are about the size of chickens.

These flightless birds are the oldest living birds. All older **species** of ratites are now extinct. However, several ratite species became extinct only recently. *Genyornis* of **Australia** survived long enough to be hunted by aborigines about 30,000 years ago. The largest bird ever found, the **elephant** bird or roc (*Aepyornis*), lived in Madagascar. A huge New Zealand moa (*Dinornis*), which became extinct only about 200 years ago, may have been as tall, but did not weigh as much as the roc. The moa had no wings at all.

Although ratites are the most ancient of the living birds, they are no more closely related to the **reptiles** from which they evolved than other birds are. In fact, they are probably descended from flying birds. Their ancestors lost the ability to fly because they did not need to fly to obtain food or escape from predators. They probably had no important enemies in their habitats.

The structure of their feathers also prevents ratites from flying. On a flying bird's feathers, the barbs, those branches that grow at an **angle** from the shaft (or quill), are fastened together by hooked structures called barbules. This design makes a smooth, flat, light surface that can push against the air during flight. The feathers of ratites, however, lack barbules. The strands that grow from the quill separate softly, allowing the air through. This quality of softness is what makes the feathers of many ratites particularly desirable. Ostrich plumes, for example, have long been used as decoration on helmets and hats.

The living flightless birds are classified into four orders and five families. The single species of ostrich is in the order Struthioniformes, family Struthionidae. The two species of rhea are in the order Rheiformes, family Rheidae. Emus and cassowaries are classified in the same order, Casuariiformes; emus belong to the family Dromaiidae, while cassowaries comprise the family Casuariidae. Kiwis belong to the order Apterygiformes, family Apterygidae. Some ornithologists consider the **tinamous** of Central and **South America** to be ratites because they seldom fly. However, tinamous are capable of flight (although they prefer to run from predators and other danger) and have keeled breastbones. **Penguins** are also flightless birds, but they are not regarded as ratites. Their powerful breast muscles are used for swimming instead of flying.

Ostriches

The ostrich is Earth's largest living bird. There is only one species, *Struthio camelus*. The specific name comes from the fact that these tall, desert-living birds have been called camel birds. They may be as much as 8 ft (2.4 m) tall and weigh up to 400 lb (181 kg). A prominent distinction among subspecies of ostrich is skin **color**. The long legs and long, straight neck show red skin in some subspecies and blue in others.

Natives of **Africa**, ostriches are found on the dry plains, where they seem more at home among big **mammals**, such as giraffes, than they do among other birds. They are currently found in three areas Western Africa, at the farthest western portion of the bulge; South Africa; and in East Africa from the Horn of Africa (Ethiopia). They were formerly found on the Arabian Peninsula, but this subspecies was hunted for sport and became extinct during the first half of the twentieth century. An effort is being made to re-introduce ostriches to this region, although of a different subspecies.

Generally, ostriches have whitish neck, head, and underparts, with a thick covering of black or dark-brown feathers crowning the entire back. The female's feathers are almost uniformly brown, while males have a black body with white wing and tail feathers. Ostrich plumage seems almost more like fur than feathers. Ostrich plumes, especially the long ones from the birds' tail and wings, have been popular for centuries, primarily for use on hats. Today, their softness makes them useful for dusting and cleaning delicate parts in machinery.

Ostriches have scaly legs and feet. There are only two toes on each foot, both of which hit the ground when the bird walks. Each toe ends in a thick, curved nail that digs into the **soil** as the ostrich runs. One toe is immense, almost the size of a human foot; the other is much smaller. Each toe has a thick, rough bottom that protects it.

There is a myth that ostriches put their heads in the **sand** when frightened, but this is not the case. In reality, ostriches can run faster than just about any **predator**, or they can stand their ground and kick with powerful slashing motions of their sharp-nailed feet. Ostriches can run at a steady pace of about 30 MPH (48 km/h), with a stride that can cover more than 20 ft (6 m). At top speed for a brief **time**, they can run almost 45 MPH (72 km/h).

There is little up and down **motion** as they run. Instead, their legs handle all of the motion and their body stays in one **plane**. Ostriches, as well as the slightly smaller rheas, use their wings rather like sails. When running, they hold them out from their bodies. This helps them balance and, by changing the level of one wing or the other, it helps them easily change direction as they run. If frightened, a running ostrich may swerve into a circular pattern that takes it nowhere.

These large birds have been farmed for more than 150 years, starting in Australia. Originally the birds were raised just for their plumes, but in recent years they have been raised for their large eggs, their skin, which tans into attractive leather, and their meat. The feathers are actually harvested, in that they are cut off before falling out as they would naturally. This harvesting keeps them from being damaged. New feathers grow in to replace the harvested ones.

Ostriches have the largest eyes of any land animal—a full 2 in (5 cm) in diameter. The eyes are shielded by eyelash-lined outer eyelids that blink from the top downward, as well as nictitating membranes that close from the bottom of the **eye** upward. This **membrane** protects the eye but is semitransparent so that the bird can still see.

Because a tall ostrich may get the first sight of approaching danger on the savannah, its alarm may alert other animals to the presence of danger. The birds are usually left undisturbed by herding mammals. The ostriches act as lookouts and the mammals stir up **insects** and other small animals with which the ostriches supplement their herbivorous diet. Actually, ostriches will eat just about anything, especially if it is shiny and attracts their attention.

During the dry season, herds containing as many as 500-600 ostriches may gather at a watering hole. The males, or cocks, tend to stay in one group, while the females, or hens, stay in their own groups. When the rainy season begins, however, they split into harem groups, consisting of one male and two to four females.

A male ostrich performs a **courtship** dance involving considerable puffing and strutting, accompanied by booming noises. At its conclusion, he kneels and undulates his body in front of his chosen female. If she has found the performance enchanting, she also bends her knees and sits down. The male's dance may be interrupted by a competing male. The two males then hiss and chase each other. Any blows are given with their beaks.

The male selects the nesting site for his several females. He prepares it by scraping a slight indentation in the soil or sand. The dominant female lays her eggs first, followed by the others, over a period of several weeks. Altogether, there may eventually be as many as 30 off-white, 8 in (20 cm) eggs in a single clutch, perhaps half of them belonging to the dominant female. However, all the eggs won't hatch because they cannot all be incubated. Abandoned eggs do not attract many scavengers because the shells are too thick to break easily.

Both the dominant female and the male take turns incubating the eggs, with their insulating feathers spread over the nest. The sitter also takes time to turn the eggs on a regular basis. The eggs hatch about six weeks after the start of incubation, but the hatching may be a slow process because the chicks are able to break only tiny bits of the tough shell at a time. The mottled-brown chicks, each about 1 ft (30.5 cm) tall, are tended by all the females in the group. The chicks are ready to feed on their own, but are protected by the adults as they grow. They develop adult feathers by their third year, and soon afterward are mature enough to mate. Ostriches can live to be more than 40-50 years old.

Rheas

Rheas are similar in appearance to ostriches, but they are smaller and live in South America instead of Africa. The two species of rheas, often called South American ostriches, vary in size and location. The common rhea (*Rhea americana*) of Argentina and Brazil stands 5 ft (1.5 m) tall, several feet shorter than an ostrich, but it is still the largest North or South American bird. Darwin's rhea (*Pterocnemia pennata*) of southern Peru to the Patagonian region of Argentina is considerably smaller and has white tips on its brown plumage. Rheas live on open grassy plains in most of South America except the Andes Mountains and the northeastern region along the Atlantic. They can usually be found in flocks of about 50 birds, often in the vicinity of cattle herds.

Rhea males attract females by swinging their heads from side to side and making a loud, roaring sound. The females are mute. Unlike ostriches, a rhea male lines the nest with leaves and assumes total responsibility for incubating the eggs. The male incubates the eggs of five or six females for about five weeks, and then takes care of the young. The eggs are dark green, almost black, in color.

Emus

The single living species of emu (*Dromaius novaehollandiae*) looks very much like an ostrich but without the long neck. This Australian bird stands between the os-

A male Masai ostrich in Amboseli National Park, Kenya. *Photograph by Michelle Burgess. Stock Market. Reproduced with permissions.*

trich and rhea in height, about 5-6 ft (1.5-1.8 m) tall. It has a black head, long, brown body feathers, and white upper legs. Its feathers are unusual in that two soft feathers grow out of only one quill. Only the emu and the related cassowary have feathers like this. The emu's plumage droops downward, as if from a central part along its back.

Emus live on the open dry plains of central Australia. They do not stay in one place, but migrate several hundred miles as the **seasons** change. Emus spend the cold, dry season in the south, and then return north when the rains start. As they travel, they communicate with each other by powerful voices that boom across the plain.

An emu male chooses only one mate. She lays a dozen or more dark green eggs, but then the male sits on them for the eight-week incubation period. The chick has lengthwise white stripes on its brown body and a speckled brown and white head. The male protects the chicks until they are about six months old and can defend themselves against predators.

Until about a hundred years ago, there were several other species of emus living on the islands near Australia. However, they were killed for their meat and are now ex-

tinct. On the mainland, emus were plentiful, so plentiful that in the 1930s, Australian farmers started a campaign to exterminate emus because they competed for grass and **water** needed for cattle and **sheep**. But the birds' ability to run or blend with the surroundings, plus some ineptness on the part of the farmers, allowed the emus to survive. Even in the early 1960s, emu hunters could still collect a payment from the government for each bird they killed. However, that changed as Australians began to value the uniqueness of this bird. Now the emu and the kangaroo are featured on Australia's coat of arms.

Cassowaries

The three species of cassowaries (*Casuarius*) are found only on the **island** of New Guinea and the nearby portion of Australia. They are about the same height as a rhea and weigh about 185 lb (84 kg). However, all resemblance ends there. The southern, or Australian, cassowary (*C. casuarius*) has a vivid blue, featherless head rising from a red-orange neck. Flaps of flesh, called wattles, hang from the neck, as on a male turkey. The wattles can be almost red, green, blue, purple, or even yellow.

The body is covered by a thick coat of shiny, black feathers. Bennett's cassowary (*C. bennetti*) is considerably smaller and lacks the wattles. The female cassowary is larger than the male. This is the only large flightless bird that lives in **forests** instead of on open plains.

On top of a cassowary's head, stretching down over the base of the beak, is a bony protuberance called a casque, which means "helmet." A cassowary thrusts its casque out in front of it when it runs through the forest. Its unusual wing feathers also help it move through the forest. The cassowary's wings are almost nonexistent, but from them grow several quills that lack barbs. These bare quills stretch out beyond the other feathers on each side and serve to help push obstructions aside. Cassowaries eat mainly fruit that has fallen from trees, along with leaves and some insects.

Cassowaries live alone instead of in flocks and are nocturnal. A male and a single female come together only at mating time, when the female lays three to eight dark green eggs. The male incubates the eggs and then takes care of the young. The young cassowaries are striped from head to tail, even more vividly than the emu young.

Kiwis

Kiwis are three species of small, forest-dwelling, flightless birds that live only in New Zealand. The body length of kiwis ranges from 14-21 in (35-55 cm), and they typically stand about 15 in (38 cm) tall. Adult birds weigh 3-9 lb (1.5-4 kg). Kiwis have a rounded body with stubby, rudimentary wings, and no tail. Their legs and feet are short but strong, and they have three forward-pointing toes as well as a rudimentary hind spur. The legs are used in defense and for scratching about in the forest floor while feeding.

The bill is long, flexible, slightly down-curved, and sensitive; it is used to probe for earthworms and insects in moist soil. Their nostrils are placed at the end of the beak. Kiwis appear to be among the few birds that have a sense of smell, useful in detecting the presence of their invertebrate **prey**. They snuffle as they forage at night, and their feeding grounds are recognized by the numerous holes left by their subterranean probings.

Kiwis have coarse feathers, which are hair-like in appearance because they lack secondary aftershaft structures, such as barbules. Their shaggy plumage is brown or gray. The sexes are similar in color, but the female is larger than the male.

Kiwis lay one to two eggs, each of which can weigh almost 1 lb (0.5 kg), or about 13% of the weight of the female. Proportionate to the body weight, no other bird lays an egg as large. The female lays the eggs in an underground burrow—a cavity beneath a **tree** root, or a fallen log. The male then incubates them. The young do not feed for the first six to twelve days after hatching, and they grow slowly thereafter. Kiwis reach sexual maturity at an age of five to six years.

Kiwis are solitary, nocturnal birds. Because of the difficulties of making direct observations of wild kiwis, relatively little is known about these extraordinary birds. Kiwis make a variety of rather simple whistles and cries. That of male birds is two-syllabic, and sounds like" *ki-wi*." Obviously, this bird was named after the sound that it makes.

The brown kiwi (*Apteryx australis*) is the most widespread species, occurring in moist and wet native forests on South and North Islands, New Zealand. The little spotted kiwi (*Apteryx oweni*) is a gray-colored bird, while the great spotted kiwi (*A. haasti*) is more chestnut-colored, and larger.

Kiwis are the national symbol of New Zealand, and New Zealanders are commonly known as "kiwis." However, because these birds are nocturnal and live in dense forest, relatively few human kiwis have ever seen the feathered variety. Unfortunately, kiwis have suffered severe population declines over much of their range. This has been caused by several interacting factors. First, like other flightless birds (such as the extinct moas of New Zealand), kiwis were commonly hunted as food by the aboriginal Maori peoples of New Zealand. The feathers of kiwis were also used to ornament the ceremonial flaxen robes of the Maori. After the European colonization of New Zealand, settlers also hunted kiwis as food, and exported their skins to **Europe** for use as curiosities in the then-thriving millinery trade.

The excessive exploitation of kiwis for food and trade led to a rapid decline in their populations, and since 1908 they have been protected by law from hunting. However, some kiwis are still accidentally killed by poisons set out for pest animals, and they may be chased and killed by domestic dogs.

Kiwis have also suffered greatly from ecological changes associated with introduced mammals, especially species of **deer**. These invasive, large mammals have severely over-browsed many forests where kiwis live, causing **habitat** changes that are unfavorable to the bird, which prefers dense woody vegetation in the understory. Deer are now regarded as **pests** in New Zealand, and if these large mammals were locally exterminated, this would markedly improve the habitat available for kiwis and other native species. Fortunately, the **conservation** efforts of the government and people of New Zealand appear to be successful in their efforts to increase numbers of kiwis. These birds are now relatively abundant in some areas.

Resources

Books

Arnold, Caroline. *Ostriches and Other Flightless Birds*. Minneapolis: Carolrhoda, 1990.

Baskin-Salzberg, Anita, and Allen Salzberg. *Flightless Birds*. New York: Franklin Watts, 1993.

Forshaw, Joseph. *Encyclopedia of Birds*. New York: Academic Press, 1998.

Green, Carl R., and William R. Sanford. *The Ostrich*. New York: Crestwood House, 1987.

Ostriches, Emus, Rheas, Kiwis, and Cassowaries. San Diego: Wildlife Education, Ltd., 1990.

Jean F. Blashfield
Bill Freedman

Flooding

Flooding, although it usually carries a negative connotation, is quite a natural process and is simply the response of a natural system (a river system) to the presence of too much **water** during an **interval** of **time**. **Rivers** and streams are governed by a simple equation, $Q = A \times V$, where Q is discharge (amount of water), A is area of the river channel, and V is **velocity**. When excess discharge is present in a river or stream, at first the water moves more quickly (V increases) and perhaps some **erosion** of the channel takes place (i.e., A increases). If discharge (Q) increases too rapidly, however, water must move out of the channel (the confining area, [A]) and out onto the surrounding area, known as the floodplain. The floodplain is the area that floods first.

Floods are caused by a variety of factors, both natural and man-made. Some obvious causes of floods are heavy rains, melting snow and **ice**, and frequent storms within a short time duration. The common practice of humans to build homes and towns near rivers and other bodies of water (i.e., within natural floodplains) has contributed to the disastrous consequences of floods. In fact, floods have historically killed more people than any other form of natural disaster. Because of this fact, humans have attempted to manage floods using a variety of methods with varying degrees of success.

Causes of floods

Many floods are directly related to changes in **weather**. The most common cause of flooding is due to rain falling at extremely high rates or for an unusually long period of time. Additionally, areas that experience a great deal of snow in winter are prone to springtime flooding when the snow and ice melt, especially if the thaw is relatively sudden. Furthermore, rainfall and snowmelt can sometimes combine to cause floods.

Sometimes, floods occur as a result of a unique combination of factors that only indirectly involve weather conditions. For instance, a low-lying coastal area may be prone to flooding whenever the **ocean** is at high tide. Exceptionally high **tides** may be attributed to a **storm** caused by a combination of factors, like low barometric **pressure** and high winds. Finally, floods sometimes can occur regardless of the climate. Examples are tsunamis (seismic waves on the sea or large lakes that are caused by earthquakes), volcanic heating and rapid melting of a snow pack atop a volcanic mountain or under a glacier, or even failures of natural or man-made **dams**.

Hydrologic cycle

An underlying influence on many floods is the **hydrologic cycle**. The hydrologic cycle is the **evaporation** of water from the oceans into the atmosphere from which it falls as rain or snow on land. The water, then, runs off the land or is absorbed by it and, after some period of time, makes its way back to the oceans. Scientists have found that the total amount of water on **Earth** has not changed in three billion years; therefore, the hydrologic cycle is said to be constant. The same water has been filtered by **soil** and **plant** use and purified by **temperature** changes over many generations. Rivers and streams may feed water into the ground, or where springs persist,

Flooding on the Salt River, Arizona. *JLM Visuals. Reproduced by permission.*

groundwater may supply water to streams (allowing them to flow even when there is a **drought**).

Although the hydrologic cycle is a constant phenomenon, it is not always evident in the same place, year after year. If it occurred consistently in all locations, floods and droughts would not exist. Thus, some places on Earth experience more than average rainfall, while other places endure droughts. It is not surprising, then, that people living near rivers eventually endure floods.

Human populations

For millennia, human populations have chosen to live near bodies of water. There are three main reasons for this practice: (1) the soil near the waters is very fertile and can be used for growing **crops**; (2) the bodies of water themselves are sources of drinking and **irrigation** water; and (3) water courses support transportation and facilitate commerce and trade.

While floods can have disastrous effects, they leave behind silt and other sediments making the land surrounding rivers and other bodies of water rich and fertile. The soil deposited by moving water is known as alluvial soil. At first, populations avoided settling directly on the low-lying land, called flood plains, surrounding the rivers and instead built their villages on terraces or bluffs close to but higher than the rivers. Examples of cities developing on such terraces are Washington, Paris, and Budapest. The advantages of building on terraces is that towns are relatively safe from floods because they are situated higher in elevation than the natural flood plain, but they are also close to fertile land so food is accessible. As populations grew, however, they needed the extra land near the rivers and, therefore, moved closer to the water.

In 1992, in the United States alone, there were almost 3,800 settlements containing 2,500 or more people, each located in an area likely to flood. Furthermore, according to another estimate, nearly 1.5 billion people worldwide still farm their crops in alluvial soil; this is almost one third of the world's population. Likewise, ever since the Mesopotamians established the "cradle of civilization" between the Tigris and Euphrates Rivers in the Middle East in about 3000 B.C., populations have been attracted to rivers for transportation and trade. Narrow stretches of rivers have always been especially attractive locations for people wanting to take advantage of the natural commerce along a trade route.

Human influence on flooding

Although human populations have been victims of natural flooding, their presence and subsequent activities near rivers has also contributed to the problem. In naturally occurring conditions, vegetation captures significant amounts of **precipitation** and returns it to the atmosphere before it has a chance to hit the ground and be absorbed by the earth; however, certain farming practices, like clear-cutting land and **animal** grazing, hamper this process. Without the natural growth of vegetation to trap the rain, the ground must absorb more moisture than it would otherwise. When the absorption limit is reached, the likelihood of flooding increases. Similarly, construction of **concrete** and stone buildings contributes to the problem of flooding. While rain is easily absorbed into **sand** and other porous materials, it is not easily absorbed by man-made building materials, such as pavement and concrete. These substances cause additional run-off which must be absorbed by the surrounding landscape.

Weight of water and force of floods

Floods are probably the strongest and most dangerous form of natural disaster on Earth. In fact, one study looked at all of the people killed in natural disasters in the 20-year period ending in 1967. During this time, over 173,000 people were killed as a direct result of river floods. At the same time, about 270,000 people were killed from 18 other categories of natural disasters, including hurricanes, earthquakes, and tornadoes.

Water, when unleashed, is virtually impossible to stop. The reason behind this is twofold: water is heavy and can move with significant speed. For instance, while a single gallon of water weighs 8.5 lb (3.2 kg), the weight of high volumes of impounded water really adds up. Hoover Dam alone holds back the waters of Lake Mead, which is about 15 mi (24 km) long and contains around 10.5 trillion gal (33.6 trillion kg) of water; **multiplication** shows that this water weighs almost 90 trillion lb (33.6 trillion kg). Added to its weight is the fact that water can travel up to 20 mi (32 km) per hour. As it picks up speed, it also picks up incredible strength. In fact, moving under certain conditions, 1 in (2.54 cm) of rain can have the same **energy** potential as 60,000 tons (54,400 metric tons) of TNT.

Flood intervention

Because of the potential of a flood to destroy life and property, men and women have, for centuries, developed ways to prepare for and fight this natural disaster. One common way to manage floodwaters is to construct dams to stop excess water from inundating dry

KEY TERMS

Alluvial soils—Soils containing sand, silt, and clay, which are brought by flooding onto lands along rivers; these young soils are high in mineral content, and are the most productive soils for agriculture.

Flood plain—A clearly defined border of flat land along a river that is composed of sediment, which was deposited by the river during periodic floods or instances of high water.

Hazard zoning—Examining historical records, geological maps, and aerial photographs to predict likely areas where flooding could occur. Used for planning the location of new settlements.

Minimizing encroachment—Carefully planning where buildings are located so that they do not restrict the flow of water.

areas. Another way is to divert floodwaters away from populated areas to planned areas of flood storage. To this end, flood control reservoirs are kept partially empty so that they can catch floodwaters when the need arises. These reservoirs then release the water at a slower **rate** than would occur under flood conditions; hence, reservoirs give the soil time to absorb the excess water. About one-third of reservoirs in the United States are used for this purpose.

Two other ways to safeguard life and property are known as "hazard zoning" flood plains and "minimizing encroachment." When hazard zoning a flood plain, planners look at such things as historical records of 40-year floods, geological maps, and aerial photographs to predict likely areas where flooding could occur. Rather than relocating populations, hazard zoning is used for planning the location of new settlements. Minimizing encroachment means carefully planning where buildings are located so that they do not restrict the flow of water or cause water to pond excessively; however, as long as people choose to live in low-lying, flood-prone areas, scientists and engineers can only do so much to protect them from the risks of floods caused by both natural conditions and human activities.

See also Alluvial systems; Hydrology; Soil conservation; Watershed.

Resources

Books

Collier, Michael, and Robert H. Webb. *Floods, Droughts, and Climate Change.* Tucson, AZ: University of Arizona Press, 2002.

Dingman, S. Lawrence. *Physical Hydrology.* 2nd. ed. Upper Saddle River, NJ: Prentiss Hall, 2002.

Parker, Sybil P. and Robert A. Corbett, eds. *McGraw-Hill Encyclopedia of Environmental Science and Engineering.* 3rd ed. New York: McGraw-Hill, Inc., 1993.

World Commission on Dams. *Dams and Development: A New Framework for Decision-Making.* Earthscan Publications, 2001.

Other

United States Geological Survey. "Floods" August 28, 2002 [cited January 15, 2003]. <http://www.usgs.gov/themes/flood.html>.

Kathryn Snavely

Flora

Flora is a word used to describe the assemblage of **plant** species that occurs in some particular area or large region. Flora can refer to a modern assemblage of plant **species**, or to a prehistoric group of species that is inferred from the fossil record. The zoological analogue is known as a **fauna**, although this word is usually used in reference to a large region. More locally, "vegetation" refers to the occurrence of groupings of plants, often called communities, in some area or region.

The word flora is also sometimes used to refer to a book that describes a taxonomic treatment of plants in some region. Floras of this sort often contain identification keys, diagrammatic, and written descriptions of the species, range maps, and descriptions of **habitat**.

Plant biogeographers have divided **Earth** and its regions into floristic units on the basis of their distinctive assemblages of plant species. The species of these large regions (sometimes called biomes, especially in the ecological context) are segregated on the basis of two complexes of factors: (1) geographic variations of environmental conditions, especially climate and to a lesser degree, **soil**, and (2) physical and ecological barriers to **migration**, which prevent the distinctive species of floras from mixing together.

In cases where regions have been physically separated for very long periods of **time**, their differences in plant species are especially great. In particular, isolated oceanic islands often have unique floras, composed of many **endemic** species of plants that occur nowhere else. For example, islands of the Hawaiian archipelago have been isolated from the nearest mainland for millions of years, and are estimated to have had an original flora of about 2,000 species of **angiosperm** plants, of which 94-98% were endemic. Unfortunately, many of these unique species have become extinct since these islands were discovered and colonized by humans, first by Polynesians and, more recently and with much greater ecological damages, by Europeans.

In cases where the physical isolation is less ancient, there can be a substantial overlap of genera and species in the floras of different regions. For example, eastern Siberia and the Alaska-Yukon region of **North America** were physically connected by a land bridge during the most recent era of glaciation, which abated about 14,000 years ago. **Reciprocal** movements of plants (and some animals, including humans) occurred across that land bridge, and this is indicated today by numerous examples of the occurrence of the same plant species in both regions. For this reason, these regions are considered to have a floristic affinity with each other.

See also Biome; Fauna.

Flounder *see* **Flatfish**

Flourescence in situ hybridization *see* **Fluorescence in situ hybridization (FISH)**

Flower

A flower is the reproductive structure of an **Angiosperm** plant. Flowers have ovaries with ovules that develop into **fruits** with **seeds**. There are over 300,000 **species** of Angiosperms, and their flowers and fruits vary significantly. Flowers and fruits are among the most useful features for the identification of plant species and determination of their evolutionary relationships.

Study of flowers throughout history

The hunter-gatherer ancestors of modern humans surely noticed that flowers gave rise to fruits which could be eaten. Because flowers signaled an anticipated harvest, it has been suggested that these early humans instinctively considered flowers attractive, an **instinct** that modern humans may also have. Many modern cultures consider flowers attractive, and scholars have been fascinated with flowers for millennia.

Dioscorides, a Greek physician in Emperor Nero's army (first century A.D.), wrote the most influential early book on plants, *De Materia Medica.* This was the first book about the medicinal uses of plants, referred to as an *herbal.* Dioscorides's book had diagrams of many plants and their flowers, and this helped other physicians to identify the species of plant to prescribe to their patients for a particular ailment.

De Materia Medica remained an important reference on plants for more than 1,500 years. However, early scholars lacked the **printing** press, so all copies had to be scripted by hand. Over **time**, the pictures of plants and their flowers in these hand-copied herbals became more romanticized and less accurate.

The 1500s were the "golden age" of herbals, when European scholars published their own books whose illustrations were based on observations of living plants, rather than upon Dioscorides's diagrams and descriptions. With the invention of the movable type printing press, these herbals became the first published scholarly works in **botany** and were widely read.

Carolus Linnaeus of Sweden revolutionized botany in the mid-1700s. He classified plant species according to the morphology of their flowers and fruits. Modern botanists continue to rely upon flowers for identification as well as the determination of evolutionary relationships.

In Linnaeus's time, many people argued in the doctrine of "Divine Beneficence," which held that all things on **Earth** were created to please humans. Thus, people believed that flowers with beautiful colors and sweet smells were created by God to please humans. Christian Konrad Sprengel of Germany disputed this view in the late 1700s. He held that the characteristics of flowers are related to their method of reproduction. Sprengel published his theory of flowers in his book *The Secret of Nature Revealed* (1793).

Sprengel's ideas were not widely accepted in his own time. However, in 1841 the English botanist Robert Brown gave Charles Darwin a copy of Sprengel's book. This book influenced Darwin's development of his theory of **evolution** by natural **selection**, which culminated in the publication of *The Origin of Species* (1859). Sprengel's work also stimulated Darwin's subsequent study of orchids, and he wrote *The Various Contrivances by Which Orchids Are Fertilized by Insects* (1862). Darwin's important studies of flowers and **pollination** supported Sprengel's view that there is a relationship between the characteristics of a flower and its method of pollination. Moreover, Darwin demonstrated that some of the highly specialized characteristics of flowers had evolved by natural selection to facilitate their pollination.

Parts of the flower

There are considerable differences among the flowers of the 300,000 species of Angiosperms. Botanists rely upon a large vocabulary of specialized terms to describe the parts of these various flowers. The most important morphological features of flowers are considered below.

Flowers can arise from different places on a plant, depending on the species. Some flowers are terminal, meaning that a single flower blooms at the apex of a stem. Some flowers are axial, in that they are borne on the axes of branches along a stem. Some flowers arise in an inflorescence, a branched cluster of individual flowers.

There are four whorls of organs in a complete flower. From the outside to the inside, one encounters sepals, petals, stamens, and carpels. The sepals are leaf-like organs, which are often green, but can sometimes be brown or brightly colored, depending on the species. The petals are also leaf-like and are brightly colored in most animal-pollinated species but dull in **color** or even absent in wind-pollinated plants.

The stamens and carpels, the reproductive organs, are the most important parts of a flower. The stamens are the male, pollen-producing organs. A stamen typically consists of an anther attached to a filament (stalk). The anther produces many microscopic pollen grains. The male sex **cell**, a sperm, develops within each pollen grain.

The carpels are the female ovule-producing organs. A carpel typically consists of an ovary, style, and stigma. The stigma is the tip of the carpel upon which pollen grains land and germinate. The style is a stalk that connects the stigma and ovary. After the pollen grain has germinated, its pollen tube grows down the style into the ovary. The ovary typically contains one or more ovules, structures which develop into seeds upon **fertilization** by the sperm. As the ovules develop into seeds, the ovary develops into a fruit, whose characteristics depend on the species.

In some species, one or more of the four whorls of floral organs is missing, and the flower is referred to as an incomplete flower. A bisexual flower is one with both stamens and carpels, whereas a unisexual flower is one which has either stamens or carpels, but not both. All complete flowers are bisexual since they have all four floral whorls. All unisexual flowers are incomplete since they lack either stamens or carpels. Bisexual flowers, with stamens and carpels, can be complete or incomplete, since they may lack sepals and/or petals.

Evolution of flowers

The flower originated as a structure adapted to protect ovules, which are borne naked and unprotected in the Gymnosperms, ancestors of the Angiosperms. Botanists are uncertain about which group of Gymnosperms is most closely related to the Angiosperms. Recently, examination of sexual fertilization in the different groups of Gymnosperms suggests that Angiosperms may be most closely related to the Gnetophyta, a small phylum with three genera (*Ephedra*, *Gnetum*, and *Welwitschia*) and about 70 species.

The Angiosperms first appeared in the fossil record in the early Cretaceous period (about 130 million years ago) and rapidly increased in diversity. Once the flowering plants had evolved, natural selection for efficient pollination by **insects** and other animals was important in their diversification. By the mid-Cretaceous, species with flowers of many different designs had evolved. These varying designs evolved as a consequence of the close association of the flowers and their **animal** pollinators, a process referred to as coevolution. In addition, many flowers became self-incompatible, in that they relied upon cross-pollination, that is, pollination by another individual of the same species. Cross-pollination increases the genetic diversity of the offspring, in general making them more fit.

Today, flowering plants are the dominant terrestrial plants in the world. There are more than 300 different families of flowering plants. The Asteraceae, with over 15,000 species, is one of the largest and most diverse families of Angiosperms, and their flowers are highly evolved. The dandelion, daisy, and sunflower are familiar species of the Asteraceae. In these species, many small individual flowers are packed closely together into a dense inflorescence called a head, which appears rather superficially like a single large flower. Each individual flower in the head has a single ovule, which produces a single seed upon fertilization. The flowers of many species of the Asteraceae, such as the dandelion, evolved highly specialized sepals, which are scale-like and are referred to as a pappus. In summer, the pappus of the dandelion expands into the furry white structure which aids the tiny attached seeds in their dispersal by the **wind**.

Induction of flowering

Many environmental cues signal a plant to produce flowers, and **light** is one of the most important of these. In many species, flowering is a photoperiodic response, in that it is controlled by the length of the light and dark periods to which the plant is exposed.

Some plants, such as Maryland mammoth tobacco, **soybean**, and **hemp**, are short-day plants, in that they flower in the spring and autumn when the days are shorter. Other plants, for example, **spinach**, mouse **ear** cress, and fuchsia, are long-day plants, in that they flower in the summer when the days are longer. Some plants, such as cucumber, corn, and garden peas, are day-neutral plants, in that they flower regardless of the daylength. Often, different varieties of the same species have different light requirements for flowering.

The dark period is as crucial as the light period for induction of flowering. In particular, when a short-day plant is exposed to short days, but given a pulse of light during the dark period, flowering is inhibited. When a long-day plant is exposed to short days but given a pulse

---

KEY TERMS

. . . .

Anther—The part of the stamen that produces pollen.

Carpel—Female reproductive organ of flowers which is composed of the stigma, style, and ovary.

Filament—Stalk of the stamen which bears the anther.

Ovary—Basal part of the carpel which bears ovules and develops into a fruit.

Ovule—Structure within the ovary which develops into a seed after fertilization.

Petal—Whorl of a flower just inside the sepals that is often colored.

Sepal—External whorl of a flower which is typically leaflike and green.

Stamen—Male reproductive organ of flowers that produces pollen.

Stigma—The part of the female organs of a plant flower (the pistil) upon which pollen lands in the first stage of fertilization.

Style—A stalk that joins the pollen-receptive surface of the stigma, to the ovary of the female organ of a plant (i.e., the pistil).

---

of light during the dark period, flowering is promoted. Phytochromes are the photoreceptive plant pigments which detect these light pulses. Phytochromes also control other stages of plant growth and development, and phytochrome genes have been cloned and sequenced in many plant species.

Regrettably, plant physiologists have made little additional progress in understanding the mechanism of flower induction in recent years. Little is known about the biochemical reactions that follow from activation of phytochrome or how plants measure photoperiod. This is an area in which future botanists may make great progress.

See also Sexual reproduction.

Resources

Books

The American Horticultural Society. *The American Horticultural Society Encyclopedia of Plants and Flowers.* New York: DK Publishing, 2002.

Judd, Walter S., Christopher Campbell, Elizabeth A. Kellogg, Michael J. Donoghue, and Peter Stevens. *Plant Systemat-*

ics: A Phylogenetic Approach. 2nd ed. with CD-ROM. Suderland, MD: Sinauer, 2002.

Kaufman, P.B., et al. *Plants: Their Biology and Importance.* New York: Harper College Press, 1990.

Periodicals

Adams, K.L., et al. "Repeated, Recent and Diverse Transfers of a Mitochondrial Gene to the Nucleus in Flowering Plants." *Nature* 408 (2000): 354-357.

Peter A. Ensminger

Fluid dynamics

Fluid dynamics is the study of the flow of liquids and gases, usually in and around solid surfaces. The flow patterns depend on the characteristics of the fluid, the speed of flow, and the shape of the solid surface. Scientists try to understand the principles and mechanisms of fluid dynamics by studying flow patterns experimentally in laboratories and also mathematically, with the aid of powerful computers. The two fluids studied most often are air and **water**. **Aerodynamics** is used mostly to look at air flow around planes and automobiles with the aim of reducing drag and increasing the efficiency of **motion**. Hydrodynamics deals with the flow of water in various situations such as in pipes, around ships, and underground. Apart from the more familiar cases, the principles of fluid dynamics can be used to understand an almost unimaginable variety of phenomena such as the flow of **blood** in blood vessels, the flight of **geese** in V-formation, and the behavior of underwater plants and animals.

Factors that influence flow

The **viscosity**, **density**, and compressibility of a fluid are the properties that determine how the liquid or gas will flow. Viscosity measures the internal **friction** or resistance to flow. Water, for instance, is less viscous than honey and so flows more easily. All gases are compressible whereas liquids are practically incompressible and cannot be squeezed into smaller volumes. Flow patterns in compressible fluids are more complicated and difficult to study than those in incompressible ones. Fortunately for **automobile** designers, air, at speeds less than 220 MPH (354 km/h) or one-third the speed of sound, can be treated as incompressible for all practical purposes. Also, for incompressible fluids, the effects of **temperature** changes can be neglected.

Reynolds number

The speed of flow is another factor that determines the nature of flow. The speed of flow is either that of a liquid or gas moving across a solid surface or, alternatively, the speed of a solid object moving through a fluid. The flow patterns in either case are exactly the same. That is why airplane designs can be tested in **wind** tunnels where air is made to flow over stationary test models to simulate the flight of actual planes moving through the air.

The speed of flow is related to the viscosity by virtue of the fact that a faster moving fluid behaves in a less viscous manner than a slower one. Therefore, it is useful to take viscosity and speed of flow into account at the same time. This is done through the Reynolds number named after the English scientist Observe Reynolds (1842-1912). This number characterizes the flow. It is greater for faster flows and more dense fluids and smaller for more viscous fluids. The Reynolds number also depends on the size of the solid object. The water flowing around a large **fish** has a higher Reynolds number than water flowing around a smaller fish of the same shape.

As long as the shape of the solid surface remains the same, different fluids with the same Reynolds number flow in exactly the same way. This very useful fact is known as the principle of similarity or similitude. Similitude allows smaller scale models of planes and cars to be tested in wind tunnels where the Reynolds number is kept the same by increasing the speed of air flow or by changing some other property of the fluid. The Ford Motor Company has taken advantage of the principle of similarity and conducted flow tests on cars under water. Water flowing at 2 MPH (3.2 km/h) was used to simulate air flowing at 30 MPH (48 km/h).

Laminar and turbulent flow

Flow patterns can be characterized as laminar or turbulent. In laminar or streamlined flow, the fluid glides along in layers which do not mix so that the flow takes place in smooth continuous lines called streamlines. This is what we see when we open a water faucet just a little so that the flow is clear and regular. If we continue turning the faucet, the flow gradually becomes cloudy and irregular. This is known as turbulent flow. This change to **turbulence** takes place at high Reynolds numbers. The critical Reynolds number at which this change occurs differs from case to case.

Bernoulli's principle

An important idea in fluid flow is that of the **conservation** of **mass**. This implies that the amount of fluid that goes in through one end of a pipe is the same as the amount of fluid that comes out through the other end. Thus the fluid has to flow faster in narrower sections or constrictions in the pipe. Another important idea, expressed by **Bernoulli's principle**, is that of the conservation of **energy**.

Daniel Bernoulli (1700-1782) was the first person to study fluid flow mathematically. He imagined a completely non-viscous and incompressible or "ideal" fluid in order to simplify the **mathematics**. Bernoulli's principle for an ideal fluid essentially says that the total amount of energy in a laminar flow is always the same. This energy has three components—potential energy due to gravity, potential energy due to **pressure** in the fluid, and kinetic energy due to speed of flow. Since the total energy is constant, increasing one component will decrease another. For instance, in a horizontal pipe in which gravitational energy stays the same, the fluid will move faster through a constriction and will, therefore, exert less pressure on the walls. In recent years, powerful computers have made it possible for scientists to attack the full mathematical complexity of the equations that describe the flow of real, viscous, and compressible fluids. Bernoulli's principle, however, remains surprisingly relevant in a variety of situations and is probably the single most important principle in fluid dynamics.

Boundary layer theory

Even though Bernoulli's principle works extremely well in many cases, neglecting viscosity altogether often gives incorrect results. This is because even in fluids with very low viscosity, the fluid right next to the solid boundary sticks to the surface. This is known as the no-slip condition. Thus, however fast or easily the fluid away from the boundary may be moving, the fluid near the boundary has to slow down gradually and come to a complete stop exactly at the boundary. This is what causes drag on automobiles and airplanes in spite of the low viscosity of air.

The treatment of such flows was considerably simplified by the boundary layer concept introduced by Ludwig Prandtl (1875-1953) in 1904. According to Prandtl, the fluid slows down only in a thin layer next to the surface. This boundary layer starts forming at the beginning of the flow and slowly increases in thickness. It is laminar in the beginning but becomes turbulent after a point determined by the Reynolds number. Since the effect of viscosity is confined to the boundary layer, the fluid away from the boundary may be treated as ideal.

Shape and drag

Moving automobiles and airplanes experience a resistance or drag due to the **force** of air sticking to the surface. Another source of resistance is pressure drag, which is due to a phenomenon known as flow separation. This happens when there is an abrupt change in the

> ## KEY TERMS
>
> **Boundary layer**—The layer of fluid that sticks to the solid surface and in which the speed of the fluid decreases.
>
> **Compressibility**—The property that allows a fluid to be compressed into a smaller volume.
>
> **Laminar**—A mode of flow in which the fluid moves in layers along continuous well-defined lines known as streamlines.
>
> **Reynolds number**—A number that characterizes a flow situation and allows flows of different fluids in different situations to be compared.
>
> **Turbulent**—An irregular, disorderly mode of flow.
>
> **Viscosity**—The internal friction within a fluid that makes it resist flow.
>
> **Wake**—The area of low pressure turbulent flow behind a moving body that causes the body to experience resistance to its forward motion.

shape of the moving object, and the fluid is unable to make a sudden change in flow direction and stays with the boundary. In this case, the boundary layer gets detached from the body, and a region of low pressure turbulence or wake is formed below it. This creates a drag on the vehicle due to the higher pressure in the front. That is why aerodynamically designed cars are shaped so that the boundary layer remains attached to the body longer, creating a smaller wake and, therefore, less drag. There are many examples in nature of shape modification for drag control. The sea anemone, for instance, continuously adjusts its form to the **ocean** currents in order to avoid being swept away while gathering food.

Resources

Books

Batchelor, G. K. *An Introduction to Fluid Dynamics*. Cambridge: Cambridge University Press, 2000.

Fox, Robert W., and Alan T. McDonald. *Introduction to Fluid Mechanics*. 5th ed. New York: John Wiley & Sons, 1998.

Ingram, Jay. *The Science of Everyday Life*. New York: Viking Penguin Inc., 1989.

The Japan Society of Mechanical Engineers, eds. *Visualized Flow-Fluid Motion in Basic Engineering Situations Revealed by Flow Visualization*. Oxford: Pergamon Press, 1988.

Kundu, Pijush K., and Ira M. Cohen. *Fluid Mechanics*. 2nd ed. San Diego: Academic Press, 2001.

Wegener, Peter P. *What Makes Airplanes Fly?* New York: Springer-Verlag, 1991.

Periodicals

Valenti, Michael. "Underwater Creatures Go With The Flow." *Mechanical Engineering* (May 1993): 130.

Sreela Datta

Fluid mechanics

Fluid mechanics is the study of gases and liquids at rest and in **motion**. Fluid statics studies the behavior of stationary fluids and tells us, for instance, how much air to put in our tires and whether a boat in a **lake** will float or sink. **Fluid dynamics** studies the flow behavior of moving fluids. Both global **weather** patterns and the flow of **water** from a faucet are governed by the laws of fluid dynamics.

A fluid at rest exerts static **pressure** on its container and on anything that is submerged in it. The pressure at any point in the fluid is the **force** that would be exerted on unit area of a submerged object. This pressure is the same in all directions. Because of gravity, the pressure in a fluid increases as one goes deeper below the surface. Marine creatures dwelling deep down in the **ocean** have to withstand greater pressures, due to the weight of the water above, than **fish** swimming near the surface. The exact pressure at different depths depends only the **density** of the fluid and the depth from the surface.

This pressure increase with depth also provides the upward buoyant force on a floating object. The pressure below a boat is greater than the pressure at higher points and, therefore, pushes the boat upwards. The upward buoyant force is equal to the weight of the **volume** of water displaced by the boat. This is known as Archimedes's principle. A heavy boat will float as long as it has a large enough volume to displace enough water to balance its weight.

External pressure exerted on a fluid is transmitted, undiminished, throughout the fluid. This is known as Pascal's principle. This principle is used in a hydraulic lever in which pressure applied on a small piston is transmitted unchanged to a large piston. Since the force exerted is equal to the pressure times the area of the piston, a small force exerted on the small piston can lift a heavy load placed on the large piston. Hydraulic **jacks**, used to lift cars, are based on this principle (Figure 1).

In a moving fluid some of this static pressure is converted into dynamic pressure due to the **velocity** of the fluid. A fluid moving faster in a pipe has more dynamic pressure and thus exerts less static pressure on the sides. The complete fluid flow equations are so complicated that only recent advances in computational capability

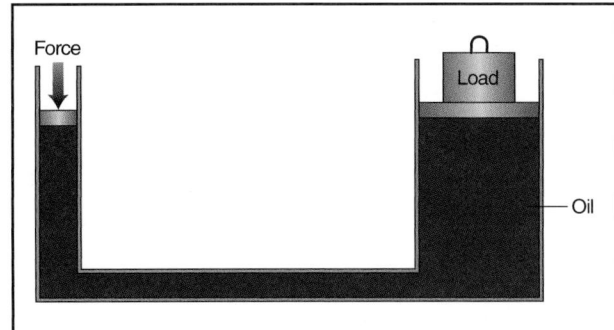

Figure 1. A hydraulic press. *Illustration by Hans & Cassidy. Courtesy of Gale Group.*

have made it possible to describe fluid flow fairly accurately in some situations. There are, however, many simplified ways to study flow mathematically and obtain a great deal of insight.

The practical science of hydraulics enabled human beings to design water clocks, irrigate **crops**, and build waterwheels long before the mathematical study of fluids was begun. Now, experiment and theory support each other in designing **dams**, underwater tunnels, and hydraulic machines, and in predicting flows in **rivers**, around airplane wings, and in the atmosphere.

See also Aerodynamics.

Fluke *see* **Flatworms**

Fluorescence

Fluorescence is the process by which a substance absorbs electromagnetic **radiation** (visible or invisible **light**) from another source, then re-emits the radiation with a wavelength that is longer than the wavelength of the illuminating radiation. It can be observed in gases at low **pressure** and in certain liquids and solids, such as the ruby gemstone. Fluorescence is the principle that is the basis of the common fluorescent lamp used for lighting; it is also a useful laboratory diagnostic tool.

Fundamentals

Matter interacts with electromagnetic radiation (such as ultraviolet and visible light) through the processes of absorption and **emission**. The internal structure of **atoms** and molecules is such that absorption and emission of electromagnetic radiation can occur only between distinct **energy** levels. If the atom is in its lowest energy level or ground state, it must absorb the exact

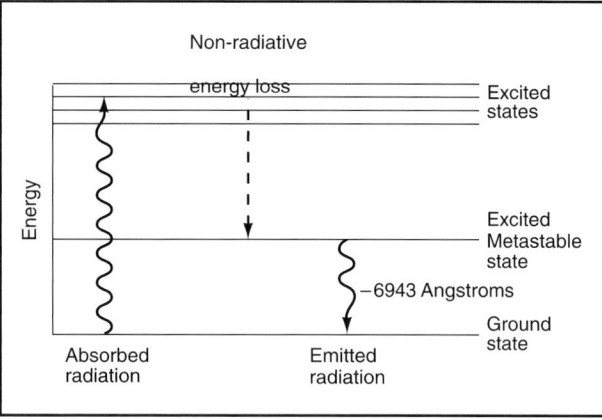

Figure 1. Fluorescence in ruby. *Illustration by Hans & Cassidy. Courtesy of Gale Group.*

amount of energy required to reach one of its higher energy levels, called excited states. Likewise an atom that is in an excited state can only emit radiation whose energy is exactly equal to the difference in energies of the initial and final states. The energy of electromagnetic radiation is related to its wavelength as follows; shorter wavelengths correspond to greater energies, longer wavelengths correspond to lower energies.

In addition to emitting radiation, atoms and molecules that are in excited states can give up energy in other ways. In a gas, they can transfer energy to their neighbors through collisions which generate **heat**. In liquids and solids, where they attached to their neighbors to some extent, they can give up energy through vibrations. The observation of fluorescence in gases at low pressure comes about because there are too few neighboring atoms or molecules to take away energy by collisions before the radiation can be emitted. Similarly, the structure of certain liquids and solids permits them to exhibit strong fluorescence.

If the wavelength of the radiation that was absorbed by the fluorescent material is equal to that of its emitted radiation, the process is called resonance fluorescence. Usually though, the atom or **molecule** loses some of its energy to its surroundings, so that the emitted radiation will have a longer wavelength than the absorbed radiation. In this process, simply called fluorescence, Stoke's Law says that the emitted wavelength will be longer than the absorbed wavelength. There is a short delay between absorption and emission in fluorescence that can be a millionth of a second or less. There are some solids, however, that continue to emit radiation for seconds or more after the incident radiation is turned off. In this case, the phenomenon is called phosphorescence.

As an example of fluorescence, consider the energy level diagram for the gemstone ruby in Figure 1. Ruby is

KEY TERMS

Angstrom—A unit of length equal to one ten-billionth of a meter.

Energy level—The internal energy state of an atom or molecule which is characterized by having only discrete, discontinuous values.

Excited state—Any energy level with an energy greater than that of the ground state.

Fluorescent efficiency—The ratio of the intensity of the fluorescent radiation to the intensity of the absorbed radiation.

Fluorescent lamp—A device that utilizes the phenomenon of fluorescence to produce light for illumination.

Ground state—The lowest energy level of an atom or molecule.

Metastable state—An energy level in which an atom or molecule can remain for a period longer than its other energy levels before returning to its ground state.

Phosphorescence—The persistent emission of radiation by a substance following the removal of the illuminating radiation.

Resonance fluorescence—Fluorescence in which the emitted radiation has the same wavelength as the absorbed radiation.

Stoke's law—In fluorescence, the emitted wavelength is always longer than the absorbed wavelength.

Ultraviolet radiation—Radiation similar to visible light but of shorter wavelength, and thus higher energy.

Visible light—Electromagnetic radiation of wavelength between 4,000 and 8,000 angstroms.

Wavelength—The distance between two consecutive crests or troughs in a wave.

a crystalline solid composed of **aluminum**, **oxygen**, and a small amount of chromium, which is the atom responsible for its reddish **color**. If blue light strikes a ruby in its ground state, it is absorbed, raising the ruby to an excited state. After losing some of this energy to internal vibrations the ruby will settle into a metastable state—one in which it can remain longer than for most excited states (a few thousandths of a second). Then the ruby will spontaneously drop to its ground state emitting red radiation whose wavelength (longer than the blue radiation) measures 6,943 angstroms. The fluorescent efficiency of the ruby—the **ratio** of the intensity of fluorescent radiation to the intensity of the absorbed radiation—is very high. For this reason the ruby was the material used in building the first **laser**.

Applications

The most well-known application of fluorescence is the fluorescent lamp, which consists of a **glass** tube filled with a gas and lined with a fluorescent material. **Electricity** is made to flow through the gas, causing it to radiate. Often mercury vapor, which radiates in the violet and ultraviolet, is used. This radiation strikes the coating, causing it to fluoresce visible light. Because the fluorescence process is used, the fluorescent lamp is more efficient and generates less heat than an incandescent bulb.

Resonance fluorescence can be used as a laboratory technique for analyzing different phenomena such as the gas flow in a **wind** tunnel. Art forgeries can be detected by observing the fluorescence of a painting illuminated with ultraviolet light. Painting medium will fluoresce when first applied, then diminish as **time** passes. In this way paintings that are apparently old, but are really recent forgeries, can be discovered.

John Appel

Fluorescence in situ hybridization (FISH)

Fluorescent in situ hybridization (**FISH**) is a powerful technique for detecting RNA or DNA sequences in cells, tissues, and tumors. FISH provides a unique link among the studies of **cell biology**, cytogenetics, and molecular **genetics**.

Fluorescent in situ hybridization is a technique in which single-stranded nucleic acids (usually DNA, but RNA may also be used) are permitted to interact so that complexes, or hybrids, are formed by molecules with sufficiently similar, complementary sequences. Through **nucleic acid** hybridization, the degree of sequence identity can be determined, and specific sequences can be detected and located on a given **chromosome**.

The method comprises of three basic steps: fixation of a specimen on a **microscope** slide, hybridization of labeled probe to homologous fragments of genomic DNA, and enzymatic detection of the tagged target hybrids. While probe sequences were initially detected with isotopic reagents, nonisotopic hybridization has become increasingly popular, with fluorescent hybridization now a common choice. Protocols involving nonisotopic probes are considerably faster, with greater signal resolution, and provide options to visualize different targets simultaneously by combining various detection methods.

The detection of sequences on the target chromosomes is performed indirectly, commonly with biotinylated or digoxigenin-labeled probes detected via a fluorochrome-conjugated detection reagent, such as an antibody conjugated with fluorescein. As a result, the direct visualization of the relative position of the probes is possible. Increasingly, nucleic acid probes labeled directly with fluorochromes are used for the detection of large target sequences. This method takes less **time** and results in lower background; however, lower signal intensity is generated. Higher sensitivity can be obtained by building layers of detection reagents, resulting in amplification of the signal. Using such means, it is possible to detect single-copy sequences on chromosome with probes shorter than 0.8 kb.

Probes can vary in length from a few base pairs for synthetic oligonucleotides to larger than one Mbp. Probes of different types can be used to detect distinct DNA types. PCR-amplified repeated DNA sequences, oligonucleotides specific for repeat elements, or cloned repeat elements can be used to detect clusters of repetitive DNA in heterochromatin blocks or centromeric regions of individual chromosomes. These are useful in determining aberrations in the number of chromosomes present in a cell. In contrast, for detecting single **locus** targets, cDNAs or pieces of cloned genomic DNA, from 100 bp to 1 Mbp in size, can be used.

To detect specific chromosomes or chromosomal regions, chromosome-specific DNA libraries can be used as probes to delineate individual chromosomes from the full chromosomal complement. Specific probes have been commercially available for each of the human chromosomes since 1991.

Any given **tissue** or cell source, such as sections of frozen tumors, imprinted cells, cultured cells, or embedded sections, may be hybridized. The DNA probes are hybridized to chromosomes from dividing (metaphase) or non-dividing (interphase) cells.

The observation of the hybridized sequences is done using epifluorescence **microscopy**. White **light** from a source lamp is filtered so that only the relevant wavelengths for excitation of the fluorescent molecules reach the **sample**. The light emitted by fluorochromes is generally of larger wavelengths, which allows the distinction between excitation and **emission** light by means of a second optical filter. Therefore, it is possible to see bright colored signals on a dark background. It is also possible to distinguish between several excitation and emission bands, thus

KEY TERMS

. . . .

Kilobase (kb)—A distance unit used in physical maps. A kilobase (kb) unit reflects 1,000 bases.

Megabase (mb)—A distance unit used in physical maps. A megabase (mb) unit reflects 1,000,000 (one million) bases.

Physical genetic map—A genetic map are based upon actual distances between genes on a chromosome. Contigs are physical maps are based on collections of overlapping DNA fragments.

between several fluorochromes, which allows the observation of many different probes on the same target.

FISH has a large number of applications in **molecular biology** and medical science, including **gene** mapping, **diagnosis** of **chromosomal abnormalities**, and studies of cellular structure and function. Chromosomes in three-dimensionally preserved nuclei can be "painted" using FISH. In clinical research, FISH can be used for prenatal diagnosis of inherited chromosomal aberrations, postnatal diagnosis of carriers of genetic **disease**, diagnosis of infectious disease, viral and bacterial disease, **tumor** cytogenetic diagnosis, and detection of aberrant gene expression. In laboratory research, FISH can be used for mapping chromosomal genes, to study the **evolution** of genomes (Zoo FISH), analyzing nuclear organization, visualization of chromosomal territories and **chromatin** in interphase cells, to analyze dynamic nuclear processes, somatic **hybrid** cells, replication, chromosome sorting, and to study tumor biology. It can also be used in developmental biology to study the temporal expression of genes during differentiation and development. Recently, high resolution FISH has become a popular method for ordering genes or DNA markers within chromosomal regions of interest.

See also Chromosome mapping; Genetic engineering; Genetic testing.

Resources

Books

Spector, D.L., R.D. Goldman, and L.A. Leinwand. *Cells: A Laboratory Manual.* Plainview, NY: Cold Spring Harbor Laboratory Press, 1998.

Periodicals

Nath, J., et al. "A Review of Fluorescence in situ Hybridization (FISH): Current Status and Future Prospects." *Biotech Histochem.* 75 (March 2000): 54-78.

Nicole D. Le Brasseur

Fluorescent light

Fluorescent **light** is the most common type of electrical light found in the United States; it is used for practically all commercial lighting, i.e. offices, factories, stores and schools, and it is estimated that there are 1.5 billion fluorescent lamps in use nationwide. Fluorescent lighting is popular due to its high efficacy, i.e. it produces between three to five times more light than an incandescent lamp consuming the same electrical power. The main reason for this is that the fluorescent lamp employs a phosphor which converts the non-visible light produced by the lamp into visible light, whereas a large fraction of the output from the incandescent lamp is infra-red light which escapes as **heat**.

Although the fluorescent lamp was first demonstrated by Becquerel in the 1860s, it was not commercially available until 1938 with the introduction of phosphors which could withstand the rigors of operation for a reasonable length of **time**. Since then improvements have been made in all aspects of the lamp: electrodes, phosphors, gas mixtures, and control circuitry. These improvements are particularly important simply because there are so many fluorescent lamps in use. Over its lifetime, a standard fluorescent lamp consumes as much **electricity** as is generated by a barrel of oil: the importance of even small increases in efficacy become apparent when one considers that even a 10% increase will result in savings of approximately 40 million barrels a year in the United States alone.

Construction and operation

The fluorescent lamp is formed from a sealed, hollow **glass** tube which is straight, although other shapes can also be used. The tube contains a low **pressure** mixture of noble gas and mercury vapor through which an AC electrical discharge is run, has electrodes located at either end, and has a coating of an inorganic phosphor on the inside surface. Each electrode acts as **cathode** and **anode** during one complete period of the AC discharge and is coated with a material of low work-function, such as **barium** oxide, which, when heated, acts as a source of electrons to feed the electrical discharge. Other electrons are created in the discharge by impact ionization of the gas mixture. The gas mixture uses a noble gas, usually krypton, to act as a **buffer** in the discharge. On excitation by electrons in the discharge, the mercury **atoms** emit light, mostly at a wavelength of 254 nm which is in the deep ultraviolet (UV). This UV light reaches the phosphor coating on the walls of the tube where it is absorbed and re-emitted at a longer wavelength in the visible. The visible light passing out of the glass envelope is used for illumination. The **color** of the emitted light is determined by the phosphor and is a particularly important characteristic of the lamp.

Starting and running the discharge

Unlike the electrical circuit for an incandescent lamp, which contains only a switch, the control circuit for a fluorescent lamp must do two things. It must first provide a high voltage spike to strike the discharge, and it must thereafter control the current and voltage once the discharge is stable. The latter is important because the discharge itself is unstable and will terminate if the current is not controlled externally.

There are several types of starter circuits which all do two things. They supply a large current to the electrodes in order to produce electrons via thermioemission (the electrons "boil off" as the electrodes heat up) and they supply a high voltage to strike the discharge. Typical examples of these include the switch start, instant-start, and rapid start. The switch start has the advantage of being actively controlled and therefore avoids the misfirings which can have the deleterious effect of removing the coating on the electrodes and thus shorten the tube's life.

The switch is initially closed, thus shorting the electrodes and allowing a large current to flow which heats the electrodes to their operating **temperature**. After a short time (1-2 seconds), the switch is opened. The large voltage spike created by the sudden reduction of current through the ballast (an inductor) then strikes the discharge and the lamp lights up. The **capacitor** reduces the reactance of the inductive ballast.

The switch used to be an argon glow tube with a bimetallic electrode, but this function has been replaced in recent years with solid state circuitry which can be actively controlled.

AC operation

Fluorescent lamps are usually operated with an AC discharge whose **frequency** is set by the power supply-60 Hz in the United States. However, it has been found that the tube has a higher efficacy if it is operated at a high frequency, for example 20-30 kHz. The reason for this increase in power is that there is less time between field reversals for the ions to collide with the electrodes, and so the **rate** of **energy** loss through electrode collision is reduced. Operation at high frequency requires a transistorized ballast, which has the added advantage that the lamp can be dimmed, unlike low frequency lamps where the current and voltage to the tube are fixed and the tube cannot be dimmed.

Phosphors and color

The phosphor converts the UV output from the mercury discharge into visible light via **fluorescence**. The mix of color emitted depends on the chemical compounds used in the phosphor. Many compounds produce what is perceived as a white light, which may indeed be a broad **emission** centered around 590 nm, as in the case of the so-called cool white and warm white halophosphates (the warm contains more red than the cool). However, recent developments in phosphors for **television** tubes have resulted in the introduction of the "triphosphor," which is a mixture of three different phosphor components emitting in the blue, green, and red. The light from a triphosphor tube distorts an object's perceived color less than that of a halophosphate tube, and changing the mix of the three components allows the lighting engineer to adapt the output of the lamp to suit certain specific purposes, for instance to better match the lighting within a building to the activities of its occupants.

Lifetime

The lifetime of a fluorescent lamp is limited primarily by the electron-emitting material on the electrodes and the phosphor. The electro-emissive material is consumed in a number of ways when the tube is used. First, the "dark space," a region of high electric field found near a cathode, accelerates ions towards the electrode, and the resulting bombardment removes the material. This effect can be alleviated by operating at high frequencies, since the bombardment is reduced as explained above. A specially shaped cathode can also be used to reduce the electric field across the dark **space**, and thus reduce impact **erosion** during normal operation. Second, the electro-emissive material suffers excess erosion when the discharge is struck due to the short-lived, high electric fields. Modern electronic control circuitry can prevent misfiring and striking the discharge when the electrodes are cold and thus reduce this erosion. The use of electronic starters can double the lifetime of a tube. The induction lamp, a commercial version of which was introduced by GE in

1994, contains no electrodes, and the discharge current is induced by a radio-frequency discharge. Since there is no erosion problem, the induction lamp has the capability of lasting for up to 60,000 hours, many times longer than standard fluorescent lamps.

The phosphor in fluorescent lamps has a finite lifetime. The older halophosphates, which were widely used before the introduction of triphosphors, exhibit a drop of fluorescent light output of 30-50% over a period of 8,000 hours. Triphosphors, however, only demonstrate a drop of 10-20% over 8,000 hours, thus extending the useful lifetime of the tube.

See also Electric circuit; Incandescent light.

Resources

Books

Cayles, M.A., and A.M. Martin. *Lamps and Lighting.* London: Edwin Arnold, 1983.

Periodicals

White, Julian. "Green Lights." *Physics World* (October 1994).

Iain A. McIntyre

Fluoridation

Fluoridation consists of adding fluoride to a substance (often drinking **water**) to reduce tooth decay. Fluoridation was first introduced into the United States in the 1940s in an attempt to study its effect on the reduction of tooth decay. Since then many cities have added fluoride to their water supply systems. Proponents of fluoridation have claimed that it dramatically reduced tooth decay, which was a serious and widespread problem in the early twentieth century. Opponents of fluoridation have not been entirely convinced of its effectiveness, are concerned by possible side effects, and are disturbed by the moral issues of personal rights that are raised by the addition of a chemical substance to an entire city's water supply. The decision to fluoridate drinking water has generally rested with local governments and communities and has always been a controversial issue.

Fluoride and tooth decay

Tooth decay occurs when food acids dissolve the protective enamel surrounding each tooth and create a hole, or cavity, in the tooth. These acids are present in food, and can also be formed by acid-producing **bacteria** that convert sugars into acids. There is overwhelming evidence that fluoride can substantially reduce tooth decay. When ingested into the body, fluoride concentrates in bones and in dental enamel which makes the tooth enamel more resistant to decay. It is also believed that fluoride may inhibit the bacteria that convert sugars into acidic substances that attack the enamel.

Fluoride is the water soluble, ionic form of the element fluorine. It is present in most water supplies at low levels and nearly all food contains traces of fluoride. When water is fluoridated, chemicals that release fluoride are added to the water. In addition to fluoridation of water supplies, toothpaste and mouthwash also contain added fluoride.

Early fluoridation studies

In 1901 Frederick McKay (1874-1959), a dentist in Colorado Springs, Colorado, noticed that many of his patients' teeth were badly stained. Curious about the cause of this staining, or dental fluorosis as it is also known, McKay concluded after three decades of study that the discolorations were caused by some substance in the city's water supply. Analysis of the water indicated high levels of fluoride, and it was concluded that the fluoride was responsible for the stained teeth. McKay also observed that although unsightly, the stained teeth of his patients seemed to be more resistant to decay. The apparent connection between fluoride and reduced decay eventually convinced H. Trendley Dean (1893-1962), of the U.S. Public Health Service (USPHS), to examine the issue more closely.

In the 1930s, Dean studied the water supplies of some 345 U.S. communities and found a low incidence of tooth decay where the fluoride levels in community water systems were high. He also found that staining was very minor at fluoride concentrations less than or equal to one part fluoride per million parts of water (or one ppm). The prospect of reducing tooth decay on a large scale by adding fluoride to community water systems became extremely appealing to many public health officials and dentists. By 1939, a proposal to elevate the fluoride levels to about one ppm by adding it artificially to water supplies was given serious consideration, and eventually several areas were selected to begin fluoridation trials. By 1950, USPHS administrators endorsed fluoridation throughout the country.

To fluoridate or not to fluoridate

The early fluoridation studies apparently demonstrated that fluoridation was an economical and convenient method to produce a 50-60% reduction in the tooth decay of an entire community and that there were no health risks associated with the increased fluoride consumption. Consequently, many communities quickly moved to fluoridate their water supplies in the 1950s. However strong opposition to fluoridation soon emerged as opponents claimed that the possible side effects of fluoride had been inadequately investigated. It was not surprising that some people were concerned by the addition of fluoride to water since high levels of fluoride ingestion can be lethal. However, it is not unusual for a substance that is lethal at high **concentration** to be safe at lower levels, as is the case with most vitamins and **trace elements**.

Opponents of fluoridation were also very concerned on moral grounds because fluoridation represented compulsory mass medication. Individuals had a right to make their own choice in health matters, fluoridation opponents argued, and a community violated these rights when fluoride was added to its water supply. Fluoridation proponents countered such criticism by saying that it was morally wrong not to fluoridate water supplies because this would result in many more people suffering from tooth decay which could have easily been avoided through fluoridation.

The issue of fluoridation had become very much polarized by the 1960s since there was no middle ground: water was either fluoridated or not. Controversy and heated debate surrounded the issue across the country. Critics pointed to the known harmful effects of large doses of fluoride that led to bone damage and to the special risks for people with kidney **disease** or those who were particularly sensitive to toxic substances. Between the 1950s and 1980s, some scientists suggested that fluoride may have a mutagenic effect (that is, it may be capable of causing human **birth defects**). Controversial claims that fluoride can cause **cancer** were also raised. Today, some scientists still argue that the benefits of fluoridation are not without health risks.

Fluoridation outside the United States

The development of the fluoridation issue in the U.S. was closely observed by other countries. Dental and medical authorities in **Australia**, Canada, New Zealand, and Ireland endorsed fluoridation, although not without considerable opposition from various groups. Fluoridation in Western **Europe** was greeted less enthusiastically and scientific opinion in some countries, such as France, Germany, and Denmark, concluded that it was unsafe. Widespread fluoridation in Europe is therefore uncommon.

Fluoridation today

Up until the 1980s the majority of research into the benefits of fluoridation reported substantial reductions (50-60% on average) in the incidence of tooth decay where water supplies had fluoride levels of about one ppm. By the end of the decade however, the extent of this

KEY TERMS

Element—A pure substance that can not be changed chemically into a simpler substance.

Fluoridation—The addition to a city's water supply of chemicals that release fluoride into the water.

Fluoride—The ionic form (negatively charged) of the element fluorine which is soluble in water.

Parts per million (ppm)—A way to express low concentrations of a substance in water. For example, 1 ppm of fluoride means 1 gram of fluoride is dissolved in 1 million grams of water.

reduction was being viewed more critically. By the 1990s, even some fluoridation proponents suggested that observed tooth decay reduction, directly as a result of water fluoridation, may only have been at levels of around 25%. Other factors, such as education and better dental hygiene, could also be contributing to the overall reduction in tooth decay levels. Fluoride in food, **salt**, toothpastes, rinses, and tablets, have undoubtedly contributed to the drastic declines in tooth decay during the twentieth century. It also remains unclear as to what, if any, are the side effects of one ppm levels of fluoride in water ingested over many years.

Although it has been argued that any risks associated with fluoridation are small, these risks may not necessarily be acceptable to everyone. The fact that only about 50% of U.S. communities have elected to adopt fluoridation is indicative of people's cautious approach to the issue. In 1993, the National Research Council published a report on the health effects of ingested fluoride and attempted to determine if the maximum recommended level of four ppm for fluoride in drinking water should be modified. The report concluded that this level was appropriate but that further research may indicate a need for revision. The report also found inconsistencies in the scientific studies of fluoride toxicity and recommended further research in this area.

See also Groundwater; Poisons and toxins; Water conservation.

Resources

Books

Martin, B. *Scientific Knowledge in Controversy: The Social Dynamic of the Fluoridation Debate.* Albany, New York: State University of New York Press, 1991.

Whitford, G.M. *The Metabolism and Toxicity of Fluoride.* Basel, New York: Karger, 1989.

Periodicals

Hileman, B. "Fluoridation of Water." *Chemistry and Engineering News* 66 (August 1, 1988): pp. 26-42.

Other

National Research Council Committee on Toxicology. *Health Effects of Ingested Fluoride.* Washington, DC: National Academy Press, 1993.

United States Department of Health and Human Services Committee to Coordinate Environmental Health and Related Programs. Ad Hoc Subcommittee on Fluoride. *Review of Fluoride Benefits and Risks: Report of the Ad Hoc Subcommittee on Fluoride.* Washington, DC: Public Health Service, Department of Health and Human Services, 1991.

Nicholas C. Thomas

Fluorine *see* **Halogens**

Flycatchers *see* **Monarch flycatchers**

Flying fish

Flying **fish** belong to the family Exocoetidae in the **bony fish** order Atheriniformes. They are close relatives of the needlefish, halfbeaks, and sauries. Flying fish are characterized by a low lateral line, soft fins without spines, and a caudal fin with the lower lobe larger than the upper lobe. The lower jaw of the young flying fish has an extended filament longer than the body, which becomes detached as the fish grows.

Flying fish have large pectoral fins almost as long as the body which serve as wings, helping the fish glide through the air when it leaves the **water**. The pectoral fin expands and stiffens while in the air for a short **distance** before the fish reenters the water. A flying fish can remain airborne for at least 30 seconds and can reach a top speed of at least 40 MPH (64 km/h) produced by the rapid movement and vibration of the tail. The tail is the first part of the fish to reenter the water, making it possible for the fish to gain speed rapidly for another thrust into the air. It is estimated that the tail fin may vibrate as rapidly as 50 times per second. By these movements the fish may make several thrusts into the air in rapid succession. Flying fish extend a flight by plunging the vibrating tail into the water to supply added **momentum**.

When gliding, flying fish barely skim over the surface of the water. Larger fish can leap to a height of 3.3 ft (1 m) above the water and glide for over 330 ft (100 m). A flying fish, however, can be carried to the topmost part of the wave possibly 15 ft (4.5 m) above the trough so that the fish may appear to be so high out of the water. It is thought that flying fish fly to escape from predators

KEY TERMS

. . . .

Caudal fin—The tail fin of a fish.

Lateral line—A line of pores along the sides of a fish containing sensory organs to detect frequency vibrations of low intensity, movements of water, and possibly pressure changes.

Pectoral—Paired fins of a fish, located close to the gill openings. In air-breathing vertebrates they become forelegs or arms.

Pelvic fins—Paired fins ventral to the pectorals and in varying positions relative to the pectorals according to the species of fish. They correspond to the hind limbs of air-breathing vertebrates.

(such as fish-eating bonitos, albacores, or blue fish), but airborne flying fish are also exposed to fish-eating **birds**.

Species of flying fish

Flying fish prefer the warm waters of the Atlantic and Pacific Oceans. Tropical flying fish such as *Exocoetus volitans* and *Hirundichthys speculiger* are found in tropical regions of the world where the water **temperature** is rarely below 68°F (20°C). The flying fish genus *Exocoetus* includes 22 **species** found in the Pacific and Atlantic Oceans.

The Atlantic flying fish (*Cypselurus heterurus*) inhabits the tropical Atlantic and Caribbean, has a black band extending through its wings, and measures less than 10 in (25.4 cm) in length. The California flying fish (*Cypselurus californicus*) is reputed to be the largest of all the flying fish, growing up to 1.6 ft (0.5 m), and is caught commercially for human consumption. This species is considered a four-winged flying fish, because its pectoral and pelvic fins resemble large wings.

The large margined flying fish (*Cypselurus cyanopterus*), the bandwing flying fish (*Cypselurus exsiliens*), and the short-winged flying fish (*Parexocoetus mesogaster*) are widely distributed throughout the tropical seas. The smallwing flying fish (*Oxyporhamphus micropterus*) is found in tropical and subtropical waters, and **flies** only short distance due to its short wings.

Resources

Books

Dickson-Hoese, H., and R.H. Moore. *Fishes of the Gulf of Mexico, Texas, Louisiana, and Adjacent Waters.* College Station: Texas A&M University Press, 1977.

Moyle, Peter B., Joseph Cech. *Fishes: An Introduction to Ichthyology.* 4th ed. New York: Prentice Hall, 1999.

Whiteman, Kate. *World Encyclopedia of Fish & Shellfish.* New York: Lorenz Books, 2000.

Nathan Lavenda

Flying foxes *see* **Bats**

FM *see* **Radio**

Focus *see* **Conic sections**

Focused ion beam (FIB)

Focused ion beams have been used since the 1960s to investigate the chemical and isotopic composition of minerals. An FIB blasts **atoms** and molecules free from the surface of a small **sample** of material; some of these free particles are also ions, and these are guided by electric fields to a **mass** spectrometer which identifies them with great precision.

Focused-ion-beam (FIB) systems are now routinely used by failure analysts and microchip engineers who require submicron imaging. In addition to diagnostic imaging, FIB techniques are now also used in rewiring microchip repair.

An ion is an atom or **molecule** with a net **electric charge**. Electric fields subject electric charges to forces; therefore, electric fields can be used to move and steer ions. A continuous stream of ions moving together is termed an ion beam; a focused ion beam (FIB) is produced by using electric fields to guide a beam of ions.

In addition to precise imaging, FIB technology can be used in a variety of manufacturing environments requiring high levels of precision and accuracy. The image above, using time-lapse, shows a computer-controlled ion beam helping shape a mirror for the Keck telescope. © *Roger Ressmeyer/Corbis. Reproduced by permission.*

In a typical FIB analysis, a narrow beam of argon, gallium, or **oxygen** ions traveling about 800,000 miles per hour (500,000 km/hr) is directed at a polished flake of the material to be analyzed. Some of the atoms and molecules in the sample are kicked loose by the beam, a process termed sputtering. Some of these sputtered particles are themselves ions and so can be collected and focused by electric fields. The sputtered ions are directed to a mass spectrometer, which sorts them by mass. Even the very slight mass differences between isotopes of a single element can be distinguished by **mass spectrometry**; thus, not only the chemical but also the isotopic composition of a sample can be determined with great precision. Very small, even microscopic, samples can be analyzed by FIB techniques.

The abundances of **trace elements** in a mineral can reveal information about the processes that formed it, helping petrologists and geochemists unravel geological history. Further, the decay of radioactive elements into isotopes of other elements acts as a built-in clock recording when the host mineral was formed. The hands of this clock are the relative **isotope** abundances in the mineral, and these can be determined by FIB analysis. **Carbon** isotope ratios also reveal whether a carbon-containing mineral was assembled by a living **organism** or by a nonliving process. Using FIB analysis, scientists have exploited this property of carbon isotopes to show that life existed on **Earth** at least 3.85 billion years ago and that certain **rocks** originating on **Mars** and recovered as meteorites lying on the Antarctic **ice** probably, despite appearances, do not contain fossils of Martian microbes.

FIB facilities are complex and expensive. Accordingly, only about 15 facilities devoted to Earth sciences exist worldwide.

See also Dating techniques; Microtechnology; Radioactive dating.

Fog

Fog is caused by the condensation of **water** at or near Earth's surface. The atmosphere is obscured—essentially by cloud formation—near the surface and fog conditions are generally characterized as existing when atmospheric visibility is reduced to about one-half mile (approximately 0.8 km).

Causes and types of fog

Fog forms either by air cooling its dew point—resulting in **radiation** fog, advection fog, or upslope fog—or by **evaporation** and mixing, when moisture is added

to the air by evaporation and then mixes with drier air to form evaporation fog or frontal fog.

Other types of fog include **ice** fog (a fog of suspended ice crystals, frequently forming in Arctic locations), acid fog (fog forming in polluted air, and turning acidic due to oxides of **sulfur** or **nitrogen**), or **smog** (fog consisting of water and smoke particles). While any type of fog can be hazardous because of the potential dangers of reduced atmospheric visibility—especially for ground and air transportation—acid fog and smog can pose additional risk to human health, causing **eye** irritations or respiratory problems.

Radiation fogs

Radiation fog (or ground fog) generally occurs at night, when radiational cooling of Earth's surface cools the shallow moist air layer near the ground to its **dew point** or below. This causes moisture in the nearby layers of air to condense into fog droplets. Radiation fog usually occurs under calm **weather** conditions, when no more than **light** winds mix the air layers near the surface. Strong winds normally mix the lower-level cold air with the higher-level dry air, thus preventing the air at the bottom from becoming saturated enough to create observable fog. The presence of **clouds** at night can also prevent fog formation of this type, because they reduce radiational cooling. Radiation fog often forms in late fall and winter nights, especially in lower areas, because cold and heavy air moves downhill to gather in valleys or other relatively low-lying areas. Accordingly, radiation fog is also called valley fog. In the morning, radiation fog usually dissipates or "burns off" when the Sun's **heat** warms the ground and air above the dew point **temperature** (i.e., the temperature at which moisture in the air condenses).

Advection fogs

Advection fog forms when warm, moist air moves horizontally over a relatively cooler surface. During such contact, the layer of air near the surface may cool to below its dew point to form advection fog. Because advection fog can form at any time, it can be very persistent. It is common along coastlines where moist air moves from over a water surface to a cooler coastal land mass. Advectional fog can also occur if an already cool air mass moves over a still colder surface (e.g. snow), so that even the reduced levels of moisture in the cold air can condense into fog as the surface continues to cool the air mass. Advection-radiation fog forms when warm, moist air moves over a cold surface that is cold as a result of radiation cooling. When warm, humid air moves over cold water, a sea fog may form.

Dense fog engulfing the Golden Gate Bridge (San Francisco, CA). *CORBIS/Kevin Schafer. Reproduced by permission.*

Upslope fog forms in higher areas, where a moist air mass is forced to move up along a mountain incline. As the air mass moves up the slope, it is cooled below the dew point to produce fog. Upslope fog formation generally requires a stronger **wind** along with warm and humid conditions at the surface. Unlike radiation fog, this type of fog dissipates as wind dissipates, and it can form more easily under cloudy conditions. Upslope fog is usually dense, and often extends to high altitudes.

Evaporation fogs

Evaporation fog forms as a result of the mixing of two unsaturated air masses. Steam fog is a type of evaporation fog that appears when cold, dry air moves over warm water or warm, moist land. When some of the water evaporates into low air layers, and the warm water warms the air, the air rises, mixes with colder air, cools, and water vapor condensation occurs to form a fog. Over oceans, this form of evaporation fog is referred to as sea smoke. Examples of cold air over warm water occur over

swimming pools or hot tubs, where steam fog easily forms. It is also common in the autumn, when winds and air fronts turn cool as water bodies remain warm.

Precipitation fog

Precipitation fog is a type of evaporation fog that happens when relatively warm rain or snow falls through cool, almost saturated air, and evaporation from the precipitation saturates the cool air. It can turn dense, persist for a long time, and may extend over large areas. Although usually associated with warm fronts, precipitation fog also occurs with slow moving fronts or stationary fronts to form frontal fogs.

See also Atmosphere, composition and structure; Atmospheric circulation; Atmospheric optical phenomena; Atmospheric temperature; Land and sea breezes; Wind.

Resources

Books

Hamblin, W. K., and E. H. Christiansen. *Earth's Dynamic Systems.* 9th ed. Upper Saddle River: Prentice Hall, 2001.
Hancock, P. L., and B. J. Skinner, eds. *The Oxford Companion to the Earth.* New York: Oxford University Press, 2000.
Keller, E. A. *Introduction to Environmental Geology.* 2nd ed. Upper Saddle River: Prentice Hall, 2002.

Agnes Galambosi

Fold

A fold is a bend in a body of rock or sediment that forms due to a change in **pressure**. Wave-like folds are composed of layers of the earth's crust that bend and buckle under enormous pressure as the crust hardens, compresses, and shortens. Folds form much the same way as a hump arises in a sheet of **paper** pushed together from both ends.

Folds may be softly rolling or severe and steep, depending on the intensity of the forces involved in the deformation and the nature of the **rocks** involved. The scale of folding may be massive, creating mile upon mile of **mountains** like the Appalachian chain, traversing eastern

An exposed fold in New Jersey. *JLM Visuals. Reproduced by permission.*

North America from Alabama to the Gulf of St. Lawrence in eastern Canada. In general, folded mountain belts represent periods of compression or squeezing during which the crust may be shortened significantly. During the formation of the European Alps, stratified rock layers that originally covered an area about 300 mi (482 km) wide were squeezed together until they had a width of less than 120 mi (193 km). Folds may also be minute, seen simply as tiny ripples a few centimeters in size.

Horizontal pressure results in two basic fold forms: anticlines, arched, upfolded **strata** that generally appear convex upward, and synclines, downfolds or reverse arches that are typically concave upward. An important and definitive characteristic of these folds is the relative position of the oldest and youngest layers within the fold. At the core of an anticline lie stratigraphically older layers. The outer most layers that make up the fold are younger in age. The opposite is true in the case of a syncline. At the core of a syncline are the youngest layers, with the oldest beds situated at the outside of the fold.

A line drawn along the points of maximum curvature of the fold is called the axis. The inclined rock that lies on either side of the axis are called the fold limbs. One limb of a downfold is also the limb of the adjacent upfold. Limbs on either side of a symmetrical fold are at relatively equal angles. A fold that has only a single limb it is known as a monocline. These often form step-like ridges rising from flat or gently sloping terrain.

As the intensity of the folding increases, the resultant folds often become more asymmetrical, i.e., one limb of an anticline dips at a steeper **angle**. In overturned folds, the angle of this limb becomes so steep that the tilted limb lies almost beneath the upper limb. Recumbent folds literally lie on their sides, with the lower limb turned completely upside-down.

In many cases the axis of the fold in not horizontal. Such folds are known as plunging folds, and are said to plunge in the direction that the axis is tilted. Folds with a curved axis are called doubly plunging folds. Domes are broad warped areas in which the plunge of the anticline is approximately equal in all directions. The corresponding synclinal structure is known as a structural **basin**.

See also Fault; Tectonics; Unconformity.

Food chain/web

A food chain is a series of organisms dependent on each other for food; a food web is an interconnected set of food chains in the same **ecosystem**. Organisms that eat similar foods are assigned to a particular trophic level, or feeding level, within a food web. Food web is a more accurate term because food chains only exist on **paper**. In nature feeding habits are complex because many organisms feed on different **trophic levels**. For example humans feed on the bottom consumer level when they eat plants but they also eat organisms from all of the higher trophic levels of the food web.

History of food web research

Food web research is an extensive area of ecological research. Charles Elton, Raymond Lindeman, Stuart Pimm, Stephen Carpenter, and James Kitchell are some of the major figures in food web research. Charles Elton was an English ecologist who first described the characteristic shape of food webs, which he called the **pyramid** of numbers. Elton observed that most food webs have many organisms on their bottom trophic levels and successively fewer on the subsequent, higher levels. His pyramid of numbers is now called the Eltonian Pyramid and is the basic model used to describe all food webs.

The American ecologist Raymond L. Lindeman published a classic paper in 1942 that examined the Eltonian pyramid in terms of **energy** flow. By using energy as the currency of the ecosystem Linderman quantified and explained that the Eltonian Pyramid was a result of successive energy losses at each trophic level. This loss is due to thermodynamic inefficiency in the transformation of energy and is referred to as ecological efficiency. Later, researchers discovered that ecological efficiency varies from 5-30% with an average of about 10%, depending on the **species** and the environment in which it lives.

Stuart Pimm published his classic book *Food Webs* in 1982. This book consolidated various aspects of food web theory and has become a reference for ecologists.

The book's many topics include food web complexity and stability and hypotheses on food chain length.

More recently, Stephen Carpenter and James Kitchell have become leaders in aquatic food web research. Their theory regarding the trophic cascade in aquatic food webs has been central to the current debate on top-down and bottom-up control of food webs.

Structure of food webs

Within food webs there are three main categories of organisms: producers, consumers, and decomposers. Producers are organisms that synthesize their own organic compounds or food using diffuse energy and inorganic compounds. Producers sometimes are called autotrophs (self-feeders) because of this unique ability. For example, green plants are autotrophs because they manufacture the compounds they need through **photosynthesis**. The process of photosynthesis is summarized below:

$$\text{solar energy} + \text{carbon dioxide} + \text{water} \rightarrow$$
$$\text{glucose (sugar)} + \text{oxygen}$$

Photosynthesis uses radiant energy from the **sun** to transform gaseous **carbon** dioxide (released as organisms respire) and water into glucose and other simple sugars which plants use as a food to survive and grow. Gaseous oxygen is released as a waste product of photosynthesis and is used by other organisms during their respiratory **metabolism**. Photosynthetic organisms are called primary producers and they are the first trophic level of the food web. Their **rate** of productivity determines how much fixed energy in the form of potential energy of **plant biomass** is available to higher trophic levels.

Above the primary producers are all of the consumers heterotrophs (other feeders). Heterotrophs feed on other organisms to obtain their energy and are classified according to the types of food they eat. Consumers that eat plants are called herbivores. Herbivores comprise the second trophic level of the food web and are called primary consumers because they are the first consumer group. Grass-eating **deer** and cows are primary consumers.

Above the primary consumers, the food web fans out to include consumers that eat other animals, carnivores, and consumers that eat both plants and animals, omnivores. Within the food web carnivores and omnivores can be on any higher trophic levels. Some are secondary consumers meaning they eat primary consumers. Wolves that eat deer (primary consumers) are secondary consumers. Other higher-level consumers are tertiary and eat further up on the food web or perhaps on many levels.

In addition to this grazing food web there is another trophic section known as the decomposer food web. There are two main types of consumers of dead biomass:

detritus feeders and decomposers. Both are called detritivores since they utilize dead plants and animals, or detritus. Detritus feeders, such as earthworms, ingest organic waste and fragment it into smaller pieces which the decomposers such as **bacteria** and **fungi** can digest. Unlike organisms in the grazing part of the food web decomposers are extremely efficient feeders. Species can rework detritus, progressively extracting more fixed energy. Eventually the waste is broken down into simple inorganic chemicals such as H_2O and CO_2 structures and **nutrients**. The nutrients then may be re-used by the primary producers in the grazing part of the food web. The decomposer food web is very active inside of compost piles and turns kitchen wastes into a **soil** conditioner. Decomposers are active in all natural ecosystems.

Contaminants in food webs

Food webs all over the world have become contaminated by **insecticides** and other manufactured chemicals. Some of these compounds are having profound effects on the reproduction and **behavior** of some wild **animal** species. These chemicals were released into the environment because it was believed that their concentrations were too small to have an effect on organisms. Now we know better. **Contamination** of very remote habitats, such as the Antarctic and the Arctic, has convincingly demonstrated that even small amounts of certain compounds can have massive effects. Some of these compounds are persistent hydrophobic (water-fearing) contaminants accumulating in the fatty tissues of organisms because they do not dissolve in water.

Some of the most important persistent, hydrophobic contaminants are PCBs and DDT. PCBs, or polychlorinated biphenyls, are a suite of about 209 different compounds, each with slight variations in their chemical structure. PCBs were widely used as insulating material in electric transformers and for other purposes. There is currently a worldwide ban on their production and use but large quantities still persist in the environment. DDT, or dichlorodiphenyltrichloroethane, is an insecticide that has been dispersed all over the world. Unfortunately, both PCBs and DDT now occur in all plant and animal tissues even in remote areas where they were never used (for example, in animals such as polar **bears** and **seals**). Humans are also contaminated, and mothers pass these chemicals to their babies in their milk, which is rich in **fat** where DDT and PCBs concentrate.

Bioaccumulation

Bioaccumulation refers to the tendency of persistent hydrophobics and other chemicals such as methyl mercury to be stored in the fatty tissues of organisms.

When these compounds are spilled into the environment they are rapidly absorbed by organisms in food webs. It is estimated that 99% of **pesticides** do not reach the target pest which means the chemical ends up in the general environment. If these pesticides are hydrophobic they build in the tissues of non-pest organisms.

Once inside the fatty tissues of an **organism**, persistent hydrophobics are not excreted easily. There are excretion mechanisms in most species. But each time the organism is exposed to the contaminant, more is taken in and deposited in the fatty tissues, accumulating progressively. Bioaccumulation is particularly acute in long-lived species because the period during which they bioaccumulate is longer. This is why some governments do not recommend consuming **fish** over a certain age or size because the older and larger they get, the more contaminated they are likely to be.

Biomagnification

Biomagnification (also called food web magnification or food web accumulation) is the progressive increase in the **concentration** of contaminants in organisms as the trophic level increases. This means lower trophic levels generally have smaller concentrations of contaminants than higher levels. This occurs because of the ecological inefficiency of food webs and persistent, hydrophobic contaminants bioaccumulating in organisms. These two factors **mean** that each trophic level has a larger concentration of contaminants dissolved in a smaller amount of biomass than the previous level. Each trophic level becomes more contaminated than those below it. For example, DDT in the Lake Ontario food web is biomagnified up the food web, so that top predators like herring **gulls** have **tissue** concentrations that are 630 times greater than primary consumers like **zooplankton**.

Dolphins have been studied by Japanese researchers as a model species for biomagnification because their migratory routes are known, they live in relatively unpolluted waters, and they live a long time (20-50 years). DDT has been found in dolphin blubber in greater concentrations (100 times greater than **sardines**) than would be expected given the small concentrations present in the water and in sardines, their favorite food. These unexpectedly large concentrations are the result of DDT biomagnification up the food web.

Biomagnification has serious consequences for all species. It is particularly dangerous for **predator** species especially if they are at the top of long food webs. Predators are usually at or near the top of their food web. This puts them at risk because the degree of biomagnification is high by the time it reaches their trophic level. Also, top predators usually consume large quantities of meat

KEY TERMS

Bioaccumulation—The tendency of substances, like PCBs and other hydrophobic compounds, to build in the fatty tissues of organisms.

Biomagnification—Tendency of organisms to accumulate certain chemicals to a concentration larger than that occurring in their inorganic, non-living environment, such as soil or water, or in the case of animals, larger than in their food.

Food chain—A sequence of organisms directly dependent on one another for food.

Food web—The feeding relationships within an ecological community, including the interactions of plants, herbivores, predators, and scavengers; an interconnection of many food chains.

Hydrophobic compounds—"Water-hating" chemical substances, such as PCBs and DDT, that do not dissolve in water and become concentrated in the fatty tissues of organisms.

Photosynthesis—The conversion of radiant energy into chemical energy that is stored in the tissues of primary producers (e.g., green plants).

Primary consumer—An organism that eats primary producers.

Primary producer—An organism that photosynthesizes.

Trophic level—A feeding level in a food web.

which has lots of fatty tissue and contaminants. Polar bears, humans, **eagles**, and dolphins are examples of top predators, and all of these organisms are vulnerable to the effects of biomagnification. Predators that consume large amounts of fish also have a high degree of risk because persistent hydrophobics are widely dispersed in aquatic food webs and are biomagnified in fish.

Current research

Currently there is much debate over what forces control the structure of food webs. Some ecologists believe that food webs are controlled by bottom-up forces referring to the strong connection between primary production and the subsequent production of consumers. For example, adding large amounts of nutrients like **phosphorus** causes rapid growth of **phytoplankton**, the primary producers in lakes which subsequently influences consumers in the food web. Other ecologists believe that food webs are controlled by top-down forces meaning the predators near or at the top of the food web. For example, in the Pacific Ocean researchers have found that when sea **otters** disappear from an area, sea urchin (the favorite food of seaotters) populations increase, and these **invertebrates** dramatically overgraze down the kelp beds. Removing top predators causes changes all the way down to the primary producers. Carpenter and Kitchell have called this type of control the trophic cascade because such food webs are controlled by forces that cascade down from the top trophic level. Understanding the roles of top-down and bottom-up forces within food webs will allow more effective management of ecosystems.

See also Autotroph; Carnivore; Herbivore; Heterotroph; Omnivore.

Resources

Books

Begon, M., J.L. Harper, and C. R. Townsend. *Ecology: Individuals, Populations and Communities.* 2nd ed. Boston: Blackwell Scientific Publications, 1990.

Bradbury, I. *The Biosphere.* New York: Belhaven Press, Pinter Publishers, 1991.

Colborn, T.E., et al. *Great Lakes: Great Legacy?* Baltimore: The Conservation Foundation and the Institute for Research on Public Policy, 1990.

Miller, G.T., Jr. *Environmental Science: Sustaining the Earth.* 3rd ed. Belmont, CA: Wadsworth Publishing Company, 1991.

Pimm, S.L. *Food Webs.* New York: Chapman and Hall, 1982.

Jennifer LeBlanc

Food irradiation

Food irradiation refers to a process where food is exposed to a type of **radiation** called **ionizing radiation**. The high-energy of the radiation, which can come from a radioactive or a non-radioactive source, breaks apart the genetic material of **microorganisms** that are on the surface of the food. Microorganisms and other surface contaminants, including **insects**, are killed as a result.

This scrutiny of food irradiation, combined with the public controversy surrounding the exposure of foods to radioactivity, has meant that the effects of irradiation on foods have been extensively studied. The consensus from these studies is that radioactive sterilization of food does not cause the food itself to become radioactive, nor does the irradiation appreciably alter the nutritional characteristics of the food.

The practice of irradiating foods is not new. Patents were issued in the United States and Britain for food irra-

diation in the first decade of the twentieth century. Scientists demonstrated in 1947 that meat and other foods could be sterilized by ionizing radiation. The military took a great interest in this development, seeing it as a way of supplying field troops with food. Military experiments on the irradiation of **fruits**, **vegetables**, milk and milk products, and various meats began in the U.S. in the 1950s.

In 1958, the U.S. Food and Drug Administration became the official government agency concerned with the evaluation and approval of irradiated foods. Congress gave the FDA authority over the food irradiation process.

The manned space program undertaken by the U.S. beginning in the 1960s gave a great boost to food irradiation technology. Astronauts have always eaten irradiated foods. In addition, in the 1960s, the United Nations established a Joint Expert Committee on Food Irradiation. The committee concluded in 1980 that the irradiation of foods posed no unique nutritional or microbiological problems.

Irradiation methods

Food irradiation can be accomplished in three different ways, using three different types of rays; gamma rays, **electron** beams, and **x rays**. Gamma rays are given off by the radioactive elements cobalt and cesium. Gamma rays are powerful, and can penetrate through several feet of material. As such, precautions against technician exposure to the radiation are necessary, and a special irradiation chamber is needed.

Electron beams are not as powerful as gamma rays. They can penetrate to depths of a few centimeters. Nonetheless, they are excellent for the sterilization of surfaces. Electron beam sterilization of medical and dental equipment has been routine for decades. Additionally, electron beams are not radioactive.

X-ray irradiation of food was introduced in the mid–1990s. X rays are a blend of the other two techniques, in that x rays are as powerful and penetrating as gamma rays. But, like electrons, x rays are not radioactive.

Foods such as solid meat and poultry, and fresh produce are well suited to irradiation sterilization. Not all foods, however, are as suited to the irradiation process. Eggs, milk, and shellfish, for example, should be treated by another process to best preserve their quality. Food irradiation alters the taste or appearance of some varieties of **grapes**, lemons, and nectarines. Irradiation is no substitute for proper cooking and storage. Even irradiated food can become contaminated if it is improperly cooked or stored.

Food irradiation sparks debate

Like **biotechnology**, food irradiation has sparked fierce public debate. Some scientists are ardent sup-

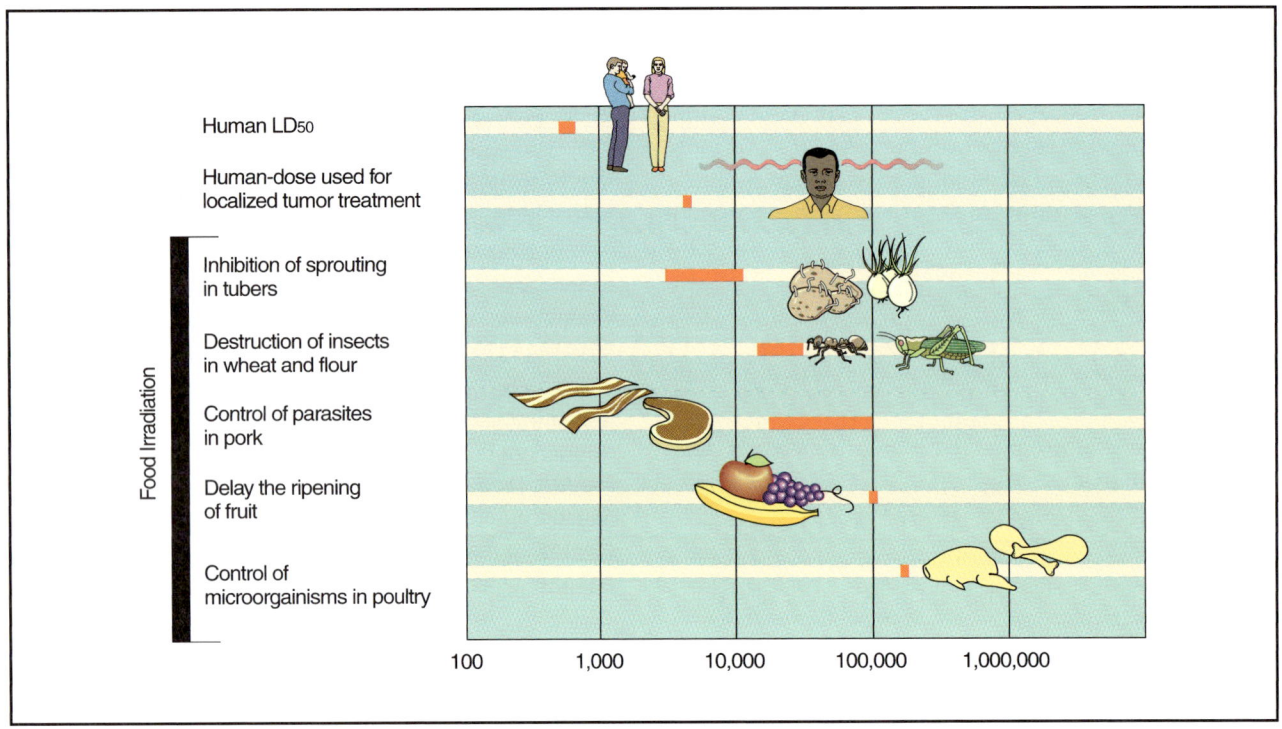

Levels of radiation (in rads) for various applications. Ranges shown for food irradiation are approved by the U.S. government. *Illustration by Hans & Cassidy. Courtesy of Gale Group.*

porters while other public groups are detractors of food irradiation. Supporters of food irradiation contend that its widespread use has the potential to reduce death and illness internationally due to food-borne microorganisms such as **salmonella** in poultry and **trichinosis** in pork.

Salmonella causes four million people to become ill and results in 1,000 deaths annually in the United States alone. **Contamination** of food products with a bacterium called *Escherichia coli* 0157 causes over 20,000 illnesses and 500 deaths a year. Internationally, as much as 30% of the world's food supply cannot be used each year because it is either spoiled or consumed by insects. Globally, there are an estimated 24,000–120,000 cases of *Salmonella* **food poisoning** and 4,900–9,800 cases of *E. coli* 0157:H7 food poisoning each year. Treatment of such food-borne illnesses and lost productivity costs an estimated US$5–6 billion each year.

Despite the weight of evidence and the need for a more effective food treatment strategy, advocates of food irradiation face public opposition. Some consumers and groups are concerned about the unforeseen reactions in food caused by the presence of high-energy particles. Others do not want the food they eat to have been exposed to radioactive substances. Ultimate-

ly, the benefits of food irradiation will be weighed against the public's wariness concerning radiation and the food supply.

As public awareness and understanding of irradiation continues, the acceptance of irradiation as a food protection strategy could prevail. The weight of evidence supports the technique. As of 2000, food irradiation is endorsed by the United States Food Protection Agency, the American Medical Association, and the World Health Association, and over 40 countries sterilize food by irradiation.

See also Bacteria; Deoxyribonucleic acid (DNA); Food preservation.

Resources

Books

Satin, Morton. *Food Irradiation: A Guidebook.* Lancaster, PA: Technomic Publishing Company, 1993.

Periodicals

Osterholm, M.T., and M.E. Potter. "Irradiation Pasteurization of Solid Foods: Taking Food Safety to the Next Level." *Emerging Infectious Disease* 3 (1997): 575–577.

Radomyski, T., E.A. Murano, D.G. Olson, et al. "Elimination of Pathogens of Significance in Food by Low-dose Irradiation: A Review." *Journal of Food Protection* 57 (1994): 73–86.

Other

Centers for Disease Control and Prevention. 1600 Clifton Road, Atlanta, GA 30333. (404) 639–3311. <http://www.cdc.gov/ncidod/dbmd/diseaseinfo/foodirradiation.htm.>.

Brian Hoyle

Food poisoning

Food poisoning refers to an illness that is caused by the presence of **bacteria**, poisonous chemicals, or another kind of harmful compound in a food. Bacterial growth in the food is usually required. Food poisoning is different from food intoxication, which is the presence of pre-formed bacterial toxin in food.

There are over 250 different foodborne diseases. The majority of these are infections, and the majority of the infections are due to contaminating bacteria, viruses, and **parasites**. Bacteria cause the most food poisonings. The United States Centers for Disease Control and Prevention estimates that 76 million Americans become ill each year from food poisoning. The cost to the economy in medical expenses and lost productivity is estimated at $5–6 billion per year. Infections with the common foodborne bacteria called *Salmonella* alone exacts about a $1 billion economic toll per year.

Aside from the economic costs, food poisoning hospitalizes approximately 325,000 Americans each year, and kills more than 5,000 Americans.

Food intoxication is technically separate from food poisoning. But, because food intoxication and food poisoning both cause foodborne illness, two noteworthy bacteria responsible for food intoxications will be mentioned.

Contamination by *Staphylococcus* is the most common cause of food poisoning. The bacteria grow readily in foods such as custards, milk, cream-filled pastries, mayonnaise-laden salads, and prepared meat.

Two to eight hours after eating, the sudden appearance of nausea, stomach cramps, vomiting, sweating, and diarrhea signal the presence of food poisoning. Usually only minor efforts need be made to ease the symptoms, which will last only a short time even if untreated. Over-the-counter preparations to counter the nausea and diarrhea may help to cut short the course of the condition. Recovery is usually uneventful.

This **syndrome** is especially prevalent in summer months when families picnic out of doors and food can remain in the warmth for hours. Bacterial growth is rapid under these conditions in lunchmeat, milk, **potato** salad, and other picnic staples. The first course of eating may be without consequences, but after the food remains at ambient **temperature** for two hours or more, the probability of an infectious bacterial presence is increased dramatically. The second course or mid-afternoon snacks can lead to an uncomfortable sequel.

A far more serious form of food intoxication results from a toxin secreted by the bacterium *Clostridium botulinum*. This **infection** is called **botulism** and is frequently fatal. The bacterium differs in that it grows under **anaerobic** conditions in food that has been improperly preserved.

Botulism is a hazard of home canning of food and can develop from commercially canned products in which the can does not maintain the sterile environment within it. Affected food has no tainted taste. Normal heating of canned products in the course of food preparation will neutralize the toxin but will not kill the bacterial spores. These will open inside the body, the bacterium will multiply, and sufficient toxin can be produced to bring about illness.

Ingestion of botulism-contaminated food does not lead to the gastric symptoms usually associated with food poisoning. Botulism toxin affects the **nervous system**, so the symptoms of botulism may involve first the eyes, with difficulty in focusing, double **vision**, or other conditions, then subsequent difficulty in swallowing and weakness of the muscles in the extremities and trunk. Death may follow. Symptoms may develop in a matter of hours if the tainted food has been consumed without heating, or in four to eight days if the food is heated and the bacterium needs the time to grow.

Diagnosis is made through observation of the symptoms and by culturing the bacterium from the suspected food source. Up to 65% of individuals infected with botulism die, usually within two to nine days after eating the affected food. Treatment of botulism usually requires the patient to be hospitalized to receive specific antitoxin therapy.

The most common foodborne bacterial infections are caused by *Campylobacter*, *Salmonella*, and a type of *Escherichia coli* designated O157:H7. The latter is the cause of "hamburger disease." A **virus** known as Calcivirus or Norwalk-like virus also is a common cause of food poisoning.

Travelers, especially those to foreign countries, often suffer pangs of gastric upset because of low sanitation levels in some areas. This type of food poisoning, which may actually stem from drinking the **water** rather than eating the food, is often called "tourista," "Montezuma's revenge," or "Delhi belly." The organisms contaminating the water can be the same as those that contaminate food (i.e., *Salmonella* and *Escherichia coli*).

Campylobacter is the most common cause of bacterial diarrhea. The bacteria live in the intestines of **birds**, and can be spread to the carcass upon slaughter. Eating undercooked chicken or food contaminated with the juices from raw chicken is a typical cause of *Campylobacter* food poisoning.

Salmonella is also found in the intestines of birds, as well as **reptiles** and **mammals**. It spreads to food because of contamination by feces; for example, by the handling of food by someone who did not washing their hands thoroughly after using the washroom. For most people, the infection is inconvenient, with cramping and diarrhea. But, for people in poor health or with malfunctioning immune systems, the bacteria can infect the bloodstream and threaten life.

Escherichia coli O157:H7 lives in the intestines of cattle. When it contaminates food or water, it can cause an illness similar to that caused by *Salmonella*. However, in a small number of cases, a much more devastating illness occurs. A condition called hemolytic uremic syndrome produces bleeding, can lead to kidney failure and, in the worst cases, can cause death.

The final common cause of foodborne illness is the Norwalk-like virus. It is also spread from feces to food, often again by handling of the food by someone who has not washed their hands. This type of foodborne illness is more difficult to diagnose, because not every testing laboratory has the equipment needed to detect the virus.

Food poisoning often affects numbers of individuals who have dined on the same meal. This enables physicians to trace the contaminated food and, if needed, determine the **species** of bacterium that is at fault.

KEY TERMS

Culturing—Growing bacteria on certain substrates such as beef broth, agar plates, or other nutrient medium.

Endotoxin—A heat-stable toxin produced in the cell wall of some bacteria.

Food poisoning is easily prevented. Proper handling of food includes washing the hands before preparing food, making certain that implements such as spoons and knives are clean, and providing proper cooling for foods that are most likely to nurture bacterial growth. Home canning must include careful and thorough heating of canned foods.

See also Membrane; Poisons and toxins.

Resources

Books

Latta, S.L. *Food Poisoning and Foodborne Diseases*. Berkeley Heights, NJ: Enslow Publishers. 1999.

Organizations

Centers for Disease Control and Prevention. 1600 Clifton Road, Atlanta, GA 30333. (404) 639–3311 [cited October 22, 2002]. <http://www.cdc.gov/ncidod/dbmd/diseaseinfo/foodborneinfections_g.htm.>.

National Institutes of Allergy and Infectious Diseases, National Institutes of Health, 31 Center Drive, MSC 2520, Bethesda, MD 20892-2520 [cited October 22, 2002]. <http://www.niaid.nih.gov/factsheets/foodbornedis.htm.>.

Brian Hoyle

Food preservation

The term food preservation refers to any one of a number of techniques used to prevent food from spoiling. It includes methods such as canning, pickling, drying and freeze-drying, irradiation, pasteurization, smoking, and the addition of chemical additives. Food preservation has become an increasingly important component of the food industry as fewer people eat foods produced on their own lands, and as consumers expect to be able to purchase and consume foods that are "out of season."

Scientific principles

The vast majority of instances of food spoilage can be attributed to one of two major causes: (1) the attack by **pathogens** (disease-causing **microorganisms**) such as **bacteria** and molds, or (2) oxidation that causes the destruction of essential biochemical compounds and/or the destruction of **plant** and **animal** cells. The various methods that have been devised for preserving foods are all designed to reduce or eliminate one or the other (or both) of these causative agents.

For example, a simple and common method of preserving food is by heating it to some minimum **temperature**. This process prevents or retards spoilage because high temperatures kill or inactivate most kinds of pathogens. The addition of compounds known as BHA and BHT to foods also prevents spoilage in another different way. These compounds are known to act as **antioxidants**, preventing **chemical reactions** that cause the oxidation of food that results in its spoilage. Almost all techniques of preservation are designed to extend the life of food by acting in one of these two ways.

Historical methods of preservation

The search for methods of food preservation probably can be traced to the dawn of human civilization. Certainly people who lived through harsh winters found it necessary to find some means of insuring a food supply during **seasons** when no fresh **fruits** and **vegetables** were available. Evidence for the use of dehydration (drying) as a method of food preservation, for example, goes back at least 5,000 years. Among the most primitive forms of food preservation that are still in use today are such methods as smoking, drying, salting, freezing, and fermenting.

Smoking

Early humans probably discovered by accident that certain foods exposed to smoke seem to last longer than those that are not. Meats, **fish**, fowl, and cheese were among such foods. It appears that compounds present in **wood** smoke have anti-microbial actions that prevent the growth of organisms that cause spoilage.

Today, the process of smoking has become a sophisticated method of food preservation with both hot and cold forms in use. Hot smoking is used primarily with fresh or frozen foods, while cold smoking is used most often with salted products. The most advantageous conditions for each kind of smoking—air **velocity**, relative **humidity**, length of exposure, and **salt** content, for example–are now generally understood and applied during the smoking process. For example, electrostatic precipitators can be employed to attract smoke particles and improve the penetration of the particles into meat or fish.

So many alternative forms of preservation are now available that smoking no longer holds the position of importance it once did with ancient peoples. More frequently the process is used to add interesting and distinctive flavors to foods.

Drying

Since most disease-causing organisms require a moist environment in which to survive and multiply, drying is a natural technique for preventing spoilage. Indeed, the act of simply leaving foods out in the **sun** and **wind** to dry out is probably one of the earliest forms of food preservation. Evidence for the drying of meats, fish, fruits, and vegetables go back to the earliest recorded human history.

At some point, humans also learned that the drying process could be hastened and improved by various mechanical techniques. For example, the Arabs learned early on that apricots could be preserved almost indefinitely by macerating them, boiling them, and then leaving them to dry on broad sheets. The product of this technique, quamaradeen, is still made by the same process in modern Muslim countries.

Today, a host of dehydrating techniques are known and used. The specific technique adopted depends on the properties of the food being preserved. For example, a traditional method for preserving **rice** is to allow it to dry naturally in the fields or on drying racks in barns for about two weeks. After this period of time, the native rice is threshed and then dried again by allowing it to sit on straw mats in the sun for about three days.

Modern drying techniques make use of fans and heaters in controlled environments. Such methods avoid the uncertainties that arise from leaving **crops** in the field to dry under natural conditions. Controlled temperature air drying is especially popular for the preservation of grains such as maize, **barley**, and bulgur.

Vacuum drying is a form of preservation in which a food is placed in a large container from which air is removed. **Water vapor pressure** within the food is greater than that outside of it, and water evaporates more quickly from the food than in a normal atmosphere. Vacuum drying is biologically desirable since some enzymes that cause oxidation of foods become active during normal air drying. These enzymes do not appear to be as active under vacuum drying conditions, however.

Two of the special advantages of vacuum drying is that the process is more efficient at removing water from a food product, and it takes place more quickly than air drying. In one study, for example, the drying time of a fish fillet was reduced from about 16 hours by air drying to six hours as a result of vacuum drying.

Coffee drinkers are familiar with the process of dehydration known as spray drying. In this process, a concentrated **solution** of coffee in water is sprayed though a disk with many small holes in it. The surface area of the original coffee grounds is increased many times, making dehydration of the dry product much more efficient.

Freeze-drying is a method of preservation that makes use of the physical principle known as sublimation. Sublimation is the process by which a solid passes directly to the gaseous phase without first melting. Freeze-drying is a desirable way of preserving food since it takes place at very low temperatures (commonly around 14°F to -13°F [-10°C to -25°C]) at which chemical reactions take place very slowly and pathogens survive only poorly. The food to be preserved by this method is first frozen and then placed into a vacuum chamber. Water in the food first freezes and then sublimes, leaving a moisture content in the final product of as low as 0.5%.

Salting

The precise mechanism by which salting preserves food is not entirely understood. It is known that salt binds with water molecules and thus acts as a dehydrating agent in foods. A high level of salinity may also impair the conditions under which pathogens can survive. In any case, the value of adding salt to foods for preservation has been well known for centuries.

Sugar appears to have effects similar to those of salt in preventing spoilage of food. The use of either compound (and of certain other natural materials) is known as curing. A desirable side effect of using salt or sugar as a food preservative is, of course, the pleasant flavor each compound adds to the final product.

Curing can be accomplished in a variety of ways. Meats can be submerged in a salt solution known as brine, for example, or the salt can be rubbed on the meat by hand. The injection of salt solutions into meats has also become popular. Food scientists have now learned that a number of factors relating to the food product and to the preservative conditions affect the efficiency of curing. Some of the food factors include the type of food being preserved, the **fat** content, and the size of treated pieces. Preservative factors include brine temperature and **concentration** and the presence of impurities.

Curing is used with certain fruits and vegetables, such as cabbage (in the making of sauerkraut), cucumbers (in the making of pickles), and olives. It is probably most popular, however, in the preservation of meats and fish. Honey-cured hams, bacon, and corned beef ("corn" is a term for a form of salt crystals) are common examples.

Freezing

Freezing is an effective form of food preservation because the pathogens that cause food spoilage are killed or do not grow very rapidly at reduced temperatures. The process is less effective in food preservation than are thermal techniques such as boiling because pathogens are more likely to be able to survive cold temperatures than hot temperatures. In fact, one of the problems surrounding the use of freezing as a method of food preservation is the danger that pathogens deactivated (but not killed) by the process will once again become active when the frozen food thaws.

A number of factors are involved in the selection of the best approach to the freezing of foods, including the temperature to be used, the **rate** at which freezing is to take place, and the actual method used to freeze the food. Because of differences in cellular composition, foods actually begin to freeze at different temperatures ranging from about 31°F (-0.6°C) for some kinds of fish to 19°F (-7°C) for some kinds of fruits.

The rate at which food is frozen is also a factor, primarily because of aesthetic reasons. The more slowly food is frozen, the larger the **ice** crystals that are formed. Large ice crystals have the tendency to cause rupture of cells and the destruction of texture in meats, fish, vegetables, and fruits. In order to deal with this problem, the technique of quick-freezing has been developed. In quick-freezing, a food is cooled to or below its freezing point as quickly as possible. The product thus obtained, when thawed, tends to have a firm, more natural texture than is the case with most slow-frozen foods.

About a half dozen methods for the freezing of foods have been developed. One, described as the plate, or contact, freezing technique, was invented by the American inventor Charles Birdseye in 1929. In this method, food to be frozen is placed on a refrigerated plate and cooled to a temperature less than its freezing point. Or, the food may be placed between two **parallel** refrigerated plates and frozen.

Another technique for freezing foods is by immersion in very cold liquids. At one time, **sodium chloride** brine solutions were widely used for this purpose. A 10% brine solution, for example, has a freezing point of about 21°F (-6°C), well within the desired freezing range for many foods. More recently, liquid **nitrogen** has been used for immersion freezing. The temperature of liquid nitrogen is about -320°F (-195.5°C), so that foods immersed in this substance freeze very quickly.

As with most methods of food preservation, freezing works better with some foods than with others. Fish, meat, poultry, and citrus fruit juices (such as frozen orange juice concentrate) are among the foods most commonly preserved by this method.

Fermentation

Fermentation is a naturally occurring chemical reaction by which a natural food is converted into another form by pathogens. It is a process in which food "goes bad," but results in the formation of an edible product. Perhaps the best example of such a food is cheese. Fresh milk does not remain in edible condition for a very long period of time. Its **pH** is such that harmful pathogens begin to grow in it very rapidly. Early humans discovered, however, that the spoilage of milk can be controlled in such a way as to produce a new product, cheese.

Bread is another food product made by the process of fermentation. Flour, water, sugar, milk, and other raw materials are mixed together with yeasts and then baked. The addition of yeasts brings about the fermentation of sugars present in the mixture, resulting in the formation of a product that will remain edible much longer than will the original raw materials used in the bread-making process.

Thermal processes

The term "thermal" refers to processes involving **heat**. Heating food is an effective way of preserving it because the great majority of harmful pathogens are killed at temperatures close to the **boiling point** of water. In this respect, heating foods is a form of food preservation comparable to that of freezing but much superior to it in its effectiveness. A preliminary step in many other forms of food preservation, especially forms that make use of packaging, is to heat the foods to temperatures sufficiently high to destroy pathogens.

In many cases, foods are actually cooked prior to their being packaged and stored. In other cases, cooking is neither appropriate nor necessary. The most familiar example of the latter situation is pasteurization. During the 1860s, the French bacteriologist Louis Pasteur discovered that pathogens in foods can be destroyed by heating those foods to a certain minimum temperature. The process was particularly appealing for the preservation of milk since preserving milk by boiling is not a practical approach. Conventional methods of pasteurization called for the heating of milk to a temperature between 145 and 149°F (63 and 65°C) for a period of about 30 minutes, and then cooling it to room temperature. In a more recent revision of that process, milk can also be "flash-pasteurized" by raising its temperature to about 160°F (71°C) for a minimum of 15 seconds, with equally successful results. A process known as ultra-high-pasteurization uses even higher temperatures—of the order of 194 to 266°F (90 to 130°C)—for periods of a second or more.

Packaging

One of the most common methods for preserving foods today is to enclose them in a sterile container. The term "canning" refers to this method although the specific container can be **glass**, plastic, or some other material as well as a **metal** can, from which the procedure originally obtained its name.

The basic principle behind canning is that a food is sterilized, usually by heating, and then placed within an air-tight container. In the absence of air, no new pathogens can gain access to the sterilized food.

In most canning operations, the food to be packaged is first prepared in some way—cleaned, peeled, sliced, chopped, or treated in some other way—and then placed directly into the container. The container is then placed in hot water or some other environment where its temperature is raised above the boiling point of water for some period of time. This heating process achieves two goals at once. First, it kills the vast majority of pathogens that may be present in the container. Second, it forces out most of the air above the food in the container.

After heating has been completed, the top of the container is sealed. In home canning procedures, one way of sealing the (usually glass) container is to place a layer of melted paraffin directly on top of the food. As the paraffin cools, it forms a tight solid seal on top of the food. Instead of or in addition to the paraffin seal, the container is also sealed with a metal screw top containing a rubber gasket. The first glass jar designed for this type of home canning operation, the Mason jar, was patented in 1858.

The commercial packaging of foods frequently makes use of tin, **aluminum**, or other kinds of metallic cans. The technology for this kind of canning was first developed in the mid-1800s, when individual workers hand-sealed cans after foods had been cooked within them. At this stage, a single worker could seldom produce more than 100 "canisters" (from which the word "can" later came) of food a day. With the development of far more efficient canning machines in the late nineteenth century, the **mass production** of canned foods became a reality.

As with home canning, the process of preserving foods in metal cans is very simple in concept. The foods are prepared and the empty cans sterilized. The prepared foods are then added to the sterile metal can, the filled can is heated to a sterilizing temperature, and the cans are then sealed by a machine. Modern machines are capable of moving a minimum of 1,000 cans per minute through the sealing operation.

Chemical additives

The majority of food preservation operations used today also employ some kind of chemical additive to reduce spoilage. Of the many dozens of chemical additives available, all are designed either to kill or retard the growth of pathogens or to prevent or retard chemical reactions that result in the oxidation of foods.

Some familiar examples of the former class of food additives are **sodium benzoate** and **benzoic acid**; **calcium**, **sodium** propionate, and propionic acid; calcium, potassium, sodium sorbate, and sorbic acid; and sodium and potassium sulfite. Examples of the latter class of additives include calcium, sodium ascorbate, and ascorbic acid (**vitamin C**); **butylated hydroxyanisole** (BHA) and **butylated hydroxytoluene** (BHT); **lecithin**; and sodium and potassium sulfite and **sulfur dioxide**.

A special class of additives that reduce oxidation is known as the sequestrants. Sequestrants are compounds that "capture" metallic ions, such as those of **copper**, **iron**, and nickel, and remove them from contact with foods. The removal of these ions helps preserve foods because in their free state they increase the rate at which oxidation of foods takes place. Some examples of sequestrants used as food preservatives are ethylenediamine-tetraacetic acid (EDTA), **citric acid**, sorbitol, and **tartaric acid**.

Irradiation

The lethal effects of **radiation** on pathogens has been known for many years. Since the 1950s, research in the United States has been directed at the use of this technique for preserving certain kinds of food. The radiation used for food preservation is normally gamma radiation from radioactive isotopes or machine-generated **x rays** or **electron** beams. One of the first applications of radiation for food preservation was in the treatment of various kinds of herbs and spices, an application approved by the U.S. Food and Drug Administration (FDA) in 1983. In 1985, the FDA extended its approval to the use of radiation for the treatment of pork as a means of destroying the pathogens that cause **trichinosis**. Experts predict that the ease and efficiency of food preservation by means of radiation will develop considerably in the future.

That future is somewhat clouded, however, by fears expressed by some scientists and members of the general public about the dangers that irradiated foods may have for humans. In addition to a generalized concern about the possibilities of being exposed to additional levels of radiation in irradiated foods (not a possibility), critics have raised questions about the creation of new and possibly harmful compounds in food that has been exposed to radiation.

Resources

Books

Considine, Glenn D. *Van Nostrand's Scientific Encyclopedia.* New York: Wiley-Interscience, 2002.

Francis, Frederick. *Wiley Encyclopedia of Food Science and Technology.* New York: Wiley, 1999.

Periodicals

Hwang, Deng Fwu. "Tetrodotoxin In Gastropods (Snails) Implicated In Food Poisoning." *Journal of Food Protection* 65, no. 8 (2002): 1341-1344.

"Preventing Food Poisoning." *Professional Nurse* 18, no. 4 (2002): 185-186.

Zurer, Pamela S. "Food Irradiation: A Technology at a Turning Point." *Chemical & Engineering News* (May 5, 1986): 46-56.

David E. Newton

KEY TERMS

Additive—A chemical compound that is added to foods to give them some desirable quality, such as preventing them from spoiling.

Antioxidant—A chemical compound that has the ability to prevent the oxidation of substances with which it is associated.

Curing—A term used for various methods of preserving foods, most commonly by treating them with salt or sugar.

Dehydration—The removal of water from a material.

Fermentation—A chemical reaction in which sugars are converted to organic acids.

Irradiation—The process by which some substance, such as a food, is exposed to some form of radiation, such as gamma rays or x rays.

Oxidation—A chemical reaction in which oxygen reacts with some other substance.

Pasteurization—A method for treating milk and other liquids by heating them to a high enough temperature for a long enough period of time to kill or inactivate any pathogens present in the liquid.

Pathogen—A diseasecausing microorganism such as a mold or a bacterium.

Food pyramid

The food pyramid was developed by the U.S. Department of Agriculture (USDA) as a **nutrition** guide for healthy persons over the age of two years. The guide stresses eating a wide variety of foods from the five major food groups while minimizing the intake of fats and sugars. The daily quantity of foods from each group is represented by the triangular shape. The pyramid is composed of four levels. The tip represents fats and sweets, the second level emphasizes foods primarily from animals (milk and meat groups), the third level emphasizes foods from plants (vegetable and fruit groups), and the bottom level emphasizes foods from grains (breads, cereals, and **rice**).

The food guide pyramid was developed in 1992 as a modification of the previously used *Basic Four* food guide. The updated guide was designed to provide nutritional information in a manner that was easily understood by the public. Also, the pyramid emphasizes fats because the American diet is too high in fats. The guide was developed following the Recommended Dietary Allowances (RDA) and recommendations by certain health organizations.

The food pyramid guidelines for healthy living are:

- Balancing diet with physical activity
- Eating a variety of foods
- Eating plenty of **vegetables**, **fruits**, and grain products
- Eating foods low in **fat**, saturated fat, and **cholesterol**
- Eating sweets in moderation
- Eating **salt** in moderation
- Limiting intake of alcohol.

Using the food pyramid

The recommended servings of each food group are expressed in ranges so that the pyramid can fit most members of a household. The number of servings chosen from each food group is based upon the number of calories a person needs. A **calorie** is the amount of **energy** obtained from food. Most persons should always have at least the lowest number of servings for each group. In general, the low to middle numbers of servings are appropriate for most women and the middle to upper numbers of servings are appropriate for most men. Servings do not need to be measured for grain products, vegetables, and fruits but should be followed carefully when eating foods that contribute a significant amount of fat (meats, dairy, and fats used in food preparation). Persons who are dieting should reduce their fat intake and increase physical activity but *not* reduce the number of servings from each group.

Sample daily diets at three calorie levels:

- Lower calorie diet. Nonactive women and some elderly persons may need a lower calorie diet (1,600 calories) comprised of: grains, six servings; vegetables, three servings; fruits, two servings; milks, two to three servings; meat, 5 oz (142 g); fat, 2 oz (53 g); and sugar, 6 teaspoons.

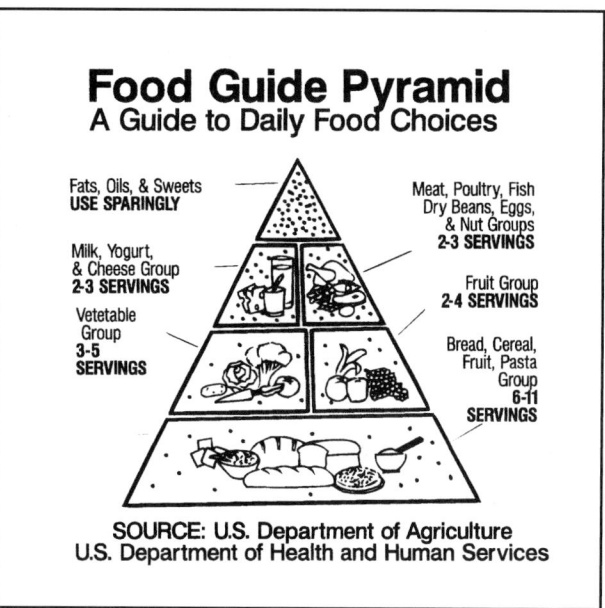

Food Guide Pyramid *Gale Group.*

- Moderate calorie diet. Children, teenage girls, active women, pregnant or breast feeding women, and nonactive men may need a moderate calorie diet (2,200 calories) comprised of: grains, nine servings; vegetables, four servings; fruits, three servings; milks, two to three servings; meat, 6 oz (171 g); fat, 2.5 oz (73 g)s; and sugar, 12 teaspoons.

- Higher calorie diet. Teenage boys, active men, and active women may need a high calorie diet (2,800 calories) comprised of: grains, 11 servings; vegetables, five servings; fruits, four servings; milks, two to three servings; meat, 7 oz (198 g); fat, 3 oz (93 g); and sugar, 18 teaspoons.

Children between two and six years of age can follow the food pyramid but with smaller serving sizes (about two thirds of a regular serving) and two cups of milk daily. Preschool children may need fewer than 1,600 calories and children under the age of two years have special dietary needs. A pediatrician should be consulted as to the appropriate diet for young children. Persons with special dietary needs (vegetarians, diabetics, etc.) can consult a dietician or nutritionist.

Food groups

Fats, oils, and sweets

Fats, oils, and sweets are at the very top of the pyramid because these foods should be used sparingly. In general, these foods provide only calories, little else nutritionally. Persons should choose lower fat foods from each group, reduce the use of fats (such as butter) and sugars (such as jelly) at the table, and reduce the intake of sweet foods (soda, candy, etc.).

Fats should not contribute more than 30% of a persons daily calories. To determine the number of grams of fat that contributes 30% of the calories multiply the total day's calories by 0.30 and divide by 9. For example, a 2,200 calorie diet should contain no more than 2.5 oz (73 g) of fat. Some fats are worse than others. The intake of saturated fats should be limited because they raise **blood** cholesterol levels which increases the risk of **heart disease**. Saturated fats are primarily found in **animal** and dairy products, and coconut, palm, and palm kernel oils. Saturated fats should not contribute more than 10% of the daily calories. Unsaturated fats are a healthier choice and include olive, peanut, canola, safflower, corn, sunflower, cottonseed, and **soybean** oils. Cholesterol is a fat-like **molecule** found only in animal products. Egg yolks and liver are especially high in cholesterol. Daily cholesterol intake should be limited to 300 mg or less.

The daily intake of sugar should be limited to 6 tsp for a diet of 1,600 calories. Sugars include white sugar, raw sugar, brown sugar, corn syrup, molasses, and honey. Naturally found sugars, those in fruits, 100% fruit juices, and milk, are not a major source of sugar in the American diet.

Fats, oils, and sweets are often found in foods from the five groups. For instance, meats contain fats and baked goods contain fats and sugars. These sources should be considered when choosing foods from each group. To reduce the intake of fats, leaner cuts of meat, low fat milk, unsaturated vegetable oils, and margarines prepared from liquid vegetable oil should be chosen.

Milk, yogurt, and cheese

The food pyramid recommends two to three servings of milk products daily. Women who are pregnant or breast feeding, teenagers, and adults up to the age of 24 years need three servings daily. Milk products are the best food source of **calcium** and also provide protein, **minerals**, and vitamins. A serving size is one cup of milk or yogurt, 2 oz (56 g) of processed cheese, or 1.5 oz (43 g) of natural cheese. To reduce the intake of fat and cholesterol, skim milk, nonfat yogurt, and low fat cheese and milk desserts should be chosen. The intake of high fat **ice** cream and cheeses should be reduced.

Meat, poultry, fish, dry beans, eggs, and nuts

The food pyramid recommends eating two to three servings (or 5-7 oz [142-198 g] of meat) from this group. Meat, **fish**, and poultry provide protein, **iron**, zinc, and B vitamins. Eggs, nuts, and dry beans supply protein, vitamins, and minerals. To help determine the serving size of meats, an average hamburger is about 3 oz. One half a cup of cooked dry beans, 2 tbsp of peanut butter, one egg, or one third a cup of nuts are all equivalent to 1 oz (28 g) of meat.

Lean meats and poultry should be chosen to reduce the intake of fat and cholesterol. Lean meats include: sirloin steak, pork tenderloin, veal (except ground), lamb leg, chicken and turkey (without skin), and most fish. The intake of nuts and **seeds**, which contain large amounts of fat, should be reduced.

Vegetable

The food guide pyramid recommends eating three to five servings of vegetables each day. Vegetables provide vitamins, minerals, and fiber, and are low in fat. A serving size of vegetable is 1 cup of raw salad greens, one half a cup of other cooked or raw vegetables, or three quarters of a cup of vegetable juice. Limit the use of toppings or spreads (butter, salad dressing, mayonnaise, etc.) because they add fat calories.

The food pyramid recommends eating a variety of vegetables because different classes of vegetables provide different **nutrients**. Vegetables classes include: dark green leafy (broccoli, **spinach**, romaine lettuce, etc.), deep yellow (sweet potatoes, carrots, etc.), starchy (corn, potatoes, peas, etc.), **legumes** (kidney beans, chickpeas, etc.), and others (tomatoes, lettuce, onions, green beans, etc.). The vegetable subgroups dark green leafy and legume should be chosen often because they contain more nutrients than other vegetables. Also, legumes can substitute for meat.

Fruit

The food guide pyramid recommends two to four servings of fruit daily. Fruits provide **vitamin** A, vitamin C, and potassium and are low in fat. A serving size of fruit is three quarters of a cup of fruit juice, one half a cup of cooked, chopped, or canned fruit, or one medium sized **banana**, orange, or apple.

The food pyramid recommends choosing fresh fruits, 100% fruit juices, and canned, frozen, or dried fruits. Intake of fruits that are frozen or canned in heavy syrup should be limited. Whole fruits are preferred because of their high fiber content. Melon, citrus, and berries contain high levels of vitamin C and should be chosen frequently. Juices that are called punch, -ade, or drink often contain considerable added sugar and only a small amount of fruit juice.

Bread, cereal, rice, and pasta

With 6-11 servings daily, this food group is the largest group, hence the bottom position on the pyramid.

KEY TERMS
. . . .

Calorie—The amount of energy obtained from food. The number of calories needed daily is based upon a persons age, gender, weight, and activity level.

Cholesterol—A fat-like substance that contains lipids; found in animal products.

Complex carbohydrate—Also called starches, complex carbohydrates are long chains of sugar molecules. Carbohydrates are used by the body as an energy source.

Saturated fats—Fats found in meat, dairy products, and palm, palm kernel, and coconut oils. Saturated fats elevate blood cholesterol levels which increases the risk of heart disease.

Unsaturated fats—Fats found in vegetable oils including canola, peanut, olive, sunflower, safflower, soybean, corn, and cottonseed. Unsaturated fats are preferable over saturated fats.

This group provides complex carbohydrates (starches), which are long chains of sugars, as well as vitamins, minerals, and fiber. Carbohydrates are the gasoline for the body's many energy-requiring systems. A serving size from this group is one slice of bread, 1 oz (27 g) of cold cereal, or one half a cup of pasta, rice, or cooked cereal.

Complex carbohydrates in and of themselves are not fattening, it is the spreads and sauces used with these foods that add the most calories. For the most nutrition, foods prepared from whole grains (whole **wheat** bread or whole grain cereals for instance) with little added fat and/or sugar should be chosen. The intake of high fat and/or high sugar baked goods (cakes, cookies, croissants, etc.) and the use of spreads (butter, jelly, etc.) should be reduced.

Resources

Books

Francis, Frederick. *Wiley Encyclopedia of Food Science and Technology.* New York: Wiley, 1999.

Other

The Food Guide Pyramid. United States Department of Agriculture Center for Nutrition Policy and Promotion, Home and Garden Bulletin, Number 252.
Shaw, Anne, Lois Fulton, Carole Davis, and Myrtle Hogbin. *Using The Food Guide Pyramid: A Resource For Nutrition Educators.* U.S. Department of Agriculture Food, Nutrition, and Consumer Services, Center for Nutrition Policy and Promotion.

Tips For Using The Food Guide Pyramid For Young Children 2 to 6 Years Old. United States Department of Agriculture Center for Nutrition Policy and Promotion, Program Aid 1647. 1999.
USDA Food Guide Pyramid [cited February 2003]. <http.www.nal.usda.gov.8001/py/pmap.htm>.

Belinda Rowland

Foods, genetically modified *see* **Genetically modified foods and organisms**

Foot and mouth disease

Foot and mouth **disease** is caused by a particular type of **virus**. The disease affects cloven hooved animals; that is, animals with hooves that are split into two main segments. Examples of domestic cloven hooved animals include cattle, **sheep**, **pigs**, and **goats**. Wild cloven hooved animals that are susceptible to foot and mouth disease include elephants, **hedgehogs**, and **rats**.

Foot and mouth disease occurs all over the world. In parts of **Asia**, **Africa**, the Middle East, and **South America**, foot and mouth disease is common to the point of being a continual occurrence among various **livestock** herds. In other areas of the world, stringent control and inspection measures have made outbreaks infrequent. For example, there has not been an outbreak of foot and mouth disease in domestic animals in the United States since 1929. Canada last experienced an outbreak in 1952, and Mexico in 1954.

Other developed countries have been less fortunate. Outbreaks have occurred in Britain periodically since 1839. An outbreak in 1864–1866 devastated cattle herds throughout Britain, prompting legislation governing the transport and export of cattle. Outbreaks in the 1910s and from 1922–1924 saw slaughter introduced as an attempt to limit the spread of the disease. This control measure has been controversial ever since its implementation, because herds that may not be infected are often ordered destroyed.

In 1967–1968, over 400,000 domestic animals were slaughtered in an attempt to limit the spread of another outbreak. The latest large outbreak occurred in England beginning on February 20, 2001. The outbreak was declared over on January 14, 2002. In between these dates, 2030 cases were confirmed, and over 3,900,000 cattle, sheep, pigs, and goats had been slaughtered in England and Western **Europe** in order to contain the outbreak.

A cow and her calf stand drooling from hoof-and-mouth disease. © Reuters NewMedia Inc./Corbis. Reproduced by permission.

The virus that is responsible for the disease is a member of the viral family called Picornaviridae. Specifically, the virus is a member of the genus called *Aphthovirus*. A genus is a more detailed grouping of organisms based on common characteristics. The virus contains **ribonucleic acid (RNA)** as its genetic material. When the virus infects host cells, the RNA is used to make **deoxyribonucleic acid (DNA)**, using the host's genetic replicating processes. The viral DNA then forms the template for the production of viral RNA, which is packaged into new viral particles and released from the host cells.

This infectious process is destructive for the host cells that have been housed the virus. Typically, an **infection** is apparent as blistering in the mouth and feet. Hence, the name of the disease. The blisters cause the feet to become very tender and sore, so that animals have difficulty walking. Often an infected **animal** will suddenly become lame. Other symptoms of infection include slobbering and smacking of the lips, fever with shivering, and reduced milk production.

Routes of infection

Foot and mouth disease is very infectious. The infection can spread quickly through a herd of cattle and sheep. Large numbers of infectious virus particles are present in the fluid from the blisters. The virus is also present in saliva, milk, feces, and, because lungs cells can also become infected, even in the air that the animals exhales. In an advanced infection, the virus can be widespread through the body.

The virus spreads from animal to animal in a number of ways. Direct contact between an infected animal and non-infected animal can spread the virus. Indirect spread is possible, for example when an animal eats food that has been contaminated by the virus (from saliva, for example). The virus can also become airborne, particularly when it has been exhaled, thus an infection in one herd can quickly become widespread over the countryside. An outbreak of foot and mouth disease causes alarm in farmers many miles away from the site of the infection.

Direct spread of the virus is aided by the fact that the appearance of symptoms does not occur until anywhere from 2 to 21 days after infection has occurred. But, the infected animals can be spreading the virus during this time. Thus, infected animals can be present in a herd, allowing an infection to become established before the farmer becomes aware of a problem.

The virus can also be spread by dogs, **cats**, poultry, and wild **birds** and other animals. Usually, there are no symptoms of infection among these animals, as they are often carriers of the virus, without themselves being harmed by the virus. This secondary route of **contamination** makes an infection difficult to control, particularly on a farm where dogs, cats, and poultry abound.

Another indirect route of virus spread is via contaminated shipping trucks, loading ramps, and market stalls. Because the virus is capable of surviving for at least a month in cold and dark environments, contaminated transport and storage facilities can be reservoirs for virus spread for a long time after becoming contaminated. Stringent cleaning and disinfection of such facilities should be done routinely, and especially during an outbreak of foot and mouth disease.

Not all disinfectants are effective against foot and mouth disease. For example, phenol- and hypochlorite-based disinfectants are insufficient to kill the virus. Disinfectants such as **sodium hydroxide**, **sodium carbonate**, and **citric acid** are effective, likely because they destroy the protective structure that surrounds the genetic material.

The death **rate** among infected animals varies depending on the animal **species** and age. For example, in pigs and sheep the death rate among adults can be only 5%, while almost 100% of infected young animals will die. Survivors can continue to carry the virus in their bodies for one to two months (sheep) up to 24 months (cattle). Surviving pigs do not continue to carry the virus.

Even though many adult animals survive, they suffer. As well, the animal's commercial value is diminished because of weight loss and reduced milk production. The economic losses can be considerable. Estimates of the losses that could result in the United States from a widespread outbreak are in the billions.

Foot and mouth disease is confirmed by the recovery of the virus from infected cells, or by detection of antibodies to the virus in host fluids.

Vaccination

There is a limited **vaccine** for foot and mouth disease. The vaccine consists of killed viruses. The viruses are unable to cause the disease, but stimulate the **immune system** to form the antibodies that will help protect a vaccinated animal from developing the disease. The full promise of a foot and mouth vaccine has not yet been fulfilled, however. This is because there are seven different types of the foot and mouth disease virus. Furthermore, these types have multiple subtle differences. As of 2002, a single vaccine that is capable of stimulating immunity to all these different versions of the virus does not exist.

Vaccination against foot and mouth disease must be accomplished each year to confer protection to the virus. The cost of an annual vaccination of the domestic cattle, sheep, and swine of any country would run to the many millions of dollars. And the vaccine protects an animal from developing symptoms, but not from acquiring the virus. Thus, a vaccinated animal could acquire the virus and pass the virus on to other animals that were not vaccinated.

For the reasons of cost and the possible contribution of vaccination to the spread of the disease, the widespread use of foot and mouth vaccine has not been sanctioned. It is conceivable that future versions of the vaccine will be modified using the tools of **biotechnology** to provide long lasting immunity.

See also DNA technology; Zoonoses.

Resources

Organizations

United States Food and Drug Administration, Veterinary Services Emergency Programs, 4700 River Road, Unit 41, Riverdale, Maryland 20737–1231. January 2002 [cited November 6, 2002]. <http://www.aphis.usda.gov/oa/pubs/fsfmd vac.html.>.

Brian Hoyle

Force

Force is the term used for an outside influence exerted by one body on another which produces a change in state of **motion** or state of configuration. This limited meaning in science compared to our everyday usage is most important because of the specific results of this outside influence.

Force producing a change in state of motion gives a body **acceleration**. If forces acting on a body produces no acceleration, the body will experience some change in configuration: a change of size (longer or shorter), a change of shape (twisted or bent), or a positional change (relative to other masses, charges, or magnets). Changes of size or shape involve elastic properties of materials.

Forces are given various names to indicate some specific character. For example, a wagon can be made to go forward by pushing from behind or pulling from the front so push or pull is more descriptive. Electrical and magnetic forces can result in attraction (tendency to come together) or repulsion (tendency to move apart) but gravitational force results only in attraction of masses. The gravitational force exerted by **Earth** on a body is called weight. A body moving, or attempting to move, over another body experiences a force opposing the motion called **friction**. When wires, cables, or ropes are stretched, they then in turn exert a force which is called tension. Specific names give information about the nature of the force, what it does, and direction of action.

See also Laws of motion.

Forensic science

Forensic science reflects multidisciplinary scientific approach to examining crime scenes and in examining evidence to be used in legal proceedings. Forensic science techniques are also used to verify compliance with international treaties and resolutions regarding weapons production and use.

Forensic science techniques incorporate techniques and principles of **biology**, **chemistry**, medicine, **physics**, computer science, **geology**, and **psychology**.

Forensic science is the application of science to matters of law. Both defense and prosecuting attorneys sometimes use information gleaned by forensic scientists in attempting to prove the innocence or guilt of a person accused of a crime.

A basic principle of forensic science is that a criminal always brings something to the scene of a crime, and he or she always leaves something behind. The "something" left behind is the evidence that detectives and criminalists (people who make use of science to solve crimes) look for. It might be fingerprints, footprints, tooth marks, **blood**, semen, hair, fibers, broken **glass**, a knife or gun, a bullet, or something less tangible such as the nature of the wounds or bruises left on the victim's body, which might indicate the nature of the weapon or the method of assault. Careful analysis of evidence left at the scene of a crime often can be used in establishing the guilt or innocence of someone on trial.

History

Archimedes, who proved that his king's crown was not pure gold by measuring its **density**, was perhaps the world's first forensic scientist. However, it was Sir Arthur Conan Doyle's fictional stories of Sherlock Holmes, written in the late nineteenth century, that first anticipated the use of science in solving crimes in the twentieth century. At about the same time, Sir Francis Galton's studies revealed that fingerprints are unique and do not change with age. As early as 1858, William Herschel, a British official in India, used imprints of inked fingers and hands as signatures on documents for people who could not write. Unknown to Herschel, contracts in Japan had been sealed by using a thumb or fingerprint for centuries.

During the 1890s, Scotland Yard, headquarters for the metropolitan police of London, began to use a system developed by a French police official named Alphonse Bertillon. The Bertillon system consisted of a photograph and 11 body measurements that included dimensions of the head, length of arms, legs, feet, hands, and so on. Bertillon claimed that the likelihood of two people having the same measurements for all 11 traits was less than one in 250 million. In 1894, fingerprints, which were easier to use and more unique (even identical twins have different fingerprints), were added to the Bertillon system.

Edmond Locard, a French criminalist, established the first laboratory dedicated to crime analysis in 1910. A decade later, crime labs had been established throughout **Europe**. The first crime lab in the United States was opened in Los Angeles in 1923, but it was 1932 before the Federal Crime Laboratory was established by the Federal Bureau of Investigation (FBI) under the direction of J. Edgar Hoover. Today, there are about 400 crime labs and nearly 40,000 people involved in forensic science in the United States alone.

Fingerprints

Although fingerprints have been used by crime investigators for more than a century, they remain one of the most sought after pieces of evidence. All human beings are born with a characteristic set of ridges on our fingertips. The ridges, which are rich in sweat pores,

form a pattern that remains fixed for life. Even if the skin is removed, the same pattern will be evident when the skin regenerates. Some of the typical patterns found in fingerprints are arches, loops, and whorls.

Oils from sweat **glands** collect on these ridges. When we **touch** something, a small amount of the oils and other materials on our fingers are left on the surface of the object we touched. The pattern left by these substances, which collect along the ridges on our fingers, make up the fingerprints that police look for at the scene of a crime. It is the unique pattern made by these ridges that motivate police to record people's fingerprints. To take someone's fingerprints, the ends of the person's fingers are first covered with ink. The fingers are then rolled, one at a time, on a smooth surface to make an imprint that can be preserved. Fingerprints collected as evidence can be compared with fingerprints on file or taken from a suspect.

Everyone entering military service, the merchant marine, and many other organizations are fingerprinted. The prints are there to serve as an aid in identification should that person be killed or seriously injured. The FBI maintains a fingerprint library with patterns taken from more than 10% of the entire United States population. Each year the FBI responds to thousands of requests to compare samples collected as evidence with those on file at their library. The process of comparison has been improved in terms of speed and effectiveness in recent years by the development of automated fingerprint identification systems (AFIS) that allows police departments with computer access to search the collection.

Many fingerprints found at crime scenes are not visible. These latent fingerprints, which are often incomplete, are obtained in various ways. The oldest and most frequently used method is to use a powder such as ninhydrin to dust the surface. The powder sticks to the oily substances on the print making the pattern visible. The print can then be photographed and lifted off the surface by using a tape to which the powder adheres. To search for fingerprints on porous materials such as **paper**, forensic technicians use fumes of iodine or cyanoacrylate. These fumes readily collect on the oils in the print pattern and can be photographed. Since 1978, argon lasers have also been used to view latent fingerprints. When illuminated by **light** from an argon **laser**, a latent print is often quite visible. Visibility under laser light can be enhanced by first dusting the print with a fluorescent fingerprint powder.

Fingerprints are not the only incriminating patterns that a criminal may leave behind. Lip prints are frequently found on glasses. Footprints and the **soil** left on the print may match those found in a search of an accused person's premises. Tire tracks, bite marks, toe prints, and prints left by bare feet may also provide useful evidence. In cases where the identity of a victim is difficult because of **tissue decomposition** or death caused by explosions or extremely forceful collisions, a victim's teeth may be used for comparison with the dental records of missing people.

Genetic fingerprints

The nuclei within our cells contain coiled, thread-like bodies called chromosomes. Chromosomes are paired, one member of each pair came from your father; the other one from your mother. Chromosomes are made of deoxyribonucleic acid, often called DNA. It is DNA that carries the "blueprint" (genes) from which "building orders" are obtained to direct the growth, maintenance, and activities that go on within our bodies.

Except for identical twins, no two people have the same DNA. However, we all belong to the same **species**; consequently, large strands of DNA are the same in all of us. The segments that are different among us are often referred to as junk DNA by biologists. It is these unique strands of DNA that are used by forensic scientists. Strands of DNA can be extracted from cells and "cut" into shorter sections using enzymes. Through chemical techniques involving **electrophoresis**, radioactive DNA, and **x rays**, a characteristic pattern can be established-the so-called genetic fingerprint. Because different people have different junk DNA, the prints obtained from different people will vary considerably; however, two samples from the same person will be identical. If there is a match between DNA extracted from semen found on the body of a rape victim and the DNA obtained from a rape suspect's blood, the match is very convincing evidence-evidence that may well lead to a conviction or possibly a confession.

Although genetic fingerprinting can provide incriminating evidence, DNA analysis is not always possible because the amount of DNA extracted may not be sufficient for testing. Furthermore, there has been considerable controversy about the use of DNA, the statistical nature of the evidence it offers, and the validity of the testing.

Genetic fingerprinting is not limited to DNA obtained from humans. In Arizona, a homicide detective found two seed pods from a paloverde **tree** in the bed of a pickup truck owned by a man accused of murdering a young woman and disposing of her body. The accused man admitted giving the woman a ride in his truck but denied ever having been near the factory where her body was found. The detective, after noting a scrape on a paloverde tree near the factory, surmised that it was caused by the accused man's truck. Using RAPD (Randomly Amplified Polymorphic DNA) markers—a technique developed by Du Pont scientists—forensic scientists were able to show that the seed pods found in the truck must have come from the scraped tree at the factory.

DNA analysis is a relatively new tool for forensic scientists, but already it has been used to free a number of people who were unjustly sent to prison for crimes that genetic fingerprinting has shown they could not have committed. Despite its success in freeing victims who were unfairly convicted, many defense lawyers claim prosecutors have overestimated the value of DNA testing in identifying defendants. They argue that because analysis of DNA molecules involves only a fraction of the DNA, a match does not establish guilt, only a probability of guilt. They also contend that there is a lack of quality control standards among laboratories, most of them private, where DNA testing is conducted. Lack of such controls, they argue, leads to so many errors in testing as to invalidate any statistical evidence. Many law officials argue that DNA analysis can provide probabilities that establish guilt beyond reasonable doubt.

Evidence and tools used in forensic science

Long before DNA was recognized as the "ink" in the blueprints of life, blood samples were collected and analyzed in crime labs. Most tests used to tentatively identify a material as blood are based on the fact that peroxidase, an **enzyme** found in blood, acts as a catalyst for the reagent added to the blood and forms a characteristic **color**. For example, when benzidine is added to a **solution** made from dried blood and **water**, the solution turns blue. If phenolphthalein is the reagent, the solution turns pink. More specific tests are then applied to determine if the blood is human.

The evidence available through blood typing is not as convincing as genetic fingerprinting, but it can readily prove innocence or increase the probability of a defendant being guilty. All humans belong to one of four blood groups–A, B, AB, or O. These blood groups are based on genetically determined antigens (A and/or B) that may be attached to the red blood cells. These antigens are either present or absent in blood. By adding specific antibodies (anti-A or anti-B) the presence or absence of the A and B antigens can be determined. If the blood cells carry the A antigen, they will clump together in the presence of the anti-A antibody. Similarly, red blood cells carrying the B antigen will clump when the anti-B antibody is added. Type A blood contains the A antigen; type B blood carries the B antigen; type AB blood carries both antigens; and type O blood, the most common, carries neither antigen. To determine the blood type of a blood **sample**, antibodies of each type are added to separate samples of the blood. The results,

Table: Testing for blood type.
A + indicates that the blood cells clump and, therefore, contain the antigen specific for the antibody added. A- indicates there is no clumping and that the blood lacks the antigen specific for the antibody added.

Antibody added to sample		Results of test indicates
anti-A	anti-B	blood type to be
—	—	O
+	—	A
—	+	B
+	+	AB

which are summarized in the table, indicate the blood type of the sample.

If a person accused of a homicide has type AB blood and it matches the type found at the crime scene of a victim, the evidence for guilt is more convincing than if a match was found for type O blood. The reason is that only 4% of the population has type AB blood. The percentages vary somewhat with race. Among Caucasians, 45% have type O, 40% have type A, and 11% have type B. African Americans are more likely to be type O or B and less likely to have type A blood.

When blood dries, the red blood cells split open. The open cells make identification of blood type trickier because the clumping of **cell** fragments rather than whole red blood cells is more difficult to see. Since the antigens of many blood-group types are unstable when dried, the FBI routinely tests for only the ABO, Rhesus (Rh), and Lewis (Le) blood-group antigens. Were these blood groups the only ones that could be identified from blood evidence, the tests would not be very useful except for proving the innocence of a suspect whose blood type does not match the blood found at a crime scene. Fortunately, forensic scientists are able to identify many blood **proteins** and enzymes in dried blood samples. These substances are also genetic markers, and identifying a number of them, particularly if they are rare, can be statistically significant in establishing the probability of a suspect's guilt. For example, if a suspect's ABO blood type matches the type O blood found at the crime scene, the evidence is not very convincing because 45% of the population has type O blood. However, if there is a certain match of two blood proteins (and no mismatches) known to be inherited on different chromosomes that appear respectively in 10% and 6% of the population, then the evidence is more convincing. It suggests that only $0.45 \times 0.10 \times 0.06 = 0.0027$ or 0.27% of the population could be guilty. If the accused person happens to have several rarely found blood factors, then the evidence can be even more convincing.

Since handguns are used in half the homicides committed in the United States and more than 60% of all homicides are caused by guns, it is not surprising that ballistic analysis has been an important part of the work performed in crime labs. Comparison microscopes, which make it possible to simultaneously view and compare two bullets, are an important tool for forensic scientists. When a bullet is fired, it moves along a **spiral** groove in the gun barrel. It is this groove that makes the bullet spin so that it will follow a straight path much like that of a spiraling football. The striations or markings on the bullet made by the groove and the marks left by the firing pin are unique and can be used to identify the gun used to fire any bullets found at the scene of a homicide. Similarly, tool marks, which are often left by burglars who pry open doors or windows, can serve as useful evidence if comparisons can be made with tools associated with a person accused of the crime. Particularly incriminating are jigsaw matches-pieces of a tool left behind that can be shown to match pieces missing from a tool in the possession of the accused.

In the event that bullets have been shattered making microscopic comparisons impossible, the fragments may be analyzed by using **neutron activation analysis**. Such analysis involves bombarding the sample with neutrons to make the **atoms** radioactive. The gamma rays emitted by the sample are then scanned and compared with known samples to determine the **concentration** of different metals in the bullet-lead. The technique can be used to compare the evidence or sample with bullet-lead associated with the accused.

Autopsies can often establish the cause and approximate time of death. Cuts, scrapes, punctures, and rope marks may help to establish the cause of death. A drowning victim will have soggy lungs, water in the stomach, and blood diluted with water in the left side of the **heart**. A person who was not breathing when he or she entered the water will have undiluted blood in the heart. Bodies examined shortly after the time of death may have stiff jaws and limbs. Such stiffness, or rigor mortis, is evident about ten hours after death, but disappears after about a day when the tissues begin to decay at normal temperatures. Each case is different, of course, and a skillful coroner can often discover evidence that the killer never suspected he or she had left behind.

Modern crime labs are equipped with various expensive analytical devices usually associated with research conducted by chemists and physicists. Scanning **electron** microscopes are used to magnify surfaces by as much as a factor of 200,000. Because the material being scanned emits x rays as well as secondary electrons in response to the electrons used in the scanning process, the **microscope** can be used together with an x ray micro analyzer to identify elements in the surface being scanned. The technique has been particularly successful in detecting the presence of residues left when a gun is fired.

The **mass** spectrometer and the gas chromatograph have been particularly effective in separating the components in illegal drugs, identifying them, and providing the data needed to track down their source and origin. Thin layer **chromatography** (TLC) has proved useful in identifying colored fibers. Although many fibers may appear identical under the microscope, they can often be distinguished by separating the component dyes used in coloring the fabric. Fusion microscopy—using changes in birefringence with temperature—has also proved useful in identifying and comparing synthetic fibers found at crime scenes. In addition to using such physical properties as density, dispersion, and refractive index to match and identify glass samples, the **plasma emission spectroscope** has proven helpful in analyzing the component elements in glass as well as distinguishing among various types of glass found in windows, bottles, and windshields.

The role and impact of forensic sciences came to the forefront of public attention during the highly publicized and televised O.J. Simpson murder trial. Lawyers for both sides offered a wide variety of forensic evidence—and disputed the validity of opposing forensic evidence—before the controversial acquittal of Simpson.

In 2002, forensic science specialists played an integral role in the tracking and eventual identification of evidence (e.g. similarities in **ballistics**, psychological profiles, etc.) that allowed investigators link a nationwide a string of crimes that culminated in several snipers attacks in the Washington-Virginia-Maryland area.

A new and emerging area of forensic science involves the reconstruction of computer data. High speed and large memory capacity computers also allow for what forensic investigators term as "virtual criminality," the ability of computer animation to recreate crime scenes.

See also Antibody and antigen; DNA technology; Pathology; Toxicology.

KEY TERMS

Birefringence—Splitting of light into two separate beams in which the light travels at different speeds.

Gas chromatograph—A device that separates and analyzes a mixture of compounds in gaseous form.

Mass spectrometer—A device that uses a magnetic field to separate ions according to their mass and charge.

Polymorphic—Distinct inherited traits that are different among members of the same species. Blood groups, for example, are polymorphic, but weight is not.

Scanning electron microscope—A device that emits a focused beam of electrons to scan the surface of a sample. Secondary electrons released from the sample are used to produce a signal on a cathode ray tube where the enlarged image is viewed.

Resources

Books

Butler, John M. *Forensic DNA Typing: The Biology and Technology Behind STR Markers* Academic Press, 2001.
Lee, Henry C., and Charles D. Gill. *Cracking Cases: The Science of Solving Crimes.* Prometheus Books, 2002.
Nordby, Jon J. *Dead Reckoning: The Art of Forensic Detection.* CRC Press, 2000.
Sachs, Jessica S. *Corpse: Nature, Forensics, and the Struggle to Pinpoint Time of Death. An Exploration of the Haunting Science of Forensic Ecology.* Perseus Publishing, 2001.
Saferstein, Richard. *Criminalistics: An Introduction to Forensic Science.* New York: Prentice-Hall, 2000.

Other

Consulting and Ducation in Forensic Science. "Forensic Science Timeline." Norah Rudin [cited March 16, 2003]. <http://www.forensicdna.com/Timeline.htm.>.

Robert Gardner

Forestry

Forestry is the science of harvesting, planting, and tending trees, within the broader context of the management of forested landscapes. Traditionally, forestry has focused on providing society with sustainable yields of economically important products, especially **wood** for the manufacturing of lumber or **paper**, or for the generation of **energy**. Increasingly, however, forestry must consider other, non-traditional goods and services provided by the forested landscape, such as populations of both hunted and non-hunted **wildlife**, recreational opportunities, aesthetics, and the management of landscapes to maintain clean air and **water**. Because not all of these values can always be accommodated in the same area, there are often conflicts between forestry and other uses of the landscape. However, the use of systems of integrated management can often allow an acceptable, working accommodation of forestry and other resource values to be achieved.

Forestry and its broader goals

Forestry is a science, but also somewhat of an art. The ultimate objective of forestry is to design and implement management systems by which forested landscapes can yield sustainable flows of a range of ecological goods and services. The most important of the resource values dealt with in forestry are products directly associated with **tree biomass**, such as lumber, paper, and fuelwood. However, non-tree resource values are also important, and these must be co-managed by foresters in conjunction with the traditional industrial products.

In many respects, forestry is analogous to agricultural science, and foresters are akin to farmers. Forestry and agriculture both deal with the harvesting and management of ecological systems, and both are seeking optimized, sustainable yields of economically important, biological commodities. However, compared with forestry, agriculture deals with a greater diversity of economic **species** and biological products, a wider range of harvesting and management systems (most of which are much more intensive than in forestry), and relatively short harvesting rotations (usually annual). Still, the goals of forestry and agriculture are conceptually the same.

Another shared feature of forestry and agriculture is that both cause substantial ecological changes to sites and the larger landscape. The various activities associated with forestry and agriculture are undertaken in particular sites. However, in aggregate these places are numerous, and therefore entire landscapes are affected. Inevitably, these activities result in substantial ecological changes, many of which represent a significant degradation of the original ecological values. For example, populations of many native species may be reduced or even extirpated, the viability of natural communities may be placed at risk through their extensive conversion to managed ecosystems, **erosion** is often caused, the environment may become contaminated with **pesticides** and **fertilizers**, and aesthetics of the landscape may be degraded. One of the most important challenges to both forestry and agriculture is to achieve their primary goals of maintaining sustainable harvests of economically important commodities, while keeping the associated environmental degradations within acceptable limits.

Resource values managed in forestry

Forested landscapes support a variety of resource values. Some of these are important to society because they are associated with natural resources that can be harvested to yield commodities and profit. Other values, however, are important for intrinsic reasons, or because they are non-valued but important ecological goods and services. (That is, their importance is not measured in monetary units, but they are nevertheless important to society and to **ecological integrity**. Some of these non-valued resources are described below.) Often, there are substantial conflicts among the different resource values, a circumstance that requires choices to be made when designing management systems. In particular, activities associated with the harvesting and management of trees for profit may pose risks to other, non-timber resources. In any cases of conflict among management objectives, societal choices must be made in order to assign emphasis to the various resource values. Sometimes timber values are judged to be most important, but sometimes not.

The most important of the resources that modern foresters consider in their management plans are the following:

(1) Traditional forest products are based on harvested tree biomass. These include large-dimension logs that are cut into lumber or manufactured into laminated products such as plywood, and more varied sizes of trees used for the production of pulp and paper, or burned to generate energy for industry or homes. These are all economically important forest products, and they are harvested to sustain employment and profits. Almost always, managing for a sustained yield of these tree-based products is the primary objective in forestry.

(2) Some species of so-called game animals are exploited recreationally (and also for subsistence) by hunters and maintenance of their populations is often a prominent management objective in forestry. The most important of the species hunted in forested lands in **North America** are large **mammals** such as **deer**, elk, **moose**, and bear; smaller mammals such as rabbit and hare; gamebirds such as **grouse**, ptarmigan, and **quail**; and sportfish such as trout and **salmon**. In some cases, forestry can enhance the abundance of these species, but in other cases forestry can damage populations of game animals, and this conflict must be managed to the degree possible.

(3) Species that sustain a commercial hunt are another common consideration in forestry. Terrestrial examples of this type of non-tree economic resource are

fur-bearing animals such as marten, fisher, weasel, beaver, bobcat, lynx, wolf, and coyote. Foresters may also be involved in the management and protection of the **habitat** of river-spawning **fish** such as salmon, which are commercially exploited in their marine habitat.

(4) So called non-game species comprise the great majority of the species of forested landscapes. Most of these elements of **biodiversity** are native species, occurring in natural communities dispersed across the ecological landscape. Although few of these species are of direct economic importance, all of them have intrinsic value. Forestry-related activities may pose important threats to many of these species and their communities, and this can engender great controversy and require difficult social choices about the priorities of resource values. For example, in North America there are concerns about the negative effects of forestry on **endangered species** such as the spotted owl and red-cockaded woodpecker, and on endangered ecosystems such as old-growth forest. To some degree, these concerns will have to be addressed by declaring ecological reserves of large tracts of natural forest, in which the commercial harvesting of timber is not allowed.

(5) Recreational opportunities are another important resource value of forested landscapes, and these may have to be maintained or enhanced through the sorts of forestry activities that are undertaken. Examples of forest recreation include wildlife observation (such as bird watching), hiking, cross-country skiing, and driving off-road vehicles. In some cases these activities are made easier through forestry which may, for example, improve access by building roads. In other cases, forestry may detract from recreational values because of the noise of industrial equipment, dangers associated with logging trucks on narrow roads, and degraded habitat qualities of some managed lands.

(6) The visual aesthetics of sites and landscapes is another important consideration in forestry. Aesthetics are important in recreation, and for intrinsic reasons such as wilderness values. Compared with natural, mature **forests**, many people consider recently clear-cut sites to have very poor aesthetics, although this value is often judged to have improved once a new forest has re-established on the site. In contrast, foresters may not share this interpretation of the aesthetics of the same sites. Clearly, aesthetics are partly in the mind of the beholder. Societal choices are required to determine the most appropriate management objectives for site or landscape aesthetics in particular regions.

(7) Non-valuated, ecological goods and services are also important considerations in forestry. Examples of these ecological values include the ability of the landscape to prevent erosion, to maintain a particular hydro-

Checking the age of a sitka spruce in Gustavus, Alaska. *Photograph by Tom Bean. Stock Market. Reproduced by permission.*

logic regime in terms of the timing and quantities of water flow, to serve as a sink for atmospheric **carbon dioxide** through the growth of vegetation, and to serve as a source of atmospheric **oxygen** through the **photosynthesis** of growing plants. As noted previously, these are all significant resource values, although their importance is not assessed in terms of dollars.

Harvesting and management

Forest harvesting refers to the methods used to cut and remove trees from the forest. Harvesting methods vary greatly in their intensity. Clear-cutting is the most intensive system, involving the harvest of all trees of economic value at the same time. The areas of clear-cuts can vary greatly, from patch-cuts smaller than a hectare in size, to enormous harvests thousands of hectares in area, sometimes undertaken to salvage timber from areas that have recently been affected by **wildfire** or an insect **epidemic**. Strip-cutting is a system involving a series of long and narrow clear-cuts, with alternating uncut strips of forest left between. A few years after the first strip-cuts were made, tree regeneration should be well established by seeding-in from the uncut strips, and the uncut strips would then be harvested. Shelter-wood cutting is a partial harvest of a stand, in which selected, large trees are left to favor particular species in the regeneration, and to stimulate growth of the uncut trees to produce high-quality sawlogs at the time of the next harvest, usually one or several decades later. In some respects, the shelterwood system can be viewed as a staged clear-cut, because all of the trees are harvested, but in several steps. The least intensive method of harvesting is the selection-tree system, in which some of the larger individual trees of desired species are harvested every ten or more years, always leaving the physical integrity of the forest essentially intact.

Usually when trees are harvested, they are de-limbed where they have fallen, the branches and foliage are left on the site, and the logs taken away for use. However, some harvest systems are more intensive in their removal of tree biomass from the site. A whole-tree harvest, usually used in conjunction with clear-cutting, involves removal of all of the above-ground biomass. A complete-tree harvest is rare, but would attempt the additional harvest of root biomass, as is possible on sites with peaty soils. These very intensive harvesting methods may be economically advantageous when trees are being harvested for the production of industrial energy, for which the quality of the biomass is not an important consideration. However, the whole-tree and complete-tree methods greatly increase the removal of **nutrients** from the site compared with stem-only harvests, and this can be a consideration in terms of impairment of fertility of the land.

Forest management refers to the activities associated with establishing new **crops** of trees on harvested sites, tending the stands as they develop, and protecting them from **insects** and diseases. As was the case with harvesting, the intensity of management activities can vary greatly. The least intensive management systems rely on natural regeneration of trees and natural stand development. Although relatively natural systems are softer in terms of their environmental impacts, the **rate** of forest productivity is often less than can be accomplished with more intensive management systems.

One natural system of regeneration utilizes the so-called advance regeneration, or the population of small individuals of tree species that occurs in many mature forests, and is available to contribute to development of the next stand after the overstory trees are harvested. Other systems of natural regeneration try to encourage the post-harvest establishment of seedlings of desired tree species after the site is harvested. For some species of

trees, the site must be prepared to encourage seedling establishment. This may require burning of the slash and surface organic **matter**, or mechanical scarification using heavy machines. Depending on the particular nature of the forest and the tree species, either of the advance regeneration or seeding-in regeneration systems might be utilized along with selective harvesting, or with clear-cutting.

If the forester believes that natural regeneration will not be adequate in terms of **density**, or that it would involve the wrong species of trees, then a more intensive system might be used to establish the next stand of trees. Often, young seedlings of desired species of trees will be planted, to establish an even-aged, usually single-species plantation. The seedlings are previously grown under optimized conditions in a greenhouse, and they may represent a narrow genetic lineage selected for desirable traits, such as rapid productivity or good growth form.

Once an acceptable regeneration of trees is established on a harvested site, the stand may require tending. Often, non-desired plants are believed to excessively compete with the trees and thereby interfere with their growth. As such, these plants are considered to be silvicultural "weeds." This management problem may be dealt with by using a herbicide, or by mechanical weeding. Similarly, once the growing stand develops a closed canopy of foliage, the individual trees may start to excessively compete among themselves, reducing the overall growth rate. This problem may be dealt with by thinning the stand, an activity in which the least productive individuals or those with poorer growth form are selectively removed, to favor productivity of the residual trees.

In some cases, the regenerating stand may be threatened by a population outbreak of an insect capable of severely reducing productivity, or even killing trees. This pest-management problem may be managed by "protecting" the stand using an insecticide. In North America, **insecticides** have most commonly been used to deal with severe defoliation caused by outbreaks of **spruce** budworm or gypsy moth, or with damage associated with **bark beetles**.

Silvicultural systems and management

Silvicultural systems are integrated activities designed to establish, tend, protect, and harvest crops of trees. Activities associated with silvicultural systems are carried out on particular sites. However, the spatial and temporal patterns of those sites on the landscape must also be designed, and this is done using a management plan appropriate to that larger scale. The landscape-scale management plan is typically detailed for the first five years, but it should also contain a 25-year forecast of objectives and activities. The design and implementation of silvicultural systems and management plans are among the most important activities undertaken by modern foresters.

The primary goal of forestry is generally to achieve an optimized, sustainable yield of economically important, tree-derived products from the landscape. In places where the mandate of forestry is focused on the economic resource of trees, the silvicultural system and management plan will reflect that priority. However, in cases where society requires effective management of a range of resource values (that is, not just trees), then integrated management will be more prominent in the system and plan.

As with the individual harvesting and management practices described in the preceding section, silvicultural systems can be quite intensive, or much less so. An example of an intensive system used in North America might involve the following series of activities, occurring sequentially, and beginning with a natural forest composed of a mixture of native species of trees: (1) whole-tree, clear-cut harvesting of the natural forest, followed by (2) scarification of the site to prepare it for planting, then (3) an evenly spaced planting of young seedlings of a narrow genetic lineage of a single species (usually a **conifer**), with (4) one or more herbicide applications to release the seedlings from the deleterious effects of **competition** with weeds, and (5) one or more thinnings of the maturing plantation, to optimize spacing and growth rates of the residual trees. Finally, the stand is (6) harvested by another whole-tree clear-cut, followed by (7) establishment, tending, and harvesting of the next stand using the same silvicultural system. If the only objective is to grow trees as quickly as possible, this system might be used over an entire landscape.

In contrast, a much softer silvicultural system might involve periodic selection-harvesting of a mixed-species forest, perhaps every decade or two, and with reliance on natural regeneration to ensure renewal of the economic resource. However, even a system as soft as this one might pose a risk for certain non-timber resource values. For example, if certain species dependent on old-growth forest were believed to be at risk, then an appropriate management plan would have to include the establishment of ecological reserves large enough to sustain that old-growth resource value, while the rest of the land is "worked" to provide direct economic benefits.

Because silvicultural systems can differ so much in their intensity, they also vary in their environmental impacts. As is the case with agriculture, the use of intensive systems generally results in substantially larger yields of the desired economic commodity (in this case, tree biomass). However, intensive systems have much greater environmental impacts associated with their activities. The challenge of forestry is to design socially acceptable

KEY TERMS

Plantation—A tract of land on which economically desired trees have been planted and tended, often as a monoculture.

Rotation—In forestry, this refers to the time period between harvests. A forestry rotation is typically 50-100 years.

Scarification—The mechanical or chemical abrasion of a hard seedcoat in order to stimulate or allow germination to occur.

Silvicultural system—A system designed to establish, tend, protect, and harvest a crop of trees.

Silviculture—The branch of forestry that is concerned with the cultivation of trees.

Weed—Any plant that is growing abundantly in a place where humans do not want it to be.

systems that sustain the economic resource, while at the same time accommodating concerns about the health of other resources, such as hunted and non-hunted biodiversity, **old-growth forests**, and ecologically important, but non-valuated goods and services that are provided by forested landscapes.

See also Deforestation.

Resources

Books

Freedman, B. *Environmental Ecology.* 2nd ed. San Diego: Academic Press, 1995.

Kimmins, H. *Balancing Act. Environmental Issues in Forestry.* Vancouver: University of British Columbia Press, 1992.

Bill Freedman

Forests

A forest is any ecological community that is structurally dominated by tree-sized woody plants. Forests occur anywhere that the climate is suitable in terms of length of the growing season, air and **soil temperature**, and sufficiency of soil moisture. Forests can be classified into broad types on the basis of their geographic range and dominant types of trees. The most extensive of these types are boreal coniferous, temperate **angiosperm**, and tropical angiosperm forests. However, there are regional and local variants of all of these kinds of forests. Old-growth tropical rainforests support an enormous diversity of **species** under relatively benign climatic conditions, and this **ecosystem** is considered to represent the acme of Earth's ecological development. Within the constraints of their regional climate, temperate and boreal forests also represent peaks of ecological development.

Types of forests

Many countries have developed national schemes for an ecological classification of their forests. Typically, these schemes are based on biophysical information and reflect the natural, large-scale patterns of species composition, soil type, topography, and climate. However, these classifications may vary greatly among countries, even for similar forest types.

An international system of ecosystem classification has been proposed by a scientific working group under the auspices of the United Nations Educational, Scientific and Cultural Organization (UNESCO). This scheme lists 24 forest types, divided into two broad classes: (i) closed-canopy forests with a canopy at least 16.5 ft (5 m) high and with interlocking **tree** crowns, and (ii) open woodlands with a relatively sparse, shorter canopy. A selection of forest types within these two broad classes is described below:

A. Tropical and Subtropical Forests.

1. Tropical rain forest. This is a species-rich forest of angiosperm tree species (sometimes known as "tropical hardwoods") occurring under conditions of high rainfall and constant, warm temperatures. Consequently, the species in this ecosystem are tolerant of neither **drought** or frost, and the forest itself is commonly in an old-growth condition. Most of Earth's terrestrial **biodiversity** occurs in this type of forest ecosystem.

2. Tropical and subtropical evergreen forest. This is also a rather species-rich forest, but occurring in regions in which there is seasonally sparse rain. Individual trees may shed their leaves, usually in reaction to relatively dry conditions. However, the various species do not all do this at the same time, so the canopy is always substantially foliated.

3. Tropical and subtropical drought-deciduous forest. This is a relatively open angiosperm forest, in which tree foliage is shed just before the dry season, which usually occurs in winter.

4. Mangrove forest. This is a relatively species-poor forest, occurring in muddy intertidal **habitat** in the tropics and subtropics. Mangrove forest is dominated by evergreen angiosperm trees that are tolerant of flooded soil and exposure to **salt**. Some genera of mangrove trees are widespread. Examples from south Florida are red mangrove (*Rhizophora mangle*) with its characteristic

stilt roots, and black mangrove (*Avicennia nitida*) with its pneumatophores, which poke out of the oxygen-poor sediment and into the atmosphere.

B. Temperate and Subpolar Forests.

5. Temperate deciduous forest. This is a deciduous forest dominated by various species of angiosperm trees growing under seasonal climatic conditions, including moderately cold winters. In eastern North American forests of this type, the common trees include species of maple, birch, hickory, ash, walnut, tulip-tree, oak, and **basswood**, among others (*Acer, Betula, Carya, Fraxinus, Juglans, Liriodendron, Quercus,* and *Tilia,* respectively).

6. Temperate and subarctic, evergreen **conifer** forest. This is a northern coniferous forest (sometimes called boreal forest), growing in regions with highly seasonal conditions, including severe winters. The dominant genera of conifer trees are fir, **spruce**, pine, cedar, and hemlock, among others (*Abies, Picea, Pinus, Thuja,* and *Tsuga,* respectively).

7. Temperate and subpolar, evergreen rain forest. This forest occurs in wet, frost-free, oceanic environments of the Southern Hemisphere, and is dominated by evergreen, angiosperm species such as southern beech (*Nothofagus* spp.) and southern pine (*Podocarpus* spp.).

8. Temperate, winter-rain, evergreen broadleaf forest. This is an evergreen angiosperm forest, growing in regions with a pronounced wet season, but with summer drought. In **North America**, this forest type occurs in coastal parts of southern California, and is dominated by evergreen species of **oaks** (*Quercus* spp.).

9. Cold-deciduous forest. This is a northern deciduous forest growing in a strongly seasonal climate with very cold winters. This forest type is typically dominated by angiosperm trees such as aspen and birch (*Populus* spp. and *Betula* spp.) or the deciduous conifer, larch (*Larix* spp.).

Forest process

Forests are among the most productive of Earth's natural ecosystems. On average, tropical rain forests have a net primary productivity of about 4.8 lb/ft^2/year (2.2 kg/m^2/yr), compared with 3.5 lb/ft^2/year (1.6 kg/m^2/yr) for tropical seasonal forests, 2.9 lb/ft^2/year (1.3 kg/m^2/yr) for temperate conifer forests, 2.6 lb/ft^2/year (1.2 kg/m^2/yr) for temperate angiosperm forests, and 1.8 lb/ft^2/year (0.8 kg/m^2 yr) for conifer subarctic forests.

Although tropical rain forests have relatively high rates of net primary productivity, their net ecosystem productivity is very small or **zero**. This occurs because these forests are typically in an old-growth condition, so that there are always some individual trees that are dying or recently dead, resulting in a relatively large number of standing dead trees and logs lying on the forest floor. These deadwood components decompose rather quickly under the warm and humid conditions of tropical rain forests, as does the rain of **leaf** litter to the forest floor. Because the **rate** of **decomposition** approximately counterbalances the positive net primary production, the net ecosystem productivity is more-or-less zero in these **old-growth forests**. Old-growth temperate rain forests are less common than tropical ones, but these also typically have a small or zero net ecosystem productivity.

Mature forests store more **carbon** (in **biomass**) than any other kind of ecosystem. This is especially true of old-growth forests, which typically contain large trees and, in temperate regions, a great deal of dead organic **matter**. Because all of the organic carbon stored in forests was absorbed from the atmosphere as **carbon dioxide** (CO_2), these ecosystems are clearly important in removing this greenhouse gas from the atmosphere. Conversely, the conversion of forests to any other type of ecosystem, such as agricultural or urbanized lands, results in a large difference in the amount of carbon stored on the site. That difference is made up by a large flux of CO_2 to the atmosphere. In fact, **deforestation** has been responsible for about one-half of the CO_2 emitted to the atmosphere as a result of human activities since the beginning of the **industrial revolution**.

Because they sustain a large biomass of foliage, forests evaporate large quantities of **water** to the atmosphere, in a hydrologic process called **evapotranspiration**. Averaged over the year, temperate forests typically evapotranspire 10-40% of their input of water by **precipitation**. However, this process is most vigorous during the growing season, when air temperature is highest and the amount of **plant** foliage is at a maximum. In fact, in many temperate forests evapotranspiration rates during the summer are larger than precipitation inputs, so that the ground is mined of its water content, and in some cases streams dry up.

Intact forests are important in retaining soil on the land, and they have much smaller rates of **erosion** than recently harvested forests or deforested landscapes. Soil eroded from disturbed forests is typically deposited into surface waters such as streams and lakes, in a process called sedimentation. The resulting shallower water depths makes flowing waters more prone to spilling over the banks of **rivers** and streams, causing **flooding**.

Forests are also important in moderating the peaks of water flow from landscapes, both seasonally and during extreme precipitation events. When this function is degraded by deforestation, the risk of flooding is further increased.

A forest of deciduous trees. *Photograph by M. Faltner. Stock Market/Zefa Germany. Reproduced by permission.*

Forests as habitat

Although trees are the largest, most productive organisms in forests, the forest ecosystem is much more than a population of trees growing on the land. Forests also provide habitat for a host of other species of plants, along with numerous animals and **microorganisms**. Most of these associated species cannot live anywhere else; they have an absolute requirement of forested habitat. Often that need is very specific, as when a bird species needs a particular type of forest, in terms of tree species, age, and other conditions.

For example, Kirtland's warbler (*Dendroica kirtlandii*) is an **endangered species** of bird that only nests in stands of jack pine (*Pinus banksiana*) of a particular age and density in northern Michigan. This songbird does not breed in any other type of forest, including younger or older stands of jack pine. Similarly, the endangered spotted owl (*Strix occidentalis*) only occurs in certain types of old-growth conifer forests in western North America. These same old-growth forests also sustain other species that cannot exist in younger stands, for example, certain species of **lichens**, mosses, and liverworts.

Usually, however, the many species occurring in forests have a broader ecological tolerance, and they may in fact require a mosaic of different habitat types. In eastern North America, for example, white-tailed **deer** (*Odocoileus virginianus*) do well in a mixture of habitats. They require access to relatively young, successional stands with abundant and nutritious food for this species, along with refuge habitat of mature forest with some conifer-dominated areas that have shallower snow depth in winter. Similarly, ruffled **grouse** (*Bonasa umbellus*) does best on a landscape that has a checkerboard of stands of various age, including mature forest dominated by trembling aspen (*Populus tremuloides*) with a few conifers mixed in.

More generally, forests provide the essential habitat for most of Earth's species of plants, animals, and microorganisms. This is especially true of tropical rain forests. Recent reductions of forest area, which since the 1950s have mostly been associated with the conversion of tropical forest into agricultural land-use, are a critical environmental problem in terms of losses of biodiversity. Deforestation also has important implications for climate change and access to natural resources.

Forests as a natural resource

The global area of forest of all kinds was about 8.4 billion acres (3.4 billion hectares) in 1990, of which 4.3 billion acres (1.76 billion ha) was tropical forest and the rest temperate and boreal forest. That global forest area is at least one-third smaller than it was prior to extensive deforestation caused by human activities. Most of the deforested land has been converted to permanent agricultural use, but some has been ecologically degraded into semi-desert or **desert**. This global deforestation, which is continuing apace, is one of the most serious aspects of the environmental crisis.

Forests are an extremely important natural resource that can potentially be sustainably harvested and managed to yield a diversity of commodities of economic importance. **Wood** is by far the most important product harvested from forests. The wood is commonly manufactured into **paper**, lumber, plywood, and other products. In addition, in most of the forested regions of the less-developed world firewood is the most important source of **energy** used for cooking and other purposes. Potentially, all of these forest products can be sustainably harvested. Unfortunately, in most cases forests have been unsustainably overharvested, resulting in the "mining" of the forest resource and widespread ecological degradation. It is critical that in the future all forest harvesting is conducted in a manner that is more responsible in terms of sustaining the resource.

Many other plant products can also be collected from forests, such as **fruits**, nuts, **mushrooms**, and latex for manufacturing rubber. In addition, many species of animals are hunted in forests, for recreation or for subsistence. Forests provide additional goods and services that are important to both human welfare and to **ecological integrity**, including the control of erosion and water

---

KEY TERMS

Conversion—A longer-term change in the character of the ecosystem at some place. When a natural forest is harvested and changed into a plantation forest, this represents an ecological conversion, as does deforestation to develop agricultural land.

Monoculture—An ecosystem dominated by a single species, as may occur in a forestry plantation.

---

flows, and the cleansing of air and water of pollutants. These are all important forest values, although their importance is not necessarily assessed in terms of dollars. Moreover, many of these values are provided especially well by old-growth forests, which in general are not very compatible with industrial **forestry** practices. This is one of the reasons why the **conservation** of old-growth forest is such a controversial topic in many regions of North America and elsewhere. In any event, it is clear that when forests are lost or degraded, so are these important goods and services that they can provide.

Resources

Books

Barnes, B.V., S. Spurr, and D. Zak. *Forest Ecology*. J. Wiley and Sons, 1998.

Begon, M., J.L. Harper, and C.R. Townsend. *Ecology. Individuals, Populations and Communities*. 2nd ed. London: Blackwell Sci. Pub., 1990.

Freedman, B. *Environmental Ecology*. 2nd ed. San Diego: Academic Press, 1995.

Hamblin, W.K., and Christiansen, E.H. *Earth's Dynamic Systems*. 9th ed. Upper Saddle River: Prentice Hall, 2001.

Kimmins, J.P. *Forest Ecology: A Foundation for Sustainable Management*. Prentice Hall, 1997.

Bill Freedman

Formula, chemical

Chemical formula is a symbolized representation of a chemical compound. It tells us the type of atom(s) (element) present in the compound and in what ratios. **Atoms** are indicated by their symbols as shown in the **periodic table**, and the number of atoms are indicated as subscripts. For example, the chemical formula for **water** is H_2O, consists of two **hydrogen** atoms (H) and one **oxygen** atom (O).

A chemical formula may be written in two ways, as an empirical formula or a **molecular formula**. The empirical formula is commonly used for both ionic compounds (compounds formed by donation and reception of electrons by participating elements, e.g. NaCl (**sodium chloride** or common **salt**) and for covalent compounds (compounds formed by sharing of electrons by participating elements, (e.g. CH_4, methane). Molecular formula is commonly used for covalent compounds (e.g., C_2H_6, ethane).

The empirical formula denotes the smallest possible **ratio** that corresponds to the actual ratio by atoms or formula unit. To construct an empirical formula for an ionic compound, one needs to write the symbol of cations (positively charged ions) first and then the **anion** (negatively charged ion). Then fulfill the **valence** requirement of each atom as well as the least possible ratio of atoms present in that compound (e.g., Al_2O_3 for **aluminum** oxide). For **carbon** containing compounds, one needs to write the carbon atom first, then hydrogen atom, followed by other atoms in alphabetical order (e.g., $CHCl_3$ for **chloroform**).

The molecular formula denotes the actual number of different atoms present in one **molecule** of that compound. In some cases a compound's molecular formula is the same as its empirical formula (e.g., water H_2O, **ammonia** NH_3, methane/natural gas CH_4) and in others it is an **integral** multiple of empirical formula (e.g., **hydrogen peroxide**, empirical formula is HO and molecular formula is H_2O_2, which is multiple of two of empirical formula). To construct the molecular formula, one needs to follow the steps as for writing empirical formulas, although the actual number of atoms not the smallest ratio is used. Molecular formulas provide the foundation of structure and the **molecular weight** of a molecule. Yet, it does not provide a complete picture of a molecule, especially for organic molecules. In almost all organic molecules, only part of the molecule (functional groups) participate in a chemical reaction. Also, for one molecular formula, it is possible to have several compounds or isomers (e.g., for C_4H_{10}, two compounds; butane and methyl propane) with totally different physical and chemical properties. Hence, organic chemists can use an expanded version of the molecular formula, called the structural formula.

A compound's structural formula consists of the actual number of atoms in the compound as well as showing where the chemical bonds are between them. It also provides information about length of chemical bond(s) and **angle** between chemical bonds. A structural formula has several representations: Lewis dot form, bond-line, stick bond notation, valence orbital notation, and projection form. Firstly, Lewis dot form is the simplest representation of communicating a chemical structure. In Lewis dot form, the atoms are represented by their corresponding symbols, and chemical bonds are represented by a pair of electrons or dots. Each **chemical bond** is

represented by a pair of electrons. Thus single bond, double bond, and triple bonds are represented by two, four, and six dots, respectively. One can easily count the sharing (involved in chemical bond formation) and unsharing electrons (not involved in chemical bond formation). Secondly, "bond-line" notation is similar to Lewis dot form except the bonding electrons are replaced by line(s). Therefore, single, double, and triple bonds are represented by one, two, and three line(s) respectively. Thirdly, "stick-bond" notation is the condensed version of bond-line notation. Each end of a open chain with a single line or a line branching out from a open chain or from a closed cyclic structure represents one methyl (-CH_3) group. Each corner in an open chain or a cyclic structure represents a methylene (-CH_2-) group. Functional groups such as **alcohol** (-OH), aldehyde (-CHO), acid (-COOH), amine (-NH_2), **ester** (-COOR), etc. are represented by their actual atomic symbols. Fourthly, valence orbital notation, in addition to the above information, reveals the shape of orbital or distribution of **electron density** around atoms. Fifthly, structure of a compound can be represented in a projected form, because atoms in any molecule occupy **space** or possess three dimensional structure. Projected form can further be represented in wedge, sawhorse, Newman projection, ball and stick, space filling molecular model, and Fischer projection forms. All these projection forms additionally enable one to see the spatial relationship between atoms and **rotation** around the connecting chemical bonds. Conceptually, projection forms are an advanced level of **learning**, but they provide almost a complete insight into structure and properties of a molecule.

See also Chemical bond; Compound, chemical; Formula, structural.

Formula, structural

A structural formula is a chemical formula that gives you a more complete picture of a compound than its **molecular formula** can. While a molecular formula, such as H_2O, shows the types of **atoms** in a substance and the number of each kind of atom, a structural formula also gives information about how the atoms are connected together. Some complex types of structural formulas can even give you a picture of how the atoms of the **molecule** are arranged in **space**. Structural formulas are most often used to represent molecular rather than ionic compounds.

There are several different ways to represent compounds in structural formulas, depending on how much detail needs to be shown about the molecule under con-

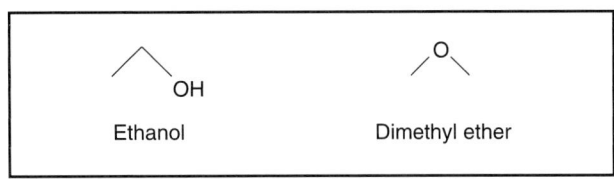

Ethanol   Dimethyl ether

Illustration by Hans & Cassidy. Courtesy of Gale Group.

sideration. We will look at complete structural formulas, condensed formulas, line formulas, and three-dimensional formulas.

Complete structural formulas

Complete structural formulas show all the atoms in a molecule, the types of bonds connecting them, and how they are connected with each other. For a simple molecule like **water**, H_2O, the molecular formula, becomes H-O-H, the structural formula. This structural formula shows that in a water molecule, the **oxygen** atom is the central atom, and it is connected by single covalent bonds to the **hydrogen** atoms. **Carbon dioxide**, CO_2, can be represented structurally as O=C=O. This structural formula tells you that in this case the **carbon** atom is the central one, and the oxygen atoms are joined by double covalent bonds to the carbon atom.

For small molecules like these, the amount of new information in a structural formula is not great, but structures become more important when we study larger molecules. Let's look at the molecular formula C_2H_6O. With some knowledge of valences for the three kinds of atoms involved, we can arrange these atoms in a complete structural formula as shown below.

$$
\begin{array}{ccc}
& H & H \\
& | & | \\
H - & C - & C - O - H \\
& | & | \\
& H & H
\end{array}
$$

This is the formula of **ethanol**, which is well-known for its intoxicant and antiseptic properties, and is also being used in reformulated gasoline. It is a liquid with a **boiling point** of 172°F (78°C). However, we can also produce another structural formula that satisfies all the bonding requirements for the atoms involved but shows a completely different molecule.

$$
\begin{array}{ccc}
H & & H \\
| & & | \\
H - C - & O - & C - H \\
| & & | \\
H & & H
\end{array}
$$

This molecule is methyl **ether**. It is a gas at room **temperature** and has very different chemical properties from ethanol.

Condensed structural formulas

After you become familiar with the rules for writing complete structural formulas, you find yourself taking shortcuts and using condensed structural formulas. You still need to show the complete molecule, but the inactive parts can be more sketchily shown. Thus the two formulas above look like this when written in condensed form:

$$CH_3CH_2OH \qquad CH_3OCH_3$$

ethyl alcohol   dimethyl ether

Line formulas

Even the condensed formulas take up a lot of space and a lot of time to write. They can be transformed still further by the shorthand of line formulas. Line formulas show the main bonds of the molecule instead of the individual atoms, and only show a particular atom if it is different from carbon or hydrogen, or if it is involved in a reaction under consideration. Our examples of condensed formulas look like this when represented by line formulas.

At each unmarked vertex of the lines, there is a carbon atom with enough hydrogen atoms to satisfy its **valence** of four. There is also a carbon atom with its accompanying hydrogen atoms at the end of any bond line that doesn't show some other atom. Compare the condensed formulas of these three compounds with the condensed formulas in order to find the atoms implied in the line formulas.

Three dimensional formulas

All of these structural formulas show you a flat molecule on a flat piece of **paper**. However, most carbon-containing molecules are three-dimensional; some of the atoms stick forward toward you from the carbon chain, and some project to the rear of the molecule. Chemists have devised special ways to show these forward- and backward-projecting atoms in order to understand how three-dimensional molecules behave. These three-dimensional structural formulas are often used when complex molecules are studied.

See also Chemical bond; Formula, chemical; Compound, chemical

Resources

Books

Carey, Francis A. *Organic Chemistry.* New York: McGraw-Hill, 2002.

Djerassi, Carl. *Steroids Made It Possible.* Washington, DC: American Chemical Society, 1990.

Mark, Herman F. *From Small Organic Chemicals to Large: a Century of Progress.* Washington, DC: American Chemical Society, 1993.

KEY TERMS

Chemical formula—A way to show the number and kind of atoms combined together in a single pure substance.

Compound—A pure substance that consists of two or more elements, in specific proportions, joined by chemical bonds. The properties of the compound may differ greatly from those of the elements it is made from.

Covalent compound—A chemical compound which uses shared electrons to form bonds between atoms. The atoms do not become electrically charged as in ionic compounds.

Ionic compound—A compound consisting of positive ions (usually, metal ions) and negative ions (nonmetal ions) held together by electrostatic attraction.

Molecule—A chemical compound held together by covalent bonds.

Valence—The combining power of an atom, or how many bonds it can make with other atoms. For the examples used in this article, carbon atoms can make four bonds, oxygen atoms can make two bonds, and hydrogen atoms can make one bond.

Mauskopf, Seymour H. *Chemical Sciences in the Modern World.* Pennsylvania: University of Pennsylvania Press, 1993.

G. Lynn Carlson

Fossa

Fossas are cat-like Madagascan carnivores in the family Viverridae, which also includes **civets**, linsangs, **genets**, and **mongooses**. Fossas are quite different from other viverrids and are the sole members of the subfamily Cryptoproctinae. They are the largest Madagascan carnivores, measuring 24-30 in (60-75 cm) long. With a number of cat-like features-including a rounded head, long whiskers, large frontal eyes, middle-sized roundish ears, and relatively short jaw-led to its original classification as a felid. However, the fossa's resemblance to **cats** is superficial and a result of convergent evolution.

The neck of the fossa is short and thickset, and its body is muscular, long, and slender. The legs are short, and its sharp, curved, retractile claws help it climb trees.

The long tail of the fossa is used for balance, but is not prehensile. The feet support well-developed hairless pads which help secure the footing. The fossa is plantigrade, meaning it walks upon the whole foot rather than just the toes, as in most viverrids. The fur is short, thick, and reddish brown on the upperside; the underside is lighter. Occasionally, the fossa is melanistic (an increased amount of nearly black pigmentation.) Fossas are nocturnal terrestrial and arboreal predators, living primarily in coastal **forests**, and are rarely seen in the central highlands of Madagascar. Their home range is several square kilometers depending on the type of country. The fossa is territorial, marking upright objects within its boundaries with oily secretions from the anal and preputial (loose skin covering the penis) **glands**. Possessing excellent **hearing**, sight, and scent, along with having no natural enemies, the fossa is the most powerful **predator** in Madagascar. Possessing 32 sharp teeth, the fossa has a varied diet including civets and young bush-pigs and other **mammals** up to the size of **lemurs**, **birds** up to the size of **guinea fowl**, eggs, lizards, **snakes**, **frogs**, and **insects**. Additionally, it will **prey** on domestic poultry, rabbits, and **sheep**.

KEY TERMS

Arboreal—Living in trees.

Convergent evolution—An evolutionary pattern by which unrelated species that fill similar ecological niches tend to develop similar morphologies and behavior. Convergence occurs in response to similar selection pressures.

Diurnal—Refers to animals that are mainly active in the daylight hours.

Felid—Of or belonging to the family Felidae which includes the lions, tigers, jaguars, and wild and domestic cats; resembling or suggestive of a cat.

Gestation—The period of carrying developing offspring in the uterus after conception; pregnancy.

Melanistic—A dark or black pigmentation; dark coloration.

Nocturnal—Active by night.

Plantigrade—Walking with the entire lower surface of the foot on the ground, as humans and bears do.

Prehensile—Adapted for seizing or holding, especially by wrapping around an object.

Retractile—Capable of being drawn back or in.

A fossa (*Cryptoprocta ferox*). © *National Audubon Society Collection/Photo Researchers, Inc. Reproduced with permission.*

Fossas are solitary, except during the breeding season in September and October, when the nocturnal habits of the fossa also change slightly and become diurnal. After a three-month gestation, two to four young are born in burrows, among **rocks**, in holes in trees, or forks at the base of large boughs. Only the female fossa will rear the young. The newborn are quite small when compared with other viverrids, and their physical development is slow. The eyes do not open for 16-25 days. The first milk tooth appears when the young fossa is ready for the inaugural venture from the nest at one and a half months. Though weaned by four months, solid food is not taken until three months of age. Climbing begins at about three and a half months. Fully grown and independent at two years, the fossa does not reach sexual maturity for another two years. Male fossa possess a penis-bone, called a baculum. Females exhibit genital **mimicry** of the males, though the mimicry is not well developed.

Grooming techniques are similar to felids, with licking and scratching with the hind feet, and face washing with the fore feet. Fossas display a variety of vocalizations: a short scream repeated five to seven times is a threat, while the mating call of the female is a long mew, lasting up to 15 seconds. The female calls the young with a sharp, long whimpering. When first beginning to

suckle, the young will growl, and both sexes mew and growl when mating.

Fossas live in low population densities and require undisturbed forests which are unfortunately rapidly disappearing on the heavily logged **island** of Madagascar. Though fossas have lived as long as 17 years in captivity, these animals are unlikely to survive as long in the wild. Recognition of fossas as **endangered species** is likely.

Resources

Books

Burton, Maurice, ed. *The New Larousse Encyclopedia of Animal Life.* New York: Bonanza Books, 1984.

Farrand, John, Jr., ed. *The Audubon Society Encyclopedia of Animal Life.* New York: Clarkson N. Potter, Inc./Publishers, 1982.

Haltenorth, Theodor, and Helmut Diller. Translated by Robert W. Hayman. *A Field Guide to the Mammals of Africa Including Madagascar.* London: William Collins & Co. Ltd., 1984.

MacDonald, David, and Sasha Norris, eds. *Encyclopedia of Mammals.* New York: Facts on File, 2001.

National Geographic Society, ed. *Book of Mammals, Volume One.* Washington, DC: National Geographic Society, 1981.

Periodicals

Sunquist, Fiona. "Of Quolls and Quokkas." *International Wildlife* (Jul/Aug 1991): 16.

Betsy A. Leonard

Fossil and fossilization

Fossils are a significant window into Earth's history and organic **evolution**. The term fossil literally means something that has been 'dug up,' but its modern meaning has been restricted to evidence of past life. Such evidence may take the form of body fossils (both **plant** and **animal**), trace fossils or ichnofossils (physical features formed in rock due to animal-sediment interaction), and chemical trace fossils (chemical evidence of life processes preserved in **minerals** within the **rocks**).

Fossilization refers to the series of postmortem processes that lead to development of a body, trace, or chemical fossil. For original hard parts (e.g., shell, skeleton, and teeth), which are composed of various kinds of organic minerals, fossilization may include replacement by new minerals, permineralization (filling open spaces with minerals). Fossil shells may be represented by external or internal (steinkern) sediment molds. Soft parts of plants or animals may also be mineralized and preserved as fossils in the process of carbonization. Soft **tissue** can be preserved as fossil material under special conditions where **bacteria** and moisture are excluded (e.g., fossils buried in glacial **ice**, anoxic peat bogs, and amber).

Fossils and their enclosing sediment (or **sedimentary rock**) are carefully studied in order to reconstruct ancient sedimentary environments and ancient ecosystems. Such analysis is called **paleoecology**, or the study of ancient ecologic systems. Fossils occur in nearly all sediments and sedimentary rock, and some volcanic rocks (e.g., ash deposits) as well. The bulk of these fossils are **invertebrates** with hard parts (e.g., clam shells). **Vertebrates**, the class that includes **reptiles** (e.g., dinosaurs) and **mammals** (e.g., mastodons), are a relatively late development, and the finding of a large, complete vertebrate fossil, with all its parts close together, is rather rare. Microfossils, on the other hand, are extremely common. Microfossils include very early bacteria and **algae**; the unicellular organisms called foraminiferans, which were common in the Tertiary periods, and fossil pollen. The study of microfossils is a specialized field called micropaleontology.

Fossils of single-celled organisms have been recovered from rocks as old as 3.5 billion years. Animal fossils first appear in Upper Precambrian rocks dating back about a billion years. The occurrence of fossils in unusual places, such as **dinosaur** fossils in **Antarctica** and **fish** fossils on the Siberian steppes, reflects both shifting of continental position by **plate tectonics** and environmental changes over **time**. The breakup of the supercontinent Pangaea during and since Triassic pulled apart areas that were once contiguous and thus shared the same floras and faunas. In particular, Earth's tectonic plates carrying the southern hemisphere continents-South America, southern **Africa**, the Indian subcontinent, **Australia**, and Antarctica-moved in different directions, isolating these areas. Terrestrial vertebrates were effectively marooned on large continental "islands." Thus, the best explanation for dinosaurs on Antarctica is not that they evolved there, but that Antarctica was once part of a much larger land mass with which it shared many life forms.

An important environmental factor influencing the kinds of fossils deposited has been radical and episodic alteration in sea levels. During episodes of high **sea level**, the interiors of continents such as **North America** and **Asia** are flooded with seawater. These periods are known as marine transgressions. The converse, periods of low sea level when the waters drain from the continents, are known as marine regressions. During transgressions, fossils of marine animals may be laid down over older beds of terrestrial animal fossils. When sea level fall occurs, thus exposing more land at the edges of continents, sediments with fossils of terrestrial animals may accumulate over older marine animals. In this way, plate **tectonics** and the occasional marine **flooding** of inland areas could result in unusual collections of fossil floras and faunas in sediments and sedimentary rocks where the living plants or animals could not exist today—such as fishes on the Siberian steppes.

Changes in sea level over the past million years or so have been related to episodes of glaciation. During glaciation, proportionately more **water** is bound up in the **polar ice caps** and less is available in the seas, making the sea levels lower. It is speculated, but not certain, that the link between glaciation and lower sea levels holds true for much of Earth's history. The periods of glaciation in turn are related to broad climatic changes that affect the entire **Earth**, with cooler **weather** increasing glaciation and with warmer temperatures causing glacial melting and a rise in sea levels. Global climatic change has had a profound effect on Earth's **fauna** and **flora** over time. This is strongly reflected in the fossil record and the record of paleoecology of Earth found in sedimentary **strata**.

The fossil clock

The principal use of fossils by geologists has been to date rock strata (layers) that have been deposited over

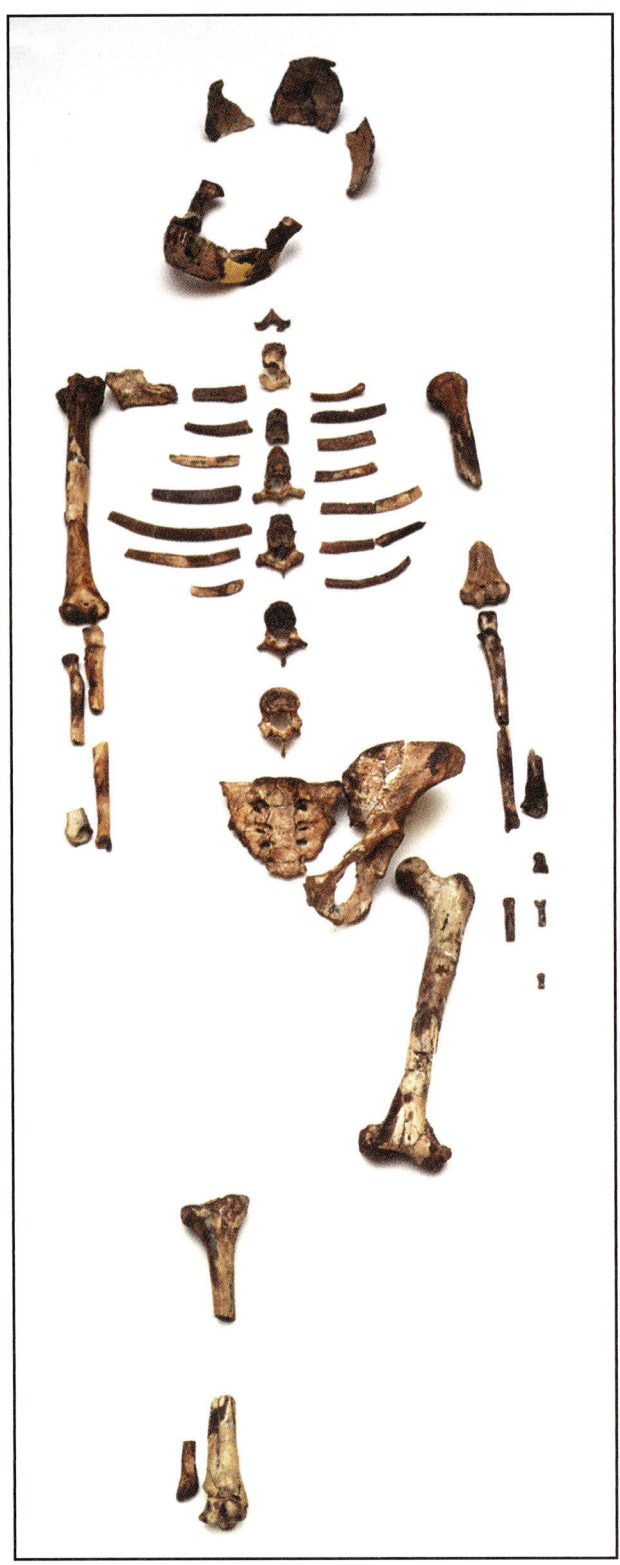

One of the most complete early hominid fossils is this *Australopithecus afarensis* specimen commonly known as "Lucy," which was found by Donald Johanson in the Afar region of Ethiopia. © *John Reader/Science Photo Library/Photo Researchers, Inc. Reproduced by permission.*

millions of years. As different episodes in Earth's history are marked by different **temperature**, aridity, and other climatic factors, as well as different sea levels, different life forms were able to survive in one locale or period but not in another. Distinctive fossilized life forms that are typically associated with given intervals of **geologic time** are known as index fossils, or **indicator species**. Over the past 200 years, paleontologists have determined an order of successive index fossils that not only allows geologists to date strata, but also is the foundation for understanding organic evolution.

The temporal relationship of the strata is relative: it is more important to know whether one event occurred before, during, or after another event than to know exactly when it occurred. Recently geologists have been able to subdivide time periods into progressively smaller intervals called epochs, ages, and zones, based on the occurrence of characteristic indicator (index fossil) **species**, with the smallest time slices about one-half million years. Radiometric dating measures that measure the decay of radioactive isotopes have also been used to derive the actual rather than relative dates of geological periods; the dates shown on the time scale were determined by radiometry. The relative dating of the fossil clock and the quantitative dating of the radiometric clock are used in combination to date strata and geological events with **accuracy**.

The fossil clock is divided into units by index fossils. Certain characteristics favor the use of one species over another as an index fossil. For example, the ammonoids (ammonites), an extinct mollusk, function as index fossils from Lower Devonian through Upper Cretaceous—an **interval** of about 350 million years. The ammonoids, marine animals with coiled, partitioned shells, in the same class (Cephalopoda) as the present-day *Nautilus*, were particularly long-lived and plentiful. They evolved quickly and colonized most of the seas on the **planet**. Different species preferred warmer or colder water, evolved characteristically sculpted shells, and exhibited more or less coiling. With thousands of variations on a few basic, easily visible features—variations unique to each species in its own time and place—the ammonoids were obvious candidates to become index fossils. For unknown reasons, this group of immense longevity became extinct during the Cretaceous-Triassic **mass extinction**. The fossils are still quite plentiful; some are polished and sold as jewelry or paperweights.

Index fossils are used for relative dating, and the geologic time scale is not fixed to any one system of fossils. Multiple systems may coexist side by side and be used for different purposes. For example, because macrofossils such as the ammonoids may break during the extraction of a core **sample** or may not be frequent enough to lie within the exact area sampled, a geologist may choose to use the extremely common microfossils as the indicator species. Workers in the oil industry may use conodonts, fossils commonly found in oil-bearing rocks. Regardless of which system of index fossils is used, the idea of relative dating by means of a fossil clock remains the same.

From biosphere to lithosphere

The likelihood that any living **organism** will become a fossil is quite low. The path from **biosphere** to lithosphere—from the organic, living world to the world of rock and mineral—is long and indirect. Individuals and even entire species may be 'snatched' from the record at any point. If an **individual** is successfully fossilized and enters the **lithosphere**, ongoing tectonic activity may stretch, abrade, or pulverize the fossil, or the sedimentary layer housing the fossil may eventually be subjected to high temperatures in **Earth's interior** and melt, or be weathered away at the Earth's surface. A fossil that has survived or avoided these events may succumb to improper collection techniques at the hands of a human.

Successful fossilization begins with the conditions of death in the biosphere. Fossils occur in sedimentary rock and are incorporated as an integral part of the rock during rock formation. Unconsolidated sediments such as **sand** or mud, which will later become the fossiliferous (fossil-bearing) sandstone or limestone, or shale, are an ideal matrix for burial. The organism should also remain undisturbed in the initial phase of burial. Organisms exposed in upland habitats are scavenged and weathered before they have an opportunity for preservation, so a low-lying **habitat** is the best. Often this means a watery habitat. The fossil record is highly skewed in favor of organisms that died and were preserved in calm seas, estuaries, tidal flats, or the deep **ocean** floor (where there are few scavengers and little disruption of layers). Organisms that died at altitude, such as on a plateau or mountainside, and are swept by **rivers** into a **delta** or estuary may be added to this death assemblage, but are usually fragmented.

A second factor contributing to successful fossilization is the presence of hard parts. Soft-bodied organisms rarely make it into the fossil record, which is highly biased in favor of organisms with hard parts—skeletons, shells, woody parts, and the like. An exception is the Precambian Burgess Shale, in British Columbia, where a number of soft-bodied creatures were fossilized under highly favorable conditions. These creatures have few relatives that have been recorded in the fossil record; this is due to the unlikelihood of the soft animals being fossilized.

From the time of burial on, an organism is technically a fossil. Anything that happens to the organism after

A fossil trilobite from the Mid-ordovician period. *Photograph by Neasaphus Rowalewkii. JLM Visuals. Reproduced by permission.*

burial, or anything that happens to the sediments that contain it, is encompassed by the term diagenesis. What is commonly called fossilization is simply a postmortem alteration in the **mineralogy** and **chemistry** of the original living organism.

Fossilization involves replacement of minerals and chemicals by predictable chemical means. For example, the shells of molluscs are made of **calcium carbonate**, which typically remineralizes to calcite or aragonite. The bones of most vertebrates are made of **calcium** phosphate, which undergoes subtle changes that increase the phosphate content, while cement fills in the pores in the bones. These bones may also be replaced by silica.

The replacement of original minerals and chemicals takes place according to one of three basic schemes. In one scheme; the skeleton is replaced one to one with new minerals. This scheme is known as replacement. In a second scheme, the hard parts have additional mineral material deposited in their pores. This is known as permineralization. In a third scheme, both hard and soft parts dissolve completely and a void is left in the host rock (which may later be filled with minerals). If in the third scenario, the sediments hardened around the hard part and took its shape before it dissolved, and the dissolved hard part was then not replaced (i.e., there is a void), a thin **space** remains between two rock sections. The rock section bearing the imprint of the interior face of the shell, let us say, is called the part, or internal **mold**, and the rock section bearing the imprint of the exterior of the shell is called the counterpart, or external mold. External molds are commonly but mistakenly discarded by amateur fossil collectors.

Because of the nature of fossilization, fossils are often said to exist in communities. A fossil community is defined by space, not time. Previously fossilized speci-

A fly in amber, 35 million years old. *JLM Visuals. Reproduced by permission.*

mens of great age may be swept by river action or carried by scavengers into young sediments that are just forming, there to join the fossil mix. For this reason, it may be difficult to date a fossil with precision on the basis of a presumed association with nearby fossils. Nevertheless, geologists hope to confirm relationships among once living communities by comparing the makeup of fossil communities.

One of the larger goals of paleontologists is to reconstruct the prehistoric world, using the fossil record. Inferring an accurate life assemblage from a death assemblage is insufficient and usually wrong. The fossil record is known for its extreme biases. For example, in certain sea environments over 95% of species in life may be organisms that lack hard parts. Because such animals rarely fossilize, they may never show up in the fossil record for that locale. The species diversity that existed in life will therefore be much reduced in the fossil record, and the proportional representation of life forms greatly altered.

To gain some idea of the likelihood of fossilization of an individual or a species, scientists have sampled the death assemblages—decaying plants and animals that have gained the security of undisturbed sediments—in modern-day harbor floors and offshore sediments, and compared those death assemblages with actual life assemblages in the overlying waters. It seems that no more than 30% of species and 10% of individuals are preservable after death. The death assemblage is still millions of years away from becoming a fossil community, however, and once such factors as consumption of the decaying organisms by scavengers, transport of the organisms out of the area, disturbance of sediments, reworking of the rock after it has formed, and **erosion** are added to the picture, the fossilization **rate** falls well below the preservation rate.

In some cases, however, a greater than usual proportion of preservable individuals in a community has fossilized in place. The result is a bed of fossils, named after the predominant fossil component, "bone bed" or "mussel bed," for example. Geologists are divided over whether high-density fossil deposits are due to reworking and condensation of fossiliferous sediments or to mass mortality events. Mass mortality—the contemporaneous death of few to millions of individuals in a given area—usually is attributed to a natural catastrophe. In North America, natural catastrophe is thought to have caused the sudden death of the dinosaurs in the bone beds at Dinosaur National Park, Colorado, and of the fossil fishes in the Green River Formation, Wyoming. These are examples of local mass mortality. When mass mortality occurs on a global scale and terminates numerous species, it is known as a mass **extinction**. The greatest mass extinctions have been used to separate strata formed during different geological eras: the Permian-Triassic extinction separates the Paleozoic era from Mesozoic; the Cretaceous-Tertiary extinction, which saw the demise of the dinosaurs and the rise of large mammalian species to fill newly available biological niches, separates Mesozoic from Tertiary. Thus, mass extinctions are recorded not only in the high-density fossil beds, but in the complete disappearance of many species from the fossil record.

From field to laboratory

A fossil identified in the field is not immediately chiseled out of its matrix. First, photographs are taken to show the relationship of the fossil fragments, and the investigator notes the rock type and age, and the fossil's orientation. Then a block of rock matrix that contains the entire fossil is cut out with a rock saw, wrapped in muslin, and wrapped again in wet plaster, a process known as jacketing. The jacketed fossils may additionally be stored in protective crates for air transport.

In the laboratory, the external wrappings are removed, exposing the fossil in its matrix. The technique used to remove the excess rock varies with the type of rock and type of fossil, but three methods are common. Needle-sharp pointed tools, such as dental drills and engraving tools, may be used under a binocular **microscope**; or pinpoint blasting may be done with a fine abrasive powder; or acid may be used to dissolve the rock. Because some fossils also dissolve in some acids, the fossil's composition must be determined before a chemical solvent is used. If the investigator wishes to see the complete **anatomy** of the fossil, the entire rocky matrix may be removed. Thin slices of the fossil may be obtained for microscopic study. If replicas are desired, the fossil may be coated with a fixative and a rubber cast made. For security purposes, most prehistoric skeletons on display in museums and public institutions are models cast from a mold, and not the original fossil itself.

The study of fossils is not limited to freeing the fossil from its matrix, looking at it microscopically, or making articulated reproductions to display in museum halls. Since about 1980, a variety of techniques developed in other fields have been used to make discoveries about the original life forms that were transformed into fossils. Immunological techniques have been used to identify **proteins** in fossilized dinosaur bones. The ability to recover DNA, not only from **insects** preserved in amber but also

KEY TERMS

Bone bed—High-density accumulation of fossilized bones.

Diagenesis—The processes to which a dead organism is exposed after burial as it becomes a fossil; for example, compaction and replacement.

External mold—Fossilized imprint of the exterior portion of the hard part of an organism, left after the fossilized organism itself has been dissolved; related to internal mold, bearing the imprint of the interior portion of the hard part, for example, of a clam shell.

Fossil record—The sum of fossils known to humans, and the information gleaned from them.

Fossiliferous—Fossil bearing; usually applied to sedimentary rock strata.

Ichnofossil—A trace fossil, or inorganic evidence of a fossil organism, such as a track or trail.

Index fossil—A distinctive fossil, common to a particular geological period, that is used to date rocks of that period; also called *indicator species*.

from fossilized fish and dinosaurs, may soon be realized. Studies of temperature-dependent **oxygen** isotopes formed during fossilization have been used to support the theory that dinosaurs were warm-blooded. And even as laboratory research is moving toward the **molecular biology** of fossilized organisms, aerial reconnaissance techniques for identifying likely locales of fossil beds are being refined. The true value of a fossil, however, is realized only when its relationships to other organisms, living and extinct, and to its environment are known.

Interpreting the fossil record

The fossil record—the sum of all known fossils— has been extremely important in developing the **phylogeny**, or evolutionary relations, of ancient and living organisms. The contemporary understanding of a systematic, phylogenetic heirarchy descending through each of the five kingdoms of living organisms has replaced earlier concepts that grouped organisms by such features as similar appearance. It is now known that unrelated organisms can look alike and closely related organisms can look different; thus, terms like "similar" have no analytical power in **biology**.

In addition to providing important information about the history of Earth, fossils have industrial uses. **Fossil fuels** (oil, **coal**, **petroleum**, bitumen, **natural gas**) drive industrialized economies. Fossil aggregates such as limestone provide building material. Fossils are also used for decorative purposes. This category of functional use should be distinguished from the tremendous impact fossils have had in supporting evolutionary theory.

See also Dating techniques; Paleontology; Stratigraphy.

Resources

Books

Donovan, Stephen K., ed. *The Processes of Fossilization*. London: Belhaven Press, 1991.

Pough, Harvey. *Vertebrate Life*. 6th ed. Upper Saddle River, NJ: Prentice Hall, 2001.

Prothero, Donald R. *Bringing Fossils to Life, an Introduction to Paleontology*. Boston: McGrawHill, 1999.

Rich, Pat Vickers, et al. *The Fossil Book: A Record of Prehistoric Life*. Mineola, NY: Dover Publishing, 1997.

Other

Edwards, Lucy E. and John Pojeta, Jr. "Fossils, Rocks, and Time" United States Geological Survey August 14, 1997 [cited January 15, 2003] <http://pubs.usgs.gov/gip/fossils/about.html>.

Marjorie Pannell

Fossil fuels

Fossil fuels are buried deposits of **petroleum, coal**, peat, **natural gas**, and other carbon-rich organic compounds derived from the dead bodies of plants and animals that lived many millions of years ago. Over long periods of **time, pressure** and **heat** generated by overlying sediments concentrate and modified these materials into valuable **energy** sources for human purposes. Fossil fuels currently provide about 90% of all commercial energy used in the world. They provide the power to move vehicles, heat living spaces, provide **light**, cook our food, transmit and process information, and carry out a wide variety of industrial processes. It is no exaggeration to say that modern industrial society is nearly completely dependent on (some would say addicted to) a continual supply of fossil fuels. How we will adapt as supplies become too limited, too remote, too expensive, or too environmentally destructive to continue to use is a paramount question for society.

The amount of fossil fuels deposited over history is astounding. Total coal reserves are estimated to be in the vicinity of ten trillion metric tons. If all this resource could be dug up, shipped to market, and burned in an economically and environmentally acceptable manner, it would fuel all our current commercial energy uses for several thousand years. Petroleum (oil) deposits are thought to have originally amounted to some four trillion barrels (600 billion metric tons), about half of which has already been extracted and used to fuel industrial society. At current rates of use the proven oil reserves will be used up in about 40 years. World natural gas supplies are thought to be at least 10 quadrillion cubic feet or about as much as energy as the original oil supply. At current rates of use, known gas reserves should last at least 60 years. If we substitute gas for oil or coal, as some planners advocate, supplies will be used up much faster than at current rates. Some unconventional **hydrocarbon** sources such as oil shales and tar sands might represent an energy supply equal to or even surpassing the coal deposits on which we now depend.

In the United States, oil currently supplies about 40% of all commercial energy use, while coal contributes about 22%, and natural gas provide about 24%. Oil and its conversion products, such as gasoline, kerosene, diesel fuel, and jet fuel are the primary fuel for internal **combustion** engines because of the ease with which they can be stored, transported, and burned. Coal is burned primarily in power plants and other large, stationary industrial boilers. Methane (natural gas) is used primarily for **space** heating, cooking, **water** heating, and industrial processes. It is cleaner burning than either oil or coal, but is difficult to store or to ship to places not served by gas pipelines.

The use of fossil fuels as our major energy source has many adverse environmental effects. Coal **mining** often leaves a devastated landscape of deep holes, decapitated mountain tops, toxic spoil piles, and rocky rubble. Acid drainage and toxic seepage from abandoned mines poisons thousands of miles of streams in the United States. Every year the 900 million tons of coal burned in the U.S. (mainly for electric power generation) releases 18 million tons of **sulfur dioxide**, five million tons of **nitrogen** oxides (the main components of **acid rain**), four million tons of **carbon monoxide** and unburned hydrocarbons, close to a trillion tons of **carbon dioxide**, and a substantial fraction of the toxic metals such as mercury, cadmium, thallium, and zinc into our air. Coal often contains **uranium** and thorium, and that most coal-fired power plants emit significant amounts of radioactivity— more, in fact, than a typical **nuclear power plant** under normal conditions. Oil wells generally are not as destructive as coal mines, but exploration, drilling, infrastructure construction, waste disposal, and transport of oil to markets can be very disruptive to wild landscapes and **wildlife**. Massive **oil spills**, such as the grounding of the *Exxon Valdez* on Prince William Sound, Alaska, in 1989, illustrate the risks of shipping large amounts of oil over great distances. Nitrogen oxides, unburned hydrocarbons, and other combustion byproducts produced by gasonine and diesel engines are the largest source of **air pollution** in many American cities.

One of the greatest concerns about our continued dependence on fossil fuels is the waste **carbon** dioxide produced by combustion. While carbon dioxide is a natural atmospheric component and is naturally absorbed and recycled by **photosynthesis** in green plants, we now burn so much coal, oil, and natural gas each year that the amount of carbon dioxide in the atmosphere is rapidly increasing. Because carbon dioxide is a greenhouse gas (it is transparent to visible light but absorbs long wavelength infrared **radiation**), it tends to trap heat in the lower atmosphere and increase average global temperatures. Climatic changes brought about by higher temperatures can result in heat waves, changes in rainfall patterns and growing **seasons**, rising **ocean** levels, and could increase the frequency and severity of storms. These potentially catastrophic effects of **global climate** change may limit our ability to continue to use fossil fuels as our major energy source. All of these considerations suggest that we urgently need to reduce our dependency on fossil fuels and turn to environmentally benign, renewable energy sources such as solar power, **wind**, **biomass**, and small-scale hydropower.

Foxes *see* **Canines**

Foxglove *see* **Snapdragon family**

Fractal

A fractal is a geometric figure, often characterized as being self-similar; that is, irregular, fractured, fragmented, or loosely connected in appearance. Benoit Mandelbrot coined the term fractal to describe such figures, deriving the word from the Latin "fractus" meaning broken, fragmented, or irregular. He also pointed out amazing similarities in appearance between some fractal sets and many natural geometric patterns. Thus, the term "natural fractal" refers to natural phenomena that are similar to fractal sets, such as the path followed by a dust particle as it bounces about in the air.

Another good example of a natural phenomenon that is similar to a fractal is a coastline, because it exhibits three important properties that are typical of fractals. First, a coastline is irregular, consisting of bays, harbors, and peninsulas. Second, the irregularity is basically the same at all levels of magnification. Whether viewed from **orbit** high above **Earth**, from a helicopter, or from land,

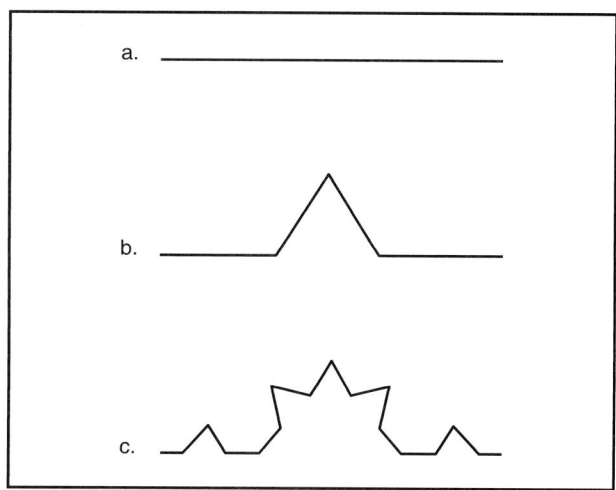

The construction of a well known self-similar figure (fractal), the triadic Koch curve. A line segment (a) has its central 1/3 removed and replaced with a segment twice as long (b). In order to accomplish this, the replacement segment is "broken" in the middle. The resulting curve has four segments each 1/3 of the length of the original line. Next, the center 1/3 of *each.* of the four segments is replaced with a "broken" line twice as long as the segment that was removed (c). Now the curve has 16 segments, each 1/9 the length of the original line. Repeating this process indefinitely results in a Koch curve, which is self-similar because any piece of it can be magnified by a factor of three and look the same as an unmagnified segment. *Illustration by Hans & Cassidy. Courtesy of Gale Group.*

whether viewed with the naked **eye**, or a magnifying **glass**, every coastline is similar to itself. While the patterns are not precisely the same at each level of magnification, the essential features of a coastline are observed at each level. Third, the length of a coastline depends on the magnification at which it is measured. Measuring the length of a coastline on a photograph taken from space will only give an estimate of the length, because many small bays and peninsulas will not appear, and the lengths of their perimeters will be excluded from the estimate. A better estimate can be obtained using a photograph taken from a helicopter. Some detail will still be missing, but many of the features missing in the space photo will be included, so the estimate will be longer and closer to what might be termed the "actual" length of the coastline. This estimate can be improved further by walking the coastline wearing a pedometer. Again, a longer measurement will result, perhaps more nearly equal to the "actual" length, but still an estimate, because many parts of a coastline are made up of **rocks** and pebbles that are smaller than the length of an average stride. Successively better estimates can be made by increasing the level of magnification, and each successive measurement will find the coastline longer. Eventually, the level of magnification must achieve atomic or even nuclear resolution to allow measurement of the irregularities in each grain of **sand**, each clump of dirt, and each tiny pebble, until the length appears to become infinite. This problematic result suggests the length of every coastline is the same.

The resolution of the problem lies in the fact that fractals are properly characterized in terms of their dimension, rather than their length, area, or **volume**, with typical fractals described as having a dimension that is not an integer. To explain how this can happen, it is necessary to consider the meaning of dimension. The notion of dimension dates from the ancient Greeks, perhaps as early as Pythagoras (582-507 B.C.) but at least from Euclid (c. 300 B.C.) and his books on **geometry**. Intuitively, we think of dimension as being equal to the number of coordinates required to describe an object. For instance, a line has dimension 1, a **square** has dimension 2, and a cube has dimension 3. This is called the topological dimension. However, between the years 1875 and 1925, mathematicians realized that a more rigorous definition of dimension was needed in order to understand extremely irregular and fragmented sets. They found that no single definition of dimension was complete and useful under all circumstances. Thus, several definitions of dimension remain today. Among them, the Hausdorf dimension, proposed by Felix Hausdorf, results in fractional dimensions when an object is a fractal, but is the same as the topological value of dimension for regular geometric shapes. It is based on the increase in length, area, or volume that is measured when a fractal object is magnified by a fixed scale **factor**. For example, the Hausdorf dimension of a coastline is defined as $D = \log(\text{Length Increase})/\log(\text{scale factor})$. If the length of a coastline increases by a factor of four whenever it is magnified by a factor of three, then its Hausdorf dimension is given by $\log(\text{Length Increase})/\log(\text{scale factor}) = \log(4)/\log(3) = 1.26$. Thus, it is not the length that properly characterizes a coastline but its Hausdorf dimension. Finally, then, a fractal set is defined as a set of points on a line, in a **plane**, or in space, having a fragmented or irregular appearance at all levels of magnification, with a Hausdorf dimension that is strictly greater than its topological dimension.

Great interest in fractal sets stems from the fact that most natural objects look more like fractals than they do like regular geometric figures. For example, **clouds**, trees, and **mountains** look more like fractal figures than they do like circles, triangles, or pyramids. Thus, fractal sets are used by geologists to model the meandering paths of **rivers** and the rock formations of mountains, by botanists to model the branching patterns of trees and shrubs, by astronomers to model the distribution of **mass** in the universe, by physiologists to model the human **circulatory**

system, by physicists and engineers to model turbulence in fluids, and by economists to model the stock market and world economics. Often times, fractal sets can be generated by rather simple rules. For instance, a fractal dust is obtained by starting with a line segment and removing the center one-third, then removing the center one-third of the remaining two segments, then the center one-third of those remaining segments and so on.

Rules of generation such as this are easily implemented and displayed graphically on computers. Because some fractal sets resemble mountains, islands, or coastlines, while others appear to be clouds or snowflakes, fractals have become important in graphic art and the production of special effects. For example, "fake" worlds, generated by computer, are used in science fiction movies and **television** series, on CD-ROMs, and in video games, because they are easily generated from a set of instructions that occupy relatively little computer memory.

Resources

Books

Peterson, Ivars. *Islands of Truth, A Mathematical Mystery Cruise*. New York: W. H. Freeman, 1990.

J. R. Maddocks

Fraction, common

Fraction is the name for part of something as distinct from the whole of it. The word itself means a small amount as, for example, when we ask someone to "move over a fraction." We mean them to move over part of the way, not all the way.

Fractional parts such as half, quarter, eighth, and so on form a part of daily language usage. When, for example, we refer to "half an hour," "a quarter pound of coffee," or "an eighth of a pie." In **arithmetic**, the word fraction has a more precise meaning since a fraction is a numeral. Most fractions are called common fractions to distinguish them from special kinds of fractions like decimal fractions.

A fraction is written as two stacked numerals with a line between them, e.g.,

$$\frac{3}{4}$$

which refers to three-fourths (also called three quarters). All fractions are read this way.

$$\frac{5}{9}$$

is called five-ninths and 5, the top figure, is known as the numerator, while the bottom figure, 9, is called the denominator.

A fraction expresses a relationship between the fractional parts to the whole. For example, the fraction

$$\frac{3}{4}$$

shows that something has been divided into four equal parts and that we are dealing with three of them. The denominator denotes into how many equal parts the whole has been divided. A numerator names how many of the parts we are taking. If we divide something into four parts and only take one of them, we show it as

$$\frac{1}{4}$$

This is known as a unit fraction.

Whole numbers can also be shown by fractions. The fraction

$$\frac{5}{1}$$

means five wholes, which is also shown by 5.

Another way of thinking about the fraction

$$\frac{3}{4}$$

is to see it as expressing the relationship between a number of items set apart from the larger **group**. For example, if there are 16 books in the classroom and 12 are

collected, then the relationship between the part taken (12) and the larger group (16) is

$$\frac{12}{16}$$

The fraction

$$\frac{12}{16}$$

names the same number as

$$\frac{3}{4}$$

Two fractions that stand for the same number are known as equivalent fractions.

A third way of thinking about the fraction

$$\frac{3}{4}$$

is to think of it as measurement or as a **division** problem. In essence the symbol

$$\frac{3}{4}$$

says: take three units and divide them into four equal parts. The answer may be shown graphically. The size of each part may be seen to be

$$\frac{3}{4}$$

To think about a fraction as a measurement problem is a useful way to help understand the operation of division with fractions which will be explained later.

A fourth way of thinking about

$$\frac{3}{4}$$

is as expressing a **ratio**. A ratio is a comparison between two numbers. For example, 3 is to 4, as 6 is to 8, as 12 is to 16, and 24 is to 32. One number can be shown by many different fractions provided the relationship between the two parts of the fraction does not change. This is most important in order to add or subtract, processes which will be considered next.

Operations with fractions

Fractions represent numbers and, as numbers, they can be combined by **addition**, **subtraction**, **multiplication**, and division. Addition and subtraction of fractions present no problems when the fractions have the same denominator. For example

$$\frac{1}{8} + \frac{5}{8} = \frac{6}{8}$$

We are adding like fractional parts, so we ignore the denominators and add the numerators. The same holds for subtraction. When the fractions have the same denominator we can subtract the numerators and ignore the denominators. For example

$$\frac{5}{6} - \frac{4}{6} = \frac{1}{6}$$

To add and subtract fractions with unlike denominators, the numbers have to be renamed. For example, the problem

$$\frac{1}{2} + \frac{2}{3}$$

requires us to change the fractions so that they have the same denominator. We try to find the lowest common denominator since this makes the calculation easier. If we write

$$\frac{1}{2} \text{ as } \frac{3}{6}$$

and

$$\frac{2}{3} \text{ as } \frac{4}{6}$$

the problem becomes

$$\frac{3}{6} + \frac{4}{6} = \frac{7}{6}$$

Similarly, with subtraction of fractions that do not have the same denominator, they have to be renamed.

$$\frac{3}{4} - \frac{1}{12}$$

needs to become

$$\frac{9}{12} - \frac{1}{12}$$

which leaves

$$\frac{8}{12}$$

Now consider:

$$(\frac{7}{6})$$

which is known as an improper fraction. It is said to be improper because the numerator is bigger than the denominator. Often an improper fraction is renamed as a mixed number which is the sum of a whole number and a fraction. Take six of the parts to make a whole (1) and show the part left over as

$$1\frac{1}{6}$$

A fraction is not changed if you can do the same operation to the numerator as to the denominator. Both the numerator and denominator of

$$\frac{8}{12}$$

can be divided by four to reduce the fraction to

$$\frac{2}{3}$$

Both terms can also be multiplied by the same number and the number represented by the fraction does not change. This idea is helpful in understanding how to do division of fractions which will be considered next. When multiplying fractions the terms above the line (numerators) are multiplied, and then the terms below the line (denominators) are multiplied, e.g.,

$$\frac{3}{4} \times \frac{1}{2} = \frac{3}{8}$$

We can also show this graphically. What we are asking is if I have half of something, (e.g., half a yard) what is

$$\frac{3}{4}$$

of that? The answer is

$$\frac{3}{8} \text{ of a yard.}$$

It was mentioned earlier that a fraction can be thought of as a division problem. Division of fractions such as

$$\frac{3}{4} \div \frac{1}{2}$$

may be shown as one large division problem

1692

$$\cfrac{\dfrac{3}{4}\ (N)}{\dfrac{1}{2}\ (D)}$$

The easiest problem in the division of fractions is dividing by one because in any fraction that has one as the denominator, e.g.,

$$\frac{7}{1}$$

we can ignore the denominator because we have 7 wholes. So in our division problem, the question becomes what can we do to get 1 in the denominator? The answer is to multiply

$$\frac{1}{2}$$

by its **reciprocal**

$$\frac{2}{1}$$

and it will **cancel** out to one. What we do to the denominator we must do to the numerator. The new equation becomes

$$\cfrac{\dfrac{3}{4}\ \text{x}\ \dfrac{2}{1}}{\dfrac{1}{2}\ \text{x}\ \dfrac{2}{1}} = \cfrac{\dfrac{6}{4}}{1} = \cfrac{\dfrac{6}{4}}{1} = 1\frac{1}{2}$$

We can also show this graphically. What we want to know is how many times will a piece of cord

$$\frac{1}{2}\ \text{inch long}$$

fit into a piece that is

$$\frac{3}{4}\ \text{inch long}$$

The answer is

$$1\frac{1}{2}\ \text{times}$$

Fractions are of immense use in **mathematics** and **physics** and the application of these to modern technolo-

KEY TERMS

Denominator—Notes the number of parts into which the whole has been divided.

Equivalent fraction—Where the value of the fraction remains the same but the form of the fraction changes.

Improper fraction—Where the numerator is the same as the denominator or greater than the denominator

Lowest common denominator—The smallest denominator which is common to all the fractional units being dealt with.

Numerator—That part of a fraction which enumerates how many of the factional parts are being dealt with.

Unit fraction—Symbol which shows that one part only is being dealt with.

gy. They are also of use in daily life. If you understand fractions you know that

$$\frac{1}{125}$$

is bigger than

$$\frac{1}{250}$$

so that shutter speed in **photography** becomes understandable. A screw of

$$\frac{3}{16}$$

is smaller than one of

$$\frac{3}{8}$$

so tire sizes shown in fractions become meaningful rather than incomprehensible. It is more important to understand the concepts than to memorize operations of fractions.

Resources

Books

Barrow, J.D. *Pi in the Sky.* New York: Oxford University Press, 1992.

Hamilton, Johnny E., and Margaret S. Hamilton. *Math to Build On: A Book for Those Who Build.* Clinton, NC: Construction Trades Press, 1993.

Savin, Steve. *All the Math You'll Ever Need.* New York: John Wiley & Sons, 1989.

Selma E. Hughes

Francium *see* **Alkali metals**

Fraunhofer lines

Fraunhofer lines are dark absorption lines in the solar **spectrum** that can be seen when sunlight is passed through a **prism** to separate it into the colors of the rainbow. They occur because cooler gas, which is higher in the Sun's atmosphere, absorbs some colors of the **light** emitted by hotter gas lower in the Sun's atmosphere. Sir Isaac Newton (1642-1727) discovered that if white light is passed through a prism, it separates into a rainbow, which is called a spectrum. While studying the spectrum that sunlight made, Joseph Fraunhofer (1787-1826) discovered some dark lines scattered among the colors. These dark lines were segments of colors missing from the complete spectrum. Fraunhofer counted 574 of these lines, which we now call Fraunhofer lines. Today, using much more sophisticated techniques, astronomers have discovered tens of thousands of Fraunhofer lines. Why doesn't the **Sun** emit these missing colors? Or, if the Sun does emit these colors, what happens to the light before it reaches **Earth**? The answer lies at the surface of the Sun.

When we look at a picture of the Sun, the surface that we see is called the photosphere. The photosphere is a region, several hundred kilometers thick, in which the Sun changes from opaque to transparent. It is not actually the outermost surface: the Sun extends for thousands of kilometers beyond the photosphere, but it is not usually visible from Earth. The photosphere is interesting because within this thin layer of the Sun (thin compared to the whole Sun, of course) sunlight is created, and some of the colors are lost almost immediately. The lower region of the photosphere has a **temperature** of about 10,000° F (about 5,500° C) and glows white-hot. Any object that glows due to a high temperature gives off a complete spectrum, that is, it has all the colors of the rainbow. As this light proceeds upwards in the Sun into a higher region of the photosphere, the temperature drops several thousand degrees. Although most of the light passes right through, some of the light is absorbed by the cooler gas. Only certain colors are removed because the chemical elements in the photosphere can only absorb certain wavelengths of light, and different wavelengths correspond to different colors. For example, **sodium** absorbs some yellow light at a wavelength of about 5.89×10^{-7}m. These absorbed colors cause the Fraunhofer lines. By measuring precisely the wavelengths of the missing colors, that is, the Fraunhofer lines, and how much light is actually absorbed, astronomers have learned much about the temperature inside the Sun and its chemical composition.

We can also learn about other stars in the sky by looking at the absorption lines in their spectra. By studying the similarities and differences that they have with the Fraunhofer lines, we can learn a lot about the similarities and differences that other stars have with our Sun.

Free radical *see* **Radical (atomic)**

Freeway

Freeways, also called superhighways, are roads specifically designed to allow for the free flow of traffic. Freeways typically feature two or more traffic lanes in each direction, medians to divide the opposing directions, full access control, a system of ramps to prevent merging and diverging traffic from interrupting the traffic flow, and grading to separate intersecting traffic on other roads.

Rise of the freeway

The advent and eventual domination of the **automobile** created a corresponding demand for roads capable of handling the increasing traffic and loads. Increasing numbers of cars began to choke the cities with traffic. The need for linking cities to one another also became apparent, especially as the truck proved its flexibility and reliability for transporting goods, materials, and products.

The freeway was first conceived as a means for reducing the crush of traffic within the cities, and for linking the cities together. The first freeway was opened in the Grunewald Forest in Berlin, Germany, in 1921. The idea for a national highway system for the United States was also developed during this time. The first United States freeways appeared in 1940, when California opened the Arroyo Seco Parkway between Pasadena and Los Angeles, and when Pennsylvania opened the first section of the Pennsylvania Turnpike.

The numbers of automobiles in use skyrocketed in the years after the World War II. With this increase came an alarming increase in traffic congestion and automobile accident fatalities. In 1956, legislation was passed creating the Federal Interstate Highway System (FIHS). This network of freeways was meant to link nearly all cities in the United States with populations greater than 50,000 people. Although the original plans called for exactly 40,000 mi (64,630 m) of road, by the 1990s nearly 45,000 mi (72,405 m) of road surface had been completed, carrying more than 20% of all traffic in this country. Freeways in the FIHS are constructed according to strict guidelines governing the materials and other elements of their design and construction.

Freeways dramatically changed the pattern of life in the United States. Access to the city by automobile allowed people to move beyond the traditional trolley and horse-drawn cart routes. The spread of people outside of the city created what is known as urban sprawl, in which the city extends farther and farther from its center. Meanwhile, the former centers of city life lost more and more manufacturers and other industries to the suburbs, draining the cities of vital resources and jobs. Although the freeway was originally meant to alleviate traffic, it actually increased traffic levels, by encouraging the use of the automobile over **mass transportation** methods such as trains and buses. The resulting increases in congestion brought problems of **pollution** and noise. What were once "walking cities" were now accessible only by car. Entire new communities, the suburbs, became so dependent on automobiles that most families found it necessary to have two or more. Meanwhile, the increased traffic on the roads brought corresponding increases in the number of traffic fatalities.

Nonetheless, the FIHS remains the most ambitious public works undertaking in American history. The FIHS has made nearly every part of the country accessible by car. It has brought a greater flexibility and choice of places for people to live, work, and travel, and a greater mobility over longer distances and safer roads.

Features of the freeway

All freeways share a number of common features. A freeway has at least four lanes, two lanes in each direction. Many freeways, however, feature more than four, and as many as ten lanes, especially as they near the cities. Lanes are required to be from 11-12 ft (3.35-3.66 m) wide. Shoulder lanes provided on each side of the driving lane for each direction of the freeway allow vehicles to safely leave the traffic stream in the event of an emergency. Shoulder lanes are generally 8-10 ft (2.4-3 m) wide. A **median**, or center strip, separates the oppos-

An aerial shot of a freeway system under construction in southern California. *Photograph by Tom Carroll. Phototake NYC. Reproduced by permission.*

ing directions of traffic. Medians may vary from 16-60 ft (4.9-18.3 m) wide. The median improves safety by preventing head-on collisions of automobiles traveling toward each other.

Freeways are called controlled access highways. This means that traffic is limited in where it may come onto or leave the freeway. These entrance and exit points are referred to as interchanges. Minor roads and driveways are diverted away from the freeway so that their traffic does not interfere with the freeway traffic flow.

Many roads, from small local roads and streets to other highways in the freeway system, intersect with a freeway. Grade separation prevents the intersection of two roads traveling crossways to each other from interrupting each others' traffic flow. Generally, one road, usually the road minor to the freeway, is raised on a grade, or slope, so that it is higher than the freeway, and allowed to cross it over a bridge. Ramps are constructed to lead the crossing road to the grade separation. Additional access ramps, often called on-ramps and off-ramps, connect the freeway to the intersecting road. They allow vehicles entering the freeway to accelerate to the proper speed before merging with the freeway traffic; the ramps allow vehicles leaving the freeway to decelerate to the slower speeds of the crossing road.

As part of the FIHS, freeways are designated by red, white, and blue signs in the shape of shields. Freeways are also numbered, with the numbering system used to indicate the direction of the road. Freeways traveling in an east-west direction are given even numbers; those traveling north-south are given odd numbers.

Construction of a freeway

Planning

With the completion of the FIHS, few new freeways may be expected to be built in the United States. Existing freeways, however, will continue to be expanded and improved. In all cases, work on a freeway must be carefully planned, its route laid out, and its impact on the environment and surrounding area thoroughly investigated. Engineers design the freeways, following government specifications. In addition, geographical and geological features are examined, including the grade, or slope of the land, and the type of **soil** found along different sections of the proposed roadway. The type of soil will affect the nature of the pavement to be laid, so soil samples are analyzed both in the field and in the laboratory.

Many questions must be answered when designing a freeway. The expected **volume** of traffic must be estimated, with minimum and maximum levels established. The expected use of the freeway is another consideration, and takes into account where people live and work and how they currently travel, and also the location and type of industry in the area, the types of goods that are produced, the markets or destinations of those goods, and how those goods have been transported in the past. These questions will affect the volume of traffic on the proposed freeway; they will also affect the type of vehicles that will use it. A freeway that will serve heavy trucks will require different surfacing, lane widths, and bridge heights than freeways serving mostly or only automobiles.

Clearing, grading, and drainage system

Work begins by clearing the right-of-way, the path, of the freeway. Vegetation will be removed, and the course for the freeway will be laid out. The use of modern construction equipment, including bulldozers and other specifically designed heavy equipment, has made this process much easier and faster than in the past. At this time, hills and valleys along the freeway route may be smoothed out, to minimize the variability of the route.

At the same time, features of the **water** drainage system—an important part of any roadway—are formed. These include the slope of the road, and ditches and culverts alongside of the road. The drainage may be the single most costly part of constructing a freeway; yet, if the water is not properly guided away from the road, the road will quickly weaken. The cleared right-of-way, including the shoulders and drainage ditches, will next be compacted, so as to provide a firm underbed for the freeway. Any **bridges** to be placed along the freeway will then be constructed, before the freeway itself is paved.

Paving

Paving a freeway may actually take place in several phases, adding layer upon layer of road surface over a long period of time, even years, until the freeway has achieved its final form. This allows weaknesses, and the effects of settling, in the roadway and drainage system to be detected and corrected.

Roads, including freeways, are generally composed of three layers: the subbed, or subgrade; the bed, or base; and the surface, or pavement or wearing course. The subbed is the soil on which the freeway is built. It is prepared by leveling and compacting the soil, and may be treated with asphalt, tar, or other substances to provide greater firmness. Next, the base is laid, consisting of crushed stone, gravel, or **concrete** pieces in a variety of sizes ranging from dust to 3 in (8 cm) **rocks** mixed in exact proportions. This allows the base to remain porous, so that moisture will not build up beneath the pavement. This course is also compacted, then sprayed with a thin, liquid layer of tar or asphalt to fill in the gaps and spaces between stones and make this surface even.

The pavement is then laid on top of the base. A layer of tar or asphalt is added, then covered with gravel or stones that are all the same size. The gravel layer is compacted into the asphalt so that they are firmly mixed together. This process, which forms the pavement, may be repeated several times, until the road surface reaches the proper thickness. Each layer is rolled with special machines until it is hard and smooth. Sudden bumps or dips in the road will make the freeway more dangerous to drive on, especially with the high speeds allowed on the freeway. The thickness of the road surface will depend on the type of traffic expected, that is, whether it is expected to be high volume, or whether it is expected to carry many heavy trucks as well as automobiles. The

KEY TERMS

. . . .

Asphalt—A substance composed primarily of hydrocarbons found in nature or produced as a by-product in petroleum refining; also refers to a road surface.

Grade—A slope or inclination in a road.

Grade separation—A crossing over or under a highway.

Interchange—An intersection of two or more highways that allows the flow of traffic to occur without stopping or crossing the other traffic streams.

Ramp—A section of roadway raising or lowering traffic from one level to second level used to allow the entrance or exiting of traffic to or from a freeway.

Right-of-way—The width and length of land through which all structures of a freeway pass.

Tar—A viscous liquid obtained by burning substances such as wood and coal that is used in the surfacing of roads and other structures requiring a waterproof seal.

pavement must be watertight, because moisture can destroy the surface as it expands or contracts with **temperature** changes. The addition of stones or gravel in the blacktop, or surface layer, allows tires to grip the surface more easily.

Safety features

Keeping the driver alert to the road is important for preventing accidents. Lighting by overhead lamps allows the driver to see the road ahead at night, even at great distances. Guardrails may be placed alongside the roadway at curves and where the land drops away suddenly beyond the shoulder. Reflectors are often placed on guardrails alongside the roadway and in the lines between lanes. Landscaping along the road and in the median helps to reduce the monotony of a long drive.

Resources

Books

Borth, Christy. *Mankind on the Move: The Story of Highways.* Automotive Safety Federation, 1969.

Davies, Richard O. *The Age of Asphalt.* J. B. Lippincott Company, 1975.

Kilareski, Walter P. *Principles of Highway Engineering and Traffic Analysis.* John Wiley & Sons, 1990.

Williams, Owen. *How Roads Are Made.* Facts on File Publications, 1989.

M. L. Cohen

Freezing *see* **States of matter**

Freezing point *see* **States of matter**

Frequency

Any process that is repetitive or periodic has an associated frequency. The frequency is the number of repetitions, or cycles, during a given time **interval**. The inverse of the frequency is called the period of the process.

Pendulums, as in a grandfather clock, also have a frequency of a certain number of swings per minute. A complete oscillation for a pendulum requires the pendulum bob to start and finish at the same location. Counting the number of these **oscillations** during one minute will determine the frequency of the pendulum (in units of oscillations/minute). This frequency is proportional to the **square root** of the **acceleration** due to gravity divided by the pendulum's length. If either of these are changed, the frequency of the pendulum will change accordingly. This is why you adjust the length of the pendulum on your grandfather clock to change the frequency, which changes the period, which allows the clock to run faster or slower.

Vibrating strings also have an associated frequency. Pianos, guitars, violins, harps, and any other stringed instrument requires a particular range of vibrational frequencies to generate musical notes. By changing the frequency, generally by changing the length of the string, you change the pitch of the note you hear.

In any type of wave, the frequency of the wave is the number of wave crests (or troughs) passing a fixed measuring position in a given time interval; and, is also equal to the wave's speed divided by the wavelength. As a wave passes by a fixed measurement point, a specific number of wave crests (or troughs) pass a fixed point in a given amount of time. In the case of waves, the frequency is also equal to the speed of the wave divided by the wavelength of the wave.

Light also exhibits the characteristics of waves; so, it too has a frequency. By changing this frequency, you also change the associated **color** of the light wave.

Frequency modulation *see* **Radio**

A freshwater mountain stream in Rocky Mountain National Park. *JML Visuals. Reproduced by permission.*

Freshwater

Freshwater is chemically defined as containing a **concentration** of less than two parts per thousand (<0.2%) of dissolved salts.

Although **water** is abundant on the surface of **Earth**, freshwater is a very limited resource. Freshwater, in all forms, makes up less than 2.8% of the world water supply. Freshwater on Earth exists in several forms. These include lakes, which represent 0.009% of the global water supply, **rivers** (0.0001%), atmospheric water including vapor, **clouds**, and **precipitation** (0.001%), shallow **groundwater** in **soil** and subterranean aquifers (0.31%), and polar icecaps and **glaciers** (2.15%). The supply of water available for human and other biological demands excludes those waters that are saline (salty), situated in the atmosphere, or frozen in icecaps and glaciers. The waters that fit into useable criteria constitute less than 0.5% of all of the water on Earth. **Pollution**, waste, population growth, and **competition** over available resources further restrict the availability of freshwater and are likely to become more acute in the future.

Most of the dissolved, inorganic chemicals in freshwater occur as ions. The most important of the positively charged ions (or cations) in typical freshwaters are **calcium** (Ca^{2+}), **magnesium** (Mg^{2+}), **sodium** (Na^+), ammonium (NH_4^+), and **hydrogen** (H^+). The most important of the negatively charged ions (or anions) are sulfate (SO_42^+), chloride (Cl^-), and nitrate (NO_3^-). Other ions are also present, but in relatively small concentrations. Some freshwaters can have large concentrations of dissolved organic compounds, known as humic substances. These can stain the water a deep-brown, in contrast to the transparent **color** of most freshwaters.

Lakes in watersheds with hard, slowly **weathering bedrock** and soils, are at the dilute end of the chemical **spectrum** of surface waters. Such lakes can have a total **salt** concentration of less than 0.002% (equivalent to 20 mg/L, or parts per million, ppm). For example, Beaverskin Lake in Nova Scotia has clear, dilute water, with the most important dissolved chemicals being chloride (4.4 mg/L), sodium (2.9 mg/L), sulfate (2.8 mg/L), calcium (0.41 mg/L), magnesium (0.39 mg/L), and potassium (0.30 mg/L). A nearby body of water, Big Red Lake, has similar concentrations of these inorganic ions. However, this lake also receives drainage from a nearby bog, and its **chemistry** includes a large concentration of dissolved organic compounds (23 mg/L), which stain the water the color of dark tea.

More typical concentrations of major inorganic ions in freshwater are somewhat larger: calcium, 15 mg/L; sulfate, 11 mg/L; chloride, 7 mg/L; silica, 7 mg/L; sodium, 6 mg/L; magnesium, 4 mg/L; and potassium, 3 mg/L.

The freshwater of precipitation is considerably more dilute than that of surface waters. For example, precipitation falling on the Nova Scotia lakes is dominated by sulfate (1.6 mg/L), chloride (1.3 mg/L), sodium (0.8 mg/L), nitrate (0.7 mg/L), calcium (0.13 mg/L), ammonium (0.08 mg/L), magnesium (0.08 mg/L), and potassium (0.08 mg/L). Because the sampling site is within 31 mi (50 km) of the Atlantic Ocean, its precipitation is significantly influenced by sodium and chloride originating with sea sprays. Locations that are more continental have much smaller concentrations of these ions in their precipitation water. For example, precipitation at a remote place in northern Ontario has a sodium concentration of 0.09 mg/L and chloride 0.15 mg/L, compared with 0.75 mg/L and 1.3 mg/L, respectively, at the maritime Nova Scotia site.

See also Groundwater; Lake; Saltwater; Water conservation.

Bill Freedman

Friction

Friction is the **force** that resists **motion** when the surface of one object slides over the surface of another. Frictional forces are always **parallel** to the surfaces in contact, and they oppose any motion or attempted motion. No movement will occur unless a force equal to or greater than the frictional force is applied to the body or bodies that can move.

While friction is often regarded as a nuisance because it reduces the efficiency of machines, it is, nevertheless, an essential force for such items as nails, screws, pliers, bolts, forceps, and matches. Without friction we could not walk, play a violin, or pick up a glass of **water**.

Gravity and friction are the two most common forces affecting our lives, and while we know a good deal about gravitational forces, we know relatively little about friction. Frictional forces are believed to arise from the **adhesive** forces between the molecules in two surfaces that are pushed together by **pressure**. The surface of a material may feel smooth, but at the atomic level it is filled with valleys and hills a hundred or more **atoms** high. Pressure squeezes the hills and valleys in the two surfaces together and the molecules adhere to one another. The actual contact area, from a microscopic perspective, is much less than the apparent area of contact as viewed macroscopically. As the weight of an object resting on a surface increases, it squeezes the two surfaces together and the actual area of contact increases. The actual contact area is believed to be proportional to the weight pushing the bodies together.

In addition to the adhesive forces between molecules, there are other factors that affect friction. They include the force needed to raise one surface over the high places of another; the fact that a rough region along a hard surface may "plough" a groove in a softer material; and electrical forces of attraction required to separate oppositely charged regions of the surfaces.

There are three laws that apply to friction. (1) The force of friction between an object and the surface on which it rests is proportional to the weight of the object. (The magnitude of the frictional force depends on the nature of the two surfaces.) (2) The force of friction between an object and the surface on which it rests is independent of the surface area of the object. (Remember, the actual contact area depends on the weight. If the weight remains constant so will the actual area of contact regardless of what the apparent area may be.) (3) The force of friction between an object and the surface on which it rests is independent of the speed at which the object moves as long as the speed is not **zero**.

The third law applies only to moving objects. Static friction, the force required to make an object at rest begin to move, is always greater than kinetic friction, which is the resistance to motion of an object moving across a surface. The reduction of friction that arises with motion is the result of fewer areas of contact once a body is in motion; the molecules are not in contact long enough to form firm bonds. Rolling friction for an object mounted on wheels or **rollers** is far less than kinetic friction. The reason that rolling friction is so small is probably the result of minimal contact area between wheel and surface, particularly if both are very hard, and because any molecular adhesions are pulled apart by vertical shearing rather than horizontal tearing.

In breaking molecular adhesions, molecular vibrations increase, causing a rise in **temperature**. You can easily verify this by simply rubbing your hands together. In machines, the adhesion and tearing of molecular bonds between surfaces causes wear. To reduce wear, we add a lubricant (often oil). The oil decreases the actual area of contact between surfaces. As a result, it reduces the wear associated with tearing surface molecules apart and thereby keeps **heat** and wear at lower levels than would otherwise be the case.

A frigate bird on its nest on Barbuda Island in the West Indies. *Photograph by Ormond Gigli. Stock Market. Reproduced by permission.*

Frigate birds

Frigate **birds** are five **species** of oceanic birds that make up the family Fregatidae. Frigate birds occur along the coasts of the tropical oceans, but also hundreds of miles out to sea.

Frigate birds typically weigh about 3 lb (1.5 kg), but the spread of their long, narrow, swept-back, pointed wings can exceed 6.5 ft (2 m). These are highly favorable wing-loading characteristics, and frigate birds are among the most skilled of the birds at flying and seemingly effortless gliding. Their tail is long, with extensive pointed forks. Their legs are short, and the small, partially webbed feet are only used for perching. Frigate birds are very ungainly on the ground and in the **water**, on which they rarely set down.

The bill of frigate birds is long, and both the upper and lower mandibles hook downwards. The plumage is a dark-brownish black, with whitish underparts in some species, but the throat is naked and colored a bright red in males. The throat sac of male frigate birds can be inflated with air and is used to impress the females, both visually and by helping to resonate the loud rattlings and yodels of courting males.

Frigate birds are highly graceful and skilled aerialists. They are excellent fliers, both in terms of the distance they can cover across the vast oceans and their extremely skilled maneuverability in flight. They feed on **fish**, **squid**, **jellyfish**, and other **invertebrates** by hovering over the surface of the **ocean** and swiftly diving to snatch **prey** at the surface, often without getting their body feathers wet. Frigate birds frequently catch **flying fish** during those brief intervals when both bird and fish are airborne. Frigate birds sometimes predate the young of other seabirds, especially **terns** and noddys.

Frigate birds also commonly swoop aggressively on **pelicans**, boobies, and **gulls**, poking them and biting their tail and wings. This pugnacious **behavior** forces these birds to drop or disgorge any fish that they have recently caught and eaten, which is then consumed by the frigate bird. Frigate birds also force other seabirds to drop scarce nesting material, which is also retrieved. This foraging strategy is known to scientists as kleptoparasitism. In view of the well-deserved, piratical reputation of frigate birds, they are known as man-o'-war birds.

Frigate birds nest in trees or on remote, rocky ledges. Females bring sticks and other appropriate materials to the nest site, where the male constructs the nest. Both sexes of the pair share in incubating the eggs and raising their babies. Frigate birds do not migrate, but they may wander extensively during their non-breeding season.

The magnificent frigate bird (*Fregata magnificens*) is a seasonally common seabird around the Florida Keys, and an occasional visitor elsewhere in the coastal southeastern and southwestern United States, ranging through the Caribbean, Gulf of Mexico, and western Mexico, and as far south as the Atlantic coast of Brazil, the Pacific coasts of Ecuador and Peru, and the Galapagos Islands. Male birds have a purplish black body and a bright-red throat pouch. The females are browner and have a white breast, and juveniles are lighter brown and have a white head and breast.

Both the great frigate bird (*Fregata minor*) and the lesser frigate bird (*F. ariel*) have pan-tropical distributions, occurring in tropical waters over most of the world (but not in the Caribbean). Much more local distributions are exhibited by the Ascension frigate bird (*F. aquila*), which breeds only on Ascension Island in the South Atlantic, while the Christmas Island frigate bird (*F. andrewsi*) only breeds on Christmas Island in the South Pacific Ocean.

Bill Freedman

Frog's-bit family

The frog's-bit or tape-grass family (Hydrocharitaceae) is a relatively small group of herbaceous, aquatic, monocotyledonous plants, occurring in fresh and marine waters. There are about 100 **species** in the frog's-bit family, distributed among 15 genera.

The flowers of members of this family are **water** pollinated, shedding their pollen into the water, which disperses it to the stigmatic surfaces of other flowers.

Several species in this family are native to **North America**. The elodea, or Canadian water-weed (*Elodea canadensis*), is a common aquatic **plant** in fertile, calcium-rich ponds, lakes, and other still waters. The elodea is a monoecious plant, meaning the male and female functions are carried out by different plants. The leaves of elodea occur opposite each other on the stem, or in whorls.

The tape-grass (*Vallisneria americana*) is also a native aquatic plant of non-acidic lakes and ponds. This species is also dioecious. The tape-grass has long, narrow, strap-like leaves, and it can form dense, perennial stands in still or slowly moving waters.

This family gets its common name from the frog's-bit (*Limnobium spongia*). This plant is a **herb** of marshes and calm waters of lakes and ponds. The frog's-bit is monoecious and has broad leaves with a well-defined petiole.

Species of plants in this family are commonly used as ornamental vegetation in fresh-water aquaria, and they are widely sold in pets shops for this purpose. These plants may also be used in horticultural ponds.

Some species in the frog's bit family have been introduced beyond their native range and have become serious weeds of ponds, lakes, and canals. The Canadian water-weed (*Hydrilla vertricillata*) is a problem in North America. An Argentinean water-weed (*Elodea densa*) has also become a pest after being introduced to North America, probably as plants that were discarded from aquaria.

Bill Freedman

Frogs

Frogs are tail-less **amphibians** (class Amphibia, order Anura). With some 3,500 living **species**, frogs are the most numerous and best known of amphibians. They are found on all continents except **Antarctica** and are common on many oceanic islands. The terms "frog" and "toad" are derived from early usage in England and northern **Europe**, where two families of the order Anura occur. One includes slender, long-legged, smooth-skinned animals that live near **water**: frogs; the other includes short-legged warty animals that live in fields and gardens: **toads**. When other kinds of animals of this group were discovered elsewhere, such as tree-frogs, fire-bellied toads, and others, it was realized that these various forms actually represented one major group. This group, the anurans, is now commonly referred to as frogs.

History and fossil record

Frogs and their ancestors are among the most ancient of terrestrial **vertebrates**. A frog-like fossil **animal** more than 240 million years old is known from early Triassic **rocks** of Madagascar. This ancient amphibian, named *Triadobatrachus*, differs from true frogs in having more vertebrae in its spinal column (14, rather than 5-9) and in having a tail made up of six additional vertebrae. For these and several other differences, it is classified in a different order, the Proanura. By Jurassic times, 208-146 million years ago, such ancestral amphibians had evolved into true frogs, whose skeletal remains are little different from those living today.

Morphology

Frogs are amphibians, a term derived from two Greek words: *amphi* meaning double and *bios* meaning life. The double life of frogs involves living in water and also on land. Because of this amphibious habit, they must have adaptations for each environment. As in other animals that have a separate larval stage and a complex life cycle, frogs have two extremely different morphologies.

Adult morphology

If frogs were not so common and familiar, they would be regarded as among the strangest of vertebrate animals. The typical frog has a broad head with an enormous mouth and protruding eyes. The body is short and plump, and there is no tail. The forelegs are rather short but normal-looking, and are used mainly for propping up the front part of the body and for stuffing food into the mouth. The hind limbs are much larger and more muscular, and have an extra joint that makes them even longer and provides extra power for jumping, which is their major mode of locomotion. Among aquatic frogs, the hind limbs also provide the propulsion for swimming.

The frog skeleton has been evolutionarily reduced. The skull is a framework of bones that hold the brain-case, eyes, internal ears, and jaws, while giving support to the jaw muscles. The vertebral column has been reduced to only 5-9 body vertebrae, and the caudal (tail) vertebrae have become fused into a single mass, the urostyle. Although the bones of the forelimbs are relatively normal-looking, those of the hindlimbs are highly modified for jumping. The tibia and fibula are fused into a single rod, and an extra joint has developed from the elongation of some of the foot bones, thus providing a jumping apparatus considerably longer than the torso.

Most frogs have a smooth, obviously moist skin. Even toads, with their warty, seemingly dry skin, have a surface cover that is moist and permeable to liquids and gases. This has advantages and disadvantages, but is necessary for frogs to carry on normal **respiration**. The lungs of amphibians are too small and simple in construction to provide adequate gas exchange, and the skin plays an important role in this regard. A significant amount of **oxygen** comes into the body via the skin, and as much as half of the **carbon dioxide** produced is released through this covering.

The internal **anatomy** is broadly similar to that of other vertebrate animals. There is a **heart** and associated **circulatory system**, a **brain** and **nervous system**, and a **digestive system** made up of esophagus, stomach, and small and large intestines, with the associated liver and other organs. The urinary system is relatively simple, having two kidneys as in most vertebrates. The **reproductive system** consists of paired ovaries or testes, with associated ducts. As in many vertebrates, the digestive, urinary, and reproductive systems empty through a common posterior chamber, the cloaca.

Larval morphology

Tadpoles, the larval stage of frogs, are adapted to a purely aquatic life. They are seemingly reduced to the essentials, which in this case includes a globular body with a muscular, finned tail. Typically, tadpoles have no bones but rather a simple cartilaginous skull and skeleton. They also have no true teeth, instead having rows of denticles and a beak of keratin (a fingernail-like substance). The globular body is mainly filled with a long, highly coiled intestine.

Ecology

The highly permeable skin of frogs might lead us to expect that they must always have access to water. This is generally true, but not always. It is true that if a common aquatic species, such as a leopard frog (*Rana pipiens*), were to escape from its cage and roam on the floor for a night, it would be little more than a dried-up mummy by the next morning. However, during the millions of years of frog **evolution**, many species have found ways of adapting to varying water availability in natural habitats. Although many frogs are aquatic, and

some never leave the water, there are also **desert** frogs, tree-frogs, and others that can withstand the drying power of tropical **heat** for a day or more.

Life history and behavior

Like other amphibians, such as **salamanders** (order Caudata) and **caecilians** (order Gymnophiona), most frogs hatch from a shell-less egg into a gilled, water-dwelling, larval stage (a tadpole). After a period of growth they metamorphose into the adult form. Most species of tadpoles are vegetarians, feeding upon **algae** and other **plant** material. All adult frogs, however, are carnivores, most of them feeding upon **insects** and other **invertebrates**.

In the temperate zones of the world the breeding season begins in the spring, but the precise **time** depends upon the species of frog. In much of temperate **North America**, for example, the beginning of springtime is proclaimed by the breeding calls of chorus frogs (*Pseudacris* spp.). Their high trills are soon followed by the calls of the spring peeper (*Hyla crucifer*). These may be followed by the rasping calls of the **wood** frog (*Rana sylvatica*), the leopard frog (*Rana pipiens*), and the green frog (*Rana clamitans*). Then the American toad (*Bufo americanus*) trills in, and when the larger ponds eventually warm up, the bullfrog (*Rana catesbeiana*) begins its booming jug-of-rum calls. As many as 16 different species of frogs have been found calling at various times at a single pond in Florida.

The males of each frog species have their own distinctive call. It has recently been found that the **ear** of the female is "tuned" to the call of her own species, so that not only is she not attracted to the calls other species, she may not even hear them! A female carrying eggs will typically approach a calling male of her choice (and of her species), and nudge him. He immediately ceases calling and grasps her around the waist. They enter the water (if they are not already in it), and as she expels eggs from her cloaca, the male sprays sperm over them. Depending upon the species, the eggs may appear in strings, in clusters, or as individual ova.

The eggs are enclosed in a protective jelly coating, and will develop over several days to a week into a tadpole. The tadpole will grow over a period of time (weeks, months, or years, depending upon the species), and ultimately sprouts legs, changes other elements of its external and internal morphology, and emerges as a small replica of the adult.

This sequence is typical of frogs living in temperate regions. In the tropics breeding is often initiated by a change in **weather** (such as dry to wet), the calling males may be on the moist forest floor or in a **tree**, the

A lesser gray treefrog (*Hyla versicolor*) at Mahn-Go-Ta-See Camp, Michigan. The lesser gray is the most widely distributed of the 13 species of North American treefrogs. *Photograph by Robert J. Huffman. Field Mark Publications. Reproduced by permission.*

eggs may be laid on foliage or beneath a rock or in a pond, and the tadpole stage may be completed inside the egg capsule, so that froglets appear directly from the egg. In other words, there is enormous variation in breeding habits, particularly in the tropics.

Classification

During the 200 million years of their existence, frogs have been evolving in response to varying environmental conditions. Common elements of their adaptations have given rise to clusters of species that share certain morphological, physiological, and behavioral traits. A system of classification has been established, mostly based on morphological features of adult frogs and their larvae. In the one presented here, two suborders, five superfamilies, and 21 families are recognized. It should be emphasized, however, that several systems of classification are recognized by scientists.

The families are often distinguished by such characters as the kind and number of vertebrae, the shape of the pectoral girdle, the presence of ribs, the kind and number of limb bones, and other elements. The structure of the pectoral girdle is an especially distinctive feature that separates large groups of otherwise similar-looking frogs.

The two genera of the family Leiopelmatidae are thought to be relics of an ancient group of frogs. They differ from all other frogs, and are also quite different from each other in habits and distribution, reflecting a long separation. Members of the genus *Leiopelma* are small terrestrial frogs of New Zealand, whereas *Ascaphus*, the "tailed frog" of western North America, inhabits streams. (The "tail" is actually an extension of the cloaca of males, and is used to place sperm into the cloaca of the female.) The Discoglossidae are another primitive group, made up of Asian and European pond frogs.

Other primitive frogs include the burrowing frog of Mexico (*Rhinophrynus dorsalis*) and a number of highly aquatic frogs, the Pipidae of **Africa** and **South America**. (One of this group, the African clawed frog [*Xenopus laevis*], has escaped from captivity and established wild populations in coastal California.) The spadefoot and parsley frogs (Ascaphidae) of North America and Europe are adapted to arid regions. They fall between the "primitive" and "advanced" frogs in structure, and show no close relationship to either. Their larvae are adapted to the rigors of desert life, and have very short periods of aquatic life.

Most of the world's frogs are included in the modern suborder Neobratrachia, with the superfamily Bufonoidea including several large families such as the Australian Myobatrachidae, the South American Leptodactylidae, and the widespread Hylidae. A number of smaller, specialized families are associated with them because of the common possession of a similarly structured pectoral girdle (known as arciferal).

The superfamily Ranoidea includes the large and widespread family Ranidae (the true or water frogs), the arboreal Rhacophoridae of **Asia** and Madagascar (flying frogs and allies), and the sedge frogs of Africa. Both of the latter appear to be derived from the ranids. The Ranoidea also includes the narrow-mouth toads, or Microhylidae. This widespread family of distinctively shaped ant-eating frogs has a so-called firmasternal pectoral girdle, and does not appear to be closely related to any other frog family, differing especially in their larval morphology. It has been placed only tentatively with the ranoid group.

In general, the species and genera of frogs in any region are relatively easy to recognize on the basis of their external features. These include the skin texture and color, the shape of the pupil of the **eye** (horizontally elliptic, vertically elliptic, or round), the amount of toe webbing, and the general body proportions, together with the geographic location and **habitat**. For example, a toad (family Bufonidae) is easily recognized throughout the world because of its warty skin. Water-dwelling frogs (Ranidae) are usually distinguished by their webbed hind feet. Tree-frogs usually have expanded toe-tips, although this can be misleading because some hylids (such as the cricket frogs of North America; genus *Acris*) have taken up a terrestrial existence and lost their climbing pads. Also, there are three quite different families of treefrogs: the Hylidae, which is primarily South American but with some members in North America and northern Eurasia; Centrolenidae, found only in the American tropics; and the Rhacophoridae of Asia, with a few species in Madagascar and Africa. The classification of the tree-frogs of **Australia** is still under consideration.

Nevertheless, the skin texture and color, the shape of the pupil of the eye (horizontally elliptic, vertically elliptic, or round), the amount of toe webbing, and the general body proportions together with the geographic location, are useful as local means of identification.

Frogs and humans

Frogs and humans have interacted for many thousands of years. Toads are referred to in ancient writings, as: a "rain of toads," the "eye of toad" as part of a witch's brew, and in many other relationships.

Frogs are also used in research, and to teach **biology**. A core element of many high school and college biology classes in the United States might involve each student dissecting a frog. Millions of leopard frogs have been utilized in this way in schools. By the 1950s, however, it was found that their numbers in the wild had decreased drastically, particularly in the midwestern U.S. This meant that frogs had to be imported from Canada and Mexico for use in teaching biology. During the past decade or so the emphasis on dissection has been much reduced, but large numbers of frogs are still used each year in physiological experiments. The frog populations of the Midwest have not recovered, and those of Canada and Mexico have also declined greatly.

Similarly, "frog-legs" used to be a prominent dish in many restaurants. American bullfrogs (*Rana catesbeiana*) of the swamps of Florida and Louisiana were the major source of this food. However, because of over-hunting it became too uncommon to be exploited by frog hunters in the United States, and imported legs of *Rana tigrina* and other species from India became the major source of frog legs.

More recently, the poison-arrow frogs of tropical America (Dendrobatidae) have become of great interest to pharmaceutical companies. Each species has an unique mix of biochemicals that may have a role to play in the treatment of human diseases. Frogs are useful to humans in various ways, although uncontrolled hunting of them can lead to serious problems for their populations.

Many people have kept pet toads or frogs, but the recent commercial market for captive frogs is primarily in exotic species such as South American horned frogs (*Ceratophrys*), the African "bullfrog" (*Pyxicephalus adspersus*), and brightly colored poison-arrow frogs. These animals are beautiful and interesting pets, but could cause ecological harm if they were to be released and develop wild populations beyond their natural range.

The future of frogs

Judging by recent observations, the prospects for many species of frogs is grim. During the 1990s, numerous species of frogs apparently vanished from nature without any obvious cause of their demise. For example, a newly described, extremely unusual Australian frog, (*Rheobatrachus silus*), could not be found in its only known habitat the following year. Numerous other Australian frogs have also disappeared. Similarly, the golden toad (*Atelopus zeteki*) of Costa Rica, which once occurred in large numbers, has apparently disappeared. The populations of the Yosemite toad (*Bufo canorus*) in the Sierra Nevada of California have plummeted. Similar re-

KEY TERMS

. . . .

Anurans—A general term for frogs and toads.

Arciferal—Having the coracoid elements of the pectoral girdle free and overlapping.

Denticles—Toothlike structures of keratin found around the mouth of tadpoles.

Firmisternal—Having the coracoid elements fused to the girdle.

Mattison, C. *Frogs and Toads of the World.* Sterling Publications, 1998.

Zug, George R., Laurie J. Vitt, and Janalee P. Caldwell. *Herpetology: An Introductory Biology of Amphibians and Reptiles.* 2nd ed. New York: Academic Press, 2001.

Herndon G. Dowling

Front (weather) *see* **Air masses and fronts**

Frost *see* **Precipitation**

ports have come from other parts of the world, and there is now an international group of biologists investigating the causes of the apparently simultaneous declines of many species of frogs.

Some biologists believe that the cause of the loss of these species may be somehow related to **pollution** caused by human activities. Emissions of chemicals known as chlorofluourocarbons, for example, may be causing the stratospheric **ozone** layer to become thinner, allowing greater amounts of ultraviolet **energy** to reach Earth's surface. There is some evidence that this environmental change may be a cause of the decline of the boreal toad (*Bufo boreas*) of the northwestern United States. This species breeds in open ponds at high altitude, and the intensified exposure to ultraviolet **light** may be killing its eggs.

Chemical pollutants may also be spread widely through the atmosphere, or be transported by surface water to places far from their original source of **emission**. Consequently, trace amounts of **pesticides** have been found in frogs living far from human populations. Although these poisons do not seem to kill the adults, they may be interfering with reproductive processes, and may be causing unusual deformities in rapidly developing tadpoles and juvenile frogs. Because frogs have such delicate, water-absorbing skin, they may be serving as environmental "canaries." Like the actual canaries that **coal** miners used to take into the mines as an early warning of the presence of toxic gas, frogs seem to be among the most sensitive indicators of ecological damage caused by toxic chemicals.

Resources

Books

Conant, Roger, et al. *A Field Guide to Reptiles & Amphibians of Eastern & Central North America (Peterson Field Guide Series).* Boston: Houghton Mifflin, 1998

Hofrichter, Robert. *Amphibians: The World of Frogs, Toads, Salamanders and Newts.* Toronto: Firefly Books, 2000.

Frostbite

Frostbite is the freezing of tissues. It occurs when body parts, most commonly the fingers, toes, and the tips of ears and the nose, are exposed for long periods to the cold. Frostbite is a direct result of limited **blood** circulation. The blood is the body's internal heating system; it carries **heat** to the body tissues. But prolonged exposure to the cold can constrict blood vessels, causing blood circulation within tissues to slow down. When tissues are deprived of the warmth of circulating blood, **ice** crystals can then form, leading to **tissue** death and loss of the affected body parts.

Stages of frostbite

There are three degrees, or stages, of frostbite: frostnip, superficial frostbite, and deep frostbite. Frostnip is the least serious form of frostbite; deep frostbite the most serious. If frostnip goes untreated, it can quickly progress to the more serious forms. Recognizing and treating the first signs of frostnip may prevent the development of the more serious forms of frostbite.

Frostnip is the "warning sign" of frostbite. In frostnip, the skin reddens then turns white. The person may also experience numbness in the affected area. The treatment for frostnip is simple; get the person out of the cold and gently warm the affected area. The warming procedure does not require special equipment; if the hands are affected, placing the hands under the armpits may be sufficient; if the fingers are affected, blowing warm air on the fingers should be enough to warm them.

The next two stages of frostbite may set in if frostnip is not treated promptly. In superficial frostbite, only the top layer of the skin is frozen. The top layer of skin is rigid, but the layers beneath the frozen layer are soft to the touch. The affected area appears white. In deep frostbite, the deeper layers of skin and tissue are frozen. The area feels rigid, and if the area is gently pressed no

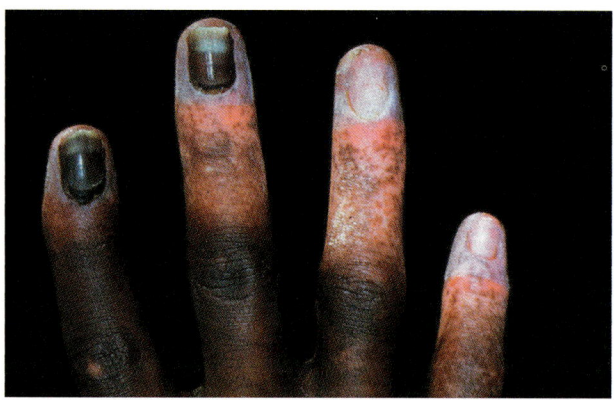

Frostbite injury of the hand. *Photograph by SIU. National Audubon Society Collection/Photo Researchers, Inc. Reproduced by permission.*

"give" or softness can be felt. The **color** of the affected area progresses from white to a grayish-yellow color and finally to a grayish blue color. In severe cases of deep frostbite, muscles, bones, and even the organs may become frozen.

Treatment for frostbite

People with superficial or deep frostbite should be taken to the hospital immediately. If transport to a hospital is delayed, the following measures can be taken to prevent further injury:

• Keep the person warm. Try to get the person indoors to a heated environment and cover with blankets, if available.

• Don't let the person smoke or drink **alcohol**. Both **nicotine** and alcohol can further constrict blood vessels and decrease blood circulation.

• Don't rub the affected area.

Most experts advise people not to rewarm or thaw frostbitten areas themselves. If thawing is not done properly, serious consequences such as loss of the affected area can result. For instance, thawing a frostbitten area by holding the affected area close to a campfire or in front of an open oven door can **burn** already-damaged tissues. Rubbing a frostbitten area with snow or ice is also not recommended. Rubbing with snow or ice will only cause more ice crystals to form in frozen tissue.

Rewarming should only be performed by trained hospital personnel. The rewarming should be gradual. The affected part of the body (or sometimes the entire person if the affected area is extensive) is submerged in a tub of warm **water**. The water is usually between 100–106°F (37.7–41.1°C). **Pain** is a sign that thawing is taking place. During the rewarming procedure, it is extremely important that the person remain as still as possible. **Motion** can cause still-frozen tissue to break into pieces which may injure delicate, newly-thawed tissues.

After rewarming, the affected area is sometimes loosely wrapped in sterile cloth. The person is carefully monitored for **infection** and for signs of restored circulation. If the area does not show renewed circulation or if the area becomes infected, **amputation** is sometimes performed. Amputation is necessary to prevent infection from spreading into other areas of the body.

In the United States, hospitals in Alaska have established guidelines for the assessment and treatment of frostbite. Many physicians and other health care personnel have also made helpful observations about the incidence of frostbite that may prevent some cases. One physician has noted a sharp increase in the incidence of frostbite affecting the ears in young male patients in the late 1970s and 1980s compared to the 1960s and early 1970s. This physician surmised that in the 1960s and 1970s, men wore their hair longer, thus protecting their ears from frostbite, but the shorter hair styles of the 1980s exposed ears to the elements. Another physician notes a correlation between snorting **cocaine** and frostbite affecting the tip of the nose. Like the cold, cocaine also leads to constriction of the mucous membranes and the **arteries** in the nose. A person using cocaine may develop frostbite faster than a person who isn't using cocaine. Yet another observation is a correlation between frostbite affecting the feet and the wearing of tennis and running shoes. In cold climates especially, proper footwear is essential in preventing frostbite.

In addition to these guidelines, other measures can be taken to prevent frostbite. When working or playing outdoors, a person should refrain from drinking alcohol or smoking. They should always wear proper clothing and take special care of the areas most vulnerable to cold exposure: the ears, tip of the nose, fingers, and toes. When any part of the body starts to feel numb, a person should immediately go indoors to warm that part. These common sense tactics should ensure safety when the mercury drops.

See also Circulatory system; Integumentary system; Nerve growth factor; Organ.

Resources

Books

Grant, Henry, et. al. *Brady Emergency Care.* 6th ed. Englewood Cliffs: Prentice Hall, 1990, pp. 566-568.

Periodicals

Mills, William J. "Summary of Treatment of the Cold Injured Patient: Frostbite." *Alaska Medicine* 35. (1): 61-66.

Fructose *see* **Carbohydrate**

Fruits

A fruit is an often edible part of a **plant** that is derived from a fertilized, ripened ovary. As a dietary staple, fruits are appreciated for their sweetness and as a rich source of **nutrients**, especially vitamins. Gardeners enjoy planting fruit-bearing plants for their usefulness as a food and also for the array of **color** and diversity they bring into the garden. In agricultural industry, fruits are grown for food consumption, such as apples and oranges, and for their use to manufacture a wide variety of drinks, jams and jellies, and flavorings, as well as for the production of wine.

Classification

Fruit-bearing plants are categorized in several ways. Among the **tree** fruits are the citruses (oranges, lemons, limes, and grapefruits) and apples, pears, peaches, plums, and figs. Most notable among fruits that grow on vines are **grapes** and kiwi fruit. Many berries are classified as bush fruits (currants and gooseberries), but some, like raspberries, blackberries, and loganberries grow on canelike shoots and are thereby referred to as cane fruit. Strawberries grow on plants that have little or no woody **tissue** and are classified as herbaceous. The bush, cane, and herbaceous fruits are also referred to as soft fruits.

Another category of fruit is the tender fruits, like pineapples, pomegranates, citrus, prickly pears, and tree tomatoes that need a warm climate to thrive. Nuts, which have a hard outer shell surrounding an inner tissue that can be eaten, are another category of fruit. An important classification of fruits is the distinction between pome fruits (a fleshy fruit surrounding a central core of **seeds**, such as apples and peaches) and stone fruits (those with a single pit or stone in the center, like avocado and cherries).

Fruit trees are also classified by their size, some coming in a standard size. Fruits that are fleshy, like berries, bananas, and grapes, are often referred to as succulent fruits. Another common grouping of fruits are the subtropical fruits, which in the United States are grown primarily in California. The subtropicals include bananas, which is also called an accessory fruit because it is not formed from the ovary of the plant's **flower** but an accessory part of it. Papayas, kiwi, and mango are some of the other subtropicals.

Growing fruits

Fruits are grown in temperate and tropical climates throughout the world. Native fruits from one region of the world have been cultivated to grow in other areas. For example, the kiwi fruit, also called the Chinese gooseberry, is native to parts of eastern **Asia** but is now also grown in **Australia** and the United States.

Climate and **soil** are particularly important to the growing of fruits. Both cold and **heat** ranges determine which fruit plants can be grown in a particular region. Because of the way fruit develops on its plant, it has specific seasonal needs. Too much rain or too little rain at certain times during the year can spoil the crop. Temperatures which are too cool can kill the flower buds of a plant, and it will not produce fruit. Each variety of fruit also needs particular amounts of **water** and summer temperatures in order for the fruit to ripen properly.

The specific location of fruiting plants is another important factor in the growing of fruits. Whether fruit plants are being grown in the garden or used for commercial purposes, **wind**, walls, hills, and other conditions that may affect the **temperature** a plant has to endure, need to be considered for good fruiting to take place. When citrus-growing regions, like Florida, are hit with late spring frosts, the citrus crop may become endangered.

Soil

All plants, edible or not, require certain soil factors such as the ability of the soil to drain excess water away from the plant's roots, since roots standing in water may become damaged. Other important considerations about soils are the amount of air in the soil and the type of nutrients. The nutrients that fruit plants need include **nitrogen**, **phosphorus**, potassium, and other micronutrients.

Nitrogen deficiency is apparent in a plant that has yellowing leaves. Nitrogen is essential for the formation of chlorophyll in a plant and also helps it produce **proteins** and **hormones**. Phosphorus is needed by fruit plants for the production of carbohydrates, and potassium helps the plant open and close its pores as it exhales moisture and inhales **carbon dioxide**. When these elements are not present in the soil in sufficient quantities, they must be added to for good fruit crop production.

Pollination and propagation

Some fruit plants are self-pollinating, which means they do not require another plant to pollinate its flowers. **Pollination** can take place when the male part of the plant, the stamen, pollinates the stigma of the plant, a female part that receives pollen from the anther of the male part of the plant. In self-pollination this takes place within the same flower. For plants that require cross-pollination, pollen is carried by the wind or by **insects** such as **bees** from the stamen of one plant to the stigma of another of the same variety. Where cross-pollination is necessary for fruiting, the plants must blossom at approximately the same time.

The propagation of fruit plants can take place by seeding, but with this method the new plants are usually different from the parent plant and from one another. This method is not preferred by fruit growers since it will take years for the new plant to produce fruit. More common propagation methods include a variety of forms of layering. In simple layering, a branch is bent and the tip is buried in the soil. After the branch has developed roots and a shoot, it is cut away and planted elsewhere. Other forms of layering are air layering, tip layering, trench layering, and mound layering.

Other propagation methods include stem or root cuttings, soft and hardwood cuttings, budding, and grafting. The advantage of these methods is that the type of fruit produced can be controlled. Budding, where a single bud is cut and placed under the **bark** of another tree, is usually done with fruit trees of the same variety. With older trees, grafting is usually done. In this method, a stem or branch from one tree is grafted onto another. The tree may produce more than one variety of fruit from this method. Micropropagation is a method of plant tissue propagation that can mass-produce plants that are identical, or clones. It also has the advantage of producing disease-free plants for fruit production.

Care of fruit plants

Some aspects of caring for fruit trees, besides ensuring they have the proper nutrients and sufficient water, is pruning and training of fruit plants, processes that help provide the proper amount of sunlight and make it easier to harvest the fruit crop. Keeping the plants **disease** and pest free is another aspect of the care of fruit plants.

Among the **pests** that can destroy a fruit crop are **aphids**, **slugs**, spider **mites**, scales, and other insect infestations. There are also a number of diseases that can afflict fruit plants, such as fireblight, brown rot, peach **leaf** curl, verticillium wilt, and powdery **mildew**. Some **wildlife species**, such as **deer**, **birds**, **moles**, **gophers**, **mice**, and rabbits can destroy the fruit crop either by stripping leaves from young trees, gnawing on the roots, or feasting on the fruit before it can be harvested. Fruit growers have developed a number of damage-control methods that include careful selection of plants, chemical deterrents, cutting away damaged parts of the plants, the use of nets to protect fruit, and wire mesh to protect the roots and bark of trees.

Economics of fruit production

Fruit harvesting has to be done either when the fruit is ripe if it is ready to be marketed or before it is ripe if it is to be transported or stored before marketing. Picking

---

KEY TERMS

Chlorophyll—Green pigment in a plant leaf that is involved in the process of photosynthesis.

Grafting—The process of attaching a branch from one tree onto another one for the purpose of propagating new fruit.

Herbaceous—A plant with the characteristics of an herb, it has little or no woody stems.

Layering—A method of growing new plants by rooting part of an older plant until it roots and forms a shoot and is then cut away and planted as a new plant.

Micropropagation—The production of new plants from plant tissue.

Pollination—The transfer of pollen from the male part of a flower to the female part.

Propagation—The production of new plants either from seed (sexual propagation) or by asexual methods like layering, grafting, and cutting.

---

fruit is, of course, labor-intensive work that requires great care in handling since the fruit can be easily damaged. While the growing and marketing of fruit such as apples, pears, peaches, and berries is done on a large scale to supply the needs of the large networks of supermarkets across the United States, many fruit **crops** are picked by the consumer. For example, in Illinois, the pick-your-own crop of strawberries has increased from 23% of the crop in 1967 to 86% in 1987. Many people grow fruit trees and berry bushes in their yards as well and benefit from the fruit they can pick by hand.

For the growers of fruit on a large scale, however, the storage of fruit is necessary to meet year-round market demands for fresh fruit. This requires that the fruit is picked before it has ripened and refrigerated until it is ready for marketing. Ideal conditions for storing apples, for example, are at 28–32°F (-2–0°C) with 90% **humidity**.

Resources

Books

Bilderback, Diane E., and Dorothy Hinshaw Patent. *Backyard Fruits & Berries.* Emmaus, PA: Rodale Press, 1984.

Brickell, Christopher, ed. *Encyclopedia of Gardening.* London: Dorling Kindersley, 1993.

Galletta, Gene J., and David G. Himelrick. *Small Fruit Crop Management.* Englewood Cliffs, NJ: Prentice Hall, 1990.

Otto, Stella. *The Backyard Orchardist.* Maple City, MI: Otto-Graphics, 1993.

Reich, Lee. *Uncommon Fruits Worthy of Attention*. Reading, MA: Addison-Wesley, 1991.

Vita Richman

Fuel cells

Fuel cells are a clean and quiet way to convert chemical-energy of fuels directly into **electricity**. Specifically, they transform **hydrogen** and **oxygen** into electric power, emitting **water** as their only waste product.

A fuel cell consists of two electrodes, an **anode** and a **cathode**, sandwiched around an **electrolyte**. (An electrolyte is a substance, usually liquid, capable of conducting electricity by means of moving ions [charged **atoms** or molecules]). The fuel—usually hydrogen—enters at the anode of the fuel cell while oxygen enters at the cathode. The hydrogen is split by a catalyst into hydrogen ions and electrons. Both move toward the cathode, but by different paths. The electrons pass through an external circuit, where they constitute electricity, while the hydrogen ions pass through the electrolyte. When the electrons return to the cathode, they are reunited with the hydrogen and the oxygen to form a **molecule** of water.

Fuel cells have several advantages: they are quiet, produce only water as a waste product, extract electricity from fuel more efficiently than combustion-boiler-generator systems. They can run on pure hydrogen—usually derived from methane by combining methane with steam at high temperature—or, in one recently developed design, on methane itself. **Biomass**, **wind**, solar power, or other renewable sources can supply **energy** to make hydrogen or other fuels for use in fuel cells, which could be installed in buildings (e.g., schools, hospitals, homes), in vehicles, or in small devices such as mobile phones or laptop computers. Fuel cells today are running on many different fuels, even gas from landfills and wastewater treatment plants.

The principles of the fuel cell were developed by Welsh chemist William Grove (1811–1896) in 1839. As early as 1900, scientists and engineers were predicting that fuel cells would be the primary source of electric power within a few years. It wasn't until the 1960s, however, when the U.S. National Aeronautics and Space Administration (NASA) chose the fuel cell to furnish power to its Gemini and Apollo spacecraft, that fuel cells received serious attention. Today, NASA still uses fuel cells to provide electricity and water (as a byproduct) for the **space shuttle**.

For years, experts predicted that fuel cells would eventually replace less-efficient gasoline engines and

Ford Motor Company zero emission fuel cell vehicle. *Ford Motor Company, Environement and Safety Public Affairs. Reproduced by permission.*

other clumsy, dirty devices for extracting energy from fuel. These predictions have yet to be fully realized, even though fuel cells are becoming more widely used. **Automobile** manufacturers are developing ways to extract hydrogen from **hydrocarbon** fuels in on-board devices, allowing a fuel-cell vehicle to run on methanol (as with Mercedes-Benz's and Toyota's prototypes) or even on gasoline, as Chrysler is proposing. DaimlerChrysler expects to produce a fuel-cell bus for the European market by 2003. The proposed 70-passenger bus will cost approximately $1.2 million and have a range of about 186 mi (300 km) and a top speed of 50 MPH (80 km/h).

Types of fuel cells

There are five basic types of fuel cells, differentiated by the type of electrolyte separating the hydrogen from the oxygen. The cells types now in use or under development are alkaline, **phosphoric acid**, **proton** exchange **membrane**, molten carbonate, and solid oxide.

Long used by NASA on space missions, alkaline cells can achieve power-generating efficiencies of up to 70%. NASA's fuel cells use alkaline potassium hydroxide as the electrolyte and the electrodes of porous **carbon**. At the anode, hydrogen gas combines with hydroxide ions to produce water vapor. This reaction results in extra electrons that are forced out of the anode to produce the **electric current**. At the cathode, oxygen and water plus returning electrons from the circuit form hydroxide ions that are again recycled back to the anode. The basic core of the fuel cell, consisting of the manifolds, anode, cathode, and electrolyte, is generally called the stack. Until recently, such cells were too costly for commercial applications, but several companies are examining ways to reduce costs and improve operating flexibility.

KEY TERMS

· ·

Anode—A positively charged electrode.

Cathode—A negatively charged electrode.

Cogeneration—The simultaneous generation of electrical energy and low-grade heat from the same fuel.

Electricity—An electric current produced by the repulsive force produced by electrons of the same charge.

Electrode—A conductor used to establish electrical contact with a nonmetallic part of a circuit.

Electrolyte—The chemical solution in which an electric current is carried by the movement and discharge of ions.

The fuel-cell type most commercially developed today is the phosphoric acid, now being used in such diverse settings as hospitals, nursing homes, hotels, office buildings, schools, utility power plants, and airport terminals. They can also be used in large vehicles such as buses and locomotives. Phosphoric-acid fuel cells generate electricity at more than 40% efficiency. If the steam produced is captured and used for heating, the efficiency jumps to nearly 85%. This compares to only 30% efficiency for the most advanced internal **combustion** engines. Phosphoric-acid cells operate at around 400°F (205°C).

Proton exchange membrane cells operate at relatively low temperatures (about 200°F [93°C]) and have high power **density**. They can vary their output quickly to meet shifts in power demand, and are suited for small-device applications. Experts say they are perhaps the most promising fuel cell for light-duty vehicles where quick startup is required.

Molten carbonate fuel cells promise high fuel-to-electricity efficiencies and the ability to consume coal-based fuels such as **carbon monoxide**. These cells, however operate at very high temperatures (1,200°F [650°C]) and therefore cannot be used in small-scale applications.

The solid oxide fuel cell could be used in big, high-power applications including industrial and large-scale central electricity generating stations. Some developers also see a potential for solid oxide use in motor vehicles. A solid oxide system usually uses a hard ceramic electrolyte instead of a liquid electrolyte, allowing operating temperatures to reach 1,800°F (980°C). Power generating efficiencies could reach 60%.

Direct methanol fuel cells (DMFC), relatively new members of the fuel cell family, are similar to the proton exchange membrane cells in that they both use a **polymer** membrane as the electrolyte. However, in the DMFC, the anode catalyst itself draws the hydrogen from the liquid methanol, eliminating the need for a fuel reformer. Efficiencies of about 40% are expected with this type of fuel cell, which would typically operate at a **temperature** between 120–190°F (50–90°C). Higher efficiencies are achieved at higher temperatures.

Regenerative fuel cells use sunlight as their energy source and water as a working medium. These cells would be attractive as a closed-loop form of power generation. Water is separated into hydrogen and oxygen by a solar-powered electrolyser. The hydrogen and oxygen are fed into the fuel cell, which generates electricity, **heat**, and water. The water is then recycled back into the system to be reused.

See also Alternative energy sources; Electric motor; Electric vehicles; Electrical conductivity; Electrical power supply.

Resources

Periodicals

"DaimlerChrysler Offers First Commercial Fuel Cell Buses to Transit Agencies, Deliveries in 2002." *Hydrogen & Fuel Cell Letter* (May 2000).

"Will Fuel Cells Power an Automotive Revolution?" *Design News* (June 22, 1998).

Other

Adam, David. "Bringing Fuel Cells Down to Earth." Nature: Science Update. March 24, 2000 [cited October 26, 2002]. <http://www.nature. com/nsu/000330/000330-3.html>.

"Beyond Batteries." Scientific American.com. December 23, 1996 [cited October 26, 2002]. <http://www.sciam.com/article.cfm?articleID=000103AE-74A1-1C76-9B81809E C588EF21>.

Raman, Ravi. "The Future of Fuel Cells in Automobiles." Penn State University, College of Earth and Mineral Sciences. May 7, 1999 [cited October 26, 2002]. <http://www.ems. psu.edu/info/explore/FuelCell.html>.

Laurie Toupin

Function

A function represents a mathematical relationship between two sets of **real numbers**. These sets of numbers are related to each other by a rule which assigns each value from one set to exactly one value in the other set. The standard notation for a function y = f(x), developed in the 18th century, is read "y equals f of x." Other representations of functions include graphs and tables. Functions are classified by the types of rules which govern their relationships including; algebraic, trigonometric, and logarithmic and exponential. It has been found by mathematicians and scientists alike that these elementary functions can represent many real-world phenomena.

History of functions

The idea of a function was developed in the seventeenth century. During this time, Rene Descartes (1596-1650), in his book *Geometry* (1637), used the concept to describe many mathematical relationships. The term "function" was introduced by Gottfried Wilhelm Leibniz (1646-1716) almost fifty years after the publication of *Geometry*. The idea of a function was further formalized by Leonhard Euler (pronounced "oiler" 1707-1783) who introduced the notation for a function, $y = f(x)$.

Characteristics of functions

The idea of a function is very important in **mathematics** because it describes any situation in which one quantity depends on another. For example, the height of a person depends on his age. The distance an object travels in four hours depends on its speed. When such relationships exist, one **variable** is said to be a function of the other. Therefore, height is a function of age and distance is a function of speed.

The relationship between the two sets of numbers of a function can be represented by a mathematical equation. Consider the relationship of the area of a **square** to its sides. This relationship is expressed by the equation $A = x^2$. Here, A, the value for the area, depends on x, the length of a side. Consequently, A is called the dependent variable and x is the independent variable. In fact, for a relationship between two variables to be called a function, every value of the independent variable must correspond to exactly one value of the dependent variable.

The previous equation mathematically describes the relationship between a side of the square and its area. In functional notation, the relationship between any square and its area could be represented by $f(x) = x^2$, where $A = f(x)$. To use this notation, we substitute the value found between the parenthesis into the equation. For a square with a side 4 units long, the function of the area is $f(4) = 4^2$ or 16. Using $f(x)$ to describe the function is a matter of tradition. However, we could use almost any combination of letters to represent a function such as $g(s)$, $p(q)$, or even $LMN(z)$.

The set of numbers made up of all the possible values for x is called the **domain** of the function. The set of numbers created by substituting every value for x into the equation is known as the range of the function. For the function of the area of a square, the domain and the range are both the set of all positive real numbers. This type of function is called a one-to-one function because for every value of x, there is one and only one value of A. Other functions are not one-to-one because there are instances when two or more independent variables correspond to the same dependent variable. An example of this type of function is $f(x) = x^2$. Here, $f(2) = 4$ and $f(-2) = 4$.

Just as we add, subtract, multiply or divide real numbers to get new numbers, functions can be manipulated as such to form new functions. Consider the functions $f(x) = x^2$ and $g(x) = 4x + 2$. The sum of these functions $f(x) + g(x) = x^2 + 4x + 2$. The difference of $f(x) - g(x) = x^2 - 4x - 2$. The product and quotient can be obtained in a similar way. A composite function is the result of another manipulation of two functions. The composite function created by our previous example is noted by $f(g(x))$ and equal to $f(4x + 2) = (4x + 2)^2$. It is important to note that this composite function is not equal to the function $g(f(x))$

Functions which are one-to-one have an inverse function which will "undo" the operation of the original function. The function $f(x) = x + 6$ has an inverse function denoted as $f^{-1}(x) = x - 6$. In the original function, the value for $f(5) = 5 + 6 = 11$. The inverse function reverses the operation of the first so, $f^{-1}(11) = 11 - 6 = 5$.

In addition to a mathematical equation, graphs and tables are another way to represent a function. Since a function is made up of two sets of numbers each of which is paired with only one other number, a graph of a function can be made by plotting each pair on an X,Y coordinate system known as the Cartesian coordinate system. Graphs are helpful because they allow you to visualize the relationship between the domain and the range of the function.

Classification of functions

Functions are classified by the type of mathematical equation which represents their relationship. Some functions are algebraic. Other functions like $f(x) = \sin x$, deal with angles and are known as trigonometric. Still other functions have logarithmic and exponential relationships and are classified as such.

Algebraic functions are the most common type of function. These are functions that can be defined using addition, **subtraction**, **multiplication**, **division**, powers, and roots. For example $f(x) = x + 4$ is an algebraic function, as is $f(x) = x/2$ or $f(x) = x^3$. Algebraic functions are called polynomial functions if the equation involves

KEY TERMS

. . . .

Dependent variable—The variable in a function whose value depends on the value of another variable in the function.

Independent variable—The variable in a function which determines the final value of the function.

Inverse function—A function which reverses the operation of the original function.

One-to-one function—A function in which there is only one value of x for every value of y and one value of y for every x.

Range—The set containing all the values of the function.

---

powers of x and constants. The most famous of these is the quadratic function (quadratic equation), $f(x) = ax^2 + bx + c$ where a, b, and c are constant numbers.

A type of function that is especially important in **geometry** is the trigonometric function. Common trigonometric functions are sine, cosine, tangent, secant, cosecant, and cotangent. One interesting characteristic of trigonometric functions is that they are periodic. This means there are an infinite number of values of x which correspond to the same value of the function. For the function $f(x) = \cos x$, the x values 90° and 270° both give a value of 0, as do 90° + 360° = 450° and 270° + 360° = 630°. The value 360° is the period of the function. If p is the period, then $f(x + p) = f(x)$ for all x.

Exponential functions can be defined by the equation $f(x) = b^x$, where b is any **positive number** except 1. The variable b is constant and known as the base. The most widely used base is an **irrational number** denoted by the letter e, which is approximately equal to 2.71828183. Logarithmic functions are the inverse of exponential functions. For the exponential function $y = 4^x$, the logarithmic function is its inverse, $x = 4^y$ and would be denoted by $y = f(x) = \log_4 x$. Logarithmic functions having a base of e are known as natural **logarithms** and use the notation $f(x) = \ln x$.

We use functions in a wide variety of areas to describe and predict natural events. Algebraic functions are used extensively by chemists and physicists. Trigonometric functions are particularly important in architecture, **astronomy**, and navigation. Financial institutions use exponential and logarithmic functions. In each case, the power of the function allows us to take mathematical ideas and apply them to real world situations.

Resources

Books

Kline, Morris. *Mathematics for the Nonmathematician.* New York: Dover Publications, 1967.
Larson, Ron. *Calculus With Analytic Geometry.* Boston: Houghton Mifflin College, 2002.
Paulos, John Allen. *Beyond Numeracy.* New York: Alfred A. Knopf Inc., 1991.

Perry Romanowski

Fundamental theorems

A fundamental **theorem** is a statement or proposition so named because it has consequences for the subject matter that are difficult to overestimate. Put another way, a fundamental theorem lies at the very heart of the subject. Mathematicians have designated one theorem in each main branch as fundamental to that branch.

Fundamental theorem of arithmetic

The fundamental theorem of **arithmetic** states that every number can be written as the product of **prime numbers** in essentially one way. For example, there are no prime factors of 30 other than 2, 3, and 5. You cannot **factor** 30 so that it contains 2s and a 7 or some other combination.

Fundamental theorem of algebra

The fundamental theorem of **algebra** asserts that every polynomial equation of degree n ≥ 1, with complex coefficients, has at least one solution among the complete numbers. An important result of this theorem says that the set of **complex numbers** is algebraically closed; meaning that if the coefficients of every polynomial equation of degree n are contained in a given set, then every solution of every such polynomial equation is also contained in that set. To see that the set of **real numbers** is not algebraically closed consider the origin of the **imaginary number** i. Historically, i was invented to provide a solution to the equation $x^2 + 1 = 0$, which is a polynomial equation of degree 2 with real coefficients. Since the solution to this equation is not a real number, the set of real numbers is not algebraically closed. That the complex numbers are algebraically closed, is of basic or fundamental importance to algebra and the solution of polynomial equations. It implies that no polynomial equation exists that would require the invention of yet another set of numbers to solve it.

Fundamental theorem of calculus

The fundamental theorem of **calculus** asserts that differentiation and integration are inverse operations, a

---

KEY TERMS

. . . .

Complex number—The set of numbers formed by adding a real number to an imaginary number. The set of real numbers and the set of imaginary numbers are both subsets of the set of complex numbers.

Composite number—A composite number is a number that is not prime.

Derivative—A derivative expresses the rate of change of a function, and is itself a function.

Integral—The integral of a function is equal to the area under the graph of that function, evaluated between any two points. The integral is itself a function.

Polynomial—An algebraic expression that includes the sums and products of variables and numerical constants called coefficients.

Prime number—Any number that is evenly divisible by itself and 1 and no other number is called a prime number.

fact that is not at all obvious, and was not immediately apparent to the inventors of calculus either. The **derivative** of a **function** is a measure of the **rate** of change of the function. On the other hand, the **integral** of a function from a to b is a measure of the area under the graph of that function between the two points a and b. Specifically, the fundamental theorem of calculus states that if F(x) is a function for which f(x) is the derivative, then the integral of f(x) on the **interval** [a,b] is equal to F(b) - F(a). The reverse is also true, if F(x) is continuous on the interval [a,b], then the derivative of F(x) is equal to f(x), for all values of x in the interval [a,b]. This theorem lies at the very heart of calculus, because it unites the two essential halves, differential calculus and integral calculus. Moreover, while both differentiation and integration involve the evaluation of limits, the limits involved in integration are much more difficult to manage. Thus, the fundamental theorem of calculus provides a means of finding values for integrals that would otherwise be exceedingly difficult if not impossible to determine.

Resources

Books

Bittinger, Marvin L., and Davic Ellenbogen. *Intermediate Algebra: Concepts and Applications.* 6th ed. Reading, MA: Addison-Wesley Publishing, 2001.

Hahn, Liang-shin. *Complex Numbers and Geometry.* 2nd ed. The Mathematical Association of America, 1996.

Immergut, Brita and Jean Burr Smith. *Arithmetic and Algebra Again.* New York: McGraw Hill, 1994.

Larson, Ron. *Calculus With Analytic Geometry.* Boston: Houghton Mifflin College, 2002.

Silverman, Richard A. *Essential Calculus with Applications.* New York: Dover, 1989.

J. R. Maddocks

Fungi

Fungi are one of the five kingdoms of organisms. Like higher plants (of the kingdom Plantae), most fungi are attached to the substrate they grow on. Unlike plants, fungi do not have **chlorophyll** and are not photosynthetic. Another key difference from plants is that fungi have **cell** walls composed of chitin, a **nitrogen** containing **carbohydrate**. All fungi have nuclei and the nuclei of most **species** are haploid at most times. Many species have two or more haploid nuclei per cell during most of the life cycle. All fungi reproduce asexually by **spore** production. Most species reproduce sexually as well.

General characteristics

The different taxonomic groups of fungi have different levels of cellular organization. Some groups, such as the yeasts, consist of single-celled organisms, which have a single nucleus per cell. Some groups, such as the conjugating fungi, consist of single-celled organisms in which each cell has hundreds or thousands of nuclei. Groups such as the **mushrooms**, consist of multicellular, filamentous organisms which have one or two nuclei per cell. These multicellular fungi are composed of branched filaments of cells called hyphae. The hyphae, in turn, often mass together to form a **tissue** called mycelium.

Mycology, the study of fungi, has traditionally included groups such as the cellular **slime molds**, plasmodial slime molds, **water** molds, chytrids, and several other groups of fungus-like organisms. Most modern biologists consider these groups as diverse assemblages of organisms unrelated to the true fungi considered here. However, it should be emphasized that biologists are very uncertain about the evolutionary relationships of these other groups and the true fungi.

Nutrition and ecology

Most species of fungi grow on land and obtain their **nutrients** from dead organic **matter**. Some fungi are symbionts or **parasites** on other organisms. The majority of species feed by secreting enzymes, which partially digest the food extracellularly, and then absorbing the partially digested food to complete digestion internally. As with animals, the major storage carbohydrate of fungi is glycogen. Fungi lack the complex vascular system found in higher plants, so their transport of food and water is less efficient.

Along with **bacteria**, fungi have an important ecological role in the **decomposition** of dead plants, animals, and other organic matter. Thus, fungi are ecologically important because they release large amounts of **carbon dioxide** into the atmosphere and recycle nitrogen and other important nutrients within ecosystems for use by plants and other organisms. Some fungi are parasitic, in that they obtain their nutrients from a living host **organism**, a relationship which usually harms the host. Such parasitic fungi usually have specialized tissues called haustoria, which penetrate the host's body. Most of the diseases which afflict agricultural plants are caused by parasitic fungi. Some examples are corn smut, black stem rust of **wheat** and **barley**, and **cotton** root rot. Some species of fungi can also parasitize animals. Even humans can be parasitized by fungi which cause diseases such as athlete's foot, ringworm, and **yeast** infections.

Evolution

The Fungi constitute a large and diverse group of organisms. Until the 1960s, fungi were considered members of the **plant** kingdom. With the advent of the five-kingdom system of biological classification, fungi were assembled into a single kingdom because of their similar ecological roles as primary decomposers of organic matter and their similar anatomical and biochemical features. Recent studies that compare the sequence of amino acids in **proteins** from fungi, plants, and animals now indicate that fungi share a closer evolutionary relationship to animals than to plants.

The evolutionary ancestry and relationships of the different fungi are not well understood. There are few fossils of fungi, presumably because their relatively soft tissues are not well preserved. There is some fossil evidence that they existed in the Precambrian era (over four billion years ago), although identification of these very early fossils is uncertain. There is definite fossil evidence for fungi in the lower Devonian (about 400 million years ago) period. Fossils of all the major groups of fungi are found in the Carboniferous period (about 300 million years ago). By the late Tertiary period (about 20 million years ago), the fossil record shows a rapid and divergent evolution of fungi. Many of the Tertiary fossils of fungi are similar to existing species.

In addition to the incomplete fossil record, there are at least two other reasons why the evolutionary relationships of fungi are not well-known: they tend to have simple morphologies and they lack embryos which follow a definite sequence of developmental stages. Biologists often use these two features to determine evolutionary relationships of animals and plants.

Another complication is that many fungi presumably evolved similar morphologies by convergent evolution. That is, unrelated species may share a common morphology because they have been subjected to similar selective pressures.

The relatively new technique of molecular systematics is particularly useful in the study of the **evolution** of fungi. This technique compares the sequence of DNA segments of different species to determine evolutionary relationships. One important finding from this new technique is that the plasmodial slime molds, cellular slime molds, and water molds are only distantly related to the true fungi (the taxonomic groups considered here). Evolutionary relationships among organisms also are being studied by comparing the sequences of ribosomal RNAs and transfer RNAs from different organisms. Although these RNAs are similar to DNA, they have structural roles in cells, rather than coding for proteins as do DNA and messenger RNAs. Therefore, the sequences of ribosomal and transfer RNAs tend to be more conserved through evolution. Comparison of these RNAs among fungi, plants, and animals also suggests that fungi are more closely related to animals than to plants. Future work in molecular systematics is expected to tell us more about the evolution and relationships of the fungi.

Classification

Biologists have estimated that over 200,000 species of fungi exist in nature, although only about 100,000 have been identified so far. Since classification schemes of organisms are usually based on evolutionary relationships, and the evolutionary relationships of fungi are not well known, biologists have proposed numerous classification schemes for fungi over the years. Below, we consider the five major phyla that nearly all mycologists would agree belong in the kingdom of Fungi.

Zygomycota, conjugating fungi

Species in this phylum reproduce sexually by forming a zygospore, a thick-walled, diploid cell which contains thousands of nuclei. There are about 600 species in this phylum. Most species are terrestrial and feed on organic matter, although there are a few parasitic species. The conjugating fungi are coenocytic, in that they have a continuous mycelium, containing hundreds or thousands of haploid nuclei, with no divisions between them. However, the

Zygomycota do have septa (cross walls) between their reproductive structures and the rest of their mycelium.

The conjugating fungi have a life cycle that includes a sexual phase and an asexual phase. In the asexual phase, thousands of spores develop inside a sporangium, a small spherical structure. The sporangium grows on the tip of a sporangiophore, a specialized aerial hypha, typically about as thin as a hair.

In the sexual phase of their life cycle, these fungi form specialized hyphae, called gametangia, which are of two different strains (sexes), plus and minus. The plus and minus strains are very similar morphologically, but differ physiologically and biochemically. Plus and minus gametangia conjugate with one another and form a structure with hundreds or thousands of nuclei from each strain.

Then, a thick-walled structure, called the zygospore, develops from the conjugated gametangia. Inside the zygospore, the many thousands of nuclei from the plus and minus strains pair off and fuse together to form thousands of diploid nuclei. The zygospore is typically spherical in shape and has a thick, dark outer wall. It usually remains dormant for several months or more before development continues.

As the zygospore germinates, it produces germsporangia which are born on germsporangiophores, structures morphologically similar to the asexual sporangium and sporangiophore (see above). The germsporangium contains thousands of haploid germspores which arose from the diploid nuclei of the zygospore by **meiosis**. Each germspore is liberated, germinates, and gives rise to a new haploid mycelium.

One of the best known of the conjugating fungi is *Phycomyces blakesleeanus*, a species which grows on **animal** feces in nature. The sporangiophores of *Phycomyces* respond to a variety of sensory stimuli. For example, they bend in response to **light (phototropism)**, gravity (gravitropism), **wind** (anemotropism), and nearby objects (avoidance response). Physiologists and biophysicists have intensively studied the response to light. One important finding is that the light sensitivity of the sporangiophore is about the same as the eyes of humans. Furthermore, like humans, the sporangiophore can adapt to a one-billion-fold change in ambient light intensity. One of the pigments involved in the extraordinary light responses of *Phycomyces* is a flavin (**vitamin** B2) bound to a special protein. This pigment is commonly called the blue light photoreceptor, since it is most sensitive to blue light.

Ascomycota, sac fungi

Species in this phylum reproduce sexually by forming a spore-filled structure called an ascus, which means

An American fly agaric (*Amanita muscaria formosa*). This mushroom is very common in all of North America, but is more slender, tinged with a salmon-like coloration, and somewhat more rare in the southern states. *Photograph by Robert J. Huffman. Field Mark Publications. Reproduced by permission.*

literally "a sac." The hyphae of the sac fungi are divided by septa with pores, that is, they have perforated walls between adjacent cells. They reproduce asexually by producing spores, called conidia, which are born on specialized erect hyphae, called conidiophores. The sac fungi are typically prolific producers of conidia.

The sac fungi also have a **sexual reproduction** phase of their life cycles. In the first step of this process, compatible hyphae fuse together by one of several different methods. Second, the nuclei from the different hyphae move together into one cell to form a dikaryon, a cell with two haploid nuclei. Third, several cell divisions occur, resulting in several cells with two different haploid nuclei per cell. Fourth, **nuclear fusion** of the two haploid nuclei occurs in one of these cells, the ascus mother cell. Fifth, the ascus mother cell develops into an ascus. Then, meiosis occurs in the diploid cells and, depending on the species, four or eight haploid ascospores form inside the ascus. In some species, such as the fleshy and edible morels, a large number of asci are massed together to form an ascocarp.

This large phylum of fungi includes many species which are beneficial to humans. For example, the yeasts are a major group of ascomycetes. Different yeasts in the genus *Saccharomyces* are employed by bakers, brewers, and vintners to make their bread, beer, or wine. Truffles are subterranean ascomycetes which grow in association with **tree** roots. Traditionally, **pigs** have been used to sniff out these underground fungi, so that French chefs could use truffles to complement their finest cuisine.

Some other ascomycetes are significant plant **pathogens**. For example, *Endothia parasitica* is an ascomycete which causes **chestnut** blight, a **disease** which virtually extirpated the American chestnut as a mature forest tree. *Ceratocystis ulmi* is a pathogenic ascomycete which causes Dutch **elm** disease, a scourge of American elm trees. *Claviceps purpurea*, the ergot fungus, infects agricultural grains, and when ingested can cause intense hallucinations or death due to the presence of LSD (D-Lysergic acid diethylamide).

Another well known ascomycete is *Neurospora crassa*, the red bread **mold**. The ordered manner in which the eight spores of this fungus align during sexual reproduction allows geneticists to construct a map of the genes on its chromosomes. Earlier in this century, biologists used *Neurospora* as a model organism to investigate some of the basic principles of **genetics** and heredity. More recently, biologists have shown that the mycelium of this species can produce spores at approximately 24 hour intervals, a circadian rhythm, in a constant environment. Many biologists are currently using *Neurospora crassa* as a model organism for investigation of circadian rhythms, which occur in a wide diversity of organisms, including humans.

Basidiomycota, club fungi

Species in this phylum reproduce sexually by forming spores on top of club-shaped structures called basidia. The club fungi are believed to be closely related to the sac fungi. Both groups have cells which are separated by septa (walls), and both have a dikaryotic phase in their life cycle; a phase with two haploid nuclei per cell. The septum of the club fungi is somewhat different from those of sac fungi and is referred to as a dolipore septum. The dolipore septum has a bagel-shaped pore in its center.

The club fungi reproduce asexually by producing asexual spores or by fragmentation of mycelium.

The sexual reproduction phase of the club fungi involves three developmental stages of the mycelium. In the primary stage, a haploid spore germinates and grows a germ tube, which develops into mycelium. The mycelium initially contains a single haploid nucleus. Then, its haploid nucleus divides and septa form between the nuclei.

A secondary mycelium forms upon conjugation of two sexually compatible hyphae. The secondary mycelium is dikaryotic, in that it has two haploid nuclei, one from each parent. As the dikaryotic mycelium grows, the cells divide and more septa are formed between the new cells.

Each of the new cells in the secondary mycelium has one haploid nucleus from each parent. This is assured by clamp connections, specialized structures unique to the club fungi. These are loop-like hyphae which connect the cytoplasm of adjacent cells and through which nuclei move during **cell division**. In particular, during cell division, one nucleus divides directly into the newly formed cell; the other nucleus divides inside the clamp connection and the two daughter nuclei migrate through the clamp connection in opposite directions to the two daughter cells.

The tertiary mycelium is simply an organized mass of secondary mycelium. It is a morphologically complex tissue and forms structures such as the typically mushroom-shaped basidiocarps commonly seen in nature.

Sexual reproduction of the club fungi begins upon fusion of two primary hyphae to form a club-shaped structure, known as a basidium. Second, the two haploid nuclei inside the basidium fuse together to form a diploid zygote. Third, the zygote undergoes meiosis to form two haploid nuclei. Fourth, these two haploid nuclei undergo **mitosis** to form a total of four haploid nuclei. These four nuclei then migrate into projections, which form on the tip of the basidium. These projections then develop into four separate haploid spores, each with a single nucleus.

In the species of club fungi which are large and fleshy, such as the mushrooms, a mass of basidia form a structure called a basidiocarp. The spores on the basidia are released from the underside of the fleshy gills of the mushroom. The **color** and shape of the basidiocarp, as well as the color of the spores are often diagnostic for species identification.

This large phylum includes species which are known as mushrooms, toadstools, earthstars, stinkhorns, puffballs, jelly fungi, coral fungi, and many other interesting common names. Some species, such as the **rusts and smuts**, are pathogens which attack agricultural grains. Other species, such as the fly agaric (*Agaricus muscaria*) and some species in the genus *Psilocybe*, produce chemical **hallucinogens** and have been used by numerous cultures in their religious ceremonies. Another species, *Agaricus bisporus*, is the common edible mushroom found in supermarkets.

An important aspect of the club fungi is the great diversity of alkaloids and other toxic and psychogenic chemicals produced by some species. For example, *Amanita virosa*, a mushroom colloquially known as "death angel," is so deadly poisonous that a small bite can kill a person. A related mushroom is *Amanita muscaria*, known as "fly agaric," which is hallucinogenic. Over the millennia, numerous cultures have eaten the fly agaric as part of their religious ceremonies. For example, R. Gordon Wasson has shown that *Amanita muscaria* is the hallucinogenic plant referred to as "Soma" throughout *Rg Veda,* the ancient religious text. According to *Rg Veda,* the ancient Aryans who invaded India about four millennia ago ingested "Soma" as a euphoriant.

While mushrooms are the best-known club fungi, many other club fungi grow underground as mycorrhizae. Mycorrhizae result from a **symbiosis** between a plant root and a fungus. In mycorrhizae, the fungus typically supplies nitrogen-containing compounds to the plant, and the plant supplies carbohydrates and other organic compounds to the fungus. Mycorrhizae are very important for the growth of orchids. One reason many orchids are difficult to grow is because they require particular fungal species to form mycorrhizae on their roots.

A recent report investigated a subterranean club fungus, *Armillaria bulbosa,* which is a pathogen on tree roots. The investigators used **molecular biology** techniques to demonstrate that a single subterranean "individual" of this species in Northern Michigan was spread out over 37 acres (15 ha) and weighed an estimated 22,000 lb (10,000 kg). Based on the estimated growth **rate** of this species, of about 0.7 ft (0.2 m) per year, this individual was about 1,500 years old.

Deuteromycota, imperfect fungi

The Deuteromycota is a heterogeneous group of unrelated species in which sexual reproduction has never been observed. Since mycologists refer to the "perfect phase" of a life cycle as the phase in which sexual reproduction occurs, these fungi are often referred to as imperfect fungi. These fungi may have lost their sexual phase through the course of evolution. Alternatively, biologists simply may not have found the appropriate environmental conditions to observe development of the sexual phase of their life cycle.

The Deuteromycota are classified as fungi for two main reasons. First, their multicellular tissue is similar to the hyphae of sac fungi and club fungi. Second, they have erect hyphae with asexual spores, called conidiophores, which are similar to those of the sac fungi and club fungi.

Most imperfect fungi are believed to be related to the sac fungi because their conidiophores closely resemble those produced by the sac fungi during their sexual phase. The imperfect fungi are not placed in the Ascomycota phylum because classification of that group is based on the morphology of sexual structures which the Deuteromycota do not have.

The best known fungus in this phylum is *Penicillium.* Some species in this genus appear as pathogenic, blue-green molds on **fruits**, **vegetables**, and cheeses. Several other species are important for the making of cheeses, such as blue cheese, Roquefort, and Camembert. Certainly the best known product from this genus is penicillin, the first widely-used antibiotic. Penicillin was first discovered in *Penicillium notatum* over 50 years ago, but is now known to be produced by many other species in this genus.

Mycophycophyta, lichens

A lichen is a symbiotic relationship between a fungus and an alga, or between a fungus and a photosynthetic cyanobacterium. They constitute a very diverse and polyphyletic group of organisms and are classified together simply because they all result from a fungus-alga symbiosis. In most **lichens**, the fungal species is in the Ascomycota phylum and the photosynthetic species is a green alga from the Chlorophyta phylum. Typically, the photosynthetic species supplies carbohydrates to the fungus and the fungus supplies nitrogen and other nutrients to the alga. The morphology of a lichen differs from its component species.

Lichens can reproduce by several methods. The fungal component of the lichen can produce spores which are dispersed, germinate, and then recombine with the algal component. Alternatively, the lichen can produce soredia, specialized reproductive and dispersal structures in which the algal component is engulfed by fungal mycelium. Typically, the soredia break off from the thallus, the main body of the lichen.

Ecologists have shown that many species of lichens are very sensitive to air pollutants, such as **sulfur dioxide**. Thus, they are often used as **indicator species** for **air pollution**; the presence of certain lichen species correlates with the cleanliness of the air.

Many lichens can inhabit harsh environments and withstand prolonged periods of desiccation. In the temperate region of **North America**, lichens often grow on tree trunks and bare **rocks** and **soil**. In the arctic and antarctic regions, lichens constitute a large proportion of the **ecosystem biomass**. Many lichens are even found growing upon and within rocks in **Antarctica**. In the arctic region, the lichen species known colloquially as reindeer mosses (*Cladonia rangifera* and several other species) are an important food for **caribou** and reindeer.

KEY TERMS

. . . .

Biomass—Total weight, volume, or energy equivalent of all living organisms within a given area.

Clamp connection—Loop-like hypha which connects the cytoplasm of adjacent cells. Characteristic feature of Basidiomycota.

Coenocytic—Lacking walls for separation of the nuclei of cytoplasm.

Cyanobacteria (singular, cyanobacterium)—Photosynthetic bacteria, commonly known as blue-green alga.

Diploid—Nucleus or cell containing two copies of each chromosome, generated by fusion of two haploid nuclei.

Haploid—Nucleus or cell containing one copy of each chromosome.

Hypha (plural, hyphae)—Cellular unit of a fungus, typically a branched and tubular filament. Many strands (hyphae) together are called mycelium.

Mycorrhiza—Subterranean symbiotic relationship between a fungus, typically a species of Basidiomycota, and a plant root.

Phylum—Broadest taxonomic category within a kingdom.

Septum—Wall that separates the cells of a fungal hypha into segments.

Symbiogenesis—Evolutionary origin of a completely new life form from the symbiosis of two or more independent species.

Symbiosis—A biological relationship between two or more organisms that is mutually beneficial. The relationship is obligate, meaning that the partners cannot successfully live apart in nature.

Studies of the symbiotic nature of lichens in the late 1800s laid an important foundation for development of the theory of symbiogenesis. This theory says that new life forms can evolve from the symbiotic relationship of two or more independent species. Nearly all modern biologists now agree that symbiogenesis of different bacteria led to the origin of eukaryotic cells, which contain many different organelles, intracellular" small organs" which are specialized for different functions.

Resources

Books

Griffin, D.H. *Fungal Physiology.* New York: Wiley-Liss, 1993.

Margulis, L., and K.V. Schwartz. *Five Kingdoms.* New York: W. H. Freeman and Company, 1988.

Soothill, E., and A. Fairhurst. *The New Field Guide to Fungi.* Transatlantic Arts, 1993.

Peter A. Ensminger

Fungicide

A fungus is a tiny plant-like **organism** that obtains its nourishment from dead or living organic **matter**. Some examples of **fungi** include **mushrooms**, toadstools, smuts, molds, rusts and **mildew**.

Fungi have long been recognized as a serious threat to plants and **crops**. They attack food both while it is growing and after it has been harvested and placed in storage. One of the great agricultural disasters of the second half of the twentieth century was caused by a fungus. In 1970, the fungus that causes southwest corn-leaf blight swept through the southern and Midwestern United States and destroyed about 15% of the nation's corn crop. **Potato** blight, **wheat** rust, wheat smut, and grape mildew are other important diseases caused by fungi.

Chestnut blight is another example of the devastation that can be caused by fungi. Until 1900, chestnut trees were common in many parts of the United States. In 1904, however, chestnut trees from **Asia** were imported and planted in parts of New York. The imported trees carried with them a fungus that attacked and killed the native chestnut trees. Over a period of five decades, the native trees were all but totally eliminated from the eastern part of the country.

It is hardly surprising that humans began looking for fungicides (substances that will kill or control the growth of fungi) early in history. The first of these fungicides was a naturally occurring substance, **sulfur**. One of the most effective of all fungicides, Bordeaux mixture, was invented in 1885. Bordeaux mixture is a combination of two inorganic compounds, **copper** sulfate and lime.

With the growth of the chemical industry during the twentieth century, a number of synthetic fungicides have been developed: these include ferbam, ziram, naban, dithiocarbonate, quinone, and 8-hydroxyquinoline. For a period of time, compounds of mercury and cadmium were very popular as fungicides. Until quite recently, for example, the compound methylmercury was widely used by farmers in the United States to protect growing plants and treat stored grains. During the 1970s, however, evidence began to accumulated about a number of adverse effects of mercury- and cadmium-based fungicides. The most serious effects were observed among **birds** and small animals who were exposed to sprays and dusting or who ate treated grain. A few dramatic incidents of methylmercury poisoning among humans, however, were also recorded. The best known of these was the 1953 disaster at Minamata Bay, Japan. At first, scientists were mystified by an **epidemic** that spread through the Minamata Bay area between 1953 and 1961. Some unknown factor caused serious nervous disorders among residents of the region. Some sufferers lost the ability to walk, others developed mental disorders, and still others were permanently disabled. Eventually researchers traced the cause of these problems to methylmercury in **fish** eaten by residents in the area.

As a result of the problems with mercury and cadmium compounds, scientists have tried to develop less toxic substitutes for the more dangerous fungicides. Dinocap, binapacryl, and benomyl are three examples of such compounds.

Another approach has been to use **integrated pest management** and to develop plants that are resistant to fungi. The latter approach was used with great success during the corn blight disaster in 1970. Researchers worked quickly to develop strains of corn that were resistant to the corn-leaf blight fungus and by 1971 had provided farmers with **seeds** of the new strain.

See also Agrochemicals; Herbicides; Pesticides.

Fusion *see* **Nuclear fusion**

G

Gadolinium *see* **Lanthanides**

Gaia hypothesis

Gaia, **Earth**, was believed by the ancient Greeks to be a living, fertile ancestor of many of their important gods. The Romans, who adopted many Greek gods and ideas as their own, also believed in this organismic entity, who they renamed *Terra*. The Gaian notion has been personified in more recent interpretations as "Mother Earth." The Gaia hypothesis is a recent and highly controversial theory that views Earth as an integrated, pseudo-organismic entity and not as a mere physical object in **space**. The Gaia hypothesis suggests that organisms and ecosystems on Earth cause substantial changes to occur in the physical and chemical nature of the environment, in a manner that improves the living conditions on the **planet**. In other words, it is suggested that Earth is an organismic planet, with homeostatic mechanisms that help to maintain its own environments within the ranges of extremes that can be tolerated by life.

Earth is the only planet in the universe that is known to support life. This is one of the reasons why the Gaia hypothesis cannot be tested by rigorous, scientific experimentation-there is only one known replicate in the great, universal experiment. However, some supporting evidence for the Gaia hypothesis can be marshaled from certain observations of the structure and functioning of the planetary **ecosystem**. Several of these lines of reasoning are described in the next section.

Evidence in support of a Gaian Earth

One supporting line of reasoning for the Gaia hypothesis concerns the presence of **oxygen** in Earth's atmosphere. It is believed by scientists that the primordial atmosphere of Earth did not contain oxygen. The appearance of this gas required the evolution of photosynthetic life forms, which were initially blue-green **bacteria** and, somewhat later, single-celled **algae**. Molecular oxygen is a waste product of **photosynthesis**, and its present atmospheric **concentration** of about 21% has entirely originated with this biochemical process (which is also the basis of all biologically fixed **energy** in ecosystems). Of course, the availability of atmospheric oxygen is a critically important environmental factor for most of Earth's **species** and for many ecological processes.

In addition, it appears that the concentration of oxygen in the atmosphere has been relatively stable for an extremely long period of time, perhaps several billions of years. This suggests the existence of a long-term equilibrium between the production of this gas by green plants, and its consumption by biological and non-living processes. If the atmospheric concentration of oxygen were much larger than it actually is, say about 25% instead of the actual 21%, then **biomass** would be much more readily combustible. These conditions could lead to much more frequent and more extensive forest fires. Such conflagrations would be severely damaging to Earth's ecosystems and species.

Some proponents of the Gaia hypothesis interpret the above information to suggest that there is a planetary, homeostatic control of the concentration of molecular oxygen in the atmosphere. This control is intended to strike a balance between the concentrations of oxygen required to sustain the **metabolism** of organisms, and the larger concentrations that could result in extremely destructive, uncontrolled wildfires.

Another line of evidence in support of the Gaian theory concerns **carbon dioxide** in Earth's atmosphere. To a substantial degree, the concentration of this gas is regulated by a complex of biological and physical processes by which **carbon** dioxide is emitted and absorbed. This gas is well known to be important in the planet's **greenhouse effect**, which is critical to maintaining the average **temperature** of the surface within a range that organisms can tolerate. It has been estimated that in the absence of this greenhouse effect, Earth's average surface temperature would be about -176°F (-116°C), much too cold for organisms and ecosystems to tolerate over the longer term. Instead, the existing greenhouse effect, caused in large part by atmospheric carbon dioxide, helps to maintain an average surface temperature of about 59°F (15°C). This is within the range of temperature that life can tolerate.

Again, advocates of the Gaia hypothesis interpret these observations to suggest that there is a homeostatic system for control of atmospheric carbon dioxide, and of climate. This system helps to maintain conditions within a range that is satisfactory for life.

Scientists agree that there is clear evidence that the non-living environment has an important influence on organisms, and that organisms can cause substantial changes in their environment. However, there appears to be little widespread support within the scientific community for the notion that Earth's organisms and ecosystems have somehow integrated in a mutually benevolent **symbiosis** (or **mutualism**), aimed at maintaining environmental conditions within a comfortable range.

Still, the Gaia hypothesis is a useful concept, because it emphasizes the diverse connections of ecosystems, and the consequences of human activities that result in environmental and ecological changes. Today, and into the foreseeable future, humans are rapidly becoming a dominant force that is causing large, often degradative changes to Earth's environments and ecosystems.

See also Biosphere; Chemical evolution; Ecological pyramids; Ecosystem; Homeostasis; Origin of life.

Resources

Books

Lovelock, J. *The Ages of Gaia: A Bibliography of Our Living Earth.* New York: Norton & Co., 1988.

Margulis, L., and L. Olendzenski. *Environmental Evolution. Effects of the Origin and Evolution of Life on Planet Earth.* Cambridge, MA: MIT Press, 1992.

Smith, L. E. *Gaia. The Growth of an Idea.* New York: St. Martin's Press, 1991.

Periodicals

Huggett, R.J. "Ecosphere, Biosphere, Or Gaia? What To Call The Global Ecosystem." *Global Ecology And Biogeography* 8, no. 6 (1999): 425-432.

Bill Freedman

Galaxy

A galaxy is a large collection of stars similar to the **Milky Way** galaxy in which our **solar system** is located.

Astronomers classify galaxies according to their shape as either **spiral**, elliptical, or irregular. Spiral galaxies are further subdivided into normal and barred spirals. Elliptical galaxies can be either giant or dwarf ellipticals, depending on their size.

Galaxies can contain anywhere from a few million stars, for dwarf ellipticals, to a few trillion stars, for giant ellipticals or spirals. Galaxies emitting far more **energy** than can easily be explained by a collection of stars are classified as active galaxies. The study of other galaxies in addition to being intrinsically interesting both helps us understand our own Milky Way galaxy and gives us clues to understanding the universe as a whole.

Outside of the galaxy

Astronomers did not recognize galaxies as separate from the Milky Way until the early part of the twentieth century. The Andromeda Galaxy, which is the nearest spiral galaxy to the Milky Way and the Large and Small Magellanic Clouds, which are the nearest irregular galaxies to the Milky Way, are visible to the naked **eye**, and have therefore been observed since antiquity. Their nature was, however, unknown.

With the development of the **telescope**, astronomers were able to discern the whorled shape of spiral galaxies, which were called spiral nebulae at the time. Until the 1920s, there was a controversy: Were these "spiral nebulae" part of our Milky Way galaxy, or were they external galaxies similar to our Milky Way? In April 1920, there was a debate on this topic between Harlow Shapley and Heber D. Curtis before the National Academy of Sciences. Curtis argued that spiral nebulae were external galaxies, Shapley that they were part of the Milky Way. Curtis did not win the debate, but **astronomy** has since proven him right—"spiral nebulae" are external galaxies similar to the Milky Way.

To settle the controversy, scientists needed an accurate method to gauge the distance to galaxies. Working at Harvard College Observatory in the early twentieth century, the American astronomer Henrietta Leavitt (1868-1921) found the required celestial yardstick. Leavitt was studying a type of **star** in the Magellanic Clouds known as a Cepheid variable, when she discovered a way to measure the distance to any Cepheid variable by comparing the star's apparent and absolute magnitudes. The distance to the variable star gave the distance to the galaxy or cluster of stars containing the Cepheid variable. Cepheid variables have since become a fundamental yardstick for measuring the distance scale of the universe.

In 1924, the American astronomer Edwin Hubble (1889-1953) used Leavitt's Cepheid variable technique to measure the distance to the Andromeda galaxy. Hubble's original distance estimates have since been refined; the modern distance to the Andromeda galaxy is about 2.2 million **light** years. (A light year is the distance light can travel in one year, about 6 trillion mi, or 9.654 trillion km). The Milky Way galaxy is however only a little over 100,000 light years in diameter. Hubble therefore conclusively proved that the Andromeda galaxy must be outside the Milky Way. Other galaxies are more distant.

With his work, Hubble launched the science of extragalactic astronomy—the study of galaxies outside the Milky Way. Hubble devised the classification scheme for galaxies that astronomers still use today. More importantly, Hubble found that more distant galaxies are moving away from us at a faster **rate**. From this observation, known as Hubble's law, he deduced that the universe is expanding. Hubble used his study of galaxies to uncover a fundamental fact about the nature of the universe. Fittingly, one of the scientific goals of the Hubble's namesake, the **Hubble Space Telescope**, is to continue this work.

Classification of galaxies

Hubble classified the galaxies he observed according to their shape. His scheme is still in use today. The basic regular shapes are elliptical and spiral. He classified galaxies with no regular shape as irregular galaxies. Galaxies that basically look like either elliptical or spiral galaxies but have some unusual feature are classified as peculiar galaxies. They are classified according to the closest match in the classification scheme then given the added designation peculiar (pec). Hubble initially thought that his classification scheme represented an evolutionary sequence for galaxies; they started as one type and gradually evolved into another type.

Modern astronomers have supplemented Hubble's original scheme with luminosity classes. The luminosity of a galaxy is its total energy output each second. Note that the luminosity refers to the intrinsic energy output of the galaxy corrected for the distance of the galaxy. Therefore a high luminosity but distant galaxy might appear fainter than a nearby low luminosity galaxy. The luminosity classes are the roman numerals I, II, III, IV, and V. The most luminous galaxies are class I, and the least luminous are V. As one might guess, the more luminous galaxies are generally larger in size and contain more stars.

How common are the various types of galaxies? In a given **volume** of **space**, about one third of all the galaxies (34%) are spirals, a little over half (54%) are irregulars, and the rest (12%) are ellipticals. However irregular and elliptical galaxies tend to be smaller and fainter on

A photograph of the Andromeda Galaxy (M31). Also seen, as the bright spot below and to the left of Andromeda, is one of its two dwarf elliptical satellite galaxies, M32 (NGC 221). Andromeda is a spiral galaxy some 2.2 million light years from our own galaxy, the Milky Way. It measures some 170,000 light years across, and as the largest of the nearby galaxies is faintly visible even to the naked eye. *Photograph by Tony Ward. Photo Researchers, Inc. Reproduced by permission.*

the average than spiral galaxies. They are therefore harder to find. Of the galaxies that we can observe the overwhelming majority (77%) are spirals and only 3% are irregular galaxies. The remaining 20% of observed galaxies are ellipticals.

Elliptical galaxies

Elliptical galaxies have a three-dimensional ellipsoidal shape, so they appear in their two dimensional projections on the sky as ellipses. In his scheme, Hubble denoted elliptical galaxies with the letter E. He further subdivided ellipticals according to the amount of elongation of the **ellipse**, using numbers from 0 to 7. An E0 galaxy appears spherical. The most elongated elliptical galaxies are E7. The E1 through E6 galaxies are intermediate.

Note that this classification is based on the appearance of a galaxy, which may be different from its true shape owing to projection effects. Since Hubble's time, astronomers have learned that some ellipticals are relatively small and others are large. We now have the additional classification of either dwarf ellipticals or giant ellipticals. For finer divisions astronomers use the luminosity classes I, for the supergiant ellipticals, down to V for the smallest dwarf ellipticals.

Dwarf elliptical galaxies tend to be fairly small. They average about 30,000 light years in diameter, but can be as small as about 10,000 light years. The diameters of galaxies are a little uncertain because galaxies do not end sharply. Instead, they tend to gradually fade out with increasing distance from the center. By contrast, Giant elliptical galaxies average about 150,000 light years in diameter. The largest supergiant ellipticals are a few million light years in diameter.

The dwarf ellipticals have masses ranging from 100,000 to 10 million times the **mass** of the **sun**, suggesting that they have about that many stars. Giant ellipticals on the other hand will typically have 10 trillion times the mass of the sun and therefore roughly that many stars. Both giant and dwarf elliptical galaxies have only old stars and very small amounts of the interstellar gas and dust that is the raw material for forming new stars, probably due to the loss of gas clouds to **star formation** during the collisions that formed the elliptical shape.

Spiral galaxies

Spiral galaxies have a disk shape with a bulging central nucleus, so that they look like an astronaut's pancake floating in midair with a fried egg in the center on both sides. Surrounding the disk is a spherical halo consisting of globular clusters—spherical clusters of roughly 100,000 stars each. The astronaut's breakfast has drops of syrup floating in a spherical distribution around the pancake.

The disk of a spiral galaxy contains the spiral arms that give class of galaxy its name. There are usually two spiral arms that **wind** around each other several times in a whorl from the nucleus to the edge of the disk. A few spiral galaxies have more than two spiral arms.

There are two types of spiral galaxies, normal spirals and barred spirals. In the normal spiral galaxies, the spiral arms wind outward from the nucleus. In barred spirals, there is a central bar structure extending out on either side of the nucleus. The spiral arms wind outward from the edge of this bar structure.

In his classification scheme, Hubble denoted normal spiral galaxies by S and barred spiral galaxies by SB. He then subclassified spirals according to how tightly the spiral arms wind around the nucleus, using a, b, and c. Galaxies denoted Sa are the most tightly wound and therefore have a relatively small disk compared to the spiral arms. Sc galaxies are the most loosely wound. They therefore extend well beyond the nucleus and have a relatively larger disk compared to the nucleus. Sb galaxies are intermediate between the Sa and Sc galaxies. Hubble used a similar scheme for barred spirals, producing the classifications SBa, SBb, and SBc.

Some galaxies have a disk surrounding a nucleus, but do not have spiral arms in the disk. Hubble classified these galaxies as SO. They are now also called lenticular galaxies. As for elliptical galaxies, modern astronomers also add luminosity classes (I, II, III, IV, V) to Hubble's classification scheme. Luminosity class I galaxies are the most luminous and are referred to as supergiant spirals. Luminosity class V galaxies are the least luminous.

The luminosity classes of spiral galaxies do not have as wide a range as elliptical galaxies, so there are no dwarf spiral galaxies. Spiral galaxies are typically about the size of the Milky Way, roughly 100,000 light years in diameter. They will typically have a mass of about 100 billion times the mass of the sum, so will contain roughly 100 billion stars. The largest supergiant spirals can have as much as several trillion times the mass of the sun.

Spiral galaxies contain fairly young stars in their disks and spiral arms and older stars in their nuclei and halos. The disks and spiral arms also contain interstellar gas and dust, which are the raw materials for forming new stars. The halos like elliptical galaxies contain very little gas and dust. This difference in the distribution of the contents of spiral galaxies tells us that they were originally spherical in shape. The rotation of these galaxies caused them to flatten out and form their disks.

Irregular galaxies

Hubble classified galaxies that do not fit neatly into his scheme of ellipticals and spirals as irregular (Irr) galaxies. Irregular galaxies as a class have no particular shape, and have no spherical or circular symmetries as the ellipticals and spirals do. There is a range of sizes, but irregulars tend to be small. They average about 20,000 light years in diameter. The smallest irregulars, dwarf irregulars, are only about 1,000 light years in diameter.

Because they are relatively small, irregular galaxies have small masses (typically about one million times the mass of the sun) and therefore relatively few stars. Astronomers now classify irregular galaxies into two groups, Irr I and Irr II. In Irr I galaxies, we can resolve young stars and evidence of ongoing star formation. In Irr II galaxies, we cannot resolve individual stars. They also have no distinct shape. Both types of irregular galaxies contain a large percentage of young stars and interstellar gas and dust.

Active galaxies

Many galaxies look almost like one of the Hubble classifications, but with some unusual feature. For example, imagine an elliptical galaxy that looks like someone sliced it through the center, pulled it apart a little bit, and displaced each half sideways. Hubble called these galaxies peculiar and added the designation, pec, to the classification. The galaxy described above might be an E0 pec galaxy. Whatever causes a galaxy to look as if it were ripped apart as described above would require large amounts of energy. Peculiar galaxies are therefore interesting because they often tend to be the active galaxies that emit large amounts of energy.

Active galaxies are galaxies that emit far more energy than normal galaxies. A galaxy is considered an active galaxy if it emits more than 100 times the energy of the Milky Way galaxy. Active galaxies often have a very compact central source of energy, much of which is emitted as **radio waves** rather than optical light. These **radio** waves are emitted by electrons moving in a helical path in a strong magnetic field at speeds near the speed of light. Active galaxies also often have a peculiar photographic appearance, which can include jets of material streaming out from the nucleus or the appearance of either explosions or implosions. They also tend to vary erratically in brightness on rapid time scales. There are a number of varieties of active galaxies, including: compact radio galaxies, extended radio galaxies, Seyfert galaxies, **BL Lacertae** objects, and quasars.

Compact radio galaxies appear photographically as ordinary giant elliptical galaxies. Radio telescopes however reveal a very energetic compact nucleus at the center. This radio nucleus is the source of most of the energy emitted by the galaxy.

Perhaps the best known compact radio galaxy is M87. This giant elliptical galaxy has both a very compact energetic radio source in the nucleus and a jet consisting of globs of material shooting out from the nucleus. Recent observations from the Hubble Space Telescope provide strong evidence that this core contains a supermassive **black hole**.

Extended radio galaxies consist of two giant lobes emitting radio waves. These lobes are on either side of a peculiar elliptical galaxy. The lobes can appear straight or curved as if the galaxy is moving through space. These lobes are the largest known galaxies and can stretch for millions of light years.

Seyfert galaxies look like spiral galaxies with a hyperactive nucleus. The spiral arms appear normal photographically, but they surround an abnormally bright nucleus. Seyfert galaxies also have evidence for hot turbulent interstellar gas.

BL Lacertae objects look like stars. In reality they are most likely to be very active nuclei of elliptical galaxies. However, BL Lacertae objects have sufficiently unusual **behavior**, including extremely rapid and erratic variations in observed properties, that their exact nature is not known for certain.

Quasars also look like stars, but they are perhaps the most distant and energetic objects in the universe known so far. Most astronomers consider them the very active nuclei of distant galaxies in the early stages of **evolution**. As for the other types of active galaxies they produce large amounts of energy in a very small volume. Most astronomers currently think that the energy source is a supermassive black hole.

Formation and evolution

For many years scientists had no ideas how galaxies formed. According to all observations at the time, galaxies formed during a single epoch very far back in the history of the universe. In the absence of direct evidence, astronomers formed two theories: the theory of accretion, in which blobs of stars came together to form galaxies; and the theory of collapse, in which galaxies were formed in the collapse of an enormous gas cloud.

In late 1996, scientists got their first view of galaxy formation, looking back in time 11 billion years to see clumps of young star clusters gradually banding together into a galaxy. It is too early to fully dismiss the gas collapse theory, however; there may be more than one way to form a galaxy.

When Hubble first devised his classification scheme, he thought that the different types of galaxies represented different evolutionary stages; they started as one type and gradually evolved into another type. We now know that his theory was true, though the phrase gradual evolution is something of a misnomer.

Elliptical galaxies are formed by the collison of two spiral galaxies. The process is slow—scientists estimate that it takes nearly half a billion years for the merging spiral galaxies to smooth into an elliptical galaxy—but can be quite violent. Although galaxies are mostly empty space, gravitational interaction between stars can cause them to explode into supernovas. More important, gravitationally induced collisions between clouds of interstel-

KEY TERMS

Active galaxy—A galaxy that emits more energy than can easily be explained, usually at least 100 times the energy output of the Milky Way.

Barred spiral galaxy—A spiral galaxy in which the spiral arms start at the end of a central bar structure rather than the nucleus.

Cepheid variable star—A type of star that varies in brightness as the star pulsates in size. Cepheid variables are important distance yardsticks in establishing the distance to nearby galaxies.

Disk—The flat disk-shaped part of a spiral galaxy that contains the spiral arms.

Elliptical galaxy—A galaxy having an elliptical shape.

Galaxy—A large collection of stars and clusters of stars, containing anywhere from a few million to a few trillion stars.

Halo—A spherical distribution of older stars and clusters of stars surrounding the nucleus and disk of a spiral galaxy.

Irregular galaxy—A galaxy that does not fit into Hubble's classification scheme of elliptical and spiral galaxies.

Light year—The distance light travels in one year, roughly 6 trillion mi, or 9,654 trillion km.

Milky Way—The galaxy in which we are located.

Nucleus—The central core of a galaxy.

Spiral arms—The regions where stars are concentrated that spiral out from the center of a spiral galaxy.

Spiral galaxy—A galaxy in which spiral arms wind outward from the nucleus.

lar **hydrogen** gas can create intense **heat** and **pressure** that can trigger the formation of new stars.

One clue to the evolution of galaxies is the distribution of different types of galaxies at different distances from us. Because light travels at a finite speed, when we look at a distant galaxy, we are seeing the galaxy as it appeared in the distant past when the light left it. Some types of active galaxies, such as quasars and BL Lax objects, occur only at great distances from us. They existed when the universe was much younger, but no longer exist. Many astronomers therefore think that active galaxies are an early stage in the evolution of galaxies. If this idea is correct, an astronomer living now on a distant

quasar might see the quasar as a normal galaxy, and the Milky Way in its earlier active stage as a quasar.

However galaxies formed and evolved, the process must have occurred quickly very early in the history of the universe. The age of the oldest galaxies appears to be not much younger than the **age of the universe**. Though astronomers now have some support for theories of galactic formation and evolution, they are still searching for more evidence and trying to understand the details.

See also Radio astronomy.

Resources

Books

Bacon, Dennis Henry, and Percy Seymour. *A Mechanical History of the Universe.* London: Philip Wilson Publishing, Ltd., 2003.

Bartusiak, Marcia. *Thursday's Universe.* Redmond, WA: Tempus Books, 1988.

Hodge, Paul. *Galaxies.* Cambridge: Harvard University Press, 1986.

Morrison, David, Sidney Wolff, and Andrew Fraknoi. *Abell's Exploration of the Universe.* 7th ed. Philadelphia: Saunders College Publishing, 1995.

Smolin, Lee. *The Life of the Cosmos.* Oxford: Oxford University Press, 1999.

Snow, Theodore P. *The Dynamic Universe: An Introduction to Astronomy.* St. Paul: West Publishing, 1991.

Periodicals

Cowen, Ron. "The Debut of Galaxies." *Astronomy* 22 (December 1994): 44-45.

Eicher, David J. "The Wonderful World of Galaxies." *Astronomy* 21 (January 1993): 60-66.

Lake, George. "Understanding the Hubble Sequence." *Sky & Telescope* 83 (May 1992): 515-21.

Paul A. Heckert

Gallium *see* **Element, chemical**

Game theory

Game theory is a branch of **mathematics** concerned with the analysis of conflict situations. It involves determining a strategy for a given situation and the costs or benefits realized by using the strategy. First developed in the early twentieth century, it was originally applied to parlor games such as bridge, chess, and poker. Now, game theory is applied to a wide range of subjects such as economics, behavioral sciences, sociology, military science, and political science.

The notion of game theory was first suggested by mathematician John von Neumann in 1928. The theory received little attention until 1944 when Neumann and economist Oskar Morgenstern wrote the classic treatise *Theory of Games and Economic Behavior*. Since then, many economists and operational research scientists have expanded and applied the theory.

Characteristics of games

An essential feature of any game is conflict between two or more players resulting in a win for some and a loss for others. Additionally, games have other characteristics which make them playable. There is a way to start the game. There are defined choices players can make for any situation that can arise in the game. During each move, single players are forced to make choices or the choices are assigned by **random** devices (such as dice). Finally, the game ends after a set number of moves and a winner is declared. Obviously, games such as chess or checkers have these characteristics, but other situations such as military battles or animal **behavior** also exhibit similar traits.

During any game, players make choices based on the information available. Games are therefore classified by the type of information that players have available when making choices. A game such as checkers or chess is called a "game of perfect information." In these games, each player makes choices with the full knowledge of every move made previously during the game, whether by herself or her opponent. Also, for these games there theoretically exists one optimal pure strategy for each player which guarantees the best outcome regardless of the strategy employed by the opponent. A game like poker is a "game of imperfect knowledge" because players make their decisions without knowing which cards are left in the deck. The best play in these types of games relies upon a probabilistic strategy and, as such, the outcome can not be guaranteed.

Analysis of zero-sum, two-player games

In some games there are only two players and in the end, one wins while the other loses. This also means that the amount gained by the winner will be equal to the amount lost by the loser. The strategies suggested by game theory are particularly applicable to games such as these, known as zero-sum, two-player games.

Consider the game of matching pennies. Two players put down a penny each, either head or tail up, covered with their hands so the orientation remains unknown to their opponent. Then they simultaneously reveal their pennies and pay off accordingly; player A wins both pennies if the coins show the same side up, otherwise player B wins. This is a zero-sum, two-player game because each time A wins a penny, B loses a penny and visa versa.

To determine the best strategy for both players, it is convenient to construct a game payoff **matrix**, which shows all of the possible payments player A receives for any outcome of a play. Where outcomes match, player A gains a penny and where they do not, player A loses a penny. In this game it is impossible for either player to choose a move which guarantees a win, unless they know their opponent's move. For example, if B always played heads, then A could guarantee a win by also always playing heads. If this kept up, B might change her play to tails and begin winning. Player A could counter by playing tails and the game could cycle like this endlessly with neither player gaining an advantage. To improve their chances of winning, each player can devise a probabilistic (mixed) strategy. That is, to initially decide on the percentage of times they will put a head or tail, and then do so randomly.

According to the minimax **theorem** of game theory, in any zero-sum, two-player game there is an optimal probabilistic strategy for both players. By following the optimal strategy, each player can guarantee their maximum payoff regardless of the strategy employed by their opponent. The average payoff is known as the minimax value and the optimal strategy is known as the solution. In the matching pennies game, the optimal strategy for both players is to randomly select heads or tails 50% of the time. The expected payoff for both players would be 0.

Nonzero-sum games

Most conflict situations are not zero-sum games or limited to two players. A nonzero-sum game is one in which the amount won by the victor is not equal to the amount lost by the loser. The Minimax Theorem does not apply to either of these types of games, but various weaker forms of a solution have been proposed including noncooperative and cooperative solutions.

When more than two people are involved in a conflict, oftentimes players agree to form a coalition. These players act together, behaving as a single player in the game. There are two extremes of coalition formation; no formation and complete formation. When no coalitions are formed, games are said to be non-cooperative. In these games, each player is solely interested in her own payoff. A proposed solution to these types of conflicts is known as a non-cooperative equilibrium. This solution suggests that there is a point at which no player can gain an advantage by changing strategy. In a game when com-

KEY TERMS

. . . .

Coalition—A situation in a multiple player game in which two or more players join together and act as one.

Game—A situation in which a conflict arises between two of more players.

Game payoff matrix—A mathematical tool which indicates the relationship between a players payoff and the outcomes of a game.

Minimax theorem—The central theorem of game theory. It states that for any zero-sum two-player game there is a strategy which leads to a solution.

Nonzero-sum game—A game in which the amount lost by all players is not equal to the amount won by all other players.

Optimal pure strategy—A definite set of choices which eads to the solution of a game.

Probabilistic (mixed) strategy—A set of choices which depends on randomness to find the solution of a game.

Zero-sum, two-player games—A game in which the amount lost by all players is equal to the amount won by all other players.

plete coalitions are formed, games are described as cooperative. Here, players join together to maximize the total payoff for the group. Various solutions have also been suggested for these cooperative games.

Application of game theory

Game theory is a powerful tool that can suggest the best strategy or outcome in many different situations. Economists, political scientists, the military, and sociologists have all used it to describe situations in their various fields. A recent application of game theory has been in the study of the behavior of animals in nature. Here, researchers are applying the notions of game theory to describe the effectiveness of many aspects of animal behavior including aggression, cooperation, hunting and many more. Data collected from these studies may someday result in a better understanding of our own human behaviors.

Resources

Books

Beasley, John D. *The Mathematics of Games.* Oxford: Oxford University Press, 1990.

Hoffman, Paul. *Archimedes' Revenge: The Joys and Perils of Mathematics.* New York: Fawcett Crest, 1988.

Newman, James R., ed. *The World of Mathematics.* New York: Simon and Schuster, 1956.

Paulos, John Allen. *Beyond Numeracy.* New York: Alfred A. Knopf Inc, 1991.

Perry Romanowski

Gamete

A gamete is a specialized reproductive **cell**. The cells usually have one half as many chromosomes in their nuclei as the majority of body cells, which are known as somatic cells. All sexually-reproducing plants, animals, and microbes produce gametes sometime during their life span.

During the second and third quarters of the nineteenth century the scientists J. L. Prevost, J. B. Dumas, T. Schwann, and R. Virchow were especially influential in the evolving consensus that sperm were cells, and that these cells united with other cells, ova or egg cells, to form a fertilized cell (a zygote) that went on to form a new **organism**.

Gametes usually form in the gonads, organs which form the sex cells. In flowering plants, the gonads are found in the flowers. The male gonads are the anthers, seen as the enlarged tips of the stamens. The anthers produce pollen (male gametes) in flowering plants. The female gametes are formed in the base of the **flower**, in the ovules, located in the ovary of the pistil.

In **vertebrates** such as **fish**, **amphibians**, **reptiles**, **birds**, and **mammals** the male gonads are the testis, where very large numbers of gametes (spermatozoa) are formed. The female gonads (ovaries) of these animals produce low numbers of gametes known as eggs, or ova. Usually the number of mature egg cells produced in females is far fewer (a handful) than the number of sperm cells formed in the males (counted in the millions).

In most organisms the gametes are produced by a special double **cell division** process, a reduction division known as **meiosis**, in which new cells (gametes) end up with half as many chromosomes as the original cell. The fusion of the egg and sperm at **fertilization** restores the normal **chromosome** number. In the case of **bees**, the fertile female queen bee is fertilized by male gametes from a male bee called a drone. But a drone develops from an unfertilized egg, which is a single gamete (ovum)! Therefore the sperm of the drone must be produced by ordinary cell division (**mitosis**) instead of the meiotic cell division which usually is involved in the formation of fertile gametes.

Currently there is much interest in manipulating the gametes of domestic animals in breeding programs to promote characteristics yielding economic advantages in agriculture. Selected female cattle are given **hormones** to cause multiple ovulation (release of ova) which are then artificially fertilized with male gametes (bull semen) in the uterus. After a few days the tiny embryos are flushed out of the waters and stored as frozen embryos for later insertion and gestation in surrogate mothers. A similar technique has been developed for use in humans.

Another new **biotechnology** that is being tried in a few human cases is intracytoplasmic sperm injection (ICSI). This technique involves a skilled technician with a micro pipette and **microscope** capturing a single sperm and injecting it directly into the cytoplasm of a female gamete, the ovum. When successful, the resulting embryo must then be implanted into a uterus prepared to receive and nourish the new offspring.

See also Sexual reproduction.

Gametogenesis

Gametogenesis is the production of haploid sex cells (in humans, ovum and spermatozoa) that each carry one-half the genetic compliment of the parents from the germ **cell** line of each parent.

The production of ovum is termed oogenesis and the production of spermatozoa is called spermatogenesis. Both oogenesis and spermatogenesis provide a mechanism through which genetic information may be passed to offspring. The fusion of spermatozoa and ova during **fertilization** results in a zygote with a fully restored diploid **genome**.

The production of male and female gametes is a highly complex and coordinated sequence of a mitotic division, two meiotic divisions, cytoplasmic apportionment (divisions) and cellular differentiation. Any chronic alteration in the sequence of morphological and biochemical transformations required to produce gametes usually results in sterility for the affected parent.

Spermatogenesis provides the haploid gametes necessary to pass on paternal genetic information. Oogenesis provides the haploid **gamete** necessary to pass on maternal genetic information and extranuclear genetic information (e.g., mitochondrial DNA).

In eukaryotic organisms the gametes are derived from primordial germ cells, which enter the gonads during early development. During embryogenesis, the primordial germ cells are determined early in development by the presence of a cytoplasmic component termed germ plasm. Once germ cells are determined they follow a different maturation and, of course, genetic function, than do the remaining somatic cells of the body. Primordial germ cells are the **stem cells** that, via **mitosis**, supply both spermatogonia and oogonia.

In humans, spermatogenesis starts with a diploid (2N) spermatogonium that carries the full genetic compliment of 46 chromosomes (22 autosomal pairs, one X and one Y sex chromosomes). The spermatogonium represents the germ cell line from which all sperm cells are derived. Sequentially, the process of spermatogenesis via mitosis produces a primary spermatocyte that is also diploid (2N) and then via **meiosis**, two secondary spermatocytes that are haploid (N). The haploid secondary spermatocytes carry 22 autosomes and either an X or a Y sex **chromosome**. The secondary spermatocytes each undergo a second meiotic division to form a total of four haploid spermatids. Subsequently, nurtured by surrounding somatic cells, through the process of cellular differentiation the four spermatids produce 4 sperm cells capable of motility and fertilization. Although there is variation between sperm cells as to the exact nature of their genetic information (i.e., what **alleles** they carry or which chromosome trace back to a maternal or parental line) in sharp contrast to female gamete production all the terminal male gametes (the sperm cells) have roughly the same cytoplasmic **volume** and contents and the same amount of genetic material.

In human females the germ cell line is represented by the diploid (2N) oogonium that carries the full female genetic compliment of 22 autosomal pairs and two X chromosomes. Mitotic division yields a diploid primary oocyte. Meiotic divisions then produce one female gamete—the ovum. In humans, the first meiotic division is suspended in the diplonema stage during embryonic development. Meiosis resumes, one ovum at a time following **puberty** and during the ovulatory period of the **menstrual cycle**. Maturation proceeds with the production of haploid (N) secondary oocytes with 22 autosomal chromosomes and an X sex chromosome (the sex chromosome must be an X chromosome because normal human females carry two X chromosomes and no Y chromosomes). Also formed is a haploid polar body that is nearly devoid of cytoplasmic contents. This is a fundamental difference between male and female Gametogenesis. In males, there is a nearly equal division of cytoplasm to the gametes, in females the cytoplasmic contents are preserved for the eventual "egg" or ovum. Extraneous genetic material is removed via polar bodies. Another meiotic division results in the production of an ootid and yet another polar body (the eventual number of polar vies associated with an ovum may equal as many as three if the first sloughed off polar body undergoes a subsequent division). Cellular differentiation of the ootid yields an ovum ready for fertilization. In many cases, however, the last maturational processes are accelerated because in human females, meiosis II is usually completed after fertilization.

During ovum maturation, there is a tremendous increase in ribosomal related component so that the cellular machinery is present to handle the tremendous amount of translation and protein synthesis required in the rapid cellular divisions that follow the formation of the zygote.

Germ cell line manipulation (e.g., **gene** "knock-outs) is a powerful potential tool to manipulate an organism's genome. Each generation of sexually-reproducing organisms is dependent upon the continuation of the germ cell line The germ line is also the vehicle of genetic transmission and alteration of the genome via mutations and recombination (i.e., **evolution**).

See also Birth defects; Embryo and embryonic development; Embryology; Evolutionary mechanisms; Genetics; Germ cells and the germ cell line; Sexual reproduction.

Resources

Books

Gilbert, Scott F. *Developmental Biology.* 6th ed. Sunderland, MA: Sinauer Associates, Inc., 2000.

Sadler, T.W., Jan Langman. *Langman's Medical Embryology.* 8th ed. New York: Lippincott Williams & Wilkins Publishers, 2000.

Periodicals

Nielsen H.I., et al. "Definitions of Human Fertilization and Preimplantation Growth Revisited." *Reprod Biomed Online.* 3(2) (2001):90–93.

Readhead C., and C. Muller-Tidow. "Genes Associated with the Development of the Male Germ Line." *Reprod Biomed Online.* 4 Suppl 1(2002):52–7.

Westphal H. "International Stem Cell Research Considerations." *C R Biol.* 325(10) (Oct 2002):1045–8.

K. Lee Lerner

Gamma ray *see* **Electromagnetic spectrum**

Gamma-ray astronomy

Gamma rays are a highly energetic form of electromagnetic **radiation**. The wavelength of a gamma ray is very short—less than the radius of an atom—the **energy** they carry can be millions of **electron** volts. Gamma rays originate in the nucleus of an atom, and are created when cosmic rays collide with **atoms** in molecules of gas. In the collision, the nucleus of the atom is destroyed, and gamma rays are emitted.

Gamma rays are emitted from a variety of sources, including **neutron** stars, black holes, supernovas, and even the **sun**. Observations at gamma-ray energies allow astronomers to study objects that are not highly visible in other spectral regions; for example Geminga, a **pulsar** located in Orion, is more visible in the gamma ray region than at any other wavelength. Because gamma rays identify locations of extreme particle **acceleration** processes, and are emitted by the interaction of interstellar gas with cosmic rays, they provide scientists with a tool to study both phenomena. Gamma rays can also help scientists learn more about **active galactic nuclei** and the process of **star formation**.

Gamma rays are as perplexing as they are informative, however. In 1979, instruments aboard several satellites recorded an ultra-high intensity burst of electromagnetic radiation passing through our **solar system**. When astronomers monitoring the satellites discovered this phenomenon, they tried to explain it. All that was known for certain was that the radiation was caused by gamma rays.

Since the 1979 incident, gamma rays have been observed occurring in short bursts several times a day as brief high-energy flashes. Most astronomers believed their origin was from within our own **Milky Way galaxy**. In 1991, NASA launched its Compton Gamma Ray Observatory **satellite**. For more than two years the Compton Observatory detected gamma ray bursts at a **rate** of nearly one a day for a total of over 600. The energy of just one of these bursts has been calculated to be more than one thousand times the energy that our sun will generate in its entire 10-billion-year lifetime.

Gamma ray bursts appear uniformly across the sky, surrounding **Earth** in a spherical shell of fireworks. Because of the shape of the Milky Way and our location within it, the bursts would appear to be concentrated in just one area in the sky if they were coming from within our galaxy. This perfectly symmetrical distribution tells us that these gamma rays originate far outside the Milky Way.

The late 1990s turned gamma ray **astronomy** on its **ear**. For years, it was accepted that gamma ray bursts never appeared in the same location twice, which led to theories that the pulses of radiation were generated by colliding neutron stars, or other catastrophic cosmic events. Then in October of 1996, the Compton observatory captured two bursts from the same region of the sky: a 100 s pulse followed 15 minutes later by a 0.9 s pulse.

Two days later, gamma rays flared again in the same spot, in a 30-s burst followed by a 23-minute burst 11 minutes afterward. Although scientists are still unclear on the cause of the radiation, many are certain that more than one of the bursts were generated by the same stellar object. If they are correct, then annihilation-based theories of **gamma ray burst** generation are invalid, and science must look elsewhere for answers to the riddle.

In 1996, an Italian and Dutch collaboration launched the Beppo-SAX orbiting observatory, designed to pinpoint the location of gamma ray bursts. In 1998, the investigators hit pay dirt—Beppo-SAX registered a burst that was determined to be larger than any other cosmic explosion yet detected, except for the big bang. At the time, though, no one was particularly excited. The intensity of the burst, as measured by the Compton observatory, appeared to be nothing unusual. As the gamma rays faded into an afterglow that included lower-energy radiation such as **x rays**, astronomers worldwide continued to monitor the output. Then two weeks after the initial burst, a faint galaxy was discovered in the spot from which the gamma ray burst emerged.

Calculations showed that the galaxy is more than 12 million light-years away from Earth. This data, combined with the burst intensity measured by the Compton observatory, allowed scientists to calculate the total energy released by the event. The numbers were stupefying—the gamma ray burst released 3×10^{53} ergs of energy, several hundred times the amount released by a **supernova**. If the calculations are accurate and the faint galaxy really was the source of the gamma ray burst, the 1998 event was the largest cosmic explosion ever detected, except for the big bang.

In January 1999, astronomers made a giant leap forward in the study of gamma ray bursts when a complex net of observatories captured a gamma ray burst as it took place. Previously, gamma ray bursts had only been observed after the fact. The Burst and Transient Source Experiment, aboard the Compton observatory, captured a burst of gamma rays, simultaneously notifying a computer at Goddard Space Flight Center in Greenbelt, Maryland. The computer passed a message across the Internet to activate an observatory in Los Alamos, New Mexico, which automatically began making observations. Meanwhile, scientists at Beppo-SAX were called in to identify the location of the gamma ray source.

NASA and the scientific community have proposed a new orbital gamma ray **telescope**. The high-sensitivity Gamma-ray Large Area Space Telescope (GLAST) will feature a wide field-of-view, high-resolution positional **accuracy**, and long-life detectors. Slated for launch in the first decade of the twenty-first century, GLAST will

KEY TERMS

Black hole—A supermassive object with such a strong gravitational field that nothing, not even light, can escape it.

Neutron star—The remnant of an extinct supernova. Next to black holes, neutron stars are the most dense objects in the universe.

Pulsar—A rapidly spinning neutron star with its magnetic axis inclined relative to its rotation axis. Radiation streams continuously from the pulsar along its magnetic axis, so if the magnetic axis passes through our line of sight as the pulsar rotates, we see a flash. The rate of the

Supernova—The final collapse stage of a supergiant star.

provide astronomers with a new tool to study gamma ray bursts, pulsars, active galactic nuclei, diffuse background radiation, and a host of other high-energy puzzles.

See also Nuclear fission; Neutron star.

Resources

Books

Bacon, Dennis Henry, and Percy Seymour. *A Mechanical History of the Universe.* London: Philip Wilson Publishing, Ltd., 2003.

Periodicals

Cowen, Ron. "Catching Some Rays." *Science News* 139 (11 May 1991).
Folger, Tim. "Bright Fires Around Us." *Discover* (August 1993).
Taubes, Gary. "The Great Annihilator." *Discover* (June 1990).

Johanna Haaxma-Jurek

Gamma ray burst

Gamma ray bursts are brief, seconds-long, blasts of **radiation** of mysterious origin that, in nature, seem to come from the depths of interstellar **space**. Bursts of gamma radiation also have been measured coming from severe thunderstorms and are a component of nuclear bomb detonation.

The accidental discovery of cosmic gamma ray bursts was confirmed in 1973. Ten years earlier, the United States Air Force had launched the first in a series of satellites that were intended to monitor the effectiveness of the Nuclear Test Ban Treaty. By signing this treaty, nations of the world agreed not to test nuclear devices in the atmosphere of **Earth** or in space. The Vela satellites (from the Spanish verb velar, which means "to watch") were part of a research and development program that had the goal of developing the technology to monitor nuclear tests from space. Along with a variety of optical and other instruments, the satellites carried x ray, gamma ray, and **neutron** detectors. The x-ray detectors were intended to sense the flash of a nuclear blast. Although most of the **energy** of a bomb detonated in space would be directly visible as an x-ray flash, a gamma ray burst at the same time would provide confirmation of a nuclear event. A further confirmation would come from the detection of neutrons. The Vela designers knew that detonating a nuclear bomb behind a thick shield or on the far side of the **moon** would effectively hide the initial flash of **x rays** from the satellites' view. But the gamma ray detectors provided a way around this because they could measure the radiation from the cloud of radioactive material blown out from a nuclear blast. This could not be completely shielded from view as it rapidly expanded outward. The Vela satellites could easily detect these gamma rays even if the detonation took place behind the moon, out of direct view of the satellites' x-ray detectors.

By 1969, a number of gamma ray bursts had been detected that were clearly not caused by nuclear explosions on or near the earth. It was concluded that these were "of cosmic origin," and more accurate observations showing the actual directions of future bursts were collected. Finally, the discovery of these cosmic gamma ray bursts was announced in 1973. It has been found that if we could see gamma rays, every day powerful explosions would illuminate the sky. They come from **random** directions and their cause is not completely understood. Until recently it was even uncertain whether they come from our own **solar system** or from as far away as the edge of the known universe.

In 1991 the National Aeronautics and Space Administration (NASA) launched the Compton Gamma Ray Observatory, carrying an instrument called the Burst and Transient Source Experiment (BATSE). Specifically designed to study gamma ray bursts, BATSE has added much new information about their origin and distribution in the universe. By studying the information BATSE can gather, scientists hope to determine what causes them. A gamma ray burst detected on January 23, 1999— GRB 990123—was the first to be observed as a visual object as well as in the gamma ray region. This has given scientists an even better look at these mysterious objects, giving clues to the structure of the explosion. Even though the object was nine billion **light** years away, the light was so bright observers on Earth could see it with a pair of binoculars.

Today, a cosmic gamma ray burst seems like a strange interstellar traffic accident, with material flowing out from the explosion at different speeds. Because of this, some material from the explosion collides with other parts of the expanding shell. This causes "pile-ups" of material that create shock waves. These, in turn, generate energy at various wavelengths. Scientists are currently trying to untangle these clues to better understand the mechanics of the explosion, but it is now clear that in addition to gamma rays, GRBs also emit light, large amounts of x rays, and other radiation.

Clint Hatchett

Gangrene

Gangrene involves the death of human **tissue**, usually due to ischemia, which is an interruption in the **blood supply** to a particular area. Loss of **blood** supply means loss of **oxygen** delivery to that tissue, as well as loss of other nutritive factors usually carried in the blood circulation. Tissue deprived in this manner will die, and often becomes infected with **bacteria** during this process.

Gangrene in the setting of atherosclerotic disease

The presence of atherosclerotic **disease** (disease in which **arteries** are stiff and hard, with fatty deposits blocking blood flow), is a major predisposing factor to gangrene, particularly of the toes, feet, and legs.

People with diabetes often have advanced, severe cases of atherosclerosis, as well as a condition called neuropathy. Neuropathy is a type of nerve disease which results in a significant decrease in sensation. Diabetics, then, may be unable to feel any **pain** from a relatively minor injury (for example, a developing blister) to their foot or leg. Because the diabetic patient does not feel the blister, due to neuropathy, and because the blood supply to the area is so severely compromised, a small initial area of damage can be extremely difficult to heal, and can rapidly spread. Furthermore, any small opening in the skin, such as a blister, can provide an entry point for bacteria (most commonly staphylococcal and/or streptococcal bacteria) The combination of tissue damage from a blister, along with lack of blood supply to the area to either help in healing the blister or in delivering immune cells to fight **infection**, can result in the ultimate development of gangrene from a seemingly insignificant injury. This can be severe enough to require **amputation** of part or all of the affected body part.

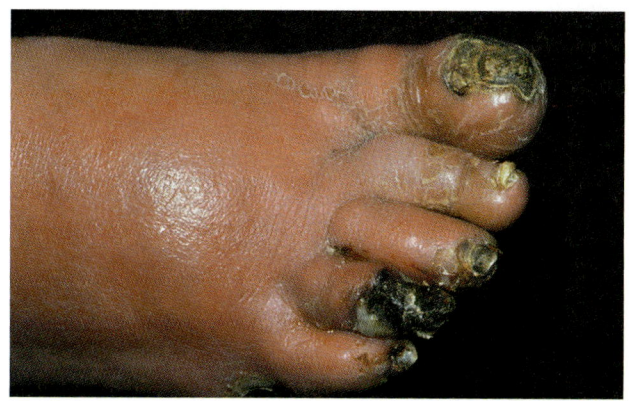

A close-up of gangrene in the toes of a diabetic patient.
Science Photo Library, National Audubon Society
Collection/Photo Researchers, Inc. Reproduced by permission.

Gas gangrene

The scenario most frequently called to mind by the word gangrene is of an extremely rapidly progressing disorder, classically affecting a leg wounded in battle, and resulting in a blackening of the limb which leads either to death of the individual or amputation of the limb to save that person's life. In fact, this scenario (courtesy of all those old war movies on late at night) is due to a very specific form of gangrene called "gas gangrene." Gas gangrene is a wound infection caused most frequently by the bacteria *Clostridium perfringens*, spores of which are present in **soil**. Individuals who suffer bullet wounds during the course of battle are very likely to have had these wounds contaminated with dirt or with shreds of their own clothing. This introduces the bacteria *C. perfringens* (or certain other clostridia cousins), into the wound. Some types of extreme injury in civilian life can also result in *C. perfringens* infection. Because *C. perfringens* bacteria sometimes reside within the gallbladder, spillage of gallbladder contents during **surgery** can result in gas gangrene of the abdominal muscles.

C. perfringens causes much of its effect due to its ability to produce toxins, or poisons. In fact, *C. perfringens* is a close cousin to the bacteria which cause **tetanus** (*C. tetani*) and **botulism food poisoning** (*C. botulinum*). These bacteria also produce their effects through the production of toxins.

Gas gangrene receives its name from another characteristic of the *C. perfringens* bacteria. These bacteria ferment (breakdown) certain chemical components of muscle, giving off gas in the process. During examination of the affected area, one can actually feel bubbles of gas which have risen up just under the layers of skin.

C. perfringens bacteria multiply so quickly that gas gangrene can develop in just a matter of hours. An individual with gas gangrene will note severe pain at the wound site, with increasing swelling of the area. The wound will begin to give off a watery, sometimes frothy fluid which has a unique sweet odor, probably due to the digestion of muscle **carbohydrate** by the bacteria. As muscle breakdown progresses, the muscle feels cooler and appears paler than normal. The muscle feels softer and more liquid, as the bacterial toxins actually work to liquefy it. Ultimately, the area turns a deep blue-black, the classic **color** of gangrenous tissue. Low blood **pressure**, kidney failure, and a state of shock (severely decreased blood circulation to all major **organ** systems) may set in. Survival time for an individual with untreated gas gangrene can be as short as a single day.

Diagnosis is by examination of tissue under a **microscope**, where the clostridia can be definitively identified. Certainly, gas gangrene has enough unique characteristics to allow a high level of suspicion based just on the appearance of the wound and the presence of gas as noted by the examiner's hand or as revealed by x ray of the area.

Treatment of gas gangrene is with massive doses of **antibiotics**, in particular Penicillin G. Surgical removal of infected tissue, with a wide margin around it, is necessary to halt the spread of infection, and gangrenous limbs may require amputation.

A fascinating type of treatment for gas gangrene is hyperbaric oxygen therapy (HBO). HBO involves placing an individual in a completely closed, carefully pressurized space, within which the patient will breathe 100% oxygen (as opposed to the 21% oxygen present in normal room air). This high level of oxygen reaches the tissues, where it slows the multiplication of the bacteria, inactivates toxin, and decreases further toxin production.

Resources

Books

Andreoli, Thomas E., et al. *Cecil Essentials of Medicine.* Philadelphia: W. B. Saunders Company, 1993.

Berkow, Robert, and Andrew J. Fletcher. *The Merck Manual of Diagnosis and Therapy.* Rahway, NJ: Merck Research Laboratories, 1992.

Cotran, Ramzi S., et al. *Robbins Pathologic Basis of Disease.* Philadelphia: W. B. Saunders Company, 1994.

Isselbacher, Kurt J., et al. *Harrison's Principles of Internal Medicine.* New York: McGraw Hill, 1994.

Kobayashi, G., Patrick R. Murray, Ken Rosenthal, and Michael Pfaller. *Medical Microbiology.* St. Louis, MO: Mosby, 2003.

Rosalyn Carson-DeWitt

Gardening *see* **Horticulture**

Garlic *see* **Lily family (Liliaceae)**

Garpike

Garpike (gar) are **bony fish** classified in the family Lepisosteidae. These **fish** are differentiated from garfish which belong to the family Belonidae. Garpike were once abundant and widely distributed, but are now rare. Some **species** of garpike are found in Mexico, Central America, and the West Indies, and in eastern **North America**. Garpike are found in shallow waters with dense weeds.

Garpike have a gas bladder, which is well supplied with **blood** and which is obtaining **oxygen**. At intervals, garpike rise to the surface to dispel waste air from the bladder and to refill its contents with fresh air. This source of fresh air helps garpike survive in polluted anoxic **water**, which would be intolerable for other fish. Garpike actually drown if caught in a net that denied access to the surface. It has been postulated that the capability of these fishes to breathe air may have been a factor in their survival to modern times. Gar spend their time either near the bottom or rising to the surface. Gar can develop considerable speed for a short period to obtain food. Garpike are cylindrical in shape like a cigar, have a long jaw equipped with many sharp teeth, and a long, flat snout. Garpike have ganoid scales, which fit together to form a hard armor or shell, rendering the fish difficult to catch. The scale surface is covered with ganoin, a substance that could be polished to a high luster, and is hard enough to protect against a fish spear. The scales of the large gars

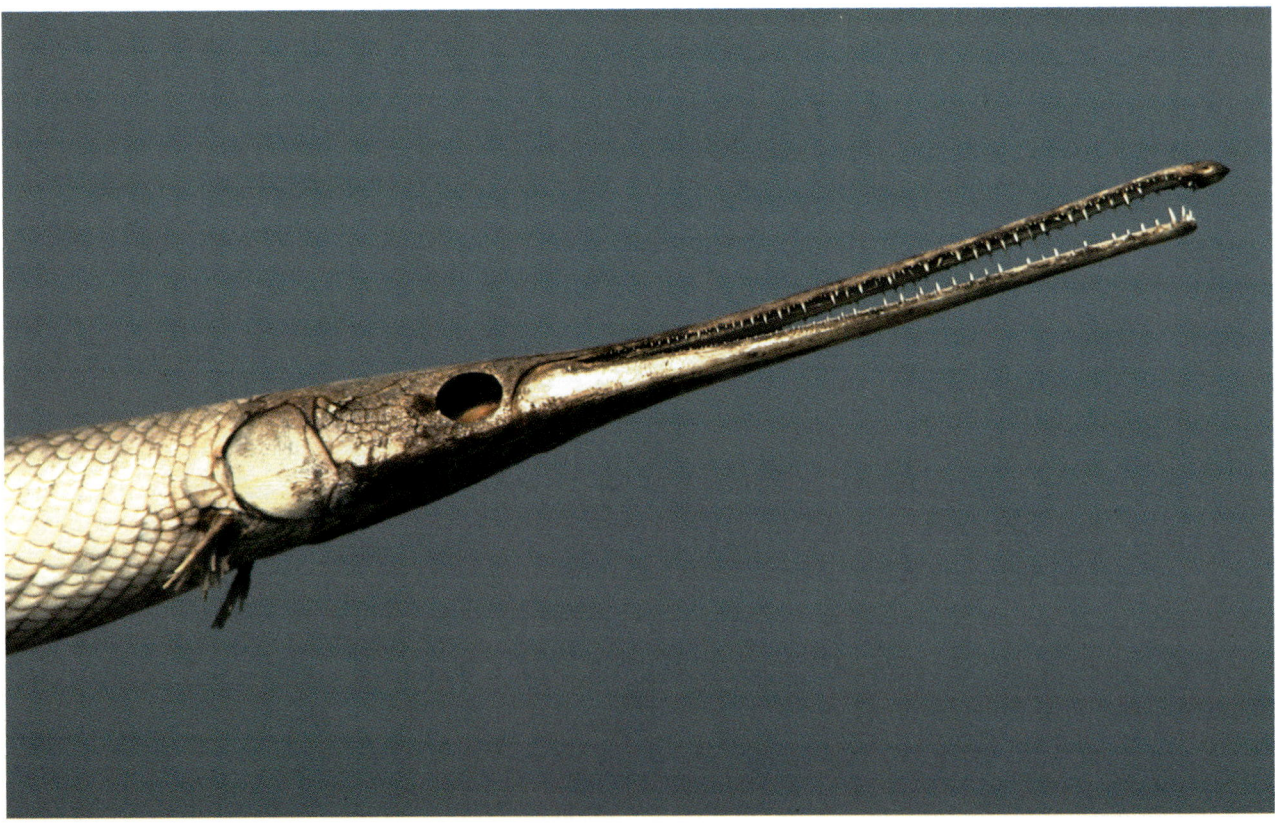

Longnose gar (*Lepisosteus osseus*). *Photograph by Robert J. Huffman. Field Mark Publications. Reproduced by permission.*

Ganoid scales—Thick scale composed of rhomboid bony plates covered with an enamel-like substance called ganoin, which is characteristically found in some primitive fishes. Its hard surface provides an excellent protective mechanism.

Gas bladder—A pouch connected to the throat provided with a blood supply. It helps the fish obtain a better supply of oxygen.

Lateral line—A row of pores on the side of the tail and trunk, enabling the fish to detect low-intensity vibrations, movement, and possibly pressure changes.

were used by native Americans for arrowheads. In pre-Columbian cultures, the shells were used for breastplates. Early farmers would at times use gar hides to cover wooden plowshares to make a hard surface.

The longnose gar (*Lepisosteus osseus*) is cylindrically shaped, and covered with small ganoid scales arranged in regular rows over its body. Its long and slender jaws are equipped with sharp teeth.

The longnose gar is found over a wide expanse of territory eastward from Montana, the Great Lakes to the St. Lawrence River, to Florida, Alabama, Texas, Mexico, and the Mississippi River drainage system. In the southern part of its range, the longnose gar prefers quiet waters with heavy vegetation, while further north these fish are found in less turbid lakes and streams.

Spawning takes place in the spring in shallow waters. Females bulging with eggs are accompanied by several males waiting to fertilize the eggs as they are laid. It is estimated that an excess of 35,000 eggs may be laid by a 3-ft (1 m) female.

The diet of the longnose gar consists mainly of live and dead fish. Gliding near their **prey** they capture it with a sudden movement. At other times the fish will lie motionless near the surface and suddenly seize an unwary fish swimming by.

Garfish have no commercial value. In some areas these fish are used for human consumption, but not considered a prized sport fish.

The shortnose gar (*L. platostomus*) resembles the longnose gar but has shorter jaws, and a short, broad snout. It is the smallest of the gars, rarely more than 2.5 ft (76 cm) long, and is found in the Mississippi River drainage **basin**.

The largest of the gars in North America is the alligator gar (*L. spatula*) found in the streams entering the Gulf

of Mexico. This species may reach a length of 10 ft (3 m) and 300 lb (136 kg) in weight, and is highly voracious and is considered especially dangerous to human beings.

Resources

Books

Dickson Hoese, H., and R.H. Moore. *Fishes of the Gulf of Mexico, Texas, Louisiana and Adjacent Waters.* College Station: Texas A&M University Press, 1977.

Migdalski, E.C., and G.S. Fichter. *The Fresh and Salt Water Fishes of the World.* New York: Greenwich House, 1983.

Moyle, Peter B., Joseph Cech. *Fishes: An Introduction to Ichthyology.* 4th ed. New York: Prentice Hall, 1999.

Whiteman, Kate. *World Encyclopedia of Fish & Shellfish.* New York: Lorenz Books, 2000.

Nathan Lavenda

Gases *see* **States of matter**

Gases, liquefaction of

Liquefaction of gases is the process by which substances in their gaseous state are converted to the liquid state. When **pressure** on a gas is increased, its molecules closer together, and its **temperature** is reduced, which removes enough **energy** to make it change from the gaseous to the liquid state.

Critical temperature and pressure

Two important properties of gases are important in developing methods for their liquefaction: critical temperature and critical pressure. The critical temperature of a gas is the temperature at or above which no amount of pressure, however great, will cause the gas to liquefy. The minimum pressure required to liquefy the gas at the critical temperature is called the critical pressure.

For example, the critical temperature for **carbon dioxide** is 304K (87.8°F [31°C]). That means that no amount of pressure applied to a **sample** of **carbon** dioxide gas at or above 304K (87.8°F [31°C]) will cause the gas to liquefy. At or below that temperature, however, the gas can be liquefied provided sufficient pressure is applied. The corresponding critical pressure for carbon dioxide at 304K (87.8°F [31°C]) is 72.9 atmospheres. In other words, the application of a pressure of 72.9 atmospheres of pressure on a sample of carbon dioxide gas at 304K (87.8°F [31°C]) will cause the gas to liquefy.

Differences in critical temperatures among gases means that some gases are easier to liquify than are others. The critical temperature of carbon dioxide is high enough so that it can be liquified relatively easily at or near room temperature. By comparison, the critical temperature of **nitrogen** gas is 126K (-232.6°F [-147°C]) and that of helium is 5.3K (-449.9°F [-267.7°C]). Liquefying gases such as nitrogen and helium obviously present much greater difficulties than does the liquefaction of carbon dioxide.

Methods of liquefaction

In general, gases can be liquefied by one of three general methods: (1) by compressing the gas at temperatures less than its critical temperature; (2) by making the gas do some kind of work against an external **force**, causing the gas to lose energy and change to the liquid state; and (3) by making gas do work against its own internal forces, also causing it to lose energy and liquefy.

In the first approach, the application of pressure alone is sufficient to cause a gas to change to a liquid. For example, **ammonia** has a critical temperature of 406K (271.4°F [133°C]). This temperature is well above room temperature, so it is relatively simple to convert ammonia gas to the liquid state simply by applying sufficient pressure. At its critical temperature, that pressure is 112.5 atmospheres, although the cooler the gas is to begin with, the less pressure is needed to make it condense.

Making a gas work against an external force

A simple example of the second method for liquefying gases is the **steam engine**. The principle on which a steam engine operates is that **water** is boiled and the steam produced is introduced into a cylinder. Inside the cylinder, the steam pushes on a piston, which drives some kind of machinery. As the steam pushes against the piston, it loses energy. That loss of energy is reflected in a lowering of the temperature of the steam. The lowered temperature may be sufficient to cause the steam to change back to water.

In practice, the liquefaction of a gas by this method takes place in two steps. First, the gas is cooled, then it is forced to do work against some external system. For example, it might be driven through a small **turbine**, where it causes a set of blades to rotate. The energy loss resulting from driving the turbine may then be sufficient to cause the gas to change to a liquid.

The process described so far is similar to the principle on which refrigeration systems work. The coolant in a refrigerator is first converted from a gas to a liquid by one of the methods described above. It then absorbs **heat** from the refrigerator box, changing back into a gas in the process. The difference between liquefaction and refrigeration, however, is that in the former process, the liquefied gas is constantly removed from the system for use in some other process, while in the latter process, the liquefied gas is constantly recycled within the refrigeration system.

Making a gas work against internal forces

In some ways, the simplest method for liquefying a gas is simply to take advantage of the forces that operate between its own molecules. This can be done by forcing the gas to pass through a small nozzle or a porous plug. The change that takes place in the gas during this process depends on its original temperature. If that temperature is less than some fixed value, known as the inversion temperature, then the gas will always be cooled as it passes through the nozzle or plug.

In some cases, the cooling that occurs during this process may not be sufficient to cause liquefaction of the gas. However, the process can be repeated more than once. Each time, more energy is removed from the gas, its temperature falls further, and, eventually, it changes to a liquid. This kind of cascade effect can, in fact, be used with either of the last two methods of gas liquefaction.

Practical applications

The most important advantage of liquefying gases is that they can then be stored and transported in much more compact form than in the gaseous state. Two kinds of liquefied gases are widely used commercially for this reason, liquefied **natural gas** (LNG) and liquefied **petroleum** gas (LPG). LPG is a mixture of gases obtained from natural gas or petroleum that has been converted to the liquid state. The mixture is stored in strong containers that can withstand very high pressures. LPG is used as a fuel in motor homes, boats, and homes that do not have access to other forms of fuel.

Liquefied natural gas is similar to LPG, except that it has had almost everything except methane removed. LNG and LPG have many similar uses.

In principle, any gas can be liquefied, so their compactness and ease of transportation has made them popular for a number of other applications. For example, liquid **oxygen** and liquid **hydrogen** are used in rocket engines. Liquid oxygen and liquid acetylene can be used in **welding** operations. And a combination of liquid oxygen and liquid nitrogen can be used in aqualung devices.

Liquefaction of gases is also important in the field of research known as **cryogenics**. Liquid helium is widely used for the study of behavior of **matter** at temperatures close to absolute zero—0K (-459°F [-273°C]).

History

Pioneer work on the liquefaction of gases was carried out by the English scientist Michael Faraday (1791-

1867) in the early 1820s. Faraday was able to liquefy gases with high critical temperatures such as **chlorine**, hydrogen sulfide, hydrogen bromide, and carbon dioxide by the application of pressure alone. It was not until a half century later, however, that researchers found ways to liquefy gases with lower critical temperatures, such as oxygen, nitrogen, and **carbon monoxide**. The French physicist Louis Paul Cailletet (1832-1913) and the Swiss chemist Raoul Pierre Pictet (1846-1929) developed devices using the nozzle and porous plug method for liquefying these gases. It was not until the end of the nineteenth century that the two gases with the lowest critical temperatures, hydrogen (-399.5°F [-239.7°C; 33.3K]) and helium (-449.9°F [-267.7°C; 5.3K]) were liquefied by the work of the Scottish scientist James Dewar(1842-1923) and the Dutch physicist Heike Kamerlingh Onnes (1853-1926), respectively.

Resources

Books

Kent, Anthony. *Experimental Low-Temperature Physics.* New York: American Institute of Physics, 1993.

McClintock, P.V.E., D.J. Meredith, and J.K. Wigmore. *Matter at Low Temperatures.* Glasgow: Blackie and Sons, 1984.

Mendelssohn, K. *The Quest for Absolute Zero: The Meaning of Low Temperature Physics.* 2nd ed. London: Taylor and Francis, 1977.

David E. Newton

Gases, properties of

The fundamental physical properties of a gas are related to its **temperature**, **pressure** and **volume**. These

Name	Formula	% Content in Atm	Color	Odor	Toxicity
Ammonia	NH_3	-	Colorless	Penetrating	Toxic
Argon	Ar	0.93	Colorless	Odorless	Non-toxic
Carbon dioxide	CO_2	0.03	Colorless	Odorless	Non-toxic
Carbon monoxide	CO	-	Colorless	Odorless	Very toxic
Chlorine	Cl_2	-	Pale green	Irritating	Very toxic
Helium	He	0.00052	Colorless	Odorless	Non-toxic
Hydrogen	H_2	0.0005	Colorless	Odorless	Non-toxic
Hydrochloric acid	HCl	-	Colorless	Irritating	Corrosive
Hydrogen sulfide	H_2S	-	Colorless	Foul	Very toxic
Krypton	Kr	0.00011	Colorless	Odorless	Non-toxic
Methane	CH_4	0.0002	Colorless	Odorless	Non-toxic
Neon	Ne	0.0018	Colorless	Odorless	Non-toxic
Nitrogen	N_2	78.1	Colorless	Odorless	Non-toxic
Nitrogen dioxide	NO_2	-	Red brown	Irritating	Very toxic
Nitric oxide	NO	-	Colorless	Odorless	Very toxic*
Ozone	O_3	Varied	Bluish	Sharp	Sharp
Oxygen	O_2	20.9	Colorless	Odorless	Non-toxic
Radon	Rd		Colorless	Odorless	Toxic
Sulfur dioxide	SO_2	-	Colorless	Choking	Toxic
Xenon	Xe	0.0000087	Colorless	Odorless	Non-toxic

*Contact with air will immediately convert nitric oxide to nitrogen dioxide— oxidation

Name of Gas	Partial Pressure in Air (mm Hg)	Percent Content in Air
N$_2$	593	78.1%
O$_2$	159	20.9%
CO$_2$	0.3	0.004%
WATER VAPOR*	17.5*	2.3%

*— This is the equilibrium vapor pressure at 20°C.

STATE	VOLUME FORM SHAPE COM-PRESSIBILITY	ARRANGEMENT AND CLOSENESS OF PARTICLES	MOTION OF PARTICLES	ATTRACTION BETWEEN PARTICLES	BOILING POINT
GAS	No definite volume, form or shape Compressible	Random; far apart	Fast	Little to none	Lower than room temperature
LIQUID	Has a definite volume, but no definite form or shape. Non-compressive tendency	Random; close	Moderate	Moderate	Higher than room temperature
SOLID	Definite volume, has own shape or form. Non-compressible	Definite; close	Slow	Strong	Much higher than room temperature

properties can be described and predicted by a set of equations, known as the gas laws. While these laws were originally based on mathematical interpretations for an ideal or perfect gas, modern atomic and kinetic theory of gases has led to a modified expression that more accurately reflects the properties of real gases.

Current understanding of gas properties came as a result of study of the interaction between volume, pressure and temperature. Robert Boyle was the first to describe the relationship between the volume and pressure of a gas. In 1660 he learned that if an enclosed amount of a gas is compressed to half its original volume while the temperature is kept constant, the pressure will double. He expressed this law mathematically as PV = constant, where P stands for pressure, V stands for volume, and the value of the constant depends on the temperature

and the amount of gas present. This expression is known as Boyle's law.

The second fundamental property of gasses was defined by Jacques Charles in 1787. He found that the temperature and volume of a gas are directly related. Charles observed that a number of gases expanded equally as **heat** was applied and the pressure was kept constant. This can be expressed mathematically as

$$\frac{V}{T} = \text{Constant (a)}$$

His ideas were expanded upon in research by others in the field, most notably Joseph Gay-Lussac who also studied the **thermal expansion** of gases. Even though Charles did not publish the results of his work, the volume/temperature relationship become known as Charles's law.

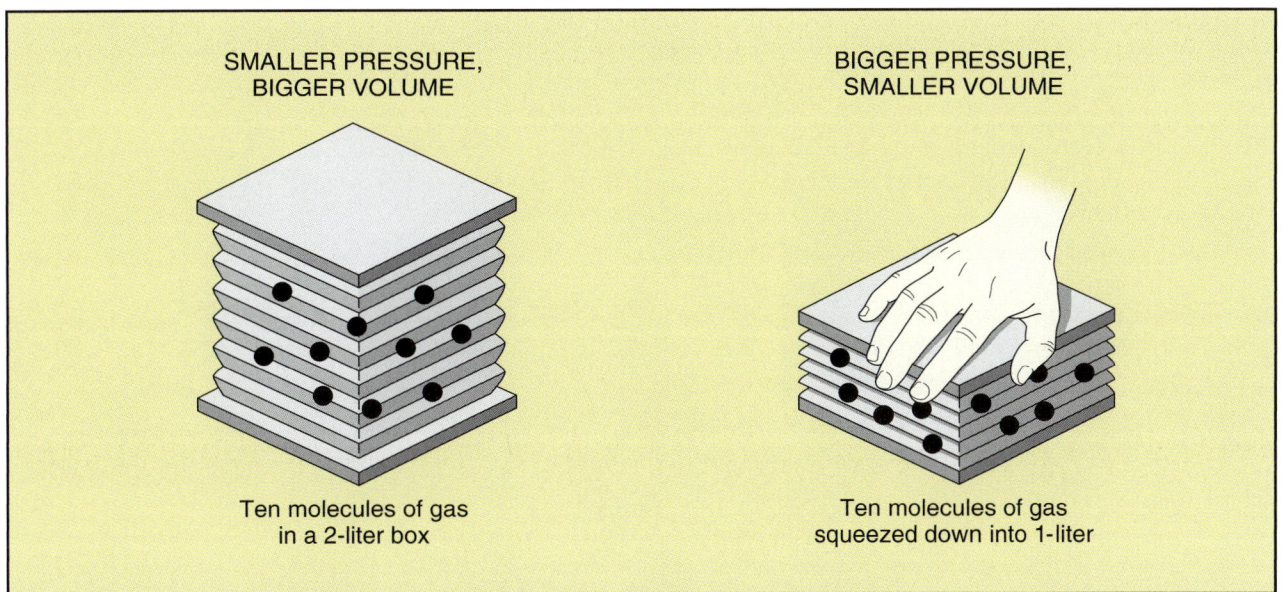

The effect of pressure change on the volume of a gas, with temperature being held constant. *Illustration by Argosy. The Gale Group.*

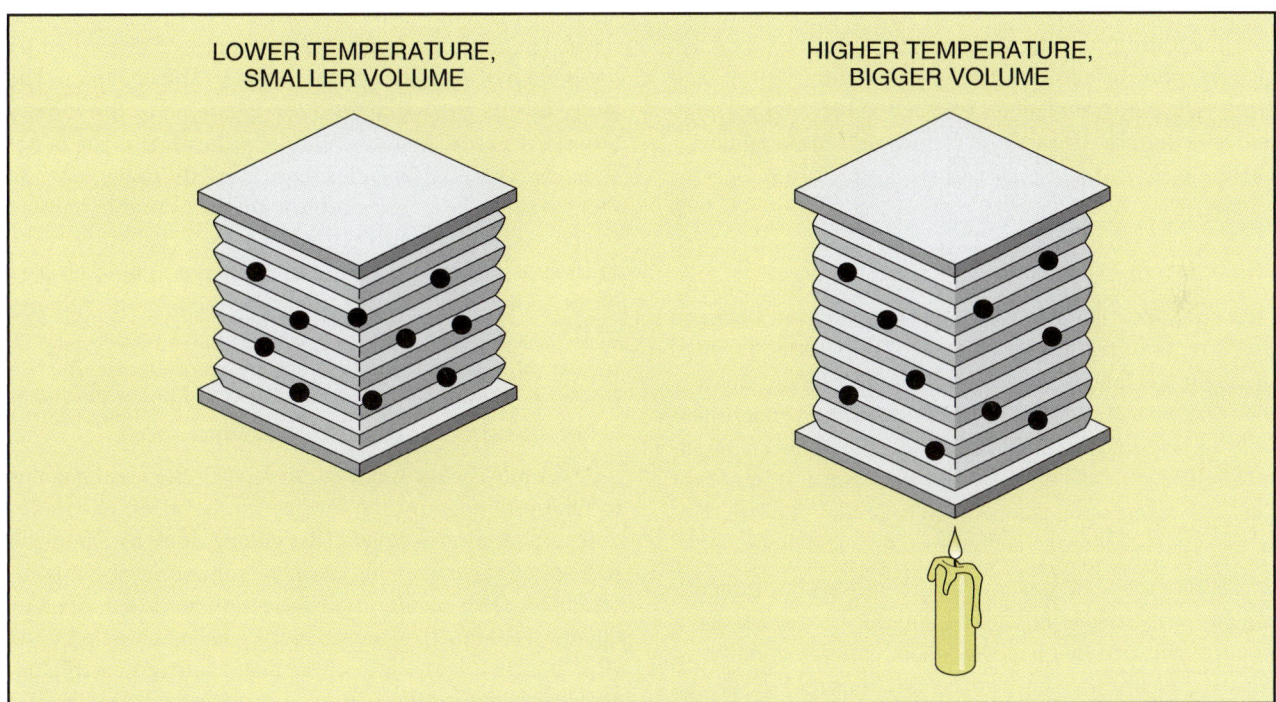

The effect of temperature change on the volume of a gas, with pressure being held constant. *Illustration by Argosy. The Gale Group.*

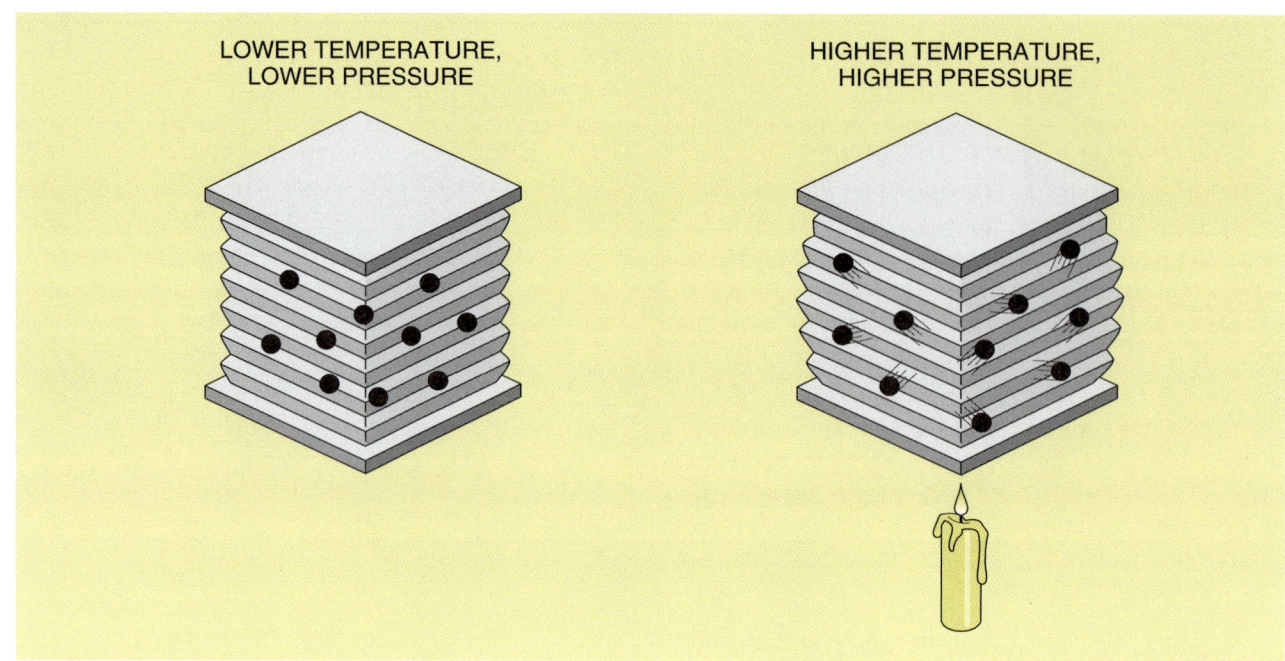

LOWER TEMPERATURE,
LOWER PRESSURE

HIGHER TEMPERATURE,
HIGHER PRESSURE

The effect of temperature change on the pressure of a gas, with volume being held constant. *Illustration by Argosy. The Gale Group.*

The third property of gases was described by Gay-Lussac who, in addition to his work with volume and temperature, researched the connection between pressure and temperature. In 1802, he formed an additional law:

$$\frac{V}{T} = \text{Constant (b)}$$

These three laws can be combined into one generalized equation that expresses the interrelation between pressure, temperature and volume. This equation, called the ideal gas law, is written as $PV = nRT$ where the R is the gas constant, which has been determined experimentally to be equal to 0.082 liter-atmospheres per Kelvin-moles. The symbol "n" stands for the number of moles of gas. This expression can be used to predict the behavior of most gasses at moderate temperatures and pressures.

While the ideal gas law works very well in predicting gas properties at normal conditions, it does not accurately represent what happens under extreme conditions. Neither does it account for the fact that real gases can undergo phase change to a liquid form. Modern **atomic theory** helps explain these discrepancies.

It describes molecules as having a certain freedom of **motion** in **space**. Molecules in a solid material are arranged in a regular lattice such that their freedom is restricted to small vibrations about lattice sites. Gas molecules, on the other hand, have no macroscopic spatial order and they can move about their containers at ran-

dom. The motion of these particles can be described by the branch of **physics** known as classical mechanics. The study of this particulate motion is known as the kinetic theory of gases. It states that the volume of a gas is defined by the position distribution of its molecules. In other words, the volume represents the available amount of space in which a **molecule** can move. The temperature of the gas is proportional to the average kinetic **energy** of the molecules, or to the square of the average **velocity** of the molecules. The pressure of a gas, which can be measured with gauges placed on the container walls, is a function of the particle **momentum**, which is the product of the **mass** of the particles and their speed.

Atomic theory was used to modify the ideal gas law to take into account the interaction between gas molecules on an atomic level. This can be done by factoring in a set of experimental parameters that describe this interaction. The resultant variation of the ideal gas law equation is known as the van der Waals equation of state: $(P + a/V2) (V - b) = RT$, where a and b are adjustable parameters determined by measuring intramolecular forces. According to this expression, a strong repulsive **force** comes into play when molecules are situated very close to one another. This force becomes mildly attractive when the molecules are at moderate distances, and its effect is not measurable at all at greater distances. **Van der Waals forces** help explain how a gas can undergo a change from a gas to a liquid state. At low temperatures (reduced molecular motion) and at high pressures

KEY TERMS

. . . .

Ideal gas law—The mathematical expression that predicts the behavior of a "perfect" gas.

Kinetic theory of gases—The physical principles that describe how gas molecules interact.

Van der Waals forces—Weak atomic forces that affect gas molecules when they are in close proximity.

or reduced volumes (reduced intermolecular spacing), the molecules in a gas come under the influence of one another's attractive force and they undergo a phase transition to a liquid. This modified gas law can be used to predict many secondary gas properties including transport properties such as thermal conductivity, the **coefficient** of **diffusion**, and **viscosity**.

Science continues to explore the basic properties of gases. For example, superconductivity, the study of **electricity** at very low temperatures, relies on super cooled gases like **nitrogen** to lower the temperature of materials to a point at which they gain special electrical properties. Furthermore, gas analysis techniques have been developed based on the discovery that the speed of sound through a given gas is a function of its temperature. These techniques rely on recently developed ultrasonic technology to analyze two-component gas mixtures that vary by as little as 1%.

Resources

Books

Dickson, T.R. *Introduction to Chemistry.* Wiley and Sons, 1991.

Holum, John R. *Fundamentals of General, Organic and Biological Chemistry.* Wiley and Sons, 1994.

Periodicals

"Gas-phase Clusters: Spanning the States of Matter." *Science,* (1 July 1988): 36.

Randy Schueller

Gasoline *see* **Hydrocarbon**

Gavials *see* **Crocodiles**

Gazelles

Gazelles are medium-sized fawn-colored antelopes found in arid parts of the world, mainly in Ethiopia, Somalia, northern **Africa** and around the Sahara Desert,

parts of the Middle East, India, and Central **Asia**. Gazelles are horned animals with a four-chambered stomach and cloven hooves. Gazelles are cud chewers (ruminants), and they lack upper canine and incisor teeth. Gazelles tear grass, foliage, buds, and shoots with a sideways **motion** of their jaws, superficially chewing and swallowing it. The food is acted on by **bacteria** in the S-shaped rumen section of the stomach, then regurgitated and chewed again.

Gazelles are grayish brown with white underbellies and rumps. They have conspicous black and white face markings and a horizontal dark-colored band along their flanks. Gazelles have slender bodies, long necks, S-shaped, ringed horns, and long legs. Their **vision** and **hearing** are well-developed. Gazelles have a distinctive way of walking, called stotting, a stiff-legged bouncing motion where all four legs hit the ground at the same time. Gazelles can be seen performing this unusual movement in moments of playfulness or when they are frightened. They have a 10-12 year life span.

Territory and social arrangements

Gazelle social arrangements vary according to the terrain they inhabit. Where food sources are abundant they are found in large herds, but in desert regions their populations are lower. In the **savanna** areas of Africa, Thomson gazelles are found in large numbers. The size of the territory ranges from 38-150 acres (15-61 hectares; Grant's gazelle, East Africa), to 250-550 acres (101-223 hectares; Edmi gazelles, Middle East), to 325-850 acres, (132-344 hectares; gerenuk or giraffe gazelle, East Africa).

Males establish territories during the mating season and routinely exclude other males. Harem herds of female gazelles with dependent young, are defended by one dominant male. Maternal herds, without a male present in the territory, and bachelor herds of male gazelles are also found. At times there are large mixed herds without a territorial male present, seen during periods of **migration**.

Gazelles mark their territories in much the same way as other ruminants do. They deposit dung heaps around the territory and they mark bushes with their scent **glands**. Glands can be found under the eyes (preorbital glands), on the hooves, shins, back and around the genital area depending on the particular **species**. When another male enters a territorial male's domain, there is no fighting as long as the intruder displays subordinate **behavior**. A subordinate male will keep his head low with his chin out and will not approach the females of the herd.

One of the smallest species is the Dorcas gazelle (*Gazella dorcas*) of North Africa (Algeria to Egypt) and Sudan, which is less than 2 ft (0.61 m) at the shoulder. The common gazelles of East Africa include Thomson's gazelle (*G. thomsoni*), with black flanks and erect horns, and Grant's gazelle (*G. granti*) which is up to 3 ft (0.915 m) at the shoulder, and the largest of all gazelles. The red-fronted gazelle (*G. rufifrons*) is found from Senegal to the Sudan.

Males within an all-male herd will frequently display intimidation behavior toward one another, but these do not often lead to attacks or injuries. Bucks will push their foreheads against one another in a display of intimidation. This may lead to interlocking horns, but they usually disengage before any serious damage occurs. Bucks will sometimes stand **parallel** to one another, head to rump, and walk around each other in circles. They may also engage in a chin-up display where they stretch their necks and bend them backwards towards one another. Within the male herds this behavior establishes dominant and submissive roles.

Mating and breeding

The mating ceremony among gazelles is ritualized. The male lowers and stretches his head and neck, following the female closely in a march-like walk, lifting his head, and prancing. The lifting of a foreleg during the mating march is also characteristic and vocal noises are made by the male. The female responds to the male's low stretch by urinating. She may walk away, circle, and make sharp turns. When she is ready for mating, she will display submissive behavior by holding her tail out.

Gestation (pregnancy) lasts around six months for gazelles. During birthing, the mother alternates between standing and lying down. Twenty minutes after **birth** a Grant's gazelle has been seen to stand up and be nursed by its mother. In its early days a fawn (newborn) spends its time between feeding & hiding out in the grass. Typically they lie in a different hiding place after each feeding. The mother will keep watch over the newborn from a distance. Many gazelles reproduce twice a year when sufficient food supplies are available.

Preservation and adaptation

In parts of Africa where national **wildlife** parks have been established, gazelles can be found in large numbers. In some parts of North Africa, Arabia, and the Near East, however, where they have not had protection, many species of gazelle have been nearly wiped out. Some gazelles that were close to **extinction** have been preserved through the efforts of particular governments or by individuals in cooperation with zoos.

A number of species of gazelle survive well in arid desert regions. Notable among them is the gerenuk, or giraffe gazelle, so called for the habit of standing up to

KEY TERMS

Bachelor herds—A group of young, nonterritorial males.

Intimidation—Threatening behavior among the same sex for the purpose of expressing dominance or for preventing intruders from entering the territory.

Maternal herd—A group of females with their dependent young.

Ruminant—A cud-chewing animal with a four-chambered stomach and even-toed hooves.

Stotting—A bounding movement where the animal will bounce and land on all four legs in response to threatening situations.

Territorial male—A male that defends its area and harem from other males.

forage for food. The gerenuk is able to balance itself on its rear legs and it has an unusually long neck. In zoos this gazelle seems never to drink **water** and has only on rare occasions been seen to drink in its natural **habitat**.

Resources

Books

Estes, Richard D. *Behavior Guide to African Mammals.* Berkeley: University of California, 1991.

Estes, Richard D. *The Safari Companion.* Post Mills, Vermont: Chelsea Green, 1993.

Haltenorth, T. and Diller, H. *A Field Guide to the Mammals of Africa.* London: Collins, 1992.

Spinage, C.A. *The Natural History of Antelopes.* New York: Facts on File, 1986.

Vita Richman

Gears

A gear is a toothed disk attached to a rotating rod or shaft that transmits and modifies rotary **motion** by working in conjunction with another gear. Usually circular in shape, the protrusions of one gear mesh into the profile of its mate to obtain a predetermined mechanical advantage. For example, if one gear wheel has ten times as many teeth as the wheel that drives it, it will make one tenth of a turn for every full turn of the latter, while simultaneously exerting ten times the **torque** or turning **force** applied to it by the driving wheel. This process converts a weak force applied to the driving wheel into a strong force delivered by the driven wheel.

An example of early gear trains is the Antikythera mechanism. This gear-driven calendar device made in Rhodes about 87 B.C. contains at least 25 gears cut in bronze. With it, the positions of the **sun** and **moon** could be predicted as well as the rising and setting of certain stars. By the first century A.D. all the simple kinds of gears were well known.

A pinion is a gear with a small number of teeth engaging with a rack or larger gear. A bevel gear is one of a pair of toothed wheels whose working surfaces are inclined to nonparallel axes. A worm gear transmits power from one shaft to another, usually at right angles. Automobiles employ a differential gear, which permits power from the engine to be transferred to a pair of driving wheels, dividing the force equally between them but permitting them to follow paths of different lengths, as when turning a corner or traversing an uneven road.

Researchers at the NASA Ames Research Center are developing molecule-sized gears and other machine parts in the hopes of producing nanostructures capable of self-repair or that could adapt to a given environment. The Ames team "built" hypothetical gears by forming tubes from fullerenes, a class of molecules consisting of 60 **carbon atoms** arranged in a ball-like lattice. They attached **benzene** molecules onto these fullerenes for "teeth." Researchers propose to turn the gears with a **laser** that will create an electronic field around the nanotube that will drag the tube around similar to a shaft turning. Although these gears presently exist only in computer simulations, the simulations predict that the gears would rotate best at about 100 billion turns per second, or six trillion rotations per minute and are virtually unbreakable.

Resources

Books

Glover, David. *Pulleys and Gears* Oxford: Heineman Library, June 1997.

Macaulay, D. *The Way Things Work* Boston: Houghton Mifflin Co. 1988.

Williams, T. *The Triumph of Invention.* London: Macdonald Orbis. 1987.

Periodicals

Hall, Alan. "A Turn of the Gear." *Scientific American* April 28, 1997.

Laurie Toupin

Geckos

Geckos are small night-lizards found in the tropics and subtropics, and number more than 650 **species** in the

A parachute gecko (*Ptychozoon kuhli*). *Photograph by Tom McHugh. The National Audubon Society Collection/Photo Researchers, Inc. Reproduced by permission.*

family Gekkonidae, divided into four subfamilies (the Diplodactylinae, the Gekkoninae, the Sphaerodactylinae, and the Eublepharinae). Only the Eublepharinae have eyelids, while members of the other three subfamilies have transparent scales protecting their eyes.

Geckos are small lizards, ranging in length from less than 2 in (5 cm), to seldom more than 1 ft (30 cm). Geckos are primarily insectivorous and nocturnal, and are unique in that they are the only lizards with a true voice. Depending on the species, geckos utter anything from a soft, high-pitched squeak to a loud **bark**. The name *gecko* arose as an attempt by humans to mimic the sound made by a common North African species (*Gekko gekko*). Geckos have a soft, scaly, often transparent skin which readily tears away, allowing the little creature to escape the jaws or beak of a **predator**. Special toe pads enable geckos to walk upside down across **rocks**, on ceilings, and up the walls of city skyscrapers. Geckos are thought to have originated in Southeast **Asia** and the western Pacific, but are now found in large numbers in the warmer parts of every **continent**, and even on isolated islands around the world. Geckos make popular house pets, since they are harmless, relatively unafraid of humans, and provide effective and natural control of insect **pests** such as the cockroach. Geckos may live as long as 15 years in their natural environment, but seldom that long in captivity.

Distribution and habitat

Geckos began their **migration** from the Pacific Rim thousands of years ago, some "stowing away" on the canoes of unsuspecting sea voyagers; others beginning colonization from eggs deposited under the bark of logs subsequently swept out to sea and washed up on a distant shore. As humans graduated from forest and land dwelling, building cities in which artificial lights illuminate the night skies attracting billions of **insects**, geckos also graduated from their original habitats to these new urban feasting grounds. Today, flicking on the **light** in the middle of the night in apartments, homes, and even tall office buildings in many parts of the world, one may interrupt the nocturnal feeding foray of one of these little creatures.

Only a small number of gecko species occur in **North America**. The tiny, two-inch leaf-toed gecko or *Phyllodactylus tuberculoses*, leaf-toed gecko, thrives in southwestern Californian among the rocks of semiarid lower mountain regions and canyon lands. The banded gecko inhabits southern California's coastal plains, rocky deserts, juniper-covered hillsides, and **sand** dunes. Several species of West Indian geckos are now established in Florida, and many different species thrive on the Hawaiian islands.

Physiology and reproduction

The texture and **color** of a gecko's skin provides excellent camouflage. Four strong legs and five specially-equipped toes on each foot provide for excellent climbing abilities; while two round eyes with vertical pupils allow sharp, nocturnal **vision**. Diurnal (daytime) geckos, such as the wall gecko (*Tarentola mauritanica*), of North **Africa**, Spain, and Croatia, have rounded pupils.

Geckos do not have a forked tongue. Geckos use their tongues to help capture **prey** and some—like the Australian naked-toed gecko and the Asian tokay gecko—use their tongues to clear their **eye** scales of dust and debris. The head is relatively large in comparison to the tubular-shaped body, and the long, sheddable tail comprises up to one half of the total body length, snapping off in sections if it is grabbed by a predator. The discarded tail wriggles around on the ground, distracting the attacker's attention and providing precious seconds for the **animal** to flee. A new tail grows back within a few months. The tail also stores **fat**, providing **nutrients** in times of food scarcity. Being cold-blooded creatures, geckos draw their body **heat** from their environment by basking in direct sunlight or on warm surfaces.

When mating, the male gecko grasps the skin at the back of the female's neck in his jaws and wraps his tail around that of the female, bringing their cloacas—the reproductive openings together. Some species of gecko reproduce asexually, when the female produces fertile eggs without mating with a male. All geckos, except some species found in New Zealand, lay eggs. Some species lay one egg in each clutch while others lay two. Eggs are deposited under rocks, **tree** bark, and even behind win-

dow shutters. Only a few species lay two clutches per year and incubation may take several months. Eggs of the banded gecko and of many other species have a leathery, parchment-like texture, while those of such species as the leaf-toed gecko have a hard, calcareous (containing **calcium**) shell, the durable nature of which has aided in the wide-spread distribution of many species, particularly the species that reproduce asexually, where just one viable egg can begin a whole new colony.

Defensive behavior

The Australian spiny-tailed gecko (*Diplodactylus williamsi*) displays the most unique defense of all lizards. When this grey, inconspicuous gecko suddenly swings opens its jaws, it displays a vivid, dark purple mouth outlined in bright blue. It may also emit a high-pitched squeak and, if attacked, shoots a thick, gooey liquid from spiny knobs on its tail, covering its enemy with a sticky weblike substance.

Although geckos in general show aggressive displays such as arching the back, stiffening the limbs to increase their height, and wagging their tails, they are relatively nonaggressive, fighting among themselves only when defending a homesite or feeding territory from a determined invader. Although small geckos will attack a foe many times their size if threatened. The Australian barking gecko (*Underwoodisaurus milii*) barks and lunges even at humans. Very few species of gecko are strong enough to break the human skin, and none are poisonous.

See also Reptiles.

Resources

Books

Bustard, Robert. *Australian Lizards*. Sydney: Collins, 1970.
Cogger, Harold G., David Kirshner, and Richard Zweifel. *Encyclopedia of Reptiles and Amphibians*. 2nd ed. San Diego, CA: Academic Press, 1998.

Conant, Roger, et al. *A Field Guide to Reptiles & Amphibians of Eastern & Central North America (Peterson Field Guide Series).* Boston: Houghton Mifflin, 1998

Periodicals

Petren, Kenneth, and Ted J. Case. "Gecko Power Play in the Pacific." *Natural History* (September 1994): 52-60.

Petren, Kenneth, Douglas T. Bolger, and Ted J. Case. "Mechanisms in the Competitive Success of an Invading Sexual Gecko over an Asexual Native." *Science* 259 (January 15 1993): 354-57.

Marie L. Thompson

Geese

Geese are large **birds** in the subfamily Anserinae of the waterfowl family Anatidae, consisting of **ducks**, geese, and **swans**.

Geese occur in many types of aquatic habitats, on all continents but **Antarctica**. Most geese breed in **freshwater** marshes, **salt** marshes, or marsh-fringed, open-water **wetlands**. Geese typically winter in those sorts of natural habitats and in estuaries, although in some regions they also use grainfields in winter, mostly for feeding. Geese are more terrestrial than either ducks or swans, and they typically feed on roots, rhizomes, and shoots of graminoid (grass-like) plants, and on **seeds** and grains, when available.

Geese are not sexually dimorphic, meaning that there are no obvious, external morphological traits that serve to distinguish between the female, properly named a goose, and the male, or gander. Ganders do tend to be somewhat larger, but size is not a reliable indicator of gender. Like other waterfowl, geese undertake a simultaneous moult of their major wing feathers, and are flightless at that time. This moult occurs during the breeding season, while the geese are taking care of their young.

Most **species** of goose undertake substantial migrations between their breeding and wintering grounds, in some cases traveling thousand of miles, twice yearly. Flocks of migrating geese commonly adopt a V-shaped formation, which is aerodynamically favorable, because it reduces resistance to passage, so less **energy** is expended in flying. Geese can be rather noisy when flying in groups, which may sometimes be heard before they are seen.

Geese of North America

The six species of goose that breed in **North America** are the Canada goose (*Branta canadensis*), brant (*B. bernicla*), black brant (*B. nigricans*), snow goose (*Chen*

caerulescens), Ross's goose (*C. rossii*), and white-fronted goose (*Anser albifrons*). Two other Eurasian species are occasionally seen during winter: the barnacle goose (*B. leucopsis*) on the northeastern coast, and the emperor goose (*Philacte canagica*) along the Alaskan coast. The two most abundant species of geese in North America are the Canada goose and the snow goose. The Canada goose is also known as the "honker" because of its resonant call, given especially enthusiastically during migratory flights and while staging. Because of geographic variations in size, morphology, and **color** patterns, the Canada goose has been divided into about 11 races. However, these races intergrade with each other, and should best be considered to represent continuous, geographic variations of a genetically polymorphic species. The largest race is the giant Canada goose (*B. c. maxima*), of which mature ganders typically weigh about 12.5 lb (5.7 kg). The giant Canada goose has become rather common in some urban and suburban areas, where it has been widely introduced and has established feral, non-migratory, breeding populations. However, because of past overhunting, this race is much less abundant than it used to be in its natural breeding range of southern Manitoba, northwestern Ontario, and Minnesota. The smallest race is the cackling Canada goose (*B. c. minima*). Males of this rather dark goose only weigh about 3.5 lb (1.5 kg). This relatively abundant race breeds in the western subarctic, especially in Alaska, and winters in the Pacific Northwest of the United States and southwestern British Columbia.

Because of its abundance and widespread migrations, the Canada goose is probably the most familiar goose to most North Americans. During their migrations, the larger-sized races of Canada goose tend to occur in relatively small, often family-sized flocks, and their calls tend to be extended, sonorous honks. Smaller-bodied races flock in much larger groups, ranging up to thousands of individuals, and often flying in large V-shaped formations. These smaller geese have calls that tend to be relatively higher pitched yelps and cackles. Wintering populations of Canada goose often occur as large, dense aggregations in the vicinity of good feeding **habitat**.

During the era of unregulated market and sport hunting of the nineteenth and early twentieth centuries, the populations of Canada geese were greatly reduced from their historical abundance. This decline was exacerbated by large losses of breeding habitat in the more southern parts of their range, largely due to the conversion of North America's prairies to agriculture, which was accompanied by the draining of many small, marsh-fringed ponds known as potholes. However, the federal governments of the United States and Canada, and the states and provinces, have since instituted effective **conservation** measures for the Canada goose and most other species of waterfowl. The most important of these actions is the regulation of hunting effort by restricting the numbers of birds that can be killed, and by limiting the hunting season to a period during the autumn and thereby eliminating the spring hunt, which killed animals before they had an opportunity to breed that year. Also very important has been the designating of a large network of protected areas, mostly to provide essential habitat and refuges from hunting for migrating and wintering waterfowl. In the case of the Canada goose, these measures have proven to be effective, and populations have recovered substantially from lows in the second decade of the twentieth century. At the end of the breeding season and during the autumn **migration**, North America now supports about three million Canada geese.

At least one million of these birds are subsequently killed by hunters, or by the insidious toxicity of ingested lead shot, or they may suffer natural mortality. Young, relatively inexperienced birds-of-the-year are most commonly killed by hunters, with the more wary adult birds tending to survive this type of predation. This is an important aspect of the hunt, because it results in the reproductive capacity of the population being left relatively intact. It appears that present-day populations of Canada geese are capable of withstanding the intense, annual mortality they are exposed to through hunting and other factors. However, it is important that this situation be continuously monitored, so that any emergent problems are quickly identified, and actions taken to prevent future population declines of this important species of **wildlife**.

The snow goose is another abundant species of goose in North America, tending to breed to the north of the major range of the Canada goose. This mostly white goose is often divided into two races, the relatively abundant and widespread lesser snow goose (*C. c. caerulescens*) and the greater snow goose (*C. c. atlantica*) of the high Arctic. The lesser snow goose has two color variants, the familiar white-bodied form with black wing-tips, and the so-called "blue" variant. The blue phase is genetically dominant over the white, and it occurs most frequently in the populations of snow geese breeding in the eastern low Arctic of Canada.

Like the Canada goose, the snow goose has been exploited heavily, and its populations were once imperilled by overhunting and habitat loss, especially of wintering habitat. However, strong conservation measures have allowed a substantial increase in the abundance of this species, which, although still hunted, may be approximately as abundant as prior to its intensive exploitation. In fact, since the 1980s, the rapidly increasing breeding populations of snow geese have caused significant degradations of parts of their habitat in the vicinity of Hudson Bay, through overgrazing of important forage species.

Brant and black brant are less common species of goose, occurring in eastern and western North America, respectively. These species are ecologically different from the other North America geese, because of their affinity for estuarine habitats, where they prefer to forage vascular plants known as eelgrass (*Zostera marina*). The brant is less abundant than it used to be, because of overhunting, and degradations of its wintering habitat, caused in part by occasional declines of its preferred forage of eelgrass. The causes of the eelgrass declines have not been determined, but they may be natural in origin, or somehow caused by human influences, possibly associated with **eutrophication**. In some years, these geese have also suffered reproductive failures due to unfavorable **weather** in their northern breeding grounds. This circumstance may also have contributed to the decline of brant.

Economic importance of geese

Like other waterfowl, wild geese have long been hunted for subsistence purposes, and more recently for sport. In recent decades, North American hunters have killed about two million geese each year, although the bag has varied depending on the annual abundance of the birds. About 75% of the geese are typically killed in the United States, and the rest in Canada. Goose hunting is an economically important activity, generating direct and indirect cash flows through spending on travel, guns and other equipment, licenses, and fees paid to hunt on private lands.

Compared with the unregulated, open-access hunts of the past, which devastated populations of all waterfowl and other animals, hunting now appears to be relatively sustainable of the avian resource. Each year the federal governments of the United States and Canada cooperate in setting bag limits on the basis of estimates of the productivity of geese in the breeding habitats. The regulation of the direct kill of geese, coupled with the development of a network of protected areas of breeding, staging, and wintering habitat, appears to be effective in maintaining populations of the most abundant species of geese, while still allowing a large sport hunt.

Two species of goose have been domesticated. The most commonly raised species is derived from the greylag goose (*Anser anser*) of Eurasia. This goose has been domesticated for about 4,000 years, and there are a number of agricultural races, most of which are white. Another, less common, domesticated species is the swan goose (*A. cygnoides*).

Like ducks and other birds, geese have increasingly attracted the interest of bird-watching, also an activity of significant economic importance.

Geese are sometimes viewed as agricultural **pests**, because they may invade fields in large numbers during

A Canada goose (*Branta canadensis*) with her goslings in the Ottawa National Wildlife Refuge, Ohio. *Photograph by Robert J. Huffman. Field Mark Publications. Reproduced by permission.*

the autumn and spring, raiding unharvested **crops** or damaging fields of winter **wheat** (which is sown in the autumn to be harvested in the following summer) and some other crops. These damages can be severe in smaller areas, but can be managed by providing the geese with alternative foods, or by scaring them away.

Factors affecting the abundance of geese

Geese are affected by many of the same environmental factors that influence populations of ducks. Some of these influences are natural. These include the effects of severe weather on the northern breeding grounds of geese, which in extreme cases can wipe out a year's breeding success. Sometimes, natural predators such as foxes and **bears** can disrupt breeding in a particular area. When they aggregate in large populations during staging or wintering, geese are also vulnerable to epidemics of diseases such as avian **cholera**. Natural degradations of staging or wintering habitats may also be important, as may be the case of eelgrass declines in estuaries used by the brant. As was noted previously for the snow goose, large populations of geese can sometimes degrade their own habitat through overgrazing.

Humans have also greatly affected goose populations. The most important of the negative influences of humans on geese have been overhunting, destruction of staging and wintering habitats, and the toxic effects of ingested lead shot. However, as with ducks, many of these negative influences are now being managed in North America, by controlling the size of annual hunts, by instituting a network of key habitat reserves, and by banning the use of lead shot. These actions have mostly been carried out by agencies of government, as well as by nongovernmental organizations, including hunter-focused groups such as Ducks Unlimited, and groups with a conservation mandate, such as the Nature Conservancies.

Humans have increasingly been undertaking activities on behalf of geese and other wildlife. However, these animals are still threatened by many human activities. The eventual balance of the positive and negative interactions of humans and wildlife remains to be determined. Hopefully, the conservation of the populations of all of the world's species of goose will become an important priority to humans, so that these creatures will always be available to be sustainably harvested, while still maintaining their populations.

Status

- Greater white-fronted goose (*Anser albifrons*). The population in North America appears to have declined in the 1970s, but to have increased since then.

- Lesser whitegfronted Goose (*Anser erythropus*). Alaskan stray. An uncommon species in its native habitat in the Old World, where it seems to be in decline. Has strayed at least once to the Aleutian Islands. Occasional sightings outside Alaska have probably been escapees from captivity.

- Snow goose (*Chen caerulescens*). The population of the greater snow goose had declined to no more than 3,000 by 1900. In 1988, the snow goose population in eastern Canada was estimated to be 2.4 million, with an annual **rate** of increase of 130,000/year. The lesser snow goose population has undergone a pronounced increase in recent decades.

- Ross's goose (*Chen rossii*). In 1983, the population in the central Canadian Arctic was estimated to be in excess of 100,000 in 30 colonies. Today, the population appears to still be increasing. This goose frequently hybridizes with the snow goose, but there is no evidence of genetic swamping by that species.

- Canada goose (*Branta canadensis*). The Aleutian Canada goose is Endangered, having almost been exterminated following the introduction of foxes to the Aleutian Islands. The population of the species as a whole is probably increasing.

- Brant (*Branta bernicla*). Decline due to disappearance of eelgrass along much the eastern seaboard since the 1930s. Eelgrass has also disappeared in England over the same period.

- Bar-headed goose (*Anser indicus*). Exotic. Native of Central **Asia**. Birds that have escaped from captivity in the United States are sometimes seen in the wild.

- Barnacle goose (*Branta leucopsis*). Eastern stray. Resident of Arctic coasts from Greenland to Siberia, wintering in northwestern **Europe**. Most strays seen in United States have escaped from captivity, though some occasionally arrive in North America from Greenland.

- Bean goose (*Anser fabalis*). Alaskan stray. A common goose in northern Asia and Europe, this bird sometimes shows up in Alaska in the spring and, more rarely, in other parts of North America.

- Chinese goose (*Anser cygnoides*). Exotic. Native of Asia. Domesticated birds in the United States sometimes abandon their home ponds.

KEY TERMS

Feral—This refers to a non-native, often domesticated species that is able to maintain a viable, breeding population in a place that is not part of its natural range, but to which it has been introduced by humans.

Graminoid—Any grass-like plant, usually referring to grasses, sedges, reeds, rushes, and other erect, monocotyledonous species.

Overhunting—The unsustainable harvesting of wild animals at a rate greater than that of recruitment of new individuals, so that the population decreases in size.

Polymorphic—Refers to genetically based variations in shape, size, color, and other traits.

Staging—A characteristic of certain migratory birds, in which individuals collect in large numbers in places with extensively appropriate habitat. Weight gain in staging habitats is important to successful completion of the subsequent arduous, long-distance migration.

- Egyptian goose (*Alopochen aegyptiacus*). Exotic. Native of **Africa**. Sometimes escapes from captivity in the United States.

- Emperor goose (*Chen canagica*). Threatened. The Alaska population, estimated at 139,000 in 1964, had declined to 42,000 in 1986. The status of the population is not well known, but the population there appears to have declined in the twentieth century.

- Graylag goose (*Anser anser*). Exotic. A native of Eurasia. Rare sightings in the United States have probably been of domesticated birds that have escaped captivity.

- Pink-footed goose (*Anser brachyrhynchus*). Eastern stray. Many of these birds nest in Greenland and Iceland, migrating to Britain and northwestern Europe where they spend the winter. Strays have been observed a couple of times in eastern Canada.

- Red-breasted goose (*Branta ruficollis*). Exotic. A native of Eurasia. Birds that have escaped from captivity have been seen in the Northeast.

Resources

Books

Bellrose, F. C. *Ducks, Geese, and Swans of North America.* Harrisburg, PA: Stackpole Books, 1976.

Ehrlich, Paul R., David S. Dobkin, and Darryl Wheye *The Birder's Handbook* New York: Simon & Schuster Inc., 1988.

Godfrey, W. E. *The Birds of Canada.* Toronto: University of Toronto Press, 1986.

Johnsgard, P. A. *Ducks in the Wild. Conserving Waterfowl and Their Habitats.* P-H Reference and Travel, 1993.

Owen, M., and J. M. Black. *Waterfowl Ecology.* London: Blackie, 1990.

Peterson, Roger Tory *North American Birds.* Houghton Miflin Interactive (CD-ROM), Somerville, MA: Houghton Miflin, 1995.

Bill Freedman
Randall Frost

Geiger counter *see* **Radiation detectors**

Gelatin

Gelatin is an edible protein made from the skin, bones and ligaments of animals. It is clear, usually colorless or pale yellow, odorless and tasteless, and dissolves in **water**. The hot **solution** is liquid, but as it cools, it "gels," forming a semi-solid, which is soft and flexible, yet firm enough to hold any shape into which it may be molded or cut. A familiar example of gelatin is the clear, sticky substance found on parts of a chicken leg after it has been cooked.

Most manufactured gelatin comes from pork skins and the skin and bones of cattle. These contain a tough, fibrous protein called **collagen**. First they are treated with either acids or bases (alkali) to dissolve hair, flesh, and other unwanted substances. Then they are cooked in hot water. The **heat** converts the collagen to gelatin, which dissolves in the water. The solution is purified and the water is removed by **evaporation**. Finally, the pure, solid gelatin is ground into flakes or powder.

Like all **proteins**, gelatin is a **polymer**. That is, its molecules are built up of smaller units called amino acids linked together by chemical bonds like beads along a string. In collagen three **amino acid** strands, each about a thousand amino acid units long, are twisted together as a sort of braid. Individual "braids" are joined by chemical bonds, making very tough web-like structures. Heating in water breaks some of the bonds. Therefore, gelatin consists of shorter strands, with fewer chemical bonds between them. These smaller fragments dissolve in water. As a hot gelatin solution cools, some of the bonds form once again, causing the solution to thicken. Cooling further, the protein strands form a three-dimensional mesh, with water filling the holes of the mesh. The resulting "gel" is soft enough to cut, yet rigid enough to hold its shape.

Gelatin cannot be a major source of protein in the human diet because it lacks tryptophan, one of the amino acids essential for human **nutrition**. Its main use in the food industry is to provide texture and shape to foods, especially desserts, candies, and dairy products. Gelatin also has many non-food uses. Made into capsules, it encloses drugs for the pharmaceutical industry and microscopic drops of ink for "carbonless" copying papers. A layer of gelatin binds light-sensitive chemicals to the surface of photographic film. Gelatin is also used as glue for objects as diverse as match-heads and the bindings of **telephone** books.

Gene

A gene is the basic structural unit of inheritance in biological organisms. It is made up of a short segment of DNA and contains the necessary information to produce a specific protein. Each gene is separated from each other by non-coding sequences that serve other functions. Genes are strung together and tightly packed into structures called chromosomes. All the genes in an **organism** are located on chromosomes in the nucleus of most cells and represent the blueprint for instructions that make up an organism. For example, genes can determine physical characteristics in humans such as height, **eye color**, skin color, or any other trait. Genes are passed from one generation to the next through sex cells (the egg and the sperm) called gametes. Maternal and paternal genes combine at **fertilization** and each contribute to the observable features of the offspring, explaining why children often look like one or both parents.

Mutations, which are changes in the structure or sequence of DNA, can cause **disease** if it involves disruption in the specific sequence of a gene. For example, if a **mutation** disrupts a gene that encodes a protein responsible for controlling **cell division**, this loss of function might cause the **cell** and the cells that arise from it to continuously divide, producing **cancer**.

History

In 1909, Wilhelm Johannsen (1857–1927), a Danish biologist, first proposed the name *gene* as the term designating the basic unit of information that is inherited. In 1944, the Canadian bacteriologist Oswald T. Avery (1887–1955) and American scientists Colin M. Macleod (1909–1972) and Maclyn McCarty demonstrated that DNA is the material responsible for a process called transformation in **bacteria**, or the transfer of genetic information from one bacterium to another. These re-

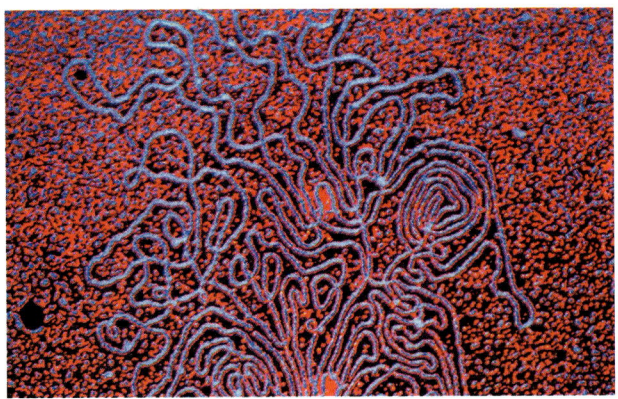

Strands of DNA. *Photograph by Howard Sochurek. Stock Market. Reproduced by permission.*

searchers had no idea how important their discoveries were until many years later when further studies demonstrated that DNA was the material responsible for the transfer of genetic information in most living organisms. James Watson, an American biochemist, and Francis Crick, a British scientist, presented in 1953 a model of DNA that resembles a twisted ladder. The sides of the ladder are composed of sugar-phosphate groups, and the rungs consist of paired nitrogenous bases. It was shown that there are bases in DNA and the arrangement of the four bases encodes the information held by genes. The DNA model explains how DNA replicates, or makes copies of itself. Later, the American biochemist Marshall W. Nirenberg, and others, used the model to work out the genetic code—the relationship between the arrangement of the DNA bases and the amino acids produced by the DNA sequences in each gene.

Gene expression

DNA is made up of four building blocks or nitrogenous bases; adenine (A), guanine (G), cytosine (C), and thymine (T). A, C, G, and T represents the DNA alphabet and different combinations of these letters mean something different. Every gene begins with a specific start sequence and ends with a stop sequence. Therefore, the specific sequences of these four bases determines whether the DNA codes for **proteins** (coding DNA), the specific protein it encodes, or whether it represents non-coding DNA that does not encode for protein. Non-coding DNA, also called junk DNA accounts for 97% of the **genome** and despite its name, it serves many purposes including the proper functioning of genes. Each gene can be converted or transcribed into a type of RNA called messenger RNA (mRNA). RNA is very similar to DNA except that instead of thymine as one of its four nitrogenous bases, uracil (U) is substituted. Gene expression can be controlled by proteins called transcription factors,

enhancers, and silencers. These regulatory proteins influence whether proteins will be expressed by binding to specific sequences of DNA or through interactions with other DNA binding proteins.

Information is passed from the DNA **molecule** to the messenger RNA (mRNA) by the pairing of complementary bases in each of the two strands. The mRNA then carries the instructions from the DNA in the nucleus to a ribosome in the cytoplasm. A molecule called transfer RNA (tRNA) transports an **amino acid** that is designated to match a codon, or a three base pair sequence in the mRNA. Amino acids strung together form a particular protein. The proteins that are produced might, for example, be important for human growth and development or represent important enzymes in physiological pathways.

Knowing the sequence of every gene found in the human genome, made possible in part by The **Human Genome Project**, will allow scientists to better understand the cause of diseases such as cancer or **cystic fibrosis** and develop new ways to treat or cure these diseases by characterizing **individual** genes as well as gene-gene interactions. The rough draft sequence of the human genome was completed and published in February 2001 in both *Nature* and *Science* scientific journals and the final sequence is expected to be completed sometime during 2003.

See also Genetic disorders; Genetics; Meiosis; Molecular biology; Mutagen; Mutagenesis.

Resources

Periodicals

The International Human Genome Mapping Consortium. "A Physical Map of the Human Genome." *Nature* 409, 934–941 (2001).
International Human Genome Sequencing Consortium. "Initial Sequencing and Analysis of the Human Genome." *Nature* 409, 860–921 (2001).

Other

National Institutes of Health. "Guide to the Human Genome" [cited October 19, 2002]. <http://www.ncbi.nlm.nih.gov/genome/guide/human/>.

Bryan R. Cobb

Gene chips and microarrays

The **Human Genome Project** began in 1990, with the goal of sequencing the complete human **genome**. Although estimates to complete the daunting project initially ranged up to forty years, with advances in technology—including **gene** chip and microarray technology—

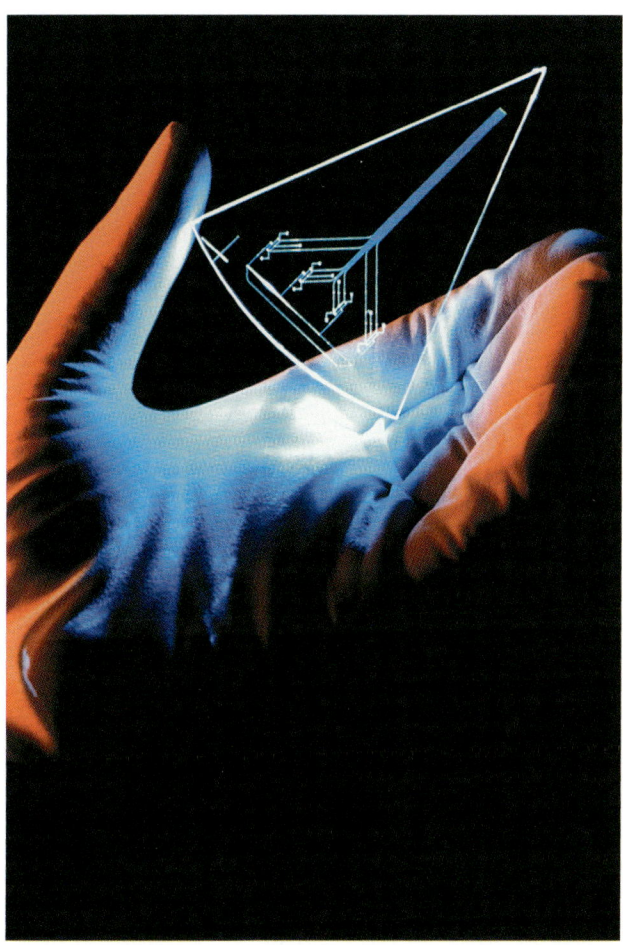

Gloved hand holding a device for rapidly analyzing samples of DNA at the scene of a crime. *Photograph by Sam Ogden. Photo Researchers, Inc. Reproduced by permission.*

by 2001 the project was nearly complete. Although researchers continue to refine data, as of 2003, the human genome is well characterized and researchers are increasingly turning their interests to linking genes with specific cellualar processes.

A gene chip is typically constructed from **glass**. Wafer-like in appearance, it resembles microtransitor chips. However, instead of transitors, a chip is contains an orderly and densely packed array of DNA **species**. The thousands of sequences of DNA are robotically spotted onto the chip. The pattern is called a microarray.

Genetic material is obtained from the biological system of interest. This can be DNA or RNA. For example, to gain insight into which genes are active at a particular time, messenger RNA is isolated. The isolated material is reacted with a fluorescent probe and then the mixture is flooded over the surface of the chip. Hybridization of the added **nucleic acid** and a piece of the tethered DNA will occur if the sequences compliment one another. The

development of **fluorescence** on the chip's surface identifies regions of binding, and the known pattern of the tethered DNA can be used to deduce the identity of the added **sample**.

Vast amounts of information are obtained from a single experiment. Up to 260,000 genes can be probed on a single chip. The analysis of this information has spawned a new science called bioinformatics, where **biology** and computing mesh. Gene chips are having a profound impact on research. Pharmaceutical companies are able to screen for gene-based drugs much faster than before. The future of chips economically is with the practicing physician. For example, a patient with a sore throat could be tested with a single-use, disposable, inexpensive gene chip in order to identify the source of the **infection** and its antibiotic susceptibility profile. Therapy could commence sooner and would be precisely targeted to the causative infectious agent.

See also Bioinformatics and computational biology; Genetic identification of microorganisms; Genetic testing; Genetics.

Gene mutation

The term *mutation* was originally coined by Dutch botanist Hugo De Vries (1848–1935) to describe a new approach to explain **evolution**, although it is quite different than the current definition. De Vries discovered new forms of the Evening Primrose (*Oenothera lamarcklana*) that were growing in a meadow. He attributed these new varieties and the method for which new **species** arise to what he called mutations. As a result of his observations, Gregor Mendel's principles of heredity were rediscovered and helped to explain variability within and between species.

Today, technological advances in **deoxyribonucleic acid (DNA)** analysis have provided scientists with tools to rapidly sequence the human **genome**. One of the main benefits of this technology is to identify mutations or alterations in the DNA sequence that might be associated with **disease**. A growing field called bioinformatics is becoming a useful field in understanding and identifying **gene** mutations by addressing the computational challenges of analyzing the large amount of **sequencing** data. DNA chips or microarrays have also recently emerged with applications that involve whole-genome scanning **mutation** detection.

There are many different types of mutations in the human genome and are either considered major gene rearrangements or point mutations, both of which are discussed in more detail below. Major gene rearrangements involve DNA sequences that have deletions, duplications, or insertions. Point mutations are single substitutions of a specific letter of the DNA alphabet (i.e. adenine, guanine, cytosine, or thymine). Alterations in the DNA sequence can result in an alteration of the protein sequence, expression, and/or function.

Genes represent the basic hereditary unit that allows species to pass its information from one generation to the next. The human gene pool is the set of all genes carried within the human population. Genetic changes (including mutations) can be beneficial, neutral or deleterious. Beneficial mutations are less common and result in a selective survival advantage for a particular gene, **cell**, or whole **organism**. Beneficial mutations can become integrated into the human gene pool, particularly when it allows an organism to live longer or to reproduce. Neutral gene mutations usually involve point mutations that do not change the **amino acid** sequence or affect transcription/translation. Deleterious mutations are gene mutations leading to alterations in gene expression or protein function that results in human disease or is fatal. Recombination, or the crossing over and exchange of information between chromosomes during **meiosis**, can lead to gene rearrangements if the chromosomes are paired inappropriately.

Point mutations within a gene can be nonsense mutations (early termination of protein synthesis), missense mutations (a mutation that results an a substitution of one amino acid for another in a protein), or silent mutations that cause no detectable change in the corresponding protein sequence. Accordingly, the effects of point mutations range from 100% lethality (usually early in fetal development) to no observable (phenotypic) change.

There are four main types of genetic rearrangements: deletions, duplications, inversions, and translocations that are often caused by chemical and radioactive agents. Deletions result in the loss of DNA or a gene. Deletions can involve either the loss of a single base or the loss of a larger portion of DNA. Duplications can result in multiple copies of genes, and are caused most commonly by unequal crossover or **chromosome** rearrangements. During crossing over in meiosis, misaligned chromosomes can result in one of the chromosomes having extra material (duplication), while the other loses the same material that is duplicated in the other chromosome (deletion). Inversions, or changes in the orientation of chromosomal regions, may cause deleterious effects if the inversion involves a gene or an important sequence involved in the regulation of gene expression. Translocations are a type of rearrangement that occurs when a portion of two different chromosomes (or a single chromosome in two different places) breaks and rejoins such that the DNA sequence or gene is lost, repeated, or interrupted. If this affects the sequence of a gene or genes, it can result in disease.

The **frequency** of a mutation in a given population may be strikingly different from another population. There are many reasons for this including gene flow, genetic drift, and natural **selection**. Gene flow occurs when individuals move from one place to another. These migrations allow the introduction of new variations of the same gene (**alleles**) when they mate and produce offspring with members of their new group. In effect, gene flow acts to increase the gene pool in the new group. Because genes are usually carried by many members of a large population that have undergone **random** mating for several generations, random migrations of individuals away from the population or group usually do not significantly change the gene pool of the group left behind.

Genetic drift is represented by fluctuations in gene frequencies and occurs by chance, usually in very small populations, or due to sampling errors. During reproduction, one allele (one form of a gene) is passed to the next generation while the other is not. The allele that is not passed on, by chance, can affect the gene frequency if the population is very small. Random genetic drift can occur as a result of sampling **error**. Genetic drift can be profoundly affected by geographical barriers, catastrophic events (i.e. natural disasters or wars that significantly affect the reproductive availability of selected members of a population), as well as other political-social factors.

Natural selection is based upon the differences in the viability and reproductive success of different genotypes with a population (differential reproductive success). If a gene mutation results in the ability of an organism to live longer by protecting it from environmental threats or allowing it to become more reproducible, than this mutation will have a survival advantage.

There are three basic types of natural selection. Directional selection occurs when an extreme phenotype is favored (high or low body **fat**). Stabilizing selection takes place when intermediate phenotype is fittest (e.g., neither too high nor too low a body fat content) and for this reason it is often referred to as normalizing selection. Disruptive selection occurs when two extreme phenotypes are better that an intermediate phenotype. In studying changes in the human genome, natural **evolutionary mechanisms** are complicated by geographic, ethnic, religious, and social groups and customs. Accordingly, the effects of various evolution mechanisms on human populations are not as easy to predict. Increasingly sophisticated statistical studies are carried out by population geneticists to characterize changes in the human genome.

See also Chromosomal abnormalities; Chromosome mapping; DNA replication; DNA synthesis; DNA technology; Evolution, convergent; Evolution, divergent; Evolution, evidence of; Evolution, parallel; Evolutionary change, rate of; Genetic disorders.

Resources

Books

Friedman, J., F. Dill, M. Hayden, B. McGillivray *Genetics*. Maryland: Williams & Wilkins, 1996.

Nussbaum, Robert L., Roderick R. McInnes, Huntington F. Willard. *Genetics in Medicine*. Philadelphia: Saunders, 2001.

Rimoin, David L. *Emery and Rimoin's Principles and Practice of Medical Genetics*. London; New York: Churchill Livingstone, 2002.

Thompson, M. *Thompson & Thompson Genetics in Medicine*. Philadelphia: Saunders, 1991.

Periodicals

Graf, W.D. "Can Bioinformatics Help Trace the Steps from Gene Mutation to Disease?" *Neurology* (August 2000): 55(3):331–3.

Other

Wesleyan University. "De Vries, Hugo 1848–1935" [cited December 13, 2001]. <http://dbeveridge.web.wesleyan.edu/wescourses/2001f/chem160/01/Who's%20Who/hugo_de_vries.htm>.

Bryan Cobb

Gene splicing

Genes are DNA sequences that code for protein. **Gene** splicing is a form of **genetic engineering** where specific genes or gene sequences are inserted into the **genome** of a different **organism**. Gene splicing can also specifically refer to a step during the processing of **deoxyribonucleic acid (DNA)** to prepare it to be translated into protein.

Gene splicing can also be applied to **molecular biology** techniques that are aimed at integrating various DNA sequences or gene into the DNA of cells. Individual genes encode specific **proteins** and it is estimated that there are approximately 50,000 genes in each **cell** of the human body. Because the cellular functions in different tissues have varying purposes, the genes undergo a complex concerted effort to maintain the appropriate level of gene expression in a **tissue** specific manner. For example, muscle cells require specific proteins to function, and these proteins differ remarkably from proteins in **brain** cells. Although the genetic information is, for the most part, the same in both cell types, the different functional purposes result in different cellular needs and

therefore different proteins are produced in different tissue types.

Genes are not expressed without the proper signals. Many genes can remain inactive. With the appropriate stimulation of gene expression, the cell can produce various proteins. The DNA must first be processed into a form that other molecules in the cell can recognize and translate it into the appropriate protein. Before DNA can be converted into protein, it must be transcribed into **ribonucleic acid (RNA)**. There are three steps in RNA maturation; splicing, capping, and polyadenylating. Each of these steps are involved in preparing the newly created RNA, called the RNA transcript, so that it can exit the nucleus without being degraded. In terms of gene expression, the splicing of RNA is the step where gene splicing occurs in this context at specific locations throughout the gene. The areas of the gene that are spliced out are represent noncoding regions that are intervening sequences also known as introns. The DNA that remains in the processed RNA is referred to as the coding regions and each coding regions of the gene are known as exons. Therefore, introns are intervening sequences between exons and gene splicing entails the excision of introns and the joining together of exons. Hence, the final sequence will be shorter than the original coding gene or DNA sequence.

In order to appreciate the role splicing plays in how genes are expressed, it is important to understand how a gene changes into its functional form. Initially, RNA is called precursor RNA (or pre-RNA). Pre-RNAs are then further modified to other RNAs called transfer RNA (tRNA), ribosomal RNA (rRNA), or messenger RNA (mRNA). mRNAs encode proteins in a process called translation, while the other RNAs are important for helping the mRNA be translated into protein. **RNA splicing** creates functional RNA molecules from the pre-RNAs.

Splicing usually proceeds in a predetermined way for each gene. Experiments which have halted transcript formation at different intervals of **time** show that splicing will follow a major pathway beginning with some intron and proceeding selectively to another, not necessarily adjacent, intron. Although other pathways can be followed, each transcript has its own primary sequence for intron excision.

Alternative splicing

A single gene can be processed to create numerous gene products, or proteins and this process is referred to as alternative splicing. In this case, a different combination of exons remain in the processed RNA. Alternate gene splicing at various intron-exon sites within a gene can be used to create several proteins from the same pre-

RNA **molecule**. Proteins are made up of multiple domains. Different exons can code for different domains. Selective splicing can remove unwanted exons as well as introns. The combination of proteins that can be produced from alternate splicing are related in structure or function but are not identical. By using a single gene to create multiple proteins, the cells DNA can be utilized more efficiently.

Alternate splicing can be tissue specific such that different proteins are made from the same original gene by two or more different cell types. Or one cell type may make multiple configurations using the same gene. For instance, a type of immune cell called a B-cell manufactures antibodies to numerous antigens. Antigens are foreign substances which trigger immune responses and antibodies bind and antigens so that they can be broken down and removed. Although an infinite number of antibodies can be produced, all antibodies fall into one of five basic subtypes. Alternate splicing is used to create these five antibody-types from the same gene.

Antibodies are made up of multiple immunoglobulin (Ig) molecules. These molecules in turn have multiple domains. A particular domain called the heavy chain constant region distinguishes the five antibody subtypes, called IgM, IgD, IgG, IgE, and IgA. The different types of antibodies serve various functions in the body and act in distinct body tissues. For example, IgAs are secreted into the gastrointestinal mucosa, and IgGs passes through the placenta. The gene encoding these heavy chain regions contains exons that direct the production of individual subtypes, and the gene is alternately spliced to yield a final mRNA transcript, which can make any one of them.

Most genes yield only one transcript; however, genes that yield multiple transcripts have numerous cellular and developmental roles. Alternate splicing controls sex determination in ***Drosophila melanogaster*** flies. And a number of proteins are differentially expressed from the same gene in various cells. Different muscle cells use alternate splicing to create cell-specific myosin proteins. And embryonic cells in varying developmental stages produce multiple forms of the protein, retinoic acid. Some transcripts differ from related transcripts in the 5' end and others can vary at the 3' end.

Spliceosomes

The molecules or molecular complexes that actually splice RNA in the cellular nucleus are called spliceosomes. Spliceosomes are made of small sequences of RNAs bound by additional small proteins. This spliceosome complex recognizes particular nucleotide sequences at the intron-exon boundary. DNA and RNA are both generally read in the 5' to 3' direction. This desig-

nation is made on the basis of the phosphodiester bonds which make up the backbone of DNA and RNA strands. Introns are first cut at their 5' end and then at their 3' end. The two adjacent exons are then bonded together without the intron. The spliceosome is an enzymatic complex which performs each of these steps along the pre-RNA to remove introns.

The small RNAs which make up the spliceosome are not mRNAs, rRNAs, or tRNAs; they are small nuclear RNAs (snRNA's). snRNAs are present in very low concentrations in the nucleus. The snRNAs combine with proteins to comprise, small nuclear ribonuclearprotein particles. Several snRNPs aggregate to form a spliceosome. This secondary structure recognizes several key regions in the intron and at the intron-exon border. In essence, snRNPs play a catalytic splicing role. The absence of individual snRNP components can inhibit splicing. snRNPs are only one of many complexes which can regulate gene expression.

In addition to snRNPs, some introns have auto (self) splicing capabilities. These introns are called group II introns. Group II introns are found in some mitochondrial genes, which come from a genome that is separate from the nucleus and is located in small compartments within the cell called mitochondria. Mitochrondria function in provide **energy** for the cells energy requirements. Although all chromosomal DNA is located in the nucleus, a few genes are located in the cells mitochondria. Group II introns form secondary structures using their internal intron region in a similar way to nuclear introns. However, these mitochondrial introns direct exon-exon rejoining by themselves without snRNPs.

Splicing out introns

Various splicing signal sequences are universal and are found within every intron site spliced, while some signal sequences are unique to individual genes. DNA is made up of bases called nucleotides, which represent the DNA alphabet. There are four bases, Adenine (A), Guanine (G), Thymine (T), and Cytosine (C). Most introns in higher life forms begin with the nucleotide sequence G-T and end with the sequence A-G. The sequences define the "left" (5') and "right" (3') borders of the intron and are described as conforming to the GT-AG rule. Mutations in any of these four positions produce introns that cannot be removed by normal splicing mechanisms. Within the intron is another highly conserved sequence that has some variability in the genes of a **species**; this region (called the branch site) is the area that connects to the 5' end of the intron as it is cut and then curls around to form a lariat shape. This lariat is a loop in the intron which is formed as it is removed from the maturing RNA.

KEY TERMS

Antibody—A molecule created by the immune system in response to the presence of an antigen (a foreign substance or particle). It marks foreign microorganisms in the body for destruction by other immune cells.

Antigen—A molecule, usually a protein, that the body identifies as foreign and toward which it directs an immune response.

Capping—A modification to the 5′ end of a mature mRNA transcript.

Cytoplasm—All the protoplasm in a living cell that is located outside of the nucleus, as distinguished from *nucleoplasm,* which is the protoplasm in the nucleus.

Deoxyribonucleic acid (DNA)—The genetic material in a cell.

Exons—The regions of DNA that code for a protein or form tRNA or mRNA.

Gene—A discrete unit of inheritance, represented by a portion of DNA located on a chromosome. The gene is a code for the production of a specific kind of protein or RNA molecule, and therefore for a specific inherited characteristic.

Genome—The complete set of genes an organism carries.

Introns—Noncoding sequences in a gene that are spliced out during RNA processing.

Mitochondria—Intracellular organelle that is separate from the nucleus, has it's own genome and is important for producing energy for various tissues.

Polyadenylation—A modification to the 3′ end of a mature mRNA transcript.

Recombinant DNA—DNA that is cut using specific enzymes so that a gene or DNA sequence can be inserted.

Splicesome—The intracellular machinery that processes RNA by removing introns from the sequence.

Other splicing events

Splicing can also involve molecules other than mRNA. tRNAs, which play a crucial role of aligning amino acids along a protein being synthesized can undergo splicing. tRNAs are encoded by DNA just like all other RNA molecules. However, tRNAs have a unique structure and function distinct from other RNA molecules in that they are responsible for matching the actual protein building blocks (amino acids) from the encoded nucleotide sequence to build a protein, or polypeptide. Since these specialized RNAs have unique conformations, enzymes that join exons after intron removal differ from those that join introns in other RNA molecules. While introns are removed, and exons are joined, the enzymatic molecules are not the same as those used for mRNA processing. Intron removal in tRNA processing is less dependent on internal intron sequences compared to other RNA introns.

Recombinant DNA technology

Advances in understanding the mechanisms that describe how gene splicing occurs has lead to the ability of scientists to cut and anneal nucleotide sequences, also called **recombinant DNA** technology. Since splice literally means the joining of separate ends, gene splicing refers to the joining of almost any nucleotide sequences to create a new gene product or to introduce a new gene sequence. Hence, just about any genetic sequence could be spliced into another sequence.

Certain enzymes called restriction enzymes (REs) are used in laboratories to splice, connect (or ligate), and remove or add nucleotides to sequences. REs are used in recombinant **DNA technology** to remove and insert genetic sequences from and into other sequences. This technology has enabled some **biotechnology** and pharmaceutical companies to manufacture large quantities of essential proteins for medical and research purposes. For example, a human **insulin** protein can be made in great supply by inserting the insulin gene into the genome of **bacteria**, for example, in order to produce large amounts of the protein. Like a photocopy machine, such sequences can produce lots of insulin for diabetics who are not able to make enough insulin on their own. These patients can then self-inject the purified insulin to treat their **disease**.

Applications of gene splicing

Using gene-splicing technology, vaccines have been produced. DNA from a **virus** can be spliced into the genome of a harmless strain of bacterial strain. When the bacteria produced the viral protein, this protein can be harvested. Since bacteria grow quickly and easily, a large amount of this protein can be extracted, purified and used as a **vaccine**. It is introduced into an individual by injection, which will elicit an immune response. When a person is infected with a virus by natural exposure, a rapid immune response can be initiated due to the initial innoculation. Another application of gene spicing technology is related to the gene involved in **Vitamin B** production. This gene has been removed from a carrots genome and spliced into the genome of **rice**. The genetically engineered recombinant rice strain therefore, is modified to produce Vitamin B. This can have many health-related benefits, particularly in third world countries that rely on rice as a major food source and do not have access to food sources rich in vitamins.

Gene splicing technology, therefore, allows researchers to insert new genes into the existing genetic material of an organisms genome so that entire traits, from disease resistance to vitamins, and can be copied from one organism and transferred another.

See also Chromosome; Gene; Genetics; Deoxyribonucleic acid (DNA); DNA replication; DNA synthesis; DNA technology; Molecular biology.

Resources

Books

Lewin B, ed. *Genes V.* New York: Oxford University Press, 1994.

Nussbaum, Robet l., Roderick R. McInnes, Huntington F. Willard. *Genetics in Medicine.* Philadelphia: Saunders, 2001.

Rimoin, David L. *Emery and Rimoin's Principles and Practice of Medical Genetics.* London; New York: Churchill Livingstone, 2002.

Louise Dickerson

Gene therapy

Gene therapy is a rapidly growing field of medicine in which genes are introduced into the body to treat diseases.

Gene therapy is the name applied to the treatment of inherited diseases by corrective **genetic engineering** of the dysfunctional genes. It is part of a broader field called genetic medicine, which involves the screening, **diagnosis**, prevention and treatment of hereditary conditions in humans. The results of genetic screening can pinpoint a potential problem to which gene therapy can sometimes offer a solution.

Genes represent the genetic material that organisms pass on from one generation to the next. Therefore, genes are responsible for controlling hereditary traits and provide the basic biological code or blueprint for living organisms. Genes produce protein such as hair and skin as well as **proteins** that are important for the proper functioning of organs. Mutated or defective genes often cause **disease**. The purpose of gene therapy is to replace a defective gene with a normal copy of the same gene in attempt to restore function. Somatic gene therapy introduces a normal gene into tissues or cells to treat an **individual** that has an abnormal gene. Germline gene therapy inserts genes into reproductive cells (the egg or the sperm) or into embryos to correct genetic defects that could be passed on to future generations. Germline gene therapy differs from somatic gene therapy in that germline integration of a gene will ideally correct every progenitor **cell** that differentiates from the germ cell. Somatic gene therapy involves integrating corrected genes into cell and tissues that are fully differentiated or mature.

Initially conceived as an approach for treating inherited diseases, like **cystic fibrosis** and Huntington's disease, the scope of potential gene therapies has grown to include treatments for cancers, **arthritis**, and infectious diseases. Although gene therapy testing in humans has rapidly advanced, in general, the field of gene therapy has proven to be problematic and complicated by a variety of ethical issues. For example, some scientists are concerned that the integrating genes into the human **genome** may cause disease. There has been evidence that randomly integrating corrected genes might disrupt other genes in the genome and if the disrupted gene is a **tumor** suppressor gene, **cancer** may develop. Others fear that germ-line gene therapy may be used to control human development in ways not connected with disease, like intelligence or appearance.

The biological basis of gene therapy

Gene therapy has grown out of the field of **molecular biology**. Life begins with a single cell, the basic building block of all multicellular organisms. Humans, for instance, are made up of trillions of cells, that make-up tissues that form into organs. Each cell type can perform a specific function. Within the cells nucleus (the center part of a cell that regulates its chemical functions) are pairs of chromosomes. These threadlike structures are made up of DNA (deoxyribonucleic acid), which carries the blueprint of life in the form of codes, or genes, that are interspersed throughout the DNA sequence.

A DNA **molecule** looks like a twisted ladder. The rungs of these represent bonds between each letter of the DNA sequence called base pairs. Base pairs are made up of nitrogenous molecules. Thousands of these base pairs, or DNA sequences, can make up a single gene, specifically defined as a segment of the **chromosome**. The gene, or combination of genes formed by these base pairs ultimately direct an organisms growth and charac-

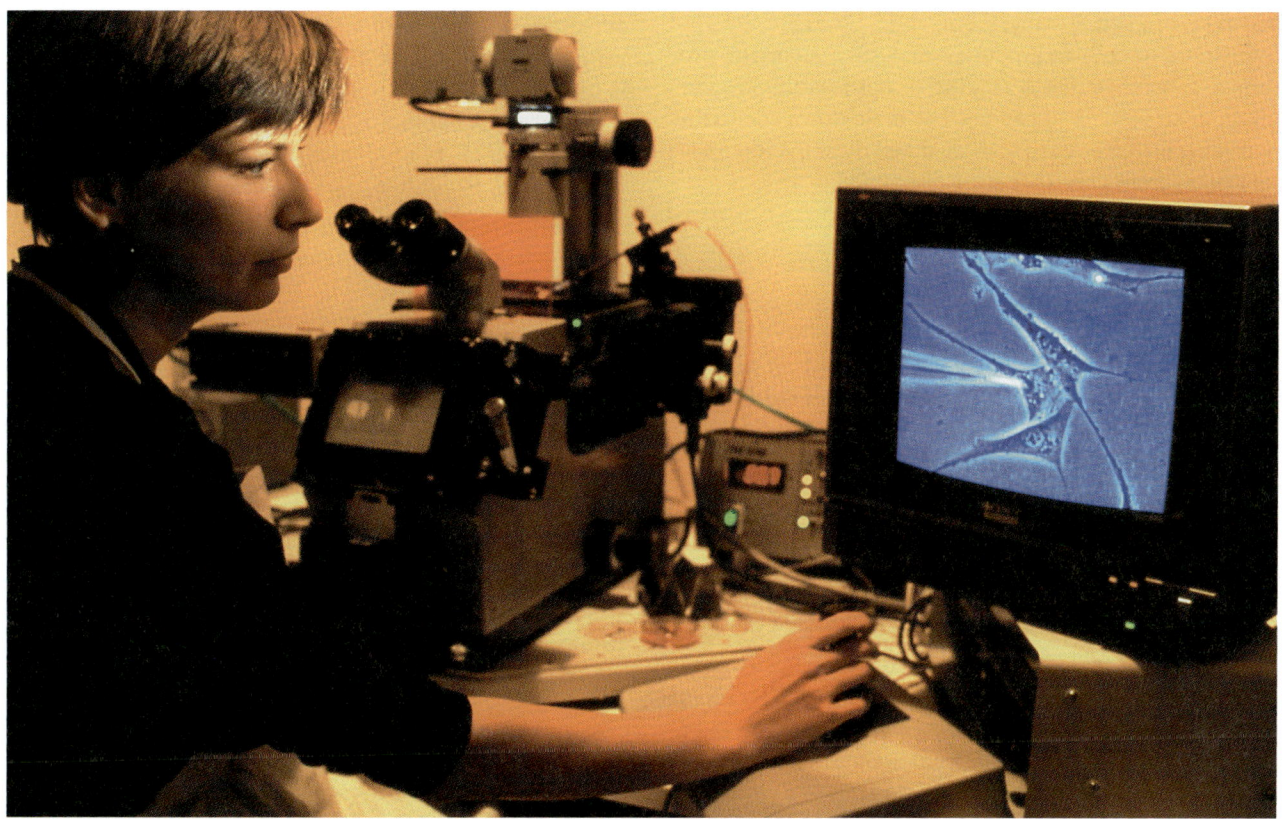

A scientist performing a microinjection of a corrective gene into a human T lymphocyte (a white blood cell). On screen can be seen the tip of the micropipette used for the injection (left) and the cell (right). The pipette is controlled by the joystick in the scientist's right hand. The procedure is being performed to treat a young girl whose immune system had collapsed because of her body's failure to produce an enzyme called adenosine deaminase (ADA) which controls the maturation of T cells. Lymphocytes collected from her are treated with the gene that expresses ADA and then reinjected into a vein. Results have been very good. *Photograph by Philippe Plailly. National Audubon Society Collection/Photo Researchers, Inc. Reproduced by permission.*

teristics through the production of certain proteins, which are important for many biochemical functions.

Scientists have long known that defects in genes present within cells can cause inherited diseases such as cystic fibrosis, sickle-cell **anemia**, and **hemophilia**. Similarly, a gain or a loss of an entire chromosome can cause diseases such as **Down Syndrome** or Turners **syndrome**. As the study of **genetics** advanced, however, scientists learned that an altered genetic sequence can also make people more susceptible to develop diseases making these individuals predisposed to having atherosclerosis, cancer, or **schizophrenia**. These diseases have a genetic component, but are also influenced by environmental factors (like diet and lifestyle). The objective of gene therapy is to treat diseases by introducing corrected genes into the body to replace a missing or dysfunctional protein. The inserted genes can be naturally- occurring genes that produce the desired effect or may be genetically engineered (or altered) genes.

Scientists have known how to manipulate the sturcture of a gene in the laboratory since the early 1970's through a process called **gene splicing**. The process involves cutting a sequence of the genome with restriction enzymes, or proteins that act like molecular sicssors. The ends where the DNA has been cut are sticky in the sense that they will easily bind to another sequence of DNA that was cut with the same **enzyme**. A DNA sequence and a gene sequence to be integrated in the DNA sequence can both be cut with the same type of enzyme and their ends will stick together. The new DNA sequence will now have the gene inserted into it. The resulting product is called genetic engineered **recombinant DNA**.

There are basically two types of gene therapy. Germ-line gene therapy introduces genes into reproductive cells (sperm and eggs) or into embryos in order to correct genetic defects that could be passed on to future generations. Most of the current research, however, has been in the applications of somatic cell gene therapy. In this type of gene therapy, therapeutic genes are inserted into **tissue** or cells to produce a naturally occurring protein or substance that is lacking or not functioning correctly in an individual patient. The main downside to this approach is that as each corrected cell dies, the therapeutic effects from gene therapy are lessened.

Viral vectors

In both types of therapy, scientists need something to transport either the entire gene or a recombinant DNA to the cells nucleus, where the DNA is located. In essence, vectors are molecular delivery trucks. One of the first and most popular vectors developed was viral vectors, or vectors made of viruses because they invade cells as part of a natural **infection** process. Viruses were originally considered the most ideal vector because they have a specific relationship with the host in that they can infect specific cell types or tissues. As a result, vectors are chosen according to their affinity for certain cells and areas of the body.

One of the first viral vectors used was the **retrovirus**. Because these viruses are easily cloned (artificially reproduced) in the laboratory, scientists have studied them extensively and learned a great deal about their biological characteristics. They have also learned how to remove the genetic information that governs viral replication, thus reducing the chances of mutliple rounds of infection. Additionally, many of the proteins from these viruses that can cause an immune response can be removed.

Retroviruses work best in actively dividing cells, but most of the cells in the body particularly those that are fully differentiated are relatively stable and do not divide often. As a result, these cells are used primarily for *ex vivo* (outside the body) manipulation. First, the cells are removed from the patient's body, and the **virus**, or vector, carrying the gene is inserted into them. Next, the cells are placed into a nutrient culture where they grow and replicate. Once enough cells are gathered, they are returned to the body, usually by injection into the **blood** stream. Theoretically, as long as these cells survive, they can have therapeutic potential.

Another class of viruses, called the adenoviruses, have proven to be good gene vectors in certain cases. These cells can effectively infect nondividing cells in the body, where the desired gene product is then expressed. These viruses, which cause respiratory tract infections, are more easily purified and more stable than retroviruses, resulting in less chance of an unwanted viral infection. These viruses live for several days in the body and can have potentially life-threatening complications related to immune cell responses. Other viral vectors include **influenza** viruses (that causes the flu), Sindbis virus, and a herpes virus that infects nerve cells. Each of these vec-

tors can be modified to minimize the risk of causing disease or immune cell responses.

Scientists have also developed nonviral vectors. These vectors rely on the natural biological process in which cells uptake (or gather) macromolecules (large molecules). One approach is to use liposomes, or globules of **fat** produced by the body and taken up by cells. Scientists are also investigating the introduction of recombinant DNA by directly injecting it into the bloodstream or placing it on microscopic beads of gold shot into the skin with a "gene-gun." Another possible vector under development is based on dendrimer molecules. This is a class of polymers or naturally occurring or artificial substances that have a high **molecular weight** and are formed by smaller molecules of the same or similar substances. They have been used in manufacturing Styrofoam, polyethylene cartons, and Plexiglass. In the laboratory, dendrimers have shown the ability to transport genetic material into human cells. They can also be designed with a high affinity for the **membrane** of a cell by attaching sugars and protein groups to it.

The history of gene therapy

In the early 1970s, scientists proposed what they called "gene surgery" for treating inherited diseases caused by defective genes. In 1983, a group of scientists from Baylor College of Medicine in Houston, Texas, proposed that gene therapy could one day be a viable approach for treating Lesch-Nyhan disease, a rare neurological disorder. The scientists conducted experiments in which an enzyme-producing gene for correcting the disease was injected into a group of cells. The scientists theorized the cells could then be injected into people with Lesch-Nyhan disease.

As the science of genetics advanced throughout the 1980s, gene therapy gained an established foothold in the minds of medical scientists as a viable approach to treatments for specific diseases. However, its promises were more than what it could deliver. One of the major impetuses in the growth of gene therapy was an increasing ability to identify the genetic abnormalities that cause inherited diseases. Interest grew as further studies showed that specific genetic defects in one or more genes occurred in successive generations of certain family members who suffered from diseases like intestinal cancer, manic-depression, **Alzheimer disease**, **heart** disease, diabetes, and many more. Although the genes may not be the sole cause of the disease in all cases, they may make certain individuals more susceptible to developing the disease because of environmental influences, such as smoking, **pollution**, and **stress**. In fact, many scientists believe that all diseases have a genetic component.

On September 14, 1990, a four-year old girl suffering from a genetic disorder that prevented her body from producing a crucial enzyme became the first person to undergo gene therapy in the United States. Since her body could not produce adenosine deaminase (ADA), she had a weakened **immune system**, making her extremely susceptible to severe, life-threatening infections. W. French Anderson and colleagues at the National Institutes of Health's Clinical Center in Bethesda, Maryland, took white blood cells (which are crucial for proper immune system functioning) from the girl, inserted ADA producing genes into them, and then transferred the cells back into the patient. Although the young girl continued to show an increased ability to produce ADA, debate arose as to whether the improvement resulted from the gene therapy or from an additional drug treatment she received.

Nevertheless, a new era of gene therapy began as more and more scientists sought to conduct clinical trials in this area. In that same year, gene therapy was tested on patients suffering from melanoma (skin cancer). The goal was to help them produce antibodies (disease fighting substances in the immune system) to battle the cancer.

These experiments have spawned a growing number of attempts to refine develop new gene therapies. For example, gene therapy for cystic fibrosis, a disease that affects the airways, is being developed. However, due to the complications involved in penetrating the natural barriers that impedes viral entry into the airways, it is unlikely that currently used vectors for cystic fibrosis gene therapy represent a plausible approach. Modifications of these vectors by adding compounds that naturally bind to areas on the outermost membranes of the lung and gain entrance into these tissues are currently being investigated. Another approach was developed for treating **brain** cancer patients, in which the inserted gene was designed to make the cancer cells more likely to respond to drug treatment. Additionally, gene therapy for patients suffering from artery blockage, which can lead to strokes, that induces the growth of new blood vessels near clogged **arteries** improving normal blood circulation is also being investigated.

In the United States, both DNA-based (*in vivo*) treatments and cell-based (*ex vivo*) treatments are being investigated. DNA-based gene therapy uses vectors (like viruses) to deliver modified genes to target cells. Cell-based gene therapy techniques remove cells from the patient, which are genetically altered and then reintroduce them to the patients body. Presently, gene therapies for the following diseases are being developed: cystic fibrosis (using adenoviral vector), HIV infection (cell-based), malignant melanoma (cell-based), kidney cancer (cell-based), Gaucher's Disease (retroviral vector), breast cancer (retroviral vector), and lung cancer (retroviral vector).

The medical has contributed to transgenic research that is supported by government funding. In 1991, the U.S. government provided $58 million for gene therapy research, with increases of $15-40 million dollars a year over the following four years. With fierce **competition** over the promise of major medical benefit in addition to huge profits, large pharmaceutical corporations moved to the forefront of transgenic research.

Diseases targeted for treatment by gene therapy

The potential scope of gene therapy is enormous. More than 4,200 diseases have been identified that result directly from defective genes. People suffering from cystic fibrosis lack a gene needed to produce a salt-regulating protein. This protein regulates the flow of chloride into epithelial cells, which cover the air passages of the nose and lungs. Without this regulation, cystic fibrosis patients suffer from a buildup of a thick mucus, which can cause lung infections and respiratory problems, which usually leads to death within the first 30 years of life. A gene therapy technique to correct this defect might employ an adenovirus to transfer a normal copy of what scientists call the cystic fibrosis transmembrane conductance regulator, or CTRF, gene. The gene is introduced into the patient by spraying it into the nose or lungs.

Familial hypercholesterolemia (FH) is also an inherited disease, resulting in the inability to process **cholesterol** properly, which leads to high levels of artery-clogging fat in the blood stream. FH patients often suffer heart attacks and strokes because of blocked arteries. A gene therapy approach to battle FH that is currently being investigated involves partially and surgically removing the patients liver (*ex vivo* transgene therapy). Corrected copies of a gene that serve to reduce cholesterol build-up are inserted into the liver sections, which are then transplanted back into the patient.

Gene therapy has also been tested on **AIDS** patients. AIDS is caused by the human immunodeficiency virus (HIV), which weakens the body's immune system to the point that sufferers are unable to fight off diseases such as **pneumonia**. An approach to treat AIDS is to insert genes into a patients bloodstream that have been genetically engineered to produce a receptor that would attract HIV and reduce its chances of replicating.

Several cancers also have the potential to be treated with gene therapy. A therapy that is currently being tested for the treatment of melanoma (a form of skin cancer), involves introducing a gene with an anticancer protein called tumor necrosis factor (TNF) into test tube samples of the patient's own cancer cells, which are then reintroduced into the patient. In brain cancer, the approach is to insert a specific gene that increases the cancer cells susceptibility to a common drug used in fighting the disease. Gene therapy can also be used to treat diseases that involve dysfunctional enzymes. For example, Gaucher's disease is an inherited disease caused by a mutant gene that inhibits the production of an enzyme called glucocerebrosidase. Gaucher patients have enlarged livers and spleens and eventually their bones fall apart. Clinical gene therapy trials focus on inserting the gene for producing this enzyme.

Gene therapy is also being considered as an approach to solving a problem associated with a surgical procedure known as angioplasty. In this procedure, a type of tubular scaffolding is used to open a clogged artery. However, in response to the trauma of the scaffold insertion, the body often initiates a natural healing process resulting in restenosis, or reclosing of the artery. The gene therapy approach to preventing this unwanted side effect is to cover the outside of the stents with a soluble gel. This gel is designed to contain vectors for genes that would reduce restenosis.

The Human Genome Project

Although great strides have been made in gene therapy in a relatively short time, its potential usefulness has been limited. For instance, it is now known that the vast majority of non-coding regions are no longer considered junk DNA anymore. In fact, these large portions of the genome are involved in the control and regulation of gene expression, and are thus much more complex than originally thought. Even so, each individual cell in the body carries thousands of genes coding for proteins many of which have not yet been identified or characterized.

To address this issue, the National Institutes of Health initiated the **Human Genome Project** in 1990. The projects 15-year goal was to map the entire human genome. A genome map would help to identify the location of all genes as well as better understand the remaining three billion base pairs. A milestone in the human genome project was completed in 1999 when the first full sequence of an entire chromosome was completed (chromosome 22). The human genome draft sequence was published by HGP and Celera scientists in February 2001 in the journals *Science* and *Nature,* respectively.

Some of the genes identified include a gene that predisposes people to **obesity**, one associated with programmed **cell death** or apoptosis, a gene that guides HIV viral reproduction, and the genes of **inherited disorders** like Huntington's disease, Lou Gehrig's disease, and some colon and breast cancers.

The future of gene therapy

There are many obstacles and ethical questions concerning gene therapy. For example, some retrovirusal vectors, can also enter normal cells and interfere with the natural biological processes, possibly leading to other diseases. Other viral vectors, like the adenoviruses, are often recognized and destroyed by the immune system so their therapeutic effects are short-lived. One of the primary limitations in gene therapy is that delivering a gene using a viral vector that can only undergo one round of infection (making it safer) may provide only temporary therapeutic value that lasts only as long as the corrected gene is expressed. As a result, some therapies need to be repeated often to provide long-lasting benefits.

One of the most pressing issues, however, involves gene regulation. Several genes may play a role in turning other genes on and off. For example, certain genes work together to stimulate **cell division** and growth, but if these are not regulated, the inserted genes could cause unregulated cell growth leading to the formation of a tumor. Another difficulty is **learning** how to make the gene be expressed in a regulated way. A specific gene should turn on, for example, when certain levels of a protein or enzyme are not sufficiently meeting cellular demands. This type of controlled regulation of gene expression for these delivered genes is very difficult to achieve.

Ethical considerations in gene therapy

While gene therapy holds promise as a revolutionary approach to treating disease, ethical concerns over its use and ramifications have been expressed. For example, it is difficult to determine the long-term effect of exposure to viral vectors and the effects these engineered viruses have on the human genome.

As the technology develops and more mainstream applications become possible, it is likely that medically unrelated genetic traits might be the target of manipulation. For example, perhaps a gene could be introduced that prevents balding in males. Or what if genetic manipulation was used to alter skin **color**, prevent homosexuality, or to enhance physical attractiveness and intelligence? Will this only be available to the rich? Gene therapy has been surrounded by more controversy and scrutiny in both scientists and the general public than many other technologies.

As with every new medical technique, there are many potential dangers and unpredictable factors with gene therapy, which make its practical application risky. Even though every precaution is taken to prevent accidents, they sometimes do occur. Jesse Gelsinger, a 17 year-old boy suffering from the disease ornithine transcarbamylase (OTC) deficiency became the first tragic victim of gene

KEY TERMS

Cells—The smallest living units of the body which together form tissues.

Chromosomes—he structures that carry genetic information in the form of DNA. Chromosomes are located within every cell and are responsible for directing the development and functioning of all the cells in the body.

Clinical trial—The testing of a drug or some other type of therapy in a specific population of patients.

Clone—A cell or organism derived through asexual reproduction, which contains the identical genetic information of the parent cell or organism.

DNA—Deoxyribonucleic acid; the genetic material in a cell.

Enzyme—Biological molecule, usually a protein, which promotes a biochemical reaction but is not consumed by the reaction.

Eugenics—A social movement in which the population of a society, country, or the world is to be improved by controlling the passing on of hereditary information through selective breeding.

Gene—A discrete unit of inheritance, represented by a portion of DNA located on a chromosome.

The gene is a code for the production of a specific kind of protein or RNA molecule, and therefore for a specific inherited characteristic.

Gene transcription—The process by which genetic information is copied from DNA to RNA.

Genetic engineering—The manipulation of genetic material to produce specific results in an organism.

Germ-line gene therapy—The introduction of genes into reproductive cells or embryos to correct inherited genetic defects that can cause disease.

Macromolecules—A large molecule composed of thousands of atoms.

Nucleus—The central part of a cell that contains most of its genetic material, including chromosomes and DNA.

Protein—Macromolecules made up of long sequences of amino acids.

Somatic gene therapy—The introduction of genes into tissue or cells to treat a genetic related disease in an individual.

Vectors—A molecular device to transport genes or DNA sequences into a cell or organ.

---

therapy and died on September 17, 1999. He had volunteered to test the potential use of gene therapy in the treatment of OTC in young babies. His therapy consisted of an infusion of corrective genes, encased in a weakened adenovirus vector. Gelsinger suffered an unexpected chain reaction that resulted in his early death from multiple **organ** system failure. The reason for his extreme reaction to the treatment is suspected to have been an overwhelming inflammatory response to the viral vector, though the reason why is not known. Subsequent investigations revealed the deaths of six other gene therapy patients, some prior to Gelsinger, who were undergoing trials for the use of gene therapy in the treatment of heart conditions. Unlike Gelsinger, these latter six victims are thought to have died from complications stemming from their underlying illnesses rather than the gene therapy itself.

See also DNA replication; DNA synthesis; DNA technology.

Resources

Books

Lemoine, Nicholas R. and Richard G. Vile. *Understanding Gene Therapy.* New York: Springer-Verlag, 2000.

Periodicals

Kaiser J. "RAC Hears a Plea for Resuming Trials, Despite Cancer Risk." *Science.* 299:5609 (2003): 991.

Kasuya H, S. Nomoto, A. Nakao. "The Potential of Gene Therapy in the Treatment of Pancreatic Cancer." *Drugs Today* 38(7) (2002):457-64.

Stephenson J. "The World in Medicine: Gene Therapy Setback." *JAMA.* 289(6) (2003): 691.

Sylven C. "Angiogenic Gene Therapy." *Drugs Today* 38(12) (2002): 819-27.

David Petechuk
Brian Cobb

Generator

A generator is a machine by which mechanical **energy** is transformed into electrical energy. Generators can be sub-divided into two major categories depending on whether the **electric current** produced is alternating current (AC) or direct current (DC). The basic principle on which both types of generator works is the same, although the details of construction of the two may differ somewhat. Generators can also be classified according to the source of the mechanical power (or *prime mover*) by which they are driven, such as **water** or steam power.

Principle of operation

The scientific principle on which generators operate was discovered almost simultaneously in about 1831 by the English chemist and physicist, Michael Faraday, and the American physicist, Joseph Henry. Imagine that a coil of wire is placed within a magnetic field, with the ends of the coil attached to some electrical device, such as a galvanometer. If the coil is rotated within the magnetic field, the galvanometer shows that a current has been induced within the coil. The magnitude of the induced current depends on three factors: the strength of the magnetic field, the length of the coil, and the speed with which the coil moves within the field.

In fact, it makes no difference as to whether the coil rotates within the magnetic field or the magnetic field is caused to rotate around the coil. The important factor is that the wire and the magnetic field are in **motion** in **relation** to each other. In general, most DC generators have a stationary magnetic field and a rotating coil, while most AC generators have a stationary coil and a rotating magnetic field.

Alternating current (AC) generators

In an electrical generator, the galvanometer mentioned above would be replaced by some electrical device. For example, in an **automobile**, electrical current from the generator is used to operate headlights, the car **radio**, and other electrical systems within the car. The ends of the coil are attached not to a galvanometer, then, but to slip rings or collecting rings. Each slip ring, in turn, is attached to a brush, through which electrical current is transferred from the slip ring to an external circuit.

As the **metal** coil passes through the magnetic field in a generator, the electrical power produced constantly changes. At first, the generated electric current moves in one direction (as from left to right). Then, when the coil reaches a position where it is **parallel** to the magnetic lines of **force**, no current at all is produced. Later, as the coil continues to rotate, it cuts through magnetic lines of force in the opposite direction, and the electrical current generated travels in the opposite direction (as from right to left).

Thus, a spinning coil in a fixed magnetic field of the type described here will produce an alternating current, one that travels in one direction for a moment of time, and then the opposite direction at the next moment of time. The **rate** at which the current switches back and forth is known as its **frequency**. The current used for most household devices, for example, is 60 hertz (60 cycles per second).

The efficiency of a generator can be increased by substituting for the wire coil described above an armature. An armature consists of a cylindrical **iron** core around which is wrapped a long piece of wire. The longer the piece of wire, the greater the electrical current that can be generated by the armature.

Commercial generators

One of the most important practical applications of generators is in the production of large amounts of electrical energy for industrial and residential use. The two most common prime movers used in operating AC generators are water and steam. Both of these prime movers have the ability to drive generators at the very high rotational speeds at which they operate most efficiently, usually no less than 1,500 revolutions per minute.

Hydroelectric power (the power provided by running water, as in large **rivers**) is an especially attractive power source since it costs nothing to produce. It has the disadvantage, however, that fairly substantial superstructures must be constructed in order to harness the mechanical energy of moving water and use it to drive a generator.

The intermediary device needed in the generation of hydroelectric power is a **turbine**. A turbine consists of a large central shaft on which are mounted a series of fanlike vanes. As moving water strikes the vanes, it causes the central shaft to rotate. If the central shaft is then attached to a very large magnet, it causes the magnet to rotate around a central armature, generating **electricity** that can then be transmitted for industrial and residential applications.

Electrical generating plants also are commonly run with steam power. In such plants, the burning of **coal**, oil, or **natural gas** or the energy derived from a **nuclear reactor** is used to boil water. The steam thus produced is then used to drive a turbine which, in turn, propels a generator.

Direct current (DC) generators

An AC generator can be modified to produce direct current (DC) electricity also. The change requires a commutator. A commutator is simply a slip ring that has been cut in half, with both halves insulated from each other. The brushes attached to each half of the commutator are arranged so that at the moment the direction of the current in the coil reverses, they slip from one half of the commutator to the other. The current that flows into the external circuit, therefore, is always traveling in the same direction.

See also Electromagnetic field; Electric current; Electrical power supply; Faraday effect.

KEY TERMS
. . . .

Alternating current—Electric current that flows first in one direction, then in the other; abbreviated AC.

Armature—A part of a generator consisting of an iron core around which is wrapped a wire.

Commutator—A split ring that serves to reverses the direction in which an electrical current flows in a generator.

Direct current (DC)—Electrical current that always flows in the same direction.

Prime mover—The energy source that drives a generator.

Slip ring—The device in a generator that provides a connection between the armature and the external circuit.

Resources

Books

Macaulay, David. *The New Way Things Work.* Boston: Houghton Mifflin Company, 1998.
McGraw-Hill Encyclopedia of Science & Technology. 6th edition. New York: McGraw-Hill Book Company, 1987, vol. 7, pp 635-37.

David E. Newton

Genetic code *see* **Deoxyribonucleic acid (DNA)**

Genetic disorders

Genetic disorders refer to medical conditions that develop as the result of abnormalities in an individuals genetic material, usually that is inherited. Inheriting or developing a genetic disorder leads to a collection of clinical manifestations known as a **syndrome**. These clinical manifestations can vary from person to person with the same genetic defect or have similar presentations.

Principles of genetic inheritance patterns

Genetic information is packaged into chromosomes that are found in thecells nucleus, or DNA containing organelle. Virtually every human **cell** has 46 chromosomes, except for the sperm and egg (reproductive cells) which each have 23 chromosomes. **Fertilization** of the egg by the sperm results in a newly formed cell called the zygote. Each zygote receives 23 chromosomes from the egg and 23 chromosomes from the sperm. All but one of the 23 chromosomes are called autosomes. The remaining **chromosome** is the sex chromosome and it is either an X or a Y. After these sex cells unite, gender is determined. Females have two X's (XX), and males have one of each (XY). Females can only pass an X to their offspring, and males can pass either an X or a Y. Therefore, the male sperm is responsible for gender **selection**.

The duplicated set of 22 autosomes, numbered 1 through 22, are called homologous pairs in that there are two chromosome number 1 with similar genes and genetic material. There are approximately 50,000 genes on all the chromosomes. Individual genes are made of **deoxyribonucleic acid (DNA)**, which makes up the genetic code or alphabet that produces specific **proteins**. Proteins can play important structural and functional roles in the body. Each **gene** has a set **locus**, or position, on a particular chromosome. Identical genes that are located on the same locus on corresponding chromosomes are called **alleles**.

A persons genotype represents the genes that they inherited. If is an autosomally inherited **disease**, the genotypes are written as a lower or an upper case letter A (such as AA, aa or Aa to represent both alleles known as the genotype), where capital letters define dominant genes and lower case letters define recessive genes. Genotypes are either homozygous or heterozygous. Having two identical alleles, such as AA or aa, makes the genotype homozygous for that locus. Having different alleles (for example, Aa) at a locus represents a heterozygous genotype. The observable features that characterize an individual are collectively called the phenotype. A phenotype can also extend to observable characteristics that can only be visualized with the help of a **microscope** or other equipment.

Types of genetic inheritance

There are many types of genetic disorders. Mendelian disorders are a group of disorders that are inherited either as an autosomal (through one of the 44 chromosomes, excluding the X or Y chromosome) or X-linked (through the X chromosome) defect in a dominant or recessive pattern. Although most of them are commonly associated with inherited defects, many of them can involve spontaneous or de novo alterations in an individuals genetic material. There are also genetic disorders that have an unknown **etiology** and the mechanisms for how a disease develops have not been clearly elucidated. Some genetic disorders can be induced by environmental factors and are collectively called multifactorial disorders. Mitochondrial disorders involve disorders in the mitochondrial **genome** and are inherited only from the mother. The mitochondria are tiny compartments found in almost every cell and has a distinct genome from the DNA found in the nucleus and is involved primarily in producing **energy** for all the tissues of the body. Finally, polygenic disorders mean that many genes act together to produce a genetic disease.

Dominant and recessive

A dominant gene means that a single allele can control whether the disease develops. If only one parents (usually affected) passes on an autosomal, defective gene which results in the child having a genetic disorder, then the disorder is called autosomal dominant. A recessive gene means that there is enough normal protein product to function properly from the normal gene and, therefore, two copies of the defective gene are necessary for the disease to develop. If both parents are unaffected and they each pass on a defective gene causing their child to be affected, then the genetic disorder is autosomal recessive. The parents are called carriers. For example, sickle-cell **anemia** is a recessive disorder characterized by abnormal hemoglobin production. The genetic defect involves a gene that produces hemoglobin. Although sickle-cell carriers produce, in part, abnormal hemoglobin, although they usually do not experience clinical manifestations since the normal hemoglobin produced from the normal gene is enough to function normally.

However, Many other genetic disorders are caused by defects related to the sex chromosomes, or the X and Y chromosomes. If a defective gene on the X-chromosome are inherited, it is called X-linked. Like autosomal disorders, X-linked genetic diseases also can be inherited by dominant and recessive mechanisms. X-linked dominant means that if the father passes on the defective gene on his only X chromosome, all his offspring (which will be females) will be affected. If he passes on his Y chromosome, none of these males will be affected. Therefore, there is no male-to-male transmission. If it is X-linked recessive, all daughters will be carriers. If the mother passes on a recessive X-linked gene, then all her sons will be affected and all her daughters will be carriers. Understanding the mechanisms by which genetic disorders are inherited are very important for interpreting recurrence risks.

A variation of Mendelian patterns of inheritance is called incomplete dominance. Incomplete dominance occurs when both alleles are expressed. An example of this is observed in Four-O-Clock **flower color**. A white and a red phenotype are neither dominant nor recessive. If a flower is heterozygous and carries genes that produce both the white color and the red color, the flower color results in pink. Another variation is codominance. In incomplete dominance, the phenotype is a blending of the two different gene effects. In codominance, both gene variants are expressed at the same time, representing a third phenotype. For example, a flower with alleles that can produce either white or red in the homozygote form will produce both colors in the heterozygote form expressed as spotted flowers.

Genetic analysis

Chromosomal analysis can be performed on cell samples from an individual, a technique called karyotyping. A karyotype involves visualizing the chromosomes using a specific dye and a high-resolution microscope. The chromosomes can be distinguished from one another in size and staining pattern. Corresponding chromosomes 1 through 22 and the sex chromosomes can be lined up and visually inspected for abnormalities. Any obvious defect can indicate a diseased state. Sometimes it is apparent that a part of one chromosome was incorrectly combined with a different chromosome during cellular division. This is called a translocation and represents a structural abnormality. Numerical abnormalities occur when an extra chromosome is present. When more than 46 chromosomes are observed in chromosome, it is called aneuploidy. Most aneuploidies that occur are called trisomies which involve three homologous chromosomes, or the presence of extra chromosome. Numerical abnormalities are typically incompatible with life, with the exception of Trisomy 21 (**Down Syndrome**) and a few other rare genetic disorders.

Dominant genetic disorders

If one parent has an autosomal dominant disease, then offspring have a 50% chance of inheriting that disease. There are roughly 2000 autosomal dominant disorders (ADDs) with effects that range from mild clinical manifestations to death. These diseases may develop early or late in life. ADDs include **Huntington disease**, **Marfan syndrome** (extra long limbs), achondroplasia (a type of dwarfism), some forms of glaucoma, most forms of porphyrias, and hypercholesterolemia (high **blood cholesterol**).

Huntington Disease is a late onset nuerodegenerative disease. It is characterized by progressive chorea (involuntary, rapid, jerky motions) and mental deterioration that often develop after the fourth decade of life, eventually leading to death approximately 15 years later. The Huntington disease gene locus is on chromosome 4, and can be identified.

Marfan syndrome is an ADD characterized by long arms, legs, and fingers. People with Marfan syndrome also have a blue sclera that represents a hallmark clinical feature that can be detected when observing the eyes. In addition, these individuals have a high incidence of **eye** and aortic **heart** problems. The **elasticity** of the vessels in the aorta are susceptible to rupture, which can cause death. Not all people with Marfan syndrome inherit it from a parent, about 15% of Marfan syndrome cases are caused by a spontaneous **mutation**.

Recessive genetic disorders

Recessive genetic disorders (RGD) result from inheriting two defective recessive alleles of a gene, one from each parent. RGD often require careful molecular or biochemical genetic analyses to determine carrier status. Hence, the **birth** of a child with a recessive disorder may surprise unknowing parents. The probability of two carrier or heterozygous parents having an affected child is 25% each time they conceive. The chance that they will have a heterozygous (carrier) child is 50% for each conception. And the chance of having an unaffected homozygous child is also 25% for each pregnancy. More than 1,000 RGDs have been identified and include: **cystic fibrosis**, **phenylketonuria** (PKU), galactosemia, retinoblastoma (Rb), **albinism**, sickle-cell anemia, thalassemia, **Tay-Sachs disease**, **autism**, growth hormone deficiency, adenosine deaminase deficiency, and Werner's syndrome (juvenile muscular dystrophy).

A number of eye disorders are RGDs and are often associated with a mutant gene on chromosome 13. The Retinoblastoma (Rb) gene was the first human gene found to cause **cancer** and the first human cancer gene in which its location on a chromosome was determined. Rb is a gene that can cause a **tumor** in the retina, called a retinoblastoma. Most retinoblastomas are inherited; however, in some cases, inheriting a mutation on one allele combined with a spontaneous mutation on the other allele can result in retinoblastoma. Environmental carcinogens (cancer causing agents) can induce a spontaneous mutation. Other recessive eye disorders include myopia (nearsightedness), albinism, day blindness, displaced pupils, and dry eyes.

Some RGDs affect people having a particular ethnic background. For example, cystic fibrosis (CF), sickle-cell anemia (SCA), and Tay-Sachs disease (TSD) all have specific mutations in the gene that causes each disorder that are preferentially found in individuals of a certain ethnic background. CF is a common autosomal recessive disease in individuals of Northern European decent and one in every 25 people in this population are carriers. SCA usually is most common in black and His-

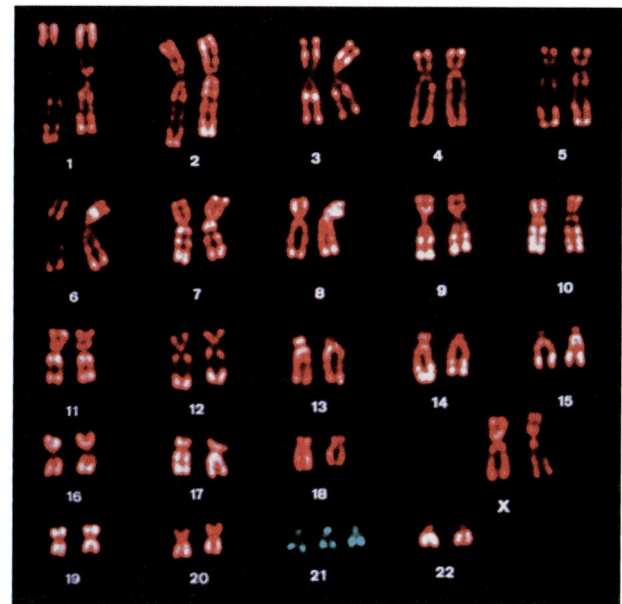

Down syndrome is a congenital disorder resulting from trisomy (three chromosomes instead of two) in pair 21. *Custom Medical Photo. Reproduced by permission.*

panic populations; however, some mutations in the gene that causes SCA are also found in Italian, Greek, Arabian, Maltese, Southern Asian, and Turkish populations. About 1 in 12 blacks are carriers for SCA gene mutations in one of the two hemoglobin genes. Hemoglobin carries **oxygen** in red blood cells to tissues and organs throughout the body. SCA patients have red blood cells that live only a fraction of the normal life span of 120 days. The abnormal blood cells have a sickled appearance and cannot transport oxygen efficiently. TSD **gene mutation** carriers are commonly found in Ashkenazi Jewish populations and approximately 1 in 30 are carriers of a mutation.

Galactosemia and PKU are examples of metabolic RGDs and are commonly called biochemical disorders. Both these disorders result from mutations in two different genes, both of which produce dysfunctional enzymes that are important in **metabolism**. Enzymes speed up **chemical reactions** and are essential for many cellular processes. People with galactosemia are **enzyme** deficient for Galactose-1-phosphate uridyl transferase, without which they can not metabolize galactose, a sugar found in milk. If milk and other galactose-containing food items are consumed, these individual cannot digest this compound properly and the result is severe developmental delay. Individuals that inherit PKU are deficient in the enzyme Phenylalanine hydroxylase, which is responsible for converting phenylalanine (an **amino acid**, a building block of protein) to tyrosine. The build-up of phenylalanine leads to severe developmental delay, and preventing clinical manifestations can be achieved by dietary modification. A phenylalanine-free diet containing sufficient amino acids is available for people diagnosed with PKU. Both of these **metabolic disorders** result from mutations in two different genes. Carriers are detected through a nationwide newborn screening programs, which differ in testing services from state to state.

Adenoside deaminase deficiency another autosomal recessive genetic disorder that is also called severe combined immunodeficiency. It is caused by a single mutation on chromosome 20 in a gene that encodes an enzyme important for proper functioning of the **immune system**. Adenosine deaminase converts breaksdown a small **molecule** called adenosine. **Gene therapy** has been successful in delivering a normal gene and prevents the damaging effects of adenosine on specific cells important for immune system function.

X-Linked genetic disorders

X-Linked genetic disorders (XLGDs) can be either dominant or recessive. Dominant XLGDs affect females and males. Dominant XLGD's include: Albright's hereditary osteodystrophy (seizures, mental retardation, stunted growth), Goltz's syndrome (mental retardation), cylindromatosis (deafness and upper body tumors), oral-facial-digital syndrome (no teeth, cleft tongue, some mental retardation), and incontinentia pigmenti (abnormal swirled skin pigmentation).

Recessive XLGDs are passed to sons through their mothers, who are carriers. Often, a carrier mother will have an affected male relative. Major XLGDs include: severe combined immune deficiency syndrome (SCID), **color blindness**, **hemophilia**, Duchenne's muscular dystrophy (DMD), some spinal ataxias, and Lesch-Nyhan syndrome. Roughly one third of these XLGDs result from a spontaneous mutation. Of these disorders, color blindness is the most benign.

Hemophilia is a more serious XLGD caused by failure of one of the clotting proteins, which serves to prevent an injured person from bleeding to death. Hemophilia A is the most severe form of this disease, and is characterized by extreme bleeding. It primarily affects males, although a few females can develop symptoms. This disease has been associated with royalty, as England's queen Victoria was a carrier and her descendants became rulers in several European countries.

Multi-factorial genetic disorders

Statistics and twin studies are often used to determine the genetic basis for multi-factorial genetic disorders (MFGDs). Because environment can play an important role in the development of these diseases, identical and fraternal twins who have been raised in different homes are ideally studied. MFGD include some disorders associated with diet and metabolism, such as **obesity**, diabetes, **alcoholism**, rickets, and high blood **pressure**. And the tendency of contracting certain infections such as measles, **scarlet fever**, and **tuberculosis** can be considered MFGDs. In addition, **schizophrenia** and some other psychological illnesses represent as well as **congenital** hip, club foot, and cleft lip are also inherited in this manner. Cancer, where the risk is associated with the environmental exposure also falls into this class of disorders. Certain breast, colon, skin, and small-cell lung cancers have been shown to have a genetic link. Certain genes predisposed people to a certain type of cancer and this risk is enhanced when there is a specific environmental exposure. This susceptibility is influenced by inherited variations in genes, which encode proteins that may be more or less functional. For example, if the protein is involved in the metabolism of a carcer causing substance and an individual inherits a variation in the gene sequence, this might affect the function of the protein it encodes. If it reduced its function, this might lead to more damaging effects of the cancer causing substance.

Aneuploidy

The two most common aneuploidies, trisomies and extra sex chromosomes, can be due to maternal or paternal factors including advanced age. A number of aneuploidies can be attributed to dispermy, or when two sperm fertilizes one egg. The resulting genetic disorders can result in a spontaneous mutation. Live-born children with autosomal aneuploidies can have trisomy 13, 18, or 21, and all have some form of developmental delay, while trisomy of any other chromosome is usually always fatal. Trisomy 13 (Patau's syndrome) is characterized by retarded growth, cleft lip, small head and chin, and often polydactyly. Trisomy 18 (Edward's syndrome) is marked by severe, variable abnormalities of the head, thumbs, ears, mouth, and feet. Trisomy 21 (Down's syndrome) occurs equally in all ethnic groups, and is closely related to increased maternal age.

Aneuploidy of the sex chromosomes can cause abnormal genital development, sterility, and other growth problems. The most common aberrations involve multiple X chromosome syndromes. Males with an XXY aneuploidy have Klinefelter's syndrome, and have small testes and typically no sperm. Hermaphrodites are individuals that have both male and female genitals, are extremely rare, and result from cell lines that have two different chromosome patterns with both XX and XY cells.

KEY TERMS

....

Allele—Any of two or more alternative forms of a gene that occupy the same location on a chromosome.

Chromosomes—he structures that carry genetic information in the form of DNA. Chromosomes are located within every cell and are responsible for directing the development and functioning of all the cells in the body.

Dominant trait—A trait which can manifest when inherited from one parent.

Gene—A discrete unit of inheritance, represented by a portion of DNA located on a chromosome. The gene is a code for the production of a specific kind of protein or RNA molecule, and therefore for a specific inherited characteristic.

Heterozygote—A person possessing two non-identical alleles.

Homozygote—A person possessing two identical alleles.

Multifactorial trait—A trait which results from both genetic and environmental influences.

Recessive trait—A trait which is not expressed in heterozygotes but is expressed if two defective genes are inherited by carrier parents.

Sclera—White of the eye.

X-Linked trait—A trait that is inherited due to mutations in genes on the X-chromosome.

Genetic screening

Genetic tests are available and can reveal varying degrees of genetic information depending on the test. Most of these tests are performed by isolating chromosomes or by measuring a protein product that is excreted in the urine to test for biochemical defects. These tests can be used prior to conception to determine a couple's risk for having an affected child, during pregnancy, at birth or later in life.

The most successful wide-spread test for genetic disorders is the newborn program that tests for diseases such as PKU. Newborn screening for hypothyroidism and galactosemia are also performed in several states. Prenatal testing in embryos and fetuses include **chorionic villus sampling (CVS)**, **amniocentesis**, and ultrasound. CVS can detect Down syndrome, hemophilia, DMD, CF, SCA, and sex chromosomal aberrations. Amniocentesis can detect Tay-Sachs disease, Down syndrome, hemophilia, **spina bifida**, and other abnormalities. Ultrasound is used to visualize the developing baby; it can detect spina bifida, anencephaly (no **brain**), and limb deformities.

Genetic counseling and testing can help people find out if they carry the gene for some disorders, or whether they will develop a late-onset genetic disorder themselves. Genetic probes can identify the genes for Huntington's disease, cystic fibrosis, Tay-Sachs, sickle-cell, thalassemia, and abnormalities associated with growth hormone. **Genetic testing** capabilities increase each year as additional genetic disorders are better characterized and the gene localization and protein function is determined.

See also ADA (adenosine deaminase) deficiency; Albinism; Birth defects; Embryo and embryonic development.

Resources

Books

Lodish, J., D. Baltimore, A. Berk, S. L. Zipursky, P. Matsudaira, and J. Darnell. *Molecular Cell Biology.* New York: Scientific American Books, Inc., 1995.

Nussbaum, Robert L., Roderick R. McInnes, and Huntington F. Willard. *Genetics in Medicine.* Philadelphia: Saunders, 2001.

David L. Rimoin *Emery and Rimoin's Principles and Practice of Medical Genetics.* London; New York: Churchill Livingstone, 2002.

Louise Dickerson
Bryan Cobb

Genetic engineering

Genetic **engineering** is the alteration of genetic material with a view to producing new substances or creating new functions. The technique became possible in the 1950s, when scientists discovered the structure of DNA molecules and learned how these molecules store and transmit genetic information. Largely as a result of the pioneering work of James Watson and Francis Crick, scientists were able to discover the sequence of **nitrogen** bases that constitute the particular DNA **molecule** codes for the manufacture of particular chemical compounds. This is the sequence that acts as an "instruction manual" for all **cell** functions. Certain practical consequences of that discovery were immediately apparent. Suppose that the base sequence T-G-G-C-T-A-C-T on a DNA molecule carries the instruction "make insulin." (The actual sequence for such a message would in reality be much longer). The DNA in the cells of the islets of Langerhans in the pancreas would normally contain that base sequence—since the islets are the region in which **insulin** is produced in **mammals**. It should be noted, however, that the base sequence carries the same message no matter where it is found. If a way could be found to insert that base sequence into the DNA of **bacteria**, for example, then those bacteria would be capable of manufacturing insulin.

Although the concept of **gene** transfer is relatively simple, its actual execution presents considerable technical obstacles. The first person to surmount these obstacles was the American biochemist Paul Berg, often referred to as the "father of genetic engineering." In 1973, Berg developed a method for joining the DNA from two different organisms: a monkey **virus** known as SV40 and a virus known as lambda phage. The accomplishment was extraordinary; however, scientists realized that Berg's method was too laborious. A turning point in genetic engineering came later that year, when Stanley Cohen at Stanford and Hubert Boyer at the University of California at San Francisco discovered an **enzyme** that greatly increased the efficiency of the Berg procedure. The gene transfer technique developed by Berg, Boyer, and Cohen forms the basis of much of contemporary genetic engineering.

This technique requires three elements: the gene to be transferred, a host cell in which the gene is to be inserted, and a vector to effect the transfer. Suppose, for example, that one wishes to insert the insulin in a bacterial cell. The first step is to obtain a copy of the insulin gene. This copy can be obtained from a natural sources (from the DNA in islets of Langerhans cells, for example), or it can be manufactured in a laboratory. The second step is to insert the insulin gene into the vector. The most common vector is a circular form of DNA known as a plasmid. Scientists have discovered enzymes that can "recognize" certain base sequences in a DNA molecule and cut the molecule open at these locations. In fact, the plasmid vector can be cleaved at almost any point chosen by the scientist. Once the plasmid has been cleaved, it is mixed with the insulin gene and another enzyme that has the ability to glue the DNA molecules back together. In this particular case, however, the insulin gene attaches itself to the plasmid before the plasmid is re-closed. The **hybrid** plasmid now contains the gene whose product (insulin) is desired. It can be inserted into the host cell, where it begins to function as a bacterial gene. In this case, however, in addition to normal bacterial functions, the host cell is also producing insulin, as directed by the inserted gene. Because of the nature of the procedure, this method is sometimes referred to as *gene splicing*; and since the genes have come from two different sources have been combined with each other, the technique is also called *recombinant DNA* (rDNA) *research*.

The possible applications of genetic engineering are virtually limitless. For example, rDNA methods now enable scientists to produce a number of products that were previously available only in limited quantities. Until the 1980s, for example, the only source of insulin available to diabetics was found in animals slaughtered for meat and other purposes, and the supply was never high enough to provide a sufficient amount of affordable insulin for diabetics. In 1982, however, the U.S. Food and Drug Administration approved insulin produced by genetically altered organisms, the first such product to become available. Since 1982, a number of additional products, including human growth hormone, alpha interferon, interleukin-2, factor VIII, erythropoietin, **tumor** necrosis factor, and **tissue** plasminogen activator have been produced by rDNA techniques.

The commercial potential of genetically products was not lost on entrepreneurs in the 1970s. A few individuals believed, furthermore, that the impact of rDNA on American technology would be comparable to that of computers in the 1950s. In many cases, the first genetic engineering firms were founded by scientists involved in fundamental research. Boyer, for example, joined the venture capitalist Robert Swanson in 1976 to form Genetech (Genetic Engineering Technology). Other early firms like Cetus, Biogen, and Genex were formed similarly through the collaboration of scientists and businesspeople.

The structure of genetic engineering (**biotechnology**) firms has, in fact, long been a source of controversy. Many have questioned the scientists' right to make a personal profit by running companies which benefit from research that had been carried out at publicly-funded universities.

The early 1990s saw the creation of formalized working relations between universities, individual researchers, and the corporations founded by these individuals. However, despite these arrangements, many ethical issues remain unresolved.

One of the most exciting potential applications of genetic engineering involves the treatment of **genetic disorders**. Medical scientists know of about 3,000 disorders that arise because of errors in individuals DNA. Conditions such as sickle-cell **anemia**, **Tay-Sachs disease**, Duchenne muscular dystrophy, Huntington's chorea, **cystic fibrosis**, and Lesch-Nyhan **syndrome** result from the mistaken insertion, omission, or change of a single nitrogen base in a DNA molecule. Genetic engineering enables scientists to provide individuals lacking a particular gene with correct copies of that gene. If and when the correct gene begins functioning, the genetic disorder may be cured. This procedure is known as *human gene therapy* (HGT).

The first approved trials of HGT with human patients began in the 1980s. One of the most promising sets of experiments involved a condition known as severe combined immune deficiency (SCID). In 1990, a research team at the National Institutes of Health (NIH) led by W. French Anderson attempted HGT on a four-year-old SCID patient, whose condition was associated with the absence of the enzyme adenosine deaminase (ADA). The patient received about a billion cells containing a genetically engineered copy of the ADA gene that his body lacked. Another instance of HGT was a procedure, approved in 1993 by NIH, to introduce normal genes into the airways of cystic fibrosis patients.

Human **gene therapy** is the source of great controversy among scientist and non-scientists alike. Few individuals maintain that the HGT should not be used. If we could wipe out sickle-cell anemia, most agree, we should certainly make the effort. But HGT raises other concerns. If scientists can cure genetic disorders, they can also design individuals in accordance with the cultural and intellectual fashions of the day. Will humans know when to say "enough" to the changes that can be made with HGT?

Genetic engineering also promises a revolution in agriculture. **Recombinant DNA** techniques enable scientists to produce plants that are resistant to **herbicides** and freezing temperatures, that will take longer to ripen, that will convert atmospheric nitrogen to a form they can use, that will manufacture a resistance to **pests**, and so on. By 1988, scientists had tested more than two dozen kinds of plants engineered to have special properties such as these. As with other aspects of genetic engineering, however, these advances have been controversial. The development of herbicide-resistant plants means that farmers will use still larger quantities of herbicides—not a particularly desirable trend, according to critics. How sure can we be, others ask, about the risk to the environment posed by the introduction of "unnatural," engineered plants?

Many other applications of genetic engineering have already been developed or are likely to be realized in the future.

See also ADA (adenosine deaminase) deficiency; Birth defects; Diabetes mellitus; Gene splicing; Genetic disorders; Genetics.

Resources

Books

Beurton, Peter, Raphael Falk, Hans-Jörg Rheinberger., eds. *The Concept of the Gene in Development and Evolution.* Cambridge, UK: Cambridge University Press, 2000.
Drlica, Karl A. *The Double-Edged Sword: The Promises and Risks of the Genetic Revolution.* Reading, MA: Addison-Wesley, 1994.
Jorde, L.B., J.C. Carey, M.J. Bamshad, and R.L. White. *Medical Genetics.* 2nd ed. New York: Year Book, Inc., 2000.
Lewin, B. *Genes.* 7th ed. Oxford: Oxford University Press, 2000.
Sylvester, Edward J., and Lynn C. Klotz. *The Gene Age: Genetic Engineering and the Next Industrial Evolution.* rev. ed. New York: Scribner's, 1987.
Tudge, Colin. *The Engineer in the Garden: Genes and Genetics: From the Idea of Heredity to the Creation of Life.* New York: Hill and Wang, 1995.

Periodicals

Amar A.R. "A Search For Justice In Our Genes." *New York Times.* May 7, 2002: A31.
Jeffords, J.M. and Tom Daschle. "Political Issues in the Genome Era." *Science* 291 (February 16, 2001): 1249-50.
Veuille, E. "Genetics and the Evolutionary Process." *C. R. Academy of Science, i III* 323, no. 12 (December 2000): 1155-65.
Yaspo, M. L. et al. "The DNA Sequence of Human Chromosome 21." *Nature* 6784 (May 2000): 311-319.

Genetic identification of microorganisms

The advent of molecular technologies and the application of genetic identification in clinical and forensic microbiology have greatly improved the capability of laboratories to detect and specifically identify an **organism** quickly and accurately.

In the wake of the 2001 **anthrax** attacks utilizing the United States mail, a great deal of investigative attention turned to identification of the source of the anthrax used in the attacks. Scientists continue to track the source of the anthrax utilizing genetic identification principles, techniques, and technologies.

The genetic identification of **microorganisms** utilizes molecular technologies to evaluate specific regions of the **genome** and uniquely determine to which genus, **species**, or strain a microorganism belongs. This work grew out of the similar, highly successful applications in human identification using the same basic techniques. Thus, the genetic identification of microorganisms has also been referred to a microbial fingerprinting.

Genetic identification of microorganisms is basically a comparison study. To identify an unknown organism, appropriate sequences from the unknown are compared to documented sequences from known organisms. Homology between the sequences results in a positive test. An exact match will occur when the two organisms are the same. Related individuals have genetic material that is identical for some regions and dissimilar for others. Unrelated individuals will have significant differences in the sequences being evaluated. Developing a database of key sequences that are unique to and characteristic of a series of known organisms facilitates this type of analysis. The sequences utilized fall into two different categories, 1) fragments derived from the transcriptionally active, coding regions of the genome, and, 2) fragments present in inactive, noncoding regions. Of the two, the noncoding genomic material is more susceptible to **mutation** and will therefore show a higher degree of variability.

Depending on the level of specificity required, an assay can provide information on the genus, species, and/or strain of a microorganism. The most basic type of identification is classification to a genus. Although this general identification does not discriminate between the related species that comprise the genus, it can be useful in a variety of situations. For example, if a person is thought to have **tuberculosis**, a test to determine if *Mycobacterium* cells (the genus that includes the tuberculosis causing organism) are present in a sputum **sample** will most likely confirm the **diagnosis**. However, if there are several species within a genus that cause similar diseases but that respond to entirely different drugs, it would then be critical to know exactly which species is present for proper treatment. A more specific test using genomic sequences unique to each species would be needed for this type of discrimination. In some instances, it is important to take the analysis one step further to detect genetically distinct subspecies or strains. Variant strains usually arise as a result of physical separation and **evolution** of the genome. If one homogeneous sample of cells is split and sent to two different locations, over time, changes (mutations) may occur that will distinguish the two populations as unique entities. The importance of this issue can be appreciated when considering tuberculosis. Since the late 1980s, there has been a resurgence of this **disease** accompanied by the appearance of several new strains with antimicrobial resistance. The use of genetic identification for rapid determination of which strain is present has been essential to protect health care workers and provide appropriate therapy for affected individuals.

The tools used for genetic studies include standard molecular technologies. Total sequencing of an organism's genome is one approach, but this method is time consuming and expensive. Southern blot analysis can be used, but has been replaced by newer technologies in most laboratories. Solution-phase hybridization using DNA probes has proven effective for many organisms. In this procedure, probes labeled with a reporter **molecule** are combined with cells in solution and upon hybridization with target cells, a chemiluminescent signal that can be quantitated by a luminometer is emitted. A variation of this scheme is to capture the target cells by hybridiza-

tion to a probe followed by a second hybridization that results in **precipitation** of the cells for quantitation. These assays are rapid, relatively inexpensive and highly sensitive. However, they require the presence of a relatively large number of organisms to be effective. Amplification technologies such as **PCR** (polymerase chain reaction) and LCR (ligase change reaction) allow detection of very low concentrations of organisms from cultures or patient specimens such as **blood** or body tissues. Primers are designed to selectively amplify genomic sequences unique to each species, and, by screening unknowns for the presence or absence these regions, the unknown is identified. Multiplex PCR has made it possible to discriminate between a number of different species in a single amplification reaction. For viruses with a RNA genome, RT-PCR (reverse transcriptase PCR) is widely utilized for identification and quantitation.

The anthrax outbreak in the Unites States in the fall of 2001 illustrated the significance of these technologies. Because an anthrax **infection** can mimic cold or flu symptoms, the earliest victims did not realize they were harboring a deadly bacterium. After confirmation that anthrax was the causative agent in the first death, genetic technologies were utilized to confirm the presence of anthrax in other locations and for other potential victims. Results were available more rapidly than would have been possible using standard microbiological methodology and appropriate treatment regimens could be established immediately. Furthermore, unaffected individuals are quickly informed of their status, alleviating unnecessary **anxiety**.

The second stage of the investigation was to locate the origin of the anthrax cells. The evidence indicated that this event was not a **random**, natural phenomenon, and that an individual or individuals had most likely dispersed the cells as an act of **bioterrorism**. In response to this threat, government agencies collected samples from all sites for analysis. A key element in the search was the genetic identification of the cells found in patients and mail from Florida, New York, and Washington, D.C. The PCR studies clearly showed that all samples were derived from the same strain of anthrax, known as the "Ames strain" since the **cell** line was established in Iowa. Although this strain has been distributed to many different research laboratories around the world, careful analysis revealed minor changes in the genome that allowed investigators to narrow the search to about fifteen United States laboratories. Total genome sequencing of these fifteen strains and a one-to-one base comparison with the lethal anthrax genome may detect further variation that will allow a unique identification to be made.

See also Bioassay; Biological warfare; Genetically modified foods and organisms; Genomics (comparative); Microbial genetics.

Resources

Books

Flint, S.J., L.W. Enquist, R.M. Krug, et al. *Principles of Virology: Molecular Biology, Pathogenesis, and Control.* Washington, DC: American Society for Microbiology Press, 1999.

Shaw, Karen Joy. *Pathogen Genomics: Impact on Human Health* Totowa, NJ: Humana Press, 2002.

Stahl, F.W. *We Can Sleep Later: Alfred D. Hershey and the Origins of Molecular Biology.* Cold Spring Harbor, NY: Cold Spring Harbor Press, 2000.

Periodicals

Fraser, C.M., J. Eisen, R.D. Fleischmann, K.A. Ketchum, and S. Peterson. "Comparative Genomics and Understanding of Microbial Biology." *Emerging Infectious Diseases.* 6, no. 5 (September-October 2000).

Other

Ronald Koopman et al. HANAA: Putting DNA Identification in the Hands of First Responder [cited January, 15 2003]. <http://coffee.phys.unm.edu/BTR/2001%20Conference/pdf/Koopman_Ronald.pdf>.

Constance Stein

Genetic mapping *see* **Chromosome mapping**

Genetic testing

The use of genetic information to predict future onset of **disease** in an asymptomatic (presymptomatic) person is called predictive genetic testing.

Every aspect of our being is influenced by both genes and environment. In the future, a strategy for influencing development may be to alter genes. At present, the environment in which genes act can sometimes be changed, and thereby moderate their impact (taking medications or avoiding specific hazards, for example). Sometimes there is no known way to change the deterministic power of a **gene**, though with increased knowledge of its workings there is always hope for future interventions. Whether or not the course of a disease can be altered, predictive information is increasingly available, and some people choose knowledge over uncertainty.

For generations, people have used family information to anticipate outcomes for themselves. Insurers consider parental age and cause of death for actuarial tables. **Evolution** in knowledge has been from information with considerable associated uncertainty to that with greater pre-

dictive capacity. **Huntington Disease** (HD) became the prototype for predictive testing and serves to illustrate.

HD is a neurological disease with onset of symptoms usually during adulthood. It is inherited as an autosomal dominant trait; someone with an affected parent has a 50/50 chance of eventually developing the disease. The HD gene was the first human gene to be linked to an otherwise anonymous DNA marker (a restriction fragment length polymorphism, called G8), and long before the gene itself was identified, this marker and others like it became powerful predictive tools. Families in which HD was segregating were studied to determine which variant of the marker was tracking with the mutant HD gene; once that relationship was established, the marker(s) could be used to test family members who wished to know their genetic status. This indirect approach to testing was associated with some probability of **error**, since the markers were only close to the gene, not within it. With discovery in 1993 of the gene responsible for HD, a direct assay was immediately possible, with or without access to samples from other family members, and results became highly predictive.

The laboratory advances made access to this information possible, but it was quickly recognized that great care would be needed in the application of such knowledge to individuals at risk. A large Canadian collaborative study of predictive testing for HD, initiated in the late 1980s, has been particularly informative for assessing the impact of such information on individuals and families and developing guidelines for the practice of predictive medicine, including the need for supportive counseling and follow-up. Lessons from experience with this relatively obscure disorder were soon applied to other late-onset diseases for which predisposing mutations were identified. Notable in this context are inherited cancers such as familial breast **cancer** or colon cancer, other neurological disorders such as spino-cerebellar ataxias (including Machado-Joseph Disease), and familial **Alzheimer disease**. Common afflictions such as **heart** disease, diabetes, and **arthritis** will eventually be amenable to similar investigations.

The **Human Genome Project** recognized the need for ethical considerations to match scientific advances, and its mandate includes significant support for research into ethical, legal and social issues. This has set new standards for the application of knowledge, respecting public concerns about the implications of new technologies. The opportunity to know ones genetic destiny has potential risks that must be mitigated in order for the benefits to be realized. Once the predictive test for HD was available, it was soon apparent that not everyone at risk wished to be tested. The right not to know is a significant issue. The genetic nature of these diseases adds complication, becasue information revealed about one **individual** may secondarily imply information about other family members, and individual choices will impact others in the family network. Acting upon respect for individual autonomy, early guidelines have advised against the testing of children for late-onset disorders in the absence of preventive options. In countries without universal health care, insurance implications of predictive testing are huge. Will people be required to submit a clean genetic bill of health in order to secure health or life insurance?

Eventually, there will be effective therapeutic interventions for diseases such as HD and Alzheimer disease, individually tailored to the needs of those at risk. Until then, there will be controversy over the practice of predictive testing, but many will continue to choose knowledge and maintain hope for the future.

See also Archaeogenetics; Chromosomal abnormalities; Medical genetics; Pharmacogenetics.

Resources

Books

Nussbaum, R.L., Roderick R. McInnes, Huntington F. Willard *Genetics in Medicine.* Philadelphia: Saunders, 2001.

Rimoin, D.L. *Emery and Rimoin's Principles and Practice of Medical Genetics.* London; New York: Churchill Livingstone, 2002.

Strachan, T., and A. Read. *Human Molecular Genetics.* New York: Bios Scientific Publishers, 1998.

Periodicals

Leparc. G.F. "Nucleic Acid Testing for Screening Donor Blood." *Infectious Medicine,* no. 17 (May 2000): 310–333.A.

Selwa, R. "Researcher Talks About Ethics of Genetic Therapy" *Macomb Daily* (2000):1A, 8A.

Other

National Human Genome Research Institute. "Ethical, Legal and Social Implications of Human Genetic Research." October 2002 [cited February 2, 2003]. <http://www.nhgri.nih.gov/ELSI/>.

Janet A. Buchanan

Genetically modified foods and organisms

While the term genetically modified organisms has arisen within the past decade, humans have for centuries been using **microorganisms** to make products like beer and cheese, and plants and animals have been carefully bred to improve the quality and quantity of the food supply. The elucidation of the structure of DNA and the de-velopment of the discipline of **molecular biology** has made possible the accurate insertion or removal specific genes into or out of the DNA of particular organisms. This enables the design organisms with specific desirable characteristics and the ability to understand which genes control the growth, reproduction, and aging and **disease** susceptibility of plants and animals.

Aside from foods, genetically modified organisms are making their way into other commercial venues. For example, the **forestry** industry is actively utilizing molecular **biology** to generate trees capable of faster and straighter growth.

The use of genetically modified organisms in agriculture has expanded at a rapid **rate** in key agricultural exporting countries in the past decade. Countries where transgenic **crops** are in advanced stages of field-testing or commercialization include the United States, Argentina, Canada, and **Australia**. The global area devoted to transgenic crops has increased from 1.7 hectares in 1996, to 27.8 hectares in 1998. In **North America**, the use of genetically modified **cotton**, **soybean** and canola now represents some 50 percent of the total acreage.

Genetically modified organisms have generated considerable debate.

Critics on one side of the debate contend that number of countries without a strong scientific infrastructure fear genetically modified foods. Others countries with advanced scientific and medical research infrastructure, (e.g., France and other European Union countries) have passed laws regulating genetically modified organisms for economic and political reasons (e.g. as a form of protectionism for their less progressive agricultural systems.) In 2001 and 2002, European countries, including France and Germany, pushed for tough European Union rules regulating the sale of genetically modified foods. The US State Department branded the news rules as "unnecessary" and without scientific merit. The US has already warned that a trade war over "biotechnology foods" might develop if the European Union fails to lift blocks to imports.

In 2002, reports surfaced that French scaremongering concerning genetically modified foods caused several African countries fighting starvation to reject genetically modified food supplements that would have reduced starvation and death rates.

On the other side of the debate, critics argue that the impact of these totally new organisms on the environment and on human health cannot presently be completely predicted. Within recent years several studies have purported to demonstrate harmful effects to monarch **butterflies** by their ingestion of pollen from Bt corn (corn modified by a bacterium called *Bacillus thuringiensis*(Bt)), and to **rats** by their ingestion of modified potatoes. The validity of these studies remains controversial. As well, the increased yields of genetically modified organisms may contribute to a decrease in crop biological diversity—genetic differences between **species**. Homogeneity may make crops more susceptible to disease. Thus, the present uncertainty about the cumulative effects in ecosystems or the food chain is making consumers wary.

Considerable controversy has arisen concerning the genetic modification of plants such that their **seeds** are not capable of growth upon planting. The commercial control and potential monopolization of food production has been decried by some. Some critics also point out that prudence on the part of France and other countries reflect warranted scientific prudence that also continues to respect closer cultural ties to food and agricultural production.

In January, 2000, The Cartagena Protocol on Biosafety was adopted in Montreal, Canada. The protocol, negotiated under the United Nations Convention on Biological Diversity, is one of the first legally binding international agreements to govern the trade or sale of genetically modified organisms of agricultural importance.

Such social, political and legal debates surrounding genetically modified organisms will likely not be resolved soon.

It is scientifically demonstrated that genetically modified crops are resistant to or tolerant to disease or insect attack. For example, a **gene** encoding an insecticidal protein from the bacterium *Bacillus thuringiensis* (Bt) has made cotton, corn and other crops resistant to attack by caterpillars. Data from several years of use of genetically modified crops in the Unites States has shown that the requirement for **pesticides** is reduced. Genetically modified crops may also permit higher yields. This may offer real hope to the estimated billion people who are chronically under-nourished and hungry, and to the many more as the global population doubles in the next 50 years. Additionally, crops that have improved nutritional value or with therapeutic value are being designed. Such nutraceuticals are driving the development of an industry whose annual sales are expected to grow to billions in the United States alone.

The direct genetic modification of foods is a modern extension of agricultural practices that have long selected genetically controlled traits of agriculturally relevant **plant** and **animal** species, so as to instills in these species beneficial genetic traits (often found in other organisms). Techniques to analyze genetic material developed within the last twenty years allow a quicker and accurate identification and propagation of superior traits, speeding the overall process of genetic improvement.

Transgenic crops (also called genetically modified crops) contain genetic material from some source other than themselves (all crops have been somewhat genetically modified from their original wild state by domestication, **selection** and controlled breeding over long periods of **time**). The inserted gene sequence, called a transgene, may come from another related species, or from a completely different species, such as a bacterial **cell**.

A significant advance in agricultural **genetics** has been the harnessing of transgenic crops to express biopesticides (also known as biological pesticides). At the end of 1998, there were approximately 175 registered biopesticide active ingredients and 700 products. The three main classes of biopesticides are microbial pesticides (microorganism as the active ingredient), biochemical pesticides (natural and non-toxic compounds as the pest control agent) and plant pesticides. Herbicide resistance is a popular transgenic trait. Plants have been engineered to be resistant to **herbicides** like glyphosate or glufosinate, which are broad **spectrum** in their activity, killing nearly all kinds of plants except those possessing the resistance transgene. Another popular biopesticide target is **insects**, such as European corn borer and the cotton bollworm, which can be killed by a protein produced by *Bacillus thuringiensis*(Bt).

Once a useful gene has been identified, isolated and copies made, it must be modified so it can be effectively inserted into the DNA of the target plant or animal. An efficient on-off switch for the expression of the gene is added to one end of the gene. A commonly used promoter sequence is CaMV35S from the cauliflower mosaic **virus**. At the other end of the gene a sequence is added which signals an end to expression. The gene can also be modified slightly to increase its expression. In the same stretch of DNA as the above construct lies a marker gene, complete with its own promoter and termination sequences. The marker gene codes for resistance to a selected antibiotic or herbicide. Development of resistance following transformation means that the inserted DNA has been expressed.

The new genetic material is inserted into the plant or animal genetic material in a process called transformation. The two main methods of transformation are the gene gun method and the Agrobacterium method. In the first, millions of DNA-coated particles are shot from a specialized gun inside the plant or animal cell. Some of the DNA will recombine with the cellular DNA. The second method takes advantage of the ability of a **soil** bacterium called Agrobacterium tumefaciens to inject, through a wound in plant cells, a specialized portion of its DNA. Following transformation, plant tissues are transferred to a growth source containing the selective antibiotic or herbicide. Cells which grow are those in which the foreign DNA has been expressed. Before the transgenic cell is ready for commercial use it must be rigorously verified and demonstrated to legislative authorities that the transgene has been stably incorporated, and does not pose harm to other plant functions, final product or the natural environment where it will reside. Often the transgenic crop will be crossed with existing parents to produce an improved variety. The improved variety is then used for several cycles of crosses to the parent to recover as much of the improved parent's genetic material as possible, with the addition of the transgene.

Development of transgene technology has been slowed by the limited knowledge of the complexities of gene expression, with the myriad of other factors, some responsive to environmental change, which control the gene's expression. Despite this limitation, the early successes of the technology have met with great commercial acceptance. As of 2000, the most popular transgenic crop and trait worldwide in terms of acreage planted is soybean (more than 55 million acres) and herbicide (more than 70 million acres). Total worldwide acreage of transgenic crops in 2000 was approximately 166 million acres. In that year, almost half of the United States' soybean crop and about 28 % of its corn crop were transgenic varieties. One attraction has been the savings in pesticide use. In the case of cotton, the use of the Bt crops has dramatically reduced the amount of chemical pesticides used.

See also Biodiversity; Biotechnology; Food chain/web; Food irradiation; Food poisoning; Food preservation; Food pyramid; Genetics.

Resources

Books

Ruse, Michael and David Castle, eds. *Genetically Modified Foods: Debating Biotechnology* Contemporary Issues Series, Amherst, NY: Prometheus Books, 2002.

Nelson, Gerald C., ed. *Genetically Modified Organisms in Agriculture: Economics and Politics* San Diego, CA: Academic Press, 2001.

Other

American Society for Microbiology. "Statement of the American Society for Microbiology on Genetically Modified Organisms" [cited February 27, 2003]. <http://www.asmusa.org/pasrc/genmodorg.htm>.

Untied States Department of Energy Office of Science. Human Genome Project Information. "Genetically Modified Foods and Organisms" [cited February 27, 2003]. <http://www.ornl.gov/hgmis/elsi/ gmfood.html>.

Brian Hoyle
K. Lee Lerner

Genetics

Genetics is the branch of **biology** concerned with the science of heredity, or the transfer of specific characteristics from one generation to the next. Genetics focuses primarily on genes, coded units found along the DNA molecules of the chromosomes, housed by the **cell** nucleus. Together, genes make up the blueprints that determine the entire development of the **species** of organisms down to specific traits, such as the **color** of eyes and hair. Geneticists are concerned with three primary areas of **gene** study: how genes are expressed and regulated in the cell, how genes are copied and passed on to successive generations, and what are the genetic basis for differences between the species. Although the science of genetics dates back at least to the nineteenth century, little was known about the exact biological makeup of genes until the 1940s. Since that time, genetics has moved to the forefront of biological research. Scientists are now on the verge of identifying the location and function of every gene in the entire human **genome**. The result will not only be a greater understanding of the human body, but new insights into the origins of **disease** and the formulation of possible treatments and cures.

The history of genetics

Although humans have known about inheritance for thousands of years, the first scientific evidence for the existence of genes came in 1866, when the Austrian monk and scientist Gregor Mendel published the results of a study of hybridization of plants—the combining of two **individual** species with different genetic make-ups to produce a new individual. Working with pea plants with specific characteristics such as height (tall and short) and color (green and yellow), Mendel bred one type of **plant** for several successive generations. He found that certain characteristics appeared in the next generation in a regular pattern. From these observations, he deduced that the plants inherited a specific biological unit (which he called factors (now called **alleles**), genes determining different forms of a single characteristic) from each parent. Mendel also noted that when factors or alleles pair up, one is dominant (which means it determines the trait, like tallness) while the other is recessive (which means it has no bearing on the trait). It is now understood that alleles may be single genes or sets of genes working together, each contributing to the final form of a physical characteristic (multiple allelism).

The period of classical genetics, in which researchers had no knowledge of the chemical constituents in cells that determine heredity, lasted well into the

Three generations of identical twins. © Gerald Davis/Phototake NYC. Reproduced with permission.

twentieth century. However, several advances made during that time contributed to the growth of genetics. In the eighteenth century, scientists used the relatively new technology of the **microscope** to discover the existence of cells, the basic structures in all living organisms. By the middle of the nineteenth century, they had discovered that cells reproduce by dividing.

Although Mendel laid the foundation of genetics, his work began to take on true significance in 1903 when Theodore Boveri and Walter Sutton independently proposed a chromosomal theory of inheritance. They discovered that chromosomes during **gamete** production behave like the so-called Mendel's particles behave. In 1910, Thomas Hunt Morgan (1866–1946) confirmed the existence of chromosomes through experiments with fruit **flies**. He also discovered a unique pair of chromosomes called the sex chromosomes, which determined the sex of offspring. Morgan deduced that specific genes reside on chromosomes from his observation that an X-shaped **chromosome** was always present in flies that had white eyes. A later discovery showed that chromosomes could mutate or change structurally, resulting in a change in characteristics which could be passed on to the next generation.

More than three decades passed before scientists began to delve into the specific molecular and chemical structures that make up chromosomes. In the 1940s, a research team led by Oswald Avery (1877–1955) discovered that **deoxyribonucleic acid (DNA)** was responsible for transformation of non-pathogenic **bacteria** into pathogenic ones. The final proof that DNA was the specific **molecule** that carries genetic information was made by Alfred Hershey and Martha Chase in 1952. They used radioactive label to differentiate between viral protein and DNA, proving that over 80% of viral DNA entered bacterial cell causing **infection**, while protein did not cause infection.

The most important discovery in genetics occurred in 1953, when James Watson and Francis Crick solved the mystery of the exact structure of DNA. The two scientists used chemical analyses and x-ray **diffraction** studies performed by other scientists to uncover the specific structure and chemical arrangement of DNA. X-ray diffraction is a procedure in which **parallel** x-ray beams are diffracted by **atoms** in patterns that reveal the atoms' **atomic weight** and spatial arrangement. A month after their double-helix model of DNA appeared in scientific journals, the two scientists showed how

DNA replicated. Armed with these new discoveries, geneticists embarked on the modern era of genetics, including efforts like **genetic engineering**, **gene therapy**, and a massive project to determine the exact location and function of all of the more than 100,000 genes that make up the human genome.

The biology of genetics

Genetic information is contained in the chromosomes, threadlike structures composed of DNA, and present in the nuclei of all cell types and are passed to daughter cells during **cell division**. Multicellular organisms contain two types of cells—body cells (or somatic cells) and germ cells (or reproductive cells). Germ cells are the ones that pass on the genetic information to the progeny. In contrast to somatic cells that contain dual copies of chromosomes in each cell, germ cells replicate through a process called **meiosis**, which ensures that the germ cells have only a single set of chromosomes, a condition called haploidy (designated as n). The somatic cells of humans have 23 pairs of chromosomes (46 chromosomes overall), a condition known as diploid (or 2n). Through the process of meiosis, a new cell, called a haploid gamete, is created with only 23 chromosomes: this is either the sperm cell of the father or the egg cell of the mother. The fusion of egg and sperm restores the diploid chromosome number in the zygote. This cell carries all the genetic information needed to grow into an embryo and eventually a full grown human with the specific traits and attributes passed on by the parents. Offspring of the same parents differ because the sperm cells and egg cells vary in their gene **sequences**, due to **random** recombination.

The somatic, or body cells are the primary components of functioning organisms. The genetic information in these cells is passed on through a process of cell division called **mitosis**. Unlike meiosis, mitosis is designed to transfer the identical number of chromosomes during cell regeneration or renewal. This is how cells grow and are replaced in exact replicas to form specific tissues and organs, such as muscles and nerves. Without mitosis, an organism's cells would not regenerate, resulting not only in **cell death**, but possible death of the entire **organism**. (It is important to note that some organisms reproduce asexually by mitosis alone.)

The genetic code

To understand genes and their biological function in heredity, it is necessary to understand the chemical makeup and structure of DNA. Although some viruses carry their genetic information in the form of **ribonucleic acid (RNA)**, most higher life forms carry genetic information in the form of DNA, the molecule that makes up chromosomes.

The complete DNA molecule is often referred to as the blueprint for life because it carries all the instructions, in the formation of genes, for the growth and functioning of most organisms. This fundamental molecule is similar in appearance to a **spiral** staircase, which is also called a **double helix**. The sides of the DNA double helix ladder are made up of alternate sugar and phosphate molecules, like links in a chain. The rungs, or steps, of DNA are made from a combination of four nitrogen-containing bases—two purines (adenine [A] and guanine [G]) and two pyrimidines (cytosine [C] and thymine [T]). The four letters designating these bases (A, G, C, and T) are the alphabet of the genetic code. Each rung of the DNA molecule is contains a combination of two of these letters, one jutting out from each side. In this genetic code, A always combines with T, and C with G to make what is called a base pair. Specific sequences of these base pairs, which are bonded together by atoms of **hydrogen**, make up the genes.

While a four-letter alphabet may seem rather small for constructing the comprehensive vocabulary that describes and determines the myriad life forms on **Earth**, the sequences or order of these base pairs are nearly limitless. For example, various sequences or rungs that make up a simple six base gene could be ATCGGC, or TAATCG, or AGCGTA, or ATTACG, and so on. Each one of these combinations has a different meaning. Different sequences provide the code not only for the type of organism, but also for specific traits like brown hair and blue eyes. The more complex an organism, from bacteria to humans, the more rungs or genetic sequences appear on the ladder. The entire genetic makeup of a human, for example, may contain 120 million base pairs, with the average gene unit being 2,000 to 200,000 base pairs long. Except for identical twins, no two humans have exactly the same genetic information.

Genetic information is duplicated during the process of **DNA replication**, which begins a few hours before the initiation of cell division (mitosis). To produce identical genetic information during mitosis, the hydrogen bonds holding together the two halves of the DNA ladder unzip, in presence of **proteins** called helicases, to expose single strands of DNA. These old strands act as templates to make new DNA molecules. Replication is initiated by this separation of DNA, and requires short DNA fragments (primers) to start synthesis of a new DNA strand by specific cellular enzymes called DNA polymerases. DNA rarely mutates during replication, as the proofreading and "repair" enzymes make sure that any errors are quickly repaired to protect the **accuracy** of the genetic information. Once completed, each new half of the DNA ladder has the identical information as the old one. This is achieved by the fact that T always combines with A and C with G, therefore if the template had a sequence ATGCTG the newly made second strand will be TACGAC. When cell mitosis is completed, each new cell contains an exact replica of the DNA.

Cells contain hundreds of different proteins and its functions are dependent on which of the thousands of types of different proteins it contains. Proteins are made up of chains of amino acids. The arrangement of the amino acids to build specific proteins is determined by the basepair sequence contained or encoded in DNA. This genetic information has to be converted to proteins building over half of all solid body tissues and control most biological processes within and among these tissues. This is achieved by using the genetic code, which is a set of 64 triplets of bases (called **codons**) corresponding to each **amino acid** and the initiation and termination signals for protein synthesis.

As the sites of protein production lie outside the cell nucleus, the instructions for making them have to be transported out of the nucleus. The messenger that carries these instructions is messenger RNA, or mRNA (a single stranded molecule that has a mirror image of the base pairs on the DNA). mRNA is made in the nucleus during a process called transcription and a single molecule of RNA carries instructions for making only one protein. After being exported out of the nucleus it is transported to **ribosomes**, which are the protein factories in the cell. In ribosomes the information from mRNA is decoded to produce a protein. This process is called translation. The flow of information is only one way from DNA to RNA and to protein. Therefore characteristics acquired during an organism's life, such as larger muscles or the ability to play the piano, cannot be inherited. However, people may have genes that make it easier for them to acquire these traits through **exercise** or practice.

Dominant and recessive traits

The expression of the products of genes is not equal, and some genes will override others in expressing themselves as an inherited characteristic. The offspring of organisms that reproduce sexually contain a set of chromosome pairs, half from the father and half from the mother. However, normally people do not have one blue **eye** and one brown eye, or half brown hair and half blond hair because most genetic traits are the result of the expression of either the dominant or the recessive genes. If a dominant and a recessive gene appear together (the heterozygous condition), the dominant will always win, producing the trait it is coded for. The only time a recessive trait (such as the one for blond hair) expresses itself is when two recessive genes are present (the homozygous condition). As a result, parents with heterozygous genes for brown hair could produce a child with blond hair if the child inherits two recessive blond-hair genes from the parents. The genes residing in the chromosome's DNA can also be present in alternative forms called alleles. It is important to note that some characteristics are a result of presence of various alleles, e.g. pink snapdragon flowers or **blood** types.

This hereditary law also holds true for genetic diseases. Neither parent may show signs of a genetic disease, caused by a defective gene, but they can pass the double-recessive combination on to their children. Some genetic diseases are dominant and others are recessive. Dominant genetic defects are more common because it only takes one parent to pass on a defective allele. A recessive genetic defect requires both parents to pass on the recessive allele that causes the disease. A few inherited diseases (such as **Down syndrome**) are caused by abnormalities in the number of chromosomes, where the offspring has 47 chromosomes instead of the normal 46.

Genetic recombination and mutations

The DNA molecule is extremely stable, ensuring that offspring have the same traits and attributes that will enable them to survive as well as their parents. However, a certain amount of genetic variation is necessary if species are to adapt by natural **selection** to a changing environment. Often, this change in genetic material occurs when chromosome segments from the parents physically exchange segments with each other during the process of meiosis. This is known as cross over or intrinsic recombination.

Genes can also change by mutations on the DNA molecule, which occur when a **mutagen** alters the chemical or physical makeup of DNA. Mutagens include ultraviolet **light** and certain chemicals. Genetic mutations in somatic (body) cells result in malfunctioning cells or a mutant organism. These mutations result from a change in the base pairs on the DNA, which can alter cell functions and even give rise to different traits. Somatic cell mutations can result in disfigurement, disease, and other biological problems within an organism. These mutations occur solely within the affected individual.

When mutations occur in the DNA of germ (reproductive) cells, these altered genes can be passed on to the next generation. A germ cell **mutation** can be harmful or result in an improvement, such as a change in body coloring that acts as camouflage. If the trait improves an individual organism's chances for survival within a particular environment, it is more likely to become a permanent trait of the species because the offspring with this gene would have a greater chance to survive and pass on the trait to succeeding generations.

Mutations are generally classified into two groups, spontaneous mutations and induced mutations. Spontaneous mutations occur naturally from errors in coding during DNA replication. Induced mutations come from outside influences called environmental factors. For example, certain forms of **radiation** can damage DNA and cause mutations. A common example of this type of mutating agent is the ultraviolet rays of the **sun**, which can cause skin **cancer** in some people who are exposed to intense sunlight over long periods of time. Other mutations can occur due to exposure to man-made chemicals. These types of mutations modify or change the chemical structure of base pairs.

Population genetics

Population genetics is the branch of genetics that focuses on the occurrence and interactions of genes in specific populations of organisms. One of its primary concerns is **evolution**, or how genes change from one generation to the next. By using mathematical calculations that involve an interbreeding population's gene pool (the total genetic information present in the individuals within the species), population geneticists delve into why similar species vary among different populations that may, for example, be separated by physical boundaries such as bodies of **water** or **mountains**.

As outlined in the previous section, genetic mutations may cause changes in a population if the mutation occurs in the germ cells. Many scientists consider mutation to be the primary cause of genetic change in successive generations. However, population geneticists also study three other factors involved in genetic change or evolution: **migration**, genetic drift, and natural selection.

Migration occurs when individuals within a species move from one population to another, carrying their genetic makeup with them. Genetic drift is a natural mechanism for genetic change in which specific genetic traits coded in alleles (alternate states of functioning for the same gene) may change by chance often in a situation where organisms are isolated, as on an **island**.

Natural selection, a theory first proposed by Charles Darwin in 1858, is a process that occurs over successive generations. The theory states that genetic changes that enhance survival for a species will come to the forefront over successive generations because the gene carriers are better fit to survive and are more likely reproduce, thus establishing a new gene pool, and eventually, perhaps, an entirely new species. One proposed mechanism of natural selection is gradualism, which predicts very slow and steady accumulation of beneficial genes. **Punctuated**

equilibrium, in contrast, depicts natural selection as occurring in brief, but accelerated periods of "survival of the fittest" with lengthy periods of relative stagnation of genetic change in populations. Some scientists hold that both processes occur and have occurred.

Genetics and the golden age of biology

More than any other biological discipline, genetics is responsible for the most dramatic breakthroughs in biology and medicine today. Scientists are rapidly advancing in their ability to engineer genetic material to achieve specific characteristics in plants and animals. The primary way to genetically engineer DNA is called gene cloning, in which a segment of one DNA molecule is removed and then inserted, into another DNA molecule. This process takes advantage of restriction enzymes to cut DNA into fragments of different lengths and ligase to re-create new molecules. Restriction enzymes act as molecular scissors, cutting larger molecules (like DNA) at specific sites. The ends of these fragments are "sticky" in that they have an affinity for complimentary ends of other DNA fragments. DNA ligase acts as a glue to join the ends of the two molecules together. This approach has applications in agriculture and medicine.

In agriculture, genetic **engineering** is used to produce transgenic animals or plants, in which genes are transferred from one organism to another. This approach has been used to reduce the amount of **fat** in cattle raised for meat, or to increase proteins in the milk of dairy cattle that favor cheese making. **Fruits** and **vegetables** have also been genetically engineered so they do not bruise as easily, or so they have a longer shelf life. On the other hand, in medicine, genetic engineering provided great advancements in production of **antibiotics**, **hormones**, vaccines, understanding disease mechanisms and in therapy. Gene therapy is currently being developed and used as it provides the opportunity to introduce specific genes into the body to either correct a genetic defect or to enhance the body's capabilities to fight off disease and repair itself. Because many inherited or genetic diseases are caused by the lack of an **enzyme** or protein, scientists hope to one day treat the unborn by inserting genes to provide the missing enzyme.

Genetic fingerprinting (DNA typing) is based on each individual's unique genetic code. To identify parentage, diagnose inherited diseases in prenatal laboratories or the presence of someone at a crime, scientists use **molecular biology** techniques such as **DNA fingerprinting** by applying restriction fragment length polymorphisms (RFLPs) analysis (identifying the characteristic patterns in DNA cut with the restriction enzymes), microsattelite analysis (looking at the small specific

KEY TERMS

Allele—Any of two or more alternative forms of a gene that occupy the same location on a chromosome.

Amino acid—An organic compound whose molecules contain both an amino group ($-NH_2$) and a carboxyl group (-COOH). One of the building blocks of a protein.

Base—A chemical unit that makes up part of the DNA molecule. There are four bases: adenine (A) and guanine (G), which are purines, and cytosine (C) and thymine (T), which are pyrimidines.

Chromosomes—he structures that carry genetic information in the form of DNA. Chromosomes are located within every cell and are responsible for directing the development and functioning of all the cells in the body.

DNA—Deoxyribonucleic acid; the genetic material in a cell. Chromosomes are made of DNA.

Dominant (dominant gene)—An allele of a gene that results in a visible phenotype if expressed in a heterozygote.

Gene—A discrete unit of inheritance, represented by a portion of DNA located on a chromosome. The gene is a code for the production of a specific kind of protein or RNA molecule, and therefore for a specific inherited characteristic.

Genetic recombination—New configurations produced when two DNA molecules are broken and rejoined together during meiosis.

Heredity—Characteristics passed on from parents to offspring.

Heterozygous—Two different forms of the same allele pair on the chromosome.

Homozygous—Two identical forms of the same allele pair on the chromosome.

Meiosis—The process of cell division in germ or reproductive cells, producing haploid genetic material.

Mitosis—The process of cell division in somatic, or body, cells, producing no change in genetic material.

Proteins—Macromolecules made up of long sequences of amino acids. They make up the dry weight of most cells and are involved in structures, hormones, and enzymes in muscle contraction, immunological response, and many other essential life functions.

Recessive—Refers to the state or genetic trait that only can express itself when two genes, one from both parents, are present and coded for the trait, but will not express itself when paired with a dominant gene. (See Dominant; Allele)

Ribonucleic acid—RNA; the molecule translated from DNA in the nucleus that directs protein synthesis in the cytoplasm; it is also the genetic material of many viruses.

Transcription—The process of synthesizing RNA from DNA.

DNA sequences), DNA hybridization, DNA sequencing or polymerase chain reaction (**PCR**). Development of PCR allows to analyse small amounts of DNA acquired from hair, semen, blood, fingernail fragments, or fetal cells by utilizing DNA polymerase enzyme (the same enzyme used naturally by cells in mitosis) to create identical copies of a DNA molecules from small samples.

One of the most exciting recent developments in genetics is the initiation of the **Human Genome Project** (HGP). This project is designed to provide a complete genetic road map outlining the location and function of the 100,000 or so genes found in human cells encoded in over three billion bases. The first human genome draft sequences were published in February 2001 by the Celera company and the HGP consortium in the journals *Science* and *Nature*, respectively. As a result, genetic researchers will have easy access to specific genes to study

how the human body works and to develop therapies for diseases. Gene maps for other species of animals are also being developed.

Future of genetics

Full sequencing of many bacterial genomes, **yeast**, *Caenorhabditis elegans*, *Drosophila*, mouse, and human genomes has brought about a new era in genetics, and a development of a new area—genomics. Availability of full DNA sequences of multiple organisms allows the comparative analysis (comparative genetics) of genomes allowing gene identification, finding of regulatory sequences and tracing evolution.

Genetic analysis proved very successful in Mendelian diseases. New challenges for genetics are the studies of common complex diseases such as **asthma**, **obesity** or

hypertension. These diseases are caused by interaction of multiple genes and also environment, making their analysis even more difficult. Geneticists analyze DNA sequence to correlate any changes with the disease (association studies). Small fragments of repetitive DNA sequence (microsatellites) or single nucleotide polymorphisms (SNPs) are analyzed. Such studies require analysis of large control (healthy) population in addition to the affected group before any conclusions can be made. Solving of the puzzle of complex traits is going to be possible by combining molecular genetics, biostatistics, further clinical and computational/bioinformatical analysis.

Ethical questions and the future of genetics

Despite the promise of genetics research, many ethical and philosophical questions arise. Many of the concerns about this area of research focus on the increasing ability to manipulate genes. There is a fear that the results will not always be beneficial. For example, some fear that a genetically re-engineered **virus** could turn out to be extremely virulent, or deadly, and may spread if there is no way to stop it.

Another area of concern is the genetic engineering of human traits and qualities. The goal is to produce people with specific traits such as better health, improved looks, or even high intelligence. While these traits may seem to be desirable on the surface, the concern arises about who will decide exactly what traits are to be engineered into human offspring, and whether everyone will have equal access to an expensive technology. Some fear that the result could be domination by a particular socioeconomic group.

Despite these fears and concerns, genetic research continues. In an effort to ensure that the science is not abused in ways harmful to society, governments in the United States and abroad have created panels and organizations to oversee genetic research. For the most part, international committees composed of scientists and ethical experts state that the benefits of genetic research for medicine and agriculture far outweigh the possible abuses.

See also Chromosomal abnormalities; Gene splicing.

Resources

Books

Beurton, Peter, Raphael Falk, Hans-Jörg Rheinberger., eds. *The Concept of the Gene in Development and Evolution.* Cambridge, UK: Cambridge University Press, 2000.

Edlin, Gordon. *Human Genetics.* Boston: Jones and Bartlett, 1990.

Jacob, François. *Logic of Life: A History of Heredity.* New York: Random House, 1982.

Thro, Ellen. *Genetic Engineering: Shaping the Material of Life.* New York: Facts On File, 1993.

Periodicals

Brookes, Anthony. "Rethinking Genetic Strategies to Study Complex Diseases." *Trends in Molecular Medicine* (November 2001): 512–6.

Brownlee, Shannon, and Joanne Silberne, "The Age of Genes." *U.S. News & World Report.* (4 November 1991): 64–72.

Guo, Sun-Wei, and Kenneth Lange, "Genetic Mapping of Complex Traits: Promises, Problems, and Prospects." *Theoretical Population Biology* (February 2000): 1–11.

Philips, Tamara J., and John K. Belknap, "Complex-trait Genetics: Emergence of Multivariate Strategies." *Nature Reviews. Neuroscience* (June 2002):478–485

Tijan, Robert. "Molecular Machines That Control Genes." *Scientific American.* (February 1995): 54–61.

Other

National Institutes of Health. "Guide to the Human Genome" [cited October 19, 2002]. <http://www.ncbi.nlm.nih.gov/genome/guide/human/>.

David Petechuk

Genets

Genets are mongoose-like **mammals** in the family Viverridae in the order Carnivora. Other members of this family include **civets**, linsangs, **mongooses** and the **fossa**. The genet genus *Genetta* has three subgenera and nine **species**. Genets are found in **Africa** south of the Sahara **desert**, in the southwestern Arabian **Peninsula**, and in southern **Europe**. Genets have a long body, short legs, a pointed snout, prominent rounded ears, short curved retractile claws, and soft dense hair. They emit a musky scent from the anal **glands**, and females have two pairs of abdominal mammae. The **color** of the fur is variable, generally grayish or yellowish, with brown or black spots on the sides, generally grayish or yellowish, with brown or black spots on the sides, sometimes arranged in rows. The tail may be black with white rings, and completely black genets are fairly common. Genets weigh 2.2-6.6 lb (1-3 kg). They have a head and body length of 16.5-22.9 in (42-58 cm), and the tail is 15.4-20.9 in (39-53 cm) long.

Genets live in savannas and in **forests**. They feed mostly on the ground hunting **rodents**, **birds**, **reptiles** and **insects**, and they climb trees to **prey** on nesting and roosting birds. Genets also eat game birds and poultry. When stalking their prey, genets crouch and seem to glide along the ground, and their bodies seem to lengthen. They can get through any opening their head can enter. Genets travel alone or in pairs. **Radio** tracking of

A large-spotted genet with a ringed tail. *Photograph by Nigel J. Dennis/The National Audubon Society Collection/Photo Researchers, Inc. Reproduced by permission.*

genets in Spain indicated a home range of 0.5 sq mi (1.4 sq km) for a three-month-old female and over 19 sq mi (50 sq km) in about five months for an adult male.

Genets communicate by vocal, olfactory, and visual signals. In Kenya, East Africa, pregnant and lactating female genets were found in May and from September to December. A captive genet (*Genetta genetta*) regularly produced two litters each year: in April/May and in July/August. Gestation lasts 56-77 days. Litter sizes range from one to four, but usually two or three young are born, which weigh 2.1-2.9 oz (61-82 g) at **birth**. The young begin to eat solid food at two months and in two years reach their adult weight. Captive genets became sexually mature at four years and produced offspring until death over 20 years later.

Genetta genetta is found in southern Europe and northwestern Africa, *G. felina* is found south of the Sa-hara and in the southwestern Arabian Peninsula, *G. tigrina* is found in South Africa and Lesotho, and *G. rubiginosa* is found elsewhere in Africa. Genets in Europe are declining in numbers because of persecution for depredation on game birds and poultry and because their winter pelts are highly valued. The subspecies *G. genetta isabelae* of Ibiza Island in the Balearics is endangered.

The aquatic genet (*Osbornictis piscivora*) occurs in northeastern Zaire. The head and body length of an adult male is 17.5 in (44.5 cm) and the tail length is 13.4 in (34 cm). Males weigh 3.1 lb (1.43 kg) and females 3.3 lb (1.5 kg). *Osbornictis piscivora* is red to dull red with a black tail, and it has elongated white spots between the eyes. The front and sides of the muzzle and sides of the head below the eyes are whitish. There are no black spots or bands on the body and the tail is not ringed. The fur is long and dense, especially on the tail, but the palms and soles are not furred as in *Genetta* and related forms. The skull is long and lightly built. Teeth are relatively small and weak, but seem adapted to catching slippery prey like **fish** and **frogs**. It is believed that the bare palms help these animals feel for fish in muddy holes. Fish are the chief diet of these rare animals which do not live in groups or families.

See also Civets; Fossa; Mongooses.

Sophie Jakowska

Genome

The genome (sometimes spelled geneome) is, in the broadest use of the term, the full set of genes or genetic material carried by a particular **organism** representing a particular **species** or population. The size of a genome is usually measured in numbers of genes or base pairs.

With the success of the **Human Genome Project** and other international genome projects and programs, by 2003, scientists have, to a great extent, constructed genetic maps delineating **individual** base sequences that constitute the basis of human genome.

A genomic sequence is the actual order of the nitrogenous bases in the DNA nucleotide sequence that, with subtle alterations that create differing **gene** forms (**alleles**), comprise an organism's genetic material.

In humans, the genome comprises one representative of each of the **chromosome** pairs of the adult diploid parent. In this sense, a genome is a single set of genetic instructions.

Not all of alleles within a genome are expressed, some are masked by the presence of dominant forms. The genomic formula is a mathematical expression of the number of subsets of genomes present in an individual **cell** or organism. One of the commonly encountered genomic formulae designations is the haploid number (n) that represents a set with a single copy of each gene. This is sometimes called the basic number. The diploid form contains two sets of genes and is designated 2n; the triploid is 3n, and the tetraploid is 4n. Genetic abnormalities, where one chromosome is missing from the genome, can be represented in the same manner. For example, a diploid organism with one chromosome missing is a monosomic cell and is represented by 2n-1. A diploid with two chromosomes missing is termed a nullisomic and is represented by 2n-2. Additions of chromosomes can also occur and are represented in the same form; for example, 2n+1 is trisomic.

In a report published in the February 2001 issue of the international scientific journal *nature* (usually spelled in the lowercase) researchers reported findings that indicated that the human genome consisted of far fewer genes than previous projected by estimation of phylogenic relationships between humans and other species. Since then additional estimates fix the size of the human genome at about 30,000 genes. By contrast, some worms carry about 22,000 genes in their genome.

As of 2003, while work still continues on the Human Genome Project scientists are also beginning a "Genomes to Life" research program designed to identify and characterize the protein complexes important in **animal**, especially human, and microbial cell reactions and to further identify the specific genes that regulate these processes.

Genomic libraries have been used in human **genetics** as part of the Human Genome Project.

A genomic library is a comprehensive collection of cloned DNA fragments derived from a genome. Each part of the genome is represented in the library several times, and the number of times it is represented on average is called the coverage of a library. The library can be screened for the presence of the sequence of interest by radioactively labelling the DNA (usually between 100 and 500 nucleotides long) and using this as a probe to identify the clone that contains the selected sequence. The clone selected can then be grown in **bacteria** to produce large amounts of clone DNA, which can be studied. If for example, the sequence of interest was part of a gene, by using this sequence as a probe, the clone containing, hopefully, the whole gene could be isolated.

See also Chromosome mapping; Evolution; Evolutionary mechanisms; Genetic engineering; Genetic identification of microorganisms; Genetic testing; Molecular biology.

Genomic fingerprinting *see* **DNA fingerprinting**

Genomic imprinting *see* **Imprinting**

Genomics (comparative)

The study of an organism's total complement of genetic material, called its **genome**, has become indispensable for shedding **light** on its **biochemistry**, **physiology**, and patterns of inheritance. Even more can be gained by comparing the genomes of multiple organisms to discern how their DNA sequences have changed over evolutionary **time**.

This technique has become increasingly valuable with the explosion of genome sequencing activity in recent years. Today, hundreds of complete or near-complete genome sequences, ranging from simple microbes to human, have been deposited in scientific databases around the world.

All life on **Earth** has a common history, reflected by its common biochemical basis in DNA. Different organisms vary in their DNA sequences, of course, but perhaps not so much as one might think. Some of the genes controlling very basic biological tasks, such as the mechanism by which DNA is transcribed into RNA to code the **proteins** that determine function, originate with the Archaea, **microorganisms** believed to be the most direct descendants of the first living things. The genome of the humble mouse is 85 % identical to our own. Our closest relatives, the **chimpanzees**, differ from us genetically by only about 1%, a testimony to the incredible power of a relatively small amount of DNA.

The degree of disparity in the genomes of different organisms reflects their phylogenetic relationship; that is, their relative distances from one another and position on the branches of life's "family tree". Evolutionary biologists use this information to determine whether organisms are descended from a common ancestor, and at what point the different lineages divided. If the same **gene** is present in two organisms, they are presumed to have a common ancestor. The more the DNA sequences have changed since that point, the longer the two **species** have been evolving independently.

An example of the use of this technique was the comparison in the late 1990s of DNA from Neanderthal remains, modern humans and chimpanzees. The analysis yielded the conclusion that modern humans almost certainly did not descend directly from Neanderthals, as had once been thought, but rather shared a common ancestor with this earlier hominid.

Although the evidence is preliminary and far from conclusive, published reports of genome analysis in late 2002 provided evidence that early migrant populations of humanoids may have been able to intermix with established or indigenous humanoid populations to a greater degree than previously believed.

Genome analysis helps to distinguish physical similarities derived from common ancestry from those that have evolved separately in response to a similar environment, a phenomenon called convergent evolution. An example of convergent **evolution** can be seen in the **fauna** of **Australia**, where **marsupials** diverged to fill ecological niches dominated by placental **mammals** on other continents. As a result, Australia has marsupial **mice**, marsupial wolves, and kangaroos, which are the marsupial equivalent of **deer** and antelopes.

Comparative genomics has a vital role to play in research contributing to human health. The mouse is a useful model **organism** for biomedical research because of the similarities of its genome to that of humans. At the same time, unlike humans, the well-studied mouse has been bred over time into genetically identical strains, and its environment may be strictly controlled. These factors combine to reduce the potential sources of uncertainty about what might be causing a given result.

Almost every human gene has an exact counterpart in the mouse, despite the fact that the chromosomes are arranged differently. The differences in the species arise primarily not from the identity of the genes, but from the exact sequences that make them up, resulting in a change in the proteins that are built when the DNA is transcribed. Sequence changes reflect mutations that may have had an effect on the organism's ability to survive and reproduce, the driving **force** of evolution by natural **selection**. When scientists find a **mutation** in a mouse gene that is associated with the trait they are studying, they look for a similar DNA sequence in humans to find the corresponding human gene.

Comparing the billions of nucleotides that make up organisms' DNA sequences to tease out sequences with similar functions requires powerful database search engines and sophisticated software. The task is complex, and fraught with the possibility of **error**. First, since the sequence of a given gene is not expected to be identical between species, scientists must determine how close a match is close enough. In many cases throughout evolutionary history, genes have become duplicated, and then their functions diverge. Researchers look for relationships between genes in such a lineage just as they seek to place related organisms in a phylogenetic **tree**. A large proportion of genetic material, called "selfish DNA", has no apparent function in the organism at all, but rather exists merely to propagate itself from one generation to the next. The rigorous requirements for sequence analysis have given rise to a specialized discipline called bioinformatics, combining high-throughput computing with an extensive knowledge of **biology**.

See also Chromosome mapping; Evolutionary mechanisms; Molecular biology.

Resources

Books

Beurton, Peter, Raphael Falk, Hans-Jörg Rheinberger., eds. *The Concept of the Gene in Development and Evolution*. Cambridge, UK: Cambridge University Press, 2000.

Lewin, B. *Genes*. 7th. ed. New York, Oxford University Press Inc., 2000

Periodicals

Fraser, C.M., J. Eisen, R.D. Fleischmann, K.A. Ketchum, and S. Peterson. "Comparative Genomics and Understanding of Microbial Biology." *Emerging Infectious Diseases* 6, no. 5 (September-October 2000).

Veuille, E. "Genetics and the evolutionary process." *C. R. Acad. Sci III* 323, no.12 (December 2000):1155–65.

Sherri Chasin Calvo

Genotype and phenotype

A genotype describes the actual set (complement) of genes carried by an **organism**. In contrast, phenotype refers to the observable expression of characters and traits coded for by those genes.

Although phenotypes are based upon the content of the underlying genes comprising the genotype, the expression of those genes in observable traits (phenotypic expression) is also, to varying degrees, influenced by environmental factors.

The exact relationship between genotype and **disease** is an area of intense interest to geneticists and physicians and many scientific and clinical studies focus on the relationship between the effects of a genetic changes (e.g., changes caused by mutations) and disease processes. These attempts at genotype/phenotype correlations often require extensive and refined use of statistical analysis.

The term genotype was first used by Danish geneticist Wilhelm Johannsen (1857 – 1927) to describe the entire genetic or hereditary constitution of an organism, In contrast, Johannsen described displayed characters or traits (e.g., anatomical traits, biochemical traits, physiological traits, etc) as an organism's phenotype.

Genotype and phenotype represent very real differences between genetic composition and expressed form. The genotype is a group of genetic markers that describes the particular forms or variations of genes (**alleles**) carried by an **individual**. Accordingly, an individual's genotype includes all the alleles carried by that individual. An individual's genotype, because it includes all of the various alleles carried, determines the range of traits possible (e.g., a individual's potential to be afflicted with a particular disease). In contrast to the possibilities contained within the genotype, the phenotype reflects the manifest expression of those possibilities (potentialities). Phenotypic traits include obvious observable traits as height, weight, **eye** color, hair **color**, etc. The presence or absence of a disease, or symptoms related to a particular disease state, is also a phenotypic trait.

A clear example of the relationship between genotype and phenotype exists in cases where there are dominant and recessive alleles for a particular trait. Using an simplified monogenetic (one **gene**, one trait) example, a capital "T" might be used to represent a dominant allele at a particular **locus** coding for tallness in a particular **plant**, and the lowercase "t" used to represent the recessive allele coding for shorter plants. Using this notation, a diploid plant will possess one of three genotypes: TT, Tt, or tt (the variation tT is identical to Tt). Although there are three different genotypes, because of the laws governing dominance, the plants will be either tall or short (two phenotypes). Those plants with a TT or Tt genotype are observed to be tall (phenotypically tall). Only those plants that carry the tt genotype will be observed to be short (phenotypically short).

In humans, there is genotypic sex determination. The genotypic variation in sex chromosomes, XX or XY decisively determines whether an individual is female (XX) or male (XY) and this genotypic differentiation results in considerable phenotypic differentiation.

Although the relationships between genetic and environmental influences vary (i.e., the degree to which genes specify phenotype differs from trait to trait), in general, the more complex the biological process or trait, the greater the influence of environmental factors. The genotype almost completely directs certain biological processes. Genotype, for example, strongly determines when a particular tooth develops. How long an individual retains a particular tooth, is to a much greater extent, determined by environmental factors such diet, dental hygiene, etc.

Because it is easier to determine observable phenotypic traits that it is to make an accurate determination of the relevant genotype associated with those traits, scientists and physicians place increasing emphasis on relating (correlating) phenotype with certain genetic markers or genotypes.

There are, of course, variable ranges in the nature of the genotype-environment association. In many cases, genotype-environment interactions do not result in easily predictable phenotypes. In rare cases, the situation can be complicated by a process termed phenocopy where environmental factors produce a particular phenotype that resembles a set of traits coded for by a known genotype not actually carried by the individual. Genotypic frequencies reflect the percentage of various genotypes found within a given group (population) and phenotypic frequencies reflect the percentage of observed expression. Mathematical measures of phenotypic variance reflect the variability of expression of a trait within a population.

See also Chromosome mapping; Evolution; Evolutionary mechanisms; Genetic engineering; Genetic disorders; Genetic identification of microorganisms; Genetic testing; Genetically modified foods and organisms; Molecular biology; Rare genotype advantage.

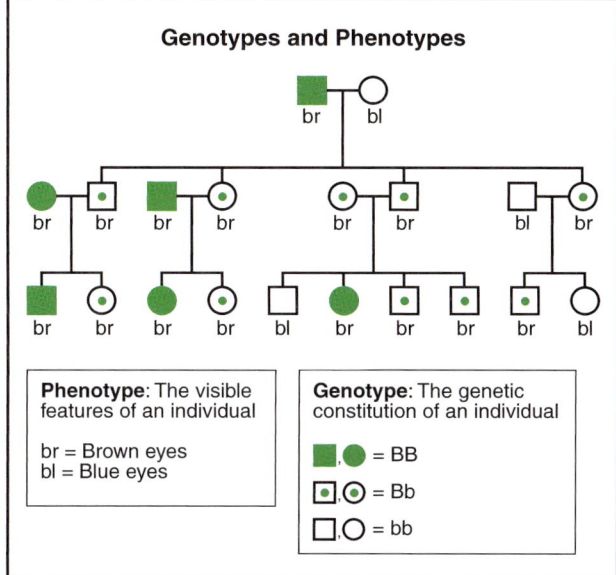

Genotypes and Phenotypes

Phenotype: The visible features of an individual

br = Brown eyes
bl = Blue eyes

Genotype: The genetic constitution of an individual

■,● = BB
▣,◉ = Bb
□,○ = bb

Pedigree analysis chart showing inheritance pattern for genotypes and phenotypes. *Illustration by Argosy. The Gale Group.*

Resources

Books

Beurton, Peter, Raphael Falk, Hans-Jörg Rheinberger., eds. *The Concept of the Gene in Development and Evolution.* Cambridge, UK: Cambridge University Press, 2000.

Gilbert, Scott F. *Developmental Biology,* 6th ed. Sunderland, MA: Sinauer Associates, Inc., 2000.

Jorde, L.B., J. C. Carey, M.J. Bamshad, and R.L. White. *Medical Genetics,* 2nd ed. St. Louis: Mosby-Year Book, Inc., 2000.

Lodish, H., et. al. *Molecular Cell Biology,* 4th ed. New York: W. H. Freeman & Co., 2000.

Periodicals

Collins F.S, and V.A. McKusick. "Implications of the Human Genome Project for Medical Science." *JAMA* 285 (7 February 2001): 540–544.

Fields, S. "Proteomics in Genomeland." *Science* 291 (16 February 2001): 1221–1224.

Venter, J.C., et al. "The Sequence of the Human Genome." *Science* 291 (2001): 1304–1351.

Veuille, E. "Genetics and the evolutionary process." *C. R. Acad. Sci III* 323, no.12 (December 2000):1155–65.

K. Lee Lerner

Geocentric theory

Rejected by modern science, the geocentric theory (in Greek, *ge* means *earth*), which maintained that **Earth** was the center of the universe, dominated ancient and medieval science. It seemed evident to early astronomers that the rest of the universe moved about a stable, motionless Earth. The **Sun**, **Moon**, planets, and stars could be seen moving about Earth along circular paths day after day. It appeared reasonable to assume that Earth was stationary, for nothing seemed to make it move. Furthermore, the fact that objects fall toward Earth provided what was perceived as support for the geocentric theory. Finally, geocentrism was in accordance with the theocentric (God-centered) world view, dominant in in the Middle Ages, when science was a subfield of theology.

The geocentric model created by Greek astronomers assumed that the celestial bodies moving about the Earth followed perfectly circular paths. This was not a **random** assumption: the **circle** was regarded by Greek mathematicians and philosophers as the perfect geometric figure and consequently the only one appropriate for celestial **motion**. However, as astronomers observed, the patterns of celestial motion were not constant. The Moon rose about an hour later from one day to the next, and its path across the sky changed from month to month. The Sun's path, too, changed with **time**, and even the configuration of constellations changed from season to season.

These changes could be explained by the varying rates at which the celestial bodies revolved around the Earth. However, the planets (which got their name from the Greek word *planetes*, meaning *wanderer* and *subject of error*), behaved in ways that were difficult to explain. Sometimes, these wanderers showed retrograde motion—they seemed to stop and move in a reverse direction when viewed against the background of the distant constellations, or fixed stars, which did not move relative to one another.

To explain the motion of the planets, Greek astronomers, whose efforts culminated in the work of Claudius Ptolemy (c. 90-168 A.D.), devised complicated models in which planets moved along circles (epicycles) that were superimposed on circular orbits about the Earth. These geocentric models were able to explain, for example, why Mercury and **Venus** never move more than 28° and 47° respectively from the Sun.

As astronomers improved their methods of observation and measurement, the models became increasingly complicated, with constant additions of epicycles. While these complex models succeeded in explaining **retrograde motion**, they reportedly prompted Alfonso X (1221-1284), king of Castile, to remark that had God asked his advice while engaging in Creation, he would have recommended a simpler design for the universe. Nonetheless, the geocentric theory persisted because it worked.

The scientific refutation of geocentrism is associated with the work of the Polish astronomer Nicolaus

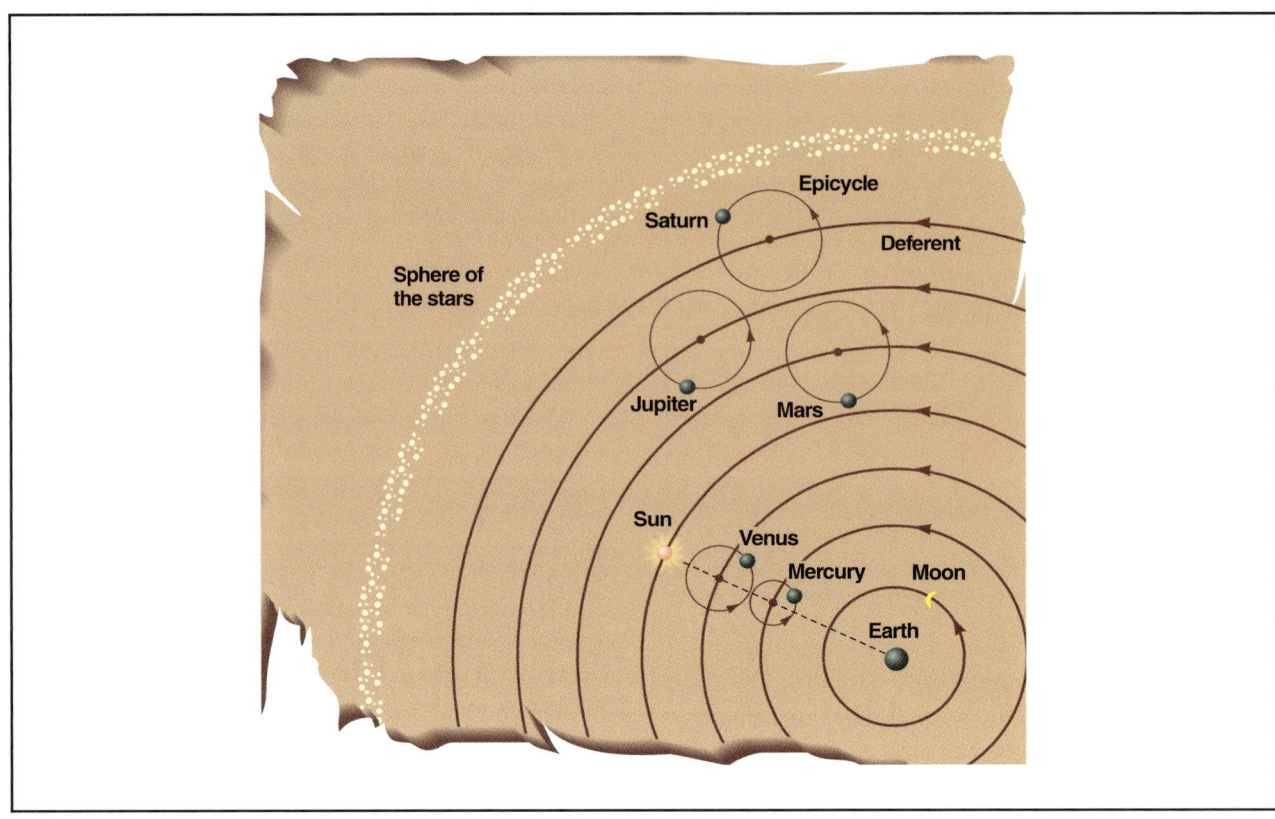

The geocentric universe. *Illustration by Hans & Cassidy. Courtesy of Gale Group.*

Copernicus (1473-1543). In *Commentariolus*, a short work composed around 1514, Copernicus suggested a replacement for the replacement for the geocentric system. According to Copernicus, who fully developed his ideas in *De revolutionibus orbium coelestium* (1543), known as *On the Revolution of the Celestial Spheres*, a **heliocentric theory** could explain the motion of celestial bodies more simply than the geocentric view. In the Copernican model, the Earth orbits the Sun along with all the other planets. Such a model can explain the retrograde motion of a **planet** without resorting to epicycles, and can also explain why Mercury and Venus never stray more than 28° and 47° from the Sun.

Copernicus's work did not spell the demise of geocentrism, however. The Danish astronomer Tycho Brahe (1546-1601), a brilliant experimental scientist whose measurements of the positions of the stars and planets surpassed any that were made prior to the invention of the **telescope**, proposed a model that attempted to serve as a compromise between the geocentric explanation and the Copernican theory. His careful observation of a comet led him to the conclusion that the comet's **orbit** could not be circular; but despite this insight, he was unable to abandon the geocentric system. Instead, he proposed a model which preserved the ancient geometric

structure, but suggested that all the planets except the Earth revolved around the Sun. The Sun, however, in accordance with the geocentric view, carrying all the planets with it, still moved about the Earth.

After Galileo (1564-1642) built a telescope and turned it toward the heavens, evidence supporting a heliocentric model started to accumulate. Through his refracting (using lenses to form images), Galileo saw that Venus and Mercury go through phases similar to those of the Moon. The geocentric model could not fully explain these changes in the appearance of the inferior planets (the planets between the Earth and the Sun). Furthermore, Galileo's observations of Jupiter's moons made it clear that celestial bodies do move about centers other than the Earth.

Around the time when Galileo began surveying the skies with his telescope, Johannes Kepler (1571-1630), a remarkable mathematician and theoretical astronomer, used Brahe's precise measurements to determine the exact paths of the planets. Kepler was able to show that the planets did not move along circular paths, but rather that each planet followed an elliptical course, with the Sun at one focus of the **ellipse**. The fact that the orbits of the planets about the Sun are ellipses became known as Kepler's first law. His second law states that for each planet,

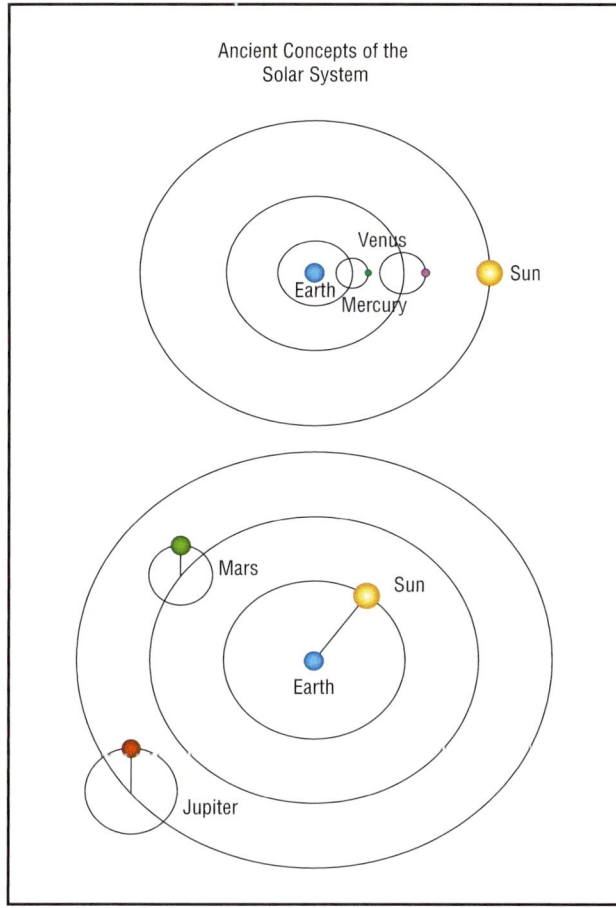

Ancient Concepts of the
Solar System

Venus
Earth
Mercury
Sun

Mars
Sun
Earth
Jupiter

In ancient geocentric theory, Earth was the center of the universe, and the body around which the Sun and planets revolved. *Illustration by Argosy. The Gale Group.*

an imaginary line connecting the planet to the Sun sweeps out equal areas in equal times; and the third law, which was later used by Isaac Newton (1642-1727) in establishing the universal law of gravitation, reveals that the **ratio** of the cube of a planet's semimajor axis to the square of its period (the time to make one revolution) is a constant; that is, the ratio is the same for all the planets. By the time Newton established the laws of motion—laws that he demonstrated to be valid for both celestial and earthly objects—there was no doubt that the workings of the solar systems clearly invalidated the geocentric model.

See also Celestial mechanics.

Resources

Books

Bacon, Dennis Henry, and Percy Seymour. *A Mechanical History of the Universe.* London: Philip Wilson Publishing, Ltd., 2003.

Hancock P.L. and Skinner B.J., eds. *The Oxford Companion to the Earth.* Oxford: Oxford University Press, 2000.

Kirkpatrick, Larry D., and Gerald F. Wheeler. *Physics: Building a World View.* Englewood Cliffs, NJ: Prentice-Hall, 1983.

Kline, Morris. *Mathematics in Western Culture.* London: Oxford University Press, 1972.

Winchester, Simon, and Soun Vannithone. *The Map That Changed the World: William Smith and the Birth of Modern Geology.* New York: Perennial, 2002.

Geochemical analysis

Geochemical analysis is the process through which scientists discover and unravel the chemical compounds that make up the **earth**, its atmosphere, and its seas. The process requires a thorough grounding in **chemistry** and the earth sciences, and an understanding of the different ways elements can interact in a given geologic situation. Geochemical analysis can predict where **petroleum**, metals, **water**, and commercially valuable **minerals** can be located—a branch of the science known as geochemical prospecting. It can also be used to predict or trace toxic leakages from waste disposal sites, and to track and understand fluctuations in the earth's climate throughout its history, a branch known as paleogeochemistry. Still another form of the science is cosmochemistry, which attempts to chart the composition of celestial bodies through the analysis of reflected **light** and other forms of **radiation**.

Branches of geochemical analysis

Geochemical analysis became important in the nineteenth and twentieth centuries, when chemists first began investigating the compounds that formed naturally in the earth, air, and water. Much of this early work was credited to a chemist named V. M. Goldschmidt, who with his students created detailed charts of the chemical breakdown of common compounds, mainly **igneous rocks**. He also created a series of guidelines known collectively as "Goldschmidt's Rules" for understanding the different ways in which elements interact to form different types of rock. Scientists have expanded on Goldschmidt's program, forming a series of disciplines that help them predict and interpret the chemical composition through **time** of this **planet**, other objects in the **solar system** (including planets), and their constituent ingredients. Goldschmidt based his analysis of chemical **behavior** on two separate items: size and electrical charge. Later scientists have added radiation to the process of geochemical analysis, grouping elements by their radioactive and stable isotopes. Isotopic analysis can give clues to the place of origin of the compound, and the environment in which it was first put together. Isotopes are also used to determine the age of a compound, and the study of the process through which they decay from one form to another is known as geochronology. Astronomers have discovered certain isotopes in compounds located in celestial bodies—like supernovae—which have relatively short half-lives, and they use these substances to help date the formation of the universe.

One of the most commercially popular subfields of modern **geochemistry** is geochemical prospecting, usually in order to locate metals like **uranium** or hydrocarbons like petroleum. The methodology for geochemical prospecting was pioneered in **Europe** and the Soviet Union during the 1930s. It was taken up by prospectors in the United States after World War II. Prospectors find that the most profitable way to search for valuable rock and mineral samples is to look in areas that have undergone extensive **weathering**, especially the beds of streams. Using their knowledge of weathering and dispersion patterns, these scientists examine samples drawn from areas where streams intersect each other and from places where **fault** lines have caused slippages of the local geography to detect the presence of valuable substances. They also can detect minerals that have undergone chemical **decomposition** by analyzing the surrounding water and **sand** and silt deposits for trace remnants, which form a characteristic spread known as a secondary dispersion halo. By examining the characteristics of these elements and comparing the results to a series of known features in areas like valency, ionic size, and type of **chemical bond**, geochemists can discover if commercial valuable minerals are present in the area. Other elements, especially volatile ones like **chlorine**, fluorine, **sulfur**, **carbon dioxide**, and water, also serve as indicators of elements that may be found in the area. Prospectors searching specifically for petroleum look for a **polymer** called kerogen, thought to be a substance falling between the original organic material that makes up petroleum or **natural gas**, and the final product, in **soil** and rock samples. The presence of **radon**, which can be detected relatively easily because of its characteristic alpha radioactivity, in the water of streams is often an indicator of uranium deposits.

Geochemists have also developed a variety of innovative and cost-saving ways of performing geochemical analysis without requiring to be in direct contact with the **rocks** and minerals they are examining. One relatively common way is to examine the surface **flora** and **fauna** for traces of chemicals or metals. Certain plants growing in contaminated areas develop characteristic diseases, such as chlorosis or nongenetic dwarfism. **Contamination** can also be detected from chemical residue collected in the internal organs of **fish**, molluscs, and **insects**. Some geochemists have even used dogs to recognize and locate minerals that are found in combination with sulfur compounds by teaching them to sniff out the gasses re-

leased in the oxidation process. Prospectors also use aerial surveys, computer mapping and modeling, and atomic absorption spectrometry to gather clues as to where the minerals they are seeking can be found. Apparatuses that can record gamma radiation are mounted in airplanes and used to locate radioactive minerals.

Geochemical analysis in other environments

By examining the chemical composition of sea water and polar **ice**, geochemists can draw conclusions and make predictions about the environment. Although natural weathering processes can take various **trace elements** into the sea or **lock** them into ice caps, scientists also find that by analyzing these compounds they can determine the impact which humans are having on the earth and possible climatological shifts, either induced by the activity of people—global warming—or the result of natural processes, such as **ice ages**. This field of study, known as low-temperature geochemistry, is a valuable diagnostic tool for understanding the impact **pollution** has on the environment.

Geochemists also make valuable contributions to understanding the history of the earth in general and human beings in particular. They perform isotopic analyses on cores drawn from rock **strata** or chemical breakdowns on ice cores to determine how the world's climate has shifted in the past. Specific events—the ash fallout of a large volcanic eruption like Mount St. Helens or Krakatoa, for instance, or records of the hydrocarbons released by factories in Europe and American during the Industrial Revolution—leave chemical traces in the sediments of sea and **lake** beds, and in the unmelting ice of the polar regions.

Resources

Books

Ingamells, C.O., and Francis F. Pitard. *Applied Geochemical Analysis.* New York, NY: Wiley, 1986.

Jungreis, Ervin. *Spot Test Analysis: Clinical, Environmental, Forensic, and Geochemical Applications.* New York, NY: Wiley, 1997.

Methods for Geochemical Analysis. Denver, CO: Dept. of the Interior, U.S. Geological Survey, 1987.

Stanton, Ronald Ernest. *Rapid Methods of Trace Analysis for Geochemical Applications.* London: Edward Arnold, 1966.

Kenneth R. Shepherd

Geochemistry

Geochemistry is the science or study of the **chemistry** of the **earth**. Geochemists who practice this science are interested in the origin of chemical elements, their **evolution**, the classes and many divisions of **minerals** and **rocks** and how they are created and changed by earth processes, and the circulation of chemical elements through all parts of the earth including the atmosphere and biological forms.

The circulation of elements in nature has many practical applications. Understanding the distribution of chemical isotopes and their stability (or instability) helps in fields as varied as age-dating in **archaeology** and medical uses of radioactive isotopes. Some significant chemical elements like **carbon**, **phosphorus**, **nitrogen**, and **sulfur** have geochemical cycles that are indicators of environmental **contamination** or the need to rotate **crops** in fields.

Geochemistry has many subdivisions. Inorganic geochemistry explains the relationships and cycles of the elements and their distribution throughout the structure of the earth and their means of moving by **thermodynamics** and kinetics. Exploration geochemistry (also called geochemical prospecting) uses geochemical principles to locate **ore** bodies, mineral fields, **groundwater** supplies, and oil and gas fields. Organic geochemistry uses the chemical indicators associated with life forms to trace human habitation as well as **plant** and **animal** activity on Earth. It has been important in understanding the **paleoclimate**, paleooceanography, and primordial life and life's evolution. Sedimentary geochemistry interprets what is known from hard rock geochemistry in **soil** and other sediments and their **erosion**, deposition patterns, and **metamorphosis** into rock. Environmental geochemistry is the newest branch of the science and came into prominence in the 1980s when environmental concerns made the tracking of chemicals in organic tissues, groundwater, surface **water**, the marine environment, soil, and rock important to scientists, engineers, and government agencies responsible for the public's well-being.

History of the science of geochemistry

Geochemistry did not come into its own as a science until the 1800s. The discoveries of **hydrogen** and **oxygen** (two of the most important elements needed for life on earth) in 1766 by Henry Cavendish and in 1770 by Joseph Priestley, respectively, opened the door for understanding chemical elements and the concept of the atom. Russia stakes claim to the founder of geochemistry, V. I. Vernadsky. Vernadsky was an expert in **mineralogy** and was the first to relate chemical elements to the formation of minerals in nature. Also in the 1800s, pioneering geochemists discovered how to produce **salt** from seawater and identified other elements and compounds like **ozone**, which helped in the understanding of the creation of life

on earth and the chemical requirements for maintaining it. The first American geochemist was F. W. Clarke, who became the chief chemist of the United States Geological Survey shortly after it was founded in 1884.

Characteristics and processes

Just as the **biochemistry** of life is centered on the properties and reaction of carbon, the geochemistry of the Earth's crust is centered upon silicon. (Si). Also important to geochemistry is oxygen, the most abundant element on earth. Together, oxygen and silicon account for 74% of the Earth's crust.

Unlike carbon and biochemical processes where the covalent bond is most common, ionic bonds are typically found in **geology**. Accordingly, silicon generally becomes a **cation** and will donate four electrons to achieve a noble gas configuration. In quartz, each silicon atom is coordinated to four oxygen **atoms**. Quartz crystals are silicon atoms surrounded by a **tetrahedron** of oxygen atoms linked at shared corners.

Rocks are aggregates of minerals and minerals are composed of elements. All minerals have a definite structure and composition. Diamonds and graphite are minerals that are polymorphs (many forms) of carbon. Although they are both composed only of carbon, diamonds and graphite have very different structures and properties. The types of bonds in minerals can affect the properties and characteristics of minerals.

Pressure and **temperature** also affect the structure of minerals. Temperature can determine which ions can form or remain stable enough to enter into **chemical reactions**. Olivine ($Fe, Mg)_2 SiO_4$), for example, is the only solid that will form at 1,800°C. According to Olivine's formula, it must be composed of two atoms of either **iron** or **magnesium**. The atoms are interchangeable because they carry the same electrical charge and are of similar size, thus, Olivine exists as a range of compositions termed a solid solution series. Depending upon the ionic substitution of iron or magnesium, Olivine is said to be either rich in iron or rich in magnesium.

The determination of the chemical composition of rocks involves the crushing and breakdown of rocks until they are in small enough pieces that **decomposition** by hot acids (hydrofluoric, nitric, hydrochloric, and perchloric acids) allows the elements present to enter into a solution for analysis. Other techniques involve the high temperature fusion of powdered inorganic reagent (flux) and the rock. After melting the **sample**, techniques such as x-ray **fluorescence** spectrometry may be used to determine which elements are present.

Chemical and mechanical **weathering** break down rock through natural processes. Chemical weathering of rock requires water and air. The basic chemical reactions in the weathering process include solution (disrupted ionic bonds), hydration, **hydrolysis**, and oxidation.

Geochemistry for the future

Of all the important work done by geochemists, improving our understanding of life on Earth may be the highest priority. In their mastery of the cycles of elements, geochemists can analyze other cycles such as the delicate balance between the atmosphere and the **hydrosphere** that produces the **biosphere**, or the portions of the earth either on land or in the water and air where life flourishes. New advances and understanding are broadening public awareness of the need to preserve the atmosphere and the oceans and our need to see the earth as a complicated mechanism.

Geode

Geodes are hollow rock masses that are lined with crystals that have grown toward the center of the cavity. Geodes are usually roughly spherical in shape, up to 12

A blue geode. © *Royalty-Free/Corbis. Reproduced by permission.*

in (30 cm) or more in diameter. Most frequently, the crystals growing within a geode are quartz, calcite, or fluorite, though occurrences of other **minerals** are found. These objects are prized by collectors for their well-formed crystals and outstanding beauty.

Geodes are typically characterized by an outer shell of chalcedony, a dense microcrystalline form of quartz. The hard outer shell of the geode can usually be separated from the enclosing rock material. Geodes are most often formed in limestone or volcanic **rocks**, though they are rarely found in mudstones. The **mineralogy** of the geode and the rock in which it formed are commonly different. Because chalcedony is often harder and more **weather** resistant than the host rock, the hollow spheres resist the effects of **erosion** and are left behind as the host rock is eroded. Collectors identify the geodes in the field based upon their shape, the characteristic chalcedony shell, and the lower **density** arising from the hollow center.

The initial requirement for the formation of a geode is the presence of a cavity in the host rock. In a volcanic rock, such voids are frequently a result of the release of gases from the molten lava. As the lava hardens, the gas bubbles are preserved as holes in the rock. In the cases of sedimentary rocks, such as limestone, a hole may develop as **groundwater** dissolves the rock itself or as a result of the decay of biologic material buried at the time of deposition of the sediments. Groundwater within the sediments then carries dissolved minerals, including silica, through the host rock and into the cavity. The chalcedony shell is formed first, only fully hardening after an extended period of time. Subsequent mineralized groundwater flow may then **deposit** additional layers of minerals within the void. These **crystal** growths form first on the walls of the shell and then grow toward the center, producing the distinctive crystalline interior of the geode. The formation of a geode with large crystals may require tens or hundreds of millions of years.

Some of the most spectacular geodes come from Brazil. Some of these may be as large as one meter in diameter and contain very large amethyst (purple quartz) crystals. Once found, the geodes are often cut open with a diamond-tipped saw blade and the cut surfaces of the **sphere** polished for display.

Geodesic

A geodesic is the shortest path between two points along a surface. On a **plane**, it is the straight line segment joining the two points. On a **sphere**, it is the shorter **arc** of a great **circle** joining the two points.

Geodesic dome

A **geodesic** dome is a spherical building in which the supporting structure is a lattice of interconnecting tetrahedrons (a **pyramid** with three sides and a base) and octahedrons (an eight sided figure—two pyramids with four sides and a base, placed base to base). The first contemporary geodesic dome on record is Walter Bauersfeld's, who realized the utility of projecting the constellations on the inner surface of an icosasphere, Omnimax-style, thereby creating a breakthrough planetarium in Jena, Germany, in 1922. However, the geodesic dome common today was invented and patented by R. Buckminster Fuller in 1947.

Geodesic domes are fractional parts of complete geodesic spheres. Actual structures range from less than 5%-100% (a full **sphere**). The Spaceship Earth Pavilion constructed by Tishman Construction for AT&T at Walt Disney World's EPCOT is the best-known example of a full sphere.

Several physical and mathematical ideas factor into building a geodesic dome. For example, a convexly curved surface is stronger than a flat one, most materials are stronger in tension than in compression, and the most rigid structure is a triangle. A hemisphere encloses the most **space** with the least amount of material while the **tetrahedron** encloses the least **volume** with the most surface. These principles make geodesic domes the strongest, lightest, most **energy** efficient buildings ever devised. Structural patterns of geodesic domes vary in complexity. Some domes have been built using simple interconnecting triangles as a support structure while others have icosahedrons as their supporting structure. An icosahedron is the geometric form having the greatest number of identical and symmetrical faces—it has 20 faces, 12 vertices, and 30 edges. The more complex the structure is, the stronger it is.

Geodesic spheres and domes come in various frequencies. The **frequency** of a dome relates to the number of smaller triangles into which it is subdivided. A high frequency dome has more triangular components and is more smoothly curved and sphere-like.

Geodesic dome structures are used as private residences, commercial buildings, places of worship, schools, sports arenas, theaters, and vacation homes. Dome homes can be found in all 50 of the United States. They can be found in many places throughout the world such as China, **Africa**, **Europe**, and the Antarctic. Some notable geodesic domes are the Climatron, a climate controlled botanical garden in St. Louis, Missouri (1960), the Houston Astrodome (1965), and the dome for the American pavilion at Expo '67 in Montreal. Humans

A geodesic radardome at Lowther Hill Civil Aviation Authority Radar Station in Scotland. The dome is made of fiberglass which, unlike concrete, is transparent to radio waves. *Photograph by John Heseltine. National Audubon Society Collection/Photo Researchers, Inc. Reproduced by permission.*

have been living in domes such as mud huts, igloos, and thatch huts for millions of years.

Geodesy *see* **Surveying instruments**

Geographic and magnetic poles

Earth's geographic poles are fixed by the axis of **Earth's rotation**. On maps, the north and south geographic poles are located at the congruence of lines of longitude. Earth's geographic poles and magnetic poles are not located in the same place – in fact they are hundreds of miles apart. As are all points on **Earth**, the northern magnetic pole is south of the northern geographic pole (located on the polar **ice** cap) and is presently located near Bathurst **Island** in northern Canada (approximately 1,000 miles (1,600 km) from the geographic North Pole. The southern magnetic pole is displaced hundreds of miles away from the southern geographic pole on the Antarctic **continent**.

Although fixed by the axis of **rotation**, the geographic poles undergo slight wobble-like displacements in a circular pattern that shift the poles approximately six meters per year. Located on shifting polar ice, the North Pole (geographic pole) is technically defined as that point 90° N latitude, 0° longitude (although, because all longitude lines converge at the poles, any value of longitude can be substituted to indicate the same geographic point. The South Pole (geographic pole) is technically defined as that point 90° S latitude, 0° longitude. Early explorers used sextants and took celestial readings to determine the geographic poles. Modern explorers reply on GPS coordinates to accurately determine the location of the geographic poles.

Earth's magnetic field shifts over **time**, eventually completely reversing its polarity. There is evidence in magnetic mineral orientation that, during the past 10–15 million years, reversals have occurred as frequently as every quarter million years. Although Earth's magnetic field is subject to constant change (periods of strengthening and weakening) and the last magnetic reversal occurred approximately 750,000 years ago, geophysicists assert that the next reversal will not come within the next few thousand years. The present alignment means that at the northern magnetic pole, a dip compass (a compass with a vertical swinging needle) points straight down. At the southern magnetic pole, the dip compass needle would point straight up or away from the southern magnetic pole.

The magnetic poles are not stationary and undergo polar wandering. The north magnetic pole migrates about 10 km per year. The magnetic reversals mean that as **igneous rocks** cool from a hot **magma**, those that contain magnetic **minerals** will have those minerals align themselves with the magnetic polarity present at the time of cooling. These volcanic **rocks** preserve a history of magnetic reversals and when found in equidistant banded patterns on either side of sites of sea floor spreading, provide a powerful paleomagnetic proof of **plate tectonics**.

Navigators using magnetic compass readings must make corrections both for the distance between the geographic poles and the magnetic poles, and for the shifting of the magnetic poles. Moreover, the magnetic poles may undergo displacements of 40–60 km from their average or predicted position due to magnetic storms or other disturbances of the ionosphere and/or Earth's magnetic field. Angular corrections for the difference between the geographic poles and their corresponding magnetic pole are expressed as magnetic declination. The values for magnetic declination vary with the observer's position and are entered into navigation calculations to relate magnetic heading to true directional heading.

See also Bowen's reaction series; Cartography; Continental drift; Earth's interior; Global Positioning System; Latitude and longitude; Magnetism.

Geologic map

Geologic maps are graphical representations of **rocks**, sediments, and other geologic features observed or inferred to exist at or beneath Earth's surface. They can be based on observations of outcrops in the field, interpretation of aerial photographs or **satellite** images, and information obtained during the drilling of exploratory boreholes. Outcrops can be obscured, particularly in areas covered by dense vegetation or thick **soil**, and borehole information is often limited. Therefore, geologic maps are in most cases interpretive rather than purely descriptive scientific documents. Geologic maps are used for a variety of purposes, including **petroleum**, mineral, and **groundwater** exploration; **land use** planning; and natural hazard studies.

The first modern geologic maps were drawn by William Smith (1769-1839), a British **canal** builder. He recognized that sedimentary rocks occurred in a consistent sequence throughout the countryside. Knowing the position of a **coal** bed within the **sedimentary rock** sequence in one location allowed Smith to predict its occurrence and depth beneath the surface in other locations. Likewise, knowledge of rock sequences allowed Smith to predict the kinds of rocks that would be encountered during canal construction.

A general geologic **map** classifies rocks primarily according to their ages and secondarily according to their formation names. Formations are rock units that have a distinctive appearance or physical properties that can be identified in the field, and must also be laterally extensive and thick enough to depict on maps of a specified scale. In the United States, the scale at which formations must be mappable is 1:24,000, which corresponds to the scale of U.S. Geological Survey topographic maps covering 7.5 minutes of **latitude and longitude**.

Formations are named according to a protocol that requires a published description of each proposed formation in a scientific journal. If a formation consists of only one rock type, that rock type is included in the formation name (Berea Sandstone). If a formation consists of different rock types, then the word Formation is appended (Morrison Formation). Formations are named after nearby landmarks and, unlike fossils, never named directly after people. They can be, however, named after places that are named after people. The **Gene** Autry Shale, for example, is named after the town of Gene Autry, Oklahoma. The town, in turn, was named after the famous American singing cowboy Gene Autry (1907-1998).

Geologic maps also contain symbols representing geometric elements that are collectively known as geologic structures. These include faults, joints (fractures across which very little movement has occurred), aligned prismatic or platy mineral crystals known as lineations or foliations, and **strata** that have been tilted or folded in response to stresses within Earth's crust.

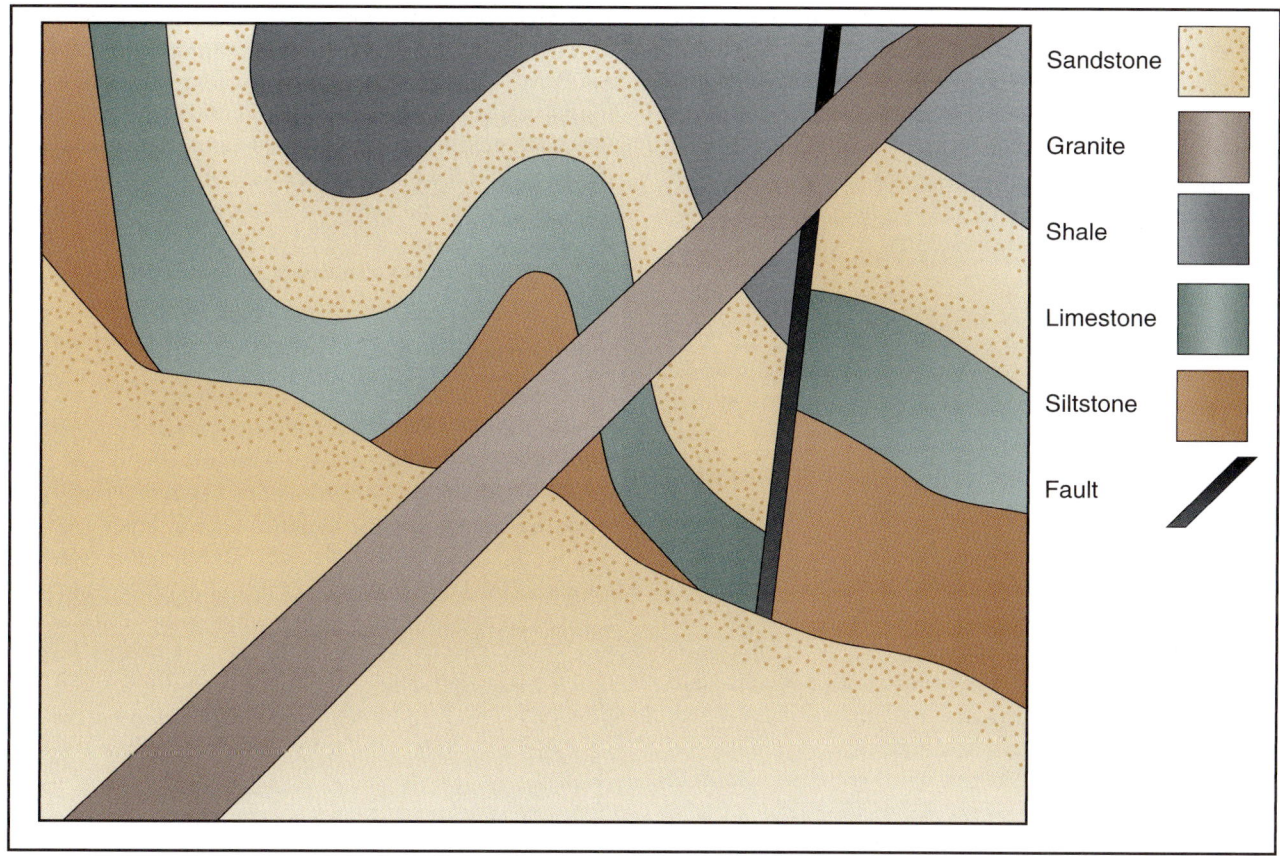

Sandstone	
Granite	
Shale	
Limestone	
Siltstone	
Fault	

A simplified geologic map. *Photo Researchers, Inc. Reproduced by permission.*

Specialized geologic maps can also be drawn. **Engineering** geologists conducting investigations for construction projects, for example, may be more concerned with rock types than ages. They may therefore depict rocks using a system that emphasizes rock type and origin over age. Petroleum geologists and hydrogeologists often prepare maps based solely on information obtained during the drilling of oil, gas, and **water** wells. Those that depict changes in the thickness of a particular formation are known as isopach maps. Structure maps show the elevation or depth of a formation that may be an important petroleum reservoir or **aquifer**.

Geologic time

Although historical **time** covers centuries, and archeological times covers millennia, geologic time describes the immense span of time—billions of years—revealed in the fossil and rock record of **Earth**. Geochronology is the science of finding out how old **rocks** and **minerals** are. Absolute time and relative time are terms used to de-

scribe the age of rocks and events used by geologists. Radiometric age determination is a method used by geologists to determine the absolute age, in years, of rocks and minerals. Knowledge of **stratigraphy**, the branch of **geology** that catalogues Earth's successions of rock layers, is essential to establish the relative ages of rock units. By finding which rock unit formed first, the order of the events in Earth history can be sorted out.

History of the concept of geologic time

Before scientific methods were used to investigate geologic time, ideas about time and Earth history came from religious theories. The Hindu and Mayan religions taught about endlessly repeating cycles of time, each lasting for billions of years. Ideas in western culture about the age of Earth were just as imprecise, and just as incorrect. In the 1650s, the Irish clergyman and scholar James Ussher (1581–1656) used apparent genealogy within the Bible's Book of Genesis to determine that the Earth was created in 4004 B.C. Ussher based his results on the only information he had available to study the question of Earth's age. Ussher's estimates have been shown to be flawed, but unfortunately have been repeated by more modern writers who oppose modern views of Earth's age. Isaac Newton also speculated on the age of Earth, using the investigative techniques of the time that could be considered archaic today. As early as the eighteenth century, scientists knew that Earth's lifetime must have been immense. But geologists were not able to measure the dimensions of Earth's history until **mass** spectrometers became available in the 1950s. The mass spectrometer is an instrument used to separate different varieties of **atoms** from each other. Before that time, educated guesses had been made by comparing the rock record from different parts of the world and estimating how long it would take natural processes to form all the rocks on Earth.

Georges Louis Leclerc de Buffon (1707–1788), for example, calculated Earth to be 74,832 years old by figuring how long it would take the **planet** to cool down to the present **temperature**. Writing around 1770, he was among the first to suggest that Earth's history can be known about by observing the planet's current state.

James Hutton (1726–1797) did not propose a date for the formation of Earth, but is famous for the statement that Earth contains "no vestige of a beginning—no prospect of an end." The German geologist Abraham Werner (1750–1817), the first scientist to make use of a stratigraphic column, a diagram of order of sedimentary layers. An original approach to geological history was suggested by the French zoologist and paleontologist Georges Cuvier (1769–1832), who observed that specific fossil animals occurred in specific rock layers, forming recognizable groups, or assemblages. William Smith (1769–1768) combined Werner's and Cuvier's approaches, using fossil assemblages to identify identical sequences of layers distant from each other, linking or correlating rocks which were once part of the same rock layer but had been separated by faulting or **erosion**.

In 1897, the physicist Lord Kelvin (1824–1907) developed a model for Earth history, which assumed that Earth has been cooling steadily since its formation. Because he did not know that **heat** moves around in currents in the earth (**convection**), or that Earth generates its own heat from the decay of radioactive minerals buried inside it, Kelvin proposed that the earth was formed from 20 to 40 million years ago.

In the late eighteenth century, geologists began to name periods of geologic time. In the nineteenth century, geologists such as William Buckland (1784–1856), Adam Sedgwick (1785–1873), Henry de la Beche (1796–1855), and Roderick Murchison (1792–1871) identified widespread rock layers beneath continental **Europe**, the British isles, Russia, and America. They named periods of time after the places in which these rocks were first described. For instance, Cambrian Period was named for Cambria (the Roman name for Wales), and Permian, for the Perm province in Czarist Russia. Mississippian and Pennsylvanian Periods widely used by American geologists were named for a U.S. state and a region around the upper reaches of a large river, respectively. By the mid-nineteenth century, most of the modern names of the periods of geologic time had been proposed; all of them are still in use.

Relative age determination

A rock layer may or may not contain evidence that reveals its age. Rock layers whose ages are defined by relationships with the dated rock units around it are examples of relative age determination. That relationship is found by observing the unknown rock layer's stratigraphic relationship with the rock layers whose ages are known. If the known rock layer is on top of the unknown layer, then the lower layer is probably the older of the two. That is based on the principle of superposition, which states that when two rock layers are stacked one above the other, the lower one was formed before the overlying one, unless the layers have been overturned.

Radiometric age determination

Every rock and mineral exists in the world as a mixture of elements, and every element exists as a population of atoms. One element's population of atoms will not all have the same number of neutrons, and so two or more kinds of the same element will have different atomic masses or atomic numbers. These different kinds of the same chemical element are called nuclides of that element. A nuclide of a radioactive element is known as a radionuclide.

The nucleus of every radioactive element spontaneously disintegrates over time. This process results in **radiation**, and is called **radioactive decay**. Losing high-energy particles from their nuclei turns the atoms of a radioactive nuclide into the daughter product of that nuclide. A daughter product is either a different element altogether, or is a different nuclide of the same parent element. A daughter product may or may not be radioactive. If it is, it also decays to form its own daughter product. The last radioactive element in a series of these transformations will decay into a stable element, such as **lead**.

While there is no way to discern whether an individual atom will decay today or two billion years from today, the **behavior** of large numbers of the same kind of atom is so predictable that certain nuclides of elements are called radioactive clocks. The use of these radioactive clocks to calculate the age of a rock is referred to as radiometric age determination. First, an appropriate ra-

KEY TERMS

Blocking temperature—Temperature below which a mineral's atomic framework is rigid enough to trap the daughter products of radioactive decay, thus starting the radioactive clock.

Concordant—Said of a rock's age when determining the absolute age of two or more rocks or minerals by radiometric methods yields the same result.

Daughter product—The element made when a radioactive element's nucleus spontaneously falls apart (or decays). The daughter product may or may not be radioactive.

Discordant—Said of a rock's age when different results come from determining the absolute age of two or more rocks or minerals by radiometric methods.

Fission-track dating—An age determination technique in which the number of trails torn by alpha particles through a zircon crystal's crystal framework.

Geochronology—The study of determining how old rocks and minerals are, in order to sort out the events of Earth history.

Half-life—The time it takes for half of the original atoms of a radioactive element to be transformed into the daughter product.

Nuclides—Different versions of the same chemical element that have slightly different numbers of subatomic particles in their nuclei, and therefore different atomic masses or in some cases different atomic numbers.

Radioactive clock—The ratio of a radionuclide and its daughter product, which accumulate within the atomic framework of a mineral. Also used to mean the radionuclide itself.

Radioactive decay—The predictable manner in which a population of atoms of a radioactive element spontaneously fall apart, and lose subatomic particles from their nuclei, becoming their daughter products.

Radiometric date—A measurement of time derived by measuring how much of a short-lived radioisotope (carbon-14) is left in a formerly living object, or how much of a long-lived radioisotope is left in a rock or mineral compared to how much of the daughter product is there into which the long-lived radioisotope decayed.

Radionuclide—Radioactive or unstable nuclide.

Stratigraphy—The study of layers of rock or soil, based on the assumption that the oldest material will usually be found at the bottom of a sequence.

dioactive clock must be chosen. The **sample** must contain measurable quantities of the element to be tested for, and its radioactive clock must tell time for the appropriate **interval** of geologic time. Then, the amount of each nuclide present in the rock sample must be measured.

Each radioactive clock consists of a radioactive nuclide and its daughter product, which accumulate within the atomic framework of a mineral. These radioactive clocks decay at various rates, which govern their usefulness in particular cases. A three-billion year old rock needs to have its age determined by a radioactive clock that still has a measurable amount of the parent nuclide decaying into its daughter product after that long. The same radioactive clock would reveal nothing about a two million year old rock, for the rock would not yet have accumulated enough of the daughter product to measure.

The time it takes for half of the parent nuclide to decay into the daughter product is called one **half-life**. The remaining population of the parent nuclide is halved again, and the population of daughter product doubled, with the passing of every succeeding half-life. The

amount of parent nuclide measured in the sample is plotted on a graph of that radioactive clock's known half-life. The absolute age of the rock, within its margin of **error**, can then be read directly from the time axis of the graph.

When a rock is tested to determine its age, different minerals within the rock are tested using the same radioactive clock—similar to questioning different witnesses at a crime scene to determine if they saw the same event happen in the same way. Ages may be determined on the same sample by using different radioactive clocks. When the age of a rock is measured in two different ways, and the results are the same, the results are said to be concordant.

Discordant ages means the radioactive clock showed different absolute ages for a rock sample, or different ages for different minerals within the rock. A discordant age result means that at some time after the rock was formed, something happened to it which reset one of the radioactive clocks back to **zero**.

For example, if a discordant result happens in the potassium-argon test, the rock may have been heated to a blocking temperature above which a mineral's atomic framework becomes active and wiggly enough to allow trapped gaseous argon-40 to escape.

Concordant ages mean that no complex sequence of events-deep burial, **metamorphism**, and mountain-building, for example has happened that can be detected by the two methods of age determination that were used.

A form of radiometric dating is used to determine the ages of organic **matter**. A short-lived radioisotope, carbon-14, is accumulated by all living things on Earth. Upon the organism's death, the carbon-14 is fixed and then begins to decay into carbon-12 at a known **rate** (its half-life is 5,730 years). By measuring how much of the carbon-14 is left in the remains, and plotting that amount on a graph showing how fast the carbon-14 decays, the approximate date of the organism's death can be known.

When **uranium** atoms decay, they emit fast, heavy alpha particles. Inside a zircon **crystal**, these **subatomic particles** tear long trails of destruction through the zircon's crystal framework. The age of a zircon crystal can be estimated by counting the number of these trails. The rate at which the trails form has been found by determining the age of rocks containing zircon crystals, and noting how torn-up the zircon crystals become over time. This age determination technique is called fission-track dating. This technique has detected the world's oldest rocks, between 3.8 billion and 3.9 billion years old, and yet older crystals, which suggest that Earth had some solid ground on it 4.2 billion years ago.

The age of Earth is deduced from the ages of other materials in the **solar system**, namely, meteorites. Meteorites are pieces formed from the cloud of dust and debris left behind by a **supernova**, the explosive death of a **star**. Through this cloud the infant Earth spun, attracting more and more pieces of matter. The meteorites that fall to Earth today have orbited the **Sun** since that time, unchanged and undisturbed by the processes that have destroyed Earth's first rocks. Radiometric ages for meteorites fall between 4.45 billion and 4.55 billion years.

The radionuclide iodine-129 is formed in nature only inside stars. A piece of solid iodine-129 will almost entirely decay into the gas xenon-129 within a hundred million years. If this decay happens in open **space**, the xenon-129 gas will float off into space, blown by the **solar wind**. Alternatively, if the iodine-129 was stuck in a rock within a hundred million years of being formed in a star, then some very old rocks should contain xenon-129 gas. Both meteorites and Earth's oldest rocks contain xenon-129. That means the star that provided the material for the solar system died its cataclysmic death less than 4.65 billion (4,650,000,000) years ago.

See also Dating techniques; Fossil and fossilization; Spectroscopy; Strata.

Resources

Books

Hartman, William, and Ron Miller. *The History of the Earth.* New York: Workman Publishing, 1991.

Press, Frank, and Raymond Sevier. *Understanding Earth.* San Francisco: Freeman, 2000.

Tarbuck, Edward. J., Frederick K. Lutgens, and Dennis Tassa, eds. *Earth: An Introduction to Physical Geology,* 7th ed. Upper Saddle River, New Jersey: Prentice Hall, 2002.

Other

Newman, William L. "Geologic Time." United States Geological Survey, July 28, 1997 [cited January 5, 2003]. <http://pubs.usgs.gov/gip/geotime/contents.html>.

Clinton Crowley

Geology

Geology is the study of **Earth**. Modern geology includes studies in seismology (**earthquake** studies), volcanology, **energy** resources exploration and development, **tectonics** (structural and mountain building studies), **hydrology** and hydrogeology (water-resources studies), geologic mapping, economic geology (e.g., **mining**), **paleontology** (ancient life studies), **soil** science, **historical geology** and **stratigraphy**, geological **archaeology**, glaciology, modern and ancient climate and **ocean** studies, atmospheric sciences, **planetary geology**, **engineering** geology, and many other subfields.

Geologists study **mountains**, valleys, plains, sea floors, **minerals**, **rocks**, fossils, and the processes that create and destroy each of these. Geology consists of two broad categories of study. Physical geology studies the Earth's materials (**erosion**, volcanism, sediment deposition) that create and destroy the materials and landforms. Historical geology explores the development of life by studying fossils (petrified remains of ancient life) and the changes in land (for example, distribution and latitude) via rocks. But the two categories overlap in their coverage: for example, to examine a fossil without also examining the rock that surrounds it tells only part of the preserved organism's history.

Physical geology further divides into more specific branches, each of which deals with its own part of Earth's materials, landforms, and/or processes. **Mineralogy** and petrology investigate the composition and origin of minerals and rocks, respectively. Sedimentologists look at sedimentary rocks—products of the accumulation of rock fragments and other loose Earth materials—to determine how and where they formed. Volcanologists tread on live, dormant, and extinct volcanoes checking lava, rocks and gases. Seismologists set up instruments to monitor and to predict earthquakes and volcanic eruptions. Structural geologists study the ways rock layers bend and break. **Plate tectonics** unifies most aspects of Physical Geology by demonstrating how and why plates (sections of Earth's outer crust) collide and separate and how that movement influences the entire **spectrum** of geologic events and products.

Fossils are used in Historical geology as evidence of the **evolution** of life on Earth. Plate tectonics adds to the story with details of the changing configuration of the continents and oceans. For years paleontologists observed that the older the rock layer, the more primitive the fossil organisms found therein, and from those observations developed evolutionary theory. Fossils not only relate evolution, but also speak of the environment in which the **organism** lived. Corals in rocks at the top of the Grand Canyon in Arizona, for example, show a shallow sea flooded the area around 290 million years ago. In addition, by determining the ages and types of rocks around the world, geologists piece together continental and oceanic history over the past few billions of years. For example, by matching fossil and tectonic evidence, geologists reconstructed the history and shape of the 200-300 million year-old supercontinent, Pangea (Pangaea).

Many other sciences also contribute to geology. The study of the **chemistry** of rocks, minerals, and volcanic gases is known as **geochemistry**. The **physics** of the Earth is known as **geophysics**. Paleobotanists study fossil plants. Paleozoologists reconstruct fossil animals. Paleoclimatologists reconstruct ancient climates.

Much of current geological research focuses on resource utilization. Environmental geologists attempt to minimize human impact on the Earth's resources and the impact of natural disasters on human kind. Hydrology and hydrogeology, two subdisciplines of environmental geology, deal specifically with **water** resources. Hydrologists study surface water whereas hydrogeologists study ground water. Both disciplines try to minimize the impact of **pollution** on these resources. Economic geologists focus on finding the minerals and **fossil fuels** (oil, **natural gas**, **coal**) needed to maintain or improve global standards of living. Extraterrestrial geology, a study in its infancy, involves surveying the materials and processes of other planets, trying to unlock the secrets of the Universe and even to locate useful mineral deposits.

Geologic employment has been traditionally dominated by the **petroleum** industry and related geologic service companies. In the modern world, this is no longer so. Mining and other economic geology occupations (e.g., prospecting and exploration), in former days plentiful, have also fallen away as major employers. En-

vironmental geology, engineering geology, and ground water related jobs are more common employment opportunities today. As these fields are modern growth areas with vast potential, this trend will likely hold true well into the future. Many modern laws and regulations require that licensed, professional geologists supervise all or part of key tasks in certain areas of engineering geologic work and environmental work. It is common for professional geologists and professional engineers to work together on such projects, including construction site preparation, waste disposal, ground-water development, engineering planning, and highway construction. Many federal, state, and local agencies employ geologists, and there are geologists as researchers and teachers in most academic institutions of higher education.

See also Earth science; Earth's interior.

Resources

Books

Hancock P.L., and Skinner B.J., eds. *The Oxford Companion to the Earth.* New York: Oxford University Press, 2000.
Tarbuck, Edward D., Frederick K. Lutgens, and Tasa Dennis. *Earth: An Introduction to Physical Geology.* 7th ed. Upper Saddle River, NJ: Prentice Hall, 2002.
Winchester, Simon. *The Map That Changed the World: William Smith and the Birth of Modern Geology.* New York: Harper Collins, 2001.

Periodicals

Hellfrich, George, and Bernard Wood. "The Earth's Mantle." *Nature.* (August 2, 2001): 501–507.

Other

Connelly, William J. "geology.com Earth Science on the Web" [cited February, 24, 2003]. <http://geology.com/>.

Geometry

Geometry, the study of points, lines, and other figures in **space**, is a very old branch of **mathematics**. Its ideas were undoubtedly used, intuitively if not formally, from earliest times. Walking along a straight line toward a particular destination is the shortest way to get there; lining an arrow up with the target is the way to hit it; sitting in a **circle** around a fire is the most equitable way to share the warmth. Early humans need not have been students of formal geometry to know and to use these ideas.

As early as 2,600 years ago the Greeks had not only discovered a large number of geometric properties, they had begun to see them as abstract ideas to be studied in their own right. By the third century B.C., they had created a formal system of geometry. Their system began with the simplest ideas and, with these ideas as a foundation, went well beyond much of what is taught in schools today.

Proof

Typically one learns **arithmetic** and **algebra** by experiment or by being told how to do it. Geometry, however, is taught logically. Its ideas are established by means of "proof." One starts with definitions, postulates, and primitive terms; then proves his or her way through the course.

The reason for this goes back to the forenamed Greeks, and in particular to Euclid. Twenty-three hundred years ago he wrote a beautiful book called the *Elements*. This book contains no exercises, no experiments, no applications, no questions—just proofs, the **proof** of one proposition after another.

For centuries the *Elements* was the basic text in geometry. Heath, in his 1925 translation of the *Elements*, quotes De Morgan: "There never has been...a system of geometry worthy of the name, which has any material departures...from the plan laid down by Euclid." Nowadays the *Elements* has been replaced with texts which do have exercises, problems, and applications, but the emphasis on proof remains. Even the most obvious fact, such as the fact that the opposite sides of a **parallelogram** are equal, is supposed to go unnoticed, or at least unused, until it has been proved. Whether or not this makes sense, the reader will have to decide for himself or herself, but sensible or not, proof is and will probably continue to be a dominant component of a course in geometry.

Proofs can vary in formality. They can be as formal as the two-column proofs used in text-books in which each statement is identified as an assumption, a definition, or the consequence of a previously proved property; they can be informal with much left for the reader to fill in; or they can be almost devoid of explanation, as in the ingenious proof of the **Pythagorean theorem** given by the Hindu mathematician Bhaskara in the twelfth century. His proof consisted of a single word "behold" and a drawing.

Constructions

Another lasting influence of Euclid's *Elements* is the emphasis which is placed on **constructions**. Three of the five postulates on which Euclid based his geometry describe simple drawings and the conditions under which they can be made. One such drawing (construction) is the circle. It can be drawn if one knows where its center and one point on it are. Another construction is drawing a line segment between two given points. A third is extending a given line segment. These are the so-called ruler-and-compass constructions upon which Euclidean geometry is based.

Modern courses in geometry are frequently based on other postulates. Some, for example, permit one to use a protractor to draw and measure angles; some allow the use of a scale to measure distances. Even so, the traditional limitations which the Euclidean postulates placed on constructions are often observed. Protractors, scales, and other drawing tools which would be easier and more accurate to use are forbidden. Constructions become puzzles, intriguing but separate from the logical structure of the course, and not overly practical.

Points, lines, and planes

Points, lines, and planes are primitive terms; no attempt is made to define them. They do have properties, however, which can be explicitly described. Among the most important of these properties are the following:

Two distinct points determine exactly one line. That line is the shortest path between the two points. Bricklayers use these properties when they stretch a string from corner to corner to guide them in laying bricks.

Two points also determine a ray, a segment, and a **distance**, symbolized for points A and B by AB (or BA when B is the endpoint), AB, and AB respectively. (Some authors use AB to symbolize all of these, leaving it to the reader to know which is meant.) Three non-collinear points determine one and only one **plane**.

The photographer's tripod exploits this to hold the camera steady; the chair on an uneven floor **rocks** back and forth between two different planes determined by two different combinations of the four legs.

If two points of a line lie in a plane, the entire line lies in the plane. It is this property which makes the plane "flat." Two distinct lines intersect in at most one point; two distinct planes intersect in at most one line. If two *coplanar* lines do not intersect, they are **parallel**. Two lines which are not coplanar cannot intersect and are called "skew" lines. Two planes which do not intersect are parallel.

A line which does not lie in a plane either intersects that plane in a single point, or is parallel to the plane.

Angles

An **angle** in geometry is the union of two rays with a common endpoint. The common endpoint is called the "vertex" and the rays are called the "sides." Angle ABC is the union of BA and BC. When there is no danger of confusion, an angle can be named by its vertex alone. It is also handy from time to time to name an angle with a letter or number written in the interior of the angle near the vertex. Thus angles ABC, B, and x are all the same angle.

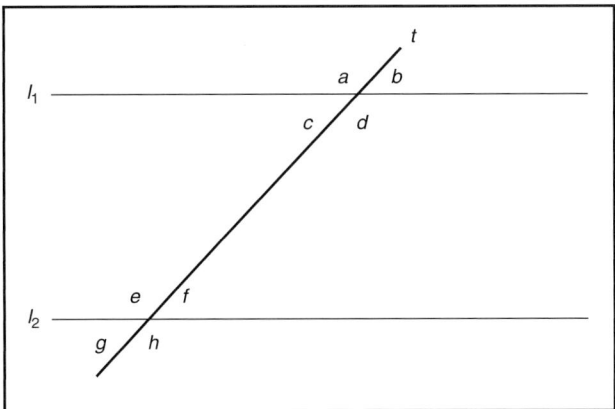

Figure 1. *Illustration by Hans & Cassidy. Courtesy of Gale Group.*

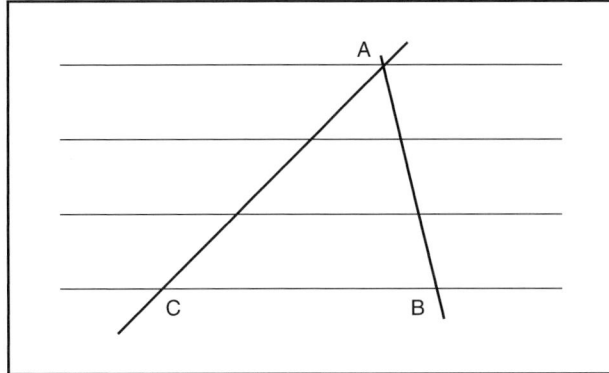

Figure 2. *Illustration by Hans & Cassidy. Courtesy of Gale Group.*

When the two sides of an angle form a line, the angle is called a "straight angle." Straight angles have a measure of 180°. Angles which are not straight angles have a measure between 180° and 0°. The "reflex" angles, whose measures exceed 180°, encountered in other branches of mathematics are not ordinarily used in geometry. If the sum of the measures of two angles is 180°, the angles are said to be "supplementary." "Right" angles have a measure of 90°. Lines which form right angles are also said to be **perpendicular**. If the sum of the measures of two angles is 90°, the angles are called "complementary." Angles which are smaller than a right angle are called "acute." Those which are bigger than a right angle but smaller than a straight angle are called "obtuse." When two lines intersect, they form two pairs of "opposite" or "vertical" angles. Vertical angles are equal.

A ray which divides an angle into two equal angles is called an angle "bisector." Points on an angle bisector are equidistant from the sides of the angle.

Parallel lines and planes

Given a line and a point not on the line, there is exactly one line through the point parallel to the line.

Two *coplanar* lines l_1 and l_2, cut by a transversal t are parallel if and only if

1) Alternate interior angles (e.g., d and e) are equal.

2) Corresponding angles (e.g., b and f) are equal.

3) Interior angles on the same side of the transversal are supplementary (see Figure 1).

These principles are used in a variety of ways. A draftsman uses 2) to rule a set of parallel lines. Number 1) is used to show that the sum of the angles of a triangle is equal to a straight angle.

If a set of parallel lines cuts off equal segments on one transversal, it cuts off equal segments on any other

transversal (see Figure 2). A draftsman finds this useful when he or she needs to subdivide a segment into parts which are not readily measured, such as thirds. If transversal AC in Figure 2 is slanted so that AC is three units, then the parallel lines through the unit points will divide AB into thirds as well.

If a set of parallel planes is cut by a plane, the lines of intersection are parallel. This property and its converse are used when one builds a bookcase. The set of shelves are, one hopes, parallel, and they are supported by parallel grooves routed into the sides.

Perpendicular lines and planes

If A is a given point and CD a given line, then there is exactly one line running through A that is perpendicular to CD. If B is the point on line CD that also resides on the line running perpedicular to CD, then that line, AB, is the shortest distance from point A to line CD.

In a plane, if CD is a line and B a point on CD, then there is exactly one line through B perpendicular to CD. If B happens to be the midpoint of CD, then AB is called the perpendicular bisector of CD. Every point on AB is equidistant from C and D.

If a line QP is perpendicular to a plane at a point P, then it is perpendicular to every line in the plane which passes through P. This property is used by carpenters when they make sure that a door frame is perpendicular to the floor. Otherwise the door will rub on the floor, as someone who lives in an old house is likely to know.

A line will be perpendicular to a plane if it is perpendicular to two lines in the plane. The carpenter, in setting up the door frame, need not check every line with his or her square; two will do.

If perpendiculars are not confined to a single plane, there will be an infinitude of lines through B perpendicular to CD, all lying in the plane which is perpendicular to

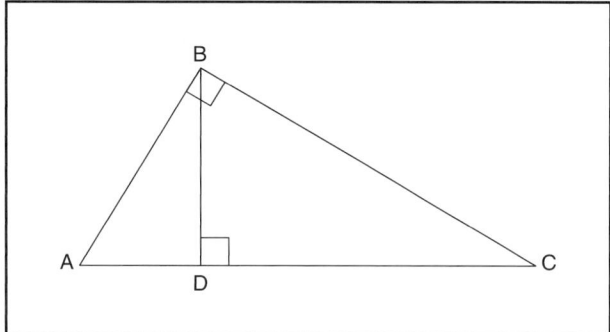

Figure 3. *Illustration by Hans & Cassidy. Courtesy of Gale Group.*

CD. If B is a midpoint, this plane will be the perpendicular-bisector plane of CD, and every point on this plane will be equidistant from C and D.

Two planes are perpendicular if one of the planes contains a line which is perpendicular to the other plane. The panels of folding screens, for example, stay perpendicular to the floor because the hinge lines are perpendicular to the floor.

Triangles

Triangles are plane figures determined by three non-collinear points called "vertices." They are made up of the segments, called sides, which join them. Although the sides are segments rather than rays, each pair of them makes up one of the triangle's angles.

Triangles may be classified by the size of their angles or by the lengths of their sides. Triangles whose angles are all less than right angles are called "acute." Those with one right angle are "right" triangles. Those with one angle larger than a right angle are "obtuse." (In a right triangle the side opposite the right angle is called the "hypotenuse" and the other two sides "legs.") Triangles with no equal sides are "scalene" triangles. Those with two equal sides are "isosceles." Those with three equal sides are "equilateral." There is no direct connection between the size of the angles of a triangle and the lengths of its sides. The longest side, however, will be opposite the largest angle; and the shortest side, opposite the smallest angle. Equal sides will be opposite equal angles.

In comparing triangles it is useful to set up a correspondence between them and to name corresponding vertices in the same order. If CXY and PST are two such triangles, then angles C and P correspond; sides CY and PT correspond; and so on.

Two triangles are "congruent" when their six corresponding parts are equal. Congruent triangles have the same size and shape, although one may be the mirror image of the other. Triangles ABC and FDE are congruent provided that the sides and angles which appear to be equal are in fact equal.

One can show that two triangles are congruent without establishing the equality of all six parts. Two triangles will be congruent whenever

1) Two sides and the included angle of one are equal to two sides and the included angle of the other (SAS congruence).

2) Two angles and the included side of one are equal to two angles and the included side of the other (ASA congruence).

3) Three sides of one are equal to three sides of the other (SSS congruence).

Triangle congruence applies not only to two different triangles. It also applies to one triangle at two different times or to one triangle looked at in two different ways. For example, when the girders of a bridge are strengthened with triangular braces, each triangle stays congruent to itself over a period of time, and does so by virtue of SSS congruence.

Two triangles can also be similar. Similar triangles have the same shape, but not necessarily the same size. They are alike in the way that a snapshot and an enlargement of it are alike. When two triangles are similar, corresponding angles are equal and corresponding sides are proportional.

One can show that two triangles are similar without showing that all the angles are equal and all the sides proportional. Two triangles will be similar when

1) Two sides of one triangle are proportional to two sides of another triangle and the included angles are equal (SAS similarity).

2) Two angles of one triangle are equal to two angles of another triangle (AA similarity).

3) Three sides of one triangle are proportional to three sides of another triangle (SSS similarity).

The properties of similar triangles are widely used. Artists, for example, use them in making smaller or larger versions of a picture. **Map** makers use them in drawing maps; and users, in reading them.

Figure 3 shows a right triangle in which an altitude BD has been drawn to the hypotenuse AC. By AA similarity, the triangles ABC, ADB, and BDC are similar to one another.

By virtue of these similarities one can write AC/BC = BC/DC and AC/AB = AB/AD. Then, using AD + DC = AC and a little algebra, one ends up with $(AB)^2 + (BC)^2 = (AC)^2$, or the Pythagorean **theorem**: "In a right triangle the sum of the squares on the legs is equal to the square on the hypotenuse." This neat proof was discovered by Bhaskara, mentioned earlier.

The altitude BD in Figure 3 is also the **mean** proportional between AD and DC. That is, AD/BD = BD/DC.

In triangle ABC, if DE is a line drawn parallel to AC, it creates a triangle similar to ABC. It therefore divides AB and BC proportionally. Conversely, a line which divides two sides of a triangle proportionally is parallel to the third side. A special case of this is a segment joining the midpoints of two sides of a triangle. It is parallel to the third side and half its length.

Each triangle has four sets of lines associated with it: medians, altitudes, angle bisectors, and perpendicular bisectors of the sides. In each set, the three lines are, remarkably, concurrent, that is, they all pass through a single point. In the case of the medians, which are lines from a vertex to the midpoint of the opposite side, the point of concurrency is the "centroid," the center of gravity. The angle bisectors are concurrent at the "incenter," the center of a circle tangent to the three sides. The perpendicular bisectors of the sides are concurrent at the "circumcenter," the center of a circle passing through all three vertices. The altitudes, which are lines from a vertex perpendicular to the opposite side, are concurrent at the "orthocenter."

Quadrilaterals

Quadrilaterals are four-sided plane figures. Various special quadrilaterals are defined in various ways. The following are typical:

Trapezoid: A **quadrilateral** with one pair of parallel sides.

Parallelogram: A quadrilateral with two pairs of parallel sides.

Rhombus: A parallelogram with four equal sides.

Kite: A quadrilateral with two pairs of equal adjacent sides.

Rectangle: A parallelogram with four right angles.

Square: A rectangle with four equal sides. It is a special kind of rhombus.

Cyclic quadrilateral: A quadrilateral whose four vertices lie on one circle.

In any quadrilateral the sum of the angles is 360°. In a cyclic quadrilateral opposite angles are supplementary.

The diagonals of any parallelograms bisect each other. The diagonals of kites and any rhombus are perpendicular to each other.

Opposite sides of parallelograms are equal.

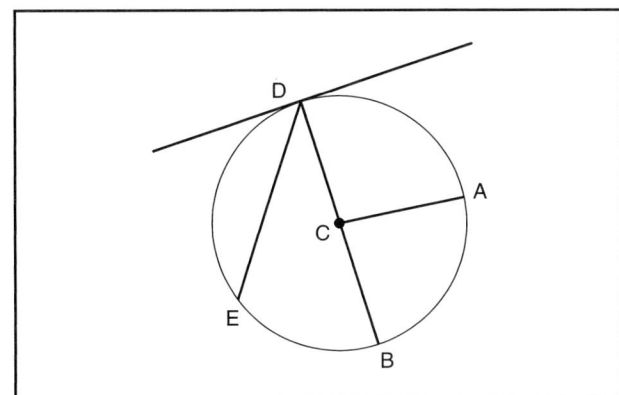

Figure 4. *Illustration by Hans & Cassidy. Courtesy of Gale Group.*

Circles

A circle is a set of points in a plane which are a fixed distance from a point called the center, C (see Figure 4). A "chord" is a segment, DE, joining two points on the circle; a radius is segment, CA, joining the center and a point on the circle; a diameter is a chord, DB, through the center. A "tangent," DF, is a line touching the circle in a single point. The words "radius" and "diameter" can also refer to the lengths of these segments.

An "arc" is the portion of the circle between two points on the circle, including the points. A major **arc** is the longer of the two arcs so determined; a minor arc, the shorter. When an arc is named it is usually the minor arc that is meant, but when there is danger of confusion, a third letter can be used, e. g. arc DAB.

All circles are similar, and because of this the **ratio** of the circumference to the diameter is the same for all circles. This ratio, called **pi** or π, was shown to be smaller than 22/7 and larger than 223/71 by the mathematician Archimedes about 240 B.C.

An arc can be measured by its length or by the central angle which it subtends. A central angle is one whose vertex is the center of the circle.

An inscribed angle is one whose vertex is on the circle and whose sides are two chords or one chord and a tangent. Angles EDB and BDF are inscribed angles. The measure of an inscribed angle is one half that of its intercepted arc. Any inscribed angle that intercepts a semicircle is a right angle; so is the angle between a tangent and a radius drawn to the point of tangency.

If X is a point inside a circle and AB and CD any two chords through X, then X divides the chords into segments whose products are equal. That is, AX XB = CX XD.

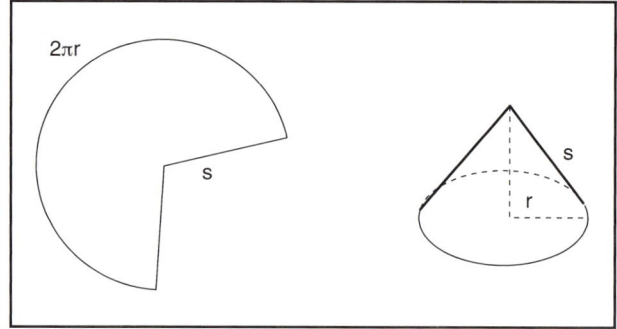

Figure 5. *Illustration by Hans & Cassidy. Courtesy of Gale Group.*

Area

Areas are expressed in terms of squares such as square inches, meters, miles, etc. Formulas for the areas of various plane figures are based upon the formula for the area of a rectangle, lw, where l is the length and w the width. The area of a parallelogram is bh, where b is the base and h the height (altitude), measured along a line perpendicular to the base. The area of a triangle is half that of a parallelogram with the same base and height, bh/2. When the triangle is equilateral, $h = \sqrt{3}\,b/2$ so the area is $\sqrt{3}\,b^2/4$. A trapezoid whose parallel sides are b_1 and b_2 and whose height is h can be divided into two triangles with those bases and altitudes. Its area is $(b_1 + b_2)h/2$.

The area of a quadrilateral with sides a, b, c, and d depends not only on the lengths of the sides but on the size of its angles. When the quadrilateral is cyclic (all four endpoints are on in a circle), its area is given by a remarkable formula discovered by the Hindu mathematician Brahmagupta in the seventh century:

$$A = \sqrt{(s-a)(s-b)(s-c)(s-d)}$$

where s is the semi-perimeter (a + b +c + d)/2. This formula includes Heron's formula, discovered in the first century, for the area of a triangle,

$$\sqrt{(s-a)(s-b)(s-c)s}$$

as a special case. By letting d = 0, the quadrilateral becomes a triangle, which is always cyclic.

The area of a circle can be approximated by the area of an inscribed regular polygon. As the number of sides of this polygon increases without **limit**, its area approaches cr/2, where c is the circumference of the circle and r the radius. Since $c = 2\pi r$, the area of the circle is πr^2.

The surface area of a **sphere** of radius r is four times the area of a circle of the same radius, $4\pi r^2$.

The lateral surface of a right circular cone can be unrolled to form a sector of a circle (see Figure 5). Its

A dodecahedron is one of Plato's five regular polyhedrons (along with the tetrahedron, hexahedron, octohedron, and icosahedron): all of its faces are identically sized regular polygons. This dodecahedron has had a pentagon shaped window removed from each of its faces to allow viewing of the interior. © *Richard Duncan, National Audubon Society Collection/Photo Researchers, Inc. Reproduced with permission.*

area is πrs, where s is the *slant* height of the cone and r the radius of its base.

Volumes

The volumes of geometric solids are expressed in terms of cubes which are one unit on a side, such as cubic centimeters or cubic yards. The **volume** of a rectangular solid (box) whose length, width, and height are l, w, and h is lwh. The volume of a **prism** or a cylinder is Bh, where B is the area of its base and h its height measured along a line perpendicular to the base. The volume of a **pyramid** or cone is one third that of a prism or cylinder with the same base and height, that is Bh/3. The volume of a sphere of radius r is $4\pi r^3/3$.

It is interesting to note that the volumes of a cylinder, a hemisphere, and a cone having the same base and height are in the simple ratio 3:2:1.

Other geometries

The foregoing is a summary of Euclidean geometry, based on Euclid's postulates. Euclid's fifth **postulate** is equivalent to assuming that through a given point not on a given line, there is exactly one line parallel to the given line. When one assumes that there is no such line, elliptical geometry emerges. When one assumes that there is more than one such line, the result is hyperbolic geometry. These geometries are called "non-Euclidean." Non-Euclidean geometries are as correct and consistent as Euclidean, but describe special spaces. Geometry can also be extended to more than three dimensions. Other special geometries include **projective geometry**, affine geometry, and **topology**.

Resources

Books

Euclid. *Elements*. translated by Heath, Sir Thomas L., New York: Dover Publishing Co., 1956.

Gullberg, Jan, and Peter Hilton. *Mathematics: From the Birth of Numbers*. W.W. Norton & Company, 1997.

Weisstein, Eric W. *The CRC Concise Encyclopedia of Mathematics*. New York: CRC Press, 1998.

J. Paul Moulton

Geomicrobiology

Geomicrobiology refers to the activities of **microorganisms** (usually **bacteria**) that live beneath the surface of the **Earth**. The field of study is also referred to as biogeochemistry and subsurface microbiology. Habitats of the organisms include the **ocean** and deep within the rock that makes up Earth's crust. The study of the identities and activities of such organisms is important from a basic science standpoint and for commercial reasons.

Microorganisms are a vital part of the cycling of **carbon**, **nitrogen**, and **sulfur** between the surface of Earth and the surrounding atmosphere. These cycles in turn support the diversity of life that exists on the **planet**. As well, microorganisms break down other compounds that are present in **water**, **soil**, and the **bedrock**.

Many of the bacteria involved in geomicrobiological activities live in environments that are extremely harsh to other life forms. For example, bacteria such as *Thermus aquaticus* thrives in boiling hot springs, where the **temperature** approaches the **boiling point** of water. Such bacteria have been dubbed "extremophiles" because of their extraordinary resilience and **adaptation** to environmental pressures of temperature, **pressure**, acidity, **salt concentration**, or **radiation**. Other extremophiles live deep in the ocean under enormous **atmospheric pressure**. The bacteria that live around hot vents at the ocean floor, for example, use the **minerals** expelled by the vent in a way that supports the development of all the other life that can exist in the vicinity of the vent. Another type of bacteria lives within rock located miles under the surface of the Earth. Indeed, bacteria have been recovered from almost two miles beneath the Earth's surface, an environment that is hostile to all other forms of life. It is presumed that the ancestors of these bacteria entered the rock through nearby oil deposits or by percolating into the rock through microscopic cracks.

These and other bacteria have adapted to live in the absence of **oxygen** and **light**. They use materials from the surrounding surface as their fuel for survival and growth. These bacteria are very different from those traditional bacteria, such as *Escherichia coli*, that use carbon as a basis for growth.

The origin of geomicrobiology dates back to the 1920s. Then, Edson Bastin, a geologist at the University of Chicago, studied the source of **hydrogen** sulfide in water from oil fields that were located far underground. Bastin found that a type of bacteria subsequently named sulfate-reducing bacteria were responsible for the production of hydrogen sulfide. Critics were skeptical, arguing that the nature of the drilling for oil had introduced the microbes into the subsurface environment. Ultimately, however, the reality of Bastin's observations were confirmed.

Geomicrobiology took on additional significance in the 1970s and 1980s, as the fragility of **groundwater** to **contamination** was realized. The activity of microbes within the surface on the Earth, particularly the use of toxic substances as food by the microbes, is important for the health of groundwater. In the United States, for example, about 40% of the nation's drinking water comes from underground. The increasing use of land for human activity is degrading this resource and has spurred research geared towards understanding the microbiology of the soil and the underlying rock.

The study of geomicrobiological processes has required the development of techniques that are not in the repertoire of conventional laboratory microbiologists. Thus, geomicrobiology has brought together microbiologists, geologists, hydrologists, geochemists, and environmental engineers to study subsurface microbiology in a multi-disciplinary fashion.

Aside from fostering a collaborative approach to science, geomicrobiology has had, and continues to have practical value both commercially and socially. For example, *Thermus aquaticus* contains an **enzyme** that forms the basis of the polymerase chain reaction (**PCR**). The use of PCR to increase the amount of genetic material so as to permit analysis or manipulation revolutionized the field of **biotechnology**. Other heat-tolerant bacterial enzymes are being exploited for use in detergents, to provide cleaning power in hot water. A third example is the use of bacteria resident in the environment to clean up spills such as oil and polychlorinated biphenols in water, soil, and other environmental niches.

Since the 1970s, the participation of bacteria in the degradation of radioactive substances has been discovered. One such microbe, a bacterium of the *Thermus sp.*, can utilize **uranium**, **iron**, chromium, and cobalt. These elements can be found in contaminated soils. Research is underway to try to harness the bacteria to detoxify soil and **radioactive waste**.

The field of geomicrobiology can yield information on the development of life on Earth. This is because many of the extremophilic bacteria that live in the Earth's surface or in the oceans are ancient forms of life. Such microbes have lived at and within the Earth's surface for about 85 percent of the planet's age. Organisms such as cyanobacteria were vital in shaping the planet's atmosphere. By understanding the structure and functioning of these microbes, more light is shed on the characteristics of the ancient Earth that spawned the organisms, and on the development of other life.

Geomicrobiology will continue to increase in importance as the number and diversity of microbes on the planet becomes known more clearly. As of 2002, microbes that live in environments that include the oceans and the subsurface are estimated to make up over half of all living **matter** on Earth. Yet, less than one percent of all the predicted microbes have been identified. Most of what is known about bacteria comes from studies that require microbial growth. Yet, it is estimated that more than 99% of all microbes cannot be grown in the lab. So, the current number of known microbes represents the limit to what is attainable using culturing techniques.

Discovering and **learning** about such microbes is difficult, since, even if a bacterium is capable of being grown, the growth conditions are not always easy to reproduce in the laboratory. For example, many laboratory detection methods rely on the rapid growth (i.e., hours to a few days) of the bacteria. Bacterial growth in the Earth's oceans and subsurface, however, can occur extremely slowly, as the metals required for growth (such as iron, manganese, zinc, nickel, and **copper**) are slowly leached from the surrounding rock by the bacteria. In the subsurface world, growth is measured in years, not days.

A promising avenue of discovery is the detection and deciphering of genetic sequences, since the growth of the microbes is not required. Indeed, in 2001, **species** of **archaebacteria** that are thought to play a major role in the cycling of **nutrients** in the ocean were discovered based on their genetic sequences. Such studies are very daunting, since the deciphering of the bacterial sequence information from all the other genetic information in the environment requires massive computer power. Improved methods of bioinformatics will make genetic studies more feasible.

Geomicrobiology is not only important for life on Earth, but may be important in identifying life on other planets. For example, the presence of **calcium carbonate** crystals in meteorites is mimicked by certain bacteria that live in the hot springs of Yellowstone National Park in the United States. Study of the bacteria could provide clues as to how microbial life might arise in hostile environments elsewhere in the **solar system**.

See also Carbon cycle; Earth's interior; Microorganisms.

Resources

Books

Fredrickson, J.K., and M. Fletcher. *Subsurface Microbiology and Biogeochemistry*. New York: Wiley, 2001.

Periodicals

Bekins, B.A., E.M. Godsy, and E. Warren. "Distribution of Microbial Physiologic Types in an Aquifer Contaminated by Crude Oil." *Microbial Ecology* 37 (1999): 263–275.

Brian Hoyle

Geophysics

Geophysics is the study of Earth's physical character, including the solid **planet**, the atmosphere, and bodies of **water**. Geophysical investigations, therefore, often draw upon information and techniques developed in scientific disciplines such as **physics**, **geology**, and **astronomy**. Major areas of modern geophysical research include seismology, volcanology and geothermal studies, **tectonics**, geomagnetism, geodesy, **hydrology**, **oceanography**, atmospheric sciences, planetary science, and mineral physics.

Geophysics has many practical applications. Some seismologists, for example, help to explore for new **petroleum** reservoirs, monitor nuclear weapon testing by other countries, and better understand the structure and **stratigraphy** of important aquifers. Others provide the information necessary to design earthquake-resistant buildings and determine the risk posed by future earthquakes. Physical oceanographers monitor changes in **ocean temperature** that give rise to **El Niño and La Niña** phenomena, resulting in better long-term **weather** forecasts, and atmospheric physicists study the conditions that can give rise to **lightning** strikes. Hydrologists study the flow of surface water and **groundwater**, including the conditions that are likely to produce destructive floods.

Aristotle (384–322 B.C.) performed some of the first known geophysical investigations and published his findings in a work entitled *Meteorologica*. That work addressed such modern topics as weather, earthquakes, the oceans, **tides**, the stars, and meteors. By the first century B.C., Chinese investigators had developed a simple device for recording earthquakes and their points of origin. However, little additional progress was made in the field of geophysics until the fifteenth century A.D., when Leonardo da Vinci (1452–1519) took up the study of gravitational attraction and wave propagation.

The 300 years following da Vinci's death were marked by steady advances in the understanding of geophysical phenomena such as **magnetism**, gravity, and earthquakes. Most of these investigations were concerned only with what could be observed with the senses. But starting in the nineteenth century, scientists began to develop much more sophisticated techniques for geophysical observation.

Seismology is a branch of geophysics that draws on the physics of wave propagation to study earthquakes and determine the physical characteristics of **Earth's interior**. Seismic wave **velocity** is proportional to rock **density**. Therefore, seismologists can infer the composition and structure of Earth's interior by calculating the velocity of seismic waves from distant earthquakes. Seismic tomography uses computer analysis of seismic wave velocities to visualize structures within **Earth** and produce images that are much the same as medical CAT scans of the human body.

Seismologists can also generate seismic waves by using explosions or vibrating devices. Specialized seismometers known as geophones record the arrival of artificially created seismic waves, some of which are reflected back to Earth's surface when they reach boundaries between different rock types. The reflected waves can help to identify areas likely to contain undiscovered petroleum reserves and locate faults that may generate future earthquakes. Like **earthquake** seismologists, exploration seismologists can use three-dimensional data collection and computer processing techniques to produce detailed images of **rocks** that are otherwise inaccessible to humans.

Several types of **energy** produced by Earth's interior vary from place to place. This variation, called a potential field, is a characteristic of magnetism, gravity, temperature, and **electrical conductivity**. Geophysicists measure potential fields to learn about the distribution, composition, and physical state of rocks beneath Earth's surface. Because the gravitational field varies with changes in Earth's density from place to place, for example, geophysicists can use gravity measurements to **map** variations in rock composition and locate faults. Studies of variations in **Earth's magnetic field** through time are fundamental to our understanding of **plate tectonics**.

Experimental geophysicists attempt to reproduce the **heat** and **pressure** present in Earth's interior to determine how rocks and **minerals** behave under extreme conditions. **Diamond** anvils, for example, can generate pressures in the laboratory that equal or exceed any in Earth's interior.

See also Earthquake; Earth's magnetic field; Earth's rotation; Global warming; Plate tectonics.

Geothermal energy *see* **Alternative energy sources**

Geotropism

Plants can sense the Earth's gravitational field. Geotropism is the term applied to the consequent orientation response of growing **plant** parts. Roots are positively geotropic, that is, they will bend and grow downwards, towards the center of the **Earth**. In contrast, shoots are negatively geotropic, that is, they will bend and grow upwards, or away, from the surface.

These geotropisms can be demonstrated easily with seedlings grown entirely in darkness. A seedling with its radicle (or seedling root) and shoot already in the expected orientation can be turned upside down, or placed on its side, while kept in darkness. The root will subsequently bend and grow downwards, and the shoot upwards. Because the plant is still in darkness, **phototropism** (a growth movement in response to **light**) can be eliminated as an explanation for these movements.

Several theories about the manner by which plants perceive gravity have been advanced, but none of them is entirely satisfactory. To account for the positive geotropism of roots, some researchers have proposed that under the influence of gravity, starch grains within the cells of the root fall towards the "bottom" of the **cell**. There they provide signals to the cell **membrane**, which are translated into growth responses. However, there have been many objections to this idea. It is likely that starch grains are in constant **motion** in the cytoplasm of living root cells, and only "sink" during the process of fixation of cells for microscopic examination. Roots can still be positively geotropic and lack starch grains in the appropriate cells.

A more promising hypothesis concerns the transport of auxin, a class of plant-growth regulating **hormones**. Experiments since 1929 have shown that auxin accumulates on the "down" side of both shoots and roots placed in a horizontal position in darkness. This gradient of auxin was believed to promote bending on that side in shoots, and to do the opposite in roots. Confirmation of the auxin gradient hypothesis came in the 1970s. When **seeds** are germinated in darkness in the presence of morphactin (an antagonist of the hormonal action of auxin), the resulting seedlings are disoriented—both the root and shoot grow in **random** directions. Auxin gradients are known to affect the expansion of plant cell walls, so these observations all support the idea that the transport of auxin mediates the bending effect that is an essential part of the directional response of growing plants to gravity.

See also Gravity and gravitation.

Gerbils

Gerbils are rat-like **rodents** in the mammalian family Muridae, which also includes **rats**, **mice**, **voles** and **lemmings**. Some authorities place the gerbils in a separate family Gerbilidae, together with the pigmy gerbils. Wild gerbils are rat-sized, long-tailed rodents with rather long hind feet. Nearly all live in self-dug burrows and forage at night feeding mostly on **seeds**.

Gerbils are probably derived from hamster-like rodents, and fossil gerbils have been consistently found since the Upper Miocene. Gerbils are found in **desert** and semi-desert areas of **Africa**, Mongolia, southern Siberia, northern China, Sinkiang, and Manchuria. There are at least 70 **species** of true gerbils. Members of the genus *Gerbillus* have yellow to **light** grayish brown long and delicate hair, which is snow-white on the belly. They appear delicate and ghost-like, with big dark eyes. *Gerbillus campestris* is common in northern Africa, and is known for the absence of hair on the soles of the hind feet. *Gerbillus gerbillus* is a small gerbil found mostly in sandy desert in Egypt, and has no hair on its soles. The Namib gerbil of the genus *Gerbillurus* found in the Kalahari region is known to survive without drinking **water**. The naked-soled gerbils of the genus *Tatera*, look very much like rats in shape and size and populate somewhat wetter habitats than do other species of gerbils. The naked-soled gerbils are found from Syria to India and Sri Lanka in Steppes and semi-deserts. *Tatera robusta* lives in Africa, from the Sahel region through eastern Africa to Tanzania.

In the United States gerbils are popular pets and are valuable as laboratory animals for scientific research. This gerbil is also known as the jird, and its taxonomic

A gerbil in the Gobi Desert, central Asia. *JLM Visuals. Reproduced by permission.*

name is *Meriones unguiculatus*. This species is somewhat different from true gerbils of the genus *Gerbillus*. *Meriones unguiculatus* was brought to the United States in the early 1950s from Mongolia for laboratory research and is often referred to as the Mongolian gerbil. They are highly adaptable and there is danger that they may become established in the wild if released.

The natural **habitat** of the Mongolian gerbil is desert or semi-desert from the Sahara to the Gobi Desert. Jirds are usually sand-colored and have tails about as long as the body. The molar teeth have a chewing surface that resembles that of burrowing mice with high crowns and small roots. The ears of the Mongolian gerbil are relatively short and their hind feet make them appear sturdier than other gerbils. They are primarily active during the night and their diet includes leaves, seeds, and **insects**. They live in small colonies and, when upset, they may drum with their hind feet like rabbits. Like other gerbils, jirds have a sebaceous gland in the center of the belly. They smear the secretion on various objects to mark their territories, and recognize each other by scent.

There is no evidence of **hibernation** or estivation, and the Mongolian gerbil may be active throughout the year, either by night or day. This species adapts to a range of temperatures from sub-zero to above 86°F (30°C). It may remain underground for long periods depending on the amount of stored food. Daily summer movements of the Mongolian gerbil may cover 0.75-1.1 mi (1.2-1.8 km). One marked **animal** moved as far as 31 mi (50 km). Its social **behavior** under laboratory conditions indicates that adults may live together, but the introduction of a stranger may result in a fight to the death. Females are as territorial and aggressive as the males. Some studies suggest that males disrupt maternal behavior and many young are lost. But monogamous pairs seem to do well and some males share in caring for the young, cleaning, grooming and warming the newborns. Fathers and juvenile males help to rear the younger animals.

Wild Mongolian gerbils breed from February to October and up to three litters may be produced each year. Captive gerbils may breed all year round. The estrous cycle lasts four to six days and may occur right after the **birth**. Gestation lasts 19-21 days although longer periods of 24-30 days were reported. There are usually 4-7 young born, but litter sizes vary from 1-12. Newborn gerbils weigh about 0.09 oz (2.5 g). They open their eyes after 16-20 days, and are weaned between 20 and 30 days. Gerbils reach sexual maturity in 65-85 days and females may reproduce for 20 months, although in the wild they may not survive for more than three or four months.

Mongolian gerbils kept as pets should be provided with a clean, comfortable, escape-proof cage. They must be protected from **cats** and from rough handling. With gentle and loving care they become quite tame and respond to the keeper. It is usually best to separate pregnant females so that birth occurs without the **interference** of other adults, especially the males, although males occasionally care for newborns. Picking up and handling newborns should be avoided because the mother may become excited and kill them. A good healthy diet must include some fresh greens and sufficient protein from a good standard gerbil diet.

See also Hamsters.

Resources

Books

Barrie, Anmarie. *The Proper Care of Gerbils*. Neptune City, NJ: TFH Pubs, 1992.
Nowak, Ronald M. *Walker's Mammals of the World*. 5th ed. Baltimore: Johns Hopkins University Press, 1991.

Sophie Jakowska

Germ cells and the germ cell line

Germ cells are one of two fundamental **cell** types in the human body. Germ cells are responsible for the production of sex cells or gametes (in humans, ovum and spermatozoa). Germ cells also constitute a cell line through which genes are passed from generation to generation.

The vast majority of cells in the body are somatic cells. Indeed, the term somatic cell encompasses all of the differentiated cell types, (e.g., vascular, muscular, cardiac, etc.) In addition, somatic cells may also contain undifferentiated **stem cells** (cells that, with regard to differentiation are still multipotential). Regardless, while the mechanism of genetic replication and **cell division** is via **mitosis** in somatic cells, in germ cells a series of meiotic divisions during **gametogenesis** produces male and female gametes (i.e., ovum and spermatozoa that upon fusion (**fertilization**) are capable of creating a new **organism** (i.e., a single celled zygote).

While somatic cell divisions via mitosis maintain a diploid chromosomal content in the daughter cells produced, germ cells—in contrast—through a series of miotic divisions produce haploid gametes (i.e., cells with one-half the normal **chromosome** compliment s—one autosomal chromosome from each homologous pair and a sex chromosome (X in females, X or Y).

Although all humans start out as single cell zygotes, the germ cells for each **individual** are set-aside early in embryogenesis (development). If the cells comprising the germ cell line are subject to **mutation** or other impairments, those mutations may be passed down to offspring. It is from the germ cell line that all spermatogonia and all oogonia are derived.

Although controversial because of ethical considerations, both germ cells and stem cell research focus on the pluripotent potential of these cells (i.e., their ability to differentiate into cells found in various tissues of the body). Stem cells are derived from the inner cell **mass** of human blastocysts, embryonic germ cells can be obtained from the primordial germ cells located in the gonadal folds, ridge, and surrounding mesenchymal cells of fetal **tissue** during the middle of the first trimester of development (e.g., four to nine weeks).

In 2000 and 2001, research using extracted embryonic germ cells grown in culture over twenty generations showed that the cells had the ability to differentiate in all three fundamental embryonic tissue types (ectoderm, mesoderm, and endoderm).

See also Cell division; Embryo and embryonic development; Embryo transfer; Embryology; Meiosis; Mitosis.

Germ theory

The germ theory is a fundamental tenet of medicine that states that **microorganisms**, which are too small to be seen without the aid of a **microscope**, can invade the body and cause certain diseases.

Until the acceptance of the germ theory, many people believed that **disease** was punishment for a person's evil **behavior**. When entire populations fell ill, the disease was often blamed on swamp vapors or foul odors from sewage. Even many educated individuals, such as the prominent seventeenth century English physician William Harvey, believed that epidemics were caused by *miasmas*, poisonous vapors created by planetary movements affecting the **Earth**, or by disturbances within the Earth itself.

The development of the germ theory was made possible by the certain laboratory tools and techniques that permitted the study of **bacteria** during the seventeenth and eighteenth centuries.

The invention of primitive microscopes by the English scientist Robert Hooke and the Dutch merchant and amateur scientist Anton van Leeuwenhoek in the seventeenth century, gave scientists the means to observe microorganisms. During this period a debate raged among biologists regarding the concept of *spontaneous generation*.

Until the second part of the nineteenth century, many educated people believed that some lower life forms could arise spontaneously from nonliving **matter**, for example, **flies** from manure and maggots from decaying corpses. In 1668, however, the Italian physician Francisco Redi demonstrated that decaying meat in a container covered with a fine net did not produce maggots. Redi asserted this was proof that merely keeping egg-laying flies from the meat by covering it with a net while permitting the passage of air into the containers was enough to prevent the appearance of maggots. However, the belief in **spontaneous generation** remained widespread even in the scientific community.

In the 1700s, more evidence that microorganisms can cause certain diseases was passed over by physicians, who did not make the connection between vaccination and microorganisms. During the early part of the eighteenth century, Lady Montague, wife of the British ambassador to that country, noticed that the women of Constantinople routinely practiced a form of **smallpox** prevention that included "treating" healthy people with pus from individuals suffering from smallpox. Lady Montague noticed that the Turkish women removed pus from the lesions of smallpox victims and inserted a tiny bit of it into the **veins** of recipients.

While the practice generally caused a mild form of the illness, many of these same people remained healthy while others succumbed to smallpox epidemics. The reasons for the success of this preventive treatment, called *variolation*, were not understood at the **time**, and depended on the coincidental use of a less virulent smallpox **virus** and the fact that the virus was introduced through the skin, rather than through its usually route of entry—the respiratory tract.

Lady Montague introduced the practice of variolation to England, where physician Edward Jenner later modified and improved the technique of variolation. Jenner noticed that milkmaids who contracted cowpox on their hands from touching the lesions on the udders of cows with the disease rarely got smallpox. He showed that inoculating people with cowpox can prevent smallpox. The success of this technique, which demonstrated that the identical substance need not be used to stimulate the body's protective mechanisms, still did not convince many educated people of the existence of disease-causing microorganisms.

Thus, the debate continued well into the 1800s. In 1848, Ignaz P. Semmelweis, a Hungarian physician working in German hospitals, discovered that a sometimes fatal **infection** commonly found in maternity hospitals in **Europe** could be prevented by simple hygiene. Semmelweis demonstrated that medical students doing autopsies on the bodies of women who died from puerperal fever often spread that disease to maternity patients they subsequently examined. He ordered these students to wash their hands in chlorinated lime **water** before examining pregnant women. Although the **rate** of puerperal fever in his hospital plummeted dramatically, many other physicians continued to criticize this practice as being useless.

In 1854, modern **epidemiology** was born when the English physician John Snow determined that the source of **cholera epidemic** in London was the contaminated water of the Broad Street pump. After he ordered the pump closed, the epidemic ebbed. Nevertheless, many physicians refused to believe that invisible organisms could spread disease.

The argument took an important turn in 1857, however, when the French chemist Louis Pasteur discovered "diseases" of wine and beer. French brewers asked Pasteur to determine why wine and beer sometimes spoiled. Pasteur showed that, while yeasts produce **alcohol** from the sugar in the brew, bacteria can change the alcohol to vinegar. His suggestion that brewers **heat** their product enough to kill bacteria but not **yeast**, was a boon to the **brewing** industry—a process called *pasteurization*. In addition, the connection Pasteur made between food spoilage and microorganisms was a key step in demonstrating the link between microorganisms and disease. Indeed, Pasteur observed that, "There are similarities between the diseases of animals or man and the diseases of beer and wine." The notion of spontaneous generation received another blow in 1858, when the German scientist Rudolf Virchow introduced the concept of *biogenesis*. This concept holds that living cells can arise only from preexisting living cells. This was followed in 1861 by Pasteur's demonstration that microorganisms present in the air can contaminate solutions that seemed sterile. For example, boiled nutrient media left uncovered will become contaminated with microorganisms, thus disproving the notion that air itself can create microbes.

In his classic experiments, Pasteur first filled short-necked flasks with beef broth and boiled them. He left some opened to the air to cool and sealed others. The sealed flasks remained free of microorganisms, while the open flasks were contaminated within a few days. Pasteur next placed broth in flasks that had open-ended, long necks. After bending the necks of the flasks into S-shaped curves that bent downward, then swept sharply upward, he boiled the contents. Even months after cooling, the uncapped flasks remained uncontaminated. Pasteur explained that the S-shaped **curve** allowed air to pass into the flask; however, the curved neck trapped airborne microorganisms before they could contaminate the broth.

Pasteur's work followed earlier demonstrations by both himself and Agostino Bassi, an amateur micro-scopist, that silkworm diseases can be caused by microorganisms. While these observations in the 1830s linked the activity of microorganisms to disease, it was not until 1876 the German physician Robert Koch proved that bacteria can cause diseases. Koch showed that the bacterium *Bacillus anthracis* was the cause of **anthrax** in cattle and **sheep**, and he discovered the **organism** that causes **tuberculosis**.

Koch's systematic methodology in proving the cause of anthrax was generalized into a specific set of guidelines for determining the cause of infectious diseases, now known as Koch's postulates. Thus, the following steps are generally used to obtain proof that a particular organism causes a particular disease:

1. The organism must be present in every case of the disease.

2. The organism must be isolated from a host with the corresponding disease and grown in pure culture.

3. Samples of the organism removed from the pure culture must cause the corresponding disease when inoculated into a healthy, susceptible laboratory **animal**.

4. The organism must be isolated from the inoculated animal and identified as being identical to the original organisms isolated from the initial, diseased host.

By showing how specific organisms can be identified as the cause of specific diseases, Koch helped to destroy the notion of spontaneous generation, and laid the foundation for modern medical microbiology.

Koch's postulates introduced what has been called the "Golden Era" of medical bacteriology. Between 1879 and 1889, German microbiologists isolated the organisms that cause cholera, **typhoid fever**, **diphtheria**, **pneumonia**, **tetanus**, **meningitis**, gonorrhea, as well the staphylococcus and streptococcus organisms.

Even as Koch's work was influencing the development of the germ theory, the influence of the English physician Joseph Lister was being felt in operating rooms. Building on the work of both Semmelweis and Pasteur, Lister began soaking surgical dressings in carbolic acid (phenol) to prevent postoperative infection. Other surgeons adopted this practice, which was one of the earliest attempts to control infectious microorganisms.

Thus, following the invention of microscopes, early scientists struggled to show that microbes can cause disease in humans, and that public health measures, such as closing down sources of **contamination** and giving healthy people vaccines, can prevent the spread of disease. This led to reduction of disease transmission in hospitals and the community, and the development of techniques to identify the organisms that for many years were considered to exist only in the imaginations of those researchers and physicians struggling to establish the germ theory.

See also Vaccine.

Marc Kusinitz

Germanium *see* **Element, chemical**

Germination

Germination is the process by which a seed begins its development into a mature **plant**. Germination begins with an increase of metabolic activity within the seed. The first visible sign of germination in angiosperms (flowering plants) is generally an enlargement of the seed, due to intake of **water** from the environment. The seed's covering may wrinkle and crack at this time. Soon afterward, the embryonic root (called the radicle) emerges from the seed and begins to grow down into the **soil**. At about this time the shoot (plumule) also emerges, and grows upward out of the soil.

In most **species**, the food reserves that provide fuel for the seed's development are contained in the fleshy part of the seed. In some **seeds**, this fleshy part is divided into two seed leaves, or cotyledons. Seeds having two seed leaves are said to be dicotyledonous; those having only one are monocotyledonous. In some plants, the growth of the shoot carries the cotyledons above the soil into the sunlight, where they become more leaf-like in appearance while continuing to provide sustenance for the growing plant. Germination that follows this pattern is called epigeal germination. In other species the cotyledons remain underground; this is known as hypogeal germination.

Germination requires the presence of suitable environmental conditions, including sufficient water, **oxygen**, and an appropriate **temperature**. However, in many species the onset of germination is preceded by a period of metabolic inactivity, known as dormancy. While dormant, seeds will not germinate even under favorable conditions, but eventually they break their dormancy and begin to develop. The processes in the seed by which dormancy is broken are known as after-ripening. Dormancy serves to give seeds a better chance of surviving unfavorable conditions and developing successfully into plants. For example, seeds produced and dispersed just before the beginning of a cold season might not survive if they germinated at once. Dormancy enables them to wait out the cold season, and to begin growth when conditions are more favorable for the mature plant, in the springtime. Typical dormancy periods of seeds vary widely from species to species (and even within the seeds of a given species), as do the mechanisms by which dormancy is broken.

Gerontology

Gerontology is a branch of sociology that studies aging among populations internationally, and monitors efforts to deal with problems arising in old age. It differs from geriatrics the same way that **psychology** is separate from **psychiatry**. A psychologist's inquiries apply to general questions about how the human **brain** and mind work. A psychiatrist is more concerned with involving patients in a particular course of therapy. Geriatrics is a specialty within medicine concerned with treating illnesses which occur most often in the aged. Gerontology, on the other hand, considers geriatrics as part of a larger **spectrum** of issues which face older people, their immediate families and society at large.

In America the past few decades has seen a shift in the **median** age of the total population. On the average there have been more and more older people than younger ones in the country as time goes by. Not all of these elderly are in the same economic bracket, and not all will remain healthy until their deaths. Gerontologists have been researching the impact which might be felt in a community, and the cost which may be incurred by the federal government, if many of them need institutional care as they grow older. The most recent literature has been produced on developing trends, like those involving **AIDS** in older patients.

Even Aristotle observed the differing life spans in the **animal** kingdom. Since the days of the Ancient Greeks, speculation about aging has gone hand in hand with the development of medicine as a science. During the 1800s, certain researchers like Lambert Quetelet of Belgium and S. P. Botkin in Russia began to study populations and social patterns of aging in a systematic fashion. During the 1930s, the International Association of Gerontology was organized. The National Institutes of Health (NIH) joined with other governmental bodies to sponsor conferences on aging during the following decade, and by 1945 the Gerontological Society of America, Inc., was established in Washington, D.C.

Aside from history, politics and economics, more personal experiences are also investigated by a gerontologist. The impact of death on a widow or widower, the interrelationships of different generations within a family, and the circulation of myths about aging are also subject to qualitative research. Qualitative studies rely less on **statistics** than on interviews and records of emerging situations within a small group of test subjects. Coping strategies and other forms of therapy are assessed in terms of their suitability and success rates. Trends in medical research are also analyzed in terms of their impact on public opinion and their contributions towards our understanding of the aging process.

See also Aging and death.

Gesnerias

Members of the Gesneria family, the Gesneriaceae, are herbs, shrubs, sometimes trees or woody vines. The Gesneriaceae is a large family composed of approximately 120 genera and 1,800 **species**. With the exception of two genera (*Haberlea* and *Ramonda*), which are native to temperate **Europe**, they are found only in the tropical and subtropical regions of the world. Although none are native to the United States, more than a dozen genera are found in Mexico.

These plants have simple leaves that are opposite or grow in rosette form around the base. The flowers are showy and borne solitary or in **flower** clusters that bloom from the center outward. The flower petals have five lobes, and are often fused at the base to form a tubular corolla. The fruit is usually a capsule, but in some species it is fleshy and berrylike. Numerous tiny **seeds** are produced. Taxonomically, the family is closely allied to the Scrophulariaceae, the **Snapdragon family**, the Bignoniaceae, the Trumpet-vine family, and the Orobanchaceae, the Broom-rape family.

Gesneria, the genus for which the family is named, contains about 50 species native to the **forests** of tropical America, found on the mainland and islands of the West Indies. The genus was named after the Swiss naturalist Conrad Gesner (1516-1565) by Charles Plumier (1646-1704), a French missionary, botanist, and explorer, who published a book about the Caribbean Islands he visited in 1703. *Gesneria* species are characterized by their long red and green tubular flowers that co-evolved with their special pollinators, **hummingbirds** and **bats**. Humming-

***Kohleria,** a genus of gesneriad.* Photograph by James Sikkema. Reproduced by permission.

birds are well adapted to extracting **nectar** from flowers by their ability to hover, their visual acuity for red, their long bill and tongue. Since flower-feeding bats are not visually sensitive to **color**, bat pollinated *Gesneria* species are green. They produce abundant nectar with a fruity odor attractive to bats.

Members of the Gesneria family are important economically as ornamentals that are grown outside in warm tropical climates, and in greenhouses in cooler climates. The genera *Ramonda* and *Haberlea* are prized plants for rock gardens in temperate regions. Popular ornamentals include the African-violet (*Saintpaulia*), gloxinias (*Sinningia*), Cape primrose (*Streptocarpus*), and others. African violets are a popular houseplant prized for their attractive leaves and profuse blooms that range in color from white to pink, lavender, and dark purple. They also come in a variety of variegated colors and different flower types, from single to double with simple or ruffled petals. They are native to the tropical lowlands of East **Africa**. Gloxinias have much larger, red or purple bell-shaped flowers, and are native to Brazil.

Geyser

A geyser is an intermittent or semi-regularly periodic spout of geothermally heated **groundwater** and steam.

Any subsurface encounter between **water** and **heat** produces a hydrothermal process. The heat is usually supplied by upwellings of **magma** from the mantle, the water by **precipitation** that percolates downward through surface **rocks**. Some oceanic water enters the mantle at subduction zones and becomes an important ingredient in upper-mantle magmas.

Most hydrothermal processes are driven by **convection**. Convection occurs because water, like most substances, expands when heated. The result is that hot water rises and cool water sinks. Convection occurs when any water-permeated part of the earth's crust is heated from below: heated water fountains upward over the **hot spot** and cool water descends around its edges. These movements occur through cracks and channels in the rock, forcing the water to move slowly and remain in constant contact with various **minerals**. Water convecting through rock is thus an effective means of dissolving, transporting, and depositing minerals. Most deposits of concentrated minerals, including large, shapely crystals, are created by hydrothermal processes.

Manifestations of hydrothermal processes can be dramatic, including the geysers and hot springs that sometimes occur where shallow magma is present. However,

The geyser, Old Faithful, located in Yellowstone National Park (Wyoming). *Photograph by John Noble. CORBIS/John Noble. Reproduced by permission.*

most hydrothermal circulation occurs inconspicuously in the vicinity of large magmatic intrusions. These can cause water to convect through the rocks for miles around.

Some geysers erupt as perdictable intervals, others irregularly; a few send jets of water and steam hundreds of feet into the air, others only a few feet. There are fewer than 700 geysers in the world, all concentrated in a few dozen fields. More than 60% of the world's geysers are in Yellowstone National Park in the northwestern United States, including the famous geyser, "Old Faithful."

The word *geyser* comes from the name of a single Icelandic geyser, Geysir, written mention of which dates back to A.D. 1294.

Geysers form only under special conditions. First, a system of underground channels must exist in the form of a vertical neck or series of chambers. The exact arrangement cannot be observed directly, and probably varies from geyser to geyser. This system of channels must vent at the surface. Second, water deep in the system—tens or hundreds of meters underground—must be in contact with or close proximity to magma. Third, this water must come in contact with some rock rich in silica (silicon dioxide, SiO_2), usually rhyolite.

Silica dissolves in the hot water and is chemically altered in **solution**. As this water moves toward the surface, it deposits some of this chemically altered silica on the inner surfaces of the channels through which it flows, coating and sealing them with a form of opal termed *sinter*. Sinter sealing allows water and steam to be forced through the channels at high **pressure**; otherwise, the pressure would be dissipated through various cracks and side-channels.

The episodic nature of geyser flow also depends on the fact that the **boiling point** of water is a function of pressure. In a **vacuum** (**zero** pressure), liquid water boils at 0°C; under high pressure, water can remain liquid at many hundreds of degrees. Water heated above 100°C but kept liquid by high pressure is said to be *superheated*.

The sequence of events in an erupting geyser follows a repeating sequence. First, groundwater seeps into the geyser's reservoirs (largely emptied by the previous eruption), where it is heated—eventually, superheated—by nearby magma. Steam bubbles then form in the upper part of the system, where the boiling point is lower because the pressure is lower. The steam bubbles eject some water onto the surface and this takes weight off water deeper in the system, rapidly lowering its pressure and therefore its boiling point. Ultimately, the deeper water flashes to steam, forcing a mixed jet of water and steam through the geyser's surface vent.

Many of the world's geysers are endangered by drilling for geothermal **energy** in their vicinity. Drilling draws off water and heat, disrupting the unusual balance of underground conditions that makes a geyser possible.

See also Bedrock.

Gibbons and siamangs

Gibbons are **species** of tropical forest **apes** in the family Pongidae. This family contains all of the anthropoid apes, which are the closest living relatives of humans (*Homo sapiens*), in terms of their **anatomy**, **physiology**, and **behavior**. Like other anthropoid **primates**, gibbons lack a tail, they have a more-or-less upright posture, and they have a well-developed **brain**. However, gibbons are generally regarded as the least intelligent of the anthropoid apes.

A hoolock gibbon (*Hylobates hoolock*). *Photograph E. Hanumantha Rao/Photo Researchers, Inc. Reproduced by permission.*

The gibbons comprise a distinctive group within the Pongidae, making up the sub-family Hylobatinae. The true gibbons are five species in the genus *Hylobates*, while the siamangs are two larger species of *Symphalangus*. (However, some taxonomists also classify the siamangs in the genus *Hylobates*, thereby treating them as large gibbons.) All of the gibbons occur in tropical **forests** of Southeast **Asia**, Malaysia, and Indonesia.

The gibbons are the smallest of the apes, being less than one meter tall, and weighing about 11-17.5 lbs (5-8 kg). Siamangs weigh 17.5-28.5 lbs (8-13 kg). Gibbons have a willowy body shape, and have much longer and more slender arms than the other anthropoid apes. In fact, the arms of gibbons are long enough to easily touch the ground as these animals walk. The hands and fingers are also elongate and slender, with a distinctively deep cleft between the thumb and the index finger. All of these are adaptations for the active life of these animals, which is mostly spent in the forest canopy.

Gibbons are highly arboreal animals—the name of their genus, *Hylobates*, is derived from Greek words for "dweller in the trees." Gibbons are extremely agile, and are sometimes referred to as the most acrobatic of all the **mammals**. They can move swiftly and gracefully through the tree-tops by swinging hand over hand from

A siamang yelling. © R. Van Nosstrand, National Audubon Society Collection/ Photo Researchers, Inc. Reproduced with permission.

branch to branch, a method of locomotion known as brachiation. Gibbons can also get about by leaping through the air until they manage to hook onto a secure branch, or land on a stable limb. These flying leaps, sometimes assisted using elastic branches, can cover a distance as great as 40 ft (12 m).

Gibbons do not often venture to the ground, but when they do, they are awkward walkers, typically holding their arms high to maintain their balance as they ambulate. Gibbons do not swim, and are in great danger of drowning if they ever fall into deep **water**.

Gibbons are typically colored in various hues of brown or black, with white body markings in some species. There is a great deal of variation in coloration of individuals within species, and even within the same family of gibbons. The young of most gibbons have white fur, and do not attain the darker, adult coloration until they are 2-5 years old, depending on the species.

Gibbons are highly vocal animals. Groups of gibbons (known as a "troop") often make very loud hootings, barks, and hollers in the early morning, known as their noisy "dawn chorus." These animals are also very vociferous during the day.

Gibbons become sexually mature at an age of 5-8 years. The gestation period is about seven months, and one baby is born. The young are weaned after about seven months, and until that time they are constantly carried by their mother. Gibbons have lived as long as 23 years in captivity.

The usual family group is a monogamous pair of an adult male and female, plus three or four pre-reproductive offspring. Sometimes, however, a mature male will live in a polygynous relationship with several mature females, with the young being raised communally. The family groups defend a territory, which can range in size from about 25 to more than 50 acres (10-20 hectares) in area.

Gibbons forage during much of the day. They typically **sleep** while sitting erect in dense vegetation at night. Gibbons are mostly herbivorous, eating a wide range of **fruits** and leaves. However, they also sometimes feed on bird eggs, nestlings, and other small animals. These agile creatures are known to capture flying **birds**, while leaping through the air, or while brachiating rapidly. Gibbons drink by dipping a hand into water, and then sucking the moisture from the fur.

The usual **habitat** of gibbons is tropical forest. This can range from lowland forest around **sea level**, to montane forest at an altitude as great as 7,900 ft (2,400 m). The siamang occurs as high as almost 9,850 ft (3,000 m) on the **island** of Sumatra, in Indonesia.

The most important natural predators of gibbons and siamangs are large **cats** such as the clouded leopard, large **snakes** such as **pythons**, and **eagles**.

Within their range, gibbons are sometimes kept as pets in villages. Some species of gibbons, especially the siamangs, are endangered by extensive losses of their natural habitat of tropical forests.

Species of gibbons and siamangs

The hoolock gibbon (*Hylobates hoolock*) occurs in Southeast Asia. Male hoolock gibbons are black, while the females are variable in **color**, ranging through black, grey, and brown, with a white band across the forehead.

The white-handed or lar gibbon (*Hylobates lar*), occurs in parts of mainland Southeast Asia, Malaya and Sumatra. The fur of this species has a basal color of black, brown, or yellow, but the upper surfaces of the hands and feet are white-colored, and the face is circled by white.

KEY TERMS

. . . .

Anthropoid apes—These consist of the gibbons, orangutan, chimpanzee, and gorilla, all of which lack a tail, have an upright posture, and a well-developed brain. Anthropoid apes are the closest living relatives of humans.

Brachiation—A method of arboreal locomotion involving hand-over-hand travelling, while holding onto branches. This is a characteristic locomotion of gibbons and some types of monkeys.

Monogamous—A breeding system in which a mature male and a mature female live as a faithful, mated pair.

Montane forests—Forests that occur relatively high on mountains, but below the open grasslands and tundra. Montane forests at low latitudes, for example in Southeast Asia, have a relatively cool and temperate climatic regime.

Polygynous—A breeding system in which a single male breeds with more than one mature female.

The dark-handed gibbon (*Hylobates agilis*) occurs on the Malayan Peninsula and Sumatra. The fur of this species varies from yellowish to dark-brown, and the upper surfaces of the hands and feet are always dark-colored.

The grey gibbon (*Hylobates moloch*) occurs in Java and Borneo. The fur of this species is light or dark grey, and the face is black.

The black gibbon (*Hylobates concolor*) occurs in Southeast Asia, particularly Vietnam, Myanmar (Laos), and Thailand. This is a dark-colored **animal**, with a distinctive, erect crest of long hair on the crown of the head, especially elongate in adult males. The black gibbon has a throat pouch, used to amplify its territorial noises, similar to the siamangs.

The siamang (*Symphalangus syndactlyus*) occurs in parts of Malaya and the island of Sumatra in Indonesia, while the dwarf siamang (*S. klossi*) is native to the Mentawei Islands off the west coast of Sumatra. Siamangs are heavier than gibbons, typically weighing 17-29 lbs (8-13 kg), with a body length as great as 36 in (90 cm), and an arm-spread of up to 5 ft (1.5 m). Siamangs have black fur, and a distinctive throat-pouch, which appears to amplify the booming and bellowing territorial noises of these animals. Siamangs are somewhat less agile than the true gibbons. Siamangs occur in montane and sub-montane forests between about 2,000-6,400 ft (600-2,800 m) in elevation. These animals defend their

foraging range, and live in social groups consisting of an adult male and an adult female, plus any babies and submature offspring that may be associated with the parents. These family groups defend a territory of about 25 acres (10 hectares) or more in area.

Resources

Books

Else, J.G., and P.C. Lee, eds. *Primate Ecology and Conservation.* Cambridge: Cambridge University Press, 1987.

Fleagle, J.G. *Primate Adaptation and Evolution.* San Diego: Academic Press, 1988.

Nowak, R.M., ed. *Walker's Mammals of the World.* 5th ed. Baltimore: Johns Hopkins University Press, 1991.

Smuts, B.B., D.L. Cheney, R.M. Seyfarth, R.W. Wrangham, and T.T. Struhsaker. *Primate Societies.* Chicago: University of Chicago Press, 1987.

Stephens, M.E., and J.D. Paterson, eds. *The Order Primates: An Introduction.* New York, 1991.

Wilson, D.E., and D. Reeder. *Mammal Species of the World.* 2nd ed. Washington, DC: Smithsonian Institution Press, 1993.

Wolfheim, J. H. *Primates of the World. Distribution, Abundance, and Conservation.* Newark, NJ: Gordon and Breach Science Publications, 1983.

Bill Freedman

Gila monster

The Gila monster (*Heloderma suspectum*) is a large, strikingly-colored venomous lizard. The gila monster and the Mexican beaded lizard (*Heloderma horridum*) are the only members of the beaded lizard family, Helodermatidae. The Gila monster occurs in rocky, semi-arid habitats from the Colorado River **basin** in the southwestern United States to the western regions of Mexico and Guatemala.

The Gila monster reaches a length of 18 in (46 cm), and is typically black or dark brown, with bright yellow markings. The body is heavy and cylindrical, ending with a thick, rounded tail, where **energy** reserves are stored against lean times. The head is relatively large, massive, and flattened, with numerous, slightly recurved teeth.

The Gila monster is a terrestrial lizard, and is most active at dawn and dusk. Since the body **temperature** of Gila monsters depends on environmental temperature, these lizards are less active during the winter in northern parts of their range. Gila monsters can live for 20 years in captivity.

The Gila monster has venom **glands** on the front part of the lower jaws, and (together with the Mexican beaded lizard) are the world's only venomous lizards.

A banded gila monster. *Photograph by Renee Lynn. The National Audubon Society Collection/Photo Researchers, Inc. Reproduced by permission.*

The venom glands synthesize and store a potent toxin that can cause paralysis of the cardiac and respiratory systems of vertebrate animals. The Gila monster bites and chews a wound in its victim, into which the venom flows by **capillary action** along deep grooves on the lizard's teeth. The Gila monster is a tenacious biter, using its venomous bite to immobilize **prey**, and to defend itself against predators. The bite of a Gila monster is painful, but rarely fatal to a human.

The Gila monster eats a wide range of small animals, bird eggs and nestlings, earthworms, and carrion. Food is swallowed whole, except for eggs, which are broken before eating. Like most lizards, the Gila monster uses its forked tongue and an associated sensory **organ** (Jacobson's organ) on the roof of the mouth for chemosensation, an important aid to finding its food.

Gila monsters lay as many as 13 eggs in a clutch. The eggs are buried, and incubate for as long as 130 days until hatching.

Because it is a potentially dangerous and unusual looking **animal**, the Gila monster is often kept as a pet. Populations of gila monsters in some areas have been de-pleted because some people fear and kill these animals, and because they are hunted for their skins and the commercial pet trade.

See also Reptiles.

Bill Freedman

Ginger

The ginger family, Zingiberaceae, includes about 50 genera and 1,300 **species** of plants, a few of which have culinary or medicinal uses. The common ginger (*Zingiber officinale*) is one of the oldest and most commonly used spices. Ginger for these uses is obtained from the tuberous **rhizome**, or underground stem of the **plant**. The common ginger is native to Southeast **Asia**, where it has been cultivated for thousands of years. It is now grown commercially throughout the tropics, and was the first oriental spice to be grown in the Americas. Jamaican ginger is considered to be the finest in the world.

Cultivated ginger (*Zingiber officinale*) in the Amazon, Peru. *JLM Visuals. Reproduced with permission.*

Zingiber officinale is a perennial, creeping plant, with reed-like, leaf-bearing stems up to 4 ft (1.3 m) high that emerge from the stout rhizome. The leaves are yellowish green, alternately arranged on the stem, lanceolate in shape (i.e., long and tapered), and 0.5-1 in (1-2 cm) wide and up to 1 ft (30 cm) long. The cone-shaped flowers are about 3 in (7 cm) long and colored yellow and purple. They occur on stalks that grow directly from the rhizome, and are about as tall as the leafy stems.

The rhizome grows relatively quickly, sprouting new above-ground shoots as it spreads. Because the rhizomes grow roots, the ginger plant can be easily propagated by taking pieces of the rhizome and planting them in the soil. Ginger grows best in rich, sandy, partially shaded places with high humidity, warm temperature, and abundant rain. Rhizomes grown specifically for drying and grinding into a powdery spice are harvested after about nine months of growth. Ginger that is to be used fresh can be harvested as soon as about one month after planting.

Ginger is used as a flavoring in tea, wine, liqueur, soft drinks, and candies. Ginger ale, once an alcoholic beverage, is now a popular, carbonated soft drink. Ginger can be purchased fresh, dried and ground, dried whole, candied, or preserved in syrup. Medicinally, ginger has been used to relieve nausea. During the Middle Ages, ginger was used as an antidote for the plague, although it did not actually work for that purpose.

Ginkgo

The ginkgo, or maidenhair tree (*Ginkgo biloba*) is an unusual species of gymnosperm, having broad leaves, and seasonally deciduous foliage that turns yellow and is dropped in autumn. The ginkgo is a dioecious plant, which means that male and female functions are performed by separate trees. The ginkgo is famous as a so-called "living fossil," because it is the only surviving member of the family Ginkgoaceae and the class Gingkoales. This is a group of gymnosperms with a fossil lineage extending back to the lower Jurassic, some 190 million years ago, and once probably having a global distribution.

In more modern times, the natural distribution of the ginkgo was apparently restricted to a small area of southeastern China. It is very likely, however, that there are no longer any truly natural, wild populations of ginkgo in forests in that region. It appears that the only reason this remarkable species still survives is because it was preserved and cultivated in small groves around a few Buddhist temples. This was apparently done because far-sighted monks recognized the ginkgo as a special, unique species of tree, and because they valued its edible, possibly medicinally useful fruits and leaves. Gingko is still being used in this way today, as a "herbal" or "folk" medicine thought to be useful in the treatment of memory loss, asthma, circulatory disorders, headaches, impotence, and a variety of other ailments.

Today, the ginkgo is no longer a rare species, and it has a virtually worldwide distribution in temperate climates. This is because the ginkgo has become commonly grown in cities and gardens as a graceful and interesting shade tree. The ginkgo has attractive, golden-yellow foliage in autumn, is easy to transplant and cultivate, and is remarkably resistant to diseases, insects, and many of the stresses of urban environments including, to some degree, air pollution. The ginkgo is also often cultivated because of its special interest to botanists and others as a living fossil.

Often, an attempt is made by horticulturists to only plant male ginkgos, because the outer flesh of the fruits

A ginkgo tree in Georgia. *JLM Visuals. Reproduced with permission.*

of female trees can have an uncomfortably foul odor, making some people nauseous, and causing skin rashes upon contact. Ginkgo trees can reach an height of about 115 ft (35 m), and can achieve a trunk diameter of more than 27 in (68.5 cm). Trees mature at about 20 years, and can live to be older than 1,000 years.

Ginseng

Ginseng refers to several **species** of plants in the genus *Panax*, family Araliaceae. Ginseng is a perennial, herbaceous **plant**, with compound leaves that grow from a starchy root. The natural **habitat** of ginseng is the understory of mature **angiosperm** forest in the temperate zones of east **Asia** and eastern **North America**.

The root of ginseng is highly valued as having many therapeutic properties by practitioners of traditional Chinese medicine, who regard it as a tonic, stimulant, aphrodisiac, and cure for some diseases. Oriental ginseng (*Panax ginseng*) is the original ginseng upon which this medicinal usage was based. Because of the insatiable demand for its roots, this Asian species has been overharvested from its natural habitat of hardwood forest in eastern Asia, and is now endangered in the wild. Although Oriental ginseng is now cultivated as a medicinal crop, it is widely believed that wild plants are of much better medicinal quality than cultivated ginseng. Consequently, virtually any wild ginseng plants that are found are harvested, because they are so valuable.

Soon after the colonization by the French of parts of eastern North America in the sixteenth century, it was realized that there was a large and profitable market in China for the roots of American ginseng (*Panax quinquifolium*), which grew abundantly in the temperate angiosperm **forests** of that region. These wild plants were initially collected in southern Quebec, and then anywhere else that ginseng could be found. For a while, ginseng root was one of the most important commodities being exported from North America. Inevitably, however, the once abundant natural resource of wild ginseng was quickly exhausted, and today these plants are extremely rare in the wild in North America. American ginseng is now considered an **endangered species** in the wild. Another, much smaller species known as dwarf ginseng (*P. trifolium*) was not over-collected, and is more common.

An agricultural system has been developed for the cultivation of ginseng, and it is now grown as a valuable cash crop in various places in North America. The plants are started from seed, which are collected from mature plants and stored in moist **sand** for one year, so that they will scarify and be capable of germinating. It can take as long as five to seven years for cultivated ginseng plants to reach their prime maturity for harvesting. However, the plants are sometimes harvested when smaller, and less valuable, because of the risk that a longer period of growth might allow a fungal **infection** to develop. Such an infection can ruin an entire crop, and devastate the result of years of patient work and investment. Agricultural ginseng is grown under a shading, plastic or wood-lattice canopy, because this species is a plant of the forest understory and does not tolerate full sunlight.

Once harvested, the largest, best-quality ginseng roots are dried, and are mostly exported to China, Korea, and Japan to be sold in traditional-medicine stores. Customers purchase their carefully selected roots, and then watch as the ginseng is prepared. Poorer-quality, thinner, cracked roots may b processed into ginseng tea and other bulk preparations.

Resources

Books

Moramarco, J. *The Complete Ginseng Handbook: A Practical Guide for Energy, Health, and Longevity.* NTC/Contemporary Publishing, 1998.

Bill Freedman

Giraffes and okapi

Giraffes are a **species** of large, long-legged, long-necked **ungulates** in the family Giraffidae, order Artiodactyla. Giraffes are the tallest living animals on **Earth**. Okapis are a close relative, but these animals do not have such long legs or neck.

The giraffe is a widespread **animal** of **grasslands** and savannas of sub-Saharan **Africa**. The okapi is a much rarer animal and occurs in tropical forest.

Both species are exclusively herbivorous, mostly browsing the foliage of woody plants. These animals are ruminants, meaning they have a complex stomach divided into four pouches. Each of these sections is responsible for a particular stage of the digestion process. **Rumination** actually specifically refers to the chewing of the cud, which is a regurgitated **mass** of pre-digested **plant matter** from one of the fore-stomachs. The cud is re-chewed in a leisurely fashion, and then swallowed one last time, to undergo further digestion. The material then passes through the alimentary system, and **nutrients** are absorbed during this final passage, which is followed by defecation.

Giraffes

The most distinctive characteristics of giraffes are their very long legs, and their enormously long neck. It is interesting that, compared with related families such as the **deer** (Cervidae), giraffes have the same number of neck vertebrae—the remarkable elongation of their neck is due solely to lengthening of the individual vertebrae. A short, dark mane runs along the top of the length of the neck.

The fore legs are slightly longer than the hind legs, but the profile of giraffes is also influenced by the extreme development of musculature on their shoulders and base of the neck. These large muscles are used to keep the heavy neck erect, and they give the animal a rather

An adult male okapi grazing. *Photograph by Tom Brakerfield. Stock Market. Reproduced by permission.*

Herd of giraffes standing in a field. *Photograph by Michael C.T. Smith. The National Audubon Society Collection/Photo Researchers, Inc. Reproduced by permission.*

hunched appearance, with a steeply sloping back. Giraffes have a rather long tail, which ends in a dark tassel.

Giraffes can run quite quickly, using a rather stiff, ambling gait because of their long legs. To drink, giraffes must stoop awkwardly to reach the **water**.

The largest male giraffes can attain a height of 19 ft (6 m). Females are somewhat shorter, by about 3 ft (1 m). Large male giraffes can weigh as much as 1,650 lb (750 kg).

The pelage of giraffes is highly variable, and several geographic races have been named on the basis of their colors and especially their patterns. The basic **color** is brownish, with a network of white lines breaking up the solid profile. Formerly, two different species of giraffes were recognized on the basis of distinctive differences in the patterns of their pelage and their non-overlapping ranges. These were the relatively widespread giraffe (*Gi-raffa camelopardalis*) and the reticulated giraffe (*G. reticulata*) of east Africa. However, further study has demonstrated that these animals are fully interfertile, and their differences are not sufficiently great to warrant their designation as full species. Today, taxonomists recognize only one species of giraffe, *Giraffa camelopardalis*.

The head of giraffes is relatively small, at least in comparison with the large body size of these animals. The head has a rather elongated profile, with a long, thin upper lip, which is prehensile and used along with the long, black, mobile tongue to dextrously grasp and tear foliage while the animal is feeding. Giraffes have large eyes, with very long eyelashes. Their ears are short, but quite mobile, and both **hearing** and **vision** are acute.

The horns of giraffes are two to five, permanent, knobby outgrowths on the forehead or top of the head, covered by skin. Both sexes have these horns. The horns are smaller than, but anatomically comparable to, the antlers of deer, except those of the giraffe are never shed and are always covered by skin.

Giraffes are social animals, but not highly so, as they do not occur in large herds. The largest herds can include as many as 20-50 animals, with several dominant male animals (or bulls) and many females (or cows) and offspring. Bull giraffes fight among themselves, using powerful swings of their knobby-topped heads, aiming at the neck or chest of their rival. Old bulls that are unable to maintain a harem live a life solitary from other giraffes. A single baby (or calf) giraffe is born after a gestation period of 14-15 months.

Often, giraffes will associate with other large herbivores such as **zebras**, gnus, and ostriches in mixed foraging groups. Giraffes commonly have ox-peckers (*Buphaga* spp.) riding on their backs. These useful **birds** feed on large insect and tick **parasites** that can be common on the hides of giraffes and other large **mammals**.

Adult giraffes are not an easy mark for their natural predators, unless they can be ambushed while in an awkward stance, such as when they are drinking. Giraffes can run quickly for a long distance, and they can inflict sharp wounds with the hooves of their front legs. The most important predators of giraffes are lions, but a pack of these large **cats** is required to kill an isolated giraffe. Young giraffes are more vulnerable, but they are generally well protected by their social group, which is very alert for the presence of nearby predators.

Giraffes are still relatively abundant in some parts of their range. However, they have become widely extirpated from large areas, equivalent to more than one-half of their original range. This substantial decline in the overall population and range of giraffes is mostly associated with conversions of their natural habitats into agriculture, as well as over-hunting of these animals.

The okapi

The okapi, or forest giraffe (*Okapia johnstoni*), did not become known to European scientists until 1900, when a native pygmy hunter showed a striped-legged skin of this species to a British zoologist in what was then the Congo in central Africa. The discovery of this unusual large animal caused a quite a sensation among European naturalists and the public. As a result, many museums and zoological gardens mounted expeditions to secure living or dead specimens of this novel, but rare animal. Wealthy big-game hunters also organized expeditions to acquire trophy heads of the "newly discovered" okapi.

By today's standards, it seems rather barbaric for scientists and hunters to have mounted those sorts of campaigns, which could only have further endangered an already rare species. However, attitudes and morality were different in late-Victorian times, when the notions of **conservation** and **ecology** were only beginning to make faint impressions on scientists, and on the broader public.

The okapi has a much shorter neck and legs than the giraffe, and the two horns of the males are pointed and uncovered by skin at their tips (female okapis do not have horns). The okapi has a fairly uniform-chestnut pelage, but distinctive, horizontal stripes on its legs. The largest okapis stand about 79 in (2 m) tall, and weigh 551 lb (250 kg).

From the first discovery of the okapi, great efforts were made to capture live specimens and transport them to European or American zoos for display and study. For many years, these efforts were quite unsuccessful. Although methods were developed for the safe capture of wild okapis (using pits dug across the paths these animals habitually use), it proved extremely difficult to transport them to the far away zoos.

Today, because of more efficient hunting by local people (some of whom have modern weapons), coupled with extensive loss of their **rainforest habitat**, the okapi is an even more rare animal than it was when Europeans first discovered it. Okapis will breed in zoos, although successes in this regard are sporadic. The survival of this unusual animal will certainly require the preservation of a large area of its natural habitat of old-growth, tropical rainforest in central Africa.

Resources

Books

Dagg, A. and J. B. Foster. *The Giraffe: Its Biology, Behavior, and Ecology.* Melbourne, FL: Krieger Pub., 1982.

Nowak, R.M., ed. *Walker's Mammals of the World.* 5th ed. Baltimore: John Hopkins University Press, 1991.

Wilson, D. E. and D. Reeder. *Mammal Species of the World.* 2nd ed. Washington, DC: Smithsonian Institution Press, 1993.

Bill Freedman

KEY TERMS

Browse—A food consisting of the foliage, twigs, and flowers of woody plants.

Harem—A group of females associated with one or several males.

Ruminant—A cud-chewing animal with a four-chambered stomach and even-toed hooves.

GIS

GIS is the common abbreviation for geographic information systems, a powerful and widely used computer database and software program that allows scientists to link geographically referenced information related to any number of variables to a **map** of a geographical area. GIS allows its users to analyze and display data using digitized maps. In addition, GIS can generate maps and tables useful to a wide-range of applications involving planning and decision-making. GIS programs allow the rapid storage, manipulation, and correlation of geographically referenced data (i.e., data tied to a particular point or **latitude and longitude** intersection on a map).

In addition to scientific studies, by 2003, GIS programs were in wide use in a number of emergency support agencies and systems (e.g., the Federal Emergency Management Agency [FEMA]). Broad in scope, GIS is has also attained a significant role in business and marketing decisions.

GIS programs allow scientists to layer information so that different combinations of data plots can be assigned to the same defined area. GIS also allows users to manipulate data plots to predict changes or to interpret the **evolution** of historical data.

GIS maps are able to convey the same information as conventional maps, including the locations of **rivers**, roads, topographical features, and geopolitical information (e.g., location of cites, political boundaries, etc.). In addition, to conventional map features, GIS offers geologists, geographers, and other scholars the opportunity to selectively overlay data tied to geographic position. By overlaying different sets of data, scientists can look for points or patterns of correspondence. For example, rainfall data can be layered over another data layer describing terrain features. Over these layers, another layer data representing **soil contamination** data might be used to identify sources of **pollution**. In many cases, the identification of data correspondence spurs additional study for potential causal relationships.

GIS software data plots (e.g., sets of data describing roads, elevations, stream beds, etc.) are arranged in layers that be selectively turned on or turned off.

In addition to scientific studies, GIS technology is increasingly used in resource management. When tied in with GPS data, GIS provides very accurate mapping. GIS provides, for example, powerful data correlation between pollution patterns monitored at specific points and **wildlife** population changes monitored by GPS tracking tags.

NASA engineers and teams of other scientists—including researchers and undergraduates from Stephen F. Austin University in Nacogdoches, Texas—employed GIS mapping to map remain found in after the break up of the **space shuttle** *Columbia* in January 2003. Debris field maps helped narrow search patterns and, by linking the location of debris, allow engineers and investigators to reconstruct critical elements of the disaster sequence. GPS data was used to construct the debris maps and to provide accurate representations of the retrogressive pattern of debris impacts.

See also Archeological mapping; Cartography; Geologic map.

Glaciers

Glaciers are flowing masses of **ice**, created by years of snowfall and cold local temperatures. Approximately one tenth of the **Earth** is covered by glaciers. Glaciers are most numerous near the poles, covering most of **Antarctica** and Greenland and parts of Iceland, Canada, Russia, and Alaska; they also exist in mountainous regions on every **continent** except **Australia**. From the air, a glacier looks deceptively smooth and pliant; in reality, it is an abrasive **mass** that can reshape the Earth. The glaciers themselves are being reshaped by human activity. Recent measurements show that glaciers have been melting worldwide since the beginning of the **Industrial Revolution** in the mid-nineteenth century (when human beings first began to add large amounts of greenhouse gasses to the atmosphere). **Water** from melting glaciers is a significant input to rising sea levels worldwide, which threaten coastal ecosystems and the approximately 100 million people who live 3.28 ft (or about 1 m) or less above **sea level**.

How glaciers form

Glaciers are created in areas where the air **temperature** never gets warm enough to completely melt snow. Snowflakes may partially melt when they come into contact with the ground; as the air temperature drops further, the partially melted snow refreezes, turning into ice. The resulting mixture of snow and ice is compacted as additional layers of snow accumulate on top. Eighty **percent** of fresh snow is air; as the weight of fresh upper layers of snow and ice increases, air is pressed out of the lower layers, producing ice that contains less than 20– air. As the years pass, snow accumulates and the slab of ice grows steadily thicker (if the glacier is in a growth phase). Eventually, the layer of ice becomes so massive that it begins to flow slowly downhill. When an ice mass begins to flow under the influence of gravity, it is considered a glacier.

Muir Inlet, a fiord in Glacier Bay National Park, Alaska, and the glacier creating it. *JLM Visuals. Reproduced by permission.*

Types of glaciers

Ice masses take on a variety of characteristics as they flow and retreat. Glaciers that pour down a valley from mountainous ground, for example, usually follow paths originally formed by **rivers** of snowmelt in the spring and summer. These glaciers, termed alpine or mountain glaciers, end either in valleys or in the **ocean** and tend to increase the sharpness and steepness of the **mountains** surrounding them by eroding them. They are thus, partially responsible for carving the high-relief mountain peaks of the Himalayas, Andes, and alpine regions of the Cascades and Northern Rocky Mountains.

Piedmont glaciers are large, gently sloping ice mounds. Piedmont glaciers are common in Alaska, Greenland, Iceland, and Antarctica.

Glaciers often form in small bowl-like valleys called cirques on the sides of mountains. Found in Norway, Iceland, Greenland, and Antarctica, glaciers within cirques usually do not move out of their basinlike areas.

The largest form of glacier is called an ice sheet or continental glacier, a huge ocean of ice that spreads slowly outward from its center. Ice sheets may cover millions of square miles and are so heavy that they cause the continental crust beneath to float lower on the Earth's mantle, like a heavy-laden barge. The largest ice sheet is found on Antarctica, where the ice is more than 2.5 mi (4 km) thick at its center, hiding entire mountain ranges (mapped using seismic waves and **radar**). The Antarctic ice sheet covers more than 5 million sq mi (12.9 million sq km), which exceeds the combined areas of the United States, Mexico, and Central America. It contains about 90% of all the world's ice and 70% of its fresh water. The Greenland ice sheet is 670,000 sq mi (1,735,000 sq km) in area, covering virtually the entire **island**. Smaller ice sheets are found in Iceland, northern Canada, and Alaska.

Glaciers' effects

While the Greenland and Antarctic ice sheets are enormous, they are only a fraction the size of the kilometers-deep ice sheets that have covered large portions of the Earth during extensive periods of glaciation, such as during an Ice Age. Geologists assume that glaciers have expanded to mammoth proportions at least six times over the past 960 million years, sweeping slowly down from the polar regions every 250 million years or so and persisting, usually, for 5–10 million years.

Hubbard Glacier calving. © *Mark Newman/Phototake NYC. Reproduced by permission.*

Most glaciers that exist today are remnants of the last glacial period, which lasted from 1.8 million to 11,000 years ago and which occurred in four periods of advance and retreat. At their maximum, the glaciers of this period covered 30– of the Earth's land surface, particularly in the Northern Hemisphere. As the glaciers advanced, they lowered sea levels by hundreds of feet, creating land **bridges** between continents. This is the most likely explanation for how humans reached **North America** from Asia—that is, glaciers probably crossed over via land that was exposed between **Asia** and Alaska.

As a glacier advances it grinds up the land beneath it, scooping up **rocks** and **soil**. These add to the glacier's weight and abrasive power; V-shaped valleys can be altered to U-shaped valleys, and mountains can go from peaky to rounded. As they melt, this burden of rock, gravel, and dirt is dropped in place. This material is termed glacial till. Glacial till, which accumulates preferentially along the leading edges of the advancing glacier, is deposited in huge mounds along glacier's edge when it ceases to advance and begins to melt, creating new hills, or moraines. Formerly placated areas are covered by 200 –1,200 ft (61–366 m) of till that was carried and dropped by glaciers. Chunks of ice buried in this till create large

depressions that later become lakes called "kettle lakes." Glaciers also scour the land to great depths, creating larger lakes such as the Great Lakes of North America.

During the last ice age, much of the Earth's surface was depressed due to the weight of the glaciers. As the glaciers retreated, the crust began to rise. This crustal rebounding, as it is called, is still occurring at in parts of North America and **Europe**.

Glaciers advance relatively slowly, moving anywhere from a few centimeters per year to a few meters per day. When ice melts under the glacier as a result of **pressure** from above and **friction** with the ground, accumulated meltwater may act like a lubricant to increase the glacier's **rate** of flow; this sudden increase in speed is termed a surge

Clues to the Earth's past and future

While the effects of glaciers—scouring, till deposits, and rebound—can tell us where they have been in the past. Scientists continue to debate the reasons why **ice ages** occur, but the consensus view is that several factors interact to produce them: (1) placement by **continental drift** of large land masses near the poles, on which glaci-

KEY TERMS

Alpine or mountain glaciers—Glaciers that form at high elevations in mountain regions and flow downhill through valleys originally created by rivers.

Cirques—Small basinlike depressions in the sides of a mountain that provide sites for circular glaciers to form.

Glacial till—Rocks, soil and other sediments transported by a glacier then deposited along its line of farthest advance.

Ice age—An extended period of time in the Earth's history when average annual temperatures were significantly lower than at other times, and polar ice sheets extended to lower latitudes.

Ice sheet—The largest form of glacier and the slowest moving, covering large expanses of a continent.

Iceberg—A large piece of floating ice that has broken off a glacier, ice sheet, or ice shelf.

Kettle lakes—Bowl-shaped lakes created by large boulders or ice blocks, which formed depressions in the Earth's surface.

Meltwater—Melted ice in the glacier's bottom layer, caused by heat that develops as a result of friction with the Earth's surface.

Moraines—Large deposits of glacial till that form hills.

Piedmont glacier—Large, gently sloping glaciers found at the feet of mountains and fed by alpine glaciers.

Surging—A sudden increase in a glacier's movement as a result of meltwater beneath the glacier that decreases friction.

ers can form; (2) **uplift** of continental plates by plate-tectonic forces, with subsequent changes in global circulations of air and water; (3) reductions in the amount of **carbon dioxide** in the atmosphere, with diminished **greenhouse effect**; and (4) long-term **oscillations** in the shape of the Earth's **orbit** and the tilt of the its poles.

Present-day glaciers are providing clues to recent and future changes in climate. **Satellite** radar and aircraft-mounted **laser** altimetry systems have recently been used to measure contemporary glaciers with great **accuracy**; the data show that many glaciers are retreating, reflecting an overall **global warming** trend. The glaciers in the Alps in Europe have lost an estimated one-third to one-half of their ice in the last century, while Alaskan glaciers losing ice thickness at an average rate of about 6 ft (2 m) per year, retreating at rates of 2 mi (3.2 km) in 20 years. By glacial standards, this is a hasty retreat. The U.S. National Academy of Sciences has predicted that, if global temperatures rise from 1.5–5°F (0.75–2.5°C) over the next century as a result of the greenhouse effect, significant portions of the Earth's ice cover could melt. This would result in **flooding** of every continent's coastlines. Indeed, sea level is already rising. Global average sea level has been rising at about .12 in (3 mm) per year for the last decade, and this rate is expected to accelerate. Alaskan glaciers—which contain for about 13% of the world's glacier area but whose melting accounts for about half of observed sea-level rise—have been thinning twice on average as fast over the last five years as during the preceding 40. In Peru, glacial melting is occurring at exponentially increasing

speed; the present rate is 33 times the rate between 1963 and 1978. There is little doubt that **global climate** change caused by human agricultural and industrial activity is contributing strongly to these effects; data from Antarctic ice cores have shown a direct correlation between warming and cooling trends and the amount of the two major greenhouse gases, **carbon** dioxide and methane, in the atmosphere. These same cores show significant increases in both gases in the past 200 years. Today, thanks to human activity, atmospheric carbon dioxide is at its highest level in at least 420,000 years.

Glaciers may offer clues about the possibility of life on other planets. In Switzerland, **bacteria** have been found living under the ice sheets. If microbes can thrive in the dark, cold environment under glaciers, the vast ice sheets that blanket Jupiter's **moon** Europa and which underlie the soil of **Mars** may have their own microscopic residents.

See also Ice age refuges.

Resources

Books

Bender, Lionel. *Glacier: The Story of the Earth.* New York: Franklin Watts, 1988.
Walker, Sally M. *Glaciers: Ice on the Move.* Minneapolis: Carolrhoda Books, 1990.

Periodicals

Bradley, Ray. "1000 Years of Climate Change." *Science.* 5470 (May 26, 2000): 1353–1355.
Meier, Mark F., and Mark B. Dyurgerov. "How Alaska Affects the World." *Science.* 5580 (July 19, 2002): 350–351.

Other

Ball, Philip. "Alaskan Glaciers Raise Sea Level." *Nature Science Update.* July 19, 2002 [cited December 16, 2002]. <http://www.nature.com/nsu/020715/020715-12.html>.

Whitfield, John. "Tropical Glaciers in Retreat." *Nature Science Update.* February 19, 2001 [cited December 16, 2002]. <http://www.nature.com/nsu/010222/010222-14.html>.

Sally Cole-Misch

Glands

Glands are aggregates of specialized cells that secrete or excrete chemical substances which are used elsewhere in the body. Glands carry out regulatory, digestive, reproductive, and other functions in the body. A gland may be an independent structure or may be incorporated into another, larger, structure that has still other functions. In addition, a gland can be endocrine, secreting its **hormones** directly into the **blood** stream without a duct; or it can be exocrine, secreting its products through a duct into the digestive tract, onto the skin, or other target areas.

The two adrenal glands, one atop each kidney, are endocrine glands that secrete various hormones, including epinephrine (adrenalin) corticosteroids, and mineralocorticoids that are part of the body's response to stressful situations.

The islets of Langerhans of the pancreas are endocrine glands that secrete the hormones **insulin** and glucagon, which lower and raise the levels of blood glucose (sugar). The pancreas, too, is an exocrine gland, for it also secretes digestive enzymes (pancreatic juice) through ducts which lead into the duodenum of the small intestine. Other endocrine glands include the thyroid gland, the parathyroid glands, the testes and the ovaries, the thymus gland, and the pituitary gland.

Other **exocrine glands** include the lacrimal glands, which manufacture and secrete tears, the salivary glands, which secrete saliva, the liver, which manufactures and secretes bile, the mammary glands, which manufacture and secrete milk, and the eccrine sweat glands of the skin, which secrete sweat to regulate body **temperature**. The kidneys are glands in that the juxtaglomerular cells of the nephrons secrete renin, which helps to regulate blood **pressure**.

Glands increase or decrease their activities in response to changes in body temperature, salinity, temperature, and other stimuli, most of which are coordinated by control centers in the **brain**.

See also Adrenals; Endocrine system.

Glass

Glass is a brittle, inorganic solid, composed mostly of inorganic oxides. The main ingredient of most glasses is silicon dioxide, SiO_2, or silica—found in nature as **sand**. Generally manufactured by heating sand, soda, lime, and other ingredients (and quickly cooling the molten **mass**), glass is a fundamental component of a variety of products, including tableware, windshields, thermometers, and **telescope** lenses. Given its durability and versatility, glass plays an important role in human culture. Glass blowing was first developed around 30 B.C.

Early peoples were likely to have discovered natural glass, which is created when **lightning** strikes sand, and were certain to have used obsidian-a dark volcanic glass-for weapons, ornaments, and money. The first manufactured glass probably took the form either of glass beads or ceramic glaze and appeared around 4000-5000 B.C. Surviving examples of Egyptian and Mesopotamian glass objects date to around 1550 B.C.

For centuries, glass, shaped by the use of molds, remained costly and difficult to produce. The invention of the blowpipe method of glass making (in which molten glass is puffed into shape with the use of a hollow tube) in about 30 B.C. made glass more commonplace. Typical uses at the time included windows as well as decorative objects.

The first four centuries after the **birth** of Christ are sometimes referred to as the First Golden Age of glass making, for during this period artisans produced a wide variety of artifacts that are now highly valued. After the decline of the Roman Empire, few developments took place in European glass making until the twelfth and thirteenth centuries, when stained glass windows (formed of pieces of colored glass outlined by **lead** strips and assembled into a narrative picture) began to appear in English and French churches. During the Crusades, Europeans were exposed to the accomplished glass making of the Near East, an influence evidenced by the growth of the craft in Italy, particularly Venice. Beginning around 1300, the Venetians ushered in the Second Golden Age of glass making; they became widely known for a particularly transparent, crystalline glass that was worked into a number of delicate objects.

In the late 1400s and 1500s the Germans and other northern Europeans were producing containers and drinking vessels that differed markedly in their utilitarian value from those produced by the Venetians. Nonetheless, Venetian glass was immensely popular during the reign of Queen Elizabeth I (1558-1603). In 1674, George Ravenscroft (1618-1681) brought fame to English glass making when he invented lead glass (now usually called lead **crystal**), an especially brilliant glass he produced accidentally when he added lead oxide to his mixture instead of lime. In colonial America, the glass made by this technique became known as flint glass, and was usually etched or cut into facets to lend it additional luster.

The first glass **plant** built in the United States was founded at Jamestown, Virginia, in 1608, but it survived for less than a year. Much later, in 1739, Caspar Wistar successfully launched the American glass industry with a plant in Salem City, New Jersey. Other prominent figures in early American glass making included Henry William "Baron" Stiegel (1729-1785) and John F. Amelung. The renowned Sandwich glass that is now much coveted by American collectors was made by the Boston and Sandwich Glass Company; the Bakewell Company of Pittsburgh was another famous glass manufacturer of the time.

The early 1800s saw a tremendous demand for glass windows, which were a symbol of affluence, particularly in the frontier communities of America. Window glass was originally made by spinning out a bubble of blown glass until it became flat; because of the bump or "crown" that was invariably left in its center, this was called crown glass. Around 1825, the cylinder process replaced the earlier method. Now the glass was blown into a cylinder shape that, when cooled, was cut down one side; when reheated, the cylinder flattened out to form a sheet. In 1842, John J. Adams invented a more sophisticated glass-flattening and tempering process that made not only plate glass but **mirrors**, showcases, and other products more widely available. During the last half of the nineteenth century, glass found wide use in medicinal containers, tableware, and kerosene lamps. Tempered glass (made exceptionally strong through a reheating process) was invented by François Royer de la Bastie in 1874, and wire glass (industrial sheet glass with **metal** mesh laminated into it) by Leon Appert in 1893. In 1895, Michael J. Owens (1859-1923) invented a bottle-making machine that allowed bottled drinks to be produced inexpensively.

The great technological advances of the twentieth century broadened the range of ingredients, shapes, uses, and manufacturing processes for glass. **Natural gas** replaced the **wood** and **coal** that had previously been used in the glass making process, and huge operations were established. One of the most common forms of glass now produced is flat glass, used for windows, doors, and furniture. Formed by flattening melted glass between **rollers**, annealing (**heat** treating) in an oven called a lehr, then cutting into sheets and grinding and polishing until smooth, this category includes sheet glass and the higher quality plate glass. The best quality of all is achieved in float glass, invented in 1952 by Alistair Pilkington. Float glass is made by floating a ribbon of liquefied glass on top of molten tin so that it forms a perfectly

even layer; the result is glass with a brilliant finish that requires no grinding or polishing. In 1980, Pilkington invented kappa float glass, which features a special, energy-efficient glaze that traps thermal heat while allowing solar heat to filter through.

Other modern forms of glass include the laminated safety glass used for **automobile** windows, which is composed of sandwiched layers of plastic and glass; nonreflecting glass (invented by Katherine Burr Blodgett and others); structural glass, used in buildings; heat-resistant cookware such as Pyrex; and fiberglass.

Resources

Books

Doremus, R. H. *Glass Science.* New York: Wiley, 1990.
Zerwick, Chloë. *A Short History of Glass.* New York: Springer-Verlag, 1994.

Global climate

The long-term distribution of **heat** and **precipitation** on Earth's surface is called global climate. Heat from the **sun** keeps the Earth's average **temperature** at about 60°F(16°C), within a range that allows for biological life and maintains the planet's life-sustaining reservoirs of liquid **water**. Astronomical variations and atmospheric shielding cause incoming solar **radiation** to fall unevenly on the Earth's surface. **Ocean currents** and winds further redistribute heat and moisture around the globe, creating climate zones. Climate zones have characteristic annual precipitation, temperature, **wind**, and ocean current patterns that together determine local, short-term **weather**, and affect development of ecologically adapted suites of plants and animals. Changes in the astronomical, oceanographic, atmospheric, and geological factors that determine global climate can lead to global climate change over **time**. The term climate is reserved for regional patterns of temperature and precipitation that persist for decades and centuries. Local atmospheric, oceanic, and temperature phenomena like storms and droughts that occur over hours, days, or **seasons**, is generally referred to as weather.

Global climate patterns

The Earth's climate zones are classified according to their average temperature and rainfall accumulation, and, in general, form latitudinal, east-west oriented bands on the Earth's surface. Average temperatures increase with latitude and decrease with altitude; temperatures are highest near the equator and near **sea level**. This pattern of uneven heating drives **convection**, or heat-driven circulation, of the oceans and atmosphere. Warm, moisture-laden air at the equator rises and flows toward the poles, cooling, releasing precipitation, and sinking as it flows. The tropical zone, which extends about 15° north and south of the equator, is extremely warm and wet. The Earth's hot semi-arid and arid zones lie beneath dry, sinking air between about 15° and 30° North and South. This vertical convection cycle of rising warm, wet air and sinking cool, dry air is called a Hadley cell. The Earth's has six Hadley cells that are responsible for the Earth's alternating wet and dry climate bands. The temperate zones, between 30° and 60° North and South, lie beneath Hadley cells with rising limbs at 60° and a sinking limbs at 30°. The polar climates form beneath sinking, dry, very cold air at the north and south poles.

The **Earth's rotation**, the global distribution of ocean basins and continents, and the location of high mountain ranges add complexity to the pattern of latitudinal climate bands. The **Coriolis effect**, a phenomenon that deflects air and water currents to the right in the northern hemisphere, and to the left in southern hemisphere, is a consequence of the Earth's eastward spin. For example, surface air flowing south in the northern equatorial Hadley cell creates the southwesterly Trade Winds instead of a direct, southerly wind. (Winds are named for the direction from which they originate; a nor'easter, for instance, blows from the northeast toward the southwest.) Belts of alternating easterly and westerly winds drive corresponding west- and east-flowing ocean currents, and distribute heat and moisture east and west within the climate bands. Air flowing across a **continent** loses moisture the farther it travels from the ocean. Consequently, the windward side of a continent is often wetter than its leeward side, and the interior a large continent is dryer than its coasts. When flowing air reaches a mountain front, it rises, cools, and releases its moisture as precipitation. Large mountain ranges thus receive heavy rain and snowfall on their upwind flank, and arid deserts and semi-arid **grasslands** form in their leeside rainshadows.

The German climatologist, Wladimir Köppen, developed the most common classification nomenclature for climatic zones in the early 1900s. The Köppen system recognizes five general types of regional climate based on average temperature and precipitation: humid tropical, dry, humid mid-latitude with mild winters, humid mid-latitude with cold winters, and polar. The system further divides the general categories into sub-types. Dry regions, for example, can be arid deserts or semi-arid steppes, and polar regions contain frozen **tundra** as well as **ice** sheets. The Köppen system has been modified over the years to include finer sub-divisions, and a sixth category for alpine environments was added, but the system remains a valuable and widely used tool for general climatic mapping.

Ecosystems of specifically adapted plants and animals inhabit each climatic zone. The climatic zones delineated by the Köppen system generally correspond to characteristic networks of **species** that have evolved to survive the region's seasonal temperature changes, precipitation fluctuations, and weather events. **Desert** plants, for example, have waxy leaves and stems that reduce the amount of water lost by **transpiration**, and many desert animals are nocturnal, an **adaptation** that has allowed them to survive in some of the hottest regions on the **planet**. Biologically productive rainforests and corral reefs flourish in the warmth and **humidity** of tropical zones. Arctic plants and animals are adapted to take advantage of the short polar summer season by reproducing and storing **nutrients** quickly before the long, dark, cold polar winter.

Global climate change

A complex group of astronomical, atmospheric, geological, and oceanographic factors account for the Earth's global climate. Many of these factors vary naturally over decades, centuries, and millennia. Furthermore, astronomical and geological variations begin a cascade of compensatory adjustments in the coupled, or linked, ocean-atmosphere system, which, in turn, require major adjustments to biological systems. These variations **force** changes in the global pattern of long-term precipitation and temperature, or global climate change. Global climate change causes permanent redistribution of climatic zones, alteration of major weather patterns, and establishment of new ecosystems. Global climate change has occurred throughout the Earth's history, and has been a major driving force in biological **evolution**; species unable to adapt to new climate regimes have become extinct, while others have flourished. Scientists predict that human activities, notably **combustion** of carbon-based **fossil fuels** like oil and **coal**, will affect the climate-regulating properties of the atmosphere, which may cause anthropogenic (human-induced) global climate change.

Astronomical factors affecting global climate change

Energy from the sun drives the Earth's climate. Changes that affect the amount of solar radiation reaching the planet, called insolation, and that alter the distribution of sunlight on its surface, can cause global climate change. Each minute, the Earth's outer atmosphere receives about two calories of energy per square centimeter of area, a value known as the solar constant. In spite of its name, the solar constant varies over time. Astronomers have, for example, observed a correlation between the solar constant and changes in the pattern of **sunspots**, or solar storms, on the Sun's surface.

The Earth's position with respect to the Sun over time affects its climate. During its annual circuit around the sun, the Earth's present elliptical **orbit** brings it closest to the sun in January (perihelion), and carries it farthest away in July (aphelion). The planet receives about 6% more solar energy in January than in July. The Earth's axis, a line through the poles, is tilted 23.4° with respect to the sun. Consequently, the Sun's rays strike the northern hemisphere most directly on June 21st, the summer **solstice**, and the southern hemisphere most directly in December 21st, the winter solstice. The equinoxes, on April 21st and September 21st, mark the dates when the Sun shines directly on the equator, and day and night are the same length around the globe. Orbital **geometry** and axial tilt together determine the Earth's pattern of seasons. Variations in this astronomical geometry would cause climatic variations.

In the 1920's, the Serbian astronomer, Milutin Milankovitch, proposed an astronomical explanation for long-term, cyclical global climate changes that caused the Pleistocene "ice ages". By observing variations in the Earth's orbital geometry and axial tilt, and calculating the time for a complete cycle of change to occur, Milankovitch predicted a pattern of varying insolation and global climate change. According to his theory, three so-called Milankovitch cycles—precession, obliquity, and eccentricity—repeat approximately every 21, 41, and 100 thousand years, respectively. The 21,000-year precession cycle occurs because the direction of the Earth's spin axis changes over time, much in the way a spinning top wobbles. This phenomenon, called the **precession of the equinoxes**, causes a particular season, northern hemisphere summer for example, to occur at different places along the Earth's orbital path, and hence, at a different time of year. During the 41,000-year obliquity cycle, the tilt **angle** of the Earth's axis changes, altering the intensity of the seasons. Changes in the shape, or eccentricity, of the Earth's orbit cause the 100,000-year Milankovitch cycle. The Earth's present orbit is almost circular, so the difference in insolation between aphelion and perihelion is fairly minor. When the orbit becomes more elliptical, the **Earth** receives more radiation at the perihelion, and less at the aphelion. The eccentricity cycle also modulates the precession and obliquity cycles; the most intense northern hemisphere summer, for example, would occur when the June solstice coincided with the perihelion of an eccentric elliptical orbit, and the axial tilt was at its highest.

Geological data from the most recent portion of the Earth's history seem to support Milankovitch theory. The pattern of insolation variations that Milankovitch predicted generally matches the pattern of polar ice sheet advance and retreat since about two million years ago. Observations of northern hemisphere glacial features,

deep sea cores that record the amount of water stored in glacial ice, and sea-level records all corroborate the timing of global cooling and warming predicted by Milankovitch theory. The correlations are more difficult to prove farther back in geologic history.

Geological factors affecting global climate

Geological changes on the Earth's surface can also affect global climate. The distribution of continental landmasses and ocean basins affects the pattern of global atmospheric and oceanographic circulation, and the shape, or topography, of the Earth's surface directs winds and ocean currents. According to the widely accepted, and well-supported theory of **plate tectonics**, the continents move, ocean basins open and close, and mountain ranges form over time. The continents have assumed new configurations on the Earth's surface throughout geologic history, and geologists know, from examination of fossil environments and organisms, that the movement of landmasses had significant climatic effects. For example, during the Cretaceous Period, about 100 million years ago, continents covered the poles, and a warm ocean called Tethys circled the equator. An intense period of volcanic activity added insulating gasses to the atmosphere. The Cretaceous was the warmest and wettest period in Earth history. There is no evidence of Cretaceous **polar ice caps**, shallow seas covered many continental interiors, and tropical plants and animals lived on all the continents. The collision of the Indian subcontinent with **Asia**, and formation of the Himalayan mountain range about 40 million years ago is another example of a plate tectonic event that caused significant climate change. The Himalayas obstruct equatorial winds and ocean currents, and contribute to major climatic phenomena, namely the **monsoon** seasons of southern Asia and the Indian Ocean, and the El Niño Southern Oscillation in the Pacific Ocean.

Changes in atmospheric composition and anthropogenic global warming

The Earth's climate is strongly affected by the way solar radiation is reflected, absorbed, and transmitted by the atmosphere. Presently, about 30% of the incoming solar energy reflects back into **space**, the atmosphere absorbs about 20%, and the remaining 50% reaches the Earth's surface. The major gaseous components of Earth's atmosphere are **nitrogen**, **oxygen**, argon, and **carbon dioxide**. Other components include relatively small amounts of neon, helium, methane, krypton, **hydrogen**, xenon and **ozone** gases, water vapor, and particulate **matter**. Except for relatively uncommon natural events, such as volcanic eruptions, the composition of the atmosphere stays constant over long periods of time.

The structure and composition of the atmosphere function to maintain the Earth's surface temperature within the phase boundaries of liquid water, and to protect organisms from damaging ultraviolet radiation. Gases, like ozone, in the outer atmosphere reflect or absorb much of the incoming short-wavelength solar radiation. Much of the sunlight that reaches the Earth's surface is re-radiated into the atmosphere as longer-wavelength infrared energy, or heat. Gases in the middle and lower atmosphere, namely **carbon** dioxide and water vapor, absorb this infrared radiation, and the temperature of the atmosphere increases, a phenomenon known as the **greenhouse effect**. This heat, trapped in the atmosphere, drives atmospheric and oceanographic circulation, keeps the oceans liquid, and maintains global climate zones. The greenhouse effect makes the Earth livable for biological organisms, including humans.

In the last century, humans have burned large quantities of fossil fuels like coal, oil, and **natural gas** to operate factories, generate **electricity**, and run **automobile** engines. Because carbon dioxide is always produced during the combustion of a carbon-based fuel, these activities have significantly increased the **concentration** of that greenhouse gas in the atmosphere. Many scientists now believe that higher concentrations of carbon dioxide will enhance the greenhouse effect, and lead to **global warming**. If global warming should occur, a number of terrestrial changes could follow. Some simulations predict melting of the polar ice caps, increasing volume of water in the oceans, and inundation of coastal cities. Models also show changes in ocean currents and wind patterns and redistribution of the Earth's major climate zones. Such events would have severe consequences for human agriculture, fishing, and civil planning, as well as for the natural environment. The complexity of the interrelated systems that create global climate, however, makes predicting the climatic effect of increased atmospheric carbon dioxide extremely difficult. The issue of anthropogenic global climate change remains a subject of heated debate among scientists and policy makers.

See also Atmospheric circulation.

Resources

Books

Ahrens, C. Donald. *Meteorology Today.* 2nd ed. St. Paul, MN: West Publishing Company, 1985.

Eagleman, Joe R. *Meteorology: The Atmosphere in Action.* 2nd ed. Belmont, CA: Wadsworth Publishing Company, 1985.

Lin, Charles. *The Atmosphere and Climate Change.* Dubuque, IA: Kendall/Hunt Publishing Company, 1993.

Lutgens, Frederick K., and Edward J. Tarbuck. *The Atmosphere: An Introduction to Meteorology.* 4th ed. Englewood Cliffs, NJ: Prentice Hall, 1989.

KEY TERMS

Anthropogenic effect—Any effect on the environment caused by human activities.

Aphelion—The point in the Earth's orbit at which it is at its greatest distance from the sun.

Axis of inclination—The angle at which the Earth's axis is tipped in relation to the plane of the Earth's orbit around the sun.

Climate—The sum total of the weather conditions for a particular area over an extended period of time, at least a few decades.

Greenhouse effect—The warming of the Earth's atmosphere as a result of the capture of heat re-radiated from the Earth by certain gases present in the atmosphere.

Ice age—An extended period of time in the Earth's history when average annual temperatures were significantly lower than at other times, and polar ice sheets extended to lower latitudes.

Perihelion—The point in the Earth's orbit when it makes its closest approach to the sun.

Solar constant—The rate at which solar energy strikes the outermost layer of the Earth's atmosphere.

Newton, David E. *Global Warming.* Santa Barbara, CA: ABC-CLIO, 1993.

Open University Course Team. *Ocean Circulation.* Oxford: Pergamon Press, 1993.

Press, Frank, and Raymond Siever *Understanding Earth. Chapter 14: Winds and Deserts* New York: W.H. Feeman and Company, 2001.

Periodicals

Jones, P. D., "The Climate of the Past 1000 Years," *Endeavour.* (Fall, 1990): 129–136.

Other

United States Naval Observatory. "The Seasons and the Earth's Orbit-Milankovitch Cycles." Astronomical Applications Department. August 21, 2000 [cited March 14, 2002]. <http://aa.usno.navy.mil/faq/docs/seasons_orbit.html>.

Laurie Duncan

Global Positioning System

Long before the space age, people used the heavens for navigation. Besides relying on the **Sun, Moon,** and Stars, the early travelers invented the magnetic compass, the **sextant**, and the seagoing chronometer. Eventually, **radio** navigation in which a position could be determined by receiving radio signals broadcast from multiple transmitters, came into existence. Improved high **frequency** signals gave greater **accuracy** of position, but they were blocked by **mountains** and could not bend over the **horizon**. This limitation was overcome by moving the transmitters into space on Earth-orbiting satellites, where high frequency signals could accurately cover wide areas.

The principle of **satellite** navigation is relatively simple. When a transmitter moves toward an observer, **radio waves** have a higher frequency, just like a train's horn sounds higher as it approaches a listener. A transmitter's signal will have a lower frequency when it moves away from an observer. If measurements of the amount of shift in frequency of a satellite radiating a fixed frequency signal with an accurately known **orbit** are carefully made, the observer can determine a correct position on **Earth**.

The United States Navy developed such a system, called Transit, in the late 1960s and early 1970s. Transit helped submarines update their on-board inertial navigation systems. After nearly 10 years of perfecting the system, the Navy released it for civilian use. It is now used in surveying, fishing, private and commercial maritime activities, offshore oil exploration, and drifting buoys. However, a major drawback to Transit was that it was not accurate enough; a user had to wait until the satellite passed overhead, position fixes required some time to be determined, and an accurate fix was difficult to obtain on a moving platform.

As a result of these shortcomings, the United States military developed another system: Navstar (Navigation Satellite for Time and Ranging) Global Positioning System. This system consists of 24 operational satellites equally divided into six different orbital planes (each containing four satellites) spaced at 60° intervals. The new system can measure to within 33 ft, (10 m), whereas Transit was accurate only to 0.1 mi (0.16 km).

With the new Global Positioning System (GPS), two types of systems are available with different frequencies and levels of accuracy. The Standard Positioning System (SPS) is used primarily by civilians and commercial agencies. As of midnight, May 1, 2000, the SPS system became 30 times more accurate when President Bill Clinton ordered that the Selective Availability (SA) component of SPS be discontinued. SA was the deliberate decrease of accurate positioning information available for commercial or civilian use. The SPS obtains information from a frequency labeled GPS L1. The United States military has access to GPS L1 and a second frequency, L2.

The use of L1 and L2 permits the transfer of data with a higher level of security. In addition to heightened security, the United States military also has access to much more accurate positioning by using the Precise Positioning System (PPS). Use of the PPS is usually limited to the U.S. military and other domestic government agencies.

Both Transit and Navstar use instantaneous satellite position data to help users travelling from one place to another. But another satellite system uses positioning data to report where users have been. This system, called Argos, is a little more complicated: an object on the ground sends a signal to a satellite, which then retransmits the signal to the ground. Argos can locate the object to within 0.5 mi (0.8 km). It is used primarily for environmental studies. Ships and buoys can collect and send data on **weather**, **currents**, winds, and waves. Land-based stations can send weather information, as well as information about hydrologic, volcanic, and seismic activity. Argos can be used with balloons to study weather and the physical and chemical properties of the atmosphere. In addition, the system is being perfected to track animals.

Use of the GPS system in our everyday lives is becoming more frequent. Equipment providing and utilizing GPS is shrinking both in size and cost, while it increases in reliability. The number of people able to use the systems is also increasing. GPS devices are being installed in cars to provide directional, tracking, and emergency information. People who enjoy the outdoors can pack hand held navigational devices that show their position while exploring uncharted areas. Emergency personnel can respond more quickly to 911 calls thanks to tracking signal devices in their vehicles and in the cell phones of the person making the call. As technology continues to advance the accuracy of navigational satellite and without the impedance of Selective Availability, the uses for GPS will continue to develop.

Global warming

Global warming refers to a long-term increase in the Earth's surface **temperature** that results in large-scale changes in **global climate**, namely redistribution of climatic zones defined by temperature, **precipitation**, and associated adapted ecosystems. Global climate changes, and episodes of global warming, have occurred throughout geologic history as a result of natural variations in incoming solar **radiation**, atmospheric **chemistry**, and oceanic and **atmospheric circulation**. Anthropogenic, or human-caused, global warming and climate change are a potential outcome of human activities during the last 150 years. Scientific data show that atmospheric concentrations of **carbon dioxide**, methane, nitrous oxide, and man-made chemicals called halocarbons are increasing as a result of emissions associated with human activities, and models predict that this environmental change may lead to global warming.

Earth's greenhouse effect

Solar radiation is the major source of **energy** to Earth's surface. Much of that incoming short-wavelength energy is absorbed by the surface where it drives atmospheric and oceanic circulation, and fuels biological processes like **photosynthesis**. The land and sea surfaces then reradiate extra longer-wavelength **heat**, or infrared, energy. If Earth's atmosphere were transparent to the emitted infrared radiation, the **planet** would cool relatively efficiently and would have an average surface temperature of about 0°F (-18°C). However, the Earth's naturally occurring "greenhouse effect" maintains the planet's average temperature at a more livable 59°F (15°C) by trapping some of the escaping heat within the atmosphere. Small concentrations of so-called "greenhouse gases," also known as radiatively active gases, absorb some of the infrared energy and thereby delay its passage to **space**. **Water** vapor (H_2O), **carbon** dioxide (CO_2), methane (CH_4), nitrous oxide (N_2O), and **ozone** (O_3) are the most concentrated and effective greenhouse gases. The **greenhouse effect** has been extremely important to the **evolution** and survival of life on **Earth**. A surface temperature of 59°F is sufficient to maintain the Earth's reservoirs of life-sustaining liquid water, and to impel climatic processes, whereas 0°F is too cold for most organisms to live or for ecological processes to function well.

Atmospheric concentrations of greenhouse gases

Prior to the modern influence of human activities on atmospheric chemistry, the naturally occurring greenhouse gases had fairly stable atmospheric concentrations: carbon dioxide about 280 ppm (or parts per million by **volume**), methane 0.7 ppm, and nitrous oxide 0.285 ppm. (Human activities do not appear to affect the **concentration** of water vapor, which varies naturally over time.) Today, the atmospheric concentration of CO_2 has increased to about 364 ppm, while that of CH_4 is 1.7 ppm, and N_2O is 0.304 ppm. The concentrations of **chlorofluorocarbons (CFCs)**, and other completely man-made, or synthetic, greenhouse gases, have increased from essentially **zero** to about 0.7 ppb (parts per billion by volume).

Atmospheric concentrations of the greenhouse gases have increased particularly quickly since the middle of the twentieth century, coinciding with rapid human population growth and intensive global industrialization. The combined effects of fossil fuel use and **deforestation** have increased the atmospheric concentration of CO_2. **Fossil fuels**, like oil, **natural gas**, and **coal** contain carbon in their chemical structure that, when liberated by **combustion**, combines with **oxygen** to create CO_2. Trees, like all plants, take in CO_2, incorporate carbon in their structure, and emit O_2 back into the atmosphere; deforestation destroys carbon "sinks" that lower the atmospheric concentration of CO_2. Fossil-fuel **mining**, **decomposition** of organic materials in human and **livestock** waste treatment facilities, and **flooding** in **rice** agriculture have led to increased emissions of CH_4. Agricultural **fertilizers**, and combustion of fossil fuels and solid wastes account for increased N_2O emissions. Industrial processes emit a variety of powerful synthetic greenhouse gases like CFCs, hydrofluorcarbons (HFCs), perfluorocarbons (PFCs) and **sulfur** hexafluoride (SF_6).

The greenhouse gases vary greatly in their ability to absorb infrared radiation. On a per-molecule basis, methane is about 25–40 times more absorptive than carbon dioxide, nitrous oxide is 200–270 times stronger, and CFCs are 3–15 thousand times more effective. CO_2, however, has by far the largest atmospheric concentration, and has experienced the greatest increases; CO_2 is responsible for about 60% of the human contribution to increased atmospheric heat retention.

Predictions and evidence of global warming

Most atmospheric scientists assume that the well-documented increase in greenhouse gases will result in an intensification of Earth's naturally occurring greenhouse effect, and to global warming. The exact climatic response to increased concentrations of radiatively active gases, and its potential effects on humans are, however, difficult to measure or predict. However, if global warming were to occur as most scientific studies predict, it would have substantial climatic, ecological, and sociopolitical consequences.

The Earth's surface is surface temperature is extremely variable from place to place, and over **time**. Furthermore, the systems that interact to maintain the planet's temperature and climate are extremely complex; cause-and-effect relationships between changes in one system, the atmosphere in this case, and results in another, global climate, are very difficult to predict, observe, and "prove." In spite of these scientific challenges, there is significant evidence that the Earth has warmed significantly during the past 150 years or so, and that global climate has responded to the temperature increase. Climate records show a 1°F increase in the average temperature of the Earth's oceans, atmosphere, and solid surface since the late 1900s. Geologic and historical studies document dramatic thinning and shrinkage of the **polar ice caps**, and retreat of Earth's alpine **glaciers**. Less conclusive, but still suggestive, data supporting anthropogenic global warming include a several centimeter increase in global sea-level since 1900, and alterations in large-scale **weather** phenomena like the southeast Indian **monsoon**, Atlantic hurricane season, El Niño Southern Oscillation, and North African **drought** cycle.

The empirical, or observed, data listed above generally agree with predictions computed by mathematical models of global climate processes. These "virtual experiments," called three-dimensional general circulation models (GCMs), simulate the complex movements of energy and **mass** involved in the global circulation of the atmosphere and oceans. Scientists use GCMs to predict the effects of a change in a specific variable, like the concentration of atmospheric CO_2, on the rest of the global climate system. Because of the complexity of the computational problem, GCMs that attempt to predict global climate change have had somewhat variable results. However, most experiments do suggest that the increased concentration of atmospheric greenhouse gases has resulted, and will continue to result, in global warming. For example, one GCM that doubles the present CO_2 concentration to about 700 ppm predicts a 2°-6°F rise in global temperature, and suggests that the warming would be 2–3 times more intense at high latitudes than in the tropics.

Other predicted consequences of warming include large-scale shifts in atmospheric and oceanographic circulation patterns, melting of the polar **ice** caps, global sea-level rise, reorganization of the Earth's climatic zones, and establishment of new large-scale weather patterns. Such changes in the distribution of heat, precipitation, and weather phenomena like storms and floods would affect the productivity and distribution of natural and managed vegetation. Animals and **microorganisms** would experience dramatic changes in their habitats, and perhaps face much higher rates of **species extinction**. Most ecologists consider that global warming, if were it to occur as predicted, would represent a serious threat to **biodiversity** and to the health of ecosystems worldwide.

The predicted climatic and biological changes associated with anthropogenic global warming could have potentially disastrous outcomes for the Earth's human population. In 1998, more than half of the world's population, some 3.2 billion people, lived with in 120 miles of the **ocean**. Even small increases in global **sea level**, and in the intensity of coastal storms and floods, would threaten the lives and property of large numbers of people. Changes in

KEY TERMS

. . . .

Global warming—A projected increase in Earth's surface temperature caused by an increase in the concentration of greenhouse gases, which absorb infrared energy emitted by Earth's surface, thereby slowing its rate of cooling.

Greenhouse effect—The warming of the Earth's atmosphere as a result of the capture of heat re-radiated from the Earth by certain gases present in the atmosphere.

regional temperature, precipitation, and weather, as well as biological health, would affect the managed agriculture, fishing, and **forestry** that provide food and shelter for the Earth's burgeoning human population.

Most scientists, and many international policy-makers, now consider global warming to be a credible threat to the Earth's natural environment and human population. However, because the specific consequences of global warming are difficult to predict, and in some cases unknown, the scientific community remains divided about the potential effects of the phenomenon. Attempts to prevent anthropogenic global warming, especially measures that require socioeconomic sacrifice, have therefore been extremely controversial. The 1992 United Nations Framework Convention on Climate Change (UNCCC), also called the Kyoto Protocol, acknowledges that human activities can alter global climate, and requires signatory nations to reduce greenhouse gas emissions. As of November 2002, 181 nations had signed, ratified, or acceded to the conditions of the Kyoto protocol. However, the United States, by far the world's largest per-capita producer of greenhouse gases, did not sign the treaty on the grounds that the science of global warming remains inconclusive, and that the economic consequences of action would be too great.

Resources

Books

Evans, C.A., and N.H. Marcus. *Biological Consequences of Global Climate Change.* University Science Books, 1996.

Freedman, B. *Environmental Ecology.* Academic Press, 1996.

Houghton, J.T. *Global Warming: The Complete Briefing.* Cambridge University Press, 1997.

Philander, S.G. *Is the Temperature Rising?: The Uncertain Science of Global Warming.* Princeton University Press, 1998.

Organizations

Intergovernmental Panel on Climate Change. United Nations Environment Program, Two UN Plaza, Room DC2–803, New York, NY 10017. (212) 963–8210. <http://www.ipcc.ch.>

Other

United Nations. "United Nations Framework Convention on Climate Change." November 10, 2002 [cited November 13, 2002]. <http://unfccc.int/index.html>.

United States Environmental Protection Agency. "Global Warming." October 2, 2002 [cited November 13, 2002]. <http://yosemite.epa.gov/oar/globalwarming.nsf/content/index.html>.

Bill Freedman
Laurie Duncan

Gluons *see* **Subatomic particles**

Glycerin *see* **Glycerol**

Glycerol

Glycerol is the common name of the organic compound whose chemical structure is $HOCH_2\text{-}CHOH\text{-}CH_2OH$. Propane-1,2,3-triol or glycerin (USP), as it is also called, consists of a chain of three **carbon atoms** with each of the end carbon atoms bonded to two **hydrogen** atoms (C-H) and a hydroxyl group (-OH) and the central carbon atom is bonded to a hydrogen atom (C-H) and a hydroxyl group (-OH). Glycerol is a trihydric **alcohol** because it contains three hydroxyl or alcohol groups. Glycerin is a thick liquid with a sweet taste that is found in fats and oils and is the primary triglyceride found in coconut and olive oil. It was discovered in 1779, when the Swedish chemist Carl Wilhelm Scheele (1742-1786) washed glycerol out of a heated a mixture of **lead** oxide (PbO) and olive oil. Today, it is obtained as a byproduct from the manufacture of soaps.

One important property of glycerol or glycerin is that is not poisonous to humans. Therefore it is used in foods, syrups, ointments, medicines, and cosmetics. Glycerol also has special chemical properties that allow it to be used where oil would fail. Glycerol is a thick syrup that is used as the "body" to many syrups, for example, cough medicines and lotions used to treat **ear** infections. It is also an additive in vanilla extracts and other food flavorings. Glycerin is added to **ice** cream to improve the texture, and its sweet taste decreases the amount of sugar needed. The base used in making toothpaste contains glycerin to maintain smoothness and shine. The cosmetic industry employs glycerin in skin conditioning lotions to replace lost skin moisture, relieve chapping, and keep skin soft. It is also added in hair shampoos to make them flow easily when poured from the bottle. The raisins found in cereals remain soft because they have been soaked in glycerol. Meat casings and food wrapping papers use glycerin to give them flexibility without brittleness. Similarly, tobacco is treated with this thick chemical to prevent the leaves from becoming brittle and crumbling during drying. It also adds sweetness to chewing tobacco. Glycerol is added during the manufacture of soaps in order to prepare shiny transparent bars. The trihydric alcohol structure of glycerin makes it a useful chemical in the manufacture of various hard foams, like those that are placed under siding in buildings and around dish washers and refrigerators for insulation and sound proofing. Analogously, the chemical structure of glycerol makes it an excellent catalyst in the microbiological production of vinegar from alcohol.

In the manufacture of foods, drugs, and cosmetics, oil cannot be employed as a lubricant because it might come in contact with the products and contaminate them. Therefore, the nontoxic glycerol is used to reduce **friction** in pumps and bearings. Gasoline and other **hydrocarbon** chemicals dissolve oil-based greases, so glycerin is used in pumps for transferring these fluids. Glycerol is also applied to **cork** gaskets to keep them flexible and tough when exposed to oils and greases as in **automobile** engines. Glycerin is used as a lubricant in various operations in the textile industry, and can be mixed with sugar to make a nondrying oil. Glycerol does not turn into a solid until it is cooled to a very low **temperature**. This property is utilized to increase the storage life of **blood**. When small amounts of glycerin are added to red blood cells, they can be frozen and maintained for up to three years.

Chemical derivatives of glycerol or propane- 1,2,3-triol are important in a wide range of applications. Nitroglycerin is the trinitrate derivative of glycerol. One application of this chemical is as the key ingredient in the manufacture of dynamite **explosives**. Nitroglycerin can also be used in conjunction with gun **cotton** or nitrocellulose as a propellant in military applications. In the pharmaceutical industry, nitroglycerin is considered a drug to relieve chest pains and in the treatment of various **heart** ailments. Another derivative, guaiacol glyceryl **ether**, is an ingredient in cough medicines, and glycerol methacrylate is used in the manufacture of soft contact lenses to make them permeable to air. Glycerol esters are utilized in cakes, breads, and other bakery products as lubricants and softening agents. They also have similar applications in the making of candies, butter, and whipped toppings. A specially designed glycerol **ester** called *caprenin* can be used as a low **calorie** replacement for cocoa butter.

The acetins are derivatives of glycerol that are prepared by heating glycerol with **acetic acid**. Monoacetin is used in the manufacture of dynamite, in tanning leather, and as a solvent for various dyes. Diacetin, another derivative of glycerol, is used as a solvent and a softening agent. Triacetin, the most useful of the acetins, is used in the manufacture of cigarette filters and as a component in solid rocket fuels. It is also used as a solvent in the production of photographic films, and has some utility in the perfume industry. Triacetin is added to dried egg whites so that they can be whipped into meringues.

See also Fat.

Resources

Books

Carey, Francis A. *Organic Chemistry.* New York: McGraw-Hill, 2002.
Newman, A.A. *Glycerol.* Cleveland: C.R.C. Press, 1968.

Andrew Poss

Glycol

A glycol is an aliphatic organic compound in which two hydroxyl (OH) groups are present. The most important glycols are those in which the hydroxyl groups are attached to adjacent **carbon atoms**, and the term glycol is often interpreted as applying only to such compounds. The latter are also called vicinal diols, or 1,2-diols. Compounds in which two hydroxyl groups are attached to the same carbon atom (geminal diols) normally cannot be isolated.

The most useful glycol is **ethylene glycol** (IUPAC name: 1,2-ethanediol). Other industrially important glycols include propylene glycol (IUPAC name: 1,2-propanediol), diethylene glycol (IUPAC name: 3-oxa-1,5-pentanediol) and tetramethylene glycol (IUPAC name: 1,4-butanediol)(Figure 1).

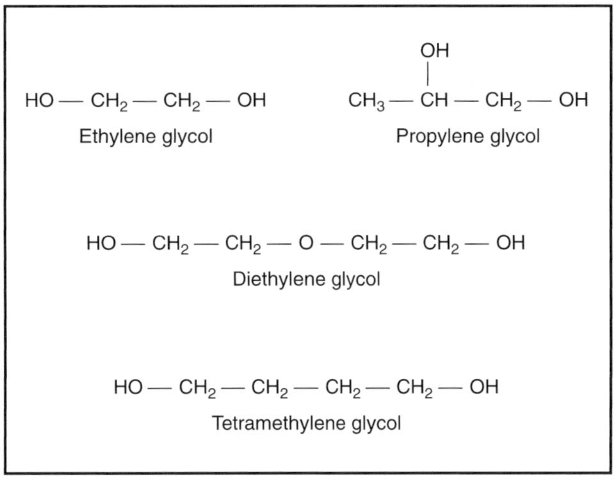

Physical properties of glycols

The common glycols are colorless liquids with specific gravities greater than that of **water**. The presence of two hydroxyl groups permits the formation of **hydrogen** with water, thereby favoring **miscibility** with the latter. Each of the glycols shown above is completely miscible with water. Intermolecular hydrogen bonding between glycol molecules gives these compounds boiling points which are higher than might otherwise have been expected; for example, ethylene glycol has a **boiling point** of 388.5°F (198°C).

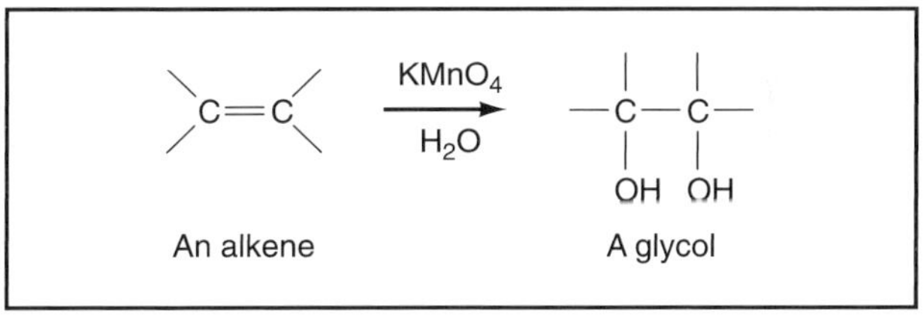

Figure 1. Structures of common glycols. *Illustration by Hans & Cassidy. Courtesy of Gale Group.*

Figure 2. Laboratory preparation of a glycol. *Illustration by Hans & Cassidy. Courtesy of Gale Group.*

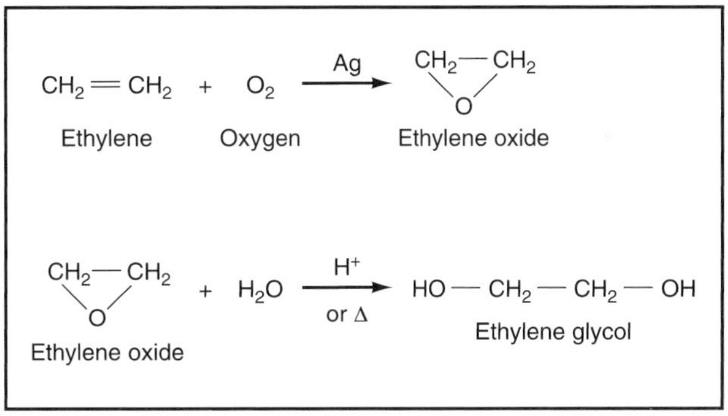

Figure 3. Industrial preparation of ethylene glycol. *Illustration by Hans & Cassidy. Courtesy of Gale Group.*

Laboratory preparation

The most convenient and inexpensive method of preparing a glycol in the laboratory is to react an alkene with cold dilute potassium permanganate, $KMnO_4$ (Figure 2).

Yields from this reaction are often poor and better yields are obtained using osmium tetroxide, OsO_4. However, this reagent has the disadvantages of being expensive and toxic.

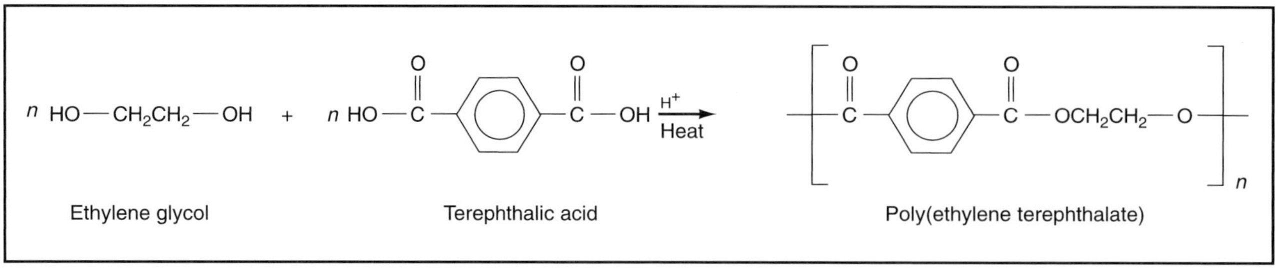

$$n \; HO-CH_2CH_2-OH \quad + \quad n \; HO-\overset{O}{\overset{\|}{C}}-\bigcirc-\overset{O}{\overset{\|}{C}}-OH \; \xrightarrow[\text{Heat}]{H^+} \quad \left[\overset{O}{\overset{\|}{C}}-\bigcirc-\overset{O}{\overset{\|}{C}}-OCH_2CH_2-O\right]_n$$

Ethylene glycol   Terephthalic acid   Poly(ethylene terephthalate)

Figure 4. Synthesis of poly(ethylene terephthalate). *Illustration by Hans & Cassidy. Courtesy of Gale Group.*

Industrial preparation

In the industrial preparation of ethylene glycol, ethylene (IUPAC name: ethene) is oxidized to ethylene oxide (IUPAC name: oxirane) using **oxygen** and a silver catalyst. Ethylene oxide is then reacted with water at high **temperature** or in the presence of an acid catalyst to produce ethylene glycol. Diethylene glycol is a useful by-product of this process (Figure 3).

Alternative methods of preparing ethylene glycol that avoid the use of toxic ethylene oxide are currently being investigated.

Uses

In the 1993 ranking of chemicals according to the quantity produced in the United States, ethylene glycol ranked 30th, with 5.23 billion lb (2.37×10^9 kg). Much of this ethylene glycol is used as antifreeze in **automobile** radiators. The addition of ethylene glycol to water causes the freezing point of the latter to decrease, thus the damage that would be caused by the water freezing in a radiator can be avoided by using a mixture of water and ethylene glycol as the coolant. An added advantage of using such a mixture is that its boiling point is higher than that of water, which reduces the possibility of boil-over during summer driving. In addition to ethylene glycol, commercial antifreeze contains several additives, including a dye to reduce the likelihood of the highly toxic ethylene glycol being accidentally ingested. Concern over the toxicity of ethylene glyco-the lethal dose of ethylene glycol for humans is 1.4 ml/kg-resulted in the introduction, in 1993, of antifreeze based on non-toxic propylene glycol.

The second major use of ethylene glycol is in the production of poly(ethylene terephthalate), or PET. This **polymer**, a polyester, is obtained by reacting ethylene glycol with terephthalic acid (IUPAC name: 1,4-benzenedicarboxylic acid) or its dimethyl **ester** (Figure 4).

Poly(ethylene terephthalate) is used to produce **textiles**, large soft-drink containers, photographic film, and overhead transparencies. It is marketed under various

trademarks including DACRON®, Terylene®, Fortrel®, and Mylar®. Textiles containing this polyester are resistant to wrinkling, and can withstand frequent laundering. Poly(ethylene terephthalate) has been utilized in the manufacture of clothing, bed linen, carpeting, and drapes.

Other glycols are also used in polymer production; for example, tetramethylene glycol is used to produce polyesters, and diethylene glycol is used in the manufacture of polyurethane and unsaturated polyester **resins**. Propylene glycol is used in the manufacture of the polyurethane foam used in car seats and furniture. It is also one of the raw materials required to produce the unsaturated polyester resins used to make car bodies and playground equipment.

See also Chemical bond; Compound, chemical; Polymer.

Resources

Books

Bailey, James E. *Ullmann's Encyclopedia of Industrial Chemistry.* New York: VCH, 2003.

Budavari, Susan, ed. *The Merck Index.* 11th ed. Rahway, NJ: Merck, 1989.

Loudon,G. Mark. *Organic Chemistry.* Oxford: Oxford University Press, 2002.

Szmant, H. Harry. *Organic Building Blocks of the Chemical Industry.* New York: Wiley, 1989.

Arthur M. Last

Glycolysis

Glycolysis, a series of enzymatic steps in which the six-carbon glucose **molecule** is degraded to yield two three-carbon pyruvate molecules, is a central catabolic pathway in plants, animals and many **microorganisms**.

In a sequence of 10 enzymatic steps, **energy** released from glucose is conserved by glycolysis in the form of **adenosine triphosphate** (ATP). So central is glycolysis to life that its sequence of reactions differs among **species** only in how its **rate** is regulated, and in the metabolic fate of pyruvate formed from glycolysis.

In **aerobic** organisms (some microbes and all plants and animals), glycolysis is the first phase of the complete degradation of glucose. The pyruvate formed by glycolysis is oxidized to form the acetyl group of acetyl-coenzyme A, while its **carboxyl group** is oxidized to CO_2. The acetyl group is then oxidized to CO_2 and H_2O by the **citric acid** cycle with the help of the **electron** transport chain, the site of the final steps of oxidative phosphorylation of **adenosine diphosphate** molecules to high-energy ATP molecules.

In some **animal** tissues, pyruvate is reduced to lactate during **anaerobic** periods, such as during vigorous **exercise**, when there is not enough **oxygen** available to oxidize glucose further. This process, called anaerobic glycolysis, is an important source of ATP during very intense muscle activity.

Anaerobic glycolysis also serves to oxidize glucose to **lactic acid** with the production of ATP in anaerobic microorganisms. Such lactic acid production by **bacteria** sours milk and gives sauerkraut its mildly acidic taste.

A third pathway for pyruvate produced by glycolysis produces **ethanol** and CO_2 during anaerobic glycolysis in certain microorganisms, such as brewer's yeast—a process called alcoholic **fermentation**. Fermentation is an anaerobic process by which glucose or other organic **nutrients** are degraded into various products to obtain ATP.

Because glycolysis occurs in the absence of oxygen, and living organisms first arose in an anaerobic environment, anaerobic **catabolism** was the first biological pathway to evolve for obtaining energy from organic molecules.

Glycolysis occurs in two phases. In the first phase, there are two significant events. The addition of two phosphate groups to the six-carbon sugar primes it for further degradation in the second phase. Then, cleavage of the doubly phosphorylated six-carbon chain occurs, breaking fructose 1,6-diphosphate into two 3-carbon isomers. These are fragments of the original six-carbon sugar dihydroxyacetone phosphate and glyceraldehyde 3-phosphate.

In the second phase, the two 3-carbon fragments of the original 6-carbon sugar are further oxidized to lactate or pyruvate.

Entry into the second phase requires the **isomer** to be in its glyceraldehyde 3-phosphate form. Thus, the dihydroxyacetone phosphate isomer is transformed into glyceraldehyde 3-phosphate before being further oxidized by the glycolytic pathway.

Glycolysis produces a total of four ATP molecules in the second phase, two molecules of ATP from each glyceraldehyde 3-phosphate molecule. The ATP is formed during substrate-level phosphorylation-direct transfer of a phosphate group from each 3-carbon fragment of the sugar to adenosine diphosphate (ADP), to form ATP. But because two ATP molecules were used to phosphorylate the original six-carbon sugar, the net gain is two ATP.

The net gain of two ATP represents a modest **conservation** of the chemical energy stored in the glucose molecule. Further oxidation, by means of the reactions of the Kreb's cycle and oxidative phosphorylation are required to extract the maximum amount of energy from this fuel molecule.

See also Adenosine triphosphate; Krebs cycle.

Resources

Books

Atkinson, D.E. *Cellular Energy Metabolism and Its Regulation.* New York: Academic, 1977.

Lehninger, A.L. *Principles of Biochemistry.* New York: Worth Publishers, Inc., 1982.

Marc Kusinitz

Gnat-eaters *see* **Antbirds and gnat-eaters**

Goats

Goats belong to the order Artiodactyla (genus *Capra*), which is made up of a number of hoofed **mammals** having an even number of toes. Goats have existed on **Earth** for at least 35 million years and, during the course of **evolution**, have undergone an incredibly wide **radiation**, both in distribution and **ecology**. Although the **taxonomy** of this group is still unclear, eight **species** are generally recognized as being true goats, with representatives found from the barren plains of central **Asia** to an altitude of 22,000 ft (6,700 m) on snow-clad peaks.

Despite such divergence, all goats have a similar design and frequently display similar behaviors. The majority are stocky, gregarious animals that live in barren habitats, often under inhospitable **weather** conditions. All goats are adapted to living in steep and often unstable terrain, and their physical appearance demonstrates several features that have evolved to cope with these conditions. The main toes of the hoof are often concave on the underside, are hard as **steel**, and can be widely splayed to spread the animal's body weight over a large area. The legs are usually short and highly muscular.

The ancestor of all domestic breeds of goats is thought to have been the wild goat of western Asia (*Capra aegagrus*). This species, which occupies a wide range of habitats at altitudes up to 13,800 ft (4,200 m), was formerly widespread throughout much of Eurasia. The wild goat reaches a height of 28-39 in (70-100 cm) at the shoulder and weighs from 55-200 lb (25-90 kg). The coat is usually a silver-white **color**, with a gray facial pattern and black or brown undersides. There is usually a distinct line of longer, darker hairs extending from the neck down along the spine. Females are usually a yellowish brown or reddish gray color. The males, which are larger than females, bear a pair of arched, scimitar-shaped horns that extend far along the back and may reach 47 in (120 cm) in length. The horns of a female are much thinner and may measure just 8-12 in (20-30 cm) in length.

Goats living at higher altitudes or under colder conditions always have a much thicker coat that is made up of several layers. More primitive forms tend to have small, pointed horns and a patterned coast, while more evolved forms have larger and more curved horns. The horns themselves are often used to classify different species and vary considerably in this group: the horns of the East Caucasian tur (*Capra cylindricornis*), for example, are curved directly back and down towards the shoulders, while those of the ibex (*Capra ibex*) are upright, curved, and heavily ridged, and those of the Kashmir markhor (*C. falconeri*) are upright and spiralled like a corkscrew. The horns, which are not shed each year, display growth patterns which enable biologists to determine the age of the animals—an important feature when management of certain threatened species are involved.

Goats are highly sociable animals that live in herds whose size and composition varies according to the species and time of year. Herds of up to 500 goats have at times been recorded but, in general, groups tend to be much smaller, often around 20-30 animals. In some species, the adult males (or billies) may either follow a solitary existence or form small groups of 3-5 animals for much of the year, while the females (or nannies) and offspring (kids) form larger, more cohesive groups. The two come together prior to the breeding season—period known as the rutting season—when males compete against one another in an attempt to mate with as many females as possible. In some herds, there will be just one dominant male, but even he will have to defend the herd of females from potential competitors from outside the herd. Within the herd, there is also a distinct hierarchy, with older animals almost always being dominant over younger ones. Females are responsible for bringing up the kids and a mother will only suckle its own kids, those of another goat being gently rebuffed.

Wild goats vary considerably in their activity patterns. Most species are active in the early morning and late afternoon, resting during the hottest part of the day to digest their food. Many species display distinctive seasonal and even daily patterns of **migration**, coming down to the lower parts of their ranges to feed and then returning to their scaly heights to rest.

All goats are herbivorous animals that feed on a vast range of plants. They are not grazing animals like the majority of herbivores, but prefer to browse on the leaves and twigs of shrubs and coarse weeds. Goats will go to any lengths to obtain a meal, and in some parts of North **Africa**, feral domestic goats are commonly seen browsing in *Acacia* trees some 23 ft (7 m) off the ground, their cleft hooves and powerful legs enabling them to jump and climb into trees. Herds of feral goats, which are widespread throughout the world, are known to cause considerable environmental destruction as they destroy natural vegetation and contribute to **erosion** and, in some drier regions, **desertification**. In some countries, feral goats have had to be exterminated because of the damage they cause to native, and often endangered, vegetation.

Goats are well-known for their aggressive **behavior** when settling territorial or reproductive disputes. The

Markhors (*Capra falconeri*) are found in various mountain ranges in India, Pakistan, Afghanistan, Tajikistan, and Uzbekistan. These goats are considered endangered throughout their range. The primary cause of the markhor's decline is excessive hunting, primarily for its horns, but also for its meat and hide. *Photograph by R. Van Nostrand/Photo Researchers, Inc. Reproduced by permission.*

most common means of settling such issues is in head-to-head combat, with both animals using their skulls and horns as offensive weapons. Among the most dramatic of these encounters are the clashes of adult male ibex: standing 10 ft (3m) apart, often on a steep precipice, the males rear up on their hind legs and charge one another, bringing their large curved horns down at the last moment to crash against its opponents. Scientists have estimated that the **force** of these blows may be as much as 60 times greater than that needed to fracture a human skull. Goats, however, have highly efficient shock absorbers built into their skulls and are able to withstand such attacks without too great an injury to their heads. In this obvious show of physical strength, the weaker **animal** usually recognizes its shortcomings at an early stage of the encounter and tries

to escape before too much damage is caused to other parts of the body.

Goats have long been domesticated, and in many parts of the world these herds may constitute an important source of food and revenue for people. A great many domestic breeds have been developed, some of which have been either deliberately released, or escaped and later bred in the wild. Some of these feral domestic goats breed with wild populations—a point of concern for some threatened wild species, as the genetic component of the original true stock might be diluted by such breeding activities.

Although goats live in almost inaccessible regions, their populations have been seriously affected by hunting and human encroachment to the foothills, where many species feed during the summer months. Wild goats have long been sought after as trophy specimens. The magnificent ibex was exterminated in the Alps during the nineteenth century, but has since been reestablished in many of its former habitats as a result of a concentrated breeding and reintroduction programme. The West Caucasian tur (*C. caucasica*) is now confined to a narrow strip of montane **habitat** in the western Caucasus, where its existence is threatened by hunters and human encroachment. Natural predators, too, take their toll on wild goat populations, and it is because of this **pressure** that goats have developed many of their behavioral patterns, such as living in small groups, and their ability to rapidly flee over rough terrain. Wolves and big **cats** such as snow leopards are among the main predators of goats, while **bears**, wild dogs, and foxes may **prey** on kids. Aerial predators such as golden **eagles** are also a threat to kids. Despite the vigilance of the adults, kids are also highly susceptible to natural causes of death as a result of their playful behavior; many engagements and mock fights can result in an inexperienced animal slipping from a rock or precipice to its death.

See also Ungulates.

David Stone

Goatsuckers

The goatsuckers, nightjars, and nighthawks number 70 **species** of **birds** in the family Caprimulgidae. These birds have a relatively large head, with a wide beak, and a large mouth with a seemingly enormous gape. The mouth is fringed by long, stiff bristles, and is an **adaptation** for catching **insects** in flight. The unusually large mouth of goatsuckers was once believed to be useful for suckling milk at night from lactating **goats**. This was, of course, an erroneous folk belief, but it is perpetuated today in the common name of the family of these birds.

Most goatsuckers and nightjars have long, pointed wings, and short, feeble feet. Most species are crepuscular, being active mostly in dim **light** around dusk. Some species are nocturnal, or active at night. The colors of these birds are subdued, mostly consisting of drab, streaky browns, blacks, and greys. This coloration makes goatsuckers and nightjars very well camouflaged, and they can be exceedingly hard to detect during the day, when they are roosting or sitting on a nest.

Species of goatsuckers and nighthawks

This family of birds is richest in species in **Africa**, and south **Asia**. There are only eight species of goatsuckers in **North America**, most of which are migratory species, breeding in North America and wintering in Central and **South America**.

One of the most familiar species of goatsuckers in North America is the whip-poor-will (*Caprimulgus vociferous*), occurring throughout the eastern United States and southeastern Canada. Unfortunately, the population of this species has declined over much of its range, due to the large loss of natural **habitat**, and the fragmentation of the remnants.

Chuck-will's-widow (*C. carolinensis*) occurs in pine **forests** of the southeastern United States while the common poorwill (*Phalaenoptilus nuttallii*) is the most common goatsucker in the western United States. The common names of the whip-poor-will, poorwill, chuck-will's-widow, and the pauraque (*Nyctidromus albicollis*) of southern Texas have all been derived from the very distinctive, loud calls made by these birds. Naming animals after the sounds that they make is known as onomatopoeia.

The common nighthawk (*Chordeiles minor*) is another relatively familar species, ranging through all of the United States and much of Canada. This species is highly aerial when hunting, swooping gracefully and swiftly on its falcon-like wings to capture its **prey** of **moths**, **beetles**, **ants**, and other flying insects. The common nighthawk tends to breed in open, rocky places, and it also accepts flat, gravelly roofs in cities as a nesting substrate. Most urban residents are not aware that breeding populations of this native bird occur in their midst, although they may have often wondered about the source of the loud "peeent" sounds that nighthawks make while flying about at dusk and dawn. For reasons that are not understood, populations of the common nighthawk appear to be declining markedly, and the species may be in jeopardy. The lesser nighthawk (*C. acutipennis*) is a smaller species that occurs in the southwestern United States.

The poorwill (*Phalaenoptilus nuttallii*) of the western United States is the only species of bird that is known to hibernate. This has not been observed many times, but poorwills have occasionally been found roosting in crevices in canyons in winter, in an obviously torpid state, and not moving for several months. These hibernating birds maintain a body **temperature** of only 95.4-97.2° F (18-19° C), compared with their normal 135-136.8° F (40-41° C).

The Puerto Rican nightjar (*Caprimulgus noctitherus*) is a rare species that only occurs on the Caribbean **island** of Puerto Rico. Only about 1,000 individuals of this **endangered species** survive. The Puerto Rican nightjar has been decimated by losses of its natural habitat, especially **deforestation**, and by depredations by introduced predators, such as the mongoose.

Resources

Books

Ehrlich, P.R., D.S. Dobkin, D. Wheye. *Birds in Jeopardy*. Stanford, California: Stanford University Press, 1992.
Forshaw, Joseph. *Encyclopedia of Birds*. New York: Academic Press, 1998.

Bill Freedman

Gobies

Gobies, belonging to the suborder Gobidioidei, are small **fish** that usually live off the coast in tropical and warm temperate regions. They spend the majority of their time resting on the bottom near protective cracks in coral reefs or burrows in the **sand**. Most **species** of this fish have fused pelvic fins which form a suction cup on their undersides. A goby uses this suction cup to cling to **rocks** so that it does not wash away with **ocean currents**.

The suborder of Gobidioidei is divided into six families. The largest family in the suborder, and indeed the largest family of all tropical fishes, is the Gobiidae. Although the count is not complete, there are approximately 212 genera and 1,900 species within the Gobiidae family worldwide; at least 500 of these species live in the Indo-Pacific Ocean. Within the Gobidioidei family, two subfamilies are distinguished: the Sleepers (Eleotrinae) and the True Gobies (Gobiinae).

General characteristics

Although some species are moderately elongated, Gobies are usually very small, compact fish. The smallest vertebrate in the world is, in fact, a goby known as

Trimmatom nanus which lives off the Philippine islands. This goby never grows larger than 8-10 mm long. Two other gobies living in the Philippines—the *Pandaka pygmaea* and the *Mistichthys luzonensis*—are among the shortest **freshwater** fishes in the world; the females of these species mature at 10-11 mm long. While most of them get no bigger than 4 in (10 cm) long, the largest range up to 19.5 in (50 cm).

One of the most unusual traits of True Gobies is the "suction cup" located on their undersides near their pelvic areas. Their pelvic bones are fused with each other; thus, their pelvic fins are united, at least at the base. In True Gobies, the fin is connected by a thin **membrane** which enables the suction cup to create a **vacuum**; gobies can use this vacuum to gain a firm hold on objects. This suction cup exists in many different variations. In some species, the pelvic fins are completely connected by a membrane; in others, the fins are partially or completely separated.

Gobies are also characterized by the presence of a two-part dorsal fin, a fin located on their backs. The first part of the dorsal fin can have up to eight unbranched rays, although sometimes these rays are completely absent. Gobies usually, but not always, have some scales; these scales are sometimes present only in specific parts of their bodies. Their mouths are usually located at the very tip of their bodies and often protrude from their faces. Their jaws contain powerful teeth which are well suited for eating meat.

In general, gobies have developed in quite diverse ways during the course of their **evolution**. While they usually live in **salt water**, they are often found in **brackish** water, and sometimes even freshwater. In fact, gobies are often the most plentiful fish in freshwater on oceanic islands. A few species even live in **rivers** in the **mountains**. They have adapted to live in widely varying habitats, living, for example, inside **sponges** and on land.

Behavior

Most goby species are bottom dwellers. Furthermore, they are not very graceful swimmers, because their movements are characteristically jerky. They propel themselves by a few strong beats of their tails and steer themselves with their pectoral fins. Gobies are carnivorous, feeding on crustaceans, small **invertebrates**, fish eggs, worms, and other small fish.

In most species, gobies' eyes are their most important sensory **organ**, especially for detecting **prey** and danger. It should be noted, however, that some species have adapted to living in caves and subsequently have no eyes. These species rely primarily on their sense of **smell**. Also, even in species with normal eyesight, smell is used to recognize members of the opposite sex. Furthermore, gobies have been proven to possess the ability to hear.

Unusual distinctions

Three goby genera—the mudskippers (*periophthalmus*), the *Boleophthalmus*, and the *Scartelaos*—act as true **amphibians**. Perhaps the most well known of these genera is the mudskippers. These gobies can move at considerable speed on land using their armlike pectoral fins. In many species of mudskippers, the pelvic fins are separate and used as independent active arms as well. Interestingly, mudskippers' eyes, which are well suited to seeing in air, are located on stalks on the tops of their heads; the fish are able to elevate and retract these stalks depending on their need.

Reproduction and longevity

Gobies breed in the spring and summer. The adult males define a territory around their chosen nests, which are often holes in the rocks, under stones or shells, or even in old shoes. After spawning occurs, the females lay the eggs in a patch on the underside of the nest roof. Male gobies guard the eggs until they hatch.

Depending on the species, gobies can live between one and 12 years. At one extreme, the *Aphia* and *Crystallogobius* species die right after their first breeding season when they are one year old. At the other extreme, the rock goby and the leopard-spotted goby do not even mature until they are two years old.

Resources

Books

Nelson, Joseph S. *Fishes of the World.* 3rd ed. New York: Wiley, 1994.

Webb, J.E. *Guide to Living Fishes.* New York: Macmillan, 1991.

Whiteman, Kate. *World Encyclopedia of Fish & Shellfish.* New York: Lorenz Books, 2000.

Kathryn Snavely

Gold *see* **Element, chemical**

Golden mole *see* **Moles**

Goldenrod *see* **Composite family**

Goldenseal

Goldenseal (*Hydrastis canadensis* L.) is a woodland **plant** belonging to the family Ranunculaceae. The plant is also known as eyebalm, eyeroot, hydrastis, orangeroot, tumeric root, and yellowroot. Mainly found in the wild, goldenseal grows to a height of about 1 ft (30 cm). It has an erect, hairy stem, and produces small, greenish-white flowers that bloom in early spring, and later turn into clusters of red berries. The plant gets its common name from its thick yellow **rhizome**.

Native Americans used goldenseal as a multi-purpose medicinal plant. The Cherokees used it as a wash to treat skin diseases and sore eyes and mixed a powder made from the root with bear **fat** for use as an insect repellent. Other uses were as a diuretic, stimulant, and treatment for **cancer**. The Catawbas used the boiled root to treat **jaundice**, an ulcerated stomach, colds, and sore mouth; they also chewed the fresh or dried root to relieve an upset stomach. The Kickapoo used goldenseal as an infusion in **water** to treat eyes irritated by smoke caused by burning the **prairie** in the autumn. Some Native American tribes made use of the plant as a source of a natural yellow dye.

Many early European settlers turned to Native American remedies to treat their ailments. In the seventeenth century, colonists in Virginia used such native plants as **ginseng** (*Panax pseudoginseng*), tobacco (*Nicotiana tabacum*), sassafras (*Sassafras albidum*), snake-root (*Echinacea angustifolio*), Collinsonia (*Collinsonia canadensis*), Sanguinaria (*Sanguinaria canadensis*), and lobelia (*Lobelia inflata*) to treat medical problems.

Goldenseal grows in high, open woods, usually on hillsides or bluffs with good drainage. It is found in its native **habitat** from the north-east border of South Carolina to the lower half of New York, and east to northern Arkansas and the south-east corner of Wisconsin, as well as in Nova Scotia. Today it is only found in abundance in Ohio, Indiana, West Virginia, Kentucky, and parts of Illinois. Goldenseal has vanished from some of its historical locations, mostly because of habitat loss. However, it has been cultivated in other places.

The roots and rhizomes of goldenseal contain a number of isoquinoline alkaloids, including hydrastine, berberine, canadine, canadaline, and l—hydrastine. It is berberine that gives the rootstock its distinctive golden **color**.

The medical uses for goldenseal are quite numerous. It is able to treat a variety of infections from **tonsillitis**, gonorrhea, and **typhoid fever**, to hemorrhages, gum **disease**, and pelvic inflammatory disease. Traditional uses of the rhizome have been as an antiseptic, astringent, diuretic, laxative, antihemorrhaging agent, digestive aid, tonic, and deworming agent. Goldenseal has also been used as an anticancer agent. Goldenseal's effectiveness against sores and inflammations is presumably due to the antiseptic properties of berberine against **bacteria** and **protozoa**, and to berberine's antimalarial and fever-reducing properties. The alkaloids hydrastine and hydrastine hydrochloride have been reported to stop uterine bleeding and prevent **infection**, and canadine acts as a sedative and muscle relaxant.

Goldenseal stimulates the liver, kidneys, and lungs, and is often used to treat **ulcers**. It has excellent antimicrobial properties that treat **inflammation** and infections of respiratory mucous membranes, the digestive tract, and urinary tract. External applications of goldenseal can be used to treat impetigo, ringworm, conjunctivitis, and gum disease.

The use of goldenseal as a "herbal medicine" is not restricted by the U.S. Food and Drug Administration (FDA), which does not regulate herbs. Consequently, goldenseal remains a popular medicinal **herb** among many practitioners of **alternative medicine**. However, some health professionals recommend not using goldenseal for medicinal purposes because of the plant's toxicity. If ingested as a fresh, raw plant, goldenseal can be very posionous. Improper preparations of goldenseal may cause serious side effects such as mouth and throat irritation, skin sensations including burning or tingling; paralysis; respiratory failure; and even death. Before using goldenseal, patients should consult with their health practitioner.

Randall Frost

Gonorrhea *see* **Sexually transmitted diseases**

Gooseberries *see* **Saxifrage family**

Gophers

Gophers are small **rodents**. Although the name is often used popularly to refer to a variety of animals, including **snakes**, in the United States gophers are the pocket gophers that live in the **grasslands** of western Canada, eastward to the Great Lakes, and down into northern **South America**. Pocket gophers (family Geomyidae) have fur-lined cheek pouches that let them carry food in large quantities. These rodents eat grain as well as underground roots, so a large population can do serious damage to farm fields.

Pocket gophers are burrowing animals with round little bodies without much visible neck. Their pouches, used only for carrying food, are located from the face back onto the shoulder region. They open on the outside, not into the mouth.

Gophers vary in size from only about 4 in (10 cm) to 14 in (35.5 cm) with a short, usually naked tail. They resemble small woodchucks, but they are not nearly as visible because they spend most of their lives underground. Their fur is colored varying shades of brown. It also occurs in varying lengths on a single **animal** because they continually molt, losing their hair in large patches.

The 25 **species** of gophers usually do not overlap very much in their ranges. The western pocket gophers (*Thomomys*) live all the way from **sea level** to perhaps an elevation of about 13,000 ft (3,965 m) in the **mountains**. Their gnawing teeth have a smooth front surface. The eastern species (*Geomys*) live in the flat plains and prairies of the southern states. Their gnawing teeth have a deep lengthwise groove, as does the third group of North American gophers, the yellow-faced pocket gophers *(Pappogeomys)*. They are found only in a small region from Colorado down into Mexico. There are no pocket gophers in the northeastern section of the United States. In that area, the name gopher is often used for the chipmunk. Additional genera of pocket gophers live in Central America.

Life underground

Gophers are well adapted to digging, with strong, large forearms and sharp claws. They have yellowish gnawing teeth that can keep digging even when their lips are closed, an aid in keeping the dirt out of their mouths while they dig. Also, they have special tear **glands** that continuously clean their eyes as they dig. Their ears can be closed against the dirt.

Gophers spend most of their lives underground. They dig shallow feeding tunnels that allow them to make their way to the juicy roots and tubers of **crops** and gardens. They also dig deeper tunnels in which they nest, rest, and store food. Their living tunnels are usually blocked at the end and are not noticeable from above except for a fan of **earth** that spreads out from where the opening would be. This fan may be as much as 6 in (15 cm) high.

These rodents do not hibernate, so their food stored during the summer must last them through the winter. They bring **plant** stems into their burrows in one of two ways. If remaining underground, they can eat the roots and then pull the plant stem down through the **soil** and carry it into their burrow. However, sometimes they go outside at night. Then they bite off plant stems and drag them back to their burrows. They also collect food in their cheek pouches. These externally opening pouches can be turned inside out for cleaning, after which a muscle pulls them back right-side-out again.

Gopher burrows do not support colonies of gophers. They are solitary animals, although so many of them can live so close to each other that they may seem to an observer to be part of a colony. This closeness allows them readily to find mates. A female takes a male into her own burrow for mating. He leaves and she remains to raise her litter. A female gives **birth** to four or five young usually only once each year, although some breed twice a year. The young are weaned and out on their own, digging their own burrows, within a month or two. Gophers rarely live more than two years.

When gophers are out of their burrows at night, they readily fall **prey** to **owls** and snakes. Their burrows may be dug up by foxes and coyotes. However, these small diggers may still be safe because they have the ability to run backward in their burrows almost as fast as they can move forward. Their sensitive tails are used in determining their direction.

Farmers tend to kill gophers because of the way they can destroy crops from the roots up. However, burrowing gophers keep the soil aerated and well-turned.

The southeastern pocket gopher (*Geomys pinetis*) of Florida, and coastal Georgia and Alabama is threatened with **extinction**. Its **habitat** has fallen prey to development.

See also Chipmunks.

Resources

Books

Caras, Roger A. *North American Mammals: Fur-Bearing Animals of the United States and Canada.* New York: Meredith Press, 1967.

Knight, Linsay. *The Sierra Club Book of Small Mammals.* San Francisco: Sierra Club Books for Children, 1993.

Jean F. Blashfield

Gorillas

Gorillas inhabit **forests** of Central **Africa**, and are the largest and most powerful of all **primates**. Adult males stand 6 ft (1.8 m) upright (although this is an unnatural position for a gorilla) and weigh up to 450 lb (200 kg), while females are much smaller. Gorillas live up to about 44 years. Mature males (older than 13 years), or silverbacks, are marked by a band of silver-gray hair on their back; the body is otherwise dark-colored.

Gorillas live in small family groups of several females and their young, led by a dominant silverback male. The females comprise a harem for the silverback,

A nine-year-old male lowland gorilla playing with a two-year-old juvenile. *Photograph by Tom McHugh/Photo Researchers, Inc. Reproduced by permission.*

who holds the sole mating rights in the group. Female gorillas produce one infant after a gestation period of nine months. The large size and great strength of the silverback are advantages in competing with other males for dominance of the group, and in defending against outside threats.

Gorillas are herbivores. During the day these ground-living **apes** move slowly through the forest, selecting **species** of leaves, fruit, and stems to eat from the surrounding vegetation. Their home range is about 9-14 square miles (25-40 sq km). At night the family group sleeps in trees, resting on platform nests that they make from branches; silverbacks usually **sleep** at the foot of the **tree**.

Gorillas belong to the family Pongidae, which also includes **chimpanzees**, orangutans, and gibbons. Chimpanzees and gorillas are the **animal** species most closely related to humans. Gorilla numbers are declining rapidly,

and only about 50,000 survive in the wild. There are three subspecies: the western lowland gorilla (*Gorilla gorilla gorilla*), the eastern lowland gorilla (*G. g. graueri*), and the mountain gorilla (*G. g. beringei*). Recent population estimates are 44,000 western lowland gorillas, 3,000-5,000 eastern lowland gorillas, and fewer than 400 mountain gorillas. All species are endangered, and the mountain gorilla critically so.

The rusty-gray, western lowland gorilla is found in Angola, Cameroon, Central African Republic, Congo, Equatorial Guinea, Gabon, Nigeria, and the Republic of Congo (formerly Zaire). The black-haired eastern lowland gorilla is found in eastern Republic of Congo. **Deforestation** and hunting are serious and intensifying threats to lowland gorillas throughout their range.

The mountain gorilla has been well-studied in the field, notably by George Schaller and Dian Fossey (the film *Gorillas in the Mist* is based on the work of Fossey). This critically endangered subspecies inhabits forest in the **mountains** of eastern Rwanda, Republic of Congo, and Uganda at altitudes up to 9,000 ft (3,000 m). Field research has shown these powerful primates to be intelligent, peaceful, shy, and of little danger to humans (unless provoked).

Other than humans, adult gorillas have no important predators, although leopards occasionally take young individuals. Illegal hunting, capture for the live-animal trade (a mountain gorilla is reputedly worth $150,000), and **habitat** loss are causing populations of all gorillas to decline rapidly. The shrinking forest refuge of these great apes is being progressively deforested to accommodate the ever-expanding human population of all countries of Central Africa. Mountain gorillas are somewhat safeguarded in the Virunga Volcanoes National Park in Rwanda, although the recent civil war there has threatened their population and status. The protection of gorillas in that park has been funded by closely controlled, small-group, gorilla-viewing **ecotourism**, existing alongside long-term field research programs, although these enterprises were seriously disrupted by the civil war.

All three subspecies of gorillas are in serious trouble. These evolutionarily close relatives of humans could easily become extinct if people do not treat them and their habitat in a more compassionate manner.

Resources

Books

Dixson, A.F. *The Natural History of the Gorilla.* New York: Columbia University Press, 1981.

Fossey, D. *Gorillas in the Mist.* Boston: Houghton Mifflin, 1983.

Fossey, D. *The Year of the Gorilla.* Chicago: University of Chicago Press, 1988.

Schaller, G.B. *The Mountain Gorilla: Ecology and Behavior.* Chicago: University of Chicago Press, 1988.

Periodicals

Gouzoules, Harold. "Primate Communication By Nature Honest, Or By Experience Wise." *International Journal of Primatology* 23, no. 4 (2002): 821-848.

Maestripieri, Dario. "Evolutionary Theory And Primate Behavior." *International Journal of Primatology* 23, no. 4 (2002): 703-705.

"Profile: Ian Redmond: An 11th-Hour Rescue for Great Apes?" *Science* 297 no. 5590 (2002): 2203.

Sheeran, L. K. " Tree Of Origin: What Primate Behavior Can Tell Us About Human Society." *American Journal off Human Biology* 14, no. 1 (2002): 82-83.

Neil Cumberlidge

Gortex *see* **Artificial fibers**

Gourd family (Cucurbitaceae)

Gourds and their relatives are various **species** of plants in the family Cucurbitaceae. There are about 750 species in this family divided among 90 genera. Some members of the gourd family include the cucumber, squash, melon, and pumpkin. Most species of gourds are tropical or subtropical, but a few occur in temperate climates. A few species in the gourd family produce large, edible **fruits**, and some of these are ancient food plants. Gourds are still economically important as foods and for other reasons.

Biology of gourds

Plants in the gourd family are herbaceous or semi-woody, climbing or trailing plants. Their leaves are commonly palmately lobed or unlobed and are arranged in an alternate fashion along the stem. Special structures known as tendrils develop in the area between the **leaf** and the stem in some species of gourds. The thin tendrils grow in a **spiral** and help to anchor the stem as it climbs or spreads over the ground.

The flowers of species in the gourd family are unisexual, containing either male stamens or female pistils, but not both. Depending on the species, **individual** plants may be monoecious and have unisexual flowers of both sexes, or dioecious, meaning only one sex is represented on the **plant**. The flowers of gourds are radially symmetric, that is, the left and right halves look identical. They can be large and trumpet-shaped in some species. The petals are most commonly yellow or white.

Strictly speaking, the fruits of members of the gourd family are a type of berry, that is, a fleshy, multi-seeded fruit. In this family, these fruits are sometimes known as pepos. The pepos of some cultivated varieties of squashes and pumpkins can be enormous, weighing as much as hundreds of pounds and representing the world's largest fruits. In many species of gourds, the fruit is indehiscent, meaning it does not open when ripe in order to disperse the **seeds**. With few exceptions, the natural dispersal mechanisms of the pepos of members of the Cucurbitaceae are animals which eat the fruit and later deposit the seeds when they defecate some distance away from the parent plant.

The seeds of plants in the Cucurbitaceae are usually rather large and flattened, and they commonly have a large **concentration** of oils.

Agricultural species of gourds

Various species in the gourd family are cultivated as agricultural **crops**. The **taxonomy** of some of the groups

Yellow crookneck squash (*Cucurbita pepo*). *JLM Visuals. Reproduced with permission.*

of closely related species is not yet understood. For example, some of the many distinctive varieties of pumpkins and squashes are treated by some taxonomists as different species, whereas other botanists consider them to be a single, variable species complex under the scientific name, *Cucurbita pepo*. This taxonomic uncertainty is also true for some of the other agricultural groups of gourds such as the melons.

The most important of the edible gourds are of two broad types-the so-called "vegetable fruits" such as cucumber, pumpkin, and squash, and the sweeter melons.

The cucumber (*Cucumis sativus*) is an annual plant, probably originally native to southern **Asia** but possibly to India. This species has been cultivated in Asia for at least 4,000 years. The cucumber grows as a rough-stemmed, climbing, or trailing plant with large and yellow flowers. The fruit of the cucumber is an elongate, usually green-skinned pepo with a fairly tough, exterior rind but a very succulent interior which is about 97% moisture. Most cucumber fruits contain many white seeds, but seedless varieties have been developed by plant breeders, for example, the relatively long, "English" cucumber. Cucumbers are most productively grown in fertile, organic-rich soils, either outdoors or in greenhouses. Cucumbers come in various agricultural varieties. The fruit of the larger cucumbers is mostly used in the preparation of fresh salads or sometimes cooked. Pickles are made from smaller-fruited varieties of the cucumber or from a close relative known as the gherkin (*Cucumis anguria*), probably native to tropical **Africa**. Cucumber and gherkin pickles are usually made in a **solution** of vinegar often flavored with garlic and dill or in a sweeter pickling solution.

The pumpkin, squash, vegetable marrow, or ornamental gourd (*Cucurbita pepo*) is an annual, climbing, or trailing species with prickly stems, large, deeply cut leaves, yellow flowers, and large fruits. This species was originally native to a broad range from Mexico to Peru. There are many cultivated varieties of this species, the fruits of which are of various shapes and sizes and with rinds of various colors. Some recently developed varieties of pumpkins and squashes can grow gigantic fruits, each weighing as much as 882 lb (400 kg) or more. The pepos of pumpkins and squashes have a relatively thick rind, and a moist, fibrous interior. These plants can be baked or steamed as a vegetable and are often served stuffed with other foods. The seeds can be extracted, roasted, and salted, and served as a snack, or they can be pressed to extract an edible oil. Some varieties of gourds have been bred specifically for their beautiful fruits, which may be displayed either fresh or dried in ornamental baskets and in decorative centerpieces for dining-room or kitchen tables.

The melon, muskmelon, winter melon, cantaloupe, or honey dew (*Cucumis melo*) is a climbing or spreading annual plant with many cultivated varieties. The species was probably originally native to southern Africa, or possibly to southeastern Asia. The large, roughly spherical fruits of this species have a yellow or orange sweet interior which can be eaten fresh. This species occurs in many varieties which are often grown in greenhouses or outside in warmer climates.

The watermelon (*Citrullus lanatus*) is a large, annual species, probably native to tropical Africa where it has long been an important food for both people and wild animals. The watermelon has been cultivated in southern **Europe** for at least 2,000 years and is now grown worldwide wherever the climate is suitable. The fruits of the watermelon are large, reaching 55 lb (25 kg) in some cases. The watermelon has a thick, green rind, and the interior flesh is red or yellow and very sweet and juicy. A variety called the citron or preserving melon is used to make jams and preserves.

Some other cultivated species in the gourd family are minor agricultural crops. The chayote (*Sechium edule*), a perennial species of tropical Central America, produces a pepo that is cooked as a vegetable. The underground **tuber** of the chayote can also be eaten as can be the young leaves and shoots. The bitter apple or colocynth (*Citrullus colocynthis*) also produces a pepo that is eaten as a cooked vegetable.

The fruits of the loofah, luffa, vegetable sponge, or dish-rag gourd (*Luffa cylindrica*) have many uses. To expose the stiff, fibrous interior of the pepos of this plant, the ripe fruits are immersed in **water** for 5-10 days after which the skin and pulp are easily washed away. The skeletonized interior of the fruit is then dried and is commonly used as a mildly abrasive material, sometimes known as a loofah sponge. This has commonly been used for scouring dishes or for bathing. Loofah material has also been used for many other purposes, including as insulation, as a packing material, and to manufacture filters.

The fruits of the white-flowered gourd or bottle-gourd (*Lagenaria siceraria*) have long been used by ancient as well as modern peoples of both the tropical and subtropical Americas and Eurasia, as far as the Polynesian Islands. The dried, hollowed fruits of this plant are used as jugs, pots, baskets, and utensils, especially as dipping spoons. In addition, varieties with long necks have been used as floats for fishing nets. Rattles are also made of these dried squashes.

More on the *Cucurbita* squashes of the Americas

The pre-Columbian aboriginal peoples of North, Central, and **South America** cultivated or otherwise used about 17 species of squashes and gourds in the genus *Cucurbita*, a genus indigenous to the Americas.

The fruits of *Cucurbita*s were used by Native Americans in many ways, and some of these practices still persist. The ripe fruits can be cooked and eaten as **vegetables**. The fruits of several species are especially useful as foods because they can be stored for several months without rotting. For even longer-term storage, the squashes can be cut into strips and dried in the **sun**. In addition, the nutritious, oil-rich seeds of these gourds can be eaten fresh or roasted, and they also store well.

The best-known species of squash is *Cucurbita pepo*, the progenitor of the important cultivated pumpkins and squashes, as well as numerous other useful cultivars. According to archaeological evidence, this species has been used by humans for as long as 7,000 years. Other cultivated species include several known as winter squash or pumpkin (*Cucurbita mixta*, *C. moschata*, and *C. maxima*) and the malabar or fig-leaf gourd (*C. ficifolia*).

The buffalo gourd or chilicote (*C. foetidissima*) is a species native to the southwestern United States and northern Mexico. This is a relatively drought-resistant, perennial species and was commonly harvested by pre-Columbian Native Americans, although they apparently did not cultivate the plant.

Additional gourds native to North America

Most species in the gourd family are tropical and subtropical in their distribution. However, a few species occur in the north-temperate zone, including several native to **North America**. These wild plants are not eaten by people.

The creeping cucumber (*Melothira pendula*) is widespread in woods in the United States and south into Mexico. The bur-cucumber (*Sicyos angulatus*) occurs in moist habitats from southeastern Canada to Florida and Arizona.

KEY TERMS

Berry—A soft, multi-seeded fruit developed from a single compound ovary.

Dioecious—Plants in which male and female flowers occur on separate plants.

Indehiscent—Refers to a fruit that does not spontaneously split along a seam when it is ripe in order to disperse the seeds.

Monoecious—A plant breeding system in which male and female reproductive structures are present on the same plant, although not necessarily in the same flowers.

Pepo—A berry developed from a single, compound ovary and having a hard, firm rind and a soft, pulpy interior.

Tendril—A spirally winding, clinging organ that is used by climbing plants to attach to their supporting substrate.

The balsam apple or squirting cucumber (*Echinocystis lobata*) is an annual, climbing plant that occurs in moist thickets and disturbed places over much of southern Canada and the United States. When the green, inflated, spiny fruits of the squirting cucumber are ripe, they eject their seeds under hydrostatic **pressure** so they are dispersed some distance away from the parent plant.

Resources

Books

Brucher, H. *Useful Plants of Neotropical Origin and Their Wild Relatives.* New York: Springer-Verlag, 1989.

Hvass, E. *Plants That Serve and Feed Us.* New York: Hippocrene Books, 1975.

Judd, Walter S., Christopher Campbell, Elizabeth A. Kellogg, Michael J. Donoghue, and Peter Stevens. *Plant Systematics: A Phylogenetic Approach.* 2nd ed. with CD-ROM. Suderland, MD: Sinauer, 2002.

Klein, R.M. *The Green World. An Introduction to Plants and People.* New York: Harper and Row, 1987.

Whitaker, T.W., and G.N. Davis. *Cucurbits. Botany, Cultivation, and Utilization.* New York: Interscience Pub., 1962.

Bill Freedman

Graft

A graft is a horticultural term for a bud or shoot of one variety or **species** of **plant** that is positioned on the stem of another, compatible plant, in such a way that integrated growth results. The recipient plant is called the stock or rootstock, and the grafted part is referred to as the scion. A simple method for stem grafting involves both stems being cut with a sharp blade at the same acute **angle**, in order to maximize the area of contact. Then the stems are joined, and the union is bandaged with waterproof tape (or tape plus wax) until the wound has healed. Variations on this method involve complementary notches and the tongue being cut, according to how sturdy the scion is. Budding is the term applied when a bud with supporting **tissue** is grafted into a slit or notch cut into the stem of the stock.

Compatibility and incompatibility

The process of wound healing is absolutely necessary for successful grafting. Healing involves the cooperative production of new cells, some of which form cambium. From the cambium, new vascular (transport) tissues develop, permitting the transfer of **water**, **nutrients**, and **hormones** (growth regulators) to and from the scion. This interaction at the cellular level requires that the scion not be rejected by the stock. Hence, grafting is most likely to succeed with plants that are very closely related: either varieties of the same species, or members of the same genus. However, not all members of the same genus are compatible with each other. Sometimes the union can only be successful if one member is always the rootstock. For example, within the genus *Prunus*, peach scions cannot be grafted onto plum rootstocks, but plums can be grafted onto peach. Surprisingly, some pears (*Pyrus* species) can be grafted onto quince (*Cydonia oblonga*), despite the generic difference. Whether a particular combination is compatible can only be discovered by testing.

Advantages of grafting

Despite being labor intensive, grafting is commonly undertaken as a means of vegetative propagation of woody plants for any or all of the following reasons: (1) to impart **disease** resistance or hardiness, contributed by the rootstock; (2) to shorten the **time** taken to first production of flowers or **fruits** by the scion, in some cases by many years; (3) to dwarf the scion, making both its height and shape more convenient for harvesting fruit, as with apples; (4) to allow scion cultivars to retain their desirable **leaf**, floral, or fruit characters, without the risk of these being lost through **sexual reproduction**; and (5) to provide the most economic use of scion material, in cases where there is some difficulty with stem cuttings producing roots.

History and important examples of grafting

The origin of grafting is uncertain. The peoples of ancient civilizations who grew fruit trees may have observed natural unions made by twigs and branches of compatible trees growing next to one another, and copied what had occurred through **wind** and abrasion. Grafting was applied routinely to apples and pears in England by the eighteenth century, and was utilized to great effect by the English plant breeder Thomas Andrew Knight. Thomas Jefferson wrote that he had "inoculated common cherry buds into stocks of large kind" in 1767, in a Garden Journal he kept for his residence Monticello in Virginia. Jefferson's record predates the work of Knight, and indicates that knowledge of grafting techniques was widespread at that time.

Disease resistance

The rescue of the European wine-grape (*Vitis vinifera*) industry from the ravages of Phylloxera disease depended on grafting European cultivars onto Phylloxera-resistant rootstocks of native American species: the northern fox grape (*Vitis labrusca*) and the southern muscadine (*V. rotundifolia*). Since 1960 another American species, *V. champini*, has been widely utilized to confer additional resistance to *V. vinifera* to root-knot nematode worms. This new rootstock also confers **salt** tolerance, and hence is particularly useful for sultana **grapes** grown under **irrigation**.

The practice of grafting onto disease-resistant stocks now extends even to annual plants like tomato. Disease-sensitive cultivars producing high quality fruit, such as Grosse Lisse, are grafted onto wilt and nematode resistant stocks of varieties that would themselves produce fruit deficient in flavor and nutrients.

Hardiness of citrus trees

Most cultivated **citrus trees** are propagated by grafting desirable types onto hardy rootstocks. For example popular lemon such as Eureka, which has few thorns, is grafted as a bud onto a thorny wild or rough lemon (all lemons are *Citrus limon*). For other types of citrus such as grapefruit and orange (*C. sinensis*), use of the rough lemon or sour orange (*C. aurantium*) as rootstock has been discontinued in favor of the wild orange (*C. trifoliata*). As a rootstock the latter species can tolerate wetter conditions than the other stocks, and its use does not diminish the quality of sweet oranges as rootstocks of rough lemon do.

Hardiness in flowering shrubs

Among cool-temperate ornamental flowering shrubs, the **lilac** (*Syringa vulgaris*) is often grafted onto privet (*Ligustrum* species), another example of rare, cross-generic compatibility. Rhododendrons, many of

KEY TERMS

Cambium—A layer of actively dividing cells, from which tissues used for conducting water and nutrients (xylem, phloem) are derived.

Graft incompatibility—The failure of a scion to establish a viable connection with a rootstock, sometimes involving active rejection by release of toxins.

Hardiness—The ability of a plant to withstand environmental stresses, such as extremes of temperature, low soil fertility, waterlogging, salinity, drought, ultraviolet light, or shade.

Hybrid—A plant derived by crossing two distinct parents, which may be different species of the same genus, or varieties of the same species.

Phylloxera—A fatal disease of grape vines caused by an infestation of the aphid *Dactylasphaera vitifoliae* in the roots.

Rootstock—The basal component of a grafted plant.

Scion—The upper or transferred component of a grafted plant.

Vegetative propagation—A type of asexual reproduction in plants involving production of a new plant from the vegetative structures—stem, leaf, or root—of the parent plant.

which have been deliberately bred for variants of **flower** size and **color**, are usually grafted onto a rootstock of *Rhododendron ponticum*. This species has pale purple flowers and is native from Spain and Portugal to Turkey. *Rhododendron ponticum* was the first rhododendron introduced to England in the mid-eighteenth century, and it is still the hardiest rootstock available, even surviving fires that destroy the above-ground scion.

See also Citrus trees; Plant breeding.

Resources

Books

Hartmann, H.T., et. al. *Plant Science: Growth, Development and Utilization of Cultivated Plants.* 2nd ed. Englewood Cliffs, NJ: Prentice-Hall, 1988.

Judd, Walter S., Christopher Campbell, Elizabeth A. Kellogg, Michael J. Donoghue, and Peter Stevens. *Plant Systematics: A Phylogenetic Approach.* 2nd ed. with CD-ROM. Suderland, MD: Sinauer, 2002.

David R. Murray

Grains *see* **Crops**

Grand unified theory

One of the major theoretical hurdles to a reachable synthesis of current theories of particles and **force** interactions into a grand unification theory (also known as Grand Unified Field Theory, Grand Unified Theory, or GUT) is the need to reconcile the evolving principles of quantum theory with the principles of general relativity advanced by German-American physicist Albert Einstein (1879-1955) nearly a century ago. The synthesis is made difficult because the unification of **quantum mechanics** (itself a unification of the laws of **chemistry** with atomic **physics**) with special relativity to form a complete quantum field theory consistent with observable data is itself not yet complete.

A grand unified theory of physics is not within the reach of our present technology and there are also theoretical obstacles to formulating a Grand unified theory.

A grand unified theory is a theory that will reconcile the electroweak force (the unified forces of **electricity** and **magnetism**) and the strong force (the force that binds **quarks** within the atomic nucleus together). A grand unified theory that could subsequently incorporate gravitational theory would, become the ultimate unified theory, often referred to by physicists as a "theory of everything" (TOE).

The technological barriers to a unified theory are a consequence of the tremendous energies required to verify the existence of the particles predicted by the theory. In essence, experimental physicists are called upon to recreate the conditions of the universe that existed during the first few millionths of a second of the Big bang - when the universe was tremendously hot, dense, and therefore energetic.

There are admittedly great difficulties and high **mountains** of inconsistency between quantum and relativity theory that may put such a "theory of everything" (TOE) far beyond our present grasp. Some scientists speculate that although a TOE is beyond our reach, we may be within reach of a grand unified theory (GUT) that, excepting quantum gravity, will unite the remaining fundamental forces.

Quantum theory was principally developed during the first half of the twentieth century through the independent **work** on various parts of the theory by German physicist Maxwell Planck (1858–1947), Danish physicist Niels Bohr (1885–1962), Austrian physicist Erwin Schrödinger (1887–1961), English physicist P.A.M. Dirac (1902–1984) and German physicist Werner Heisenberg (1901–1976). Quantum mechanics fully describes wave particle duality, and the phenomena of su-

KEY TERMS

Electroweak force—A unification of the fundamental forces of electromagnetism (that light is carried by quantum packets termed photons manifested by alternating fields of electricity and magnetism) and the weak force.

Field theory—A concept first advanced Scottish physicist James Clerk Maxwell (1831–1879) as part of his development of the theory of electromagnetism to explain the manifestation of force at a distance without an intervening medium to transmit the force. Einstein's general relativity theory is also a field theory of gravity.

Fundamental forces—The forces of electromagnetism (light), weak force, strong force, and gravity. Aptly named, the strong force is the strongest force, but acts over only the distance of the atomic nucleus. In contrast, gravity is 10^{39} times weaker than the strong force and acts at infinite distances.

Gravitational force—A force dependent upon mass and the distance between objects. The English physicist and mathematician Sir Isaac Newton (1642–1727) set out the classical theory of gravity in his *Philosophiae Naturalis Principia Mathematica (Mathematical Principles of Natural Philosophy)*. According to classical theory, gravitational force, always attractive between two objects, increases directly and proportionately with mass of the objects but is inversely proportional to the square of the distance between the objects. According to general

relativity, gravity results from the bending of fused space-time. According to modern quantum theory, gravity is postulated to be carried by a vector particle termed a graviton.

Local gauge invariance—In physics, a concept that asserts that all field equations ultimately contain symmetries in space and time. Gauge theories depend on difference in values as opposed to absolute values.

Strong force (or Strong interactions)—A force that binds quarks together to form protons and neutrons and hold protons and neutrons—and to hold together the electrically repelling positively charged protons within the atomic nucleus.

Unified field theory—In physics, a theory describing how a single set of particles and fields can become (or underlie) the observable fundamental forces of the electroweak force (electromagnetism and weak force unification) and the strong force.

Virtual particles—A particle that is emitted and then reabsorbed by particles involved in a force interaction (e.g., the exchange of virtual photons between charged particles in involved in electromagnetic force interactions).

Weak force—The force that causes transmutations of certain atomic particles. For example, weak force interactions in beta decay change neutrons and protons allowing Carbon-14 to decay into Nitrogen at a predictable rate useful in Carbon-14 dating.

perposition in term of probabilities. Quantum field theory describes and encompasses **virtual particles** and renormalization.

In contrast, special relativity describes space-time **geometry** and the relativistic effects of different inertial reference frames (i.e., the relativity of describing **motion**) and general relativity describes the nature of gravity. General relativity fuses the dimensions of **space** and **time**. The motion of bodies under apparent gravitational force is explained by the assertion that, in the vicinity of **mass**, space-time curves. The more massive the body the greater is the curvature or "force of gravity."

Avoiding the mathematical complexities, a fair simplification of the fundamental incompatibility between quantum theory and relativity theory may be found in the difference between the two theories with respect to the nature of the gravitational force. Quantum theory depicts a quantum

field with a carrier particle for the gravitational force—that although not yet discovered—is termed a graviton. As a force carrier particle, the graviton is analogous to the **photon** that acts as the boson or carrie of the **electromagnetism** (i.e., **light**). In stark contrast, general relativity theory does away with the need for the graviton by depicting gravity as a consequence of the warping or bending of space-time by **matter** (or, more specifically, mass).

Although both quantum and relativity theories work extremely well in explaining the universe at the quantum and cosmic levels respectively, the theories themselves are fundamentally incompatible and hence the search for unification theories.

Such unifications are not trivial mathematical or rhetorical flourishes; they evidence an unswerving trail back towards the beginning of time and the creation of the universe in the big bang. What the electroweak unification reveals is that at higher levels of **energy**, (e.g., the energies associated with the big bang), the forces of electromagnetism and the weak force are really one in the same. It is only at the more modest present state of the universe, far cooler and less dense, that the forces take on the characteristic differences of electromagnetism and the weak force.

Experiments at high energy levels have revealed the existence of a number of new particles. According to modern field theory and the **Standard model**, particles are manifestations of field and particles interact (exert forces) through fields. Accordingly, for every particle (e.g., quarks and leptons—one form of a lepton is the **electron**) there must be an associate field. Forces between particles result from the exchange of particles that are termed virtual particles. Electromagnetism depends upon the exchange of photons (QED theory). The weak force depends upon the exchange of W^+, W^-, and Z_o particles. Eight different forms of gluons are exchanged in a gluon field to produce the strong force. Regardless, the energy requirements required to identify the particles associated with a unified field required by a grand unified theory are greater than present technologies can achieve. Most mathematical calculations involving quantum fields indicate that unification of the fields may require 1016 GeV. Some models allow the additional fusion of the gravitational force at 1018 GeV.

The higher energies needed are not simply a question of investing more time and money in building larger **accelerators**. Using our present technologies, the energy levels achievable by a particle accelerator are proportional to the size of the accelerator (specially the diameter of the accelerator). Alas, to archive the energy levels required to find the particles of a grand unified force would require an accelerator larger than our entire **solar system**.

Although a quantum explanation of gravity is not required by a grand unification theory that seeks only to reconcile electroweak and strong forces, it is important to acknowledge that the unification of force and particle theories embraced by the Standard model is not yet complete. Further, it may not be possible to rule out gravity and develop a unified theory of electroweak and strong forces that ignores gravity.

See also Atomic theory; Electromagnetic spectrum; Feynman diagrams; Gravity and gravitation; Particle detectors; Relativity, general; Subatomic particles.

Resources

Books

Feynman, Richard and Steven Weinberg *Elementary Particles and the Laws of Physics.* Cambridge, UK: Cambridge University Press, 1987.

Greene, Brian. *The Elegant Universe: Superstrings, Hidden Dimensions, and the Quest for the Ultimate Theory.* New York: Vintage books, 2000.

Gribbin, John. *Q is for Quantum: An Encyclopedia of Particle Physics.* New York: The Free Press, 1998.

Hawking, Stephen. *The Illustrated Brief History of Time, Updated and Expanded.* New York: Bantam, 2001.

Klein, Etienne, et al. *The Quest for Unity: The Adventure of Physics.* Oxford, UK: Oxford University Press, 2000.

Mohapatra, Rabindra. *Unification and Supersymmetry.* Oxford, UK: Oxford University Press, 2002.

Periodicals

Weinberg, Steven. "A Unified Physics by 2050?" *Scientific American.* December, 1999.

Other

Particle Data Group. Lawrence Berkeley National Laboratory. "The Particle Adventure: The Fundamentals of Matter and Force" [cited February, 5, 2003]. <http://particleadventure.org/particleadventure/>.

K. Lee Lerner

Grapefruit tree *see* **Citrus trees**

Grapes

Grapes are various **species** of woody vines in the genus *Vitis*, family Vitaceae. This family contains about 700 species most of which occur in tropical and subtropical climates, although some occur in temperate habitats. The genus *Vitis* has about 50 species. Grapes are ecologically important as food for **wildlife**. They are also cultivated by humans in large quantities, mostly for the production of table grapes, raisins, and wines.

Biology of grapes

Grapes are perennial, woody vines. They often form thickets along **rivers** and other naturally open habitats, and often drape trees in open **forests** or at forest edges.

Grape leaves are entire, and they often have three distinct lobes. The leaves are alternately arranged along the stem. Opposite most leaves are structures known as tendrils which grow in a **spiral** fashion and are important in anchoring the vine to its supporting structure.

Grapes have small, inconspicuous flowers arranged in clusters. The flowers have associated nectaries which are important in attracting the **insects** that are the pollinators of grapes. The fruit of grapes is an edible, two-seeded berry, usually purple in **color**. Grapes are avidly

Fredonia grapes. *Photograph by James Sikkema. Reproduced by permission.*

eaten by **birds** and **mammals**. The grape seed passes intact through the gut of these animals and is deposited into the ground with feces. The edible fruit of grapes is an **adaptation** for dispersal by **animal** vectors.

Cultivated varieties, or cultivars, of grapes are usually propagated by grafting shoots of the desired type onto the root of a relatively hardy **plant**. In this way, the desirable traits of the cultivar will be displayed by the grafted shoot, while the grape grower can also take advantage of the adaptation of the rootstock to the local environment.

Native grapes of North America

Various species of grapes are native to **North America**. Some of the more widespread species are the muscadine-grape (*V. rotundifolia*), the fox-grape (*V. labrusca*), the summer-grape (*V. aestivalis*), the forest-grape (*V. vulpina*), and the river-bank grape (*V. riparia*). Most of these are species of moist sites, often growing luxuriantly along forest edges and in riparian habitats.

Wild grapes provide a nutritious and seasonally important food for many species of birds and mammals. Wild grapes also contribute to the pleasing aesthetic of some habitats, for example, when they luxuriously drape the edges of forests beside rivers and lakes.

Agricultural grapes

By far the most common species of cultivated grape is the wine grape (*Vitis vinifera*), probably native to southwest **Asia**, possibly in the vicinity of the Black Sea. This species may have been cultivated for as long as 7,000 years. The wine grape now occurs in hundreds of cultivated varieties and is planted in temperate climates in all parts of the world. The **fruits** of this species can be blue, yellow, or green in color, and they contain one to

four **seeds**. Ripe wine grapes typically contain 70% of their weight as juice and 20-24% as sugar. This species is widely grown in warm-temperate regions of **Europe**, especially in France and Italy, and to a lesser degree in Germany, Spain, and elsewhere. Other notable centers for the cultivation of wine grapes are California, Chile, **Australia**, Portugal, Russia, Algeria, and South **Africa**.

Two North American species of grapes are also cultivated for the production of wine. These are the fox-grape (*V. labrusca*) and, less commonly, the summer-grape (*V. aestivalis*). The skin of the fruits of the fox-grape separate quite easily from the interior pulp, which makes it easy to distinguish agricultural varieties of this species from the wine grape.

Grapes are often eaten fresh as a tasty and nutritious table fruit. Grapes can also be crushed to manufacture a highly flavorful juice. Grapes can also be preserved by drying, usually in the **sun**. Most dried grapes are called raisins, but dried seedless grapes are known as sultanas.

Wine is an alcoholic beverage that is produced by a careful **fermentation** of grape juice. The fermentation is carried out by the wine **yeast** (*Saccharomyces ellipsoides*), a microscopic fungus that occurs naturally on the surface of grapes. However, specially prepared strains of the wine yeast are generally used by commercial vintners in order to help ensure a constant, predictable fermentation and final product.

Wine yeast ferments the sugar content of the juice of pressed grapes into **carbon dioxide** and **ethanol**, a type of **alcohol**. The yield of alcohol is about 1% for every 2% of sugar in the juice, but the final alcohol **concentration** cannot exceed 12%, because this is the upper limit of tolerance of the yeast to alcohol in its growth medium. (Actually, there are wines with an alcohol concentration greater than 12%, but these are prepared by adding pure ethanol, a process known as *fortifying*.) The initial grape juice is prepared by pressing the ripe grapes. Originally, this was done by barefoot people stomping about in large wooden tubs. Today, however, the grapes are usually pressed using large machines. Red wines are obtained when the skins of blue grapes are left in with the fermenting juice. White wines are obtained when the skins are removed prior to the fermentation, even if the juice was pressed from red grapes.

The quality of the resulting wine is influenced by many factors. The variety, sugar content, and other aspects of the grapes are all important, as is the strain of wine yeast that is used. The **soil** conditions and climate of the growing region are also highly influential. The incubation **temperature** during the fermentation is important as is the sort of container that is used during this process. In addition, once the fermentation is stopped,

KEY TERMS

Cultivar—A distinct variety of a plant that has been bred for particular, agricultural or culinary attributes. Cultivars are not sufficiently distinct in the genetic sense to be considered to be subspecies.

Fermentation—This is a metabolic process during which organic compounds are partially metabolized, often producing a bubbling effervescence. During the fermentation of sugar, this compound is split into carbon dioxide and an alcohol.

Grafting—This is a method by which woody plants can be propagated. A shoot, known as a scion, is taken from one plant, and then inserted into a rootstock of another plant and kept wrapped until a callus develops. The genetically based, desirable attributes of the scion are preserved, and large numbers of plants with these characteristics can be quickly and easily propagated by grafting.

Raisin—A grape that has been preserved by drying.

Riparian—A moist habitat that occurs in the vicinity of streams, rivers, ponds, and lakes.

Sultana—A raisin produced by drying a seedless grape.

Tendril—A spirally winding, clinging organ that is used by climbing plants to attach to their supporting substrate.

Vine—A plant, usually woody, that is long and slender and creeps along the ground or climbs upon other plants.

the period of **time** during which the wine is stored can be important. However, a storage which is too long can be detrimental because the alcohol in the wine may be spoiled by a further **metabolism** of the ethanol into **acetic acid**, or vinegar.

Grapes in horticulture

Some species of grapes are occasionally used in **horticulture**. The desired utilization is generally as a wall covering and sometimes for the visual aesthetics of the foliage in the autumn. Species commonly grown for these horticultural purposes are *Vitis vinifera* and *V. coignetiae*. The Virginia creeper (*Parthenocissus quinquefolia*) is a closely related native species that is also often used for these purposes as is the introduced Boston ivy (*P. tricuspidata*).

See also Graft.

Resources

Books

Judd, Walter S., Christopher Campbell, Elizabeth A. Kellogg, Michael J. Donoghue, and Peter Stevens. *Plant Systematics: A Phylogenetic Approach.* 2nd ed. with CD-ROM. Suderland, MD: Sinauer, 2002.

Klein, R. M. *The Green World. An Introduction to Plants and People.* New York: Harper and Row, 1987.

Raven, Peter, R. F. Evert, and Susan Eichhorn. *Biology of Plants.* 6th ed. New York: Worth Publishers Inc., 1998.

Bill Freedman

Graphite *see* **Carbon**

Graphs and graphing

In **mathematics**, a graph is a geometric representation, a picture, of a **relation** or function. A relation is a subset of the set of all ordered pairs (x,y) for which each x is a member of some set X and each y is a member of another set Y. A specific relationship between each x and y determines which ordered pairs are in the subset. A function is a similar set of ordered pairs, with the added restriction that no two ordered pairs have the same first member. A graph, then, is a pictorial representation of the ordered pairs that comprise a relation or function. At the same time, it is a pictorial representation of the relationship between the first and second elements of each of the ordered pairs.

Representing ordered pairs

In 1637, René Descartes (1594-1650), the French mathematician and philosopher, published a book entitled *Géométrie*, in which he applied algebraic methods to the study of **geometry**. In the book, Descartes described a system (now called the rectangular coordinate system or the Cartesian coordinate system) for using points in a **plane** to represent ordered pairs. Given any two sets X and Y, the Cartesian product (written X × Y) of these two sets is the set of all possible ordered pairs (x,y) formed by choosing an element x from the set X and pairing it with an element y from the set Y. A relation between two sets X and Y, is a subset of their Cartesian product. To graph a relation, it is first necessary to represent the Cartesian product geometrically. Then, the graph of a particular relation is produced by highlighting that part of the representation corresponding to the points contained in the relation. Geometrically, the Cartesian product of two sets is represented by two **perpendicular** lines, one horizontal, one vertical, called axes. The point where the axes intersect is called the origin. Members of the set X are represented in this picture by associating each member of X with points on the horizontal axis (called the x-axis). Members of the set Y are represented by associating each member of Y with points on the vertical axis (called the y-axis). It is interesting to note that this picture is easily extended to three dimensions by considering the Cartesian product of the sets (X × Y) and Z. Z is then represented by a third axis perpendicular to the plane that represents the ordered pairs in the set (X × Y). Having established a picture of the set of all possible ordered pairs, the next step in producing a graph is to represent the subset of ordered pairs that are contained in a given relation. This can be done in a number of ways. The most common are the bar graph, the scatter graph and the line graph.

Bar graphs

A bar graph is used to picture the relationship between a relatively small number of objects, such as information listed in tabular form. Tables often represent mathematical relations, in that they consist of ordered pairs (listed in rows) for which the first and second elements of each pair (listed in separate columns) are related in a specific way. For example, a department store receipt is a relation defined by a table. It lists each item purchased together with its retail price. The first element of each ordered pair is the item, the second element is that item's purchase price. This type of relation lends itself well to the bar graph, because it contains information that is not strictly numeric, and because there are relatively few ordered pairs. In this example the prices are represented by points on the vertical axis, while the items purchased are represented by short line segments centered about the first few positive **integers** on the horizontal axis. To create the graph, the price of each item is located on the vertical axis, and a bar of that height is filled in directly above the location of the corresponding item on the horizontal axis. The advantage of the bar graph is that it allows immediate comparison of the relative purchase prices, including identification of the most expensive and least expensive items. It also provides a visual means of estimating the average cost of an item, and the total amount of money spent.

Scatter graphs

The scatter graph is similar to the bar graph in that it is used to represent relations containing a small number of ordered pairs. However, it differs from the bar graph in that both axes can be used to represent sets of **real numbers**. Since it is not feasible to represent pairs of real numbers with bars that have some width, ordered pairs of the relation are plotted by marking the corresponding point with a small symbol, such as a **circle**, or **square**. Since the scatter graph represents relations between sets

of real numbers, it may also include **negative** as well as positive numbers. In producing a scatter graph, the location of each point is established by its horizontal **distance** from the y-axis and its vertical distance from the x-axis. Scatter graphs are used extensively in picturing the results of experiments. Data is generated by controlling one variable (called the independent variable) and measuring the response of a second variable (called the dependent variable). The data is recorded and then plotted, the independent variable being associated with the x-axis and the dependent variable with the y-axis.

Line graphs

Very often, a function is defined by an equation relating elements from the set of real numbers to other elements, also from the set of real numbers. When this is the case the function will usually contain an infinite number of ordered pairs. For instance, if both X and Y correspond to the set of real numbers, then the equation $y = 2x + 3$ defines a function, specifically the set of ordered pairs (x, 2x + 3). The graph of this function is represented in the rectangular coordinate system by a line. To graph this equation, locate any two points in the plane, then connect them together. As a check a third point should be located, and its position on the line verified. Any equation whose graph is a straight line, can be written in the form $y = mx + b$, where m and b are constants called the slope and y-intercept respectively. The slope is the **ratio** of vertical change (rise) to horizontal change (run) between any two points on the line. The y-intercept is the point where the graph crosses the y-axis. This information is very useful in determining the equation of a line from its graph. In addition to straight lines, many equations have graphs that are curved lines. **Polynomials**, including the **conic sections**, and the trigonometric functions (sine, cosine, tangent, and the inverse of each) all have graphs that are curves. It is useful to graph these kinds of functions in order to "picture" their **behavior**. In addition to graphing equations, it is often very useful to find the equation from the graph. This is how mathematical models of nature are developed. With the aid of computers, scientists draw smooth lines through a few points of experimental data, and deduce the equations that define those smooth lines. In this way they are able to model natural occurrences, and use the models to predict the results of future occurrences.

Practical applications

There are many practical applications of graphs and graphing. In the sciences and **engineering**, sets of numbers represent physical quantities. Graphing the relationship between these quantities is an useful tool for understanding nature. One specific example is the graphing of current versus voltage, used by electrical engineers, to

KEY TERMS

Cartesian product—The Cartesian product of two sets X and Y is the set of all possible ordered pairs (x, y) formed by taking the first element of the pair from the set X and the second element of the pair from the set Y.

Function—A function is a relation for which no two ordered pairs have the same first element.

Ordered pair—A pair of elements (x,y) such that the pair (y,x) is not the same as (x,y) unless x = y.

Relation—A relation between two sets X and Y is a subset of all possible ordered pairs (x,y) for which there exists a specific relationship between each x and y.

Variable—A variable is a quantity that is allowed to have a changing value, or that represents an unknown quantity.

picture the behavior of various circuit components. The rectangular coordinate system can be used to represent all possible combinations of current and voltage. Nature, however, severely limits the allowed combinations, depending on the particular electrical device through which current is flowing. By plotting the allowed combinations of current and voltage for various devices, engineers are able to "picture" the different behaviors of these devices. They use this information to design circuits with combinations of devices that will behave as predicted.

See also Variance.

Resources

Books

Bittinger, Marvin L, and Davic Ellenbogen. *Intermediate Algebra: Concepts and Applications.* 6th ed. Reading, MA: Addison-Wesley Publishing, 2001.
Larson, Ron. *Calculus With Analytic Geometry.* Boston: Houghton Mifflin College, 2002.
McKeague, Charles P. *Elementary Algebra.* 5th ed. Fort Worth: Saunders College Publishing, 1995.
Tobias, Sheila. *Succeed with Math.* New York: College Entrance Examination Board, 1987.

J. R. Maddocks

Grasses

Grasses are monocotyledonous plants in the family Poaceae (also known as Gramineae). There are as many as 10,000 **species** of grasses distributed among more than 600 genera. The richest genera of grasses are the panic-grasses (*Panicum* spp.) with 400 species, the blue-grasses (*Poa* spp.) and love-grasses (*Eragrostis* spp.) with 300 species each, and the needle-grasses (*Stipa* spp.) with 200 species.

Species of grasses occur worldwide in virtually any habitats that are capable of supporting vascular plants. Grasses are the dominant species in some types of natural vegetation such as prairies and steppes, and they are an important source of forage for many species of herbivorous animals. Some species of grasses are grown as agricultural **crops**, and these are among the most important foods for humans and domestic **livestock**. The most important of the agricultural grasses are maize, **wheat**, **rice**, **sorghum**, **barley**, and sugar cane.

Biology of grasses

Most grasses are annual plants or are herbaceous perennials that die back to the ground surface at the end of the growing season and then regenerate the next season by shoots developing from underground **rhizome** or root systems. A few species, such as the bamboos, develop as shrub- and tree-sized, woody plants.

The shoots of grasses typically have swollen nodes, or bases, and they are often hollow between the nodes. The leaves are usually long and narrow and have parallel **veins**. A specialized **tissue** called a ligule is usually present at the location where a **leaf** sheaths to the stem. The flowers of grasses are typically small, monoecious or dioecious, and are called florets. The florets have various specialized tissues, and often contain a long bristle called an awn, which can be quite prominent in some species. The florets are generally arranged into an inflorescence, or cluster, which can be quite large in some species. **Pollination** of grasses occurs when grass pollen is shed to the **wind** and carried opportunistically to other grasses. The **fruits** of grasses are known as a caryopsis or grain, are one-seeded, and can contain a large **concentration** of starch.

Native grasses of North America

Hundreds of species of grasses are native to **North America**. Native grasses are present in virtually all habitats, and they are among the most dominant plants in prairies, some types of marshes, and similar, herbaceous types of vegetation. In addition, many species of grasses have been introduced by humans from elsewhere, especially from western Eurasia.

Although many rich varieties of form and function are represented by the native grasses of North America, only a few of the most prominent species of selected, grass-dominated habitats will be briefly mentioned.

The temperate prairies of North America are dominated by herbaceous perennial plants, many of which are species of grasses. In the tall-grass prairies, some of the grasses can grow as high as 6.5 ft (2 m). Examples of these tall species include the big blue-stem (*Andropogon gerardi*), indian grass (*Sorghastrum nutans*), dropseed (*Sporobolus asper*), needle grass (*Stipa spartea*), panic grass (*Panicum virgatum*), wild rye (*Elymus virginicus*), and others. Somewhat drier sites support mixed-grass prairies containing shorter species, for example, little blue-stem (*Andropogon scoparius*), grama grass (*Bouteloua gracilis*), wheat grass (*Agropyron smithii*), and green needlegrass (*Stipa viridula*). The driest habitats support semi-arid, short-grass prairies with species such as grama grasses (*Bouteloua dactyloides and* B. gracilis), dropseed (*Sporobolus cryptandrus*), muhly grass (*Muhlenbergia torreyana*), and Junegrass (*Koerlia comata*).

Some species of grasses can be abundant in marshes, including the reed (*Phragmites communis*) which can reach a height greater than 13 ft (4 m) and is North America's tallest grass. The reed is a very widespread species, occurring in marshes on all of the continents. Some seaside habitats can also develop perennial **grasslands**. Sandy habitats are typically dominated by species of grasses such as the beach grass (*Ammophila breviligulata*), sand-reed (*Calamovilfa longifolia*), and beach rye (*Elymus mollis*). Salt-marshes are **brackish**, estuarine habitats that are typically dominated by cord grasses, such as *Spartina alterniflora* and *S. patens*, two species which segregate within the same salt-marshes on the basis of salinity and moisture gradients.

Although it was actually introduced to North America from **Europe**, the so-called Kentucky blue-grass (*Poa pratensis*) is now a very widespread species. Kentucky blue-grass is one of the most common species in lawns, and it also occurs widely in disturbed habitats.

Grasses in agriculture

In terms of the economic and nutritional values of foods provided for humans and domestic livestock, no other **plant** family is as important as the grasses. All of the important cereals and grains are members of the grass family, and some of these agronomic species have been cultivated for thousands of years. There are useful cereal species available for all of the climatic zones in which humans commonly live, and this has been one of the most important reasons why our species has been able to develop such large and prosperous populations during the past several thousand years.

Wheats

The bread wheat (*Triticum aestivum* or *T. vulgare*) is a very important grain species. The origins of the modern bread wheat are somewhat uncertain, because this species occurs in many **hybrid** varieties which have been selectively bred over time by complex, unrecorded hybridizations of various species of *Triticum*. Some botanists believe that the major progenitor species was an ancient cultivated wheat known as emmer (*T. dicoccum*) which was grown in the Middle East at least 5,000 years ago. Other ancient wheats which have also contributed to the genetic make-up of the modern bread wheat include einkorn (*T. monococcum*) from southwestern **Asia** and spelt (*T. spelta*) and durum (*T. durum*) from the Mediterranean region.

The numerous varieties of wheat have been bred for various purposes and climatic regimes. The flowering heads of wheats can have long awns as in the so-called "bearded" wheats, or they can be awnless. Wheat can be sown in the spring or in the previous autumn, known as winter wheat. Winter wheat generally has larger yields than spring wheat because it has a longer growing season. The so-called "soft" wheats are mostly used for baking breads and pastries, while the "hard" or durham wheats are used to prepare pastas and other types of noodles.

Wheat is rarely grown in subtropical or tropical climates because it is too susceptible to fungal diseases under warm and humid conditions. The best climatic regime for growing wheat involves a temperate climate with **soil** moisture available during the spring and summer while the plants are actively growing, and drier conditions later on while the **seeds** are ripening and when the crop is being harvested.

Certain landscapes of the temperate zones that used to support natural prairies and steppes are now the best regions for the cultivation of wheat. These include the mixed-grass and short-grass prairies of North America and similar zones in the pampas of **South America**, the steppes of western Russia and Ukraine, parts of central China and **Australia**, and elsewhere. Winter wheat tends to be the favored type grown in places where the environmental regime is more moderate, while spring wheats are sown under more extreme climatic conditions.

Wheat grains are manufactured into various edible products. Most important is flour, finely milled wheat, which is mostly used to bake breads, sweetened cakes, and pastries, and also for manufacturing into pastas and noodles. Wheat is also used to manufacture breakfast cereals, such as puffed wheat, shredded wheat, and fiber-rich bran flakes. Wheat grains are fermented in a mash to produce beer and other alcoholic beverages and also industrial **alcohol**. Wheat straw and hay are sometimes

Grasses grow in virtually any habitat that is capable of supporting plant life. *Photograph by Robert J. Huffman. Field Mark Publications. Reproduced by permission.*

used as fodder for animals or as stuffing, although the latter use is now uncommon because so many synthetic materials are available for this purpose.

Maize or corn

Maize, corn, or mealies (*Zea mays*) is derived from grasses native to Central America, probably from Mexico. Maize has a very distinctive, flowering structure, with a tassel of male flowers perched above the larger clusters of female flowers. Each of the several female **flower** clusters is an elongated, head-like structure known as a cob or **ear**, enclosed within sheathing leaves or bracts, known as husks. Each ear contains as many as several hundred female flowers, each of which may produce a seed known as a kernel. During the time that they are ripe for pollination, and the stigmas of the female flowers are borne outside of the sheathing leaves of the cob on very long styles known as corn silk.

As with wheat, maize occurs in a wide range of cultivated varieties bred for particular uses and climates. The maize plant has been so highly modified by selective breeding for agriculture that it is now incapable of repro-

ducing itself without the aid of humans. The ripe grains of the plant are no longer able to detach from their husk or cob (this is known as shattering). Moreover, the ripe kernels are tightly enclosed within their sheathing husks so that they are trapped by those leaves when they germinate. Modern maize can only be propagated if humans remove the leaves and grains from the cob and sow the ripe seeds.

Some of the presumed, wild ancestors of maize still occur in natural habitats in Mexico. One of these is teosinte (*Zea mexicana*), a wild grass that does not form a cob encased in husks. Another possible progenitor of maize is a grass called *Tripsacum mexicanum* that does not look much like corn but will readily hybridize with it. The wild relatives of maize are of enormous importance because they contain genetic variation that no longer is present in the highly inbred races of maize that exist today, particularly the varieties that are used widely in modern, industrialized agriculture. As such, some of the genetic information in the remaining wild species that participated in the cultural **evolution** of the modern maize plant may prove to be incredibly important in the future breeding of **disease** resistance, climatic tolerance, and other useful attributes of this critical food plant for humans.

Maize grows well under a hot and moist climatic regime, and it can be cultivated in both the tropical and temperate zones. Maize is used in many forms for direct consumption by humans. During the harvest season, much sweet corn is eaten after boiling or steaming. Maize is also eaten as a cooked porridge made of ground meal (in the southeastern United States, this food is known as grits). Other foods include canned or frozen cooked kernels, corn flakes, tortillas, corn chips, and popcorn. The small, unripe cobs of maize can also be steamed or boiled and eaten as a nutritious vegetable. Corn seeds can also be pressed to manufacture an edible oil.

Much of the maize crop in North America is fed to livestock. The nutritional value of the maize is greatly enhanced if the plants are chopped up and subjected to a **fermentation** process before being used for this purpose. This type of preparation which can also be prepared from other grasses and from mixed-species hay is known as silage.

In some regions such as the midwestern United States, much of the maize production is utilized to manufacture **ethanol** for use as a fuel in automobiles, usually blended with liquid **petroleum** hydrocarbons as a mixture known as gasohol. Other products made from maize include corn starch, corn syrup, and alcoholic beverages, such as some types of whiskey.

Rice

Rice (*Oryza sativa*) is probably a native of south Asia, and it has been cultivated on that **continent** for more than 5,000 years. The natural **habitat** of rice is tropical marshes, but it is now cultivated in a wide range of subtropical and tropical habitats.

If rice is being cultivated under flooded conditions, its seeds are germinated, grown until they are about 6-12 in (15-30 cm) tall, and then out-planted into the sediment in shallow **water**. In Asia, this cultivation system is known as paddy. A variant of this system is also used in the southern United States where fields are flooded to plant and grow the crop and then drained for optimal ripening and the harvest. Rice can also be cultivated under drier conditions, called "upland" rice, although the soil must be kept moist because the species is intolerant of **drought**. On moist, fertile sites in some parts of tropical Asia, two to four rice crops can be harvested each year, although this eventually could deplete the soil of its nutrient capital.

Rice is mostly eaten steamed or boiled, but it can also be dried and ground into a flour. Like most grains, rice can be used to make beer and liquors. Rice straw is used to make **paper** and can also be woven into mats, hats, and other products.

Other important agricultural grasses

Barley (*Hordeum vulgare*) is a relatively ancient crop species, having been grown in northeastern **Africa** and the Middle East for as long as 6,000 years. The environmental conditions favorable to the growth of barley are similar to those for wheat, although barley can be cultivated in somewhat cooler conditions and therefore farther to the north in Eurasia. Most barley is used as feed for domestic animals, but it is also used as a malt in **brewing** ale and other alcoholic beverages.

Rye (*Secale cereale*) is an agricultural grass that originated in Asia. This species is mostly cultivated in north-temperate regions of central Asia and Europe. The flour is used to make rye breads and crisp breads, and it is sometimes used in a mash to prepare rye whisky.

Oats (*Avena sativa*) probably originated in western Asia, and they have been cultivated for more than 2,000 years. Unlike most of the temperate, agricultural grasses, oats are relatively tolerant of late-summer and autumn rains. Oats are mostly used as fodder for cattle and **horses**, but they are also used to prepare breakfast cereals, such as rolled oats and oatmeal porridge. The Turkish oat (*A. orientalis*) and short oat (*A. brevis*) are relatively minor cultivated species.

Sorghum (*Sorghum bicolor*) is a small-grained cultivated species. Sorghum has been grown in Africa for at least 4,000 years, and it is still probably the most important crop for the making of bread flour on that continent. Sorghum is also widely used in Africa to prepare a mash for the brewing of beer. Some varieties of sorghum, known as broom-corns, are used to manufacture brushes, while others are used as forage crops.

Various other small-grain grasses are commonly known as millet. The most important species is the proso millet (*Panicum miliaceum*), which originated in tropical Africa or Asia, and has been cultivated for more than 5,000 years. This species is relatively tolerant of drought, and it is most commonly cultivated under drier climatic regimes in Africa and Asia. Proso millet is commonly eaten as a cooked porridge, and it is also an important ingredient in commercial birdseeds. More minor species of millets include pearl millet (*Pennisetum glaucum*), foxtail millet (*Setaria italica*), Japanese millet (*Echinochloa frumentacea*), shama millet (*E. colona*), barnyard millet (*E. crus-galli*), and ragi millet (*Eleusine coracana*).

Wild rice (*Zizania aquatica*) is a North American grass that grows naturally in shallow waters of temperate lakes and ponds, and has long been collected from the wild, usually by beating the ripe grains off their heads into a canoe, using a paddle. During the past several decades, however, this species has also been cultivated on farms in

Bamboo on Avery Island, Louisiana. *JLM Visuals. Reproduced with permission,*

the southwestern United States. This grain is relatively expensive, and is mostly used as an epicurean food and served with fine meals often mixed with *Oryza* rice.

Sugar cane

Sugar cane (*Saccharum officinarum*) is a very tall, tropical grass which can grow as high as 23 ft (7 m), most likely derived from wild plants that grew in marshes in India. The stems or canes of this species can be as thick as 2 in (5 cm), and they have a sweet pith that typically contains 20% of a sugar known as sucrose. The concentration of sugar varies greatly during the life cycle but is greatest when the cane is flowering, so this is when the harvest typically occurs. Sugar cane is propagated by planting sections of stems with at least one node, known as cuttings.

Sugar cane is grown widely in the subtropics and tropics; for example, in southern Florida, Cuba, and Brazil. Most of the harvest is manufactured into refined sucrose, or table sugar. Increasingly, however, sugar cane is used to manufacture alcohol as a fuel for vehicles.

Pasture grasses

Pasture grasses are species that are cultivated as nutritious fodder for agricultural animals such as cattle,

sheep, horses, and **goats**. These grasses are often grown in combination with fodder **legumes** to provide better **nutrition** for the livestock. The pasture foods may be eaten directly by the grazing animals, or they may be harvested, baled, dried, and used later as hay. In recent decades, there has been a great increase in the use of hay silage in which harvested pasture materials are stored under moist, oxygen-poor conditions for some time while **microorganisms** ferment some of the materials and develop a more nutritious product for the livestock.

Some of the pasture grasses that are commonly grown in North America include cock's-foot (*Dactylis glomerata*), timothy (*Phleum pratense*), meadow fox-tail (*Alopecurus pratensis*), and rye-grasses (*Lolium perenne* and *L. multiflorum*).

Other economic products obtained from grasses

The bamboos (*Bambusa* spp.) are fast-growing, woody species of grasses. The largest species of bamboos can grow taller than 131 ft (40 m) and can have a diameter of 12 in (30 cm). The most important genera of the larger bamboos are *Arundinaria, Bambusa, Dendrocalamus, Gigantochloa,* and *Phyllostachys*. These tree-sized grasses occur in **forests** and in cultivation in subtropical and tropical parts of the world. Bamboo stalks are woody and strong and are widely used as a building and scaffolding material, especially in Asia. Bamboo canes are also split and used for thatching and for many other purposes. The young shoots can be steamed or boiled and eaten as a vegetable.

Some tropical species of grasses have essential oils in their tissues, and these can be extracted and used in the manufacturing of perfumes. Oil of citronella is distilled from the foliage of citronella grass (*Cymbopogon nardus*) and is used as a scent and as an insect repellent. The lemongrass (*C. citratus*) and ginger-grass (*C. martinii*) also yield aromatic oils which are used as scents and in medicine.

Sweet grass (*Hierochloe odorata*) is an aromatic grass that grows in temperate regions of North America. This grass has long been used by Native Americans for basket weaving, and it is also smoked in culturally significant "sweetgrass" ceremonies.

The Job's tears (*Coix lachryma-jobi*) of southeast Asia produces large, white, lustrous seeds that can be eaten but are mostly used to make attractive necklaces, rosaries, and other decorations, often dyed in various attractive colors.

Grasses in horticulture

Some species of grasses are grown in **horticulture** as attractive foliage plants. Some varieties have been developed with variegated leaves, that is, with foliage that is mottled with green or white areas. Examples include reed canary-grass (*Phalaris arundinacea*) and bent-grass (*Agrostis stolonifera*). Various species of bamboos, both large and small, are also cultivated in gardens in climates where the winters are not severe. The pampas grasses (*Cortaderia* spp.) are tall, herbaceous grasses that are cultivated for their large, whitish fruiting heads.

Of course, grasses are also the most commonly cultivated plants to develop lawns around homes, public buildings, parks, and golf courses. Various species are favored as so-called turf-grasses, depending on the soil type, climate, amount of shading that the site has, and the type of use that the lawn is likely to receive. Commonly used species include Kentucky bluegrass (*Poa pratensis*), meadowgrass (*P. palustris*), Canada bluegrass (*P. compressa*), bent-grass (*Agrostis tenuis*), red-top (*a. alba*), creeping red fescue (*Festuca rubra*), tall fescue (*F. arundinacea*), and ryegrasses (*Lolium perenne* and *L. multiflorum*).

Grasses as weeds

Some people have developed allergies to grass pollen which can be very abundant in the atmosphere at

---

KEY TERMS

Awn—A sometimes long, bristle-like structure that extends from the tip of a leaf or floral part.

Caryopsis—A dry, one-seeded fruit in which the seed is tightly connected to its sheathing pericarp, a tissue derived from the ovary wall. (Also known as a grain.)

Essential oil—These are various types of volatile organic oils that occur in plants and can be extracted for use in perfumery and flavoring.

Inflorescence—A grouping or arrangement of florets or flowers into a composite structure.

Ligule—In grasses, this is a small hair- or scale-like tissue that develops where the leaf blade, leaf sheath, and stem all meet.

Malt—This is a preparation in which grain is soaked in water and allowed to germinate, and then fermentation by yeast is encouraged by removing the supply of oxygen. Malts are used in the preparation of ales, and they may be distilled to prepare a malt liquor or to manufacture pure grain alcohol, or ethanol.

Pith—A diffuse, spongy tissue that occurs inside of the stems of most herbaceous plants and is mostly used for storage of energy-rich nutrients such as carbohydrates.

Rhizome—This is a modified stem that grows horizontally in the soil and from which roots and upward-growing shoots develop at the stem nodes.

Tassel—A terminal, spike-like inflorescence of male flowers, usually with one or more inflorescences of female flowers located beneath. The flowering structures of maize plants have this arrangement.

Weed—Any plant that is growing abundantly in a place where humans do not want it to be.

---

times when these plants are flowering. Although many wind-pollinated plants contribute to hay fever, grasses are among the most important causes during the early and mid-summer **seasons** in temperate climates.

Some species of grasses may be deemed to be weeds for other reasons. Crabgrasses (*Digitaria* spp.), for example, are unwanted in lawns, and for that reason they are considered to be important, aesthetic weeds. Other grasses interfere with the productivity of agricultural crops, and they may be weeds for that reason. The barnyard grass (*Echinochloa crus-galli*), for example, can be abundant in fields of cultivated rice, causing losses of economic yield in the form of rice grains. Another example is ilang-ilang (*Imperata cylindrica*), a weed of various types of cultivated lands in tropical Asia. This grass can be such an aggressive plant that it is sometimes referred to as the world's worst weed.

Other weed grasses are non-native species that have been introduced beyond their original range and have become seriously invasive in their new habitats. Sometimes these species can become dominant in natural communities and thereby seriously degrade the habitat for native plants and animals. In North America, for example, the reed canary-grass (*Phalaris arundinacea*) and giant manna-grass (*Glyceria maxima*) have invaded some type of **wetlands**, causing serious ecological damages in terms of habitat availability for native species. In semi-arid parts of the Great Plains of the western United States, the downy brome-grass (*Bromus tectorum*), a Eurasian species, has become abundant. The highly flammable, late-season **biomass** of this grass has encouraged frequent fires in this habitat. This too-frequent disturbance regime has converted the naturally shrub-dominated **ecosystem** into a degraded system dominated by the brome-grass, which supports few of the original, native species of plants and animals.

Clearly, the grass family contains species that are extraordinarily important to the welfare of humans and other creatures. Some of these grasses are consequential because they are such important sources of food. Others species are important because they have been able to take advantage of ecological opportunities provided for them by human activities and disturbances. Especially important in this respect has been the dispersal of some species of grasses far beyond their native ranges. In their new, colonized habitats the productivity and fecundity of these invasive grasses are not limited by the natural constraints that they experience in their original range such as diseases and herbivory. This is how these plants become invasive weeds.

Resources

Books

Barbour, M.G., and W.D. Billings, eds. *North American Terrestrial Vegetation.* New York: Cambridge University Press, 1988.

Hvass, E. *Plants That Serve and Feed Us.* New York: Hippocrene Books, 1975.

Judd, Walter S., Christopher Campbell, Elizabeth A. Kellogg, Michael J. Donoghue, and Peter Stevens. *Plant Systematics: A Phylogenetic Approach.* 2nd ed. with CD-ROM. Suderland, MD: Sinauer, 2002.

Klein, R.M. *The Green World. An Introduction to Plants and People.* New York: Harper and Row, 1987.

Bill Freedman

Grasshoppers

Grasshoppers are plant-eating **insects** characterized by long hind legs designed for locomotion by jumping. Like all insects, the body of grasshoppers is divided into three main parts: head, thorax, and abdomen. On the head are two antennae for feeling and detecting scent, and two compound eyes comprised of many optical units called facets, each of which is like a miniature **eye**. The chewing mouthparts comprise two sets of jaws which move from side to side. The sides of the mouth have two palps, tiny appendages for feeling and detecting chemicals, which aid in food selection. There are three pair of legs and two pairs of wings attached to the thorax, although some **species** are wingless. At the tip of the abdomen are two appendages called cerci, and the external reproductive organs. Females have an ovipositor at the end of the abdomen through which the eggs are laid. Grasshoppers develop by incomplete **metamorphosis**, passing from egg, to a small wingless larval stage through several molts, to the mature adult.

Classification, distribution, and habitat

Grasshoppers belong to the insect order Orthoptera and the suborder Caelifera. The family Acrididae includes more than 8,000 species of grasshoppers and locusts distributed worldwide. Grasshoppers are found in almost all types of **habitat** including the tropics, temperate grassland, **rainforest**, **desert**, and **mountains**. If adverse conditions prevail, some species migrate in huge numbers to maximize survival. Grasshoppers feed on grass, leafy plants, and bushes. Some species eat only particular food plants, but most species broaden their food base following depletion of their preferred food.

Maintaining appropriate moisture content in the body is achieved primarily through food selection. All species of grasshopper consume both wet and dry food; however, a hydrated insect will choose leaves with low **water** content, while a dehydrated one selects leaves higher in moisture. Captive grasshoppers will drink water directly when food moisture drops below about 50%.

Leaping

Leaping is so advantageous that some grasshopper species have lost the ability to fly. Grasshoppers can repeatedly jump many times their body length without tiring, attaining speeds up to ten times greater than the speed of a running insect. The muscular back legs of grasshoppers allow powerful propulsion. The legs have a muscular femur (thigh), a long, slender tibia (shin), and a five-jointed foot with claws. Before jumping, the grass hopper flexes its rear legs and projects itself through the air with an explosive kick, sometimes using its wings to help it glide. Grasshoppers mainly move by leaping and seldom fly long distances.

Size and color

Male grasshoppers are smaller than females, and size varies greatly between species—from a length of 0.4 in (1 cm) to more than 5.9 in (15 cm). The large Costa Rican grasshopper (*Tropidacris cristatus*) has a 9.9 in (25 cm) wing-span and weighs more than 1 oz (30 g). Colors range from the drab shades of the field dwellers to the brilliant hues of some rainforest species. In some instances, males and females are colored differently.

Body temperature

Although grasshoppers have a body **temperature** that ranges with the environmental temperature, the actual body temperature is important since it can affect movement, digestion, food consumption, water retention, egg/nymph survival **rate**, life expectancy, mating, and habitat selection. The preferred temperature range is 86-112°F (30-44°C). Because grasshoppers normally produce little body **heat**, they thermoregulate (maintain appropriate body temperature) by using heat gained from the environment. Long, thin species increase body heat by exposing their sides to the **sun**. Broad, flat grasshoppers turn their back **perpendicular** to the sun's rays. Crouching allows heat absorption from a warm surface into the abdomen, while stilting (extending the legs) cools the insect by lifting it off a warm surface and permitting air to circulate around its body.

Defense

Grasshoppers are eaten upon by a number of vertebrate and arthropod predators. Defense mechanisms include leaping and camouflage (blending in with their environment). For example, the grass-dwelling *Cylindrotettix* of Brazil changes the **color** of its body from straw-tone in the dry season to green after the rains. Larger species such as *Agriacris trilineata* of Peru's rainforests may use physical defense, kicking predators with powerful hind legs ominously equipped with long spines that can draw **blood**. Other species use startle tactics. The Mexican species *Taeniopoda auricornis*, a tiny black-and-white grasshopper, flashes glorious crimson wings to startle and scare off predators.

Chemical deterrents, such as the regurgitation and defecation of sticky, obnoxious-smelling fluids, are employed by many species of grasshoppers. A few species produce a stinking glandular excretion which effectively repels predators as large as **geckos**, jays, domestic **cats**, and **monkeys**. Certain species sequester toxic chemicals from their **plant** food and predators ingesting them become ill. Most of the toxic species of grasshoppers have conspicuous vivid warning colors which predators learn to avoid. Some nontoxic species of grasshopper mimic the color of toxic species so that predators also avoid them.

Courtship and mating

Grasshoppers have an amazing ability to identify their mates. Each species has its **individual** song, produced by rubbing or flicking the lower back legs on the forewings to create either a chirping or a clicking sound (this is known as stridulation). Females sing more softly than males, facilitating differentiation between both sex and species. Species that make no sound rely on sight and scent to find a mate. Males emit **pheromones**, external **hormones** which attract females, while other species use their excellent eyesight to enable identification by color. The tiny, wingless grasshopper *Drymophilacris bimaculata* of Costa Rica has a brilliant green body with glimmering gold accents on its head, thorax, and genital areas. The male of this species searches out its mate by drumming its hind legs on its preferred food plant. The female drums back, and the pair identify each other by their unique coloring.

Elaborate **courtship** routines are performed by males in some species. The American grasshopper *Syrbula admirabilis* displays 18 individual poses using its wings, legs, and palps. Males of other species may wave brilliantly colored wings when wooing the female, while other species forego courtship altogether.

Mating occurs when the male lights on the female's back and may last anywhere from 45 minutes to well over a day. In the species *Extatosoma tiaratum*, a female mates with several males. Most of the sperm in her genital tract from the first suitor is replaced by the sperm of her next mate. Males therefore mate many times with the same partner and other females to gain the maximum opportunity to pass on their genes. Males of some species die shortly after mating. The females die after egg-laying, which may last until cold **weather** begins.

Reproduction and development

Female grasshoppers **deposit** fertilized eggs in batches in the ground, on the ground, or less commonly, on grass or plant stems. When burying eggs, the female uses four horn-like appendages at the tip of the abdomen, and twists her body and forces her ovipositor into the ground. The desert species *Locusta migratoria* extends her abdomen from its normal length of 1 in (2.5 cm) to 3.2 in (8 cm) in order to bury her eggs as deep as possible.

Chapman, R.F., and A. Joern, eds. *Biology of Grasshoppers.* New York: John Wiley & Sons, Inc., 1990.

Helfer, Jacques R. *How to Know the Grasshoppers and Their Allies.* Toronto: Dover Publications, 1987.

Preston-Mafham, Ken. *Grasshoppers and Mantids of the World.* London: Blandford, 1990.

Marie L. Thompson

KEY TERMS

Cerci—A pair of "feelers" at the tip of the abdomen.

Diapause—A period of delayed development.

Ovipositor—Egg-laying organ on the tip of a female insect's abdomen.

Palps—Tiny appendages near the mouth sensitive to touch and chemical detection or taste.

Pheromone—Hormonal chemical excretion used to attract a mate.

Stridulation—Chirping, clicking or other audible sounds made by certain insects by rubbing body parts together.

In tropical species the eggs hatch after three or four weeks, whereas in temperate climates eggs usually undergo diapause (suspended development) over the winter. Eventually, tiny larvae hatch and burrow to the surface, molting immediately to emerge as undeveloped miniatures of the adult (nymphs). These nymphs may undergo as many as six molts before reaching maturity at an average age of three months.

Grasshoppers and the environment

Swarming grasshoppers and locusts can be extremely destructive to vegetation. A single swarm of African locusts (*Schistocerca gregaria*) can contain 50 billion individuals, and consume as much food in one day as the daily food intake of all the people in New York, London, Paris, and Los Angeles combined. Clearly, such immense irruptions are capable of causing tremendous damage to agriculture. **Insecticides** and the introduction of pathogenic **fungi** deadly to the insects are methods used to try to control such plagues, but this is not always successful. Sometimes, less conventional methods prove effective. In Thailand, Mexico, parts of **Africa**, and other countries, grasshoppers are edible delicacies, providing important dietary protein. During a locust plague in Thailand, government authorities encouraged citizens to catch the swarming masses. Domestic and commercial **crops** were saved from complete destruction and billions of grasshopper bodies were sold to restaurants and market places for seasoning, stir frying, and consumption by many a delighted connoisseur.

Resources

Books

Carde, Ring, and Vincent H. Resh, eds. *Encyclopedia of Insects.* San Diego, CA: Academic Press, 2003.

Grasslands

Grasslands are environments in which herbaceous **species**, especially **grasses**, make up the dominant vegetation. Natural grasslands, commonly called **prairie**, pampas, shrub steppe, palouse, and many other regional names, occur in regions where rainfall is sufficient for grasses and forbs but too sparse or too seasonal to support **tree** growth. Such conditions occur at both temperate and tropical latitudes around the world. In addition, thousands of years of human activity—clearing pastures and fields, burning, or harvesting trees for materials or fuel—have extended and maintained large expanses of the world's grasslands beyond the natural limits dictated by climate.

Precipitation in temperate grasslands (those lying between about 25° and 65° latitude) usually ranges from approximately 10-30 in (25-75 cm) per year. At tropical and subtropical latitudes, annual grassland precipitation is generally between 24-59 in (60-150 cm). Besides its relatively low **volume**, precipitation on natural grasslands is usually seasonal and often unreliable. Grasslands in **monsoon** regions of **Asia** can receive 90% of their annual rainfall in a few weeks; the remainder of the year is dry. North American prairies receive most of their moisture in spring, from snow melt and early rains that are followed by dry, intensely hot summer months. Frequently windy conditions further evaporate available moisture.

Grasses (family Gramineae) can make up 90% of grassland **biomass**. Long-lived root masses of perennial bunch grasses and sod-forming grasses can both endure **drought** and allow **asexual reproduction** when conditions make reproduction by seed difficult. These characteristics make grasses especially well suited to the dry and variable conditions typical of grasslands. However, a wide variety of grass-like plants (especially **sedges**, Cyperaceae) and leafy, flowering forbs contribute to species richness in grassland **flora**. Small shrubs are also scattered in most grasslands, and **fungi**, mosses, and **lichens** are common in and near the **soil**. The height of grasses and forbs varies greatly, with grasses of more humid regions standing 7 ft (2 m) or more, while arid land grasses may be less than 1.6 ft (0.5 m) tall. Wetter grasslands may also contain scattered trees, especially in

Lush grassland growth in autumn in the Mallee region of New South Wales, Australia. *Photograph by Bill Bachman. National Audubon Society Collection/Photo Researchers, Inc. Reproduced by permission.*

low spots or along stream channels. As a rule, however, trees do not thrive in grasslands because the soil is moist only at intervals and only near the surface. Deeper tree roots have little access to **water**, unless they grow deep enough to reach **groundwater**.

Like the plant community, grassland **animal** communities are very diverse. Most visible are large herbivores—from American **bison** and elk to Asian **camels** and **horses** to African kudus and wildebeests. Carnivores, especially wolves, large **cats**, and **bears**, historically preyed on herds of these herbivores. Because these carnivores also threatened domestic herbivores that accompany people onto grasslands, they have been hunted, trapped, and poisoned. Now most wolves, bears, and large cats have disappeared from the world's grasslands. Smaller species compose the great wealth of grassland **fauna**. A rich variety of **birds** breed in and around ponds and streams. **Rodents** perform essential roles in spreading **seeds** and turning over soil. **Reptiles**, **amphibians**, **insects**, **snails**, worms, and many other less visible animals occupy important niches in grassland ecosystems.

Grassland soils develop over centuries or millennia along with regional vegetation and according to local climate conditions. Tropical grassland soils, like tropical forest soils, are highly leached by heavy rainfall and have moderate to poor nutrient and **humus** contents. In temperate grasslands, however, generally **light** precipitation lets **nutrients** accumulate in thick, organic upper layers of the soil. Lacking the acidic **leaf** or pine needle litter of **forests**, these soils tend to be basic and fertile. Such conditions historically supported the rich growth of grasses on which grassland herbivores fed. They can likewise support rich grazing and crop lands for agricultural communities. Either through **crops** or domestic herbivores, humans have long relied on grasslands and their fertile, loamy soils for the majority of their food.

Along a moisture gradient, the margins of grasslands gradually merge with moister savannas and woodlands or with drier, **desert** conditions. As grasslands reach into higher latitudes or altitudes and the climate becomes to cold for grasses to flourish, grasslands grade into **tundra**, which is dominated by mosses, sedges, willows, and other cold-tolerant plants.

See also Savanna.

Resources

Books

Coupland, R.T., ed. *Grassland Ecosystems of the World: Analysis of Grasslands and Their Uses.* London: Cambridge University Press, 1979.

Cushman, R.C., and S.R. Jones. *The Shortgrass Prairie.* Boulder, CO: Pruett Publishing Co., 1988.

Mary Ann Cunningham

Gravitational lens

Gravitational lenses are accidental natural arrangements of gravity, **light**, and distant astronomical objects that create altered images of the those objects. Commonly, a **lens** is a piece of **glass** shaped so as to bend light passing through it. In the process, it alters the image of the light source as observed through the lens. A gravitational lens bends light using gravity rather than glass. Gravitational lensing is a useful tool for astronomers, allowing them to accurately determine the **mass** of distant galaxies and clusters of galaxies, including non-radiating (but gravitating) **matter** that cannot be observed directly.

Gravitational lensing is predicted by Einstein's theory of general relativity, which states that a gravitational field will bend the path of a ray of light. (Newton's older theory, according to which light is a stream of material particles, also predicted that light would be influenced by gravity; however, Einstein predicted a bending effect twice as great as Newton's, and has been confirmed by observation.) This bending effect is generally slight. Therefore, to produce significant lensing (image focusing) a comparatively large mass, such as a **black hole**, **galaxy**, cluster of galaxies, or the like, is required. What is more, gravitational lensing requires not only a lensing mass, but also a light source behind the lensing mass. Quasars, for example, are among the most distant objects in the Universe. If by chance a **quasar** is aligned with a galaxy (as seen from **Earth**), the galaxy may act as a gravitational lens and alter the image of the quasar.

General relativity was dramatically confirmed in 1919 when its prediction that starlight would be bent by passing near the **Sun** was verified. However, gravitational lensing of an entire image was not observed until 1979, when astronomers noticed that the two quasars, designated 0957+561A and 0957+561B, are unusually close together in the sky. (The designation numbers refer to the quasars' position in the sky, while the A and B distinguish the two nearby objects.) Investigating further, astronomers found that these quasars have nearly identical properties, as if they were a double image of the same quasar. Detailed photographs of the region revealed a fuzzy area near one of the quasar images. This fuzz, it turned out, was the faint image of an elliptical galaxy. This galaxy acts as a gravitational lens that bends the light from a *single* quasar, almost directly behind it as seen from Earth, to produce a double image. Since this initial discovery, dozens of other gravitational lenses have been discovered. Two of the most famous have been dubbed Einstein's Ring and Einstein's Cross. Einstein's Ring is observed by **radio** telescopes to be a near perfect ring-image of a quasar. The **Hubble Space Telescope** reveals Einstein's Cross as four images of a quasar, arranged in a cross pattern around a central image of the lensing galaxy.

Objects other than single galaxies can also serve as gravitational lenses. Images of some clusters of galaxies show bright arcs in their vicinity, the gravitationally lensed images of more distant galaxies. By studying these arcs, astronomers can determine the total mass of the lensing cluster. It turns out that only 10% of the total mass of the cluster of galaxies can be accounted for by the visible galaxies in the cluster; the other 90% of the mass is unseen "dark matter," one of the standing mysteries of modern **cosmology**. Astronomers do not know what **dark matter** is, (or if it is matter at all, rather than "dark energy," a currently favored theory) but have observed that it seems to constitute 90% of the mass of the Universe.

One possible component of dark matter is massive compact halo objects (MACHOs). MACHOs are faint or nonradiating objects that may exist in large numbers in a spherical halo surrounding each galaxy (including ours). An otherwise invisible MACHO passing in front of a **star** in a nearby galaxy such as the Andromeda galaxy will produce a small gravitational-lens effect. Because MACHOs are in rapid **motion** relative to the Earth, such a microlensing event would produce a transient brightening of the distant star rather then a drastic, semipermanent distortion like that produced by a galactic lens. Numerous MACHO-type microlensing events have been observed, but their low **rate** shows that there are not enough MACHOs to account for the Universe's dark matter.

See also Gravity and gravitation.

Resources

Periodicals

Glanz, James, "In the Dark Matter Wars, Wimps Beat Machos." *New York Times.* (February 29, 2000).

Gravity *see* **Gravity and gravitation**

Gravity and gravitation

Gravity is a **force** of attraction that exists between every pair of objects in the Universe. This force is proportional to the **mass** of each object in each pair, and inversely proportional to the square of the **distance** between the two; thus,

$$F = Gm_1m_2/r^2,$$

where m_1 is the mass of the first object, m_2 is the mass of the second object, r is the distance between their centers, and G is a fixed number termed the gravitational constant. (If m_1 and m_2 are given in kilograms and r in meters, then $G = 6.673 \times 10^{-11}$ N m^2/kg^2.)

The history of gravity

The Greek philosopher Aristotle (384–322 B.C.) posed, following earlier traditions, that the material world consisted of four elements: **earth**, **water**, air, and fire. For example, a rock was mostly earth with a little water, air, and fire, a cloud was mostly air and water with a little earth and fire. Each element had a natural or proper place in the Universe to which it spontaneously inclined; earth belonged at the very center, water in a layer covering the earth, air above the water, and fire above the air. Each element had a natural tendency to return to its proper place, so that, for example, **rocks** fell toward the center and fire rose above the air. This was one of the earliest explanations of gravity: that it was the natural tendency for the heavier elements, earth and water, to return to their proper positions near the center of the Universe. Aristotle's theory was for centuries taken as implying that objects with different weights should fall at different speeds; that is, a heavier object should fall faster because it contains more of the center-trending elements, earth and water. However, this is not correct. Objects with different weights fall, in fact, at the same **rate**. (This statement still only an **approximation**, however, for it assumes that the Earth is perfectly stationary, which it is not. When an object is dropped the Earth accelerates "upward" under the influence of their mutual gravitation, just as the object "falls," and they meet somewhere in the middle. For a heavier object, this meeting *does* take place slightly sooner than for a light object, and thus, heavier objects actually do fall slightly faster than light ones. In practice, however, the Earth's movement is not measurable for "dropped" objects of less than planetary size, and so it is accurate to state that all *small* objects fall at the same rate, regardless of their mass.)

Aristotle's model of the Universe also included the **Moon**, **Sun**, the visible planets, and the fixed stars. Aris-

totle assumed that these were outside the layer of fire and were made of a fifth element, the **ether** or quintessence (the term is derived from the Latin expression *quinta essentia*, or *fifth essence*, used by Aristotle's medieval translators). The celestial bodies circled the Earth attached to nested ethereal spheres centered on Earth. No forces were required to maintain these motions, since everything was considered perfect and unchanging, having been set in **motion** by a Prime Mover—God.

Aristotle's ideas were accepted in **Europe** and the Near East for centuries, until the Polish astronomer Nicolaus Copernicus (1473–1543) developed a heliocentric (Sun-centered) model to replace the geocentric (Earth-centered) one that had been the dominant cosmological concept ever since Aristotle's time. (Non-European astronomers unfamiliar with Aristotle, such as the Chinese and Aztecs, had developed geocentric models of their own; no heliocentric model existed prior to Copernicus.) Copernicus's model placed the Sun in the center of the Universe, with all of the planets orbiting the Sun in perfect circles. This development was such a dramatic change from the previous model that it is now called the Copernican Revolution. It was an ingenious intellectual construct, but it still did not explain why the planets circled the Sun, in the sense of what caused them to do so.

While many scientists were trying to explain these celestial motions, others were trying to understand terrestrial mechanics. It seemed to be the common-sense fact that heavier objects fall faster than light ones of the same mass: drop a feather and a pebble of equal mass and see which hits the ground first. The fault in this experiment is that air resistance affects the rate at which objects fall. What about another experiment, one in which air resistance plays a smaller role: observing the difference between dropping a large rock and a medium rock? This is an easy experiment to perform, and the results have profound implications. As early as the sixth century A.D. Johannes Philiponos (c. 490–566) claimed that the difference in landing times was small for objects of different weight but similar shape. Galileo's friend, Italian physicist Giambattista Benedetti (1530–1590), in 1553, and Dutch physicist Simon Stevin (1548–1620), in 1586, also considered the falling-rock problem and concluded that rate of fall was independent of weight. However, the individual most closely associated with the falling-body problem is Italian physicist Galileo Galilei (1564–1642), who systematically observed the motions of falling bodies. (It is unlikely that he actually dropped weights off the Leaning Tower of Pisa, but he did write that such an experiment might be performed.)

Because objects speed up (accelerate) quickly while falling, and Galileo was restricted to naked-eye observation by the technology of his day, he studied the slower motions of pendulums and of bodies rolling and sliding down incline. From his results, Galileo formulated his Law of Falling Bodies. This states that, disregarding air resistance, bodies in free fall speed up with a constant **acceleration** (rate of change of **velocity**) that is independent of their weight or composition. The acceleration due to gravity near Earth's surface is given the symbol g and has a value of about 32 feet per second per second (9.8 m/s^2) This means that 1 second after a release a falling object is moving at about 10 m/s; after 2 seconds, 20 m/s; after 10 seconds, 100 m/s. That is, after falling for 10 seconds, it is dropping fast enough to cross the length of a football field in less than one second. Writing v for the velocity of the falling body and t for the time since commencement of free fall, we have $v = gt$.

Galileo also determined a formula to describe the distance d that a body falls in a given time:

$$d = \tfrac{1}{2}gt^2$$

That is, if one drops an object, after 1 second it has fallen approximately 5m; after 2 seconds, 20m; and after 10 seconds, 500 meters.

Galileo did an excellent job of describing the effect of gravity on objects on Earth, but it wasn't until English physicist Isaac Newton (1642–1727) studied the problem that it was understood just how universal gravity is. An old story says that Newton suddenly understood gravity when an apple fell out of a **tree** and hit him on the head; this story may not be exactly true, but Newton did say that a falling apple helped him develop his theory of gravity.

Newtonian gravity

Newton's universal law of gravitation states that all objects in the Universe attract all other objects. Thus the Sun attracts Earth, Earth attracts the Sun, Earth attracts a book, a book attracts Earth, the book attracts the desk, and so on. The gravitational pull between small objects, such as molecules and books, is generally negligible; the gravitational pull exerted by larger objects, such as stars and planets, organizes the Universe. It is gravity that keeps us on the Earth, the Moon in **orbit** around the Earth, and the Earth in orbit around the Sun.

Newton's law of gravitation also states that the strength of the force of attraction depends on the masses of the two objects. The mass of an object is a measure of how much material it has, but it is not the same as its weight, which is a measure of how much force a given mass experiences in a given gravitational field; a given rock, say, will have the same mass anywhere in the universe but will weight more on the Earth than on the Moon.

We do not feel the gravitational forces from objects other than the Earth because they are weak. For example,

The antenna (cylindrical bar, bottom center) of the gravitational wave detector Auriga under construction at Legnaro, Italy. Auriga is one of the first ultracryogenic antennas in the world. Gravitational waves will be detected by the 10 ft (3 m) long bar which will be suspended in the shell seen in the background on the left. The shell will shield it from external vibrations. The bar will also be cooled to a temperature of -459.2°F (-272.9°C [0.1K]) to minimize its own atomic vibrations. Auriga will be detecting gravitational waves from supernova explosions within the galaxies of the Local Group. *Photograph by Tommaso Guicciardini. Photo Researchers, Inc. Reproduced by permission.*

the gravitational force of attraction between two friends weighing 100 lb (45.5 kg) standing 3 ft (1 m) apart is only about $3 = 10^{-8}$ N = 0.00000003 lb, which is about the weight of a bacterium. (Note: the pound is a measure of weight—the gravitational force experienced by an object—while the kilogram is a measure of mass. Strictly speaking, then, pounds and kilograms cannot be substituted for each other as in the previous sentence. However, near Earth's surface weight and mass can be approximately equated because Earth's gravitational field is approximately constant; treating pounds and kilograms as proportional units is therefore standard practice under this condition.)

The gravitational force between two objects becomes weaker if the two objects are moved apart and stronger if they are brought closer together; that is, the force depends on the distance between the objects. If we take two objects and double the distance between them, the force of attraction decreases to one fourth of its former value. If we triple the distance, the force decreases to one ninth of its former value. The force depends on the inverse square of the distance.

All these statements are derived from one simple equation: for two objects having masses m_1 and m_1 respectively, the magnitude of the force of gravity acting on each object is given by:

$$F = Gm_1m_2/r^2,$$

where r is the distance between the objects' centers and G is the gravitational constant (6.673×10^{-11}N m²/kg².) Note that the gravitational constant is an extremely small number; this explains why we only feel gravity when we are near a large mass (e.g., the Earth).

Newton also explained how bodies respond to forces (including gravitational forces) that act on them. His Second Law of Motion states that a net force (i.e., force not canceled by a contrary force) causes a body to accelerate. The amount of this acceleration is inversely proportional to the mass of the object. This means that under the influence of a given force, more massive objects accelerate more slowly than less massive objects. Alternatively, to experience the same acceleration, more massive objects require more force. Consider the gravitational force exerted by the Earth on two rocks, the first with a mass of 2 lb (1 kg) and a second with a mass of 22 lb (10 kg). Since the mass of the second is 10 times the mass of the first, the gravitational force on the second will be 10 times the force on the first. But a 22-lb (10-kg) mass requires 10 times more force to accelerate it, so both masses accelerate Earthward at the same rate. Ignoring the Earth's acceleration toward the rocks (which is extremely small), it follows that equal falling rates for small objects are a natural consequence of Newton's law of gravity and second law of motion.

What if one throws a ball horizontally? If one throws it slowly, it will hit the ground a short distance away. If one throw sit faster, it will land farther. Since the Earth is round, the Earth will **curve** slightly away from the ball before it lands; the farther the throw, the greater the amount of curve. If one could throw or launch the ball at 18,000 mi/h (28,800 km/h), the Earth would curve away from the ball by the same amount that the ball falls. The ball would never get any closer to the ground, and would be in orbit around the Earth. Gravity still accelerates the ball at 9.8 m/s² toward the Earth's center, but the ball never approaches the ground. (This is exactly what the Moon is doing.) In addition, the orbits of the Earth and other planets around the Sun and all the motions of the stars and galaxies follow Newton's laws. This is why Newton's law of gravitation is termed "universal;" it describes the effect of gravity on all objects in the Universe.

Newton published his **laws of motion** and gravity in 1687, in his seminal *Philosophiae Naturalis Principia Mathematica* (Latin for *Mathematical Principles of Natural Philosophy*, or *Principia* for short). When we need to solve problems relating to gravity, Newton's laws usually suffice. There are, however, some phenomena that they cannot describe. For example, the motions of the **planet** Mercury are not exactly described by Newton's laws. Newton's theory of gravity, therefore, needed modifications that would require another genius, Albert Einstein, and his Theory of General Relativity.

General relativity

German physicist Albert Einstein (1879–1955) realized that Newton's theory of gravity had problems. He knew, for example, that Mercury's orbit showed unexplained deviations from that predicted by Newton's laws. However, he was worried about a much more serious problem. As the force between two objects depends on the distance between them, if one object moves closer, the other object will feel a change in the gravitational force. According to Newton, this change would be immediate, or instantaneous, even if the objects were millions of miles apart. Einstein saw this as a serious flaw in Newtonian gravity. Einstein assumed that nothing could travel instantaneously, not even a change in force. Specifically, nothing can travel faster than light in a **vacuum**, which has a speed of approximately 186,000 mi/s (300,000 km/s). In order to fix this problem, Einstein had not only to revise Newtonian gravity, but to change the way we think about **space**, time, and the structure of the Universe. He stated this new way of thinking mathematically in his general theory of relativity.

Einstein said that a mass bends space, like a heavy ball making a dent on a rubber sheet. Further, Einstein contended that space and time are intimately related to each other, and that we do not live in three spatial dimensions and time (all four quite independent of each other), but rather in a four-dimensional space-time continuum, a seamless blending of the four. It is thus not "space," naively conceived, but space-time that warps in reaction to a mass. This, in turn, explains why objects attract each other. Consider the Sun sitting in space-time, imagined as a ball sitting on a rubber sheet. It curves the space-time around it into a bowl shape. The planets orbit around the Sun because they are rolling across through this distorted space-time, which curves their motions like those of a ball rolling around inside a shallow bowl. (These images are intended as analogies, not as precise explanations.) Gravity, from this point of view, is the way objects affect the motions of other objects by affecting the shape of space-time.

Einstein's general relativity makes predictions that Newton's theory of gravitation does not. Since particles of light (photons) have no mass, Newtonian theory predicts that they will not be affected by gravity. However, if gravity is due to the curvature of space-time, then light should be affected in the same way as **matter**. This proposition was tested as follows: During the day, the Sun is too bright to see any stars. However, during a total solar eclipse the Sun's disk is blocked by the Moon, and it is possible to see stars that appear in the sky near to the Sun. During the total solar eclipse of 1919, astronomers measured the positions of several stars that were close to the Sun in the sky. It was determined that the measured positions were altered as predicted by general relativity; the Sun's gravity bent the starlight so that the stars appeared to shift their locations when they were near the Sun in the sky. The detection of the bending of starlight by the Sun was one of the great early experimental verifications of general relativity; many others have been conducted since.

Another surprising prediction made by general relativity is that waves can travel in gravitational forces just as waves travel through air or other media. These gravitational waves are formed when masses move back and forth in space-time, much as **sound waves** are created by the **oscillations** of a speaker cone. In 1974, two stars were discovered orbiting around each other, and scientists found out that the stars were losing **energy** at the exact rate required to generate the predicted gravity waves; that is, they were steadily radiating energy away in the form gravitational waves. So far, gravitational waves have not been detected directly, but new detectors will be completed in the U.S., Japan, and Europe in 2003 and it is expected that these devices will detect gravitational waves produced by violent cosmic events such as supernovae. Scientists have already verified that changes in gravitation do propagate at the speed of light, as predicted by Einstein's theory but not by Newton's.

Of all the predictions of general relativity, the strangest is the existence of black holes. When a very massive **star** runs out of fuel, the gravitational self-attraction of the star makes it shrink. If the star is massive enough, it will collapse it to a point having finite mass but infinite **density**. Space-time will be so distorted in the vicinity of this "singularity," as it is termed, that not even light will be able to escape; hence the term "black hole." Astronomers have been searching for objects in the sky that might be black holes, but since they do not give off light directly, they must be detected indirectly. When material falls into a **black hole**, it must **heat** up so much that it glows in **x rays**. Astronomers look for strong x-ray sources in the sky because these sources may be likely candidates to be black holes. Numerous black holes have been detected by these means, and it is now believed that many or most galaxies contain a supermassive black hole at their center, having a mass millions or billions of times greater than that of the Sun.

The greatest remaining challenge for gravity theory is unification with **quantum mechanics**. Quantum theory describes the **physics** of phenomena at the atomic and subatomic scale, but does not account for gravitation. General relativity, which employs continuous variables,

KEY TERMS

Acceleration—The rate at which the velocity of an object changes over time.

Force—Influence exerted on an object by an outside agent which produces an acceleration changing the object's state of motion.

General relativity—Einstein's theory of space and time, which explains gravity and the shape of space.

Mass—A measure of the amount of material in an object.

Velocity—The speed and direction of a moving object.

Weight—The gravitational force pulling an object toward a large body, e.g., the Earth, that depends both on the mass of the object and its distance from the center of the larger body.

does not describe the **behavior** of objects at the quantum scale. Physicists therefore seek a theory of "quantum gravity," a unified set of equations that will describe the whole range of known phenomena.

See also Geocentric theory; Heliocentric theory; X-ray astronomy; Relativity, general; Relativity, special.

Resources

Books

Hartle, James B. *Gravity: An Introduction to Einstein's General Relativity* Boston: Addsion-Wesley, 2002.

Hawking, Stephen W. *A Brief History of Time: From the Big Bang to Black Holes.* New York: Bantam Books, 1988.

Thorne, Kip S. *Black Holes and Time Warps: Einstein's Outrageous Legacy.* New York: W. W. Norton, 1994.

Periodicals

"Einstein Was Right on Gravity's Velocity." *New York Times.* (January 8, 2003).

Jim Guinn

Great Barrier Reef

The Great Barrier Reef lies off the northeastern coast of **Australia** and is both a scientific wonder and an increasingly popular tourist attraction. It has been described as "the most complex and perhaps the most productive biological system in the world." The Great Barrier Reef is the largest structure ever made by living organisms including human beings, consisting of the skeletons of tiny coral polyps and hydrocorals bounded together by the soft remains of coralline **algae** and **microorganisms**.

Location and extent

The Great Barrier Reef is over 1,250 mi (2,000 km) long and is 80,000 sq mi (207,000 sq km) in surface area, which is larger than the **island** of Great Britain. It **snakes** along the coast of the **continent** of Australia, roughly paralleling the coast of the State of Queensland, at distances ranging from 10-100 mi (16-160 km) from the shore. The Reef is so prominent a feature on **Earth** that it has been photographed from satellites. The Reef is located on the **continental shelf** that forms the perimeter of the Australian land mass where the **ocean water** is warm and clear. At the edge of the continental shelf and the Reef, the shelf becomes a range of steep cliffs that plunge to great depths with much colder water. The coral polyps require a **temperature** of at least 70°F (21°C), and the water temperature often reaches 100°F (38°C).

Formation

The tiny coral polyps began building their great Reef in the Miocene Epoch which began 23.7 million years ago and ended 5.3 million years ago. The continental shelf has subsided almost continually since the Miocene Epoch so the Reef has grown upward with the living additions to the Reef in the shallow, warm water near the surface; live coral cannot survive below a depth of about 25 fathoms (150 ft or 46 m) and also depend on the **salt** content in sea water. As the hydrocorals and polyps died and became cemented together by algae, the spaces between the skeletons were filled in by wave action that forced in other debris called infill to create a relatively solid mass at depth. The upper reaches of the Reef are more open and are riddled with grottoes, canyons, caves, holes bored by molluscs, and many other cavities that provide natural homes and breeding grounds for thousands of other **species** of sea life. The Great Barrier Reef is, in reality, a string of 2,900 reefs, cays, inlets, 900 islands, lagoons, and shoals, some with beaches of **sand** made of pulverized coral.

Discovery and exploration

The aborigines (the native people of Australia) undoubtedly were the first discoverers of the Great Barrier Reef. The Chinese probably explored it about 2,000 years ago while searching for marine creatures like the sea cucumber that are believed to have medicinal properties. During his voyage across the Pacific Ocean in 1520, Ferdinand Magellan missed Australia and its Reef. Captain James Cook, the British explorer credited with discovering Australia, also found the Great Barrier Reef by sudden impact. His ship, the Endeavour, ran aground on the Reef on June 11, 1770. Cook's crew unloaded ballast (including cannon now imprisoned in the coral growth) and, luckily, caught a high tide that dislodged the ship from the Reef. After extensive repairs, it took Cook and his crew three months to navigate through the maze-like construction of the Great Barrier Reef. These obstacles did not discourage Cook from exploring and charting the extent of the Reef and its cays, passages, and other intricacies on this first of three expeditions of discovery he undertook to the Reef.

In 1835, Charles Darwin's voyage of scientific discovery on the British ship the *Beagle* included extensive study of the Reef. Mapping the natural wonder continued throughout the nineteenth century, and, in 1928, the Great Barrier Reef Expedition was begun as a scientific study of coral lifestyles, Reef construction, and the **ecology** of the Reef. The Expedition's work concluded in 1929, but a permanent marine laboratory on Heron Island within the Reef was founded for scientific explorations and environmental monitoring. The Reef is also the final resting place of a number of ships that sank during World War II.

Biology

The Reef is the product of over 350 species of coral and red and green algae. The number of coral species in the northern section of the Reef exceeds the number (65) of coral species found in the entire Atlantic Ocean. Polyps are the live organisms inside the coral, and most are less than 0.3 in (8 mm) in diameter. They feed at night by extending frond-like fingers to wave **zooplankton** toward their mouths. In 1981, marine biologists discovered that the coral polyps spawn at the same time on one or two nights in November. Their eggs and sperm form an orange and pink cloud that coats hundreds of square miles of the ocean surface. As the polyps attach to the Reef, they secrete lime around themselves to build secure turrets or cups that protect the living organisms. The daisy-or feather-like polyps leave limestone skeletons when they die. The creation of a 1 in (2.5 cm)-thick layer of coral takes five years.

The coral is a laboratory of the living and once-living; scientists have found that coral grows in bands that can be read much like the rings in trees or the icecaps in polar regions. By drilling cores 25 ft (7.6 m) down into the coral, 1,000 years of lifestyles among the coral can be interpreted from the **density**, skeleton size, band thickness, and chemical makeup of the formation. The drilling program also proved that the Reef has died and revived at least a dozen times during its 25-million-year history, but it should be understood that this resiliency predated human activities. The Reef as we know it is about 8,000 years old and rests on its ancestors. In the early 1990s, study of the coral cores has yielded data about temperature ranges, rainfall, and other climate changes; in fact, rainfall data for design of a dam were extracted from the wealth of information collected from analysis of the coral formation.

Coral also shows considerable promise in the field of medicine. Corals produce chemicals that block ultraviolet rays from the **sun**, and the Australian Institute has applied for a patent to copy these chemicals as potential **cancer** inhibitors. Chemicals in the coral may also yield analgesics (**pain** relievers) and anti-AIDS medications.

Animal life forms flourish on and along the Reef, but plants are rare. The Great Barrier Reef has a distinctive purple fringe that is made of the coralline or encrusting algae *Lithothamnion* (also called stony seaweed), and the green algae *Halimeda discodea* that has a creeping form and excretes lime. The algae are microscopic and give the coral its many colors; this is a symbiotic relationship in which both partners, the coral and the algae, benefit. Scientists have found that variations in water temperature stress the coral causing them to evict the resident algae. The loss of **color** is called coral bleaching, and it may be indicate **global warming** or other effects like El Niño.

Other animal life includes worms, **crabs**, prawns, **crayfish**, **lobsters**, anemones, **sea cucumbers**, **starfish**, gastropods, **sharks**, 22 species of whales, dolphins, eels, sea snakes, **octopus**, **squid**, dugongs (sea cows), 1,500 species of **fish** including the largest black marlin in the world, and **birds** like the shearwater that migrates from Siberia to lay its eggs in the hot coral sand. The starfish *Acanthaster planci*, nicknamed the crown-of-thorns, is destructive to the Reef because it eats the live coral. The starfish ravages the coral during periodic infestations then all but vanishes for nearly 20 years at a time. The crown-of-thorns has lived on the Great Barrier Reef for ages (again according to the history shown in the drilling cores), but scientists are concerned that human activities may be making the plague-like infestations worse. Giant clams that grow to more than 4 ft (1.2 m) across and 500 lb (187 kg) in weight are the largest molluscs in the world. Of the seven species of sea **turtles** in the world, six nest on Raine Island within the Reef and lay over 11,000 eggs in a single reproductive night.

This **biodiversity** makes the Reef a unique **ecosystem**. Fish shelter in the Reef's intricacies, find their food there, and spawn there. Other marine life experience the same benefits. The coastline is protected from waves and the battering of storms, so life on the shore also thrives.

Tourism and environmental hazards

In 1990, Conservation, Education, Diving, Archaeology, and Museums International (CEDAM International) gathered the opinions of the world's most respected marine experts and selected "seven underwater wonders of the world" including the Great Barrier Reef. Of course, the idea was inspired by the seven wonders of the ancient world, which were all manmade and of which only the Great Pyramid survives. The underwater wonders give people points of interest and focus for preserving our planet's vast oceans.

Education of the public is needed if the Great Barrier Reef is to survive. Over 1.5 million visitors per year visit the tropical paradise, and development along the Australian coast to accommodate the tourists was largely uncontrolled until 1990. In the 1980s, the island resort of Hamilton was built following the dredging of harbors, leveling of hills, construction of hotels and an airport, and the creation of artificial beaches. About 25 resorts like this dot the Reef. Fishing has also decimated local fish populations; fish that are prized include not only those for food but tropical fish for home aquariums. Fishing nets, boat anchors, and waste from fishing and pleasure boats all do their own damage. Greedy prospectors have mined the coral itself because it can be reduced to lime for manufacture of cement and for **soil** improvement in the sugar cane fields. Reefs in other parts of the world are near collapse, thanks to such irresponsibility.

Environmental hazards like **oil spills** have seriously threatened the Reef. The maze of reefs includes a narrow, shallow shipping channel that is used by oil and chemical tankers and that has a high accident **rate**. Over 2,000 ships per year navigate the channel, and the environmental organization Greenpeace is campaigning to ban the oil traffic through this vulnerable channel. Lagoons have collected waste runoff from towns, agriculture, and tourist development; and the waste has allowed algae (beyond the natural population) to flourish and strangle the live coral. **Pesticides** and **fertilizers** also change the balance between the coral and algae and zoo- and phyto-plankton, and the coral serves as an indicator of chemical damage by accumulating PCBs, metals, and other contaminants. Sediment also washes off the land from agricultural activities and development; it **clouds** the water and limits **photosynthesis**. A thousand other hazards inflict unknown damage on the Reef. Periodic burning off of the sugar cane fields fills the air with smoke that settles on Reef waters, overfishing of particular species of fish shifts the balance of power in the un-

KEY TERMS

. . . .

Aborigines—The native people of the continent of Australia.

Algae—A group of aquatic plants (including seaweed and pond scum) with chlorophyll and colored pigments.

Biodiversity—The biological diversity of an area as measured by the total number of plant and animal species.

Cay—A low-lying reef of sand or coral.

Continental shelf—A relatively shallow, gently sloping, submarine area at the edges of continents and large islands, extending from the shoreline to the continental slope.

Ecotourism—Ecology-based tourism, focused primarily on natural or cultural resources.

El Niño—The phase of the Southern Oscillation characterized by increased sea water temperatures and rainfall in the eastern Pacific, with weakening trade winds and decreased rain along the western Pacific.

Phytoplankton—Minute plant life that lives in water.

Polyp—The living organism in coral with an attached end and an open end with a mouth and fine tentacles.

Symbiosis—A biological relationship between two or more organisms that is mutually beneficial. The relationship is obligate, meaning that the partners cannot successfully live apart in nature.

Zooplankton—Minute animal life that lives in water.

dersea world, and shells and coral are harvested (both within and beyond legal limits) and sold to tourists.

The Government of Australia has declared the Great Barrier Reef a national park, and activities like explorations for gold and oil and spearfishing were permanently banned with the Reef's new status. The United Nations Educational, Scientific, and Cultural Organization (UNESCO) has named it a world heritage site in attempts to encourage awareness and protect the area. Despite many threats, the marine park is one of the best protected in the world, thanks to citizens who recognize the worth of this treasure and visitors who are willing to practice **ecotourism**, and thanks to an extensive body of protective laws.

Resources

Books

Care, Patricia. *The Struggle for the Great Barrier Reef.* New York: Walker and Company, 1971.

McGregor, Craig. *The Great Barrier Reef.* Amsterdam: Time-Life Books, 1975.

Reader's Digest Guide to the Great Barrier Reef. Sydney, Australia: Reader's Digest, 1988.

Periodicals

Belleville, Bill, and David Doubilet. "The Reef Keepers." *Sea Frontiers* (Mar-Apr. 1993): 50+.

Drogin, Bob. "Trouble Down Under." *Los Angeles Times Magazine* (Sept. 19, 1993):16+.

FitzGerald, Lisa M. "Seven Underwater Wonders of the World." *Sea Frontiers* (Dec. 1990): 8-21.

Organizations

Great Barrier Reef Marine Park Authority [cited April 2003]. <www.gbrmpa.gov. au>.

Other

National Gerographic Society. "Virtual World: Great Barrier Reef" [cited April 2003]. <http://www.nationalgeographic.com/earthpulse/ reef/reef1_flash.html>.

Gillian S. Holmes

Greatest common factor

The greatest common **factor** (or *greatest common divisor*) of a set of **natural numbers** is the largest natural number that divides each member of the set evenly (with no remainder). For example, 6 is the greatest common factor of the set because 1246 = 2, 1846 = 3, and 3046 = 5.

Similarity, the greatest common factor of a set of **polynomials** is the polynomial of highest **degree** that divides each member of th set with no remainder. For example, $3(x+2)^3 (x-4)^2$, $12(x+2)^4 (x-4)^3 (x2+x+5)$, and $6(x+2)^2 (x-4)$ have $3(x+2)^2 (x-4)$ for the highest common factor. Polynomials is the polynomial of highest degree that divides each member of the set with no remainder. For example, $3(x+2)^3 (x-4)^2$, $12(x+2)^4 (x-4)^3 (x2+x+5)$, and $6(x+2)^2 (x-4)$ have $3(x+2)^2 (x-4)$ for the highest common factor.

Grebes

Grebes are aquatic **birds** that make up the family Podicipedidae. This is the only family in the order Podicipitiformes, a rather unique group of birds that is not

Western grebes (*Aechmophorus occidentalis*) displaying across the water. *Photograph by Phil Dotson/The National Audubon Society Collection/Photo Researchers, Inc. Reproduced by permission.*

closely related to other living orders, and has a fossil lineage extending back 70 million years. The 20 **species** of grebes range in size from the least grebe (*Podiceps dominicus*), with a body length of 9.9 in (25 cm) and weight of 4 oz (115 g), to the great crested grebe (*Podiceps cristatus*), 18.9 in (48 cm) long and weighing 3.1 lbs (1.4 kg). The wintertime **color** of grebes is brown, grey, or black on top and white below, but during the breeding season most species develop a rather colorful plumage, especially around the head and neck.

Grebes are well adapted to swimming, with feet placed far back on the body, and paddle-like, lobed toes that provide a greater surface area for propulsion and a very short tail. The dense plumage of these birds provides waterproofing and grebes are strong, direct flyers. However, once they are settled in a particular place for breeding or feeding, grebes tend not to fly much.

Grebes breed on **freshwater** lakes and marshes on all of the continents except **Antarctica**. Some species winter in coastal marine waters, or on large lakes. Grebes have a noisy **courtship behavior**, often accompanied by a spectacular display. For example, courting western grebes (*Aechmophorus occidentalis*) run in tandem over the **water** surface, each bird striking a symmetric, ritualized pose known as the penguin dance.

The nests of most grebes are made of anchored, piled-up mounds of vegetation in shallow water. The young chicks often ride on the back of their parents, where they are brooded. The **prey** of these birds includes **fish** and aquatic **invertebrates**.

Species of grebes

Most species of grebes are found in the Americas, especially in Central and **South America**. Three species are flightless and confined to single lakes, these being the short-winged grebe (*Rollandia micropterum*) of Lake Titicaca in Bolivia and Peru, the Junin grebe (*Podiceps taczanowskii*) of Lake Junin in Peru, and the giant pied-billed grebe (*Podilymbus gigas*) of Lake Atitlan in Guatemala. The latter species may now be extinct as a result of hunting and development activities around its lake.

Six species of grebes occur regularly in **North America**. The largest species is the western grebe (*Aechmophorus occidentalis*) of the western United States and southwestern Canada. The western grebe breeds on lakes and marshes, and winters in near-shore waters of the Pacific, and on some large lakes. This is the only species of grebe that spears its prey of fish with its sharp beak. Other grebes catch their food by grasping with the mandibles.

The red-necked grebe (*Podiceps grisegena*) breeds in northwestern North America, and winters on both the Atlantic and Pacific coasts. The horned grebe (*Podiceps auritus*) also breeds in the northwest and winters on both coasts. A similar looking species, the eared grebe (*Podiceps caspicus*), breeds in southwestern North America. The pied-billed grebe (*Podilymbus podiceps*) has the widest distribution of any grebe in North America, breeding south of the boreal forest and wintering in Mexico and further south.

Conservation of grebes

Small species of grebes are not often hunted, because their meat is not very tasty, but the larger grebes have been hunted for their plumage. "Grebe fur" is the patch of breast skin with plumage attached, which can be stripped from the dead bird. Grebe fur from the western grebe and great crested grebe was used to make hand muffs, capes, and hats for fashionable ladies, while that of the short-winged grebe was used locally around Lake Titicaca to make saddle blankets.

Grebe populations also suffer from **pollution**. Species that winter in coastal waters are highly vulnerable to **oil spills**, and grebes can be killed in large numbers when this type of pollution occurs.

Grebes may also be affected by **pesticides**. One of the earliest, well documented examples of birds being killed by exposure to chlorinated-hydrocarbon **insecticides** occurred at Clear Lake, California. This lake is important for recreational use, but there were numerous complaints about a non-biting midge (a tiny, aquatic fly) that could sometimes be extremely abundant. In 1949 this perceived problem was dealt with by applying the insecticide DDT to the lake. This chemical was used again in 1954, and soon afterward about 100 western grebes were found dead on the lake. It took several years of study to determine that the grebes had been killed by the insecticide, which they had efficiently accumulated from the residues in their diet of fish, achieving unexpectedly large, toxic concentrations in their bodies. This case study proved to be very important in allowing ecologists and toxicologists to understand the insidious effects that persistent **chlorinated hydrocarbons** could achieve through food-web accumulation.

Status of North American Grebes

• Western grebe (*Aechmophorus occidentalis*). Plume hunters devastated the population in the beginning of the twentieth century. The species has apparently recovered, taking up residence in areas not historically used. The population in Mexico may be declining.

• Clark's grebe (*Aechmophorus clarkii*). Plume hunters contributed greatly to the decline in population. Past population counts are unreliable because of confusion of this bird with the Western Grebe. The population in Mexico may be declining due to loss of nesting **habitat** (i.e., tules on lakes).

• Red-necked grebe (*Podiceps grisegena*). Declines in population have resulted from damage to eggs and eggshells by pesticides and PCBs, and by raccoon predation. This species continues to be vulnerable to polluted wintering areas along the coast. The population status today is not well known.

• Horned grebe (*Podiceps auritus*). Population is apparently declining, though hard numbers are lacking.

• Eared grebe (*Podiceps nigricollis*). Feathers were once used for hats, capes, and muffs; and eggs gathered for food. Today the populations appear stable, but the species is considered vulnerable because large numbers depend on a very few lakes at certain **seasons** (for example, the Great **Salt** Lake, Mono Lake, and the Salton Sea).

• Pied-billed grebe (*Podilymbus podiceps*). This species has proven adaptable, and is now found in developed areas. Surveys suggest a population decline in recent decades, however.

• Least grebe (*Tachybaptus dominicus*). Normally found in southern Texas in the United States. Sometimes killed by exceptionally cold Texas winters.

See also Biomagnification.

Resources

Books

Ehrlich, Paul R., David S. Dobkin, and Darryl Wheye. *The Birder's Handbook.* New York: Simon & Schuster Inc., 1988.

Forshaw, Joseph. *Encyclopedia of Birds.* New York: Academic Press, 1998.

Freedman, B. *Environmental Ecology.* 2nd ed. San Diego: Academic Press, 1994.

Peterson, Roger Tory. *North American Birds.* Houghton Miflin Interactive (CD-ROM), Somerville, MA: Houghton Miflin, 1995.

Bill Freedman
Randall Frost

Greenhouse effect

The greenhouse effect is the retention by the Earth's atmosphere in the form of **heat** some of the **energy** that arrives from the **Sun** as **light**. Certain gases, including **carbon dioxide** (CO_2) and methane (CH_4), are transparent to most of the wavelengths of light arriving from the Sun but are relatively opaque to infrared or heat **radiation**; thus, energy passes through the Earth's atmosphere on arrival, is converted to heat by absorption at the surface and in the atmosphere, and is not easily re-radiated into space. The same process is used to heat a solar greenhouse, only with **glass**, rather than gas, as the heat-trapping material. The greenhouse effects happens to maintain the Earth's surface **temperature** within a range comfortable for living things; without it, the Earth's surface would be much colder.

The greenhouse effect is mostly a natural phenomenon, but its intensity, according to a majority of climatologists, may be increasing because of increasing atmospheric concentrations of CO_2 and other greenhouse gases. These increased concentrations are occurring because of human activities, especially the burning of **fossil fuels** and the clearing of **forests** (which remove CO_2 from the atmosphere and store its **carbon** in **cellulose**, $[C_6H_{10}O_5]_n$). A probable consequence of an intensification of Earth's greenhouse effect will be a significant warming of the atmosphere. This in turn would result in important secondary changes, such as a rise in **sea level** (already occurring), variations in the patterns of **precipitation**. These, in turn, might accelerate the **rate** at which **species** are already being to **extinction** by human activity, and impose profound adjustments on human society.

The greenhouse effect

The Earth's greenhouse effect is a reasonably well-understood physical phenomenon. Scientists believe that in the absence of the greenhouse effect, Earth's surface temperature would average about -0.4°F (-18°C), which is below the freezing point of **water** and more frigid than life on the surface of the **Earth** could tolerate over the longer term—except, perhaps, organisms deriving their energy from hot deep-sea vents. The greenhouse effect maintains Earth's surface at an average temperature of about 59°F (15°C). This is about 59.5°F (33°C) warmer than it would otherwise be.

The energy budget

To understand the greenhouse effect, Earth's energy budget must be known. An energy budget is an account of all of the energy coming into and leaving a system and of any energy that is stored in (or produced by) the system itself. Almost all of the energy coming to Earth from space has been radiated by the closest **star**, the Sun. The Sun emits electromagnetic energy at a rate and spectral quality determined by its surface temperature. In this it resembles all bodies having a temperature greater than **absolute zero** (i.e., -459°F or -273°C). Fusion reactions occurring in the core the Sun give it a high surface temperature, about 10,800°F (6,000°C). As a consequence, about one-half of the Sun's emitted energy is visible radiation with wavelengths between 0.4 and 0.7 æm, so called because this is the range of electromagnetic wavelengths that the human **eye** can perceive. Most of the remainder is in the near-infrared wavelength range, between about 0.7 and 2.0 æm. The Sun also emits radiation in other parts of the **electromagnetic spectrum**, such as ultraviolet and **x rays**; however, these wavelengths convey relatively insignificant amounts of energy away from the Sun.

At the average distance of Earth from the Sun, the rate of input of solar energy to the Earth's surface is about 2 calories per minute per square centimeter, a value termed the solar constant. There is a nearly perfect energetic balance between this quantity of energy incoming to Earth and the amount that is eventually dissipated to outer space. The myriad ways in which the incoming energy is reflected, dispersed, transformed, and stored make up Earth's energy budget.

REFLECTION. On average, one-third of incident solar radiation is reflected back to space by the Earth's atmosphere or its surface. Earth's local reflectivity (**albedo**) is strongly dependent on cloud cover, the **density** of tiny particulates in the atmosphere, and the nature of the surface, especially vegetation and **ice** and snow.

Atmospheric absorption and radiation

Another one-third of incoming solar radiation is absorbed by certain gases and vapors in Earth's atmosphere, especially water vapor and carbon dioxide. Upon absorption, the solar electromagnetic energy is transformed into thermal kinetic energy (i.e., heat or energy of molecular vibration). The warmed atmosphere then reradiates energy in all directions as longer-wavelength (7–14 æm) infrared radiation. Much of this reradiated energy escapes to outer space.

ABSORPTION AND RADIATION AT THE SURFACE.
Much of the solar radiation that penetrates to Earth's surface is absorbed by living and nonliving materials. This results in a transformation to thermal energy, which increases the temperature of the absorbing surfaces and of air in contact with those surfaces. Over the medium term (days) and longer term (years) there is little net storage of energy as heat; almost all of the thermal energy is re-radiated by the surface as electromagnetic radiation of a longer wavelength than that of the original, incident radiation. The wavelength **spectrum** of typical, reradiated electromagnetic energy from Earth's surface peaks is within the long-wave infrared range.

EVAPORATION AND MELTING OF WATER. Some of the electromagnetic energy that penetrates to Earth's surface is absorbed and transformed to heat. Much of this thermal energy subsequently causes water to evaporate from **plant** and open-water surfaces, or melts ice and snow.

WINDS, WAVES, AND CURRENTS. A small amount (less than 1%) of the absorbed solar radiation causes mass-transport processes to occur in the oceans and lower atmosphere, which disperses of some of Earth's unevenly distributed thermal energy. The most important of these physical processes are winds and storms, water **currents**, and waves on the surface of the oceans and lakes.

PHOTOSYNTHESIS. Although small, an ecologically critical quantity of solar energy, averaging less than 1% of the total, is absorbed by plant pigments, especially **chlorophyll**. This absorbed energy is used to drive **photosynthesis**, the energetic result of which is a temporary storage of energy in the interatomic bonds of certain biochemical compounds. This energy is released when plant material is digested or burned.

Now we are ready to explain the greenhouse effect. If the atmosphere was transparent to the long-wave infrared energy that is reradiated by Earth's atmosphere and surface, then that energy would travel unobstructed to outer space. However, so-called radiatively active gases (or RAGs; also known as "greenhouse gases") in the atmosphere are efficient absorbers within this range of infrared wavelengths, and these substances thereby slow the radiative cooling of the **planet**. When these atmospheric gases absorb infrared radiation, they develop a larger content of thermal energy, which is then dissipated by a reradiation (again, of a longer wavelength than the electromagnetic energy that was absorbed). Some of the secondarily reradiated energy is directed back to Earth's surface, so the net effect of the RAGs is to slow the rate of cooling of the planet.

This process has been called the "greenhouse effect" because its mechanism is analogous to that by which a glass-enclosed space is heated by solar energy. That is, a greenhouse's glass and humid atmosphere are transparent to incoming solar radiation, but absorb much of the re-radiated, long-wave infrared energy, slowing down the rate of cooling of the structure.

Water vapor (H_2O) and CO_2 are the most important radiatively active constituents of Earth's atmosphere. Methane (CH_4), nitrous oxide (N_2O), **ozone** (O_3), and **chlorofluorocarbons (CFCs)** play lesser roles. On a per-molecule basis, all these gases differ in their ability to absorb infrared wavelengths. Compared with CO_2, methane is 11–25 times more effective at absorbing infrared, nitrous oxide is 200–270 times, ozone 2,000 times, and CFCs 3,000–15,000 times.

Other than water vapor, the atmospheric concentrations of all of these gases have increased in the past century because of human activities. Prior to 1850, the **concentration** of CO_2 in the atmosphere was about 280 parts per million (ppm), while 2002 it was over 360 ppm. During the same period, CH_4 increased from 0.7 ppm to 1.7 ppm, N_2O from 0.285 ppm to 0.304 ppm, and CFCs from nothing to 0.7 parts per billion. These increased concentrations are believed by climatologists to contribute to a significant increase in the greenhouse effect. Overall, CO_2 is estimated to account for about 60% of this enhancement of the greenhouse effect, CH_4 for 15%, N_2O for 5%, O_3 for 8%, and CFCs for 12%.

The greenhouse effect and climate change

The physical mechanism of the greenhouse effect is conceptually simple, and this phenomenon is acknowledged by scientists as helping to keep Earth's temperature within the comfort zone for organisms. It is also known that the concentrations of CO_2 and other RAGs have increased in Earth's atmosphere, and will continue to do so. However, it has proven difficult to demonstrate that the observed warming of Earth's surface or lower atmosphere has been caused significantly by a stronger greenhouse effect rather than by some still-unknown process of natural climate change.

Since the beginning of instrumental recordings of surface temperatures around 1880, almost all of the warmest years on record have occurred since the late 1980s. Typically, these warm years have averaged about 1.5–2.0°F (0.8-1.0°C) warmer than occurred during the decade of the 1880s. Overall, Earth's surface air temperature has increased by about 0.9°F (0.5°C) since 1850.

However, the temperature data on which these apparent changes are based suffer from some important deficiencies, including: (1) air temperature is **variable** in time and space, making it difficult to determine statistically significant, longer-term trends; (2) older data are

An atmosphere with natural levels of greenhouse gases (left) compared with an atmosphere of increased greenhouse effect (right). *Illustration by Hans & Cassidy. Courtesy of Gale Group.*

generally less accurate than modern records; (3) many **weather** stations are in urban areas, and are influenced by "heat island" effects; and (4) climate can change for reasons other than a greenhouse response to increased concentrations of CO_2 and other RAGs, including albedo-related influences of volcanic emissions of **sulfur dioxide**, sulfate, and fine particulates into the upper atmosphere. Moreover, it has long been thought that the **interval** 1350 to 1850, known as the Little Ice Age, was relatively cool, and that **global climate** has been generally warming since that time period. (The data one which this claim was based, however, have recently been called into question; no instrumental or global data at all are available from the period in question.)

However, some studies have provided evidence for linkages between historical variations of atmospheric CO_2 and surface temperature. Important evidence comes, for example, from a core of Antarctic glacial ice that represents a 160,000–year period. Concentrations of CO_2 in the ice are determined by analysis of air bubbles in ice layers of known age (determined by counting annual snowfall layers back from the present), while changes in air temperature are inferred from ratios of **oxygen** isotopes in the ancient ice. (Because **atoms** of various isotopes differ in weight, their rates of **diffusion** are affected by temperature differently; differences in diffusion rate, in turn, affect their relative abundance in the ice). Because changes in CO_2 and surface temperature are positively correlated, a greenhouse mechanism is sug-

gested. However, this study could not determine causal direction—that is, whether increased CO_2 might have resulted in warming through an intensified greenhouse effect, or whether, conversely, warming (caused by something unknown) could have accelerated CO_2 release from ecosystems by increasing the rate of **decomposition** of **biomass**, especially in cold regions.

Because of the difficulties in measurement and interpretation of climatic change using real-world data, computer models have been used to predict potential climatic changes caused by increases in atmospheric RAGs. The most sophisticated simulations are the so-called "three-dimensional general circulation models" (GCMs), which are run on supercomputers. GCM models simulate the extremely complex mass-transport processes involved in **atmospheric circulation** and the interaction of these processes with other variables that contribute to climate. To perform a simulation "experiment" with a GCM model, components are adjusted to reflect the probable physical influence of increased concentrations of CO_2 and other RAGs.

Many simulation experiments have been performed using a variety of GCM models. Their results have, of course, varied according to the specifics of the experiment. However, a central tendency of experiments using a common CO_2 scenario (i.e., a doubling of CO_2 from its recent concentration of 360 ppm) is an increase in average surface temperature of 1.8–7.2°F (1–4°C). This warming is predicted to be especially great in polar regions, where temperature increases could be two or three times greater than in the tropics.

One of the best-known models was designed by the International Panel on Climate Change (IPCC). This GCM model makes assumptions about population and economic growth, resource availability, and management options that result in increases or decreases of RAGs in the atmosphere. Scenarios were developed for emissions of CO_2, other RAGs, and sulfate **aerosols**, which may cool the atmosphere by increasing its albedo and by affecting cloud formation. For a simple doubling of atmospheric CO_2, the IPCC estimate was a 4.5°F (2.5°C) increase in average surface temperature. The estimates of more advanced IPCC scenarios (with adjustments for other RAGs and sulfate) were similar, and predicted a 2.7–5.4°F (1.5–3°C) increase in temperature by the year 2100, compared with 1990. Thus, theoretical studies tend to back the claim that CO_2 can cause **global warming**, whether or not the reverse process may also occur.

Effects of climatic change

It is likely that the direct effects of climate change caused by an intensification of the greenhouse effect would be substantially restricted to plants. The temperature changes might cause large changes in the quantities, distribution, or timing of precipitation, and this would have a large effect on vegetation. There is, however, even more uncertainty about the potential changes in rainfall patterns than of temperature, and effects on **soil** moisture and vegetation are also uncertain. Still, it is reasonable to predict that any large changes in patterns of precipitation would result in fundamental reorganizations of vegetation on the terrestrial landscape.

Studies of changes in vegetation during the warming climate that followed the most recent, Pleistocene, glaciation, suggest that plant species responded in unique, individualistic ways. This results from the differing tolerances of species to changes in climate and other aspects of the environment, and their different abilities to colonize newly available **habitat**. In any event, the species composition of plant communities was different then from what occurs at the present time. Of course, the vegetation was, and is, dynamic, because plant species have not completed their post-glacial movements into suitable habitats.

In any region where the climate becomes drier (for example, because of decreased precipitation), a result could be a decreased area of forest, and an expansion of **savanna** or **prairie**. A landscape change of this character is believed to have occurred in the New World tropics during the Pleistocene glaciations. Because of the relatively dry climate at that time, presently continuous rainforest may have been constricted into relatively small refugia (that is, isolated patches). These forest remnants may have existed within a landscape matrix of savanna and grassland. Such an enormous restructuring of the character of the tropical landscape must have had a tremendous effect on the multitude of rare species that live in that region. Likewise, climate change potentially associated with an intensification of the greenhouse effect would have a devastating effect on Earth's natural ecosystems and the species that they sustain.

There would also be important changes in the ability of the land to support crop plants. This would be particularly true of lands cultivated in regions that are marginal in terms of rainfall, and are vulnerable to **drought** and **desertification**. For example, important **crops** such as **wheat** are grown in regions of the western interior of **North America** that formerly supported natural shortgrass prairie. It has been estimated that about 40% of this semiarid region, measuring 988 million acres (400 million hectares), has already been desertified by agricultural activities, and crop-limiting droughts occur there sporadically. This climatic handicap can be partially managed by **irrigation**. However, there is a shortage of water for irrigation, and this practice can cause its own environmental problems, such as salinization. Clearly, in many areas substantial changes in climate would place the present agricultural systems at great risk.

Patterns of **wildfire** would also be influenced by changes in precipitation regimes. Based on the predictions of climate models, it has been suggested that there could be a 50% increase in the area of forest annually burned in Canada, presently about 2.5-4.9 million acres (1-2 million hectares) in typical years.

Some shallow marine ecosystems might be affected by increases in seawater temperature. Corals are vulnerable to large increases in water temperature, which may deprive them of their symbiotic **algae** (called zooxanthellae), sometimes resulting in death of the colony. Widespread coral "bleachings" were apparently caused by warm water associated with an El Niño event in 1982-83.

Another probable effect of warming could be an increase in sea level. This would be caused by the combination of (1) a **thermal expansion** of the **volume** of warmed seawater, and (2) melting of polar **glaciers**. The IPCC models predicted that sea level in 2100 could be 10.5-21 in (27-50 cm) higher than today. Depending on the rate of change in sea level, there could be substantial problems for low-lying, coastal agricultural areas and cities.

Most GCM models predict that high latitudes will experience the greatest intensity of climatic warming. Ecologists have suggested that the warming of northern ecosystems could induce a positive feedback to climate change. This could be caused by a change of great expanses of boreal forest and arctic **tundra** from sinks for atmospheric CO_2, into sources of that greenhouse gas. In this scenario, the climate warming caused by increases in RAGs would increase the depth of annual thawing of frozen soils, exposing large quantities of carbon-rich organic materials in the **permafrost** to microbial decomposition, and thereby increasing the **emission** of CO_2 to the atmosphere.

Reducing atmospheric RAGs

It is likely that an intensification of Earth's greenhouse effect would have large climatic and ecological consequences. Clearly, any sensible strategy for managing the causes and consequences of changes in the ggreenhouse effect will requir substantial reductions in the emissions of CO_2 and other RAGs.

It is important to recognize that any strategy to reduce these emissions will require great adjustments by society and economies. Because such large quantities of CO_2 are emitted through the burning of fossil fuels, there will be a need to use different, possibly new, technologics to generate energy, and there may be a need for large decreases in total energy use. The bottom line, of course, will be a requirement to add considerably smaller quantities of RAGs to the atmosphere. Such a strategy of mitigation will be difficult, especially in industrialized countries, because of the changes required in economic systems, resource use, investments in technology, and levels of living standards. The implementation of those changes will require enlightened and forceful leadership.

Under the auspices of the United Nations Environment Program, various international negotiations have been undertaken to try to get nations to agree to decisive actions to reduce their emissions of RAGs. The most recent major agreement came out of a large meeting held in Kyoto, Japan, in 1997. There, most of the world's industrial countries agreed to reduce their CO_2 emissions to 5.2% below 1990 levels by the year 2012. The United States, which has about 5% of the world's population but produces 24% of its CO_2 emissions, signed the Kyoto protocol in 1998 (that is to say, its ambassador to the United Nations signed the plan) but never ratified it as a binding treaty; shortly after taking office in 2000, President George W. Bush repudiated the protocol entirely. (China, with about 23% of the world's population, is the second-biggest CO_2 producer, at 14% of total emissions.)

A complementary way to balance the emissions of RAGs would be to remove some atmospheric CO_2 by increasing its fixation by growing plants, especially through the planting of forests onto agricultural land. Similarly, the prevention of **deforestation** will avoid

KEY TERMS

Albedo—Refers to the reflectivity of a surface.

Carbon reserve—An ecosystem, such as a forest, that is managed primarily for its ability to store large quantities of organic carbon, and to thereby offset or prevent an emission of carbon dioxide to the atmosphere.

Desertification—A climatic change involving decreased precipitation, causing a decreased or destroyed biological productivity on the landscape, ultimately leading to desert-like conditions.

Electromagnetic energy—A type of energy, involving photons, which have physical properties of both particles and waves. Electromagnetic energy is divided into spectral components, which (ordered from long to short wavelength) include radio, infrared, visible light, ultraviolet, and cosmic.

Energy budget—A physical accounting of the various inputs and outputs of energy for some system, as well as the quantities and locations where energy is internally stored.

Radiatively active gases (RAGs)—Within the context of the greenhouse effect, these gases absorb long-wave infrared energy emitted by Earth's surface and atmosphere, and thereby slow the rate of radiative cooling by the planet.

large amounts of CO_2 emissions through the conversion of high-carbon forests into low-carbon agro-ecosystems.

The development and maintenance of ecosystems that store large quantities of carbon to offset industrial emissions would require very large areas of land. These carbon reserves would preclude other types of economically important uses of the land. This strategy would therefore require a substantial commitment by society; however, so would any other possible means of decreasing greenhouse gases, and so would a decision to do nothing at all (or to keep researching the problem indefinitely, which amounts to much the same thing). There are no easy solutions to problems of this type and magnitude.

See also Air pollution; Energy budgets; Hydrochlorofluorocarbons; Ozone layer depletion.

Resources

Books

Hamblin, W.K., and Christiansen, E.H. *Earth's Dynamic Systems.* 9th ed. Upper Saddle River: Prentice Hall, 2001.

Hancock P.L. and Skinner B.J., eds. *The Oxford Companion to the Earth.* New York: Oxford University Press, 2000.

Periodicals

Kerr, R.A. American Association for the Advancement of Science (AAAS). "Clearing the Air—Global warming: Rising Global Temperature, Rising Uncertainty. Greenhouse Warming Passes One More Test. *Science.* 292 (2001): 267.

Schneider, S. H. "The Changing Climate." *Scientific American.* 261 (1989): 70-79.

Other

Nebehay, Stephanie. "2002 Second Hottest as Global Warming Speeds, Says WMO." Reuters. December 18, 2002 [cited January 6, 2003]. <http://www.enn.com/news/wire-stories/2002/12/121820 02/reu_49197.asp>.

Bill Freedman
Larry Gilman

Grosbeaks *see* **Cardinals and grosbeaks**

Groundhog

The groundhog or woodchuck (*Marmota monax*) is a husky, waddling rodent in the squirrel family Sciuridae, order Rodentia. The groundhog is a type of marmot (genus *Marmota*), and is also closely related to the ground **squirrels** and **gophers**. The natural **habitat** of the groundhog is forest edges and **grasslands**, ranging from the eastern United States and Canada through much of the Midwest, to parts of the western states and provinces. However, the groundhog is also a familiar **species** in agricultural landscapes within its range, occurring along roadsides, fence-rows, pastures, the margins of fields, and even in some suburban habitats.

The groundhog is a rather large marmot, typically weighing about 6.6-13.2 lb (3-6 kg). One captive **animal**, however, managed to achieve a enormous 37.4 lb (17 kg) just prior to its wintertime **hibernation**, when these animals are at their heaviest. The fur is red or brown, with black or dark brown feet.

Groundhogs have a plump body, a broad head, and small, erect ears. The tail and legs are short, while the fingers and toes have strong claws, useful for digging. When frightened, groundhogs can run as fast as a person, but they are normally slow, waddling animals, tending to stay close to the safety of their burrows. Groundhogs can climb rather well, and are sometimes seen feeding while perched in the lower parts of trees or shrubs.

Groundhogs are enthusiastic diggers, and they spend much of their time preparing and improving their bur-

KEY TERMS
. . . .

Hibernation—A deep, energy-conserving sleep that some mammals enter while passing the wintertime. In groundhogs, hibernation is characterized by a slowed metabolic rate, and a decrease in the core body temperature.

rows and dens. Woodchucks dig their burrow complexes in well-drained, sandy-loam soils, generally on the highest ground available. Their sleeping dens are lined with hay-like materials, both for comfort, and to provide insulation during the winter. There are separate chambers for sleeping and defecation.

Groundhogs are social animals, sometimes living in open colonies with as many as tens of animals living in a maze of interconnected burrows. Groundhogs are not very vocal animals, but they will make sharp whistles when a potential **predator** is noticed. This loud sound is a warning to other animals in the colony.

Groundhogs are herbivorous animals, eating the foliage, stems, roots, and tubers of herbaceous plants, and sometimes the buds, leaves, flowers, and young shoots of woody species. Groundhogs also store food in their dens, especially for consumption during the winter. Groundhogs are very **fat** in the autumn, just prior to hibernation. If they are living in a colony, groundhogs snuggle in family groups to conserve **heat** during the winter. They occasionally waken from their deep **sleep** to feed. However, groundhogs lose weight progressively during their hibernation, and can weigh one-third to one-half less in the springtime than in the autumn.

Groundhogs have a single mating season each year, usually beginning shortly after they emerge from their dens in the spring. After a gestation period of 30-32 days, the female usually gives **birth** to four or five young, although the size of the litter may vary from one to nine. Born blind and naked, young groundhogs acquire a downy coat after about two weeks. Soon the mother begins to bring soft **plant** stems and leaves back to the den for them to eat. Young groundhogs follow their mother out of the burrow after about a month and are weaned about two weeks later.

Groundhogs are sometimes perceived to be **pests**. They can cause considerable damage by raiding vegetable gardens, and can also consume large quantities of ripe grain and other **crops**. In addition, the excavations of groundhogs can be hazardous to **livestock**, who can break a leg if they step unawares into a groundhog hole, or if an underground burrow collapses beneath their

A woodchuck sitting on its hind legs to view the surrounding area. *Photograph by Leonard Lee Rue, III. Photo Researchers, Inc. Reproduced by permission.*

weight. For these reasons, groundhogs are sometimes hunted and poisoned. However, groundhogs also provide valuable ecological benefits as **prey** for a wide range of carnivorous animals, and because these interesting creatures are a pleasing component of the outdoors experience for many people.

A Midwestern American folk myth holds that if a woodchuck comes out of its burrow on February 2, also known as Groundhog Day, and sees its shadow, then the cold wintertime **weather** will soon be over. However, if that day is cloudy and the woodchuck does not see its shadow, it goes back into hibernation, and the winter weather will last a while longer. Of course, there is no basis in natural history to this belief. Nor, contrary to common wisdom, do woodchucks chuck **wood**.

See also Marmots; Rodents.

Resources

Books

Banfield, A.W.F. *The Mammals of Canada.* Toronto: University of Toronto Press, 1974.

Barash, D. *Marmots. Social Behavior and Ecology.* Stanford: Stanford University Press, 1989.

Bill Freedman

Groundwater

Groundwater occupies the void space in a geological **strata**. It is one element in the continuous process of moisture circulation on **Earth**, termed the **hydrologic cycle**.

Almost all groundwater originates as surface **water**. Some portion of rain hitting the earth runs off into streams and lakes, and another portion soaks into the **soil**, where it is available for use by plants and subject to **evaporation** back into the atmosphere. The third portion soaks below the root zone and continues moving downward until it enters the groundwater. **Precipitation** is the major source of groundwater. Other sources include the movement of water from lakes or streams and contributions from such activities as excess **irrigation** and seepage from canals. Water has also been purposely applied to increase the available supply of groundwater. Water-bearing formations called aquifers act as reservoirs for storage and conduits for transmission back to the surface.

The occurrence of groundwater is usually discussed by distinguishing between a zone of saturation and a zone of aeration. In the zone of saturation, the pores are entirely filled with water, while the zone of aeration has pores that are at least partially filled by air. Suspended water does occur in this zone. This water is called vadose, and the zone of aeration is also known as the vadose zone. In the zone of aeration, water moves downward due to gravity, but in the zone of saturation it moves in a direction determined by the relative heights of water at different locations.

Water that occurs in the zone of saturation is termed groundwater. This zone can be thought of as a natural storage area or reservoir whose capacity is the total **volume** of the pores of openings in **rocks**.

An important exception to the distinction between these zones is the presence of ancient sea water in some sedimentary formations. The pore spaces of materials that have accumulated on an **ocean** floor, which has then been raised through later geological processes, can sometimes contain **salt** water. This is called connate water.

Formations or strata within the saturated zone from which water can be obtained are called aquifers. Aquifers must yield water through wells or springs at a **rate** that can serve as a practical source of water supply. To be considered an **aquifer** the geological formation must contain pores or open spaces filled with water, and the openings must be large enough to permit water to move through them at a measurable rate. Both the size of pores and the total pore volume depends on the type of material. **Individual** pores in fine-grained materials such as clay, for example, can be extremely small, but the total volume is large. Conversely, in coarse material such as **sand**, individual pores may be quite large but total volume is less. The rate of movement for fine-grained materials, such as clay, will be slow due to the small pore size, and it may not yield sufficient water to wells to be considered an aquifer. However, the sand is considered

an aquifer, even though they yield a smaller volume of water, because they will yield water to a well.

The water table is not stationary but moves up or down depending on surface conditions such as excess precipitation, **drought**, or heavy use. Formations where the top of the saturated zone or water table define the upper limit of the aquifer are called unconfined aquifers. The hydraulic **pressure** at any level with an aquifer is equal to the depth from the water table, and there is a type known as a water-table aquifer, where a well drilled produces a static water level which stands at the same level as the water table.

A local zone of saturation occurring in an aerated zone separated from the main water table is called a perched water table. These most often occur when there is an impervious strata or significant particle-size change in the zone of aeration, which causes the water to accumulate. A confined aquifer is found between impermeable layers. Because of the confining upper layer, the water in the aquifer exists within the pores at pressures greater than the atmosphere. This is termed an artesian condition and gives rise to an artesian well.

Groundwater can be pumped from any aquifer that can be reached by modern well-drilling apparatus. Once a well is constructed, hydraulic pumps pull the water up to the surface through pipes. As water from the aquifer is pulled up to the surface, water moves through the aquifer towards the well. Because water is usually pumped out of an aquifer more quickly than new water can flow to replace what has been withdrawn, the level of the aquifer surrounding the well drops, and a cone of depression is formed in the immediate area around the well.

Groundwater can be polluted by the spilling or dumping of contaminants. As surface water percolates downward, contaminants can be carried into the aquifer. The most prevalent sources of **contamination** are waste disposal, the storage, transportation and handling of commercial materials, **mining** operations, and nonpoint sources such as agricultural activities. Two other forms of groundwater **pollution** are the result of pumping too much water too quickly, so that the rate of water withdrawal from the aquifer exceeds the rate of aquifer recharge. In coastal areas, salty water may migrate towards the well, replacing the fresh water that has been withdrawn. This is called salt water intrusion. Eventually, the well will begin pulling this salt water to the surface; once this happens, the well will have to be abandoned. A similar phenomenon, called connate ascension, occurs when a **freshwater** aquifer overlies a layer of sedimentary rocks containing connate water. In some cases, overpumping will cause the connate water to migrate out of the sedimentary rocks and into the freshwater aquifer. This results in a **brackish**, briney contamination similar to the effects of a salt water intrusion. Unlike salt water intrusion, however, connate ascension is not particularly associated with coastal areas.

Groundwater has always been an important resource, and it will become more so in the future as the need for good quality water increases due to urbanization and agricultural production. It has recently been estimated that 50% of the drinking water in the United States comes from groundwater; 75% of the nation's cities obtain all or part of their supplies from groundwater, and rural areas are 95% dependent upon it. For these reasons every precaution should be taken to protect groundwater purity. Once contaminated, groundwater is difficult, expensive, and sometimes impossible to clean up.

See also Water pollution.

Resources

Books

Collins, A.G., and A.I. Johnson, eds. *Ground-Water Contamination: Field Methods.* Philadelphia: American Society for Testing and Materials, 1988.

Davis, S.N., and R.J.M. DeWiest. *Hydrogeology.* New York: Wiley, 1966.

Fairchild, D.M. *Ground Water Quality and Agricultural Practices.* Chelsea, MI: Lewis, 1988.

Freeze, R.A., and J.A. Cherry. *Ground Water.* Englewood Cliffs, NJ: Prentice-Hall, 1979.

Ground Water and Wells. St. Paul: Edward E. Johnson, 1966.

James L. Anderson

Group

A group is a simple mathematical system, so basic that groups appear wherever one looks in **mathematics**. Despite the primitive nature of a group, mathematicians have developed a rich theory about them. Specifically, a group is a mathematical system consisting of a set G and a binary operation * which has the following properties:

[1] $x*y$ is in G whenever x and y are in G (closure).

[2] $(x*y)*z = x*(y*z)$ for all x, y, and z in G (**associative property**).

[3] There exists and element, e, in G such that $e*x=x*e=x$ for all x in G (existence of an **identity element**).

[4] For any element x in G, there exists an element y such that $x*y=y*x=e$ (existence of inverses).

Note that commutativity is not required. That is, it need not be true that $x*y=y*x$ for all x and y in G.

One example of a group is the set of **integers**, under the binary operation of **addition**. Here the sum of any two integers is certainly an integer, 0 is the identity, $-a$ is the inverse of a, and addition is certainly an associative operation. Another example is the set of positive fractions, m/n, under **multiplication**. The product of any two positive fractions is again a positive fraction, the identity element is 1 (which is equal to 1/1), the inverse of m/n is n/m, and, again, multiplication is an associative operation.

The two examples we have just given are examples of commutative groups. (Also known as Abelian groups in honor of Niels Henrik Abel, a Norwegian mathematician who was one of the early users of group theory.) For an example of a non-commutative group consider the permutations on the three letters a, b, and c. All six of them can be described by

$$\begin{pmatrix} a\,b\,c \\ a\,b\,c \end{pmatrix} \begin{pmatrix} a\,b\,c \\ a\,c\,b \end{pmatrix} \begin{pmatrix} a\,b\,c \\ b\,a\,c \end{pmatrix} \begin{pmatrix} a\,b\,c \\ b\,c\,a \end{pmatrix} \begin{pmatrix} a\,b\,c \\ c\,a\,b \end{pmatrix} \begin{pmatrix} a\,b\,c \\ c\,b\,a \end{pmatrix}$$

$$\text{I} \qquad \text{P} \qquad \text{Q} \qquad \text{R} \qquad \text{S} \qquad \text{T}$$

I is the identity; it sends a into a, b into b, and c into c. P then sends a into a, b into c, and c into b. Q sends a into b, b into a, and c into c and so on. Then P*Q=R since P sends a into a and Q then sends that a into b. Likewise P sends b into c and Q then sends that c into c. Finally, P sends c into b and Q then sends that b into a. That is the effect of first applying P and then Q is the same as R.

Following the same procedure, we find that Q*P=S which demonstrates that this group is not commutative. A complete "multiplication" table is as follows:

	I	P	Q	R	S	T
I	I	P	Q	R	S	T
P	P	I	R	Q	T	S
Q	Q	S	I	T	P	R
R	R	T	P	S	I	Q
S	S	Q	T	I	R	P
T	T	R	S	P	Q	I

From the fact that I appears just once in each row and column we see that each element has an inverse. Associativity is less obvious but can be checked. (Actually, the very nature of permutations allows us to check associativity more easily.) Among each group there are *subgroups*-subsets of the group which themselves form a group. Thus, for example, the set consisting of I and P is a subgroup since P*P=I. Similarly, I and T form a subgroup.

Another important concept of group theory is that of *isomorphism*. For example, the set of permutations on three letters is isomorphic to the set of *symmetries* of an equilateral triangle. The concept of isomorphism occurs in many places in mathematics and is extremely useful in that it enables us to show that some seemingly different systems are basically the same.

The term "group" was first introduced by the French mathematician Evariste Galois in 1830. His work was inspired by a proof by Abel that the general equation of the fifth degree is not solvable by radicals.

Resources

Books

Bell, E.T. *Men of Mathematics.* Simon and Schuster, 1961.
Grossman, Israel, and Wilhelm Magnus. *Groups and Their Graphs.* Mathematical Association of America, 1965.

Roy Dubisch

Grouse

Grouse (and ptarmigan) are medium-sized **birds** in the family Tetraonidae, order Galliformes. Grouse and ptarmigan are often hunted for food and sport, and are sometimes broadly referred to as upland gamebirds because they are not hunted in **wetlands**, as are **ducks** and **geese**.

Grouse are ground-dwelling birds with a short, turned-down bill. They have long, heavy feet with a short elevated fourth toe behind the short, rounded wings. Grouse have feathered ankles, and most grow fringes of feathers on their toes in the winter. In addition, the nostrils are feathered, and some **species** have a bright colored patch around the eyes.

Grouse are found throughout the temperate and more northerly zones of Eurasia and **North America**. There are 10 species of grouse in North America: blue grouse (*Dendragapus obscurus*), **spruce** grouse (*Canachites canadensis*), ruffed grouse (*Bonasa umbellus*), sharp-tailed grouse (*Pedioecetes phasianellus*), sage grouse (*Centrocercus urophasianus*), greater **prairie chicken** (*Tympanuchus cupido*), lesser **prairie** chicken (*T. pallidicinctus*), willow ptarmigan (*Lagopus lagopus*), rock ptarmigan (*L. mutus*), and white-tailed ptarmigan (*L. leucurus*). These grouse utilize most of the major **habitat** types of North America, with particular species being adapted to **tundra**, boreal forest, temperate forest, heathlands, or **grasslands**. The capercallie (*Tetrao urogallus*) is found in coniferous **forests** in **Europe** and **Asia**.

Throughout their range, grouse are hunted intensively. Fortunately, they have a high reproductive capability, and if conserved properly, can be sustainably harvested. In most areas, the most important threats to grouse are not from hunting, but from the more insidious effects of habitat loss. This effect on grouse is primarily associated with the conversion of their natural and semi-natural habitat to agricultural or urban use, or to extensive practice of plantation **forestry**. As a result, grouse and other **wildlife** are displaced.

Wildlife biologists have been able to develop management systems that can accommodate many types of agricultural and forestry activities, as well as the needs of most species of grouse. In North America, for example, economically productive forestry can be conducted in ways that do not degrade, and in fact can enhance, the habitat of certain species of grouse. Systems of co-management for ruffed grouse and forestry are especially well known.

The ruffed grouse is the most commonly hunted upland gamebird in North America, with about six million birds being harvested each year; an additional two million individuals of other species of grouse and ptarmigan are also killed annually. Ruffed grouse prefer a temperate forest mosaic, with both mature stands and younger brushy habitats of various ages, with a great deal of edge habitat among these types. Ruffed grouse can utilize a wide range of habitat types, but they do best in hardwood-dominated forests with some conifers mixed in. The most favored variety of forest is dominated by poplars (especially trembling aspen, *Populus tremuloides*) and birches (especially white birch, *Betula papyrifera*), but stands of various age are required. In Minnesota, it has been found that clear-cuts of aspen forest develop into suitable breeding habitat for ruffed grouse after 4-12 years of regeneration. These maturing clear-cuts are utilized as breeding habitat for 10-15 years. Older, mature aspen stands are also important to ruffed grouse, especially as wintering habitat. In general, to optimize habitat for ruffed grouse over much of its eastern range, a forest can be managed to create a mosaic of stands of different ages, each less than about 25 acres (10 hectares) in size.

In some circumstances, grouse hunting can be a more important use of the land than agriculture or forestry. In such instances, the needs of these birds are the primary consideration for landscape managers. This is the case where grouse hunting on large estates is a popular sport, for example, in Britain and some other European countries. In Scotland, upland heaths of red grouse (known as the willow ptarmigan in North America) are periodically burned by wildlife managers. This treatment stimulates the flowering and sprouting of fresh shoots of heather (*Calluna vulgaris*), an important food of the red grouse. The burnt patches are arranged to create a larger habitat mosaic that includes recently burned areas, older burns, and mature heather.

Although some species of grouse can be effectively managed for sustainable hunting, and the effects of many types of forestry and agricultural practices can be mitigated, it should be pointed out that other species of grouse have fared less well. In North America, the greater prairie chicken was once abundant in tall-grass and mixed-grass prairies and coastal heathlands. However, this species is now rare and endangered over its remaining, very-much contracted range, because most of its original habitat has been converted to intensively managed agriculture. One subspecies, known as the heath hen (*T. c. cupido*), was once abundant in coastal grasslands and heath barrens from Massachusetts to Virginia. However, largely because of habitat loss in combination with overhunting, the heath hen became extinct in 1932. Another subspecies, Attwater's greater prairie chicken (*T. c. attwateri*), was formerly abundant in coastal prairies of Texas and Louisiana, but this endangered bird is now restricted to only a few isolated populations.

Bill Freedman

Growth and decay

Growth and decay refers to a class of problems in **mathematics** that can be modeled or explained using increasing or decreasing sequences (also called series). A sequence is a series of numbers, or terms, in which each successive **term** is related to the one before it by precisely the same formula. There are many practical applications of sequences. One example is predicting the growth of human populations. Population growth or decline has an impact on numerous economic and environmental issues. When the population grows, so does the **rate** at which waste is produced, which in turn affects growth rate of land fill sites, nuclear waste dumps, and other sources of **pollution**. Various other growth rates also affect our lives. For instance the growth rates of our investments and savings accounts, affect our economic well-being. Understanding the mathematics of growth is very important. For example, predicting the rate at which renewable resources, including the **forests**, marine life, and **wildlife**, naturally replenish themselves, helps prevent excessive harvesting that can lead to population declines and even **extinction**.

Arithmetic growth and decay

Arithmetic growth is modeled by an arithmetic sequence. In an arithmetic sequence each successive term is obtained by adding a fixed quantity to the previous term. For example, an investment that earns simple interest (not compounded) increases by a fixed percentage of the principal (original amount invested) in each period that interest is paid. A one-time investment of $1,000, in

an account that pays 5% simple interest per year, will increase by $50 per year. The growth of such an investment, left in place for a 10 year period, is given by the sequence, where the first entry corresponds to the balance at the beginning of the first year, the second entry corresponds to the balance at the beginning of the second year and so on. A sequence that models growth is an increasing sequence, one that models decay is a decreasing sequence. For instance, some banks require depositors to maintain a minimum balance in their checking accounts, or else pay a monthly service charge on the account. If an account, with a required minimum of $500, has $50 in it, and the owner stops using the account without closing it, then the balance will decrease arithmetically each month, by the amount of the monthly service charge, until it reaches **zero**.

Geometric growth and decay

Geometric growth and decay are modeled with geometric sequences. A geometric sequence is one in which each successive term is multiplied by a fixed quantity. In general, a geometric sequence is one of the form, where $P_1 = cP_0$, $P_2 = cP_1$, $P_3 = cP_2$,..., $P_n = cP_{n-1}$, and c is a constant called the common **ratio**. If c is greater than 1, the sequence is increasing. If c is less than 1, the sequence is decreasing. The rate at which an investment grows when it is deposited in an account that pays compound interest is an example of a geometric growth rate. Suppose an initial deposit of P_0 is made in a bank paying a fixed interest rate that is compounded annually. Let the interest rate in decimal form be r. Then, the account balance at the end of the first year will be $P_1 = (P_0 + r P_0) = (1+r) P_0$. At the end of the second year, the account balance will be $P_2 = (P_1 + rP_1) = (1+r)P_1$. By continuing in this way it is easy to see that the account balance in any given year will be equal to (1+r) times the previous years balance. Thus, the growth rate of an initial investment earning compound interest is given by the geometric sequence that begins with the initial investment, and has a common ratio equal to the interest rate plus 1.

This same compounding model can be applied to population growth. However, unlike the growth of an investment, population growth is limited by the availability of food, **water**, shelter, and the prevalence of **disease**. Thus, population models usually include a variable growth rate, rather than a fixed growth rate, that can take on negative as well as positive values. When the growth rate is negative, a declining population is predicted. One such model of population growth is called the logistic model. It includes a variable growth rate that is obtained by comparing the population in a given year to the capacity of the environment to support a further increase. In this model, when the current population exceeds the

. . . .

Limit—A limit is a bound. When the terms of a sequence that are very far out in the series grow ever closer to a specific finite value, without ever quite reaching it, that value is called the limit of the sequence.

Mathematical model—A mathematical model is the expression of a physical law in terms of a specific mathematical concept.

Rate—A rate is a comparison of the change in one quantity with the simultaneous change in another, where the comparison is made in the form of a ratio.

Sequence—A sequence is a series of terms, in which each successive term is related to the one before it by a fixed formula.

capacity of the environment to support the population, the quantity in parentheses becomes negative, causing a subsequent decline in population.

Still another example of a process that can be modeled using a geometric sequence is the process of **radioactive decay**. When the nucleus of a radioactive element decays it emits one or more alpha, beta or gamma particles, and becomes stable (nonradioactive). This decay process is characteristic of the particular element undergoing decay, and depends only on **time**. Thus, the probability that one nucleus will decay is given by: Probability of Decay = λt, where λ depends on the element under consideration, and t is an arbitrary, but finite (not infinitesimally short), length of time. If there are initially N_0 radioactive nuclei present, then it is probable that $N_0\lambda t$ nuclei will decay in the time period t. At the end of the first time period, there will be $N_1 = (N_0 - N_0\lambda t)$ or $N_1 = N_0 (1-\lambda t)$ nuclei present. At the end of the second time period, there will be $N_2 = N_1 (1-\lambda t)$, and so on. Carrying this procedure out for n time periods results in a sequence similar to the one describing compound interest, however, λ is such that this sequence is decreasing rather than increasing. In order to express the number of radioactive nuclei as a continuous function of time rather than a sequence of separated times, it is only necessary to recognize that t must be chosen infinitesimally small, which implies that the number of terms, n, in the sequence must become infinitely large. To accomplish this, the common ratio is written $(1 - \lambda t/n)$, where t/n will become infinitesimally small as n becomes infinitely large. Since a geometric sequence has a common ratio, any term can be written in the form $T_{n+1} = c^n T_0$, where T_0 is the initial term, so that the number of radioactive nuclei

at any time, t, is given by the sequence $N = N_0(1 - \lambda t/n)^n$ when n approaches **infinity**. It is well known that the limit of the expression $(1 + x/n)^n$ as n approaches infinity equals e^x, where e is the base of the natural **logarithms**. Thus, the number of radioactive nuclei present at any time, t, is given by $N = N_0e^{\lambda t}$, where N_0 is the number present at the time taken to be t = 0.

Finally, not all growth rates are successfully modeled by using arithmetic or geometric sequences. Many growth rates are patterned after other types of sequences, such as the **Fibonacci sequence**, which begins with two 1s, each term thereafter being the sum of the two previous terms. Thus, the Fibonacci sequence is. The population growth of male honeybees is an example of a growth rate that follows the Fibonacci sequence.

See also Fibonacci sequence.

Resources

Books

Bittinger, Marvin L, and Davic Ellenbogen. *Intermediate Algebra: Concepts and Applications.* 6th ed. Reading, MA: Addison-Wesley Publishing, 2001.
Garfunkel, Soloman A., ed. *For All Practical Purposes, Introduction to Contemporary Mathematics.* New York: W. H. Freeman, 1988.
Tobias, Sheila. *Succeed With Math.* New York: College Entrance Examination Board, 1987.

James Maddocks

Growth hormones

Several **hormones** play important roles in human growth. The major human growth hormone (hGH), or somatotropin, is a protein made up of 191 amino acids secreted by the anterior pituitary and coordinates normal growth and development. Human growth is characterized by two spurts, one at **birth** and the other at **puberty**. HGH plays an important role at both of these times. Normal individuals have measurable levels of hGH throughout life. Yet, levels of hGH fluctuate during the day and are affected by eating and **exercise**. Receptors that respond to hGH exist on cells and tissues throughout the body. The most obvious effect of hGH is on linear skeletal development. But, the metabolic effects of hGH on muscle, liver, and **fat** cells are critical to its function. Humans have two forms of hGH, and the functional difference between the two is unclear. They are both formed from the same **gene**, but one lacks the amino acids in positions 32–46.

Additional hormones that affect growth are the somatomedins, thyroid hormones, androgens, estrogens, glucocorticoids, and **insulin**. Somatomedins are small **proteins** produced in the liver in response to stimulation by hGH. The two major somatomedins are insulin-like growth factor I and II (IGF-I and IGF-II). IGF-I causes increased cartilage growth and **collagen** formation, and its **plasma** levels peak between the ages of 13 and 17. IGF-II is important during fetal development and is present at constant levels in adult brains; however, its neuronal role is unclear. IGH-II increases protein synthesis as well as RNA and **DNA synthesis**. Levels of all these hormones are measured in the plasma, which is the liquid, cell-free, portion of **blood**.

Normal growth

Normal growth is regulated by hormones, but is also greatly influenced by genetic makeup and **nutrition**. Parental stature and growth patterns are usually indicative of the same in their offspring. Poor nutrition will negatively affect the growth process. This nutritional boost to growth occurs at conception, and must continue through embryonic and fetal development.

Newborn babies have high hGH levels, which continue through early infancy. Baseline plasma levels of hGH are normal, however, through childhood until puberty, when the resting plasma hGH level increases. Metabolically, hGH functions to increase the **rate** of protein synthesis in muscle, increase the rate of fat breakdown in fatty **tissue**, and decrease the rate of glucose use by tissues, resulting in an increase in glucose output by the liver. In the gastrointestinal tract, (GI) growth hormone increases absorption of **calcium**, an increase in metabolic rate, and a decrease in **sodium** and potassium excretion. The sodium and potassium are thought to be diverted to growing tissues. In essence, hGH frees up **energy** to build up tissues. hGH creates an increase in both **cell** sizes and numbers.

hGH is produced in the anterior portion of the pituitary gland under the control of hormonal signals in the hypothalamus. Two hypothalamic hormones regulate hGH; they are growth hormone-releasing hormone (GHRH) and growth hormone-inhibiting hormone (GHIH). When blood glucose levels fall, GHRH triggers the secretion of stored hGH. As blood glucose levels rise, GHRH release is turned off. Increases in blood protein levels trigger a similar response. As a result of this hypothalamic feedback loop, hGH levels fluctuate throughout the day. Normal plasma hGH levels are 1–3 ng/ml with peaks as high as 60 ng/ml. In addition, plasma glucose and **amino acid** availability for growth is also regulated by the hormones: adrenaline, glucagon, and insulin.

Most hGH is released at night. Peak spikes of hGH release occur around 10 p.m., midnight, and 2 a.m. The logic behind this night-time release is that most of hGH's effects are mediated by other hormones, including the somatomedins, IGH-I and IGH-II. As a result, the effects of hGH are spread out more evenly during the day. There is also evidence that GH secretion in humans follows a sexually dimorphic pattern, meaning that secretion patterns and levels of hormone are different in males and females.

Other fluctuations in growth occur naturally or because of illness. Growth slows in sick children, so that resources are channeled to heal. However, most children experience a catch-up **acceleration** of growth after a sick period. This growth can be as much as 400–above normal, but resumes normal levels once the child has caught up. Children given long treatments with steroids may experience hindered growth, as steroids stop growth.

Factors influencing hGH secretion include diet (nutrition) and stressors. Inhibition of hGH secretion occurs with high blood glucose levels, steroid use, and during REM **sleep**. HGH secretion increases with ingestion of a protein meal, deep sleep, low blood glucose levels, fasting, exercise, physical **stress** (such as **infection** or trauma), and psychological stress.

A second major growth spurt occurs at puberty with the coupled effect of sex hormones on growth. Puberty usually occurs earlier in girls (around the ages of age 10–12) than in boys (a few months later). During puberty, the epiphyseal ends of long bones begin to close, signaling the end of length growth. This closure is usually completed by the age of twenty.

Abnormal growth

A number of hormonal conditions can lead to excessive or diminished growth. Because of its critical role in producing hGH and other hormones, an aberrant pituitary will often yield altered growth. Dwarfism (very small stature) can be due to lack or under-production of hGH, lack of IGH-I, abnormalities in GH receptor or other changes leading to irresponsiveness of the target tissues to GH. Overproduction of hGH or IGH-I or an exaggerated response to these hormones can lead to giagantism or acromegaly both characterized by a very large stature.

Short stature can result not only from total absence of hGH, but also from GH deficiency. Children deficient in hGH have normal size at birth, but their postnatal growth is decreased leading to short stature, delayed bone maturation and delayed puberty. In contrast, absence of hGH can cause smaller birth lengths.

Giagantism is the result of hGH overproduction in early childhood leading to a skeletal height up to 8 ft (2.5 m) or more. Another condition, called acromegaly results from overproduction of hGH in adulthood. In this condition, the epiphyseal plates of the long bones of the body do not close, and they remain responsive to additional stimulated growth by hGH leading to increased bone thickness and length. People diagnosed with acromegaly develop increasingly enlarged and exaggerated facial bones. Also, acromegalic patients have pronounced, enlarged joints such as in the hands, feet, and spine.

Simple variation in height are due, in part, to a range of hGH levels due to factors already mentioned. However, parents concerned about their child's growth should discuss this with a pediatrician. hGH levels can be evaluated, and hormone therapy using synthetic hGH is a possibility. The hGH that is therapeutically used is now produced using **genetic engineering**. Recombinant human growth hormone (rhGH) is a protein produced from genetically altered cells. General growth patterns vary normally, however, and bone age is actually a more accurate reflection of regular development than age itself. Some children can be as much as a year off of average development. These children's growth rates may catch up or slow down compared to their peers. Growth patterns often follow family trends, such that if a boy's father was relatively small until puberty, during which time he outgrew his peers, then the boy may repeat this path. Some debate exists over the use of genetically engineered hGH to treat small stature. Some have debated that using hGH simply to make people taller is unethical and an example of science tampering with processes best left untouched.

Aging and growth hormone therapy in adults

Studies on mouse models aberrant in GH signaling indicate that dwarf **mice** live longer than normal or oversized ones. It appears that the impairment of GH signaling reduces IGF-1 levels dramatically, and implicates IGF-1 as regulator of aging. However, in presence of some other hormonal abnormalities in some of the mouse models it is impossible to dissect out the effects of GH and IGF-1 alone.

Independent of the **animal** studies, the use of GH for fighting off **obesity**, increasing energy, and as an anti-aging hormone increases. As people age, levels of GH decrease and the market for drugs to prolonging youth increases. Some people have a medical condition known as growth hormone deficiency. Defining it is problematic as there is not one universal definition. Growth hormone deficiency occurring in adults is associated with increased abdominal fatness, reduced muscle mass and strength, increased risk for cardiovascular **disease**, **memory** difficulties, and psychological problems. Treatment of adults with GH deficiency by hormone replacement is not universally accepted and is still being tested for efficacy and long-term side effects. The trials up to date indicate that patients treated with GH had reduced fat-mass and some improvements in quality of life.

KEY TERMS

Amino acid—An organic compound whose molecules contain both an amino group (-NH$_2$) and a carboxyl group (-COOH). One of the building blocks of a protein.

Epiphyseal closure—Closure of the epiphyses, the cartilaginous stretch next to a bone's end, which signifies the end of linear growth.

Hypothalamus—A region of the brain comprised of several neuronal centers, one of which regulates human growth hormone production in the pituitary.

Plasma—The non-cellular, liquid portion of blood.

Somatotrophs—Cells in the anterior pituitary which produce somatotropin, human growth hormone (hGH).

See also Physiology.

Resources

Books

Ganong, W., ed. *Review of Medical Physiology.* 15th ed. Norwalk, CT: Appleton & Lange, 1991.

Rhoads, R., and R. Pflanzer. *Human Physiology* 2nd ed. New York: Saunders College, 1992.

Periodicals

Carter, Christy S., Ramsey, Melinda M., and William E. Sonntag, "A Critical Analysis of the Role of Growth Hormone and IGF-1 in Aging and in Lifespan." *Trends in Genetics* (June 2002): 295–301

Cuneo, Ross C. et al., "The Australian Multicenter Trial of Growth Hormone (GH) Treatment in GH-deficient Adults." *Journal of Clinical Endocrinology and Metabolism* (January 1998): 107–116

Partridge, Linda, and David Gems, "Mechanisms of Aging: Public Or Private?" *Nature Reviews: Genetics* (March 2002): 165–175.

Louise Dickerson

Guanacos *see* **Camels**

Guava *see* **Myrtle family (Myrtaceae)**

Guenons

Guenons are small to medium-sized **monkeys** widespread throughout sub-Saharan **Africa**. These **primates** are classified in the infraorder of Old World simian primates

(Cataffhina) and the family Cercopithecidae. Their genus, *Cercopithecus*, is large, very diverse, and successful.

The Cercopithecidae family consists of two subfamilies: the omnivorous Cercopithecinae (including guenons, talapoin, and **baboons** from Africa) and the vegetarian Colobinae. There are approximately 14 **species** of monkeys in the genus *Cercopithecus* and a total of over 70 subspecies. These species are: the grass monkey, L'Hoest's monkey, the diademed guenon, the diana monkey, De-Brazza's monkey, the mona monkey, the crowned guenon, the lesser white-nosed guenon, the red-bellied guenon, the greater white-nosed guenon, the mustached guenon, the owl-faced guenon, the dwarf guenon, and the swamp guenon. The red guenon has been classified in another genus altogether, *Erythrocebus*.

General characteristics

While guenons vary greatly in coloring and facial characteristics, there are certain characteristics that they have in common. Generally, the guenon is a medium to large monkey, slender in build, measuring (head and body) 13-27.5 in (32.5-70 cm). Its tail is considerably longer than its body, ranging in length from 19.5-34 in (50-87.5 cm). Guenons are most active during the day, can be either terrestrial or arboreal, and walk on four feet, sometimes with slightly lifted wrists and ankles. While their jaws are short, they have well-developed cheek pouches. These pouches allow them to forage for food in open areas and then return to shelter to chew and swallow their food.

All guenons are easily recognized by their colorful fur. They are usually grayish green on their backs and the tops of their heads, and get much lighter down their sides and on their stomachs and undersides—usually light gray to white. However, as many of their species' names indicate, their coloring varies a great deal between species. Many species have ornamentation around their heads, such as white noses, white mustaches, white beards, prominent side whiskers, white throats, and/or white or brown brow bands. Their legs can also be decorated, sometimes with white or beige stripes on their thighs or bright coloring on the insides of their legs. Furthermore, the coloring of the young is occasionally different from that of adults.

The size and territory of guenon troops varies from species to species. Some species, such as the mustached guenons, stay within fairly strict territorial boundaries. Other species do not observe these boundaries at all, and roam freely. Within guenon troops, the ranking system found with baboons is only very loosely established.

The manner in which guenons feed is closely related to their **habitat**. For example, arboreal guenons primarily eat leaves and **tree** fruit. Reportedly, guenons in the wild will also eat eggs, although this has not been verified. In their natural habitats, they eat a wide variety of foods, such as **fruits** and **vegetables**, nuts, **insects**, **birds**, lizards, and other small animals.

Breeding

While guenons breed throughout the year, in some species, the births of the young are concentrated during specific times of the year. Pregnant for about seven months, guenon mothers, usually bear one baby, although twins sometime occur. The newborn clings to the mother's stomachs, supported by the mother's hand. According to scientists, some species of guenon newborns can walk after only a few days. At one or two months, they are active, and eat their first solid food. At four months, they can take care of themselves. Guenons mature sexually at the age of four years.

Habitat

While some guenons stray into the temperate climates found in southern Africa or high altitudes, they mostly thrive in tropical conditions. To survive, these monkeys need a **temperature** of at least 70°F (21°C). They are most comfortable when the temperature is between 75–85°F (24–29°C).

Guenons live primarily in the tropical **rainforest** belt in Africa to the south of the Sahara Desert, but a few species have adapted to the **forests** along principle African **rivers**. Furthermore, one species—the grass-monkey-thrives in the open **savanna** and spends a significant amount of **time** on the ground.

Overall, the forest-dwelling species prefer to live deep within the shelter of forests. However, they often prefer to inhabit different types of forests. For example, some species like to live high in the canopy and rarely come down to the lower branches or the forest floor. Other species are very active near the ground and commonly leave the trees. Interestingly, guenon distribution is heavily influenced by varying conditions. Population shifts sometimes cause migrations to areas previously uninhabited by guenons.

The savanna-dwelling species of guenon, the grass monkey, inhabits areas near the boundaries of rainforests. They live only near running **water** and tend to avoid open areas and forest interiors. They prefer certain trees, and are reportedly fond of the fruit of the wild fig tree. When they are pursued, they have been seen pressing themselves against tree branches for camouflage. Typically, they live in groups of 20-50, each group usually limiting its activities to a specific territory.

Activity

Like most monkeys, guenons are most active during the daytime. During the hottest part of the day, around noon, they rest, and groom themselves and each other. This is their only form of social contact. In the morning and late afternoon, they spend their time looking for food. While foraging, they communicate with each other through a series of peaceful calls. If a member of the troop encounters a dangerous situation, he emits a sharp barking sound, which is repeated by the troop members. On these occasions, all of the monkeys in the troop climb higher to get a better view of the danger. After doing so, the monkeys flee by running and jumping through the tree canopy. By extending their bodies vertically, guenons can leap very long distances from branch to branch.

Guenon relatives

The genus *Cercopithecus* has three offshoots that have adapted to other environmental conditions and evolved morphological traits that distinguish them from other guenons. Consequently, some scientists have assigned them separate generic status. The first of these relatives, is the dwarf guenon. Known formally as *Cercopithecus talapoin*, the dwarf monkey has been classified by some scientists in a separate genus called *Miopithecus*. Other scientists place these guenons in a subgenus by the same name. This monkey is significantly smaller than all other guenons and has morphological traits (in body structure) are directly related to its reduced size. It is found is swampy forests and mangrove swamps near the coast. It eats plants, nuts, insects, and, on occasion, small animals.

The second of these relatives is the swamp guenon (*Cercopithecus nigroviridis*). Like the dwarf guenon, it prefers to live in swampy forests. This monkey's skull and other anatomical characteristics are similar to those of baboons. These differences, combined with significant behavioral and vocal differences, have caused some scientists to classify them in a separate genus (*Allenopithecus*). Other scientists, who allow a wide range of guenon characteristics, classify swamp guenons in the subgenus *Allenopithecus* within the genus *Cercopithecus*. Very little is known about these monkeys in the wild. In captivity, they are very agile and tireless. It is believed that they live in small groups and eat a vegetarian diet.

The third relative of the guenon is the red guenon or dancing monkey, classified by all in a completely separate genus (*Erythrocebus*). It is the only species of guenon that lives primarily in semi-arid savanna, avoiding forests even when threatened. Thus, it has made several adaptations distinguishing it from all of its relatives.

These characteristics are: a rough coat, long and slender arms and legs, short hands and feet, and whiskers and mustaches on adult males. These guenons live in troops of 7-15, containing only one male. The male acts as a sentry, and is always looking for potential enemies. If something threatens the troop, the male red guenon distracts the **animal** while the others flee. Red guenons feed on plants, insects, and small animals. There are two subspecies of red guenon: the patas monkey and the Nisnas monkey.

In captivity

Often referred to as "organ grinder's" monkeys, young guenons make gentle, trusting pets. If treated well, they usually have pleasing dispositions and like attention. However, as they mature or if they are mistreated, they are large enough to become a threat. The dispositions of adult guenons can be unpredictable, sometimes bordering on aggressive. The males can inflict serious bites with their sharp canine teeth. Therefore, it is inadvisable to keep them as house pets.

In zoos, guenons are generally a public favorite. They are kept in family groups or pairs, and are fed a mixed diet of fruits, nuts, and vegetables. While guenons do not breed in captivity as readily as some other types of monkeys, breeding is not impossible. In fact, some zoos have been successful at interbreeding various guenon species.

Rhesus monkeys have long been the staple to most scientist performing animal experiments. However, these monkeys have been getting more and more difficult to secure. Consequently, guenons have increasingly been used for medical and pharmaceutical experiments.

Guenons live a long time in captivity; some guenons have reportedly lived to be more than 20 years old in zoos. Indeed, one mona monkey lived to be 26 years old in a United States zoo. While their life span may be as high as 25-30 years, guenons in the wild probably do not live to such an old age.

Resources

Books

Hill, W.C. Osman. *Evolutionary Biology of the Primates.* New York: Academic Press, 1972.

Jolly, Alison. *The Evolution of Primate Behavior.* New York: Maccmillan, 1972.

Preston-Mafham, Rod, and Ken Preston Mafham. *Primates of the World.* London: Blanford, 1992.

Walker, Ernest P. *The Monkey Book.* New York: Macmillan, 1954.

Kathryn Snavely

Guillain-Barre syndrome

Guillain-Barre **syndrome** (GBS) is a cause of progressive muscle weakness and paralysis which evolves over days or weeks, and resolves over the next several weeks or months. About 85% of patients recover completely, with no residual problems.

Causes

The classic scenario in GBS involves a patient who has just recovered from a typical, seemingly uncomplicated viral **infection**. The most common preceding infection is a Herpes infection (caused by cytomegalovirus or **Epstein-Barr virus**), although a gastrointestinal infection with the **bacteria** *Campylobacter jejuni* is also common. About 5% of GBS patients have a surgical procedure as a preceding event, and patients with lymphoma or systemic lupus erythematosus have a higher than normal risk of GBS. In 1976-1977, a hugely increased number of GBS cases occurred, with the victims all patients who had been recently vaccinated against the Swine flu. The reason for this phenomenon has never been identified, and no other flu **vaccine** has caused such an increase in GBS cases.

The cause of the weakness and paralysis of GBS is demyelination of the nerve pathways. Myelin is an insulating substance that is wrapped around nerves in the body, serving to speed conduction of nervous impulses. Without myelin, nerve conduction slows or ceases. GBS is considered an acute inflammatory demyelinating polyneuropathy (acute: having a short, severe course; inflammatory: causing symptoms of **inflammation**; demyelinating: destructive of the myelin sheath; polyneuropathy: disturbance of multiple nerves).

The basis for the demyelination is thought to be autoimmune (meaning that components of the patient's own **immune system** go out of control, and direct themselves not against an invading **virus** or bacteria, but against parts of the body itself). Next to nothing is understood about why certain viruses, surgical events, or predisposing conditions cause a particular patient's system to swing into autoimmune overdrive.

Symptoms

Symptoms of GBS begin five days to three weeks following the seemingly ordinary viral infection (or other preceding event), and consist originally of weakness of the limbs (legs first, then arms, then face), accompanied by prickly, tingling sensations (paresthesias). Symptoms are symmetric (affecting both sides of the body simultaneously), an important characteristic which helps distinguish GBS from other causes of weakness and paresthesias. Normal reflexes are first diminished, then lost. The weakness ultimately affects all the voluntary muscles, eventually resulting in paralysis. Paralysis of the muscles of **respiration** necessitates mechanical ventilation, occurring about 30% of the time. Very severely ill GBS patients may have complications stemming from other **nervous system** abnormalities which result in problems with fluid balance in the body, and **blood pressure** and **heart** rhythm irregularities.

About 5% of all GBS patients die, most from cardiac rhythm disturbances. While the majority of patients recover fully, there are some patients (particularly children) who have some degree of residual weakness, or even permanent paralysis. About 10% of GBS patients begin to improve, then suffer a relapse. These patients suffer chronic GBS symptoms.

Diagnosis

Diagnosis of GBS is made by virtue of the cluster of symptoms (ascending muscle weakness and then paralysis) and by examining the fluid which bathes the **brain** and spinal canal (cerebrospinal fluid or CSF). This fluid is obtained by inserting a needle into the lumbar (lower back) region. When examined in a laboratory, the CSF of a GBS patient will reveal an increased amount of protein over normal, with no increase over the normal amount of white blood cells usually present in CSF.

Treatment

Treatment of GBS is usually only supportive in nature, consisting of careful monitoring of the potential need for mechanical assistance in the event of paralysis of the muscles of respiration, as well as attention to the patient's fluid and cardiovascular status.

Plasmapheresis, performed early in the course of GBS, has been shown to shorten the course and severity of GBS, and consists of withdrawing the patient's blood, passing it through a **cell** separator, and returning all the cellular components (red and white blood cells, platelets) along with either donor **plasma** or a manufactured replacement solution. This is thought to rid the blood of the substances which are attacking the patient's myelin.

Fairly recently, it has been shown that the use of high doses of immunoglobulin given intravenously (by drip through a needle in a vein) may be just as helpful as plasmapheresis. Immunoglobulin is a substance naturally manufactured by the body's immune system in response to various threats. It is interesting to note that corticosteroid medications (such as prednisone), often the mainstay of anti-autoimmune **disease** treatment, are

KEY TERMS

Autoimmune—Immune response in which lymphocytes mount an attack against normal body cells.

Inflammatory—Having to do with inflammation. Inflammation is the body's response to either invading foreign substances (such as viruses or bacteria) or to direct injury of body tissue.

Myelin—The substance which is wrapped around nerves, and which is responsible for speed and efficiency of impulses traveling through those nerves. Demyelination is when this myelin sheath is disrupted, leaving bare nerve, and resulting in slowed travel of nerve impulses.

not only unhelpful, but may in fact be harmful to patients with GBS.

Resources

Books

Andreoli, Thomas E., et al. *Cecil Essentials of Medicine.* Philadelphia: W.B. Saunders Company, 1993.

Isselbacher, Kurt J., et al. *Harrison's Principles of Internal Medicine.* New York: McGraw-Hill, 1994.

Rosalyn Carson-DeWitt

Guinea fowl

Guinea fowl are seven **species** of medium-sized terrestrial **birds** in the family Phasianidae, order Galliformes, which also includes other fowl-like birds, such as the **grouse**, ptarmigan, turkey, **quail**, **peafowl**, and **pheasants**.

The natural range of guinea fowl is sub-Saharan **Africa**, the Arabian Peninsula, and Madagascar. However, these birds have been introduced to some other places, and are commonly kept in aviculture. The usual habitats of guinea fowl are open **forests**, savannas, and **grasslands**.

The range of body lengths of guinea fowl is 17-29 in (43-75 cm). Their head and the upper part of their neck are devoid of feathers, but the skin is brightly colored in hues of blue, red, yellow, or grey. Some species have a bony structure known as a casque on the top of their head, while others have a wattle or other types of colored protuberances. Their bill is short but stout, the wings

rather short and rounded, and the legs and feet are large and used for running and scratching in litter for their food of **insects**, **seeds**, roots, and rhizomes. The plumage is dark colored, but patterned with white spots and bars. The sexes are similar in shape and **color**.

Guinea fowl are terrestrial birds. They are powerful fliers, but only over a short distance. Guinea fowl generally prefer to run swiftly from danger rather than fly. Guinea fowl do not migrate.

The nests are crude scrapes on the ground, containing 2-20 eggs, which are brooded by the female. Both sexes care for the hatched young. Guinea fowl are highly gregarious birds, occurring in large flocks, especially during the non-breeding season, when they may also wander extensively. These flocks scatter readily when any bird perceives danger and utters an alarm call. The flock re-assembles later, as soon as one of the older, more-experienced males sounds an all-clear call.

The largest and most ornamentally plumaged species is the vulturine guinea fowl (*Acryllium vulturinum*) of central and east Africa. This is a relatively tall species, with long legs, an elongate neck, blue-skinned head, long downward hanging neck feathers known as hackles, a cobalt-blue colored breast, and a black body with white spots and stripes.

The helmeted or domestic guinea fowl (*Numida meleagris*) is originally from a wide range in sub-Saharan Africa, where it is commonly hunted as a game bird, as are other wild species of guinea fowl. However, the helmeted guinea fowl has also been domesticated. This species has long been kept in domestication in Africa and now more widely in tropical and south-temperate climates. Wild, naturalized populations also occur beyond the original range of this species, probably including the wild birds of Madagascar and smaller islands in the Indian Ocean, but also in Central America. The domestic guinea fowl is kept as a source of meat and eggs, although it is used for these purposes much less commonly than the domestic chicken (*Gallus gallus*, family Phasianidae). The domestic guinea fowl is also commonly kept as a pet.

Guinea pigs and cavies

Guinea **pigs**, or cavies, are about 20 **species** of **rodents** in the family Caviidae. Guinea pigs are native to **South America**, occurring from Colombia and Venezuela in the north, to Brazil and northern Argentina. These animals occur in rocky habitats, savannas, forest edges, and swamps, and can be rather common within their preferred **habitat**.

Guinea pigs have a stout body, with a relatively large head, short limbs and ears, and a vestigial tail that is not visible externally. The molars of guinea pigs grow continuously, an **adaptation** to the tooth wear associated with their vegetarian diet. Guinea pigs weigh 16-25 oz (450-700 g), and have a body length of 9-14 in (22-36 cm). They have four digits on their fore feet, and three on the hind, with naked soles, and small but sharp nails. Females have a single pair of teats, or mammae. The fur of wild guinea pigs is coarse, long, dense, and generally brown or grayish. However, the fur of domesticated varieties can be quite variable in **color** and form—the fur can be short or angora-long, dense or thick, straight or curly, and colored white, brown, gray, red, or black, occurring in monotones or in piebald patterns.

These nocturnal animals walk on four legs and are quick but not particularly fast runners. They live in burrows in small groups of several to ten individuals, and are rather vocal, emitting loud squeals, especially when warning the group of danger. Guinea pigs excavate their own dens, or they may take over diggings abandoned by other animals. They are vegetarian, eating a wide variety of plants. They tend to follow well-worn paths when they forage away from their burrows, and will quickly flee to the safety of their den at the first sign of potential danger.

Guinea pigs give **birth** to one to four babies after a gestation period of 60-70 days. Wild animals can breed as often as twice each year, but pet guinea pigs will breed more frequently than this. The young are mobile within hours of birth, and are sexually mature after about 55-70 days. Guinea pigs have a potential longevity of eight years, but they average less than this in the wild, largely because they are important food for a wide range of **predator** species, including humans.

The wild cavy (*Cavia aperea tschudii*) of Chile is believed to be the ancestor of the domestic guinea pig (*Cavia aperea porcellus*). The wild cavy is a montane and **tundra** species, living in habitats as high as 13,100 ft (4,000 m). Other relatively common species include the Amazonian wild cavy (*C. fulgida*), the southern mountain cavy (*Microcavia australis*), and the rock cavy (*Kerodon rupestris*). The most widespread species is the aperea (*Cavia aperea*).

The guinea pig was domesticated by Peruvian Incas in prehistoric times, and cultivated as a source of meat. They are still widely eaten in the highlands of Andean South America. The common name of the domestic guinea pig is likely derived from their piggish squeals, their plump and round body form, and the likelihood that the first animals to arrive in **Europe** in the late sixteenth century were transported by an indirect route from the coast of West **Africa** (where Guinea is located), probably in slave-trading ships.

Guinea pigs are widely used in medical and biological research, and are commonly kept as pets. They are gentle animals that are easily tamed, do not bite, and fare well in captivity as long as they are kept warm. Guinea pigs live easily in an open-topped box (because they do not climb), with newspaper or sawdust as bedding and to absorb their feces, a sleeping box, and a feeding bowl. If fed with a variety of grains and **vegetables**, guinea pigs do not need **water** to drink. Guinea pigs are social animals, and are happiest when kept with at least one other guinea pig. However, mature males will fight with each other, and are best maintained in separate cages.

Bill Freedman

Gulls

Gulls are 43 **species** of seabirds, in the subfamily Larinae of the family Laridae, which also includes the **terns**. Gulls occur in a wide range of coastal habitats, ranging from inland lakes, **rivers**, and **wetlands**, to marine shores and estuaries. Their distribution is virtually world-wide, but most species occur in the Northern Hemisphere.

Species of gulls range in body length from 8-32 in (20-81 cm). Their wings are long and pointed, and gulls have a short squared tail. The legs are short and stout, and the feet have webbing between the toes, useful for swimming. The bill is rather stout and hooked at the end.

Gulls are typically white-colored, with the wings and back, known as the "mantle," being colored gray or black. Some species have a black head during the breeding season. The sexes are alike in **color** and shape. Immature **birds** are usually much darker colored than the adults, but in a few species they are whiter.

Gulls are strong fliers, and they can undertake long-distance movements for purposes of feeding or during their migrations. Gulls often soar and glide effortlessly, whenever possible using the **wind** and updrafts to transport them where they want to go. Gulls are gregarious animals, both during the breeding season when they nest in loose colonies and during the non-breeding season when they often occur in large foraging and roosting flocks.

Gulls are highly omnivorous and opportunistic animals, eating a wide range of foods, depending on availability. However, they mostly feed on **animal biomass**, and less commonly on vegetation, especially **fruits**. Gulls are capable fishers, aerially spotting a **fish** as it

A laughing gull (*Larus atricilla.*) in flight. Laughing gulls have been known to perch on the head of a pelican and steal food when it opens its bill to shift its catch for swallowing. *Photograph by Robert J. Huffman. Field Mark Publications. Reproduced by permission.*

swims near the surface and catching it in their beak. This is usually done either by picking the food off the surface of the **water**, or sometimes by catching the **prey** after a head-long, shallow plunge into the water. Gulls also predate on the young of other seabirds when the opportunity presents itself. In addition, they scavenge carrion whenever it is available. Many species also scavenge the edible refuse of humans, near garbage dumps, fishing boats, fish-processing factories, and similar sorts of places.

Gulls nest in loosely structured colonies, generally building a mound-like nest out of **grasses** and seaweeds. Most species nest on the ground, but a few nest on ledges on cliffs. Gulls lay one to four greenish, speckled eggs, which are incubated by both sexes of the pair, which also share the raising of the young. Depending on the species, gulls can take as long as four to five years to reach sexual maturity. Some species of gulls are long-lived, and leg-ringed individuals have reached ages greater than 40 years.

Gulls in North America

The name "sea-gull" does not really apply to any particular species of bird. However, this name would be most appropriately used to describe the herring gull (*Larus argentatus*), which is the world's most widely distributed species of gull. The herring gull breeds extensively on the coasts of large lakes, rivers, and the oceans of **North America** and Eurasia. The herring gull spends its non-breeding season in the southern parts of its breeding range, and as far into the tropics as the equatorial coasts of **Africa**, the Americas, and Southeast **Asia**. The **taxonomy** of herring gulls has engendered some controversy among ornithologists due to confusion about the identity of subspecies and whether some of these should

be considered as distinct species. Herring gulls breed freely with other seemingly distinct gulls, including the Iceland gull (*L. glaucoides*), Thayer's gull (*L. thayeri*), and even the considerably larger, glaucous gull (*Larus hyperboreus*).

The world's largest gull is the greater black-backed gull (*L. marinus*). This large, black-mantled species breeds on the north Atlantic coasts of both North America and **Europe**.

The glaucous-winged gull (*L. glaucescens*) is an abundant species on the west coast of North and Central America. The western gull (*L. occidentalis*) is a black-backed species of the west coast of North America and is rather similar to the lesser black-backed gull (*L. fuscus*) of Europe, which sometimes strays to North America during the non-breeding season.

The ring-billed gull (*L. delawarensis*) is a common and widespread breeding gull, particularly on **prairie** lakes and on the Great Lakes, migrating to winter on the Atlantic and Pacific coasts.

The California gull (*L. californicus*) breeds in large colonies on prairie lakes and winters along the Pacific coast. This is the species of gull that "miraculously" descended on the grasshopper-infested fields of the first Mormons in Utah, helping to save their new colony. When **grasshoppers** are abundant, these gulls will gorge themselves so thoroughly with these **insects** that they are temporarily unable to fly.

The laughing gull (*L. atricilla*) breeds on the coast of the Gulf of Mexico and the Atlantic states. Like other black-headed gulls, this species has a white head, with some black spots, during the non-breeding season. Franklin's gull (*L. pipixcan*) is another black-headed species, breeding on small inland lakes, potholes, and marshes of the prairies, and migrating to the west coast of **South America** to spend the winter. Bonaparte's gull (*L. philadelphia*) breeds beside lakes and in other wetlands in the subarctic taiga and muskeg. This species commonly builds its nests in short **spruce** trees.

The black-headed gull (*Larus ridibundus*) is a small, widespread European species, which has recently begun to breed in small numbers in eastern Canada, particularly in Newfoundland.

Almost all species of gulls are in the genus *Larus*. One exception in North America is the kittiwake (*Rissa tridactyla*), a subarctic, highly colonial, cliff-nesting marine species that lacks the hind toe found in other species of gulls. The kittiwake breeds in large colonies in various places in the Canadian Arctic as well as in northern Eurasia. This species spends its non-breeding season feeding pelagically at sea, as far south as the tropics.

Another non-*Larus* species is Sabine's gull (*Xema sabini*), a fork-tailed gull that breeds in the arctic **tundra** of northern Canada, Greenland, Spitzbergen, and Siberia, and migrates down both the Atlantic and Pacific coasts to winter at sea off Peru and eastern South Africa. The ivory gull (*Pagophila eburnea*) is a rare, all-white species that only breeds in a few small colonies in the High Arctic of Canada and Siberia.

Ross's gull (*Rhodostethia rosea*) is another rare gull of the Arctic, breeding in a few places in eastern Siberia and, very rarely, at Hudson Bay in Canada. Ross's gull is a particularly beautiful small-sized gull, with bright-red legs and subtly pink breast and face plumage. On rare occasions, individuals of Ross's gull will wander to more southerly regions of North America, to the great excitement of many bird watchers.

Gulls and people

Because they are both omnivorous and opportunistic in their feeding habits, some species of gulls have benefitted greatly from certain human activities. In particular, gulls often feed on an amazing repertoire of foods at garbage dumps, especially if the daily refuse has not been covered over with a layer of dirt (as it would be in a sanitary **landfill**). Gulls also follow fishing boats, feeding on offal and by-catch as it is discarded overboard. In addition, gulls frequently patrol recently plowed agricultural land, where they feed on worms and other **invertebrates** that have been exposed by disturbance of the **soil**.

These and other opportunities provided to gulls by humans have allowed a tremendous increase in the populations and ranges of some species. Gulls whose populations in North America have shown especially large increases include the herring gull, great black-baked gull, ringed-bill gull, laughing gull, and glaucous-winged gull, among others.

In places where they are common, gulls are often considered to be a significant nuisance. Gulls are most commonly regarded as **pests** at and near solid-waste disposal sites, where they generally pick over the garbage. The can also be considered a problem in parks and stadiums, where they forage for left-over foods. In cities and towns where municipal drinking water is stored in open reservoirs, the presence of large numbers of gulls can result in fecal **contamination** of the water as a result of their copious defecations. Gulls are also a hazard to airplane navigation because of the risks of collisions. A single gull taken into a **jet engine** can easily ruin the machine and has resulted in airplane crashes. However, some species of gulls benefit humans by feeding on large numbers of insects that might otherwise damage **crops**.

KEY TERMS

Offal—Wastes from the butchering of fish, mostly consisting of the head, spinal column with attached muscles, and the guts. Fishing boats that process their catch at sea commonly dispose of the offal by throwing it into the water.

Pelagic—Refers to an animal that spends time at sea, far away from land.

The larger species of gulls, such as the herring and great black-backed gulls can be formidable predators of the young of smaller seabirds. The increased populations of these predatory gulls have severely affected the breeding success and populations of some smaller species, especially terns. This is a serious **conservation** problem in many areas, and it may only be resolved by killing adult gulls with guns or poisons. The alternative to this unsavory control strategy would likely be the local extirpation, and perhaps even global endangerment of, the prey species.

In many places gulls eggs are regarded as a delicacy and are collected as a subsistence food or to sell. To ensure freshness, all of the eggs in a colony are generally smashed on the first visit to the breeding site. Consequently, the age of any eggs that are collected on the second or subsequent visits is known. Adult or young gulls are also sometimes eaten by people, though this is not very common.

In spite of some of the problems with gulls, they are a favored group among bird-watchers. Numerous species of gulls can be seen in some places, especially during the non-breeding season. Birders often undertake field trips to those avian hot-spots, with the specific goal of identifying as many rare species of gulls as possible.

Resources

Books

Brooke, M., and T. Birkhead. *The Cambridge Encyclopedia of Ornithology*. Cambridge, UK: Cambridge University Press, 1991.

Croxall, J.P., ed. *Seabirds: Feeding Biology and their Role in Marine Ecosystems*. Cambridge, UK: Cambridge University Press, 1987.

Grant, P. *Gulls: A Guide to Identification*. London, UK: Poyser Pubs, 1986.

Harrison, C.J.O., ed. *Bird Families of the World*. New York: H. N. Abrams Pubs, 1978.

Harrison, P. *Seabirds: An Identification Guide*. Beckenham, UK: Croom Helm, 1991.

Richards, A. *Seabirds of the Northern Hemisphere*. New York: Dragonsworld, 1990.

Bill Freedman

Guppy

One of the most popular **species** of **freshwater** topical **fish** is the guppy. The first specimens were brought to the British Museum in London for description in 1859 by R. J. L. Guppy, a biologist from Trinidad (West Indies) after whom the fish is named. The species originally possessed the scientific name *Lebistes reticulatus*, but in 1963, the Latin name was changed to *Poecilia reticulata* and remains so today. While characteristically quite small in size, guppies display a wide range of colors and patterns, and many forms of elaborate fin and tail shapes. The modifications in **color** and fins create numerous varieties of the species. The brightest and most ornate are termed *fancy* varieties and are highly prized as aquarium pets. Although colorful, male guppies average only 1.2 in (3 cm) in length, while the larger females average a mere 3 in (8 cm) long when mature.

Guppies belong to the phylum Chordata within the kingdom Animalia. Like many evolutionarily advanced organisms (including **birds**, **reptiles**, and **mammals**), guppies are **vertebrates**, possessing spinal cords protected by a vertebral column. Guppies also belong to the class Osteichthyes, which is the group of organisms defined as "bony" fish. As members of this large group, guppies have a skeleton made of bone (unlike **sharks** which have skeletons made of cartilage), homocercal tails (having equally sized upper and lower lobes), skin with embedded scales and mucus **glands**, and a swim bladder that helps control buoyancy. Guppies are also among the most evolutionarily advanced group of **bony fish**, known as Teleosts. Teleosts have characteristically complex, extensible mouth to aid in **prey** capture.

Guppies are tropical fish that prefer warm **water** temperatures between 73-83°F (22-28°C). They are a freshwater species native to Trinidad, Barbados, and Venezuela where they inhabit slow moving streams and relatively shallow lakes. In contrast to other aquatic organisms, guppies give **birth** to live offspring. Displaying a characteristic termed ovoviviparity, female guppies release live immature progeny (called fry) that developed from yolky eggs within the mother rather than deriving nourishment directly from the mother herself. Captive guppies feed readily on worms, small crustaceans, small **insects**, and **plant matter** in addition to commercially available flake food. The relative ease of care and breeding of guppies, in combination with their stunning varieties, make them very popular tropical tank species.

Terry Watkins

Gutenberg discontinuity

The Gutenberg discontinuity occurs within **Earth's interior** at a depth of about 1,800 mi (2,900 km) below the surface, where there is an abrupt change in the seismic waves (generated by earthquakes or explosions) that travel through **Earth**. At this depth, primary seismic waves (P waves) decrease in **velocity** while secondary seismic waves (S waves) disappear completely. S waves shear material, and cannot transmit through liquids, so it is believed that the unit above the discontinuity is solid, while the unit below is in a liquid, or molten, form. This distinct change marks the boundary between two sections of the earth's interior, known as the lower mantle (which is considered solid) and the underlying outer core (believed to be molten).

The molten section of the outer core is thought to be about 1,292°F (700°C) hotter than the overlying mantle. It is also denser, probably due to a greater percentage of **iron**. This distinct boundary between the core and the mantle, which was discovered by the change in seismic waves at this depth, is often referred to as the core-mantle boundary, or the CMB. It is a narrow, uneven zone, and contains undulations that may be up to 3-5 mi (5-8 km) wide. These undulations are affected by the heat-driven **convection** activity within the overlying mantle, which may be the driving force of plate tectonics-motion of sections of Earth's brittle exterior. These undulations in the core-mantle boundary are also affected by the underlying eddies and **currents** within the outer core's iron-rich fluids, which are ultimately responsible for **Earth's magnetic field**.

The boundary between the core and the mantle does not remain constant. As the **heat** of the earth's interior is constantly but slowly dissipated, the molten core within Earth gradually solidifies and shrinks, causing the core-mantle boundary to slowly move deeper and deeper within Earth's core.

The Gutenberg discontinuity was named after Beno Gutenberg (1889-1960) a seismologist who made several important contributions to the study and understanding of the Earth's interior. It has also been referred to as the Oldham-Gutenberg discontinuity, or the Weichhert-Gutenberg discontinuity.

See also Earthquake; Tectonics.

Gutta percha

Gutta percha is a rubberlike gum obtained from the milky sap of trees of the Sapotaceae family, found in Indonesia and Malaysia. Once of great economic value, gutta percha is now being replaced by **plastics** in many items, although it is still used in some electrical insulation and dental work. The English natural historian John Tradescant (c. 1570-1638), introduced gutta percha to **Europe** in the 1620s, and its inherent qualities gave it a slow but growing place in world trade. By the end of World War II, however, many manufacturers switched from gutta percha to plastics, which are more versatile and cheaper to produce.

Sumatra, one of the largest islands of Indonesia, is the world's leading producer of gutta percha; the **island** is home to many plantations of *Palaquium oblongifolia* and *Palaquium javense* trees. The trees reach 66-81 ft (20-25 m) in height; the lance-shaped leaves, usually 6 in (15 cm) in length, have feather-like vein patterns called pinnate venation. The greenish flowers, about 0.4 in (1 cm) wide, contain pollen-bearing stamens and seed-bearing pistils. The **seeds** contain a butter-like **fat** that is used as food.

Gutta percha sap is extracted from the leaves— unlike rubber, which is collected by producing incisions on the **tree** trunk. The leaves are ground up and boiled in **water**, and the gum is removed. At room **temperature**, the resulting gum forms a hard brown substance that can be molded if softened by **heat**; the melting point of gutta percha is 148°F (64°C). It is dielectric, which means that it can sustain an electric field but will not conduct electrical currents; this property, combined with its resistance to alkalies and many acids, made it a good insulator for underseas cables until better synthetic insulators were developed in the 1940s. Its resistance to acids also made it a good material for acid containers, but plastics have also replaced gutta percha in these products.

The primary use of gutta percha now is in the manufacture of golf ball covers, for which hard, resilient qualities are desired to withstand the golfer's strikes without shattering or chipping. However, plastics may soon replace gutta percha in this product as well, as the Dunlop Rubber Company has produced a plastic golf ball cover that is almost identical to the covers using gutta percha.

Gymnosperm

Gymnosperms are one of the two major groups of plants that produce **seeds**; the other is the angiosperms. Gymnosperm literally means "naked seed," which refers to the development of seeds exposed on a flat structure, that is, not within an ovary as in the angiosperms.

Gymnosperms became common about 290 million years ago and although many of the earlier types are now extinct, four kinds remain alive: the conifers, **cycads**, gnetophytes, and **ginkgo**, the maidenhair **tree**. Conifers are the most familiar, widespread, and abundant of the gymnosperms. Most conifers have needle- or scale-like leaves that persist for more than one year, that is, are evergreen. The conifers include **species** of pine, **spruce**, fir, hemlock, cedar, **juniper**, and redwood. The latter is the largest **plant**, exceeding 328 ft (100 m) in height. Conifers dominate boreal **forests** of high latitudes and mature forests of high altitudes, and are extremely valuable commercially. Their **wood** is used to make houses, furniture, and **paper** products. Resin, an organic secretion of conifers, has various uses ranging from turpentine for thinning paint, to resin blocks for making violin bows sticky. Some **pines** produce edible nuts and juniper berries are used to give gin its distinctive taste.

The other kinds of living gymnosperms are much less abundant, more geographically restricted, and less valuable than conifers. These gymnosperms are highly diverse, however, and sometimes quite strange. The cycads, or Sago **palms**, look like palms but are not (palms are flowering plants). Cycads are unusual among plants in that each individual is either a male or female, as in most animals. Most seed-producing plants are bisexual. The gnetophytes as a group are the oddest of the gymnosperms, and the oddest of these is *Welwitschia*. This species lives in sandy deserts of southwestern **Africa**. *Welwitschia* has a saddle-like, central core that produces only two leaves during the life of the plant. These grow continuously, frequently splitting along their length to give a mop-headed appearance.

Lastly, the ginkgo or maidenhair tree (*Ginkgo biloba*) is the only living species of a group that was most common about 170 millon years ago. The ginkgo has distinctive, fan-shaped leaves that are sometimes cleft in the middle and, unlike the leaves of most gymnosperms, fall each autumn. Ginkgo trees also are either male or female. Because of its unusual and attractive leaves, this species is widely planted as an ornamental. Curiously, the ginkgo appears to be extinct in the wild, only having survived through cultivation by Buddhist monks in China.

See also Angiosperm; Conifer; Firs; Sequoia.

Gynecology

Gynecology, from the Greek meaning "the study of women," is a medical specialty dealing with the health of a woman's genital tract. The genital tract is made up of the reproductive organs including the vagina, cervix, uterus, ovaries, fallopian tubes, and their supporting structures.

Marked changes occur in a woman's reproductive organs upon her reaching menarche (the age at which she begins to menstruate) and again during any pregnancy that occurs in her life. Later, at the stage known as **menopause**, she experiences still other changes. It is the specialty of the gynecologist to guide women through these alterations and to ensure that they retain their health throughout each stage.

Maturity of the reproductive organs has to do with hormonal regulation of the organs centering on the pituitary gland in the **brain**. This gland, the master endocrine gland, stimulates the ovaries to produce other **hormones** that encourage the maturity of an ovum (egg). The egg is released from the ovary, is carried down to the uterus (womb), and if the egg is not fertilized the woman has her "period" or menses. This is the sloughing off of the lining of the uterus which is rebuilt each month in preparation to accept a fertilized ovum.

This cycle occurs approximately once a month or so if the woman is not pregnant. Thus, each month the uterus and the ovaries go through a cycle of preparation and dissolution and rebuilding far more profound than do any organs in the male body.

History

Until the late nineteenth century, physicians linked the female menstrual cycle to the phases of the **moon**. Of course, if that were so, every female would have her menstrual period at the same time. It was late in the nineteenth century that researchers attributed menstrual changes to hormones. Not until the early twentieth century were those hormones isolated in pure form and named. Female hormones as a group are called estrogens.

The menstrual cycle

Hormonal interaction during the **menstrual cycle** includes hormones from the pituitary, the ovaries, and the uterus itself. In a complicated, interwoven pattern the hormones become dominant and retiring in turn, allowing ovulation (release of the ovum), **fertilization**, implantation (lodging of the fertilized egg on the wall of the womb), or menstruation, and then beginning over again.

The female reproductive organs are very susceptible to pathologic changes—those that constitute **disease**. Hormonal disruption can alter the cycle or stop it and other, as yet unknown causes can change **cell** development to a cancerous lesion. Also, at approximately 50 years of age, the woman undergoes what is commonly called the "change of life," or menopause. Here the hormonal pattern changes so that eggs no longer are produced and the menstrual cycle no longer takes place.

Again, at this stage the woman is susceptible to long-term pathologic changes leading to **osteoporosis** (thinning of the bones), which renders her more likely to suffer fractures.

Testing

The gynecologist can monitor a woman's stage in life and administer tests to determine whether her reproductive organs are healthy. Removing, staining, and studying cells from the vagina and cervix each year can help to detect **cancer** early, when it is curable. This test, commonly called the Pap test, is named after the physician who developed it in the mid-twentieth century-George Papanicolaou. He learned that by scraping cells from the vaginal walls at a certain stage in the woman's cycle and staining the cells for viewing under a **microscope**, he could determine whether any abnormal cells were present that could be forerunners of cancer.

Gynecologists can also investigate why a woman is unable to become pregnant. She may have obstructed fallopian tubes or a hormonal imbalance that prevents maturity and release of the ovum or prevents implantation of the fertilized ovum onto the uterine wall. In each case, steps can be taken to correct or bypass the problem so the woman can bear children.

Gynecology has advanced to the point that the physician can force the ovaries to produce eggs, which can then be removed and fertilized in a dish (called in-vitro fertilization) and then implanted in the uterus. This technique is not guaranteed to produce an infant, but in many cases the implanted ovum will mature into the desired offspring—often into more than one baby. The science of gynecology continues to make advances against the **pathology** that may deny a woman the ability to have babies.

See also Puberty; Reproductive system.

Gyroscope

A gyroscope is heavy disk placed on a spindle that is mounted within a system of circles such that it can turn freely. When the disk, called a flywheel, is made to spin, the gyroscope becomes extremely resistant to any change in its orientation in **space**. If it is mounted in gimbals, a set of pivot and frame mountings that allow it freedom of **rotation** about all three axes, a fast-spinning gyroscope will maintain the same position in space, no matter how the frame is moved. Once the flywheel is set spinning, the spindle of a gyro in a gimbal mount can be aimed toward true north or toward a **star**, and it will continue to point that direction as long as the flywheel continues spinning, no matter what kind of turning or tilting the surface bearing it experiences. This stability has allowed the gyroscope to replace the magnetic compass on ships and in airplanes.

An interesting aspect of gyroscope **motion** occurs when a the flywheel is set rotating and one end of the spindle is set on a post using a frictionless mount. Intuitively, it would appear that the gyroscope should fall over, but it instead describes a horizontal **circle** about the post, flywheel still spinning. In apparent defiance of natural laws, it is simply obeying one of the simplest laws of **physics**, that of **conservation** of angular **momentum**. To understand the motion of a gyroscope, you must first understand angular momentum and **torque**. Angular momentum can be thought of as a rigid body's tendency to turn. Specifically, it tells us how much a given bit of the gyroscope flywheel contributes toward turning the flywheel about the spindle at any instant in **time**. For a small rotating object, angular momentum L is defined as

$$L = r \times mv$$

where m is the **mass** of the object, r is the distance of between the mass and the origin of rotation, and v is the instantaneous **velocity** of the mass. For the gyroscope, the angular momentum is obtained by considering it as composed of tiny bits, and adding up the contributions of each piece. Vectors have magnitude and a direction.

Torque, on the other hand, can be thought of as the rotational analog of **force**. Its effect depends upon the distance it is applied from the pivot point. Torque (τ) is defined as $\tau = r \times f$, where f is the applied force and r is the distance between the pivot and the point at which the force is applied. In this case Newton's second law permits us to state that force equals mass times **acceleration** (f = ma), so and we can define torque as $\tau = r \times ma = r \times mg$, where g is acceleration due to gravity. When you hold a weight out horizontally with your arm fully extended, you feel the torque that the weight mg is applying at r, the length of your arm from your shoulder, the pivot point.

How does this apply to the gyroscope? When the flywheel spins at a high **rate**, the angular momentum vector is pointing straight along the spindle. The vector r points along the spindle also, until it reaches the flywheel, the center of mass m. On Earth's surface, gravity g is acting on the flywheel, pulling it downward. According to the righthand rule, the fingers of your righthand point in the direction of r, then bend to point in the direction of g, and your thumb will point in the direction of the torque vector t. Notice that t is in a horizontal direction, the same direction that the gyroscope describes as it turns. To conserve angular momentum, the gyroscope will pivot about the support post, or precess, in an effort to align the angular momentum vector with the torque vector.

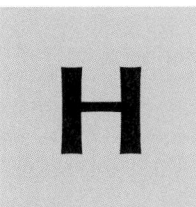

Habitat

The term habitat refers to the type of environment in which an **organism** or **species** occurs. For plants, habitat is mostly defined by its physical attributes (e.g., rainfall, **temperature**, topographic position, **soil** texture and moisture) and its chemical properties (e.g., soil acidity, concentrations of **nutrients** and toxins, oxidation reduction status). For terrestrial animals, the habitat is defined by the physical structure (e.g., grassland, shrub-dominated, or forested) and the **plant** species composition. Habitat for aquatic organisms is mostly determined by physical factors (such as running versus still **water**, depth, and **light** availability) as well as climatic and chemical components (especially nutrients). Broad ecological characteristics may also be included in the characterization of habitat, for instance forest, **prairie**, or **tundra** habitats on land, and littoral or pelagic ones in water. Within a given habitat there may be various micro-habitats, for example the hummocks and hollows in bogs, or patches of disturbance in **forests**.

Hafnium *see* **Element, chemical**

Hagfish

A primitive group of **fish**, hagfish (order Hyperotreti, family Myxinidae) resemble eels in their external appearance. These fish lack a backbone, jaws, true fins, and scales. Their body is tubelike and often covered in a slimy substance that is secreted from abundant **glands** in the skin. The body is often a pale fleshy pink, but is occasionally brown-gray above and pink below. They may reach up to 2 ft (60 cm) in length, but most measure about 1.5 ft (40 cm); females are often larger than males.

Hagfish are bottom-dwelling fish of soft, muddy substrates, living at a depth range of 65.5-1,968 ft (20-600 m). They feed on bottom-dwelling **crustacea** and polychaete worms, but may also scavenge on dead fish. The mouth is a simple slit surrounded by a ring of fleshy barbels that have a sensory role; other barbels are located around the nostrils and probably fulfill a similar role. The tongue is serrated and consists of two plates of gristle with many horny teeth that are continuously replaced as they wear out with use.

Hagfish have an unusual but simple system of obtaining **oxygen** from the surrounding **water**. Unlike most fish that have a complex arrangement of gills, hagfish have a simplified set of gills in a series of paired pouches that open to the pharynx and the exterior. Water enters the body through the snout, is compressed in the body and expelled through these breathing pouches. Under normal swimming conditions this does not present a problem, but when the fish is feeding, the flow of water is seriously reduced, if not completely cut off. This has led to the suggestion that hagfish may be able to tolerate temporary periods of oxygen deficiency and are later able to expel all metabolic body wastes when normal breathing resumes.

Fifteen **species** of hagfish have been recognized to date, all of which are marine-dwelling with the majority living in temperate oceans; one species has been recorded from the tropical waters off Panama. Hagfish are completely blind, but have a very keen sense of **smell**, which is used to locate food. They are active hunters (although poor swimmers), and also attack sick and dead fish, latching onto their **prey** and absorbing the flesh of the other species through continued rasping motions of their tongues. While hagfish play an important role in eliminating weak or ill fish, they can sometimes be a nuisance to commercial fisheries, as they are known to attack captured fish on long line fisheries.

Once they have reached sexual maturity—usually by the time they grow to 10-11 in (25-28 cm)—hagfish may breed throughout the year. Some species are hermaphrodites (each fish has both male and female reproductive organs), but others are either male or female.

Hagfish lay relatively large yolky eggs in a horny shell with hooked filaments at each end. The young are free-living and resemble the adults when they hatch.

Hail *see* **Precipitation**

Hail *see* **Thunderstorm**

Half-life

The half-life of a process is an indication of how fast that process proceeds—a measure of the **rate** or rapidity of the process. Specifically, the half-life is the length of **time** that it takes for a substance involved in that process to diminish to one-half of its initial amount. The faster the process, the less time it will take to use up one-half of the substance, so the shorter the half-life will be.

The rates of some biological processes, such as the elimination of drugs from the body, can be characterized by their half-lives, because it takes the same amount of time for half of the drug to disappear no matter how much there was to begin with. Processes of this kind are called first-order processes. On the other hand, the speeds of many **chemical reactions** depend on the amounts of the various substances that are present, so their rates cannot be expressed in terms of half-lives; more complicated mathematical descriptions are necessary.

Half-lives are most often heard of in connection with radioactive decay—a first-order process in which the number of **atoms** of a radioisotope (a radioactive **isotope**) is constantly diminishing because the atoms are transforming themselves into other kinds of atoms. (In this sense, the word "decay" does not mean to rot; it means to diminish in amount.) If a particular radioisotope has a half-life of one hour, for example, then at 3 P.M. there will be only half as many of the original species of radioisotope atoms remaining as there were at 2 P.M.; at 4 P.M., there will be only half as many as there were at 3 P.M., and so on. The amount of the radioactive material thus gets smaller and smaller, but it never disappears entirely. This is an example of what is known as exponential decay.

The half-life of a radioisotope is a characteristic of its nuclear instability, and it cannot be changed by ordinary chemical or physical means. Known radioisotopes have half-lives that range from tiny fractions of a second to quadrillions of years. Waste from the reprocessing of **nuclear reactor** fuel contains radioisotopes of many different half-lives, and can still be at a dangerously high level after hundreds of years.

The mathematical equation which describes how the number of atoms, and hence the amount of radioactivity, in a **sample** of a pure radioisotope decreases as time goes by, is called the **radioactive decay** law. It can be expressed in several forms, but the simplest is this: $\log P = 2 - 0.301 \, t/t_{12}$. In this equation, P is the percentage of the original atoms that still remain after a period of time t, and t_{12} is the half-life of the radioisotope, expressed in the same units as t. In other words: To get the logarithm of the percentage remaining, divide t by the half-life, multiply the result by 0.301, and subtract that result from 2.

See also Radioactive waste.

Robert L. Wolke

Halibut *see* **Flatfish**

Halide, organic

Organic halides are organic compounds containing a halogen atom bonded to a **carbon** (C) atom. Fluorine (F), **chlorine** (Cl), bromine (Br), and iodine (I) are all types of halogen **atoms**. A compound that contains a carbon atom bonded to a fluorine atom (C-F) is called an organofluoride. If the carbon atom is part of a chain of carbon atoms, the organofluoride compound is referred to as an alkyl fluoride. If the carbon atom is contained in a **benzene** or phenyl ring, the organofluoride is called an aryl fluoride. Other halide compounds are named in a similar fashion.

The reactivity of organic halides depends on the halogen atom that is bonded to the carbon atom in the particular compound. Organoiodides are the most reactive and can be converted into many other compounds. Organobromides are less reactive than organoiodides but more reactive than organochlorides. Organofluorides are the least reactive of the organic halides.

Organofluorides

Organofluorides are very stable compounds that are nonflammable, have very limited toxicity, and do not react with other chemicals. Perfluorocarbons (PFCs) are alkyl fluorides that consist of chains of carbon atoms bonded only to fluorine atoms. In 1966, scientists demonstrated that large amounts of **oxygen** could be dissolved in PFCs. Two years later, scientists replaced the **blood** in a laboratory rat with a **solution** of oxygen dissolved in PFCs. The **animal** lived and scientists began researching the use of PFCs as artificial blood. In 1990, the Food and Drug Administration (FDA) approved the use of Fluosol-DA, a PFC solution licensed to Green

Cross Corporation, as an oxygen carrier during the medical process of cleaning **heart arteries** with a **balloon**. PFCs are also used to temporarily replace **eye** fluid during **surgery** on the eye. PFC chains are the foundation of many products used to repel **water**, oil, and dirt from carpets and upholstery. Many aryl fluorides are important pharmaceutical and agricultural products. The anti-inflammatory agent dislunisal, the tranquilizer haloperidol, and the sedative flurazepam hydrochloride are examples of drugs that are also aryl fluorides. Fluometuron, an aryl fluoride herbicide, is used to kill weeds in grain and **cotton** fields. Flutriafol, another aryl fluoride, is a **fungicide** used to stop diseases on the grains used in the manufacture of various cereals.

Organochlorides

Because of their low chemical reactivity, alkyl chlorides are useful in dissolving other chemicals, greases, and oils. They are used as solvents for dry cleaning, removing oil from **metal** parts, and running **chemical reactions**. Methylene chloride (CH_2Cl_2) is an alkyl chloride with a low **molecular weight** that is used in many paint and varnish removers. It is also used as a solvent for removing **caffeine** from coffee. The well-known pesticide, DDT, is an aryl chloride that was first used in 1939 to kill the **mosquitoes** that transmitted **malaria**. This chemical was beneficial in eliminating the spread of malaria throughout the world. However, DDT is also poisonous to **fish** and **birds**; as a result, the Environmental Protection Agency (EPA) stopped its use in 1972. Aryl chlorides, such as chlozolinate and quintozene, are used to stop the growth of fungus on **fruits** and **vegetables**.

Chlorofluorocarbons

Chlorofluorocarbons (CFCs) are compounds that contain both chlorine atoms and fluorine atoms bonded to carbon atoms. These compounds are very stable and are usually gases at room **temperature**. CFCs are often called Freons (trademark of E.I. du Pont de Nemours & Co.) because they are used in refrigerators and air conditioners. CFCs are also employed in the manufacture of various hard foams that are used under the siding in buildings and around dish washers and refrigerators for insulation and sound proofing. Since CFCs are excellent at dissolving oil and grease, they are a primary component of dry cleaning solutions. They are also used to remove oil and grease from electronic parts. The chlorofluorocarbon, dichlorodifluoromethane (CCl_2F_2), is not poisonous and is employed as the carrier gas in **asthma** and **allergy** inhalers. This compound is mixed with ethylene oxide, and the resulting gas is used to sterilize medical equipment and materials that are sent into outer **space**.

In 1971, scientists determined that CFCs were accumulating in the atmosphere; they later showed that this "build up" was destroying the **ozone** layer, a level of the stratosphere that absorbs much of the harmful ultraviolet rays from the **Sun**. The Antarctic ozone hole was discovered in 1985, prompting the international community to sign the Montreal Protocol. The agreement between 24 nations limited the production of CFCs with the intention of ceasing production by the year 2000. The Protocol went into effect on January 1, 1989, and by July 1992, 81 countries had signed the agreement to save the ozone layer.

Organobromides

Organobromides form highly reactive compounds when mixed with metals such as **magnesium** (Mg) or **aluminum** (Al); for this reason, they are used extensively in the manufacture of dyes, drugs, and other chemicals. The alkyl bromide, bromotrifluoromethane ($CBrF_3$) is not poisonous and will not burn. It is used in portable fire extinguishers and in airplanes to stop engine fires while in flight. Halothane ($CF_3CHClBr$), another organobromide, is used as a general medical anesthetic. Aryl bromides are colored and are used extensively as dyes and colorants. Alizarine Pure Blue B is an aryl bromide used to dye wool. The orange **color** in lipsticks is often D & C Orange No. 5, another member of the organobromide family.

Organoiodides

The most chemically reactive of the organic halides are those that contain carbon atoms bonded to iodine atoms. Organoiodides are not used as extensively as organobromides or chlorides because they are expensive. Alkyl iodides react with metals such as **lithium** (Li) or mercury (Hg) to make useful chemicals in the manufacture of pharmaceutical and organic intermediates. The aryl iodide, thyroxin, is a thyroid hormone used to stimulate human **metabolism**. Erythrosin, or FD&C; Red No. 3, was used to add red color to maraschino cherries. This dye was removed from the market when researchers found that it caused **cancer** in laboratory animals.

See also Dyes and pigments; Halogens; Halogenated hydrocarbons; Ozone layer depletion.

Resources

Books

Hudlicky, Milos. *Chemistry of Organic Fluorine Compounds.* Englewood Cliffs, NJ: Prentice Hall, 1992.
Kirk-Othmer Encyclopedia of Chemical Technology. "Bromine Compounds," vol. 4, p. 567; "Chlorocarbons and Chlorohydrocarbons," vol. 5, p. 1017; "Fluorine Compounds, Or-

KEY TERMS

Chlorofluorocarbons (CFCs)—Compounds that contain both chlorine atoms and fluorine atoms bonded to carbon atoms.

Halide, organic—An organic functional group that consists of a halogen atom bonded to a carbon atom.

Montreal Protocol—An agreement initially between 24 nations to limit and eventually stop the production of ozonedepleting chlorofluorocarbons.

Organobromide—A compound that contains a carbon atom bonded to a bromine atom (C-Br).

Organochloride—A compound that contains a carbon atom bonded to a chlorine atom (C-Cl).

Organofluoride—A compound that contains a carbon atom bonded to a fluorine atom (C-F).

Organoiodide—A compound that contains an carbon atom bonded to a iodine atom (C-I).

Perfluorocarbons (PFCs)—Compounds that consist of chains of carbon atoms bonded exclusively to fluorine atoms.

ganic," vol. 11, p. 467; and "Iodine and Iodine Compounds," vol. 13, p. 667. New York: John Wiley and Sons, 1991.
McMurry, J. *Organic Chemistry*. 5th ed. Pacific Grove, CA: Brooks/Cole Publishing Company, 1999.
Patai, S., ed. *The Chemistry of the Carbon-Halogen Bond*. New York: John Wiley, 1973.

Andrew Poss

Hall effect

A current-carrying body placed in a magnetic field with the current direction unaligned with the field experiences a **force** leading to a transient sidewise drift of the charge carriers of the current. This drift continues until the force is balanced by an electric field produced by the charge accumulating at points on the body's surface in the direction of the drift. At points on the body's surface opposite the direction of the drift, there will clearly be an equal depletion of charge, which is equivalent to an accumulation of charge of opposite sign. The electric field created by this transient **behavior** is called the Hall field and results in a potential difference between corresponding points on the two oppositely charged surfaces. Which

of the two surfaces is at the higher potential is determined by the sign of the charge carriers. If the carriers are positive, the surface in the direction of their drift will be at the higher potential; if the carriers are negative, the surface in the direction of their drift will be at the lower potential. The phenomenon thus described is called the Hall effect after E. H. Hall, who discovered it in 1879. A little over a century later, it was discovered by Klaus von Klitzingthat the Hall potential in a semiconducting material experiences quantum jumps as the magnetic field is increased when subjected to temperatures far below room **temperature**. This remarkable discovery has made it possible to measure an important constant of **physics**, called the fine structure constant, to a heretofore unattainable **accuracy**. Also, it provides scientists with a readily achieved standard for making accurate determinations of conductivity. For this discovery, von Klitzing was awarded the Nobel prize in 1985.

Importance of Hall effect

Of monumental importance to today's technology is a class of materials whose ability to conduct **electric current** increases with temperature and whose charge carriers can be either positive or negative, depending on the impurity introduced into them. These materials are called semiconductors, prime examples of which are the elements silicon and germanium. When these elements are given traces of the appropriate impurity element, they can be made into either p-type (containing positive carriers called holes) or n-type (containing negative carriers called electrons). The Hall effect is then used to confirm which type of material one is dealing with. Furthermore, by measuring the Hall potential, the current, the magnetic field, and the sample **geometry**, it is easy to calculate the number of charge carriers per unit **volume** in the material tested. In the 1940s, it was found that junctions could be formed with these two different types of semiconductors across which current could flow only in one direction. Devices of this kind are called rectifiers or diodes and are vital for converting alternating current to direct current, adding or removing audio and video signals from their carrier waves, and many other applications. It was also found that more than two junctions could be formed in one device, and these were called transistors. These devices were capable of being employed in **amplifier** and oscillator circuits in radios and TVs. Previously, rectifiers, amplifiers, and oscillators used **vacuum** tubes as their essential components, which were generally bulky, used lots of power, and burned out frequently. The new semiconductor devices had none of these problems. In the 1950s and 1960s, it was learned how to create many **transistor** circuits on a small chip using integrated circuitry. Without this new technology, the powerful com-

KEY TERMS

Electric current—The flow of charge carriers, like electrons and holes, whose direction is defined as that of the carriers if positive and opposite that of the carriers if negative.

Electric field—The force per unit charge acting on a charged body when placed in the vicinity of electric charges. The direction of the field is the same as that of the force on a positive charge or the opposite of that of the force on a negative charge.

Hole—An electron vacancy in the lattice structure of a semiconductor caused by an impurity atom with one less electron than needed for complete bonding with the structure. Holes can drift through the lattice under the action of an electric field as though they were positively charged particles.

Magnetic field—The force per unit pole strength acting on a magnetic pole when placed in the vicinity of a magnet. The direction of the field is the same as that of the force on a north pole or the opposite of that of the force on a south pole. To provide a test pole to measure the field, a long magnet must be used so that its north and south poles are far enough apart to consider them isolated.

Potential difference—The energy per unit charge that is necessary to move a positive charge from a point of low potential to a point of high potential.

Semiconductor—A material whose electrical conductivity is midway between that of a conductor and an insulator and which increases with temperature, such as silicon or germanium.

puters that were used in our **space** program and are now found universally in the form of compact personal computers would not have been possible. Even more importantly, without the discovery of the Hall effect and its use in the scientific investigation of semiconducting materials, this sequence of developments could not have even begun. Finally, with the discovery of its large-scale quantum behavior, the future role of the Hall effect in the advancement of science and technology may eventually prove to be even greater than its past role.

Resources

Books

Serway, Raymond A., *Physics: For Engineers and Scientists with Modern Physics.* 3rd ed., Philadelphia: Saunders College Publishing, 1992.

Frederick L. Culp

Halley's comet

Halley's comet, a periodic comet usually appearing every 76 years, is named after English astronomer Edmond Halley (1656-1742), the first person to accurately predict the return of a comet. This famous comet follows a retrograde (east-west), elliptical **orbit**, providing a magnificent, astronomical spectacle. In 1910, **Earth** passed through its brilliant, fan-shaped tail which soared 99 million mi (160 million km) into **space**. During its last apparition (appearance) in 1986, space probes and ground-based technology gathered valuable scientific data on its size, shape, and composition. In 2024, the comet will reach aphelion (furthest point from the **Sun**) millions of miles outside Neptune's orbit, make a U-turn, and begin its thirty-first observed return to perihelion (point nearest the Sun) inside the orbit of **Venus**, arriving in 2061. Observed by Chinese astronomers in 240 B.C. and maybe even 466 B.C., Halley's Comet may make 3,000 more revolutions and live another 225,000, if recent estimates calculated from data collected by the **space probe** Giotto are correct.

Edmond Halley's prediction

In the late seventeenth century, **comets** were believed to follow parabolic (U-shaped) orbits and appear only once. The gregarious, outgoing Edmond Halley boldly suggested to his reclusive but genius friend, Isaac Newton, that comets may travel in an **ellipse** and appear more than once. Newton initially rejected the idea, even though his **laws of motion** and gravitation clearly allowed for such orbits. Later Newton accepted the possibility that comets can follow elliptical paths, orbiting the Sun repeatedly. In 1695, basing his work on Newton's laws of cometary **motion**, Halley computed the orbits of two-dozen comets, including the comet of 1682. He suggested the comets of 1531, 1607, and 1682 were one and the same, even venturing to predict its return in 1758. He was also the first to consider the perturbative (disruptive) effect of planets on a comet's orbit. Allowing for Jupiter's influence, he narrowed the comet's return to late 1758 or early 1759. Astronomers around the world anxiously watched the sky, aspiring to be the first to recover (find) the comet. On Christmas eve, 1758, German farmer and amateur astronomer, Johann Palitzch, spotted the comet which would forever bear Halley's name.

Ancient and modern perspectives

Throughout history, comets were viewed as omens. Halley's comet is no exception, and almost every apparition is linked to a major world event: in 11 B.C. to Agrip-

Halley's comet as seen from Peru on April 21, 1910. *U.S. National Aeronautics and Space Administration (NASA).*

pa's death; in A.D. 451 to Atilla the Hun's only defeat; in A.D. 1066 to William of Normandy's conquest of England. Even in 1910, people panicked, believing the comet's tail contained poisonous gas which would exterminate all life on Earth.

A different picture preceded Halley's 1986 apparition. Astronomers worldwide trained their telescopes on the heavens and the "International Halley Watch" became the largest international scientific cooperative ever. Ironically, the comet was first seen by a California Institute of Technology graduate student, David Jewitt, and staff astronomer, Edward Danielson, who "borrowed" a few hours' viewing time through the 200-in (508-cm) **telescope** on Palomar Mountain in California. Also, six spacecraft soared to probe Halley's secrets, collecting data which confirmed Fred Whipple's 1950 theory of a solid nucleus composed of **ice** and **rocks** and providing new information. Giotto came to within 370 mi (596 km) of Halley's nucleus, capturing for the first time fascinating images of a potato-shaped, $9 \times 5 \times 5$ mi ($15 \times 8 \times 8$ km) core with an irregularly shaped, dark surface crust. Only about 4% of the ices were exposed, the vapors of which emit gas and dust which create the gigantic, glowing **coma** and tail.

Cometary dust particles consist primarily of silicates-silicon, **magnesium**, and **iron**; and CHON particles-carbon, **hydrogen**, **oxygen**, and **nitrogen**, which were undetected until the VEGA and Giotto space missions. CHON particles suggest organic **matter** in the nucleus and, although providing no proof, the discovery renewed speculation that cometary molecules may have provided the **stimulus** for living organisms on Earth.

Gas analysis suggests that about 78% of Halley's nucleus is ice from **water**; 13% from **carbon monoxide**; 2% **carbon** dioxide-undetected until VEGA 1; 1-2% **ammonia** and methane-undetected until Giotto; while

hydrocyanic acid, **sulfur**, and other gases combine for less than 1%. Giotto may also have detected the unexpected presence of polymers, created by formaldehyde molecules. The comet's basic chemical composition is similar to other **solar system** bodies.

Resources

Books

Asimov, Isaac. *Asimov's Guide to Halley's Comet.* New York: Walker and Company, 1985.

Bailey, M.E., S.V.M. Clube, and W.M. Napier. *The Origin of Comets.* Oxford: Pergamon Press, 1990.

Lancaster-Brown, Peter. *Halley & His Comet.* Poole, England: Blandford Press, 1985.

Yeomans, Donald K. *Comets: A Chronological History of Observation, Science, Myth, and Folklore.* New York: John Wiley & Sons, Inc., 1991.

Marie L. Thompson

Hallucinogens

Hallucinogens are substances that alter the user's thought processes or mood to the extent that he perceives objects or experiences sensations that in fact have no reality. Many natural and some manmade substances have the ability to bring about hallucinations. In fact, because of the ready market for such chemicals, they are manufactured in illegal chemical laboratories for sale as hallucinogens. LSD and many so-called designer drugs have no useful clinical function.

Hallucinogens have long been a component in the religious rites of various cultures, both in the New and

Old Worlds. The tribal shaman or medicine man swallowed the hallucinogenic substance or inhaled fumes or smoke from a burning substance to experience hallucinations. They believed that such a state, separated from reality, enabled them to better communicate with the gods or their ancestors. In fact, such rituals remain a central part of life for many peoples whose culture has been handed down from one century to the next. Of course, these hallucinogens were natural substances or derivatives from them. Among the oldest are substances from **mushrooms** or **cactus** that have been in use in Native American rites since before recorded **time**. Some Native American tribes have established the legality of their use of such compounds, which still form a central part of tribal ritual.

In recent years hallucinogens have been discovered and embraced by a subculture that cannot claim tribal history. The so-called Hippies, a movement that burgeoned in the 1960s, adopted hallucinogens as a part of their culture. Artists, poets, and writers of the time believed that the use of hallucinogens enhanced their creative prowess. The use of these substances as recreational drugs resulted in a great number of psychological casualties because of the accumulation of substances in the user's body or because of unforeseen adverse side effects such as "flashbacks," which occurred after the user had ceased using the drug.

True hallucinogens must be differentiated from other, less potent drugs such as the psychedelics. The latter can alter reality to some degree and may in certain circumstances push the user into experiencing hallucinations, but their primary effect is one of inducing euphoria, relaxation, stimulation, relief from **pain,** or relief from **anxiety.** Probably the most commonly used of the psychedelics is **marijuana**, which is available worldwide and constitutes one of the primary illegal money **crops** in the United States. Opiates such as heroin or **morphine**, phencyclidine (PCP), and certain **tranquilizers** such as diazepam (Valium) also can have such a psychedelic effect. These drugs are not considered true hallucinogens, though they remain a substantive part of the drug subculture **ecology.**

LSD

LSD (lysergic acid diethylamide) is a synthetic (not naturally occurring) substance first synthesized in 1938 by Dr. Albert Hofmann, a Swiss chemist who was seeking a headache remedy. He first isolated lysergic acid from the ergot fungus that grows on **wheat**. In the laboratory he manipulated the **molecule** to add the diethylamide molecule to the base compound. His initial tests on animals failed to elicit any outward sign that the sub-

stance was having any effect. Convinced that it was inactive he stored the chemical on his laboratory shelf.

In 1943, Hofmann decided to work with the LSD again, but in the process of using it he ingested a small, unknown quantity. Shortly afterwards he was forced to stop his work and go home. He lay in a darkened room and later recorded in his diary that he was in a dazed condition and experienced "an uninterrupted stream of fantastic images of extraordinary plasticity and vividness...accompanied by an intense kaleidoscope-like play of colors." Three days later Hofmann purposely took another dose of LSD to verify that his previous experience was the result of taking the drug. He ingested what he thought was a small dose (250 micrograms), but which in fact is about five times the amount needed to induce pronounced hallucinations in an adult male. His hallucinatory experience was even more intense than what he had experienced the first time. His journal describes the symptoms of LSD toxicity: a metallic taste, difficulty in breathing, dry and constricted throat, cramps, paralysis, and visual disturbances.

American chemists, **hearing** of Hofmann's experiences, imported LSD in 1949. Thereafter began a series of **animal** experiments in which the drug was given to **mice**, spiders, **cats**, dogs, **goats**, and an **elephant**. All of the animals showed dramatic outward changes in **behavior**, but few symptoms of toxicity. This led to an extension of research into the use of human subjects in an effort to find some therapeutic use for LSD. In the 1950s, such use of human subjects in drug experimentation was not under strict controls so scientists could give the drug as they wished.

Early experiments on humans involved using LSD for the treatment of various psychiatric disorders such as **schizophrenia**, **alcoholism**, and **narcotic addiction**. The rationale was that LSD induced major changes in **brain** function and behavior and that the patient might better be able to gain insight into his illness or addiction while under the influence of the drug. After only a short time, however, it became evident that this line of research was fruitless and it was abandoned.

LSD as a recreational drug

Lysergic acid diethylamide is one of the most potent hallucinogens known. That is, a dramatic effect can be elicited by only a tiny amount of the drug. The usual dose for an adult is 50-100 micrograms. A microgram is a millionth of a gram. Higher doses will produce more intense effects and lower doses will produce milder effects. The so-called "acid trip" can be induced by swallowing the drug, smoking it (usually with marijuana), injecting it, or rubbing it on the skin. Taken by mouth, the drug will take about 30 minutes to have any effect and up to an hour for its full effect to be felt, which will last 2-4 hours.

Physiologically the user will experience blurred **vision**, dilation of the pupils of the **eye**, muscle weakness and twitching, and an increase in **heart rate**, **blood pressure**, and body **temperature**. He may also salivate excessively and shed tears, and the hair on the back of his arm may stand erect. Women who are pregnant and who use LSD or other of the hallucinogens may have a miscarriage because these drugs cause the muscles of the uterus (womb) to contract. Such a reaction in pregnancy would expel the fetus.

To the observer, the user usually will appear to be quiet and introspective. Most of the time the user will be unwilling or unable to interact with others, to carry on a conversation, or engage in intimacies. At times LSD will have profoundly disturbing effects on an individual even at moderate doses. Although the physiologic effects will be approximately the same, the psychological result is a terrifying series of events. The distortions in reality, exaggeration of **perception,** and other effects can be horrifying, especially if the user is not aware that he has been given the drug. This constitutes what is called the "bad trip." The psychological effects reported by LSD users consist of depersonalization, the separation from one's self, yet with the knowledge that the separated entity is one's self and is observing the passing scene. A confused body image in which the user cannot tell where his own body ends and the surroundings begin also is common. Removal from reality is the third most common experience. In this, the user's perception of colors, distance, shapes, and sizes is totally distorted and constantly changing. Hallucinations in the form of perceiving objects that are not present or forms that have no substance also occur. He may be able to taste colors or smell sounds, a mixing of the senses called synesthesia. Sounds, colors, and taste are all greatly enhanced, though they may be an unrealistic and constantly changing tableau.

The user often talks endlessly on social subjects, history, current events, philosophy, or other areas, often babbling meaningless phrases. On the other hand, the user may become silent and unmoving for long periods of time as he listens to music or contemplates a **flower** or his thumb. As well, he may become hyperactive and talk unendingly for long periods. Mood swings are frequent, with the user alternating between total euphoria and complete despair with no reason for doing so.

Some users will exhibit symptoms of paranoia. They become suspicious of persons around them and tend to withdraw from others. They become convinced that other people are talking about them and plotting against them. This mood may be one of many temporary responses to the drug that the user will experience or it may be the only response. Feelings of anxiety can come to the fore when the user is removed from a quiet environment and placed in an active one. His feelings of inability to cope can be elicited by no more than standing in line with other people or being taken for a walk down a city sidewalk.

All of these effects can be hazardous to the LSD consumer. With his distorted sense of reality and his belief that he is removed from everyday events he may feel invincible. Users have been known to jump off buildings or walk in front of moving trucks, with fatal consequences, because their grasp on reality is gone.

How LSD and the other hallucinogens produce these bizarre effects remains unknown. The drug attaches to certain chemical binding sites widely spread through the brain, but what ensues thereafter has yet to be described. A person who takes LSD steadily with the doses close together can develop a tolerance to the drug. That is, the amount of drug that once produced a pronounced "high" no longer is effective. A larger dose is required to achieve the same effect. However, if the individual keeps increasing his drug intake he will soon pass over the threshold into the area of toxicity. His experiences no longer will be perceived as pleasurable.

Curiously, when an LSD user has attained a high threshold of tolerance for LSD he also has one for other hallucinogens. He cannot change to psilocybin or peyote and be able to attain the desired high at a low dose. This indicates that the hallucinogenic drugs occupy the same receptors in the brain and must bring about their effects in a similar manner.

Discontinuing LSD or the other hallucinogens, especially after having used them for an extended period of time, is not easy. The residual effects of the drugs produce toxic symptoms and "flashbacks," which are similar to an LSD "trip." Many LSD users do not take the drug at close intervals, but use it on weekends or other occasions.

Currently, the most common form of LSD administration is by licking the back of a stamp torn from a perforated sheet of homemade stamps. The design on the front of the stamp is unique to an individual LSD chemist and is a form of guarantee that the LSD is pure. The drug is coated on the back of the sheet of stamps or is deposited as a colored dot on the **paper**. Removing one stamp, the user places it on his tongue and allows the LSD to dissolve in his saliva.

Some marijuana is sold with LSD mixed with it to enhance the psychedelic effects of the **plant**. Because LSD can produce such a potent reaction with a very small dose, the drug can be administered unbeknownst to the victim by placing it in a drink or other means by which it may be ingested. The person who does not know he is being given the drug may experience a terrifying series of events over the next few hours.

Of course, LSD is an illegal drug and is sold on the street in various forms. LSD is produced by a chemical process, so the buyer is trusting that the seller knows how to manufacture the drug. The purity of such a product cannot be guaranteed, of course, and the impurities or other drugs present in the LSD can cause serious side effects or even death. The subculture of steady users, called acid heads, remains a part of civilization in developed countries. Though the middle 1960s were the years of greatest use of LSD and the consumption of the drug dropped off somewhat thereafter, a fairly constant number of users has formed a market for LSD and other hallucinogens since then.

Not everyone can consume LSD or other hallucinogens and experience a moderate and short-lived response. Some people have a reaction far beyond what would be expected at a moderate dose of LSD for reasons unknown. There is no way to determine who will have such a reaction prior to his consuming the hallucinogen, so the first-time user may provide a frightening experience for those around him as well as for himself. Perhaps these people have more numerous receptor sites than do other people so they experience a more intense effect from the drug because it affects a greater portion of the brain. The explanation remains undetermined as yet. Once given, the LSD cannot be countered with any other drug. The user must simply endure the next several hours of alteration of his consciousness. Those who experience a bad trip can be helped through it by calm reassurance, but for individuals whose grasp on reality is completely gone, even that modest form of therapy is ineffective.

Mushrooms

Among the many **species** of mushrooms, edible, poisonous, and others, are certain species known to bring about hallucinations. Their usage far predates that of LSD or other modern hallucinogens. In fact, artifacts remaining from pre-Columbian eras often were sculpted with mushrooms surrounded by human figures. These small statues were the first indication that mushrooms were a part of any kind of tribal rite. The significance of such a figure remained obscure for many years. Not until the twentieth century were scientists aware of the existence of hallucinogenic mushrooms. Efforts were then made to collect them and analyze their content.

In 1936, an ethnologist named Roberto Weitlaner collected some mushrooms said to have hallucinogenic properties and sent them to a commercial laboratory, but they were decomposed beyond the point of usefulness. Scientists collected the same mushrooms, preserved them carefully, and sent them for identification. They proved to be *Panaeolus campanulatis*. The first description of these **fungi** was published in 1939, alleging to their prowess as hallucinogens.

Not until the 1950s was another mushroom, *Psilocybe mexicana*, discovered. In 1957, a dried specimen of the mushroom was sent to Sandoz Pharmaceuticals in Switzerland for analysis. An **alkaloid** in the mushroom was isolated, but its use in animals proved unequivocal. One of the laboratory chemists consumed 0.08 oz (2.4 g) of the dried fungus, a moderate dose by standards of the Indians who regularly used it. He experienced vivid hallucinations. The active ingredient was named psilocybin. Additional analysis disclosed that its chemical structure is similar to serotonin, a **neurotransmitter** in the brain. A neurotransmitter is a chemical that provides the means of communication from one brain **cell** (**neuron**) to another.

Yet another species of hallucinogenic mushroom was found in 1973 on the campus of the University of Washington. It was named *Psilocybe stuntzii*.

Hallucinogenic mushrooms have been used for centuries in rites of medicine men to foresee the future or communicate with the gods. The privilege of using the mushroom may or may not be passed on to the other tribal members. The mushroom is consumed by eating it or by drinking a steeped beverage in which the mushroom has been boiled. The effects are similar to those experienced by an LSD user-enhancement of colors and sounds, introspective interludes, perception of objects or persons who are not present, and sometimes terrifying visions that predict dire circumstances to come.

Peyote

Another ancient, natural hallucinogenic substance is derived from any of a number of Mexican cacti of the genus *Lophophora*. Relics dating back hundreds of years depict animals with a peyote button in their mouths. The part of the cactus used is the flowering head which contains a potent alkaloid called mescaline. The uses of peyote parallel those of the hallucinogenic mushrooms. The peyote flower was used to induce a state of intoxication and happiness in the user. American Indians of the southwest often employed the cactus in their tribal rites.

Other hallucinogens

A number of other plant species produce hallucinogenic substances. Some also have uncomfortable side effects that go along with the hallucinations, so they seldom are used.

KEY TERMS

Acid trip—The description for the sensations experienced by a user of LSD. The trip may be a pleasant one, a good trip, or a terrifying experience, a bad trip.

Alkaloid—A nitrogen-based chemical, usually of plant origin, also containing oxygen, hydrogen, and carbon. Many are very bitter and may be active if ingested. Common alkaloids include nicotine, caffeine, and morphine.

Recreational drug—A substance used socially for artificially enhancing mood or feeling, but not for the treatment of any medical condition. LSD and marijuana are two of the most common such drugs.

Synesthesia—A mixing of the senses so that one who experiences it claims to have tasted color or heard taste or smelled sounds. It is a common phenomenon among users of hallucinogens.

Trees of the Barbados cherry family (Malpighiaceae), which grow in the tropics contain certain alkaloids or beta-carboline. The **bark** of the **tree** is boiled or is squeezed and twisted in cold **water** and the water extract of the bark is drunk. The resulting liquid is bitter and, along with hallucinations, brings on pronounced nausea. It is seldom used because of its unpleasant side effects.

The **seeds** of two species of morning glory of the family Convolvulaceae contain lysergic acid amide, a substance closely related to LSD. Chewing the seeds releases the hallucinogen. Here again, however, some morning glory seeds are poisonous instead of hallucinogenic.

Even some members of the bean family can produce hallucinogens. Two species within the genus *Anadinanthera* contain tryptamines or beta-carbolins, which are hallucinogenic. Some of the 500 species of the genus *Mimosa* produce a hallucinogen used in ancient tribal rites. The tribes that used these plants no longer exist and the secret of extracting the hallucinogen has gone with them.

The belladonna plant, or deadly **nightshade**, produces hyoscymamine and to a lesser extent scapolamine, both of which are hallucinogens. Belladonna is a dangerous plant that can cause death, hence its nickname. Greatly diluted purified extracts from it have been used in clinical medicine.

Hallucinogens, then, have been known and used for centuries. They may be found in surprising sources, though the effective dose of a hallucinogen may closely border the lethal dose and the ability to select the proper source to extract the active drug is something to be left to experts. Hallucinogens may leave a legacy of long-lasting toxicity that may permanently alter brain functions and render the user helpless. These potent substances are not as harmless or recreational as one is led to believe by those in the drug subculture.

Resources

Periodicals

Fernandes, B. "The Long, Strange Trip Back." *World Press Review* 40 (September 1993): 38-39.

Porush, D. "Finding God in the Three-Pound Universe: The Neuroscience of Transcendence." *Omni* 16 (October 1993): 60-62+.

Larry Blaser

Halo *see* **Atmospheric optical phenomena**

Halogenated hydrocarbons

Halogenated hydrocarbons are derivatives of hydrocarbons (that is, organic compounds that only contain **carbon** and **hydrogen atoms**) which include some halogen atoms within their chemical structure. The most commonly encountered **halogens** in halogenated hydrocarbons are fluorine and **chlorine**, but sometimes bromine or iodine occur, or combinations of any of these.

Some halogenated hydrocarbons occur naturally, being synthesized by halogenation reactions occurring during **combustion** of **biomass** containing the constituent atoms (that is, carbon, hydrogen, and halogens). For example, these syntheses occur commonly but at low rates during forest fires. However, most **species** of halogenated hydrocarbons are synthetic, and are manufactured by humans as industrially useful materials, or are incidentally produced as a by-product during industrial **chemical reactions**, or during the **incineration** of municipal waste.

Chlorinated hydrocarbons are a well known group, with a wide variety of uses. A number of these chemicals have been used as **insecticides**, including DDT, DDD, lindane, **chlordane**, aldrin, and dieldrin. Others have been used as **herbicides**, especially 2,4-D and 2,4,5-T. Polychlorinated biphenyls or PCBs have been widely used as dielectric fluids in electrical transformers and for other purposes. Dioxins, including the deadly TCDD, are trace contaminants synthesized during the manufacture of other chlorinated hydrocarbons and in spontaneous **chlo-**

rination reactions in incinerators and pulp mills. Chlorinated hydrocarbons are associated with some well known environmental problems, because most of these chemicals are persistent in the environment, and they accumulate in organisms, sometimes causing toxicity.

Chlorofluorocarbons or freons are another group of halogenated hydrocarbons that have been used extensively in refrigeration, air conditioning, and for cleaning electrons. After their use these chemicals are often discharged to the atmosphere, where they are very persistent, and appear to be involved in ozone-destroying reactions occurring in the stratosphere. This is an important environmental problem, because ozone is critical in screening life on Earth's surface from the deleterious effects of exposure to solar ultraviolet radiation, which can cause skin cancers, cataracts, and other problems. In recognition of the environmental problems associated with these chemicals, the manufacturing and use of chlorofluorocarbons are rapidly being curtailed through international agreements.

See also Bioaccumulation; Hydrocarbon; Hydrochlorofluorocarbons; Ozone layer depletion.

Bill Freedman

Halogens

The halogens are a group of chemical elements that includes fluorine, chlorine, bromine, iodine, and astatine. Halogen comes from Greek terms meaning "produce sea salt." None of the halogens occur naturally in the form of elements, but, except for astatine, they are very widespread and abundant in chemical compounds where they are combined with other elements. Sodium chloride, common table salt, is the most widely known.

All of the halogens exist as diatomic molecules when pure elements. Fluorine and chlorine are gases. Bromine is one of only two liquid elements, and iodine is a solid. Astatine atoms exist only for a short time and then decay radioactively. Fluorine is the most reactive of all known elements. Chemical activity, the tendency to form chemical compounds, decreases with atomic number, from fluorine through iodine. Simple compounds of these elements are called halides. When one of the elements becomes part of a compound its name is changed to an -ide ending, e.g., chloride.

Chlorine

Chlorine was the first halogen to be separated and recognized as an element. It was named in 1811 by Humphry Davy from a Greek term for its greenish yellow color. Huge deposits of solid salt, mostly sodium chloride, and salts dissolved in the oceans are vast reservoirs of chloride compounds. "Salt" is a general term for a metal and nonmetal combination; there are many different salts. To obtain chlorine an electrical current is applied to brine, a water solution of sodium chloride. Chlorine gas is produced at one electrode. The chlorine must be separated by a membrane from the other electrode, which produces sodium hydroxide.

Chlorine gas itself is toxic. It attacks the respiratory tract and can be fatal. For this reason it was used as a weapon during World War I. Chlorine in solutions has been used as a disinfectant since 1801. It was very effective in hospitals in the 1800s, particularly in an 1831 cholera epidemic in Europe. Chlorine bleaches are employed in most water treatment systems in the United States as well as over much of the rest of the world and in swimming pools.

Chlorine will combine directly with almost all other elements. Large amounts are used yearly for making chlorinated organic compounds, bleaches, and inorganic compounds. Organic compounds, ones which have a skeleton of carbon atoms bonded to each other, can contain halogen atoms connected to the carbon atoms. Low molecular weight organic chlorine compounds are liquids and are good solvents for many purposes. They dissolve starting materials for chemical reactions, and are effective for cleaning such different items as computer parts and clothing ("dry cleaning"). These uses are now being phased out because of problems that the compounds cause in Earth's atmosphere.

Chlorine-containing organic polymers are also widely employed. Polymers are large molecules made of many small units that hook together. One is polyvinyl chloride (PVC), from which plastic pipe and many other plastic products are made. Neoprene is a synthetic rubber made with another chlorine-containing polymer. Neoprene is resistant to the effects of heat, oxidation, and oils, and so is widely used in automobile parts.

Many medicines are organic molecules containing chlorine, and additional chlorine compounds are intermediate steps in the synthesis of a variety of others. Most crop protection chemicals, herbicides, pesticides, and fungicides have chlorine in them. Freon refrigerants are chlorofluorocarbons (CFCs). These perform well because they are volatile, that is they evaporate easily, but they are not flammable. Freon 12, one of the most common, is CCl_2F_2, two chlorine atoms and two fluorine atoms bonded to a carbon atom.

Chlorine is part of several compounds, such as the insecticide DDT, that are soluble in fats and oils rather than in water. These compounds tend to accumulate in the fatty tissues of biological organisms. Some of these compounds are carcinogens, substances that cause **cancer**. DDT and other pesticides, **polychlorinated biphenyls (PCBs)**, and dioxins are substances that are no longer manufactured. However, they are still present in the environment, and disposal of materials containing these compounds is a problem.

Bromine

The next heaviest element in the halogen family is bromine, named from a Greek word for stink, because of its strong and disagreeable odor. It was first isolated as an element in 1826. Bromine is a reddish brown liquid that vaporizes easily. The vapors are irritating to the eyes and throat. Elemental bromine is made by oxidation, removal of electrons from bromide ions in brine. Brines in Arkansas and Michigan, in the United States, are fairly rich in bromide. Other world-wide sources are the Dead Sea and **ocean** water.

There are a variety of applications for bromine compounds. The major use at one time was in ethylene dibromide, an additive in leaded gasoline. This need has declined with the phase-out of leaded fuel. Several brominated organic compounds have wide utilization as pesticides or disinfectants. Currently the largest **volume** organic bromine product is methyl bromide, a fumigant. Some medicines contain bromine, as do some dyes.

Halons, or halogenated carbon compounds, have been utilized as flame retardants. The most effective contain bromine, for example, halon 1301 is $CBrF_3$. Inorganic bromine compounds function in water sanitation, and silver bromide is used in photographic film. Bromine also appears in quartz-halide **light** bulbs.

Iodine

The heaviest stable halogen is iodine. Iodine forms dark purple crystals, confirming its name, Greek for violet colored. It was first obtained in 1811 from the ashes of seaweed. Iodine is purified by heating the solid, which sublimes, or goes directly to the gas state. The pure solid is obtained by cooling the vapors. The vapors are irritating to eyes and mucous membranes.

Iodine was obtained commercially from mines in Chile in the 1800s. In the twentieth century brine from wells has been a better source. Especially important are brine wells in Japan, and, in the United States, in Oklahoma and Michigan.

Iodine is necessary in the diet because the thyroid gland produces a growth-regulating hormone that contains iodine. Lack of iodine causes goiter. Table salt usu-

ally has about 0.01% of sodium iodide added to supply the needed iodine. Other compounds function in chemical analysis and in synthesis in a **chemistry** laboratory of organic compounds. Iodine was useful in the development of **photography**. In the daguerreotype process, an early type of photography, a silver plate was sensitized by exposure to iodine vapors.

Astatine

Astatine could be described as the most rare element on **Earth**. All isotopes, atoms with the same number of protons in the nucleus and different numbers of neutrons, are radioactive; even its name is Greek for "unstable." When an atom decays its nucleus breaks into smaller atoms, **subatomic particles**, and **energy**. Astatine occurs naturally as one of the atoms produced when the **uranium** 235 **isotope** undergoes **radioactive decay**. However, astatine does not stay around long. Most of its identified isotopes have half-lives of less than one minute. That is, half of the unstable atoms will radioactively decay in that time.

Astatine was first synthesized in 1940 in **cyclotron** reactions by bombarding bismuth with alpha particles. The longest-lived isotope has a **half-life** of 8.3 hours. Therefore, weighable amounts of astatine have never been isolated, and little is known about its chemical or physical properties. In a **mass** spectrometer, an instrument that observes the masses of very small samples, astatine behaves much like the other halogens, especially iodine. There is evidence of compounds formed by its combining with other halogens, such as AtI, AtBr, and AtCl.

Fluorine

Fluorine was the most difficult halogen to isolate because it is so chemically reactive. H. Moissan first isolated elemental fluorine in 1886, more than seventy years after the first attempts. Moissan received the 1906 Nobel Prize for Chemistry for this work. The technique that he developed, **electrolysis** of potassium fluoride in anhydrous liquid **hydrogen** fluoride, is still used today, with some modifications. The name fluorine comes from the mineral fluorspar, or calcium fluoride, in which it was found. Fluorspar also provided the term "fluorescence," because the mineral gave off light when it was heated. Hydrofluoric acid has been used since the 1600s to etch **glass**. However, it, as well as fluorine, must be handled with care because it causes painful skin burns that heal very slowly. Fluorine and fluoride compounds are toxic.

Fluorine is so reactive that it forms compounds with the noble gases, which were thought to be chemically inert. Fluorine compounds have been extremely important in the twentieth century. Uranium for the first atomic bomb and for nuclear reactors was enriched in the 235 isotope, as compared to the more abundant 238 isotope, by gaseous **diffusion**. Molecules of a uranium atom with six fluorine atoms exist as a gas. Less massive gases will pass through a porous barrier faster than more massive ones. After passage through thousands of barriers the uranium hexafluoride gas was substantially enriched in the 235 isotope.

Fluoride ions in low concentrations have been shown to prevent cavities in teeth. Toothpaste may contain "stannous fluoride," and municipal water supplies are often fluoridated. However, too high a **concentration** of fluoride will cause new permanent teeth to have enamel that is mottled. Chlorofluorocarbons were developed and used as refrigerants, blowing agents for polyurethane foam, and propellants in spray cans. Their use became widespread because they are chemically inert. Once the active fluorine is chemically bound the resulting **molecule** is generally stable and unreactive. The polymer polytetrafluoroethylene is made into Teflon, a non-stick coating.

Unexplored sources and problems

Most of the organic halogen compounds mentioned are made synthetically. However, there are also natural sources. In 1968 there were 30 known naturally occurring compounds. By 1994 around 2,000 had been discovered, and many biological organisms, especially marine **species**, those in the oceans, had not been looked at as yet. Halogenated compounds were found in ocean water, in marine **algae**, in corals, jelly fish, **sponges**, terrestrial plants, **soil** microbes, **grasshoppers**, and ticks. Volcanoes are another natural source of halogens, and they release significant amounts into the air during eruptions. Chlorine and fluorine are present in largest quantities, mostly as **hydrogen chloride** and hydrogen fluoride.

In the 1980s depletion of the layer of **ozone** (O_3) high in Earth's atmosphere was observed. Ozone absorbs much of the high energy ultraviolet **radiation** from the **Sun** that is harmful to biological organisms. During September and October, in the atmosphere over the Antarctic, ozone concentration in a roughly circular area, the "ozone hole," drops dramatically.

Chlorine-containing compounds, especially CFCs, undergo reactions releasing chlorine atoms, which can catalyze the conversion of ozone to ordinary **oxygen**, O_2. Bromine and iodine-containing carbon compounds may also contribute to ozone depletion. Countries signing the Montreal Protocol on Substances that Deplete the Ozone Layer have pledged to eliminate manufacture and use of halocarbons. However, natural sources, such as volcanic

---

KEY TERMS

Chemical activity—The tendency to form chemical compounds. Active elements are not usually found in elemental form because a more active elements will replace a less active element in a compound.

Compound—A pure substance that consists of two or more elements, in specific proportions, joined by chemical bonds. The properties of the compound may differ greatly from those of the elements it is made from.

Formula—A shorthand description for chemical substances. The number of atoms of each element is given as a subscript following the element symbol (except for 1, which is understood). For example, HF, O_3, CCl_2F_2.

Oxidation—A chemical process that removes electrons from a reacting substance.

Radioactive—The nucleus of an atom that is not stable. It falls apart to lighter atoms, subatomic particles, and energy.

Salt—A solid that is made from a combination of positive and negative ions but has no net charge itself.

Synthesize—To prepare through human activity, in contrast to preparation in some naturally-occurring process.

---

eruptions and fires, continue to add halogen compounds to the atmosphere. Finding substitutes that work as well as the banned compounds and do not also cause problems is a current chemical challenge.

See also Elements, formation of; Halogenated hydrocarbons.

Resources

Books

CRC Handbook of Chemistry and Physics. Boston: CRC Press, Inc., published yearly.

Greenwood, N.N., and A. Earnshaw. *Chemistry of the Elements.* 2nd ed. Oxford: Butterworth-Heinneman Press, 1997.

Kirk-Othmer Encyclopedia of Chemical Technology. 4th ed. Suppl. New York: John Wiley & Sons, 1998.

Periodicals

Gribble, G.W. "Natural Organohalogens." *Journal of Chemical Education* 71, no.11 (1994): 907-911.

Patricia G. Schroeder

Halosaurs

A halosaur is a thin, elongated **fish** resembling an eel. The largest of halosaurs grows to about 20 in (51 cm) long. Unlike the eel, the halosaur has a backbone composed of many vertebrae. It has somewhat large scales, numbering fewer than 30 horizontal rows on each of its sides. This fish lives close to or on the bottom of the sea and is thus referred to as a benthic fish. It feeds on the **ocean** floor; like many bottom-feeding fish, its mouth is inferior, meaning that its jaw is positioned under its projecting snout. The halosaur's eyes, like those of the eel, are covered with transparent skin, called spectacles. It is believed that this **membrane** serves to protect the fish's eyes while it feeds on the bottom. This fish has a single dorsal fin composed of 9-13 soft rays, pelvic fins on its abdomen, and a long anal fin that extends to the tip of its tail. It has no caudal (tail) fin.

Scientists differ in their classification of the halosaur. Some scientists classify halosaurs in the order Albuliformes, the suborder Notacanthoidei, and the family Halosauridae. However, other scientists classify halosaurs in the order Notacanthiformes. All fish in this order have pectoral fins placed high on their sides, pelvic fins positioned on their abdomens, and anal fins that are long and tapering into their tails. All are deep **water** fish, inhabiting depths of between 656-17,062 ft (200-5,200 m). The order is distributed world wide and contains 20 **species** and six genera. According to this classification, the order Notacanthiformes has three families; the most notable of which are the Halosauridae (halosaurs) and the Notacanthidae (**spiny eels**).

Within the family Halosauridae, there are three genera with 15 species. The eight species in the genus *Halosaurus* live in the Atlantic, Indian, and Pacific Oceans, usually near the continental shelves. The six species in the genus *Aldrovandia* occur throughout the Atlantic and Indian Oceans as well as in the central and western Pacific Ocean. There is only one species in the third genus, referred to as *Halosauropsis macrochir*, and it lives in the western Pacific, Atlantic, and Indian Oceans.

Like its close relative, the spiny eel, the halosaur commonly moves slowly over the ooze covering the deep-sea floor in search of food. Because it has a long tapering tail which ends without a fin, it has modified its **mode** of locomotion. Like other deep-sea fish with its body type, it is believed that the halosaur moves by rolling its long anal fin or by using quick strokes of its pectoral fins. Furthermore, it may accomplish locomotion by undulating its long body.

This fish has been caught swimming at up to 5,200 ft (1,585 m) below the ocean's surface. Because it lives at such extreme depths, it is rarely seen, and little is known about its habits.

Kathryn Snavely

Hamsters

Hamsters are small **rodents** with dense fur, a short tail, and large cheek pouches. They belong to the mammalian family Muridae, which also includes **rats**, **mice**, **gerbils**, **voles**, and **lemmings**.

During foraging trips, hamsters use their cheek pouches to carry **seeds** and grains back to underground food stores that are sometimes quite large. Hamsters mostly eat **plant matter**, especially seeds, nuts, soft **fruits**, tubers, and roots. However, they will also opportunistically predate on **insects**, small **reptiles**, bird eggs and nestlings, and even other small **mammals**.

Hamsters are aggressive animals. They are not very social, and generally live a solitary life. Soon after mating occurs, the male hamster is driven away by the female. Once the offspring are weaned, they are likewise driven away by their mother.

Species of hamsters

There are about 16 **species** of hamsters, all of which are found in the Old World. The common or black-bellied hamster (*Cricetus cricetus*) is an aggressive, solitary, burrowing **animal**. This species lives in grassy steppes and cultivated areas of temperate **Europe** and western **Asia** south to Iraq. The common hamster has a body length of about 12 in (30 cm), and can weigh as much as a pound. The common hamster has a reddish coat with bold, white markings, and its fur is sometimes used by furriers. The underground burrows of this species include a relatively large, central chamber, with radiating galleries used to store food, or as a toilet. Remarkably, the winter burrows of the common hamster contain separate storage chambers for each type of food.

The common hamster is an inveterate hoarder, and if the opportunity presents itself, it will store food far in excess of its actual needs. Stores weighing as much as 200 lb (90 kg) have been found. People sometimes dig up the large winter hoardings of the common hamster to retrieve the grain they contain, usually for use as chicken feed. Like other hamsters, this species carries small items of food in its large cheek pouches, although some items, such as large tubers, are carried in the teeth. The pouches are stuffed, and emptied, using the fore paws. The common hamster hibernates in winter, when it blocks up the entrances to its burrow, and sleeps lightly in a bed of straw.

The common hamster is sometimes considered an important pest of agriculture, partly because of its enthusiastic storing of food in amounts far beyond its requirements. As a result, farmers often try to kill these animals using poison, by digging or **flooding** them out of their burrows, or using dogs.

The golden hamster (*Mesocricetus auratus*) is a very rare animal that is found in only a few places in the Middle East. For about a century, the golden hamster was only known from a single specimen, collected in 1839. It was not seen again until 1930, when a single family of golden hamsters was discovered in their den in Syria. Three individuals from that group were taken into captivity, and were used as breeding stock for zoos. They were later used as laboratory animals and for the pet trade. It is likely that all of the golden hamsters presently in captivity are descended from that small, original, founder group.

The head and body length of the golden hamster is about 6 in (17-18 cm) and it weighs about 4 oz (97-113 g). The golden hamster breeds quickly. It has a gestation period of only 15 days, and becomes sexually mature after only 8-11 weeks of age. This is the shortest gestation period of any non-marsupial mammal.

Golden hamsters are also not very social or friendly animals, and in the wild they are thought to live in a solitary fashion. However, these animals can be tamed by frequent handling from an early age, and the golden hamster has become quite popular as a pet. Although this species is quite abundant in captivity and is not in danger of **extinction**, its little-known wild populations are endangered.

The dwarf hamsters (*Phodopus roborovskii* and *P. sungorus*) inhabit the deserts and semi-deserts of southern Siberia, Manchuria, and northern China. They have a head and body length of 2-4 in (5-10 cm) and are virtually tailless. When in captivity, they tame easily and are sometimes kept as pets.

The rat-like or gray long-tailed hamsters in the genus *Cricetulus* inhabit dry agricultural fields and deserts in Eurasia. The head and body length of these hamsters is 3-10 in (8-25 cm) and the tail is 1-4 in (2.5-11 cm) long. Like other hamsters, the seven *Cricetulus* species sometimes have large, underground stores, and these stores are excavated by people to retrieve the grain in some areas.

The mouse-like hamster (*Calomyscus bailwardi*) is another long-tailed hamster, occurring in rocky habitats in the **mountains** of western Asia south of the Caspian Sea. This species has a head and body length of 3-4 in (6-10 cm), a tail slightly longer than its body, and a

A golden hamster. *Photo Researchers, Inc. Reproduced by permission.*

weight of 0.5-1 oz (15-30 g). Its upper parts are buff, sandy brown, or grayish brown, its underparts and paws are white, and its tail is thickly haired and tufted. The mouse-like hamster has prominent ears and no cheek pouches.

Bill Freedman

Hand tools

Hand tools can be as easily found as made, and the earliest tools used by people included sticks and **rocks** picked up and used as projectiles, or to pound or dig. The earliest fashioned hand tools date back to the Stone Age. Currently new technologies make hand tools that are battery-powered, so they are still portable, yet easier to use than their precursors.

Tools are an extension of human limbs and teeth, and were first inspired by human limitations. Things which would be torn by an **animal** with its teeth required less well-equipped humans to use sharp rocks or sticks as knife edges. Sticks could also dig out what human hands could not pull out. They could be used as noise-makers or be thrown at intruders as an intimidation tactic. Even today **monkeys** and **apes** use found objects in these ways, so it is not hard to imagine early humans exhibiting this same ingenuity.

Earliest stone and metal tools

Technology begins in human history when the first stone flints or spear tips were deliberately cut, which are known as Oldowan tools or eoliths. It is very difficult for archaeologists to prove that the sharpened edges of some stone artifacts are the work of human hands rather than the result of the shearing of one stone against another over eons. However, certain improvised tools such as pebbles and animal bones, show clear signs of the wear and tear associated with deliberate use. Chipped quartz tools are identified as such because of the situation in which they were unearthed, accompanying human remains in areas clearly definable as settlements.

About one and a half million years ago, an improvement was made upon the basic carved tool, with the aid of better raw materials. The newer tools fall into three categories of standardized designs; mainly handaxes, picks, and cleavers. These Acheulian tools are the work of humans with larger brains than previous incarnations of the genus *Homo*. They first appeared during the Paleolithic or early Stone Age period. Handaxes from this period are flaked on both sides and often shaped carefully into teardrops. Picks are long tools, with either one sharp edge or two. Cleavers are smoothed into U-shapes with two sharp points on one side. With these inventions, humans began to consider how an object would fit the hand, and how it might be designed for optimum impact.

Acheulean tools were made in great numbers across much of **Africa** and **Europe**, as well as India and the Near East. They were produced over thousands of years but led to no modern counterparts. Archaeologists therefore have a long list of possible uses for these artifacts, which may have served more than one purpose. Butchering animals, digging for roots or **water** sources, and making other tools are the most common suggestions. More inventive ones include the "killer Frisbee" projectile, a use for disc-shaped objects proposed by two researchers at the University of Georgia. **Iceman**, a fully preserved human over 5,000 years old, was found with articles of clothing and tools and weapons on his person. This fortunate occurrence has given archaeologists a chance to theorize about the uses of particular tools, rather than piecing together scattered remains and surmising about possible uses for artifacts.

The later periods of the Neolithic and Bronze Ages saw further developments in **metallurgy** and design.

Axes were made in two pieces, a head and a shaft bound together by **plant** or animal fibers. **Metal** alloys like bronze were deliberately crafted to improve the durability and efficiency of hand tools. Smithing was an art as well as a science, well into the Iron Age. Handcrafted knives were important for nomadic peoples who hunted to survive, and swords especially became crucial tools in warfare. The invention of the metal plow brought agriculture a huge step forward, since it made systematic planting over wide areas possible. This was a great improvement over digging holes one at a **time**.

Development of modern tools

Some hand tools have gone out of style or are used only rarely, but not all. The cobbler used to make shoes by hand, but now people buy mass produced shoes and only take them to a repair shop to be worked on by hand. However, a sewing needle has not changed in centuries, and is still a common household object. Even though people now have access to big sewing machines, it is still easier to fix a button or darn a small tear with a plain needle. During colonial times only the metal parts of an implement would be sold to a user, who would then make his own handle out of **wood** to fit in his hand perfectly. Many things made with metal nowadays, like nails and shovels, were fashioned from wood instead. This is why older buildings and tools have aged well, without problems like rusting or damage to adjacent materials.

Modern technology

Simple hand tools, which cut or pound or assemble, may now be sold with attached metal or plastic handles, but their basic designs and operations have not changed over time. The plane and the file smooth down metal or wood surfaces. Drills and saws are now primarily electric, to save time and **energy**. Hammers come in all sizes, from the rock-breaking sledgehammer to the tiny jeweler's model, which is used to stamp insignias into soft metals like sterling or gold. Screwdrivers attach screws and wrenches tighten nuts and bolts together in areas where larger tools would not reach as easily. Measuring tools are also included under the category of hand tools, since they include tape or folding measures which may be carried on a tool belt. Squares and levels now measure inclines and angles with liquid **crystal** digital displays, but they otherwise look and feel like their old-fashioned counterparts.

Current research and development applies computer-aided design (CAD) programs to simulate models as if under stress of actual use, in order to test possible innovations without the expense of building real prototypes. Lightweight alloys, **plastics**, and engineered woods are used to improve versatility and convenience. Poisonous

KEY TERMS

Acheulean—A term for the tools made by *Homo erectus*, which are recognizably standard designs. The name comes from a Paleolithic site discovered in St. Acheul, France during the 1800s.

Cobbler—An old term for a shoemaker.

Cooper—An old term for a barrel maker.

Eolith—Chipped stones and flints made by humans, which give the Eolithic or Stone Age period its formal name. This period is further divided into the Paleolithic or Old Stone Age and the Neolithic or New Stone Age.

Iceman—The body of a Stone Age man dug up in the Tyrolean Alps, preserved in ice.

Oldowan—A term for tools made during the earliest several hundred thousand years of the Stone Age by *Homo habilis*, items which follow no distinct patterns.

Smith—Someone who works with metals or who makes things. A blacksmith uses iron primarily, while a gunsmith specializes in weaponry.

heavy metals are being replaced with safer plating materials, and nickel-cadmium batteries may soon be replaced with rechargeable units that are easier to recycle.

Resources

Books

Schick, Kathy D., and Nicholas Toth. *Making Silent Stones Speak: Human Evolution and the Dawn of Technology.* New York: Simon & Schuster, 1993.

Periodicals

"How Designs Evolve." *Technology Review* (January 1993).
"Iceman's Stone Age Outfit Offers Clues to a Culture." *New York Times* (June 21, 1994): B7.
"Recreating Stone Tools to Learn Makers' Ways." *New York Times* (December 20, 1994): B5.
"The Technology of Tools." *Popular Science* (September 1993).
"Tool Training at the Chimp Academy." *New Scientist* (May 11, 1991).

Jennifer Kramer

Hantavirus infections

Hantavirus infections are infections of the lungs caused by hantaviruses. There are five known types of hantaviruses, which differ only slightly from one another. These types are: Hantaan, Seoul, Puumala, Prospect Hill, and Sin Nombre. The Sin Nombre **virus** was the cause of the 1993 outbreak in the Southwestern United States, which led to a greater understanding of the virus and its transmission to humans.

The hantavirus are named for the Hantaan River in Korea. In 1976, a virus found near this river was shown to be the cause of a deadly **disease**, which was dubbed the Hantaan River disease. This same type of virus was likely responsible for a disease that appeared in United Nations troops stationed in Korea in 1951. Indeed, a 1990 study that examined the serum that has been collected and saved from Korean War victims found that over 90% of the sera contained antibodies to hantavirus.

Until the early 1990s, reports of hantavirus infections were confined to the Far East. Then, in 1993, an illness outbreak occurred in the United States Southwest, where the states of Colorado, Arizona, New Mexico, and Utah meet (an area known as the Four Corners). A disease that initially appeared similar to the flu quickly progressed to a life-threatening illness within 24 hours to a few days. Lung function dramatically reduced as fluid accumulated in the lungs. Kidney failure also occurred in several victims. At least seven people died from hantavirus infections in the early stages of the Four Corners outbreak.

After state health departments and Indian Health Services in the Four Corners area tested the victims for all known disease agents, the Special **Pathogens** branch of the United States Centers for Disease Control (CDC) assisted with the intense public health investigation into the 1993 Four Corners outbreak The cause of the outbreak was found to be a hantavirus dubbed Sin Nombre virus (from Spanish, meaning *no name*). The lung **infection** became known as hantavirus pulmonary **syndrome**. The virus was shown to live naturally in **rodents**, particularly the **deer mouse**. Mouse feces, urine, and saliva can contain the virus.

The 1993 outbreak is suspected to have arisen because of a period of heavy rain that occurred in the Four Corners region. The wet conditions produced an explosion in the deer mouse population. The virus could then be spread from **mice** to humans more easily.

Dusty environments are particularly important in the spread of hantavirus. The virus particles left behind upon the drying of feces or saliva can be distributed into the air and inhaled into the lungs.

Hantavirus pulmonary syndrome has also occurred in **South America**. Indeed, it is more common in South America than in **North America**. Additionally, the hantavirus types found in North and South America cause a monary Syndrome." *Emerging Infectious Diseases* 6 (March-April 2000): 238–247.

Kreeger, K.Y. "Stalking the Deadly Hantavirus: A Study in Teamwork." *The Scientist* 8 (July 1994): 1–4.

Nicjol, S.T., C.F. Spinopoulou, S. Morzunov, et al. "Genetic Identification of a Hantavirus Associated with an Outbreak of Acute Respiratory Illness." *Science* 262 (1993): 2615–2618.

Brian Hoyle

KEY TERMS

Hantavirus—A virus carried by rodents, especially the deer mouse, that is responsible for the disease hantavirus pulmonary syndrome.

Hantavirus pulmonary syndrome—A serious febrile illness associated with respiratory compromise or failure and caused by a hantavirus that is usually transmitted through inhalation of aerosolized rodent droppings.

Hemodialysis—A method of mechanically cleansing the blood outside of the body, used when an individual is in relative or complete kidney failure, in order to remove various substances which would normally be cleared by the kidneys.

Outbreak—The appearance of new cases of a disease in numbers greater than the established incidence rate, or the appearance of even one case of an emergent or rare disease in an area.

more serious disease than the hantavirus types that are found in the Far East.

Treatment of hantavirus pulmonary syndrome is mostly supportive and can be difficult. One reason is because the patient deteriorates so fast that **diagnosis** and hospitalization must occur very quickly. A second reason is that viral diseases are not treatable using **antibiotics**. Treatment mainly consists of clearing fluid from the lungs to preserve lung function, maintaining **blood pressure** and, if necessary, initiating kidney **dialysis** (hemodialysis). For those who survive, recovery is almost as rapid as was the progression of the infection.

Currently, an antiviral drug called ribavirin is being evaluated as a hantavirus treatment. This drug has shown some potential in the treatment of infections caused by the human immunodeficiency virus (HIV). For the present **time**, the best defense against hantavirus is to avoid environments where exposure to rodent droppings could occur.

See also Physiology; Respiratory diseases; Zoonoses.

Resources

Periodicals

Englethaler, D., D. Mosley, R. Bryan, et al. "Investigation of Climatic and Environmental Patterns in Hantavirus Pulmonary Syndrome Cases in the Four Corners States." *Emerging Infectious Diseases* 5 (September-October 1999): 87–94.

Glass, G.E., J.E. Cheek, J.A. Patz, et al. "Using Remotely Sensed Data to Identify Areas at Risk for Hantavirus Pul-

Hard water

Hard **water** is water that contains large amounts of **calcium**, **magnesium**, or **iron** ions. Hard water is undesirable since it often has an unpleasant taste, interferes with the ability of soaps to dissolve (although some synthetic detergents dissolve well in hard water), and can cause scaling (the building up of insoluble precipitates) in pipes and hot water systems.

Water hardness is most commonly the result of dissolved calcium or magnesium ions, often caused by limestone or dolomite dissolving slightly when acidic water containing **carbon dioxide** runs through these **minerals**. These dissolved minerals lead to an increase in the amounts of calcium and magnesium ions. Water hardness where the **negative** ion (**anion**) is bicarbonate (as in the cases above) is sometimes called temporary hardness, since the unwanted ions can be reduced by boiling the water. If the anion is not bicarbonate, but is instead sulfate or chloride, then permanent hardness is said to result, and this condition can not be remedied by merely boiling the water.

In either case, the calcium and carbonate ions (or calcium and sulfate ions) may **deposit** along the inside of pipes and water heating systems, leading to boiler scale. This scaling can significantly reduce the efficiency of a heating system and can build up to such an extent that the entire pipe is plugged, often leading to overheating of the boiler.

Hard water can be treated either by boiling the water (a method effective only for small quantities) or by precipitating the calcium or magnesium ions from the water (this method is also not practical for large quantities of water). A more efficient method is to use ion exchangers, in which the unwanted calcium and magnesium ions are exchanged or traded for **sodium** ions that do not form any insoluble precipitates and thus do not cause scaling. Most water softeners work by the **ion exchange** method. The soft water produced is not free of ions, only of undesirable ions. Other methods are available for removing ions, including reverse **osmosis** and magnetic water conditioning. Reverse osmosis removes almost 100% of undesirable materials from water, including the hard water ions. This method uses **pressure** to force water to flow from a **solution** of concentrated minerals to one of dilute mineral content, the reverse direction of natural osmosis. The water flows through a semi-permeable **membrane**, which allows the water molecules to pass while filtering out unwanted molecules. This procedure requires several steps and is not as common in the home as ion exchange. Magnetic water conditioning occurs when electromagnets are attached to water pipes. These electromagnets create a strong magnetic field within the pipe, which keeps the hard water minerals from precipitating into the plumbing. This method has not been scientifically proven, and it is unknown if and how it actually works. Electromagnets have been installed in thousands of homes in the United States, but are not as common or reliable as ion exchange water softeners. One drawback to the ion exchange method is that the water produced is slightly acidic and contains a large amount of sodium ions. The acidic water can damage **metal** pipes over **time**, and there is an established link between sodium consumption and **heart disease**.

See also Ion and Ionization.

Hares *see* **Lagomorphs**

Harmonics

What makes a note from a musical instrument sound rich? The **volume** of the sound is determined by the amplitude of the **oscillations** in a sound wave, the distance individual molecules oscillate. A larger amplitude produces a louder sound and transmits more **energy**. The pitch of a note is the **frequency** or number of oscillations per second. A higher frequency produces a higher pitched note. The richness or quality of a sound is produced by the harmonics.

A pure note consisting entirely of one frequency will sound boring. A musical instrument that only produced such pure notes would not sound pleasing. The harmonics are missing. The harmonics are integer multiples of the fundamental frequency. The first harmonic is the fundamental frequency, 264 cycles per second for middle C. The second harmonic will be twice this frequency, 528 cycles per second, which is an octave higher. The third harmonic will be three times the fundamental frequency, 792 cycles per second, and so on. These harmonics are also called overtones—the second har-

monic is the first overtone, the third harmonic the second overtone, and so on.

The violin, piano, and guitar all produce sounds by vibrating strings. Playing the same note, say middle C, will produce a tone with a fundamental frequency of 264 cycles per second. Yet all three instruments sound different because they have different harmonics. The amount of each harmonic present is what gives each musical instrument its own unique sound. A well made instrument will sound richer than a poorly made one because it will have better harmonics. An instrument with no harmonics will sound like a tuning fork with only one fundamental frequency present.

For reasons that we do not completely understand, sounds composed of harmonics whose frequencies are integer multiples of each other sound pleasing to the human **ear**. They are music. On the other hand, sounds composed of frequencies that are not integer multiples of each other are dissonant noise to the human ear.

Hartebeests

Hartebeests are even-toed hoofed antelopes in the family Bovidae, which are found throughout **Africa** south of the Sahara. Included among the grazing antelopes are the reedbuck, **waterbuck**, rhebok, addax, **oryx**, bluebuck, gemsbok, and roan and sable antelopes. More closely related to hartebeests are gnus, impala, topi, wildebeest, and bontebok. These are medium to large antelopes that forage for food in the **grasslands** and woodlands of Africa.

Both males and females have characteristic hook-like horns ringed with ridges. Hartebeests range from a tan to a reddish brown **color** with distinctive markings denoting the different **species**. Females are slightly smaller than males. Hartebeests have long faces, raised high shoulders with strong legs in front, and a steep sloping back. Their legs are thin and they canter for long distances, which is made possible by their long forelegs.

Social groups and behavior

Hartebeestes graze in herds and are commonly seen with wildebeestes, **gazelles,** and **zebras**. The home ranges of hartebeest can be from 800 to 1,400 acres (234 to 567 hectares). Within this area a number of different relationships exist. Small groups within the home range may occupy only the few acres that a male can defend. The female groups roam over many of the male-dominated smaller territories. Young hartebeests remain with

A small herd of hartebeests in Kenya. *JLM Visuals. Reproduced by permission.*

their mothers, who may have several offspring of different ages following her. Males leave around the age of two and a half years old and join bachelor herds.

Females are sexually mature at two years of age. Pregnancy lasts about eight months and hartebeests give **birth** to one offspring at a time. Newborns lie out in the grass for about two weeks, then join the maternal herd. Mothers will defend young males from threatening older males that claim the territory.

Hartebeest males mark their territories with dung piles. They will also advertise their territorial claim by standing on mounds within the territory and marking grass with their preorbital **glands**, which are located in front of their eyes. They also have scent glands on their front hooves.

Hartebeests may settle territorial differences by fighting or by ritualized **behavior**. This may include defecation, pawing the ground, and scratching and cleaning their heads and necks. Fighting can include something that looks like neck wrestling. One of the difficulties hartebeests encounter in maintaining control of their territories is their need for **water**. If one leaves to drink, on his return he may find that another bull has claimed his territory.

Land competition

While hartebeests once occupied a large area over much of the African **continent**, their range has diminished because of expanded farming in some of the areas they had once inhabited. Since domestic cattle graze on the same **grasses** that hartebeests prefer, the growth in cattle raising in Africa has resulted in a general decline in hartebeest populations. The most numerous species is the Kongoni or Coke's hartebeest (*Alcelaphus buscelaphus cokei*) of Kenya, while the Cape hartebeest (*A. caagma*) survives in protection on farms. The bastard hartebeests (*Damaliscus*) are smaller than the *Alcelaphus* species, and include the topi (*D. korrigum*) of East Africa and the sassaby (*D. lunatus*) of South Africa.

Particularly vulnerable has been the hirola, or Hunter's hartebeest. In a five year period from 1973 to 1978, the hirola population in Kenya declined from 10,000 to a little over 2,000. The bubal hartebeest (*Alcelaphus buscelaphus buscelaphus*) became extinct in 1940 and in 1969 the Lake Nakuru hartebeest was lost to the continent. The Swayne's hartebeest was abundant in the early part of this century and is now considered the most vulnerable to **extinction**. The kaama has been res-

KEY TERMS

Forage—Vegetation that is suitable for grazing or browsing animals.

Home range—The full territory that an animal occupies throughout its lifetime.

Land competition—When two or more animal groups use the same natural growth on a land area and one population grows while the other declines.

cued, replenished on farms and in game parks, and released again on natural ranges.

In addition to the competition for land, hartebeests face threats from predators. They are particularly vulnerable to lions, leopards, cheetahs, and hyenas. Young animals are also vulnerable to attacks from jackals, **pythons**, and **eagles**. Hartebeests get their name from the South-African, Dutch-derived language of Afrikaans. It means "tough beast." The early Dutch settlers of South Africa found them to be good runners that could not be easily overtaken by a horse.

See also Antelopes and gazelles.

Resources

Books

Estes, Richard D. *Behavior Guide to African Mammals.* Berkeley: University of California, 1991.

Estes, Richard D. *The Safari Companion.* Post Mills, Vermont: Chelsea Green, 1993.

Grzimek, Bernhard. *Encyclopedia of Mammals.* New York: McGraw-Hill, 1990.

Haltenorth, T., and H. Diller. *A Field Guide to the Mammals of Africa.* London: Collins, 1992.

MacDonald, David, and Sasha Norris, eds. *Encyclopedia of Mammals.* New York: Facts on File, 2001.

Vita Richman

Hassium *see* **Element, transuranium**

Hawks

Hawks (family Accipitridae) are one of the major groups of predatory **birds** that are active during the day. They are members of the order Falconiformes, which also includes the **falcons**, **vultures**, and osprey, and like the other Falconiformes, they have the characteristic sharp, strong claws and hooked beak suited for catching and tearing up **prey**.

Found on all continents but **Antarctica**, hawks are a diverse group. There are 26 **species** in **North America** alone that have been breeding successfully in recent times. They include four species of **eagles**, five species of **kites**, and 17 species called hawks. These North American hawks vary from the small, 3-8 oz (85-227 g) sharp-shinned hawk, with a wingspan of about 2 ft (0.6 m), to the ferruginous hawk (*Buteo regalis*), with a wingspan of 4.5 ft (1.5 m). Eagles are different primarily because of their huge size; they may weigh from 8-20 lb (4-9 kg), with wingspans up to 8 ft (2.4 m).

Besides the hooked beak and strong claws already described, the hawks share several characteristics. Their wings are generally broad and rounded, well-suited for flying over land (kites' wings are different, more like a falcon). Their nostrils are oval or slit-like, and open in the soft skin, the cere, where the upper mandible joins the head, which is round. The neck is short and strong. The large eyes are usually yellow, orange, red, or brown, and turn little in their sockets. Hawks move their heads to direct their **vision**, which is both monocular and **binocular** (especially when hunting).

Hawks' plumage is subdued, usually mottled browns and grays on the back and lighter, often barred or streaked, below. **Color** phases have been found in many species: albinos in 10 species, melanism (a black phase) in five, and erythrism (a red phase) in one.

The North American hawks fall into four groups: the buteos, the accipiters, the kites, and the harriers.

Buteos

The buteos are like the eagles, but smaller. They have broad, rounded wings, which are stubbier than those of the eagles, which help them cruise long distances over land searching for prey. Common prey items include **mice** and rabbits, for the buteos generally feed on **mammals**. A small prey item, such as a mouse, is swallowed whole. A larger item is brought to a secluded spot, held down with the feet, and pulled apart with the sharp beak. Representative buteos include the red-tailed hawk (*Buteo jamaicensis*), the rough-legged hawk (*B. lagopus*), and Swainson's hawk (*B. swainsoni*).

Old World buteos include:

• Common buzzard (*Buteo buteo*). Resident of Eurasia, with some wintering in **Africa**.

• African mountain buzzard (*Buteo oreophilus*). Resident of the **mountains** of east and southern Africa.

- Madagascar buzzard (*Buteo brachypterus*). Resident of Madagascar.

- Rough-legged buzzard (*Buteo lagopus*). Besides residing in North America, this bird also makes it home in northern and arctic Eurasia.

- Long-legged buzzard (*Buteo rufinus*). Resident of southeastern **Europe**, North Africa, and Central **Asia**.

- African red-tailed buzzard (*Buteo auguralis*). Resident of West and Central Africa.

- Jackal buzzard (*Buteo rufofuscus*). Resident of Africa, south of the Sahara.

North American buteos and their status are as follows:

- Crane hawk (*Geranospiza caerulescens*). Southwestern stray, normally resident of the tropical woodlands.

- Common black-hawk (*Buteogallus anthracinus*). Rare and apparently declining in the United States due to disturbance and loss of **habitat**. Today there are possibly 250 pairs left in the United States.

- Harris' hawk (*Parabuteo unicinctus*). Has disappeared from some former areas, such as the lower Colorado River Valley. Declining in parts of its range, but recently re-introduced in some areas. Has been threatened by illegal poaching for falconry.

- Gray hawk (*Buteo nitidus*). It is estimated that no more than 50 pairs nest north of Mexico. It is vulnerable to loss of its lowland stream forest habitat, though it remains common and widespread in the tropics.

- Roadside hawk (*Buteo magnitostris*). No information available.

- Red-shouldered hawk (*Buteo lineatus*). Declining or now stabilized at low numbers. Accumulates organochlorine **pesticides** and PCBs, however, loss of habitat is the major threat. Although this bird is today far less numerous than historically in some areas, including the upper Midwest and parts of the Atlantic Coast, current populations are believed to be stable in most regions.

- Broad-winged hawk (*Buteo platypterus*). In the early years of the twentieth century, large numbers were sometimes shot during **migration**. Now legally protected, and their numbers appear stable.

- Short-tailed hawk (*Buteo brachyurus*). May be threatened by destruction of breeding grounds (mature cypress swamps and riparian hardwoods). Today, this bird is very uncommon in Florida (with a population probably no larger than 500), but its numbers appear stable. The population may be increasing in Mexico.

- Swainson's hawk (*Buteo swainsoni*). Current status is unclear. Many have been shot while perched along roads. But expanding cultivation has increased breeding opportunities, especially in the Great Plains. The population has declined seriously in California, for reasons that are not well understood.

- White-tailed hawk (*Buteo albicaudatus*). Marked decline from 1930s to 1960s largely due to loss of habitat. Significant eggshell thinning has been observed since 1947. Its decline in Texas from the 1950s to the 1970s has been attributed to the use of pesticides, but the population in that state now appears to be stable. Numbers may be declining in Mexico due to overgrazing of its habitat.

- Zone-tailed hawk (*Buteo albonotatus*). This bird has disappeared from some of its former nesting areas. Loss of nesting sites such as tall cottonwoods near streams may have contributed to its decline.

- Red-tailed hawk (*Buteo jamaicensis*). Greatly reduced in the east by early bounties. Continued decline due to human persecution and loss of habitat. Some egg thinning. The population has increased in some areas since the 1960s. Today the population is stable or increasing.

- Ferruginous hawk (*Buteo regalis*). Currently rare in many parts of its range. Many have been shot while perched along roadsides. Today this bird is a threatened species. The current population may be less than 4,000. The decline in population is due to hunting and to loss of habitat.

- Rough-legged hawk (*Buteo lagopus*). Inadvertently poisoned by bait intended for mammals. Often shot when feeding off road kills in the winter. Local populations in the Arctic rise and fall with the rodent population there. The overall numbers appear healthy.

Accipiters

The accipiters are generally smaller than the buteos. Their shorter, rounded wings and long tails make them agile hunters of birds, which they catch on the wing. Familiar accipiters include Cooper's hawk (*Accipiter cooperii*) and the sharp-shinned hawk (*A. striatus*).

Old World accipiters include:

- Japanese sparrowhawk (*Accipiter gularis*). Japan, China, and the eastern parts of the former Soviet Union.

- Besra (*Accipiter virgatus*). Resident of the Himalayas, southeast Asia, and the East Indies.

- African goshawk (*Accipiter tachiro*). Resident of Africa, south of the Sahara.

- Crested goshawk (*Accipiter trivirgatus*). Resident of southern Asia, the Philippines, and Borneo.

- Australian goshawk (*Accipiter fasciatus*). Resident of **Australia**, New Guinea, Flores, Timor, and Christmas Island.

- France's sparrowhawk (*Accipiter francessi*). Resident of Madagascar.

North American accipiters and their status are as follows:

- Sharp-shinned hawk (*Accipiter striatus*). Dramatic decline in the eastern United States in the early 1970s. Between 8-13% of the eggs showed shell thinning. Their numbers recovered somewhat through the early 1980s, but more recently, the numbers in the east have declined.

- Cooper's hawk (*Accipiter cooperi*). A serious decline underwent a slight reversal after the ban of DDT in 1972. Their numbers appear to be stable in most areas.

- Northern Goshawk (*Accipiter gentilis*). Population formerly declined in the north, while expanding in the southeast. Eggshell thinning was reported in some areas in the early 1970s. Today the range is expanding in the northeast, but populations in the southwestern mountains may be threatened by loss of habitat.

Kites

More graceful in flight than either the buteos or the accipiters are the kites. Although they are hawks, the kites have long, pointed wings similar to those of falcons, and long tails. Found in warm areas, kites have shorter legs and less powerful talons than other members of the hawk family, but are adept at catching prey such as **frogs**, **salamanders**, **insects**, and snails—in fact, the Everglade kites (*Rostrhamus sociabilis*) prey solely on **snails** of the genus *Pomacea*. Also found in North America is a single species of harrier, the hen or marsh hawk (*Circus cyaneus*), which is common in Europe and in Asia, too. This slender little hawk (maximum weight, 1.25 lb (0.5 kg) eats mice, **rats**, small birds, frogs, **snakes**, insects, and carrion.

Old World kites include:

- Black-breasted buzzard kite (*Hamirostra melanosternon*). Resident of Australia.

- Brahminy kite (*Hamirostra indus*). Resident of Southern Asia, East Indies, New Guinea, northern Australia, and the Solomon Islands.

- Black kite (*Milvus migrans*). Resident of Europe, Asia, Africa, and Australasia.

- Black-shouldered kite (*Elanus caeruleus*). Resident of Spain, Africa, and southern Asia.

Kites found in North America and their status are as follows:

A red-tailed hawk (*Buteo jamaicensis.*) at the Kellogg Bird Sanctuary, Michigan. The red-tail is North America's most common hawk. *Photograph by Robert J. Huffman. Field Mark Publications. Reproduced by permission.*

- Hook-billed kite (*Chondrohierax uncinatus*). Decline in population with clearing of woods. Subspecies on Grenada and Cuba have been listed as endangered.

- American swallow-tailed kite (*Elanoides forficatus*). Marsh drainage, **deforestation**, and shooting have reduced the population and range. Formerly more widespread in the southeast, and north as far as Minnesota. Current population appears stable.

- White-tailed kite (*Elanus leucurus*). The population has been increasing since the 1930s, and settling in places not known historically. Has also spread to American tropics with clearing of forest land.

- Snail kite (*Rostrhamus sociabilis*). **Endangered species.** The population in Florida had been reduced to 20 by 1964, due to marsh draining and shooting. By 1983, they were recovering (with an estimated population of 700). But today the Florida population is endangered due to disruption of **water** flow (and impact on habitat and snail population). Although widespread in the tropics, the species there is vulnerable to loss of habitat.

- Mississippi kite (*Ictinia mississippiensis*). Increasing since 1950s. Breeding range has expanded westward, possibly due to **tree** planting for **erosion** control. Since about 1950, the population in some areas (such as the southern Great plains) has greatly increased. The range has extended to parts of the Southwest, where the species was previously unknown.

- Black-shouldered kite (*Elanus caeruleus*). Range has greatly expanded since 1960. This kite is probably the only raptor to have benefited from agricultural expansion. Its expansion has been aided by its ability to adapt to habitat disruption and an increase in the number of rodents.

Harriers

Old World harriers include:

- Spotted harrier (*Circus assimilis*). Found throughout most of Australia, and sometimes in Tasmania.

- European marsh harrier (*Circus ranivorus*). Resident of northern Kenya, Uganda, eastern Zaire, and Angola.

- Marsh harrier (*Circus aeruginosus*). Resident of western Europe, central Asia, and Japan; winters in Africa and southern Asia.

- Black harrier (*Circus maurus*). Resident of southern Africa.

Harriers in North America and their status are as follows:

- Northern harrier (*Circus cyaneus*). Has disappeared from many of its former nesting areas. Decline attributed to loss of habitat and effects of pesticides. In 1970, 20% of the eggs examined were found to exhibit shell thinning. Today the population appears to be declining in parts of North America.

Characteristics and behavior

Generally, hawks kill their prey with their claws, unlike the falcons, which catch prey with the claws but kill with a blow of their beak. However, despite their fierce reputations, some hawks are quiet and gentle. In addition to their familiar scream, hawks' vocalizations include a high plaintive whistle like the wood pewee (broad-shouldered hawk); a musical kee-you, kee-you (red-shouldered hawk);and a high-pitched squeal (short-tailed hawk).

Hawks are unusual among birds in that the female is generally larger than her mate. In some species, this difference—called sexual dimorphism—can be as great as the female being twice the size of the males, as in the accip-iters. Some researchers have found a correlation between the size difference between the sexes and the diet of the species. For example, among Falconiformes like vultures, which eat carrion, the sexes are similarly sized. However, moving from there through the diets of insects, **fish**, mammals and birds, the sexual dimorphism increases. So many other factors correlate with sexual dimorphism, it is difficult to say which is the major contributing factor. For instance, another hypothesis holds that a larger female bird of prey is better equipped to protect herself during contact with the potentially dangerous and certainly well-armed male. Yet another theory suggests that size is related to the vulnerability of the prey pursued. That is, the more agile the prey, the less likely the success of each hunt. Or, perhaps the secret to sexual dimorphism lies in a simpler explanation: that the larger female is better at catching some prey, and the male is better at catching others.

Courtship among the hawks is among the most spectacular of all animals. In the case of the red-tailed hawk, for example, the pair soar, screaming at each other; then the male dives at the female, who may roll in the air to present her claws to him in mock combat. The male marsh hawk flies in a series of graceful U's over the marsh from where the female is watching. Hawks generally mate for life, and are strongly attached to their nesting territory; one pair of red-shouldered hawks (and their offspring) used the same area for 45 years.

Hawks usually build their nests high in trees. The nests are quite large, up to about 3-4 ft (0.9-1.2 m) across, and consist mostly of sticks, with twigs, **bark**, **moss**, and sprigs of evergreen. Nests are often used year to year, with the bird abandoning it only at death or when the nest has grown so large that it breaks the boughs it is built upon.

Generally, the pair will defend their territory against all who approach, but some species, including the ferruginous hawk, will abandon their nest if disturbed by humans. Some hawks will dive at humans who approach too near their nests, as in the case of a pair of red-tailed hawks nesting in a park in Boston, who injured several curious passers-by before park officials removed the **raptors** and their eggs to a more secluded spot. The territory defended can range from 650 ft (198 m) between nests in small hawks to up to 18.5 mi (29.8 km) in larger ones. Some species, including the kites, are more gregarious and nest in loose colonies of about 10 pairs.

Female hawks lay between two and five eggs. Depending on the species, the female either incubates them alone or with the help of her mate. Incubation lasts about 28 days. The young hawks fledge at about 40 days of age.

Some young hawks may remain with their parents for a while after fledging, and these family groups have been observed hunting as a team. Generally, hunting buteos circle high in the air, watching the ground for any movement of prey. They then **fold** their wings and dive upon their prey. Accipiters are more likely to pursue their avian prey on the wing, darting thickets and woods during the chase. Some accipiters are decried for their impact on the populations of songbirds; in fact, in the past some ornithologists considered the sharp-shinned hawk a "harmful" species because it preyed on "beneficial" songbirds. Such human prejudice is at the root of most human-raptor conflict.

After the breeding season ends, many hawk species conduct spectacular migrations. The most spectacular is that of the Swainson's hawk. Huge flocks of these birds will travel overland from their North American summer range to their wintering grounds in **South America**, a total distance of 11,000-17,000 mi (17,699-27,353 km) annually. The broad-winged hawk (*B. platypterus*) is also noted for its large migrations: in one day (September 14, 1979), 21,448 broad-winged hawks passed over Hawk Mountain, Pa. Besides Hawk Mountain, other good sites to watch hawk migrations include Cape May, NJ; Duluth, MN.; Port Credit and Amherstburg, Ont.; and Cedar Grove, WI.

Hawks and humans

Although more humans are enjoying watching these migrations and learning to appreciate these raptors, hawks still face persecution. Many are shot each year. Others die in traps set for fur-bearing animals. Still others are killed when they alight on high-voltage power lines. Most species of hawks, like all other raptors, were hard hit by the effects of the pesticide DDT. Considered a miracle pesticide when it was introduced in the 1940s, DDT pervaded the environment, and became concentrated higher up in the food chain. The effect on the raptors was the production of eggs that were too thin-shelled to be incubated: when the female moved to sit on them, the eggs collapsed beneath her, killing the chicks inside. Recovery has been slow.

All hawks are protected by federal and state laws. Some, like the red-tail, are successfully adjusting to living in urban areas. Hawks have been known to live almost 20 years.

Resources

Books

Ehrlich, Paul R., David S. Dobkin, and Darryl Wheye. *The Birder's Handbook.* New York: Simon & Schuster Inc., 1988.

Peterson, Roger Tory. *North American Birds.* Houghton Mifflin Interactive (CD-ROM). Somerville, MA: Houghton Mifflin, 1995.

F. C. Nicholson
Randall Frost

Hazardous wastes

Hazardous wastes are by-products of human activities that could cause substantial harm to human health or the environment if improperly managed. The United States Environmental Protection Agency (EPA) classifies liquid, solid, and gaseous discarded materials and emissions as hazardous if they are poisonous (toxic), flammable, corrosive, or chemically reactive at levels above specified safety thresholds. In the United States, the term hazardous waste generally refers to potentially dangerous or polluting chemical compounds; other potentially hazardous industrial, military, agricultural, and municipal byproducts, including biological contaminants and **radioactive waste**, are regulated by other government agencies than the EPA's hazardous waste division.

The handling of hazardous wastes became a major political issue in the late 1970s in the United States and other industrialized nations when a number of high-profile human health and environmental **pollution** crises focused public attention on the problem. Since then, many governments have greatly expanded regulation of hazardous **waste management**, disposal practices, and clean-up. In the United States, the EPA oversees hazardous waste regulations that attempt to prevent new cases of environmental and human **contamination**, as well as the so-called "Superfund" program that addresses clean-up of sites contaminated in the past.

Sources of hazardous wastes

Hazardous wastes can be solids, gases, liquids, or semi-liquids like **mining** sludge and drilling mud. Most of the wastes listed by the EPA are liquids or semi-liquids. Thousands of waste materials are considered hazardous. These include familiar items like used motor oil and mercury, agricultural **pesticides**, and industrial materials such as **asbestos** and **polychlorinated biphenyls (PCBs)**. United States industries, farms, mines, military facilities, cities, and small businesses generate roughly 200 million tons of hazardous wastes each year. Furthermore, the EPA estimates that there are presently 6,500 facilities in the United States that require hazardous waste clean-up under the directives of the 1976 Resource Conservation and Reclamation Act (RCRA) and its 1984 Hazardous and Solid Waste Amendments (HSWA).

Hazardous waste management is also an international issue. Each year, industrialized nations with strict environmental regulations export more than two million tons of hazardous waste for disposal in poorer developing nations with less stringent waste disposal oversight. Developed nations also locate large corporate, industrial, and military facilities in countries that have lax environmental restrictions.

Hazardous wastes often cause problems for years after their disposal. Many industrial waste disposal sites were established, filled, and buried long before establishment of present-day standards for management and disposal of hazardous chemicals. Toxic, flammable, corrosive, and reactive chemicals are often long-lived, and sometimes the dangers they posed to the environment and to human health were unknown at the time of their disposal. The industries responsible for many pre-1970 hazardous waste sites are no longer in business, and sometimes the sites themselves are difficult to locate. Even modern legislation gives industries fairly broad leeway to produce chemicals, police their own waste disposal practices, and to contest cases of possible environmental or human health damage. It is often extremely difficult to prove a scientific link between an incident of drinking **water** poisoning, or a human **disease** cluster, and a facility that improperly handles industrial chemicals.

Industrial hazardous wastes

Four types of industry account for about 90% of industrial hazardous wastes generated in the United States: chemical manufacturing, primary **metal production**, **metal** fabrication, and **petroleum** processing. Large chemical plants and petroleum refineries, and other "large quantity generators" that produce more than 2,200 lb (1,000 kg) of hazardous wastes per month, are the most visible and heavily regulated facilities in the United States. However, businesses of all sizes generate dangerous chemicals; the EPA currently lists more than 250,000 facilities as "small-quantity generators" of hazardous waste. These diverse, smaller producers account for about 10% of the potentially harmful substances produced each year.

Pesticides like malathion, DDT, and diazanon are hazardous chemicals; some of them have been banned, but many are still manufactured and used in the United States. Pesticides are designed to kill pest **insects**, plants, and other organisms that threaten agricultural **crops**, destroy municipal and residential landscaping, and carry human diseases. Most pesticides are dangerous chemicals themselves, and their manufacture produces additional hazardous waste. The EPA's Hazardous Waste division regulates handling, disposal, and clean-up of pesticides during their production, but environmental pollution and human health effects caused by pesticides after application are not included in hazardous waste regulations. (The EPA's Office of Pesticide Programs oversees pesticide use and handles cases where pesticides in agricultural or landscaping runoff pollute air and water or compromise human health.)

Other sources

Other types of hazardous wastes are associated with military bases, mines, residential communities, and small businesses. Though large industry produces the majority of hazardous waste in the United States, the small quantity generators (SQGs) that produce between 220 and 2,200 lbs (100–1,000 kg) of hazardous waste per month present particular regulatory challenges: (1) The chemicals used by auto garages, dry cleaners, construction companies, scientific labs, photo developers, printers, large offices, and farmers are often toxic. (2) Hazardous wastes generated by SQGs are much more varied than those produced by large companies. Each chemical, be it a month's supply of dry cleaning fluid or a house-worth of residential insulation, requires its own handling and disposal strategy. (3) SQGs, who do not have the legal and administrative support common at large companies, often have difficulty deciphering hazardous waste regulations. Noncompliance can result from simple ignorance of a small business's responsibility to follow environmental laws.

United States military bases have some of the most serious hazardous waste problems in the nation, an issue only recently addressed by government and private environmental agencies. About 19,000 sites at 1,800 military installations show some degree of **soil** or **groundwater** pollution. More than 90 military bases have been on the EPA's Superfund list of high-priority, hazardous waste cleanup sites. Moreover, a law passed in 1992 allows federal and state regulatory agencies to levy fines against the military if their hazardous wastes are not properly managed. Prior to this, the armed forces were not subject to state or federal environmental laws. Consequently, the military now has a range of programs to clean up hazardous waste problems at its bases.

Mining waste, a type of industrial waste, often includes hazardous substances. Mining operations commonly use hazardous chemicals, and sometimes naturally toxic substances are released into the environment during mining and the disposal of its waste materials. For example, gold mining in the Amazon Basin of **South America** results in the release of 90–120 tons of mercury into **rivers** every year. This has resulted in elevated levels of mercury in **fish** and humans in the region. Mercury poisoning results in severe **birth defects**, neurological disorders, kidney failure, and a number of other serious health effects. Chemical separation of **ore minerals** like **lead**, **iron**, and zinc from their host **rocks** creates so-called acid-mine drainage that contains both the toxic chemicals used in the separation process like arsenic and **sulfuric acid**, and poisonous heavy metals like lead and mercury. Acid-mine drainage from metal mining in the American West has contaminated drinking water and caused serious ecological damage since the mid-1800s.

Household hazardous wastes are discarded products used in the home, which contain dangerous substances. Examples include paint, motor oil, antifreeze, drain cleaner, and pesticides. In the 1980s, many local governments in the **North America** began to set up regular collection programs for household hazardous wastes, to ensure that they are properly disposed or recycled. Local or state/provincial governments usually pay the costs of such programs. However, a system used in British Columbia, Canada, requires consumers to pay an "eco-fee" on paint they buy. This, along with funds provided by the paint industry, helps pay for a collection program for waste paint from households.

Protection from hazardous wastes

Beginning in the 1970s, a number of highly-publicized hazardous waste crises and advances in environmental science led the American people and public health authorities to recognize hazardous wastes as a significant threat to health and the environment. Today, there is a public and political debate between those who believe that public **perception** of waste hazards is worse than the actual danger, and that adequate safeguards exist to protect people from significant exposures, and those who insist that government and industry need to do a better job of managing hazardous wastes, considering the harm that can be caused by these chemicals.

The case of chemical dumping by the Hooker Chemical Company at Love Canal in Niagara Falls, New York was a catalyst that dramatically increased public concern over hazardous wastes. The Love Canal community was built at the turn of the twentieth century as a residential subdivision centered on a small hydro-power canal. The original developer never completed the canal, and the Hooker Company used the half-finished ditch as a dump for more than 20,000 tons of chemical wastes during the 1940s and 1950s. In 1953, the Hooker Company covered the dumpsite with soil and sold it to the town of Love Canal for a dollar. By 1976, residents and scientists had linked a series of public health problems including **birth** defects and childhood **leukemia** to teratogenic (birth defect-causing) and carcinogenic (cancer-causing) liquids, sludge, and gases visually seeping from the dumpsite.

The media reported extensively on the problems at Love Canal. The resulting wave of public outrage at **television** pictures of black sludge seeping from the ground, and children suffering from **cancer**, triggered a political response. In 1978, President Jimmy Carter declared Love Canal a federal disaster area. Two years later, the U.S. Congress passed "Superfund" legislation, which established a national cleanup program for hazardous waste sites.

A hazardous waste dump site. The barrels are filled with chemical wastes. *Photograph by Nancy J. Pierce. Photo Researchers, Inc. Reproduced by permission.*

Activist groups such as Greenpeace and the Citizen's Clearinghouse for Hazardous Wastes seek to increase public awareness of hazardous waste issues. Such groups frequently oppose government and industry policies and projects related to hazardous wastes. One outgrowth of the publicity surrounding hazardous waste is that it has become difficult to find locations for new treatment facilities because of local opposition. This is called the NIMBY, or "not in my backyard" **syndrome**. Civil rights groups in the United States have also called attention to the unequal distribution of hazardous waste dumpsites and handling facilities in poor and minority-dominated communities. Studies have shown that a disproportionately large fraction of African Americans and Hispanic Americans—three out of five—live in communities with hazardous waste sites.

Government management strategies

A complex web of federal agencies and legislation oversee and regulate storage, transportation, disposal, **recycling**, and use of hazardous wastes n the United States. State and local governments also have hazardous waste regulations. The private environmental consulting industry helps government agencies, industrial manufacturers, cities, and businesses of all sizes assess their hazardous waste practices and compliance with the increasingly long list of federal, state and local hazardous waste laws.

There are two main U.S. federal hazardous waste laws: the 1976 Resource Conservation and Recovery Act (RCRA), and the 1980 Comprehensive Environmental Response, Compensation and Liability Act (CERCLA), also known as the Superfund law.

RCRA legislation focuses mainly on disposal of non-hazardous solid waste, and was enacted mainly to deal with unsightly garbage disposal practices. Hazardous waste disposal was a minor issue in the mid-1970s, but enough concern existed that Congress included a section on hazardous wastes in RCRA. Prior to the passage of RCRA, factories and plants typically dumped hazardous wastes in ponds, lagoons, or streams near their facilities. Many smaller waste generators sent their chemical by-products to outdated municipal landfills that where they leaked into ground and surface water reservoirs.

RCRA mandated creation of a system to track and monitor hazardous wastes from production to disposal, or from "cradle to grave." Legislators also designed RCRA to regulate existing hazardous waste sites, and to improve hazardous waste management overall. RCRA's goals have been partly accomplished, but problems have occurred along the way. For example, EPA has been slow to put some of the changes into effect. Some industrial polluters have discovered that it is less expensive to ignore the hazardous waste disposal recommendations, and to use their financial and legal resources to contest claims of environmental damage. Also, some of the legislation expected private industry to build expensive treatment facilities, hire environmental consultants to assess their practices, and to pay clean-up costs. In many cases, companies balked at the cost of self-regulation, and failed to meet the requirements. Community opposition to local siting also delayed or prevented construction of many waste treatment and disposal facilities.

The focus of RCRA has changed over the years. Amendments and enactment of related laws have moved the EPA's focus from management and disposal practices to waste prevention. There is a growing consensus that it is less expensive, and much less dangerous to prevent a spill, leak, or poisoning than it is to clean one up. Regulations now encourage industries to produce fewer hazardous wastes, to produce wastes that are less hazardous, and to develop alternative methods that do not require dangerous materials.

In contrast to RCRA, which attempts to manage waste production, management, and treatment, CERCLA was designed to clean up sites that are already contaminated. The law established a National Priority List

of the United States' worst hazardous waste sites, and set up a fund, nicknamed Superfund, to augment remediation costs. CERCLA requires that the EPA, which enforces the law, try to find the parties, usually businesses or individuals, responsible for the hazardous waste problems, and make them pay for the cleanups. If responsible parties cannot be found, or if additional money is needed for a proper cleanup, then the governmental Superfund money may be used. The fund was started with $1.6 billion in 1980, and increased to $8.5 billion in 1986. Most of the money in the fund comes from a federal tax on chemical and petroleum companies, the industries responsible for many of the listed sites. Although the amount of money in the Superfund seems huge, cleanup costs are also enormous. The average cost to clean up a Superfund site is $30 million. There were 1,235 sites on the Superfund National Priority List in 2001.

The Superfund project and CERCLA have not been as effective as was initially hoped. Because of the technical difficulty, expense, and legal ramifications of cleanup, fewer than 100 sites have been completely cleaned up and removed from the National Priority List. CERCLA has also been widely criticized because of its liability provisions that require a "potentially responsible party" to pay cleanup costs. This party could, for example, be a business that transported waste materials to a dumpsite years ago, even if the site was not considered a problem at that time, and even if the company did not break any relevant laws. Because businesses often object to the CERCLA liability provisions, these matters frequently end up in court, slowing up the cleanup process.

Many developed nations have environmental regulations similar to RCRA and CERCLA. Some countries, Japan and Denmark for example, rely on partnerships of government and private industry to manage hazardous wastes. In both of these countries, industries receive subsidies or incentives to try new, innovative methods of handling their wastes. Ironically, the nations with the strictest environmental regulations end up exporting large quantities of hazardous wastes for recycling or disposal. Germany, for example, exported more than 500,000 tons of hazardous wastes to other countries each year in the 1980s. Non-governmental environmental groups have campaigned against the export of hazardous wastes by industrialized countries. The United Nation Environment Programme's (UNEP) 1989 Basel Convention attempts to restrict international transport of hazardous wastes and to encourage less developed nations to resist the economic temptation to take hazardous waste from developed nations. In 2002, 135 nations and the European Union had signed the Basel Convention. The convention, however, does not include the United States, one of the world's largest hazardous waste producers.

Treatment and disposal technologies

Hazardous wastes that need treatment or disposal may be freshly generated from an industrial operation, they may be old stored chemicals, or they may have been sitting in a dumpsite for many years. At a dumpsite, component chemicals difficult to identify, they are likely to have reacted with one another, and they may have already affected the surrounding soil and water. Land disposal and **incineration** are two main dumpsite remediation methods. Types of waste treatment include physical, biological, and chemical **neutralization** or stabilization. Some treated hazardous wastes can even be reclaimed or recycled.

Industries in the United States dispose of about 60% of their hazardous waste using a land disposal method called deep well injection. Liquid wastes are injected into wells located in impervious rock formations that supposedly keep the waste isolated from groundwater and surface water. Unfortunately, hydrogeologists now predict that groundwater flow actually does occur in most previously-designated impervious rock formations, and injected waste often migrates into groundwater reservoirs called aquifers. Other underground burial locations for hazardous wastes include deep mines, natural caverns, and man-made deep pits.

Landfilling is the other primary land disposal method for hazardous waste disposal in the United States. Hazardous waste landfills are similar to regular solid waste landfills, but they must meet much higher standards for safety and environmental protection. The EPA requires that most hazardous wastes be treated before being discarded in properly-designed, approved landfills and burial sites.

Incineration, or burning, is a controversial, but still common, method of handling hazardous wastes. The EPA estimates that five million tons of hazardous wastes are burned each year in the United States. Various incineration technologies exist for a variety of types of waste. For example, volatile chemicals like paint thinners, oils, and solvents are destroyed by **combustion** at cement plants whose furnaces, called kilns, reach temperatures of 2,700°F (1,500°C). Needless to say, residents living near cement plants and other hazardous waste incinerators often have concerns about **air pollution**. In 1993, the EPA tightened its regulations on emissions from most hazardous waste incinerators, including cement kilns, after discovering that the emissions contained like dioxins, furans and other substances that cause cancer and other health problems in humans. Another recent EPA study noted that medical waste incinerators that many hospitals use to burn hazardous wastes also emit dioxins.

Some hazardous wastes, including certain tars, drilling muds, and mining sludges, are relatively well-suited for incineration. Some other wastes, however, should not be burned, such as those that contain heavy metals. Burning does not destroy the metals, and they end up in the incinerator ash. Ash from hazardous waste incinerators that contains high concentrations of metals is a dangerous material in its own right, and requires careful disposal.

Stabilization, also called solidification, is a physical treatment method sometimes used on incinerator ash and other hazardous wastes before landfilling or underground burial. In this method, additives are combined with the waste material to make it more solid, or to prevent **chemical reactions**. Other physical treatment methods include soil washing at hazardous waste dumpsites, filtering hazardous waste solids out of liquids, and **distillation**.

Various biological treatments utilize microbes to break down wastes through a series of organic chemical reactions. Through these methods, substances that could cause damage to humans or the environment can be rendered harmless. New substances created by microbial reactions may be suitable for reuse or recycling. Research in **genetic engineering**, though controversial, could lead to breakthroughs in biological treatment. In chemical treatment, materials are added to or removed from the hazardous waste to produce new, less hazardous chemicals. Chemical neutralization, for example, involves mixing a corrosive acid with carbonate lime or another high-pH material until it is no longer acidic.

Waste prevention

In the 1990s, government regulators and others recognized the strengths of waste prevention as a tool for managing hazardous wastes. Waste prevention means using smaller quantities of potentially harmful materials or products, or using materials that are less toxic. The obstacles encountered by CERCLA underscored a need to manage hazardous waste by preventing its creation in the first place. Waste prevention is less expensive than treatment or disposal because it does not require transportation, processing or cleanup. It also can save on product production costs because fewer resources are needed. Furthermore, much of the environmental and human health damage caused by hazardous waste contamination is irreversible. The chemical dump at Love Canal, for example, has been cleaned up, and the site is no longer contaminated, but there is no financial remedy for a person who lost his or her life to cancer, or lived with a deformity caused by the contamination.

Businesses can prevent hazardous waste problems in a number of ways: they can reuse hazardous chemicals, improve storage and transportation methods, substitute less dangerous chemicals for more dangerous one, re-

KEY TERMS

....

Superfund—A fund created by the U.S. Congress to help clean up hazardous waste dumpsites.

Toxic waste—A type of hazardous waste that is capable of killing or injuring living creatures.

Waste prevention—A waste management method that involves preventing waste from being created, or reducing waste.

design production methods to eliminate the need for hazardous materials, and improve record-keeping and labeling of materials. Prevention measures often carry a significant up-front expense, but such waste prevention projects usually pay for themselves. Sometimes the financial benefit of such an "ounce of prevention" takes years to become apparent, but often the gain is almost immediate. Exxon Corporation, for example, spent about $140,000 to redesign several chemical storage tanks. The improvement allowed the company to reduce its chemical use by 700,000 lb (318,000 kg), and to save more $200,000 a year. Lower disposal, treatment, and shipping costs aren't the only benefits to companies that instate waste prevention practices. More efficient record-keeping, reduced legal liability, safer employee work conditions, and improved public image all promote a business's economic success.

Industrialized society will always generate some hazardous waste. However, prevention has emerged as the key to environmentally and social responsible hazardous waste management. As the costs of hazardous waste treatment and disposal continue to rise, waste prevention makes even more economic sense. The combination of national and international waste clean-up efforts, safe handling regulations, and prevention measures is expected to reduce the present threat that hazardous wastes pose to human and environmental health in coming decades.

See also Bioremediation; Chlordane; Landfill.

Resources

Books

Miller, E. Willard, and Ruby Miller. *Environmental Hazards: Toxic Waste and Hazardous Material: A Reference Handbook*. Santa Barbara: ABC-CLIO, 1991.

Harte, John, et al. *Toxics A to Z: A Guide to Everyday Pollution Hazards*. Berkeley: University of California Press, 1991.

Mazmanian, Daniel, and David Morell. *Beyond Superfailure: America's Toxics Policy for the 1990s*. Boulder: Westview Press, 1992.

Page, G.W. *Contaminated Sites and Environmental Cleanup*. London: Academic Press, 1997.

Other

United Nations Environmental Programme. "Secretariat of the Basel Convention." August 1, 2002 [cited October 22, 2002]. <http://www.basel.int/>.

United States Environmental Protection Agency. "Hazardous Wastes." October 4, 2002 [cited October 22, 2002]. <http://www.epa.gov/epaoswer/osw/hazwaste.htm>.

Tom Watson
Laurie Duncan

Hazel

Hazels or filberts are shrub-sized woody plants in the **birch family (Betulaceae)** found in temperate **forests** of **North America** and Eurasia. Hazels have simple, coarse-toothed, hairy leaves that are deciduous in the autumn.

Hazel **species** native to North America include the American hazel (*Corylus americana*) of the east and beaked hazel (*C. cornuta*) of a wider distribution. The giant filbert (*C. maxima*) is a European species that is sometimes cultivated as an ornamental.

The nuts of all of the wild hazels can be gathered and eaten raw or roasted. The hazel or cobnut (*C. avellana*) of Eurasia is grown commercially in orchards for the production of its **fruits**. These nuts can be eaten directly, or their oil may be extracted for use in the manufacture of perfumes and oil-based paints.

Y-shaped, forked branches of various species of hazel have long been used to find underground **water** by a folk method known as dowsing, or water witching. The dowser walks slowly about holding two ends of the Y in his or her hands. The place where the free end of the dowsing rod is attracted mysteriously downwards is believed to be a good location to dig or drill a well. One of the common names of the American hazel is witch hazel and is presumably derived from the use of the species to find accessible **groundwater**.

HDTV *see* **Television**

Hearing

Hearing is the ability to collect, process and interpret sound. Sound vibrations travel through air, **water,** or solids in the form of **pressure** waves. When a sound wave hits a flexible object such as the eardrum it causes it to vibrate, which begins the process of hearing. The process of hearing involves the conversion of acoustical **energy** (**sound waves**) to mechanical, hydraulic, chemical, and finally electrical energy where the signal reaches the **brain** and is interpreted.

Sound

The basis of sound is simple: there is a vibrating source, a medium in which sound travels, and a receiver. For humans the most important sounds are those which carry meaning, for example **speech** and environmental sounds. Sounds can be described in two ways, by their **frequency** (or pitch), and by their intensity (or loudness).

Frequency (the number of vibrations or sound waves per second) is measured in Hertz (Hz). A sound that is 4,000 Hz (like the sound the letter "F" makes) has 4,000 waves per second. Healthy young adults can hear frequencies between 20 and 20,000 Hz. However, the frequencies most important for understanding speech are between 200 and 8,000 Hz. As adults age, the ability to hear high frequency sounds decreases. An example of a high frequency sound is a bird chirping, while a drum beating is a low frequency sound.

Intensity (loudness) is the amount of energy of a vibration, and is measured in decibels (dB). A **zero** decibel sound (like leaves rustling in the **wind**), can barely be heard by young healthy adults. In contrast, a 120 dB sound (like a **jet engine** at 7 m [20 ft]) is perceived as very loud and/or painful. Extremes in both loudness and/or pitch may seriously damage the human **ear** and should be avoided.

The difference between frequency (pitch) and intensity (loudness) can be illustrated using the piano as an analogy. The piano keyboard contains 88 keys which represent different frequencies (or notes). The low frequencies (**bass** notes) are on the left, the higher frequencies (treble notes) are on the right. Middle C on the keyboard represents approximately 256 Hz. The intensity or loudness of a note depends on how hard you hit the key. A light touch on middle C may produce a 30 dB, 256 Hz note, while a hard strike on middle C may produce a 55 dB, 256 Hz note. The frequency (or note) stays the same, but the intensity or loudness varies as the pressure on the key varies.

Animal hearing

The difference between hearing in humans and animals is often visible externally. For example some animals (e.g. **birds**) lack external ears/pinnas, but maintain similar internal structures to the human ear. Although birds have no pinnas they have middle ears and inner ears similar to humans, and like humans, hear best at the frequencies around 2,000 to 4,000 Hz. All **mammals** (the animals most closely related to the human) have outer ears/pinnas. Many mammals have the ability to move the pinna to help with localization of sounds. Foxes, for example, have large bowl shaped pinnas which can be moved to help locate distant or faint sounds. In addition to sound localization, some animals are able to manipulate their pinnas to regulate body **temperature**. Elephants do this by using their huge pinnas as fans and for **heat** exchange.

Human hearing

Human hearing involves a complicated process of energy conversion. This process begins with two ears located at opposite sides of the human head. The ability to use two ears for hearing is called binaural hearing. The primary advantages to binaural hearing are the increased ability to localize sounds and the increased ease of listening in background noise. Sound waves from the world around us enter the ear and are processed and relayed to the brain. The actual process of sound transmission differs in each of the three parts of the human ear (the outer, middle and inner ears).

Outer ear and hearing

The pinna of the outer ear gathers sound waves from the environment and transmits them through the external auditory **canal** and eardrum to the middle ear. In the process of collecting sounds, the outer ear also modifies the sound. The external ear, or pinna, in combination with the head, can slightly amplify (increase) or attenuate (decrease) certain frequencies. This amplification or attenuation is due to individual differences in the dimensions and contours of the head and pinna.

A second source of sound modification is the external auditory canal. The tube-like canal is able to amplify specific frequencies in the 3,000 Hz region. An analogy would be an opened, half filled soda bottle. When you blow into the bottle there is a sound, the frequency of which depends on the size of the bottle and the amount of **space** in the bottle. If you empty some of the fluid and blow into the bottle again the frequency of the sound will change. Since the size of the human ear canal is consistent the specific frequency it amplifies is also constant. Sound waves travel through the ear canal until they strike the tympanic **membrane** (the eardrum). Together, the head, pinna and external auditory canal amplify sounds in the 2,000 to 4,000 Hz range by 10-15 dB. This boost is needed since the process of transmitting sound from the outer ear to the middle ear requires added energy.

Middle ear and hearing

The tympanic membrane or eardrum separates the outer ear from the middle ear. It vibrates in response to

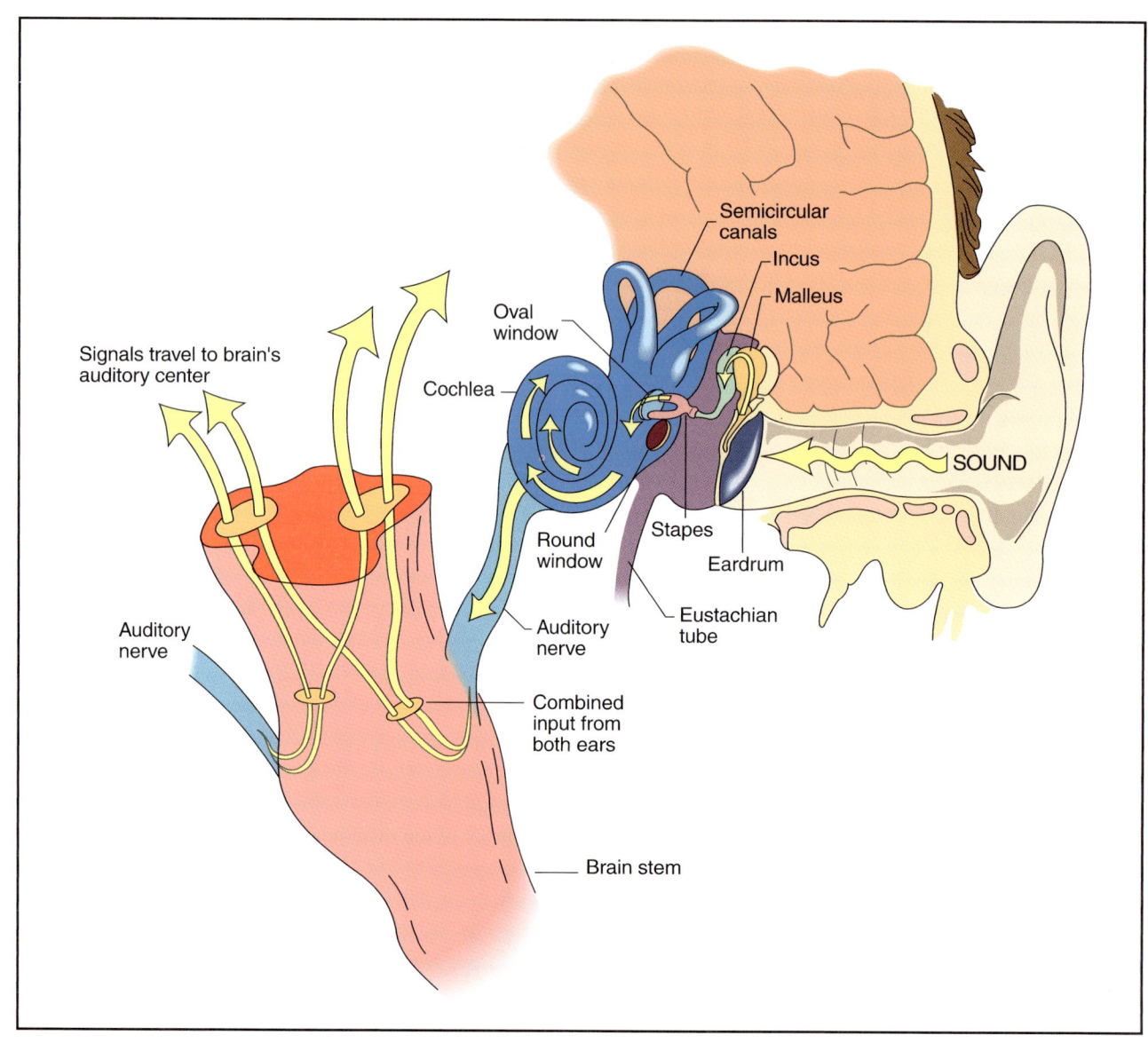

The hearing process. *Illustration by Hans & Cassidy. Courtesy of Gale Group.*

pressure from sound waves traveling through the external auditory canal. The initial vibration causes the membrane to be displaced (pushed) inward by an amount equal to the intensity of the sound, so that loud sounds push the eardrum more than soft sounds. Once the eardrum is pushed inwards, the pressure within the middle ear causes the eardrum to be pulled outward, setting up a back-and-forth **motion** which begins the conversion and transmission of acoustical energy (sound waves) to mechanical energy (bone movement).

The small connected bones of the middle ear (the ossicles—malleus, incus, and stapes) move as a unit, in a type of lever-like action. The first bone, the malleus, is attached to the tympanic membrane, and the back-and-

forth motion of the tympanic membrane sets all three bones in motion. The final result of this bone movement is pressure of the footplate of the last (smallest) bone (the stapes), on the oval window. The oval window is one of two small membranes which allow communication between the middle ear and the inner ear. The lever-like action of the bones amplifies the mechanical energy from the eardrum to the oval window. The energy in the middle ear is also amplified due to the difference in surface size between the tympanic membrane and the oval window, which has been calculated at 14 to 1. The large head of a thumbtack collects and applies pressure and focuses it on the pin point, driving it into the surface. The eardrum is like the head of the thumb tack and the oval window is the pin point. The overall amplification in the middle ear is approximately 25 dB. The conversion from mechanical energy (bone movement) to hydraulic energy (fluid movement) requires added energy since sound does not travel easily through fluids. We know this from trying to hear under water.

Inner ear and hearing

The inner ear is the site where hydraulic energy (fluid movement) is converted to chemical energy (hair **cell** activity) and finally to electrical energy (nerve transmission). Once the signal is transmitted to the nerve, it will travel up to the brain to be interpreted.

The bone movements in the middle ear cause movement of the stapes footplate in the membrane of the oval window. This pressure causes fluid waves (hydraulic energy) throughout the entire two and a half turns of the cochlea. The design of the cochlea allows for very little fluid movement, therefore the pressure at the oval window is released by the interaction between the oval and round windows. When the oval window is pushed forward by the stapes footplate, the round window bulges outward and vice versa. This action permits the fluid **wave motion** in the cochlea. The cochlea is the fluid filled, snail shell-shaped coiled **organ** in the inner ear which contains the actual sense receptors for hearing. The fluid motion causes a corresponding, but not equal, wave-like motion of the basilar membrane. Internally, the cochlea consists of three fluid filled chambers: the scola vestibuli, the scola tympani, and the scala media. The basilar membrane is located in the scala media portion of the cochlea, and separates the scala media from the scala tympani. The basilar membrane holds the key structure for hearing, the organ of Corti. The physical characteristics of the basilar membrane are important, as is its wave-like movement, from base (originating point) to apex (tip). The basilar wave motion slowly builds to a peak and then quickly dies out. The distance the wave takes to reach the peak depends on the speed at which the oval window is moved. For example, high frequency sounds have short wavelengths, causing rapid movements of the oval window, and peak movements on the basilar membrane near the base of the cochlea. In contrast, low frequency sounds have long wavelengths, cause slower movements of the oval window, and peak movements of the basilar membrane near the apex. The place of the peak membrane movements corresponds to the frequency of the sound. Sounds can be "mapped" (or located) on the basilar membrane; high frequency sounds are near the base, middle frequency sounds are in the middle, and low frequency sounds are near the apex. In addition to the location on the basilar membrane, the frequency of sounds can be identified based on the number of nerve impulses sent to the brain.

The organ of Corti lies upon the basilar membrane and contains three to five outer rows (12,000 to 15,000 hair cells) and one inner row (3,000) of hair cells. The influence of the inner and outer hair cells has been widely researched. The common view is that the numerous outer hair cells respond to low intensity sounds (quiet sounds, below 60 dB). The inner hair cells act as a booster, by responding to high intensity, louder sounds. When the basilar membrane moves, it causes the small hairs on the top of the hair cells (called stereocilia) to bend against the overhanging tectorial membrane. The bending of the hair cells causes chemical actions within the cell itself creating electrical impulses (action potentials) in the nerve fibers attached to the bottom of the hair cells. The nerve impulses travel up the nerve to the temporal lobe of the brain. The intensity of a sound can be identified based on the number of hair cells affected and the number of impulses sent to the brain. Loud sounds cause a large number of hair cells to be moved, and many nerve impulses to be transmitted to the brain.

Each of the separate nerve fibers join and travel to the lowest portion of the brain, the brainstem. Nerves from the vestibular part (balance part) of the inner ear combine with the cochlear nerves to form the VIII cranial nerve (auditory or vestibulocochlear nerve). Once the nerve impulses enter the brainstem, they follow an established pathway, known as the auditory pathway. The organization within the auditory pathway allows for a large amount of cross-over. "Cross-over" means that the sound information (nerve impulses) from one ear do not travel exclusively to one side of the brain. Some of the nerve impulses cross-over to the opposite side of the brain. The impulses travel on both sides (bilaterally) up the auditory pathway until they reach a specific point in the temporal lobe called Heschl's gyrus. Crossovers act like a safety net. If one side of the auditory pathway is blocked or damaged, the impulses can still reach Heschl's gyrus to be interpreted as sound.

See also Neuron.

Resources

Books

Mango, Karin. *Hearing Loss.* New York: Franklin Watts, 1991.

Martin, Frederick. *Introduction to Audiology.* 6th ed. Boston: Allyn and Bacon, 1997.

Moller, Aage R. *Sensory Systems: Anatomy and Physiology.* New York: Academic Press, 2002.

Rahn, Joan. *Ears, Hearing and Balance.* New York: Antheneum, 1984.

Simko, Carole. *Wired for Sound.* Washington, DC: Kendall Green Publications, 1986.

Sundstrom, Susan. *Understanding Hearing Loss and What Can Be Done.* Illinois: Interstate Publishers, 1983.

Periodicals

Mestel, Rosie. "Pinna To the Fore." *Discover* 14 (June, 1993): 45-54.

Kathryn Glynn

Hearing disorders *see* Deafness and inherited hearing loss

Heart

A heart is a means to circulate **blood** through the body of an **animal**. Among the lower **species** such as **insects**, **arachnids**, and others, the heart may simply be an expanded area in a blood vessel and may occur a number of times. The earthworm, for example, has 10 such "hearts." These areas contract rhythmically to force the blood through the aorta, or blood vessel.

Not until the **evolution** of the higher **vertebrates** does the heart achieve its ultimate form, that of a chambered **organ** with differentiated purposes. Even the lower **chordates**, such as amphioxus, possess hearts not more advanced than those in the earthworm. It is simply a pulsating blood vessel that moves blood through the body.

Blood

In a complex **organism** such as a vertebrate, with multiple **cell** layers and complex organ systems, blood serves to distribute nourishment and **oxygen** to the cells and remove waste products. Specialized cells within the blood, such as the red blood cells (erythrocytes) and white blood cells (leukocytes) serve specialized functions. The red cells hold and distribute oxygen to release in the cells and return **carbon dioxide** to the lungs to be eliminated. The white cells carry out immune functions to destroy invading **bacteria** and other foreign material. Still other components of the blood are involved in forming clots when a blood vessel is opened. The liquid medium, the **plasma**, carries vitamins and other **nutrients** throughout the body.

Animals require a means for the blood to obtain oxygen, whether through gills or lungs, and a means to propel the blood through those structures. The heart is central to that purpose.

The multiform heart

The heart is a pulsating organ that pushes a liquid medium throughout the body. It may be as simple as one chamber or as complex as four chambers, as in the higher **mammals**. In all animals, however, it is an organ that must function day after day without pause to keep the blood moving.

In general, blood that returns from the body or from the oxygen exchanging structures returns to an atrium, which is simply a holding chamber. The atrium (plural is atria) empties into another chamber called the ventricle, a muscular chamber that contracts rhythmically to propel the blood through the body. Movement of the blood between chambers and in and out of the heart is controlled by valves that allow movement only in one direction.

The lower vertebrates such as the **hagfish** and other **fish** have two-chambered hearts. The ventricle pumps blood forward through the gills to obtain oxygen and dispose of carbon dioxide. From there the blood enters the dorsal aorta and is carried through the body. The blood returns to the heart by means of the sinus venosus, which empties into the auricle or atrium. From there it is passed into the ventricle and the cycle begins again.

Terrestrial vertebrates such as **amphibians**, have three-chambered hearts and a more complex **circulatory system**. The third chamber is another auricle or atrium. Unoxygenated blood from the body of the animal returns to the right auricle and the oxygenated blood from the lungs goes into the left auricle. Both auricles contract simultaneously and empty their contents into the single ventricle. The oxygenated and unoxygenated blood does not mix to any degree because of specialized muscle strands in the ventricle. Also, the ventricle contracts immediately after the auricles so the bloods do not have time to mix. When the ventricle contracts, the unoxygenated blood is forced from the heart first and enters the pulmocutaneous vessels leading to the lungs and skin for oxygen exchange. The oxygenated blood enters the truncus arteriosus and flows to the head, arms, body, and hind legs.

Reptiles demonstrate a further step in heart development—a divided ventricle. The wall in the ventricle dividing left from right is incomplete except in **crocodiles** and alligators. Blood from the body enters the right auricle from which it passes into the right side of the ventricle and is pumped through the lungs. Oxygenated blood from the lungs enters the left auricle and passes into the left side of the ventricle. From there it moves out through the two aortic arches to the body.

Among the crocodilia, however, a peculiar structure of the aortae allows mixing of arterial and venous blood. The left aortic arch rises from the right side of the heart, as does the pulmonary artery. Thus, unoxygenated blood pumped from the right ventricle goes to the lungs through the pulmonary artery and also into the aortic arch and out into the body where it will mix with oxygenated blood. The right aortic arch rises from the left ventricle and carries only oxygenated blood.

Birds also possess four-chambered hearts, in this case with complete separation of the two ventricles. They have only one aortic arch, however, and that is the right arch.

Mammals, including humans, also have a four-chambered heart, but the aortic arch curves up and to the left as it leaves the heart. Here, as with birds, there is no mixing of venous and arterial blood under normal circumstances. Though the heart appears to be a simple organ, it requires a complex series of nerve stimulations, valve openings, and muscle contractions to adequately achieve its purpose.

The human heart

Located in the thoracic cavity, the heart is a four-chambered muscular organ that serves as the primary pump or driving force within the circulatory system. The heart contains a special form of muscle, appropriately named cardiac muscle, that has intrinsic contractility (i.e., is able to beat on its own, without **nervous system** control).

The Chinese were aware more than 2,000 years ago that the heart is a pump that forces blood through a maze of **arteries**. The Greeks, however, believed the blood did

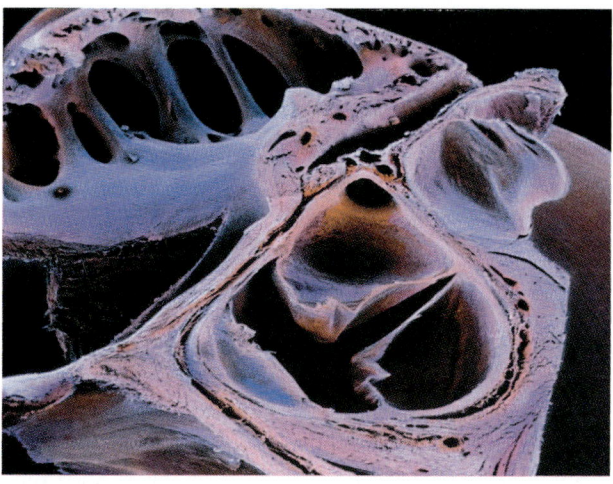

A scanning electron micrograph (SEM) of the aortic valve. It lies between the left ventricle of the heart and the aortic arch and consists of a fibrous ring with three semilunar pockets attached to it. *Photograph by Prof. P. Motta/G. Macchiarelli University "La Sapienza," Rome/Science Photo Library, National Audubon Society Collection/ Photo Researchers, Inc. Reproduced by permission.*

not circulate at all, and their ideas dominated medical science until the seventeenth century. The Greek physician Galen published a great deal on the human body and its functions, much of which was incorrect, but his doctrines held sway for hundreds of years. Not until the 1600s did William Harvey, through human and animal experiments, discover that the heart circulates the blood. He published his findings in 1628, thus bringing Western medical science in line with that of the ancient Chinese.

The human heart on the average weighs about 10.5 oz (300 g). It is a four-chambered, cone-shaped organ about the size of a closed fist that lies in the mid-thorax, under the breastbone (sternum). Nestled between the lungs, the heart is covered by a fibrous sac called the pericardium. This important organ is protected within a bony cage formed by the ribs, sternum, and spine.

In its ceaseless work, the heart contracts some 100,000 times a day to drive blood through about 60,000 mi (96,000 km) of vessels to nourish each of the trillions of cells in the body. Each contraction of the ventricles forces about 2.5 oz (0.075 l) of blood into the circulation, which adds up to about 10 pt (4.7 l) of blood every minute. On average, the heart will pump about 2,500 gal (9,475 l) of blood in a day, and that may go up to as much as 5,000 gal (18,950 l) with exertion. In a lifetime the heart will pump about 100 million gal of blood.

The chambers of the human heart are divided into two upper (superior) atrial chambers and two thicker-walled, heavily muscular inferior ventricular chambers.

The right and left sides of the heart are divided by a thick septum. The right side of the heart is on the same side of the heart as is the right arm of the patient. The atrial and ventricular chambers on each side of the septum constitute separate collection and pumping systems for the pulmonary (right side) and systemic circulation (left side). The coronary sulcus or grove separates the atria from the ventricles. The left and right side atrial and ventricular chambers each are separated by a series of one way valves that, when properly functioning, allow blood to move in one direction, but prohibit it from regurgitating (flowing back through the valve).

Deoxygenated blood—returned to the heart from the systemic circulatory venous system—enters the right atrium of the heart through the superior and inferior vena cava. Auricles lie on each atrium and are most visible when the atria are drained and deflated. The auricles (so named because they resembled **ear** flaps) allow for greater atrial expansion. Pectinate muscles on the auricles assist with atrial contraction. Small contractions within the right atrium, and **pressure** differences caused by evacuation of blood in the lower (inferior) right ventricle, cause this deoxygenated blood to move through the tricuspid valve during diastole (the portion of the heart's contractile cycle between contractions, and a period of lower pressure as compared to systole) into the right ventricle. When the heart contracts, a sweeping wave of pressure forces open the pulmonic semilunar valve that allows blood to rush from the right ventricle into the pulmonary artery where it is travels to the lungs for oxygenation and other gaseous exchanges.

Freshly oxygenated blood returns to the heart from the pulmonary circulation through the pulmonary vein the empties into the left atrium. During diastole, the oxygenated blood moves from the left atrium into the left ventricle through the mitral valve. During systolic contraction, the oxygenated blood is pumped under high pressure through the semilunar aortic valve into the aorta and thus, enters the systemic circulatory system.

As the **volume** and pressure rise during the filling of the right and left ventricles, the increased pressure snaps shut the flaps of the atrioventricular valves (tricuspid and mitral valves) anchored by fibrous connection to the left and right ventricles. The pressure in the ventricles seals the valves and as the pressure increases during systole, the valves seal becomes further compressed. A prolapse in one of the valves (a pushing through of one of the cusps) leads to blood flow back through the valve. The cusps are held against prolapse by the chordae tendineae, thin cords that attach the cusps to papillary muscles.

The heart and great vessels attached to it are encased within a multi-layered pericardium. The outer layer is fibrous and covers a double membraned inner sac-like structure termed the pericardial cavity that is filled with pericardial fluid. The pericardial fluid acts to reduce friction between the heart, the pericardial membranes, and the thoracic wall as the heart contracts and expands during the **cardiac cycle**.

The heart muscle is composed of three distinct layers. The outermost layer, the outer epicardium, is separated from the inner endocardium by the middle pericardium. The outer epicardium is continuous and in some places the same as the visceral pericardium. Epicardium protects the heart and is invested with **capillaries**, nerves, and lymph vessels. The middle myocardium is a think layer of cardiac muscle. The innermost endocardium contains **connective tissue** and Purkinje fibers. The endocardium is continuous with the lining of the great vessels attached to the heart and it lines all valve and cardiac inner surfaces.

Heart muscle does not directly take up oxygen from the blood it pumps. A specialized set of vessels (e.g., the left and right coronary arteries and their branches) supply oxygenated blood to the heart muscle and constitute the coronary circulation. A heart attack occurs whenever blood flow is occluded (blocked).

The **fossa** ovalis is a remnant or the embryonic foramen ovale that allows blood to flow between the left and right atria in the developing fetus.

Regulation of the heart

Various intrinsic, neural, and hormonal factors act to influence the rhythm control and impulse conduction within the heart. The rhythmic control of the cardiac cycle and its accompanying heartbeat relies on the regulation of impulses generated and conducted within the heart. Regulation of the cardiac cycle is also achieved via the autonomic nervous system. The sympathetic and parasympathetic divisions of the autonomic system regulate heart rhythm by affecting the same intrinsic impulse conducting mechanisms that lie within the heart in opposing ways.

Cardiac muscle is self-contractile because it is capable of generating a spontaneous electrochemical signal as it contracts. This signal induces surrounding cardiac muscle **tissue** to contract and a wave-like contraction of the heart can result from the initial contraction of a few localized cardiac cells.

The cardiac cycle describes the normal rhythmic series of cardiac muscular contractions. The cardiac cycle can be subdivided into the systolic and diastolic phases. Systole occurs when the ventricles of the heart contract and diastole occurs between ventricular contractions when the right and left ventricles relax and fill. The sino-

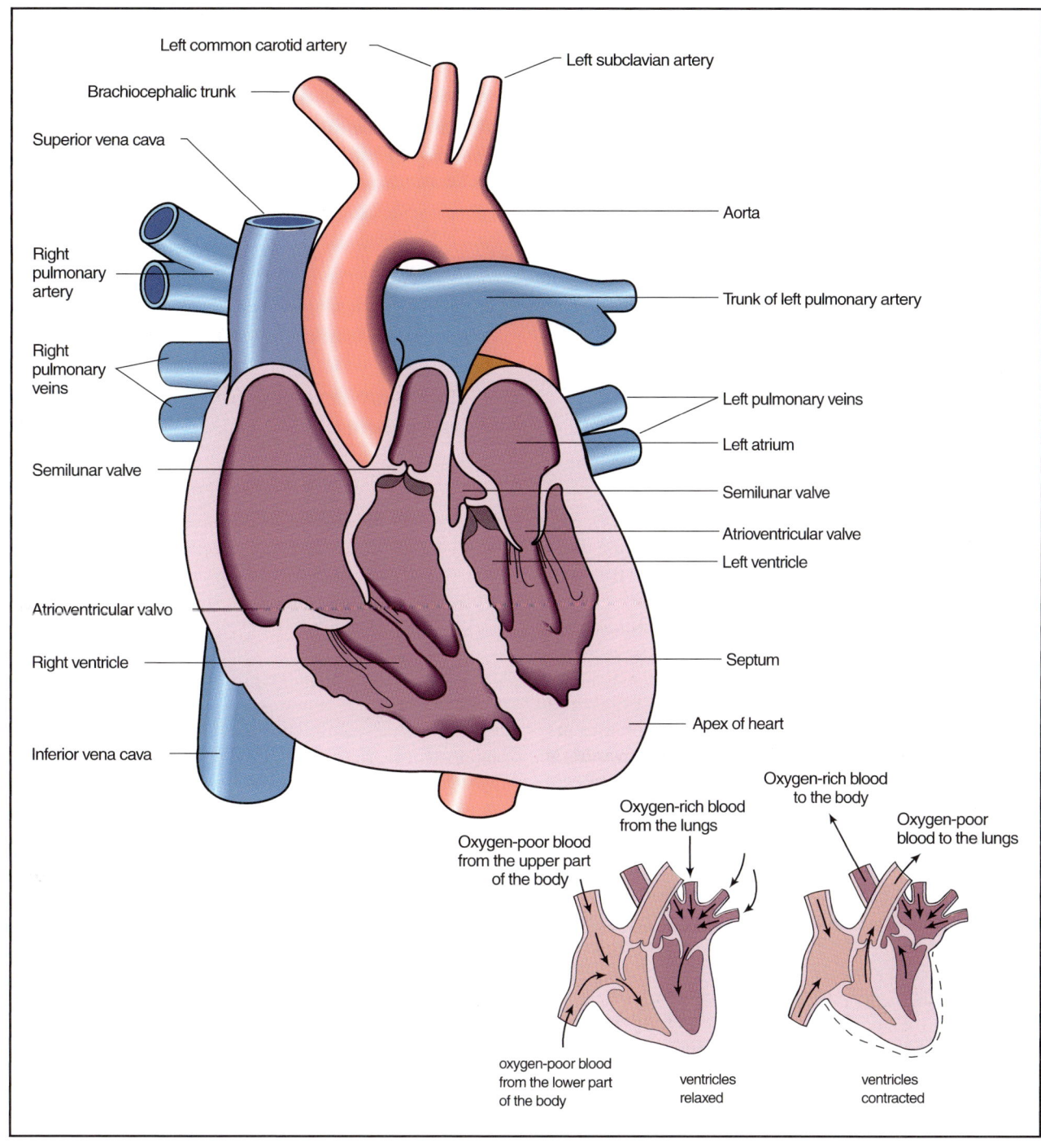

A cutaway view of the anatomy of the heart. *Illustration by Hans & Cassidy. Courtesy of Gale Group.*

atrial node (S-A node) and atrioventricular node (AV node) of the heart act as pacemakers of the cardiac cycle.

The contractile systolic phase begins with a localized contraction of specialized cardiac muscle fibers within the sino-atrial node. The S-A node is composed of nodal tissue that contains a mixture of muscle and neural cell properties. The contraction of these fibers generates an electrical signal that then propagates throughout the surrounding cardiac muscle tissue. In a contractile wave originating at the S-A node, the right atrium muscle contracts (forcing blood into the right ventricle) and then the left atrium contracts (forcing blood into the left ventricle).

Intrinsic regulation is achieved by delaying the contractile signal at the atrioventricular node. This delay also allows the complete contraction of the atria so that the ventricles receive the minimum amount of blood to make their own contractions efficient. A specialized type of neuro-muscular cells, named Purkinje cells, form a system of fibers that covers the heart and which conveys the contractile signal from S-A node (which is also a part of the Purkinje system or subendocardial plexus). Because the Purkinje fibers are slower in passing electrical signals (action potentials) than are neural fibers, the delay allows the atria to finish their contractions prior to ventricular contractions. The signal delay by the AV node lasts about a tenth (0.1) of a second.

The contractile signal then continues to spread across the ventricles via the Purkinje system. The signal travels away from the AV node via the bundle of His before it divides into left and right bundle branches that travel down their respective ventricles.

Extrinsic control of the heart **rate** and rhythm is achieved via autonomic nervous system (ANS) impulses (regulated by the medulla oblongata) and specific **hormones** that alter the contractile and or conductive properties of heart muscle. ANS sympathetic stimulation via the cervical sympathetic chain ganglia acts to increase heart rate and increase the force of atrial and ventricular contractions. In contrast, parasympathetic stimulation via the vagal nerve slows the heart rate and decreases the vigor of atrial and ventricular contractions. Sympathetic stimulation also increases the conduction **velocity** of cardiac muscle fibers. Parasympathetic stimulation decreases conduction velocity.

The regulation in impulse conduction results from the fact that parasympathetic fibers utilize **acetylcholine**, a **neurotransmitter** hormone that alters the transmission of an **action potential** by altering **membrane** permeability to specific ions (e.g., potassium ions [K^+]). In contrast, sympathetic postganglionic neurons secrete the neurotransmitter norepinephrine that alters membrane permeability to **sodium** (Na^+) and **calcium** ions (Ca^{2+}).

The ion permeability changes result in parasympathetic induced hypopolarization and sympathetic induced hyperpolarization.

Additional hormonal control is achieved principally by the adrenal **glands** (specifically the adrenal medulla) that release both epinephrine and norepinephrine into the blood when stimulated by the sympathetic nervous system. As part of the fight or flight **reflex**, these hormones increase heart rate and the volume of blood ejected during the cardiac cycle.

The electrical events associated with the cardiac cycle are measured with an electrocardiogram (EKG).

Disruptions in the impulse conduction system of the heart result in arrhythmias.

Variations in the electrical system can lead to serious, even dangerous, consequences. When that occurs an artificial electrical stimulator, called a **pacemaker**, must be implanted to take over regulation of the heartbeat. The small pacemaker can be implanted under the skin near the shoulder and long wires from it are fed into the heart and implanted in the heart muscle. The pacemaker can be regulated for the number of heartbeats it will stimulate per minute. Newer pacemakers can detect the need for increased heart rate when the individual is under exertion or **stress** and will respond.

Embryonic development of the human heart

The developing fetal heart accounts for a large percentage of the volume of the early thorax. About 20 days after **fertilization**, the heart develops from the fusion of paired endothelial tubes into a single tube. Heart growth subsequently involves the growth, expansion, and partitioning of this tube into four chambers separated by thickened septa of cardiac muscle and valves. Atrial development is initially more advanced than ventricular development. The left and right atria develop while the primitive ventricle remains a single chamber. As atrial separation nears completion, the left and right ventricles begin to form, then continue until the heart consists of its fully developed four-chambered structure.

Although the majority of the heart develops from mesoderm (splanchnic mesoderm) near the neural plate and sides of the embryonic disk, there are also contributions from neural crest cells that help form the valves.

Three systems initially return venous blood to the primitive heart. Regardless of the source, this venous blood returns to sinus venosus. Vitelline **veins** return blood from the yolk sac; umbilical veins return oxygenated blood from the placenta. The left umbilical vein enlarges and passes through the embryonic liver before continuing on to become the inferior vena cava that fuses with a common chambered sinus venosus and the right atrium of the heart. Especially early in development, venous return also comes via the cardinal system. The anterior cardinals drain venous blood from the developing head region. Subcardinal veins return venous blood from the developing renal and urogenital system, while supracardinals drain the developing body wall. The anterior veins empty into the common cardinals that terminate in the sinus venosus.

Movement of blood through the early embryonic vascular system begins as soon as the primitive heart tubes form and fuse. Contractions of the primitive heart begin early in development, as early as the initial fusion of the endothelial channels that fuse to form the heart.

The heart and the atrial tube that form the aorta develop by the compartmentalization of the primitive cardiac tube. Six separate septae are responsible for the portioning of the heart and the development of the walls of the atria and ventricles. A septum primum divides the primitive atria into left and right chambers. The septum secundum (second septum) grows along the same course of the primary septum to add thickness and strength to the partition. There are two holes in these septae through which blood passes, the foramen secundum and the foramen ovale. Specialized endocardinal tissue develops into the atrioventricular septum that separates the atrium and ventricles. The mitral and tricuspid valves also develop from the atrioventricular septum.

As development proceeds, the interventricular septum becomes large and muscular to separate the ventricles and provide strength to these high-pressure contractile chambers. The interventricular septum also has a membranous portion.

Initially, there is only a common truncus arteriosus as a channel for ventricular output. The truncus eventually separates into the pulmonary trunk and the ascending aorta.

Blood oxygenated in the placenta returns to the heart via the inferior vena cava into the right atrium. A valve-like flap in the wall at the juncture of the inferior vena cava and the right atrium directs the majority of the flow of oxygenated blood through the foramen ovale, then allows blood to flow from the right atrium to the left. Although there is some mixing with blood from the superior vena cava, the directed flow of oxygenated blood across the right atrium caused by the valve of the inferior vena cava means that deoxygenated fetal blood returning via the superior vena cava still ends up moving into the right ventricle.

While in the uterus, the lungs are non-functional. Accordingly, another shunt, the ductus arteriosis (also spelled ductus arteriosus) provides a diversionary channel that allows fetal blood to cross between the pulmonary artery and aorta and thus largely bypass the rudimentary pulmonary system.

Because only a small amount of blood returns from the pulmonary circulation, almost all of the blood in the fetal left atrium comes through the foramen ovale. The relatively oxygen-rich blood then passes through the mitral value into the left ventricle. Contractions of the heart, whether in the single primitive ventricle or from the more developed left ventricle, then pump this oxygenated blood into the fetal systemic arterial system.

In response to inflation of the lungs and pressure changes within the pulmonary system, both the foramen ovale and the ductus arteriosis normally close at **birth** to establish the normal adult circulatory pattern whereby blood flows into the right atrium, though the tricuspid valve into the right ventricle. The right atrium pumps blood into the pulmonary artery and pulmonary circulation for oxygenation in the lungs. Oxygenated blood returns to the left atrium by pulmonary veins. After collecting in the left atrium, blood flows through the mitral value into the left atrium where it is then pumped into the systemic circulation via the ascending aorta.

See also Angiography; Artificial heart and heart valve; Heart diseases; Heart-lung machine; Thoracic surgery; Transplant, surgical.

Resources

Books

Gilbert, Scott F. *Developmental Biology.* 6th ed. Sunderland, Massachusetts: Sinauer Associates, Inc., 2000.

Gray, Henry. *Gray's Anatomy.* Philadelphia: Lea & Febiger, 1992.

Guyton, Arthur C., and Hall, John E. *Textbook of Medical Physiology.* 10th ed. Philadelphia. W.B. Saunders Co., 2000.

Kandel, E.R., J.H. Schwartz, and T.M. Jessell. eds. *Principles of Neural Science.* 4th ed. New York: Elsevier, 2000.

Larsen, William J. *Human Embryology.* 3rd. ed. Philadelphia: Elsevier Science, 2001.

Martini, Frederic H., et al. *Fundamentals of Human Anatomy & Physiology.* Upper Saddle River, NJ: Prentice Hall, 2001.

Netter, Frank H., and Sharon Colacino. *Atlas of the Human Body.* Teterboro, NJ: Icon Learning Systems, 2003.

Sadler, T.W., and Jan Langman. *Langman's Medical Embryology.* 8th ed. New York: Lippincott Williams & Wilkins Publishers, 2000.

Thibodeau, Gary A., and Kevin T Patton . *Anatomy & Physiology,* 5th ed. St. Louis: Mosby, 2002.

Other

Intellimed, Inc. "Human Anatomy Online-Innerbody.com" [cited February, 5, 2003]. <http://www.innerbody.com/htm/body.html> .

Klabunde, R.E. "Cardiac Cycle." Cardiovascular Physiology Concepts. January 17, 2003 [cited January 22, 2003]. <http://www.cvphysiology.com/Heart%20Disease/HD00.htm>.

Murray Jensen College, University Of Minnesota. "Web Anatomy" [cited February 5, 2003]. <http://www.gen.umn.edu/faculty_staff/jensen/1135/webanatomy/>.

Brenda Wilmoth Lerner
K. Lee Lerner
Larry Blaser

Heart attack *see* **Heart**

Heart diseases

Heart diseases (cardiovascular **disease**) is any abnormal organic condition of the heart or the heart and circulation. A number of conditions can lead to the development of heart disease, including angina, atherosclerosis, cardiac arrhythmia, cardiomyopathy, chronic venous insufficiency, diabetes, heart attack, high **cholesterol**, high homocysteine, high **triglycerides**, **hypertension**, **insulin** resistance **syndrome**, mitral valve prolapse, and **stroke**.

Coronary artery disease (CAD), which involves atherosclerosis (hardening of the **arteries**) that supply the heart with **blood** is the most common cause of heart attacks and is a leading killer in the United States. The primary risk factors for CAD are diabetes, male gender, family history of coronary disease at an early age, smoking, elevated blood **pressure** (hypertension), high LDL cholesterol, and low HDL cholesterol. The control of diabetes and blood pressure has resulted in a small benefit in preventing heart attacks. Proper ranges of cholesterol are effective in the prevention of heart attack or stroke. Total blood cholesterol above 200 mg/dl, LDL cholesterol above 130 mg/dl, HDL cholesterol below 35 mg/dl; and lipoprotein(a) level greater than 30 mg/dl are indicators of problematic cholesterol. Cholesterol is not actually a damage mechanism but is more an indicator of compromised liver function, and increased risk of heart attack. These factors mentioned above, however, do not fully account for all of the risks for heart disease since some patients without any of the above risk factors can develop heart attacks.

Throughout history, diseases of the heart have captured the concern and interest of investigators. Ancient Greek and Roman physicians observed the serious and often fatal consequences of heart disease. But effective treatment for heart disease was limited to rest and painkillers until the eighteenth-century discovery of the therapeutic properties of the foxglove **plant**, whose dried **leaf** is still used to make the medicine **digitalis**.

While the heart was once considered a part of the body that could never be improved surgically, the twentieth century has seen a revolution in surgical treatment for heart disease. Blocked coronary arteries can be bypassed using new **tissue** and failing hearts can be transplanted. Yet heart disease remains the primary cause of death in the United States. Preventive health measures, such as improved diet and regular **exercise**, have become fundamental tools in the battle against heart disease.

Early knowledge

Early man knew that the heart was important and powerful. As early as 1550 B.C., a passage in the so-called Ebers Papyrus of the ancient Egyptians reported that evidence of pains in the arm and the breast on the side of the heart suggested that death was approaching. Suggested treatment for such problems included beer taken with herbs.

Evidence of heart disease is also present in mummies. A. R. Long described in 1931 the condition of the fragile heart he found in a mummy dating from approximately 1,000 B.C. Further research found evidence of scarring of the heart muscle and of endocarditis, an **inflammation**, on the mitral valve.

Observational knowledge about the heart increased with the flowering of ancient Greece. According to P. E. Baldry, the earliest description of the **circulatory system** was developed in 500 B.C. by Alcmaeon, a pupil of the mathematician and scientist Pythagoras. Alcmaeon wrote that the breath, or the spirit, was sent around the body by blood vessels. Hippocrates (460-375 B.C.) and his students made many important observations about the heart. They noted that sharp pains irradiating towards the breast bone and the back were fatal; that those who were **fat** were more likely to die than those who were thin; and that those in **pain** should rest immediately. In the second century A.D., another Greek author, Aretaeus, described various ways to treat heart pain while it was occurring, including the offering of wine, the bleeding of the patient, and the encouragement of the physician.

Greek knowledge of the heart was limited by the general prohibition on human dissection. The dissection of humans was allowed in the ancient culture in Alexandria, however, where it enabled such advances as a detailed study of the way the blood vessels worked and the **rate** of the arterial pulse, conducted by Herophilus in 300 B.C. But the practice of human dissection was prohibited by the ancient Romans and throughout the medieval era in **Europe**.

The heart held special fascination for Galen (A.D. 130-200), a Greek who practiced medicine in Rome. Galen's extensive writing about the way the heart worked was respected throughout the Middle Ages. Through clinical practice with humans and careful observation of dissected animals, Galen observed that lungs were responsible for expelling waste material and that the heart was responsible for the pulse. He was known for the observation that a young woman's pulse quickened when the name of the man she loved was spoken.

Some of Galen's statements about the heart were wrong. He erroneously believed that some vessels could be used for blood flowing in two directions, and, while he knew there were chambers in the heart, he believed there were invisible pores in the tissue separating the right and left ventricles through which blood could flow.

Galen wrote that these pores enabled blood to mix with air in the left ventricle. Due to the popularity of his work, these errors were passed down for centuries.

The Middle Ages

The emphasis in medieval Europe on suffering as an experience of spiritual growth did not bode well for research concerning the heart. However, advances were made by Arab scholars, whose culture encouraged scholarly research. The medical writing of Ibn Sina, known as Avicenna (980-1037), included a rich sampling of astute observations about heart disease. This book was translated widely in the East and West and was highly influential for centuries. Although he repeated some of Galen's errors about the heart, Avicenna also distinguished between many types of heart disease, including those caused by a wound or **abscess**, those caused by collapse of the heart, and those caused by an obstruction in the heart. In the thirteenth century, Ibn Nafis challenged some of Galen and Avicenna's incorrect assumptions about the heart, particularly Galen's belief that invisible pores allowed blood to pass between the left and right ventricles. Nafis correctly believed that blood was mixed with air in the lungs.

Medieval healers and magicians had many cures for heart pain, deriving from scholarship and folk medicine as well as quackery. One popular cure called for serving the individual a radish with **salt** while he sat in a vapor bath. While folk remedies may not have been effective by current standards, they may have provided patients with a sense of calm and well-being, still considered to be of value in the battle against heart disease.

The artful heart

As the prohibition against human dissection was abandoned in the Renaissance era, knowledge of **anatomy** and the heart grew significantly. Fascination with the heart led the artist and inventor Leonardo da Vinci (1452-1519) to create models and numerous finely detailed drawings of the **organ**. Da Vinci was one of many Renaissance artists who used dissection of the dead as a tool in the understanding of human life. His drawings clearly show the way the heart works as a pump and document the changes of aging blood vessels. But Da Vinci's drawings had little influence on contemporary medicine because they were held privately and were not seen by many physicians.

Andreas Vesalius (1514-1564) was far more influential among healers. His classic **physiology** text featured the first accurate descriptions and drawings of the heart to be publicized. Vesalius also challenged Galen's "hidden passage" idea, arguing that this was not possible given the physiology of the heart. His description of au-

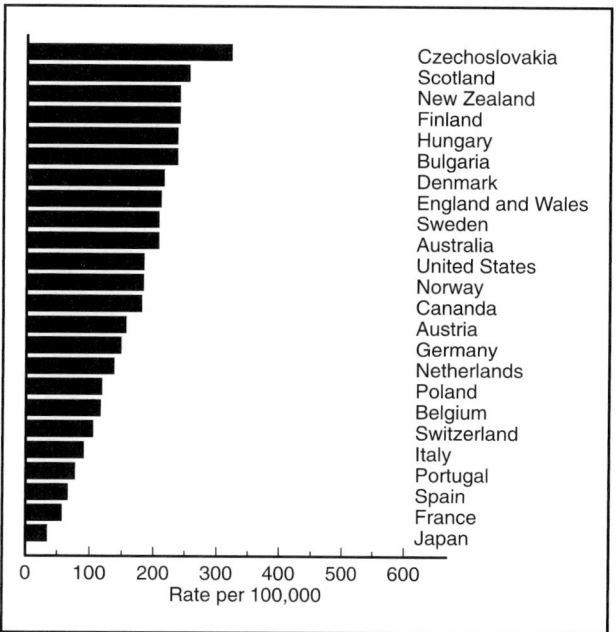

Age-standardized death rate attributed to ischemic heart disease (narrowing or blockage of the coronary arteries) for various countries. *Illustration by Hans & Cassidy. Courtesy of Gale Group.*

topsy reports revealed increasing understanding of the diversity of heart problems. One report described a huge mass of flesh in the left ventricle of a man's heart that weighed almost two pounds. Such an obstruction is known by contemporary physicians as a thrombus, a mass of blood tissue which can block blood vessels.

Another great influence in the development of knowledge about the heart was William Harvey (1578-1657), a physician whose findings about the blood and the circulatory system changed medicine profoundly. He was the first to show that blood traveled in a circle through the body. Harvey also understood the rhythmic nature of the heart's work. His work described the way blood was expelled from the heart with each contraction and entered the heart with every relaxation.

Though Harvey revealed the principles of blood circulation, he and his contemporaries did not understand the purpose of the lungs in the circulatory system. True understanding about the functioning of the lungs did not occur until the nineteenth century, when knowledge of **chemistry** advanced and researchers gained knowledge about the lungs' role in oxygenating blood for the tissues of the body.

Explosion of knowledge

Researchers in the eighteenth and nineteenth centuries developed effective treatment for some types of heart disease and greatly expanded their diagnostic knowledge. One important finding was the discovery that the purple foxglove plant contained a substance that was an effective medicine for some types of heart disease.

In 1775, William Withering (1741-1799), a British physician and botanist, was called to evaluate a folk remedy for dropsy, a serious condition involving an accumulation of fluid in the body that can affect the heart, the liver, and other organs. The remedy had 20 or more herbs, and Withering determined that foxglove was the active ingredient. Using his poor patients to test the remedy, Withering found the drug to be helpful in dropsy and in heart disease.

Contemporary physicians use foxglove, now called digitalis, to boost the strength of heart contractions and to lower the heart rate. It is often used in cases of congestive heart failure but can be used for other types of heart disease as well. Withering realized the potential danger in using too much foxglove and warned in 1785 that too much of the medicine could cause illness and death. Nevertheless, the medicine was used in excessive quantities by his contemporaries, leading to the death of patients and the eventual shunning of the medication. By the end of the eighteenth century, foxglove was no longer used widely for heart disease. The medication was reintroduced at the end of the nineteenth century, when its therapeutic properties were reassessed.

Other eighteenth century findings found more immediate acceptance, such as the discovery by Austrian Joseph Leopold Auenbrugger (1722-1809) that one could detect heart disease by tapping the chest in different places. A skilled examiner can use this technique (called percussion) to detect areas in the heart or lungs which have too much fluid.

Another advance which changed medicine was the invention of the stethoscope by René-Théophile-Hyacinthe Laënnec (1781-1826). Earlier physicians, such as Harvey, described the sounds of the heart. In 1816, Laënnec realized he could amplify those sounds. He first created a **paper** cylinder and then began using a wooden instrument. Laënnec used the stethoscope to expand knowledge about the heart, diagnosing narrowed valves and heart murmurs. The stethoscope enabled physicians to diagnose heart disease earlier in its course. The device also showed physicians that heart disease was not invariably fatal.

During the first half of the nineteenth century, physicians learned to distinguish between different types of heart murmur and the different types of valve damage they suggested. British physician James Hope conducted extensive experiments in which he used the poison **curare** to conduct **surgery** to examine the heart and other organs in animals. This experimentation led Hope to draw and describe, in 1839, two widely seen problems of the mitral valve, which is located between the left atrium and the left ventricle. Mitral valve incompetence occurs when the valve does not fit tightly, while mitral valve stenosis takes place when the valves do not open properly.

Another important finding about the heart was made in 1838, when Italian physicist Carlo Matteucci discovered that the heart muscle generates **electricity**. This electrical **force** enables the healthy heart to beat steadily and regulate its own activity. This finding cleared the way for electronic measurement of the heart, a technological advance which remains central to contemporary **diagnosis** of heart problems. The electrocardiogram was developed by William Einthoven (1860-1927), a German professor of physiology. Einthoven's electrocardiogram, which he first described in 1903, documented contraction and relaxation of different parts of the heart.

The death from heart disease of a young woman in labor motivated physician James Mackenzie (1835-1925) to begin an exhaustive study of heart disease early in his career. Mackenzie monitored the hearts of pregnant women and others using a polygraph he developed to detect and document the pulse in the neck. Through careful observation of patients over many years, Mackenzie realized that patients with certain types of irregular, extra heart beats were normal and could live normal lives. This contradicted contemporary wisdom, which advised the confinement of children and adults with abnormal heart beats. Mackenzie's 1908 textbook on heart disease also described auricular fibrillation (now known as atrial fibrillation), a type of irregular heart beat characterized by the ineffectual movement of the auricles, or atrial heart muscles, which can result in heart failure.

The critical arteries

Mackenzie's observations about coronary arteries reflected growing interest in the blood vessels leading to and from the heart. It had long before been observed by Da Vinci and others that a hardening occurred in blood vessels in some people. But the first well-documented report of coronary artery disease was presented by William Heberden, a British physician, in 1772. He named the condition angina pectoris, drawing the term from the Greek *agkhone* for strangling. He said that the condition tended to get worse and that patients often experienced it when walking. Heberden noted that when patients with this condition died, their aortas resembled bone or a bony-like substance. Contemporary researchers have found that coronary artery disease cuts down blood flow to the heart, causing pain. This pain can be triggered by emotional strain.

Another risk to individuals with thickened coronary arteries is coronary **thrombosis**, the most common cause of heart attack. This occurs when a blood clot forms, preventing blood flow and potentially causing death. This condition was identified by Dr. Adam Hammer, a German-born American, in 1878. Hammer suspected that the heart of one of his patients had been stopped by an obstruction and found upon autopsy that the heart was clogged by a jelly-like plug.

The nineteenth and twentieth centuries also saw advancing knowledge in the diagnosis of **congenital** heart disease. Currently, about eight per 1,000 infants are born with some sort of heart abnormality, including many that do not need to be treated. While most instances of congenital disorder occur for unknown reasons, congenital heart disease can also be caused by **genetic disorders**, such as Down's syndrome, or through maternal exposure to disease.

Cogenital heart disease, the atrial septal defect, was first described in 1900 by George Gibson of Edinburgh. This problem, which occurs when there is an opening in the wall (or septum) between two atria, can cause the right ventricle to be overwhelmed with blood, a condition that eventually leads to heart failure. In some cases, however, the holes are small and do not cause problems. For years, there was little physicians could do to help children with the problem. The development of successful surgical procedures to repair atrial septal defects was one of a multitude of dramatic modern advances in heart surgery.

Twentieth-century advances

In the twentieth century, physicians have acquired the tools to prevent heart disease in some cases and treat it effectively in many others. Major medical advances, such as the development of antibiotic therapy in the 1940s, have dramatically reduced heart disease due to syphilis and **rheumatic fever**. Developments in surgery, new drugs, diagnostic skill, and increasing knowledge about preventive medicine have also greatly reduced deaths from heart disease. Between 1980 and 1990, the death rate from heart disease dropped 26.7% in the United States, according to the American Heart Association (AHA).

The open heart

For most of medical history, the heart was seen as untouchable, limited by the difficulty of operating on the organ that kept the body alive. An 1896 book about chest surgery by Stephen Paget noted that "surgery of the heart has probably reached the limits set by nature to all surgery." But even as Paget cautioned doctors against trying unproven surgery on the heart, the effort was already being made. In 1882, German physician M. H.

Block described his successful suturing of rabbit hearts, and by 1896, German physician Ludwig Rehn successfully repaired a lacerated heart using sutures.

Obstacles to more ambitious heart surgery took some time to overcome. High death rates marked a series of operations performed in the 1920s to correct mitral stenosis, the narrowing of the area where the mitral valve is located. One of the first successful operations was the 1939 operation on a child by Robert Gross of Boston to correct patent ductus arteriosus, an abnormality in which the circulatory pattern used by the fetus is not converted over to the type of circulation necessary for survival outside of the womb.

New types of surgery were made possible with a series of technological advances. In 1934, the American John H. Gibbon developed a machine that allowed the heart to stop beating during surgery while the blood was oxygenated outside the body. Gibbon spent nearly 20 years testing the machine on animals. In 1953, Gibbon became the first surgeon to operate on an open heart when he repaired an 18-year-old girl's atrial septal defect.

With new access to the heart, the treatment of heart disease changed dramatically. The development of the first electric **pacemaker** in 1950 enabled doctors to correct many arrhythmias and numerous types of heart block. In 1992, a total of 113,000 pacemakers were implanted in the United States, according to the American Heart Association.

A significant change in the treatment of coronary heart disease was the development of coronary bypass surgery in 1967 in the United States. The surgery uses blood vessels taken from elsewhere in the body, often the leg, to pass around diseased tissue. In 1992, about 468,000 coronary artery bypass grafts were performed in the United States, according to the AHA.

Another commonly performed procedure for individuals with coronary heart disease is angioplasty, during which narrowed arteries are stretched to enable blood to flow more easily. The surgery involves threading a tube through the body and stretching the artery by using a plastic **balloon** that is inflated when the tube is in the coronary artery. A total of 399,000 angioplasty procedures were performed in 1992, according to the AHA.

The most dramatic change in treatment of heart disease was the development of methods to replace the most damaged hearts with healthy human hearts or even **animal** hearts. The first successful human heart transplant was performed by South African surgeon Christiaan Barnard in 1967. The patient, however, died in 18 days. Though many surgeons tried the operation, success was limited, most patients dying after days or months, until the early 1980s, when effective drugs were developed to fight organ rejection. By 1993, a total of 2,300 heart transplants were performed in the United States, where the one-year survival rate is 81.6%, according to the AHA.

A healthier life

Physician William Osler observed in 1910 that certain types of people were most likely to develop coronary artery disease, particularly individuals who were "keen and ambitious." Contemporary efforts to prevent heart disease focus on identifying types of **behavior** and activity that increase the risk of heart disease and on encouraging individuals to adopt healthier lifestyles.

The massive body of evidence linking various types of risk factors to heart disease derives from a series of ambitious twentieth-century studies of heart disease in large groups of people over a long period of time. One of the best known of these efforts is the Framingham study, which has traced thousands of residents since 1949. This and other studies have led to findings that individuals are at a greater risk of heart disease if they have high levels of certain types of cholesterol in the blood, if they smoke cigarettes, if they are obese, if they have high blood pressure, and if they are male. Blood cholesterol, a fat-like substance found in all human and animal tissue, is a primary focus of efforts to prevent heart disease. High cholesterol levels are shaped, in part, by diet and can be lowered. Experts suggest limiting consumption of foods high in saturated fats, such as cream, meat, and cheese. Such a diet reduces the risk of high levels of low-density lipoprotein, or LDL, the type of cholesterol which increases the risk of heart disease.

Exercise has also been promoted as a protection against heart disease. Numerous studies have shown that individuals who do not exercise are more likely to develop coronary heart disease. Exercise reduces blood pressure and eases blood flow through the heart. In addition, people who exercise are less likely to be overweight.

Individuals who burn more calories are also more likely to have higher levels of what has been called the "good" cholesterol—high-density lipoprotein, or HDL. This type of cholesterol is believed to reduce the risk of heart disease. Other activities that boost the level of HDL include maintaining average weight and not smoking cigarettes.

New drug and diagnostic therapies

The conventional treatment for cardiovascular disease includes specific therapy for any underlying causes and may also include drugs such as ACE inhibitors (e.g., captopril, enalapril, lisinopril), blood thinners (e.g., aspirin, warfarin), the combination of hydralazine and isosorbide dinitrate, digitalis, nitroglycerin, diuretics, and **beta-blockers** (e.g., propranolol). The last few decades of the twentieth century have also seen the introduction of numerous drugs which prolong life and activity for individuals with heart disease. Beta blockers are used to treat angina, high blood pressure, and arrhythmia. They are also given to individuals who have had heart attacks. These drugs block the neurohormone norepinephrine from stimulating the organs of the body. This makes the heart beat more slowly and slows the dilation of certain blood vessels.

Another important class of drugs for the treatment of heart disease is the vasodilators, which cause blood vessels to dilate, or increase in diameter. These drugs, including the so-called ACE inhibitors, are used to ease the symptoms of angina by easing the work of the heart, to forestall complete congestive heart failure, and to prolong life in people who have had heart attacks.

A third important type of drug reduces cholesterol in the blood. The process by which these drugs eliminate cholesterol from the blood varies, but several work by preventing the reabsorption of bile salts by the body. Bile salts play a role in digestion, and they contain cholesterol.

Diagnostic advances have also made a difference in the treatment of heart disease. Cardiac catheterization enables doctors to see how the heart works without surgery. The process, which was first explored in humans in 1936, involves sending a tube through an existing blood vessel and filling the tube with a contrast material that can be tracked as it circulates through the heart. In 1992, a total of 1,084,000 of these procedures were performed to diagnose heart problems.

Future challenges

Though knowledge about the treatment and prevention of heart disease has expanded dramatically, heart disease remains an immense threat. A total of 925,000 Americans die each year of cardiovascular disease, a general category which includes heart disease, stroke, and high blood pressure, all of which are linked. The biggest killer is coronary heart disease, which claimed 480,170 U.S. deaths in 1992. Stroke and hypertension together killed about 180,000 and artery diseases killed 40,730.

Many more people suffer than die from heart disease. For example, a total of 5.6 million Americans had angina pectoris in 1992, and 1,290 people died from it, according to the AHA. As many as 11.2 million Americans have a history of heart attack, chest pain, or both. While 39,206 Americans died of heart failure in 1991, more than 800,000 Americans were discharged from the hospital after treatment for the problem in 1992.

Approximately 250,000 Americans die sudden and unexpected deaths due to heart disease each year, the

AHA reports. Such findings support the need for educational efforts about heart disease and for the expansion of emergency care for heart attack victims.

Genetic therapy for heart disease is considered a fertile area for progress. In 1994, surgeons performed a procedure on a woman who had a genetic defect that prevented her liver from removing adequate amounts of LDL cholesterol. She had suffered a heart attack at age 16. The procedure, which took place in Michigan, involved the insertion of genetically modified cells in her liver, to enable the organ to remove LDL cholesterol properly. With the new cells, her heart should no longer be threatened by high levels of cholesterol.

Though much is known about risk factors for heart disease, new theories will continue to be tested in the future. For example, various studies have shown that individuals who eat large amounts of **fish** (especially containing particular oils called omega-3 **fatty acids**) or who consume **vitamin** E have a lower than average rate of coronary heart disease. But more ambitious studies are needed to confirm this information.

Researchers are also looking carefully at women and heart disease, a topic which has been overshadowed by research concerning men and heart disease in the past. Certain estrogen/ progestin supplements that were thought to reduce the risk of heart disease in post-menopausal women were found to actually increase the number of heart attacks in a long-term study published in 2002 by the NIH (National Institutes of Health).

Additionally, scientists are studying the correlation between levels of a protein in the blood known as known as C-reactive protein (CRP) and heart disease. Blood levels of C-reactive protein increase in the presence of systemic (throughout the body) inflammation, and increased CRP levels have been linked to heart attack and stroke. With elevated levels of CRP indicating blood vessel inflammation, the white blood cells are stimulated and may cause fatty cholesterol deposits to break from the vessel walls and clog arteries. Inflamed artery walls may also release greater portions of the weakened plaques, causing stroke. Elevated CRP is considered a predictor of heart disease even in the absence of other risk factors such as **obesity**, high blood pressure, or smoking, and may indicate heart disease even before symptoms are present. A blood test is becoming widely available to detect CRP levels.

In 1998, the American Heart Association published its third list of what it considers to be the most promising research areas in heart disease. These included: **gene therapy** which could potentially encourage the growth of new blood vessels to and from the heart, thus bypassing diseases vessels; the discovery of new "super aspirins"

KEY TERMS

Angina pectoris—Chest pain that occurs when blood flow to the heart is reduced, causing a shortage of oxygen. The pain is marked by a suffocating feeling.

Auricular fibrillation—Now known as atrial fibrillation; a condition marked by the irregular contraction of the atrial heart muscle.

Coronary artery disease, or ischemic heart disease—Common cause of death due to heart attack, which results from narrowing or blockage of the coronary arteries. The disease can also cause angina.

Coronary thrombosis—A potentially fatal event in which the coronary artery is blocked by a thrombus, a clot of platelets and other blood factors which can prevent the passage of blood.

Electrocardiogram (ECG)—A picture of the operation of the heart, obtained by measuring the electric potential of the heart muscle.

Heart block—Impairment of the electrical signal which controls the heart's activity.

Heart failure—A clinical syndrome which takes place when the heart's activity no longer meets the body's needs. Congestive heart failure is marked by water retention and breathlessness.

Stenosis—The narrowing of a canal or duct.

which seem to have even greater protective effects for both heart attack and stroke; more data to support the association between inflammation and heart attacks; better techniques for early detection of obstructed vessels in the heart (using magnetic **resonance** imaging, or MRI); hope that damaged left ventricular muscle can regain better functioning, if a mechanical device called a left ventricular assist device (LVAD) takes over the work of the left ventricle for a time; further evidence that tobacco is a crucial risk factor in the development of heart disease, as evidenced by research which showed that as few as 10 cigarettes a day shortens life; more research supporting the importance of diet and exercise on levels of cholesterol in the blood; efforts to encourage people to seek treatment more quickly when a heart attack is suspected; the association between non-responsiveness to nitric oxide, and the development of high blood pressure.

Over the past century, researchers have made huge advances in the understanding, treatment, and prevention of heart disease. For the first time in history, medicine has acquired the tools to provide many individuals who suffer from heart disease with active, full lives. Being born with a faulty heart is no longer a reason to live a sedentary, shortened life. Yet even as the medical profession has gained skill in healing, preserving, and even replacing the failing heart, this crucial organ remains vulnerable.

See also Angiography; Transplant, surgical.

Resources

Books

Bulpitt, C.J. *Epidemiology of Hypertension. Series: Handbook of Hypertension.* Vol. 20. Amsterdam: Elsevier, 2000.

Crawford, M.H., and J.P. DiMarco, ed. *Cardiology.* London: Mosby Limited Ltd., 2000.

Grubb, N., and D. Newby. *Churchill's Pocketbook of Cardiology.* London: Churchill Livingstone, 2000

Grundy, S.M., ed. *Cholesterol-Lowering Therapy—Evaluation of Clinical Trial Evidence.* New York: Dekker, 2000

Katz, A.M. *Physiology of the Heart.* 3rd ed. Philadelphia: Lippincott Williams & Wilkins, 2000

Periodicals

Avezum Jr., Alvaro, Marcus Flather, and Salim Yusuf. "Recent Advances and Future Directions in Myocardial Infarction." *Cardiology* 84 (1994): 391-407.

Detjen, Jim. "U.S. Scientists Successfully Treat Patient with Gene Therapy." *Knight-Ridder-Tribune News Service* (March 31, 1994).

"A Heart Disease Checkup." *Tufts University Diet and Nutrition Letter* (September 1993): 2.

Patricia Braus

Heart, embryonic development and changes at birth

The developing fetal **heart** accounts for a large percentage of the **volume** of the early thorax. About 20 days after **fertilization**, the heart develops from the fusion of paired endothelial tubes into a single tube. Heart growth subsequently involves the growth, expansion, and partitioning of this tube into four chambers separated by thickened septa of cardiac muscle and valves. Atrial development is initially more advanced than ventricular development. The left and right atria develop while the primitive ventricle remains a single chamber. As atrial separation nears completion, the left and right ventricles begin to form, then continue until the heart consists of it's fully developed four-chambered structure.

Although the majority of the heart develops from mesoderm (splanchnic mesoderm) near the neural plate and sides of the embryonic disk, there are also contributions from neural crest cells that help form the valves.

Three systems initially return venous **blood** to the primitive heart. Regardless of the source, this venous blood returns to sinus venosus. Vitelline **veins** return blood from the yolk sac; umbilical veins return oxygenated blood from the placenta. The left umbilical vein enlarges and passes through the embryonic liver before continuing on to become the inferior vena cava that fuses with a common chambered sinus venosus and the right atrium of the heart. Especially early in development, venous return also comes via the cardinal system. The anterior cardinals drain venous blood from the developing head region. Subcardinal veins return venous blood from the developing renal and urogenital system, while supracardinals drain the developing body wall. The anterior veins empty into the common cardinals that terminate in the sinus venosus.

Movement of blood through the early embryonic vascular system begins as soon as the primitive heart tubes form and fuse. Contractions of the primitive heart begin early in development, as early as the initial fusion of the endothelial channels that fuse to form the heart.

The heart and the atrial tube that form the aorta develop by the compartmentalization of the primitive cardiac tube. Six separate septae are responsible for the portioning of the heart and the development of the walls of the atria and ventricles. A septum primum divides the primitive atria into left and right chambers. The septum secundum (second septum) grows along the same course of the primary septum to add thickness and strength to the partition. There are two holes in these septae through which blood passes, the foramen secundum and the foramen ovale. Specialized endocardinal **tissue** develops into the atrioventricular septum that separates the atrium and ventricles. The mitral and tricuspid valves also develop from the atrioventricular septum.

As development proceeds, the interventricular septum becomes large and muscular to separate the ventricles and provide strength to these high-pressure contractile chambers. The interventricular septum also has a membranous portion.

Initially, there is only a common truncus arteriosus as a channel for ventricular output. The truncus eventually separates into the pulmonary trunk and the ascending aorta.

Blood oxygenated in the placenta returns to the heart via the inferior vena cava into the right atrium. A valve-like flap in the wall at the juncture of the inferior vena cava and the right atrium directs the majority of the flow of oxygenated blood through the foramen ovale, then allows blood to flow from the right atrium to the left. Although there is some mixing with blood from the superior vena cava, the directed flow of oxygenated blood across the right atrium caused by the valve of the inferior vena cava means that deoxygenated fetal blood returning via the superior vena cava still ends up moving into the right ventricle.

While in the uterus, the lungs are non-functional. Accordingly, another shunt, the ductus arteriosis (also spelled ductus arteriosus) provides a diversionary channel that allows fetal blood to cross between the pulmonary artery and aorta and thus largely bypass the rudimentary pulmonary system.

Because only a small amount of blood returns from the pulmonary circulation, almost all of the blood in the fetal left atrium comes through the foramen ovale. The relatively oxygen-rich blood then passes through the mitral value into the left ventricle. Contractions of the heart, whether in the single primitive ventricle or from the more developed left ventricle, then pump this oxygenated blood into the fetal systemic arterial system.

In response to inflation of the lungs and **pressure** changes within the pulmonary system, both the foramen ovale and the ductus arteriosis normally close at **birth** to establish the normal adult circulatory pattern whereby blood flows into the right atrium, though the tricuspid valve into the right ventricle. The right atrium pumps blood into the pulmonary artery and pulmonary circulation for oxygenation in the lungs. Oxygenated blood returns to the left atrium by pulmonary veins. After collecting in the left atrium, blood flows through the mitral value into the left atrium where it is then pumped into the systemic circulation via the ascending aorta.

See also Action potential; Birth defects; Cardiac cycle; Circulatory system; Embryo and embryonic development; Embryology; Heart diseases; Heart, rhythm control and impulse conduction.

Resources

Books

Gilbert, Scott F. *Developmental Biology.* 6th ed. Sunderland, MA: Sinauer Associates, Inc., 2000.

Sadler, T.W., and Jan Langman. *Langman's Medical Embryology.* 8th ed. New York: Lippincott Williams & Wilkins Publishers, 2000.

Mohrman, David E., and Lois Jane Heller. *Cardiovascular Physiology.* 5th ed. New York: McGraw-Hill, 2002.

Thibodeau, Gary A., and Kevin T. Patton. *Anatomy & Physiology.* 5th ed. St. Louis: Mosby, 2002.

Other

Abdulla, Ra-id. "Embryology." Rush Children's Heart Center. [cited January 17, 2003]. <http://www.rchc.rush.edu/rma webfiles/Embryology.htm>.

Klabunde, R.E. "Cardiac Cycle." Cardiovascular Physiology Concepts. January 17, 2003 [cited January 22, 2003]. <http://www.cvphysiology.com/Heart%20Disease/HD002. htm> .

Hill, Mark. "Vascular Development Circulation Changes at Birth." Vertebrate Development [cited January 21, 2003]. <http://anatomy.med.unsw.edu.au/teach/anat2310/2002/ Lecture01Heart(view).pdf>.

Brenda Wilmoth Lerner

Heart-lung machine

The heart-lung machine is a device used to provide **blood** circulation and oxygenation while the **heart** is stopped. It is a means of keeping a patient alive while his heart is stopped or even removed from his body. Usually called the heart-lung machine, the device also is referred to as cardiopulmonary bypass, indicating its function as a means to substitute for the normal functions of the heart (cardio) and lungs (pulmonary).

It is the function of the heart to provide circulation of blood at all times. It pushes blood out into the body and through the lungs. It must function every minute of every day of life to maintain the health of the tissues throughout the body.

The heart malfunctions at times and requires **surgery** to correct the problem. Surgeons searched for a means to stop the heart so they could correct defects yet keep the patient alive by circulating blood by another means. For many years no such means could be found. Some heart surgery was carried out while the **organ** still pumped, making delicate surgery virtually impossible. Surgeons then discovered that they could stop the heart by lowering the patient's body **temperature**, a condition called **hypothermia**, and by **flooding** the heart with a cold **solution**. In its state of artificial **hibernation** the body needed less blood circulation, but at best that gave surgeons only a few brief moments to carry out the surgery. They were still limited as to the procedures they could do because of the severe time constraints.

At the turn of the century, German scientists were studying isolated **animal** organs such as the liver and kidney and the effects that various drugs had on them. To do this they required the organ to be kept alive, meaning supplied with blood. They attempted various contrivances using syringes and pumps to maintain the viability of the organs. They experienced severe problems with blood clotting and changes in blood composition when the blood cells were damaged by the pumps. The researchers searched vainly for a means to provide oxygenated blood to their organ preparations. They filtered the blood through various screens and membranes and even pumped it through the lungs of dogs or **monkeys**, but their problem was not to be solved for decades, though this may be considered the beginning of research into a heart-lung device.

In 1953, at Jefferson Medical College in Philadelphia, Dr. John Gibbon connected the **circulatory system** of an 18-year-old female to a new machine, stopped the woman's heart, and for 26 minutes he performed surgery to close a hole in the wall of the heart between the left and right atria. It was the first successful use of a heart-lung machine and the beginning of a new era in cardiac surgery. The machine was not a sudden inspiration by anyone, but rather was the culmination of many years of dedicated research in many laboratories to find the means to oxygenate the blood and circulate it through the body.

That early machine, while functional, still was open to improvement. For one thing, it required many pints of blood to prime the machine and it was bulky and took up much of the room in the operating room. Since then, the size of the machine has been reduced and the need for blood to prime the machine has been dramatically reduced to only a few pints.

To function, the heart-lung machine must be connected to the patient in a way that allows blood to be removed, processed, and returned to the body. Therefore, it requires two hook-ups. One is to a large artery where fresh blood can be pumped back into the body. The other is to a major vein where "used" blood can be removed from the body and passed through the machine.

In fact, connections are made on the right side of the heart to the inferior and superior vena cavae (singular: vena cava). These vessels collect blood drained from the body and head and empty into the right atrium. They carry blood that has been circulated through the body and is in need of oxygenation. Another connection is made by shunting into the aorta, the main artery leading from the heart to the body, or the femoral artery, a large artery in the upper leg. Blood is removed from the vena cavae, passed into the heart-lung machine where it is cooled to lower the patient's body temperature, which reduces the tissues' need for blood. The blood receives **oxygen** which forces out the **carbon dioxide** and it is filtered to remove any detritus that should not be in the circulation such as small clots. The processed blood then goes back into the patient in the aorta or femoral artery.

During surgery the technician monitoring the heart-lung machine carefully watches the temperature of the blood, the **pressure** at which it is being pumped, its oxygen content, and other measurements. When the surgeon nears the end of the procedure the technician will increase

the temperature of the **heat** exchanger in the machine to allow the blood to warm. This will restore the normal body heat to the patient before he is taken off the machine.

See also Respiratory system; Thoracic surgery.

Heat

Heat exchange reflects and drives changes in **energy** state between two objects—or more generally systems—in thermal contact due to a difference in **temperature**. Heat flows from a system at higher temperature to one at lower temperature until both systems are at the same temperature. Systems at the same temperature are said to be in thermal equilibrium.

The term "heat" is sometimes used, incorrectly, to refer to a form of energy that a system contains. Heat is a form of energy-in-transit; it is not energy-in-residence. The energy contained in a system (exclusive of energy depending on external factors) is called internal energy and, unlike heat, is a property of a system like the **volume** or **mass**.

The first law of **thermodynamics** states that the internal energy of a system can change only if "energy" flows into or out of the system. This flow, or energy-in-transit, appears as heat or as work (or a combination), and the change in internal energy is equal to the total of heat and work appearing during the change. After the change, however, the system contains neither heat nor work; it contains internal energy.

Units of heat are units of energy. One classical unit, the **calorie**, was defined as the amount of energy required to raise the temperature of one gram of **water** one degree Celsius. A more precise definition recognizes that this energy depends slightly on the temperature of the water, so the **interval** was specified as 14.5–15.5°C (58.1–59.9°F). The dietary Calorie (capital C) is a kilocalorie (1000

TABLE 1	
SUBSTANCE	*SPECIFIC HEAT (J/g °C) at 25°C*
water	4.18
iron	0.45
mercury	0.14
ethyl alcohol	2.46

calories). The energy available from the **metabolism** of a given amount of food is commonly given in Calories.

In the International System of Units (SI or extended **metric system**) the joule is the unit of energy. Although based on mechanical rather than thermal considerations, the joule is now the preferred energy unit for both mechanical and thermal applications. The joule is about 1/4 of a calorie and now formally defines the calorie. One calorie is by definition exactly 4.184 joules, although the practical difference between this definition and the original one is negligible.

The specific **heat capacity**, or specific heat, is the heat required to raise the temperature of one gram of substance one degree Celsius. The specific heats of a few substances in joules per gram per degree Celsius are listed above.

For example, to raise the temperature of equal amounts of all four of these substances, the water would require considerably more heat than the others (over 9 times as much as the **iron**, for example, because 4.18 divided by 0.45 is 9.3). Or if you added the same amount of heat to equal amounts of all four of these substances, the temperature of the water would rise least. In short, it is more difficult to change the temperature of water than most other substances. This is one of the main reasons coastal climates usually have smaller seasonal temperature variations than inland climates. Because of its relatively high specific heat, water is a good thermal moderator.

Heat capacity

Heat capacity (often abbreviated Cp) is defined as the amount of **heat** required to raise the **temperature** of a given **mass** of a substance by one degree Celsius. Heat capacity may also be defined as the **energy** required to raise the temperature of one **mole** of a substance by one degree Celsius (the molar heat capacity) or to raise one gram of a substance by one degree Cel-

sius (the specific heat capacity). Heat capacity is related to a substance's ability to retain heat and the **rate** at which it will heat up or cool. For example, a substance with a low heat capacity, such as **iron**, will heat and cool quickly, while a substance with a high heat capacity, such as **water**, heats and cools slowly. This is why on a hot summer day the water in a **lake** stays cool even though the air above it (which has a low heat capacity) heats quickly, and why the water stays warm at night after the air has cooled.

Heat capacity and calorimetry

Calorimetry is the study of heat and heat energy. A **calorie** is a unit of heat energy in the British system of measurement. In the **metric system**, energy is measured in joules, and one calorie equals 4.184 joules. When any substance is heated, the amount of heat required to raise its temperature will depend on the mass of the object, the composition of the object, and the amount of temperature change desired. It is the temperature change and not the individual starting and final temperatures that matters. The equation that relates these quantities is

$$q = m\ Cp\ \Delta T$$

where q is the quantity of heat (in joules), m is the mass of the object (usually in grams), Cp is the heat capacity (usually in joules/gram degree,) and DELTAT is the change in temperature (in degrees Celsius). The amount of heat required depends on the mass to be heated (i.e., it takes more heat energy to warm a large amount of water than a small amount), the identity of the substance to be heated (water, for example, has a high heat capacity and heats up slowly, while metals have low heat capacities and heat up quickly), and the temperature change (it requires more energy to heat up an object by 60 degrees than by 20 degrees).

Heat capacity and the law of conservation of energy

Calculations using heat capacity can be used to determine the temperature change that will occur if two objects at different temperatures are placed in contact with each other. For example, if a 50 g piece of **aluminum metal** (Cp = 0.9 J/g C) at a temperature of 100°C is put in 50 g of water at 20°C, it is possible to calculate the final temperature of the aluminum and water. The aluminum will cool down and the water will warm up until the two objects have reached the same temperature. All of the heat lost by the aluminum as it cools will be gained by the water. This is a result of law of **conservation** of energy, which states that energy can neither be created or destroyed. The heat lost by the metal will be

$$q_{lost} = (50 \text{ grams}) \times (0.9 J/g°C) \times (100\text{-}T)$$

and the heat gained by the water will be

$$q_{gained} = (50 \text{ grams}) \times (4.184 J/g°C) \times (T\text{-}20)$$

These two equations are equivalent since heat lost equals heat gained; the final temperature of the mixture will be 27. 8°C. This final temperature is much closer to the initial temperature of the water because water has a high heat capacity and aluminum a low one.

Significance of the high heat capacity of water

Water has one of the highest heat capacities of all substances. It takes a great deal of heat energy to change the temperature of water compared to metals. The large amount of water on **Earth** means that extreme temperature changes are rare on Earth compared to other planets. Were it not for the high heat capacity of water, our bodies (which also contain a large amount of water) would be subject to a great deal of temperature variation.

See also Thermodynamics.

Resources

Books

Goldstein, Martin, and Inge Goldstein. *The Refrigerator and the Universe: Understanding the Laws of Energy.* Harvard University Press, 1993.

Pitts, Donald R., and Leighton E. Sissom. *Schaum's Outline of Heat Transfer.* 2nd ed. Whitby, Ontario: McGraw-Hill Trade, 1998.

Periodicals

Hendricks, Melissa. "Plant Calorimeter May Pick Top Crops." *Science News* 134 (September 17, 1988): 182.

Louis Gotlib

Heat of combustion *see* **Combustion**

Heat of fusion *see* **States of matter**

Heat index

The heat index is a measure of how warm an average person feels as a consequence of moisture in the air compared to the actual **temperature** measured by a **thermometer** at the same time and location. Generally speaking, the higher the relative **humidity**, the warmer the temperature will seem to be to a person. The reason for this relationship is that the human body normally loses **heat** through the process of perspiration. As the relative humidity rises, the **rate** at which perspiration occurs decreases, and the body loses heat less efficiently. Therefore, at high relative humidities, the outside temperature appears to be higher than it actually is.

The heat index can be expressed as a graph with the true air temperature charted on the vertical axis and the relative humidity on the horizontal axis. On this graph, the apparent temperature perceived by a person is expressed as a sloping line that decreases from left to right (from 0% to 100% relative humidity). As an example, a true temperature of 100°F (38°C) is likely to be perceived by the human body as 90°F (32°C) at a relative humidity of 0%. But at 50% relative humidity, that same true temperature is perceived to be about 120°F (49°C), and at 100% relative humidity, well over 140°F (60°C).

One use for the heat index is as an early warning system for possible heat problems. The National Weather Service has defined four categories on the heat index graph with increasingly serious health consequences. The lowest category (IV) covers the range of perceived temperatures from 81–90°F (27–32°C). In this range, the average human can expect to experience some fatigue as a result of prolonged exposure and physical activity. In category I, covering temperatures greater than 130°F (54.4°C), heat **stroke** or sunstroke is regarded as imminent.

Heat transfer

Heat transfer is the net change in **energy** as a result of **temperature** differences. This energy is transferred in the direction of decreasing temperature until thermal equilibrium (equality of temperatures) is achieved. The basic mechanisms involved in this process include **radiation** (the transfer of energy in the form of electromagnetic waves) and conduction (the transfer of kinetic energy). Heat transfer in fluids can occur at a faster **rate**, because large masses of a fluid can be displaced and can mix with other fluid masses of different temperatures. This process is considered a distinct mechanism called **convection**. In many heat transfer processes, radiation and convection or conduction work together, although one is often dominant.

Radiation

Every object emits electromagnetic radiation in a wave **spectrum** related to its own temperature. An object cooler than its surroundings will absorb more energy in the form of radiation than it emits. This radiation can pass through both free **space** and transparent media. Heat transfer by radiation helps sustain life on Earth—energy received from the **Sun** is an example of this process.

The electromagnetic radiation associated with heat transfer is sometimes referred to as **blackbody radiation** (where "blackbody" is an ideal emitter and absorber) or as thermal radiation. Thermal radiation is often associated with infrared radiation, although more thermal energy is received from the visible potion of the Sun's spectrum than from the infrared portion.

Conduction and convection

The molecules of a hotter material move faster and therefore have higher kinetic energy than the molecules of a cooler material. When molecules collide with slower neighboring molecules, kinetic energy is transferred from one **molecule** to another. The rate of heat transfer is high for metals (which, therefore, are said to have higher conductivity) and quite low for gases like air.

The process of convection occurs when groups of molecules are displaced to the vicinity of slower or faster molecules and mix with them. Forced convection occurs when hotter or cooler parts of a fluid are moved by way of forces other than gravity, such as a pump. Natural or free convection occurs when fluids are heated from below (like a pot on a kitchen stove) or cooled from above (like a drink with **ice** cubes on top). Hotter portions of the fluid expand, become lighter, and move upwards, while cooler, heavier portions descend. Convection can be many times faster than conduction alone. Vertical and horizontal convection plays a major role in the distribution of heat on **Earth** through the movements of atmospheric and oceanic masses.

See also Thermodynamics.

Heat of vaporizaton *see* **States of matter**

Heath family (Ericaceae)

The heath family, or Ericaceae, contains about 100-125 genera of vascular plants comprising 3,000-3,500 **species**. These plants are widespread in North and **South America**, Eurasia, and **Africa**, but are rare in Australasia. Species of heaths are most diverse and ecologically prominent in temperate and subtropical regions.

The most species-rich genus in the heath family are the rhododendrons (*Rhododendron* spp.), of which there are 850-1,200 species. The exact number is not known because species are still being discovered in remote habitats, and because the **taxonomy** of these plants is quite difficult and somewhat controversial among botanists. The "true" heaths (*Erica* spp.) are also diverse,

containing 500-600 species. The blueberries and cranberries (*Vaccinium* spp.) include about 450 species, a few of which are cultivated for their **fruits**.

Plants in the heath family are woody shrubs, trees, or vines. Their leaves are simple, usually arranged in an alternate fashion along the stem, and often dark-green colored. The foliage of many species is sometimes referred to as "evergreen," meaning it persists and remains functional in **photosynthesis** for several growing **seasons**. Other species have seasonally deciduous foliage. The flowers are radially symmetric, and are perfect, containing both staminate and pistillate organs. The fused petals (most commonly five in number) of many species give their flowers an urn- or bell-shaped appearance. The flowers may occur singly, or as inflorescences of numerous flowers arranged along the stem. The flowers of most species of heaths produce **nectar** and pleasant scents, and are pollinated by **insects** such as **bees** and **flies**. The fruits are most commonly a multi-seeded berry, a single-seeded drupe, or a capsule.

Species of heaths typically grow in acidic, nutrient-impoverished soils. Habitats range from closed to open **forests**, shrub dominated communities, bogs, and tundras. All species of heaths have a heavy reliance on mycorrhizal **fungi** to aid in the acquisition of mineral **nutrients**, especially **phosphorus**.

Species in North America

Species in the heath family are prominent in some types of habitats in **North America**, particularly in forests, shrubby places, bogs, and alpine and arctic tundras. The most important of the North American heaths are described below.

The most diverse group is the blueberries and cranberries (*Vaccinium* spp.), the delicious fruits of which are often gathered and eaten fresh or used in baking and jams. Widespread species include the blueberry (*V. angustifolium*), hairy blueberry (*V. myrtilloides*), tall blueberry (*V. corymbosum*), farkelberry (*Vaccinium arboreum*), deerberry (*V. stamineum*), bog-bilberry (*V. uliginosum*), (*V. macrocarpon*), and small cranberry or lingonberry (*V. oxycoccos*). Huckleberries also produce edible, blueberry-like fruits, including the black huckleberry (*Gaylussacia baccata*), dangleberry (*G. frondosa*), and dwarf huckleberry (*G. dumosa*).

Various species of rhododendrons occur in North America, especially in moist or wet forests, heathy shrublands, and bogs in the eastern part of the **continent**. The white laurel or rose bay (*Rhododendron maximum*) grows as tall as 32.8 ft (10 m), and has beautiful, large-sized, white or rose-colored flowers. The red laurel or rose bay (*R. catawbiense*) grows to 19.7 ft (6 m), and has

A rhododendron. *JLM Visuals. Reproduced with permission.*

beautiful, lilac or purple flowers. Shorter species include the pinksterflower (*R. nudiflorum*), mountain-azalea (*R. roseum*), swamp-azalea (*R. viscosum*), flame-azalea (*R. calendulaceum*), and rhodora (*R. canadense*). The California mountain laurel (*R. californicum*) is native to coastal forests of the western United States.

Various species of the heath family are commonly known as "wintergreens," because their foliage stays green through the winter, becoming photosynthetic again in the following spring. The shinleafs or wintergreens include *Pyrola americana* and *P. elliptica*, and occur in forests. One-flowered **wintergreen** (*Moneses uniflora*) occurs in damp forests and bogs. The spotted wintergreen (*Chimaphila maculata*) and pipsissewa or prince's pine (*C. umbellata*) occur in dry woods, especially on sandy soils. The common wintergreen (*Gaultheria procumbens*) is a common, creeping species of the ground vegetation of coniferous and mixed-wood forests of eastern North America, while shallon (*G. shallon*) is a shrub of Pacific forests.

The May-flower or trailing arbutus (*Epigaea repens*) is a low-growing, attractive, fragrant wildflower, and one of the first species to bloom in the springtime. The bearberry (*Arctostaphylos uva-ursi*) is a low-growing, evergreen shrub, especially common in open, sandy woods.

Labrador-tea (*Ledum groenlandicum*) is a shrub that grows in northern forests, tundras, and bogs of North America and Eurasia. Laurels are shrubs with attractive flowers, including the mountain-laurel (*Kalmia latifolia*) and sheep-laurel (*K. angustifolia*). The madrone or arbutus (*Arbutus menziesii*) is a beautifully red-barked **tree** of the Pacific coast.

The Indian pipe (*Monotropa uniflora*) and pinesap (*M. hypopithys*) occur widely in rich woods in North America, and also in Eurasia. These species lack **chlorophyll**, and their tissues are a waxy, whitish yellow or sometimes pinkish in **color**, and the plants are incapable of feeding themselves through photosynthesis. Instead, they are parasitic on their mycorrhizal fungus, which provides these plants with organic nutrients through the saprophytic food web, which derives its flow of fixed **energy** from the decay of organic litter and detritus. Other chlorophyll-lacking, parasitic species include pine-drops (*Pterospora andromeda*) and sweet pinesap (*Monotropsis odorata*).

Several Eurasian species have been introduced as horticultural plants and have established wild, self-maintaining populations, although none of these has become extensively invasive in North America. These include the purple-flowered Scotch heather (*Calluna vulgaris*) and several species of true heath, including *Erica tetralix*.

Economic importance

Some species in the heath family are cultivated as ornamentals in **horticulture**. The most commonly grown genera are the madrone or arbutus (*Arbutus* spp., including *A. menziesii* of North America), heather (*Calluna* spp.), heath (*Erica* spp.), and rhododendron (*Rhododendron* spp.). Cultivated rhododendrons include the white laurel or rose bay (*Rhododendron maximum*) and red laurel (*R. catawbiense*), native to eastern North America, and California mountain laurel (*R. californicum*), as well as the Asian azalea (*R. indicum*) and garden azalea (*R. sinense*).

The fruits of most species of blueberries and cranberries (*Vaccinium* spp.) are important **crops** in some areas, as are huckleberries (especially *Gaylussacia baccata* of eastern North America). Any of these may be gathered from the wild, or they may be intensively cultivated in monocultures.

Various species of blueberries are cultivated in agriculture, including the so-called lowbush blueberries

KEY TERMS

. . . .

Mycorrhiza—A "fungus root" or mycorrhiza (plural: mycorrhizae) is a fungus living in a mutually beneficial symbiosis (or mutualism) with the roots of a vascular plant.

Perfect—In the botanical sense, this refers to flowers that are bisexual, containing both male and female reproductive parts.

(*Vaccinium angustifolium, V. canadense, V. pennsylvanicum,* and *V. vacillans*), and the taller, high-bush blueberries (*V. atrococcum* and *V. corymbosum*). These are typically grown on acidic, nutrient-poor, sandy soils, and the fields are burned every several years in order to stimulate the sprouting of new twigs and branches, which then **flower** profusely. Blueberry fields may also be fertilized, but only at a relatively small **rate**. This is because agricultural weeds usually respond more vigorously to nutrient addition than do blueberries, so that excessive **fertilization** can cause problems. When they are ripe, the fruits are usually picked with a hand-held implement called a rake, which is a scoop-like device with numerous prongs on its underside, which can harvest the blueberries without collecting excessive quantities of leaves.

Cranberries are also cultivated, usually on sandy, wet, acidic soils. The most commonly grown species is *Vaccinium macrocarpon*. During the autumn harvest, cranberry fields are often flooded, and when the berries float to the surface, the fields provide a spectacularly red vista. Cranberries are also harvested using a rake-like device.

The mountain cranberry or cowberry (*Vaccinium vitis-idaea*) is collected in the wild, and is used in Scandinavia to make jams and a distinctive wine and liquor. All other cranberries and blueberries may be used to make jams, pies, and other cooked foods.

The common wintergreen or checkerberry (*Gaultheria procumbens*) is a natural source of oil-of-wintergreen (or methyl salicylate), which can be distilled from the leaves of this **plant**, and also from the twigs and inner **bark** of some species of birches (especially *Betula lenta* of eastern North America). Oil-of-wintergreen is commonly used as a flavoring for gums, candies, and condiments. This substance is also sometimes applied by massage as an analgesic for sore muscles. Oil-of-wintergreen is apparently pleasantly sweet to drink, which is unfortunate, because this material is highly toxic if ingested in large quantities, so that children have been killed by drinking this medicinal product. The smaller doses obtained from drinking a pleasant-tasting tea, made by boiling a small quantity of leaves, is said to relieve certain pains and discomforts of rheumatism.

A relatively minor use of a member of the heath family is that of briar **wood** (*Erica arborea*) of **Europe**, the wood of which has been used to make pipes for smoking tobacco.

See also Mycorrhiza.

Resources

Books

Judd, Walter S., Christopher Campbell, Elizabeth A. Kellogg, Michael J. Donoghue, and Peter Stevens. *Plant Systematics: A Phylogenetic Approach.* 2nd ed. with CD-ROM. Suderland, MD: Sinauer, 2002.
Klein, R. M. *The Green World: An Introduction to Plants and People.* New York: Harper and Row, 1987.

Bill Freedman

Heather *see* **Heath family (Ericaceae)**

Heavy water *see* **Deuterium**

Hedgehogs

Hedgehogs are small, often spine-covered members of the **insectivore** family Erinaceidae. The spiny hedgehogs are 13 **species** in subfamily Erinaceidae. Most famed is the European hedgehog, *Erinaceus europaeus*, which is also a resident of New Zealand, where it was introduced. Not all members of the hedgehog family have tough spines. The moonrats, or gymnures, of Southeast **Asia** have coarse hair instead of spines.

The common European hedgehog looks rather like a large pine cone with eyes. Its rounded body is covered from nose to tail with inch-long spines, up to 5,000 of them. The animal's size is variable, from 5-12 in (12-36 cm) long. While the spines of a hedgehog are often sharp, they are not nearly so dangerous as those of a porcupine. They lack the barbs on the end that can catch in an animal's flesh, anchoring there. Instead, the spines just provide a tough covering that very few predators are willing to try to penetrate. Some animals have learned that hedgehogs have soft underbellies that can be attacked. In defense against this, hedgehogs can curl up into a ball, protecting their softer, more vulnerable parts. If attacked, a hedgehog will fight quite noisily, screaming in fury.

Both hedgehogs and moonrats are geared strictly for eating **insects** and other small **invertebrates**, especially earthworms. Their narrow pointed snouts and their strong claws are useful for digging insects out of the ground. However, the Daurian hedgehog (*Hemiechinus dauuricus*) of the Gobi Desert has taken to eating small **rodents**. This species has longer ears than other hedgehogs.

Some species live in deserts, where they normally breed only once a year. Inhabitants of moister areas, such as the European hedgehog, normally breed twice a year. Except when mating, hedgehogs are solitary animals. The female produces usually three or four young after a gestation period of about four or five weeks. When they are born, the blind babies' spines are white and soft. They harden as they grow during their first three weeks. By that time, they can see and they are beginning to explore beyond the burrow with the mother.

Hedgehogs, like all insectivores, are very active animals that require a lot of food to survive. When confronted by food scarcity or cold **weather**, many hedgehogs will enter a period of dormancy, or even genuine **hibernation** in order to conserve **energy**. They retreat into a protected section of the burrow in which they normally live. During a mild winter, however, they may not **sleep** at all. Residents of tropical areas do not hibernate. And even European hedgehogs, when transplanted to warmer countries, do not hibernate.

Hedgehogs are active at night, eating earthworms, various insects, and even **snakes**. This makes them popular in gardens, where they often eat insect **pests**. Perhaps one of the reasons they are called hedgehogs is that these animals snuffle and snort, pig-like, as they go after their food. During the day in warm weather, they rest in a small temporary nest of leaves at the base of the hedges they frequent. During colder weather, they burrow into the ground.

European hedgehogs burrow under hedges all over **Europe**, even in busy cities, and they have long been chosen as pets. However, the animals are so infested by **fleas** that only tiny babies raised in captivity can become flea-free pets.

Hedgehogs perform a curious activity that has been described as self-anointing. They find a fluid substance, such as sap, and then lick on it enough to develop saliva. Then, moving from one side to the other, they lick the surface of their spines. It has been suggested that somehow the use of any irritating substance to self-anoint might keep predators at bay.

If hedgehogs could shed their spines, they would look like the other members of the family: gymnures, also called moonrats and thought of as spineless hedgehogs. These small, little-known animals of Asia occur in five species.

The Philippine gymnure, or wood shrew (*Podogymnura truei*), of the Philippine **island** of Mindanao lives at high altitudes. Less than 7 in (18 cm) long, it is regarded as threatened because its forest **habitat** is being destroyed. The females of the much larger moonrat of Malaysia (*Echinosorex gymnurus*) are larger than the males. This is the largest insectivore in the hedgehog family. It can be more than 2 ft (66 cm) long from head to tail. Unusually it is all white. The three species of lesser gymnures (*Hylomys*) are so small that they are easily mixed up with **shrews**, which live exactly the same kind of lifestyles.

Resources

Books

Hedgehog. Racine, WI: Western Publishing, 1993.

Kerrod, Robin. *Mammals: Primates, Insect-Eaters and Baleen Whales.* Encyclopedia of the Animal World series. New York: Facts on File, 1988.

Jean F. Blashfield

Heisenberg uncertainty principle

The Heisenberg uncertainty principle first formulated by German physicist Werner Heisenberg (1901–1976), has broad implications for quantum theory. The principle asserts that it is physically impossible to measure both the exact position and the exact **momentum** of a particle (like an **electron**) at the same **time**. The more precisely one quantity is measured, the less precisely the other is known.

Heisenberg's uncertainty principle, which also helps to explain the existence of **virtual particles**, is most commonly stated as follows: It is impossible to exactly and simultaneously measure both the momentum p (**mass** times **velocity**) and position x of a particle. In fact, it is not only impossible to *measure* simultaneously the exact values of p and x; they do not *have* exact, simultaneous values. There is always an uncertainty in momentum (Δp) and an uncertainty in position (Δx), and these two uncertainties cannot be reduced to **zero** together. Their product is given by $\Delta p \times \Delta x > h/4PI$, where h is **Planck's constant** (6.63×10^{-34} joules ċ second). Thus, if $\Delta p \to 0$, then $\Delta x \to \infty$, and vice versa.

Heisenberg's uncertainty principle is *not* equivalent to the statement that it is impossible to observe a system without perturbing it at least slightly; this is a true, but is not uniquely true in **quantum mechanics** (it is also true in Newtonian mechanics) and is not the source of Heinseberg's uncertainty principle.

Heisenberg's uncertainty principle applies even to particles that are not interacting with other systems, that is are not being "observed."

One consequence of Heisenberg's uncertainty principle is that the **energy** and duration of a particle are also characterized by complementary uncertainties. There is always, at every point in **space** and time, even in a perfect **vacuum**, an uncertainty in energy ΔE and an uncertainty in duration Δt, and these two complementary uncertainties, like Δp and Δx, cannot be reduced to zero simultaneously. Their product is given by $\Delta E \times \Delta t > h/4\text{PI}$.

Electrons and other **subatomic particles** exist in a dual particle and wave state and so one can only speak of their positions in terms of probability as to location when their velocity (energy state) is known.

Resources

Books

Barnett, R. Michael, Henry Mühry, and Helen R. Quinn. *The Charm of Strange Quarks.* New York: Springer-Verlag, 2000.

Gribbin, John. *Q is for Quantum: An Encyclopedia of Particle Physics.* New York: The Free Press, 1998.

Ne'eman, Yuval, and Yoram Kirsh. *The Particle Hunters.* Cambridge, UK: Cambridge University Press, 1996.

Silverman, Mark. *Probing the Atom* Princeton, NJ: Princeton University Press, 2000.

Other

Kalmus, P.I.P. "Particle Physics at the Turn of the Century." *Contemporary Physics* 41 (2000):129–142.

Lambrecht, Astrid. "The Casimir Effect: A Force From Nothing." PhysicsWeb. Sep. 2002 [cited Feb. 14, 2003]. <http://physicsweb.org/article/world/15/9/6>.

K. Lee Lerner
Larry Gilman
Terry Watkins

Heliocentric theory

The heliocentric theory argues that the **Sun** is the central body of the **solar system** and perhaps of the universe. Everything else (planets and their satellites, asteroids, **comets**, etc.) revolves around it.

The first evidence of the theory is found in the writings of ancient Greece. Greek philosopher-scientists deduced by the sixth century B.C. that **Earth** is round (nearly spherical) from observations that during **eclipses** of the **Moon**, Earth's shadow on the Moon is always a circle of about the same radius wherever the Moon is on the sky. Only a round body can always cast such a shadow.

The prevailing theory of the universe at that time was a geocentric (Earth-centered) one, in which all celestial bodies were believed to revolve around Earth. This was seen as a more plausible theory than the heliocentric one because to a casual observer, all celestial bodies seem to move around a motionless Earth at the center of the universe.

Over 200 years later Aristarchus of Samos (310-230 B.C.) attempted to measure the distance of the Sun from Earth in units of Earth-Moon distance by measuring intervals of the Moon. Observing the new Moon to the first quarter and the first quarter to full Moon, then using **geometry** and several assumptions, Aristarchus used the difference of the intervals of time to calculate the Sun's distance from Earth. The smaller the difference between the intervals, the more distant the Sun. From this value he determined the Sun's distance and the relative sizes of Earth, the Moon (about 1/4 that of Earth), and of the Sun. Aristarchus concluded that the Sun was several times larger than Earth. Aristarchus thought it reasonable that the smaller Earth revolved around the larger Sun.

Because the stars are all located on an enormous celestial **sphere** (the entire sky) centered on the Sun, not Earth, Earth's yearly **motion** around the Sun shows up in observations of the stars. The stars most likely to show the effect of this yearly motion are those in Gemini. The two brightest stars in Gemini, Castor, and Pollux, are about 4.56° apart and are close to the ecliptic, the Sun's yearly path among the stars. In the heliocentric theory, the ecliptic is the projection of Earth's **orbit** onto the sky. If one views the heliocentric model from the North Ecliptic Pole in Fig. 1 we see the Sun, the Earth (E) in several positions in its orbit, Castor (C), and Pollux (P) on the celestial sphere. If Castor and Pollux are fixed on the celestial sphere, then the distance CP between them is a fixed length.

Because they are fixed objects, the distance CP in this case appears largest when closest, and smallest when most distant. This effect was not detected with even the best astronomical instruments during the time of the ancient Greeks.

The Copernican revival of the heliocentric theory

Nicholas Copernicus (1472-1543) revived the heliocentric theory in the sixteenth century, after hundreds of years of building on Claudius (c. A.D. 90-168) Ptolemy's geocentric cosmological model (proving Earth is at the center of the universe). In his book, *On the Revolution of the Spheres of the Universe,* he placed the Sun at the center of the universe with the planets revolving around it on epicycles (a circle around which a **planet** moves) and deferents (the imaginary circle around Earth in whose periphery moves the epicycle). He argued that the planets in order from the Sun are Mercury, **Venus**, Earth (with the Moon orbiting it), **Mars**, **Jupiter**, and **Saturn**. The celestial sphere with the stars is far beyond Saturn's orbit. The apparent daily westward **rotation** of the celestial sphere, the Sun, Moon, and of the planets is the result of Earth's daily eastward rotation around its axis. If one assumes that the orbital velocities decrease with increasing distance from the Sun, then retrograde (the appearance of moving backward) motion of the planets on the zodiac could be explained by Earth overtaking Mars, Jupiter and Saturn near opposition (when they are 180° from the Sun on the zodiac), and by it being overtaken by the faster moving Mercury and Venus when they pass between the Sun and Earth (inferior conjunction). Copernicus's heliocentric model achieved a simpler **cosmology** than did the modified Ptolemaic geocentric model that existed in the sixteenth century, although not more accurate. The major advantage in the eyes of Copernicus was the aesthetic appearance of a system of concentric orbits with ever-widening separations and, ironically, the return to some of the fundamentals of the ancient Greeks, including purely circular motions. Copernicus's heliocentric model of the solar system did not represent accurately the observed planetary motions over many centuries. His model had many critics and was generally not accepted. An interesting variant of a geocentric model of the solar system was developed at the end of the sixteenth century by the Danish planetary observer Tycho Brahe (1546-1601). Earth was the primary center for the motions of all celestial bodies in his model but Mercury and Venus revolved around the Sun, which in turn revolved with them around Earth.

The triumph of the heliocentric theory

Johannes Kepler's (1571-1630) work enabled the heliocentric solar system model to accurately match and predict planetary positions on the zodiac for many centuries. After trying many geometric curves and solids in Copernicus's heliocentric model to match earlier observations of planetary positions, Kepler found that the model would match the observed planetary positions if the Sun is placed at one focus of elliptical planetary obits. This is Kepler's First Law of Planetary Motion. Kepler's three laws of planetary motion allow accurate matches and predictions of planetary positions.

Almost simultaneously, Galileo Galilei (1564-1642) built a small refracting **telescope** and began astronomical observations in 1609. Several of his observations lent support to Kepler's heliocentric theory:

1. Galileo discovered the four satellites of Jupiter (Io, Europa, Ganymede, and Callisto in order of increasing distance from Jupiter) in 1610. Their orbits around Jupiter showed that Jupiter and Earth were centers of orbital motion for celestial bodies (**geocentric theory** assumed that celestial bodies revolve only around Earth).

2. Galileo observed the disks of at least several planets. His observations of Venus's disk were especially important for determining whether the geocentric or heliocentric model was correct for the solar system. Ptolemy's geocentric model predicts that Venus's disk will show only the new Moon (dark) and crescent phases as it orbits Earth on its epicycle(s) and deferent (see Fig. 2). Kepler's modified Copernican heliocentric model predicts that Venus's disk will show all the phases of the Moon (including the half-moon, gibbous, and full Moon phases; see Fig. 3) as Venus and Earth both orbit the Sun. Galileo observed the second possibility for Venus's disk, which supported the heliocentric theory. The enormous variations in the angular size of Mars could not be explained by a circular orbit about Earth, but were easily understood if Mars orbits the Sun instead, thus varying its distance from Earth by a factor of five from the closest approach to the most distant retreat.

On the basis of these observational discoveries, Galileo began to teach the modified Copernican heliocentric model of the solar system as the correct one. He even used **Kepler's laws** to calculate orbital parameters for the orbits of the satellites revolving around Jupiter. However, direct proof that Earth moves around the Sun was still lacking. Furthermore, the Catholic Church considered Galileo's heliocentric theory to be heretical. It placed Copernicus's book on its Index of Restricted Books and tried Galileo before the Inquisition. Galileo was forced to recant the heliocentric theory and was placed under house arrest for the last eight years of his life.

The next major development was the generalization of Kepler's laws in 1687 by Isaac Newton (1642-1727). His generalized form of Kepler's laws showed that the Sun and planets all revolve around the solar system's center of **mass**. Telescopic observations of solar system objects gave indications of their size and when used in the generalized Kepler's laws, soon showed that the Sun is much larger and more massive than even Jupiter (the largest and most massive planet). Thus the center of the solar system, around which Earth revolves, is always in or near the Sun. Earth orbits the Sun much more than the Sun orbits Earth.

Another demonstration of Earth's orbital motion is the aberration of starlight. Astronomical observations and **celestial mechanics** indicate that Earth should have a 16-19 mi/sec (25-30 km/sec) orbital **velocity** around the solar system's center which continuously changes its direction due to the gravitational effect of the Sun. James Bradley's (1693-1762) attempt to determine the parallax-

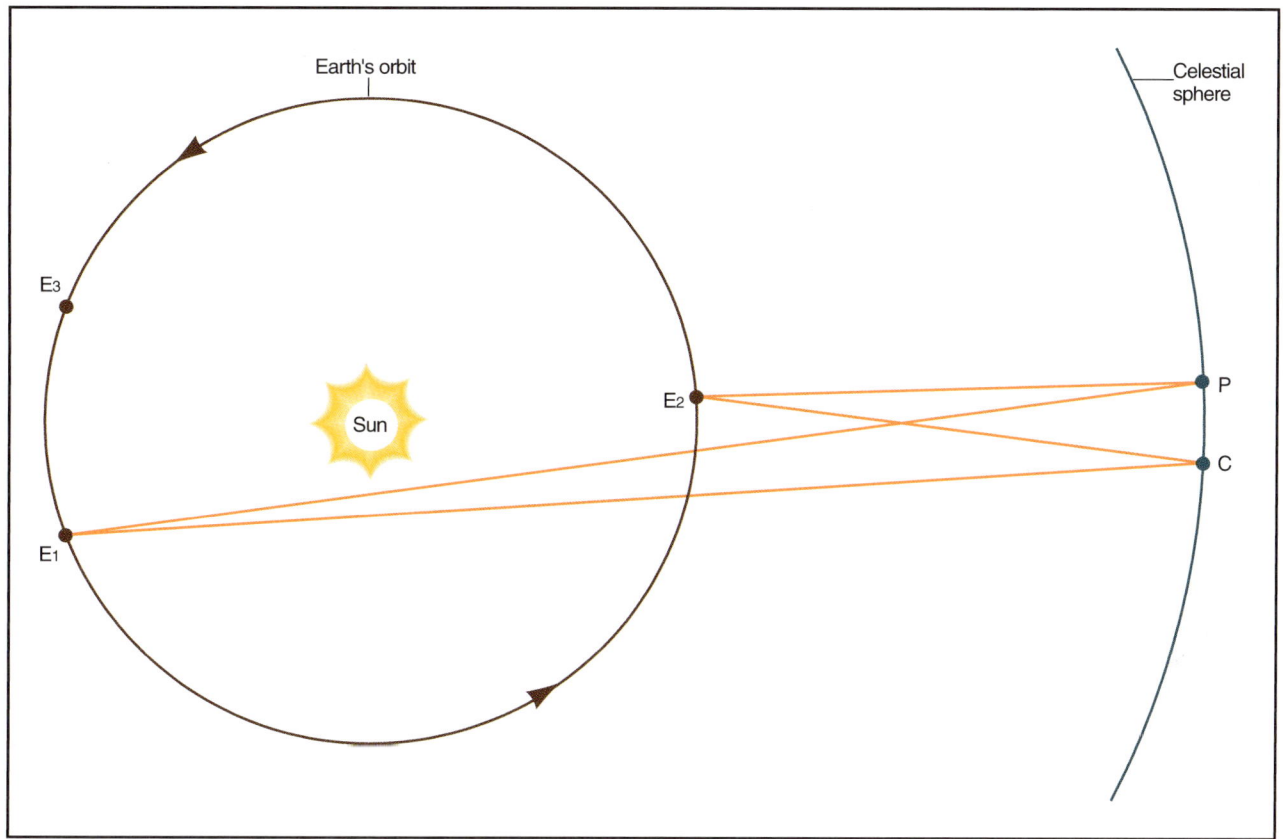

Figure 1. Illustration of how the angle PEC "opens up" as Earth approaches Castor (C) and Pollux (P) as it moves from position E_1 in its orbit around the Sun in September to E_2 in January, then PEC "closes up" as the Earth moves from E_2 to its position E_3 in early June. The distance PC between Castor and Pollux is assumed to be fixed on the celestial sphere. *Illustration by Hans & Cassidy. Courtesy of Gale Group.*

es of stars starting in 1725 with a telescope rigidly fixed in a chimney soon found that the apparent positions of the stars shifted along elliptical paths. These ellipses were 90° out of phase with the **parallax ellipse** for a nearby **star** on a distant background that is expected to be produced by Earth's motion around the Sun. Moreover the ellipses' semi-major axes were always 20.5", with no variation from the different distances of the stars. These same size ellipses were soon understood to be the yearly paths of the aberrations of the apparent positions of the stars caused by the addition of Earth's constantly changing orbital velocity to the **vacuum** velocity of the **light** arriving from the stars (whose true positions are at the centers of the aberrational ellipses). These ellipses show that Earth does indeed have the expected orbital velocity around the solar system's center of mass.

Final proof of the heliocentric theory for the solar system came in 1838, when F.W. Bessel (1784-1846) determined the first firm trigonometric parallax for the two stars of 61 Cygni (Gliese 820). Their parallax (difference in apparent direction of an object as seen from two dif-

ferent points) ellipses were consistent with orbital motion of Earth around the Sun.

In 1835 the Catholic Church removed Copernicus's book from its index of restricted books. It is fitting to mention Copernicus's book here. Bessel's successful measurement of a parallax ellipse established the Sun as the central body of the solar system, but it was not certain that the Sun was at or near the center of the universe.

The heliocentric theory and the universe

Astronomers seem to have been of differing opinions on this aspect of the heliocentric theory. Thomas Wright (1711-86) and William Herschel (1738-1822) thought that the Sun was at or near the center of the **Milky Way** which most astronomers believed to comprise most or all of the universe. Herschel arrived at this conclusion by making star counts in different directions (parts of the sky) but he did not allow for the absorption of starlight by interstellar dust. J.H. Lambert (1728-77) concluded that the Sun was somewhat away from its cen-

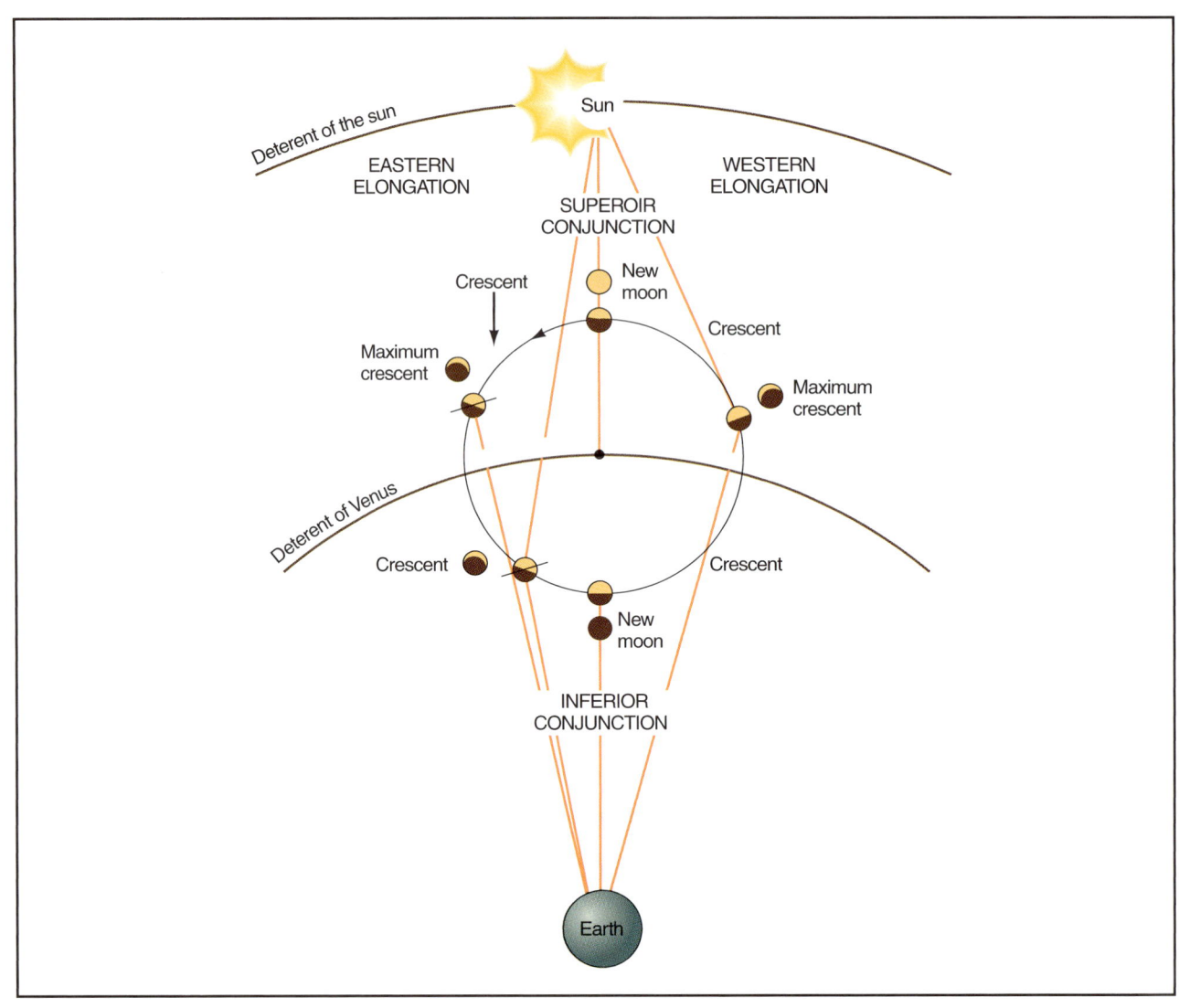

Figure 2. Diagram showing the phases of the disk of Venus in the Ptolemaic geocentric model of the solar system. Notice that only the new Moon (dark disk) or bright crescent phases are possible for Venus's disk. *Illustration by Hans & Cassidy. Courtesy of Gale Group.*

ter on the basis of the Milky Way's geometry. As long as there seemed to be evidence that the Sun was at or near the Milky Way's center and the Milky Way comprised most of the universe, a case could be made that the Sun was at or near the center of the universe.

Immanuel Kant (1724-1804) suggested that some of the nebulae seen in deep **space** were other Milky Ways, or "island universes," as he termed them. If his speculation proved to be correct, this would almost certainly mean that the Sun is nowhere near the center of the universe.

Astrometry also showed that the Sun and the other stars are moving relative to each other. The Sun is not at rest relative to the average motions of the nearby stars, but is moving relative to them at about 12 mi/sec (20

km/sec) towards the constellations Lyra and Hercules. These facts indicated that the Sun is only one of perhaps billions of ordinary stars moving through the Milky Way.

Harlow Shapley (1887-1972) postulated the first fairly correct idea about the Sun's location in the Milky Way. He found that the system of the Milky Way's globular star clusters is arranged in a halo around the Milky Way's disk (within which the Sun is located). These clusters are concentrated towards its nucleus and center, which are beyond the stars of the **constellation** Sagittarius. He found about 100 globular clusters in the hemisphere of the celestial sphere centered on the direction to the center of the Milky Way in Sagittarius, while there were only about a dozen globular clusters in the opposite hemisphere centered in the constellation Auriga. Shapley

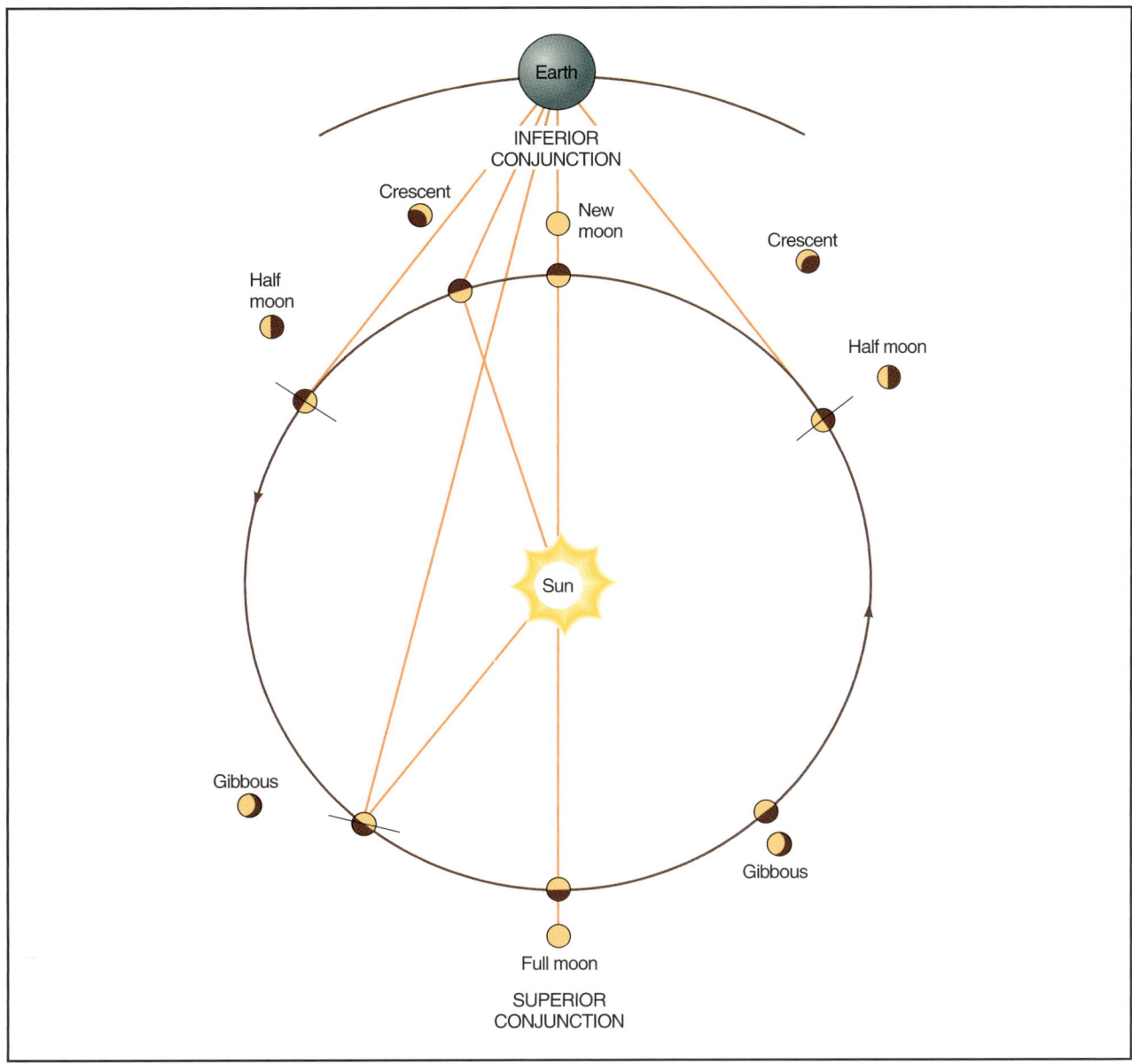

Figure 3. Diagram of the phases of Venus's disk as it gains on Earth in their orbits around the Sun in the heliocentric model of the solar system. Notice that, in theory, Venus's disk should show all the phases of the Moon. This is what Galileo observed with his telescope. *Illustration by Hans & Cassidy. Courtesy of Gale Group.*

reported this research in 1918 and estimated that the Sun is about 2/3 from the Milky Way's center to the edge of its disk which is very far from its center.

Edwin Hubble (1889-1953) confirmed Kant's hypothesis that the **spiral** and elliptical nebulae are other galaxies similar to the Milky Way in 1924. He also discovered that all the distant galaxies have spectra whose **spectral lines** are Doppler shifted towards the red end of their visible spectra, indicating that all distant galaxies are moving away from the Milky Way and its neighboring galaxies. Furthermore, the more distant such a

galaxy seems to be, the faster it seems to be receding. This indicates that our universe seems to be expanding. One result of this discovery has been to make the concept of a "center of the universe" questionable, perhaps meaningless, in a universe with three spatial dimensions.

Present estimates indicate that the Sun is between 25,000 to 30,000 light years from the Milky Way's center. The Sun is revolving around this center with an orbital velocity of about 155 mi/sec (250 km/sec). One revolution around the Milky Way's center takes about 200,000,000 years. The Sun is only one star among

KEY TERMS

Celestial sphere—The entire sky on which are situated the Sun, Moon, planets, stars, and all other celestial bodies for a geocentric observer.

Constellation—A region of the celestial sphere (sky). There are 88 officially recognized constellations over the entire celestial sphere.

Parallax (parallactic shift)—The apparent shift of position of a relatively nearby object on a distant background as the observer changes position.

Semi-major axis—The longest radii of an ellipse.

Zodiac—The zone 9° on each side of the ecliptic where a geocentric observer always finds the Sun, Moon, and all the planets except Pluto.

100,000,000,000 or more other ordinary stars which revolve around the Milky Way's center.

Heliocentric theory is valid for our solar system but its relevance extends only a few light-years from the Sun to the vicinity of the three stars of the Alpha Centauri system (Gliese 551, Gliese 559A, and Gliese 559B).

See also Astronomy; Celestial coordinates; Doppler effect.

Resources

Books

Bacon, Dennis Henry, and Percy Seymour. *A Mechanical History of the Universe.* London: Philip Wilson Publishing, Ltd., 2003.

Beer, A., ed. *Vistas in Astronomy: Kepler.* Vol. 18. London: Pergamon Press, 1975.

Morrison, David, and Sidney C. Wolff. *Frontiers of Astronomy.* Philadelphia: Sanders College Publishing, 1990.

Frederick R. West

Helium *see* **Rare gases**

Hematology

Hematology is the study of **blood** and its basic biological components, including red blood cells (erythrocytes), white blood cells (leukocytes), and blood platelets (erythrocytes). Hematologists study and help treat a variety of hematological malfunctions and diseases, one of the primary being the various anemias. Anemias, like

sickle cell anemia, result in a loss of erythrocytes, which reduces the blood's ability to transport **oxygen** to tissues.

The vital importance of blood to life has been known since at least 400 B.C. However, there were many early misconceptions concerning blood's functions and activities. It was once thought that disturbances in blood "humors" caused diseases. As a result, bloodletting was thought to eliminate contaminated fluids from the body and became a primary, though misguided, therapy for almost every **disease**. It was not until the seventeenth century that the **microscope** was invented and the science of hematology moved into the modern era. With this technology, the cellular components of blood were first discovered. In 1852, Karl Vierordt quantitatively analyzed blood cells, which led to correlations between blood cells counts and various diseases.

Blood helps the body to function in many ways. It is the main connection between different body tissues, transferring substances produced by one **organ** for use by other tissues. As a result, blood is a primary component in proper organ functioning. For example, it transports oxygen from the lungs and releases it in various tissues, providing the essential nourishment for **tissue** survival and growth.

One of the primary areas of study within hematology is hematopoiesis, or the formation and development of blood cells. Blood is formed by various hematopoietic tissues or organs depending on the stage of life. In humans, the embryo yolk sack begins producing blood usually within 20 days after **fertilization**. In the third month of embryonic life, the liver takes over with help from the spleen, kidney, thymus, and lymph nodes. Although lymph nodes continue to play a role in blood formation throughout life, bone marrow becomes the primary source of blood production in the embryo at about six months and continues this role after **birth**.

In contrast to the early theory of blood's relationship to disease, hematologists came to understand that changes in specific blood components are the result, not the cause, of disease. Anemias, for example, are the result of other body organ or tissue malfunctions. In addition to studying and treating various types of anemias, hematologists are concerned with a variety of other blood disorders, like leukemias (cancerous malignancies that occur in the blood and lymph nodes) and blood clotting, which can lead to an **embolism** (obstruction of a blood vessel).

See also Anemia; Circulatory system.

Hemophilia

Hemophilia is an inheritable disorder of the mechanism of **blood** clotting. Depending on the degree of the disorder present in an individual, excess bleeding may occur only after specific, predictable events (such as **surgery**, dental procedures, or trauma), or may occur spontaneously, with no initiating event.

Normal blood clotting

The normal mechanism for blood clotting is a complex series of events involving the interaction of the injured blood vessel, blood cells called platelets, and over 20 different **proteins** which also circulate in the blood.

When a blood vessel is injured in a way to cause bleeding, platelets collect over the injured area, and form a temporary plug to prevent further bleeding. This temporary plug, however, is too disorganized to serve as a long-term solution, so a series of chemical events result in the formation of a more reliable plug. The final plug involves tightly woven fibers of a material called fibrin. The production of fibrin requires the interaction of a variety of chemicals, in particular a series of proteins which are called clotting factors. At least 13 different clotting factors have been identified.

The clotting cascade, as it is usually called, is the series of events required to form the final fibrin clot. The cascade uses a technique called amplification to rapidly produce the proper sized fibrin clot from the small number of molecules initially activated by the injury.

The defect in hemophilia

In hemophilia, certain clotting factors are either decreased in quantity, absent, or improperly formed. Because the clotting cascade uses amplification to rapidly plug up a bleeding area, absence or inactivity of just one clotting factor can greatly increase bleeding time.

Hemophilia A is the most common type of bleeding disorder, and involves decreased activity of factor VIII. Three levels of factor VIII deficiency exist, and are classified based on the percentage of normal factor VIII activity present. Half of all people with hemophilia A have severe hemophilia. This means that their factor VIII activity level is less than 1% of the normal level. Such individuals frequently experience spontaneous bleeding, most frequently into their joints, skin, and muscles. Surgery or trauma can result in life-threatening hemorrhage, and must be carefully managed. Individuals with 1-5% of normal factor VIII activity level have moderate hemophilia, and are at risk for heavy bleeding after seemingly minor traumatic injury. Individuals with 5-40% of normal factor VIII activity level have mild hemophilia, and must prepare carefully for any surgery or dental procedures.

Individuals with hemophilia B have very similar symptoms, but the deficient factor is factor IX. Hemo-

philia C is very rare, and much more mild than hemophilias A or B; it involves factor XI.

How hemophilia is inherited

Hemophilia A and B are both caused by a genetic defect present on the X **chromosome** (hemophilia C is inherited in a different fashion). About 70% of all individuals with hemophilia A or B inherited the **disease**. The other 30% have hemophilia because of a spontaneous genetic **mutation**.

In order to understand the inheritance of these diseases, a brief review of some basic human **genetics** is helpful. All humans have two chromosomes which determine their gender: females have XX, males have XY. When a trait is carried only on the X chromosome, it is called sex-linked.

Because both factors VIII and IX are produced by direction of the X chromosome, hemophilia A and B are both sex-linked diseases. Because a female child always receives two X chromosomes, she nearly always will receive at least one normal X chromosome. Therefore she will be capable of producing a sufficient quantity of factors VIII and IX to avoid the symptoms of hemophilia. If, however, she has a son who receives her flawed X chromosome, he (not having any other X chromosome) will be unable to produce the right quantity of factors VIII or IX, and he will suffer some degree of hemophilia.

In the rare event of a hemophiliac father and a carrier mother, the right combination of parental chromosomes will result in a hemophiliac female child. This situation, however, is extraordinarily rare. The vast majority of individuals with either hemophilia A or B, then, are male.

As mentioned earlier, about 30% of all individuals with hemophilia A or B are the first members of their families to ever present with the disease. These individuals had the unfortunate occurrence of a spontaneous genetic mutation, meaning that early in development, some **random** genetic accident befell their X chromosome, resulting in the defect causing hemophilia A or B.

Symptoms of hemophilia

In the case of severe hemophilia, the first bleeding event usually occurs prior to 18 months of age. In fact, toddlers are at particular risk, because they fall so frequently.

Some of the most problematic and frequent bleeds occur into the joints, particularly into the knees and elbows. Repeated bleeding into joints can result in permanent deformities. Mouth injuries can result in compression of the airway, and therefore can be life-threatening. A blow to the head, which might be totally insignificant in a normal individual, can result in bleeding into the skull and **brain**. Because the skull has no room for expansion, the hemophiliac individual is at risk for brain damage due to blood taking up space and exerting **pressure** on the delicate brain **tissue**.

Diagnosis

Various tests are available to measure, under very carefully controlled conditions, the length of time it takes to produce certain components of the final fibrin clot. Tests can also determine the percentage of factors VIII and IX present compared to known normal percentages. Families with a positive history of hemophilia can have tests done during a pregnancy to determine whether the fetus is hemophiliac.

Treatment

Various types of factors VIII and IX are available to replace a patient's missing factors. These are administered intravenously (directly into the patient's **veins** by needle). These factor preparations may be obtained from a single donor, by pooling the donations of as many as thousands of donors, or by laboratory creation through highly advanced genetic techniques.

The frequency of treatment with factors depends on the severity of the individual patient's disease. Patients with relatively mild disease will only require treatment in the event of injury, or to prepare for scheduled surgical or dental procedures. Patients with more severe disease will require regular treatment to avoid spontaneous bleeding.

While appropriate treatment of hemophilia can both decrease suffering and be life-saving, complications of treatment can also be quite serious. About 20% of all patients with hemophilia A begin to produce chemicals within their bodies which rapidly destroy infused factor VIII. The presence of such a chemical may greatly hamper efforts to prevent or stop a major hemorrhage.

Individuals who receive factor prepared from pooled donor blood are at risk for serious infections which may be passed through blood. **Hepatitis**, a severe and potentially fatal viral liver **infection**, is frequently contracted from pooled factor preparations. Most frighteningly, pooled factor preparations in the early 1980s were almost all contaminated with Human Immunodeficiency Virus (HIV), the **virus** which causes **AIDS**. Currently, careful methods of donor testing, as well as methods of inactivating viruses present in donated blood, have greatly lowered this risk, but not before huge numbers of he-

mophiliacs were infected with HIV. In fact, some **statistics** show that, even today, the leading cause of death among hemophiliacs is AIDS.

Resources

Books

Andreoli, Thomas E., et al. *Cecil Essentials of Medicine.* Philadelphia: W. B. Saunders Company, 1993.

Berkow, Robert, and Andrew J. Fletcher. *The Merck Manual of Diagnosis and Therapy.* Rahway, NJ: Merck Research Laboratories, 1992.

Hay, William W., et al. *Current Pediatric Diagnosis and Treatment.* Norwalk, CT: Appleton & Lange, 1995.

Rosalyn Carson-DeWitt

Hemorrhagic fevers and diseases

Hemorrhagic diseases are caused by **infection** with certain viruses or **bacteria**. Viruses cause virtually all the hemorrhagic diseases of microbiological origin that arise with any frequency. The various viral diseases are also known as viral hemorrhagic fevers. Bacterial hemorrhagic **disease** does occur, but rarely. One example of a bacterial hemorrhagic disease is scrub **typhus**.

Copious bleeding is the hallmark of a hemorrhagic disease. The onset of a hemorrhagic fever or disease can produce mild symptoms that clear up quickly. However, most hemorrhagic diseases are infamous because of the speed that some infections take hold, and the ferocity of their symptoms. Such hemorrhagic maladies, such as Ebola, have high mortality rates.

Viral types and characteristics

Four main groups of viruses exist that cause hemorrhagic disease or fever: arenaviruses, filoviruses, bunyaviruses, and flaviviruses. Arenaviruses cause Argentine hemorrhagic fever, Bolivian hemorrhagic fever, Sabia-associated hemorrhagic fever, Lymphocytic choriomeningitis, Venezuelan hemorrhagic fever, and Lassa fever. Members of the filovirus group cause Ebola hemorrhagic fever and Marburg hemorrhagic fever. Bunyaviruses cause Crimean-Congo hemorrhagic fever, Rift Valley fever, and Hantavirus pulmonary **syndrome**. Lastly, Flaviviruses cause tick-borne **encephalitis**, **yellow fever**, Dengue hemorrhagic fever, Kyasanur Forest disease, and Omsk hemorrhagic fever.

These viruses differ in structure and in the severity of the symptoms they can cause. They all, however, share common features. All hemorrhagic viruses contain **ribonucleic acid (RNA)** as their genetic material. The RNA is protected and confined in a **membrane** called the viral envelope. The envelope is typically made of **lipid**. Another feature of hemorrhagic viruses, and indeed of all viruses, is the requirement for a host in which to live and produce new viral particles. Hemorrhagic viruses can live in some non-human **mammals**, such as **primates**, and in **insects**. The primates and insects are described as being natural reservoirs of the particular **virus**. Humans are not a natural reservoir. Epidemiologists (disease trackers) suspect that initial infections of humans occurs only accidentally when humans and the primate or insect come into close contact.

In contrast to the reservoir host, the presence of the hemorrhagic virus in humans typically produces a devastating illness. The symptoms can progress from mild to catastrophic very rapidly (i.e., in only hours). While catastrophic for the victims and difficult to treat, the rapid nature of the outbreaks has an advantage. Because victims succumb quickly, the transmission of the virus from human to human is limited. An outbreak can appear, ravage a local population, and fade away within days or a few weeks.

The viruses that cause the various hemorrhagic fevers and diseases do not survive in the host following the disease (the Human Immunodeficiency Virus, in contrast, is able be latent in the host, and survive for prolonged periods of time before symptoms of infection appear). However, people who are recovering from infections caused by Hantavirus and Argentine hemorrhagic fever can excrete infectious viruses in their urine.

The sporadic appearance of hemorrhagic outbreaks and the fact that they often occur in geographically isolated regions (e.g., interior of **Africa**) has made the study of the diseases difficult. It is known that there is not any timetable to the appearance of a hemorrhagic fever, such as in one season of the year relative to another season. The only factor that is known clearly is that the viruses are passed from the natural host to humans. How this transfer occurs and why it occurs sporadically are not known.

The viruses do not damage their primate or insect hosts as much as they do a human who acquires the **microorganisms**. The reasons for this difference are unknown. Researchers are attempting to discover the basis of the natural resistance, as this would help in finding an effective treatment for human hemorrhagic diseases.

The speed at which hemorrhagic fevers appear and end in human populations, combined with their frequent occurrence in relatively isolated areas of the globe has made detailed study difficult. Even though some of the diseases, such as Argentine hemorrhagic fever, have been known for almost 50 years, knowledge of the molecular basis of the disease is lacking. For example, while it is apparent that some hemorrhagic viruses can be transmitted through the air as **aerosols**, the pathway of infection once the microorganism has been inhaled is still largely unknown.

Hemorrhagic diseases are zoonotic diseases; ones that occur by the transfer of the disease causing agent from a non-human to a human. For some of the hemorrhagic viruses, the reservoir host is known. They include the cotton rat, **deer mouse**, house mouse, arthropod ticks, and **mosquitoes**. However, for viruses such as the Ebola and Marburg viruses, the natural host still is not known. Outbreaks with these two viruses have involved transfer of the virus to human via primates. Whether the primate is the natural reservoir host, or whether primates acquire the virus as the result of contact with the true natural reservoir host, is yet another aspect of hemorrhagic diseases that is not clear.

As mentioned, hemorrhagic fevers can rapidly spread through a human population. This is due to human-to-human transmission. This transmission occurs easily, often via body fluids that accidentally contact a person who is caring for the afflicted person. Funeral practices of handling and washing the bodies of the deceased have contributed to human-to-human transmission of Ebola during outbreaks in Sub-Saharan Africa.

Hemorrhagic diseases typically begin with a fever, a feeling of tiredness, and a generalized aching of muscles. In rare instances, symptoms may not progress any further, in which case recovery is rapid. For unknown reasons, however, more serious damage often occurs. Here, symptoms include copious bleeding from the mouth, eyes, and ears. Internal bleeding also occurs, as organs are attacked and destroyed by the infection. Death is typically the result of the overwhelming damage to the organs, and from the failure of the **nervous system**. Often, victims have seizures and lapse into a **coma** prior to death.

Hemorrhagic diseases are difficult to treat. One reason is because of the rapid progression of the disease. Another reason is because vaccines exist for only a few of the diseases (i.e., yellow fever and Argentine hemorrhagic fever). For the remaining diseases, supportive care such as keeping the infected person hydrated is often the only course of action.

To prevent outbreaks, the most effective policy is to curb human interaction with the natural reservoir of the microbe. For example, in the case of hantavirus pulmonary syndrome, scientists discovered in the 1990s that the responsible virus was resident in rodent populations, and that these populations exploded in numbers after rainy periods. Thus, limiting contact with places where the **rodents** live (i.e., barns), particularly after a rainy period, is a wise practice. Insect vectors are controlled by a spraying and common sense steps, such as use of insect repellent, proper clothing, insect netting over sleeping areas.

Hemorrhagic fevers are significant, not only because of the human suffering they cause, but because the viral agents could be exploited as bioweapons. For these reasons, a great deal of research effort is devoted towards understanding the origins and behaviors of the viruses.

Ebola and other hemorrhagic diseases

The best-studied hemorrhagic fever is Ebola. The **Ebola virus** is named after a river located in the Democratic Republic of the Congo, where the virus was discovered in 1976. This outbreak occurred in the western part of the African nation of Sudan and in nearby Zaire. In 1979, another outbreak occurred in Zaire. In 1995, an outbreak that involved 316 people occurred in Kikwit, Zaire. Outbreaks have also occurred in the African regions of Gabon and the Ivory Coast.

There are four **species** of Ebola virus. These differ in their arrangement of their genetic material and in the severity of the infection they cause. Ebola-Zaire, Ebola-Sudan, and Ebola-Ivory Coast cause disease in humans. The fourth species, Ebola-Reston, causes disease in primates.

Ebola Reston inspired great public awareness and terror of hemorrhagic diseases. This infamous virus is named for the United States military primate research facility where the virus was isolated during a 1989 outbreak of the disease in research primates. At that time, there was fear that Ebola fever could spread to neighboring Washington, DC. Study of the cause of this outbreak determined that Ebola viruses could remain infectious after becoming dispersed in the air. Whether inhalation of the virus plays a major role in the development of the

hemorrhagic fever is not clear. The current consensus is that airborne transmission is possible, but is not the principle route of infection.

Other hemorrhagic viruses can be spread by air. These include the Marburg, Lassa, Congo-Crimean, and Hantaviruses.

The Junin virus causes the hemorrhagic fever known as Argentine hemorrhagic fever. The virus was discovered in 1955, during a disease outbreak among corn harvesters in Argentina. It was later determined that the virus was spread to the workers by contact when rodent feces that had dried in the cornfields. The same route of transmission is used by the Machupo virus, which causes Bolivian hemorrhagic fever.

Congo-Crimean hemorrhagic fever is transmitted to people by ticks. The tick is likely not the natural reservoir host of the virus, but acquires the virus when it feeds on the natural reservoir host. This identity of the host is not known. This hemorrhagic fever occurs in the Crimea and in regions of Africa, **Asia**, and **Europe**.

Another hemorrhagic fever called Rift Valley fever occurs mainly in Africa. Like Ebola, it cause explosive outbreaks of disease.

Hantavirus disease was first described around the time of World War II, in Manchuria. United Nations troops stationed in Korea during the Korean War in the 1950s were sickened with the disease. A lung infection caused by the virus, which can rapidly progress to death, became prominent because of an outbreak in the southwestern region of the United States in the mid-1990s. Like some of the other hemorrhagic fevers, Hantavirus Pulmonary Syndrome is caused by inhalation of dried rodent feces.

Many of the above hemorrhagic fevers were discovered only in the past 50 to 75 years. Other hemorrhagic fevers have a longer history. For example, yellow fever was discovered in the first decade of the twentieth century, when a disease outbreak occurred among workers who were constructing the Panama Canal.

The **diagnosis** of hemorrhagic fevers often requires knowledge of the recent travel of the patient. This helps to clarify what natural hosts the patient may have come in contact with.

Vaccine and treatment

As of 2003, the only licensed **vaccine** for a hemorrhagic fever is that available for yellow fever. The vaccine consists of live virus particles that have been modified so as not to be capable of growth or of causing an infection. The virus is capable of stimulating the **immune system** to produce antiviral antibodies. The vaccine must be taken by those who are traveling to areas of

KEY TERMS

. . . .

Hemorrhagic—Involving life-threatening bleeding.

Reservoir host—The animal or organism in which the virus or parasite normally resides.

Vector—Any agent, living or otherwise, that carries and transmits parasites and diseases.

Zoonoses—The transmission of disease to humans from an animal.

Tao, H. *Atlas of Hemorrhagic Fever with Renal Syndrome.* Thousand Oaks, CA: Science press, 1999.

Organizations

Centers for Disease Control and Prevention, Special Pathogens Branch, National Center for Infectious Diseases, MS A–26, 1600 Clifton Road, Atlanta, GA 30333. (404) 639-1510. February 8, 2002 [cited November 12, 2002] <http://www.cdc.gov/ncidod/dvrd/spb/mnpages/dispages/vhf.htm>.

Brian Hoyle

the world where yellow fever is actively present (areas of Africa and **South America**). The vaccine may have some potential in protecting people from the virus that causes Bolivian hemorrhagic fever.

Vaccines to Rift Valley fever are under development. But these are still undergoing testing and so are not publicly available. Vaccines have not been developed to the other hemorrhagic fevers. An antiviral drug called ribavirin shows potential against Lassa fever. Unfortunately, the drug has caused mutations in test animals. Thus, its use on humans carries a risk. In determining whether or not to administer ribavirin, the risk of its use is weighed against the urgency of the illness.

At the present time, the best treatments for hemorrhagic fevers are isolation of the infected patient and care when handling the patient. For example, health care workers should be dressed in protective clothing, including gloves and protective facemask. Also, any material or equipment that comes into contact with a patient should be sterilized to kill any virus that may have adhered to the items.

The devastating infection caused by the hemorrhagic viruses is remarkable given the very small amount of genetic material that the viruses contain. For example, Ebola viruses can produce less than 12 **proteins**. How the viruses are able to evade the host immune responses, and establish infections is unknown. The virus may commandeer the host's genetic material to produce proteins that it is unable to produce. Or, hemorrhagic viruses may be exquisitely designed infection machines, containing only the resources needed to evade the host and establish an infection. Sequencing of the genetic material of hemorrhagic viruses will help distinguish between these two possibilities.

See also Immunology; Zoonoses.

Resources

Books

Specter, S.C., R.L. Hodinka, and S.A. Young. *Clinical Virology,* 3rd ed. Washington, DC: American Society for Microbiology Press, 2000.

Hemp

Hemp, or *Cannabis sativa*, is a tall, annual **plant** that thrives in temperate and subtropical climates. It is native to central and western **Asia**, and is one of the oldest cultivated plants. The word "hemp" is derived from the old English word "hanf," and refers to both the plant and the long fibers that are processed from its stems. The most common use of hemp has been as a source of fiber for manufacturing rope, canvas, other **textiles**, and **paper**. Hemp contains more than 400 biochemicals, and has been used for medicinal purposes for at least 3,000 years. Even today, it is useful as a treatment for **cancer** and **AIDS** patients, because its stimulatory effect on the appetite can help victims of these diseases to avoid weight loss. During the twentieth century hemp gained notoriety as the source of **marijuana**, a psychoactive drug banned in most countries.

Hemp is a dioecious plant, meaning there are separate male and female plants. It is an annual, herbaceous plant that can grow as tall as 10–20 ft (3–6 m). Hemp can be cultivated in a wide range of climates having adequate amounts of **sun** and moisture during the summer. It has a relatively short growing season, and, in the Northern Hemisphere, is planted in May and harvested in September. As hemp grows it improves **soil** quality somewhat, and reduces the abundance of weeds by casting a dense shade over the ground surface.

Hemp has been grown for at least 5,000 years to obtain its stem fibers for weaving cordage and textiles. Its fibers can be used for manufacturing rope, canvas, and other materials. Its **seeds** can be pressed for oil, which is used for making paint, heating and lubricating oils, **animal** feed, and pharmaceutical products. The plant also produces a sap rich in silica, which can be used for making **abrasives**.

For centuries, hemp was the largest cash crop in the world. As recently as 1941, U.S. farmers were encouraged by the federal government to grow hemp, because of the need for its fibers to make rope, parachutes, backpacks, hoses, and other necessities during World War II.

The cultivation of hemp has been outlawed in the United States and many other countries because its flowering buds, and to a lesser degree its foliage, are the source of the drug marijuana. The buds produce a yellow resin that contains various cannabinoid chemicals. Of these, delta-9-tetrahydrocannabinol, or THC, has the most psychoactive activity. THC combines with receptor sites in the human **brain** to cause drowsiness, increased appetite, giddiness, hallucinations, and other psychoactive effects. Although the causative mechanisms are not fully known, current research indicates that THC ingestion results in THC binding to receptor sites associated with measurable **memory** loss. Other studies correlate THC binding to receptors in the cerebellum and correlated decreases in motor coordination and/or the ability to maintain balance.

Plant breeders have now produced varieties of hemp with concentrations of THC that are too low for the plants to be used as a medicine or recreational drug. In Canada and many other countries, permits are being granted to allow farmers to grow low-THC hemp as a source of valuable fiber. Although United States Federal law prohibits hemp growth, as of 2002 eight states had passed state legislation authorizing industrial hemp research.

In 2002, a United States federal appeals court blocked a Drug Enforcement Administration (DEA) rule that attempted to ban food made with hemp. Prior to the ruling the DEA—relying on the Controlled Substances Act—banned food products containing tetrahydrocannabinol (THC). At press, an appeal was pending.

See also Natural fibers.

Resources

Books

Bosca, I., and M. Karus. *The Cultivation of Hemp: Botany, Varieties, Cultivation, and Harvesting.* Hemptech Pub., 1998.

Henna

Henna, **species** *Lawsonia intermis* of the family Lythracea, is a perennial shrub that grows wild in northern **Africa** and southern **Asia**. The name henna, which comes from the Arabic word *al kenna*, refers to both the **plant** and the dye that comes from the leaves. The henna plant has narrow, grayish green leaves and small, sweet smelling clustered flowers that are white, yellow, or rose in **color**.

One of the oldest known hair dyes, henna is still used worldwide. The leaves are dried, pulverized, mixed with hot **water**, and then made into a paste. The paste is applied to the hair, and later rinsed out, leaving a reddish tint. Women in Muslim countries use henna to color their nails, hands, feet, and cheeks. In India, some brides use henna to stain a beautifully intricate design on their hands. The Berbers of North Africa believe henna represents **blood** and fire, and that it links humankind to nature. Henna is used in Berber marriage ceremonies because it is thought that henna has special seductive powers and that it symbolizes youth. Henna dye is used to stain leather and horses' hooves and manes. Some mummies have been found wrapped in henna-dyed cloth.

The active dye ingredient in henna is hennotannic acid, or lawsone. Henna powder is available commercially; the quality depends on where the plant was grown and what part of the plant was used to make the dye. Sometimes henna is mixed with other plant dyes, such as indigo or coffee, to obtain other hues. The henna plant also produces an aromatic oil used as perfume. Henna is now grown commercially in Morocco, China, and **Australia**.

See also Dyes and pigments.

Hepatitis

Hepatitis is **inflammation** of the liver, a potentially life-threatening **disease** most frequently caused by viral infections but which may also result from liver damage caused by toxic substances such as **alcohol** and certain drugs. Hepatitis viruses identified to date occur in five types: hepatitis A (HAV), hepatitis B (HBV), hepatitis C (HCV), hepatitis D (HDV), and hepatitis E (HEV). All types are potentially serious and, because clinical symptoms are similar, positive identification of the infecting strain is possible only through serologic testing (analyzing the clear, fluid portion of the **blood**). Symptoms may include a generalized feeling of listlessness and fatigue, perhaps including mental **depression**, nausea, vomiting, lack of appetite, dark urine and pale feces, **jaundice** (yellowing of the skin), **pain** in the upper right portion of the abdomen (where the liver is located), and enlargement of both the liver and the spleen. Severe cases of some types of hepatitis can lead to scarring and fibrosis of the liver (**cirrhosis**), and even to **cancer** of the liver. Epidemics of liver disease were recorded as long ago as Hippocrates' time and, despite major advances in **diagnosis** and prevention methods over the past two decades, viral hepatitis remains one of the most serious global health problems facing humans today.

Hepatitis A virus

The incidence and spread of HAV is directly related to poor personal and social hygiene and is a serious problem not only in developing countries where sanitation and **water** purification standards are poor, but also in developed, industrialized nations—including the United States, where it accounts for 30% of all incidences of clinical hepatitis. Except in 1% to 4% of cases where sudden liver failure may result in death, chronic liver disease and serious liver damage very rarely develop, and "chronic carrier state," in which infected people with no visible symptoms harbor the **virus** and transfer the disease to non-infected individuals, never occurs. Also, re-infection seldom develops in recovered HAV patients because the body eventually develops antibodies, cells which provide a natural immunity to the specific virus attacking the host. Although HAV is self-limiting (after time, ends as a result of its own progress), there is as yet no effective treatment once it is contracted.

Symptoms and transmission

Apart from the symptoms described above, HAV commonly produces a medium-grade fever, diarrhea, headaches, and muscle pain. The primary route of HAV transmission is fecal-oral through ingestion of water contaminated with raw sewage, raw or undercooked shell-fish grown in contaminated water, food contaminated by infected food handlers, and close physical contact with an infected person. Heterosexual and homosexual activities with multiple partners, travel from countries with low incidences to countries with high rates of infected population, and, less frequently, blood transfusions and illicit intravenous drug use also spread **infection**.

During the infectious stage, large numbers of viruses are eliminated with the stool. Although HAV infection occurs in all age groups, high rates of disease transmission occur in day-care centers and nursery schools where children are not yet toilet trained or able to wash their hands thoroughly after defecating. The disease may then be transmitted to day-care workers and carried home to parents and siblings. In areas of the world where living quarters are extremely crowded and many people live in unhygienic conditions, large outbreaks of HAV threaten people of all ages. Because during the viruses' incubation period—from 14 to 49 days—no symptoms are observable, and because symptoms seldom develop in young children, particularly those under the age of two, the disease is often unknowingly but readily transmitted before infected people can be isolated.

Prevention and control

A **vaccine** against HAV is available. It appears to provide good protection, if the first immunization has been received at least four weeks prior to exposure. For adults, two immunizations about six months apart are recommended; for children, three immunizations are necessary (two a month apart, and the third six months later). High-risk groups who should receive HAV vaccine include child care workers, military personnel, Alaskan natives, frequent travelers to HAV **endemic** areas, laboratory technicians where HAV is handled, and people who work with **primates**. The immunization lasts for 20 years.

If someone who is unimmunized is exposed to HAV, or if a traveler cannot wait four weeks prior to departure for an HAV endemic area, then immune globulin may be utilized to avoid infection. Immune globulin is a naturally-occurring substance harvested from the **plasma** in human blood, then injected into an individual exposed to the HAV. Immune globulin prevents disease development in 80% to 90% of cases in clinical trials. It also seems to be effective in reducing the number of cases normally expected after outbreaks in schools and other institutions. As yet, the most effective control mechanisms are public education regarding the importance of improved personal hygiene, which in many instances is as simple as washing hands thoroughly after using the toilet and before handing food, and concerted worldwide efforts to purify water supplies (including **rivers** and oceans) and improve sanitation methods.

Hepatitis B virus

Acute HBV is the greatest cause of viral hepatitis throughout the world. World Health Organization figures released in 1992 indicate that as many as 350 million people worldwide carry the highly infectious HBV. Because of its severity and often lengthy duration, coupled with the lack of any effective treatment, 40% of those carriers—possibly as many as two million per year—will eventually die from resultant liver cancer or cirrhosis. HBV-related liver cancer deaths are second only to tobacco-related deaths worldwide. Infected children who survive into adulthood may suffer for years from the damage caused to the liver. In the United States alone, as many as 300,000 people become infected with HBV every year, medical costs amount to more than $1 million per day, and the death **rate** over the last 15 or so years has more than doubled in the U.S.A. and Canada.

If serology tests detect the presence of HBV six months or more from time of initial diagnosis, the virus is then termed "chronic." Chronic persistent hepatitis may develop following a severe episode of acute HBV. Within a year or two, however, this type usually runs its course and the patient recovers without serious liver damage. Chronic active hepatitis also may follow a severe attack of acute HBV infection, or it may simply develop almost unnoticed. Unlike persistent hepatitis, the chronic active type usually continues until fatal liver damage occurs. In long-term studies of 17 patients with chronic active hepatitis, 70% developed cirrhosis of the liver within two to five years. Fortunately, this type of hepatitis is rarely seen in children. Several modes of treatment—including the use of steroids—have been relatively unsuccessful, and treatment with corticosteroids, while appearing at first to have some positive benefit, actually cause additional liver damage.

Symptoms and transmission

Symptoms are similar to those manifested by HAV and may include weight loss, muscle aches, headaches, flu-like symptoms, mild **temperature** elevation, and constipation or diarrhea. By the time jaundice appears, which is often quite noticeable and prolonged in older women, the patient may feel somewhat better overall but the urine becomes dark, stools light or yellowish, the liver and possibly the spleen enlarged and painful, and fluid may accumulate around the abdominal area. Early in the disease's life, however, symptoms may be very slight or even virtually nonexistent—particularly in children—facilitating infection of others before isolation is implemented.

The incubation period for HBV varies widely—anywhere from four weeks to six months. Primary routes of transmission are blood or blood product transfusion; body fluids such as semen, blood, and saliva (including a bite by an infected human); **organ** and/or **tissue** transplants; contaminated needles and syringes in hospitals or clinical settings; contaminated needles or syringes in illegal intravenous drug use; and "vertical" transmission-from mother to baby during pregnancy, **birth**, or after birth through breast milk. Even though they may not develop symptoms of the disease during childhood, and will remain healthy, almost all infected newborns become "chronic carriers," capable of spreading the disease. Many of these infected yet apparently healthy children—particularly the males—will develop cirrhosis and liver cancer in adulthood. Where the incidence of the disease is relatively low, the primary **mode** of transmission appears to be sexual and strongly related to multiple sex partners, particularly in homosexual men. In locations where disease prevalence is high, the most common form of transmission is from mother to infant.

Prevention and control

Controlling HBV infection is an overwhelming task. In spite of the development of safe and effective vaccines capable of preventing HBV in uninfected individuals, and regardless of programs designed to vaccinate adults in high-risk categories such as male homosexuals, prostitutes, intravenous drug users, health-care workers, and families of people known to be carriers, the disease still remains relatively unchecked, particularly in developing countries.

Although effective vaccines have been available since the mid-1980s, the cost of mass immunization world-wide, and particularly in developing countries, was initially prohibitive, while immunizing high-risk adult populations did little to halt the spread of infection. Authorities now believe the most effective disease control method will be immunization of all babies within the first week following birth. Concerted efforts of researchers and health authorities worldwide, including the foundation in 1986 of an International Task Force for Hepatitis B Immunization are investigating various avenues for providing cost-effective, mass vaccination programs. These include incorporating HBV vaccination into the existing Expanded Program of Immunization controlled by the World Health Organization. Methods of cost containment, storing the vaccine, and distribution to midwives in remote villages (60% of the world's births occur at home), have been designed and are continually being refined to ultimately attain the goal of universal infant immunization. This will not only drastically decrease the number of babies infected through vertical transmission (which constitutes 40% of all HBV transmission in **Asia**), preventing them from becoming adult carriers, it provides immunity throughout adulthood.

Finding an effective treatment for those infected with HBV presents a major challenge to researchers—a challenge equal to that posed by any other disease which still remains unconquered. And HBV may present yet another challenge: mutant forms of the virus seem to be developing in resistance to the current vaccines, thus finding a way to survive, replicate and continue its devastating course. Necessary measures in disease control include: education programs aimed at health care workers to prevent accidental HBV transfer—from an infected patient to an uninfected patient, or to themselves; strict controls over testing of blood, blood products, organs, and tissue prior to transfusion or transplantation; and the "passive" immunization with immunoglobulin containing HBV antibodies as soon as possible after exposure to the active virus.

Hepatitis C and E viruses

These relatively recently discovered viruses, often called non-A, non-B hepatitis, exist in more than 100 million carriers worldwide, with 175,000 new cases developing each year in the U.S. and **Europe**.

Hepatitis C virus

Not until 1990 were tests available to identify HCV. Research since then has determined that HCV is distributed globally and, like HBV, is implicated in both acute and chronic hepatitis, as well as liver cancer and cirrhosis. Eighty-five **percent** of all transfusion-related hepatitis is caused by HCV, and mother-baby and sexual transmission are also thought to spread the disease. Symptoms are similar but usually less severe than HBV; however, it results in higher rates of chronic infection and liver disease.

Prevention and control

Control and prevention of HCV is a serious problem. First, infected people may show no overt symptoms and the likelihood that infection will become chronic means that many unsuspecting carriers will transmit the disease. Second, HCV infection does not appear to stimulate the development of antibodies, which not only means infected people often become reinfected, it creates a major challenge in the development of an effective vaccine. Third, HCV exists in the same general high-risk populations as does HBV. Combined, these factors make reducing the spread of infection extremely difficult. On a positive note, the development of accurate blood screening for HCV has almost completely eliminated transfusion-related spread of hepatitis in developed countries. Immunoglobulin injections do not protect people who have been exposed to HCV; the search is on for an adequate immunization, although this effort is hampered by characteristics of HCV, which include rapid **mutation** of the virus.

Hepatitis E virus

Undiscovered until 1980, HEV is believed to transmit in a similar fashion to HAV. HEV is most prevalent in India, Asia, **Africa**, and Central America. Contaminated water supplies, conditions which predispose to poor hygiene (as in developing countries), and travel to developing countries all contribute to the spread of HEV. Symptoms are similar to other hepatitis viruses and—like HAV—it is usually self-limiting, does not develop into the chronic stage, and seldom causes fatal liver damage. It does seem, however, that a higher percentage of pregnant women (from 10%-20%) die from HEV than from HAV.

Prevention and control

Research into the virus was slow because of the limited amounts which could be isolated and collected from both naturally infected humans and experimentally infected primates. Recently, successful genetic cloning (artificial duplication of genes) is greatly enhancing research efforts. Surprisingly, research found that antibodies exist in between 1%-5% of people who have never been infected with hepatitis. Until an effective vaccine is developed, sanitation remains the most important factor in preventing the spread of HEV.

KEY TERMS

....

Carrier—An individual who has a particular bacteria present within his/her body, and can pass this bacteria on to others, but who displays no symptoms of infection.

Coinfection—Infecting together requiring at least one other infectious organism for infection.

Self-limiting—Runs its course, ends or dies out as a result of its own progress.

Hepatitis D virus

Because it is a "defective" virus requiring "coinfection" with HBV in order to live and reproduce, HDV alone poses no threat in the spread of viral hepatitis. It also poses no threat to people vaccinated against HBV. However, when this extremely infectious and potent virus is contracted by unsuspecting carriers of HBV, rapidly developing chronic and even fatal hepatitis often follows. The coexistent requirements of HDV as yet remain unclear. Research into development of an effective vaccine is ongoing, and genetic cloning may aid in this effort.

Hepatitis G virus

Little is currently known about a relatively recently discovered hepatitis virus, G. HGV appears to be passed through contaminated blood, as is HCV. In fact, many infections with HVG occur in people already infected with HCV. HGV, however, does not seem to change the disease course in people infected with both HCV and HGV. In cases of isolated HGV infection, little liver injury is noted, and there does not appear to be a risk of chronic liver injury. Much more information must be sought about this particular hepatitis virus, and its risks.

See also Epstein-Barr virus; Tuberculosis; Vaccine.

Resources

Books

Kurstak, E. *Viral Hepatitis-Current Status and Issues.* New York: Springer-Verlag, 1993.

Nishioka, K., et al. *Viral Hepatitis and Liver Disease.* New York: Springer- Verlag, 1993.

Marie L. Thompson

Herb

An herb is an aromatic **plant** that is used by people most commonly in cooking, but sometimes for medicinal purposes, as an insect repellant, as a source of dye, and sometimes for their attractive aesthetics. Herbs are not necessarily plants that are taxonomically related to each other—what these plants share is a usefulness to humans, not an evolutionary lineage.

In general, herbs are non-woody plants that are grown from seed, and they can be annual, biennial, or perennial **species**. Plants that grow from bulbs, such as the species of crocus (*Crocus sativus*) that saffron is derived from, are not considered to be herbs. Nor are aromatic woody plants, such as the sweet bay (*Laurus nobilis*) or common **pepper** (*Piper nigrum*), which are considered to be spices.

There is a wide variety of herbs that are commonly cultivated. A few of the ones that are frequently used as foods are briefly described below.

The parsley (*Petroselinum hortense*) is a biennial plant in the carrot family (Umbelliferae or Apiaceae). The original range of this species was the Mediterranean region, from Spain to Greece. This aromatic plant is commonly used to flavor cooked meals, and as an attractive garnish of other foods. A variety known as the turnip-parsley (*P. h. tuberosum*) is cultivated for its thick, aromatic root, which is used in soups and stews.

Dill (*Anethum graveolus*) is another member of the carrot family, also native to the Mediterranean region. It is an annual plant, and is used to flavor a wide range of cooked dishes, as well as pickled cucumbers and other **vegetables**.

Caraway (*Carum carvi*) is a biennial umbellifer. The **seeds** of caraway are mostly used to flavor cheeses and breads, and also a liqueur known as kummel. The seeds of anise or aniseed (*Pimpinella anisum*) are used to flavor foods, to manufacture candies, and a liqueur known as anisette.

A number of herbs are derived from species in the **mint family** (Menthaceae). The common mint (*Mentha arvensis*), spearmint (*M. spicata*), and peppermint (*M. piperita*) are used to flavor candies, chewing gum, and toothpaste, and are sometimes prepared as condiments to serve with meats and other foods. Sweet marjoram (*Origanum majorana*) is used to flavor some cooked meats and stews. Common sage (*Salvia officinalis*) is used to flavor cooked foods, and in toothpaste and mouthwash.

Other herbs are derived from plants in the **mustard family (Brassicaceae)**. The seeds of mustard (*Brassica alba*), garden cress (*Lepidium sativum*), and white mustard (*Sinapis alba*) are ground with vinegar to produce spicy condiments known as table mustard. The root of horse radish (*Cochlearia armoracea*) is also ground with vinegar to produce a sharp-tasting condiment, often served with cooked meats.

Although they may be nutritious in their own right, most herbs are too strong tasting to be eaten in large quantities. However, these plants provide a very useful service by enhancing the flavor of other foods. Many people are great fans of the use of herbs, and they may grow a diversity of these plants in their own herb gardens, to ensure a fresh supply of these flavorful and aromatic plants.

Bill Freedman

Herbal medicine

Modern medicine has provided many breakthrough treatments for serious diseases. Some conditions, however, have eluded the healing grasp of contemporary western medicine, which emphasizes rigorous scientific investigation of therapies. In addition, rising costs of some treatments have placed modern healthcare beyond the reach of many. The drugs that routinely fill pharmacy shelves of post-industrialized nations remain inaccessible to the majority of the people in the world. Instead, populations in many areas of the globe use herbal medicine, also called botanical medicine or phytotherapy, as the principal means of healthcare. Herbal medicine is the use of natural **plant** substances to treat illness. Based upon hundreds, even thousands of years of experience, herbal medicine provides an alternative to modern medicine, making healthcare more available. In fact, the majority of the world's population uses **herb** products as a primary source of medicine. While some regulating authorities fear the consequences of unrestricted herbal remedy use, herbal medicine offers a degree of hope to some patients whose **disease** states do not respond favorably to modern pharmaceuticals. More often, however, herbal remedies are used to treat the common ailments of daily living like indigestion, sleeplessness, or the common cold. A resurgence in interest in herbal medicine has occurred in the United States as medical experts have begun to recognize the potential benefit of many herbal extracts. So popular has herbal medicine become that scientific clinical studies of the effectiveness and proper dosing of some herbal medicines are being investigated.

Herbal medicine recognizes the medicinal value of plants and plant structures such as roots, stems, **bark**, leaves, and reproductive structures like **seeds** and flowers. To some, herbal medicine may seem to be on the fringes of medical practice. In reality, herbal medicine has been in existence since prehistoric time and is far more prevalent in some countries than is modern health-

Woman herbalist preparing Chinese herbal medication.
© 1995 Eric Nelson. Custom Medical Stock Photo, Inc.
Reproduced by permission.

care. The use of herbs ground into powders, filtered into extracts, mixed into salves, and steeped into teas has provided the very foundation upon which modern medicine is derived. Indeed, herbal medicine is the history of modern medicine. Many modern drugs are compounds that are derived from plants whose pharmacological effects on humans had been observed long before their mechanisms of action were known. A common example is aspirin. Aspirin, or **acetylsalicylic acid**, is a compound found in the bark of the willow **tree** belonging to the taxonomic genus Salix. Aspirin, now sold widely without prescription, is an effective analgesic, or **pain** reliever, and helps to control mild swelling and fever. While aspirin is synthetically produced today, willow bark containing aspirin was used as an herbal remedy long before chemical synthesis techniques were available. Similarly, the modern cardiac drug **digitalis** is derived from the leaves of the purple foxglove plant, *Digitalis purpurea*. Foxglove was an herbal known to affect the **heart** long before it was used in modern scientific medicine.

A prime example of the prevalence of herbal medicine in other cultures is traditional Chinese medicine. Herbal remedies are a central aspect of traditional Asian medical practices that have evolved from ancient societies. The philosophical and experimental background of Chinese herbal medicine was established more than two thousand years ago. Large volumes of ancient Chinese medical knowledge, largely concerning herbs, have been preserved which chronicle wisdom gathered throughout periods of history. Some of the information is dated to about 200 B.C. One Chinese legend tells of how Shen Nung, the ancient Chinese father of agriculture, tested hundreds of herbs for medical or nutritional value. Many herbs from Chinese traditional medicine have documented pharmacological activity. Ma Huang, also called Chinese ephedra, is an example. This herb, *Ephedra sinica*

has a potent chemical within its structures called ephedrine. Ephedrine is a powerful stimulant of the sympathetic **nervous system**, causing widespread physiological effects such as widening of breathing passages, constriction of **blood** vessels, increased heart **rate**, and elevated blood **pressure**. Ephedrine, whether from Ma Huang or modern medication preparations, mimics the effects of adrenaline on the body. Modern medicine has used ephedrine to treat **asthma** for years. Chinese traditional herbal medicine has been using Ma Huang to treat disease for many hundreds of years.

The term **alternative medicine** is often used to describe treatments for disease that do not conform to modern medical practices, including herbal medicine. Alternative medicine includes things such as apitherapy, the use of bee stings to treat neurological diseases. Apitherapy is used by some to treat multiple sclerosis, a degenerative nerve disease that can cripple or blind its victims. Also, alternative medicine includes scientifically unfounded therapies such as kinesiology (the healing properties of human **touch**), **acupuncture**, aromatherapy, meditation, massage therapy, and homeopathy. Aromatherapy and homeopathy are closely related to herbal medicine because they both use botanical, or plant, extracts. Aromatherapy uses the strong odors from essential oils extracted from plants to induce healing and a sense of well being. Homeopathy is the art of healing the sick by using substances capable of causing the same symptoms of a disease when administered to healthy people. Many homeopathic remedies are herbal extracts. Homeopathic medicine has been practiced for over 200 years. The German physician, Samuel Hahnemann, began the practice of homeopathy using herbs in 1796. The philosophy behind this form of herbal medicine is to induce the body to heal itself. The use of herbals in homeopathic treatment follows the unscientific principle of "Let likes be cured by likes."

Homeopathic remedies, and herbal remedies in general, are primarily used in alleged self-care, without the help of a physician. Because many remedies have genuine effects, the United States government regulates the sale of homeopathic substances. The Homeopathic Pharmacopoeia of the United States (HPUS) is the official list of accepted remedies that the law uses as standard. Along with the United States Pharmacopoeia and National Formulary (USP/NF) that lists all regulated drugs and drug products, the HPUS is the legal source of information for the Federal Food, Drug, and Cosmetics Act. Standards for manufacture, purity, and sale of drugs are listed in these documents, enforced by law. Many people are concerned that herbal medicine products that are currently widely available are a danger to public health, safety, and welfare because an official federal pharmacopoeia for herbals does not yet exist. Therefore, few legal requirements exist

KEY TERMS

Aromatherapy—The use of odorous essential oils from herbs to heal and induce feelings of well-being.

Homeopathy—A system in which diluted plant, mineral, or animal substances are given to stimulate the body's natural healing powers. Homeopathy is based upon three principles: the law of similars, the law of infinitesimal dose, and the holistic medical model.

Pharmacopoeia—An official, and legal listing of approved drugs, drug manufacture standards, and use enforced by legislation.

for the manufacture, dose standardization, labeling, and sale of preparations for herbal medicines. Yet, herbal remedies are the fastest growing segment of the supplemental health product industry. Such problems with purity and dosage only add to skepticism regarding the therapeutic value of many herbals. Most of the health claims made by advertisements have not been evaluated scientifically.

Examples of herbal medicine products in wide use today are St. Johns Wort for **depression**, Echinacea for increased immune function, Saw Palmetto for prostate gland problems in men, and **ginkgo** biloba for improved mental functioning and headaches. Other forms of herbal medicine in popular culture include herbal teas, like Chamomile tea used to help people who have trouble sleeping and peppermint tea to calm stomach and digestive problems.

Resources

Books

Barney, D. Paul. *Clinical Applications of Herbal Medicine.* Woodland Publishing, 1996.

O'Neil, Maryadele J. *Merck Index: An Encyclopedia of Chemicals, Drugs, & Biologicals.* 13th ed. Whitehouse Station, NJ: Merck & Co., 2001.

Selby, Anna. *The Ancient and Healing Art of Chinese Herbalism.* Ulysses Press, 1998.

Sravesh, Amira A. *The Alchemy of Health: Herbal Medicine and Herbal Aromatherapy.* Amira Alchemy, 1998.

Taylor, Leslie. *Herbal Secrets of the Rainforest: Over 50 Powerful Herbs and Their Medicinal Uses.* Rocklin, CA: Prima Publishing, 1998.

Wood, Matthew. *The Book of Herbal Wisdom: Using Plants as Medicine.* North Atlantic Books, 1997.

Organizations

Rainforest Alliance. <http://www.rainforest-alliance.org> (March 2003).

Terry Watkins

Herbicides

A herbicide is a chemical used to kill or otherwise manage certain **species** of plants considered to be **pests**. **Plant** pests, or weeds, compete with desired crop plants for **light**, **water**, **nutrients**, and space. This ecological interaction may decrease the productivity and yield of crop plants, thereby resulting in economic damage. Plants may also be judged to be weeds if they interfere with some desired aesthetic effect, as is the case of weeds in lawns.

Clearly, the designation of plants as weeds involves a human judgment. However, in other times and places weeds may be judged to have positive values. For example, in large parts of **North America**, the red raspberry (*Rubus strigosus*) is widely considered to be one of the most important weeds in **forestry**. However, this species also has positive attributes. Its **fruits** are gathered and eaten by people and **wildlife**. This vigorously growing plant also provides useful ecological services. For example, it binds **soil** and helps prevent **erosion**, and takes up nutrients from the soil, which might otherwise be leached away by rainwater because there are so few plants after disturbance of the site by clear-cutting or **wildfire**. These ecological services help to maintain site fertility.

Still, it is undeniable that in certain situations weeds exert a significant interference with human purposes. To reduce the intensity of the negative effects of weeds on the productivity of desired agricultural or forestry **crops**, fields may be sprayed with a herbicide that is toxic to the weeds, but not to the crop species. The commonly used herbicide 2,4-D, for example, is toxic to many broad-leaved (that is, dicotyledonous) weeds, but not to **wheat**, maize or corn, **barley**, or **rice**, all of which are members of the grass family (Poaceae), and therefore monocotyledonous. Consequently, the pest plants are selectively eliminated, while maintaining the growth of the desired plant species.

Modern, intensively managed agricultural systems have an intrinsic reliance on the use of herbicides and other **pesticides**. Some high-yield varieties of crop species are not very tolerant of **competition** from weeds. Therefore, if those crops are to be successfully grown, herbicides must be used. Many studies have indicated the shorter-term benefits of herbicide use. For example, studies of the cultivation of maize in Illinois have demonstrated that the average reduction of yield was 81% in unweeded plots, while a 51% reduction was reported in Minnesota. Yields of wheat and barley can be reduced by 25-50% as a result of competition from weeds. To reduce these important, negative influences of weeds on agricultural productivity, herbicides are commonly applied to agricultural fields. As noted above, the herbicide must be toxic to the weeds, but not to the crop species.

Types of herbicides

The most important chemical groups of herbicides are chlorophenoxy acids such as 2,4-D and 2,4,5-T; triazines such as atrazine, hexazinone, and simazine; organic **phosphorus** chemicals such as glyphosate; **amides** such as alachlor and metolachlor; thiocarbamates such as butylate; dinitroanilines such as trifuralin; chloroaliphatics such as dalapon and trichloroacetate; and inorganic chemicals such as various arsenicals, cyanates, and chlorates. The first three of these groups are described in more detail below.

Chlorophenoxy acid herbicides

Chlorophenoxy acid herbicides cause toxicity to plants by mimicking their natural hormone-like auxins, and thereby causing lethal growth abnormalities. These herbicides are selective for broad-leaved or **angiosperm** plants, and are tolerated by monocots and conifers at the spray rates normally used. These chemicals are moderately persistent in the environment, with a **half-life** in soil typically measured in weeks, and a persistence of a year or so. The most commonly used compounds are 2,4-D (2,4-Dichlorophenoxyacetic acid); 2,4,5-T (2,4,5-Trichlorophenoxyacetic acid); MCPA (2-Methyl-4-chlorophenoxyacetic acid); and silvex [2-(2,4,5-Trichlorophenoxy)-propionic acid].

Triazine herbicides

Triazine herbicides are mostly used in corn agriculture, and sometimes as soil sterilants. These chemicals are not very persistent in surface soils, but they are mobile and can cause a **contamination** of **groundwater**. Important examples of this class of chemicals are: atrazine [2-Chloro-4-(ethylamino)-6-(isopropylamino)-s-triazine]; cynazine [2-(4-Chloro-6-ethylamino -5-triazin-2-ylamino)-2-methylpropionitrile]; hexazinone [3-Cyclohexyl-6-(dimethyl-amino)-1-methyl-1,3,5-triazine-2,4(1H,3H)-dione]; metribuzin [4-Amino-6-tert-butyl-3-(methylthio)-as-triazin-5(4H)-one]; and simazine [2-chloro-4,6-bis-(ethyl-amino)-s-triazine].

Organic phosphorus herbicides

Organic phosphorus herbicides are few, but they include the commonly used chemical, glyphosate (N-phosphonomethyl-glycine). Glyphosate has a wide range of agricultural uses, and it is also an important herbicide in forestry. To kill plants, glyphosate must be taken up and transported to perennating tissues, such as roots and rhizomes, where it interferes with the synthesis of certain amino acids. Because glyphosate can potentially damage many crop species, its effective use requires an understanding of seasonal changes in the vulnerability of both weeds and crop species to the herbicide. Glyphosate is not mobile in soils, has a moderate persistence, and is not very toxic to animals. Recently, varieties of certain crops, notably the oilseed canola, have been modified through **genetic engineering (transgenics)** to be tolerant of glyphosate herbicide. Previously, there were no effective herbicides that could be applied to canola crops to reduce weed populations, but now glyphosate can be used for this purpose. However, this has become controversial because many consumers do not want to eat foods made from transgenic crops.

Use of herbicides

In 1990, a total of about 290 herbicidal chemicals were available for use. Many of these chemicals are used in various types of formulations, each of which is a specific combination of the herbicidal active ingredient, a solvent such as water or kerosene, and various chemicals intended to enhance the efficacy of the herbicide, for example, by increasing the ability of the spray to adhere to foliage, or to spread freely on **leaf** surfaces. In addition, different companies often manufacture and sell the same formulations under different names, so the number of commercial products is larger than the number of actual formulations.

The United States accounts for about one-third of the global use of pesticides, much more than any other country. In 1989, herbicides accounted for about 61% of the 1,100 million lb (500 million kg) of pesticides used domestically in the United States. During recent years, eight of the ten most commonly used pesticides in the United States have been herbicides. Listed in order of decreasing quantities used, these herbicides are: alachlor (100 million lb [45 million kg] used per year), atrazine (100 million lb [45 million kg]), 2,4-D (53 million lb [24 million kg]), butylate (44 million lb [20 million kg]), metolachlor (44 million lb [20 million kg]), trifluralin (31 million lb [14 million kg]), cynazine (20 million lb [9 million kg]), and metribuzin (13 million lb [6 million kg]). During the mid-1980s, amide herbicides accounted for 30% of the herbicides used in the United States, triazines 22%, carbamates 13%, N-anilines 11%, and phenoxys 5%. These data reflect a large decrease in the usage of phenoxy herbicides, which were used much more commonly prior to the 1980s. For example, during the mid-1970s, about 50-80% of the small-grain acreage in North America was treated with phenoxy herbicides, mostly with 2,4-D. Since the mid 1990s, the use of glyphosate has increased tremendously.

In terms of quantities applied, by far the major usage of herbicides is in agriculture. Intensive systems of cultivation of most major species of annual crops requires the use of herbicides. This is especially true of crops in the grass family. For example, at least 83% of the North American acreage of maize (or corn) cultivation involves treatment with herbicides. In part, herbicide use is important in maize cultivation because of the common use of **zero** tillage systems. Zero tillage involves direct seeding into unploughed soil, a system that has great benefits by reducing erosion and saving fuel, because tractors are used much less. However, one of the most important agricultural benefits of ploughing is the reduction of weeds that results. Consequently, zero tillage systems would be not be practical if they were not accompanied by the use of herbicides. This is only one example—most of the areas of grain crops cultivated in North America and other industrialized countries receive herbicide treatments.

Herbicides are also widely used in landscaping, mostly to achieve grassy lawns that are relatively free of broad-leaved weeds, which many people find unattractive. Herbicides are commonly used in this way by individual landowners managing the lawns around their home, and by authorities responsible for maintaining lawns around public buildings, along roadways, and in parks. Golf courses rely heavily on intensive use of herbicides. This is particularly true of putting greens, where it is important to have a very consistent lawn. In fact, the intensity of pesticide use on golf-course putting greens is greater than in almost any other usage in agriculture.

Forestry also uses herbicides. Usually, silvicultural herbicide use is intended to achieve a greater productivity of the desired **conifer** trees, by reducing the abundance of unwanted weeds. However, in most regions forestry usage of herbicides is much smaller than agriculture and lawn uses, typically less than 5% of the total use.

Herbicides were used extensively by the U.S. military during the Vietnam War. Large quantities of these chemicals were sprayed in Vietnam as a military tactic intended to deprive enemy forces and their supporters of agricultural production and forest cover. So-called "Agent Orange," a 1:1 mixture of 2,4,5-T and 2,4-D, was the most commonly used herbicide. Because the intention was to destroy **forests** and agricultural productivity, herbicides were used at about ten times the **rate** typically used in forestry for management purposes. This tactical strategy of war was labeled "ecocide" by its opponents, because of the severe damage that was caused to natural and agricultural ecosystems, and possibly to people (there is ongoing debate about whether scientific evidence actually demonstrates the latter damage). For these reasons, and also because **Agent Orange** was significantly contaminated by a very toxic chemical in the **dioxin** family, called TCDD, the military use of herbicides in Vietnam was extremely controversial.

Environmental effects of herbicide use

As has been suggested above, some substantial benefits can be gained through the use of herbicides to manage unwanted vegetation. Compared with alternative means of weed control, such as mechanically weeding by hand or machine, herbicides are less expensive, often safer (especially in forestry), faster, and sometimes more selective.

However, if herbicides are not used properly, damage may be caused to crop plants, especially if too large a dose is used, or if spraying occurs during a time when the crop species is sensitive to the herbicide. Unintended but economically important damage to crop plants is sometimes a consequence of the inappropriate use of herbicides.

In addition, some important environmental effects are associated with the use of herbicides. These include unintended damage occurring both on the sprayed site, and offsite. For example, by changing the vegetation of treated sites, herbicide use also changes the **habitat** of animals such as **mammals** and **birds**. This is especially true of herbicides use in forestry, because biodiverse, semi-natural habitats are involved. This is an indirect effect of herbicide use, because it does not involve toxicity caused to the **animal** by the herbicide. Nevertheless, the effects can be severe for some species. In addition, not all of the herbicide sprayed by a tractor or **aircraft** deposits onto the intended spray area. Often there is drift of herbicide beyond the intended spray site, and unintended, offsite damages may be caused to vegetation. There are also concerns about the toxicity of some herbicides, which may affect people using these chemicals during the course of their occupation (i.e., when spraying pesticides), people indirectly exposed through drift or residues on food, and wildlife. For these and other reasons, there are many negative opinions about the broadcast spraying of herbicides and other pesticides, and this practice is highly controversial.

The intention of any herbicide treatment is to reduce the abundance of weeds to below some economically acceptable threshold, judged on the basis of the amount of damage that can be tolerated to crops. Sometimes, this objective can be attained without causing significant damage to non-target plants. For example, some herbicides can be applied using spot applicators or injectors, which minimize the exposure to non-pest plants and animals. Usually, however, the typical method of herbicide application is some sort of broadcast application, in which a large area is treated all at once, generally by an aircraft or a tractor-drawn apparatus.

An important problem with broadcast applications is that they are non-selective—they affect many plants and animals that are not weeds—the intended target of the treatment. This is especially true of herbicides, because they are toxic to a wide variety of plant species, and not just the weeds. Therefore, the broadcast spraying of herbicides results in broad exposures of non-pest species, which can cause an unintended but substantial mortality of non-target plants. For example, only a few species of plants in any agricultural field or forestry plantation are abundant enough to significantly interfere with the productivity of crop plants. Only these competitive plants are weeds, and these are the only target of a herbicide application. However, there are many other, non-pest species of plants in the field or plantation that do not interfere with the growth of the crop plants, and these are also affected by the herbicide, but not to any benefit in terms of vegetation management. In fact, especially in forestry, the non-target plants may be beneficial, by providing food and habitat for animals, and helping to prevent erosion and **leaching** of nutrients.

This common non-target effect of broadcast sprays of herbicides and other pesticides is an unfortunate consequence of the use of this non-selective technology to deal with pest problems. So far, effective alternatives to the broadcast use of herbicides have not been discovered for the great majority of weed management problems. However, there are a few examples that demonstrate how research could discover pest-specific methods of controlling weeds that cause little non-target damage. These mostly involve weeds introduced from foreign countries, and that became economically important pests in their new habitats. Several weed species have been successfully controlled biologically, by introducing native herbivores of invasive weeds. For example, the klamath weed (*Hypericum perforatum*) is a European plant that became a serious pasture weed in North America, but it was specifically controlled by the introduction of two species of herbivorous leaf **beetles** from its native range. In another case, the prickly pear **cactus** (*Opuntia* spp.) became an important weed in **Australia** after it was introduced there from North America, but it has been successfully controlled by the introduction of a moth whose larvae feed on the cactus. Unfortunately, few weed problems can now be dealt with in these specific ways, and until better methods of control are discovered, herbicides will continue to be used in agriculture, forestry, and for other reasons.

Most herbicides are specifically plant poisons, and are not very toxic to animals. (There are exceptions, however, as is the case with the herbicide paraquat.) However, by inducing large changes in vegetation, herbicides can indirectly affect populations of birds, mammals, **insects**, and other animals through changes in the nature of their habitat.

For example, studies in Britain suggest that since the 1950s, there have been large changes in the populations

of some birds that breed on agricultural land. These changes may be partly caused by the extensive use of herbicides, a practice that has changed the species and abundance of non-crop plants in agroecosystems. This affects the structure of habitats, the availability of nest sites, the food available to granivorous birds, which mostly eat weed **seeds**, and the food available for birds that eat **arthropods**, which rely mainly on non-crop plants for nourishment and habitat. During the time that herbicide use was increasing in Britain, there were also other changes in agricultural practices. These include the elimination of hedgerows from many landscapes, changes in cultivation methodologies, new crop species, increases in the use of **insecticides** and fungicides, and improved methods of seed cleaning, resulting in fewer weed seeds being sown with crop seed. Still, a common opinion of ecologists studying the large declines of birds, such as the gray partridge (*Perdix perdix*), is that herbicide use has played a central but indirect role by causing habitat changes, especially by decreasing the abundance of weed seeds and arthropods available as food for the birds.

Similarly, the herbicides most commonly used in forestry are not particularly toxic to animals. Their use does however, cause large changes in the habitat available on clear-cuts and plantations, and these might be expected to diminish the suitability of sprayed sites for the many species of **song birds**, mammals, and other animals that utilize those habitats.

Modern, intensively managed agricultural and forestry systems have an intrinsic reliance on the use of herbicides and other pesticides. Unfortunately, the use of herbicides and other pesticides carries risks to humans through exposure to these potentially toxic chemicals, and to ecosystems through direct toxicity caused to non-target species, and through changes in habitat. Nevertheless, until newer and more pest-specific solutions to weed-management problems are developed, there will be a continued reliance on herbicides in agriculture, forestry, and for other purposes, such as lawn care.

Resources

Books

Briggs, S.A. *Basic Guide to Pesticides: Their Characteristics and Hazards.* Washington, DC: Taylor & Francis, 1992.

Freedman, B. *Environmental Ecology.* 2nd ed. San Diego: Academic Press, 1995.

Hayes, W.J., and E.R. Laws, eds. *Handbook of Pesticide Toxicology.* San Diego: Academic Press, 1991.

Periodicals

Pimentel, D., et al. "Environmental and Economic Costs of Pesticide Use." *Bioscience* 41 (1992): 402-409.

Bill Freedman

Herbivore

An herbivore is an **animal** that eats plants as its primary source of sustenance. Examples of herbivores include large **mammals** such as cattle, **deer**, **sheep**, and kangaroos, as well as smaller creatures such as leaf-eating **insects** and crustaceans that graze upon aquatic **algae**. However, many animals are not exclusively herbivorous. In addition to feeding mostly upon live plants, omnivorous animals such as **pigs** and **bears** may also kill and eat other animals, opportunistically feed upon dead creatures, or eat dead **plant biomass**.

In the language of trophic **ecology**, herbivores are known as heterotrophic creatures, which means that they must ingest biomass to obtain their **energy** and **nutrition**. In contrast, autotrophs such as green plants are capable of assimilating diffuse sources of energy and materials, such as sunlight and simple inorganic molecules,

and using these in biosynthetic reactions to manufacture complex biochemicals. Herbivores are known as primary consumers, because they feed directly on plants. Carnivores that feed on herbivores are known as secondary consumers, while predators of other carnivores are tertiary consumers.

A fact of ecological energetics is that within any **ecosystem**, herbivores are always much less productive than the green plants that they feed upon, but they are much more productive than their own predators. This ecological reality is a function of the pyramid-shaped structure of productivity in ecological food webs, which is itself caused by thermodynamic inefficiencies of the transfer of energy between levels.

However, this ecological law only applies to production, and not necessarily to the quantity of biomass (also known as standing crop) that is present at a particular time. An example of herbivores having a similar total biomass as the plants that they feed upon occurs in the open-ocean, planktonic ecosystem, where the **phytoplankton** typically maintains a similar biomass as the small animals, called **zooplankton**, that graze upon these microscopic plants. In this case, the phytoplankton cells are relatively short-lived, but their biomass is regenerated quickly because of their productivity. Consequently, the phytoplankton has a much larger total production than the longer-lived zooplankton, even though at any particular time their actual biomasses may be similar. Similarly, the densities of animals are not necessarily less than those of the plants that they eat, as occurs, for example, if insects are the major herbivores in a forest of large trees.

Following further along the above line of reasoning, because herbivores eat lower in the ecological food web, there is a relatively large quantity of food resource available to sustain them, compared with what is available to sustain carnivores. This fact has implications for humans, which can choose to sustain themselves by eating various ratios of food obtained directly from plants, or from animals that feed upon plants (such as cows, pigs, sheep, or chickens). In a world in which food for humans is often present in a supply that is less than the demand, at least in some regions, many more herbivorous (or vegetarian) people could be sustained than if the predominant feeding strategy was carnivorous.

See also Carnivore; Ecological productivity; Food chain/web; Heterotroph; Omnivore.

Bill Freedman

Heredity *see* **Genetics**

Hermaphrodite

A hermaphrodite is any **organism** with both male and female reproductive organs which produce both male gametes (sperm) and female gametes (ova). In some animals, the simultaneous hermaphrodites, both male and female organs are functional at the same time. In other animals, the sequential hermaphrodites, one sex develops at one time, which later develops into the other sex. Examples of both strategies are found naturally, especially in the **invertebrates**, and for many creatures, hermaphrodism is the only method of reproduction.

Simultaneous hermaphrodism

Sponges, **sea anemones**, tapeworms, **snails**, and earthworms are all simultaneous hermaphrodites possessing both male and female reproductive organs at the same time. These animals are either sedentary **species** (remaining in one place all their life) or they are mobile, but do not range widely. These habits present problems for **sexual reproduction** and mating, for individuals rarely meet others of their own species. When two simultaneous hermaphrodites, such as two **slugs**, meet and mate, each one can fertilize the eggs of the other.

A common misconception about hermaphroditic reproduction is that organisms fertilize their own eggs with their own sperm. In fact, most species do not self-fertilize, and many are physically incapable of self-fertilization. (Bisexual plants are the exception, and some do self-fertilize.) For example, sponges fertilize externally, but they release eggs and sperm into the **plankton** at different times, so that gametes encounter, and are likely to be fertilized by, or to fertilize, gametes from other individuals. The reproductive organs of earthworms are positioned at different ends of their bodies, so that **fertilization** of the eggs is only possible when the worms are aligned in opposite directions. Some simultaneous hermaphrodites even have alternative reproductive methods. The larvae of digenetic trematodes, such as the liver fluke, are simultaneous hermaphrodites, and reproduce asexually, while the adults reproduce sexually.

Sequential or serial hermaphrodites

A few species of **vertebrates** (mostly **fish**), and some species of crustaceans (**barnacles** and shrimps) change sex during their lifetime.

Sequential hermaphroditic fish, such as the bluehead wrasse, undergo protogyny, changing sex from female to male. Other species, such as the sea **perch**, *Pagellus acurne*, undergo protandry, changing sex from male to female. A third group of fish, such as the sea **bass**, un-

dergoes both protogyny and protandry, and can do so repeatedly. One of the most dramatic sights in nature is the mating of the sea bass *Serranus subligarins*. When two female sea bass meet to mate, one undergoes protogyny, changing **color** from a deep blue to bright orange with a white stripe. After fertilization, both fish then switch sex (and coloring) and then mate again.

The sex changes of sequential hermaphrodites depend on social factors. For example, bluehead wrasse live in large colonies where only the largest fish are males. The others must remain female until the males die. Only then can some of the females (usually the largest) change from female to male. Other factors influencing protogyny and protandry are hunger, the amount of **salt** in the **water** (salinity), social behaviors, and the **ratio** of males to females in the population.

Hermaphrodism in humans

True hermaphroditic humans do not exist, but pseudohermaphrodism does, where an **individual** has both male and female external genital organs, sometimes at the same time. Female embryos exposed to high levels of androgens (the male **hormones**) develop female internal reproductive organs but male external genitalia. Alternately, genetic defects cause children to be born with female external genital organs, which change at **puberty**, with the development of a penis and the closure of the false vagina.

Resources

Books

Campbell, N., J. Reece, and L. Mitchell. *Biology.* 5th ed. Menlo Park: Benjamin Cummings, Inc. 2000.

Elia, Irene. *The Female Animal.* New York: Henry Holt, 1988.

Jorde, L.B., J.C. Carey, M.J. Bamshad, and R.L. White. *Medical Genetics* 2nd ed. New York: Year Book, Inc., 2000.

Stern, Kingsley R. *Introductory Plant Biology.* Dubuque, IA: Wm. C. Brown, 1991.

Periodicals

Berreby, D. "Sex and the Single Hermaphrodite." *Discover* 13 (1992): 88-93.

David L. Brock

Hernia

A hernia occurs when an **organ** or **tissue** bulges out of its proper location. Hernias can occur in numerous locations throughout the body, including the **brain**, abdomen, groin, spine, and chest. The most common types will be discussed below.

Hernias can be either **congenital** or acquired. Congenital means that the **individual** was born with an abnormal opening, allowing the hernia to occur. Such a hernia may show up immediately after **birth**, may cause no symptoms for years to come, or may never result in symptomatology. An acquired hernia is one which was not present at birth, but which occurred later, either due to some other anatomical abnormality, or due to weakening of an area with use and aging.

Hernias which involve loops of intestine, or other abdominal contents, may be reducible, meaning that the individual can push on the bulging area with a hand to move the intestine back into the abdomen. When a hernia cannot be reduced, it is said to be incarcerated. The greatest risk with any hernia containing intestine is strangulation, in which the hernia is incarcerated, and **blood supply** to the intestine is cut off. This is a medical emergency, and without surgical intervention, an area of the intestine may well die off.

Groin hernias

Most people, when they hear or use the word hernia, are referring to an inguinal hernia (the inguinal area, also called the groin, is that area where the thigh and the abdomen meet), in which a loop of the intestine has passed through a weak muscular area. There are three main types of groin hernias: direct inguinal, indirect inguinal, and femoral. These are classified according to the anatomical route that the hernia takes. In men, a herniated loop of intestine may enter the scrotum. This is the type of hernia which the physician is testing for when giving the dreaded command, "Cough!" Pregnancy, **obesity**, heavy lifting, and medical conditions which increase the **pressure** within the abdomen (**emphysema** or other lung conditions causing frequent coughing; constipation; swelling of prostate causing difficulty urinating) can all predispose to hernia, or worsen an already existing hernia. Most physicians favor surgical repair of groin hernia, even those which are reducible, to avoid future incarceration and strangulation, which can lead to dangerous complications.

Abdominal hernias

Abdominal hernias include umbilical hernias, hernias through the scar left by a previous surgical incision, and hernias through the muscles of the abdominal wall. All of these types of hernias involve abdominal contents (often a loop of intestine) which pop through a weakened area. Some umbilical hernias are present at birth, particularly in premature infants, and are due to incomplete closure of an area called the umbilical ring, which should close before birth. Most of these umbilical hernias do not require **surgery**, because the ring usually decreases in size and then closes on its own within the first two to four years of life.

Hiatal hernia

A hiatal hernia occurs when a portion of the stomach protrudes above the diaphragm (the diaphragm is the large, sheet-like muscle which should separate the contents of the chest from the contents of the abdomen). The majority of hiatal hernias (90%) are of a type causing reflux, which occurs when the acidic contents of the stomach wash up the esophagus (the esophagus is the tube which should only carry swallowed substances down into the stomach). Presence of these acidic contents burn the esophagus, resulting in the symptom commonly referred to as heartburn. Most of these types of hernias do not require surgical repair. Symptoms are helped by various medications which decrease the acidity of the stomach contents, and thwarting the effects of gravity which can exacerbate the problem at night (patients should go to **sleep** propped up on an angle by a sufficient number of pillows). This other type of hiatal hernia more frequently requires surgical repair, because its complications include hemorrhage (massive bleeding), incarceration, and strangulation (which can result in death of stomach tissue). Furthermore, in this more serious type of hiatal hernia, other abdominal contents (intestine, spleen) may also protrude into the chest cavity, and pressure from crowding in the chest cavity can result in **heart** problems.

Diaphragmatic hernia

Diaphragmatic hernias can be congenital, or acquired through trauma (for example, a knifing). Congen-

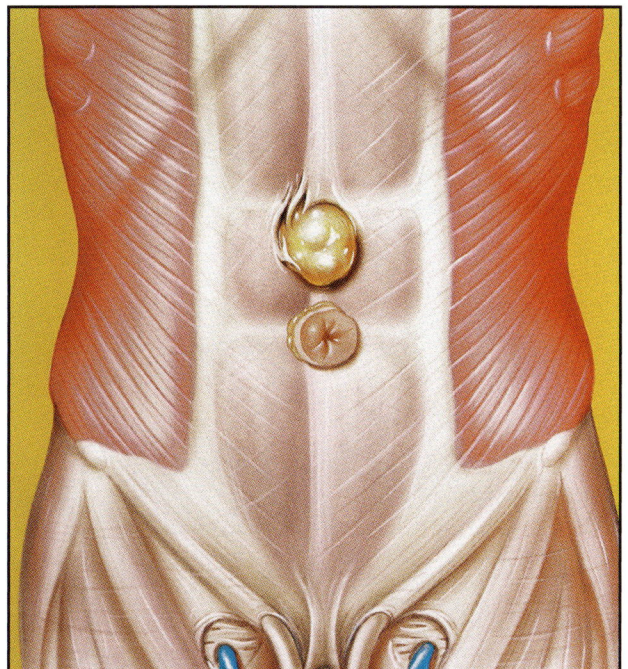

An illustration of an epigastric (abdominal) hernia in an adult male. The torso is shown with its skin removed. Epigastric hernia is caused commonly by a congenital weakness in muscles of the central upper abdomen; the intestine bulges out through the muscle at a point between the navel and breastbone. *Illustration by John Bavosi. National Audubon Society Collection/ Photo Researchers, Inc. Reproduced by permission.*

KEY TERMS

Congenital—A condition or disability present at birth.

Diaphragm—The sheet-like muscle that is supposed to separate the contents of the abdomen from the contents of the chest cavity. The diaphragm is a major muscle involved in breathing.

Incarcerated—Refers to a hernia which gets trapped protruding through an abnormal opening. Most frequently refers to loops of the intestine which cannot be easily replaced in their normal location.

Inguinal—Referring to the groin area, that area where the upper thigh meets the lower abdomen.

Invertebral disc—The cartilaginous disc located between each of the vertebral bones of the spine. This disc provides cushioning and insulation.

Reduce—The ability to put a displaced part of the anatomy (in particular, the loops of intestine present in a hernia) back in their correct location by simply pushing on the bulging area.

Strangulation—A situation which occurs when the blood supply to an organ is cut off, resulting in death of that tissue.

ital diaphragmatic hernias occur during development before birth, when the tissues making up the diaphragm do not properly close off the area between the abdomen and chest cavities. These abnormal contents, especially the intestine and spleen, push into the chest cavity, applying pressure to the heart, and sometimes preventing adequate development of the lungs. A baby born with such a defect usually experiences extreme respiratory distress, and requires immediate surgery.

Brain herniation

Herniation of brain tissue can occur when an expanding mass (**tumor**) begins to take up **space** within the finite area of the skull (for example, if there is high pressure in the skull from fluid accumulation, and a test called a spinal tap, or lumbar puncture, is performed). Displacement of brain tissue in this way results in compression of various areas of the brain, and greatly compromised vital functions (vital functions are those brain-directed functions necessary for the basics of human life, for example, breathing and heart **rate**). Herniation of brain tissue usually results either in death, or in massive and permanent brain damage.

Disc herniation

The spine is made up of individual bones, the vertebrae, separated from each other by a disc to provide insulation and cushioning. Disc herniation, or a slipped disc, occurs when the interior area of the disc breaks through the outer area of the disc, and pushes into the spinal canal, or when the entire disc becomes displaced from its normal positioning. Disc herniation occurs most commonly in the neck area, and in the lower back, and can be a result of wear-and-tear from aging or from trauma.

Problems due to disc herniation occur because the displaced disc presses on the spinal cord or the nerves leaving the spinal cord. This can result in problems ranging from tingling in the hands, feet, or buttocks; weakness of a limb; back, leg, or arm **pain**; loss of bladder control; loss of normal reflexes (for example, normally tapping the knee with an exam hammer results in an involuntary kicking out the foot; disc herniation may make it impossible to elicit this foot kick, as well as other reflexes); or in very extreme cases, paralysis.

Cases of disc herniation with less extreme symptomatology can be treated with such measures as a neck brace or back brace, medications to reduce swelling (nerve roots experiencing pressure from the protruding disc may swell, further compromising their function), **heat**, and pain medications. When pain is untreatable, or loss of function is severe or progressive, surgery may be required to relieve or halt further progression of the symptoms.

Resources

Books

Abernathy, Charles, and Brett Abernathy. *Surgical Secrets.* Philadelphia: Hanley and Belfus, Inc., 1986.

Berkow, Robert, and Andrew J. Fletcher. *The Merck Manual of Diagnosis and Therapy.* Rahway, NJ: Merck Research Laboratories, 1992.

Way, Lawrence. *Current Surgical Diagnosis and Treatment.* Los Altos, CA: Lange Medical Publications, 1983.

Rosalyn Carson-DeWitt

Heroin *see* **Addiction**

Herons

Herons, egrets, and **bitterns** are large, slender wading **birds** in the family Ardeidae, order Ciconiiformes (which also includes anhingas, **storks**, spoonbills, and **ibises**). Most of the **species** in the heron family have long legs, necks, and bills. These characteristics are all adaptive to hunting their **prey** of **fish**, **amphibians**, snakes, small **mammals**, and other animals living in the shallow waters of **wetlands**. The prey is generally caught by grasping it firmly in the mandibles, and is then killed by beating it against the ground, branches, or another hard substrate. The food is usually then rinsed, and swallowed head-first.

Herons have an unusual articulation of the sixth vertebra that is adaptive to swallowing large prey. This feature causes the neck of herons to adopt a distinctive, S-shape when they are in flight or resting, although their neck can be extended while grooming or to give greater reach while attempting to catch prey.

Herons also have an unusual type of filamentous feathers, known as powder-down. These feathers are very friable, and disintegrate into a powder that the bird rubs over the major body feathers to cleanse them of slime from its food of fish.

Herons primarily occur along the edges of lakes and other shores, and in marshes, swamps, and other relatively productive wetlands. Many species in the heron family are colonial nesters, generally on islands, if possible.

A snowy egret (*Egretta thula*) on Estero Island, Florida. This species, once hunted to near extinction by the millinery trade, became a symbol for the early conservation movement in the United States and remains the emblem of the National Audubon Society. *Photograph by Robert J. Huffman. Field Mark Publications. Reproduced by permission.*

These birds typically build platform nests of sticks in trees, sometimes with many nests in a single, large **tree**.

Species of herons

Sixty species are included in the heron family, occurring worldwide, except in **Antarctica** and arctic **North America** and Eurasia. Twelve species of herons breed regularly in North America. One of the most familiar is the great blue heron (*Ardea herodias*), occurring over most of the temperate and more-southern regions of North America, as well as in parts of Latin America. The great white heron used to be considered a separate species (under *A. occidentalis*), but it is now regarded as a **color** variety of the great blue heron that only occurs in the Florida Keys and nearby parts of Florida Bay. The great blue heron breeds in colonies of various size, usu-

A yellow-crowned night heron (*Nyctanassa violacea*) at the Ding Darling National Wildlife Refuge, Florida. *Photograph by Robert J. Huffman. Field Mark Publications. Reproduced by permission.*

ally nesting in trees. The great blue heron is very similar to the grey heron (*A. cinerea*), which has a widespread distribution in Eurasia and **Africa**. Further studies may conclude that these are, in fact, the same species.

Smaller species of herons include the Louisiana or tricolored heron (*Hydranassa tricolor*), and the little blue heron (*Florida caerulea*), found in the wetlands of the coastal plain of southeastern North America, the Caribbean, and the Pacific coasts of Mexico. The green-backed heron (*Butorides virescens*) is a relatively small and attractive species with a wide distribution in southern North America. Some individuals of this species have learned to "fish," using floating bits of material, such as small twigs, to attract **minnows**. These birds will deliberately drop their "bait" into the **water** and may retrieve it for re-use if it floats away.

The black-crowned night heron (*Nycticorax nycticorax*) is widely distributed in colonies throughout much of the United States and a small region of southern Canada. The yellow-crowned night heron (*Nyctanassa violacea*) is more southeastern in its distribution than the preceding species, and it tends to occur more frequently in the vicinity of **saltwater**.

The largest of the several species of egrets in North America is the common or American egret (*Casmerodius albus*), ranging widely over the southern half of the **continent**. The snowy egret (*Leucophyx thula*) is a smaller, more southern species. Most species of herons and egrets are patient hunters, which quietly stalk their prey or lie in wait for food to come within their grasp. However, the relatively uncommon reddish egret (*Egretta rufescens*) is an active hunter on saline mudflats of the southernmost states, where it runs boisterously about in active pursuit of its food of small fishes and **invertebrates**.

The American bittern (*Botaurus lentiginosus*) inhabits marshes over much of temperate North America and further south. This species has a resounding, "onk-a-tson-ck" call that can be heard in the springtime when male birds are establishing breeding territories and attempting to attract a mate. The least bittern (*Ixobrychus exilis*) is the smallest North American heron. Both of these bitterns are very cryptic in their reedy, marshy habitats. When they perceive that they are being observed by a potential **predator**, these birds will stand with their neck and bill extended upright, with the striped breast plumage facing the intruder, and they will even wave their body sinuously in concert with the movement of the surrounding vegetation as it is blown by the **wind**.

The cattle egret (*Bubulcus ibis*) is a naturalized species in the Americas, having apparently colonized naturally from Africa in the present century. This species was first observed in Argentina, but it has since spread widely and now occurs in suitable **habitat** throughout South, Central, and North America. The cattle egret commonly follows cattle in pastures, feeding on the **arthropods** and other small animals that are disturbed as these large animals move about.

Conservation of herons

Most species in the heron family, and many other types of birds, were unsustainably hunted during the nineteenth century to provide feathers for use in the millinery trade, mostly as decorations on ladies' hats and other clothing. Many millions of herons and egrets were killed for this reason, and their populations declined precipitously in most regions. The outcry among conservationists over the slaughter of so many birds for such a trivial purpose led to the formation of the National Audubon Society in the United States in 1886 and the Royal Society for the Protection of Birds in Great Britain in 1889. These were the first important, non-government organizations that took up the **conservation** and protection of natural **biodiversity** as their central mandate.

Today, habitat losses are the most important threat to species in the heron family and to other birds of lakes, shores, and wetlands. These habitat types are suffering world-wide declines from **pollution**, drainage, conversion to agriculture or urban development, and other stressors associated with human activities. As a result, the populations of herons, egrets, and bitterns are declining in North America and in many other regions, as are other wild life with which these birds share their wetland habitats.

Birds in the heron family are large, attractive, and sometimes relatively tame. Consequently, they are popular among birders; however, the numbers of these beautiful and charismatic birds have declined as a result of continuing influences of humans, especially damage caused to wetlands. In the future, the populations of these birds can only be sustained if sufficiently large areas of their natural habitats are preserved.

Resources

Books

Ehrlich, P.R., D.S. Dobkin, and D. Wheye. *Birds in Jeopardy.* Stanford, Cal.: Stanford University Press, 1992.
Forshaw, Joseph. *Encyclopedia of Birds.* New York: Academic Press, 1998.
Hancock, J., and J. Kushlan. *The Herons Handbook.* London: Croom Helm, 1984.
Marquis, M. *Herons.* London: Colin Baxter, 1993.

Bill Freedman

Herpes *see* **Sexually transmitted diseases**

Herpetology

Herpetology is the scientific study of **amphibians** and **reptiles**. The term "herpetology" is derived from the Greek and refers to the study of creeping things. **Birds** and **mammals**, for the most part, have legs that lift their bodies above the surface of the ground. Amphibians (class Amphibia) and reptiles (class Reptilia), with the exception of **crocodiles** and lizards, generally have legs inadequate to elevate their bellies above the terrain, thus they creep.

Both Amphibia and Reptilia are within the phylum Chordata, which also includes several classes of fishes, reptiles, birds and mammals. Amphibia include the anurans, which are **frogs** and **toads**; the urodeles, which include **salamanders** and sirens; and the gymnophioma, which are peculiar worm-like legless **caecilians**. Larval amphibians (tadpoles) respire with gills whereas adults breathe with lungs. Amphibian skin is ordinarily scaleless. Reptilia includes lizards, **snakes**, **turtles**, and crocodiles. They have scaly skin and respire with lungs. Extinct reptiles are of great scientific and popular interest and include dinosaurs, pterosaurs, and ichthyosaurs.

Some scientists are both herpetologists and ecologists. They study **habitat**, food, population movements, reproductive strategies, life expectancy, causes of death, and a myriad of other ecological problems. Their studies have significance not only to survival of the animals that they study but also to humans. Amphibians and reptiles manage their **metabolism** of xenobiotic (foreign to the body) toxic substances in much the same way as humans do, by metabolic change in the liver and other organs that permits rapid excretion. It becomes a notable concern when amphibians and/or reptiles cannot survive in an altered environment. Amphibians in a number of countries have been reported to be found in diminishing numbers and many are anatomically abnormal. Because of their similarity in managing toxic substances, whatever is causing the population perturbations and anomalous **anatomy** in the lower creatures may be of equal concern to humans.

The studies of amphibians and reptiles relating to **pathology** and medicine is less well known than similar studies with higher organisms. Herpes viruses are now recognized as being microbial agents related to **animal** and human **cancer**. Burkitt's lymphoma and Kaposi's sarcoma are two human cancers with an established link with herpes viruses. The first cancer of any type known to be causally associated with a herpes **virus** was the Lucké renal adenocarcinoma of the leopard frog, *Rana pipiens*. Virologists working with the frog cancer can perform a multiplicity of experiments with the herpes virus and frog cancer. The frog experiments would be very difficult to perform on other animals, and would be precluded for ethical reasons from human experimentation.

The feasibility of vertebrate cloning was first demonstrated in the frog, *R. pipiens*, in Philadelphia in 1952, and later in the South African clawed toad, *Xenopus laevis*, in England in 1958. Prior to the frog experiments, it was generally thought that cloning was a "fantastical" dream. Cloning has since been achieved with **sheep**, cows, and other mammals.

As economic resources, turtle meat and crocodile (raised on farms for that purpose) hides have a significant role in the Louisiana economy. Further, many amphibians and reptiles are collected for scientific study. Only **rodents** exceed in number frogs used for biomedical research.

Herrings

One of the most important fisheries in the world is provided by the true herrings, which belong to the **bony fish** family Clupeidae. This family contains a wide variety of fishes with distinctive habits. Although most of the **species** are marine, a few are anadromous—that is, they spend their lives in the sea and enter **rivers** to spawn. Other species remain permanently in **freshwater**.

Herrings are small, silvery **fish** with a deeply forked tail. They rarely grow over 11 lb (5 kg) in weight. Herrings have a ridge of scales on the belly midline, which is sharp-edged, and they have no visible lateral line.

Herrings contribute greatly to the economy of some countries—wars have even been fought for rights to important fishing grounds, which are widely distributed, except for extremely cold parts of the Arctic and Antarctic Oceans.

Since herrings tend to migrate in enormous schools, they can be caught readily be commercial fishers. They are also key parts of the diet of some species of whales, **seals**, **gulls**, and predatory fish. Herrings eat **plankton** that they strain from the **water** with their gill rakers, trapping these organisms as water passes across their gills.

The Atlantic herring (*Clupea harengus*) may be the most plentiful pelagic (or open-ocean) fish, and is found on both sides of the North Atlantic Ocean. Due to intensive overfishing, however, the population of herrings has been markedly reduced.

Spawning times vary but most often occurs in the fall and occasionally in the spring or summer. Each female may deposit 25,000-40,000 eggs, which are heavy and sink to the bottom. On the way down a thick covering of mucus causes the eggs to stick to anything they encounter. It takes up to two weeks for the eggs to hatch, the time depends on such variables as depth and **temperature**. There is no parental care. In the first year the young may reach a size of 5 in (13 cm), reaching 10 in (25 cm) after two years. In their third year they may have acquired enough **fat** to be harvested as a source of oil. Herrings become sexually mature in their fourth year.

The term sardine is generally applied to small herrings. It also is applied to such forms as the Pacific sardine (*Sardinops sagax*). The sprat or brisling (*C. sprattus*) from the European side of the Atlantic is considerably smaller than the Atlantic herring.

The Atlantic menhaden or mossbunker (*Brevoortia tyrannus*) is the most numerous of all fish in the mid-Atlantic waters of **North America**. It has a stubby shape and generally weighs under a pound. Due to its heavy oil content it is not palatable, but makes an excellent fertilizer, fishmeal, and oil. Traveling in massive schools near the surface, these fish can cause a swirling **motion** of the water. Schools of menhaden may be located by the presence of flocks of seabirds feeding on them.

The American shad (*Alosa sapidissima*) is considered to be the largest herring, since it has an average weight of 3 lb (1.5 kg) and can reach 12 lb (6 kg). It is found in the Atlantic Ocean from the St. Lawrence River south to Florida. Toward the end of the nineteenth century, shad were introduced into the Pacific Ocean and this species now ranges from Alaska to California. The shad is an anadromous fish, in that it spends its adult life in the ocean but swims up the rivers to spawn.

When spawning, the sexes separate with the males entering the river first, followed by the females, known as roe shad. Each female carries a tremendous number of eggs, estimated at 30,000 on average, although larger females can carry several times that number. As with the other herrings, the eggs are dropped at **random** since they are sticky and heavy, readily sinking to the bottom but tending to adhere to objects. The young shad remain in the streams until strong enough to enter the sea. Males are sexually mature at about their fifth year, at which time they return to spawn. Females may take a bit longer to mature and reenter the rivers to spawn. Shad are caught as they are traveling upstream when they are energetic. They are caught commercially as well as for sport, and they are highly prized for human consumption, especially the roe.

The alewife (*Alosa pseudoharengus*) is a close relative of the American shad but is smaller in size. Ocean-moving alewives are anadromous. Some populations in the eastern United States are landlocked and are found in great abundance in the Great Lakes. Tons of alewives die during some summers, resulting in an intolerable, smelly nuisance on beaches. Alewives are caught commercially in seines and nets, and are used for fishmeal and fertilizer.

Resources

Books

Dickson Hoese, H., and R.H. Moore. *Fishes of the Gulf of Mexico, Texas, Louisiana, and Adjacent Waters.* College Station and London: Texas A&M; University Press, 1977.
Whiteman, Kate. *World Encyclopedia of Fish & Shellfish.* New York: Lorenz Books, 2000.

Hertzsprung-Russell diagram

A Hertzsprung-Russell diagram, or H-R diagram, is a graph of stellar temperatures (plotted on the horizontal axis) and luminosities, or brightnesses (plotted on the vertical axis). H-R diagrams are valuable because they reveal important information about the stars plotted on them. After constructing an H-R diagram for a group of stars, an astronomer can make estimates of many important stellar properties including diameter, **mass**, age, and evolutionary state. Our understanding of the processes at work in the stars depends on knowing these parameters, so H-R diagrams have been essential tools in twentieth-century astronomical research.

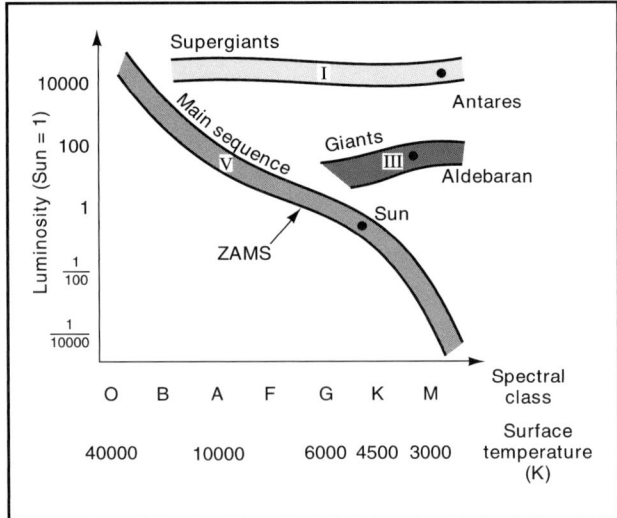

Figure 1. The spectral classes and corresponding surface temperatures are given at bottom, while the luminosities are given at left. (The luminosities are in solar units, meaning that "1" equals the luminosity of the Sun, while "10" means ten times the luminosity of the Sun, and so forth.) Clearly the stars are not randomly distributed on this graph. They fall in several well-defined areas, with most stars on a narrow strip running from upper left to lower right. This graph, the H-R diagram, was a fundamental advance in astronomy.
Illustration by Hans & Cassidy. Courtesy of Gale Group.

Stellar classification and the H-R diagram

The nineteenth century saw the development of a powerful technique called *spectroscopy*. This technique involves the use of an instrument called a *spectrograph*, which disperses **light** passing through it into its component colors in the same way that an ordinary **prism** does. Indeed, many spectrographs in use today have prisms as one or more of their components.

When sunlight or starlight passes through a spectrograph and is dispersed, the resulting **spectrum** has many narrow, dark lines in it. These lines are called *absorption lines*. A line occurs only at a certain wavelength and is caused by the presence of a specific element in the star's atmosphere. They are called absorption lines because they are caused when elements in the star's atmosphere absorb some of the light radiating outward from the star's surface. Less light escapes from the star's atmosphere where there is a line than in other portions of the spectrum, so the line looks dark.

Different stars have different patterns of absorption lines, and the pattern present in a particular **star** depends on the star's surface **temperature**. For example, **hydrogen**, the most common element in stars, produces several very strong absorption lines in the visual part of the spectrum—but only if the star's temperature is about

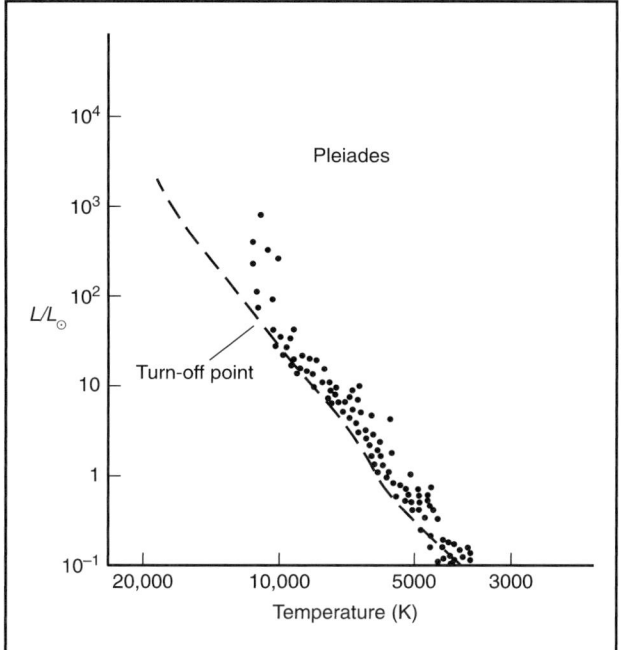

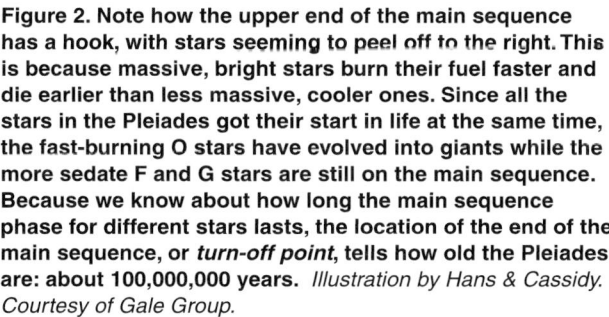

Figure 2. Note how the upper end of the main sequence has a hook, with stars seeming to peel off to the right. This is because massive, bright stars burn their fuel faster and die earlier than less massive, cooler ones. Since all the stars in the Pleiades got their start in life at the same time, the fast-burning O stars have evolved into giants while the more sedate F and G stars are still on the main sequence. Because we know about how long the main sequence phase for different stars lasts, the location of the end of the main sequence, or *turn-off point*, tells how old the Pleiades are: about 100,000,000 years. *Illustration by Hans & Cassidy. Courtesy of Gale Group.*

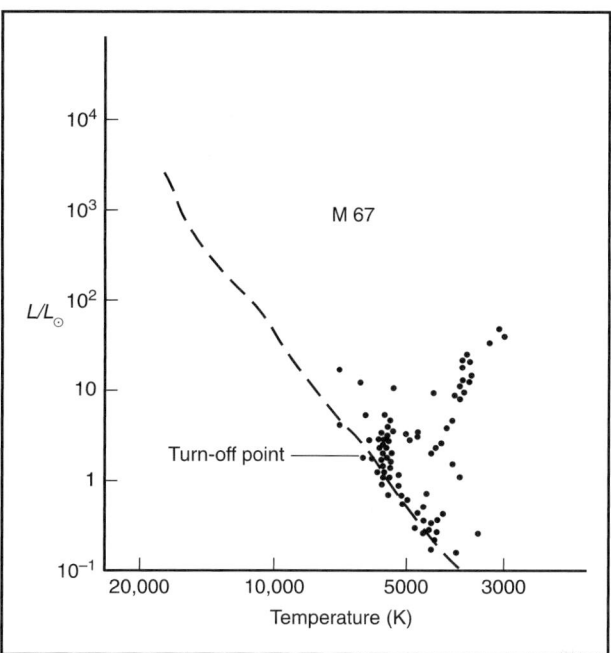

Figure 3. The cluster M67 however, is much older than the Pleiades. Its O, B, A, and F stars are already well into giant-hood. It takes F stars several billion years to burn all their hydrogen, so M67 must be around five billion years old. *Illustration by Hans & Cassidy. Courtesy of Gale Group.*

10,000K (17,541°F [9,727°C]). If the star is much hotter, say 20,000K (35,541°F [19,727°C]), the hydrogen **atoms** can no longer absorb as much light in the visual spectrum, so the lines are weaker. Very cool stars also have weaker hydrogen lines.

In the early 1900s, a group of astronomers led by Annie Jump Cannon at the Harvard Observatory began to classify stellar spectra. They grouped stars into *spectral classes*, with all the stars in a given spectral class having similar patterns of lines. This is just like the way that biologists classify animals into groups such as families and **species**. Spectral classes are denoted by letters, and the main ones, in order of decreasing surface temperature, are O, B, A, F, G, K, and M. You can remember this by the mnemonic "Oh Be A Fine Girl (or Guy), Kiss Me!" Because stars have many elements in their atmospheres (hydrogen, helium, **calcium**, **sodium**, and **iron**, to name only a few), their spectra can have thousands of lines. To accommodate this complexity, the spectral classes are each divided into 10 subclasses, de-

noted by numbers. For example, there are F0 stars, F1 stars, and so on until F9; the next class is G0. The **Sun**, with a surface temperature of 5,800K (9,981°F [5,527°C]), is a G2 star.

The first H-R diagrams were created independently in the early 1900s by the astronomers Ejnar Hertzsprung and Henry Norris Russell. Russell's graph had spectral class plotted along the x-axis and a quantity related to luminosity (or brightness) plotted along the y-axis. Figure 1 is such a graph.

The nature of the H-R diagram

Figure 1 shows all the important features of the H-R diagram. The stars fall into several relatively narrow strips which W. W. Morgan, another famous classifier of stellar spectra, called *luminosity classes*. Luminosity classes are denoted by Roman numerals.

The main sequence

Luminosity class V is the long, narrow strip running diagonally across the diagram, and it is called the *main sequence*. The Sun lies on the main sequence, as do 90% of all stars. Stars on the main sequence are stable and healthy, shining as a result of **nuclear fusion** reactions in their cores that convert their hydrogen to helium. Stars

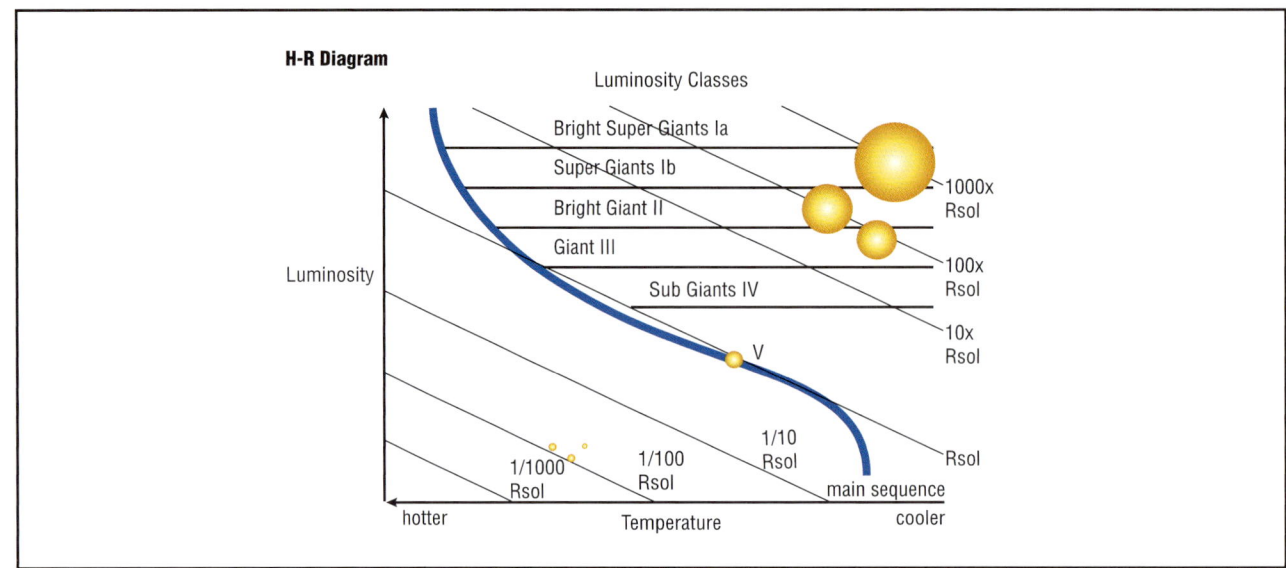

H-R Diagram

Luminosity Classes

Bright Super Giants Ia

Super Giants Ib

Bright Giant II

Giant III

Sub Giants IV

V

Luminosity

1000x
Rsol

100x
Rsol

10x
Rsol

1/10
Rsol

Rsol

1/100
Rsol

1/1000
Rsol

main sequence

hotter   Temperature   cooler

On the H-R diagram, the Sun is a main sequence star. Main sequence of stars run from extremely bright, hot stars in upper left-hand corner to faint, cool stars in lower right-hand corner. *Illustration by K. Lee Lerner and Argosy. The Gale Group.*

spend most of their lives on the main sequence, so it is not surprising that most stars are found there.

The main sequence slopes from upper left to lower right on the H-R diagram. Therefore, the hotter main sequence stars are, the brighter they are. Main sequence O stars, or O V stars (using the luminosity class numeral), are extremely hot and blaze away with the brightness of 10,000 or more Suns. At the other end of the main sequence are the little M V stars, shining with a dull glow, only 1% as bright as the Sun.

For main sequence stars, there are also relationships between surface temperature, radius, mass, and lifetime. Hotter main sequence stars are both larger (greater radius) and more massive than cooler ones. So not only are O V stars brighter than the Sun, they are also physically larger and may be 20 or more times as massive. M V stars may be only a tenth as massive as the Sun. However, the brilliant O stars have to consume their hydrogen fuel thousands of times faster than their cooler cousins. Therefore, they live for a very short time—no more than a few million years—while stars like the Sun may remain on the main sequence for 10 billion years. And the tiny, faint M stars, though not very impressive, will remain shining faintly on for hundreds of billions of years.

Giant stars

Main sequence stars are, by definition, normal. The other luminosity classes, of which the main ones are III and I, contain stars that are very different.

Consider class III stars. They are fairly cool since they lie near the right side of the H-R diagram. But they

are also much brighter than any normal K and M star should be—perhaps 100 times as luminous as the Sun. We know that luminosity depends on temperature. Normally cool stars would not be as bright as hot stars, just as a glowing ember in a campfire gradually gets dimmer as it cools off. However, luminosity also depends on the *size* of an object. Imagine a glowing ember the size of a marble and another one, equally hot, the size of a beach ball. Clearly the larger one will be brighter, simply because there is more of it. Therefore, class III stars must be huge to be so bright and yet so cool.

For this reason, stars in luminosity class III are called *giant* stars. For example, Aldebaran, a bright K5 III star in the **constellation** Taurus (the Bull), has a diameter roughly 100 times greater than the Sun's. Aldebaran and many of the other bright but reddish stars you can see with the unaided **eye** are giants. If they were small main-sequence stars, they would be too faint to see.

Now consider luminosity class I, lying at the very top of the H-R diagram. If red stars 100 times brighter than the Sun are large, red stars 10,000 times brighter must be monstrous indeed. And they are: Antares, the M1 I star in the constellation Scorpio (the Scorpion), is so large that astronomers have been able to measure its diameter directly. Antares, it turns out, is about 400 times larger than the Sun. If placed at the center of the **solar system**, Antares would extend past the **orbit** of **Mars**. All four inner planets, including **Earth**, would be swallowed in a 4,000K (6,741°F [3,727°C]) inferno. Stars like this are called *supergiants*, and Antares as well as hotter supergiants like Rigel (the foot of Orion, spec-

KEY TERMS

Giant—A star that has exhausted nearly all of its hydrogen fuel and is using heavier elements as fuel to sustain itself against its own gravity. The processes occurring in its interior have forced it to expand until it is 10 to 100 times the diameter of the Sun.

Luminosity—The amount of energy a star emits in a given amount of time. More massive stars are more luminous less massive ones, and they do not live as stable stars for as long.

Luminosity class—One of several well-defined bands of stars on the H-R diagram. The main luminosity classes are denoted by the Roman numerals I, II, III, IV, and V, and stars belonging to them are called supergiants, bright giants, giants, subgiants, and dwarfs (or main sequence stars), respectively.

Main sequence—The narrow strip of stars running from upper left to lower right on the H-R diagram. Main sequence stars are those that are shining stably and without any dramatic changes in their size or surface temperature. About 90% of all stars are main sequence stars, including the Sun.

Spectral class—A classification category containing stars with similar patterns of absorption lines in their spectra. The spectral classes are denoted by the letters O, B, A, F, G, K, M, and represent a temperature sequence. The hottest stars are type O, while the coolest are type M.

Supergiant—A star of extraordinary size and luminosity, belonging to luminosity class I. These are massive stars (five to 30 times as massive as the Sun) that have exhausted the hydrogen fuel in their cores and are burning heavier elements like helium and oxygen to sustain themselves.

Turn-off point—The upper end of the main sequence in an H-R diagram of a star cluster. Since more massive (hotter) stars evolve off the main sequence faster than less massive (cooler) ones, the turnoff point gradually "moves down" the main sequence as the cluster ages. The location of the turn-off point reveals the current age of the cluster.

tral type B8 I) are among the largest, most luminous, and most massive stars in the **galaxy**.

The H-R diagram and stellar evolution

One of the most important properties of the H-R diagram is that it lets us trace the lives of the stars. A ball of gas officially becomes a star at the moment that nuclear fusion reactions begin in its core, converting hydrogen to helium. At the point the star is a brand-new main sequence object, and lies at the lower boundary of the main sequence strip. Sensibly enough, this is called the *zero-age main sequence*, or ZAMS.

As a star ages, it gradually gets brighter. This means the star moves upward on the H-R diagram, because it is getting more luminous. That is why the main sequence is a band and not just a line: different stars of a given spectral type are different ages and have slightly different luminosities.

When a star runs out of hydrogen, many bizarre and fascinating things begin to happen. With its hydrogen nearly exhausted, the star has to begin fusion of heavier elements like helium, **carbon**, and **oxygen** to keep its interior furnace going. This causes the surface of the star to expand greatly, and it becomes very luminous, moving to the upper parts of the H-R diagram.

Giant stars, therefore, are dying beasts. They are stars that have run out of hydrogen and are now burning heavier elements in their cores. Many giant stars are unstable and pulsate, while others shine so fiercely that **matter** streams away from them in a *stellar wind*. All these are important evolutionary states and occur in stars in specific parts of the H-R diagram.

Nowhere is **stellar evolution** more dramatically illustrated than in a **star cluster** H-R diagram. Clusters are large groups of stars that all formed at the same time. Figures 2 and 3 show the H-R diagram for two clusters, the Pleiades and M67.

These are only a few of the ways in which H-R diagrams reveal the essential properties of stars. The power and elegance of the H-R diagram in improving our understanding of stars and how they evolve has made its invention one of the great advances in twentieth-century **astronomy**. More importantly, it demonstrates how careful classification, often considered mundane or even boring work, can reveal the beautiful patterns hidden in nature and reward humans with a clearer understanding of the universe of which they are such a small part.

See also Spectral classification of stars; Spectroscopy; Stellar magnitudes; Stellar wind.

Resources

Books

Introduction to Astronomy and Astrophysics. 4th ed. New York: Harcourt Brace, 1997.

Meadows, A.J. *Stellar Evolution.* 2nd ed. Oxford: Pergamon, 1978.

Shu, F. *The Physical Universe: An Introduction to Astronomy.* Chap 8-9. University Science Books, 1982.

Jeffrey C. Hall

Heterotroph

A heterotroph is a creature that must ingest **biomass** to obtain its **energy** and **nutrition**. In direct contrast, autotrophs are capable of assimilating diffuse, inorganic energy and materials and using these to synthesize biochemicals. Green plants, for example, use sunlight and simple inorganic molecules to photosynthesize organic **matter**. All heterotrophs have an absolute dependence on the biological products of autotrophs for their sustenance—they have no other source of nourishment.

All animals are heterotrophs, as are most **microorganisms** (the major exceptions being microscopic **algae** and blue-green **bacteria**). Heterotrophs can be classified according to the sorts of biomass that they eat. Animals that eat living plants are known as herbivores, while those that eat other animals are known as carnivores. Many animals eat both plants and animals, and these are known as omnivores. **Animal parasites** are a special type of **carnivore** that are usually much smaller than their **prey**, and do not usually kill the animals that they feed upon.

Heterotrophic microorganisms mostly feed upon dead plants and animals, and are known as decomposers. Some animals also specialize on feeding on dead organic matter, and are known as scavengers or detritivores. Even a few vascular plants are heterotrophic, parasitizing the roots of other plants and thereby obtaining their own nourishment. These plants, which often lack **chlorophyll**, are known as saprophytes.

Humans, of course, are heterotrophs. This means that humans can only sustain themselves by eating plants, or by eating animals that have themselves grown by eating plants. All of these foods must be specifically grown for human consumption in agricultural ecosystems, or be gathered from natural ecosystems. If humans and their societies are to be sustained over the long term, it can only be through the wise use of the **species** and ecosystems that sustain them. This is a fact, and a consequence of the inextricable links of humans with other species and with the products and services of ecosystems. The intimate dependency of humans on other creatures is a biological and ecological relationship that can be difficult for modern people to remember as they purchase their food in stores, and do not directly participate in its growth, harvesting, and processing.

See also Autotroph; Herbivore; Omnivore; Scavenger.

Bill Freedman

Hibernation

Hibernation is a state of inactivity, or torpor, in which an animal's **heart rate**, body **temperature**, and breathing rate are decreased in order to conserve **energy** through the cold months of winter. A similar state, known as estivation, occurs in some **desert** animals during the dry months of summer. Hibernation is an important **adaptation** to harsh climates, because when food is scarce, an **animal** may use up more energy maintaining its body temperature and in foraging for food than it would receive from consuming the food. Hibernating animals use 70-100 times less energy than when active, allowing them to survive until food is once again plentiful.

Many animals **sleep** more often when food is scarce, but only a few truly hibernate. Hibernation differs from sleep in that a hibernating animal shows a drastic reduction in **metabolism**, or its rate of energy usage, and arouses relatively slowly, while a sleeping animal decreases its metabolism only slightly, and can wake up almost instantly if disturbed. Also, hibernating animals do not show periods of rapid **eye** movement (REM), the stage of sleep associated with dreaming in humans. **Bears**, which many people think of as the classic hibernating animals, are actually just deep sleepers, and do not significantly lower their metabolism and body temperature. True physiological hibernation occurs only in small **mammals**, such as **bats** and woodchucks, and a few **birds**, such as poorwills and nighthawks. Some **species** of insect show periods of inactivity where growth and development are arrested and metabolism is greatly reduced: this state is generally referred to as diapause, although when correlated with the winter months it would also fit the definition of hibernation.

Preparing for hibernation

Animals prepare for hibernation in the fall by storing enough energy to last them until spring. **Chipmunks** accomplish this by filling their burrows with food, which they consume during periodic arousal from torpor throughout the winter. Most animals, however, store en-

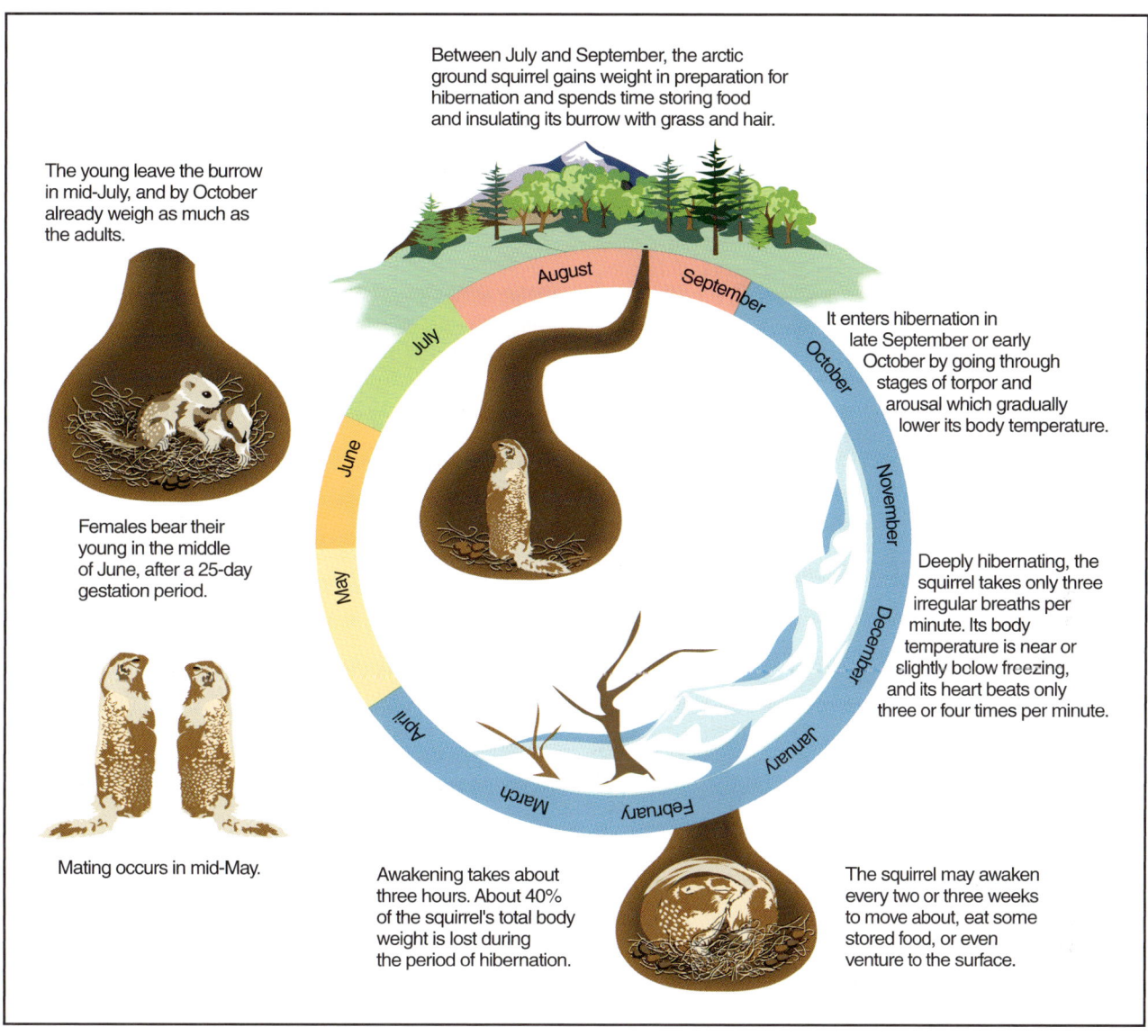

Between July and September, the arctic ground squirrel gains weight in preparation for hibernation and spends time storing food and insulating its burrow with grass and hair.

The young leave the burrow in mid-July, and by October already weigh as much as the adults.

Females bear their young in the middle of June, after a 25-day gestation period.

Mating occurs in mid-May.

It enters hibernation in late September or early October by going through stages of torpor and arousal which gradually lower its body temperature.

Deeply hibernating, the squirrel takes only three irregular breaths per minute. Its body temperature is near or slightly below freezing, and its heart beats only three or four times per minute.

The squirrel may awaken every two or three weeks to move about, eat some stored food, or even venture to the surface.

Awakening takes about three hours. About 40% of the squirrel's total body weight is lost during the period of hibernation.

A year in the life of a female arctic ground squirrel. *Illustration by Hans & Cassidy. Courtesy of Gale Group.*

ergy internally, as **fat**. A woodchuck in early summer may have only about 5% body fat, but as fall approaches changes occur in the animal's **brain chemistry** which cause it to feel hungry and to eat constantly, which results in an increase to about 15% body fat. In other animals, such as the **dormouse**, fat may comprise as much as 50% of the animal's weight by the time hibernation begins. A short period of fasting usually follows the feeding frenzy, to ensure that the digestive tract is completely emptied before hibernation begins.

Many hibernators also produce a layer of specialized fat known as brown fat (brown adipose **tissue**) which lies between the shoulder blades of the animal. Brown fat is capable of rapidly producing large amounts

of **heat** when it is metabolized, which raises the animal's body temperature and brings about the eventual arousal of the animal from hibernation.

Entering hibernation

Going into hibernation is a gradual process. Over a period of days, the heart rate and breathing rate drop slowly, reaching slow steady rates of just a few times per minute. The body temperature also plummets from mammalian levels of 101.5–103.5°F (38.6–39.7°C) to 50–68°F (10–20°C). The lowered body temperature is regulated about the new set point, and therefore makes fewer demands on metabolism and food stores.

Electrical activity in the brain almost completely ceases during hibernation, although some areas remain active. These are primarily areas which respond to external stimuli such as **light**, temperature, and noise, so that the hibernating animal can be aroused under extreme conditions.

Arousal

Periodically, perhaps every two weeks or so, the hibernating animal will arouse and take a few deep breaths to refresh its air supply, or in the case of the chipmunk, to grab a bite to eat. If it is a particularly mild winter day, some animals may venture above ground. These animals, including chipmunks, **skunks**, and **raccoons**, are sometimes called "shallow" hibernators.

Arousal begins with an increase in the heart rate. **Blood** vessels dilate, particularly around the heart, lungs, and brain, leading to an increased breathing rate. Blood also flows into the layer of brown fat, increasing activity there and causing a rise in body temperature. Eventually, the increase in circulation and metabolic activity spreads throughout the body, reaching the hindquarters last. It usually takes several hours for the animal to become fully active.

The importance of understanding hibernation

Scientists are interested in discovering the mechanisms which control hibernation and arousal, and the means by which animals survive such critically low metabolic activity. Many researchers hope to discover ways of placing human beings into a state of hibernation, thus allowing them to survive medical operations which cut off much of the supply of blood to the brain, or even to embark on long **space** voyages. Other researchers look at the changes in brain chemistry of hibernators as a way of understanding **obesity** in humans, or as a way to unravel the mysteries of sleep and the functioning of the human brain.

Resources

Books

Stidworthy, John. *Hibernation.* New York: Gloucester Press, 1991.

Periodicals

Kenzie, Aline. "Seeking the Mechanisms of Hibernation." *BioScience* 40 (June 1990).

David E. Fontes

Hickory *see* **Walnut family**

High-temperature superconductivity *see* **Superconductor**

Himalayas, geology of

Early mountaineers from India named the Himalayas "snow abode" based on two Sanskrit words *hima* and *laya*. These early climbers were attracted to the **mountains** by the same features that, today, challenge climbers from all over the world. The range includes the highest peaks in the world, notably Mount Everest; glaciated valleys and snow that never melts create unsurpassable vistas; and the scenery and dangers inspire myths and religious contemplation.

The range is 1,550 mi (2,500 km) long from west to east, and it encompasses all of Nepal and Bhutan and parts of Afghanistan, India, Pakistan, and China. The north-south width varies from 125-250 mi (200-400 km), and they cover 229,500 sq mi (594,400 sq km) of Earth's area. In height, the range rises to the top of Everest at 29,028 ft (8,848 m); much of the area is at an elevation of 2.5 mi (4 km) above **sea level**. The Himalayan Mountains are actually part of a band of ranges that cross the globe from North **Africa** to Asia's Pacific Coast. They are bordered by the Karakoram and Hindu Kush ranges as well as the high Tibetan Plateau.

Ranges and origin

The Himalayas are also made up of four distinct ranges. The northernmost Trans-Himalayas, the Greater or Tibetan Himalayas, the Lesser or Lower Himalayas, and the southernmost Outer Himalayas parallel each other in long belts from west to east. Each has a different geologic history depending on how, where, and when (in **geologic time**) the tremendous plates that make up Earth's crust collided and pushed up the ranges. **Plate tectonics** is a geologic theory that describes the crust of the **Earth** as a collection of plates floating on the molten mantle; scientists believe the movement (**tectonics**) is the planet's effort to keep itself cool.

Before the Jurassic Period (180 million years ago), India, **South America**, Africa, **Australia**, and **Antarctica** were united as one giant, southern "super continent" called Gondwanaland or Gondwana. In Jurassic times, this super **continent** began to break into fragments that moved away from each other. India began to move northward toward Eurasia, but, between the Eurasian Plate and the Indian Plate, was the Tethys Trench which was a deep **ocean**. The Indian Plate moved to the north over the course of 130 million years; in the Tertiary Period (50 million years ago), it finally collided with Eurasia. Collisions like this between continents typically take millions of years and involve volcanism, seismic activity, **metamorphism** (changes) of **rocks** due to intense **pressure**, and episodes of mountain-building. Scientists have been able to use the metamorphic rocks in the Himalayas to date these events by measuring radioactivity remaining in the rocks.

Mountain building

As India pushed relentlessly toward Eurasia, the Tethys Trench was compressed, folded, and faulted. The base of the trench consisted of sedimentary rocks that were weak to begin with. When they were broken by the colliding plates, weaknesses in these overlying materials allowed basalt and granite to intrude upward from Earth's underlying mantle. These materials were fresh and hard; when the Indian plate encountered the Tethys trench, the plate was sheared under, or subducted (curved and sucked down into the mantle) under the trench. The trench rose in elevation as it was pushed up by the subducting plate and the compressive forces, and the **water** drained away. The flat ocean bottom became today's Tibetan Plateau. The Trans-Himalayan Range formed the southern edge of the Plateau, and, as the mountains rose, new **rivers** were created, and their drainage changed the climate and the downslope topography. At this point in the development of the Himalayas, the mountains were impressive but had not reached the monumental elevations we know now.

From about 50-23 million years ago, the subduction began to slow (the Indian plate of rock was too buoyant to be drawn down into the mantle), and the plate corner intersected **Asia** and began to slide under Asia. During the Miocene Epoch (23 million years ago), the compression of the plates intensified and continued into the Pliocene Epoch (1.6 million years ago). As the Indian plate slid under the Asian plate, its upper layers were stripped off and curled back on the subcontinent. These layers, called nappes, were older **metamorphic rock** from the ancient Gondwana. As the mountains rose, the rivers steepened, the runoff and **erosion** increased, deposition of sediments similarly increased, and the weight of the sediments forced receiving basins downward so they could hold still more alluvium.

The creation of the nappes left a core zone. In the Pliestocene Epoch about 600,000 years ago, serious mountain building began in a **time** that is relatively recent in geologic terms. New granite and gneiss intrusions pushed up and rose to the astounding height of Everest. As Everest itself was uplifted, the crystalline rock dragged evidence of the older sediments including those from the Tethys Basin to the summit so the most ancient fossils in the region are present on the newest mountain tops. These **uplift** episodes again changed the climate and blocked rains from moving to the north. The mountains on the north (the Trans-Himalayas) and the Tibetan Plateau became deserts. Heavy rains to the south changed the line of the crest and shifted the direction of rivers to create a high midlands between the Greater Himalayas and the Lesser Himalayas to the south. High valleys filled with sediment to form lush valleys like the Kathmandu Valley. The Outer Himalayas including the Shiwalik Hills form the southern line of the Himalayas, and the Gangetic Plain (draining toward the Ganges River) lies below it for the full extent of the Indian subcontinent and Bangladesh.

Seismic activity

The northeastern end of the Gangetic Plain has experienced four "great" earthquakes with a Richter magnitude exceeding 8.0 in the past 100 years, beginning with the Assam **earthquake** in 1897. Over 30,000 people perished in these quakes. The origin of the earthquakes is the same tectonic, plate-moving action that welded the Indian subcontinent to Eurasia and formed the Himalayan Mountain Range as a kind of massive suture. Along the line where the outer Himalayas border the Gangetic Plain, seismic gaps store the strain from tectonic movement. One of these, called the Central Seismic Gap, has not released its strain in the form of an earthquake in an estimated 745 years (since a great quake appeared to have killed the king of Nepal in 1255). The Central Seismic Gap is about 500 mi (800 km) long and lies between the regions struck by great earthquakes in 1905 and 1934. Northern India and Nepal will experience significant devastation if a great earthquake occurs in this region with a population of 100 million.

The faults along which earthquakes occur were generated by the pressure of the Indian subcontinent. After the period when it was subducted under the Eurasian plate, the plate's direction shifted, and it pushed toward Tibet, compressing the edge of Tibet. When the resisting rock had folded as much as possible, it began to tear and the faults were born. Geologists know plate movement is continuing from the earthquake activity but also from the continuing formation of hills along the southern limits of the Himalayas. The longest **fault**, called the Main Detachment Fault, is as long as the Himalayan Range from west to east. If the fault does rupture, it is most likely to occur where the greatest strain has accumulated at the Central Seismic Gap. Although the prospect of such an earthquake is frightening, study of seismic activity has helped geologists to better understand the complex processes that have formed Earth's most upstanding mountain range.

KEY TERMS

. . . .

Alluvium—Particles of soil and rock that are moved as sediments by the downslope flow of water.

Gondwanaland—An ancestral supercontinent that broke into the present continents of Africa, South America, Antarctica, and Australia as well as the Indian subcontinent.

Metamorphism—The process of changing existing rock by increased temperature or pressure.

Nappes—Enormous folds of rock strata that become flat lying.

Plate tectonics—The motion of large sections of Earth's crust.

Seismic gap—A length of a fault, known to be historically active, that has not experienced an earthquake recently and may be storing strains that will be released as earthquake energy.

Seismology—The study of earthquakes.

Subduction—In plate tectonics, the movement of one plate down into the mantle where the rock melts and becomes magma source material for new rock.

Trilobites—Extinct crustaceans that lived from the Cambrian through the Permian Periods. Their extensive fossil evidence helps geologists and paleontologists understand early rocks and early life on Earth.

In analyzing geologic processes and earthquake hazards, geologists have used technology to measure movements in areas that are remote, frigid, and impassable. An array of 24 satellites bounce **radio** signals over the earth's surface in the **Global Positioning System** (GPS). Hikers can tune into the GPS signals and determine their own exact locations, and geologists can similarly find the locations of mountaintops and continents. Comparisons of data over time show relative movements. By routinely using GPS data to survey a line of reference points, scientists are understanding **geophysics**, geomechanics, and the convergence of continents. They have found that India is shifting to the northeast toward Asia at a **rate** of about 2.5 in (6 cm) per year. Studies of **paleontology** (fossils) in the Himalayas have also added pieces to the puzzle of the explanation of the range's geologic and seismic history. Comparison of fossilized trilobites (ancient crustaceans) found in different locations in the Himalayas and the places where they were known to have lived helps superimpose the geologic timetable on the components of the comparatively young Himalayas.

Resources

Books

Coxall, Michelle, and Paul Greenway. *Indian Himalaya: A Lonely Planet Travel Survival Kit.* Hawthorn, Australia: Lonely Planet Publications, 1996.

Hamblin, W.K., and E.H Christiansen. *Earth's Dynamic Systems.* 9th ed. Upper Saddle River: Prentice Hall, 2001.

Hancock, P.L., and B J. Skinner, eds. *The Oxford Companion to the Earth.* Oxford: Oxford University Press, 2000.

MacDougall, J.D. *A Short History of Planet Earth: Mountains, Mammals, Fire, and Ice.* New York: John Wiley & Sons, Inc., 1996.

Nicolson, Nigel. *The Himalayas.* Amsterdam: Time-Life Books, Inc., 1975.

Periodicals

Hughes, Nigel C. "Trilobite Hunting in the Himalaya." *Earth* (June 1996): 52+.

Pendick, Daniel. "Himalayan High Tension." *Earth* (Oct. 1996): 46+.

Gillian S. Holmes

Hippopotamus

The common or river hippopotamus (*Hippopotamus amphibius*) is a huge, even-toed hoofed **herbivore** that lives in bodies of **freshwater** in central and southern **Africa**. A second **species**, the pygmy hippopotamus (*Choeropsis liberiensis*), lives in **water** bodies in Western African rainforests. Both species are included in the family Hippopotamidae.

The name hippopotamus means "river horse" but hippos are only distantly related to **horses**. Horses are odd-toed hoofed animals, while hippos are even-toed in the class Artiodactyla. Hippos have four hoofed toes on each foot. The common hippo has webbing between its toes, while the pygmy hippo has less webbing.

Fossil finds indicate that hippos are formerly found throughout much of Eurasia, but today hippos are found only in the tropical regions of Africa. The common hippo is abundant in the **rivers**, lakes, and swamps of most of sub-Saharan Africa, while the pygmy hippo is limited to forested areas in West Africa.

The common hippo

The common hippo is barrel-shaped, measuring 14 ft (4 m) long, 4.5 ft (1.5 m) high and weighing about 2 tons (1,800 kg). Large males have been known to reach 4.5 tons (3,800 kg). The common hippo is slate brown in **color**, shading to either a lighter or darker color on the underside.

Common hippos have relatively short legs for their vast girth, but hippos spend most of the day under water, with only the top of their head visible. Their eyes are sited on top of their head, and sometimes only the eyes, the flicking ears, and a small mound of back are all that is visible.

The hippo's teeth and its diet

Hippos have a huge mouth, measuring up to 4 ft (1.2 m) across, and a pair of huge incisors in each jaw. Only a few teeth are immediately visible, mainly the curved lower canine teeth (which are a source of ivory) on the outer part of the jaw. Like tusks, these teeth continue to grow and can reach a length of 3 ft (1 m). Hippos are herbivores, grinding up vegetation with their big, flat molars at the back and the mouth. Hippos die when their molars have worn down too much to grind food.

After sunset, hippos come onto land in search of succulent **grasses** and **fruits**. The path to an individual bull's foraging ground may be marked by a spray of excrement that warns other hippos away. The cows do not mark their paths. Individual animals may wander as far as 20 mi (32 km) during the night to find food. It takes almost 150 lb (68 kg) of food each night to satisfy a hippo's appetite.

Hippo in water

A hippo's eyes, ears, and nostrils are all positioned in a single **plane** that can stay above water when the rest of the **animal** is submerged. Both the ears and the nostrils can close, at least partially, when in water. Hippos do not see well either on land or in the water; instead, they depend on their acute **hearing** to warn them of danger and their good sense of **smell** to find food. When alarmed, a hippo may quietly submerge or it may attack, especially if it is threatened by people in boats. Several hundred people are killed by hippos each year in Africa.

Hippos spend most of the day in groups in the water. They prefer water about 5 ft (1.5 m) deep, just deep enough to swim if they want to or to walk on the bottom. Hippos can stay completely submerged for about six minutes, but they generally rise to breathe again after only two or three minutes. They can control the **rate** of their rising and sinking in the water by changing the **volume** of air in their lungs by movements of the diaphragm. Hippos can walk on the bottom of the river or **lake** at a rate of about 8 mi (13 km) per hour. On land, they can run at up to 20 mi (32 km) an hour.

Reproduction

A single herd of hippos may include up to 100 animals. The herd's location, foraging, and movement are

A hippopotamus. *Photograph by William & Marcia Levy. The National Audubon Society Collection/Photo Researchers, Inc. Reproduced by permission.*

controlled by a group of mature females. The females and their young inhabit the center of a herd's territory, called the crèche. The male's individual territories, called refuges, are spaced around the crèche. A bull will defend his territory against another bull. If roars and open mouths do not scare off the challenger, they attack each other with open mouths, trying to stab their canine teeth into each other's head or **heart**.

The animals breed as the dry season is ending, with the females selecting their mates. Hippos mate in water. Gestation lasts about eight months, and the calves are occasionally born in the water at the height of the rainy season when the most grass is available. A new calf is about 3 ft (1 m) long and weighs about 60 lb (27 kg) when born. On land, it can stand very quickly. It will be several weeks, however, before the mother and her infant rejoin the group.

Once taken into the crèche, the young hippos are tended by all the females. Although adult hippos have few enemies, the calves are small enough to be taken by lions and **crocodiles**. Until young hippos start to swim by themselves, the young may ride on their mothers' backs when in the water. Once they can swim, the calves may nurse, eat, and even nap under water. They automatically come up to the surface to breathe every few minutes.

Young females are sexually mature at three to four years old, but usually do not mate until they are seven or eight years old. Male hippos are mature at about five years old, but do not successfully challenge the dominant males for the right to mate until they are much older. A cow with a young calf will usually have another calf when the first one is two or three years old. Because an adolescent hippo is not ready to go out on its own until

Green, Carl R., and William R. Sanford. *The Hippopotamus.* Wildlife Habits & Habitats Series. New York: Crestwood House, 1988.

Hippos. Zoobooks Series. San Diego: Wildlife Education, 1988.

Knight, Linsay. *The Sierra Club Book of Great Mammals.* San Francisco: Sierra Club Books for Children, 1992.

Lavine, Sigmund A. *Wonders of Hippos.* New York: Dodd, Mead & Company, 1983.

Stidworthy, John. *Mammals: The Large Plant-Eaters.* Encyclopedia of the Animal World. New York: Facts On File, 1988.

Jean F. Blashfield

KEY TERMS

Crèche—The central group in a herd of hippos, including mature females and calves.

about four years of age, a cow may be taking care of two calves at once. In the wild, hippos live for about 30 years, while in captivity they can live past 40 years old.

The pygmy hippo

Pygmy hippos were discovered relatively recently in 1913, when an agent for a German animal collector caught several specimens and sent them back to **Europe**.

The smaller pygmy hippo is proportioned more like a pig than the common hippo. Pygmy hippos reach a height of only about 3 ft (1 m), a length of 5 ft (1.5 m), and weigh only about 500 lb (227 kg). The oily black skin has a greenish tinge, with lighter colors, even yellow-green, on its underparts.

Unlike the common hippo, the pygmy hippo's eyes do not bulge out and it has only one set of incisors. The skin contains **glands** that give off an oil that looks reddish in sunlight, a characteristic which prompted sideshow claims that the pygmy hippo sweated **blood**. The oil keeps the animal's skin from drying out. Pygmy hippos' skin dries out very easily, so they live within an easy stride of water. Pygmy hippo calves are born after a seven month gestation, weigh less than 10 lb (4.5 kg), and have to be taught to swim.

Hippos and people

Hippos herds greatly benefit the rivers and lakes where they live, their excrement fertilizing the vegetation of the **habitat**. As a result, all animals in the food chain benefit, and fishing is usually very good in hippo areas. When the supply of nearby vegetation in areas near hippo pools became scarce, however, these huge animals sometimes feed in farm fields, where many have been shot. Also, hippos are hunted for their meat, hide, and ivory tusks.

The numbers of pygmy hippos left in the wild is uncertain because they are so rarely seen, but it is likely that they are an **endangered species**. Luckily, pygmy hippos breed well in zoos, and it may one day be possible to restock the wild habitats.

Resources

Books

Arnold, Caroline. *Hippo.* New York: Morrow Junior Books, 1989.

Histamine

Histamines are chemicals released by cells of the **immune system** during the inflammatory response, which is one of the body's defenses against **infection**. For instance, the inflammatory response helps neutralize **bacteria** that enter the body when the skin is accidentally cut with a knife. In addition, the sneezing, runny nose, and itchy eyes of allergies are actually "small-scale" inflammatory responses initiated by allergens such as dust, **mold**, and pollen. Histamines play a prominent role in both kinds of reactions.

Histamines are contained within two types of immune cells, basophils and mast cells. Basophils are free-floating immune cells, while mast cells are fixed in one place. When basophils and mast cells are activated by other immune cells—such as in response to invasion of the body by bacteria—they release histamines into body tissues.

Once histamines are released into the tissues, they exert a variety of effects. Histamines dilate **blood** vessels, stimulate gland secretion, and prompt the release of **proteins** from cells. These effects, in turn, help the body rid itself of foreign invaders. The dilation of blood vessels increases the circulation of blood to the injured area, washing away harmful bacteria. The release of proteins from cells attracts other immune cells to the area, such as macrophages, which engulf and destroy bacterial invaders. In response to these activities within the body, the injured area becomes red, swollen, and painful. These symptoms of **inflammation** signal that the body's inflammatory response is activated.

Histamines also play a role in allergic responses. Instead of responding to bacterial or viral invaders, mast cells and basophils bind to allergens and then release histamines and a special kind of antibody called IgE. Histamines released from mast cells in the nasal passages, lungs, and throat in response to allergens prompt inflammatory responses in these organs, leading to allergic symptoms, such as a running nose, coughing, sneezing, and watery eyes.

An effective way to control allergic symptoms is to disable the histamines with **antihistamines** which prevent the allergen from exerting their effects on the tissues. Antihistamines are the active ingredients in many **allergy** medications, and work by binding to the released histamines, effectively inactivating them. Until recently, antihistamines had an inconvenient side effect: they caused drowsiness in a small percentage of the population. Newer antihistamines do not cause drowsiness, and most people can tolerate these antihistamines without side effects.

Historical geology

Historical geology is the study of changes in **Earth** and its life forms over **time**. It includes sub-disciplines such as **paleontology**, paleoclimatology, and paleoseismology. In addition to providing a scientific basis for understanding the **evolution** of Earth over time, historical geology provides important information about ancient climate changes, volcanic eruptions, and earthquakes that can be used to anticipate the sizes and frequencies of future events.

Scientific interpretation of Earth's history requires an understanding of currently operating geologic processes. According to the doctrine of actualism, most geologic processes operating today are similar to those that operated in the past. The rates at which the processes occur, however, may be different. By studying modern geologic processes and their products, geologists can interpret **rocks** that are the products of past geologic processes and events. For example, the layering and distribution of different grain sizes within a sandstone layer may be similar to those in a modern beach, leading geologists to infer that the sandstone was deposited in an ancient beach environment. There have been some past geologic events, however, that are beyond the range of human experience. Evidence of catastrophic events such asteroid impacts on Earth has led geologists to abandon the doctrine of **uniformitarianism**, which holds that all of the geologic past could be explained in terms of currently observable processes, in favor of actualism.

Rocks preserve evidence of the events that formed them and the environments in which they were formed. Fossils are an especially useful type of biological evidence preserved in sedimentary rocks (they generally do not occur in igneous or metamorphic rocks). Organisms thrive only in those conditions to which they have become adapted over time. Therefore, the presence of particular fossils in a rock provides paleontologists with insights into the environment in which the fossilized organisms lived. Sediments and sedimentary rocks also preserve a variety of tracks, trails, burrows, and footprints known as trace fossils. Information about **tree** ring widths and changes in the isotopic composition of some sedimentary rocks and glacial **ice** over time have been used to reconstruct patterns of past climate changes over millennial time scales. These patterns, in turn, provide important information about the magnitude and **frequency** of future climate changes.

Any study of Earth's history involves the element of time. Relative **geologic time** considers only the sequence in which geologic events occurred. For example, rock A is older than rock B, but younger than rock C. Relative geologic time is based largely on the presence or absence of index fossils that are known to have existed over limited ranges of geologic time. Using the concept of relative geologic time, geologists in the nineteenth century correlated rocks around the world and developed an elaborate time scale consisting of eons, eras, periods, and epochs. The development of radiometric **dating techniques** during the second half of the twentieth century allowed geologists to determine the absolute ages of rocks in terms of years and assign specific dates to the relative time boundaries, which had previously been defined on the basis of changes in fossil content.

See also Fossil and fossilization; Geochemical analysis; Geochemistry; Geophysics; Stratigraphy.

Hoatzin

The hoatzin (*Opisthocomus hoazin*) is one of the world's most peculiar bird **species**. It is the sole member of its family, Opisthocomidae. It is peculiar enough to have defied taxonomists' best efforts for years.

This bird lives only in the rainforests of northern **South America**. Its feathers are dark brown on the back and lighter below, and chestnut-colored on its sides. The skin around its red eyes is a startling electric blue. Its head is topped by a crest of long chestnut-colored feathers.

The hoatzin builds its next of sticks in trees or large bushes, usually on boughs that overhang the **water**. Into this nest the female lays two or three (though sometimes as many as five) buff-colored eggs speckled with blue or brown. After an incubation period of 28 days, out of these eggs hatch some of most remarkable chicks in the order Aves.

Hoatzin chicks are naked, and proof of the old aphorism "a face only a mother could love." But beauty does not count: what does count is the tiny claws on the chick's wings. The claws help the chick to hold on as it moves through the branches. But even if it should tumble off the branch and fall into the water below, the hoatzin chick can swim to the nearest branch or **tree** trunk and climb back up the tree into the nest.

Some people have called the hoatzin a living fossil and equated these claws with those of *Archaeopteryx*, the ancestor of modern **birds** that lived 150 million years ago. However, the claws are not unique among birds: some species of **geese** retain spurs on their wings into adulthood, and young European coots have a single claw on each wing that helps them climb back to the nest as well. It is more likely that the hoatzin's claws are not a relic, but a recent **adaptation** to its rather precarious nesting site.

Hoatzins are remarkable for still another reason. Their diet consists strictly of vegetable matter—leaves, flowers, and **fruits**. It is the only tree-dwelling bird that feeds its young on leaves. To handle this fibrous diet, the hoatzin has evolved a very large crop, or gizzard, in which it grinds up the tough **cellulose** fibers of the leaves. The crop is so large that it accounts for about one-third of the adult bird's 28 oz (793 g) body weight.

Hodgkin's disease

Hodgkin's disease is a type of **cancer** involving tissues of the **lymphatic system**. The lymphatic system is a network of organs, tissues, and ducts in the human body. The lymphatic system maintains the fluid balance in the body by coordinating the draining of fluid from cells and tissues back into the bloodstream. Also, the lymphatic system aids in fighting infections caused by **microorganisms**, by supplying the body with white **blood** cells.

A variety of cancers called lymphomas affect the lymph tissues. Hodgkin's disease represents a specific type of lymphoma. Its cause is unknown, although some interaction between individual genetic makeup, environmental exposures, and infectious agents is suspected.

Hodgkin's lymphoma can occur at any age. The majority of all cases of Hodgkin's lymphoma occur in people between the ages of 15 and 34, and those who are older than 60 years of age.

The lymphatic system

The lymphatic system is part of the body's **immune system**. It consists of a number of elements. First, there is a network of vessels. The vessels drain **tissue** fluid from all the major organs of the body, including the skin, and from all four limbs. These vessels pass through lymph nodes on their way to empty their contents into major **veins** at the base of the neck and within the abdomen.

The lymph nodes are clusters of specialized cells that serve to filter the lymph fluid. In this capacity, they trap foreign substances such as viruses, **bacteria**, cancer cells, as well as any other encountered debris. For example, the examination of lymph nodes form people who live in cities typically detects gritty, dark material, which is not present in the lymph nodes of people who live in rural settings. This is because the lymph nodes of the city dweller have received fluid from the lungs, which contain debris from polluted city air.

Another component of the lymphatic system are lymphocytes. Lymphocytes are cells of the immune system. They are produced within bone marrow, lymph nodes, and spleen, and circulate throughout the body in both blood and lymph fluid. These cells work to identify and rid the body of any invaders that threaten health.

Still another component of the lymphatic system are clusters of scavenger-like immune cells. These exist in major organs, and provide immune surveillance on location. These include the tonsils and adenoids in the throat/pharynx, Kupffer cells in the liver, Peyer's patches in the intestine, and other specialized immune cells stationed in the lungs and the **brain**.

Cancer

Cancer is a general term that refers to a condition in which a particular type of **cell** within the body begins to multiply in an out-of-control fashion. This may mean that cancer cells multiply more quickly, or it may mean that cancer cells take on abnormal characteristics. For example, at a very early stage in embryonic development (development of a fetus within the uterus), generic body cells begin to differentiate. Cells acquire specific characteristics which ultimately allow liver cells to function as liver cells, blood cells as blood cells, brain cells as brain cells, and so on. Thus, cancer can be considered to be a process of "de-differentiation." In other words, a specialized type of cell loses whatever controls govern the expression of its individual characteristics and instead revert to a more embryonic cell. Such cells also lose their sense of organization and no longer position themselves appropriately within their resident tissue.

Cancer cells can also acquire the ability to invade other tissues. Normally, for example, breast cells are found only in breast tissue. However, cancerous breast cells can invade into other tissue spaces, so that breast cancer can spread to bone, liver, brain, etc.

Lymphoma is a cancer of the lymph system. Depending on the specific type, a lymphoma can have any or all of the characteristics of cancer. These characteristics include rapid multiplication of cells, abnormal cell types, loss of normal arrangement of cells with respect to each other, and invasive ability.

Causes and symptoms of Hodgkin's lymphoma

Hodgkin's lymphoma usually begins in a lymph node. This node enlarges, but may or may not cause the **pain** that typically results when lymph nodes enlarge as a consequence of an **infection** by a microorganism. Hodgkin's lymphoma progresses in a fairly predictable way, traveling from one group of lymph nodes on to the next. More advanced cases of Hodgkin's include involvement of the spleen, the liver, and bone marrow.

Constitutional symptoms (symptoms which affect the whole body) are common, and include fever, weight loss, heavy sweating at night, and itching. Some patients note pain after drinking alcoholic beverages.

As the lymph nodes swell, they may push on other nearby structures. This **pressure** produces other symptoms. These symptoms include pain from pressure on nerve roots, as well as loss of function of specific muscle groups served by the compressed nerves. Kidney failure may result from compression of the ureters, the tubes that carry urine from the kidneys to the bladder. The face, neck, or legs may swell due to pressure slowing the flow in veins that should drain blood from those regions (superior vena cava **syndrome**). Pressure on the spinal cord can result in paralysis of the legs. Compression of the trachea and/or bronchi (airways) can cause wheezing and shortness of breath. Masses in the liver can cause the accumulation of certain chemicals in the blood, resulting in **jaundice** (a yellowish discoloration of the skin and the whites of the eyes).

As Hodgkin's lymphoma progresses, a patient's immune system becomes less and less effective at fighting infection. Thus, patients with Hodgkin's lymphoma become increasingly more susceptible to both common infections caused by bacteria and unusual (opportunistic) infections caused by viruses, **fungi**, and **protozoa**.

The exact cause of Hodgkin's disease is not known. Viruses, particularly the **Epstein-Barr virus** (a herpes **virus** that causes infectious mononucleosis), are found in tissues of 20-50% of people with Hodgkin's disease. However, a link between the virus and Hodgkins disease has not been established.

Another suggested cause is socio-economic conditions. Studies have demonstrated that Hodgkin's disease is more prevalent in wealthier people in the developed

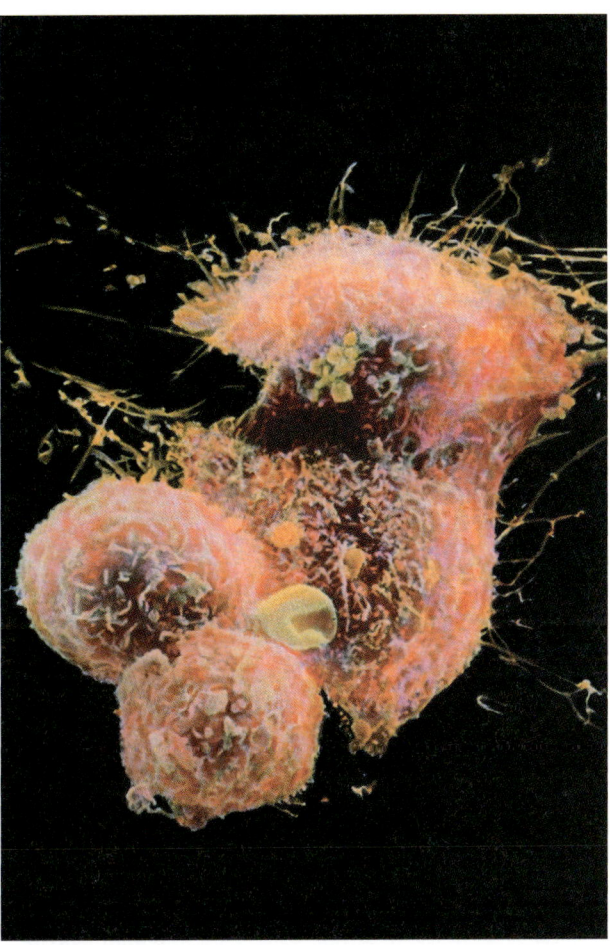

A scanning electron micrograph (SEM) image of dividing Hodgkin's cells from the pleural effusions (abnormal accumulations of fluid in the lungs) of a 55-year-old male patient. *Photograph by Dr. Andrejs Liepins. National Audubon Society Collection/Photo Researchers, Inc. Reproduced by permission.*

world. It has been speculated that the hygienic conditions that most of these people grow up in does not **stress** their immune systems in a way that is healthy for them. Other suggested causes include exposure to chemicals, and a genetic disposition (including the activity of cancerous genes known as oncogenes).

Diagnosis

As with many forms of cancer, **diagnosis** of Hodgkin's disease has two important components. First is the identification of Hodgkin's lymphoma as the cause of the patient's **disease**. Second is the staging of the disease; that is, an attempt to identify the degree of spread of the lymphoma.

Diagnosis of Hodgkin's lymphoma requires removal of a **sample** of a suspicious lymph node (biopsy) and careful examination of the tissue under a **microscope**. In Hodgkin's lymphoma, certain characteristic cells, which are called Reed-Sternberg cells, must be present in order to confirm the diagnosis. These cells usually contain two or more nuclei. The nucleus is the oval, centrally located structure within a cell that houses the genetic material of the cell. Reed-Sternberg cells also have other unique characteristics, which cause them to appear under the microscope as "owl's eyes" or yin-yang cells. In addition to the identification of these Reed-Sternberg cells, other cells in the affected tissue sample are examined. The characteristics of these other cells help to classify the specific subtype of Hodgkin's lymphoma present.

Once Hodgkin's disease has been diagnosed, staging is the next important step. This involves computed tomography scans (CT scans) of the abdomen, chest, and pelvis, to identify areas of lymph node involvement. In rare cases, a patient must undergo abdominal **surgery** so that lymph nodes in the abdominal area can be biopsied (staging laparotomy). Some patients have their spleens removed during this surgery, both to help with staging and to remove a focus of the disease. Bone marrow biopsy is also required unless there is obvious evidence of vital **organ** involvement. Some physicians also order lymphangiograms (a radiograph of the lymphatic vessels).

Staging is important because it helps to determine what kind of treatment a patient should receive. It is important to understand the stage of the disease so that the treatment chosen is sufficiently strong to provide the patient with a cure. All available treatments, however, have potentially serious side effects. The goal of staging, then, is to allow the patient to have the type of treatment necessary to achieve a remission, but to minimize the severity of short and long-term side effects from which the patient may suffer.

Treatment

Treatment of Hodgkin's lymphoma has become increasingly effective over the years. The type of treatment used for Hodgkin's depends on the information obtained by staging, and may include chemotherapy (treatment with a combination of drugs), and /or radiotherapy (treatment with **x rays** which kill cancer cells).

Both chemotherapy and radiotherapy often have side effects. Chemotherapy can result in nausea, vomiting, hair loss, and increased susceptibility to infection. Radiotherapy can cause sore throat, difficulty swallowing, diarrhea, and growth abnormalities in children. Both forms of treatment, especially in combination, can result in sterility (the permanent inability to produce offspring), as well as **heart** and lung damage.

The most serious negative result of the currently available treatments for Hodgkin's disease is the possible development in the future of another form of cancer. This phenomenon is referred to as second malignancy. Examples of second cancers include **leukemia** (cancer of a blood component), breast cancer, bone cancer, or thyroid cancer. A great deal of cancer research is devoted to preventing these second malignancies.

Prognosis

Hodgkin's is one of the most curable forms of cancer. Current treatments are quite effective. Children have a particularly high **rate** of cure from the disease, with about 75% still living cancer-free 20 years after the original diagnosis. Adults with the most severe form of the disease have about a 50% cure rate.

See also Genetic disorders.

Resources

Books

Jaffe, E.S., N.L. Harris, H. Stein, et al. *Pathology and Genetics of Tumours of the Haematopoietic and Lymphoid Tissues.* Lyon: IARC Press, 2001.

Organizations

The Leukemia and Lymphoma Society. 1311 Mamaroneck Ave. White Plains, NY 10605 (914) 949–5213 [cited November 19, 2002]. <http:// www.leukemia-lymphoma.org>.

Brian Hoyle

Holly family (Aquifoliaceae)

Members of the holly family (Aquifoliaceae) are shrubs and trees with small, white or pale green, unisexual flowers. The family consists of four genera with 419 **species**, of which 400 species are members of the holly genus, *Ilex*. The family Aquifoliaceae is a member of the class Magnoliopsida (dicotyledons), division Magnoliophyta (the angiosperms, or flowering plants).

Characteristics of holly

Most hollies are dioecious, meaning a **plant** is either a male (staminate) or a female (pistillate). Holly flowers have radial **symmetry**, and are four-merous, that is, the flowers are round and floral parts occur in fours. Holly flowers have four sepals, four petals, four stamens (male flowers), and an ovary made up of four fused carpels, called a pistil (female flowers).

The fruit is a drupe, which is similar to a berry, but with hard **seeds** instead of soft seeds, and may be red, orange, yellow, or black in **color**. The drupes of *Ilex* spp. are eaten by **wildlife**, especially **birds**. Bird droppings ef-

Blue angel holly. *Photograph by James Sikkema. Reproduced by permission.*

fectively disperse holly seeds, which pass through the bird's digestive tract undamaged. Indeed, hollies are often seen sprouting along fence rows and under other places where birds roost. Holly leaves may also be a source of food for wildlife, as they are sometimes grazed by **deer**.

Hollies may be evergreen or deciduous, depending on whether a species retains its foliage throughout the year or loses its leaves in the fall. One of the more spectacular of the deciduous hollies is winterberry (*Ilex verticillata*). Occurring in the eastern United States and southeastern Canada, winterberry is well known and loved for its crimson colored berries, which provide a stark contrast to the white landscape of winter. Holly leaves are alternate, occurring one at a time on alternating sides of a branch. Leaves are simple (as opposed to compound), and **leaf** margins may be entire, wavy, or spiny.

Distribution and ecology of hollies

The holly family occurs in most temperate and tropical regions, except **Australia** and **Africa**. About 12 species of *Ilex* occur in **North America**. Sarvis holly (*Ilex amelanchier*), gallberry (*Ilex glabra*), large gallberry (*Ilex coriacea*), myrtle-leaf holly (*Ilex myrtifolia*), and winterberry inhabit swamps, bogs, and floodplains. Possum haw (*Ilex decidua*) occurs in floodplains and second-growth **forests**. Some hollies inhabit coastal areas, such as **sand** holly (*Ilex ambigua*) and dahoon holly (*Ilex cassine*). The American holly (*Ilex opaca*) is found in moist forests. In Florida, the scrub holly, a variety of American holly, inhabits sandy, oak scrub. Yaupon holly (*Ilex vomitoria*) occurs in coastal areas, scrub, and second-growth forests. The mountain holly, (*Nemopanthus mucronata*), occurs in the eastern region of North America.

Although most hollies are small trees or shrubs, some have reached a substantial size. For example, the largest dahoon holly, normally about 33 ft (10 m) tall, can reach a height of 79 ft (24 m). Yaupon holly, typically a small or shrubby **tree** can measure 49 ft (15 m) tall.

Ilex guianensis of Central America stands out from the other, relatively shorter hollies. This species can reach a height of 141 ft (43 m). Like the American hollies, the Asian hollies also occur in a variety of sizes and habitats. The tarago (*Ilex latifolia*) of Japan is a handsome tree, with large shiny leaves and red berries. This species may grow to 66 ft (20 m). The smaller *Ilex integra* of Japan is cultivated in tranquil temple gardens.

Uses by humans

Several species of *Ilex* are planted by homeowners for their attractive foliage and berries. Among the best known of the horticultural varieties are the American holly, yaupon holly, and winterberry. Because of their colorful berries which ripen by fall and winter, many hollies are used for indoor decorating, especially during the Christmas season. Holly boughs and wreaths are popular for this purpose. English holly (*Ilex aquifolium*) is commonly used for its attractive berries, but creative decorators will also collect leaves and berries from the native, North American hollies.

Long before the **horticulture** industry discovered the hardiness and attractiveness of hollies, Native Americans used the yaupon holly for medicinal and religious purposes. A dark tea was brewed from the leaves of yaupon holly, which, when consumed, induced sweating, excitation, bowel movement, and vomiting. Only men were allowed to consume this purgative tea. The compound in the holly leaves responsible for the reaction from the tea is the stimulant **caffeine**, also found in coffee. Of the 400 species of *Ilex* worldwide, only about 60 species are known to contain caffeine.

In **South America**, Paraguay tea or yerba mate, is brewed from the leaves of *Ilex paraguariensis*. This brew, like yaupon tea, is a stimulating drink containing caffeine. Although Paraguay tea can induce sweating and urination, it is widely used on a regular basis by many South Americans, just as North Americans drink coffee.

The **bark** of winterberry was used by native American Indians to brew a refreshing tonic, but direct consumption of the berries may induce vomiting. Winterberry bark has also been made into an astringent for the skin, and also as an antiseptic.

In addition to their usage as ornamentals and teas, hollies are also used as a source of **wood**. Because hollies are primarily of small stature, holly wood is used mainly for decorative inlays in furniture and in wood sculptures.

Resources

Books

Everett, T. H. *Living Trees of the World.* New York: Doubleday, 1968.
Raven, Peter, R. F. Evert, and Susan Eichhorn. *Biology of Plants.* 6th ed. New York: Worth Publishers Inc., 1998.

Elaine L. Martin

Holmium *see* **Lanthanides**

Hologram and holography

Holography is defined as a method of producing a three-dimensional (3-D) impression of an object. The recording and the image it brings to life are each referred to as holograms.

This impression is taken by splitting a beam of coherent (that is, uniform over distance as well as over **time**) **radiation** along two paths. One is known and stays undisturbed, to act as a reference. Another strikes the object and is diffracted in an unpredictable fashion along the object's contours. This can be compared to throwing one rock into a pool of **water**, which creates a regular pattern of rings, and then scattering smaller

stones afterwards, to see what kind of design appears where the expanding rings intersect with each other. Likewise, intersections of radiation waves hold crucial information. The aim is to track and record the pattern of interference of the split rays.

The surface of the hologram acts as a **diffraction grating** by alternating clear and opaque strips. When you view a common optical hologram, this grating replicates the action of ordinary illumination, capturing the phase and amplitude of the **light** beam and its interference pattern, in an additive fashion. You can not only see how bright a jewel is, you can see how the light sparkles on each facet if you shift your own position.

Inventions and variations

Holograms were being produced by the 1960s in the East and West, but developments in each area followed different paths.

In Britain, Dr. Dennis Gabor's intention was to improve the resolution of **electron** microscopes. He wrote on his efforts to tackle the problem in 1948, but since no stable source of coherent light was available, his work excited little interest as an imaging technique. T. A. Mainman at Hughes Aircraft in the United States was the first to demonstrate a ruby **laser** in 1960. After two other researchers, E. N. Leith and J. Upatnieks, used the laser to make 3-D images in the early 1960s, Gabor was awarded the Nobel for his research in 1971.

In 1958, Yuri Denisyuk had no idea what Gabor had done. He was fond of science fiction, and came across a reference in Efremov's story "Star Ships" to a mysterious plate, which could show a face in natural dimensions with animated eyes. The Russian researcher was inspired to try to make something just like that, which he referred to as a "wave photograph." Denisyuk's hologram could be seen under white light, because the plate doubled as a **color** filter.

Materials and techniques

There are many sorts of holograms, classified by their differences in material (amplitude, thick/thin, absorption), diffraction (phase), orientation of recording (rainbow, transmission and reflection, image **plane**, Fresnel, Fraunhofer), and optical systems (Fourier and lensless Fourier). The hologram is usually defined as a record of an interference pattern in a chemical medium, but the pattern does not have to be produced by a light source, nor must the hologram be stored on photographic film. Sonic, x ray, and microwaves are used as well, and computers can generate ones just by using mathematical formulas.

Researchers have been experimenting with aspects of the holographic process all along, and new tests are always being devised, in order to explore novel ways to improve the resolution and vibrancy of the images. The most common differences among these methods involve the mechanical setup of the exposure, the **chemistry** of the recording medium, and the means of displaying the final product. Full color holograms can be made by creating three masters in red, green and blue, after painting the object in grayscale tones, according to a separation technique already used in art **printing**. Different shades of gray are interpreted by a combination of the masters as different colors. Fiber optic delivery systems can insure proper illumination and eliminate aberrations which arise during long exposures. Multiplex or multiple-exposure holograms can be in planar or cylindrical form, showing a 360-degree view or even apparent movement.

Holograms versus photographs

Ordinary **photography** only accounts for the intensity of light. The only consideration is whether or not the light is too bright. You can usually see the grains in a photographic image, but the features in the fringe pattern of a hologram measure the same as each wavelength of light (1/2000 of a millimeter), recording amplitude in their depth of modulation and phase in their varying positions.

Older "3-D" imagery constructed from photographs is known as stereoscopy. This method reproduces a single viewpoint with the aid of two images. The two are superposed to recreate the **parallax** between your left eye's view and your right eye's view, but that is where your options stop. Holography allows for a full range of parallax effects: you can see around, over and even behind objects in a hologram.

Flashbulbs can be uncomfortable, but holograms use laser technology. Direct physical contact with a low-power laser cannot harm you unless you look directly into the beam, but remove all potentially reflective surfaces from the area, in order to prevent an accident.

Current usage and future prospects

The most common holograms are now an everyday occurrence. Embossed holograms are mass produced on mylar—foil and plastic—and can be viewed under the kind of diffused light which renders higher-quality holograms blurry. These can be seen on a variety of consumer goods, but they are also used on credit or identification cards as security measures. Holographic optical elements (HOEs) do not generate images themselves, but are employed to regulate the pattern of a scanning light

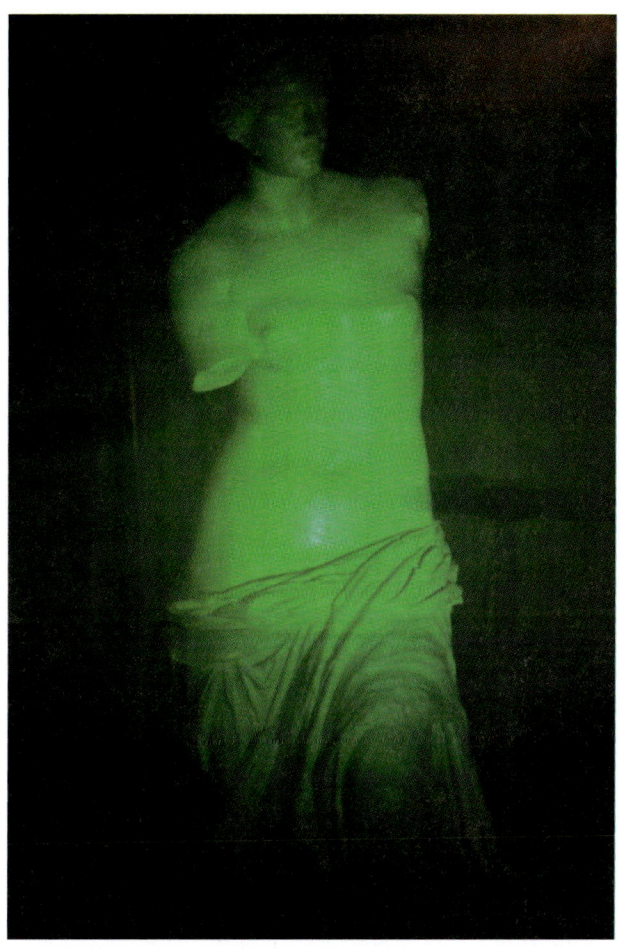

A hologram of the Venus de Milo. It was produced by an optical laboratory in Besancon, France. At 5 ft (1.5 m) tall it is one of the largest holograms in the world. *Photograph by Phillippe Plailly. National Audubon Society Collection/Photo Researchers, Inc. Reproduced by permission.*

beam. Supermarket checkout scanners are built out of a collection of HOEs mounted on a spinning disc, which can read a UPC code from any angle.

Holographic memory is an emerging technology, which aims to preserve data in a format superior to currently used magnetic ones. Binary computer code (patterns of ones and zeros) could be represented as light and dark spots. Part of a hologram can be defective or destroyed, while the remaining part will still retain all the data intact. Creative use of multiplexing can layer information, recorded from different positions.

Computer-aided design (CAD) imagery would be made more accessible to the average viewer if the full-scale plan appeared in apparent 3-D, instead of requiring that a series of linear plots be deciphered visually, which is the current practice. Holograms can be used as visualization aids and screening devices in aviation and automotives as well, since they can be viewed from a particular angle, but not others.

X rays can show detail where an electron **microscope** would only show dark undifferentiated circles, and would render less damage to a living thing or **tissue** than electronic bombardment. Subatomic or light-in-flight experiments could be recorded in fully-dimensional imagery, in real time.

Jennifer Kramer

Homeostasis

Homeostasis (a Greek term meaning same state), is the maintenance of constant conditions in the internal environment of the body despite large swings in the external environment. Functions such as **blood pressure**, body **temperature**, **respiration rate**, and blood glucose levels are maintained within a range of normal values around a set point despite constantly changing external conditions. For instance, when the external temperature drops, the body's homeostatic mechanisms make adjustments that result in the generation of body **heat**, thereby maintaining the internal temperature at constant levels.

Negative feedback

The body's homeostatically cultivated systems are maintained by negative feedback mechanisms, sometimes called negative feedback loops. In negative feedback, any change or deviation from the normal range of function is opposed, or resisted. The change or deviation in the controlled value initiates responses that bring the function of the **organ** or structure back to within the normal range.

Negative feedback loops have been compared to a thermostatically controlled temperature in a house, where the internal temperature is monitored by a temperature-sensitive gauge in the **thermostat**. If it is cold outside, eventually the internal temperature of the house drops, as cold air seeps in through the walls. When the temperature drops below the point at which the thermostat is set, the thermostat turns on the furnace. As the temperature within the house rises, the thermostat again senses this change and turns off the furnace when the internal temperature reaches the pre-set point.

Negative feedback loops require a receptor, a control center, and an effector. A receptor is the structure that monitors internal conditions. For instance, the human body has receptors in the blood vessels that monitor the **pH** of the blood. The blood vessels contain receptors that measure the resistance of blood flow against the vessel walls, thus monitoring blood pressure. Receptors sense changes in function and initiate the body's homeostatic response.

These receptors are connected to a control center that integrates the information fed to it by the receptors. In most homeostatic mechanisms, the control center is the **brain**. When the brain receives information about a change or deviation in the body's internal conditions, it sends out signals along nerves. These signals prompt the changes in function that correct the deviation and bring the internal conditions back to the normal range.

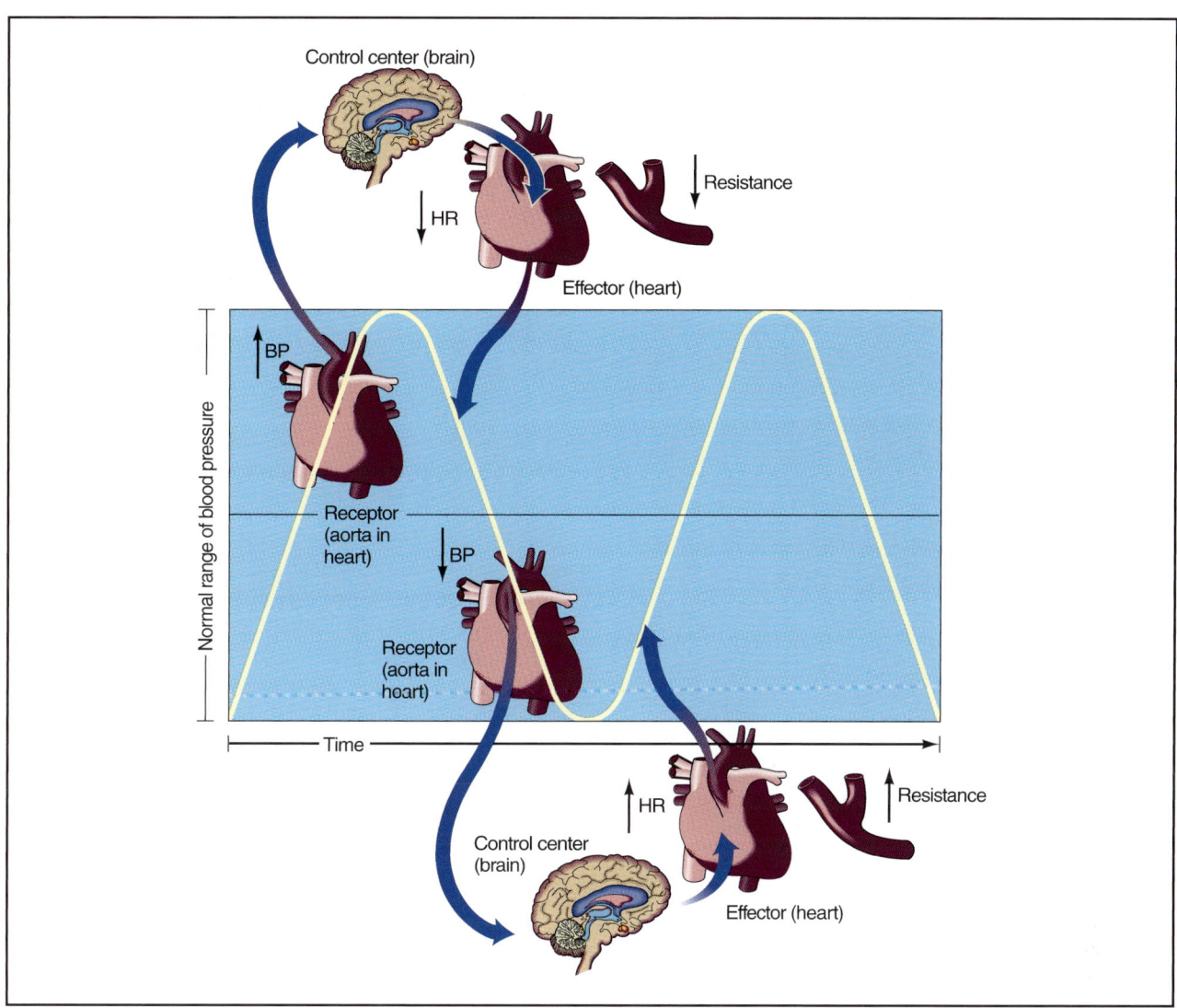

Figure 1. A negative feedback loop helps regulate blood pressure. *Illustration by Hans & Cassidy. Courtesy of Gale Group.*

Effectors are muscles, organs, or other structures that receive signals from the brain or control center. When an effector receives a signal from the brain, it changes its function in order to correct the deviation.

An example of a negative feedback loop is the regulation of blood pressure (Figure 1). An increase in blood pressure is detected by receptors in the blood vessels that sense the resistance of blood flow against the vessel walls. The receptors relay a message to the brain, which in turn sends a message to the effectors, the **heart** and blood vessels. The heart rate decreases and blood vessels increase in diameter, which cause the blood pressure to fall back within the normal range or set point. Conversely, if blood pressure decreases, the receptors relay a message to the brain, which in turn causes the heart rate to increase, and the blood vessels to decrease in diameter.

Some set points become "reset" under certain conditions. For instance, during **exercise**, the blood pressure normally increases. This increase is not abnormal; it is the body's response to the increased demand of **oxygen** by muscle tissues. When the muscles require more oxygen, the body responds by increasing the blood flow to muscle tissues, thereby increasing blood pressure. This resetting of the normal homeostatic set point is required to meet the increased demand of oxygen by muscles.

Similarly, when the body is deprived of food, the set point of the metabolic rate can become reset to a lower-than-normal value. This lowering of the metabolic rate is the body's attempt to stave off starvation and keep the body functioning at a slower rate. Many people who periodically deprive themselves of food in attempts to lose weight find that after the initial weight loss it becomes increasingly diffi-

KEY TERMS

Control center—The center that receives messages from receptors about a change in the body's internal conditions and relays messages to effectors to change their function to correct the deviation; in most homeostatic mechanisms, the control center is the brain.

Effector—A muscle or organ that receives messages from the control center to change its function in order to correct a deviation in the body's internal conditions.

Hormone—Chemical regulator of physiology, growth, or development which is typically synthesized in one region of the body and active in another and is typically active in low concentrations.

Negative feedback loop—A homeostatic mechanism that opposes or resists a change in the body's internal conditions.

Positive feedback loop—A mechanism that increases or enlarges a change in the body's internal conditions.

Receptor—A structure that monitors the body's internal functions and conditions; detects changes in the body's internal environment.

Set point—The range of normal functional values of an organ or structure.

cult to lose more pounds. This difficulty stems from the lowering of the metabolic set point. Exercise may counteract some of these effects by the increasing metabolic demands.

See also Physiology.

Resources

Books

Marieb, Elaine Nicpon. *Human Anatomy & Physiology*. 5th ed. San Francisco: Benjamin/Cummings, 2000.

Reinhardt, H. Wolfgang, Paul P. Leyssac, and Peter Bie, eds. *Mechanisms of Sodium Homeostasis: Sodium and Water Excretion in Mammals; Haemodynamic, Endocrine, and Neural Mechanisms*. Boston: Blackwell Scientific Publishers, 1990.

Periodicals

Kozak, Wieslaw. "Fever: A Possible Strategy for Membrane Homeostasis During Infection." *Perspectives in Biology and Medicine* 37 (Autumn 1993): 1.

Skorupski, Peter, et al. "Integration of Positive and Negative Feedback Loops in a Crayfish Muscle." *Journal of Experimental Biology* 187 (February 1994): 305.

Kathleen Scogna

Honeycreepers

Honeycreepers are 14 living **species** of **birds** in the family Drepanididae, which occur only on the Hawaiian and Laysan Islands and nearby islands in the equatorial Pacific Ocean. Unfortunately, a further eight species of honeycreepers have recently become extinct as a result of ecological changes that humans have caused to the habitats of these birds. In addition, at least half of the surviving species of honeycreepers are perilously endangered, as are some of the distinctive subspecies that occur on various islands.

Most species of honeycreepers breed in native forest and shrubby habitats in the Hawaiian Islands. They are resident in those habitats and do not migrate elsewhere during their non-breeding season.

The honeycreepers are small birds, ranging in body length from 4-8 in (11-20 cm). Their bills are extremely varied, depending on the diet of the species. Some honeycreepers have small, thin beaks, ideal for gleaning **arthropods** from **tree** foliage. Other species have longer, curved beaks, adaptive to feeding on **nectar** or on **insects** deep in **bark** crevices. The beaks of yet other species are heavier and more conical and are used to feed on **plant seeds**.

This extreme diversification of species with various bill shapes within such a closely related group of birds is a famous example of speciation. In the case of the honeycreepers, the speciation was driven by natural **selection** in favor of birds having adaptations favorable to taking advantage of specific ecological opportunities, which occurred in a wide variety on the Hawaiian Islands. Evolutionary biologists consider the adaptive **radiation** of the Hawaiian honeycreepers to be one of the clearest illustrations of the phenomenon of **evolution**.

Undoubtedly, all of the many species of honeycreepers evolved from a single, probably quite small founder group that somehow arrived on the Hawaiian Islands by accident in the distant past. Because few other types of birds were present, a variety of ecological niches were unfilled or were utilized by generalist organisms. Under the pervasive influence of natural selection, the original honeycreepers slowly evolved a repertoire of differing bill shapes and other useful adaptations. Eventually, the specialized populations of birds became reproductively isolated. Ultimately, they diversified into different species that were better adapted to feeding and living in specialized ways.

The honeycreepers are also highly variable in **color**, which ranges from a relatively drab gray to brown, olive, yellow, red, and black. Some species are dimorphic, with the male being larger than the female.

Honeycreepers build their cup-shaped nests in trees. They typically lay two to four eggs, which are incubated by the female. Both of the parents share the duties of raising their babies.

Species of honeycreepers

The smallest of the living honeycreepers is the anianiau (*Loxops parva*), only 4 in (11 cm) long. The largest species is the 8 in (20 cm) long Kauai akialoa (*Hemignathus procerus*). This species, and the closely related akailoa (*H. obscurus*), have long, downward-curving bills that are about one third of the total body length. The akiapolaau (*H. wilsoni*) has an especially strange bill, with the upper mandible being strongly down-curved, but the lower being straight, and only half the length of the upper mandible. This species uses the lower mandible to pry loose bark off trees, and the upper to probe and impale their food of insects.

The mamo (*Drepanis pacifica*) and the crested honeycreeper (*Palmeria dolei*) have relatively shorter, downward-curving beaks, useful in sipping nectar from flowers.

The liwi (*Vestiaria coccinia*) is a beautiful, crimson-colored bird with black wings. This species is particularly prized by aboriginal Hawaiians, who use the red feathers in the preparation of traditional garments.

The grosbeak finch (*Psittirostra kona*) has a massive bill, useful in cracking hard seeds to extract the edible **matter** inside.

Humans and honeycreepers

The Hawaiian honeycreepers have become endangered through a variety of interacting ecological stressors. **Habitat** losses have been important, especially those associated with the conversion of their limited areas of natural-forest habitats to agricultural and urban land-uses, which do not support these native birds. Introduced herbivores, such as **goats** and **pigs**, have caused serious damage to honeycreeper habitat, greatly changing the nature of the vegetation, even in remote places. Introduced diseases, such as avian **malaria**, and introduced predators such as **rats**, **mongooses**, and pigs have also caused significant damage to honeycreepers.

Today, the Hawaiian and U.S. governments have designated many of the most important remaining refuges of natural habitat as parks and ecological reserves. Some of these refuges are being managed to maintain their **ecological integrity** as much as possible. For example, some large areas have been fenced, and the populations of feral goats and pigs have been eliminated or reduced. Unfortunately, these sorts of ecological interventions are required today, and will also be needed in

the future if the extraordinary Hawaiian honeycreepers are to survive in their changed and changing world.

KEY TERMS

Adaptive radiation—An evolutionary phenomenon in which a single, relatively uniform population gives rise to numerous, reproductively isolated species. Adaptive radiation occurs in response to natural selection, in environments in which there are diverse ecological opportunities, and little competition to filling them.

Endemic—This refers to a species (or genus, family, etc.) with a restricted geographic range. For example, the honeycreepers only occur on the Hawaiian Islands and nearby islands, and are therefore endemic to that relatively small region. Some are endemic to single islands.

Founder group—A small population of original immigrants to a habitat previously not known to the species. Following the successful colonization, the population may increase, and under the influence of natural selection may diversify into various species.

Speciation—The divergence of evolutionary lineages, and creation of new species.

Resources

Books

Ehrlich, P.R., D.S. Dobkin, and D. Wheye. *Birds in Jeopardy*. Stanford: Stanford University Press, 1992.

Forshaw, Joseph. *Encyclopedia of Birds*. New York: Academic Press, 1998.

Pratt, H.D., P.L. Bruner, and D.G. Berrett. *The Birds of Hawaii and the Tropical Pacific*. Princeton, NJ: Princeton University Press, 1987.

Periodicals

Freed, L.A., S. Conant, and R.C. Fleischer. "Evolutionary Ecology and Radiation of Hawaiian Passerine Birds." *Trends in Ecology and Evolution* 2 (1987): 196-203.

Bill Freedman

Honeyeaters

As their name suggests, honeyeaters are often found near flowering plants feeding on **nectar**. All of the Honeyeaters have slender, pointed bills with a long, brushlike tongue that is used to sip the nectar. However, there are many variations of the bill shape, depending on the specific diet of each **species**. Species with longer bills are usually feeding from tubular flowers, while those with shorter bills often feed on the flowers that are more accessible. All honeyeaters supplement their diets to varying degrees with **insects** and fruit.

A symbiotic (mutually beneficial) relationship has developed between the honeyeaters and the flowers they use for food. As the honeyeater is feeding on nectar, pollen is placed on the bird's forehead by the stamen. The pollen is then deposited on the stigma of the next **flower** while feeding. Thus, the bird obtains food while the flower is pollinated.

In addition to the bill, there can be a great variety in the appearance of a honeyeater. Anatomical differences include a wattle, ear-tufts, fleshy helmets, and the length of the tail. Both the male and female are usually drab brown, gray, or green in **color**. However, there are species in which the male is brightly colored compared to the drab female. Most range in size from 4-18 in (10-46 cm) long, with strong, short legs, an **adaptation** to the tree-climbing lifestyle.

Placed in the family Meliphagidae, the 169 species of honeyeater are found in **Australia**, New Guinea, the Celebes, the Moluccas, and other smaller islands of the Western Pacific. Some species were originally found as far east as the Hawaiian Islands, but they have since been driven to **extinction** by loss of habitat and hunting; the feathers were once sought by the native Hawaiians.

Hoopoe

The hoopoe (*Upupa epops*) is the only **species** in its family, the Upupidae. This species breeds in northwestern **Africa**, on Madagascar, throughout the Middle East, and in southern **Europe** and southern **Asia**. Its usual habitats are open **forests**, savannas, **grasslands**, and some types of cultivated lands and parks. Some populations of hoopoes are sedentary, while others are migratory.

Hoopoes have a body length of 12 in (30 cm), with broad, rounded wings, a long tail, and a long, thin, slightly downward-curving beak. The head is strongly and distinctively crested. The upper body has a light brown, pinkish coloration, and the rest of the body and wings are strongly barred with black and white. The female hoopoe is slightly smaller and duller in coloration than the male.

Hoopoes feed mostly on the ground on **invertebrates**, although they sometimes also catch **insects** in the air. They commonly **perch** and roost in trees. The flight of hoopoes is erratic, and strongly undulating.

Their call is a loud "*hoop-hoop-hoop*," and is the origin of the species' name.

Hoopoes nest in a cavity in a **tree** or sometimes in a hole in a wall or building. The female incubates the five to eight eggs, and is fed by the male during her confinement. Both sexes care for the young. Unlike most **birds**, hoopoes are not fastidious and do leave their nest to defecate, nor do they remove the fecal packets of the young. Consequently, their nest becomes quite fouled with excrement and disgustingly smelly.

Hoopoes are an unusual and distinctive species and have been important in some cultures. Hoopoes are depicted in Egyptian hieroglyphics, and they are mentioned in some classical literature from Greece. They are still appealing today and are a choice sighting for bird-watchers and other naturalists.

Horizon

A **soil** horizon is a horizontal layer of soil with physical or chemical characteristics that separate it from layers above and below. More simply, each horizon contains chemicals, such as rust-like **iron** oxides, or soil particles that differ from adjacent layers. Soil scientists generally name these horizons (from top to bottom) "O," "A," "B," "C," and" R," and often subdivide them to reflect more specific characteristics within each layer. Considered together, these horizons constitute a soil profile.

Horizons usually form in residual soils: soils not transported to their present location by **water**, **wind**, or **glaciers** but formed "in place" by the **weathering** of the **bedrock** beneath them. It takes many thousand to a million years to achieve a mature soil with fully developed horizons.

The O horizon (sometimes known as the A_0) consists of freshly dead and decaying organic matter—mostly plants but also small (especially microscopic) animals or the occasional rigid cow. A gardener would call this organic **matter** (minus the cow) "compost" or "humus." Below the O lies the A horizon, or topsoil, composed of organic material mixed with soil particles of **sand**, silt, and clay. Frolicking earthworms, other small animals, and water mix the soil in the A horizon. Water forced down through the A by gravity carries clay particles and dissolved **minerals** (such as iron oxides) into the B horizon in a process called "leaching;" therefore, the A is known as the Zone of leaching. These tiny clay particles zigzag downward through the spaces (pores) between larger particles like balls in a Japanese pachinko game.

Sometimes the lower half of the A horizon is called the E ("Eluvial") horizon meaning it is depleted of clay and dissolved minerals, leaving coarser grains.

The leached material ends up in the B horizon, the Zone of Accumulation. The B horizon, stained red by iron oxides, tends to be quite clayey. If the upper horizons erode, **plant** roots have a tough time penetrating this clay; and rain which falls on the exposed clay can pool on the surface and possibly drown plants or flood basements.

Sometimes the top of the B horizon develops a dense layer called a fragipan—a claypan (compacted by vehicles) or a hardpan (cemented by minerals). In arid climates, intense **evaporation** sucks water and its dissolved minerals upward. This accumulation creates a hardpan impenetrable to any rain percolating (sinking) downward, resulting in easily evaporated pools or rapid runoff. If the hardpan is composed of the calcium-rich mineral calcite, it is called "caliche." If composed of iron oxides, it is called an "ironpan." Fragipans are extremely difficult for crop roots and water to penetrate.

Partially weathered bedrock composes the C horizon. Variously sized chunks of the rock below are surrounded by smaller bits of rock and clay weathered from those chunks. Some of the original rock is intact, but other parts have been chemically changed into new minerals.

The R layer (D horizon) is the bedrock or, sometimes, the sediment from which the other horizons develop. Originally, this rock lay exposed at the surface where it weathered rapidly into soil. The depth from the surface to the R layer depends on the interrelationships between the climate, the age of the soil, the slope, and the number of organisms. Most people do not consider the R layer as soil, but include it in the profile anyway, since the weathering of this bedrock usually produces the soil above it.

In a perfect world, all soils demonstrate these horizons, making the lives of soil scientists and soil students blissful. In reality, however, some soils, like transported soils (moved to their present locations by water, wind, or glaciers), lack horizons because of mixing while moving or because of youth. In other soils, the A and B rest on bedrock, or **erosion** strips an A, or other complicated variations. Around the world, scientists classify soils by these horizonal variations.

Hormones

Hormones are biochemical messengers that regulate physiological events in living organisms. More than 100 hormones have been identified in humans. Hormones are secreted by endocrine (ductless) **glands** such as the hypothalamus, the pituitary gland, the pineal gland, the thyroid, the parathyroid, the thymus, the **adrenals**, the pancreas, the ovaries, and the testes. Hormones are secreted directly into the **blood** stream from where they travel to target tissues and modulate digestion, growth, maturation, reproduction, and **homeostasis**. The word hormone comes from the Greek word, *hormon*, to stir up, and indeed excitation is characteristic of the adrenaline and the sex hormones. Most hormones produce an effect on specific target tissues that are sited at some distance from the gland secreting the hormone. Although small **plasma** concentrations of most hormones are always present, surges in secretion trigger specific responses at one or more targets. Hormones do not fall into any one chemical category, but most are either protein molecules or steroid molecules. These biological managers keep the body systems functioning over the long term and help maintain health. The study of hormones is called endocrinology.

Mechanisms of action

Hormones elicit a response at their target **tissue**, target **organ**, or target **cell** type through receptors. Receptors are molecular complexes which specifically recognize another molecule-in this case, a particular hormone. When the hormone is bound by its receptor, the receptor is usually altered in some way that it sends a secondary message through the cell to do something in response. Hormones that are **proteins**, or peptides (smaller strings of amino acids), usually bind to a receptor in the cell's outer surface and use a second messenger to relay their action. Steroid hormones such as cortisol, testosterone, and estrogen bind to receptors inside cells. Steroids are small enough to and chemically capable of passing through the cell's outer **membrane**. Inside the cell, these hormones bind their receptors and often enter the nucleus to elicit a response. These receptors bind DNA to regulate cellular events by controlling **gene** activity.

Most hormones are released into the bloodstream by a single gland. Testosterone is an exception, because it is secreted by both the adrenal glands and by the testes. Plasma concentrations of all hormones are assessed at some site which has receptors binding that hormone. The site keeps track of when the hormone level is low or high. The major area which records this information is the hypothalamus. A number of hormones are secreted by the hypothalamus which stimulate or inhibit additional secretion of other hormones at other sites. The hormones are part of a negative or positive feedback loop.

Most hormones work through a negative feedback loop. As an example, when the hypothalamus detects high levels of a hormone, it reacts to inhibit further production. And when low levels of a hormone are detected, the hypothalamus reacts to stimulate hormone production or secretion. Estrogen, however, is part of a positive feedback loop. Each month, the Graafian follicle in the ovary releases estrogen into the bloodstream as the egg develops in ever increasing amounts. When estrogen levels rise to a certain point, the pituitary secretes luteinizing hormone (LH) which triggers the egg's release of the egg into the oviduct.

Not all hormones are readily soluble in blood (their main transport medium) and require a transport **molecule** that will increase their **solubility** and shuttle them around until they get to their destination. Steroid hormones, in particular, tend to be less soluble. In addition, some very small peptides require a carrier protein to deliver them safely to their destination, because these small peptides could be swept into the wrong location where they would not elicit the desired response. Carrier proteins in the blood include albumin and prealbumin. There are also specific carrier proteins for cortisol, thyroxin, and the steroid sex hormones.

Major hormones

The concentrations of several important biological building blocks such as amino acids are regulated by more than one hormones. For example, both calcitonin and parathyroid hormone (PTH) influence blood **calcium** levels directly, and other hormones affect calcium levels indirectly via other pathways.

The hypothalamus

Hormones secreted by the hypothalamus modulate other hormones. The major hormones secreted by the hypothalamus are corticotrophin releasing hormone (CRH), thyroid stimulating hormone releasing hormone (TRH), follicle stimulating hormone releasing hormone (FSHRH), luteinizing hormone releasing hormone (LRH), and growth hormone releasing hormone (GHRH). CRH targets the adrenal glands. It triggers the adrenals to release adrenocorticotropic hormone (ACTH). ACTH functions to synthesize and release corticosteroids. TRH targets the thyroid where it functions to synthesize and release the thyroid hormones T3 and T4. FSH targets the ovaries and the testes where it enables the maturation of the ovum and of spermatozoa. LRH also targets the ovaries and the testes, and its receptors are in cells which promote ovulation and increase progesterone synthesis and release. GHRH targets the anterior pituitary to release **growth hormones** to most body tissues, increase protein synthesis, and increase blood glucose. Hence, the hypothalamus plays a first domino role in these cascades of events.

The hypothalamus also secretes some other important hormones such as prolactin inhibiting hormone (PIH), prolactin releasing hormone (PRH), and melanocyte inhibiting hormone (MIH). PIH targets the anterior pituitary to inhibit milk production at the mammary gland, and PRH has the opposite effect. MIH targets skin pigment cells (melanocytes) to regulate pigmentation.

The pituitary gland

The pituitary has long been called the master gland because of the vast extent of its activity. It lies deep in the **brain** just behind the nose. The pituitary is divided into anterior and posterior regions with the anterior portion comprising about 75% of the total gland. The posterior region secretes the peptide hormones vasopressin, also called anti-diuretic hormone (ADH), and oxytocin. Both are synthesized in the hypothalamus and moved to the posterior pituitary prior to secretion. ADH targets the collecting tubules of the kidneys, increasing their permeability to **water**. ADH causes the kidneys to retain water. Lack of ADH leads to a condition called diabetes insipidus characterized by excessive urination. Oxytocin targets the uterus and the mammary glands in the breasts. Oxytocin begins labor prior to **birth** and also functions in the ejection of milk. The drug, pitocin, is a synthetic form of oxytocin and is used medically to induce labor.

The anterior pituitary (AP) secretes a number of hormones. The cells of the AP are classified into five types based on what they secrete. These cells are somatotrophs, corticotrophins, thyrotrophs, lactotrophs, and gonadotrophs. Respectively, they secrete growth hormone (GH), ACTH, TSH, prolactin, and LH and FSH. Each of these hormones is either a polypeptide or a glycoprotein. GH controls cellular growth, protein synthesis, and elevation of blood glucose **concentration**. ACTH controls secretion of some hormones by the adrenal cortex (mainly cortisol). TSH controls thyroid hormone secretion in the thyroid. In males, prolactin enhances testosterone production; in females, it initiates and maintains LH to promote milk secretion from the mammary glands. In females, FSH initiates ova development and induces ovarian estrogen secretion. In males, FSH stimulates sperm production in the testes. LH stimulates ovulation and formation of the corpus luteum which produces progesterone. In males, LH stimulates interstitial cells to produce testosterone. Each AP hormone is secreted in response to a hypothalamic releasing hormone.

The thyroid gland

The thyroid lies under the larynx and synthesizes two hormones, thyroxine and tri-iodothyronine. This gland takes up iodine from the blood and has the highest iodine level in the body. The iodine is incorporated into the thyroid hormones. Thyroxine has four iodine **atoms** and is called T4. Tri-iodothyronine has three iodine atoms and is called T3. Both T3 and T4 function to increase the metabolic **rate** of several cells and tissues. The brain, testes, lungs, and spleen are not affected by thyroid hormones, however. T3 and T4 indirectly increase blood glucose levels as well as the insulin-promoted uptake of glucose by **fat** cells. Their release is modulated by TSH-RH from the hypothalamus. TSH secretion increases in cold infants. When **temperature** drops, a metabolic increase is triggered by TSH. Chronic **stress** seems to reduce TSH secretion which, in turn, decreases T3 and T4 output.

Depressed T3 and T4 production is the trademark of hypothyroidism. If it occurs in young children, then this decreased activity can cause physical and mental retardation. In adults, it creates sluggishness—mentally and physically—and is characterized further by weight gain, poor hair growth, and a swollen neck. Excessive T3 and T4 cause sweating, nervousness, weight loss, and fatigue. The thyroid also secretes calcitonin which serves to reduce blood calcium levels. Calcitonin's role is particularly significant in children whose bones are still forming.

The parathyroid glands

The parathyroid glands are attached to the bottom of the thyroid gland. They secrete the polypeptide parathyroid hormone (PTH) which plays a crucial role in monitoring blood calcium and phosphate levels. About 99% of the body's calcium is in the bones, and 85% of the **magnesium** is also found in bone. Low blood levels of calcium stimulate PTH release into the bloodstream in two steps. Initially, calcium is released from the fluid around bone cells. And later, calcium can be drawn from bone itself. Although, only about 1% of bone calcium is readily exchangeable. PTH can also increase the absorption of calcium in the intestines by stimulating the kidneys to produce a **vitamin** D-like substance which facilitates this action. High blood calcium levels will inhibit PTH action, and magnesium (which is chemically similar to calcium) shows a similar effect.

Calcium is a critical element for the human body. Even though the majority of calcium is in bone, it is also used by muscles, including cardiac muscle for contractions, and by nerves in the release of neurotransmitters. Calcium is a powerful messenger in the immune response of **inflammation** and blood clotting. Both PTH and calcitonin regulate calcium levels in the kidneys, the gut, bone, and blood. Whereas calcitonin is released in conditions of high blood calcium levels, PTH is released when calcium levels fall in the blood. Comparing the two, PTH causes an increase in calcium absorption in the kidneys, absorption in the intestine, release from bone, and levels in the blood. In addition, PTH decreases kidney phosphate absorption. Calcitonin has the opposite effect on each of these variables. PTH is thought to be the major calcium modulator in adults.

PTH deficiency can be due to autoimmune diseases or to inherited parathyroid gland problems. Low PTH capabilities cause depressed blood calcium levels and neuromuscular problems. Very low PTH can lead to tetany or muscle spasms. Excess PTH can lead to weakened bones because it causes too much calcium to be drawn from the bones and to be excreted in the urine. Abnormalities of bone mineral deposits can lead to a number of conditions including **osteoporosis** and rickets. Osteoporosis can be due to dietary insufficiencies of calcium, phosphate, or vitamin C (which has an important role in formation of the bone matrix). The end result is a loss of bone mass. Rickets is usually caused by a vitamin D deficiency and results in lower rates of bone matrix formation in children. These examples show how important a balanced nutritious diet is for healthy development.

The adrenal glands

The two adrenal glands, one on top of each kidney, each have two distinct regions. The outer region (the medulla) produces adrenaline and noradrenaline and is under the control of the sympathetic **nervous system**. The inner region (the cortex) produces a number of steroid hormones. The cortical steroid hormones include mineralocorticoids (mainly aldosterone), glucocorticoids (mainly cortisol), and gonadocorticoids. These steroids are derived from **cholesterol**. Although cholesterol receives a lot of bad press, some of it is necessary. Steroid hormones act by regulating gene expression, hence, their presence controls the production of numerous factors with multiple roles. Aldosterone and cortisol are the major human steroids in the cortex. However, testosterone and estrogen are secreted by adults (both male and female) at very low levels.

Aldosterone plays an important role in regulating body fluids. It increases blood levels of **sodium** and water and lowers blood potassium levels. Low blood sodium levels trigger aldosterone secretion via the renin-angiotensin pathway. Renin is produced by the kidney, and angiotensin originates in the liver. High blood potassium levels also trigger aldosterone release. ACTH has a minor promoting effect on aldosterone. Aldosterone targets the kidney where it promotes sodium uptake and potassium excretion. Since sodium ions influence water retention, the result is a net increase in body fluid **volume**.

Blood cortisol levels fluctuate dramatically throughout the day and peak around 8 A.M. Presumably, this early peak enables humans to face the varied daily stressors they encounter. Cortisol secretion is stimulated by physical trauma, cold, burns, heavy **exercise**, and **anxiety**. Cortisol targets the liver, skeletal muscle, and adipose tissue. Its overall effect is to provide amino acids and glucose to meet synthesis and **energy** requirements for normal **metabolism** and during periods of stress. Because of its anti-inflammatory action, it is used clinically to reduce swelling. Excessive cortisol secretion leads to **Cushing syndrome** which is characterized by weak bones, **obesity**, and a tendency to bruise. Cortisol deficiency can lead to **Addison disease** which has the symptoms of fatigue, low blood sodium levels, low blood **pressure**, and excess skin pigmentation.

The adrenal medullary hormones are epinephrine (adrenaline) and nor-epinephrine (nor-adrenaline). Both of these hormones serve to supplement and prolong the fight or flight response initiated in the nervous system. This response includes the neural effects of increased **heart** rate, peripheral blood vessel constriction, sweating, spleen contraction, glycogen conversion to glucose, dilation of bronchial tubes, decreased digestive activity, and lowered urine output.

The condition of stress presents a model for reviewing one way that multiple systems and hormones interact. During stress, the nervous, endocrine, digestive, urinary, respiratory, circulatory, and immune response are all tied together. For example, the hypothalamus sends nervous impulses to the spinal cord to stimulate the fight or flight response and releases CRH which promotes ACTH secretion by the pituitary. ACTH, in turn, triggers interleukins to respond which promote immune cell functions. ACTH also stimulates cortisol release at the adrenal cortex which helps **buffer** the person against stress. As part of a negative feedback loop, ACTH and cortisol receptors on the hypothalamus assess when sufficient levels of these hormones are present and then inhibit their further release. De-stressing occurs over a period of **time** after the stressor is gone. The systems eventually return to normal.

The pancreas

The pancreas folds under the stomach, secretes the hormones **insulin**, glucagon, and somatostatin. About 70% of the pancreatic hormone-secreting cells are called beta cells and secrete insulin; another 22%, or so, are called alpha cells and secrete glucagon. The remaining gamma cells secrete somatostatin, also known as growth hormone inhibiting hormone (GHIH). The alpha, beta, and gamma cells comprise the islets of Langerhans which are scattered throughout the pancreas.

Insulin and glucagon have **reciprocal** roles. Insulin promotes the storage of glucose, **fatty acids**, and amino acids, whereas, glucagon stimulates mobilization of these constituents from storage into the blood. Both are relatively short polypeptides. Insulin release is triggered by high blood glucose levels. It lowers blood sugar levels by binding a cell surface receptor and accelerating glucose transport into the cell where glucose is converted into glycogen. Insulin also inhibits the release of glucose by the liver in order to keep blood levels down. Increased blood levels of GH and ACTH also stimulate insulin secretion. Not all cells require insulin to store glucose, however. Brain, liver, kidney, intestinal, epithelium, and the pancreatic islets can take up glucose independently of insulin. Insulin excess can cause hypoglycemia leading to convulsions or **coma**, and insufficient levels of insulin can cause **diabetes mellitus** which can be fatal if left untreated. Diabetes mellitus is the most common endocrine disorder.

Glucagon secretion is stimulated by decreased blood glucose levels, **infection**, cortisol, exercise, and large protein meals. GHIH, glucose, and insulin inhibit its secretion. Protein taken in through the digestive tract has more of a stimulatory effect on glucagon than does injected protein. Glucagon stimulates glycogen breakdown in the liver, inhibits glycogen synthesis, and facilitates glucose release into the blood. Excess glucagon can result from tumors of the pancreatic alpha cells; and a mild diabetes seems to result. Some cases of uncontrolled diabetes are also characterized by high glucagon levels suggesting that low blood insulin levels are not always the only cause in some diabetes cases.

It was the study of glucagon and its action by Sutherland in 1961 that led to the concept of the second messenger system. Glucagon activates the intracellular molecule cyclic AMP, cAMP. Since this discovery, a number of other molecules have been found which modulate cellular activity via this second messenger.

The female reproductive organs

The female reproductive hormones arise from the hypothalamus, the anterior pituitary, and the ovaries. Although detectable amounts of the steroid hormone estrogen are present during fetal development, at **puberty** estrogen levels rise to initiate secondary sexual characteristics. Gonadotropin releasing hormone (GRH) is released by the hypothalamus to stimulate pituitary release of LH and FSH. LH and FSH propagate egg development in the ovaries. Eggs (ova) exist at various stages of development, and the maturation of one ovum takes about 28 days and is called the ovarian or **menstrual cycle**. The ova are contained within follicles which are support organs for ova maturation. About 450 of a female's 150,000 germ cells mature to leave the ovary. The hormones secreted by the ovary include estrogen, progesterone, and small amounts of testosterone.

As an ovum matures, rising estrogen levels stimulate additional LH and FSH release from the pituitary. Prior to ovulation, estrogen levels drop, and LH and FSH surge to cause the ovum to be released into the fallopian tube. The cells of the burst follicle begin to secrete progesterone and some estrogen. These hormones trigger thickening of the uterine lining, the endometrium, to prepare it for implantation should **fertilization** occur. The high progesterone and estrogen levels prevent LH and FSH from further secretion-thus hindering another ovum from developing. If fertilization does not occur, eight days after ovulation the endometrium deteriorates resulting in menstruation. The falling estrogen and progesterone levels which follow trigger LH and FSH, starting the cycle all over again.

Although estrogen and progesterone have major roles in the menstrual cycle, these hormones have receptors on a number of other body tissues. Estrogen has a protective effect on bone loss which can lead to osteoporosis. And progesterone, which is a competitor for androgen sites, blocks actions that would result from testosterone activation. Estrogen receptors have even been found in the forebrain indicating a role in female neuronal function or development.

Hormones related to pregnancy include human chorionic gonadotrophin (HCG), estrogen, human chorionic somatomammotrophin (HCS), and relaxin. HCG is released by the early embryo to signal implantation. Estrogen and HCS are secreted by the placenta. And relaxin is secreted by the ovaries as birth nears to relax the pelvic area in preparation for labor.

The male reproductive organs

Male reproductive hormones come from the hypothalamus, the anterior pituitary, and the testes. As in females, GRH is released from the hypothalamus which stimulates LH and FSH release from the pituitary. In males, LH and FSH facilitate spermatogenesis. The steroid hormone testosterone is secreted from the testes and can be detected in early embryonic development up until shortly after birth. Testosterone levels are quite low until puberty. At puberty, rising levels of testosterone stimulate male reproductive development including secondary characteristics.

LH stimulates testosterone release from the testes. FSH promotes early spermatogenesis, whereas testosterone is required to complete spermatogenic maturation

---

KEY TERMS

Amino acid—An organic compound whose molecules contain both an amino group (-NH$_2$) and a carboxyl group (-COOH). One of the building blocks of a protein.

Androgen—Any hormone with testosterone-like activity (i.e. it increases male characteristics).

Homeostasis—A condition of chemical and physical equilibrium in the human body.

Plasma—The non-cellular, fluid portion of blood in which the concentration of most molecules is measured.

---

to facilitate fertilization. In addition to testosterone, LH, and FSH, the male also secretes prostaglandins. These substances promote uterine contractions which help propel sperm towards an egg in the fallopian tubes during sexual intercourse. Prostaglandins are produced in the seminal vesicles, and are not classified as hormones by all authorities.

See also Biological rhythms; Cell; Endocrine system; Exocrine glands; Glands; Growth hormones; Reproductive system.

Resources

Books

Burnstein, K. L. *Steroid Hormones and Cell Cycle Regulation.* Boston: Kluwer Academic Publishers, 2002.

Engelking, L. R. *Metabolic and Endocrine Physiology.* Jackson, WY:, Teton NewMedia, 2000.

Goffin, V., P. A. Kelly. *Hormone Signaling.* Boston: Kluwer Academic Publishers, 2002.

Griffin, J. E., and S. R. Ojeda *Textbook of Endocrine Physiology.* New York: Oxford University Press, 2000.

Kacsoh, B. *Endocrine Physiology.* New York: McGraw-Hill Health Professions Division, 2000.

Louise Dickerson

Hornbills

Hornbills are medium- to large-sized, large-billed, long-tailed **birds** of tropical **forests**, savannas, and **grasslands**, comprising the family Bucerotidae. The 45 **species** of hornbills are distributed widely through the tropical regions of **Africa** and **Asia**. Most hornbills live in forests, and nest in holes in trees, while the species of open habitats nest in cavities in hollow trees or in holes in cliffs.

Ground hornbills in Kenya. *JLM Visuals. Reproduced with permission.*

Most hornbills are brightly colored, especially around the head and bill. The smallest hornbill is the 15 in (38 cm) long, red-billed dwarf hornbill (*Tockus camurus*) of West Africa, while the largest species are the 4 ft (1.2 m) long great hornbill (*Buceros bicornis*) of India and Southeast Asia, and the 5 ft (1.6 m) long helmeted hornbill (*Rhinoplax vigil*) of Malaysia and Indonesia.

The most distinctive characteristic of hornbills is their very unusual beak, which has a complex structure known as a casque sitting on top of the upper mandible. Remarkably, the specific function of the casque has not yet been discovered. Some species have seemingly enormous casques, as is the case of the rhinoceros hornbill (*Buceros rhinoceros*) of Malaysia and Indonesia, and the black-casqued hornbill (*Ceratogymna atrata*) of West Africa. Although bulky, the casque is light, being filled with a sponge-like matrix that is mostly air cavities. An exception is the helmeted hornbill of Southeast Asia, which has a solid casque, known as hornbill ivory. This is a valuable natural commodity in Southeast Asia and China.

Hornbills are rather conspicuous birds, because they make a wide range of loud noises, and usually fly in flocks, especially during the non-breeding season.

Most hornbills are omnivorous, typically depending on fruit as the major component of their diet. However, most hornbills are opportunistic predators, and will readily eat small animals if they can catch them. Some hornbills commonly feed upon relatively dangerous animals, such as poisonous **snakes** and scorpions. Hornbills handle their **prey** with great skill, using the very tip of their seemingly ungainly, but in fact highly dexterous, bill. Hornbills are also adept at manipulating and peeling bulky fruit, again using their large bill.

Hornbills have a remarkable breeding **biology**. The female of almost all species is sealed into the nesting chamber in the **tree** during the breeding season. She remains there, laying and brooding her eggs and hatched young, with only a narrow slit-like opening to the outside. The female builds a wall across the entrance to the nesting cavity using her excrement, which cures to a very hard consistency. Sometimes the male bird assists with the building of this wall, using moist clay. Presumably, the walled-in female and nestlings are kept relatively safe from nest predators. However, both she and the developing nestlings must be fed faithfully by the male. In some species, the female breaks out of the nesting cavity once the chicks are partially grown. The cavity is then re-walled, and the female assists the male in gathering food for the hungry young hornbills.

Most cavity-nesting birds keep the nest clean by routinely disposing the fecal sacs of the young birds outside of the nest. Obviously, the walled-in hornbills cannot do this. The female deals with her individual sanitary problem by defecating at high speed through the small, slit-like opening in the wall, and as the young birds grow they also learn to do this. Other detritus is left for scavenging **beetles**, **ants**, and other **insects** to clean up.

Hornbills are relatively large birds, and they are sometimes hunted by humans as food. The helmeted hornbill is also hunted specifically for its valuable hornbill ivory. Hornbills are also commonly caught and kept in captivity as interesting pets. Overhunting for any of these purposes can easily cause local depletions of wild hornbill populations in the vicinity of human settlements. These birds also suffer as their tropical forest habitats are converted to agriculture or other purposes. For these reasons, the populations of all species of hornbills are decreasing rapidly, and many species are endangered.

Resources

Books

Forshaw, Joseph. *Encyclopedia of Birds.* New York: Academic Press, 1998.

Kemp, A. *The Hornbills (Bucerotidae).* Vol. 1, *Bird Families of the World.* Oxford: Oxford University Press, 1994.

Bill Freedman

An Ohio buckeye (*Aesculus glabra*). *JLM Visuals. Reproduced by permission.*

Horse chestnut

The horse chestnut and buckeyes (*Aesculus* spp.) are various **species** of **angiosperm** trees in the family Hippocastaneae. There are about 20 species of trees and shrubs in this family, occurring widely in temperate, angiosperm **forests** of **Europe**, **Asia**, and **North America**.

The horse chestnut and buckeyes have seasonally deciduous, oppositely arranged, palmately compound leaves, which means that the five to seven leaflets all originate from the same place at the far end of the petiole. The margins of the leaflets are coarsely toothed. The horse chestnut and buckeyes have attractive, whitish flowers, occurring in showy clusters. The flowers of *Aesculus* species develop in the springtime from the large, over-wintering, sticky stem-bud, before the years' leaves have grown. The flowers produce large quantities of **nectar**, and are insect pollinated. The **fruits** of the horse-chestnut and buckeyes are greenish, leathery, spiny capsules containing one or two large, attractive, chestnut-brown **seeds**. These seeds are not edible by humans.

The horse chestnut (*Aesculus hippocastanum*) is a **tree** that can grow as tall as about 115 ft (35 m), and is native to Asia and southeastern Europe. The horse chestnut has been widely planted in North America as an ornamental tree, especially in cities and other residential areas. This species sometimes escapes from cultivation and becomes locally invasive, displacing native species from woodlands.

Several species of buckeyes are native to North America. Most species occur in hardwood forests of the eastern and central United States. The yellow buckeye (*A. octandra*) grows as tall as 98 ft (30 m) and can attain a diameter of almost 3.3 ft (1 m). The Ohio buckeye (*A. glabra*) and Texas buckeye (*A. arguta*) develop a characteristic, unpleasant odor when the leaves or twigs are crushed. The painted buckeye (*A. sylvatica*) and red buckeye (*A. pavia*) are relatively southeastern in distribution. The California buckeye (*A. californica*) is a shrub or small tree of drier foothill areas of the west coast of the United States.

These various, native species are of relatively minor economic importance for their **wood**, which has been used to manufacture boxes, furniture, musical instruments, and other products.

The fruits of the horse chestnut and buckeyes are eaten by various species of wild animals and some species of **livestock**. However, these fruits contain a chemical known as aesculin that is poisonous to humans if eaten in large quantities, and can cause death. The seeds of horse chestnut and buckeyes should not be confused with those of the true chestnuts (*Castanea* spp.), which are edible. (True chestnuts or sweet chestnuts are classified in the beech family, the Fagaceae.) However, it is reported that boiling or roasting the seeds of horse chestnut and buckeyes can remove or disable the aesculin, to provide a starchy food.

Some people attribute medicinal qualities to the fruits and flowers of the horse chestnut and buckeyes. In Appalachia, it is believed by some people that the seeds of buckeyes will help to prevent rheumatism, if carried around in your pockets. Various preparations of the seeds, flowers, and **bark** have also been used as folk medicines to treat hemorrhoids, **ulcers**, rheumatism, neuralgia, and fever, and as a general tonic.

Sometimes, children will collect the seeds of horse chestnut or buckeyes, drill a hole through the middle, and tie them to a strong string. The game of "conkers" involves contests in which these tethered seeds are swung at each other in turn, until one of the horse chestnuts breaks. Each time a particular conker defeats another, it is said to gain a "life." However, there are many variations of the rules of this game.

Bill Freedman

Horsehair worms

Horsehair or gordian worms are unusual **invertebrates** in the phylum Nematomorpha. These very long, thin creatures have a superficial resemblance to animated horse hairs, hence their common name. Often, horsehair worms occur in seemingly inextricable tangles of two or more individuals, especially during the breeding season, which is generally in the springtime. The second common name of these animals originates with these breeding aggregations and refers to the legendary "Gordian knot." This was a very complicated knot devised by King Gordius of Phrygia, that could not be solved and untied, and eventually had to be cut with a sword.

Mature horsehair worms are typically 4-28 in (10-70 cm) long. However, they are only 0.01-0.1 in (0.3-2.5 mm) in diameter, a characteristic that changes little over the length of their body. The mouth is at one end, and the cloacal aperture or vent is at or near the other end.

Adult horsehair worms live in fresh waters of all types. Immature stages are parasites of various types of terrestrial **insects**, most commonly **crickets**, **grasshoppers**, and **beetles**, and sometimes aquatic insects. The adult animals move about by slow undulations and tangles, which is not a very efficient means of locomotion. Consequently, these animals tend to live in static or slow-moving waters and are not generally found in more energetic aquatic habitats.

Male horsehair worms die soon after they impregnate a female during the breeding season. The female lays a stringy egg mass that can contain several million ova, and she then dies. It has not yet been discovered how the larvae manage to parasitize their host insects. It is thought that the larvae may encyst on vegetation or organic debris in shallow **water**, which becomes exposed later in the growing season, when water levels drop. Presumably, insects ingest these tiny cysts when feeding, and if the insect is an appropriate host for that **species** of horsehair worm, it is thereby parasitized by the larva. The life cycle is completed if the host insect falls into water, so the adult horsehair worm can emerge into an appropriate **habitat**. Adults that emerge into a terrestrial habitat are unable to survive for long.

Horsehair worms are widely distributed, occurring from the tropics to the **tundra**. About 110 species of horsehair worms are known, of which only one species is marine, being a parasite of small crustaceans. All of the other specis occur in **freshwater** habitats.

Horses

Horses are members of the family Equidae, which includes the wild **asses** of **Africa** and **Asia** and the **zebras** of African plains and **mountains**. The origins of horse-like **mammals** have been traced back some 55 million years to a small dog-sized, plant-eating **animal** known as *Hyracotherium*. More recently, during the Pliocene and Miocene periods (which ended some 1.5-2 million years ago) horses and their relatives as we know them today were probably the most abundant medium-sized grazing animals in the world. Since then, every **species** has experienced a major reduction in population size.

One wild horse, the tarpan, a small, shy, grey species lived on the Russian steppes of Eurasia until some time in the eighteenth century, when it became extinct because of overhunting and cross-breeding with domesticated species. Almost nothing is known about this animal apart from scant information in a few museums. The only other true wild horse, the slightly larger Przewalski's horse (*Equus przewalskii*), is now also thought to have gone extinct in the wild as recently as the mid 1960s. Some members of this species were, however, preserved in captivity so at least some representatives of this ancient lineage remain.

Horses are grazing animals of wide open plains, where constant vigilance is necessary in order to avoid predators such as lions, tigers, leopards, and wild canids. Apart from their keen senses of **vision**, **hearing**, and **smell**, horses are well equipped to outrun most potential attackers. Wild horses also undergo extensive seasonal migrations in search of optimal feeding and watering **habitat**. The feet of these hoofed animals (perissodactyls) are modified for agility and rapid movement. Horses have light feet with just one toe and, when moving, the hoof is the only part of the foot to touch the ground. Horses are also characterized by their long, slender legs, capable of a steady, prolonged movement or a long, striding gait. A deep chest allows for their large lungs, as well as the animal's large stomach, which is important for digesting the great amounts of relatively bulky **plant** materials.

Grasses and herbs form a major part of the diet. While these materials are relatively abundant, they are often not very nutritious, being low in protein and difficult to digest. Horses eat large quantities of plant materials each day and must be able to transform this into **energy** and **nutrition**. Plant cells are composed of **cellulose**, which the **digestive system** of few mammals is capable of breaking down. To assist with this process horses and their relatives rely on **microorganisms** within the large intestine and colon to break down and ferment their bulky diet. In contrast to ruminating animals such as **deer** and cattle, horses have a small and relatively simple stomach in which **proteins** are digested and absorbed. The digestive system of horses is far less efficient than that of a cow, for example, which means that the former must eat considerably more of the same materials in order to acquire a similar amount of energy.

Przewalski's horse is closely related to the domestic species (*Equus caballus*), but is distinct in its appearance. Reaching more than 7 ft (2 m) at the shoulder, and with a length of almost 8 ft (2.5 m), these horses are a dark bay-dun **color** with a much lighter underside and muzzle patch. The dark mane narrows to a single, narrow dorsal stripe along the back, ending in a black tail. Early Stone Age **cave** paintings feature many illustrations of horses that closely resemble this species. It was formerly

A Przewalski's horse (*Equus przewalskii*). *Photograph by J. Gordon Miller. Reproduced by permission.*

widespread in steppe and semiarid habitats of Kazakhstan, Sinkiang, Mongolia, and parts of southern Siberia. The Przewalski's horse first became known to Western science in 1879, when it was discovered by a Polish explorer after whom the horse is named. Although there are no known estimates of the initial population size, by the early twentieth century it was already rare and found only in parts of southern China and Mongolia.

These animals were once highly prized by Mongolian people for their stamina. The wild herds once also provided semi-nomadic tribes with an essential supply of milk, meat, and hides, the latter being used for clothing as well as construction materials for their hut-like homes. Although the species is now extinct in its native habitat, sufficient animals are kept in zoological collections to enable a systematic program of captive breeding to take place. As a result of these efforts, there are now more than 1,000 individuals in captivity in many parts of the world. Apart from the hopes of conservationists to see this horse returned to its natural habitat, there is also a strong national desire amongst people in Mongolia to see these animals returned to the plains of its rightful heritage.

In their natural habitat, wild horses live in herds that consist of a number of mares, a single stallion, and foals and colts of a wide age span. The stallion is responsible for leading the herd to safe watering and feeding grounds and for protecting the females and young from predators. Stallions are extremely protective of their herds, and fights with other males who attempt to overthrow the stallion are common. Male horses fight with their hooves and teeth, especially the enlarged **canines** of the lower jaw—a prominent feature on mature males. A wide range of facial and other expressions are used to help avoid conflicts or to ensure that these are of short duration, as animals risk injury in such sparring events. Baring the teeth and curling the lips, while at the same time flattening the ears, is one of the most aggressive threats, while a number of vocalizations and stomping movements with the feet are also used to enhance the meaning of the gestures.

Almost everything we know about the social life of these animals is based on observations of semi-wild Przewalski's horses and feral populations of domestic horses. In the Przewalski's horse, young are born from April to June, following a gestation period of about 330 days. Mares usually bear a single foal which, shortly after **birth**, is able to stand up and follow its mother—an essential ability if the foal is not to fall **prey** to ever-vigilant predators. Foals remain close to their mothers for the first few weeks of life and do not become independent until they are almost two years old. Following this, they remain with the herd for several more years until they mature. In a natural situation, males are driven away from the herd as they reach sexual maturity. These solitary males usually join with other males to form small bachelor herds. Females, in contrast, may remain with the herd they were born into and will, in time, breed with the dominant male of the herd.

The precise origins of the domestic horse are not known but they likely arose from either the tarpan or Przewalski's horse. The earliest records of domestication are unclear and it is possible that this took place simultaneously in different parts of the world. Some reports suggest that it was attempted as early as 4000 B.C. in Mesopotamia and China, while evidence suggests that by 2000 B.C. domesticated horses were in use in China. Since then, horses have been bred for a number of purposes and there are now thought to be more than 180 different breeds. The powerful Shire horses were bred as draught animals in England, while most modern thoroughbreds, bred for their speed, stamina, and grace, are derived from breeding other species with primarily Arabian horses. The increasing spread of agriculture almost certainly played an important role in the use of domesticated species for draught purposes, but others were also bred and crossbred for their hardiness in extreme climates. Horses have also featured heavily in warfare, and many battles have been won and empires taken by mounted warriors.

Wild horses have suffered considerably since the arrival of humans on **Earth**. Horses and asses were once widely harvested for their meat and skins, particularly in parts of Asia. Elsewhere, the integrity of true wild species became diluted as domestic species interbred with wild animals. Natural changes may also have had some role to play in the demise of the wild horse, but it is more likely that human encroachment on the great plains of Asia, with spreading agriculture, has had the greatest and most long-term effect.

It is now too late to protect the last true wild horses, but considerable efforts are required to ensure that the last member of this ancient lineage, Przewalski's horse, and its natural habitat are protected in a manner that would enable this species to be reintroduced to its native habitat. Consideration should also be given to the preservation of wild stocks of domesticated varieties, such as the mustangs of **North America**, the Dartmoor and Exmoor ponies of Great Britain, and the brumbies of **Australia**, where these species have a role to play in maintaining the **ecology** of their respective habitats. In some countries, however, feral horses have caused considerable destruction to local plants and control programs are required to limit herd size so that they do not cause irreversible damage to fragile ecosystems. In other regions, feral horses play a useful role in cropping long coarse grasses, which helps keep the **ecosystem** open for other smaller, more fastidious grazing animals and plants. Some plants are known to germinate only when their **seeds** have passed through a horse's digestive system, as many of these plants may have evolved at a time when large herds of wild horses roamed the plains and acted as natural seed dispersers.

See also Livestock.

Horseshoe crabs

Often referred to as a living fossil, the horseshoe crab has changed very little in over 400 million years. Related to spiders, this **animal** is easily identified by the large greenish brown, helmet-like dorsal plate, called either the cephalothorax or prosoma. A separate plate covers its abdomen. A long tail spine, referred to as the caudal spine or telson, extends from its abdomen. Measured from the front of its dorsal plate to the tip of its tail spine, the horseshoe crab can reach a length of 60 cm. Its mouth and six body segments lie underneath its dorsal plate; a pair of limbs is attached to each segment. Today's horseshoe crab populations are rather sporadically distributed. One species—*Limulus polyphemus*—lives off the coast of the eastern United States, and four **species** live in the marine waters of southeast **Asia**.

The phylum Arthropoda is the largest phylum in the animal kingdom, containing more than one million species. Within this phylum, the subphylum Chelicerata includes spiders and their relatives. This subphylum can be broken down into three classes: (1) class Arachnida (Otherwise known as **arachnids**, this class includes true spiders and scorpions); (2) class Pantopoda (also known as **sea spiders**); and (3) class Merostomata (referred to as Merostomates). Within the Merostomata class, there are two orders. One extinct order, the order Eurypterida, contained sea scorpions; the other order, Xiphosura, includes only horseshoe **crabs**. There is one family, Limulidae, and three genera within this family—*Limulus, Tachypleus,* and *Carcinoscorpius.* In total, there are four species.

Evolution

Although fossils confirm that chelicerates developed in the sea, there is some debate over their evolutionary history. Some research suggests that animals in this subphylum are descendants of trilobites, the earliest known **arthropods** which lived 570 million years ago during the Cambrian period; other research suggests that **segmented worms** are their true ancestors. Whatever the case, the history of the horseshoe crab can definitely be traced back to the Ordovician period, about 500-440 million years ago.

Ancestors and relatives of the horseshoe crab include very diverse animals. For instance, horseshoe crabs are related to **mites** that never surpass.04 in (1 mm) long and to the biggest segmented animal that ever lived—the giant sea scorpion (*Pterygotus rhenanus*)—which grew to over 6 ft (180 cm) long. Members of this subphylum have adapted to nearly every **habitat** on land and sea, and most have retained their primitive behaviors.

Physical characteristics

The horseshoe crab's body is composed of two parts: the cephalothorax and the abdomen. The cephalothorax is basically the crab's head and thorax fused together. Under the cephalothorax, there are six body segments, each equipped with a pair of limbs. Under the abdominal shell is located the circulatory, respiratory, reproductive, and nervous systems. Further, the abdomen houses part of the crab's **digestive system** and an abundant number of **glands**.

Like all members of the subphylum Chelicerata—but unlike other anthropods—the horseshoe crab does not have antennae. Instead, it uses its first pair of appendages (called cheliceras), located in front and to the sides of its mouth, to feed itself. The cheliceras, and all of their appendages except for their walking legs, are equipped with pinchers (called chelas) with which the animal grabs food from the sea floor. The second pair of legs (called the pedipalp) evidently used to be used for walking, but, over **time**, evolved more specialized functions. Currently, the second pair of legs are used in different ways, depending on the species; basically, these legs can be used for gripping, chewing, or sensing.

While the horseshoe crab does not have a conventional jaw, its four pairs of walking legs have special equipment attached to them. Known as gnathobases, these are primitive devices that the crab uses to manipu-

A horseshoe crab moving along the waters edge. *Horseshoe crab, photograph. © John M. Burnley/The National Audubon Society Collection/Photo Researchers, Inc. Reproduced by permission.*

late and shred food before passing it to its mouth. The last pair of walking legs can be used to break shells and to crush tough food. Because the crab often swallows **sand** and shell fragments, its gizzard is quite powerful and can grind up almost anything it consumes.

The horseshoe crab has two sets of eyes. The first pair are large and compound, meaning that they are composed of numerous simple eyes clustered tightly together. These large eyes are located far apart on the front side of the dorsal plate. Much less noticeable, the two small, simple eyes are located fairly close to each other at the anterior of the crab's back. Little is known about the animal's other senses.

Behavior

The horseshoe crab lives in shallow, coastal waters, usually partially covered by mud or sand. It covers itself

KEY TERMS
. . . .

Caudal spine—Also called the telson, this appendage extends from the crab's abdomen and resembles a tail; it is often as long as the crab's body. It is used by the crab to right itself if it falls on its back; the crab flaps it against the abdomen when it swims.

Cephalothorax—The head and thorax (upper part of the body) combined.

Chelas—Pincers on the last pair of walking legs with which the crab grabs food from the sea floor.

Chelicerae—Feeding appendages.

Compound eyes—Two large eyes appearing widely separated on the anterior of the dorsal plate. They are actually composed of numerous simple eyes clustered together.

Gnathobases—Attached to the legs, these spiny devices function like jaws, shredding and manipulating food before passing it to the mouth.

Pedipalp—Their second pair of legs, highly specialized, depending on the species.

Prosoma—See cephalothorax.

Salinity—The amountdissolved salts in water.

Simple eyes—Located fairly close to each other at the anterior of the crab's back. Easy to overlook.

Thorax—The area just below the head and neck; the chest.

like this by driving the front of its round dorsal plate forward and downward into the **earth**. This crab is a sturdy creature, tolerating wide swings in salinity and **temperature**. As a **scavenger**, it spends much of its life feeding on all types of marine animals, including small **fish**, crustaceans, and worms. Interestingly, it swims through the **water** with its dorsal plate facing the bottom (on its back) by flapping its tail spine into its abdomen.

Horseshoe crabs mature sexually when they are between nine and 12 years old. Typically, when they breed, horseshoe crabs congregate in large numbers in shallow coastal waters. At such times, the male climbs onto the female's back, holding the sides of her dorsal plate. (The male is significantly smaller than the female.) She carries him around, sometimes for days, until spawning takes place. When ready to lay her eggs, she digs holes about 5.9 in (15 cm) deep in a tidal area and lays up to 1,000 eggs in each hole. While she lays these eggs, the male fertilizes them. In approximately

six weeks, the eggs hatch into free-swimming larvae that look a lot like their parents, but their tail spines are missing. Because of the inflexibility of their dorsal plates, it is difficult for these animals to grow within their shells; thus they molt several times before their growth stops at sexual maturity.

Uses to humans

When a horseshoe crab is wounded, its **blood** cells release a special protein to clot the bleeding. The same thing happens when certain toxins are introduced to stop invading **bacteria**. (Horseshoe crabs are a favorite host of flatworms.) Thus, hospitals sometimes use extracts of their blood when diagnosing human bacterial diseases and checking for toxins in intravenous solutions.

Resources

Books

Bonaventura, Joseph, Celia Bonaventura, and Shirley Tesh, eds. *Physiology and Biology of Horseshoe Crabs: Studies on Normal and Environmentally Stressed Animals.* New York: Alan R. Liss, Inc., 1982.

Grzimek, H. C. Bernard, ed. *Grzimek's Animal Life Encyclopedia.* New York: Van Nostrand Reinhold Company, 1993.

The New Larousse Encyclopedia of Animal Life. New York: Bonanza Books, 1987.

Pearl, Mary Corliss, Ph.D. Consultant. *The Illustrated Encyclopedia of Wildlife.* London: Grey Castle Press, 1991.

Pearse, John and Vicki, and Mildred and Ralph Buchsbaum. *Living Invertebrates.* Palo Alto, California: Blackwell Scientific Publications; Pacific Grove, California: The Boxwood Press, 1987.

Kathryn Snavely

Horsetails

Horsetails are a group of relatively primitive, vascular plants in the genus *Equisetum*, family Equisetaceae, subdivision Sphenophytina. The sphenophytes have an ancient evolutionary lineage occurring as far back as the Devonian period. These plants were most abundant and diverse in **species** about 300 million years ago, during the late Devonian and early Carboniferous periods. Fossils from that time suggest that some of these plants were as large as 8 in (20 cm) in diameter and at least 49 ft (15 m) tall.

Today, however, this group is represented by 29 species of small, herbaceous plants all in the genus *Equisetum*. Horsetails are very widespread, although they do not occur naturally in the Amazon **basin** or in **Australia** and New Zealand. These plants are characterized by their

Field horsetails (*Equisetum arvense*) at the Long Point Wildlife Sanctuary, Ontario. *Photograph by Robert J. Huffman. Field Mark Publications. Reproduced by permission.*

conspicuously jointed stems and their reduced, scale-like leaves, which are arranged in whorls around the stem. The stems of horsetails contain deposits of silica which give the plants a coarse, grainy feel when crushed. The silica-rich horsetails are often used by campers to clean their dishes and pots, giving rise to another of their common names, the "scouring rushes." Horsetails are perennial plants, and they grow from underground systems of rhizomes. Horsetails develop specialized structures known as a strobilus (plural: strobili), containing sporangiophores which develop large numbers of spores (or sporangia). In some species the strobilus develops at the top of the green or vegetative shoot. In other so-called dimorphic species of horsetails, the strobilus occurs at the top of a specialized, whitish shoot which develops before the green shoots in the early springtime.

The woodland horsetail (*Equisetum sylvaticum*) occurs throughout the northern hemisphere in boreal and north-temperate **forests**. The common horsetail (*E. ar-*

vense) is a very widespread species occurring almost worldwide, often in disturbed habitats. This species is dimorphic, producing its whitish, fertile shoots early in the springtime and its green, vegetative shoots somewhat later. The scouring rush (*E. hyemale*) occurs widely in the northern hemisphere in wet places. The **water** horsetail (*E. fluviatile*) occurs in a wide range of aquatic habitats in boreal and north-temperate regions of **North America** and Eurasia. The dwarf scouring rush (*E. scirpoides*) is a small species of **wetlands** and moist shores, occurring widely in arctic and boreal habitats of the northern hemisphere.

See also Rushes.

Bill Freedman

Horticulture

The word horticulture comes from Latin and refers to the cultivation of gardens. There are three main branches of the science of growing plants: **forestry**, **agronomy**, and horticulture. Forestry is concerned with the cultivation of stands of trees for their commercial and ecological uses. Agronomy involves the large-scale cultivation of **crops**, such as **wheat**, **cotton**, **fruits**, and **vegetables**. Horticulture involves growing plants for their aesthetic value (e.g., in floriculture; the cultivation of flowers), or on a very local scale as food (as in a home garden).

In addition to home gardening, horticulturists are involved in the landscaping and maintenance of public gardens, parks, golf courses, and playing fields. Seed growers, **plant** growers, and nurseries are the major suppliers of plants and supplies for use in horticulture. Among the important specialists working in horticulture are plant physiologists, who work on the nutritional needs of plants, and plant pathologists, who are engaged in protecting plants from diseases and insect damage.

For the amateur home gardener, the rewards of horticulture are both recreational and emotional. Gardening is one of the most popular pastimes for many people—for those living in suburbs, as well as city dwellers who plant window boxes, grow house plants, or develop a garden in a vacant lot.

Plant needs

Whether plants are being grown on a large scale for commercial purposes or for the pleasures of having a garden, they have fundamental needs that include a suitable regime of **water**, **soil**, and climate.

A garden of perennial plants and flowers in bloom. *Photograph by Alan & Linda Detrick. Photo Researchers, Inc. Reproduced by permission.*

Climatic factors

The climatic factors that have the greatest effects on plant growth are **temperature**, **precipitation**, **humidity**, **light**, and **wind**. In deciding what plant **species** can be grown in a particular location, the horticulturist must consider whether the seasonal ranges of temperature can be tolerated. Many plants will die if exposed to temperatures as low at 28°F (-2.2°C), although others are frost hardy and can be grown in places much colder than this. While some plants die from frost, others may only die back and then recover when warmer **weather** returns. Conversely, many plants need exposure to seasonally cold temperatures, as occurs during the wintertime.

Another climatic factor affecting plants is precipitation. The amount of moisture that plants require varies greatly. **Desert** plants can survive on little water, and may perish if over-watered. Other plants need continuously moist growing conditions. Plants of some coastal habitats receive benefits from **fog** and moist air blowing in from over the water. However, too much dew can damage some plants, by predisposing them to fungal diseases. In many regions, trees overburdened by heavy, freezing rain are subject to broken branches.

The amount of sunlight plants receive also affects their growth. The intensity and duration of light controls the growth and flowering of plants. Insufficient light results in the **rate** of **photosynthesis** being insufficient to allow the plant to grow and **flower**. Wind is another important factor, which can cause damage by increasing the rate of water loss, and if extreme by breaking off plant parts. Strong wind blowing from oceans can deposit harmful salts on sensitive plants.

All these climatic factors must be considered by horticulturists when planning a garden or landscaping project. These factors determine the possible selection of plants for a particular ecological context.

Consideration of the quality of the soil is also important. If the soil does not have the proper combination of **nutrients**, organic **matter**, and moisture, plants will not grow well.

All types of soils need management for optimum plant growth. Some soils are rich with clay, while others are sandy or rocky. Clay soils are heavy and tend to drain water poorly, which may cause plant roots to become waterlogged and oxygen-starved. Sandy or rocky soils, on the other hand, drain water rapidly and may have to be irrigated to grow plants well. Usually, the preferable garden soil is loamy, meaning it consists of a balanced mixture of clay, **sand**, and organic matter. Organic matter in soil is important because it helps develop larger pore spaces and allows water and air to penetrate. This helps roots to grow well and absorb nutrients for use by the plant.

Other important soil factors include the acidity or alkalinity of the soil, the presence of beneficial or harmful microscopic organisms, and the composition and structure of the soil layers (topsoil and subsoil). The addition of mineral nutrients and organic matter to soil being prepared for planting is a common practice in horticulture. This may include the addition of **fertilizers** that meet the **nitrogen**, **phosphorus**, potassium, **calcium**, **magnesium**, **sulfur**, and trace-element needs of the plants.

Horticultural plants

Thousands of plant species are available for use in horticulture. Many of these have been domesticated, selectively bred, and hybridized from the original, wild, parent stocks, and are now available in large numbers of cultivated varieties (or cultivars). Consider, for example, the numerous varieties of roses, tulips, geraniums and many other common horticultural plants that can be obtained from commercial outlets.

In most places, almost all of the horticultural plants that are widely grown in parks and gardens are not indigenous to the region (that is, their natural habitats are far away, usually on another **continent**). This widespread cultivation of non-native plants has resulted in some important ecological problems, caused when the horticultural species "escape" to the wild and displace native plants. Because of this kind of severe ecological damage, many environmentalists are advocating the cultivation of native species of plants in horticulture. If this sensible "naturalization" is practiced, there are fewer problems with invasive aliens, and much better **habitat** is provided for native species of animals. This means that horticulture can achieve important ecological benefits, in addition to the aesthetic ones.

KEY TERMS

Agronomy—The application of agricultural science to the production of plant and animal crops, and the management of soil fertility.

Floriculture—The cultivation of flowers.

Photosynthesis—The synthesis of carbohydrates by green plants, which takes place in the presence of light.

Resources

Books

Bennett, Jennifer. *Our Gardens Ourselves.* Ontario, Canada: Camden House, Camden East, 1994.

Jackson, Ron S. *Wine Science: Principles and Applications.* San Diego: Academic Press, 1994.

Jones, Hamlyn G. *Plants and Microclimate.* 2nd ed. Cambridge, England: Cambridge University Press, 1992.

Larson, Roy A., and Allan M. Armitage. *Introduction to Floriculture.* San Diego: Academic Press, 1992.

Rice, Laura Williams, and Robert P. Rice. *Practical Horticulture.* Englewood Cliffs, New Jersey: Prentice-Hall, 1986.

Smith, Geoffrey. *A Passion for Plants.* North Pomfret, VT: Trafalgar Square Publishing, 1990.

Vita Richman

Hot spot

Hot spots are a common term for plumes of **magma** welling up through the crust (Earth's outermost layer of rock) far from the edges of plates.

To understand what hot spots are and why they are important, some understanding of the theory of **plate tectonics** is necessary. This widely accepted theory proposed by Alfred Wegener in 1912 states that the crust is composed of huge plates of rock that drift over Earth's mantle. Where the plates separate, magma from the mantle approaches the surface and encounters decreased **temperature** and **pressure**, allowing it to solidify into new rock. At the edges of plates that crash together, trenches form, in which one plate slides under the other. In some places, such as the San Andreas Fault in California, the plates slide by each other. Most volcanoes and earthquakes occur near the edges of these plates.

Some volcanoes, however, are far from the plate margins. These volcanoes tend to be very high, in the center of raised areas, and the rock produced there is alkaline, chemically different than the theoleiite rock produced at the margins. Moreover, there are several dotted lines of extinct volcanoes (such as the chain of the Hawaiian islands) that are arranged oldest-to-youngest in a line, tipped by a young active **volcano**.

These are explained, in a theory proposed by J. Tuzo Wilson in 1963, as fixed spots in Earth's mantle, from which thermal plumes penetrate the crust. The lines of extinct volcanoes do not indicate that the plume is moving: rather that the plate is moving relative to the mantle. Therefore, the hot spots can be used to deduce the direction in which a plate is moving. In the case of the Hawaiian ridge, the most recent volcano (Kilauea) is southeast of the older volcanoes. The oldest volcano in this line dates back to about 40 million years ago. From this, scientists have deduced that the Pacific plate is moving northwest at about 3.9 in (10 cm) each year. However, a line of even older extinct volcanoes, the Emperor **seamounts**, trail northward from the end of Hawaiian ridge: the youngest are southernmost and the oldest (about 70 million years old) are northernmost. From this, we can deduce that the Pacific plate changed direction sometime between 40 and 50 million years ago.

In addition to the volcanoes, hot spots have other effects on the areas around them: they lift the areas around them and represent areas of high **heat** flow.

The number of hot spots in the world is uncertain, with numbers ranging from a few dozen to over a hundred. They range in age from a few tens of millions of years in age (like the Hawaii-Emperor hot spot) to hundreds of millions of years old. Some appear to be extinct.

Hovercraft

A hovercraft is a vehicle that can be used to journey over **water** and land. Unlike a boat, which floats on the water, a hovercraft is suspended above the water on a cushion of air. This also allows a hovercraft to move over land and float over small depressions such as a ditch or over waves. A powerful and specially designed fan creates the air cushion that is part of the hovercraft. For this reason a hovercraft is also called an Air Cushion Vehicle or ACV.

Englishman Christopher Cockerell invented the hovercraft in 1956. For this accomplishment and his other efforts, which included being part of the research team that developed the **radar**, he was later knighted in 1969.

Cockerell's main idea involved a vehicle designed to that float on a cushion of air, with another power source that would move the vehicle horizontally over the surface. The feasibility of the idea was tested initially using tin cans and the nozzle end of a **vacuum** cleaner. Initially, cans of different sizes were modified so that air could be blown down through the sealed end. Cockerell found that single cans did not produce sufficient air **pressure**. However, positioning one can inside another and forcing the air through the narrow cylinder of space between the two cans created a zone of high air pressure created in the region between these cans. It was this basic design that was used for the first commercial hovercraft in 1959, the Saunders Roe Nautical One (SRN1), and for subsequent versions of the hovercraft.

In the SRN1 and other models of hovercraft, the narrow cylindrical space between the tin can test system is referred to as a plenum chamber. The word plenum is from the Latin word meaning full. Large fans, which are analogous to the vacuum cleaner nozzle on Cockerall's tabletop developmental model, blow air down through the plenum chamber.

The fan used on hovercraft differs from the standard propeller type of fan. Propeller blades generate backpressure as they rotate. In hovercraft, backpressure would decrease the efficiency of the air cushion.

The fan of a hovercraft is called a centrifugal lift fan. The fan appears like an inverted funnel positioned inside a donut-like chamber that has angled slats around the outside edge of the chamber. When the funnel shaped assembly rotates at high speed, air is sucked into the chamber and is expelled out through the slats. This design creates a more powerful airflow than does a conventional propeller.

The airflow must be constant and powerful in order to compensate for air that escapes from the edge of the hovercraft. The airflow must also be even around the edge of the hovercraft, to prevent one region of the hovercraft from lifting higher off the surface of the ground or water than other parts of the hovercraft.

The edge of a hovercraft contains a flexible curtain of material. This material, which is known as a skirt, helps prevent the escape of air from the plenum chamber, which in turn lessens the mount of **energy** that is needed to generate the suspending airflow. The skirt must be durable and flexible to accommodate the different heights of terrain or waves that the hovercraft passes over. Additionally, the skirt must be **light**, yet resistant to flapping. Originally, a skirt was a single piece of material; now it is typically made of rubber. When one region of the old-design skirt wore out, the entire skirt had to be replaced. The cost of this replacement, often in the millions of dollars, prompted a redesign. Nowadays, damaged rubber portions of a skirt can be removed and replaced without the necessity of replacing the entire structure.

As a hovercraft operates, the rubber skirt is inflated outward by the air pressure generated by the centrifugal lift fan. This effect produces an air cushion that is about 3 meters in depth beneath the hovercraft. Propellers that blow air out horizontally provide the power that moves the hovercraft over the surface of the ground or water. The blades of these propellers can be moved, or pitched, to control the speed of the hovercraft. When the propellers have a **zero** pitch, the hovercraft is not moving. Positive pitch moves a hovercraft forward and negative pitch is the braking system that slows the hovercraft down.

The hovercraft has proven to be useful for applications where passengers or cargo are transported over water and where the loading and unloading can be done on land. For example, hovercraft have been used for decades to ferry people back and forth across the English Channel between the United Kingdom and France.

See also Ocean.

Resources

Books

Amyot, J.R. *Hovercraft Technology, Economics and Applications* (Studies in Mechanical Engineering, No 11). New York: Elsevier Science Ltd., 1990.

Other

Flexitech LLC, PO Box 412, Germantown, MD 20875-0412. April 10, 2002 [cited November 3, 2002]. <http://www.hovercraftmodels.com/How_ a_Hovercraft_works.htm >.

Hubble constant *see* **Cosmology**

Hubble Space Telescope

Floating in **orbit** approximately 380 miles (612 km) above the **earth**, the 12.5-ton Hubble Space Telescope has peered farther into the Universe than any **telescope** before it. The Hubble, which was launched in 1990, has produced images with unprecedented resolution at visible, near-ultraviolet, and near-infrared wavelengths since its originally faulty **optics** were corrected in 1993. Although ground-based technology is finally starting to catch up—the European Southern Observatory's Very Large Telescope atop Cerro Paranal, Chile, can now produce narrow-field images even sharper than Hubble's—the Hubble continues to produce a stream of unique observations. Over the last decade, the Hubble has revolutionized **astronomy**.

The Hubble was the first of the four great observatories planned by the United States. National Aeronautics and Space Administration (NASA). This series of orbital telescopes also includes the Compton Gamma Ray Observatory (launched 1991), the Chandra X-Ray Observatory (launched 1999), and the Space Infrared Telescope Facility (scheduled for launch in 2003). Together, the light-sensing abilities of the Great Observatories span much of the **electromagnetic spectrum**. They are designed to do so because each part of the **spectrum** conveys different astronomical information.

Above the turbulent atmosphere

The twinkling of stars is a barrier between astronomers and the information they wish to gather. In reality, stars do not twinkle but burn steadily; they only appear to ground observers to twinkle because atmospheric **turbulence** distorts their **light** waves en route to us. Although telescopes on Earth's surface incorporate enormous **mirrors** to gather starlight and sophisticated instruments to minimize atmospheric distortion, the images gathered still suffer from some image degradation. Recently much progress has been made in the use of adaptive optical systems. These systems aim lasers along a telescope's line of sight to measure atmospheric turbulence. This information is fed to computers, which calculate and apply an ever-changing counter-warp to the surface of the telescope's mirror (or mirrors) to undo the effect of the turbulence in real **time**. Adaptive optics are starting to overcome some of the problems caused by atmospheric turbulence. However, the fact that the Earth's atmosphere absorbs much of the electromagnetic spectrum cannot be overcome from the ground; only space-based telescopes can make observations at certain wavelengths (e.g., the infrared).

Scientists first conceived of an orbital telescope in the 1940s. The observatory proposed at that time was called, optimistically, the Large Space Telescope. By the 1970s, the concept had coalesced into an actual design, less "large" thanks to political backlash against the huge space-exploration budgets of the 1960s. In 1990, after a decade of development and years of delay caused by the Challenger shuttle disaster of 1986, the **space shuttle** *Discovery* deployed the Hubble Space Telescope into an orbit approximately 380 mi (612 km) above Earth. The way we see the universe was about to be changed—but not for another three years, due to a design flaw in the main mirror.

The design

The Hubble Space Telescope is a large cylinder sporting long, rectangular solar panels on either side like the winding stems of a giant toy. Almost 43 ft (13 m) long and more than 14 ft (4.2 m) in diameter, this cylinder houses a large mirror to gather light and a host of instruments designed to analyze the light thus gathered.

The telescope itself is a Ritchey-Chretien Cassegrain type that consists of a concave primary mirror 8 feet (2.4 m) in diameter and a smaller, convex secondary mirror 1 foot (.3 m) in diameter that is mounted facing the primary. This pair of mirrors is mounted deep within the long tube of the Hubble's housing, which prevents unwanted light from degrading the image.

Light follows a Z-shaped path through the telescope. First, light from the target travels straight down the tube to the primary mirror. This reflects the light, focusing it on the secondary mirror. The secondary mirror reflects the light again and further focuses it, aiming it through a small hole in the center of the primary at the telescope's focal **plane**, which is located behind the primary. The focal plane is where the light gathered by the telescope is formed into a sharp image. Here, the focused light is directed to one of the observatory's many instruments for analysis. All data collected by the Hubble is radioed to Earth in digital form.

The Hubble's original complement of instruments, since replaced by a series of space-shuttle service missions, included the Wide Field/Planetary Camera (WF/PC1), the Faint Object Spectrograph (FOS), the High Resolution Spectrograph (HRS), and the High Speed Photometer (HSP). WF/PC1 was designed to capture spectacular photos from space. The FOS, operating from ultraviolet to near-infrared wavelengths, did not create images, but analyzed light from stars and galaxies spectroscopically, that is, by breaking it into constituent wavelengths. The FOS contained image intensifiers that amplify light, allowing it to view very faint, far away objects. The HRS also analyzed light spectroscopically, but was limited to ultraviolet wavelengths. Although it could not study very faint stars as the FOS could, the HRS operated at comparatively high precision. The HSP provided quantitative data on the amount of light emanating from different celestial objects.

Every aspect of the Hubble had to be designed for operation in space. For example, the Hubble is designed to function under radical **temperature** extremes. Although the **vacuum** of space itself has no temperature, at the Earth's distance from the **Sun**, an object in deep shadow cools to a temperature of $-250°F$ ($-155°C$) while an object in direct Sunlight can be heated to hundreds of Fahrenheit degrees above **zero**. The Hubble itself orbits the Earth every 97 minutes, spending 25 minutes of that time in Earth's shadow and the rest in direct sunlight. It thus passes, in effect, from an extreme deep freeze to an oven and back again about 15 times a day, and must be effectively insulated to keep its instruments and mirrors stable.

Another aspect of the Hubble that had to be specially designed for its orbit situation is its pointing system. Because astronomical observations often require minutes or hours of cumulative, precisely-aimed viewing of the target, the Hubble—which rotates with respect to the fixed stars an average of once every 97 minutes—must turn itself while making observations in order to keep its target in view and unblurred; ground-based telescopes must cope with a similar problem, but rotate with respect to the fixed stars at much slower keeps the Hubble aligned while it is observing a target, checking for movement 40 times per second.

Another problem for any space vehicle is the supply of electrical power. In the Hubble's case, a pair of 40 ft $\times$ 8 ft (12 m $\times$ 2.4 m) solar arrays provide power for the observatory, generating up to 2400 W of **electricity**. Batteries supply power while the telescope is in the earth's shadow.

Hubble's blurry vision

After the Hubble's launch in 1990, astronomers eagerly awaited its first observations. When they saw the test images, however, it quickly became clear that something was seriously wrong: the Hubble had defective **vision**. Scientists soon realized that the primary mirror of the space telescope suffered from a spherical aberration, an error in its shape that caused it to focus light in a thin slab of space rather than at a sharply defined focal plane. In the focal plane, therefore, a star's image appeared as a blurred disk instead of a sharp point.

The fabrication of a large astronomical mirror such as the Hubble's primary is a painstaking task. The mirror is first cast in the rough and must be ground and polished down to its precise final shape. The computer-controlled tools used for this process remove **glass** from the rough cast one micron at a time. After each grinding or polishing step, the mirror is re-measured to determine how closely it approximates the desired shape. With these measurements in hand, engineers can tell the computer how much glass to remove in the next grinding or polishing pass and where the glass must be removed. This cycle of grind, polish, measure, and re-grind, a single round of which can take weeks, must be repeated dozens of times before the mirror's final shape is achieved.

During the metrology (measuring) step, a repeated or systematic error caused the manufacturers to produce a mirror with a shape that was slightly more flat around the edges than specified. The error was small—the thickness of extra glass removed was a fraction of the width of a human hair—but it was enough to produce a significant spherical aberration. Although useful science could still be performed with the telescope's spectroscopic instruments, the Hubble was unable to perform its imaging mission.

Endeavor to the rescue

The design and manufacture of a space telescope like the Hubble is a large project that takes many years; of necessity, the design must be finalized early on. As a result, by the time the observatory reaches orbit its scientific instruments rarely represent the state of the art. Having this constraint in mind, the telescope engineers designed the Hubble's instruments as modular units that could be easily swapped out for improved designs. The Hubble was thus, engineered for periodic servicing missions by space shuttle crews over the course of its planned 15-year lifetime (since extended to 20 years). Its housing or outer shell is studded with a host of handholds and places for astronauts to secure themselves, bolt heads are large-sized for easy manipulation by astronauts wearing clumsy gloves, and more than 90 components are designed to be replaced in orbit. The Hubble's housing also includes a fixture that enables the shuttle's robot arm to seize it and draw the Hubble and shuttle together. The shuttle's cargo bay includes a servicing platform to hold the telescope while the bay doors are open, and astronauts can affect repairs while standing on small platforms nearby.

One benefit of the primary mirror's precision fabrication was that despite the error imparted by the systematic metrology error, the mirror's shape—error and all—was precisely known. Its surface is so smooth that if the mirror were the width of the United States, its largest variation in surface height would be less than 3 ft (1 m). Once scientists understood what was wrong, therefore, they knew the exact correction required. Replacing the primary mirror would have required bringing the Hubble back to Earth, re-building it, and re-launching it, much too expensive to be feasible; instead, designers developed an add-on optics module to compensate for the focusing error. This module would correct the "vision" of the telescope to the level originally designed for, much as a pair of glasses corrects for defective eyesight.

This module—the Corrective Optics Space Telescope Axial Replacement (COSTAR)—contained five mirrors that would refocus light gathered by the primary and secondary mirrors and relay it to the instruments. The challenge was to build the module to fit into the compact interior of a telescope that was, and would remain, in orbit, and which had never been designed for such a fix. Engineers also produced an improved version of the Wide Field/Planetary Camera, the WF/PC2, that included its own corrective optics to allow it to capture images of the clarity that astronomers had originally hoped for.

In addition to the flaw in its optics, the observatory was experiencing difficulties with its pointing stability and with its solar arrays, which turned out to be prone to wobbling due to thermal stress created during the transi-tion from sun to shadow. This wobbling further degraded observation quality. NASA planned an ambitious repair mission that would attempt to correct all the Hubble's problems at once.

In December, 1993, the space shuttle *Endeavor* took off to rendezvous with the Hubble Space Telescope. During the course of the mission, astronauts performed a total of five space walks. They captured the Hubble with the shuttle arm, repaired some of the pointing gyroscopes, replaced the wobbling solar arrays, and installed the WF/PC2 and COSTAR.

The mission was a success; the contrast between the images taken before and after the repairs was stunning. Suddenly the Hubble was dazzling the world and astronomers were lining up for observing time. Since the 1993 repair, the Hubble's available observing time has invariably been booked for years in advance; in fact, it is so over-subscribed that only one out of every ten proposals for observing time can be accepted.

Daily operations

Making observations with an orbital telescope is not a simple process. The telescope must be instructed where to point to acquire a new target, how to move in order to avoid light **contamination** from the Sun and **Moon**, how long to observe and with what instruments, what data format to use for transmission of result, how to orient its **radio** antennas to send and receive future commands, and so forth. All commands must be written in computer code and relayed to the Hubble by radio during a point in its orbit where it can communicate with antennas on the ground.

How does the Hubble know where to find a given target object? Like a person trying to find his or her way in unfamiliar territory, the telescope searches for stellar landmarks termed guide stars. The position of any **star**, **planet**, or **galaxy** can be specified in terms of particular guide stars—bright, easily found stars located near the object of interest. (The guide stars are not literally close to the objects they are used to locate, but appear to be near them in the sky.) Sky surveys performed by ground-based telescopes have mapped many of these stars, so the Hubble merely points itself to the appropriate coordinates, then uses the guide stars to maintain its position.

In early 1997, astronauts aboard the space shuttle *Discovery* performed another servicing mission, this time to swap out instruments. The HRS was replaced by the Space Telescope Imaging Spectrograph (STIS). Unlike the older instrument, the STIS collects light from hundreds of points over a target area instead of just one point. The servicing crew removed the FOS and in its place installed the Near Infrared Camera and Multi-Ob-

KEY TERMS

Guide star—Bright star used as landmark to identify other stellar objects.

Metrology—The process of measuring mirrors and lenses precisely during the fabrication process

Spectrograph—Instrument for dispersing light into its spectrum of wavelengths then photographing that spectrum.

Spectroscopy—A technique in which light is spread out into its constituent wavelengths (colors, for visible light). The presence of energy at certain wavelengths in the light emitted by a star or galaxy indicates the presence of certain elements or processes in that star or galaxy.

Spherical aberration—A distortion in the curvature of a lens or mirror. When spherical aberration is present in a mirror, light from different radial sections of the mirror focuses at different distances rather than all at the same point. The image produced is thus blurred, or aberrated.

ject Spectrometer (NICMOS), which allows the telescope to gather images and spectroscopic data in the infrared spectral region (0.8 and 2.5 micrometers), which in effect allows the Hubble to see through interstellar **clouds** of gas and dust that block visible light.

The crew also made repairs to the telescope's electrical, data storage, computer, and pointing systems, as well as to its battered thermal insulation blanket, which had been severely damaged by collisions with small bits of space debris. The final task of the repair mission was to nudge the observatory to an orbit six miles higher than previously, to enhance its longevity and stability. Altitude affects longevity because the orbit of any near-Earth object, including the Hubble, is degrading all the time due to friction with outlying traces of the Earth's atmosphere. Therefore, unless it is boosted out of Earth orbit or brought back to Earth by a space shuttle, the Hubble will eventually burn in the atmosphere. Because the Hubble is so massive, it would not vaporize entirely on reentry, but would shower some part of the Earth's surface with chunks of **metal** and glass. NASA is presently debating whether to (a) retrieve the Hubble intact after it is scheduled to go out of service in 2010, (b) guide it to a chosen crash zone on Earth, or (c) push it right out of Earth orbit with a specially-built rocket.

The Hubble Space Telescope has revolutionized astronomy by bringing a whole new understanding of the Universe to mankind. The following list highlights a few of the Hubble's achievements:

- Imaged comet Shoemaker-Levy 9 crashing into **Jupiter** in 1994.
- Showed that protoplanetary dust disks are common around young stars.
- Proved that Jupiter-size planets are uncommon in globular clusters.
- Shown that quasars reside in galaxies, many of which are colliding with each other.
- Shown that supermassive black holes reside at the centers of many galaxies.
- Permitted more accurate measurement of the Universe's **rate** of expansion than ever before.
- Observed distant supernovae which give evidence that the expansion of the Universe is actually accelerating, prompting a major revision of cosmological thought.
- Imaged large numbers of very distant galaxies distances with its Deep Field study, greatly increasing our estimate of how many galaxies there are in the Universe.

The Hubble will eventually be decommissioned. Work is already under way on its replacement, the James Webb Space Telescope (JWST, named for a former NASA administrator), due for launch in 2010. Unlike the Hubble, which travels around Earth in a moderately low orbit, the JWST will be located some 930,000 mi (1.5 million km) away, to avoid glare from the Earth. The JWST will make observations only at near- and mid-infrared wavelengths, seeking to study the early history of the Universe. Optical and ultraviolet wavelengths will not be observed by the new telescope.

See also Space probe; Spectral classification of stars.

Resources

Periodicals

Lawler, Andrew, "Glimpsing the Post-Hubble Universe." *Science* (February 22, 2002): 1448–1451.

Leary, Warren, "NASA Starts Planning Hubble's Going-Away Party." *New York Times.* September 17, 2002.

Other

National Aeronautics and Space Administration. "Hubble's Parts." August 8, 2002. [cited November 23, 2002]. <http://hubble.nasa.gov/technology/parts.html#optics>.

Kristin Lewotsky

Human artificial chromosomes

An artificial **chromosome** is a **deoxyribonucleic acid (DNA)** containing structure that is assembled from many different components of naturally occurring chromosomes.

Human artificial chromosomes and gene therapy

Chromosomes are located in an organelle called the nucleus that is found in almost every **cell**. Chromosomes contain DNA tightly packaged in order to conserve **space**. Chromosomes are unwound during **gene** expression, which produces **proteins**. Recently human artificial chromosomes (HAC) have come into the forefront of **gene therapy**. Gene therapy—the transfer of corrected gene to cells with an endogenously defective gene—has had many setbacks toward becoming a medically routine therapeutic approach. Gene transfer often has a low efficiency targets, limited specific cell type targets. In addition, once transferred, gene expression is poorly regulated and this leads to a reduced therapeutic value.

Many currently used vectors can only package small genes, while HACs lack size restrictions. In fact, these constructs might be useful in delivering large genes, such as the genes that cause muscular dystrophy or **cystic fibrosis**. It will also be applicable to delivery of multiple genes such as anticancer genes. Using HACs as vectors for transferring genes might also lead to reducing life threatening immune-related complications observed with other vectors, and improve regulation of gene expression due to its very similar construction, modeled after normal human chromosomes. Preliminary studies also demonstrate HACs to be more stable.

In addition to being structurally similar to normal chromosomes, HAC can be designed to carry less non-gene related DNA than other vectors for gene therapy. Because the type of genetic material used to construct human artificial chromosomes can be regulated similarly compared to how normal human chromosomes are regulated, geneticists argue that HACs will take on an increasingly important role in gene therapy. The ability to regulate gene expression from artificial chromosomes allows scientists and clinicians the ability to introduce genes that ultimately produce specific therapeutic proteins needed to treat specific genetic diseases in a more controlled way.

The key to the HAC, the centromere

Human artificial chromosomes must contain the same essential functional and stabilizing regions as do normal chromosomes. They must, for example, contain telomeric regions at the end of each the chromosome strand. Telomeres consist of DNA and associated proteins that function to protect chromosomes from breaks and other forms of damage. Another important element that must be present on every HAC is a functioning centromere that allows for the proper separation and assortment of chromosomes during **cell division**. As telomeres are located at the ends of chromosomes, centromeres are usually in the middle. Both regions contain repetitive DNA, or sequences that are repeated throughout the **genome**. These sequences are important regulatory regions and play a role in maintaining the integrity of the chromosome.

In contrast to normal chromosomes, HACs contain far less extraneous non-functional genetic material. Accordingly, the use of HACs gives researchers the ability to limit the genetic complexity by reducing the number of genes present on a chromosome. In addition to being able to control which genes are present, the construction of HACs offers researchers an opportunity to study less complex systems of gene interaction that are similar to natural chromosomes.

HACs are capable of self-assembly. When the required and proper genetic elements are introduced into cells, (e.g., telomeres, centromeric DNA, gene carrying DNA, etc.), smaller versions of chromosomes (microchromosomes) can be created. These resulting microchromosomes are what makes up a HACs.

In gene therapy, HACs have the ability to function as additional accessory chromosomes to natural chromosomes. The ability to construct artificial chromosomes that can remain stable through the cellular division offers an alternative to the use of viruses (viral vectors) to introduce therapeutic genes into natural chromosomes. The key to this design in terms of stability relied on the application of centromeres, which were shown to be critical for dividing the chromosome when the cell replicates its DNA and divides into two new cells. Additionally, the construction of a HAC carrying desired therapeutic genes eliminates potential damage to natural chromosomes often associated with the introduction of genes by viruses.

Neocentromeres

The importance of centromeres was discovered by Australian scientist Andy Choo from the Murdoch Childrens Research Institute in Melbourne, **Australia** while he was studying the genome of a developmentally delayed 5-year-old child. He observed that the tip of chromosome 7 had been broken off in all the cells he studied. Normally, fragmented DNA broken off from chromosomes gets lost or extruded from the cell. Interestingly, he also noticed that this broken off fragment remained in the nucleus and did not get extruded because it had somehow developed a new centromere called a neocentromere. By using this neocentromere, Choo and his colleagues were able to produce an HAC approximately one-hundredth the size of a normal human chromosome.

Earlier attempts to create HACs failed because such artificial chromosomes lacked fully functional centromeres. Without a functional centromere, these early

KEY TERMS

Cells—The smallest living units of the body which together form tissues.

Chromosomes—The structures that carry genetic information in the form of DNA. Chromosomes are located within every cell and are responsible for directing the development and functioning of all the cells in the body.

Deoxyribonucleic acid (DNA)—The genetic material in a cell

Enzyme—Biological molecule, usually a protein, which promotes a biochemical reaction but is not consumed by the reaction.

Gene—A discrete unit of inheritance, represented by a portion of DNA located on a chromosome. The gene is a code for the production of a specific kind of protein or RNA molecule, and therefore for a specific inherited characteristic.

Genome—The complete set of genes an organism carries.

HACs would not properly divide during cell division and thus, would not remain intact or stable for more than a few cell cell divisions. In 1997, research scientists at Case Western Reserve University and Athersys, Inc., (a private company that conducts research into the development of therapeutic and diagnostic products, including research into the stability of chromosome structure and function) announced the creation of the first stable HAC. Functional HAC centromeres were constructed from alpha **satellite** DNA, a type of highly repetitive DNA found in and surrounding normal chromosomal centromeres. Alpha satellite DNA is difficult to sequence and might not be practical clinically due to regulatory requirements mandating knowledge of the exact sequence of any vector used for gene therapy. Choo's HAC, however, does not have alpha satellite DNA and is therefore more easily sequenced.

Another report of a DNA-based HAC that has been developed came from a joint venture between Chromos Molecular Systems Inc. of Canada and the Biological Research. These HACs might potentially provide scientists with the alternative, low risk vector for gene therapy that researchers pursue. This vector has been shown to be stable, and expresses DNA in a reproducible manner. This method allows geneticists to insert genes into human cells without the risk of disrupting other genes because it is a distinct chromosome itself and does not integrate directly with the human genome.

Resources

Other

"Human Minichromosomes." SCIENCE NOW. May 4, 2001. [cited February 15, 2003] <http://bric.postech.ac.kr/science/97now/01_5now/010504c.html.>.

"Chromosome Research." MURDOCH CHILDREN'S RESEARCH INSTITUTE. August 20, 2002 [cited February 15, 2003] <http://murdoch.rch.unimelb.edu.au/pages/lab/chromosome_research/overview.html.>.

"Scientific Issues." GENETICS AND PUBLIC POLICY CENTER. February 15, 2003 [cited February 15, 2003] <http://www.dnapolicy.org/genetics/transfer.jhtml.>.

Bryan Cobb

Human chorionic gonadotropin

Human chorionic gonadotropin (HCG) is a glycoprotein hormone produced by the extraembryonic **tissue** of the early human embryo. After **fertilization**, the human zygote undergoes cleavage followed by the formation of a blastocyst. The blastocyst is a hollow **sphere** constructed of an inner **cell mass**, which becomes the embryo proper, and a trophoblast, which is embryonic tissue that will contribute to the formation of the placenta. The portion of the trophoblast that is invasive into the maternal uterus is known as the syncytiotrophoblast. The syncytiotrophoblast is the tissue of origin of HCG. The hormone is produced early in pregnancy and increases in **rate** of production until about the tenth week of pregnancy. Thereafter it decreases. The function of HCG is to stimulate the production of progesterone by the corpus luteum. This assures a continual supply of ovarian progesterone until the placenta develops a supply of progesterone around seven weeks of gestation. Progesterone prepares the uterine lining, the endometrium, for implantation and maintainence of the embryo.

The presence of HCG in the urine of a woman is indicative of pregnancy. Actually, the test reveals the presence of trophoblast cells and does not in any way indicate the health of the fetus. Early on, there were **mice** (Aschheim- Zondek) and rabbit (Friedman) tests for the presence of HCG in urine. However, these were expensive. Later, the leopard frog, *Rana pipiens*, was shown to be much less expensive as a biological test **organism**. A male leopard frog will release living sperm in an hour after receiving an injection of morning urine containing HCG. Somewhat similarly, female African clawed **toads**, *Xenopus laevis*, will release eggs after receiving an injection of HCG-containing urine. These tests have now been replaced with even more sensitive clinical tests, one of which will reveal pregnancy prior to the first missed period.

Cryptorchidism is a condition where the testes do not descent into the scrotum of a newborn baby. This is a serious condition because abdominal testes are vulnerable to testicular **cancer** at a much higher incidence than normal testes. Further, abdominal testes are generally sterile. Some infants respond to HCG treatment of this condition. HCG enhances maturation of the external genitals and often causes the undescended testes to move into the scrotum.

Human cloning

An oocyte is an unfertilized egg. Oocytes and spermatozoa are called gametes, and represent different cells that fuse their genes to form a new **cell**, the fertilized egg. The fertilized cell is called a zygote, and it rapidly divides into several totipotent cells (cells capable of developing into any cell type) called blastomers. Totipotent cells can be considered the opposite of differentiated cells (cells that are biochemically and morphologically specialized to perform a specific function), and it is worth noting how differentiated cells (gametes) can produce totipotent cells. As the fertilized egg continues to divide, totipotent cells become more differentiated and specialize into nerve cells, **blood** cells, muscular cells, and the many other cells that are required in order to produce a complete new **individual**. In the laboratory, this biological process can be modified. If an unfertilised egg is enucleated (the nucleus is removed) and fused with a somatic cell (any cell other than germ cells that produce gametes) from an adult individual, the resulting cell will have inside the nucleus only the genes from the adult individual that donated the somatic cell with his relative nucleus. Thus, a new "twin" individual is theoretically generated. Once an embryo is generated, it can be implanted in the uterus of a surrogate mother. This method defines cloning, i.e., the creation of a new being by nuclear transfer from a somatic (differentiated) cell. The first successfully cloned mammal was the **sheep** Dolly. Dolly was created in this manner using a mammary cell. Dolly, however, was not as identical as a naturally occurring twin because some of the mitochondrial DNA from the oocyte was present in the resulting zygote. The mitochondria provide **energy** needed by the cell and may play other roles as well, possibly even storing information in neurons and thus, playing a role in **memory**. Cloning is a process with a low **rate** of success; hundreds of experiments are needed to clone a single **animal**. Furthermore, in cloned animals, a higher rate of malformations and genetic **disease**, as well as signs of early aging have been observed.

Benefits of animal cloning

Pharmaceutical proteins and nutraceuticals

The possibility of deriving live animals from cultured cells provides an efficient way of producing transgenic farm animals. Furthermore, in normal transgenic breeding, successive generations often loose the incorporated **gene**. Once a transgenic animal is made, cloning makes sure that its genetic variation remains through successive generations. In this way, human **proteins** can be produced avoiding purifications from blood, an expensive process associated with risk of **contamination** of viruses such as **AIDS** and **hepatitis** C. Target proteins can be purified from milk of transgenic animals as well as sheep, **goats**, and cattle with relatively low costs. For example alpha-1-antitrypsin and factor IX can be produced and used to treat **cystic fibrosis** and haemophilia, respectively. Again, human serum albumin, which is in high demand for treatment of burns and other trauma, can be produced in transgenic cows by substitution with the human albumin gene for its bovine equivalent. By altering the nutritional content of cows' milk it is possible to insert genes for human proteins in order make high-**nutrition** milk for premature infants, for example, or to create milk without the specific proteins responsible for allergic immune responses or lactose intolerance.

Xenotransplantation organ source animals

Xenotransplantation is the use of animal organs for human **organ** transplantation. Recent advances in understanding of organ rejection and in animal genetic modification and cloning have made it possible to consider animals as a viable source of organs for transplant into humans. This need stems form the worldwide shortage of donated human organs for transplant. It is estimated that in the United States, about 1,200 patients die each year on **heart** and lung transplant waiting lists. Research into xenotransplantation has concentrated on the use of **pigs**. The prospect of xenotransplantation presents a whole new set of risks for consideration as well as the so-called xenozoonoses. This neologism refers to animal diseases that may be transmitted to the recipient of a xenotransplant. Some zoonotic **pathogens** are known to scientists and screening protocols to detect them have been developed, but it is likely that others exist that have not been identified. The use of pigs as a source for donor organs seems to reduce the risk of unusual infections, as pigs and humans, for the most part, share the same environment.

Animal models of human disease

Cloning can produce genetically identical laboratory animals that can be used as models for the study of human disease. The most commonly used laboratory animal, the mouse, reproduces rapidly and its **genetics** have been well studied for the discovery of new treatments for disease. Several other **mammals** have served as scientific models. **Cats**, for example, aided research on human AIDS using the feline AIDS (FIV) as a model. Rabbits have proved valuable for studying human cardiovascular diseases, and **primates** have been models for studying human diseases such as viral hepatitis.

Possible role of cloning in stem cell therapy

Stem cell therapy is a revolutionary new way to treat disease and injury, by transplant of new cells able to repair damaged tissues or organs. The creation of Dolly demonstrated that the normal developmental process of cellular differentiation could be reversible, since differentiated cells can be converted into all of the other cell types that make up a whole animal. This suggests a radical new approach to the problem of **tissue** incompatibility. Perhaps in the near future, when cells would be needed for transplants, it could be possible to obtain them by collecting skin fibroblasts or other cells and allowing them to proliferate before being converted into the specific cell type needed for the disease being treated. When these cells were returned to the patient, they would not be rejected because they have the patient's identical immune profile. At present, the only way to achieve such a transformation would be to collect a human egg (from which the nucleus had been removed) and incubate the resulting human embryo for six to seven days before recovery of pluripotent **stem cells**. Incubation of these cells with specific growth factors would then be used to obtain the desired cell type.

Conditions such as **Alzheimer disease**, **Parkinson disease**, diabetes, heart failure, degenerative joint disease, and other problems may be made curable from pluripotent stem cell technology. Several ethical problems have been raised regarding the use of human embryos for such a scope. In fact, before the stem cells are extracted from embryos, the embryos could potentially be implanted into the uterus and develop into a fetus. For these reasons, it has been proposed to use the frozen advanced embryos already in existence from *in vitro* **fertilization** protocols. However, even if all those embryos would be able to give proper stem cells, their number would never be sufficient to cover all the potential needs for stem cells.

Aging and reproduction

The so-called somatic mutations that occur after several cell divisions in normal individuals are thought to contribute to the ageing process, as well as to the increased incidence of **cancer** as the individual ages. Some leading proponents of human cloning suggest that the technology may someday be possible to reverse the aging process, and for many more infertile couples to have children than ever before. Infertile couples have a higher baseline risk of a sick offspring when conceived using the *in vitro* fertilization processes; a hypothetical cloning could make the risk even higher.

The topic of human cloning is a **matter** of continuing discussion. Given that a healthy cloned human could be produced someday, a clone will never be exactly like the original, as many factors including epigenetic controls, the environment, and the extranucleus (or mithocondrial) DNA are not identical in the clone. In December 2002, a private company, Clonaid, announced the **birth** of the first cloned human, a 7lb (3.2-kg) girl nicknamed Eve. The announcement was met with scepticism in the scientific community, as Clonaid is funded by the Raelians, a religious sect whose tenants hold that humans were initially cloned from extra-terrestrial visitors to **Earth**. Despite Clonaid claims of forthcoming scientific evidence explaining the successful human clone, no evidence was presented to the international scientific community, and the claim was dismissed as a hoax. The Clonaid incident sparked legislative efforts in several countries to ban human cloning, especially for reproductive purposes. As of 2003, despite backing by United States President George W. Bush, the Unites States has no federal laws banning human cloning as **asexual reproduction**. Federal funds may not be used for human cloning, however, and further legislation is pending. Several states have clarified their laws, or banned human cloning outright.

See also Clone and cloning; Embryo and embryonic development; Embryology; Gene therapy; Genetic disorders; Genetic engineering; Genetics.

Resources

Books

Kass, Leon R. *Human Cloning and Human Dignity: The Report of the President's Council on Bioethics.* Public Affairs, 2002.

Peat, David F. *Scientific and Medical Aspects of Human Reproductive Cloning.* National Academy Press, 2003.

Periodicals

Bosch, Xavier. "United Nations Debates Human Cloning Ban." *Lancet* 360, no. 9345 (November 16, 2000):1574.

Malakoff, David. "Human Cloning: New Players, Same Debate in Congress." *Science* 299, no. 5608 (February 7, 2003):799.

Veeck, L. "National Academy of Sciences Report Reaffirms Human Cloning for Stem-cell Therapy but Condemns Human Cloning for Reproductive Purposes." *Reprod Biomed Online.* 4, no. 2 (March-April 2002):198.

Other

PBS Online and WGBH/FRONTLINE. "Making Babies. Human Cloning: How Close Is It?" 1999 [cited March 11,

2003].<http://www.pbs.org/wgbh/pages/frontline/shows/fertility/etc/cloning.html> (March 11, 2003).

Antonio Farina

Human ecology

Human ecology is the study of the **reciprocal** interactions of humans with their environment. Key aspects of human ecology are demographics, resource use, environmental influences on health and society, and environmental impacts of human activities. All of these subjects are intimately linked, because increasing populations of humans require more resources, the exploitation and use of which cause increasing environmental damages. However, certain patterns of use and abuse of resources and environmental quality are clearly more destructive than others. An important goal of human ecology is to discover the causes of pathological interactions between humans and the environment that sustains them and all other species. Once this destructive **syndrome** is clearly understood, it will be possible to design better pathways towards the development of sustainable human societies.

Human demographics is the study of changes in human populations, and the factors that cause those changes to occur. The central focus of this important topic is the remarkable increase that has occurred in the size of the human population during the past several millennia, but especially during the past several centuries. The population of humans exceeded six billion in 1999; this is probably more individuals than any other **species** of large **animal** has ever been able to maintain. The growth of the human population has been made possible by technological and cultural innovations that have allowed a more efficient exploitation of environmental resources, along with advances in medicine and sanitation that have reduced death rates associated with **epidemic** diseases.

Humans and their societies have an absolute dependence on environmental resources to provide **energy**, food, and materials. Some resources, such as metals and **fossil fuels**, can only be mined because they are present in a finite supply that is diminished as they are used. Other resources, such as **forests**, hunted animals, agricultural **soil** capability, and clean air and **water**, are potentially renewable, and if sensibly used they could support sustainable economies and societies over the longer term. However, humans commonly overexploit potentially renewable natural resources, that is, they are mined as if they were nonrenewable resources. This common syndrome of resource degradation is one of the most important aspects of the environmental crisis, and it is a formidable obstacle to the achievement of a sustainable human economy.

An important activity of human ecologists is to discover the reasons for this habitual overexploitation, so that potentially renewable resources could be utilized in more sensible ways. Mostly, it appears that resource degradation is caused by the desires of individuals, corporations, and societies to gain shorter-term profits and wealth, even if this occurs at the expense of longer-term, sometimes irreversible damage caused to resources and environmental quality. The problem is complicated by the nature of ownership of certain resources, in particular common-property resources from which self-interested individuals or companies can reap short-term profit through overexploitation, while the costs of the resulting damage to the resource and environmental quality are borne by society at large.

Human ecologists are also concerned with other environmental effects of human activities, such as **pollution**, **extinction** of species, losses of natural ecosystems, and other important problems. These damages are critical because they indirectly affect the availability of resources to humans, while degrading the quality of life in various other ways. Just as important is the damage caused to other species and ecological values, which have intrinsic (or existence) value regardless of any perceived value that they may have to humans.

Human ecologists are attempting to understand the various linkages between humans and the ecosystems that sustain them. This is being done in order to understand the causes of damage caused by human activities to the environment and resources, and to find ways to mitigate or prevent this degradation before the scale and intensity of the environmental crisis becomes truly catastrophic.

See also Biodiversity; Population, human.

Resources

Books

Bates, D.G. *Human Adaptive Strategies: Ecology, Culture, and Politics.* Allyn and Bacon, 1997.

Bill Freedman

Human evolution

The history of how the human **species**, *Homo sapiens sapiens*, evolved is reconstructed by evidence gathered by paleontologists, anthropologists, archeologists, anatomists, biochemists, behavioral scientists, and many other professions. The evidence comes from the record left by fossils and by extrapolation from modern **primates** and human hunter-gatherer tribes. Fossils are evidence of past life. In practice, human fossils are mostly bones and teeth, which are the parts of the human body least likely to decompose. Most types of fossils are rare; it is extremely unusual for bodies to be subjected to all of the favorable conditions necessary for fossilization. Scientists date fossils by one of several techniques; carbon-14 dating, which measures the **ratio** of radioactive **carbon** to stable carbon, and potassium-argon dating, which measures the ratio of a radioactive form of the element potassium to its breakdown product, argon. Before these methods were available, index fossils of a particular geologic period were used to give an approximate date to other fossils. More recent dating methods include thermoluminescence, **electron** spin **resonance**, and fission track dating.

Paleontologists try to recreate the entire **animal** from sparse bone fragments by comparing the fossil fragments with similar animals, both now living or fossil, of which more information is known. Since complete fossils are rarely found, anatomists recreate the entire skeleton by comparing it with other individuals from the same species or with closely related species. Muscles are reconstructed over the skeleton based on a knowledge of **anatomy**, and the animal is positioned based upon how a similar living animal would move.

Studies of the DNA of humans and the great **apes** indicate that the closest living relatives of humans are **chimpanzees** and **gorillas**. Humans are not thought to be direct descendants of apes, rather we have descended from a common ancestor. Initial studies comparing chimpanzee and human DNA estimated that the similarity is 98.5%. However, recent studies showed that this similarity is more likely to be lower and is estimated at 95%. The final verdict will be delivered in a few years when the chimpanzee **genome** project undertaken by the Riken Institute, is finished. Despite being closely related and having some things in common (number of bones) there are distinct differences between humans and chimps. These include the human's larger **brain**, ability to speak due to a differently-built larynx, ability to walk upright on two legs instead of swinging or knuckle-walking, and greater manual dexterity, due to the opposable thumb that enables humans to manipulate small tools with precision. The faces of humans are flattened, or reduced compared to the apes. The human skeleton is similar to that of a chimp or a gorilla, but is modified for walking upright on two legs. At some point in our development, humans began to rely more on learned **behavior** (which creates culture) than on genetically fixed or instinctive behavior. This cultural development might be indicated by remains other than bones or teeth, including objects such as stone

tools. The first appearance of those traits in the fossil record indicate that those animals were nearly as human as us, which makes them a possible ancestor.

Determining when a fossil find is an early human

What is it that makes us essentially human? Our name, *Homo sapiens*, means "wise man." Intelligence is the quality most widely seen as making humans unique. Fossil evidence of intelligence is based upon brain size measured in **volume** (cubic centimeters). Human brains are three times larger than any comparable primate of a similar weight. Although they grow after **birth** at a **rate** that is average for a mammal, they continue to grow for much longer than other animals. Our brains also have different proportions than other primates. Particular areas of the human brain have developed in unique ways, especially the parts of the brain responsible for **speech**. The other physical traits that we have uniquely acquired include an upright posture, walking on two feet, and an opposable thumb. Finally, human young are cared for over a longer period of **time** than any other primate.

Many paleontologists argue that evidence of culture, such as making fire and tools, using spoken language, and having self awareness are some of the less tangible but important qualities that differentiated human ancestors from other animals. Other animals besides humans use tools, such as chimpanzees that fashion twigs into devices to poke **termites** from a termite mound. However, humans make tools with anticipation of using them in the future. In addition, we have become advanced in our tool making capabilities that we can extend our powers of observation beyond our senses. Some species of animals communicate using complex sounds or show evidence of aiding another, such as dolphins and whales. Although other primates do not use symbolic language, where the meaning of words is learned, they are capable of leaning our system of symbols; chimpanzees and gorillas have been taught to use and understand American Sign Language. Humans are unique in having developed written languages.

There is no direct evidence of when or how self-awareness first arose in humans, but some indirect evidence is available. It appears that only humans exhibit awareness that someday they will die. This death awareness leads humans to bury their dead, and intentional burial leaves a trace in the fossil record.

The hominid fossil record

The first pre-human fossil to be named was *Australopithecus africanus*, meaning the southern ape of **Africa**. The fossils were found at a site called Taung in South

Africa by Raymond Dart, who recognized it as being intermediate between apes and humans. The fossils are dated at three million years old. Additional fossils of *A. africanus* were discovered at Sterkfontein and at Makapanskat in South Africa. The bones from other animals found along with *A. africanus* were interpreted as meaning that our ancestors were hunters. Other scientists determined that those bones were actually the leftover meals from leopards and hyenas. It is now believed that *A. africanus* was primarily a vegetarian, and probably did little, if any, hunting. Teeth wear patterns indicate that *A. africanus* ate fruit and foliage. No stone tools were found with any of these fossils, so there is no evidence that *Australopithecus* made or used tools, or used fire.

In 1912, William Dawson discovered pieces of a skull and jaw along with stone tools and index fossils at Piltdown in England. The jaw was ape-like, but the skull was humanlike. British anthropologists at the time judged the find to be authentic, perhaps because it appeared to support a cherished belief that humans had first developed a big brain, and then later developed other human characteristics. It was subsequently discovered that Piltdown man was a hoax, composed of a human braincase and the jaw from an orangutan, modified to look old.

Australopithecus afarensis, the southern ape of the Afar region in Ethiopia, was discovered more recently and found to be the oldest known humanlike animal to have walked upright. The most famous of these fossils, nicknamed "Lucy," was found near Hadar, Ethiopia, by a team of anthropologists led by Dr. Donald Johanson. Lucy lived about 3.5 million years ago, and had a skull, knees and a pelvis more similar to ours than to the apes. Her brain size was about 350 cc, which was less than one third of the brain size of modern humans (1,400 cc), yet larger than any ape-like ancestor to have come before. She would have stood at a height of about 3.5 ft (1 m) tall, with long arms, a v-shaped jaw, and a large projecting face.

Fossils of several male and female *Australopithecus* have been found together. There is some uncertainty as to whether these are *A. afarensis* or another closely related species. This group find gives evidence that they were social animals. Two of these early humanlike ancestors also left a trail of footprints at Laetoli in East Africa in what was then volcanic ash that later became fossilized. These were discovered by Mary Leakey, the wife of the pioneer paleontologist, Louis Leakey. The fossil footprints look very similar to modern human prints and add further proof that our ancestors walked upright.

The reasons our ancestors started to walk upright are not known. Possibly, it was a response to environmental changes; as tropical **forests** were beginning to shrink, walking might have been a better way to cross the **grasslands** to get to nearby patches of forest for food. We can get some ideas of possible advantages of upright posture to our ancestors by studying modern apes. When chimpanzees or gorillas become excited, they stand in an upright posture and shake a stick or throw an object. By standing upright, they appear bigger and more impressive in size than they normally are. This would be useful to help protect the group against predators. Also, the ability to stand up and get a wider view of the surroundings gives an animal an advantage in the tall **grasses**. Walking upright frees up the hands to carry objects, such as tools.

Two other species of *Australopithecus* are *A. robustus* and *A. boisei*. *Australopithecus robustus*, from South Africa, was named for its massive jaws and large flat chewing teeth. This species also had a bony ridge along the top of its skull (the sagittal crest) similar to that of an adult male gorilla, which served as a site of attachment for massive jaw muscles. Its skull had the brain capacity of 500 cc. Living about 1.9–1.5 million years ago, the diet of *A. robustus* probably consisted of tough gritty foods, such as **plant** tubers. *Australopithecus robustus* was probably not a direct ancestor of modern humans. The other *Australopithecus* species, *A. boisei*, was discovered by Louis Leakey at Olduvai Gorge in Tanzania, a site that has been famous for hominid fossils for more than 60 years. Sediments and fossils are exposed in the walls of the gorge that represent almost two million years of evolutionary history. *Australopithecus boisei* had huge flat grinding teeth, a very long face, and a large elongated cranium, with a brain capacity of 530 cc and a sagittal crest atop the skull.

The record of animals that were ancestral to *Australopithecus* is poor. An ape-like animal (*Ramapithecus*), lived in Africa some 12 million years ago and is thought to have been the first representative of the line leading to humans. *Ramapithecus* lived on the forest fringe, near **rivers** and lakes, and began to make the transition to life on the more open **savanna**. Very few remains of *Ramapithecus* have been found, only fragments of upper and lower jaws and teeth. Its dental pattern was unique among other fossil finds from that time. The canine teeth were fairly small, indicating that its diet may have included **seeds** and other tough plant material that required being torn apart before eaten. A five-million-year gap in the fossil record between the time of *Ramapithecus* and *Australopithecus* has been recently partially filled by new finds in Africa, although it is not yet clear where exactly on the human **evolution tree** these fossils will be placed. Remains of a hominid from six million years ago were found at Kapsomin by a French and Kenyan anthropological group led by Martin Pickford and Brigitte Senut in 2001. It was named *Orrorin tugenensis*. However, in July 2002, Professor Michael Brunet with an international group of scientists found an even older (seven million-year-old) skull in Chad, called *Sahelanthropus tchadensis*, nicknamed Toumai. The opinions on whether it is the skull of a hominid or an ancient gorilla are divided. Independent of the final verdict, the fact that the skull was nearly intact is very important for further comparative analysis.

Australopithecus is similar enough to humans to be considered an ancestor, but different enough to be assigned to a separate genus. On the other hand, *Homo habilis*, which lived about 1.5–2 million ago, is similar enough to modern humans be included in the genus *Homo*. The braincase of *H. habilis* was appreciably larger than that of *Australopithecus*, with a brain capacity of 750 cc. *Homo habilis* individuals were short and made stone tools from pebbles about 5 in (12.7 cm) long, formed from flakes of rock. The flakes had been broken off the pebble by blows from another stone and were probably used for cutting.

Homo erectus is generally thought to have been our direct ancestor. *Homo erectus* lived about 1.7 million years ago, and had a brain capacity of 950 cc. The first fossil of *Homo erectus* was found in Java; it was nicknamed Java man. Similar fossils found in China were dubbed Peking man. Recently, an entire skeleton of a closely related species, *Homo ergaster*, was found in Kenya. Walking with a fully upright posture, tall and slender, the fossils were found with sophisticated stone tools. They were probably hunters and also scavengers. Bones found along with the fossils have been studied closely; they carry the remains of tooth marks from predators, like leopards, as well as hominid tooth marks. *Homo erectus* probably scavenged from kills made by large predators, breaking bones to eat the rich marrow. The presence of charcoal provides evidence that *H. erectus* used fire, probably to scare off predators.

H. erectus was thought to be the first hominid to leave Africa. This notion was recently (2001) shaken by David Lordkipanidze's group finds in the Georgian village of Dmanisi. The skulls found were much smaller (estimated brain size 600–780 cc) than those of *H. erectus* but had enough similarities to be classified in the same species. The fact that they were dated to 1.7–1.8 million years ago challenges the notion that long-legged, large-brained *H. erectus* left Africa around one million years ago. A hypothesis was made by Vekua and colleagues that Dminisi hominids might have evolved from *H. habilis* outside Africa. Confirmation of such a hypothesis, however, will require further fossil evidence.

Neanderthal man (*Homo sapiens neanderthalensis*) was the first human fossil to be found. It was discovered in 1856 in Germany's Neander Valley and is the source of the caveman stereotype. Neanderthals first appeared 300,000 years ago in what is now **Europe**, lived throughout the **ice ages**, and were thought to disappear about 35,000 years ago, but recently remains from 28,000 years ago were found in Croatia. Neanderthals had a large brain capacity about (1,500 cc), a strong upper body, a bulbous nose, and a prominent brow ridge. There is disagreement as to whether or not Neanderthals were the direct ancestors of modern humans. The controversy has been fueled even further by disagreement between genetic and morphological evidence. In 1997, Krings and colleagues based on the analysis of mitochondrial DNA from Neanderthal remains concluded that modern humans and Neanderthals were not related. However, just a year later in 1998 professor Erik Trinkaus discovered a skeleton of a four-year-old boy from 25,000 years ago that showed a mixture of Neanderthal and modern features suggesting that the two species interbred.

Recent excavations in Israel, Portugal, and Croatia show clearly that Neanderthals were contemporary with modern *Homo sapiens sapiens*. The two hominids apparently survived independently of each other for tens of thousands of years. Some anthropologists see this as evidence that Neanderthals were not our direct ancestors; other anthropologists speculate the two types of humans may have interbred and Neanderthal became genetically absorbed by more modern humans. We do not know why Neanderthals died out, nor what the nature of their interaction with modern *Homo sapiens sapiens* might have been.

Neanderthal man made a number of crafted flint tools with many different uses. Judging from the hearths found at many sites, Neanderthals had mastered the art of making fire. Fossil bones show signs of old injuries that had healed, indicating the victim had been cared for. Some Neanderthal caves contain burial sites, indicating that Neanderthals were probably self-aware. The Shanidar **cave** in Iraq held the remains of a Neanderthal buried 60,000 years, with bunches of flowers. Several of the flowers discovered were species used today as herbal medicines. It is therefore possible that Neanderthals had an elaborate culture, were aware of the medicinal properties of plants, and ritually buried their dead. One anthropologist in Israel found what he believed to be evidence that Neanderthals had the capacity for speech, a fossil bone from the throat (the hyoid), which anchors the muscles connected to the larynx and tongue, and which permit speech in modern humans.

Appearance of modern-looking humans

Although all of the ancestors described thus far first evolved in Africa, there is uncertainty as to where modern humans, *Homo sapiens sapiens*, first appeared. There are two theories to explain this process. The first is the "multi-regional" model, which proposes that *Homo sapiens sapiens* evolved in Europe, **Asia**, and **Australia** from *Homo erectus* after the latter left Africa about one million years ago. The second model, called *Out of Africa*, suggests that modern humans evolved only once, in Africa, leaving there within the last 200,000 years in a rapid global expansion. They replaced other populations of older human forms in Europe and Asia, including the Neanderthals. Variations and combinations of these two theories have also been proposed.

The oldest fossils of modern human beings, *Homo sapiens sapiens*, are 125,000-100,000 years old, appearing at the time of the first of the great ice ages. *Homo sapiens sapiens* are identified by a large brain (1,400 cc), a small face in proportion to the size of the skull, a small chin, and small teeth. In addition, they were tall and relatively slender in build.

The first fossil of modern *Homo sapiens sapiens* was found at Cro Magnon in France, which gave that name to all early *Homo sapiens sapiens*. Cro Magnon remains have been found along with the skeletons of **woolly mammoth**, **bison**, reindeer, and with tools made from bone, antler, ivory, stone and **wood**, indicating that Cro Magnon hunted game of all sizes. Cro Magnons also cooked their food in skin-lined pots heated with stones. Pieces of amber from the Baltic found in southern Europe together with Cro Magnon fossils indicate these humans traded material over vast distances. Cro Magnon humans buried their dead with body ornaments such as necklaces, beaded clothing, and bracelets.

Cro Magnon humans lived at the mouth of caves under shelters made of skins, or in huts made of sticks, saplings, stones, animal skins or even bones. A mammoth-bone hut 15,000 years old has been excavated at a site in the Ukraine. Anthropologists interpret some of the fossil findings of early *Homo sapiens sapiens*, by making comparisons with present-day hunter-gatherer tribes such as the Kalahari or Kung bushmen. These nomadic people live in relatively small bands of about 25 people. A larger group of about 20 bands makes up a community of people who all speak the same dialect and occasionally gather in large groups. The groups disperse into smaller bands during the wet season, and establish clusters around permanent **water** holes in the dry season. The men hunt in cooperative bands when the game is plentiful. The women gather plant material; about two thirds of their diet is made up of plant food. Since only a small portion of time is spent hunting or gathering, there is plenty of time for visiting, entertaining, and sewing. The same might be true for the hunting people living in Europe before 12,000 years ago.

Ice age humans were artists, producing hauntingly beautiful cave art. Carefully rendered pictures of animals, human and mythical representations, and geometric shapes and symbols were created using charcoal and other pigments. The remains of stone lamps found deep within these caves suggested that the caves were visited often. Carvings of stone, ivory and bone have also been discovered in these caves, including female figures. We cannot know what the significance of this art was to them, other than that it was a reflection of how early humans perceived the world around them.

The end of the ice ages brought changes in climate and ecosystems. In Europe, the vast grasslands were replaced by forests, and animal populations shifted from reindeer and bison to red **deer** and boar. The focus of cultural innovation shifted from Europe to the Middle East, where settled cultivation began.

There remain many unanswered questions. For example, *Homo sapiens sapiens* is the only species of hominid now existing. What happened to the preceding older species after the newer human form appeared and replaced it? How will *Homo sapiens sapiens* evolve in the future? Can humans consciously help to shape future evolution? Humans are the only species with the potential to consciously direct future evolution.

See also Dating techniques; Fossil and fossilization; Genetics.

Resources

Books

Dobzhansky, T., and E. Boesiger. *Human Culture: A Moment in Evolution.* New York: Columbia University Press, 1983.

Johanson, Donald L., and Maitland B. Eddy. *Ancestors: The Search for Our Human Origins.* New York, Random House, 1994.

Leakey, Richard, and Roger Lewin. *Origins: What New Discoveries Reveal About the Emergence of Our Species and its Possible Future.* New York: Viking Penguin, 1991.

Tattersall, Ian. *The Human Odyssey; Four Million Years of Human Evolution.* Englewood Cliffs, NJ: Prentice Hall, 1993.

Willis, Delta. *The Hominid Gang: Behind the Search for Human Origins.* New York: Viking Penguin, 1991.

Periodicals

Brunet, M, et al. "A New Hominid from the Upper Miocene of Chad, Central Africa." *Nature* (July 2002): 145–151.

Duarte, Mauricio, C., J. P.B. Pettitt, P. Souto, E. Trinkaus, H. Plicht van der, J. Zilhao, "The Early Upper Paleolithic Human Skeleton from Abrigo do Lugar Velho (Portugal) and Modern Human Emergence in Iberia." *Proc.Natl.Acad.Sci. USA* (June 1999): 7604–7609.

Krings, M., A. Stone, W. Schmitz, H. Krainitzki, M. Stoneking, S. Paabo. "Neandertal DNA Sequences and the Origins of Modern Humans" *Cell* (July 1997):19–30.

KEY TERMS

. . . .

Hominid—A primate in the family Homidae, which includes modern humans.

Primate—An animal of the order Primata, which includes lemurs, monkeys, apes, and humans.

Senut, Brigitte, Martin Pickford, Dominique Gommery, Pierre Mein, Kiptalam Cheboi, Yves Coppens. "First Hominid from the Miocene (Lukeino Formation,Kenya)." *C.R.Acad.Sci.Paris, Earth Planet Sci.* (January 2001) 332: 137–144.

Vekua, A., D. Lordkipanidze, G.P. Rightmire, J. Agusti, R. Ferring, G. Maisuradze, et al. "A New Skull of Early Homo from Dmanisi, Georgia." *Science* (July 2002): 85–89.

Wood, Bernard, and Mark Collard. "The Human Genus." *Science* (April 1999): 65–71.

Marion Dresner

Human Genome Project

The United States Human Genome Project (HGP) is an initiative formally launched in 1990 by the National Institutes of Health (NIH) and the U.S. Department of Energy (DOE) to better understand all aspects related to human genetic material, or **deoxyribonucleic acid (DNA)**. DNA represents a genetic alphabet and the specific sequences that are part of DNA called genes code for various **proteins** by virtue of the DNA sequence that makes up an organism's **genome**. The DNA alphabet consists of four letters (A for adenine, T for thymine, C for cytocine, and G for guanine) called nucleotides. This DNA sequence is found in the nucleus of almost every **cell** in the body. The initiative of the HGP has been to completely sequence the human genome, create databases to categorize this information, and use it for medical, research, and educational purposes.

Around the time that the HGP was formally introduced, it was an issue of debate whether it would be more important to know the complete sequence of the genome, or whether known sequences should be annotated (functionally characterized) before further sequences were determined. The scientific approach to identifying and defining the function of genes and to determine how genes interact is a field of **genetics** called functional genomics. Structural genomics is a field of genetics focused on determining the location of a **gene** by an approach called genetic mapping, or the localization of

genes with respect to each other. In the end, functional genomics became secondary to sequencing the human genome, but functional genomics is now the focus of what is called the post-genomic era. The issue was resolved in this manner mainly because functional genomic-based studies are time consuming and require a more challenging experimental design compared to direct DNA sequencing.

The goals of the Human Genome Project

The HGP outlined several targeted goals to better assist scientists in understanding the human genome. One goal was to identify the estimated 30,000 genes. Because it is estimated that there are roughly 3.9 billion nucleotide bases that makeup the human genome, identifying ways to store this information on publicly accessible databases was an important HGP goal and a challenge to computational biologists. Another goal was to improve analytical tools related to data acquisition so that sequences of the human genome could be compared to sequences from a database that includes gene sequence information from many different organisms. With only a small percentage of the human genome comprised of genes, another goal was to determine the entire DNA sequence of the human genome, including sequences that are interspersed between genes. Finally, the last objective was to address the inevitable ethical, legal, and social implications (ELSI) of having access to an individual's genetic information. This component of the HGP was strongly urged due to the nature of the information that would become available and the potential for negative impacts related to using this information inappropriately.

DNA sequencing methodology

DNA is packaged into structures called chromosomes that unwind when genes are given the appropriate cues to produce protein. Chromosomes are important structures for organizing the long stretches of DNA and provide a platform for which this material can be replicated and separated so that as the cell divides, both of the two new cells will have the appropriate amount of genetic material. Chromosomes can range in size between 50 to 250 million nucleotide bases. In order to sequence these bases, they must first be broken into fragments. Each fragment is used to produce a collection of smaller sized fragments that differ in length by only a single base and are amplified. The amplified set is separated by a gel matrix, and an electrical field is created that separates the DNA fragments in the matrix based on size and charge. These fragments are used as a template for the DNA sequencing reaction. Since DNA is double stranded where A binds to T and C binds to G (and vice versa), a single-stranded template can be used for adding fluorescent-labeled nucleotides of complementary sequence. Using current technology, up to 700 bases can be sequenced.

The timeline

The first decade

In as early as 1983, scientists at Los Alamos National Laboratory (LANL), a Department of Energy Laboratory, and Lawrence Berkeley National Laboratory (LBNL) were working to begin the production of what are called DNA libraries. DNA libraries allow scientists to categorize different DNA sequences so that they can piece together the continuous sequence for each **chromosome**. Only two years later, the feasibility of the Human Genome Initiative was carefully being considered. In 1986, the Department of Energy and the Office of Health and Environmental Research announced a $5.3 million pilot project to begin the Human Genome Initiative in order to develop resources and technologies that would improve this effort. In 1987, the Health and Environmental Research Advisory Committee recommended a 15-year goal to map and sequence the entire human genome, the first undertaking ever to be made.

In 1988, the Human Genome Organization was founded in order to provide international collaborative opportunities for scientists. In 1989, the ELSI Task Force was created. An official and formal, five-year joint agreement between NIH and DOE was presented to Congress in 1990 along with a 15-year goal to sequence the entire human genome. Already artificial chromosomes were being created that would give scientists the ability to insert large DNA sequences into these constructs. In particular, bacterial artificial chromosomes (BACs) were being produced that allowed larger fragments to be inserted, accelerating sequencing efforts. Inserts of human DNA into BACs represent a type of DNA library. In 1991, a repository called the Genome Database was created, marking the first major computational effort to begin teasing out the complex genetic material that separates humans from other organisms. In 1992, only two years after the HGP formally began, the first crude map of the human genome was published using sequence data acquired by linking various genes together based on known locations (or markers) along a chromosome. This gave the research community a glimpse into the human genome map.

The next ten years: public and private contributions

In 1993, an international consortium was established to sort out sequences derived from expressed genes and efforts to map theses sequences ensued. This consortium was called the Integrated Molecular Analysis of Gene Expression (IMAGE) Consortium and it paved the way for structural and functional genomics. Novel sequencing methodology was being developed almost as rapidly as the DNA sequences were elucidated. A new artificial chromosome vector called YAC (**yeast** artificial chromosome) was introduced providing a construct with an even larger DNA insert capacity.

In 1994, the HGP announced the completion of the five-year goal of producing the genetic map of the human genome one year earlier than proposed. Each chromosome had an expanding DNA library resource. In the same year, the first legislation to be passed initiated by the U.S. HGP and called the Genetic Privacy Act was designed to control how DNA is collected, analyzed, stored, and used.

The physical maps of chromosomes 16 and 19 were announced in 1995, followed by the publication of moderate-resolution maps of chromosomes 3, 11, 12 and 22. During this time, the HGP was not the only **species** that was being sequenced. Already, the genome from the **bacteria** that causes the flu (*Haemophilus influenzae*) was completely sequenced, followed by the yeast genome (*Saccharomyces cerevisiae*) a year later. Concerns over discrimination based on genetic information elicited a amendment to the Health Care Portability and Accountability Act that included a clause that prohibits healthcare insurance companies to use of genetic information in certain cases to determine eligibility. This was an important legislative initiative, helping to mitigate some of the immediate concerns related to genetic discrimination of the healthcare industry.

In January, 1997 the NIH declared that the National Human Genome Research Institute (NHGRI) would be a recognized collaborative institute. Following this decree, physical maps of chromosomes X and 7 were announced. GeneMap of 1998 was released allowing scientists the ability to use the mapped location of approximately 30,000 markers for genetic studies. It was also in 1998 that American geneticist Craig Ventor formed Celera Genomics, a company that would significantly contribute to the sequencing effort using many resources provided by the HGP. Celera Genomics, equipped with high-speed state of the art sequencing capabilities, became a leader in the race to sequence the human genome. Only nine years after the HGP was formally initiated, chromosome 22 was considered to be completely sequenced, meaning that although the 56 million bases that are estimated to makeup the entire sequence of chromosome 22, only 33.5 million bases were actually sequenced by the HGP. The remaining sequences, roughly 22.5 million bases, represent regions at the ends of chromosomes (called telomeres) and the center of chromosomes (centromeres) are comprised of repeated sequences that prevent them from being cloned into BACs or any other construct. There are few genes, if any, in most of these sequences. The sequenced portion of chromosome 22 represents 97% of regions that are rich in genes.

Using the sequencing data, a public database created by major pharmaceutical companies called the Single Nucleotide Polymorphism (SNP) Consortium was introduced in order to provide information about inherited variations in the human genome that might provide insight into health and **disease**. For example, inherited variations in genes that metabolize carcinogens might have inherited variations in some individuals that makes them susceptible to developing **cancer** when they are exposed to environmental contaminants. People without these variations are therefore less likely to develop cancer. Identifying these individuals has important implications for reducing cases of cancer.

The success of the HGP was celebrated in the year 2000 when the draft of the human genome was announced. An executive order issued by President Bill Clinton mandated that federal agencies were prohibited from using genetic information for employment decisions or staff promotions. Also in this year, the second chromosome to be sequenced, chromosome 21, was announced and the draft of 5,16, 19 were also finished followed by chromosome 20 in 2001. The working draft of the HGP was published by the journals nature and science. The *Science* article depicted work performed by Celera Genomics and the nature article represented data derived from the efforts of the public sector. Less than a year later, the Mouse Genome Sequencing Consortium published its own draft sequence of the mouse genome on December 5, 2002 in the journal *Nature*. In January 2003, chromosome 14 became the fourth chromosome to be entirely sequenced. Having the sequence of both mouse and humans helps scientist understand human diseases by developing mouse models and identifying genes that are homologous (the same) and might have similar functions in both organisms.

The draft sequence

When the draft sequence was published in February 2001, scientists sequenced each chromosome four to five times to be certain of the **accuracy** of their nucleotide base calling. It was called a draft because the chromosomal locations are roughly approximated. In order to determine the order of the sequence, DNA is cut into fragments using restriction enzymes. These fragments on the order of approximately 10kb (or 10,000 base pairs) are cloned into vectors (for example, BACs) by cutting open the circular bacterial DNA vector using the same enzymes that cut the DNA, ligating the fragment to the end of the vec-

tor, growing up the BACs in culture, and sequencing the inserts. Overlapping fragments in which the DNA sequences matched were carefully pieced together. The ends of BACs, therefore, were used as markers that were found in roughly every 3,000 to 4,000 bases throughout the entire genome called sequence tag connectors (STCs). In this large-scale sequencing effort, STCs provided a compass for knowing which specific BAC clones had to be sequenced to fill in gaps between STCs. The next step was to produce a higher quality sequence of approximately 95.8% of the human genome sequence that was projected to be completed sometime in 2003. This version involves an error **rate** of only one base per 10,000 bases, requiring additional sequencing and filling in of any gaps with coverage of up to 9 times base calling. Final sequences are publicly available in databases such as GenBank.

The DNA sequence: is it informative?

The DNA sequence represents a reference, not an exact match, for a geneticist to use as a guideline. It does not mean that every **individual** has the same exact sequence. Although there are also highly conserved regions (or regions that are the same between people) throughout the genome representing genes and regulatory regions important for controlling gene expression, there are also regions throughout the genome that are variable from person to person. In fact, although 99% of the genome is conserved, 1% is variable. This means that roughly 3,000,000 bases differ between individuals. These differences are significant because there are nucleotides that are variable called polymorphisms and are found in greater then 1% of the population. These single nucleotide polymorphisms (SNPs) have become very useful in the field of pharmacogenomics, or the area of research that involves understanding how an individual's genes affect the body's response to drugs. For example, it is particularly significant if an SNP occurs in a gene that encodes a protein important for metabolizing a certain pharmaceutical agent. The protein, in this case an **enzyme**, might function at a reduced rate if the SNP causes an alteration in conformation of the protein. A reduction in function might lead to adverse drug reactions for this individual because the enzyme is capable of metabolizing only a small amount of the drug. Therefore, knowing if a person has the SNP is important for determining the appropriate drug to use.

The post-genomic era

The race to completely sequence the human genome has lead to several concerns related to what to do with this information. Although the ELSI task force was created to address these concerns, many issues continue to arise resulting from the dissemination of this informa-

KEY TERMS

. . . .

Clone—A cell or organism derived through asexual reproduction, and which contain the identical genetic information of the parent cell or organism.

Deoxyribonucleic acid (DNA)—The genetic material in a cell.

Eugenics—A social movement in which the population of a society, country, or the world is to be improved by controlling the passing on of hereditary information through selective breeding.

Gene—A discrete unit of inheritance, represented by a portion of DNA located on a chromosome. The gene is a code for the production of a specific kind of protein or RNA molecule, and therefore for a specific inherited characteristic.

Genome—The complete set of genes an organism carries.

Polymorphism—An variation in an individuals genetic material that is inherited in greater than 1% of the population and does not represent a spontaneous mutation.

tion. These issues include implications that impact the development of threatening biological weapons, genetic discrimination, **eugenics**, and **human cloning** to name a few. The magnitude of ramifications related to HGP achievements is just beginning to be realized. The next challenge in the post-genomic era is to annotate, or functionally characterize genes and to build on our understanding of gene-gene interactions, gene expression, and protein-protein interaction, and apply this knowledge to better understand life.

See also Chromosomal abnormalities; Chromosome mapping; Codons; DNA replication; DNA synthesis; Forensic science; Gene chips and microarrays; Gene mutation; Gene splicing; Gene therapy; Genetic engineering; Genetic identification of microorganisms; Genetic testing; Genetically modified foods and organisms; Genomics (comparative); Genotype and phenotype; Meiosis; Mendelian genetics; Shotgun cloning.

Resources

Books

Nussbaum, Robert L., Roderick R. McInnes, and Huntington F. Willard. *Genetics in Medicine.* Philadelphia: Saunders, 2001.

Rimoin, David L. *Emery and Rimoin's Principles and Practice of Medical Genetics.* New York: Churchill Livingstone, 2002.

Periodicals

International Human Genome Mapping Consortium. "A Physical Map of the Human Genome." *Nature* 409 (2001): 934–941.

International Human Genome Sequencing Consortium. "Initial Sequencing and Analysis of the Human Genome." *Nature* 409 (2001): 860–921.

Lennon, G., C. Auffray, M. Polymeropoulos,. M.B. Soares. "The I.M.A.G.E. Consortium: An Integrated Molecular Analysis of Genomes and Their Expression." *Genomics* 33 (1996): 1512.

Other

National Institutes of Health. "The National Human Genome Research Institute: Advancing Human Health Through Genetic Research." NHGRI. February 2003 [cited February 28, 2003]<http:// www.genome.gov>.

Bryan Cobb

Humidity

Humidity is a measure of the quantity of **water** vapor in the air. There are different methods for determining this quantity and those methods are reflected in a variety of humidity indexes and readings.

The humidity reading in general use by most meteorologists is relative humidity. The relative humidity of air describes the saturation of air with water vapor. Given in terms of **percent** humidity (e.g., 50% relative humidity), the measurement allows a comparison of the amount of water vapor in the air with the maximum amount water vapor that—at a given temperature—represents saturation. Saturation exists when the phase state changes of **evaporation** and condensation are in equilibrium.

Approximately 1% of Earth's total water content is suspended in the atmosphere as water vapor, **precipitation**, or **clouds**. Humidity is a measure only of the vapor content.

Because water vapor exerts a **pressure**, the presence of water vapor in the air contributes **vapor pressure** to the overall **atmospheric pressure**. Actual vapor pressures are measured in millibars. One atmosphere of pressure (1 atm) equals 1013.25 mbar.

In contrast to the commonly used value of relative humidity, absolute humidity is a measure of the actual **mass** of water vapor in a defined **volume** of air. Absolute humidity is usually expressed in terms of grams of water per cubic meter.

Specific humidity is a measure of the mass of water vapor in a defined volume of air relative to the total mass of gas in the defined volume.

The amount of water vapor needed to achieve saturation increases with **temperature**. Correspondingly, as temperature decreases, the amount of water vapor needed to reach saturation decreases. As the temperature of a parcel of air is lowered, it will eventual reach saturation without the **addition** or loss of water mass. At saturation (**dew point**), condensation or precipitation forms. This is the fundamental mechanism for cloud formation as air moving aloft is cooled. The level of cloud formation is an indication of the humidity of the ascending air because, given the standard temperature lapse **rate**, a parcel of air with a greater relative humidity will experience condensation (e.g., cloud formation) at a lower altitude than a parcel of air with a lower relative humidity.

The differences in the amount of water vapor in a parcel of air can be dramatic. A parcel of air near saturation may contain 28g of water per cubic meter of air at 30° F (-1° C), but only 8g of water per cubic meter of air at 10°F (-12° C).

An increasingly popular measure of comfort, especially in the hotter summer months, is the **heat index**. The heat index is an integrated measurement of relative humidity and dry air temperature. The measurement is useful because higher humidity levels retard evaporation from the skin (perspiration) and lower the effectiveness of physiological cooling mechanisms.

Absolute humidity may be measured with a sling cyclometer. A hydrometer is used to measure water vapor content. Water vapor content can also be can be expressed as grains/cubic ft. Grains, a unit of weight, equals 1/7000 of a pound.

See also Atmospheric temperature; Hydrologic cycle; Weather forecasting; Weather.

Hummingbirds

Hummingbirds are small, often tiny **birds** of the Americas, named after the noise made by their extremely rapid wingbeats. There are 320 **species** of hummingbirds, which make up the family Trochilidae.

Hummingbirds are spectacularly beautiful birds, because of the vivid iridescence of their feathers. They are such accomplished fliers that they can aggressively drive away much larger, predatory birds. Invariably, people who have had the opportunity to watch hummingbirds regularly develop a loyal admiration for these lovely sprites.

Hummingbirds are widely distributed in the Americas, occurring from Tierra del Fuego in the south, to the subarctic of Alaska and Canada in the north. However, the greatest richness of hummingbird species occurs in the tropics, especially in **forests** and associated, disturbed habitats where flowers may be relatively abundant.

Biology of hummingbirds

Hummingbirds have small, weak legs and feet, which are used only for perching, and not for walking. Hummingbirds only move about by flying, and they are extremely capable aerial acrobats. Although diminutive, hummingbirds can fly quickly over short distances, at up to 31-40 miles per hour (50-65 kph). Hummingbirds can fly forwards, backwards, and briefly, upside down. These birds are also very skilled at hovering, which they typically do when feeding on **nectar** from flowers. Hovering is accomplished using a figure-eight movement of the wings, and a relatively erect posture of the body.

Because of their small size, the tiniest hummingbirds must maintain an extraordinarily rapid wingbeat **rate** of 70 beats per second to stay aloft. However, the larger species can fly with only about 20 beats per second. The flight muscles of hummingbirds typically account for 25-30% of the body weight, compared with an average of 15% for other birds.

Also as a direct result of their small body size, hummingbirds have a high rate of **heat** loss from their bodies. This is because small objects, including small organisms, have a relatively large surface area to **volume ratio**, and lose heat more rapidly from their surface than do larger-bodied animals. This relatively high rate of heat loss, combined with the fact that hummingbirds are very active animals, means that they have a very high rate of **metabolism**. Consequently, hummingbirds must feed frequently, and relatively voraciously, to fuel their high-energy life style.

When **weather** conditions make it difficult for them to forage, for example, during intense rain or cool temperatures, hummingbirds may enter a state of torpor. This involves becoming inactive, and reducing and maintaining a relatively low body **temperature**, as a means of conserving **energy** until environmental conditions improve again.

Most hummingbirds feed on nectar, obtained from flowers. Feeding is usually done while hovering. All hummingbird species have long, slender bills that are specialized for this **mode** of feeding. The bill of the swordbill hummingbird (*Ensifera ensifera*) is straight, and is as long as the bird's body and tail, about 4 in (10 cm). This sort of extremely developed feeding device is adapted to extracting nectar from the base of long, tubular flowers, in particular, certain species of **passion flower** (*Passiflora* spp.) that have corolla tubes about 4.3 in (11 cm) long. A few species of hummingbirds, for example, the white-tipped sicklebird (*Eutoxeres aquila*),

A hummingbird sipping nectar from a flower. *Zefa Germany/Stock Market. Reproduced by permission.*

have downward curving beaks that are useful for probing flowers of other shapes.

Hummingbirds also have a long tongue that can be extended well beyond the tip of their bill. The tongue of hummingbirds has inrolled edges, that can be used to form a tube for sucking nectar.

Some tropical plants occur in a mutualistic relationship with one or several species of hummingbird, that is, a **symbiosis** in which both species receive a benefit. The advantage to the hummingbirds occurs through access to a predictable source of nectar, while the **plant** benefits by **pollination**. While feeding, the hummingbird will typically have its forehead dusted with pollen, some of which is then transferred to the receptive stigmatic surfaces of other flowers of the same plant species as the bird moves around while foraging. Hummingbird-pollinated flowers are usually red in **color**, and they have a tubular floral structure, with nectar-secreting organs at the base.

While nectar is the primary food of hummingbirds, they are also opportunistic predators of small **insects**, spiders, and other **arthropods**. This **animal** food is an important source of protein, a nutrient that is deficient in sugar-rich nectar.

The largest species of hummingbird is the giant hummingbird (*Patagona gigas*) of montane habitats in the Andes, up to 0.7 oz (20 g) in body weight and 8.5 in (22 cm) long, although about one-half of the length is the elongated tail of the bird. The smallest hummingbird is the bee hummingbird (*Mellisuga helenae*) of Cuba, with a body length of only 1 in (2.5 cm) and a weight of 0.07 oz (2 g). This is also the world's smallest species of bird. Most other hummingbirds are also small, typically about 2 in (5 cm) long.

Hummingbirds can be spectacularly colored, especially the males. Most of the coloration is not due to the presence of pigments, but to iridescence. This is a physical effect associated with prism-like microstructures of the feathers of hummingbirds and some other birds. These break **light** into its spectral components, which are selectively segregated into brilliant reds, pinks, blues, purples, and greens through absorption processes, and by the angle of incidence of light. The most vivid colors of hummingbirds are generally developed by the feathers of the head and throat, which are prominently displayed by the male to the female during **courtship** flights. The male birds of some species of hummingbirds also develop crests, and intricately long tail feathers.

Even the smallest hummingbirds can be quite aggressive against much larger birds. Because of their extraordinary mobility, hummingbirds can successfully chase away potential predators, such as small **hawks** and crows. Hummingbirds are also aggressive with other hummingbirds, of the same or different species. This can result in frequent and rowdy fights as territorial claims are stated and defended at good feeding stations.

North American hummingbirds

Compared with the tropics, relatively few species of hummingbirds breed in **North America**. By far the most widespread and common species is the ruby-throated hummingbird (*Archilochus colubris*). This hummingbird occurs over most of eastern and central North America south of the boreal forest, and is the only species in the east. This species can be fairly common around gardens and other disturbed habitats where wildflowers are abundant, especially red-colored flowers, as is the case of most hummingbirds. The ruby-throated hummingbird is a migratory species, which spends the winter from the southern tip of Florida to Central America.

Many other species of hummingbirds occur in southwestern North America. The most common and widespread of these species are the broad-tailed hummingbird (*Selasphorus platycercus*), rufous hummingbird (*S. rufus*), Anna's hummingbird (*Calypte anna*), and black-chinned hummingbird (*Archilochus alexandri*). Several other species also occur in more southwestern parts of the United States, some of them barely penetrating north from Mexico.

Conservation of hummingbirds

In the past, hummingbirds were hunted in large numbers for their beautiful, iridescent feathers, which were used to decorate the clothing of fashionable women. Sometimes entire, stuffed birds were used as a decoration on hats and as brooches. Fortunately, this gruesome use of hummingbirds in fashion has long passed, and these birds are now rarely hunted.

Today, the greatest risks to hummingbirds occur through losses of their natural **habitat**. This is an especially important problem for the many species of hummingbirds that breed in mature tropical forests. This **ecosystem** type is being rapidly diminished by **deforestation**, mostly to create new agricultural lands in tropical countries.

In most places where they occur, hummingbirds are highly regarded as beautiful creatures and pleasant birds to have around gardens and other places that are frequented by people. The presence of these lovely birds is

often encouraged by planting an abundance of the red, nectar-rich flowers that hummingbirds favor. These birds will also avail themselves of artificial nectar, in the form of sugary solutions made available at specially designed feeders that are hung around homes and gardens. Of course, only those species of hummingbirds that frequent relatively open, disturbed habitats will benefit from this type of management. The many species that only breed in forests can only be sustained by preserving extensive tracts of that natural ecosystem type.

Resources

Books

Brooke, M., and T. Birkhead, eds. *The Cambridge Encyclopedia of Ornithology.* Cambridge: Cambridge University Press, 1991.

Forshaw, Joseph. *Encyclopedia of Birds.* New York: Academic Press, 1998.

Greenewalt, C.H. *Hummingbirds.* New York: Dover Press, 1991.

Tyrell, E.Q., and R.A. Tyrell. *Hummingbirds: Their Life and Behaviour.* New York: Crown, 1985.

Bill Freedman

Humus

Humus is an amorphous, dark brown, organic material that is formed by the incomplete **decomposition** of **biomass**. Strictly speaking, humus is composed of organic residues that are sufficiently fragmented and decomposed by microbial and other decomposition processes that the original source of the biotic materials is no longer recognizable.

Humus is mostly composed of a very complex mixture of large organic molecules, known as humic compounds, which are resistant to further biological oxidation by **microorganisms** and are therefore relatively persistent in the environment. Humus is the major component of the organic **matter** of **soil**. Soluble humic

---

Iridescence—A non-pigmented coloration caused by the physical dissociation of light into its spectral components. The feathers of some birds, including hummingbirds, can develop spectacular, iridescent colors. These colors disappear if the physical structure of the feathers is destroyed by grinding.

---

substances also occur in ground **water** and surface waters, sometimes giving lakes and **rivers** a dark, tea-colored appearance.

Humic substances are divided into three functional classes on the basis of their **solubility** in aqueous solutions of various **pH**. Humic acids are soluble in strongly alkaline solutions, while fulvic acids are soluble in both alkaline and strongly acidic solutions, and humins are insoluble in either. However, apart from their solubility in these solutions, these fractions of polymeric humic substances cannot be easily differentiated or characterized in terms of chemical structure. All of these humic substances are effective at absorbing water and in binding a wide range of organic and inorganic chemicals. Most of the favorable qualities of humus in soil are associated with these properties of humic substances.

Humus is a very important aspect of soil quality. Some of the most beneficial attributes of humus are associated with its ability to make small, inorganic particles adhere together as loose, friable aggregates. The resulting, relatively coarse physical structure allows **oxygen** to penetrate effectively into the soil, which is an important benefit in terms of supporting microbial processes related to decay and nutrient cycling, as well as providing for the **respiration** of **plant** roots. Humus also improves the water-holding capacity of soils, which helps to mitigate **drought** because rainwater does not drain rapidly to depths below the penetration of plant roots. Humus is also important in binding ionic forms of **nutrients**, and in serving as a nutrient reservoir of organically bound nutrients, which are slowly released for plant uptake by microbial nutrient cycling processes.

Because of these characteristics of humus, agricultural and horticultural soils that are composed of a mixture of humus and inorganic **minerals** usually have a substantially greater capability for supporting a vigorous and healthy growth of plants. Compared with soils that are lacking in humus, such substrates are better aerated and have an improved water and nutrient holding capacity, and they are generally more fertile. These important benefits are why one of the highest priority objectives of organic methods of agriculture is to improve the **concentration** of humus in soil.

See also Organic farming.

Huntington disease

Huntington disease is a rare, incurable genetic **disease** that results in the progressive degeneration of both physical and mental abilities. Huntington disease was formly known as Huntington chorea since the most obvious symptoms involve uncontrollable body movements known as chorea. As the disease progresses, its symptoms worsen and patients eventually die of respiratory failure or complications related to the neurodegenerative progression of the disease. Huntington disease is a late onset disorder, where affected individuals usually become symptomatic after 40 years of age. A genetic test for the disease is available, and its use brings to the forefront ethical and social issues related to the clinical **diagnosis**, particularly in the absence of a cure.

History

Huntington disease is named for physician George Huntington, who described the illness in an 1872 paper titled "On Chorea." Huntington practiced medicine on the eastern tip of Long Island, New York. His description of the disease was drawn from his familiarity with several affected families in his community. Both Huntington's father and grandfather had practiced medicine in the same area. Their encounters with the disease gave Huntington an appreciation of the heredity aspect of the illness.

Historically, the mental and emotional deterioration that marks the illness has frequently led to the confinement of Huntington disease patients to psychiatric hospitals. Some historians speculate that a few of the women accused of witchcraft in Salem, Massachusetts may have exhibited the involuntary twitches and turns that are hallmarks of the disease.

Symptoms

The symptoms in Huntington disease begin with noticeable behavioral changes including aggression, paranoia, and irritability. Affected individuals may seem restless, with tapping feet or odd twitches. Patients begin to suffer from impaired judgment and an inability to be organized. They become forgetful and their I.Q. declines, coinciding with the deterioration of the **brain**. Emotionally, they may suffer from psychiatric disorders and even suicidal thoughts or actions. They may drop things and become less efficient in their usual activities. **Depression**, **anxiety**, and apathy are also common experiences in the beginning stages of the disease. As the illness progresses, the chorea worsens. The entire body moves in uncoordinated, jerky movements.

Although wide variations in clinical manifestations exist, the illness typically lasts 13–16 years. In later stages of the illness, patients cannot walk or care for themselves. They may barely speak and may fail to recognize friends and family. They eventually require full time nursing care. Eating is quite difficult and death is very frequently caused by choking or by the **pneumonia** that results after accidentally inhaling bits of food. Although some symptoms can be treated with medication, currently no cure exists to delay the onset of Huntington disease or to slow its course.

Huntington disease is a hereditary disease caused by a dominant **gene** and, therefore, follows an autosomal dominant pattern of inheritance. This means that only one copy of the gene is necessary to cause the disease. It is transmitted from one generation to the next. The child of a mother or father suffering from Huntington disease has a 50% chance of inheriting the disease gene and, thus, of contracting the disease later in life. About 30,000 Americans suffer from Huntington disease and another 150,000 are at risk for developing it. One of the best-known disease victims was an American folk singer Woody Guthrie (1912–1967), who died of the disease.

Genetic defect responsible for disease

In 1993, scientists discovered the genetic defect that causes Huntington disease. A gene located on the **chromosome** 4 normally contains a sequence of three nucleotide bases (the alphabet of the genetic code) that repeats several times. The sequence is cytosine, adenine, and guanine, or CAG, which codes for the **amino acid** glutamine that is a building block for protein synthesis. In Huntington disease, patients have too many repeats. While unaffected individuals normally have 11-24 repeats, a person with Huntington disease may have anywhere from 36-100 or more repeats. Clinical research studies have demonstrated that the greater the number of repeats, the earlier the disease will develop and the clinical manifestations will be more severe. If the expanded trinucleotide sequence is passed from the father to the offspring, the offspring that inherit this expansion can have an earlier age of onset of the disease. This phenomenon is called anticipation and is paternal in origin if the father inherited the disease gene from his mother. There are also other diseases characterized by expansion of a repetitive sequence in the DNA and developmental delay, such as fragile X.

Despite the discovery of the Huntington disease gene, scientists were baffled by how this genetic defect produces such a devastating disease course. The Huntington disease gene codes for a large protein with no similarities to known **proteins**. It has been named the huntintin protein. It is important for normal development of the **nervous system** and interacts with many other proteins. Through autopsy, it was shown that an abnormality in the huntingtin protein caused the destruction of brain cells in the basal ganglia, a region of the brain with unknown functions. Using **genetic engineering**, scientists have developed strains of **mice** that express the

Huntington disease gene. These mice display the symptoms of the disease. It has been found that the huntington protein, normally present in the cytoplasm (internal fluid-like content) of cells, collects in the brain **cell** nuclei, forming masses that kill the cell. This dominant-negative effect explains why clinically asymptomatic patients develop progressive neurodegeration of the brain in the fourth decade of life.

The quest for the Huntington disease gene

The quest for the Huntington disease gene was made possible by a new era in medicine and **biology**. The researcher who found the first genetic marker for the disease used a novel scientific approach. Much of the credit for the discovery of the gene belongs to Nancy Wexler, a American clinical psychologist who organized and championed the gene hunt with unflagging enthusiasm, in part due to the fact that she had a positive family history of the disease.

In 1968, at the age of 23, Nancy and her sister, Alice, learned from their father, Milton Wexler, that their mother had been diagnosed with Huntington disease. With their mother's diagnosis, Nancy and Alice had a 50% risk of developing the disease themselves. Milton Wexler, a lawyer and psychoanalyst, later founded the Hereditary Disease Foundation. The foundation worked to attract scientists to the study Huntington disease. It formed a board of scientific advisors, held conferences, and funded workshops particularly for younger scientists. It successfully urged Congress to appropriate money for the study of the disease. Nancy Wexler, a graduate student, became increasingly involved in her father's foundation and eventually became president.

Lake Maracaibo

In 1972, Wexler learned of several large, interrelated families affected with Huntington disease who lived in small villages along Lake Maracaibo in Venezuela. Wexler realized that this was a unique and valuable resource due to the large family pedigree. The larger the family **tree**, the easier it is to find genes by linking their location on the chromosome to specific DNA markers within the **genome**. In 1979, she began making annual trips to Lake Maracaibo. With the help of a team of investigators, she created a genealogy of the families and, beginning in 1981, took **blood** samples from both sick and the healthy family members. Wexler was convinced that the key to Huntington disease lay locked in the DNA of these families.

In 1983, James Gusella, a young scientist at Massachusetts General Hospital, began applying a new technique of **molecular biology** to the blood samples from Venezuela. He was looking for patterns that were present in the DNA sequences of people with Huntington disease but absent in the DNA of people without the disease. If one particular pattern of DNA was always associated with the illness in a given family, then it could be used as a marker for the disease gene. For instance, if it were true that people who developed the disease always had green eyes, then scientists could say that the gene that is responsible for green eyes located in a position along the DNA strand that is close to the gene that causes the disease. This genetic evaluation is called linkage analysis.

Eventually, Gusella found a genetic marker for Huntington disease, and remarkably it was almost immediate. Although the gene itself was still unknown, the discovery of this genetic marker made it possible to create a genetic test for the disease (linkage analysis), in the following year. By studying blood samples from several family members, persons who had a parent die of Huntington disease could be told whether or not they had inherited the genetic marker linked to the disease in their family. Other scientific teams also began using linkage analysis to search for disease genes as a result of these studies.

Even though the discovery of the marker indicated to scientists the general location of the gene itself, that gene hunt proceeded slowly. At Wexler's urging, and in a break from usual scientific practice, a consortium of six scientific teams worked together to find the gene. Finally, on March 26, 1993, in the scientific journal *Cell,* the 58 members of the Huntington's Collaborative Research Group announced to the world the discovery of the gene that causes Huntington disease.

Ethical questions

The genetic test for Huntington disease raises profound ethical questions. It offers people who are at risk the opportunity to know whether they inherited the gene. Yet many people at risk choose not to be tested. Currently, no treatment existed to cure Huntington disease or even to delay the onset of the disease. Given this reality, many people would rather live with uncertainty than take the chance of **learning** that they will develop an incurable, fatal illness. Additionally, an ethical dilemma arises in cases where a grandson or granddaughter desires testing but their parent does not. If a grandparent is affected and the grandchild is affected, then by default the parent that is biological related to both is affected. Other concerns related to **genetic testing** of Huntington disease involves guilt associated with not having the disease gene when a sibling is a carrier. Guilt is also commonly experienced by the parent that is responsible for passing the disease gene to their offspring. These emotional experiences can have a profound effect on the family dynamics.

KEY TERMS

Dominant gene—An allele of a gene that results in a visible phenotype if expressed in a heterozygote.

Genetic marker—DNA segment that can be linked to an identifiable trait, although it is not the gene for that trait.

Nucleotides—Building blocks of DNA: a phosphate and a sugar attached to one of the bases, adenine (A), cytosine (C), guanine (G), or thymine (T).

RFLP (restriction fragment length polymorphism)—A variation in the DNA sequence, identifiable by restriction enzymes.

Prenatal testing, now offered for several genetic diseases, is also available to parents whose fetus is at risk for Huntington disease. Genetic testing also raises the right to privacy. Do employers, health insurers, or the government have the right to know whether a person at risk has been tested, or the right to know the results of the test? Most researchers and ethicists, including Wexler, promote the need for privacy. These ethical questions are not unique to Huntington disease.

As the genetic components of other illnesses are discovered, especially for late-onset illnesses like **Alzheimer disease** and certain cancers, these questions will become more relevant and pressing. In many ways, the implementation of the genetic test for Huntington disease may serve as a model for how genetic testing is used in medicine and impacts society.

See also Genetics; Gene therapy.

Resources

Books

Nussbaum, Robert L., Roderick R. McInnes, and Huntington F. Willard *Genetics in Medicine.* Philadelphia: Saunders, 2001.

Rimoin, David L. *Emery and Rimoin's Principles and Practice of Medical Genetics.* London; New York: Churchill Livingstone, 2002.

Periodicals

Revkin, Andrew. "Hunting Down Huntington's." *Discover* (December 1993).

Other

Online Mendelian Inheritance in Man. "143100 HUNTINGTON DISEASE; HD." December 17, 2002 [cited January 10, 2003]. <http://www. ncbi.nlm.nih.gov/htbin-post/Omim/dispmim?143100>.

Liz Marshall

Hurricane *see* **Tropical cyclone**

Hyacinth *see* **Lily family (Liliaceae)**

Hybrid

A hybrid is an offspring between two different **species**, or the offspring between two parents of the same species that differ in one or more heritable characteristics. An example of the first kind of hybrid is a mule, a cross between a female horse (*Equus caballus*) and a male donkey (*E. asinus*). An example of the second kind is the offspring from a cross between true-breeding red- and white-flowered garden peas (*Pisum sativum*).

Hybrids between species are often sterile because they fail to produce viable reproductive cells that is, eggs, sperm, or spores. These cells develop improperly because the chromosomes from one species do not pair correctly during **meiosis** with the chromosomes from the other species.

Despite their sterility, hybrids may thrive and expand their ranges by reproducing asexually. For example, in the eastern United States and adjacent Canada, there are hundreds of distinctive hybrids of hawthorn (*Crataegus*) and blackberry (*Rubus*) that are not interfertile. Yet these hybrids may be common because they are able to set seed from asexually produced embryos, a special form of propagation called apomixis (Greek *apo*, away from, and *mixis*, mix, union—referring to the lack of **fertilization**). Also, in blackberries, the first-year stems are able to root at the tip, a form of propagation called vegetative reproduction.

Some hybrids become fertile by doubling their **chromosome** number, a process called polyploidy. Hybridization followed by polyploidy has been extremely important in **plant** evolution, especially among **ferns** and **grasses**. Examples are the wheats used to make bread and pastas, and species of wood fern (*Dryopteris*) and spleenworts (*Asplenium*).

Hybrids are generally infrequent in nature. Nevertheless, once formed they may be important for **evolution** because of the way they combine the characteristics of their parents. Especially in changing or disturbed habitats, hybrids that contain new genetic combinations may be better adapted to the new environments than either of the parents. Thus they may be able to colonize new habitats where neither parent can grow.

See also Asexual reproduction; Genetics.

Hydra

Hydra are solitary animals of the phylum Coelenterata that measure from just a few millimeters in size to more than 3.5 ft (1 m) in length. They are all thin animals that rarely measure more than 0.4 in (1 cm) in diameter. Most are cylindrical in shape, with a broadened basal disk that serves to attach the **animal** to some firm substrate. Most **species** are sessile but some can, if conditions require, move over short distances by repeatedly looping the body over onto the substrate. Longer-range movements may be accomplished by releasing their grip on the substrate and rising into the **water** current.

The main body stalk is a simple, erect tube-like arrangement, at the top of which is the mouth. This is surrounded by a ring of tentacles whose length varies according to the species in question. The body stalk is not encased in a hard protective layer, and the animal is therefore able to flex and bend. The bulk of the body is taken up with the large intestinal cavity.

The tentacles contain a large number of specialized cells called cnidocytes, which contain stinging structures known as nematocysts. The latter vary in shape according to their required purpose. Most have an oval shaped base, attached to a long threadlike structure. When the animal is feeding or is alarmed, the cnidocytes are triggered to release the nematocysts. When the animals is feeding, most of the nematocysts that are released are hollow and elongate, their purpose being to trap and entangle **prey**. Once this has been completed, the captured prey—often small crustaceans—are grasped by the tentacles and passed down towards the mouth.

In other situations, for example in defense, the nematocysts may be shorter and often bear small spines; some may also contain a toxic substance which is injected into the attacking animal to deter or stun it.

Most hydras reproduce by asexual means through a simple system of "budding-off." In this process, a small extension of the parent animal forms on the body wall. As this grows, a separate mouth and set of tentacles develops until eventually a replicate daughter **cell** of the parent hydra is produced. When the young animal has fully developed, the two separate and the young hydra drifts off in the current to become established elsewhere. In certain circumstances, particularly where seasonal **drought** is a regular feature, some species may also practice **sexual reproduction** which involves the production of a fertilized embryo enclosed in a toughened outer coating. In this state, the young hydra is able to withstand periods of drought, cold, food shortages, or **heat**. Once conditions normal resume, the outer casing dissolves and the embryo recommences life.

Hydrangea *see* **Saxifrage family**

Hydrocarbon

A hydrocarbon is any chemical compound whose molecules are made up of nothing but **carbon** and **hydrogen atoms**.

Carbon atoms have the unique ability to form strong bonds to each other, atom after atom. Every hydrocarbon **molecule** is built upon a skeleton of carbon atoms bonded to each other either in the form of closed rings or in a continuous row like links in a chain. A chain of carbon atoms may be either straight or branched. In every case, whether ring or chain, straight or branched, all the carbon bonds that have not been used in tying carbon atoms together are taken up by hydrogen atoms attached to the carbon skeleton. Because there is no apparent limit to the size and complexity of the carbon skeletons, there is in principle no limit to the number of different hydrocarbons that can exist.

Hydrocarbons are the underlying structures of all organic compounds. All organic molecules can be thought of as being derived from hydrocarbons by substituting other atoms or groups of atoms for some of the hydrogen atoms and occasionally for some of the carbon atoms in the skeleton.

Carbon's chemical bonding

The carbon atom has four electrons in its outer, or *valence*, shell. This means that every carbon atom can form four, and only four, covalent (electron-pair-sharing) bonds by pairing its four **valence** electrons with four electrons from other atoms. This includes forming bonds to other carbon atoms, which can form bonds to still other carbon atoms, and so on. Thus, extensive skeleton structures of dozens or hundreds of carbon atoms can be built up.

A carbon atom does not form its four bonds all in the same direction from the nucleus. The bonding **electron** pairs being all negatively charged tend to repel one another, and they will try to get as far apart as possible. The bonds will therefore stick out in four equally spaced directions. In two dimensions, four equally spaced directions from a point would aim at the four corners of a **square**. But in three-dimensional **space**, four equally spaced directions from a point (the carbon atom's nucleus) aim at the four corners of a *tetrahedron*.

On two-dimensional **paper**, the formation of a covalent bond between two carbon atoms can be depicted as follows, where the dots indicate valence electrons and

the C's indicate the rest of the atoms (nucleus plus inner electrons):

$$\cdot \overset{\cdot}{\underset{\cdot}{C}} \cdot \ + \ \cdot \overset{\cdot}{\underset{\cdot}{C}} \cdot \ \longrightarrow \ \cdot \overset{\cdot}{\underset{\cdot}{C}} : \overset{\cdot}{\underset{\cdot}{C}} \cdot$$

The carbon atoms still have unused bonds shown by the unpaired dots, and they can join to third and fourth carbon atoms and so on, building up longer and longer chains:

$$\cdot \overset{\cdot}{\underset{\cdot}{C}} : \overset{\cdot}{\underset{\cdot}{C}} \cdot \ + \ \cdot \overset{\cdot}{\underset{\cdot}{C}} \cdot \ \longrightarrow \ \cdot \overset{\cdot}{\underset{\cdot}{C}} : \overset{\cdot}{\underset{\cdot}{C}} : \overset{\cdot}{\underset{\cdot}{C}} \cdot$$

and

$$\cdot \overset{\cdot}{\underset{\cdot}{C}} : \overset{\cdot}{\underset{\cdot}{C}} : \overset{\cdot}{\underset{\cdot}{C}} \cdot \ + \ \cdot \overset{\cdot}{\underset{\cdot}{C}} \cdot \ \longrightarrow \ \cdot \overset{\cdot}{\underset{\cdot}{C}} : \overset{\cdot}{\underset{\cdot}{C}} : \overset{\cdot}{\underset{\cdot}{C}} : \overset{\cdot}{\underset{\cdot}{C}} \cdot$$

Instead of lining up in straight or *normal* chains, the carbon atoms may also bond in different directions to form *branched chains*.

In all of these skeletons, there are still some carbon valence electrons that are not being used for carbon-to-carbon bonding. The remaining bonds can be filled by hydrogen atoms to form hydrocarbon molecules:

$$\cdot \overset{\cdot}{\underset{\cdot}{C}} : \overset{\cdot}{\underset{\cdot}{C}} \cdot \ + \ 6 \cdot H \ \longrightarrow \ H : \overset{H \ \ H}{\underset{H \ \ H}{C : C}} : H$$

$$C_2H_6$$

$$\cdot \overset{\cdot}{\underset{\cdot}{C}} : \overset{\cdot}{\underset{\cdot}{C}} : \overset{\cdot}{\underset{\cdot}{C}} \cdot \ + \ 8 \cdot H \ \longrightarrow \ H : \overset{H \ \ H \ \ H}{\underset{H \ \ H \ \ H}{C : C : C}} : H$$

$$C_3H_8$$

Hydrogen is a particularly good candidate for bonding to carbon because each hydrogen atom has only one valence electron; it can pair up with one of the carbon atom's valence electrons to form a bond in one of carbon's four possible directions without interfering with any of the other three because hydrogen is such a tiny atom. (In addition to its valence electron, a hydrogen atom is nothing but a proton.) Hydrocarbons are divided into two general classes: *aromatic hydrocarbons*, which contain **benzene** rings in their structures, and *aliphatic hydrocarbons*, which are all the rest.

Aliphatic hydrocarbons

The carbon-atom skeletons of aliphatic hydrocarbons may consist of straight or branched chains, or of (non-benzene) rings. In addition, all of the carbon atoms in the skeletons may be joined by sharing single pairs of electrons (a single bond, represented as C:C or C-C), as in the examples above, or there may be some carbon atoms that are joined by sharing two or three pairs of electrons. Such bonds are called double and triple bonds

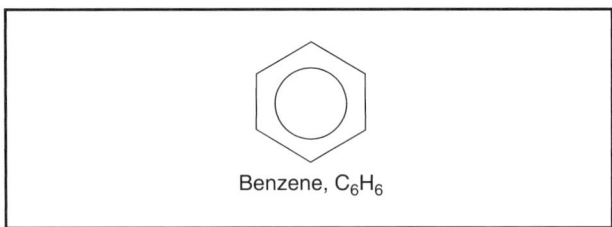

Benzene, C_6H_6

Illustration by Hans & Cassidy. Courtesy of Gale Group.

and are represented as C::C or C=C and C:::C or C≡C, respectively.

Thus, there can be three kinds of aliphatic hydrocarbons: those whose carbon skeletons contain only single bonds, those that contain some double bonds, and those that contain some triple bonds. These three series of aliphatic hydrocarbons are called *alkanes*, *alkenes*, and *alkynes*, respectively. (There can also be " hybrid" hydrocarbons that contain bonds of two or three kinds.)

Alkanes

The alkanes are also called the *saturated hydrocarbons*, because all the bonds that are not used to make the skeleton itself are filled to their capacity—saturated—with hydrogen atoms. They are also known as the *paraffin* hydrocarbons, from the Latin *parum affinis*, meaning "little affinity," because these compounds are not very chemically reactive.

The three smallest alkane molecules, containing one, two, and three carbon atoms, are shown in three ways.

The *structural formulas* are one way in which simple organic molecules can be depicted in two dimensions on paper; each line indicates a single covalent bond-a shared pair of electrons. The three-dimensional *ball-and-stick models* and *space-filling models*, in which the balls represent the carbon and hydrogen atoms (roughly to scale) and the sticks represent the bonds, are used by chemists to study the shapes of molecules.

The names and formulas of the first eight normal (not branched) alkanes are: Methane (CH_4); Ethane (C_2H_6); Propane (C_3H_8); Butane (C_4H_{10}); Pentane (C_5H_{12}); Hexane (C_6H_{14}); Heptane (C_7H_{16}); Octane (C_8H_{18}).

While the first four alkanes were named before their structures were known, the rest have been named with Greek roots that tell how many carbon atoms there are in the chain: *pent* = five, *hex* = six, and so on, all ending in the "family name," *-ane*. The **chemical formula** of an alkane hydrocarbon can be obtained quickly from the number of carbon atoms, *n*, in its skeleton: the formula is C_nH_{2n+2}. This method works because every carbon atom has two hydrogen atoms attached except for the two end carbon atoms which have two extra ones. As an example, the formula for pentane is C_5H_{12}.

The branched alkanes are named by telling what kinds of branches—methyl, ethyl or propyl groups, etc.—are attached to the main chain and where. For example,

$$H_3C\text{-}CH\text{-}CH_2\text{-}CH_2\text{-}CH_3$$
$$|$$
$$CH_3$$

is named 2-methyl pentane; the 2 indicates that the **methyl group** (-CH_3) branches off the second carbon atom from the nearest end of the pentane chain.

The four lightest normal alkanes, having the smallest (lowest **molecular weight**) molecules, are gases at room **temperature** and **pressure**, while the heavier ones are oily liquids, and still heavier ones are waxy solids. Alkanes, which are the major constituents of crude oil, do not mix with **water** and float on its surface. The wax that we call paraffin and make candles from is a mixture of alkanes containing between 22 and 27 carbon atoms per molecule.

All hydrocarbons burn in air to form **carbon dioxide** and water. Methane, CH_4, as the major constituent of **natural gas**, is widely used as a heating fuel. Also known as marsh gas, methane occurs naturally in marshes and swamps, being produced by **bacteria** during the **decomposition** of **plant** and **animal matter**. It can form explosive mixtures with air, however, and is therefore a hazard when present in **coal** mines. On the positive side, bacteria-produced methane has prospects for being developed as a commercial source of fuel.

Propane, C_3H_8, and butane, C_4H_{10}, are compressed into tanks, where they liquefy and can be used as portable fuels for such applications as barbecue grills, mobile-home cooking, and disposable cigarette lighters. Because these compounds are pure and burn cleanly, they are being explored as fuels for non-polluting **automobile** engines. They are often referred to as LPG-liquefied **petroleum** gas.

Cycloalkanes are alkanes whose carbon atoms are joined in a closed loop to form a ring-shaped molecule. Cyclopropane which contains three carbon atoms per molecule has molecules that are in the shape of a three-membered ring, or triangle. Cyclohexane, with six carbon atoms, has hexagonal molecules; it is used as a good solvent for many organic compounds.

Alkenes

The alkenes, sometimes called *olefins*, are hydrocarbons that contain one or more double bonds per molecule. Their names are **parallel** to the names of the alkanes, except that the family ending is *-ene*, rather than *-ane*. Thus, the four smallest molecule alkenes containing two, three, four, and five carbon atoms are *ethene* (also called *ethylene*), *propene* (also called *propylene*), *butene* (also called *butylene*), and *pentene*. (There can be no "methene," because there must be at least two carbon atoms to form a double bond.) A number preceding the name indicates the location of the double bond by counting the carbon atoms from the nearest end of the chain. For example, 2-pentene is the five-carbon alkene with the structure

$$H_3C-CH=CH-CH_2-CH_3$$

The locations of branches are similarly indicated by numbers. For example, 3-ethyl 2-pentene has the structure

$$H_3C-CH=C-CH_2-CH_3$$
$$|$$
$$C_2H_5$$

The lightest three alkenes, ethylene, propylene, and butene, are gases at room temperature; from there on, they are liquids that boil at higher and higher temperatures. The chemical formula of an alkene containing only one double bond per molecule can be obtained from the number of carbon atoms in its molecules: if n is the number of carbon atoms, the formula is C_nH_{2n}. Thus, the formula for pentene is C_5H_{10}.

Alkenes are called *unsaturated* hydrocarbons; if there is more than one double bond in an alkene molecule it is said to be *polyunsaturated*. In principle, two more hydrogen atoms could be added to each double bond to "saturate" the compound, and in fact this does happen quite easily when hydrogen gas is added to an alkene in the presence of a catalyst. This process is called **hydrogenation**.

Other elements, such as the **halogens** and hydrogen halides, can also be added easily to the double bonds in alkenes. The resulting **halogenated hydrocarbons** are very useful but are often toxic or environmentally damaging. Trichloroethylene is a useful solvent, **chlorinated hydrocarbons** have been used as **insecticides**, and chlorofluorocarbons (CFCs or Freons) are used as refrigerants but have been shown to damage the earth's **ozone** layer.

Alkynes

Alkynes are hydrocarbons that contain one or more triple bonds per molecule. Their names are parallel to the names of the alkanes except that the family ending is -*yne*. Thus, the four smallest-molecule alkynes are *ethyne* (more usually called *acetylene*), *propyne*, *butyne,* and *pentyne*. Alkynes containing one triple bond have chemical formulas given by C_nH_{2n-2}, where n is the number of carbon atoms in the molecule. Thus, the formula for pentyne is C_5H_8. Acetylene, propyne, and butyne are gases at room temperature; the rest are liquids.

The most famous of the alkynes is the first member of the series: acetylene, C_2H_2. It forms explosive mixtures with air or **oxygen**, but when mixed with oxygen in a controlled way in an oxyacetylene torch it burns with a very hot flame—up to 6,332°F (3,500°C) which is hot enough to cut and weld **steel**. Because acetylene is explosive when compressed into liquid form, the tanks of acetylene that welders use contain acetylene dissolved in **acetone**.

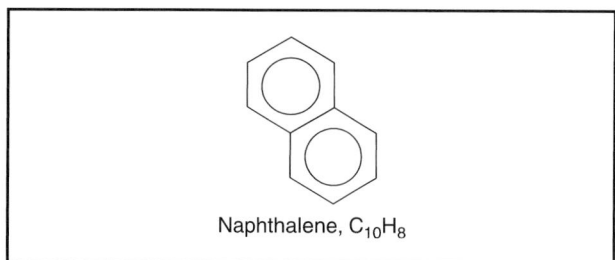

Naphthalene, $C_{10}H_8$

Illustration by Hans & Cassidy. Courtesy of Gale Group.

Other important alkenes are styrene, $C_6H_5-CH=CH_2$, from which the plastic polystyrene is made, and isoprene, $CH_2=C(CH_3)CH=CH_2$, which is the **monomer** of natural rubber. (In this shorthand structural formula for isoprene, the parentheses indicate that the CH_3 group within them is a branch attached to the preceding carbon atom.)

Aromatic hydrocarbons

An aromatic hydrocarbon is any hydrocarbon that contains one or more benzene rings in its molecule. The name "aromatic" is historical in origin, and does not at all imply that these compounds have pleasant aromas. Aromatic hydrocarbons are the basis of many *aromatic compounds* containing other atoms such as oxygen and **nitrogen** in addition to the carbon and hydrogen that are of extreme biological and industrial importance.

The simplest aromatic hydrocarbon is benzene itself, C_6H_6, whose molecule is a hexagonal ring of six CH groups. Various carbon-and-hydrogen groups can be substituted for any or all of the hydrogen atoms in benzene to form *substituted benzenes*. Benzene's own phenyl groups, C_6H_5, can bond to each other end to end, to form *polycyclic* (multiple-ring) hydrocarbons, or they can fuse together along the hexagons' sides to form *condensed ring* or *fused ring* hydrocarbons.

In this figure, the bonds leading to all the hydrogen atoms are omitted for simplicity as is the usual practice among chemists. Also, the benzene rings are drawn with alternating double and single bonds between the carbon atoms. In reality, however, **resonance** makes all the carbon-carbon bonds equal at an intermediate value between single and double. Chemists therefore usually draw the benzene ring simply as a hexagon with a circle inside:

TABLE 1. TYPICAL HYDROCARBON MIXTURES OBTAINED FROM THE FRACTIONAL DISTILLATION OF PETROLEUM.

Boiling-temperature range[a]	Name of fraction	Number of carbon atoms in molecule[a]	Uses
Below 36°C	Natural gas	1-5	Fuel; starting material for making plastics
40-60°C	Petroleum ether	5-6	Solvent
70-90°C	Naphtha	6-7	Solvent; lighter fuel
69-174°C	Gasoline	6-10	Fuel for engines, industrial solvent
174-288°C	Kerosene (coal oil)	10-16	Fuel for lamps, heaters, tractors, jet airplanes
250-310°C	Fuel oil (gas oil)	15-18	Heating oil; diesel fuel
300-370°C	Lubricating oils	16-20	Lubrication
Melts at 40-55°C	Petrolatum (petroleum jelly)	17-30	Lubrication; ointments
Melts at 50-60°C	Paraffin wax	23-29	Candles; waterproof coatings
Above 515°C	Pitch, tar	Over 39	Paving, roofing

[a] The exact temperature ranges and numbers of carbon atoms differ in different refineries, and according to various legal definitions in various states and countries.

The hexagon represents the six carbon atoms and their attached hydrogen atoms, while the circle represents all the bonding electrons as if they were everywhere in the molecule at once. In chemists' shorthand, then, naphthalene would be depicted as shown in the above figure.

Among the important substituted benzenes are methyl benzene, commonly known as toluene, and dimethyl benzene, commonly known as xylene. They are both powerful solvents for organic compounds and are used as starting materials for the synthesis of drugs, dyes, **plastics**, and **explosives**. Treatment of toluene with nitric and sulfuric acids produces the explosive trinitrotoluene, or TNT.

Among the important condensed ring aromatic hydrocarbons are naphthalene and anthracene whose molecules consist of two and three hexagonal benzene rings, respectively, fused together along one side.

Both are derived from coal tar and are used as starting materials for the synthesis of many useful compounds. Naphthalene is a crystalline solid with a strong, pungent odor; it is used as a moth repellant and a deodorant-disinfectant.

Petrochemicals

Our primary source of hydrocarbons is petroleum or crude oil, that thick, black liquid that we find in the **earth**.

Petroleum consists almost entirely of a mixture of alkanes with some alkenes and smaller amounts of aromatic hydrocarbons. When petroleum is distilled at a series of different temperatures, the lowest molecular-weight hydrocarbons boil off at the lowest temperatures and the higher-molecular-weight ones boil off at successively higher temperatures. This process, called *fractional distillation*, is used to separate the complex mixture of compounds. The table shows the various hydrocarbon mixtures ("fractions") that distill off in various temperature ranges.

In addition to harvesting the hydrocarbons that occur naturally in petroleum, oil refineries use a variety of processes to convert some of them into other more desirable hydrocarbons.

A vast number of synthetic (man-made) organic chemicals, including drugs, plastics, paints, adhesives, fibers, detergents, synthetic rubber, and agricultural chemicals, owe their existence to *petrochemicals*: chemicals derived from petroleum. The top six petrochemicals produced in the United States are ethylene, propylene, benzene, xylene, butadiene (the four-carbon-atom alkene with two double bonds), and toluene. From these, hundreds of other chemicals are manufactured.

Gasoline

Probably the most important product of the fractional **distillation** of petroleum is gasoline, a mixture of alkanes containing six to ten carbon atoms in their molecules: hexane (C_6H_{14}), heptane (C_7H_{16}), octane (C_8H_{18}), nonane (C_9H_{20}), and decane ($C_{10}H_{22}$), plus small amounts of higher-molecular weight alkanes. More than six trillion gallons of gasoline are burned each year in the United States.

Gasoline must have certain properties in order to work well in automobile engines. If the gasoline-air mixture does not explode smoothly when ignited by the spark in the cylinder, that is, if it makes a fast, irregular explosion instead of a fast but gentle burn, then the explosive **force** will hit the piston too soon, while it is still trying to move down into the cylinder. This clash of ill-timed forces jars the engine, producing a metallic clanking noise called a *knock*, which is especially audible when the engine is laboring to climb a hill. Extensive knocking can lead to serious engine damage, so gasolines are formulated to minimize this effect.

Of all the hydrocarbons that can be in gasoline, normal (straight-chain) heptane, C_7H_{16}, has been found to make auto engines knock worst. It has been assigned a value of **zero** on a scale of gasoline desirability. The hydrocarbon that knocks least is a branched-chain form of octane, C_8H_{18}, called iso-octane. It has been rated 100. Every gasoline blend is assigned an *octane rating* be-tween zero and 100, according to how much knocking it produces under standard test conditions. Most automobile fuels sold have octane ratings above 85. High-octane gasolines that are even better than iso-octane because of anti-knock additives can have ratings above 100.

The C_6 to C_{10} hydrocarbons make up only about 20-30% of crude oil, which is far from enough to supply the world's appetite for gasoline. But even if there were enough of it, the natural mixture has an octane rating of only about 40 to 60—not good enough for modern engines. Refineries therefore modify the natural mixture of molecules by breaking down big molecules into smaller ones (*cracking*) and by reshaping some of the smaller molecules into forms that knock less (*reforming*).

By the time gasolines get to the pump, they are no longer pure hydrocarbon mixtures; they have been blended with additives. Lead-containing antiknock compounds such as tetraethyl lead, $Pb(C_2H_5)_4$, are no longer used because lead is a toxic air pollutant; methyl-tert-butyl **ether** (MBTE) is used instead. Other additives remove harmful engine deposits, prevent gum formation, inhibit rusting, prevent icing, clean the carburetor, lubricate the cylinders, and dye the gasoline distinctive colors for identification purposes.

See also Chemical bond; Formula, structural

Resources

Books

Amend, John R., Bradford P. Mundy, and Melvin T. Armold. *General, Organic and Biological Chemistry.* Philadelphia: Saunders, 1990.

Jahn, F., M. Cook, and M. Graham. *Hydrocarbon Exploration and Production. Developments in Petroleum Science.* Vol. 46. The Netherlands: Elsevier Science, 2000.

Loudon, G. Mark. *Organic Chemistry.* Oxford: Oxford University Press, 2002.

Schobert, Harold H. *The Chemistry of Hydrocarbon Fuels.* Boston: Butterworth's, 1990.

Sherwood, Martin, and Christine Sutton. *The Physical World.* New York: Oxford University Press, 1991.

Robert L. Wolke

Hydrocephalus

Hydrocephalus, which means literally, "water on the brain," is a condition in which excessive fluid collects inside the skull. The fluid is a naturally produced liquid that normally is found in the **brain**. Accumulation of excessive amounts of the fluid may build **pressure** to levels that cause brain damage and subsequent disability.

The brain rests within the natural bony vault of the cranium. There it is protected by the skull and by layers of fibrous material that help to stabilize it and to contain the fluid that surrounds it. The brain itself is a very soft, gelatinous material that requires substantial protection. Three layers of **connective tissue** line the skull and surround the brain. The pia mater (which means literally "tender mother") lies directly on the brain, following its contours and continuing along the spinal cord as it descends through the spine. The second layer is the arachnoid (like a spider's web), a very thin, fibrous **membrane** without **blood** vessels. It, too, lies close to the brain but does not follow its every bump and wrinkle. The space between the pia mater and the arachnoid, called the subarachnoid space, contains the **arteries** and **veins** that circulate blood to the brain and the cerebrospinal fluid that bathes the nervous tissues. The outermost layer, the dura mater ("hard mother") is a two-layered, leathery, tough membrane that adheres closely to the inside of the skull. The inner layer is contoured to the brain to support it. The outer layer lies against the cranium and continues into the spinal **canal**.

The fluid that bathes the brain and spinal cord—cerebrospinal fluid or CSF—is manufactured and secreted in the brain by a structure called the choroid plexus. Cerebrospinal fluid is a colorless, clear fluid that contains **oxygen**, some **proteins**, and glucose (a form of sugar). Normally the fluid will circulate through the cranium and down the spinal column. It will be absorbed by special structures called villi in the arachnoid **tissue** or it will drain from one of several outlets. Excessive fluid accumulates because the brain is manufacturing too much CSF or the drainage routes are blocked and the fluid cannot drain properly.

The capacity of the ventricles in the brain and the space around the spinal cord is approximately 0.5 c (125 ml). The choroid plexus manufactures from 2-3 c (500-750 ml) of fluid each day. The pressure of the CSF within the **nervous system**, therefore, is related to the **rate** of manufacture versus the rate of drainage of the fluid. Fluid pressure can be measured by inserting a needle between two of the lumbar vertebrae into the spinal canal. The needle is connected to a meter that indicates the fluid pressure.

The choroid plexus is composed of specialized cells that line the ventricles of the brain. The ventricles are four small, naturally formed cavities in the brain that act as reservoirs for CSF. Overproduction of fluid or its failure to drain can enlarge the ventricles and press the brain against the bony vault of the skull.

Newborn babies who have hydrocephalus often will develop grossly swollen heads. The bones of the skull have not fused and the pressure of the fluid inside the skull can expand the disconnected bony plates.

KEY TERMS

Congenital—A condition or disability present at birth.

Hydro—Reference to water.

Lumbar—Reference to the lower back; the vertebrae below the ribs or thorax.

Meninges—Collectively, the three membranes that cover the brain and line the skull; the pia mater, arachnoid, and dura mater.

Tumor—An uncontrolled growth of tissue, either benign (noncancerous) or malignant (cancerous).

Ventricle—An opening in the brain that forms a reservoir for the cerebrospinal fluid.

Two types of hydrocephalus

The two types of hydrocephalus are called communicating and noncommunicating. Communicating hydrocephalus is caused by overproduction of fluid by the choroid plexus. The fluid, which overwhelms the absorption capacity of the arachnoid, collects inside the ventricles as well as outside the brain. This is the most common form of hydrocephalus occurring in adults and is the result of injury or **infection** such as **encephalitis**. At the onset of the condition the patient will become clumsy in walking and appear tired. Other signs will develop indicating a brain injury. To diagnose communicating hydrocephalus the physician will review the patient's recent history to determine whether an infection or head injury has occurred. In addition, such diagnostic measures as a magnetic **resonance** image (MRI) of the skull can reveal the presence of excess fluid. This condition is readily treatable.

Noncommunicating hydrocephalus is the most common form of the condition in childhood. Usually it will be diagnosed immediately after **birth**, when signs such as a swollen cranium are seen. Here the problem lies in a narrowing of a drainage **aqueduct** which inhibits passage of the CSF out of the cranium. The ventricles enlarge greatly and the fluid pressure begins to push the brain against the skull. In this case a drain can be implanted in the skull to drain the fluid into a vein to relieve the pressure.

This form of hydrocephalus also is associated with a **congenital** condition called meningomyelocele. A newborn with this condition is born with the spinal cord and its superficial coverings exposed. The spinal canal, the opening through which the spinal cord passes, has not fused, so the cord can protrude through the open side. Almost always, the surgical repair of the meningomyelocele will result in hydrocephalus, which will in turn require surgical correction.

This form of hydrocephalus also can occur in an adult and generally is the result of the formation of a **tumor** that blocks the drainage area.

All forms of hydrocephalus can be treated surgically, so it is important that **diagnosis** be made as soon as possible after the condition develops. With excessive fluid pressure inside the skull brain damage can occur, leading to various forms of disability. That can be avoided if treatment is timely.

See also Birth defects; Edema.

Resources

Books

Ziegleman, David. *The Pocket Pediatrician.* New York: Doubleday, 1995.

Larry Blaser

Hydrochloric acid *see* **Hydrogen**

Hydrochlorofluorocarbons

Hydrochlorofluorocarbons (HCFCs) are compounds made up of **hydrogen**, **chlorine**, fluorine, and **carbon atoms**. HCFCs and their cousins, hydrofluorocarbons (HFCs), were created in the 1980s as substitutes for **chlorofluorocarbons (CFCs)** for use in refrigeration and a wide variety of manufacturing processes. Because all three of these classes of compounds either destroy the stratospheric **ozone** layer essential to life on **Earth**, and/or contribute to an unnatural warming of the planet's climate, international agreements have been signed to eliminate their production and use by either the year 2000 (CFCs) or 2040 (HCFCs and HFCs).

Why HCFCs?

Thomas Midgley, an organic chemist working at the Frigidaire division of General Motors, created chlorofluorocarbons in 1928 as a safe and inexpensive coolant for use in refrigerators and air conditioners. CFCs are non-flammable, non-toxic, non-corroding gases. In addition to their widespread use as coolants, they were used in the manufacturing of hundreds of products, such as contact lenses, telephones, artificial hip joints, foam for car seats and furniture, and computer circuit boards. CFCs have also been used as a propellant of aerosol products.

By 1974, however, researchers discovered that CFCs emitted to the atmosphere slowly traveled to the upper-altitude layer known as the stratosphere, higher than about 15 mi (25 km) above Earth's surface. The CFCs are degraded in the stratosphere by solar ultraviolet **radiation**, and this releases chlorine radicals that attack ozone molecules. Although ozone in the lower atmosphere is a harmful pollutant, in the stratosphere it acts to shield organisms at the surface of Earth from the harmful effects of solar ultraviolet radiation.

When ultraviolet radiation in the stratosphere degrades CFCs or HCFCs, the chlorine released acts to consume ozone molecules, which contain three **oxygen** atoms, into separate chlorine-oxygen and two-oxygen molecules (the latter is known as oxygen gas). Because the chlorine atoms can persist in the stratosphere for more than a century, they are recycled through the ozone-degrading reactions; one chlorine atom can destroy up to 100,000 molecules of stratospheric ozone.

The use of CFCs as aerosol propellants was banned in the United States, Canada, Switzerland, and the Scandinavian countries in 1978, as the dangers posed by their use were increasingly understood. By the early 1980s, companies such as DuPont, the world's largest manufacturer of CFCs, were creating alternate, less damaging compounds, including HCFCs and HFCs.

The good news and the bad news

HCFC compounds react differently from CFCs. This is because the HCFCs contain a hydrogen atom, which causes these chemicals to decompose photochemically before they reach the stratosphere. HFCs do not contain chlorine and thus do not attack the ozone layer. HCFCs and HFCs survive in the atmosphere for two to 40 years, compared with about 150 years for CFCs.

As a result of their shorter persistence and different molecular composition, HCFC and HFC compounds are expected to replace CFCs in most major uses, including the production of foams for insulation, furniture, and vehicle seats, and as a coolant in refrigerators and air conditioners.

HCFCs and HFCs are much more expensive to manufacture than CFCs, and they still negatively affect Earth's atmosphere to some degree. Although HCFCs destroy 98% less ozone in the stratosphere than do CFCs, HCFCs and HFCs are still greenhouse gases that may contribute to **global warming**. In comparison to **carbon dioxide**, a more common greenhouse gas, CFCs are about 4,100 times more efficient in their global warming potential, while HFCs are 350 times more effective.

The future of HCFCs

CFCs and HCFCs have contributed to our quality of life, particularly as valuable components in refriger-

KEY TERMS

Chlorofluorocarbons (CFCs)—Chemical compounds containing chlorine, fluorine and carbon. CFCs were a key component in the development of refrigeration, air conditioning, and foam products.

Greenhouse gases—Gases that contribute to the warming of the earth's atmosphere. Examples include carbon dioxide, HCFCs, CFCs, and HFCs.

Hydrofluorocarbons (HFCs)—Chemical compounds that contain hydrogen, fluorine, and carbon atoms.

Montreal Protocol on Substances that Deplete the Ozone Layer—An agreement signed by 43 countries in 1987, and amended and signed by 90 nations in 1990, to eliminate the production and use of compounds that destroy the ozone layer.

Ozone—A gas made up of three atoms of oxygen. Pale blue in color, it is a pollutant in the lower atmosphere, but essential for the survival of life on

Earth's surface when found in the upper atmosphere because it blocks dangerous ultraviolet solar radiation.

Ozone layer—A layer of ozone in the stratosphere that shields the surface of Earth from dangerous ultraviolet solar radiation.

Stratosphere—A layer of the upper atmosphere above an altitude of 5–10.6 mi (8–17 km) and extending to about 31 mi (50 km), depending on season and latitude. Within the stratosphere, air temperature changes little with altitude, and there are few convective air currents.

Troposphere—The layer of air up to 15 mi (24 km) above the surface of the earth, also known as the lower atmosphere.

Ultraviolet radiation—Radiation similar to visible light but of shorter wavelength, and thus higher energy.

ation and computer technology. However, their impact on the atmosphere has prompted several countries to agree to stop producing them. The Montreal Protocol on Substances that Deplete the Ozone Layer was signed by 43 countries in 1987 to limit and eventually eliminate the production and use of CFCs. When additional evidence emerged that the ozone layer was being damaged more quickly than originally thought, more than 90 countries signed an amendment to the Montreal Protocol in 1990. In the year 2000, CFCs were banned from use, and guidelines included new phaseouts for HCFCs and HFCs by the year 2020 if possible, and no later than 2040.

Other research is increasing the need to develop acceptable alternatives to HCFCs. In laboratory tests, male **rats** exposed to 5,000 parts per million (ppm) of HCFCs over a two-year period (equivalent to what humans working occupationally with the compound might experience over 30-40 years) developed tumors in the pancreas and testes. The tumors were benign and did not result in death for the tested rats. Nevertheless, this research resulted in the recommended eight-hour occupational exposure levels to HCFCs for humans being reduced from 100 ppm to 10 ppm.

Two possible alternatives to HCFCs are already being used successfully. Refrigerators that use propane gas, **ammonia**, or **water** as coolants are being tested in research laboratories, and are using up to 10% less **energy** than typical models using CFCs as

a coolant. **Telephone** companies are experimenting with crushed orange peels and other materials to clean computer circuit boards, as substitutes for another important use of CFCs and HCFCs. Certain **microorganisms** are also being tested that degrade HCFCs and HFCs, which could help in controlling emissions of these compounds during manufacturing processes involving their use.

See also Greenhouse effect; Ozone layer depletion.

Resources

Books

Duden, Jane. *The Ozone Layer.* New York: Crestwood House, 1990.

Fisher, David E. *Fire and Ice.* New York: HarperCollins, 1990.

Fisher, Marshall. *The Ozone Layer.* New York: Chelsea House Publishers, 1992.

Gay, Kathlyn. *Air Pollution.* New York: Franklin Watts, 1991.

Jahn, F., M. Cook, and M. Graham. *Hydrocarbon Exploration and Production. Developments in Petroleum Science.* Vol. 46. The Netherlands: Elsevier Science, 2000.

Periodicals

MacKenzie, Debora. "Cheaper Alternatives for CFCs." *New Scientist* (June 30, 1990): 39-40.

Wallington, Timothy J., et al. "The Environmental Impact of CFC Replacement-HFCs and HCFCs." *Environmental Science & Technology* 28 (1994): 320A-326A.

Sally Cole-Misch

Hydrofoil

The hydrofoil is very similar to the **hovercraft**, because it moves in the boundary between air and **water**. It avoids drag by lifting itself out of the water, using wing-shaped structures called hydrofoils that extend into the water from the craft. These hydrofoils function like the wings on a plane, creating lift and flying the hull above the surface of the water.

The first person to work on this idea was a French priest, Ramus, in the mid-1800s. However, there was no engine that could supply sufficient thrust. In the 1890s, another Frenchman, the Count de Lambert, tried and failed to make a working model using a gasoline engine.

The first successful hydrofoil boats were created in the early 1900s. Enrico Forlanini, an Italian **airship** designer, built a small boat with hydrofoils in 1905. He showed Alexander Graham Bell a later model that impressed the famous American. Bell built one himself, based on Forlanini's patented design and set a water-based speed record of 71 mph (114 kph) with it in 1918. This record stood until the 1960s.

Although there were small improvements made over the next few decades, hydrofoils did not see commercial use until the 1950s, when Hans von Schertel, a German scientist, developed his designs for passenger hydrofoils. Italy created their Supramar boats, and Russia and the United States developed hydrofoils with both commercial and military applications.

There have been experiments with various types of foils and different types of engines, including the gas **turbine**, diesel, gasoline, and jet engines.

The foils themselves have two distinct shapes. The surface-piercing models are V-shaped, so that part of the foil stays out of the water. This type is good for calm surfaces like **rivers** and lakes. The other foil is completely submerged. It usually consists of three foils extending straight down beneath the boat. Hydrofoils with this configuration need autopilots to keep them level. Whenever the boat shifts to one side, sensors send messages to flaps on the foils, which then adjust automatically to bring the boat back to a normal position.

Hydrofoils today are used by commuter services, fishery patrols, fire fighters, harbor control, water police, and air-sea rescues. For the military, hydrofoils can be excellent small **submarine** chasers and patrol craft.

Hydrogen

Hydrogen is the chemical element of **atomic number** 1. Its symbol is H, it has an **atomic weight** of 1.008,

its specific gravity at 32°F (0°C) is 0.0000899, and it melts at -434.7°F (-259.3°C). The **boiling point** of hydrogen is -423.2°F (-252.9°C), just above **absolute zero**. Boiling liquid hydrogen is the coldest substance known, with the exception of liquid helium. At room **temperature**, hydrogen is a colorless, odorless, tasteless gas. It consists of two stable isotopes of **mass** numbers 1 and 2.

Hydrogen is "number one" among the chemical elements. That is, it is the element whose atomic number is one. Its **atoms** are the simplest and lightest of all. A hydrogen atom contains only one **electron**, and it has a nucleus that consists of nothing but a **proton**. (A small percentage of hydrogen nuclei also contain one or two neutrons; see below.) In the **periodic table**, it is in a class by itself; there are no other members of its exclusive "group." It is usually placed at the top, all by itself.

Hydrogen's name is a clue to its most important position among the world's elements. It comes from the Greek *hydro*, meaning **water**, and *genes*, meaning born or formed. Hydrogen is a substance that gives birth to water (with a little help from **oxygen**). The name was coined in 1783 by the French chemist Antoine Lavoisier (1743-1794) in honor of the fact that when hydrogen burns in air it reacts with oxygen to form water, H_2O.

Hydrogen is everywhere

There are roughly 170 million billion tons of hydrogen tied up in the earth's supply of water. Hydrogen is therefore the most abundant of all elements on **Earth**. (Remember, there are twice as many hydrogen atoms in water as there are oxygen atoms.) Because the stars are mostly made of hydrogen, it is also the most abundant element in the universe, making up about 93% of all the atoms, and about three-quarters of the mass of the entire universe. Closer to home, 61% of all the atoms in the human body are hydrogen atoms.

Every one of the 13 million known organic compounds contains hydrogen. Hydrocarbons—compounds that contain nothing but hydrogen and **carbon** atoms—are the foundation upon which the vast world of organic chemicals is built. The **proteins**, carbohydrates, fats and oils, **acids and bases** that make up all plants and animals are organic, hydrogen-containing compounds. **Petroleum** and **coal**, which are made from ancient plants and animals, are vast deposits of hydrocarbons.

Hydrogen is the source of most of the **energy** of the **sun** and stars. At the 10-million-degree temperatures of the interiors of stars, not only are hydrogen molecules separated into atoms, but each atom is ionized—separated into an electron and a nucleus. The nuclei, which are simply protons, fuse together, forming nuclei of helium atoms and giving off a great deal of energy in the process. By a series of such reactions, all of the heavier elements have been built up from hydrogen in the stars.

The element

Hydrogen gas consists of *diatomic* (two-atom) molecules, with the formula H_2. It is the lightest of all known substances. There is only about 0.05 part per million of hydrogen gas in the air. It rises to the top of the atmosphere and is lost into **space**. It is continually being replaced by volcanic gases, by the decay of organic **matter**, and from coal deposits, which still contain some of the hydrogen from when they were decaying organic matter.

There are three isotopes of hydrogen, two stable and one radioactive. Like all isotopes, they have the same number of protons in the nucleus (in this case, one) but differing numbers of neutrons. Hydrogen is the only element whose isotopes go by their own names: protium (used only occasionally, when it is necessary to distinguish it from the others), **deuterium**, and **tritium**. Their mass numbers are one, two, and three, respectively. Protium, the most common hydrogen **isotope**, constitutes 99.985% of all hydrogen atoms; it has no neutrons in its nuclei. Deuterium, the other stable isotope, has one **neutron** in its nucleus; it constitutes 0.015% of all hydrogen atoms—that's about one out of every 6,700 atoms. Water made out of deuterium instead of protium is called heavy water; it is used as a moderator—a slower of neutrons—in nuclear reactors. Tritium has two neutrons and is radioactive, with a **half-life** of 12.33 years. In spite of its short lifetime, it remains present in the atmosphere in very tiny amounts because it is constantly being produced by cosmic rays. Tritium is also produced artificially in nuclear reactors. It is used as a radioactive tracer and as an ingredient of luminous paints and hydrogen bombs.

Discovery and preparation

Hydrogen is so easy to make by adding a **metal** to an acid that it was known as early as the late fifteenth century. Paracelsus (1493?-1541) made it by adding **iron** to **sulfuric acid**, but it wasn't until 1766 that it was recognized as a distinct substance, different from all other gases, or what were then called "airs." Henry Cavendish (1731-1810), an English chemist, gets the credit for this realization and hence for the discovery of hydrogen. Only in modern times, however, were isotopes of elements discovered. In 1932 Harold Urey (1893-1981) discovered deuterium by separating out the small amounts of it that are in ordinary water. This was the first separation of the isotopes of any element.

Hydrogen can be prepared in several ways. Many metals will release bubbles of hydrogen from strong acids such as sulfuric or hydrochloric acid. Hot steam (H_2O) in contact with carbon in the form of coke reacts to produce a mixture of hydrogen and **carbon monoxide** gases. Both of these products are flammable, and this so-called" water gas" mixture is sometimes used as a fuel, although it is dangerous because carbon monoxide is poisonous. Passing an **electric current** through water—electrolysis—will break it down into bubbles of oxygen gas at the **anode** (positive electrode) and hydrogen gas at the **cathode** (**negative** electrode).

Uses of hydrogen

Hydrogen and **nitrogen** gases can react to form **ammonia**:

$$N_2 \quad + \quad 3H_2 \quad \rightarrow \quad 2NH_3$$

| nitrogen gas | hydrogen gas | ammonia gas |

This reaction, called the Haber process, is used to manufacture millions of tons of ammonia every year in the United States alone, mostly for use as fertilizer. The Haber process converts nitrogen from the air, which plants cannot use, into a form (ammonia) that they can use. In order to get the biggest yield of ammonia, the reaction has to be carried out at a high **pressure** (500 times normal **atmospheric pressure**) and a high temperature (842°F [450°C]). To make it go faster, a catalyst is also used. More than two-thirds of all the hydrogen produced in the world goes into making ammonia.

A lot of hydrogen is used to make methyl alcohol—about 4 million tons of it a year in the U.S.:

$$2H_2 \quad + \quad CO \quad \rightarrow \quad CH_3OH$$

| hydrogen gas | Carbon monoxide gas | methyl alcohol gas |

Methyl **alcohol** is a flammable, poisonous liquid that is used as a solvent and in the manufacture of paints, cements, inks, varnishes, paint strippers, and many other products. It is what burns in the camping fuel, Sterno.

Another major use of hydrogen is in the *hydrogenation* of unsaturated fats and oils. If the molecules of a **fat** contain some double bonds between adjacent carbon atoms, as most **animal** fats do, they are said to be unsaturated. Treating them with hydrogen gas "fills up" or saturates the double bonds: the hydrogen atoms add themselves to the molecules at the double bonds, converting them into single bonds. Saturated (all-single-bond) fats have higher melting points; they're not as soft, they're more stable, and they stand up to **heat** better in frying. That's why "hydrogenated vegetable oil" on many food labels. Saturated fats raise people's **blood cholesterol** and increase the risk of **heart disease**.

In the oxyhydrogen torch, the potentially violent reaction between hydrogen and oxygen is controlled by feeding the gases gradually to each other, thereby turning a potential explosion into a mere **combustion**. The resulting flame is extremely hot and is used in **welding**.

Hydrogen disasters

Liquid hydrogen, combined with liquid oxygen, is the fuel that sends space shuttles into **orbit**. The reaction between hydrogen and oxygen to form water gives off a large amount of energy. They are useful as a rocket fuel because in their liquid forms, large quantities of them can be stored in a small space. They are very dangerous to handle, however, because unless they are kept well below their boiling points (hundreds of degrees below **zero**), they will boil and change into gases. Under certain conditions, hydrogen gas in the air can explode, while oxygen gas can feed the slightest spark into a fiery inferno if there is anything combustible around. Mixed together, they make a highly explosive mixture. These sobering facts turned into disaster on January 28, 1986, when the Challenger **space shuttle** exploded shortly after liftoff, killing all seven astronauts aboard. A rubber seal had failed, spilling the explosive gases out into the jet of flame that resulted in the explosion of the center fuel tank.

An earlier flying tragedy caused by hydrogen was the explosion on May 6, 1937 of the German zeppelin (a dirigible, or blimp), *Hindenburg*. At that time, hydrogen was used as the lighter-than-air filling in dirigibles. The *Hindenburg* caught fire while mooring at Lakehurst, New Jersey after a transatlantic flight, and 36 people were killed. Ever since then, nonflammable helium gas has been used instead of hydrogen as the filling in dirigibles. It is not as buoyant, but it is completely safe.

Reactions of hydrogen

Having only one electron in each of its atoms, hydrogen has two options for combining chemically with another atom. For one thing it can pair up its single electron with an electron from a non-metal atom to make a shared-pair covalent bond. Examples of such compounds are H_2O, H_2S and NH_3 (water, hydrogen sulfide and ammonia) and virtually all of the millions of organic compounds. Or, it can take on an extra electron to become the negative ion H^-, called a hydride ion, and combine with a metallic positive ion. Examples are **lithium** hydride LiH and **calcium** hydride CaH_2, but these compounds are unstable in water and decompose to form hydrogen gas.

Hydrogen reacts with all the **halogens** to form hydrogen halides, such as **hydrogen chloride** HCl and hydrogen fluoride HF. These compounds are acids when dissolved in water, and are used among other things to dissolve metals and, in the case of HF, to etch **glass**.

With **sulfur**, hydrogen forms hydrogen sulfide, H_2S, a highly poisonous gas. Fortunately, hydrogen sulfide has such a strong and disagreeable odor that people can smell very tiny amounts of it in the air and take steps to put some distance between it and them.

Hydrogen as a clean fuel

When hydrogen burns in air, it produces nothing but water vapor. It is therefore the cleanest possible, totally nonpolluting fuel. This fact has led some people to propose an energy economy based entirely on hydrogen, in which hydrogen would replace gasoline, oil, **natural gas**, coal, and **nuclear power**. The idea is that hydrogen would be prepared by the **electrolysis** of sea water in remote coastal areas and sent to the cities in pipelines similar to the pipeline that brings natural gas from Alaska to the lower states. In addition to being used as a fuel, the hydrogen could be used in factories to produce a variety of useful chemicals (see above). The problems, however, are that hydrogen is a dangerous gas, and piping it around the country has its hazards. A more serious problem is that hydrogen is currently expensive, both in money and in energy cost. After all, where is the **electricity** supposed to come from in the first place, to electrolyze the sea water? It would have to be produced by burning coal or oil, which are hardly nonpolluting, or by nuclear power. In any energy-production scheme, the entire process must be considered, from beginning to end, with all of its ramifications. Only then can we decide whether or not there would be a net saving of energy or a reduction in overall **pollution**.

Resources

Books

Brady, James E., and John R. Holum. *Fundamentals of Chemistry.* New York: Wiley, 1988.

Greenwood, N. N., and A. Earnshaw. *Chemistry of the Elements.* 2nd ed. Oxford: Butterworth-Heinneman Press, 1997.

Kirk-Othmer Encyclopedia of Chemical Technology. 4th ed. Suppl. New York: John Wiley & Sons, 1998.

Lide, David R., ed. *Handbook of Chemistry and Physics.* 73rd ed. CRC Press, 1992-3, page 4-14.

Parker, Sybil P., ed., *McGraw Hill Encyclopedia of Chemistry.* 1993.

Sherwood, Martin, and Christine Sutton, eds., *The Physical World.* New York, Oxford University Press, 1991.

Umland, Jean B. *General Chemistry.* St. Paul: West Publishing, 1993.

Robert L. Wolke

Hydrogen bond *see* **Chemical bond**

Hydrogen chloride

Hydrogen chloride is a chemical compound composed of the elements **hydrogen** and **chlorine**. It readily dissolves in **water** to produce a **solution** called hydrochloric acid. Both substances have many important industrial applications, including those in **metallurgy**, and the manufacture of pharmaceuticals, dyes, and synthetic rubber. Hydrochloric acid is found in most laboratories since its strong acidic nature makes it an extremely useful substance in analyses and as a general acid. Because hydrogen chloride and hydrochloric acid are so closely related, they are usually discussed together.

Properties

Hydrogen chloride is represented by the **chemical formula** HCl. This means that a **molecule** of hydrogen chloride contains one atom of hydrogen and one atom of chlorine. At room **temperature** (about 77°F [25°C]) and at a **pressure** of one atmosphere, hydrogen chloride exists as a gas. Consequently it is generally stored under pressure in **metal** containers. A much more convenient way to use hydrogen chloride is by dissolving it in water to form a solution. Hydrogen chloride is very soluble in water, the latter dissolving hundreds of times its own **volume** of hydrogen chloride gas. The resulting solution is known as hydrochloric acid and this also is generally given the chemical formula HCl. Commercial hydrochloric acid usually contains 28-35% hydrogen chloride by weight, and is generally referred to as concentrated hydrochloric acid. When smaller amounts of hydrogen chloride are dissolved in water, the solution is known as dilute hydrochloric acid.

Hydrogen chloride is a colorless, nonflammable gas with an acrid odor. The gas condenses to a liquid at -121°F (-85°C) and freezes into a solid at -173.2°F (-114°C). Hydrochloric acid is a colorless, fuming liquid having an irritating odor. Both hydrogen chloride and hydrochloric acid are corrosive, and so must be treated with great care. Both substances strongly irritate the eyes and are highly toxic if inhaled or ingested. Exposure to hydrogen chloride vapor can damage the nasal passages and produce coughing, **pneumonia**, headaches and rapid throbbing of the **heart**, and death can occur from exposure to levels in air greater than about 0.2%. Concentrated hydrochloric acid solutions cause burns and **inflammation** of the skin. Chemists always wear protective rubber gloves and safety glasses when using either hydrogen chloride or hydrochloric acid, and generally work in a well ventilated area to reduce exposure to fumes.

While dry hydrogen chloride gas is fairly unreactive, moist hydrogen chloride gas (and hydrochloric acid solutions) react with many metals. Consequently, dry hydrogen chloride gas can be stored in metal containers, whereas solutions of highly corrosive hydrochloric acid must be handled in acid-proof materials such as **ceramics** or **glass**. When hydrochloric acid reacts with metals, hydrogen gas and compounds known as metal chlorides are usually generated. Metal chlorides are formed when a metal displaces the hydrogen from the hydrogen chloride. For example, zinc metal dissolves in hydrochloric acid to form hydrogen gas and zinc chloride. Both moist hydrogen chloride and hydrochloric acid also react with many compounds including metal oxides, hydroxides, and carbonates. These are all examples of basic compounds, which neutralize hydrochloric acid, and form metal chlorides.

Obviously, hydrochloric acid is acidic. Like most acids, hydrogen chloride forms hydrogen ions in water. These are positively charged **atoms** of hydrogen that are very reactive and are responsible for all acids behaving in much the same way. Because all the hydrogen atoms in hydrogen chloride are converted into hydrogen ions, hydrochloric acid is called a strong acid. Nitric and sulfuric acids are other examples of strong acids.

Early discovery of hydrogen chloride

The alchemists of medieval times first prepared hydrogen chloride by heating ordinary **salt** (**sodium chloride**) with **iron** sulfate. The German chemist Johann Glauber (1604-1668) made hydrogen chloride by the reaction of salt with **sulfuric acid**, and this became the common method for conveniently preparing hydrogen chloride in the laboratory. By passing hydrogen chloride gas into water, hydrochloric acid is produced. Because hydrogen chloride was first prepared from salt, hydrochloric acid was originally referred to as *spirits of salt.* Commercially, it was also commonly called muriatic acid, from the Latin *muria,* meaning brine, or salt water. Hydrochloric acid dissolves many substances, and the alchemists found this acid to be very useful in their work. For example, it was used to dissolve insoluble ores thereby simplifying the methods of chemical analysis to determine the metal content of the ores. A mixture of hydrochloric acid and **nitric acid** (known as *aqua regia*) also became very useful since it was the only acid that will dissolve gold.

Preparation and uses

Hydrogen chloride can be prepared on an industrial scale from the reaction of salt with sulfuric acid. It is also formed rapidly above 482°F (250°C) by direct combination of the elements hydrogen and chlorine, and it is generated as a by-product during the manufacture of chlorinated **hydrocarbons**. Hydrochloric acid is obtained by passing hydrogen chloride gas into water.

Both hydrogen chloride and hydrochloric acid have many important practical applications. They are used in the manufacture of pharmaceutical hydrochlorides (water soluble drugs that dissolve when ingested), chlorine, and various metal chlorides, in numerous reactions of organic (**carbon** containing) compounds, and in the **plastics** and **textiles** industries. Hydrochloric acid is used for the production of **fertilizers**, dyes, artificial silk, and paint pigments; in the refining of edible oils and fats; in electroplating, leather tanning, refining, and **concentration** of ores, **soap** production, **petroleum** extraction, cleaning of metals, and in the photographic and rubber industries.

Small quantities of hydrochloric acid occur in nature in emissions from active volcanos and in waters from volcanic mountain sources. The acid is also present in digestive juices secreted by **glands** in the stomach wall and is therefore an important component in gastric digestion. When too much hydrochloric acid is produced in the **digestive system**, gastric **ulcers** may form. Insufficient secretion of stomach acid can also lead to digestion problems.

See also Acids and bases.

Resources

Books

Emsley, John. *Nature's Building Blocks: An A-Z Guide to the Elements.* Oxford: Oxford University Press, 2002.

Heiserman, D.L. *Exploring the Chemical Elements and Their Compounds.* Blue Ridge Summit, PA: Tab Publications, 1992.

Mahn, W.J. *Academic Laboratory Chemical Hazards Guidebook.* New York: Van Nostrand Rheinhold, 1991.

Salzberg, H.W. *From Caveman to Chemist.* Washington, DC: American Chemical Society, 1991.

Sittig, M. *Handbook of Toxic and Hazardous Chemicals and Carcinogens.* 3rd ed. Park Ridge, NJ: Noyes Publications, 1991.

Nicholas C. Thomas

Hydrogen peroxide

Hydrogen peroxide, H_2O_2, is a colorless liquid that mixes with **water** and is widely used as a disinfectant and a bleaching agent. It is unstable and decomposes (breaks down) slowly to form water and **oxygen** gas. Highly concentrated solutions of hydrogen peroxide are powerful oxidizing agents and can be used as rocket fuel.

Hydrogen peroxide is most widely found in homes in brown bottles containing 3% solutions (3% hydrogen peroxide and 97% water). The **decomposition** of hydrogen peroxide happens much faster in the presence of **light** so that an opaque bottle helps slow this process down. The decomposition of hydrogen peroxide can be summarized by the chemical equation:

$$2H_2O_2 \rightarrow 2H_2O + O_2 + \textbf{heat}$$

which states that two molecules of hydrogen peroxide break down to form two molecules of water and one **molecule** of oxygen gas, along with heat **energy**. This process happens slowly in most cases, but once opened a bottle of hydrogen peroxide will decompose more rapidly because the built-up oxygen gas is released. A totally decomposed bottle of hydrogen peroxide consists of nothing but water. Old unopened bottles of hydrogen peroxide often bulged out from the **pressure** of the oxygen gas that has built up over time. Some bottles have been known to "pop" from that pressure of the oxygen gas.

The most common uses of hydrogen peroxide are as a bleaching agent for hair and in the bleaching of pulp for **paper** manufacturing, and as a household disinfectant. As a **bleach**, hydrogen peroxide is an oxidizing agent (a substance that accepts electrons from other molecules). It is becoming more widely used than **chlorine** bleaches in industries because the products of its decomposition are water and oxygen while the decomposition of chlorine bleaches produces poisonous chlorine gas.

As a disinfectant, hydrogen peroxide is widely used on cuts and scrapes, and produces bubbling (caused by the formation of oxygen gasmolecules). The bubbling is quite rapid on cuts because of the presence of an **enzyme** (a protein catalyst—or molecule that speeds up a reaction) in **blood**, known as catalase. A similar bubbling can be observed if a small amount of hydrogen peroxide is put on a raw sliced **potato**, as the enzyme catalase is also found in potatoes.

Hydrogenation

Hydrogenation is a chemical reaction in which **hydrogen atoms** add to carbon-carbon multiple bonds. In order for the reaction to proceed at a practical **rate**, a catalyst is almost always needed. Hydrogenation reactions are used in many industrial processes as well as in the research laboratory, and occur also in living systems. We will look at a few examples in each category in this article.

The hydrogenation reaction

Hydrogen gas, H_2, can react with a **molecule** containing carbon-carbon double or triple bonds. In its simplest form, a molecule with one double bond would react with one molecule of hydrogen gas. An example is shown below.

$$H_2C = CH_2 \xrightarrow[\text{catalyst}]{H_2} H_3C - CH_3$$

Many **carbon** compounds have triple bonds, and in a case such as that, two molecules of hydrogen are necessary to completely saturate the carbon compound with hydrogen.

$$HC \equiv CH \xrightarrow[\text{catalyst}]{H_2} H_2C = CH_2 \xrightarrow[\text{catalyst}]{H_2} H_3C - CH_3$$

Hydrogenation of a double or triple carbon-carbon bond will not occur unless the catalyst is present. Scientists have developed many catalysts for this kind of reaction. Most of them include a heavy **metal**, such as platinum or palladium, in finely divided form. The catalyst adsorbs both the carbon compound and the hydrogen gas on its surface, in such a way that the molecules are arranged in just the right position for addition to occur. This allows the reaction to proceed at a fast enough rate to be useful.

Because at least one of the reagents (hydrogen) is a gas, often the reaction will occur at an even faster rate if it occurs in a pressurized container, at a **pressure** several times higher than **atmospheric pressure**.

Hydrogenation in the research laboratory

The hydrogenation reaction is a useful tool for a scientist trying to determine the structure of a new molecule. The **molecular formula**, showing the exact number of each kind of atom, can be determined in several ways, but discovering the arrangement of these atoms requires a large amount of detective work.

Sometimes, for example, a new substance is isolated from a **plant**, and a chemist needs to determine what the structure of this substance is. One method of attack is to find out how many molecules of hydrogen gas will react with one molecule of the unknown substance. If the **ratio** is, for example, two molecules of hydrogen to one of the unknown, the scientist can deduce that there are two carbon-carbon double bonds, or else one carbon-carbon triple bond in each molecule. Other kinds of chemical clues lead to the rest of the structure, and help the scientist to decide where in the unknown molecule the multiple bonds are.

One of the simplest uses of hydrogenation in the research laboratory is to make new compounds. Almost any organic molecule that contains multiple bonds can undergo hydrogenation, and this sometimes leads to compounds that were unknown before. In this way scientists have synthesized and examined many molecules not found in nature, or not found in sufficient quantity. These newly synthesized molecules are of use to humanity in a variety of ways.

Hydrogenation in industry

Many of the carbon compounds found in crude **petroleum** are of little use. These compounds may contain multiple bonds, but can be converted to saturated compounds by catalytic hydrogenation. This is one source for much of the gasoline that we use today. Other chemicals besides gasoline are made from petroleum, and for these, too, the first step from crude oil may be hydrogenation.

Another commercial use of the hydrogenation reaction is the production of fats and oils in more useful forms. Fats and oils are not hydrocarbons, like the simple molecules we have been looking at, since they contain **oxygen** atoms, too. But they do contain long chains of carbon and hydrogen, joined together in part by carbon-carbon double bonds. Partial hydrogenation of these molecules, so that some, but not all of the double bonds react, gives compounds with different cooking characteristics, more satisfactory for consumers in some situations than the original oils. This is the source of the "partially hydrogenated vegetable oil" on the grocery shelf.

Biological hydrogenation

Many **chemical reactions** within the body require the addition of two atoms of hydrogen to a molecule in order to maintain life. These reactions are much more complex than the ones described above, because hydrogen gas is not found in the body. These kinds of reactions require "carrier" molecules, which give up hydrogen atoms to the one undergoing hydrogenation. The catalyst in biological hydrogenation is an **enzyme**, a complex protein that allows the reaction to take place in the **blood**, at a moderate **temperature**, and at a rate fast enough for **metabolism** to continue.

Hydrogenation reactions can happen to many other types of molecules as well. However, the general features for all of the reactions are the same. Hydrogen atoms add to multiple bonds in the presence of a catalyst, to product a new compound, with new characteristics. This new compound has different properties than the original molecule had.

Resources

Books

Bettelheim, Frederick A., and Jerry March. *Introduction to General, Organic, and Biological Chemistry.* 3rd ed. Fort Worth: Saunders College Publishing, 1991.

KEY TERMS

. . . .

Addition—A type of chemical reaction in which two molecules combine to form a single new molecule.

Adsorb—To attach to the surface of a solid. The more finely divided the solid is, the more molecules can absorb on its surface.

Catalyst—Any agent that accelerates a chemical reaction without entering the reaction or being changed by it.

Fat—A solid ester of glycerol and long-chain carboxylic acids.

Le Châtelier's principle—A statement describing the behavior of mixtures undergoing a chemical reaction. This principle states that in a chemical reaction at its steady state, addition of more of a reactant or product will cause the readjustment of concentrations to maintain the steady state.

Oil—A liquid ester of glycerol and long-chain carboxylic acids.

Organic—A term used to describe molecules containing carbon atoms.

Saturation—A molecule is said to be saturated if it contains only single bonds, no double or triple bonds.

Carey, Francis A. *Organic Chemistry*. New York: McGraw-Hill, 2002.

Cross, Wilbur. *Petroleum*. Chicago: Children's Press, 1983.

Other

Chemicals from Petroleum. London: Audio Learning, 1982. 35mm Film strip.

G. Lynn Carlson

Hydrologic cycle

The hydrologic, or **water**, cycle is the continuous, interlinked circulation of water among its various compartments in the environment. Hydrologic budgets are analyses of the quantities of water stored, and the rates of transfer into and out of those various compartments.

The most important places in which water occurs are the **ocean**, **glaciers**, underground aquifers, surface waters, and the atmosphere. The total amount of water among all of these compartments is a fixed, global quantity. However, water moves readily among its various compartments through the processes of **evaporation**, **precipitation**, and surface and subsurface flows. Each of these compartments receives inputs of water and has corresponding outputs, representing a flow-through system. If there are imbalances between inputs and outputs, there can be significant changes in the quantities stored locally or even globally. An example of a local change is the **drought** that can occur in **soil** after a long period without replenishment by precipitation. An example of a global change in **hydrology** is the increasing **mass** of continental **ice** that occurs during glacial epochs, an event that can remove so much water from the oceanic compartment that **sea level** can decline by more than 328 ft (100 m), exposing vast areas of **continental shelf** for the development of terrestrial ecosystems.

Major compartments and fluxes of the hydrologic cycle

Estimates have been made of the quantities of water that are stored in various global compartments. By far the largest quantity of water occurs in the deep **lithosphere**, which contains an estimated 27×10^{18} tons (27-billion-billion tons) of water, or 94.7% of the global total. The next largest compartment is the oceans, which contain 1.5×10^{18} tons, or 5.2% of the total. Ice caps contain 0.019×10^{18} tons, equivalent to most of the remaining 0.1% of Earth's water. Although present in relatively small quantities compared to the above, water in other compartments is very important ecologically because it is present in places where biological processes occur. These include shallow **groundwater** (2.7×10^{14} tons), inland surface waters such as lakes and **rivers** [0.27×10^{14} ton], and the atmosphere [0.14×10^{14} tons]).

The smallest compartments of water also tend to have the shortest turnover times, because their inputs and outputs are relatively large in comparison with the mass of water that is contained. This is especially true of atmospheric water, which receives annual inputs equivalent to 4.8×10^{14} tons as evaporation from the oceans (4.1×10^{14} tons/yr) and terrestrial ecosystems (0.65×10^{14} tons/yr), and turns over about 34 times per year. These inputs of water to the atmosphere are balanced by outputs through precipitation of rain and snow, which **deposit** 3.7×10^{14} tons of water to the surface of the oceans each year, and 1.1×10^{14} tons/yr to the land.

These data suggest that the continents receive inputs of water as precipitation that are 67% larger than what is lost by evaporation from the land. The difference, equivalent to 0.44×10^{14} tons/yr, is made up by 0.22×10^{14} tons/yr of runoff of water to the oceans through rivers, and another 0.22×10^{14} tons/yr of subterranean runoff to the oceans.

The movements of water in the hydrologic cycle are driven by gradients of **energy**. Evaporation occurs in response to the availability of thermal energy and gradients of **concentration** of water vapor. The ultimate source of energy for virtually all natural evaporation of water on **Earth** is solar electromagnetic **radiation**. This solar energy is absorbed by surfaces, increasing their **heat** content, and thereby providing a source of energy to drive evaporation. In contrast, surface and ground waters flow in response to gradients of gravitational potential. In other words, unless the flow is obstructed, water spontaneously courses downhill.

Hydrologic cycle of a watershed

The hydrological cycle of a defined area of landscape is a balance between inputs of water with precipitation and upstream drainage, outputs as evaporation and drainage downstream or deep into the ground, and any internal storage that may occur because of imbalances of the inputs and outputs. Hydrological budgets of landscapes are often studied on the spatial scale of watersheds, or the area of terrain from which water flows into a stream, river, or **lake**.

The simplest watersheds are so-called headwater systems that do not receive any drainage from watersheds at higher altitude, so the only hydrologic input occurs as precipitation, mostly as rain and snow. However, at places where **fog** is a common occurrence, windy conditions can effectively drive tiny atmospheric droplets of water vapor into the forest canopy, and the direct deposition of cloud water can be important. This effect has been measured for a foggy **conifer** forest in New Hampshire, where fogwater deposition was equivalent to 33 in (84 cm) per year, compared with 71 in (180 cm) per year of hydrologic input as rain and snow.

Vegetation can have an important influence on the **rate** of evaporation of water from watersheds. This hydrologic effect is especially notable for well-vegetated ecosystems such as **forests**, because an extensive surface area of foliage supports especially large rates of **transpiration**. **Evapotranspiration** refers to the combined rates of transpiration from foliage, and evaporation from nonliving surfaces such as moist soil or surface waters. Because transpiration is such an efficient means of evaporation, evapotranspiration from any well vegetated landscape occurs at much larger rates than from any equivalent area of non-living surface.

In the absence of evapotranspiration an equivalent quantity of water would have to drain from the **water**-shed as seepage to deep groundwater or as streamflow. Studies of forested watersheds in Nova Scotia found that evapotranspiration was equivalent to 15-29% of the hydrologic inputs with precipitation. Runoff through streams or rivers was estimated to account for the other 71-85% of the atmospheric inputs of water, because the relatively impervious **bedrock** in that region prevented significant drainage to deep ground water.

Forested watersheds in seasonal climates display large variations in their rates of evapotranspiration and streamflow. This effect can be illustrated by the seasonal patterns of hydrology for a forested watershed in eastern Canada. The input of water through precipitation is 58 in (146 cm) per year, but 18% of this arrives as snow, which tends to accumulate on the surface as a persistent snowpack. About 38% of the annual input is evaporated back to the atmosphere through evapotranspiration, and 62% runs off as river flow. Although there is little seasonal variation in the input of water with precipitation, there are large seasonal differences in the rates of evapotranspiration, runoff, and storage of groundwater in the watershed. Evapotranspiration occurs at its largest rates during the growing season of May to October, and runoff is therefore relatively sparse during this period. In fact, in small watersheds in this region forest streams can literally dry up because so much of the precipitation input and soil water is utilized for evapotranspiration, mostly by trees. During the autumn, much of the precipitation input serves to recharge the depleted groundwater storage, and once this is accomplished stream flows increase again. Runoff then decreases during winter, because most of the precipitation inputs occur as snow, which accumulates on the ground surface because of the prevailing sub-freezing temperatures. Runoff is largest during the early springtime, when warming temperatures cause the snowpack to melt during a short period of time, resulting in a pronounced flush of stream and river flow.

Influences of human activities on the hydrologic cycle

Some aspects of the hydrologic cycle can be utilized by humans for a direct economic benefit. For example, the potential energy of water elevated above the surface of the oceans can be utilized for the generation of **electricity**. However, the development of hydroelectric resources generally causes large changes in hydrology. This is especially true of hydroelectric developments in relatively flat terrain, which require the construction of large storage reservoirs to retain seasonal high-water flows, so that electricity can be generated at times that suit the peaks of demand. These extensive storage reservoirs are essentially artificial lakes, sometimes covering enormous areas of tens of thousands of hectares. These types of hydroelectric developments cause great changes in river hydrology, especially by evening out the variations of flow, and sometimes by unpredictable spillage of water at times when the storage capacity of the reservoir is full. Both of these hydrologic influences have significant ecological effects, for example, on the **habitat** of **salmon** and other aquatic biota. In one unusual case, a large water spillage from a reservoir in northern Quebec drowned 10,000 **caribou** that were trapped by the unexpected cascade of water during their **migration**.

Where the terrain is suitable, hydroelectricity can be generated with relatively little modification to the timing and volumes of water flow. This is called run-of-the-river hydroelectricity, and its hydrologic effects are relatively small. The use of geologically warmed ground water to generate energy also has small hydrological effects, because the water is usually re-injecting back into the **aquifer**.

Human activities can influence the hydrologic cycle in many other ways. The volumes and timing of river flows can be greatly affected by channeling to decrease the impediments to flow, and by changing the character of the watershed by paving, compacting soils, and altering the nature of the vegetation. Risks of **flooding** can be increased by speeding the rate at which water is shed from the land, thereby increasing the magnitude of peak flows. Risks of flooding are also increased if **erosion** of soils from terrestrial parts of the watershed leads to siltation and the development of shallower river channels, which then fill up and spill over during high-flow periods. Massive increases in erosion are often associated with **deforestation**, especially when natural forests are converted into agriculture.

The quantities of water stored in hydrologic compartments can also be influenced by human activities. An important example of this effect is the **mining** of groundwater for use in agriculture, industry, or for municipal purposes. The best known case of groundwater mining in **North America** concerns the enormous Ogallala aquifer of the southwestern United States, which has been drawn down mostly to obtain water for **irrigation** in agriculture. This aquifer is largely comprised of "fossil water" that was deposited during earlier, wetter climates, although there is some recharge capability through rain-fed groundwater flows from mountain ranges in the watershed of this underground reservoir.

Sometimes industrial activities lead to large emissions of water vapor into the atmosphere, producing a local hydrological influence through the development of low-altitude **clouds** and fogs. This effect is mostly associated with electric power plants that cool their process water using cooling towers.

KEY TERMS

Evapotranspiration—The evaporation of water from a large area, including losses of water from foliage as transpiration, and evaporation from non-living surfaces, including bodies of water.

Hydrology—The study of the distribution, movement, and physical-chemical properties of water in Earth's atmosphere, surface, and near-surface crust.

Precipitation—The deposition from the atmosphere of rain, snow, fog droplets, or any other type of water.

Watershed—The expanse of terrain from which water flows into a wetland, waterbody, or stream.

A more substantial hydrologic influence on evapotranspiration is associated with large changes in the nature of vegetation over a substantial part of a watershed. This is especially important when mature forests are disturbed, for example, by **wildfire**, clear-cutting, or conversion into agriculture. Disturbance of forests disrupts the capacity of the landscape to sustain transpiration, because the amount of foliage is reduced. This leads to an increase in streamflow volumes, and sometimes to an increased height of the groundwater table. In general, the increase in streamflow after disturbance of a forest is roughly proportional to the fraction of the total foliage of the watershed that is removed (this is roughly proportional to the fraction of the watershed that is burned, or is clear-cut). The influence on transpiration and streamflow generally lasts until regeneration of the forest restores another canopy with a similar area of foliage, which generally occurs after about 5-10 years of recovery. However, there can be a longer-term change in hydrology if the ecological character of the watershed is changed, as occurs when a forest is converted to agriculture.

Resources

Books

Freedman, B. *Environmental Ecology.* 2nd ed. San Diego: Academic Press, 1995.

Herschy, Reginald, and Rhodes Fairbridge, eds. *Encyclopedia of Hydrology and Water Resources.* Boston: Kluwer Academic Publishing, 1998.

Ricklefs, R.E. *Ecology.* 3rd ed. New York: Freeman, 1990.

Periodicals

Berbery, Ernesto Hugo. "The Hydrologic Cycle of the La Plata Basin in South America." *Journal of Hydrometeorology* 3, no. 6 (2002): 630-645.

"Temperature And Rainfall Tables: July 2002." *Journal of Meteorology* 27, no. 273 (2002): 362.

Bill Freedman

Hydrology

Hydrology is the science of **water**. It is concerned with the occurrence and circulation of water on and within **Earth**, the physical and chemical properties of bodies of water, the relationship between water and other parts of the environment, and societal or economic aspects of water resources. Hydrology is an interdisciplinary field of study, and hydrologists have academic backgrounds that include **geology**, **engineering**, **biology**, **chemistry**, geography, **soil** science, economics, and **mathematics**.

The two major sub-disciplines of hydrology are surface water hydrology, which is concerned with water on or at Earth's surface, and **groundwater** hydrology (sometimes referred to as hydrogeology or geohydrology), which his concerned with the water beneath Earth's surface. Surface water hydrology includes the analysis and prediction of floods (as well as the meteorological events that produce them); the transfer of water from Earth's surface to its atmosphere by **evaporation**, **transpiration**, and sublimation; the study of sediment **erosion**, transportation, and deposition by flowing water; and investigations of water quality in lakes and streams. Groundwater hydrology involves the study of soils and **rocks** that comprise **aquifer** systems, the exploration for new groundwater resources using geological and geophysical methods, monitoring groundwater flow directions and velocities, and the remediation of contaminated groundwater.

Atmospheric water, surface water, and groundwater are linked together by the **hydrologic cycle**, which describes the continuous movement of water in its various forms and phases. The hydrologic cycle has no beginning or end. Surface water is transformed from a liquid to atmospheric water vapor as it is evaporated from open bodies of water and transpired by plants, or from solid to vapor by sublimation of snow at high elevations. **Atmospheric pressure** and **temperature** changes then transform the water vapor into liquid or solid water that falls to Earth's surface as rain or snow. A portion of the rain and snow is returned to Earth's surface by **rivers** and streams. Another portion seeps into Earth's crust to become groundwater via a process known as infiltration. Groundwater can be pumped from aquifers by humans or flow naturally into surface water bodies by way of

seeps and springs. In some cases, particularly in arid and semi-arid regions, rain can be returned to the atmosphere by evaporation before it reaches the ground. This cycle, with many variations, occurs continuously as water is recycled through the environment. Therefore, one of the principal activities of hydrologists is the development of water balances that quantify the different components of the water cycle in a particular region.

Hydrologists rely on many techniques to collect the data they need; some are simple and straightforward, such as the measurement of snow depth and the discharge of rivers and streams. Others are more elaborate, such as the use of remote-sensing techniques to assess the quantity and quality of water resources. The development and application of computer models that simulate hydrologic systems is also an important aspect of hydrology.

See also Alluvial systems; Aqueduct; Dams; Environmental impact statement; Evaporation; Flooding; Hydrolysis; Water conservation; Water microbiology; Water pollution; Water treatment; Watershed.

Hydrolysis

Whenever **water** reacts with another chemical compound, the process is called hydrolysis. Hydrolysis differs somewhat from hydration, although the two can occur together. Hydration is the bonding of whole water molecules to an ion (a charged atom or **molecule**), usually a **metal** ion. Hydrolysis, on the other hand, involves an actual chemical reaction of the water molecule itself with another reactant. **Aluminum** ion, for example, can bond with six water molecules to form the hydrated aluminum ion. In water, however, the hydrated ion can undergo hydrolysis; some of the hydrated molecules contribute a **hydrogen** ion to the **solution**, making the solution acidic.

Solutions of non-hydrated ions often become either acidic or basic because of hydrolysis, too. In general, **negative** ions (anions) form basic solutions if they hydrolyze, because the negative charge on the ion attracts the positively charged hydrogen ion (H+) away from water, leaving the basic hydroxide ion (OH-) behind. Similarly, positive ions (cations) form acidic solutions if they hydrolyze, because the positive charge on the ion attracts the negatively charged hydroxide ion away from water, leaving the acidic hydrogen ion behind. Hydrolysis of these ions only occurs, however, if the ion originally came from a weak acid or base, or the **salt** of a weak acid or base. (A salt is an ionic chemical compound derived from an acid or base, often as the result of a **neu-**

tralization reaction.) Ions do not hydrolyze if they are from strong acids or bases—such as chloride ion from hydrochloric acid or **sodium** ion from **sodium hydroxide** (a base)—or their salts.

In **biochemistry**, hydrolysis often involves the **decomposition** of a larger molecule. If a **fat** undergoes hydrolysis, for example, it reacts with water and decomposes to **glycerol** and a collection of **fatty acids**. Similarly, complex sugars can hydrolyze to smaller sugars, and nucleotides can hydrolyze to a five **carbon** sugar, a nitrogenous base, and **phosphoric acid**.

Hydroponics

There are several early examples of hydroponics, or soil-free agriculture, including the hanging gardens of Babylon and the floating gardens of China and Aztec Mexico. Early Egyptian paintings also depict the growing of plants in **water**.

In 1600, the Belgian Jan Baptista van Helmont (1579-1644) demonstrated that a willow shoot kept in the same **soil** for five years with routine watering gained 160 lb (73 kg) in weight as it grew into a full-sized **plant** while the soil in the container lost only 2 oz (57g). Clearly, the source of most of the plant's **nutrition** was from the water, not the soil.

During the 1860s, German scientist Julius von Sachs (1832-1897) experimented with growing plants in water-nutrient solutions, calling it *nutriculture.*

In 1929, W. F. Gericke of the University of California first coined the term hydroponics, which literally means "water labor." Gericke demonstrated commercial applications for hydroponics and became known for his twenty-five-foot tomato plants. Hydroponics has been shown to double crop yield over that of regular soil. It can be categorized into two subdivisions: *water culture,* which uses the Sachs water-nutrient **solution**, with the plants being artificially supported at the base; and *gravel culture,* which uses an inert medium like **sand** or gravel to support the plants, to which the water-nutrient solution is added.

Hydroponics was used successfully by American troops stationed on non-arable islands in the Pacific Ocean during the 1940s. It has also been practiced to produce fresh produce in arid countries like Saudi Arabia. In the 1970s, researcher J. Sholto Douglas worked on what he called the *Bengal hydroponics system.* He sought to simplify the methods and equipment involved in hydroponics so it could be offered as a partial solution for food shortages in India and other developing countries.

Successfully adopted in certain situations, hydroponics will remain in limited use as long as traditional farming methods in natural soil can support the world's population.

Hydrosphere

Hydrosphere refers to that portion of **Earth** that is composed of **water**. The hydrosphere represents one component of Earth's system, operating in conjunction with the solid crust (**lithosphere**) and the air that envelopes the **planet** (atmosphere). The derivation of the term hydrosphere, from the Greek words for water and ball, is truly descriptive of our world, as it reflects the abundance and importance of water on Earth.

On Earth, water exists in the three primary **states of matter**; liquid, solid, and gas. The distance of Earth from the **Sun**, by fortunate coincidence, is such that the amount of **energy** arriving at the surface of most of the planet is sufficient to elevate the ambient **temperature** to levels above the freezing point of water, yet insufficient to cause all of the water to evaporate into the gaseous state. The capacity of water to store large quantities of **heat** energy heavily influences the nature of the **global climate**. The presence of large bodies of liquid water and the atmosphere restrict the range of temperature fluctuations on Earth. These conditions have allowed the existence of the fourth component of Earth's system, the **biosphere**.

Water is constantly being cycled through its various manifestations and through the components of Earth's systems by means of the **hydrologic cycle**. Driven by solar energy, water is evaporated from the **ocean** surface and distributed over the earth as water vapor. **Precipitation** returns the water, in liquid and solid forms, to other parts of the globe. Throughout the cycle, water may exist in a number of forms, interact with the atmosphere and lithosphere, or may be utilized by organisms within the biosphere.

One commonly cited statistic asserts that 71% of the surface area of our planet is covered by water, with the largest part covered by oceans. The total **volume** of seawater, amounting to 97.2% of all the water on the planet, is 295,000,000 mi^3 (1,230,000,000 km^3). Usable **freshwater** constitutes less than 0.5% of all water on Earth. Water in all **rivers**, lakes and streams totals only 29,800 mi^3 (124,200 km^3). The amount of **groundwater** that is within 0.5 mi (0.8 km) of the surface is 960,000 mi^3 (4,000,000 km^3). Water also exists on Earth in the solid form as icecaps and **glaciers**, occupying a volume of

6,900,000 mi³ (28,600,000 km³). Straddling the division between hydrosphere and atmosphere is water vapor. A volume of 3,000 mi³ (12,700 km³) of water can be found in the atmosphere.

Hydrothermal vents

Hydrothermal vents are places where hot fluids (up to 752°F [400°C]) related to volcanic activity are released from the **ocean** floor. Because of the high **pressure** exerted by the **water** at depth on the sea floor, hydrothermal fluids can exceed 212°F (100°C) without boiling. The most visible indications of on-going volcanic activity are the plumes of hot fluids issuing from hydrothermal vents, which have been directly observed by scientists in deep-sea submersible vessels. Oceanographer Jack Corliss is credited with discovering the seafloor geysers in volcanic ridges in the Pacific Ocean in 1977.

These vents can occur as cracks in the top of cones of basalt (a dark, fine-grained rock that makes up most of the earth's crust). Or, the vents can issue from chimney-like structures that extend upward from the ocean floor. Some vents have lower fluid temperatures and release light-colored precipitates of silica; these vents are called "white smokers." But often, the fluids are black due to the presence of very fine sulfide mineral particles that precipitate out as the fluids cool. The sulfides present in these "black smokers" may contain amounts of **iron**, **copper**, zinc, and other metals that have been dissolved from underlying fresh basalt and concentrated in the hot solutions. These **minerals** can accumulate around the vents as sulfide deposits in mounds or chimney shapes up to 148 ft (45 m) high.

Hydrothermal vents usually occur along mid-ocean ridges where erupting basalt cools and creates new sea floor. The exact locations of the vents are controlled by cracks and faults in the basaltic rock. Isolated hydrothermal vents have also been found on **seamounts** and in Lake Baikal in Siberia.

Along the mid-ocean ridges, the **heat** of the **magma** that rises continuously from the mantle to form new oceanic crust causes water to convect through the top mile or two (2–3 km) of oceanic crust over many thousands of square miles. Down-convected ocean water encounters hot **rocks** at depth, is heated, yields up its dissolved **magnesium**, and leaches out manganese, copper, **calcium**, and other metals. This hot, chemically altered brine then convects upward to the ocean floor, where it is cooled and its releases most of its dissolved minerals as solid precipitates. This process makes the concentrations

of vanadium, cobalt, nickel, and copper in recent seafloor sediments near mid-ocean ridges 10–100 times greater than those elsewhere, and has formed many commercially important ores.

Two of the metals transported in large quantities by sea-floor circulation (i.e., calcium and magnesium) are important controllers of the **carbon dioxide** (CO_2) balance of the ocean and thus of the atmosphere. A **volume** of water approximately equal to the world's oceans passes through the hydrothermal mid-ocean ridge cycle every 20 million years.

In the late 1980s, a mysterious illumination coming from some hydrothermal vents not visible to human eyes was discovered, and it has yet to be explained. Scientists at first thought the **light** was thermal **radiation** from the hot water, but other explanations have been proposed including crystalloluminescence (**salt** in the water responding to the heat) or chemiluminescence (from **energy** released during **chemical reactions** in the water). The faint glow is certainly important to the life forms around the vents.

The vents support living communities called ecotones that are transition zones between the hot vent water and the surrounding cold ocean water. The unusual forms of sea life that surround the hydrothermal vents include giant clams, tube worms, and unique types of **fish** that thrive on the energy-rich chemical compounds transported by hydrothermal fluids from the vents. This is the only environment on **Earth** supported by a food chain that does not depend on the energy of the **sun** or **photosynthesis** and lives by chemosynthesis instead. If the light source is sufficient to cause photosynthesis on the ocean floor, this is the only known photosynthesis not initiated by the Sun. Scientists have also found an apparently blind **species** of **shrimp** around the vents; instead of eyes, the shrimp has light-sensing patches on its back suggesting that **evolution** adapted the creature to the faint light source. Microbes called hyperthermophiles have also been found in vent water. The heat from the vents and the unusual life forms have prompted speculation that life on Earth originated on the sea floor near the vents or repopulated the **planet** after asteroid impacts. In fact, astrobiologists greatly interested in research on the origins of life prompted by these deep-sea finds.

See also Bedrock; Hot spot; Volcano.

Hydrozoa

Hydrozoa (phylum Coelenterata, class Hydrozoa) are coelenterates that are closely related to the **hydra**,

Medusoid—The generative bud of a sessile hydrozoa that resembles a Medusa's head.

Pelagic—Refers to the open ocean.

Polyp—Mature hydrozoa distinguished by a cylindrical body that has an oral opening surrounded by tentacles and an arboreal end that may be fixed in substrate.

Sessile—Unable to move about.

sea anemones, corals, and **jellyfish**. Although a large majority of these **species** are common and widespread, they are often overlooked, as they are all small animals. The vast majority are marine species, but several **freshwater** hydrozoans have also been identified, for example, *Cordylophora lacustris* and *Craspedacusta sowerbyi*.

There is considerable variation in the structure and appearance of hydrozoans. All species, however, have a stalk-like arrangement known as a polyp, which bears a number of tentacles, as well as the mouth. In addition, two main types of hydrozoa exist: a polypoid structure which is sessile, remaining in the same place, and a medusoid form which is free-swimming. Many polypoid hydrozoans, however, may have a medusoid larval phase which eventually settles onto some substrate. Some species are solitary, but the majority are colonial. In the latter, the colony arises from a single basal root which rests on the substrate and from which individual polyps arise. Both colonial and individual species lack a hard outer skeleton.

A special feature among colonial species is the presence of individual polyps that fulfil separate roles. Some polyps are, for example, specialized for feeding (gastrozooids), while others are responsible for reproduction (gonozooids) or defense. While most colonial sessile species are small and feed by filtering **zooplankton** from the surrounding **water currents**, some of the medusoid forms are quite large and capable of feeding on small **fish**. Floating pelagic species such as *Porpita* and *Velella*, which resemble small jellyfish, may reach 1.5-2.5 in (4-6 cm) in diameter. These are colonial species made up of large numbers of gastrozooids and gonozooids; the body is modified into a flattened structure with a float on the upper surface to provide buoyancy. Some species, such as *Velella*, or the by-the-wind-sailor, as it is commonly known, have an additional small "sail" on the upper surface to catch the **wind** and assist further with dispersal.

Hyena

Hyenas (or hyaenas) are African and Asian dog-like carnivores in the family Hyaenidae, order Carnivora, which also includes the dogs, **cats**, **seals**, and **bears**. Hyenas are very powerfully built animals with a stout head, a short snout, short ears, and powerful jaws with strong teeth, useful for crushing bones to get at the **nutrients** contained inside. The neck of the hyena is rather elongate, and the hind legs are somewhat smaller and lower than the forelegs. The four-toed paws have nonretractable claws, used for digging their burrows, or for gripping the ground while tearing away at the carcass of a dead **animal**. The fur is coarse and mostly comprised of guard hairs, and some **species** have a relatively long mantle on the back of the neck, which can be erected as an aggressive display.

Hyenas commonly feed on carrion, and on the remains of the kills of larger predators. Hyenas find the kills of others by using their keen sense of **smell**, and by observing the movements of large predators and scavenging **birds**, such as **vultures**. Spotted hyenas are particularly capable predators, often hunting in packs, and can take down **prey** that is considerably larger than themselves, even as large as domestic cattle. Striped hyenas have been reported to scavenge human bodies that are not buried deeply enough. When wild food is scarce, the larger spotted hyena will sometimes attack and kill people.

Hyenas are social animals, living in groups and hunting and scavenging in packs. Mature female hyenas can give **birth** once each year, to a litter of three or four brown cubs.

Hyenas tend to be nocturnal prowling animals, usually resting near their rocky lairs or burrow openings during the day. However, they are sometimes active during the day, although not when it is intensely hot. As they move about, especially at night, hyenas commonly make diverse noises such as barking and hysterical laughing, which are unsettling to many people.

Many people consider hyenas to be ugly and offensive creatures, because of their outward appearance—a crooked-legged stance, strange face, raucous noises, rough pelage, apparent cowardice, smelliness, and scavenging habits. Hyenas are commonly considered to be **pests**, because they sometimes kill **livestock**, they occasionally attack people, and because they are just generally disliked. For these reasons, hyenas are often killed by shooting or poisoning. However, hyenas play an important role as scavengers and cullers-of-the-weak in their **ecosystem**. Hyenas should be respected for their provision of this valuable service, as well as for their intrinsic value as wild animals.

Pack of hyenas feeding on a zebra. *Photograph by Leonard Lee Rue, III. Photo Researchers, Inc. Reproduced by permission.*

Species of hyena

The spotted or laughing hyena (*Crocuta crocuta*) is the largest species of hyena, occurring in open habitats throughout sub-Saharan **Africa**. Adults of this species typically weigh 130-180 lb (59-82 kg), and have a gray-red coat with numerous dark brown spots. The spotted hyena does not have a mane on the back of its neck. This animal is said to have the strongest jaws and teeth in proportion to its size of any animal; it can crack the largest bones of even cow-sized animals to get at the nutritious marrow inside. The spotted hyena lives in large, territory-holding groups. These animals are sometimes solitary hunters, but they usually hunt in packs.

The spotted hyena was long thought to mostly eat carrion, but closer study of the habits of this nocturnal **predator** has shown that it is an efficient hunter. Spotted hyenas are often seen waiting, in accompaniment with vultures and jackals, while lions eat their fill of a recently killed animal. However, in many of these cases it is likely that the hyenas actually did the killing at night, but were then quickly chased off the carcass by the lions. In these situations, the hyenas must patiently wait until the lions finished their scavenging of the hyena kill, before the actual hunter can eat.

Interestingly, in many places spotted hyenas and lions have developed a deep, mutual enmity for each other. Lions can be quite intolerant of nearby hyenas, and will often chase and sometimes kill these animals, although they do not eat them. There are also cases in which groups of hyenas have cornered individual or pairs of lions, forcing them to climb a **tree** for refuge, and sometimes killing the large cats. **Reciprocal**, lethal rivalry of this sort is rare among wild animals, but considering the feeding relationships of these particular species, it is not too surprising that this unusual **behavior** has developed.

It is virtually impossible to determine the sex of a spotted hyena, because the external genitalia of the female mimic those of the male, and are virtually identical in form. A traditional myth about this animal described how any hyena could act as a male or female (that is, as a **hermaphrodite**), and this was why these animals laugh so raucously.

The striped hyena (*Hyaena hyaena*) ranges from central Africa to southwestern **Asia** and Asia Minor. It weighs about 59-119 lb (27-54 kg), and has a gray coat with darker stripes and a dark mane. The brown hyena (*Hyaena brunnea*) is a closely related, similar-sized species with a rather long, heavy, uniformly brown coat and mane. The brown hyena occurs in southern Africa. This species is becoming quite rare because it is being exterminated by farmers, who erroneously believe that the brown hyena preys on livestock.

Both of these species are considerably smaller and less powerful than the spotted hyena, and they are not very social animals, foraging as individuals or in groups of two. The striped and brown hyenas are nocturnal, and feed mostly on carrion, crushing the bones to obtain nourishment from even the most picked-over carcass. They also occasionally prey upon small **mammals**. The aardwolf (*Proteles cristatus*) is an uncommon, hyena-like animal of the **grasslands** and savannas of southern and eastern Africa. This species is much smaller and less powerful than the true hyenas. The jaws and teeth of the aardwolf are relatively small and quite unlike the bone-crushing apparatus of the hyenas, but useful in feeding on small, soft prey such as the **insects** that are the main food of this animal. The fur of the aardwolf is relatively long and yellow-gray, with dark stripes, an erectile mane, and a bushy, black-tipped tail. The aardwolf lives in burrows that it usually excavates itself, a trait that is likely the origin of its common name, which is derived from the Afrikaans words for "earth wolf." The aardwolf largely feeds nocturnally on **termites** and the larvae of other insects, although it sometimes feeds on carrion, and may occasionally prey on small animals. If cornered by a threatening predator, the aardwolf emits a foul-smelling scent from its anal **glands** as a means of self-defense.

See also Carnivore; Scavenger.

Resources

Books

Ewer, R. F. *The Carnivores*. Ithaca, NY: Cornell University Press, 1985.

Grzimek, B., ed. *Grzimek's Encyclopedia of Mammals*. New York: McGraw Hill, 1990.

Jeweel, P.A., and G. Maloiy, eds. *The Biology of Large African Mammals and Their Environment*. Oxford: Oxford University Press, 1989.

Kruuk, H. *The Spotted Hyena. A Study of Predation and Social Behaviour*. Chicago: University Of Chicago Press, 1972.

Nowak, R.M., ed. *Walker's Mammals of the World*. 5th ed. Baltimore: Johns Hopkins University Press, 1991.

Wilson, D.E., and D. Reeder, comp. *Mammal Species of the World*. 2nd ed. Washington, DC: Smithsonian Institution Press, 1993.

Bill Freedman

Hyperbola

A hyperbola is a **curve** formed by the intersection of a right circular cone and a **plane** (see Figure 1). When the plane cuts both nappes of the cone, the intersection is a hyperbola. Because the plane is cutting two nappes, the curve it forms has two U-shaped branches opening in opposite directions.

Other definitions

A hyperbola can be defined in several other ways, all of them mathematically equivalent:

1. A hyperbola is a set of points P such that $PF_1 - PF_2 = \pm C$, where C is a constant and F_1 and F_2 are fixed points called the "foci" (see Figure 2). That is, a hyperbola is the set of points the difference of whose distances from two fixed points is constant. The positive value of $\pm C$ gives one branch of the hyperbola; the **negative** value, the other branch.

2. A hyperbola is a set of points whose distances from a fixed point (the "focus") and a fixed line (the "directrix") are in a constant **ratio** (the "eccentricity"). That is, $PF/PD = e$ (see Figure 3). For this set of points to be a hyperbola, e has to be greater than 1. This definition gives only one branch of the hyperbola.

3. A hyperbola is a set of points (x,y) on a **Cartesian coordinate plane** satisfying an equation of the form

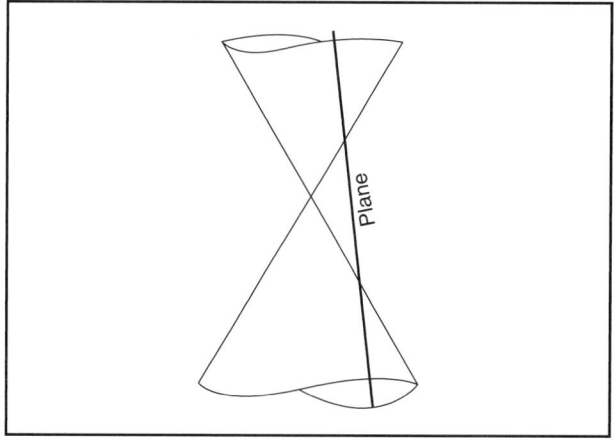

Figure 1. *Illustration by Hans & Cassidy. Courtesy of Gale Group.*

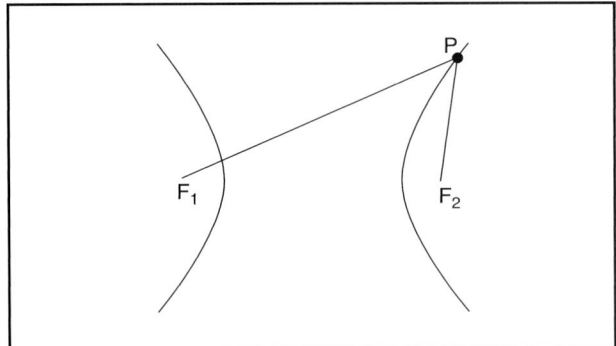

Figure 2. *Illustration by Hans & Cassidy. Courtesy of Gale Group.*

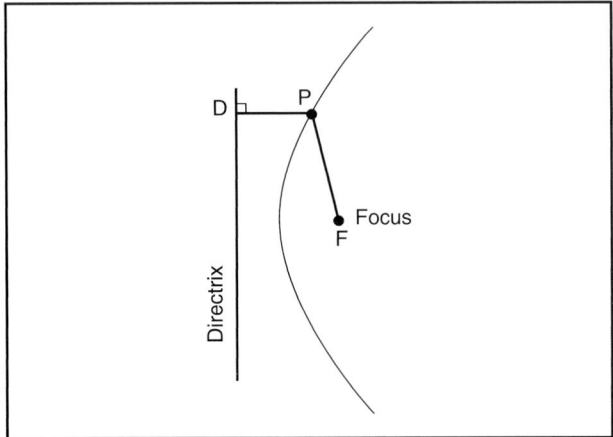

Figure 3. *Illustration by Hans & Cassidy. Courtesy of Gale Group.*

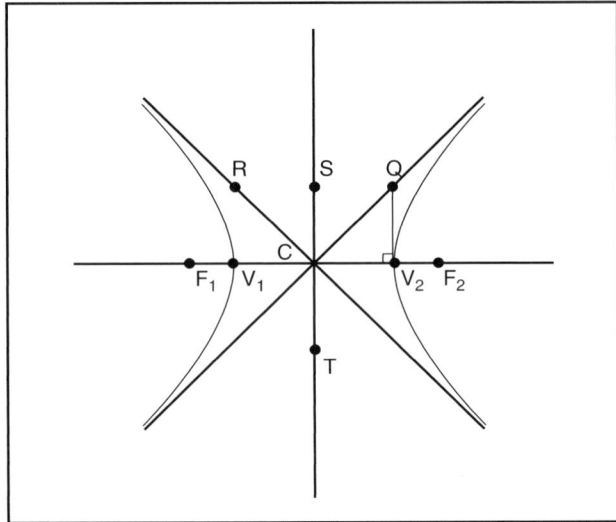

Figure 4. *Illustration by Hans & Cassidy. Courtesy of Gale Group.*

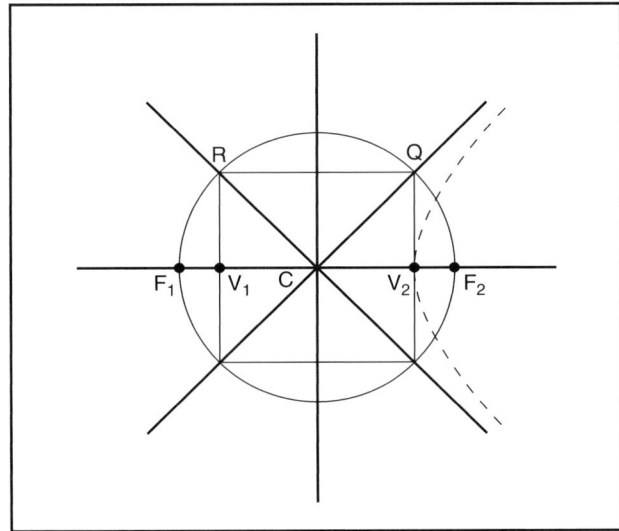

Figure 5. *Illustration by Hans & Cassidy. Courtesy of Gale Group.*

$x^2/A^2 - y^2/B^2 = \pm 1$. The equation $xy = k$ also represents a hyperbola, but of eccentricity not equal to 2. Other second-degree equations can represent hyperbolas, but these two forms are the simplest. When the positive value in ± 1 is used, the hyperbola opens to the left and right. When the negative value is used, the hyperbola opens up and down.

Features

When a hyperbola is drawn as in Figure 4, the line through the foci, F_1 and F_2, is the "transverse axis." V_1 and V_2 are the "vertices," and C the "center." The transverse axis also refers to the **distance**, V_1V_2, between the vertices.

The ratio CF_1/CV_1 (or CF_2/CV_2) is the "eccentricity" and is numerically equal to the eccentricity e in the focus-directrix definition.

The lines CR and CQ are asymptotes. An asymptote is a straight line which the hyperbola approaches more and more closely as it extends farther and farther from the center. The point Q has been located so that it is the

vertex of a right triangle, one of whose legs is CV_2, and whose hypotenuse CQ equals CF_2. The point R is similarly located.

The line ST, **perpendicular** to the transverse axis at C, is called the "conjugate axis." The conjugate axis also refers to the distance ST, where $SC = CT = QV_2$.

A hyperbola is symmetric about both its transverse and its conjugate axes.

When a hyperbola is represented by the equation $x^2/A^2 - y^2/B^2 = 1$, the x-axis is the transverse axis and the y-axis is the conjugate axis. These axes, when thought of

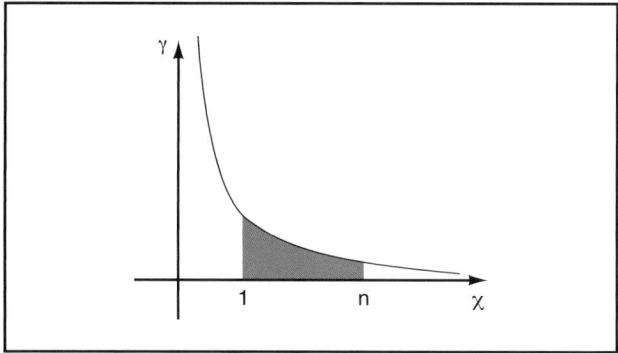

Figure 6. *Illustration by Hans & Cassidy. Courtesy of Gale Group.*

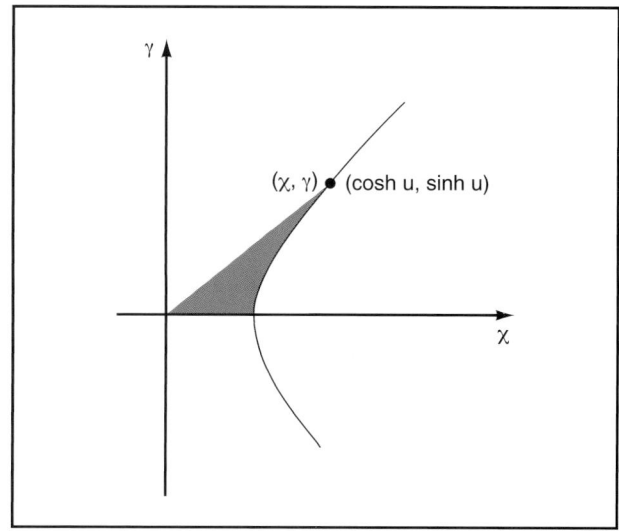

Figure 7. *Illustration by Hans & Cassidy. Courtesy of Gale Group.*

as distances rather than lines, have lengths 2A and 2B respectively. The foci are at

$$\sqrt{A^2 + B^2}, 0 \text{ and } \sqrt{A^2 + B^2}, 0;$$
the eccentricity is
$$\frac{\sqrt{A^2 + B^2}}{A}$$

The equations of the asymptotes are y = Bx/A and y = -Bx/A. (Notice that the constant 1 in the equation above is positive. If it were -1, the y-axis would be the transverse axis and the other points would change accordingly. The asymptotes would be the same, however. In fact, the hyperbolas $x^2/A^2 - y^2/B^2 = 1$ and $x^2/A^2 - y^2/B^2 = -1$ are called "conjugate hyperbolas.") Hyperbolas whose asymptotes are perpendicular to each other are called "rectangular" hyperbolas. The hyperbolas xy = k and $x^2 - y^2 = \pm C^2$ are rectangular hyperbolas. Their eccentricity is ± 2. Such hyperbolas are geometrically similar, as are all hyperbolas of a given eccentricity.

If one draws the **angle** F_1PF_2 the tangent to the hyperbola at point P will bisect that angle.

Drawing hyperbolas

Hyperbolas can be sketched quite accurately by first locating the vertices, the foci, and the asymptotes. Starting with the axes, locate the vertices and foci. Draw a **circle** with its center at C, passing through the two foci. Draw lines through the vertices perpendicular to the transverse axis. This determines four points, which are corners of a **rectangle**. These diagonals are the asymptotes.

Using the vertices and asymptotes as guides, sketch in the hyperbola as shown in Figure 5. The hyperbola approaches the asymptotes, but never quite reaches them. Its curvature, therefore, approaches, but never quite reaches, that of a straight line.

If the lengths of the transverse and conjugate axes are known, the rectangle in Figure 5 can be drawn without using the foci, since the rectangle's length and width are equal to these axes.

One can also draw hyperbolas by plotting points on a coordinate plane. In doing this, it helps to draw the asymptotes, whose equations are given above.

Uses

Hyperbolas have many uses, both mathematical and practical. The hyperbola y = 1/x is sometimes used in the definition of the natural logarithm. In Figure 6 the logarithm of a number n is represented by the shaded area, that is, by the area bounded by the x-axis, the line x = 1, the line x = n, and the hyperbola. Of course one needs **calculus** to compute this area, but there are techniques for doing so.

The coordinates of the point (x,y) on the hyperbola $x^2 - y^2 = 1$ represent the hyperbolic cosine and hyperbolic sine functions. These functions bear the same relationship to this particular hyperbola that the ordinary cosine and sine functions bear to a unit circle:

$$x = \cosh u = (e^U + e^{-u})2$$
$$y = \sinh u = (e^U - e^{-u})\, 2$$

Unlike ordinary sines and cosines, the values of the hyperbolic functions can be represented with simple exponential functions, as shown above. That these representations work can be checked by substituting them in the equation of the hyperbola. The parameter u is also related to the hyperbolas. It is twice the shaded area in Figure 7.

The definition $PF_1 - PF_2 = \pm C$, of a hyperbola is used directly in the **LORAN** navigational system. A ship at P receives simultaneous pulsed **radio** signals from sta-

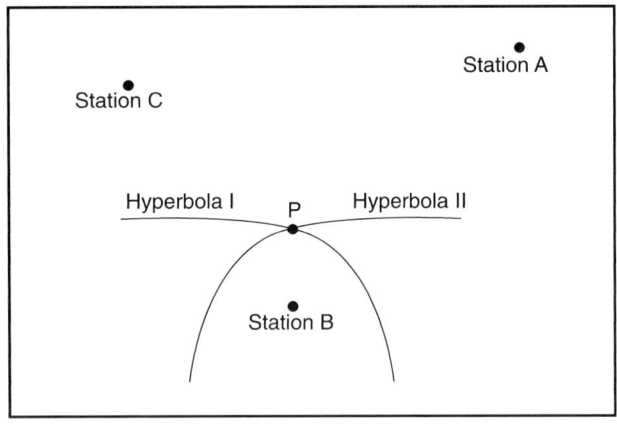

Figure 8. *Illustration by Hans & Cassidy. Courtesy of Gale Group.*

Hilbert, D., and S. Cohn-Vossen. *Geometry and the Imagination.* New York: Chelsea Publishing Co. 1952.

Larson, Ron. *Calculus With Analytic Geometry.* Boston: Houghton Mifflin College, 2002.

J. Paul Moulton

> ### KEY TERMS
>
> **Foci**—Two fixed points on the transverse axis of a hyperbola. Any point on the hyperbola is always a fixed amount farther from one focus than from the other.
>
> **Hyperbola**—A conic section of two branches, satisfying one of several definitions.
>
> **Vertices**—The two points where the hyperbola crosses the transverse axis.

tions at A and B. It cannot measure the **time** it takes for the signals to arrive from each of these stations, but it can measure how much longer it takes for the signal to arrive from one station than from the other. It can therefore compute the difference PA - PB in the distances. This locates the ship somewhere along a hyperbola with foci at A and B, specifically the hyperbola with that constant difference. In the same way, by timing the difference in the time it takes to receive simultaneous signals from stations B and C, it can measure the difference in the distances PB and PC. This puts it somewhere on a second hyperbola with B and C as foci and PC - PB as the constant difference. The ship's position is where these two hyperbolas cross (Figure 8). Maps with grids of crossing hyperbolas are available to the ship's navigator for use in areas served by these stations.

Resources

Books

Gullberg, Jan, and Peter Hilton. *Mathematics: From the Birth of Numbers.* W.W. Norton & Company, 1997.

Hahn, Liang-shin. *Complex Numbers and Geometry.* 2nd ed. The Mathematical Association of America, 1996.

Hypertension

Hypertension is high **blood pressure**. Blood pressure is the **force** of blood pushing against the walls of **arteries** as it flows through them. Arteries are the blood vessels that carry oxygenated blood from the **heart** to the body's tissues.

As blood flows through arteries it pushes against the inside of the artery walls. The more pressure the blood exerts on the artery walls, the higher the blood pressure will be. The size of small arteries also affects the blood pressure. When the muscular walls of arteries are relaxed, or dilated, the pressure of the blood flowing through them is lower than when the artery walls narrow, or constrict.

Blood pressure is highest when the heart beats to push blood out into the arteries. When the heart relaxes to fill with blood again, the pressure is at its lowest point. Blood pressure when the heart beats is called systolic pressure. Blood pressure when the heart is at rest is called diastolic pressure. When blood pressure is measured, the systolic pressure is stated first and the diastolic pressure second. Blood pressure is measured in millimeters of mercury (mm Hg). For example, if a person's systolic pressure is 120 and diastolic pressure is 80, it is written as 120/80 mm Hg. The American Heart Association considers blood pressure less than 140 over 90 normal for adults.

Hypertension is a major health problem, especially because it has no symptoms. Many people have hypertension without knowing it. In the United States, about 50 million people age six and older have high blood pressure. Hypertension is more common in men than women and in people over the age of 65 than in younger persons. More than half of all Americans over the age of 65 have hypertension. It is also more common in African-Americans than in white Americans.

Hypertension is serious because people with the condition have a higher risk for heart **disease** and other medical problems than people with normal blood pressure. Serious complications can be avoided by getting regular blood pressure checks and treating hypertension as soon as it is diagnosed.

If left untreated, hypertension can lead to the following medical conditions:

• **arteriosclerosis**, also called atherosclerosis

• heart attack

• **stroke**

• enlarged heart

• kidney damage

Arteriosclerosis is hardening of the arteries. The walls of arteries have a layer of muscle and elastic **tissue** that makes them flexible and able to dilate and constrict as blood flows through them. High blood pressure can make the artery walls thicken and harden. When artery walls thicken, the inside of the blood vessel narrows. **Cholesterol** and fats are more likely to build up on the walls of damaged arteries, making them even narrower. Blood clots can also get trapped in narrowed arteries, blocking the flow of blood.

Arteries narrowed by arteriosclerosis may not deliver enough blood to organs and other tissues. Reduced or blocked blood flow to the heart can cause a heart attack. If an artery to the **brain** is blocked, a stroke can result.

Hypertension makes the heart work harder to pump blood through the body. The extra workload can make the heart muscle thicken and stretch. When the heart becomes too enlarged it cannot pump enough blood. If the hypertension is not treated, the heart may fail.

The kidneys remove the body's wastes from the blood. If hypertension thickens the arteries to the kidneys, less waste can be filtered from the blood. As the condition worsens, the kidneys fail and wastes build up in the blood. **Dialysis** or a kidney transplant are needed when the kidneys fail. About 25% of people who receive kidney dialysis have kidney failure caused by hypertension.

Causes and symptoms

Many different actions or situations can normally raise blood pressure. Physical activity can temporarily raise blood pressure. Stressful situations can make blood pressure go up. When the **stress** goes away, blood pressure usually returns to normal. These temporary increases in blood pressure are not considered hypertension. A **diagnosis** of hypertension made only when a person has multiple high blood pressure readings over a period of time.

The cause of hypertension is not known in 90-95% of the people who have it. Hypertension without a known cause is called primary or essential hypertension.

When a person has hypertension caused by another medical condition, it is called secondary hypertension. Secondary hypertension can be caused by a number of different illnesses. Many people with kidney disorders have secondary hypertension. The kidneys regulate the balance of **salt** and **water** in the body. If the kidneys can-

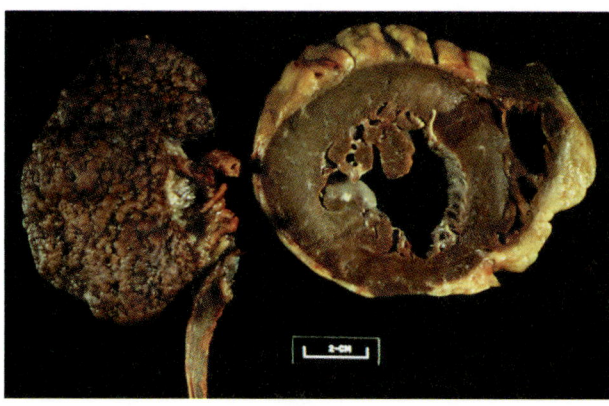

Hypertension, on the heart and kidney. *Photograph by Dr. E. Walker. Photo Researchers, Inc. Reproduced by permission.*

not rid the body of excess salt and water, blood pressure goes up. Kidney infections, a narrowing of the arteries that carry blood to the kidneys, called renal artery stenosis, and other kidney disorders can disturb the salt and water balance.

Cushing syndrome and tumors of the pituitary and adrenal **glands** often increase levels of the adrenal gland **hormones** cortisol, adrenalin and aldosterone, which can cause hypertension. Other conditions that can cause hypertension are blood vessel diseases, thyroid gland disorders, some prescribed drugs, **alcoholism,** and pregnancy.

Even though the cause of most hypertension is not known, some people have risk factors that give them a greater chance of getting hypertension. Many of these risk factors can be changed to lower the chance of developing hypertension or as part of a treatment program to lower blood pressure.

Risk factors for hypertension include:

• age over 60

• male sex

• race

• heredity

• salt sensitivity

• obesity

• inactive lifestyle

• heavy **alcohol** consumption

• use of oral contraceptives

Some risk factors for getting hypertension can be changed, while others cannot. Age, male sex, and race are risk factors that a person cannot do anything about. Some people inherit a tendency to get hypertension. People with family members who have hypertension are more likely to develop it than those whose relatives are not hypertensive. A person with these risk factors can avoid or eliminate the other risk factors to lower their chance of developing hypertension.

Diagnosis

Because hypertension does not cause symptoms, it is important to have blood pressure checked regularly. Blood pressure is measured with an instrument called a sphygmomanometer. A cloth-covered rubber cuff is wrapped around the upper arm and inflated. When the cuff is inflated, an artery in the arm is squeezed to momentarily stop the flow of blood. Then, the air is let out of the cuff while a stethoscope placed over the artery is used to detect the sound of the blood spurting back through the artery. This first sound is the systolic pressure, the pressure when the heart beats. The last sound heard as the rest of the air is released is the diastolic pressure, the pressure between heart beats. Both sounds are recorded on the mercury gauge on the sphygmomanometer.

Normal blood pressure is defined by a range of values. Blood pressure lower than 140/90 mm Hg is considered normal. A blood pressure around 120/80 mm Hg is considered the best level to avoid heart disease. A number of factors such as **pain**, stress, or **anxiety** can cause a temporary increase in blood pressure. For this reason, hypertension is not diagnosed on one high blood pressure reading. If a blood pressure reading is 140/90 or higher for the first time, the physician will have the person return for another blood pressure check. Diagnosis of hypertension usually is made based on two or more readings after the first visit.

Systolic hypertension of the elderly is common and is diagnosed when the diastolic pressure is normal or low, but the systolic is elevated, e.g., 170/70 mm Hg. This condition usually co-exists with hardening of the arteries (atherosclerosis).

Blood pressure measurements are classified in stages, according to severity:

• normal blood pressure: less than 130/85 mm Hg
• high normal: 130–139/85–89 mm Hg
• mild hypertension: 140–159/90–99 mm Hg
• moderate hypertension: 160–179/100–109 mm Hg
• severe hypertension: 180–209/110–119
• very severe hypertension: 210/120 or higher

A typical physical examination to evaluate hypertension includes:

• medical and family history
• physical examination
• ophthalmoscopy: Examination of the blood vessels in the eye
• chest x-ray

• electrocardiograph (ECG)
• blood and urine tests

The medical and family history help the physician determine if the patient has any conditions or disorders that might contribute to or cause the hypertension. A family history of hypertension might suggest a genetic predisposition for hypertension.

The physical exam may include several blood pressure readings at different times and in different positions. The physician uses a stethoscope to listen to sounds made by the heart and blood flowing through the arteries. The pulse, reflexes, height, and weight are checked and recorded. Internal organs are palpated, or felt, to determine if they are enlarged.

Because hypertension can cause damage to the blood vessels in the eyes, the eyes may be checked with a instrument called an ophthalmoscope. The physician will look for thickening, narrowing, or hemorrhages in the blood vessels.

A chest x ray can detect an enlarged heart, other vascular (heart) abnormalities, or lung disease.

An **electrocardiogram (ECG)** measures the electrical activity of the heart. It can detect if the heart muscle is enlarged and if there is damage to the heart muscle from blocked arteries.

Urine and blood tests may be done to evaluate health and to detect the presence of disorders that might cause hypertension.

Treatment

There is no cure for primary hypertension, but blood pressure can almost always be lowered with the correct treatment. The goal of treatment is to lower blood pressure to levels that will prevent heart disease and other complications of hypertension. In secondary hypertension, the disease that is responsible for the hypertension is treated in addition to the hypertension itself. Successful treatment of the underlying disorder may cure the secondary hypertension.

Treatment to lower blood pressure usually includes changes in diet, getting regular **exercise**, and taking antihypertensive medications. Patients with mild or moderate hypertension who do not have damage to the heart or kidneys may first be treated with lifestyle changes.

Lifestyle changes that may reduce blood pressure by about 5-10 mm Hg include:

• reducing salt intake
• reducing **fat** intake
• losing weight
• getting regular exercise

- quitting smoking
- reducing alcohol consumption
- managing stress

Patients whose blood pressure remains higher than 139/90 will most likely be advised to take antihypertensive medication. Numerous drugs have been developed to treat hypertension. The choice of medication will depend on the stage of hypertension, side effects, other medical conditions the patient may have, and other medicines the patient is taking.

Patients with mild or moderate hypertension are initially treated with monotherapy, a single antihypertensive medicine. If treatment with a single medicine fails to lower blood pressure enough, a different medicine may be tried or another medicine may be added to the first. Patients with more severe hypertension may initially be given a combination of medicines to control their hypertension. Combining antihypertensive medicines with different types of action often controls blood pressure with smaller doses of each drug than would be needed for monotherapy.

Antihypertensive medicines fall into several classes of drugs:

- diuretics
- **beta-blockers**
- **calcium** channel blockers
- angiotensin converting **enzyme** inhibitors (ACE inhibitors)
- alpha-blockers
- alpha-beta blockers
- vasodilators
- peripheral acting adrenergic antagonists
- centrally acting agonists

Diuretics help the kidneys eliminate excess salt and water from the body's tissues and the blood. This helps reduce the swelling caused by fluid buildup in the tissues. The reduction of fluid dilates the walls of arteries and lowers blood pressure.

Beta-blockers lower blood pressure by acting on the **nervous system** to slow the heart **rate** and reduce the force of the heart's contraction. They are used with caution in patients with heart failure, **asthma**, diabetes, or circulation problems in the hands and feet.

Calcium channel blockers block the entry of calcium into muscle cells in artery walls. Muscle cells need calcium to constrict, so reducing their calcium keeps them more relaxed and lowers blood pressure.

ACE inhibitors block the production of substances that constrict blood vessels. They also help reduce the build-up of water and salt in the tissues. They are often

KEY TERMS

Arteries—Blood vessels that carry blood to organs and other tissues of the body.

Arteriosclerosis—Hardening and thickening of artery walls.

Cushing's syndrome—A disorder in which too much of the adrenal hormone, cortisol, is produced; it may be caused by a pituitary or adrenal gland tumor.

Diastolic blood pressure—Blood pressure when the heart is resting between beats.

Hypertension—High blood pressure.

Renal artery stenosis—Disorder in which the arteries that supply blood to the kidneys constrict.

Sphygmomanometer—An instrument used to measure blood pressure.

Systolic blood pressure—Blood pressure when the heart contracts (beats).

Vasodilator—Any drug that relaxes blood vessel walls.

Ventricles—The two lower chambers of the heart; also the main pumping chambers.

given to patients with heart failure, kidney disease, or diabetes. ACE inhibitors may be used together with diuretics.

Alpha-blockers act on the nervous system to dilate arteries and reduce the force of the heart's contractions.

Alpha-beta blockers combine the actions of alpha and beta blockers.

Vasodilators act directly on arteries to relax their walls so blood can move more easily through them. They lower blood pressure rapidly and are injected in hypertensive emergencies when patients have dangerously high blood pressure.

Peripheral acting adrenergic antagonists act on the nervous system to relax arteries and reduce the force of the heart's contractions. They usually are prescribed together with a diuretic. Peripheral acting adrenergic antagonists can cause slowed mental function and lethargy.

Centrally acting agonists also act on the nervous system to relax arteries and slow the heart rate. They are usually used with other antihypertensive medicines.

Prognosis

There is no cure for hypertension. However, it can be well controlled with the proper treatment. Therapy with a combination of lifestyle changes and antihypertensive medicines usually can keep blood pressure at levels that will not cause damage to the heart or other organs. The key to avoiding serious complications of hypertension is to detect and treat it before damage occurs. Because antihypertensive medicines control blood pressure, but do not cure it, patients must continue taking the medications to maintain reduced blood pressure levels and avoid complications.

Prevention

Prevention of hypertension centers on avoiding or eliminating known risk factors. Even persons at risk because of age, race, or sex or those who have an inherited risk can lower their chance of developing hypertension.

The risk of developing hypertension can be reduced by making the same changes recommended for treating hypertension.

Resources

Books

Bellenir, Karen, and Peter D. Dresser, eds. *Cardiovascular Diseases and Disorders Sourcebook.* Detroit: Omnigraphics, 1995.

Texas Heart Institute. *Heart Owner's Handbook.* New York: John Wiley and Sons, 1996.

Toni Rizzo

Hypertonic *see* **Osmosis**

Hypotenuse *see* **Pythagorean theorem**

Hypothermia

Hypothermia is the intentional or accidental reduction of core body **temperature** to below 95°F (35°C) which, in severe instances, is fatal. Because humans are *endothermic*—warm-blooded creatures producing our own body heat—our core body temperature remains relatively constant at 98.6°F (37°C), even in fluctuating environmental temperatures. However, in extreme conditions, a healthy, physically fit person's core body temperature can rise considerably above this norm and cause **heat stroke** or fall below it far enough to cause hypothermia.

Intentional hypothermia

Intentional hypothermia is used in medicine in both regional and total-body cooling for **organ** and **tissue** protection, preservation, or destruction. Interrupted **blood** flow starves organs of **oxygen** and may cause permanent organ damage or death. The body's metabolic **rate** (the rate at which cells provide **energy** for the body's vital functioning) decreases 8% with each 1.8°F (1°C) reduction in core body temperature, thus requiring reduced amounts of oxygen. Total-body hypothermia lowers the body temperature and slows the metabolic rate, protecting organs from reduced oxygen supply during the interruption of blood flow necessary in certain surgical procedures. In some procedures, like **heart** repair and organ transplantation, individual organs are preserved by intentional hypothermia of the organ involved. In open heart **surgery**, **blood supply** to the chilled heart can be totally interrupted while the surgeon repairs the damaged organ. Organ and tissue destruction using extreme hypothermia -212 to -374°F (-100 to -190°C) is utilized in retinal and glaucoma surgery and to destroy pre-cancerous cells in some body tissue. This is called cryosurgery.

History of medical use of intentional hypothermia

Intentional hypothermia is not a modern phenomenon. With it, ancient Egyptians treated high fevers; as did Hippocrates—who also understood its analgesic (pain-killing) properties; the Romans; and Europeans of the Middle Ages. Napoleon's surgeon general used cryoanalgesia when performing amputations. He discovered that packing a limb in **ice** and snow not only killed most of the **pain**, it also helped prevent bleeding. Today, intentional hypothermia is most commonly used in heart surgery.

Accidental hypothermia

Accidental hypothermia is potentially fatal. It can happen as simply as falling off a log. Falling into icy **water**, or exposure to cold **weather** without appropriate protective clothing, can quickly result in death. Hypothermia is classified into four states. In mild cases 95–89.6°F (35–32°C), symptoms include feeling cold, shivering (which helps raise body temperature), increased heart rate and desire to urinate, and some loss of coordination. Moderate 87.8–78.8°F (31–26°C) hypothermia causes a decrease in, or cessation of, shivering; weakness; sleepiness; confusion; slurred **speech**; and lack of coordination. Deep hypothermia 77–68°F (25–20°C) is extremely dangerous as the body can no longer produce heat. Sufferers may behave irrationally, become comatose, lose the ability to see, and often cannot follow commands. In profound cases 66–57°F (19–14°C), the sufferer will become rigid and may even appear dead, with dilated pupils, extremely low blood **pressure**, and barely perceptible heartbeat and breathing. This state usually requires complete, professional cardiopulmonary resuscitation for survival.

Causes of accidental hypothermia

Although overexertion in a cold environment causes most accidental hypothermia, it may occur during **anesthesia**, primarily due to central **nervous system** depression of the body's heat-regulating mechanism; and in babies, elderly, and ill people whose homes are inadequately heated. The human body loses heat to the environment through conduction, **convection**, **evaporation** and **respiration**, and **radiation**. It generates heat through the metabolic process.

Conduction occurs when direct contact is made between the body and a cold object, and heat passes from the body to that object. Convection is when cold air or water make contact with the body, become warm, and move away to be replaced by more of the same. The cooler the air or water, and the faster it moves, the faster the core body temperature drops.

Evaporation through perspiration and respiration provides almost 30% of the body's natural cooling mechanism. Because cold air contains little water and readily evaporates perspiration; and because physical exertion produces sweating, even in extreme cold; heat loss through evaporation takes place even at very low temperatures. The dry air we inhale attracts moisture from the lining of our nose and throat so quickly that, by the time the air reaches our lungs, it is completely saturated. Combined, evaporation and convection from wet clothes will reduce the body temperature dangerously quickly.

When the body is warmer than its environment, it radiates heat into that environment. Radiation is the greatest source of heat loss, and the colder the environment, the greater the potential for heat loss. Most clothing is of little help because body heat radiates into clothing and from clothing into the atmosphere or object with which it comes into contact.

Preventing accidental hypothermia

Although profound hypothermia can be reversed in some instances, even mild states can quickly lead to death. However, through knowledge and common sense, hypothermia is avoidable. Two factors essential in preventing accidental hypothermia are reducing loss of body heat and increasing body heat production.

Appropriate clothing, shelter, and diet are all essential. Apart from new, synthetic fabrics which allow undergarments to "wick" perspiration away from the body while remaining dry and warm, and outergarments which "breathe" while keeping **wind** and moisture out,

natural wool fibers contain millions of air pockets which act as excellent insulators. Even when saturated, wool maintains 80% of its dry insulation value. Because it provides no insulation and becomes extremely cold when wet, **cotton** is often called "killer cotton" by experienced outdoors people. As 60% of body heat is lost by radiation from the head, hats can be lifesavers. Fingers, toes, hands, and feet lose heat quickly, and excellent quality boots, gloves, and mittens are a must.

Regular consumption of high-energy food rich in carbohydrates aids the body in heat production, while 13-17 c (3-4L) of water a day prevents rapid dehydration from evaporation. Exercising large muscles-like those in the legs—is the best **generator** of body heat; however, overexertion must be avoided as it will only speed the onset of hypothermia.

Resources

Books

Auerbach, Paul S. *Medicine for the Outdoors: A Guide to Emergency Medical Procedures and First Aid.* U.S.A./Canada: Little, Brown & Company, Limited.

Schönbaum, E., and Peter Lomax, ed. *Thermoregulation: Pathology, Pharmacology, and Therapy.*

Wilkerson, James A. *Hypothermia, Frostbite, and Other Cold Injuries.* Vancouver: Douglas & McIntyre, Ltd., 1986.

Marie L. Thompson

Hypotonic *see* **Osmosis**

Hyraxes

Hyraxes are rabbit-sized, hoofed African **mammals** that surprisingly share a common ancestry with elephants and manatees, or seacows. Hyraxes were originally thought to be **rodents**, and were later grouped with rhinoceroses. They are now placed in an order of their own, the Hyracoidea, since they share many common features of primitive **ungulates**. The fossil record indicates that hyraxes were the most prevalent medium-sized browsing and grazing **animal** 40 million years ago, ranging in size from that of contemporary hyraxes to that of a tapir. As competition with the bovid family (African and Asian antelope, **bison**, **sheep**, **goats**, and cattle) increased, hyraxes retreated to the more peripheral habitats with **rocks** and more trees. Rock hyraxes (*Procavia capeasis*) are dependent upon suitable **habitat** in rocky outcrops (kopjes) and cliffs, but nevertheless the five **species** of rock hyraxes have the widest geographical and altitudinal distribution in **Africa**. **Tree** hyraxes, (*Dendrohyrax arboreus*) prefer arboreal habitats and are found in Zaire, East Africa, and South Africa. Another species of hyrax (*D. dorsalis*) is found in West Africa.

Small and compact with short rounded ears, rock hyraxes have only a tiny stump of a tail. Their coat is coarse and thick, ranging in **color** from light gray to black. Males and females are approximately the same size and show little sexual dimorphism. The feet of hyraxes have naked rubbery pads with numerous sweat **glands**. There are four toes on the forefoot and three on the hind foot. All the digits have flattened nails, except the inner toe of the hind foot which has a sharp-edged nail that is used for grooming. Hyraxes have grinding teeth, like those found in rhinoceroses and a pair of incisors that are sharp and dagger-like.

The social organization of rock hyraxes consists of stable family groups composed of one adult territorial male and three to seven related adult females, commonly known as a harem. Females breed once per year producing litters of one to four young after a gestation period of seven-and-one-half to eight months. Hyraxes can live ten years or longer. They are gregarious animals, and may often be seen sunbathing on rocks or in cooler periods, huddling close together. Hyraxes regulate their body **temperature** poorly and have a low metabolic **rate** for a mammal.

Rock hyraxes feed mostly on **grasses**, and supplement their diet with herbage, leaves, berries, fruit, and the **bark** of trees during the dry season. Hyraxes have a tolerance for eating highly poisonous plants. They also need little **water** to survive, by virtue of efficient kidneys obtaining moisture from their food. In spite of their compact build, hyraxes are agile in their movements and relatively good jumpers. Both **hearing** and sight are excellent. Their main **predator** is the Verreaux eagle. Other enemies are martial and tawny **eagles**, leopards, lions, servals, caracals, jackals, large **civets**, spotted hyenas, and several snake species. To guard against predators, an alarm whistle is sounded. Hyraxes may also growl with gnashing teeth and give long-drawn piercing screams.

Rock hyraxes habitually defecate in the same spot, creating a pile of dried, hardened excrement which contains hyraceum, a substance used to make perfumes. Although tree hyraxes are heavily hunted and suffer from habitat destruction, rock hyraxes appear to be stable in their population numbers with little threat to their habitat. Rock hyraxes are also known as rock-rabbits, coneys, dassies, and kupdas, while tree hyraxes are also known as tree **bears**. The Syrian hyrax (*Procavia syriaca*) is similar to the rock hyrax and is the "coney" referred to in the Bible.

Betsy A. Leonard

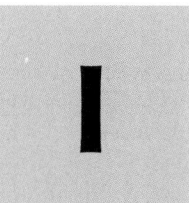

Ibises

Ibises are grouped together with large wading **birds** such as **storks**, **herons**, **flamingos**, and spoonbills, in the order Ciconiiformes. Ibises, like most birds in this order, have long legs and a long bill for feeding on **fish** and aquatic animals in shallow **water**. They also have broad wings, a short tail, and four long toes on each foot. The 26 **species** of ibis share the family Threskiornithidae with the spoonbills. Ibises have a large body, long legs, and a characteristic thin, downward-curving bill. The plumage of male and female ibises is alike, but the females are generally smaller. The heavy body of ibises means that they must flap their wings rapidly when in flight. They fly with their neck extended, often in a characteristic V-formation.

In **North America**, ibises are represented by the white ibis (*Eudocimus albus*), glossy ibis (*Plegadis flacinellus*), and white-faced ibis (*P. chici*). The wood ibis (*Mycteria americana*) is not actually an ibis, but a stork (Ciconidae). The scarlet ibis (*Guara rubra*) is the most spectacular ibis of **South America** and the Caribbean. The sacred ibis, the glossy ibis, and the hadada ibis are the principal species found in **Africa**.

Habitat and behavior

Ibises are found on shores and marshes worldwide, mainly in tropical habitats, but some are found in south-temperate regions. Ibises feed in flood plains, marshes, and swamps, and along streams, ponds, and lakes. Their diet is varied, consisting of aquatic **invertebrates**, **insects**, **snails**, fish, **reptiles**, **amphibians**, and even small **mammals**. Generally, ibises feed in large groups of up to 100 birds, and the flock may include other species of waders. Ibises usually feed by wading through shallow water and grabbing available **prey** with their beak.

Some species of ibis are solitary in their nesting habits within a prescribed territory, but most nest in large colonies of up to 10,000 pairs. Within the colony there may be several different species of ibis. These birds tend to be monogamous (faithful to a single mate) during a breeding season, but observers have also noted promiscuous mating within the large colonies.

Both male and female ibises build the nest, protect it from intruders, incubate the eggs (from two to six at a time), feed the fledglings, and care for them for about a month after they are hatched. Before mating, there is a **courtship** period, involving displays and the enhancement of the **color** of the face, legs, bill, and exposed parts of the bird's skin.

The series of courtship behaviors that ibises display (preening, shaking, and bill popping) are ritualized, beginning when the birds gather near secluded nesting areas. The male birds display, the females are attracted to them, and mating follows. Males may behave aggressively in defending their nesting site from other males, but they can also act aggressively towards females not selected as the mate.

Display preening involves pretending to preen the front or back feathers. Display shaking involves shaking loose wings up and down, and bill popping involves snapping the bill up and down with a popping sound. Ibises also have a sleeping display, in which they pretend to be asleep. The head rub during courtship is a sign for the female to enter the nesting area, where she performs a bowing display, keeping her head and body low as she comes near the male. The male may pretend to be aggressive before he finally allows the female to enter his nesting area. Head shaking is one of the displays ibises perform after mating as a greeting and acceptance. The intimacy of their relationship can be seen in mutual preening of one another, shaking, and the rubbing of their heads against each other.

Historical references

There are references to ibises in the Bible. Moses was told by God not to eat them, and they were also referred to

A white ibis (*Eudocimus albus*) at Estero Island, Florida. *Photograph by Robert J. Huffman. Field Mark Publications. Reproduced by permission.*

as birds of doom in other parts of the Bible. The ancient Egyptians considered the sacred ibis (*Threskiornis aethiopicus*) to be a sacred bird. Drawings, statues, and mummified ibises have been found in abundance in cemeteries dedicated to them. At a location near Memphis, Egypt, 1.5 million mummified birds were found. However, the sacred ibis has been absent from Egypt for well over 100 years, because of excessive hunting and **habitat** loss.

The Egyptian god of wisdom and knowledge, Thoth, is depicted in ancient Egyptian artifacts as a man with the head of an ibis. During the fourth and fifth centuries B.C., ibises were engraved on Greek coins. During the Middle Ages, Austrian nobles ate ibis as a delicacy. They were first described scientifically by a European naturalist, Konrad Gesner, in the sixteenth century. By the middle of the seventeenth century, they disappeared from central **Europe**, and the bird that Gesner described and painted was not noted again until it was seen in 1832 near the Red Sea. In the nineteenth century, a society of British ornithologists named their journal for the ibis.

A number of species of ibises are endangered. Among these is the formerly widespread Waldrapp ibis, which is reduced to present-day populations in Turkey.

There are 800 Waldrapp ibises in zoos all over the world and efforts are being made to reintroduce them into their former habitat. Other **endangered species** are the bald ibis of southern Africa, the dwarf olive ibis of the **island** of Sao Tome in West Africa, the oriental crested ibis of **Asia**, the giant ibis of Vietnam, and the white-shouldered ibis of Vietnam and Borneo. As with many other animals, the destruction of natural habitat, especially wetland drainage, is the primary threat to these wading birds. However, they are also hunted as food.

Resources

Books

Bildstein, Keith L. *White Ibis: Wetland Wanderer.* Washington, DC: Smithsonian Institution Press, 1993.

Boylan, P. *Thoth: The Hermes of Egypt.* 1922. Reprint. Chicago: Ares Publishers, 1987.

Campbell, N., J. Reece, and L. Mitchell. *Biology.* 5th ed. Menlo Park: Benjamin Cummings, Inc. 2000.

Hancock, James A., James A. Kushlan, and M. Philip Kahl. *Storks, Ibises and Spoonbills of the World.* London: Academic Press, 1992.

Vita Richman

Ice

Ice is the solid state of **water**. The great abundance of water on the surface of **Earth** includes a great quantity of ice in the Polar Regions and high elevations. The relative proportion of each of the three states of water on Earth is a delicately balanced equilibrium controlled by the amount of incoming solar **energy** and the amount of reflection, known as Aledo, from **clouds**, water, ice caps, etc. The amount of ice at any one location on Earth varies seasonally, over the long term with climatic change, and even with movements of tectonic plates. One of the most abundant of Earth's substances, ice is especially familiar to residents of high-latitude and alpine regions, and manifests itself in a variety of forms including snow, hail, **glaciers**, **icebergs**, and sea ice, along with the artificially produced ice cube. Despite being so familiar, ice and water are anomalous in a variety of respects and behave differently from other materials in a number of important ways. The study of ice, in all forms, and its related processes is known as glaciology.

Structure of ice

Because they share a common composition with their liquid state, ice molecules also consist of the same 2 to 1 **ratio** of **hydrogen** and **oxygen atoms**, the well-known H_2O **molecule**. The shape of this molecule, the oxygen atom at the center with the two hydrogen atoms separated by an angle of 104.52°, dictates the structure of the solid, crystalline ice. All naturally occurring ice crystals are hexagonal in shape and all snowflakes reflect this basic six-sided **crystal** habit. The crystal lattice consists of linked hexagonal rings of water molecules with considerable open space in the center of the ring.

Under artificial laboratory conditions of very high pressures and low temperatures, ice can be forced to crystallize in a number of allotropic forms that are stable only under those particular conditions. Crystallization can occur in these laboratory situations in one of several non-hexagonal forms. This is similar to the way that **carbon** atoms may crystallize to form graphite or, under more extreme conditions, **diamond**. The conditions under which the alternate forms might be created do not occur naturally on Earth. They may, however, be present on other bodies in space.

The crystalline structure of ice may be deformed by stress, such as the weight of overlying ice on the deeper portions of a glacier. One type of deformation involves shearing of the crystal lattice along **parallel** planes. Recrystallization, on the other hand, entails the change in the shape and orientation of crystals within the solid. Both of these processes produce the phenomenon known as creep, responsible for the flowing **motion** of massive ice bodies such as glaciers.

Physical properties of ice

Pure liquid water is transformed to its solid state, ice, at a **temperature** of 32°F (0°C) when the **pressure** is at one atmosphere. Interestingly, the **density** of liquid water at the freezing point is 62.418 lb/ft^3 (0.99984 g/cm^3) but decreases to 57.23 lb/ft^3 (0.9168 g/cm^3) when that water organizes itself into crystalline ice at 32°F (0°C). This density difference is due the large open spaces within the crystal lattice of ice. The increased **volume** of the solid lattice causes pure water to expand by approximately 9% upon freezing, resulting in ruptured pipes or damaged engines when the process occurs in a closed vessel. Ice is one of a very few solid substances that is lower in density than the corresponding liquid state. Surface ice floating on a **lake** or pond helps to insulate the water below, reduces mixing, and can prevent the water body from freezing solid. This fact has often been cited as an important factor in the development and **evolution** of life in **freshwater**.

The freezing point of water containing dissolved solids is proportionately reduced below 32°F (0°C) depending on the quantity of solutes. As the salinity of the water increases, the freezing temperature is lowered. This is the principle behind the practice of road salting. The **salt** causes the freezing point of the water to be lowered, hopefully below the ambient temperature, and the ice or snow is forced to melt.

When pressure is exerted on ice crystals at temperatures near the melting point, the edges of those crystals may melt. When that pressure is released, the water refreezes. This process, called regelation, may be familiar to those that have formed snowballs. The loose snowflakes are partially melted by the pressure of the hands. When the pressure is released, the refreezing water hardens and causes the cohesion of the flakes into a ball. On very cold days, however, the pressure that can be exerted by the hands is insufficient to cause melting, and the snowball is more difficult to form.

Natural ice occurrence

The vast majority of the natural ice on Earth is situated at the extreme latitudes; the Greenland ice sheet and sea ice at the North Pole, the Antarctic ice sheet in the South. Sea ice, massive ice sheets, valley and mountain glaciers all combine to form the **polar ice caps**. Enormous areas of the polar and subpolar regions are underlain by **permafrost**. Polar ice caps and glaciers contain a large proportion of Earth's freshwater resource. Over

KEY TERMS

Albedo—The fraction of sunlight that a surface reflects. An albedo of zero indicates complete absorption, while an albedo of unity indicates total reflection.

Allotropic—Said of substances that take multiple forms, such as graphite and diamond, usually in the same phase.

Freezing point—The temperature at which a liquid solidifies, 32°F (0°C) for water.

Glaciology—The study of all aspects of ice and its associated processes.

Hexagonal crystal system—One of six crystal systems. Characterized by one axis that is of unequal length to three identical perpendicular axes, commonly displaying three- or six-fold symmetry.

Melting point—The temperature at which a solid becomes liquid, 32∞F (0∞C) for ice.

Permafrost—Permanently frozen soil or subsoil.

Recrystallization—The formation of new crystals, while in the solid state.

Regelation—A two-fold process involving the melting of ice under pressure and the refreezing of the melt water upon the release of that pressure.

75% of all freshwater, or 2.15% of all water on Earth, presently exists in the form of ice. This proportion was significantly greater during past glacial epochs.

These vast stores of ice are particularly sensitive indicators of climatic change. The rapid retreat of mountain glaciers has been cited as evidence of **global warming**. If all ice at the poles and in glaciers melted, **sea level** would rise approximately 260 ft (80 m).

Ice is known to occur extensively on a variety of bodies in space. The origin of water on Earth has been postulated to be a result of collisions with **comets** and/or meteors containing a significant quantity of ice. The presence of ice has been confirmed at the poles of the **Moon** and the **planet** Mars. The existence of ice on **Mars** may be an indicator of the potential for the existence of life forms in the warmer and wetter past of that planet. The rings of **Saturn** and even nebulae outside our **solar system** are believed to contain ice. Europa, a moon of **Jupiter**, is thought to have a liquid-water **ocean** beneath a crust of ice. Scientists also assume that ice on such bodies might be utilized to supply the water needs of manned missions to these bodies, as well as being split into its component gases and used for fuel.

Current glaciology research

Much of the research currently being conducted in glaciology is focused on reducing the impact that ice has on modern society. Ice causes damage to pipes in homes, damages **crops**, restricts ability to travel, breaks power lines and other property, interferes with the function of airplanes and ships, along with other human considerations, such as contributing to accidental injuries. Engineers study ice to better prepare to build structures that interact with it, such as airplanes, ships and even oil platforms on the ocean. Climatologists and environmental scientists are working to understand the effects of global warming on the polar ice caps. Meteorologists study the formation of ice in the atmosphere. Other scientists are looking for improved methods by which ice can be controlled on roads. Biologists work to develop methods of protecting crops from frost damage. Physicists and engineers try to improve understanding of the properties of ice in order to improve the performance of sports equipment such as snow skis and ice **skates**. Geologists are studying the formation of ice volcanoes along the shores of the Great Lakes. Also, space scientists are looking for additional ice in our solar system and beyond, and planning new techniques and equipment that will allow man to someday utilize that ice in the exploration of other worlds.

See also Ice ages; Icebergs.

Resources

Books

Lock, G. S. H. *The Growth and Decay of Ice.* New York: Cambridge University Press, 1990.

Petrenko, Victor F., and Robert W. Whitworth. *Physics of Ice.* New York: Oxford University Press, 1999.

Pounder, Elton R. *The Physics of Ice.* New York: Pergamon Press, 1965.

Other

Dolan, Michael, and Paul Kimberly. "Ice Volcanoes of Lake Superior's South Shore." [cited January 10, 2003]. <http://www.geo.mtu.edu/volcanoes/ ice/>

NASA. "Found It! Ice on Mars." May 28, 2002 [cited January 10, 2003]. <http://science.nasa.gov/headlines/y2002/28 may_marsice.htm>.

NASA National Space Science Data Center. "Ice on the Moon." December 3, 2002 [cited January 10, 2003]. <http://nssdc. gsfc.nasa.gov/planetary/ice/ice_moon.html>.

David B. Goings

Ice age refuges

The series of **ice ages** that occurred between 2.4 million and 10,000 years ago had a dramatic effect on the climate and the life forms in the tropics. During each glacial period the tropics became both cooler and drier, turning some areas of tropical rain forest into dry seasonal forest or savanna. For reasons associated with local topography, geography, and climate, some areas of forest escaped the dry periods, and acted as refuges for forest biota. During subsequent interglacials, when humid conditions returned to the tropics, the **forests** expanded and were repopulated by plants and animals from the species-rich refuges.

Ice age refuges today correspond to present day areas of tropical forest that typically receive a high rainfall and often contain unusually large numbers of **species**, including a high proportion of **endemic** species. These species-rich refuges are surrounded by relatively species-poor areas of forest. Refuges are also centers of distribution for obligate forest species such as the gorilla, with a present day narrow and disjunct distribution best explained by invoking past episodes of **deforestation** and reforestation. The location and extent of the forest refuges have been mapped in both **Africa** and **South America**. In the African rain forests there are three main centers of species richness and endemism recognized for **mammals**, **birds**, **reptiles**, **amphibians**, **butterflies**, freshwater **crabs**, and flowering plants. These centers are in Upper Guinea, Cameroon and Gabon, and the eastern rim of the Zaire **basin**. In the Amazon basin more than 20 refuges have been identified for different groups of animals and plants in Peru, Columbia, Venezuela, and Brazil.

The precise effect of the ice ages on **biodiversity** in tropical rain forests is currently a **matter** of debate. Some have argued that the repeated fluctuations between humid and arid phases created opportunities for the rapid **evolution** of certain forest organisms. Others have argued the opposite—that the climatic fluctuations resulted in a net loss of species diversity through an increase in the **extinction rate**. It has also been suggested that refuges owe their species richness not to past climate changes but to other underlying causes such as a favorable local climate, or **soil**.

The discovery of centers of high biodiversity and endemism within the tropical rain forest **biome** has profound implications for **conservation** biology. A refuge rationale has been proposed by conservationists, whereby ice age refuges are given high priority for preservation, since this would save the largest number of species, including many unnamed, threatened, and **endangered species**, from extinction.

Since refuges survived the past dry-climate phases, they have traditionally supplied the plants and animals for the restocking of the new-growth forests when wet conditions returned. Modern deforestation patterns, how-

ever, do not take into account forest history or biodiversity, and both forest refuges and more recent forests are being destroyed equally. For the first time in millions of years, future tropical forests which survive the present mass deforestation episode could have no species-rich centers from which they can be restocked.

Resources

Books

Collins, Mark, ed. *The Last Rain Forests*. London: Mitchell Beazley Publishers, 1990.

Sayer, Jeffrey A., et al., eds. *The Conservation Atlas of Tropical Forests*. New York: Simon and Schuster, 1992.

Whitmore, T. C. *An Introduction to Tropical Rain Forests*. Oxford, England: Clarenden Press, 1990.

Periodicals

Bard, E. "Ice Age Temperatures and Geochemistry." *Science* no. 284 (May 1999): 1133-1134.

Other

Prance, Ghillean T., ed. *Biological Diversification in the Tropics*. Proceedings of the Fifth International Symposium of the Association for Tropical Biology, at Caracas, Venezuela, February 8-13, 1979. New York: Columbia University Press, 1982.

Ice ages

The ice ages were periods in Earth's history during which significant portions of Earth's surface were covered by **glaciers** and extensive fields of **ice**. Scientists sometimes use more specific terms for an ice "age" depending on the length of **time** it lasts. It appears that over the long expanse of **Earth** history, seven major periods of severe cooling have occurred. These periods are often known as ice eras and, except for the last of these, are not very well understood.

What is known is that Earth's average annual **temperature** varies constantly from year to year, from decade to decade, and from century to century. During some periods, that average annual temperature has dropped to low enough levels for fields of ice to grow and cover large regions of Earth's surface. The seven ice eras have covered an average of about 50 million years each.

The most recent ice era

The ice era that scientists understand best (because it occurred most recently) began about 65 million years ago. Throughout that long period, Earth experienced periods of alternate cooling and warming. Those periods during which the annual temperature was significantly less than average are known as ice epochs. There is evidence for the occurrence of six ice epochs during this last of the great ice eras.

During the 2.4 million year lifetime of the last ice epoch, about two dozen ice ages occurred. That means that Earth's average annual temperature fluctuated upwards and downwards to a very significant extent about two dozen times during the 2.4 million year period. In each case, a period of significant cooling was followed by a period of significant warming—an interglacial period—after which cooling once more took place.

Scientists know a great deal about the cycle of cooling and warming that has taken place on the earth over the last 125,000 years, the period of the last ice age cycle. They have been able to specify with some degree of precision the centuries and decades during which ice sheets began to expand and diminish. For example, the most severe temperatures during the last ice age were recorded about 50,000 years ago. Temperatures then warmed before plunging again about 18,000 years ago.

Clear historical records are available for one of the most severe recent cooling periods, a period now known as the Little Ice Age. This period ran from about the fifteenth to the nineteenth century and caused widespread crop failure and loss of human life throughout **Europe**. Since the end of the Little Ice Age, temperatures have continued to fluctuate with about a dozen unusually cool periods in the last century, interspersed between periods of warmer **weather**. No one is quite certain as to whether the last ice age has ended or whether we are still living through that period.

Evidence for the ice ages

A great deal of what scientists know about the ice ages they have learned from the study of mountain glaciers. For example, when a glacier moves downward out of its mountain source, it carves out a distinctive shape on the surrounding land. The "footprints" left by continental glaciers formed during the ice ages are comparable to those formed by mountain glaciers.

The transport of materials from one part of the earth's surface to another part is also evidence for continental glaciation. **Rocks** and fossils normally found only in one region of the the earth may be picked up and moved by ice sheets and deposited elsewhere. The "track" left by the moving glacier provides evidence of the ice sheets movement. In many cases, the moving ice may actually leave scratches on the rock over which it moves, providing further evidence for changes that took place during an ice age.

Causes of the ice ages

Scientists have been asking about the causes of ice ages for more than a century. The answer (or answers) to that question appears to have at least two main parts, astronomical factors and terrestrial factors. By astronomical factors, scientists mean that the way the earth is oriented in **space** can determine the amount of **heat** it receives and, hence, its annual average temperature.

One of the most obvious astronomical factors about which scientists have long been suspicious is the appearance of **sunspots**. Sunspots are eruptions that occur on the sun's surface during which unusually large amounts of solar **energy** are released. The number of sunspots that occur each year changes according to a fairly regular pattern, reaching a maximum about every 11 years or so. The increasing and decreasing amounts of energy sent out during sunspot **maxima and minima**, some scientists have suggested, may contribute in some way to the increase and decrease of ice fields on the earth's surface.

By the beginning of the twentieth century, however, astronomers had identified three factors that almost certainly are major contributors to the amount of solar **radiation** that reaches the earth's surface and, hence, the earth's average annual temperature. These three factors are the earth's angular tilt, the shape of its **orbit** around the **sun**, and its axial precession.

The first of these factors, the planet's angular tilt, is the angle at which its axis is oriented to the **plane** of its orbit around the sun. This angle slowly changes over time, ranging between 21.5 and 24.5 degrees. At some angles, the earth receives more solar radiation and becomes warmer, and at other angles it receives less solar radiation and becomes cooler.

The second factor, the shape of Earth's orbit around the sun, is important because, over long periods of time, the orbit changes from nearly circular to more elliptical (flatter) in shape. As a result of this variation, the earth receives more or less solar radiation depending on the shape of its orbit. The final factor, axial precession, is a "wobble" in the orientation of Earth's axis to its orbit around the sun. As a result of axial precession, the amount of solar radiation received during various parts of the year changes over very long periods of time.

Between 1912 and 1941, the Yugoslav astronomer Milutin Milankovitch developed a complex mathematical theory that explained how the interaction of these three astronomical factors could contribute to the development of an ice age. His calculations provided rough approximations of the occurrences of ice ages during the earth history.

Terrestrial factors

Astronomical factors provide only a broad general background for changes in the earth's average annual temperature, however. Changes that take place on the earth itself also contribute to the temperature variations that bring about ice ages.

Scientists believe that changes in the composition of the earth's atmosphere can affect the planet's annual average temperature. Some gases, such as **carbon dioxide** and nitrous oxide, have the ability to capture heat radiated from the earth, warming the atmosphere. This phenomenon is known as the **greenhouse effect**. But the composition of Earth's atmosphere is known to have changed significantly over long periods of time. Some of these changes are the result of complex interactions of biotic, geologic and geochemical processes. Humans have dramatically increased the concentration of **carbon** dioxide in the atmosphere over the last century through the burning of **fossil fuels** (**coal**, oil, and **natural gas**). As the concentration of greenhouse gases, like carbon dioxide and nitrous oxide, varies over many decades, so does the atmosphere's ability to capture and retain heat.

Other theories accounting for atmospheric cooling have been put forth. It has been suggested that **plate tectonics** are a significant factor affecting ice ages. The **uplift** of large continental blocks resulting from plate movements (for example, the uplift of the Himalayas and the Tibetan Plateau) may cause changes in global circulation patterns. The presence of large land masses at high altitudes seems to correlate with the growth of ice sheets, while the opening and closing of **ocean** basins due to

Drumlins, like this one near West Bend, Wisconsin, are composed of glacial till. Although their formation is not well understood, some geologists believe they are shaped when a glacier advances over its own end moraine. *JLM Visuals. Reproduced by permission.*

tectonic movement may affect the movement of warm **water** from low to high latitudes.

Since volcanic eruptions can contribute to significant temperature variations, it has been suggested that such eruptions could contribute to atmospheric cooling, leading to the lowering of Earth's annual temperature. Dust particles thrown into the air during an eruption can reflect sunlight back into space, reducing heat that would otherwise have reached Earth's surface. The eruption of Mount Pinatubo in the Philippine Islands in 1991 is thought to have been responsible for a worldwide cooling that lasted for at least five years. Similarly, the earth's annual average temperature might be affected by the impact of meteorites on Earth's surface. If very large meteorites had struck Earth at times in the past, such collisions would have released huge volumes of dust into the atmosphere. The presence of this dust would have had effects similar to the eruption of Mount Pinatubo, reducing Earth's annual average temperature for an extended period of time and, perhaps, contributing to the development of an ice age.

The ability to absorb heat and the reflectivity of the earth's surface also contribute to changes in the annual average temperature of Earth. Once an ice age begins, sea levels drop as more and more water is tied up in ice sheets and glaciers. More land is exposed, and because land absorbs heat less readily than water, less heat is retained in the earth's atmosphere. Likewise, pale surfaces reflect more heat than dark surfaces, and as the area covered by ice increase, so does the amount of heat reflected back to the upper atmosphere.

Whatever the cause of ice ages, it is clear that they can develop as the result of relatively small changes in

KEY TERMS

. . . .

Axial precession—The regular and gradual shift of the earth's axis, a kind of "wobble," that takes place over a 23,000 year period.

Interglacial period—A period of time between two glacial periods during which the earth's average annual temperature is significantly warmer than during the two glacial periods.

the earth's average annual temperature. It appears that annual variations of only a few degrees Celsius can result in the formation of extensive ice sheets that cover thousands of square miles of the earth's surface.

See also Geologic time.

David E. Newton

Icebergs

An iceberg is a large mass of free-floating **ice** that has broken away from a glacier. Beautiful and dangerous, icebergs wander over the **ocean** surface until they melt. Most icebergs come from the **glaciers** of Greenland or from the massive ice sheets of **Antarctica**. A few icebergs originate from smaller Alaskan glaciers. Snow produces the glaciers and ice sheets so, ultimately, icebergs originate from snow. In contrast, "sea ice" originates from freezing **saltwater**. When fragments break off of a glacier, icebergs are formed in a process called calving. Icebergs consist of **freshwater** ice, pieces of debris, and trapped bubbles of air. The combination of ice and air bubbles causes sunlight shining on the icebergs to refract, coloring the ice spectacular shades of blue, green, and white. **Color** may also indicate age; blue icebergs are old, and green ones contain **algae** and are young. Icebergs come in a variety of shapes and sizes, some long and flat, others towering and massive.

An iceberg floats because it is lighter and less dense than salty seawater, but only a small part of the iceberg is visible above the surface of the sea. Typically, about 80-90% of an iceberg is below **sea level**, so they drift with ocean **currents** rather than **wind**. Scientists who study icebergs classify true icebergs as pieces of ice which are greater than 16 ft (5 m) above sea level and wider than 98 ft (30 m) at the water line. Of course, icebergs may be much larger. Smaller pieces of floating ice are called "bergy bits" (3.3-16 ft or 1-5 m tall and 33-98 ft or 10-30

m wide) or "growlers" (less than 3.3 ft or 1 m tall and less than 33 ft or 10 m wide). The largest icebergs can be taller than 230 ft (70 m) and wider than 738 ft (225 m). Chunks of ice more massive than this are called ice islands. Ice islands are much more common in the Southern Hemisphere, where they break off the Antarctic ice sheets.

Because of the unusual forms they may take, icebergs are also classified by their shape. Flat icebergs are called tabular. Icebergs which are tall and flat are called blocky. Domed icebergs are shaped like a turtle shell, rounded, with gentle slopes. Drydock icebergs have been eroded by waves so that they are somewhat U-shaped. Perhaps the most spectacular are the pinnacle icebergs, which resemble mountain tops, with one or more central peaks reaching skyward.

The life span of an iceberg depends on its size but is typically about two years for icebergs in the Northern Hemisphere. Because they are larger, icebergs from Antarctica may last for several more years. Chief among the destructive forces that work against icebergs are wave action and **heat**. Wave action can break icebergs into smaller pieces and can cause icebergs to knock into each other and fracture. Relatively warm air and water **temperature** gradually melt the ice. Because icebergs float, they drift with water currents towards the equator into warmer water. Icebergs may drift as far as 8.5 mi (14 km) per day. Most icebergs have completely melted by the time they reach about 40 degrees latitude (north or south). There have been rare occasions when icebergs have drifted as far south as Bermuda (32 degrees north latitude), which is located about 900 mi (1,400 km) east of Charleston, South Carolina. In the Atlantic Ocean, they have also been found as far east as the Azores, islands in the Atlantic Ocean off the coast of Spain.

An iceberg struck and sank the R.M.S. *Titanic* on April 14, 1912, when the great ship was on her maiden voyage; more than 1,500 people lost their lives in that disaster, which occurred near Newfoundland, Canada. As a result of the tragedy, the Coast Guard began monitoring icebergs to protect shipping interests in the North Atlantic sea lanes. Counts of icebergs drifting into the North Atlantic shipping lanes vary from year to year, with little predictability. During some years, no icebergs drift into the lanes; other years are marked by hundreds or more—as many as 1,572 have been counted in a single year. Many ships now carry their own **radar** equipment to detect icebergs. In 1959, a Danish ship equipped with radar struck an iceberg and sank, resulting in 95 deaths. Some ships even rely on infrared sensors from airplanes and satellites. Sonar is also used to locate icebergs.

An iceberg in Disko Bay on the western coast of Greenland. *Photograph by Tom Stewart The Stock Market. Reproduced by permission.*

Modern iceberg research continues to focus on improving methods of tracking and monitoring icebergs, and on **learning** more about iceberg deterioration. In 1995, a huge iceberg broke free from the Larsen ice shelf in Antarctica. This iceberg was 48 mi (77 km) long, 23 mi (37 km) wide, and 600 ft (183 m) thick. The monster was approximately the size of the country of Luxembourg and isolated James Ross Island (one of Antarctica's islands) for the first time in recorded history. The megaberg was monitored by airplanes and satellites to make sure it did not put ships at peril. According to some scientists, this highly unusual event could be evidence of **global warming**. Surges in the calving of icebergs known as Heinrich events are also known to be caused by irregular motions of **Earth** around the **Sun** that cause ocean waters of varying temperatures and salinity to change their circulation patterns. These cycles were common during the last glacial period, and glacial debris was carried by "iceberg armadas" to locations like Florida and the coast of Chile. Scientists have "captured" icebergs for study including crushing to measure their strength. During World War II, plans were made to make floating airfields from flat-topped bergs (but this never got past the planning stage). Some people have proposed towing icebergs to drought-stricken regions of the world to solve water shortage problems; however, the cost and potential environmental impact of such an undertaking have so far discouraged any such attempts.

Resources

Books

Colbeck, S. C. *Dynamics of Snow and Ice Masses.* New York: Academic Press, 1980.

Lewis, E.O., B.W. Currie, and S. Haykin. *Detection and Classification of Ice.* Letchworth, England: Research Studies Press, 1987.

Sharp, R. P. *Living Ice: Understanding Glaciers and Glaciation.* Cambridge: Cambridge University Press, 1988.

Periodicals

Ballard, R. D. "A Long Last Look at the Titanic." *National Geographic* (December 1986): 698-727.

Dane, M. "Icehunters." *Popular Mechanics* (October 1993): 76-79.

Monastersky, R. "Satellite Radar Keeps Tabs on Glacial Flow." *Science News* (December 1993): 373.

Nicklin, F. "Beneath Arctic Ice: Life at the Edge." *National Geographic* (July 1991): 2-31.

KEY TERMS

· ·

Calving—The process in which huge chunks of ice or icebergs break off from ice shelves and sheets or glaciers to form icebergs.

Ice island—A thick slab of floating ice occupying an area as large as 180 sq mi (460 sq km).

Ice sheet—Glacial ice covering at least 19,500 sq mi (50,000 sq km) of land and obscuring the landscape below it.

Ice shelf—That section of an ice sheet that extends into the sea a considerable distance and which may be partially afloat.

Sea ice—Ice that forms from the freezing of salt water; as the saltwater freezes, it ejects salt, so sea ice is fresh, not salty. Sea ice forms in relatively thin layers, usually no more than 3–7 ft (1–2 m) thick, but it can cover vast areas of the ocean surface at high latitudes.

Raney, R. K. "Probing Ice Sheets with Imaging Radar." *Science* 262 (1993):1521-1522.

Steger, W. "Six across Antarctica: Into the Teeth of the Ice." *National Geographic* (November 1990): 67-95.

Vogt, P. R., and K. Crane. "Megabergs Left Scars in Arctic." *Science News* (August 1994): 127.

Elaine Martin

Iceman

The Iceman is an intact, 5,300-year-old mummy discovered September 19, 1991, in a melting glacier within the Italian Alps near Austria. The oldest human discovered in **Europe**, he is one of the most complete, naturally mummified humans ever found.

Age determination

Tools found near the body accompanied the Iceman to Innsbruck, and one of these tools, an axe with a **metal** blade, gave scientists a clue to the Iceman's age. The axe's distinctive shape, similar to those of the Early Bronze Age (2,200-1,000 B.C.), suggested the body is approximately 4,000 years old. Repeated radiocarbon dating of a bone **sample**, performed at two different laboratories, instead indicated the Iceman is 5,300 years old. The axe blade was analyzed and found to be **copper**, supporting the radiocarbon dating.

Therefore, scientists believe the Iceman lived during the Stone Age, or Neolithic Period. More specifically, he lived during the Copper Age, which occurred in central Europe between 4,000-2,000 B.C.

Significance of discovery

Studying the Iceman is important to many branches of science, including **archaeology**, **biology**, **geology**, and **pathology**. The Iceman, stored at 21°F (-6°C) and 98% relative **humidity**, is removed for observation or sample collection for no longer than 20 minutes at a time.

The body was found naturally "freeze dried" at an altitude of about 10,500 ft (3,200 m). Prior to the find, archaeologists had never excavated for evidence of human activity at such high altitudes in Europe. Geologists wondered how the Iceman was spared the grinding forces of glacial **ice**, and why he was not transported down the mountain within an ice flow. Fortunately, the corpse lay in a rock-rimmed depression below a ridge. The Iceman remained entombed in a stable ice pocket within this depression, undisturbed as the glacier flowed overhead.

The Iceman's axe, flint knife, bow and arrows, leather pouch, grass cape, leather shoes, and other accessories provide a glimpse of everyday life during Europe's Copper Age. The Iceman's leather clothing is rare indeed. The only evidence of leather workmanship typically recovered at an archeological site is a leather scraper.

Scientists have analyzed the Iceman's bone, **blood**, DNA, and stomach contents to assess the presence of diseases, his social status, occupation, diet, and general health. Studies of his teeth suggest a diet of coarse grain; studies of his hair suggest a vegetarian diet at the time of his death. Analyses indicate he was 5ft 2in (1.5 m), 110 lb (50 kg), and approximately 25-35 years of age at death. Scientists speculate that the Iceman died of exposure.

Biologists identified slowberries in his birch **bark** container, suggesting he died in the autumn, when the berries ripen. Speculation that he belonged to an agricultural community is based on the grains of ancient **wheat** found with the corpse.

Although errors made in handling and preserving the body destroyed our chances to answer certain questions, there is much the Iceman will teach us about life 5,000 years ago. For example, the Iceman's tattoos are 2,500 years older than any seen before. Placed on his body in locations not easily observed and thought to correspond with **acupuncture** points, they raise the question that acupuncture may have been practiced earlier than thought and possibly began in Eurasia not east **Asia**.

Ideal gas *see* **Gases, properties of**

Identity element

Any mathematical object that, when applied by an operation, such as **addition** or **multiplication**, to another mathematical object, such as a number, leaves the other object unchanged is called an identity element. The two most familiar examples are 0, which when added to a number gives the number, and 1, which is an identity element for multiplication.

More formally, an identity element is defined with respect to a given operation and a given set of elements. For example, 0 is the identity element for addition of **integers**; 1 is the identity element for multiplication of **real numbers**. From these examples, it is clear that the operation must involve two elements, as addition does, not a single element, as such operations as taking a power.

Sometimes a set does not have an identity element for some operation. For example, the set of even numbers has no identity element for multiplication, although there is an identity element for addition. Most mathematical systems require an identity element. For example, a **group** of transformations could not exist without an identity element that is the transformation that leaves an element of the group unchanged.

See also Identity property; Set theory.

Identity property

When a set possesses an **identity element** for a given operation, the mathematical system of the set and operation is said to possess the identity property for that operation. For example, the set of all functions of a **variable** over the **real numbers** has the identity element, or identity function, $I(x) = x$. In other words, if $f(x)$ is any function over the real numbers, then $f(I(x)) = I(f(x)) = f(x)$.

See also Function.

Igneous extrusions *see* **Igneous rocks**

Igneous intrusions *see* **Igneous rocks**

Igneous rocks

Igneous **rocks** are formed by the cooling and hardening of molten material called **magma**. The word igneous comes from the Latin word *igneus*, meaning fire. There are two types of igneous rocks: intrusive and extrusive. Intrusive igneous rock forms within Earth's crust; the molten material rises, filling voids in the crust, and eventually hardens. Intrusive rocks are also called plutonic rocks, named after the Greek god **Pluto**, god of the underworld. Extrusive igneous rocks form when the magma, called lava once it reaches the surface, pours out onto the earth's surface. Extrusive rocks are also known as volcanic rocks.

Igneous rocks are classified according to their texture and mineral or chemical content. The texture of the rock is determined by the **rate** of cooling. The slower the rock cools, the larger the crystals form. Because the magma chamber is well insulated by the surrounding country rock, intrusive rocks cool very slowly and can form large, well developed crystals. Rapid cooling results in smaller, often microscopic, grains. Some extrusive rocks solidify in the air, before they hit the ground. Sometimes the rock mass starts to cool slowly, forming large crystals, and then finishes cooling rapidly, resulting in rocks that have larger crystals surrounded by a fine-grained **matrix**. This is known as a porphyritic texture. Other extrusive rocks cool before the chemical constituents of the melt are able to arrange themselves into any crystalline form. These are said to have glassy texture and include the rocks obsidian and pumice.

The chemistry of the magma determines the **minerals** that will crystallize and their relative abundance in the rocks that form. Light-colored igneous rocks are likely to contain high proportions of light colored minerals, such as quartz and feldspars and are called felsic. Dark rocks will contain **iron** and magnesium-rich minerals like pyroxene and olivine and are known as mafic rocks. Those rocks with a **color** falling between the two are said to have an intermediate composition.

Once the basic composition and texture of the rock are determined, they are combined to establish the name of the rock. For example, a coarse-grained, felsic rock is called granite and a fine-grained felsic rock is called rhyolite. These two rocks are composed of the same minerals, but the slow cooling history of the granite has allowed its crystals to grow larger. These are some of the most familiar igneous rocks because continental portions of the crust are built largely of rock that is similar in composition to these felsic rocks. Coarse-grained and fine-grained mafic rocks are called gabbro and basalt, respectively. Each of these is easily recognized by their dark color. In general, oceanic crustal plates are primarily mafic in chemistry. Diorite and andesite are the respective names for coarse- and fine-grained rocks of intermediate composition. While geologists sometimes use more detailed classification systems, this basic method is used for preliminary differentiation of crystalline igneous rocks.

Certain igneous rocks are named on the basis of particular features. Fragmental rocks like tuff and volcanic

Rope-like, twisted lava like this is called pahoehoe. *JLM Visuals. Reproduced by permission.*

breccia are named on the basis of the size of particles of volcanic material ejected during an eruption. Tuff is composed of fine particles of volcanic ash, while breccia includes larger pieces. Obsidian, pumice, and often scoria have a non-crystalline, glassy texture that can be distinguished on the basis of the quantity of trapped gas. Obsidian contains no such gas and pumice has so many gas bubbles that it will sometimes actually float on **water**.

Earth's crustal plates are continually shifting, being torn open by faults, and altered by earthquakes and volcanoes. As old plates are drawn downward into the mantle, old rock material is recycled through melting. New igneous material is continually added to the crust along plate margins and other locations through igneous intrusions and volcanic activity. Igneous rocks represent both the ancient history of the formation of the **earth** and modern episodes of regeneration. Associated igneous processes are evidence of the continuing activity of **Earth's interior** and the form and composition of each of the igneous rocks give clues as to the conditions and processes under which they formed.

See also Plate tectonics; Volcano.

Iguanas

Iguanas are large, ancient, herbivorous lizards with a stocky trunk, long, slender tail, scaly skin, and a single row of spines from the nape of the neck to the tip of the tail. On either side of the head is an **eye** with a round pupil and with moveable lids. The well-defined snout has two nostrils, the mouth houses a short, thick tongue, and dangling beneath the chin is a "dewlap," or throat fan. Iguanas are well equipped for speed and climbing with four short, thick, powerful legs, each with five long thin

toes tipped with strong claws. Iguanas are found in warm, temperate, and tropical zones and, depending on the **species**, live in trees, holes, burrows, and among **rocks**. Iguanas are **oviparous** (egg-laying), diurnal (active during the day), and ectothermic (cold-blooded), thermoregulating by basking in the **sun** or sheltering in the shade. Iguanas are found only in the New World, and were completely unknown in the Old World until European explorers discovered the Americas.

Classification and characteristics

The 30 species of iguanas belong to the subfamily Iguaninae, of the family Iguanidae. Iguanas are assigned to seven genera, their common names being banded iguanas (*Brachylophus*), land iguanas (*Conolophus*), spiny-tailed iguanas (*Conolophus*), ground iguanas (*Cyclura*), **desert** iguanas (*Dipsosaurus*), green iguanas (*Iguana*), marine iguanas (*Amblyncus*), and chuckwallas (*Sauromalus*).

Size, weight, and longevity vary between species. Large land iguanas of the Galápagos Islands range in weight from a hefty 26.5 lb (12 kg) to less than 11 lb (5 kg), while the tiny ground iguana of the Bahamas and West Indies weighs scarcely 2 lb (1 kg).

Distribution and diet

Iguanas are believed to be monophyletic, that is, they have evolved from a single ancestral type dating back to the **extinction** of the dinosaurs. Except for the banded iguana of the Fijian islands, all species are found exclusively in the Western Hemisphere. Marine iguanas are the only living lizard which spends time in the **ocean**, exclusively existing on **algae** gathered from rocks either by diving or foraging on tidally exposed reefs. Only the banded iguana and the green iguana of **South America** are found in wet tropics, while all other species inhabit dry environments.

Iguanas are strictly herbivorous—with the exception of the spiny-tailed iguana whose young eat **insects** and a few species which occasionally eat readily available meat. Iguanas are selective about their diet, preferring easily digested fruit, **flower** buds, and tender young leaves.

Reproduction

After reaching sexual maturity, iguanas reproduce annually until death. Green iguanas mature during their second or third year and live to be 10 or 12 years old, while the large land iguana attains adulthood around 10 years, and may live to age 40. Adult males establish mating territories and are selected by females who prefer larger males. Females may court several males before choosing a mate,

A chuckwalla (*Sauromalus obesus*) at the Arizona Sonora Desert Museum, Arizona. This lizard basks in the sun during the day to reach its preferred body temperature of 100°F (38°C). *Photograph by Robert J. Huffman. Field Mark Publications. Reproduced by permission.*

and one male may be chosen by several females, all of which take up residence in the male's territory.

Several weeks after mating, the female selects a nesting site where she digs a burrow, creates a special chamber, and lays her single clutch of eggs. Seven to 12 weeks after mating, the green iguana lays 20 to 30 eggs, each about 1.5 in (4 cm) long. The banded iguana lays three to six eggs, each about 1 in (3 cm) long, approximately six weeks after mating.

After laying her eggs, the female exits and fills in her burrow, leaving an air pocket in the chamber for the hatchlings, which appear three to four months later at the onset of the rainy season, when food is abundant. The banded iguana is unique in that egg incubation takes an unusually long five to eight months. The young hatch simultaneously and dig to the surface. In most species, only a small percentage of hatchlings reach maturity.

Display patterns as attractions and deterrents

Males of most species use head bobbing, pushups, and expansion of the dewlap to attract a mate. More threatening postures, such as opening the mouth, tongue-flicking, and snorting, are added when defending territories or warding off rivals. Banded iguanas also puff up their torso and their green bands become much darker, increasing the contrast with their pale blue-green bands. Physical aggression is rare, and the occasional clash results in head-thrashing, tail-swinging, and sometimes biting, with the loser creeping quietly away. Females usually only show aggression when contesting for, or defending, nesting sites. Each species has a distinct display pattern which seems to aid in recognition.

Popularity and extinction

The green iguana is the largest, most prolific, and best-known species in the Americas, and is in great demand in the United States where proud owners can be seen parading this gentle green lizard on their shoulders, restrained in specially designed harnesses. This arboreal (tree-dwelling) lizard naturally inhabits the periphery of rainforests from Mexico to the tip of South America. Green iguanas live in groups near **rivers** and **water** holes, and lie along **tree** limbs high above the water,

KEY TERMS

. . . .

Dewlap—A loose fold of skin that hangs from beneath the chin.

Diurnal—Refers to animals that are mainly active in the daylight hours.

Ectothermic—A cold-blooded animal, whose internal body temperature is similar to that of its environment. Ectotherms produce little body heat, and are dependent on external sources (such as the sun) to keep their body temperature high enough to function efficiently.

Monophyletic—Evolving from a single ancestral type.

Thermoregulate—Regulate and control body temperature.

basking in the sun as still as statues, prepared to plunge if danger approaches. Green iguanas are excellent swimmers, and can remain submerged for 30 minutes, often surfacing in a safer location.

Few iguanas escape the skilled, professional human hunter, however, for apart from their value in the pet trade, their eggs are dietary delicacies, as is their flesh, which is often called "gallina de palo," or "tree chicken," and is credited with medicinal properties which supposedly cure such conditions as impotency.

While all iguanas have natural predators such as **snakes**, carnivorous **birds**, and wild **canines**, most species are in danger of extinction from human actions—direct capture, **habitat** destruction, introduction of domestic and feral **mammals**, **pesticides**, and firearms. Fortunately, green iguanas are now being successfully bred in captivity for both the food and pet trades. Some **conservation** efforts for this and other species have been implemented in the form of protective legislation, **wildlife** reserves, and public awareness campaigns. However, much effort is still necessary to prevent the rapidly increasing destruction of these ancient, docile herbivores.

See also Herbivore.

Resources

Books

Burghardt, Gordon M., and A. Stanley Rand, eds. *Iguanas of the World, Their Behavior, Ecology, and Conservation.* Park Ridge, NJ: Noyes Data Corp, 1982.

Harris, Jack C. *A Step-by-Step Book about Iguanas.* Neptune, NJ: T.F.H. Publications, 1990.

Periodicals

Leal, Jose H. "Iguanas as Island Hoppers." *Sea Frontiers* (June 1992): 11-12.

Marie L. Thompson

Imaginary number

The number $i = \sqrt{-1}$ is the basis of any imaginary number, which, in general, is any real number times i. For example, 5i is an imaginary number and is equivalent to $-1 \div 5$. The **real numbers** are those numbers that can be expressed as terminating, repeating, or nonrepeating decimals; they include positive and **negative** numbers. The product of two negative real numbers is always positive. Thus, there is no real number that equals -1 when multiplied by itself—that is, no real number satisfies the equation $x^2 = -1$ in the real number system. The imaginary number i was invented to provide a solution to this equation, and every imaginary number represents the solution to a similar equation (e.g., 5i is a solution to the equation $x^2 = -25$).

In addition to providing solutions for algebraic equations, the imaginary numbers, when combined with the real numbers, form the **complex numbers**. Each complex number is the sum of a real number and an imaginary number, such as (6 + 9i). The complex numbers are very useful in mathematical analysis, the study of **electricity** and **magnetism**, the **physics** of **quantum mechanics**, and in the practical **field** of electrical **engineering**. In terms of the complex numbers, the imaginary numbers are equivalent to those complex numbers for which the real part is **zero**.

See also Square root.

Immune system

The immune system protects the body from disease-causing **microorganisms**. It consists of two levels of protection, the non-specific defenses and the specific defenses. The non-specific defenses, such as the skin and mucous membranes, prevent microorganisms from entering the body. The specific defenses are activated when microorganisms evade the non-specific defenses and invade the body.

The human body is constantly bombarded with microorganisms, many of which can cause **disease**. Some of these microorganisms are viruses, such as those that cause colds and **influenza**; other microorganisms are

bacteria, such as those that cause **pneumonia** and **food poisoning**. Still other microorganisms are **parasites** or **fungi**. Usually, the immune system is so efficient that most of us are unaware of the battle that takes place almost everyday, as the immune system rids the body of harmful invaders. However, when the immune system is injured or destroyed, the consequences are severe. For instance, Acquired Immune Deficiency **Syndrome** (**AIDS**) is caused by a virus—Human Immunodeficiency **Virus** (HIV)—that attacks a key immune system **cell**, the helper T-cell lymphocyte. Without these cells, the immune system cannot function. People with AIDS cannot fight off the microorganisms that constantly bombard their bodies, and eventually succumb to infections that a healthy immune system would effortlessly neutralize.

Organs of the immune system

The organs of the immune system either make the cells that participate in the immune response or act as sites for immune function. These organs include the lymphatic vessels, lymph nodes, tonsils, thymus, Peyer's patch, and spleen. The lymph nodes are small aggregations of tissues interspersed throughout the **lymphatic system**. White **blood** cells (lymphocytes) that function in the immune response are concentrated in the lymph nodes. Lymphatic fluid circulates through the lymph nodes via the lymphatic vessels. As the lymph filters through the lymph nodes, foreign cells of microorganisms are detected and overpowered.

The tonsils contain large numbers of lymphocytes. Located at the back of the throat and under the tongue, the tonsils filter out potentially harmful bacteria that may enter the body via the nose and mouth. Peyer's patches are lymphatic tissues which perform this same function in the **digestive system**. Peyer's patches are scattered throughout the small intestine and the appendix. They are also filled with lymphocytes that are activated when they encounter disease-causing microorganisms.

The thymus gland is another site of lymphocyte production. Located within the upper chest region, the thymus gland is most active during childhood when it makes large numbers of lymphocytes. The lymphocytes made here do not stay in the thymus, however; they migrate to other parts of the body and concentrate in the lymph nodes. The thymus gland continues to grow until **puberty**; during adulthood, however, the thymus shrinks in size until it is sometimes impossible to detect in x-rays.

Bone marrow, found within the bones, also produces lymphocytes. These lymphocytes migrate out of the bone marrow to other sites in the body. Because bone marrow is an integral part of the immune system, certain bone **cancer** treatments that require the destruction of

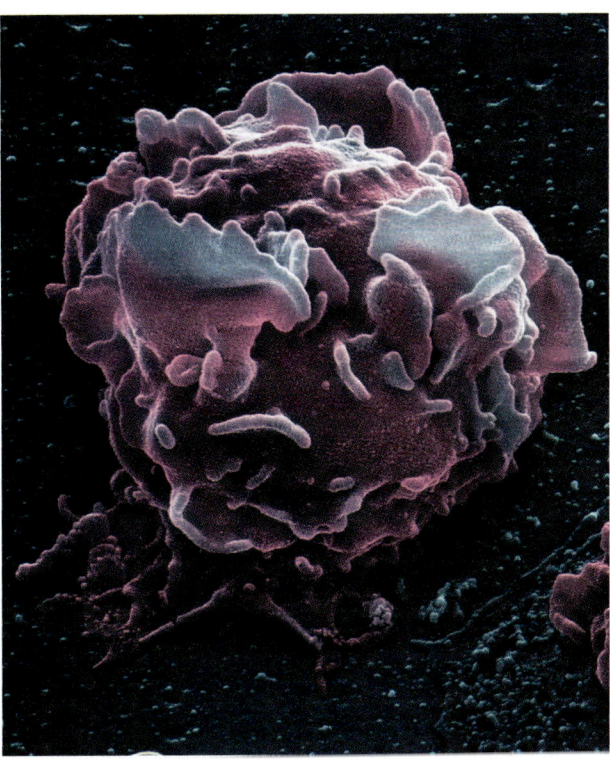

Colored scanning electron micrograph of a white blood cell (pink and white) attacking a *Staphylococcus aureus* bacterium (yellow). *Photograph by Juergen Berger. Max-Planck Institute/Science Photo Library/Photo Researchers, Inc. Reproduced by permission.*

bone marrow are extremely risky, because without bone marrow, a person cannot make lymphocytes. People undergoing bone marrow replacement must be kept in strict isolation to prevent exposure to viruses or bacteria.

The spleen acts as a reservoir for blood and any rupture to the spleen can cause dangerous internal bleeding, a potentially fatal condition. The spleen also destroys worn-out red blood cells. Moreover, the spleen is also a site for immune function, since it contains lymphatic **tissue** and produces lymphocytes.

Overview of the immune system

For the immune system to work properly, two things must happen: First, the body must recognize that it has been invaded by foreign microorganisms. Second, the immune response must be quickly activated before many body tissue cells are destroyed by the invaders.

How the immune system recognizes foreign invaders

The cell **membrane** of every cell is studded with various **proteins** that protrude from the surface of the

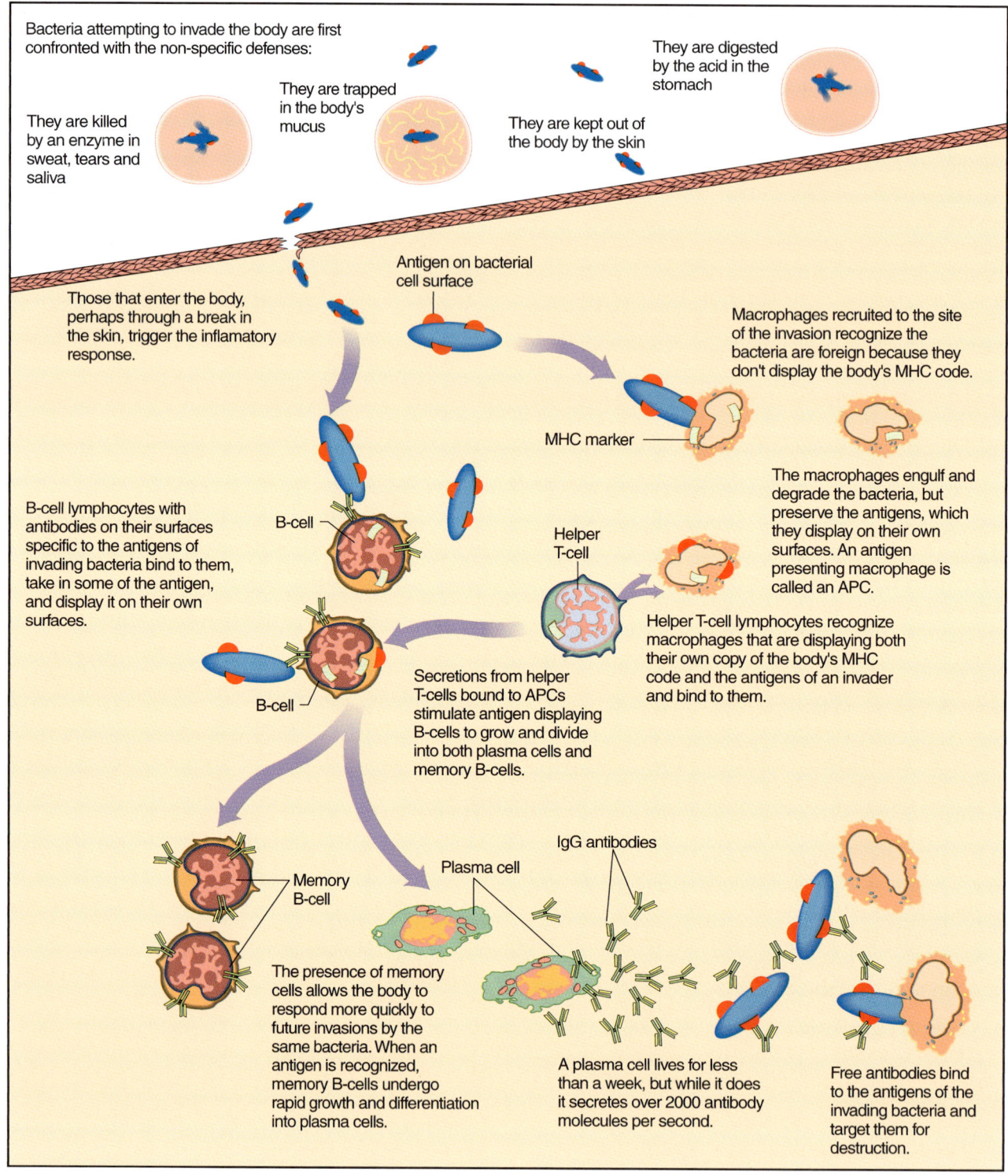

Bacteria attempting to invade the body are first confronted with the non-specific defenses:

They are killed by an enzyme in sweat, tears and saliva

They are trapped in the body's mucus

They are kept out of the body by the skin

They are digested by the acid in the stomach

Those that enter the body, perhaps through a break in the skin, trigger the inflamatory response.

Antigen on bacterial cell surface

Macrophages recruited to the site of the invasion recognize the bacteria are foreign because they don't display the body's MHC code.

MHC marker

B-cell lymphocytes with antibodies on their surfaces specific to the antigens of invading bacteria bind to them, take in some of the antigen, and display it on their own surfaces.

B-cell

Helper T-cell

The macrophages engulf and degrade the bacteria, but preserve the antigens, which they display on their own surfaces. An antigen presenting macrophage is called an APC.

Helper T-cell lymphocytes recognize macrophages that are displaying both their own copy of the body's MHC code and the antigens of an invader and bind to them.

B-cell

Secretions from helper T-cells bound to APCs stimulate antigen displaying B-cells to grow and divide into both plasma cells and memory B-cells.

IgG antibodies

Memory B-cell

Plasma cell

The presence of memory cells allows the body to respond more quickly to future invasions by the same bacteria. When an antigen is recognized, memory B-cells undergo rapid growth and differentiation into plasma cells.

A plasma cell lives for less than a week, but while it does it secretes over 2000 antibody molecules per second.

Free antibodies bind to the antigens of the invading bacteria and target them for destruction.

The antibody-mediated immune response. *Illustration by Hans & Cassidy. Courtesy of Gale Group.*

membrane. These proteins are a kind of name tag called the Major Histocompatibility Complex (MHC). They identify all the cells of the body as belonging to the "self." An invading microorganism, such as a bacterium, does not have the "self" MHC on its surface. When an immune system cell encounters this "non-self" cell, it alerts the body that it has been invaded by a foreign cell. Every person has their own unique MHC. For this reason, **organ** transplants are often unsuccessful because the immune system interprets the transplanted organ as "foreign," since the transplanted organ cells have a "non-self" MHC. Organ recipients usually take immunosuppressant drugs to suppress the immune response, and every effort is made to transplant organs from close relatives, who have genetically similar MHCs.

In addition to a lack of the "self" MHC, cells that prompt an immune response have foreign molecules (called antigens) on their membrane surfaces. An antigen is usually a protein or polysaccharide complex on the outer layer of an invading microorganism. The antigen can be a viral coat, the cell wall of a bacterium, or the surface of other types of cells. Antigens are extremely important in the identification of foreign microorganisms. The specific immune response depends on the ability of the immune lymphocytes to identify the invader and create immune cells that specifically mark the invader for destruction.

How the two defenses work together

The immune system keeps out microorganisms with non-specific defenses. Non-specific defenses do not involve identification of the antigen of a microorganism; rather the non-specific defenses simply react to the presence of a "non-self" cell. Oftentimes, these non-specific defenses effectively destroy microorganisms. However, if they are not effective and the microorganisms manage to infect tissues, the specific defenses are activated. The specific defenses work by recognizing the specific antigen of a microorganism and mounting a response that targets the microorganism for destruction by components of the non-specific system. The major difference between the non-specific defenses and the specific defenses is that the former impart a general type of protection against all kinds of foreign invaders, while the specific defenses create protection that is tailored to match the particular antigen that has invaded the body.

The non-specific defenses

The non-specific defenses consist of the outer barriers, the lymphocytes, and the various responses that are designed to protect the body against invasion by any foreign microorganism.

Barriers: skin and mucous membranes

The skin and mucous membranes act as effective barriers against harmful invaders. The surface of the skin is slightly acidic which makes it difficult for many microorganisms to survive. In addition, the **enzyme** lysozyme, which is present in sweat, tears, and saliva, kills many bacteria. Mucous membranes line many of the body's entrances, such as those that open into the res-

piratory, digestive, and uro-genital tract. Bacteria become trapped in the thick mucous layers and are thus prevented from entering the body. In the upper respiratory tract, the hairs that line the nose also trap bacteria. Any bacteria that are inhaled deeper into the respiratory tract are swept back out again by the cilia—tiny hairs—that line the trachea and bronchii. One reason why smokers are more susceptible to respiratory infections is that hot **cigarette smoke** disables the cilia, slowing the movement of mucus and bacteria out of the respiratory tract. Within days of quitting smoking, the cilia regenerate and new quitters then cough and bring up large amounts of mucus, which eventually subsides.

Non-specific immune cells

Non-specific lymphocytes carry out "search and destroy" missions within the body. If these cells encounter a foreign microorganism, they will either engulf the foreign invader or destroy the invader with enzymes. The following is a list of non-specific lymphocytes:

Macrophages are large lymphocytes which engulf foreign cells. Because macrophages ingest other cells, they are also called phagocytes (*phagein*, to eat + *kytos*, cell).

Neutrophils are cells that migrate to areas where bacteria have invaded, such as entrances created by cuts in the skin. Neutrophils phagocytize microorganisms and release microorganism-killing enzymes. Neutrophils die quickly; pus is an accumulation of dead neutrophils.

Natural killer cells kill body cells infected with viruses, by punching a hole in the cell membrane, causing the cell to lyse, or break apart.

The inflammatory response

The inflammatory response is an immune response confined to a small area. When a finger is cut, the area becomes red, swollen, and warm. These signs are evidence of the inflammatory response. Injured tissues send out signals to immune system cells, which quickly migrate to the injured area. These immune cells perform different funcions: some engulf bacteria, others release bacteria-killing chemicals. Other immune cells release a substance called **histamine**, which causes blood vessels to become wider (dilate), thus increasing blood flow to the area. All of these activities promote healing in the injured tissue.

An inappropriate inflammatory response is the cause of allergic reactions. When a person is "allergic" to pollen, the body's immune system is reacting to pollen (a harmless substance) as if it were a bacterium and an immune response is prompted. When pollen is inhaled it stimulates an inflammatory response in the nasal cavity and sinuses. Histamine is released which dilates blood

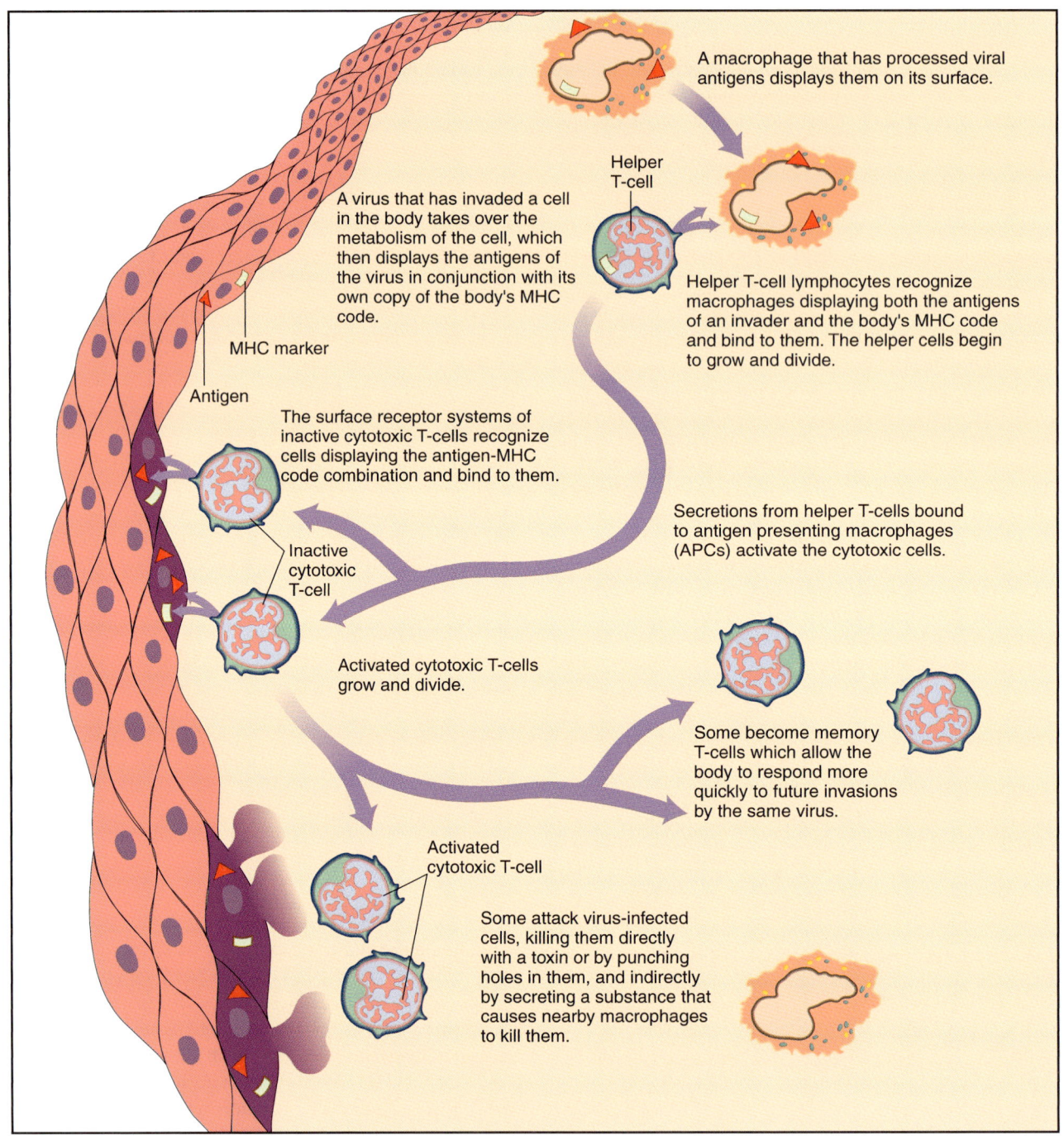

A macrophage that has processed viral antigens displays them on its surface.

A virus that has invaded a cell in the body takes over the metabolism of the cell, which then displays the antigens of the virus in conjunction with its own copy of the body's MHC code.

Helper T-cell

Helper T-cell lymphocytes recognize macrophages displaying both the antigens of an invader and the body's MHC code and bind to them. The helper cells begin to grow and divide.

MHC marker

Antigen

The surface receptor systems of inactive cytotoxic T-cells recognize cells displaying the antigen-MHC code combination and bind to them.

Secretions from helper T-cells bound to antigen presenting macrophages (APCs) activate the cytotoxic cells.

Inactive cytotoxic T-cell

Activated cytotoxic T-cells grow and divide.

Some become memory T-cells which allow the body to respond more quickly to future invasions by the same virus.

Activated cytotoxic T-cell

Some attack virus-infected cells, killing them directly with a toxin or by punching holes in them, and indirectly by secreting a substance that causes nearby macrophages to kill them.

The cell-mediated immune response. *Illustration by Hans & Cassidy. Courtesy of Gale Group.*

vessels, and also causes large amounts of mucous to be produced, leading to a "runny nose." In addition, histamine stimulates the release of tears and is responsible for the watery eyes and nasal congestion typical of allergies.

To combat these reactions, many people take drugs that deactivate histamine. These drugs, called **antihista-** **mines**, are available over the counter and by prescription. Some allergic reactions, involve the production of large amounts of histamine which impairs breathing and necessitates prompt emergency care. People prone to these extreme allergic reactions must carry a special syringe with epinephrine (adrenalin), a drug that quickly counteracts this severe respiratory reaction.

Complement

The complement system is a group of more than 20 proteins that "complement" other immune responses. When activated, the complement proteins perform a variety of functions: they coat the outside of microorganisms, making them easier for immune cells to engulf; they stimulate the release of histamine in the inflammatory response; and they destroy virus-infected cells by puncturing the **plasma** membrane of the infected cell, causing the cell to burst open.

Specific immune defenses

The specific immune response is activated when microorganisms evade the non-specific defenses. Two types of specific defenses destroy microorganisms in the human body: the cell-mediated response and the antibody response. The cell-mediated response attacks cells which have been infected by viruses. The antibody response attacks both "free" viruses that haven't yet penetrated cells, and bacteria, most of which do not infect cells. However, some bacteria, such as the Mycobacteria that cause **tuberculosis**, do infect cells.

Specific immune cells

Two kinds of lymphocytes operate in the specific immune response: T lymphocytes and B lymphocytes, (T lymphocytes are made in the thymus gland, while B lymphocytes are made in bone marrow). The T and B lymphocytes migrate to other parts of the lymphatic system, such as the lymph nodes, Peyer's patches, and tonsils. Non-specific lymphocytes attack *any* foreign cell, while B and T lymphocytes are individually configured to attack a specific antigen. In other words, the blood and lymph of humans have T-cell lymphocytes that specifically target the **chickenpox** virus, T-cell lymphocytes that target the **diphtheria** virus, and so on. When T-cell lymphocytes specific for the **chickenpox virus** encounter a body cell infected with this virus, the T cell multiplies rapidly and destroying the invading virus.

Memory cells

After the invader has been neutralized, some **T cells** remain behind. These cells, called memory cells, impart immunity to future attacks by the virus. Once a person has had chickenpox, memory cells quickly stave off subsequent infections. This secondary immune response, involving memory cells, is much faster than the primary immune response.

Some diseases, such as **smallpox**, are so dangerous that it is better to artificially induce immunity rather than to wait for a person to create memory cells after an **infection**. Vaccination injects whole or parts of killed viruses or bacteria into the bloodstream, prompting memory cells to be made without a person developing the disease.

Helper T cells

Helper T cells are a subset of T-cell lymphocytes which play a significant role in both the cell-mediated and antibody immune responses. Helper T cells are present in large numbers in the blood and lymphatic system, lymph nodes, and Peyer's patches. When one of the body's macrophage cells ingests a foreign invader, it displays the antigen on its membrane surface. These antigen-displaying-macrophages, or APCs, are the immune system's distress signal. When a helper T cell encounters an APC, it immediately binds to the antigen on the macrophage. This binding unleashes several powerful chemicals called cytokines. Some cytokines, such as interleukin I, stimulate the growth and division of T cells. Other cytokines play a role in the fever response, another non-specific immune defense. Still another cytokine, called interleukin II, stimulates the division of cytotoxic T cells, key components of the cell-mediated response. The binding also "turns on" the antibody response. In effect, the helper T cells stand at the center of both the cell-mediated and antibody responses.

Any disease that destroys helper T cells destroys the immune system. HIV infects and kills helper T cells, so disabling the immune system and leaving the body helpless to stave off infection.

B cells and the antibody response

B-cell lymphocytes, or B cells, are the primary players in the antibody response. When an antigen-specific B cell is activated by the binding of an APC to a helper T cell, it begins to divide. These dividing B cells are called plasma cells. The plasma cells, in turn, secrete antibodies, proteins that attach to the antigen on bacteria or free viruses, marking them for destruction by macrophages or complement. After the infection has subsided, a few memory B cells persist that confer immunity.

A closer look at antibodies

Antibodies are made when a B cell specific for the invading antigen is stimulated to divide by the binding of an APC to a helper T cell. The dividing B cells, called plasma cells, secrete proteins called antibodies. Antibodies are composed of a special type of protein called immunoglobin (Ig). An antibody **molecule** is Y-shaped and consists of two light chains joined to two heavy chains. These chains vary significantly between antibodies. The **variable** regions make antibodies antigen-specific. Con-

KEY TERMS

Antibody response—The specific immune response that utilizes B cells to neutralize bacteria and "free viruses."

Antigen-presenting cell (APC)—A macrophage that has ingested a foreign cell and displays the antigen on its surface.

B lymphocyte—Immune system white blood cell that produces antibodies.

Cell-mediated response—The specific immune response that utilizes T cells to neutralize cells that have been infected with viruses and certain bacteria.

Complement system—A series of 20 proteins that "complement" the immune system; complement proteins destroy virus-infected cells and enhance the phagocytic activity of macrophages.

Cytotoxic T cell—A T lymphocyte that destroys virus-infected cells in the cell-mediated immune response.

Helper T lymphocyte—The "lynch pin" of specific immune responses; helper T cells bind to APCs (antigen-presenting cells), activating both the antibody and cell-mediated immune responses.

Inflammatory response—A non-specific immune response that causes the release of histamine into an area of injury; also prompts blood flow and immune cell activity at injured sites.

Lymphocyte—White blood cell.

Macrophage—An immune cell that engulfs foreign cells.

Major Histocompatibility Complex (MHC)—The proteins that protrude from the surface of a cell that identify the cell as "self."

Memory cell—The T and B cells that remain be-

hind after a primary immune response; these cells swiftly respond to subsequent invasions by the same microorganism.

Natural Killer cell—An immune cell that kills infected tissue cells by punching a hole in the cell membrane.

Neutrophil—An immune cell that releases a bacteria-killing chemical; neutrophils are prominent in the inflammatory response.

Non-specific defenses—Defenses such as barriers and the inflammatory response that generally target *all* foreign cells.

Phagocyte—A cell that engulfs another cell.

Plasma cell—A B cell that secretes antibodies.

Primary immune response—The immune response that is elicited when the body first encounters a specific antigen.

Secondary immune response—The immune response that is elicited when the body encounters a specific antigen a second time; due to the presence of memory cells, this response is usually much swifter than the primary immune response.

Specific defenses—Immune responses that target specific antigens; includes the antibody and cell-mediated responses.

Suppressor T cell—T lymphocytes that deactivate T and B cells.

T cells—Immune-system white blood cells that enable antibody production, suppress antibody production, or kill other cells.

Vaccination—Inducing the body to make memory cells by artificially introducing antigens into the body.

---

stant regions, on the other hand, are relatively the same between antibodies. All antibody molecules, whether made in response to a chickenpox virus or to a *Salmonella* bacterium, have some regions that are similar.

How antibodies work to destroy invaders

An antibody does not itself destroy microorganisms. Instead, the antibody that has been made in response to a specific microorganism binds to the specific antigen on its surface. With the antibody molecule bound to its antigen, the microorganism is targeted by destructive immune cells like macrophages and NK cells. Antibody-

tagged microorganisms can also be destroyed by the complement system.

T cells and the cell-mediated response

T-cell lymphocytes are the primary players in the cell-mediated response. When an antigen-specific helper T cell is activated by the binding of an APC, the cell multiplies. The cells produced from this division are called cytotoxic T cells. Cytotoxic T cells target and kill cells that have been infected with a specific microorganism. After the infection has subsided, a few memory T cells persist, so conferring immunity.

How is the immune response "turned off?"

Chemical signals activate the immune response and other chemical signals must turn it off. When all the invading microorganisms have been neutralized, special T cells (called suppressor T cells) release cytokines that deactivate the cytotoxic T cells and the plasma cells.

See also Allergy; Antibody and antigen; Cyclosporine; Immunology; Inflammation; Vaccine.

Kathleen Scogna

Resources

Books

Richman, D. D., and R. J. Whitley. *Clinical Virology.* 2nd ed. Washington: American Society for Microbiology, 2002.

Schindler, Lydia Woods. *The Immune System: How It Works.* Bethesda, MD: U.S. National Institutes of Health, 1993.

Periodicals

Engelhard, Victor H. "How Cells Process Antigens." *Scientific American* 271 (August 1994): 54.

Kedzierski, Marie. "Vaccines and Immunization (sic)." *New Scientist* 133 (8 February 1992): S1.

Kisielow, Pavelrod. "Self-Nonself Discrimination by T Cells." *Science* 248 (June 15, 1990): 1369.

"Life, Death, and the Immune System." Special Issue, *Scientific American* 269 (September 1993).

Miller, Jacques. "The Thymus: Maestro of the Immune System." *BioEssays* 16 (July 1994): 509.

Radesky, Peter. "Of Parasites and Pollens." *Discover* 14 (September 1993): 54.

Travis, John. "Tracing the Immune System's Evolutionary History." *Science* 261 (July 9, 1993): 164.

Immunology

Immunology is the study of how the body responds to foreign substances and fights off **infection** and other **disease**. Immunologists study the molecules, cells, and organs of the human body that participate in this response.

History of immunology

No one knows when humans first noticed that they are better at fighting a disease the second time they get it; Chinese documents from 5,000 B.C. mention the fact. In 430 B.C., the Greek historian Thucydides (?-411 B.C.) mentioned the great plague that swept through Athens, and how those who survived it (including Thucydides himself) could tend to the sick without worrying about catching it again.

But the beginnings of our understanding of immunity date to 1798, when the English physician Edward Jenner (1749-1823) published a report that people could be protected from deadly **smallpox** by sticking them with a needle dipped in the pus from a cowpox boil. The great French biologist and chemist Louis Pasteur (1822-1895) theorized that such immunization protects people against disease by exposing them to a version of a microbe that is harmless but is enough like the disease-causing **organism**, or pathogen, that the **immune system** learns to fight it. Modern vaccines against diseases such as measles, polio, and chicken pox are based on this principle.

In the late nineteenth century, a scientific debate was waged between the German physician Paul Ehrlich (1854-1915) and the Russian zoologist Elie Metchnikoff (1845-1916). Ehrlich and his followers believed that **proteins** in the **blood**, called antibodies, eliminated **pathogens** by sticking to them; this phenomenon became known as humoral immunity. Metchnikoff and his students, on the other hand, noted that certain white blood cells could engulf and digest foreign materials: this cellular immunity, they claimed, was the real way the body fought infection.

Modern immunologists have shown that both the humoral and cellular responses play a role in fighting disease. They have also identified many of the actors and processes that form the immune response.

Friend or foe?

The immune response recognizes and responds to pathogens via a network of cells that communicate with each other about what they have "seen" and whether it "belongs." These cells patrol throughout the body for infection, carried by both the blood stream and the lymph ducts, a series of vessels carrying a clear fluid rich in immune cells.

The antigen presenting cells are the first line of the body's defense, the scouts of the immune army. They engulf foreign material or **microorganisms** and digest them, displaying bits and pieces of the invaders—called antigens—for other immune cells to identify. These other immune cells, called T lymphocytes, can then begin the immune response that attacks the pathogen.

The body's other cells can also present antigens, although in a slightly different way. Cells always display antigens from their everyday proteins on their surface. When a **cell** is infected with a **virus**, or when it becomes cancerous, it will often make unusual proteins whose antigens can then be identified by any of a variety of cytotoxic T lymphocytes. These "killer cells" then destroy the infected or cancerous cell to protect the rest of the body. Other T lymphocytes generate chemical or other signals that encourage multiplication of other infection-fighting cells. Various types of T lymphocytes are a central part of the cellular immune response; they are also involved in the humoral response, encouraging B lymphocytes to turn into antibody-producing **plasma** cells.

Selecting disease fighters

The body cannot know in advance what a pathogen will look like and how to fight it, so it creates millions and millions of different lymphocytes that recognize **random** antigens. When, by chance, a B or T lymphocyte recognizes an antigen being displayed by an antigen presenting cell, the lymphocyte divides and produces many offspring that can also identify and attack this antigen. The way the immune system expands cells that by chance can attack an invading microbe is called clonal **selection**.

Some researchers believe that while some B and T lymphocytes recognize a pathogen and begin to mature and fight an infection, others stick around in the bloodstream for months or even years in a primed condition. Such memory cells may be the basis for the immunity noted by the ancient Chinese and by Thucydides. Other immunologists believe instead that trace amounts of a pathogen persist in the body, and their continued presence keeps the immune response strong over time.

Advances in immunology—monoclonal antibody technology

Substances foreign to the body, such as disease-causing **bacteria**, viruses, and other infectious agents (known as antigens), are recognized by the body's immune system as invaders. The body's natural defenses against these infectious agents are antibodies—proteins that seek out the antigens and help destroy them. Antibodies have two very useful characteristics. First, they are extremely specific; that is, each antibody binds to and attacks one particular antigen. Second, some antibodies, once activated by the occurrence of a disease, continue to confer resistance against that disease; classic examples are the antibodies to the **childhood diseases chickenpox** and measles.

The second characteristic of antibodies makes it possible to develop vaccines. A **vaccine** is a preparation of killed or weakened bacteria or viruses that, when introduced into the body, stimulates the production of antibodies against the antigens it contains.

It is the first trait of antibodies, their specificity, that makes monoclonal antibody technology so valuable. Not only can antibodies be used therapeutically, to protect against disease; they can also help to diagnose a wide variety of illnesses, and can detect the presence of drugs, viral and bacterial products, and other unusual or abnormal substances in the blood.

Given such a diversity of uses for these disease-fighting substances, their production in pure quantities has long been the focus of scientific investigation. The conventional method was to inject a laboratory **animal** with an antigen and then, after antibodies had been formed, collect those antibodies from the blood serum (antibody-containing blood serum is called antiserum). There are two problems with this method: It yields antiserum that contains undesired substances, and it provides a very small amount of usable antibody.

Monoclonal antibody technology allows the production of large amounts of pure antibodies in the following way. Cells that produce antibodies naturally are obtained along with a class of cells that can grow continually in cell culture. The **hybrid** resulting from combining cells with the characteristic of "immortality" and those with the ability to produce the desired substance, creates, in effect, a factory to produce antibodies that works around the clock.

A myeloma is a **tumor** of the bone marrow that can be adapted to grow permanently in cell culture. Fusing myeloma cells with antibody-producing mammalian spleen cells, results in hybrid cells, or hybridomas, producing large amounts of monoclonal antibodies. This product of cell fusion combined the desired qualities of the two different types of cells: the ability to grow continually, and the ability to produce large amounts of pure antibody. Because selected hybrid cells produce only one specific antibody, they are more pure than the polyclonal antibodies produced by conventional techniques. They are potentially more effective than conventional drugs in fighting disease, since drugs attack not only the foreign substance but the body's own cells as well, sometimes producing undesirable side effects such as nausea and allergic reactions. Monoclonal antibodies attack the target **molecule** and only the target molecule, with no or greatly diminished side effects.

Goals for the future

While researchers have made great gains in understanding immunity, many big questions remain. Future research will need to identify how the immune response is coordinated. Other researchers are studying the immune systems of non-mammals, trying to learn how our immune response evolved. **Insects**, for instance, lack antibodies, and are protected only by cellular immunity and chemical defenses not known to be present in higher organisms.

Immunologists do not yet know the details behind **allergy**, where antigens like those from pollen, poison ivy, or certain kinds of food make the body start an uncomfortable, unnecessary, and occasionally life-threatening immune response. Likewise, no one knows exactly why the immune system can suddenly attack the body's tissues—as in autoimmune diseases like rheumatoid **arthritis**, juvenile diabetes, systemic lupus erythematosus, or multiple sclerosis.

The hunt continues for new vaccines, especially against parasitic organisms like the **malaria** microbe that trick the immune system by changing their antigens. Some researchers are seeking ways to start an immune response that prevents or kills cancers. A big goal of immunologists is the search for a vaccine for HIV, the virus that causes **AIDS**. HIV knocks out the immune system—causing immunodeficiency—by infecting crucial T lymphocytes. Some immunologists have suggested that the chiefly humoral response raised by conventional vaccines may be unable to stop HIV from getting to lymphocytes, and that a new kind of vaccine that encourages a cellular response may be more effective.

Researchers have shown that transplant rejection is just another kind of immune response, with the immune system attacking antigens in the transplanted organ that are different from its own. Drugs that suppress the immune system are now used to prevent rejection, but they also make the patient vulnerable to infection. Immunologists are using their increased understanding of the immune system to develop more subtle ways of fooling the immune system into accepting transplants.

See also Antibody and antigen.

Resources

Books

Joneja, Janice M. Vickerstaff, and Leonard Bielory. *Understanding Allergy, Sensitivity, and Immunity: a Comprehensive Guide.* New Brunswick: Rutgers University Press, 1990.

Paul, William E., ed. *Immunology Recognition and Response.* New York: W. H. Freeman and Company, 1991.

Porter, Roy, and Marilyn Ogilvie, eds. *The Biographical Dictionary of Scientists.* Vol. 2. Oxford: Oxford University Press, 2000.

Richman, D.D., and R.J. Whitley. *Clinical Virology.* 2nd ed. Washington: American Society for Microbiology, 2002.

Rose, N.R. *Manual of Clinical Laboratory Immunology.* 4th ed. Washington: American Society for Microbiology, 2002.

Periodicals

Cimons, M. "New Prospects on the HIV Vaccine Scene." *ASM News* no. 68 (January 2002): 19-22.

Erickson, Deborah. "Industrial immunology: Antibodies May Catalyze Commercial Chemistry." *Scientific American* (September 1991): 174-175.

KEY TERMS

. . . .

Autoimmunity—An aberrant immune response that attacks the body's own tissues.

Cellular immunity—The arm of the immune system that uses cells and their activities to kill pathogens, infected cells, and cancer cells.

Clonal selection—The process whereby one or a few immune cells that by chance recognize an antigen multiply when the antigen is present in the body.

Humoral immunity—The arm of the immune system that uses antibodies and other chemicals to clear pathogens from the body and to kill infected or cancerous cells.

Immunodeficiency—A condition where the immune response is weak or incomplete, allowing pathogens to cause disease more easily. AIDS is a kind of immunodeficiency.

"Life, Death and the Immune System." Special issue, *Scientific American* (September 1993): 52-144.

Kenneth B. Chiacchia

Impact crater

The impact crater is typically the most common type of **landform** seen on the surface of most of the rocky and icy planets and satellites in our **solar system**. Impact craters form when a minor planetary body (meteoritic fragment, asteroid, or comet) strikes the surface of a larger body or major **planet**. A physical scar is excavated on the surface and much **energy** is dispersed in the process.

Most impact craters are generally circular, although elliptical impact craters are known from very low-angle or obliquely impacting projectiles. In addition, some impact craters have been tectonically deformed and thus are no longer circular. Impact craters may be exposed, buried, or partially buried. Geologists distinguish an impact crater, which is rather easily seen, from an impact structure, which is an impact crater that may be in a state of poor preservation. A meteorite crater is distinguished from other impact craters because there are fragments of the impacting body preserved near the crater. Typically, a meteorite crater is rather small feature under 1 km (0.62 mi) in diameter.

An aerial view of Meteor Crater (near Winslow, Arizona). ©
Francois Gohier/Photo Researchers. Reproduced by permission.

Volcanic activity can also produce circular depression (which are also sometimes called "craters," but these are not impact craters). Impact craters bear the evidence of hypervelocity impact (most cosmic objects are moving in a solar **orbit** at several dozen km/sec). These features include meteoritic fragments (as small craters) and shocked and shock-melted materials within and about the impact crater.

Impact craters are obliterated or covered over by younger materials where rates of volcanic activity are very high (e.g., **Venus** and Io) and where **weathering**, **erosion**, and sedimentation are highly active (e.g., **Earth** and parts of **Mars**). At present, there are about 1000 suspected impact craters on Venus and perhaps one or two on Io. On Earth, 150 to 200 impact craters and impact structures have been scrutinized sufficiently to prove their origin. There are several hundred other possible impact features that also have been identified. Given Earth's rather rapid weathering and tectonic cycling of crust, this is a relatively large preserved crater record. On Mars, there are several thousand impact craters in various stages of degradation. Even though preserved craters are rare on Earth, there is no reason to suspect that Earth has been bombarded any less intensively than the **Moon** (which has millions of impact craters), and thus the vast majority of Earth's impact features must have been erased.

Impact craters of Earth are subdivided into three distinctive groups based upon their shape, which are in turn related to crater size. The simple impact crater is a bowl-shaped feature (usually less than 1.2 mi [2 km] in diameter) with relatively high depth to diameter **ratio**. The complex impact crater has a low depth to diameter ratio and possesses a central **uplift** and a down-faulted and terraced rim structure. Complex impact craters on Earth range from the upper limit of simple impact craters to approximately 62.1 mi (100 k) in diameter. Multi-ring craters (also called multi-ring basins) are impact craters with depth to diameter ratios like complex impact craters, but they possess at least two outer, concentric rings (marked by normal faults with downward **motion** toward crater center). Earth has five known multi-ring impact basins, but many more are known on the Moon and other planets and satellites in the solar system, where they range from several hundred kilometers up to 4000 km (2485.5 mi) in diameter. The gravity of a planet or **satellite** and the strength of the surface material determine the transition diameter from simple to complex and complex to multi-ring impact crater morphology.

Impact craters go through three separate stages during formation. The contact and compression stage comes first. Contact occurs when the projectile first touches the planet or satellite's surface. Jetting of molten material from the planet's upper crust can occur at this stage and initial penetration of the crust begins (this is the origin of most tektites or impact **glass** objects). During compression, the projectile is compressed as it enters the target crustal material. Depending upon relative strength of the target and projectile, the projectile usually penetrates only a few times its diameter into the crust. Nearly all the vast kinetic energy of the projectile is imbued into the surrounding crust as shock-wave energy. This huge shock wave propagates outward radially into the crust from the point of projectile entry. At the end of compression, which lasts a tiny fraction of a second to two seconds at most (depends upon projectile size), the projectile is vaporized by a shock wave that bounces from the front of the projectile to the back and then forward. At this point, the projectile itself is no longer a factor in what happens. The subsequent excavation stage is driven by the shock wave propagating through the surrounding target crust. The expanding shock wave moves material along curved paths, thus ejecting debris from the continually opening crater cavity. This is the origin of the transient crater cavity. It may take several seconds to a few minutes to open this transient crater cavity. Material cast out of the opening crater during this phase forms an ejecta curtain that extends high above the impact area. This ejected material will fall back thus forming an ejecta blanket in and around the impact crater (usually extending out about 3 crater radii). During the final modification stage, gravity takes over and causes crater-rim collapse in simple impact craters. In complex and multi-ring impact craters, there is central peak or peak-ring uplift and coincident gravitational collapse in the rim area. Lingering effects of the modification stage may go on for many years after impact.

Impact crater densities are used in **planetary geology** to gauge the age of surfaces that have been exposed for long periods of geological time and have not been covered by volcanic flows or sediments. Impact crater sizes are also used to gauge age because the average size of impacting bodies has declined, on average, over time since the end of planetary accretion (early in our solar system's history). Sharpness or "freshness" of craters on some planetary surfaces is also a descriptive gauge of age of the crater itself. Crater studies on old planetary surfaces, like those of airless worlds like Mercury, the Moon, some icy satellites of the outer planets, and some asteroids allow age comparisons to be made between the body's surfaces. Also, crater studies allow estimates to be made of the change in **density** or "flux" of impacting bodies over time in the solar system.

On Earth (and perhaps early Mars), it is thought that impact events related to craters greater than 62.1 mi [100 km] in diameter likely had globally devastating effects. These effects, which may have led to global **ecosystem** instability or collapse, included: gas and dust discharge into the upper atmosphere (blocking sunlight and causing greenhouse effects); heating of the atmosphere due to re-entry of ballistic ejecta (causing extensive wildfires); seismic sea waves (causing tsunamis); and acid-rain production (causing damage to soils and oceans). There is much research currently underway to about the effect of cosmic impact events upon life on Earth during the geological past.

Resources

Books

French, B.M. *Traces of Catastrophe.* Houston: Lunar and Planetary Institute, 1999.

Melosh, H.J. *Impact Cratering. A Geologic Process.* New York: Oxford University Press, 1989.

Montanari, A., and C. Koeberl. *Impact Stratigrapy, the Italian Record.* Berlin: Springer-Verlag, 2000.

David T. King Jr.

Imprinting

Imprinting is a term used to describe two very distinct processes. Genomic imprinting is an epigenetic chromosomal modification that describes the preferential expression of a specific parental form of a **gene** (allele). Imprinting is also a term used in the behavioral science to describe a **learning** process during which a younger **animal** identifies with, and adopts behaviors exhibited by, other animals, usually of the same **species**.

Genomic imprinting

Genomic imprinting is a normal but complex genetic phenomenon, that is difficult to define. As of March, 2003, no adequate explanation has been found for why genomic imprinting exists.

With genomic imprinting only one type of gene (allele) is expressed while the other allele remains genetically silent. Which allele is expressed and which remains silent depends on from which parent the genes are inherited.

A small subset of approximately 50 genes exhibit characteristics of genomic imprinting.

Following **fertilization** of a mammalian embryo most of the genes contributed by each parent begin to function equally. When a gene is expressed, the copy inherited from the mother (maternal allele), and the copy inherited from the father (paternal allele), are transcribed equally (bi-alleleic expression) and the RNA is translated into the protein product.

In contrast, with imprinting one allele is transcribed while the other is silent (i.e., imprinted). For example, in humans the insulin-like growth factor 2 gene (IGF2), which is an important fetal growth factor, is only expressed from the paternally inherited allele while the maternal allele is imprinted and never normally expressed. Similarly, the H19 gene, which is located adjacent to IGF2, is normally only expressed from the maternally inherited allele, while the paternal allele is silent.

The genetic mechanism of genomic imprinting remains uncertain but research indicates that some form of reversible genetic modification (epigenetic modification) such as DNA methylation is involved.

Impact of genomic imprinting

In most cases genomic imprinting is a normal process and has no affect on the normal **individual**. However, imprinted genes are involved in the development of some **genetic disorders** and in **cancer**.

Imprinted genes are involved in the development of some cancers. The imprinted fetal growth factor gene, IGF2, is commonly expressed in cancers such as Wilms **tumor** of the kidney, and cancers of the breast, lung, liver, and colon. In these cancers the maternal IGF2 imprint has been lost and both gene **alleles** are expressed (bi-allelic expression), this is termed "relaxation of imprinting."

There are many theories for why genomic imprinting exists. One of the most favored (in accord with the most current data), proposed by David Haig (the Haig Hypothesis), suggests that imprinting is a form of genetic reproductive conflict between the sexes each vying for a different reproductive outcome. Males desire large offspring males, so they over-express growth factors such as the paternally expressed fetal growth factor IGF2. However, females needing to limit fetal growth to ensure their successful **birth** have repressed growth factor expression by imprinting the gene.

Behavioral imprinting

With behavioral imprinting—a form of which is termed parental imprinting—a newly hatched or newborn animal is able to recognize its own parents from among other individuals of the same species. This process helps to ensure that the young will not become separated from their parents, even among large flocks or herds of similar animals.

Imprinting occurs during a sensitive period shortly after hatching, corresponding to a time when the chicks are near the nest and unlikely to encounter adults other than their parents. Many behavioral scientists assert that once an animal has imprinted on an object, it is never forgotten and the animal cannot imprint on any other object. Thus even when the chicks begin to encounter other animals they remain with their parents.

Imprinting was first studied in depth by Austrian zoologist Konrad Lorenz (1903–1989), who observed the process in **ducks** and **geese**. Lorenz found that a chick will learn to follow the first conspicuous moving object it sees after hatching. Normally, this object would be the mother bird, but in various experiments, ducklings and goslings have imprinted on artificial models of **birds**, bright red balls, and even human beings. In 1973, Lorenz's work earned a share of the Nobel Prize for Physiology and Medicine.

The effects of the imprinting process carry over into the adult life of the animal as well. In many cases it has been shown that the object imprinted upon as a hatchling determines the mating and **courtship** behaviors of the adult. Many species will avoid social contact with animals dissimilar to the one to which they have imprinted. Under normal circumstances, this helps prevent breeding between different species. Under artificial conditions, an animal which has imprinted on an individual of a different species will often attempt to court a member of that species later in life.

Imprinting in animals is most thoroughly studied in birds, although it is believed to be especially important in the hoofed **mammals**, which tend to congregate in large herds in which a young animal could easily be separated from its mother. Imprinting also occurs in humans to at least some extent. An infant separated from its mother for a prolonged period during its first year may develop serious mental retardation. Irreparable damage and even death may result from a separation of several months.

See also Behavior; Genetics.

In vitro fertilization (IVF)

In vitro fertilization (IVF) is a procedure in which eggs (ova) from a woman's ovary are removed, fertilized with sperm in a laboratory procedure, and then the resulting fertilized egg (embryo) is returned to the woman's uterus. Human **fertilization** *in vivo* (in the living body) occurs in oviducts (fallopian tubes) of the female reproductive tract.

IVF is a procedure of assisted reproductive techniques (ART) in which eggs (ova) from a woman's ovary are removed. Ova are fertilized with sperm in a laboratory procedure If fertilization occurs, a fertilized ovum, after undergoing several **cell** divisions, is transferred to the mother for normal development in the uterus, or frozen for later implantation.

IVF is one of several assisted reproductive techniques (ART) used to help infertile couples to conceive a child. If after one year of having sexual intercourse without the use of **birth** control a woman is unable to get pregnant, **infertility** is suspected. IVF is used to treat couples with unexplained infertility of long duration who have failed with other infertility treatments. Some of the reasons for infertility are damaged or blocked fallopian tubes, hormonal imbalance, or endometriosis in the woman. In the man, low sperm count or poor quality sperm can cause infertility.

IVF is one of several possible methods to increase the chance for an infertile couple to become pregnant. Its use depends on the reason for infertility. IVF may be an option if there is a blockage in the fallopian tube or endometriosis in the woman or low sperm count or poor quality sperm in the man. There are other possible treatments for these conditions, such as **surgery** for blocked tubes or endometriosis, which may be tried before IVF.

IVF will not work for a woman who is not capable of ovulating or a man who is not able to produce at least a few healthy sperm.

Other similar types of assisted reproductive technologies are also used to achieve pregnancy. A procedure called intracytoplasmic sperm injection (ICSI) uses a manipulation technique that uses a **microscope** to inject a single sperm into each egg. The fertilized eggs can then be returned to the uterus, as in IVF. In **gamete** intrafallopian tube transfer (GIFT) the eggs and sperm are mixed in a narrow tube and then deposited in the fallopian tube, where fertilization normally takes place. Another variation on IVF is zygote intrafallopian tube transfer (ZIFT). As in IVF, the fertilization of the eggs occurs in a laboratory dish. And, similar to GIFT, the embryos are placed in the fallopian tube (rather than the uterus as with IVF).

Precautions

The screening procedures and treatments for infertility can become a long, expensive, and sometimes disappointing process. Each IVF attempt takes at least an entire **menstrual cycle** and can cost $5,000-$10,000, which may not be covered by health insurance. The **anxiety** of dealing with infertility can challenge both individuals and their relationship. The added **stress** and expense of multiple clinic visits, testing, treatments, and surgical procedures can become overwhelming. Couples may want to receive counseling and support through the process.

Description

In vitro fertilization is a procedure where the joining of egg and sperm takes place outside of the woman's body. A woman may be given fertility drugs before this procedure so that several eggs mature in the ovaries at the same time. Eggs (ova) are removed from a woman's ovaries using a long, thin needle. The physician gets access to the ovaries using one of two possible procedures. One procedure involves inserting the needle through the vagina (transvaginally). The physician guides the needle to the location of the ovaries with the help of an ultrasound machine. In the other procedure, called laparoscopy, a small thin tube with a viewing **lens** is inserted through an incision in the navel. This allows the physician to see inside the patient, and locate the ovaries, on a video monitor.

Once the eggs are removed, they are mixed with sperm in a laboratory dish or test tube. (This is where the term *test tube baby* comes from.) The eggs are monitored for several days. Once there is evidence that fertilization has occurred and the cells begin to divide, they are then returned to the woman's uterus.

In the procedure to remove eggs, enough may be gathered to be frozen and saved (either fertilized or unfertilized) for additional IVF attempts.

Preparation

Once a woman is determined to be a good candidate for in vitro fertilization, she will generally be given "fertility drugs" to stimulate ovulation and the development of multiple eggs. These drugs may include gonadotropin releasing hormone agonists (GnRHa), Pergonal, Clomid, or **human chorionic gonadotropin** (hcg). The maturation of the eggs is then monitored with ultrasound tests and frequent **blood** tests. If enough eggs mature, the physician will perform the procedure to remove them. The woman may be given a sedative prior to the procedure. A local

anesthetic agent may also be used to reduce discomfort during the procedure.

Aftercare

After the IVF procedure is performed the woman can resume normal activities. A pregnancy test can be done approximately 12-14 days later to determine if the procedure was successful.

Risks

The risks associated with in vitro fertilization include the possibility of multiple pregnancy (since several embryos may be implanted) and ectopic pregnancy (an embryo that implants in the fallopian tube or in the abdominal cavity outside the uterus). There is a slight risk of ovarian rupture, bleeding, infections, and complications of **anesthesia**. If the procedure is successful and pregnancy is achieved, the pregnancy would carry the same risks as any pregnancy achieved without assisted technology. However because many IVF patients are of advanced maternal age, and thererfore have an increased risk for conceiving a child with **Down syndrome** or other abnormalities, in IVF programmes it would be better test ovocytes before implantation in order to detect potential chromosomal aneuploidies, thus avoiding the transfer of embryos affected by aneuploid oocytes.

Normal results

Success rates vary widely between clinics and between physicians performing the procedure. A couple has about a 10% chance of becoming pregnant each time the procedure is performed. Therefore, the procedure may have to be repeated more than once to achieve pregnancy.

IVF has been used successfully since 1978, when the first child to be conceived by this method was born in England. Over the past 20 years, thousands of couples have used IVF or other assisted reproductive technologies to conceive.

Abnormal results

An ectopic or multiple pregnancy may abort spontaneously or may require termination if the health of the mother is at risk.

See also Embryo and embryonic development; Embryo transfer; Embryology; Clone and cloning.

Resources

Books

Nussbaum, R.L., Roderick R. McInnes, Huntington F. Willard. *Genetics in Medicine*. Philadelphia: Saunders, 2001.

KEY TERMS

. . . .

Fallopian tubes—In a woman's reproductive system, a pair of narrow tubes that carry the egg from the ovary to the uterus.

GIFT—Gamete intrafallopian tube transfer. This is a process where eggs are taken from a woman's ovaries, mixed with sperm, and then deposited into the woman's fallopian tube.

ICSI—Intracytoplasmic sperm injection. This process is used to inject a single sperm into each egg before the fertilized eggs are put back into the woman's body. The procedure may be used if the male has a low sperm count.

ZIFT—Zygote intrafallopian tube transfer. In this process of in vitro fertilization, the eggs are fertilized in a laboratory dish and then placed in the woman's fallopian tube.

Rimoin, D.L. *Emery and Rimoin's Principles and Practice of Medical Genetics*. London; New York: Churchill Livingstone, 2002.

Sadler, T.W., and Jan Langman. *Langman's Medical Embryology*. 8th ed. Lippincott Williams & Wilkins Publishers; 2000.

Periodicals

Alper, M., P. Brinsden, M. Wikland, and R. Fischer. "International Standard for IVF Centres." *Human Reproducion* 18, no. 2 (2003): 461.

Foote RH. "Fertility estimation: a review of past experience and future prospects." *Animal Reproduction Science* 75 no. 1-2 (2003): 119-39.

Squires, J., A. Carter,. P. Kaplan. "Developmental Monitoring of Children Conceived by Intracytoplasmic Sperm Injection and In Vitro Fertilization." *Fertility and Sterility* 79, no. 2 (2003): 453-4.

Antonio Farina
Brenda Wilmoth Lerner

In vitro and *in vivo*

The definition of *in vitro* and *in vivo* research depends on the experimental model used. *In vitro* research is generally referred to as the manipulation of organs, tissues, cells, and biomolecules in a controlled, artificial environment. The characterization and analysis of biomolecules and biological systems in the context of intact organisms is known as *in vivo* research.

The basic unit of living organisms is the **cell**, which in terms of scale and dimension is at the interface between the molecular and the microscopic level. The living cell is in turn divided into functional and structural domains such as the nucleus, the cytoplasm, and the secretory pathway, which are composed of a vast array of biomolecules. These molecules of life carry out the **chemical reactions** that enable a cell to interact with its environment, use and store **energy**, reproduce, and grow. The structure of each biomolecule and its subcellular localization determines in which chemical reactions it is able to participate and hence what role it plays in the cell's life process. Any manipulation that breaks down this unit of life, that is, the cell into its non-living components is, considered an *in vitro* approach. Thus, *in vitro,* which literally means "in glass," refers to the experimental manipulation conducted using cell-free extracts and purified or partially purified biomolecules in test tubes. Most of the biochemical and molecular biological approaches and techniques are considered genetic manipulation research. Molecular cloning of a **gene** with the aim of expressing its protein product includes some steps that are considered *in vitro* experiments such as the **PCR** amplification of the gene and the ligation of that gene to the expression vector. The expression of that gene in a host cell is considered an *in vivo* procedure. What characterizes an *in vitro* experiment is in principle the fact the conditions are artificial and are reconstructions of what might happen *in vivo*. Many *in vitro* assays are approximate reconstitutions of biological processes by mixing the necessary components and reagents under controlled conditions. Examples of biological processes that can be reconstituted in vitro are enzymatic reactions, folding and refolding of **proteins** and DNA, and the replication of DNA in the PCR reaction.

Microbiologists and **yeast** geneticist working with single cells or cell populations are conducting *in vivo* research while an immunologist who works with purified lymphocytes in **tissue** culture usually considers his experiments as an *in vitro* approach. The *in vivo* approach involves experiments performed in the context of the large system of the body of an experimental **animal**. In the case of *in vitro* **fertilization (IVF)**, physicians and reproductive biologists are manipulating living systems, and many of the biological processes involved take place inside the living egg and sperm. This procedure is considered an *in vitro* process in order to distinguish it from the natural fertilization of the egg in the intact body of the female.

In vivo experimental research became widespread with the use **microorganisms** and animal models in genetic manipulation experiments as well as the use of animal models to study drug toxicity in pharmacology. Geneticists have used prokaryotic, unicellular eukaryotes like yeast, and whole organisms like *Drosophila*, **frogs**, and **mice** to study **genetics**, **molecular biology** and **toxicology**. The function of genes has been studied by observing the effects of spontaneous mutations in whole organisms or by introducing targeted mutations in cultured cells. The introduction of gene cloning and *in vitro* **mutagenesis** has made it possible to produce specific mutations in whole animals thus considerably facilitating *in vivo* research. Mice with extra copies or altered copies of a gene in their **genome** can be generated by transgenesis, which is now a well established technique. In many cases, the function of a particular gene can be fully understood only if a mutant animal that does not express the gene can be obtained. This is now achieved by gene knock-out technology, which involves first isolating a gene of interest and then replacing it *in vivo* with a defective copy.

Both *in vitro* and *in vivo* approaches are usually combined to obtain detailed information about structure-function relationships in genes and their protein products, either in cultured cells and test tubes or in the whole **organism**.

See also Embryo transfer; Stem cells.

Resources

Books

Lodish, J., D. Baltimore, A. Berk, S. L. Zipursky, P. Matsudaira, J. Darnell. *Molecular Cell Biology.* New York: Scientific American Books, Inc., 1995.

Abdel Hakim Nasr

Incandescent light

Incandescent light is given off when an object is heated until it glows. To emit white **light**, an object must be heated to at least 1,341°F (727°C). White-hot **iron** in a forge is incandescent, as is red lava flowing down a **volcano**, as are the red burners on an electric stove. The most common example of incandescence is the white-hot filament in the light bulb of an incandescent lamp.

History of incandescent lamps

In 1802, Sir Humphry Davy showed that **electricity** running through thin strips of **metal** could **heat** them to temperatures high enough that they would give off light; this is the basic principle by which all incandescent lamps work. In 1820, De La Rue demonstrated a lamp made of a coiled platinum wire in a **glass** tube with brass endcaps. When the current was switched on, electricity ran through the endcaps and through the wire (the filament). The wire was heated by its resistance to the current until it glowed white-hot, producing light. Between this time and the 1870s, the delicate lamps were unreliable, short-lived, and expensive to operate. The lifetime was short because the filament would burn up in air. To combat the short lifetime, early developers used thick low-resistance filaments, but heating them to incandescence required large currents—and generating large currents was costly.

Thomas Edison is well-known as "the inventor of the light bulb," but he was, in fact, only one of several researchers that created early electrical incandescent lamps in the 1870s. These researchers include Joseph Swan, Frederick DeMoyleyns, and St. George Lane-Fox in England, as well as Moses Farmer, Hiram Maxim, and William Sawyer in the United States.

Edison's contribution was an understanding of the necessary electrical properties for lamps. He knew that a system for delivering electricity was needed to make lamps practical; that it should be designed so that the lamps are run in **parallel**, rather than in series; and that the lamp filament should have high, rather than low, resistance. Because voltage in a circuit equals the current times the resistance, one can reduce the amount of current by increasing the resistance of the load. Increasing the resistance also reduces the amount of **energy** required to heat the filament to incandescence.

Edison replaced low-resistance **carbon** or platinum filaments with a high-resistance carbon filament. This lamp had electrical contacts connected to a **cotton** thread that had been burned to char (carbonized) and placed in a glass container with all the air pumped out. The **vacuum**, produced by a pump developed only a decade earlier by Herman Sprengel, dramatically increased the lifetime of the filament. The first practical version of the electric light bulb was lit on October 19, 1879, which burned for 40 hours, and produced 1.4 lumens per watt of electricity.

An incandescent non-electric lamp still in use is the Welsbach burner, commonly seen in camping lanterns. This burner, invented in 1886 by Karl Auer, Baron von Welsbach, consists of a mantle made of knit cotton soaked in oxides (originally nitrates were used), that is burned to ash the first time it is lit. The ash holds its shape and becomes incandescent when placed over a flame—and is much brighter than the flame itself.

Design

Incandescent lamps come in a huge variety of shapes and sizes, but all share the same basic elements as

De La Rue's original incandescent lamp. Each is contained by a glass or quartz envelope. Current enters the lamp through a conductor in an airtight joint or joints. Wires carry current to the filament, which is held up and away from the bulb by support wires. Changes in the specifics of incandescent lamps have been made to increase efficiency, lifetime, and ease of manufacture.

Although the first common electric lamps were incandescent, many lamps used today are not: Fluorescent lamps, neon signs, and glow-discharge lamps, for example, are not incandescent. Fluorescent lamps are more energy-efficient than incandescent lamps, but may not offer a desired **color** output.

Basic structure

Today, filaments are made of coiled tungsten, a high-resistance material that can be drawn into a wire and has both a high melting point of 6,120°F (3,382°C) and a low **vapor pressure**, which keeps it from melting or evaporating too quickly. It also has the useful characteristic of having a higher resistance when hot than when cold. If tungsten is heated to melting, it emits 53 lumens per watt. (Lamp filaments are not heated as high to keep the lamp lifetime reasonable, but this gives the upper limit of light available from such a filament.) The filament shape and length are also important to the efficiency of the lamp. Most filaments are coiled, and some are double and triple coiled. This allows the filament to lose less heat to the surrounding gas as well as indirectly heating other portions of the filament.

Most lamps have one screw-type base, through which both wires travel to the filament. The base may be sealed by a flange seal (for lamps 0.8 in [20 mm] or larger) or a low-cost butt seal for lamps smaller than 0.8 (20 mm) in diameter with smaller wires that carry 1 amp or less. The bases are cemented to the bulbs. In applications that require precise positioning of the filament, two-post or bayonet-type bases are preferred.

The bulb may be made from either a regular **lead** or lime glass or a borosilicate glass that can withstand higher temperatures. Even higher temperatures require the use of quartz, high-silica, or aluminosilicate glasses. Most bulbs are chemically etched inside to diffuse light from the filament. Another method of diffusing the light uses an inner coating of powered white silica.

Lower wattage lights have all the atmosphere pumped out, leaving a vacuum. Lights rated at 40 W or more use an inert fill gas that reduces the **evaporation** of the tungsten filament. Most use argon, with a small percentage of **nitrogen** to prevent arcing between the lead-in wires. Krypton is also occasionally used because it increases the efficiency of the lamp, but it is also more ex-

pensive. **Hydrogen** is used for lamps in which quick flashing is necessary.

As the bulb ages, the tungsten evaporates, making the filament thinner and increasing its resistance. This reduces the wattage, the current, the lumens, and the luminous efficacy from the lamp. Some of the evaporated tungsten also condenses on the bulb, darkening it and resulting in more absorption at the bulb. (You can tell whether a bulb has a fill gas or is a vacuum bulb by observing the blackening of an old bulb: Vacuum bulbs are evenly coated, whereas gas-filled bulbs show blackening concentrated at the uppermost part of the bulb.) Tungsten-halogen lamps are filled with a halogen (bromine, **chlorine**, fluorine, or iodine) gas and degrade much less over their lifetimes. When tungsten evaporates from the filament, instead of being deposited on the bulb walls, it forms a gaseous compound with the halogen gas. When this compound is heated (near the filament), it breaks down, redepositing tungsten onto the filament. The compactness and lifelong performance of such lamps is better than regular lamps. The **temperature** is higher (above 5,121°F [2,827°C]) in these lamps than in regular lamps, thus providing a higher percentage of visible and ultraviolet output. Linear tungsten-halogen bulbs may be coated with filters that reflect infrared energy back at the filament, thus raising the efficiency dramatically without reducing the lifetime.

Color temperature

The efficiency of the light is determined by the amount of visible light it sheds for a given amount of energy consumed. **Engineering** the filament material increases efficiency. Losses come from heat lost by the filament to the gas around it, loss from the filament to the lead-in wires and supports, and loss to the base and bulb.

Most of the output of the lamp is in the infrared region of the **spectrum**, which is fine if you want a heat lamp, but not ideal for a visible light source. Only about 10% of the output of a typical incandescent lamp is visible, and much of this is in the red and yellow parts of the spectrum (which are closer to the infrared region than green, blue, or violet). One way of providing a color balance more like daylight is to use a glass bulb with a blue tinge that absorbs some of the red and yellow. This increases the color temperature, but reduces the total light output.

Tradeoffs in design

Temperature is one of several tradeoffs in the design of each lamp. A high filament temperature is necessary, but if it is too high then the filament will evaporate quickly, leading to a short lifetime. Too low a temperature and little of the **radiation** will be visible. For tungsten-halogen lamps, the temperature must be at least 500°F (260°C) to insure operation of the regenerative cycle. Also, although the filament must be hot, the bulb and base have temperature limits, as does the cement that binds them. Many bulbs have a heat button that acts as a heat shield between the filament and the base. The position of the bulb (base-down for a table lamp, but base-up for a hanging ceiling lamp) also changes the amount of heat to which the base is exposed, which alters the lifetime of the bulb.

If the voltage at which the bulb is operating changes, this changes the filament resistance, temperature, current, power consumption, light output, efficacy (and thus color temperature), and lifetime of the bulb. In general, if the voltage increases, all the other characteristics increase—except for lifetime, which decreases. (None of these relationships are linear.)

Applications

With so many different parameters to be balanced in each lamp, it is no wonder that thousands of different lamps are available for a myriad of purposes. Large lamps (including general purpose lamps), miniature lamps (such as Christmas **tree** lights), and photographic lamps (such as those for shooting movies) cover the three major classes of lamp.

General service lamps are made in ranges from 10 W to 1500 W. The higher-wattage lamps tend to be more efficient at producing light, so it is more energy-efficient to operate one 100-W bulb than two 50-W bulbs. On the other hand, long-life bulbs (which provide longer lifetimes by reducing the filament temperature) are less efficient than regular bulbs but may be worth using in situations where changing the bulb is a bother or may a hazard.

Spotlights and floodlights generally require accurately positioned, compact filaments. Reflectorized bulbs, such as those used for car headlights (these are tungsten-halogen bulbs) or overhead downlights (such as those used in track lighting) are made with reflectors built into the bulb: The bulb's shape along one side is designed so that a reflective coating on that inner surface shapes the light into a beam.

Lamps used for color **photography** have to provide a good color balance, keep the same balance throughout their lives, and interact well with the film's sensitivity. These lamps tend to be classified according to their color temperatures, which range from 5,301°F (2,927°C) for photography, 5,571°F (3,077°C) for professional movies, to 8,541°F (4,727°C) for "daylight blue" lamps, and even some "photographic blue" lamps that approximate sunshine and have a color temperature of 9,441°F (5,227°C).

Resources

Books

Rea, Mark. S., ed. *IES Lighting Handbook: Reference & Application, 8th ed.* New York: Illuminating Engineering Society of North America, 1993.

Yvonne Carts-Powell

Incineration

Incinerators are industrial facilities used for the controlled burning of waste materials. The largest incinerators are used to burn municipal solid wastes, often in concert with a technology that utilizes the **heat** produced during **combustion** to generate **electricity**. Smaller, more specialized incinerators are used to burn medical wastes, general chemical wastes such as organic solvents, and toxic wastes such as polychlorinated biphenyls and other **chlorinated hydrocarbons**.

Municipal solid wastes

Municipal solid waste comes from a wide range of sources in cities and suburban areas, including residences, businesses, educational and government institutions, industries, and construction sites. Municipal solid waste is typically composed of a wide range of materials, including food wastes, **paper** products, **plastics**, metals, **glass**, demolition debris, and household **hazardous wastes** (the latter assumes that hazardous wastes from industries, hospitals, laboratories, and other institutions are disposed as a separate waste stream).

Depending on the municipality, some of this solid waste may be recycled, reused, or composted. More typically, however, most of the wastes are disposed in some central facility, generally some sort of sanitary **landfill**. These are regulated, engineered disposal sites to which the wastes are hauled, dumped on land, compacted, and covered with **earth**. The **basin** of a modern sanitary landfill is generally lined with an impermeable material, such as heavy plastic or clay. This allows the collection of **water** that has percolated through the wastes, so it can be treated to reduce the concentrations of pollutants to acceptable levels, prior to discharge to the environment.

However, in many places large, sanitary landfills are no longer considered a preferable option for the disposal of general solid wastes. In some cases, this is because land is locally scarce for the development of a large landfill. More usually, however, local opposition to these facilities is the constraining factor, because people living in the vicinity of operating or proposed disposal sites object to these facilities. These people may be variously worried about odors, local **pollution**, truck traffic, poor aesthetics, effects on property values, or other problems potentially associated with large, solid-waste disposal sites.

Everyone, including these people, recognizes that municipalities need large facilities for the disposal of solid wastes. However, no one wants to have such a facility located in their particular neighborhood. This popularly held view about solid waste disposal sites, and about other large, industrial facilities, is known as the "not in my back yard" or NIMBY **syndrome**, and sometimes as the "locally unacceptable land use" or LULU syndrome.

Municipal incinerators

Incinerators are an alternative option to the disposal of general municipal garbage in solid-waste disposal sites. Municipal incinerators accept organic wastes and combust them under controlled conditions. The major benefit of using incinerators for this purpose is the large reductions that are achieved in the mass and **volume** of wastes.

In addition, municipal incinerators can be engineered as waste-to-energy facilities, which couple incineration with the generation of electricity. For example, a medium-sized waste-to-energy facility can typically take 550 tons (500 tonnes) per day of municipal solid wastes, and use the heat produced during combustion to generate about 16 megawatts of electricity. About 2-3 megawatts would be used to operate the facility, including its **energy** demanding air-pollution control systems, and the rest could be sold to recover some of the costs of waste disposal.

Among the major drawbacks of incinerators is the fact that these facilities have their own problems with NIMBY, mostly associated with the fears of people about exposures to air pollutants. As is discussed in the next section, incinerators emit a wide range of potentially toxic chemicals to the environment.

In addition, municipal incinerators produce large quantities of residual materials, which contain many toxic chemicals, especially metals. The wastes of incineration include bottom ash that remains after the organic **matter** in the waste stream has been combusted, as well as finer fly ash that is removed from the waste gases of the incineration process by **pollution control** devices. These toxic materials must be disposed in sanitary landfills, but the overall amounts are much smaller than that of the unburned garbage.

Incinerators are also opposed by many people because they detract from concerted efforts to reduce the amounts of municipal wastes by more intensive reducing, **recycling**, and reusing of waste materials. Incinerators require large quantities of organic garbage as fuel, especially if they are waste-to-energy facilities that are contracted to deliver certain quantities of electricity. As a result of the large fuel demands by these facilities, it can be difficult to implement other mechanisms of refuse management. Efforts to reduce the amounts of waste produced, to recycle, or to compost organic debris can suffer if minimal loads of fuels must be delivered to a large incinerator to keep it operating efficiently. These problems are best met by ensuring that incinerators are used within the context of an integrated scheme of solid **waste management**, which would include vigorous efforts to reduce wastes, reuse, recycle, and compost, with incineration as a balanced component of the larger system.

Emissions of pollutants

Incinerators are often located in or near urban areas. Consequently, there is intense concern about the emissions of chemicals from incinerators, and possible effects on humans and other organisms that result from exposure to potentially toxic substances. Consequently, modern incinerators are equipped with rigorous pollution control technologies to decrease the emissions of potentially toxic chemicals. The use of these systems greatly reduces, but does not eliminate the emissions of chemicals from incinerators. Also, as with any technology, there is always the risk of accidents of various sorts, which in the case of an incinerator could result in a relatively uncontrolled **emission** of pollutants for some period of time.

Uncertainty about the effects of potentially toxic chemicals emitted from incinerators is the major reason for the intense controversy that accompanies any plans to build these facilities. Even the best pollution-control systems cannot eliminate the emissions of potentially toxic chemicals, and this is the major reason for incinerator-related NIMBY. In fact some opponents of incinerators believe that the technology is unacceptable anywhere, a syndrome that environmental regulators have dubbed by the acronym **BANANA**, for "build absolutely nothing near anybody or anything." During the incineration process, small particulates are entrained into the flue gases, that is, the stream of waste gases that vents from the combustion chamber. These particulates typically contain large concentrations of metals and organic compounds, which can be toxic in large exposures.

To reduce the emissions of particulates, the flue gases of incinerators are treated in various ways. There are three commonly used systems of particulate removal. Electrostatic precipitators are devices that confer an electrical charge onto the particulates, and then collect them at a charged electrode. A baghouse is a physical filter, which collects particulates as flue gases are forced through a fine fabric. Cyclone filters cause flue gases to swirl energetically, so that particles can be separated by physical impaction at the periphery of the device. For incinerators located in or near urban areas, where concerns about emissions are especially acute, these devices may be used in series to achieve especially efficient removals, typically greater than 99% of the particulate mass. Virtually all particulates that are not removed by these systems are very tiny, and therefore behave aerodynamically as gases. Consequently, these emitted particulates are widely dispersed in the environment, and do not **deposit** locally in significant amounts.

The most important waste gases produced by incinerators are **carbon dioxide** (CO_2), **sulfur dioxide** (SO_2), and oxides of **nitrogen** (NO and NO_2, together known as NO_x). The major problem with carbon dioxide is through its contribution to the enhancement of Earth's **greenhouse effect**. However, because incinerators are a relatively small contributor to the total emissions of carbon dioxide from any municipal area, no attempts are made to reduce emissions from this particular source.

Sulfur dioxide and oxides of nitrogen are important in the development of urban **smog**, and are directly toxic to vegetation. These gases also contribute to the deposition of acidifying substances from the atmosphere, for example, as acidic **precipitation**. Within limits, sulfur dioxide and oxides of nitrogen can be removed from the waste gases of incinerators. There are various technologies for flue-gas desulfurization, but most rely on the reaction of sulfur dioxide with finely powdered limestone ($CaCO_3$) or lime [$Ca(OH)_2$] to

form a sludge containing gypsum ($CaSO_4$), which is collected and discarded in a solid-waste disposal site. This method is also effective at reducing emissions of **hydrogen chloride** (HCl), an acidic gas. Emissions of oxides of nitrogen can be controlled in various ways, for example, by reacting this gas with **ammonia**. Because urban areas typically have many other, much larger sources of atmospheric emissions of sulfur dioxide and oxides of nitrogen, emissions of these gases from incinerators are not always controlled using the technologies just described.

Various solid wastes can contain substantial concentrations of mercury, including thermometers, electrical switches, batteries, and certain types of electronic equipment. The mercury in these wastes is vaporized during incineration and enters the flue-gas stream. Pollution control for mercury vapor can include various technologies, including the injection of fine activated carbon into the flue gases. This material absorbs the mercury, and is then removed from the waste gases by the particulate control technology.

One of the most contentious pollution issues concerning incinerators involves the fact that various chlorinated hydrocarbons are synthesized during the incineration process, including the highly toxic chemicals known as dioxins and furans. These are formed during combustions involving chlorine-containing organic materials, at a **rate** influenced by the **temperature** of the combustion and the types of material being burned, including the presence of metallic catalysts. The synthesis of dioxins and furans is especially efficient at 572–932°F (300–500°C), when **copper**, **aluminum**, and **iron** are present as catalysts. These reactions are an important consideration when incineration is used to dispose of chlorinated plastics such as polyvinyl chloride (PVC, commonly used to manufacture piping and other rigid plastic products) and **polychlorinated biphenyls (PCBs)**.

Attention to combustion conditions during incineration can greatly reduce the rate of synthesis of dioxins and furans. For example, temperatures during incineration are much hotter, typically about 1,742–2,102°F (950–1,150°C), than those required for efficient synthesis of dioxins and furans. However, the synthesis of these chemicals cannot be eliminated, so emissions of trace quantities of these chemicals from incinerators are always a concern, and a major focus of NIMBY and BANANA protests to this technology.

Specialized incinerators

Relatively small, specialized incinerators are used for the disposal of other types of wastes, particularly

KEY TERMS

. . . .

Flue gas—The waste gases of a combustion. These may be treated to reduce the concentrations of toxic chemicals, prior to emission of the flue gases to the atmosphere.

Incinerator—An industrial facility used for the controlled burning of waste materials.

NIMBY—Acronym for "not in my back yard."

hazardous wastes. For example, hospitals and research facilities generally use incinerators to dispose of biological tissues, blood-contaminated materials, and other medical wastes such as disposable hypodermic needles and tubing. These are all considered to be hazardous organic wastes, because of the possibilities of spreading pathogenic **microorganisms**.

Incinerators may also be used to dispose of general chemical wastes from industries and research facilities, for example, various types of organic solvents such as **alcohol**. More specialized incinerators are used to dispose of more toxic chemical wastes, for example, chlorinated hydrocarbons such as PBCs, and various types of synthetic **pesticides**. For these latter purposes, the incineration technology includes especially rigorous attention to combustion conditions and pollution control. However, emissions of potentially toxic chemicals are never eliminated.

The role of incinerators

Industrialized and urbanized humans have a serious problem with solid wastes. These materials must be dealt with by society in a safe and effective manner, and incineration is one option that should be considered. However, incinerators have some drawbacks, including the fact that they invariably emit some quantities of potentially toxic chemicals. The role of incinerators in waste disposal would best be determined by an objective consideration of the best available scientific information.

Environmental damages have been caused in the past by the use of less efficient technologies to dispose of the wastes of society, including incinerators without modern combustion and pollution-control systems. In large part, these damages were associated with industries, politicians, and societies that were not sufficiently aware of the potential environmental damages, or did not care about them to the degree that is common today.

See also Air pollution.

Resources

Books

Dennison, R.A., and J. Rushton, eds. *Recycling and Incineration: Evaluating the Choices.* Washington, DC: Island Press, 1990.

Freedman, B. *Environmental Ecology.* 2nd ed. San Diego, Academic Press, 1994.

Hemond, H.F. and E.J. Fechner. *Chemical Fate and Transport in the Environment.* San Diego Academic Press, 1994.

McConnell, Robert, and Daniel Abel. *Environmental Issues: Measuring, Analyzing, Evaluating.* 2nd ed. Englewood Cliffs, NJ: Prentice Hall, 2002.

Bill Freedman

Indicator, acid-base

An acid-base indicator is not always a synthetic chemical. It is often a **complex** organic dye that undergoes a change in **color** when the **pH** of a **solution** changes over a specific pH range. Many **plant** pigments and other natural products are good indicators, and synthetic ones like phenolphthalein and methyl red are also available and widely used. **Paper** dipped in a mixture of several indicators and then dried is called pH paper, useful for obtaining the approximate pH of a solution. Blue litmus paper turns red in acidic solution, and red litmus paper turns blue in basic solution.

The pH at which the color of an indicator changes is called the transition **interval**. Chemists use appropriate indicators to signal the end of an acid-base **neutralization** reaction. Such a reaction is usually accomplished by titration—slowly adding a measured quantity of the base to a measured quantity of the acid (or vice versa) from a **buret**. (A buret is a long tube with **volume** markings for precise measurement and a stopcock at the bottom to control the flow of liquid.) When the reaction is complete, that is, when there is no excess of acid or base but only the reaction products, that is called the endpoint of the titration. The indicator must change color at the pH which corresponds to that endpoint.

The indicator changes color because of its own neutralization in the solution. Different indicators have different transition intervals, so the choice of indicator depends on matching the transition interval to the expected pH of the solution just as the reaction reaches the point of complete neutralization. Phenolphthalein changes from colorless to pink across a range of pH 8.2 to pH 10. Methyl red changes from red to yellow across a range of pH 4.4 to pH 6.2. Those are the two most common indicators, but others are available for much higher and lower pH values. Methyl violet, for example, changes from yellow to blue at a transition interval of pH 0.0 to pH 1.6. Alizarin yellow R changes from yellow to red at a transition interval of pH 10.0 to pH 12.1. Other indicators are available through most of the pH range, and can be used in the titration of a wide range of weak **acids and bases**.

Indicator species

Indicator species are plants and animals that, by their presence, abundance, lack of abundance, or chemical composition, demonstrate some distinctive aspect of the character or quality of an environment.

For example, in places where metal-rich **minerals** occur at the **soil** surface, indicator species of plants can be examined to understand the patterns of naturally occurring **pollution**, and they can even be a tool used in prospecting for potential **ore** bodies. Often, the indicator plants accumulate large concentrations of metals in their tissues. Nickel concentrations as large as 10% have been found in the tissues of indicator plants in the mustard family (*Alyssum bertolanii* and *A. murale*) in Russia, and a concentration as large as 25% occurs in the blue-colored latex of *Sebertia acuminata* from the Pacific **island** of New Caledonia. Similarly, *Becium homblei*, related to mint, has been important in the discovery of **copper** deposits in parts of **Africa**, where it is confined to soils containing more than 0.16 oz/lb (1,000 mg/kg) of copper, because it can tolerate more than 7% copper in soil. So-called copper mosses have been used by prospectors as botanical indicators of surface mineralizations of this **metal** in Scandinavia, Alaska, Russia, and elsewhere.

Plants are also used as indicators of serpentine minerals, a naturally occurring soil constituent that in large concentrations can render the substrate toxic to the growth of most plants. The toxicity of serpentine influenced soils is mostly caused by an imbalance of the availability of **calcium** and **magnesium**, along with the occurrence of large concentrations of toxic nickel, chromium, and cobalt, and small concentrations of potassium, **phosphorus**, and **nitrogen**. Serpentine soils are common in parts of California, where they have developed a distinctive **flora** with a number of indicator species, many of which are **endemic** to this **habitat** type (that is, they occur nowhere else). A genus in the mustard family, *Streptanthus*, has 16 species endemic to serpentine sites in California. Three **species** have especially narrow distributions: *Streptanthus batrachopus, S. brachiatus,* and *S. niger,* only occur at a few sites. *Streptanthus glandulosus, S. hesperidis,* and *S. polygaloides*

maintain wider distributions, but they are also restricted to serpentine sites.

Indicator plants also occur in many semiarid areas on soils containing selenium. Some of these plants can accumulate this element to large concentrations, and they can be poisonous to **livestock**, causing a **syndrome** known as "blind staggers" or "alkali disease." The most important selenium-accumulating plants in **North America** are in the genus *Astragalus*, of the legume family. There are about 500 species of *Astragalus* in North America, 25 of which can accumulate up to 15 thousand ppm (parts per million) of selenium in foliage. These species of *Astragalus* can emit selenium-containing chemicals to the atmosphere, which gives the plants a distinctive and unpleasant odor.

Sometimes indicator species are used as measures of habitat or **ecosystem** quality. For example, animals with a specialized requirement for **old-growth forests** can be used as an indicator of the integrity of that type of ecosystem. Old-growth dependent **birds** in North America include the spotted owl (*Strix occidentalis*), red-cockaded woodpecker (*Picoides borealis*), marbled murrelet (*Brachyramphus marmoratus*), and pine marten (*Martes americana*). If the area and quality of old-growth forest in some area is sufficient to allow these indicator animals to maintain viable populations, this suggests something positive about the health of the larger, old-growth ecosystem. In contrast, if a proposed forest-harvesting plan is considered to pose a threat to the populations of these species, this also indicates a challenge to the integrity of the old-growth forest more broadly.

Indicator species can also be used as measures of environmental quality. For example, many species of **lichens** are very sensitive to toxic gases, such as **sulfur dioxide** and **ozone**. These "species" (actually, lichens are a **symbiosis** between a fungus and an alga) have been monitored in many places to study **air pollution**. Severe damage to lichens is especially common in cities with chronic air pollution, and near large point sources of toxic gases, such as metal smelters.

Similarly, aquatic **invertebrates** and **fish** have commonly been surveyed as indicators of **water** quality and the health of aquatic ecosystems. If a site has populations of so-called "sewage worms" or tubificids (Tubificidae), for example, this almost always suggests that water quality has been degraded by inputs of sewage or other oxygen-consuming organic **matter**. Tubificid worms can tolerate virtually anoxic water, in contrast with most of the animals of unpolluted environments, such as **mayflies** (Ephemeroptera) and **stoneflies** (Plecoptera), which require well-oxygenated conditions.

Often, the lacking presence of an indicator species is indicative of environmental change or **contamination**. For instance, the nymphs of stoneflies mentioned above, if absent from a stream where they would normally be expected to reside, might indicate a lack of oxygenation or the presence of a pollutant. Caddisfly larvae, mayfly nymphs, and stonefly nymphs are often used to evaluate water quality and the presence of acid mine drainage in western Pennsylvania, where **coal mining** is prevalent and can affect nearby watersheds.

Another current example involves **frogs** and **salamanders** as indicator species. Populations of **amphibians** are declining on a global scale. Their decline is thought to be an indicator of tainted environments. Therefore, the numbers of amphibians worldwide are being closely monitored. In a related example, the eggs of certain bird species are tested for the presence of organic **pesticides**.

Much research is being done by governments to accurately establish which species of plants and animals can act as sentinels of particular environmental contaminants. Here, the indicator species shows directly the persistence of hazardous chemicals in the environment. Through the use of indicator species, then, it is hoped that potential environmental problems may be identified before they result in irrevocable damage.

See aaalso Ecological monitoring; Water pollution.

Bill Freedman

Indium *see* **Element, chemical**

Individual

An individual, in the sense of evolutionary **biology**, is a genetically unique **organism**. An individual has a complement of genetic material, encoded in its DNA (deoxyribonucleic acid), that is different from other members of its **species**. At the level of populations and species, this variation among individuals constitutes genetic **biodiversity**.

Phenotype, genotype, plasticity, and evolution

Morphology, **physiology**, and **behavior** are attributes of individual organisms that can be observed. These attributes are known as the phenotype. Two factors that influence the phenotype are: (1) the specific genetic information of the individual (its genotype), and (2) environmental influences on the expression of the individual's genetic potential. The term phenotypic plasticity refers to the **variable** growth, physiology, and behavior that an individual organism displays, depending on environmental conditions experienced during its lifetime.

Because organisms vary in character (phenotype), they also differ in their abilities to cope with environmental stresses and opportunities. Under certain conditions, an individual with a particular phenotype (and genotype) may be relatively successful, compared with other individuals. In evolutionary biology, the "success" of an individual is measured by how many offspring it has produced, and whether those progeny go on to reproduce. This is similar to fitness, or the genetic contribution of an individual to all the progeny of its population. A central element of evolutionary theory is that individuals seek to maximize their fitness, and thereby to optimize their genetic influence on future generations.

Biologists believe natural **selection** is the most important means by which **evolution** occurs. Natural selection can only proceed if: (a) there is genetically based variation among individuals within a population, and (b) some individuals are better adapted to coping with the prevailing environmental conditions. Better-fit organisms tend to be more successful having offspring, and they have a greater influence on the evolution of subsequent generations. Individuals themselves do not evolve, however, they are capable of phenotypic plasticity.

Unusual individuals

In virtually all species, individuals differ genetically. However, there are a few interesting exceptions to this generalization. Populations of some plants may have no genetic variability because the species propagates by non-sexual (or vegetative) means. In such plants, genetically uniform populations (or clones) may develop. These represent a single genetic "individual." For example, extensive clones of trembling aspen (*Populus tremuloides*) can develop when new trees sprout from underground stems (or rhizomes). Such aspen clones can cover more than 100 acres (40 ha) and consist of tens of thousands of trees. In terms of total **biomass**, such aspen clones may represent the world's largest "individual" organisms. Another case involves the **duckweed**, *Lemna minor*, a tiny aquatic **plant** that grows on **water** surfaces. Duckweed propagates by growing buds on the edge of a single **leaf**. These buds grow and break off to produce "new" plants genetically identical to the parent. These interesting cases of asexual propagation are exceptional because most populations and species contain a great deal of genetic variation amongst their individuals.

Bill Freedman

Indoor air quality

The chemical, physical, and biological characteristics of the atmosphere inside of dwellings and in commercial and institutional buildings are influenced in numerous ways. Sometimes, effects on indoor air quality can be sufficient to cause people to experience significant discomfort, and even to become physically ill.

People vary greatly in their sensitivity to **air pollution**, both inside and outside of buildings. People also differ in the sorts of symptoms that they develop in response to deterioration of air quality. Consequently, it has proven difficult for scientists to characterize the dimensions of indoor air quality, and to precisely define the nature of the subsequent environmental illnesses that some people appear to develop. This has led to a great deal of environmental and medical controversy, concerning the extent and intensity of a **syndrome** of air-quality related illnesses, known as the "sick building syndrome."

Factors influencing indoor air quality

Air quality inside of buildings is related to a diverse range of chemical, physical, and biological factors. In any situation, the importance of these many influences can vary greatly, depending on the **emission** rates of various chemicals, the **frequency** with which inside air is exchanged with ambient air, the efficiency of **atmospheric circulation** within the building, and numerous other factors.

In response to the need to conserve **energy** (and money), modern buildings are well insulated to retain their **heat** in winter and their coolness in summer. Such buildings receive almost all of their inputs of relatively clean, outside air through their carefully designed, ventilation system. Such systems have only a few, discrete intakes of ambient air, and outputs of "used" air back to the outside, as well as particular, internal-circulation characteristics. It is not possible, for example, to open any windows in many modern office buildings, because this would interfere with **pressure** gradients and upset the designed balance of the ventilation system. Of course, the ventilation characteristics of many recently constructed modern buildings have a substantial influence on the quality of the internal atmosphere of the structure.

When ventilation systems are operated with a view to saving energy, there are relatively few exchanges of indoor air with relatively clean, ambient air. Sometimes, too much attention to the efficiency of energy use in air-tight buildings can lead to the build-up of excessive concentrations of indoor air pollutants, because of on-going emissions of chemicals within the building.

In addition, in some cases the intake pipes for ambient air to buildings are located too close to ducts that exhaust contaminated air from the same or a nearby building. This faulty design can lead to the intake of poor-quality outside air, impairing atmospheric quality within the building. Similarly, sinks and other **water** drains installed without proper systems to prevent the back-up of sewer gases can lead to incursions of noxious smells and chemicals into buildings. In other cases, the poor faulty design or operation of internal ventilation systems can lead to the development of local zones of restricted air circulation, which can develop into areas of degraded air quality within the building.

Clearly, the appropriate design and operation of air-handling systems in modern, air-tight buildings is a critical factor affecting indoor air quality.

Emission rates of chemicals and dusts within buildings are affected by many factors. The sorts of materials of which the building or its furnishings are constructed may be important in this regard. For example, **minerals** contained in cement or in stone may emit gaseous **radon**, or may slowly degenerate to release fine, inhalable dusts. The oxidation of materials in humidification systems and ventilation duct works can also generate large quantities of fine, metallic dusts, as can the wear of painted surfaces. Many composite **wood** products, such as plywood and particle boards, emit gaseous formaldehyde, as do many types of synthetic fabrics.

Chemicals may also be emitted to the internal air from laboratories that do not have adequate fume hoods to vent noxious vapors and gases to the atmosphere. Similarly, industrial processes involving chemicals may be an important source of emissions in some buildings. The use of some kinds of solvents, detergents, and other substances during cleaning and sanitation of the building may also be important.

Even the human occupants of buildings emit large quantities of gases and vapors that affect air quality, for example, **carbon dioxide**. Also, although the practice is increasingly being restricted, many people smoke tobacco inside of buildings, releasing diverse chemicals to the atmosphere. More than 2,000 chemicals have been identified in tobacco fumes, including various carcinogens such as benzo(a)pyrene and nickel carbonyl, as well as many other toxic chemicals.

These are just a few of the diverse sources of emissions of gases, vapors, and particulates inside of modern buildings. All of these sources of emissions contribute to the degradation of the quality of the indoor atmosphere.

Some buildings can develop indoor-air problems associated with **fungi** and other microbes that grow in damp places, and whose spores or other so-called bioaerosols become spread within the building through the ventilation system. This microbial problem can develop in systems designed to humidify the indoor air, in places where stagnant water accumulates within the air-circulation system, or in other damp places. Some people may be allergic to these spores, or in rare cases the **microorganisms** may be **pathogens**. The latter is the case of **Legionnaires' disease**, a rare condition involving pathogenic **bacteria** spread through the ventilation system of buildings.

Aspects of indoor air quality

Indoor air quality has many components, some of which are physical, others chemical, and a few biological. The most significant of these are briefly described below.

The most important physical aspects of indoor air quality are air **temperature** and **humidity**. Air temperatures that are too warm or cool for human comfort can be caused by improper placement or adjustment of thermostats, and by an inability of the heating or air-conditioning system to compensate for extremes of outdoor **weather**, or to adequately deal with heat generated by machinery or large numbers of people. Excessive or insufficient humidity can be caused by similar problems, including poorly operating or non-existent humidity-control mechanisms within the ventilation system.

Carbon dioxide (CO_2) is a normal constituent of the ambient atmosphere, occurring in a **concentration** of about 350 parts per million (ppm, on a volumetric basis). However, there are many sources of emission of carbon dioxide inside of buildings, including potted plants and their **soil**, **respiration** by humans, and stoves or space heaters fueled by kerosene, propane, or methane. Consequently, the concentrations of carbon dioxide are typically relatively large inside of buildings, especially in inadequately ventilated rooms that are crowded with people. Commonly measured concentrations of this gas are about 600-800 ppm, but in some situations concentrations of thousands of ppm can be achieved. Longer-term exposure to concentrations of carbon dioxide greater than about 5,000 ppm is not recommended. Symptoms of excessive exposure to carbon dioxide include drowsiness, dizziness, headaches, and shortage of breath.

Carbon monoxide (CO) is a product of the incomplete oxidation of organic fuels. Indoor emissions are mostly associated with stoves or space heaters fueled by kerosene or **natural gas**, with **cigarette smoke**, or with poorly vented emissions from automobiles in garages or loading docks. Longer-term exposures to carbon monoxide concentrations greater than nine ppm should be avoided, as should shorter-term (about one-hour) exposures greater than 35 ppm. Carbon monoxide is a relatively

toxic gas because it combines strongly with the hemoglobin of **blood**, thereby restricting the ability of the circulation system to transport an adequate supply of **oxygen** to the various parts of the body. Excessive exposures to carbon monoxide under poorly ventilated conditions can cause headaches, drowsiness, nausea, fatigue, impaired judgement, and other symptoms of insufficient oxygen supply. Anoxia and death can ultimately be caused.

Formaldehyde is a pungent, organic vapor that can be detected by smell at a concentration greater than about 0.2 ppm. There are diverse sources of emission of formaldehyde, including poorly sealed plywoods and particle boards, urea-formaldehyde foam insulation, and many fabrics, carpets, glues, and copy papers. Some people are quite sensitive to formaldehyde, developing symptoms that can include a dry or sore throat, headaches, fatigue, nausea, and stinging sensations in the eyes. Most people can tolerate formaldehyde concentrations of less than 0.5 ppm without developing these sorts of symptoms, but other, hypersensitive people may be adversely affected at concentrations as small as 0.01 ppm. In general, exposures to formaldehyde exposure in work areas should be less than 0.1 ppm.

Volatile organic compounds (VOCs) are a wide range of molecular **species** that vaporize at normally encountered temperatures. Common examples of volatile organic compounds found in buildings include (in alphabetical order): **acetone**, butyl acetate, dichlorobenzene, dichloromethane, hexane, octane, toluene, trichloroethane, and xylene. These organic chemicals have diverse sources, including synthetic materials used to manufacture carpets and fabrics, paints, solvents, adhesives, cleaning solutions, perfumes, hair sprays, and cigarette smoke. All of the common VOCs and many others have recommended indoor-exposure limits, which vary depending on the toxicity of the particular chemical, and on the length of the exposure. Human responses to large concentrations of volatile organic compounds include dizziness, fatigue, drowsiness, tightness of the chest, numbness or tingling of the extremities, and skin and **eye** irritation. Some people are hypersensitive to specific compounds or groups of VOCs.

The gases nitric oxide (NO), nitrogen dioxide (NO_2), and **sulfur dioxide** (SO_2) may also be important pollutants of the indoor atmosphere, especially where there are fuel-burning appliances or stoves used for cooking or space heating. These gases can be irritating to the eyes and upper **respiratory system** of people exposed to large concentrations.

Radon is a radioactive gas emitted by a wide range of geological sources, including mineral-containing building materials and ground water. Many poorly vented homes and some commercial buildings become significantly contaminated by radon, a gas that carries a risk of causing human toxicity through the development of cancers, especially lung **cancer**.

Particulates are various sorts of solid or liquid materials that are small enough to be suspended in the atmosphere as fine dusts or **aerosols**. Particulate emissions inside of buildings are associated with smoke, physical-chemical deterioration of ducts, insulating materials, walls, ceiling tiles, and paints, fibers from clothing and other fabrics, and many other sources. Particulates may also be drawn into buildings along with unfiltered, ambient air. Particulates are aggravating to many people, who may develop irritations of the upper respiratory tract, such as **asthma**. Some chemicals contained in particulates, especially certain metals and **polycyclic aromatic hydrocarbons**, are widely regarded as toxic substances, and unnecessary exposures should generally be avoided. The particulate size range of 0.004-0.4 in (0.1-10 mm) is of particular importance in terms of human exposures, because this size range is efficiently retained in the deepest parts of the lungs. Particulates smaller than 0.004 in are generally re-exhaled, while particles larger than 0.4 in are trapped in the upper respiratory system and have little toxic effect.

Sometimes, microbial **matter** (or bioaerosols) can be an indoor-air problem. Usually, this involves spore-producing fungi that occur in damp places in the ventilation system, carpets, or other places. Many people have allergies to fungal spores, and can be made ill by excessive exposures to these bioaerosols in indoor air. Bioaerosols of other microbes such as **yeast**, bacteria, viruses, and protozoan may also be important problems in the atmosphere of buildings. On a rare occasion, pathogenic bacteria such as the *Legionella* associated with pneumonia-like Legionnaires' disease, can be spread through the ventilation system of buildings. Other potential pathogens in the inside air of buildings include the fungi *Aspergillus fumigatus* and *Histoplasma capsulatum*.

Sick building syndrome

The "sick building syndrome" exists. However, it has proven very difficult for scientists to characterize the causes, treatment, or human responses to the sick building syndrome. This is because of the extremely **variable** natures of both the exposures to environmental stressors in buildings, and the responses of individual people, a small fraction of whom appear to be hypersensitive to particular aspects of the indoor atmosphere.

The effects of the sick building syndrome on people range from drowsiness and vague feelings of discomfort, with subsequent decreases in productivity, to the devel-

KEY TERMS

Bioaerosols—Spores or actual microorganisms that occur suspended in the atmosphere.

Hypersensitivity—The occurrence of extreme sensitivity to chemicals or pathogens in a small fraction of a larger human population. Hypersensitivity may be related to an extreme allergic response, or to a deficiency of the immune system.

Sick building syndrome—A condition in which people frequently complain about a number of ailments while they are in a particular building, but feel relief when they go outside.

Ventilation rate—This refers to the amount of outside or ambient air that is combined with re-circulating inside or return air, and is then supplied to the interior space of a building. This may also apply to some part of a building, such as a particular room.

opment of actual illnesses. In many cases it may be necessary for the afflicted people to leave the building for some length of **time**. Sometimes, sensitive people must give up their jobs, because they find the indoor air quality to be intolerable.

As a result of the difficult-to-define nature of the sick building syndrome, important medical and environmental controversies have developed. Some scientists suggest that people who display building-related illnesses are imagining their problems. It is suggested that these people may have developed so-called psychosomatic responses, in which clinical illnesses are caused by non-existent factors that the victim believes are important. Increasingly, however, scientists are convinced that the relatively sensitive physiologies of severely afflicted people are direct responses to physical, chemical, or biological stressors in the poorly ventilated, enclosed spaces where they live or work. Increasingly, indoor air quality issues are being taken seriously by private individuals, commercial property owners, health organizations, and federal, state, and local governments.

Further research and monitoring will be required before a better understanding of the sick building syndrome can be achieved. This knowledge is required in order to design sensible systems of avoiding or treating the problems of poor-quality indoor air of buildings, and to better protect people who are exposed to this type of **pollution**. Federal and state agencies are working with home owners, developers and building maintenance professionals to develop plans and programs for dealing with indoor air quality programs. Particular attention is being paid to schools, because the relatively more-sensitive physiologies of children make them particularly susceptible to health threats from poor indoor air quality.

Resources

Books

Indoor Air Quality in Office Buildings: A Technical Guide. Ottawa: Health Canada, 1993.

Indoor Allergens: Assessing and Controlling Adverse Health Effects. Washington, DC: National Academy Press, 1993.

Bill Freedman

Indri *see* **Lemurs**

Industrial minerals

Industrial **minerals** is a term used to describe naturally occurring non-metallic minerals that are used extensively in a variety of industrial operations. Some of the minerals commonly included in this category include **asbestos**, barite, boron compounds, clays, corundum, feldspar, fluorspar, phosphates, potassium salts, **sodium chloride**, and **sulfur**. Some of the mineral mixtures often considered as industrial minerals include construction materials such as **sand**, gravel, limestone, dolomite, and crushed rock; **abrasives** and refractories; gemstones; and lightweight aggregates.

Asbestos

Asbestos is a generic term used for a large group of minerals with complex chemical composition that includes **magnesium**, silicon, **oxygen**, **hydrogen**, and other elements. The minerals collectively known as asbestos are often sub-divided into two smaller groups, the serpentines and amphiboles. All forms of asbestos are best known for an important common property—their resistance to **heat** and flame. That property is responsible, in fact, for the name *asbestos* (Greek), meaning "unquenchable." Asbestos has been used for thousands of years in the production of heat resistant materials such as lamp wicks.

Today, asbestos is used as a reinforcing material in cement, in vinyl floor tiles, in fire-fighting garments and fire-proofing materials, in the manufacture of brake linings and clutch facings, for electrical and heat insulation, and in **pressure** pipes and ducts.

Prolonged exposure to asbestos fibers can block the **respiratory system** and **lead** to the development of asbestosis and/or lung **cancer**. The latency period for these disorders is at least 20 years, so men and women who mined the mineral or used it for various construction purposes during the 1940s and 1950s were not aware of their risk for these diseases until late in their lives. Today, uses of the mineral in which humans are likely to be exposed to its fibers have largely been discontinued.

Barite

Barite is the name given to a naturally occurring form of **barium sulfate**, commonly found in Canada, Mexico, and the states of Arkansas, Georgia, Missouri, and Nevada. One of the most important uses of barite is in the production of heavy muds that are used in drilling oil and gas wells. It is also used in the manufacture of a number of other commercially important industrial products such as **paper** coatings, **battery** plates, paints, linoleum and oilcloth, **plastics**, lithographic inks, and as a filler in some kinds of **textiles**. **Barium** compounds are also widely used in medicine to provide the opacity that is needed in taking certain kinds of **x rays**.

Boron compounds

Boron is a non-metallic element obtained most commonly from naturally occurring minerals known as borates. The borates contain oxygen, hydrogen, **sodium**, and other elements in addition to boron. Probably the most familiar boron-containing mineral is borax, mined extensively in **salt** lakes and alkaline soils.

Borax was known in the ancient world and used to make glazes and hard **glass**. Today, it is still an important ingredient of glassy products that include heat-resistant glass (Pyrex), glass wool and glass fiber, enamels, and other kinds of ceramic materials. Elementary boron also has a number of interesting uses. For example, it is used in nuclear reactors to absorb excess neutrons, in the manufacture of special-purpose alloys, in the production of semiconductors, and as a component or rocket propellants.

Corundum

Corundum is a naturally occurring form of **aluminum** oxide that is found abundantly in Greece and Turkey and in New York State. It is a very hard mineral with a high melting point. It is relatively inert chemically and does not conduct an electrical current very well.

These properties make corundum highly desirable as a refractory (a substance capable of withstanding very high temperatures) and as an abrasive (a material used for cutting, grinding, and polishing other materials). One of the more mundane uses of corundum is in the preparation of toothpaste, where its abrasive properties help in keeping teeth clean and white.

In its granular form, corundum is known as emery. Many consumers are familiar with emery boards used for filing finger nails. Emery, like corundum, is also used in the manufacture of cutting, grinding, and polishing wheels.

Feldspar

The feldspars are a class of minerals known as the aluminum silicates. That is, they all contain aluminum, silicon, and oxygen, as sodium, potassium, and **calcium**. In many cases, the name feldspar is reserved for the potassium aluminum silicates. The most important commercial use of feldspar is in the manufacture of pottery, enamel, glass, and ceramic materials. The hardness of the mineral also makes it desirable as an abrasive.

Fluorspar

Fluorspar is a form of calcium fluoride that occurs naturally in many parts of the world including **North America**, Mexico, and **Europe**. The compound gets its name from one of its oldest uses, as a flux. In Latin, the word *fluor* means "flux." A flux is a material that is used in industry to assist in the mixing of other materials or to prevent the formation of oxides during the refining of a **metal**. For example, fluorspar is often added to an open hearth **steel** furnace to react with any oxides that might form during that process. The mineral is also used during the smelting of an **ore** (the removal of a metal form its naturally occurring ore).

Fluorspar is also the principal source of fluorine gas. The mineral is first converted to hydrogen fluoride which, in turn, is then converted to the element fluorine. Some other uses of fluorspar are in the manufacture of paints and certain types of cement, in the production of emery wheels and **carbon** electrodes, and as a raw material for phosphors (a substance that glows when bombarded with **energy**, such as the materials used in **color television** screens).

Phosphates

The term phosphate refers to any **chemical compound** containing a characteristic grouping of **atoms**, given by the formula PO_4, or comparable groupings. In the field of industrial minerals, the term most commonly refers to a specific naturally occurring phosphate, calcium phosphate, or phosphate rock.

By far the most important use of phosphate rock is in agriculture, where it is treated to produce **fertilizers** and **animal** feeds. Typically, about 80% of all the phosphate rock used in the United States goes to one of these agricultural applications.

Phosphate rock is also an important source for the production of other phosphate compounds, such as sodium, potassium, and ammonium phosphate. Each of these compounds, in turn, has a very large variety of uses in everyday life. For example, one form of sodium phosphate is a common ingredient in dishwashing detergents. Another, ammonium phosphate, is used to treat cloth to make it fire retardant. Potassium phosphate is used in the preparation of baking powder.

Potassium salts

As with other industrial minerals mentioned here, the term potassium salts applies to a large group of compounds, rather than one single compound. Potassium chloride, sulfate, and nitrate are only three of the most common potassium salts used in industry. The first of these, known as sylvite, can be obtained from salt **water** or from fossil salt beds. It makes up roughly 1% of each deposit, the remainder of the deposit being sodium chloride (halite).

Potassium salts are similar to phosphate **rocks** in that their primary use is in agriculture, where they are made into fertilizers, and in the chemical industry, where they are converted into other compounds of potassium. Some compounds of potassium have particularly interesting uses. **Potassium nitrate**, for example, is unstable and is used in the manufacture of **explosives**, fireworks, and matches.

Sodium chloride

Like potassium chloride, sodium chloride (halite) is found both in sea water and in underground salt mines left as the result of the **evaporation** of ancient seas. Sodium chloride has been known to and used by humans for thousands of years and is best known by its common name of salt, or table salt. By far its most important use is in the manufacture of other industrial chemicals, including **sodium hydroxide**, hydrochloric acid, **chlorine**, and metallic sodium. In addition, sodium chloride has many industrial and commercial uses. Among these are in the preservation of foods (by salting, pickling, corning, curing, or some other method), highway de-icing, as an additive for human and other animal foods, in the manufacture of glazes for **ceramics**, in water softening, and in the manufacture of rubber, metals, textiles, and other commercial products.

Sulfur

Sulfur occurs in its elementary form in large underground deposits from which it is obtained by traditional **mining** processes or, more commonly, by the Frasch process. In the Frasch process, superheated water is forced down a pipe that has been sunk into a sulfur deposit. The heated water melts the sulfur, which is then forced up a second pipe to the earth's surface.

The vast majority of sulfur is used to manufacture a single compound, **sulfuric acid**. Sulfuric acid consistently ranks number one in the United States as the chemical produced in largest quantity. Sulfuric acid has a very large number of uses, including the manufacture of fertilizers, the refining of **petroleum**, the pickling of steel (the removal of oxides from the metal's surface), and the preparation of detergents, explosives, and synthetic fibers.

A significant amount of sulfur is also used to produce **sulfur dioxide** gas (actually an intermediary in the manufacture of sulfuric acid). Sulfur dioxide, in turn, is extensively used in the pulp and paper industry, as a refrigerant, and in the purification of sugar and the bleaching of paper and other products.

Some sulfur is refined after being mined and then used in its elemental form. This sulfur finds application in the **vulcanization** of rubber, as an insecticide or **fungicide**, and in the preparation of various chemicals and pharmaceuticals.

Resources

Books

Greenwood, N.N., and A. Earnshaw. *Chemistry of the Elements.* 2nd ed. Oxford: Butterworth-Heinneman Press, 1997.

Klein, C. *The Manual of Mineral Science.* 22nd ed. New York: John Wiley & Sons, Inc., 2002.

David E. Newton

Industrial Revolution

Industrial Revolution is the name given by the German socialist author Friedrich Engels in 1844 to changes that took place in Great Britain during the period from roughly 1730 to 1850. In general, those changes involved the transformation of Great Britain from a largely agrarian society to one dominated by industry. In a broader context, the term has also been applied to the transformation of the Trans-Atlantic economy, including continental **Europe** and the United States in the nineteenth century.

Without question, the Industrial Revolution involved some of the most profound changes in human society in history. However, historians have long argued over the exact nature of the changes that occurred during this period, the factors that brought about these changes, and the ultimate effects the Revolution was to have on Great Britain and the world.

Most of the vast array of changes that took place during the Industrial Revolution can be found in one of three major economic sectors—textiles, **iron**, and **steel**, as well as transportation. These changes had far-flung effects on the British economy and social system.

The textile industry

Prior to the mid-eighteenth century, textile manufacture in Great Britain (and the rest of the world) was an activity that took place almost exclusively in private homes. Families would obtain thread from wholesale outlets and then produce cloth by hand in their own houses. Beginning in the 1730s, however, a number of inventors began to develop machines that took over one or more of the hand-knitting operations previously used in the production of **textiles**.

For example, John Kay invented the first flying shuttle in 1733. This machine consisted of a large frame to which was suspended a series of threads through which a shuttle carrying more thread could be passed. Workers became so proficient with the machine that they could literally make the shuttle "fly" through the thread framework as they wove a piece of cloth.

Over the next half century, other machines were developed that further mechanized the weaving of cloth. These included the spinning jenny, invented by James Hargreaves in 1764; the **water** frame, invented by Richard Arkwright in 1769; the spinning mule, invented by Samuel Crompton in 1779; the power loom, invented by Edmund Cartwright in 1785; and the **cotton** gin, invented by Eli Whitney in 1792. (Dates for these inventions may be in dispute because of delays between actual inventions and the issuance of patents for them.) One indication of the **rate** at which technology was developing during this period is the number of patents being issued. Prior to 1760, the government seldom issued more than a dozen patents a year. By 1766, however, that number had risen to 31 and, by 1783, to 64. By the end of the century, it was no longer unusual for more than 100 new patents to be issued annually.

At least as important as the invention of individual machines was the organization of industrial operations for their use. Large factories, powered by steam or water, sprang up throughout the nation for the manufacture of cloth and clothing.

The development of new technology in the textile industry had a ripple effect on society, as is so often the case with technological change. As cloth and clothing became more readily available at more modest prices, the demand for such articles increased. This increase in demand had the further effect, of course, of encouraging the expansion of business and the search for even more efficient forms of technology.

Technological change also began to spread to other nations. By the mid-nineteenth century, as an example, the American inventor Elias Howe had applied the principles of the Industrial Revolution to hand sewing. He invented a machine that, in a demonstration contest in 1846, allowed him to sew a garment faster than five women sewing by hand.

Iron and steel manufacture

One factor contributing to the development of industry in Great Britain was that nation's large supply of **coal** and iron **ore**. For many centuries, the British had converted their iron ores to iron and steel by heating the raw material with charcoal, made from trees. By the mid-eighteenth century, however, the nation's timber supply had largely been decimated. Iron and steel manufacturers were forced to look elsewhere for a fuel to use in treating iron ores.

The fuel they found was coal. When coal is heated in the absence of air it turns into coke. Coke proved to be a far superior material for the conversion of iron ore to iron and then to steel. It was eventually cheaper to produce than charcoal and it could be packed more tightly into a blast furnace, allowing the heating of a larger **volume** of iron.

The conversion of the iron and steel business from charcoal to coke was accompanied, however, by a number of new technical problems which, in turn, encouraged the development of even more new inventions. For example, the use of coke in the smelting of iron ores required a more intense flow of air through the furnace. Fortunately, the **steam engine** that had been invented by James Watt in 1763 provided the means for solving this problem. The Watt steam engine was also employed in the **mining** of coal, where it was used to remove water that collected within most mines.

By the end of the eighteenth century, the new approach to iron and steel production had produced dra-matic effects on population and industrial patterns in Great Britain. Plants were moved or newly built in areas close to coal resources such as Southern Wales, Yorkshire, and Staffordshire.

Transportation

For nearly half a century, James Watt's steam engine was used as a power source almost exclusively for stationary purposes. The early machine was bulky and very heavy so that its somewhat obvious applications as a source of power for transportation were not readily solved. Indeed, the first forms of transport that made use of steam power were developed not in Great Britain, but in France and the United States. In those two nations, inventors constructed the first ships powered by steam engines. In this country, Robert Fulton's steam ship *Clermont*, built in 1807, was among these early successes.

During the first two decades of the nineteenth century, a handful of British inventors solved the host of problems posed by placing a steam engine within a carriage-type vehicle and using it to transport people and goods. In 1803, for example, Richard Trevithick had built a "steam carriage" with which he carried passengers through the streets of London. A year later, one of his steam-powered locomotives pulled a load of ten tons for a distance of almost 10 mi (16 km) at a speed of about 5 MPH (8 km/h).

Effects of the Industrial Revolution

The Industrial Revolution brought about dramatic changes in nearly every aspect of British society, including demographics, politics, social structures and institutions, and the economy. With the growth of factories, for example, people were drawn to metropolitan centers. The number of cities with populations of more than 20,000 in England and Wales rose from 12 in 1800 to nearly 200 at the close of the century. As a specific example of the effects of technological change on demographics, the growth of coke smelting resulted in a shift of population centers in England from the south and east to the north and west.

Technological change also made possible the growth of capitalism. Factory owners and others who controlled the means of production rapidly became very rich. As an indication of the economic growth inspired by new technologies, purchasing power in Great Britain doubled and the total national income increased by a factor of ten in the years between 1800 and 1900.

Such changes also brought about a revolution in the nation's political structure. Industrial capitalists gradually replaced agrarian land owners as leaders of the nation's economy and power structure.

Working conditions were often much less than satisfactory for many of those employed in the new factory systems. Work places were often poorly ventilated, overcrowded, and replete with safety hazards. Men, women, and children alike were employed at survival wages in unhealthy and dangerous environments. Workers were often able to afford no more than the simplest housing, resulting in the rise of urban slums. Stories of the unbelievable work conditions in mines, textile factories, and other industrial plants soon became a staple of Victorian literature.

One consequence of these conditions was that action was eventually taken to protect workers—especially women and children—from the most extreme abuses of the factory system. Laws were passed requiring safety standards in factories, setting minimum age limits for young workers, establishing schools for children whose parents both worked, and creating other standards for the protection of workers. Workers themselves initiated activities to protect their own interests, the most important of which may have been the establishment of the first trade unions.

Overall, the successes of the technological changes here were so profound internationally that Great Britain became the world's leading power, largely because of the Industrial Revolution, for more than a century.

David E. Newton

Inequality

In **mathematics**, an inequality is a statement about the relative order of members of a set. For instance, if S is the set of positive **integers**, and the symbol < is taken to mean less than, then the statement 5 < 6 (read "5 is less than 6") is a true statement about the relative order of 5 and 6 within the set of positive integers. The comparison that is symbolized by < is said to define an ordering **relation** on the set of positive integers. An inequality is often used for defining a subset of an ordered set. The subset is also the solution set of the inequality.

Ordered sets

A set is ordered if its members obey three simple rules. First, an ordering relation such as "less than" (<) must apply to every member of the set, that is, for any two members of the set, call them a and b, either a < b or b < a. Second, no member of the set can have more than one position within the ordering, in other words, a < a has no meaning. Third, the ordering must be **transitive**, that is, for any three members of the set, call them a, b, and c, if a < b, and b < c, then a < c. There are many examples of ordered sets. The alphabet, for instance, is an ordered set whose members are letters. An encyclopedia is an ordered set whose members are entries that are ordered alphabetically. The **real numbers** and subsets of the real numbers are also ordered. As a consequence, any set that is ordered can be associated on a one-to-one basis with the real numbers, or one of its subsets. The **algebra** of inequalities, then, is applicable to any set regardless of whether its members are numbers, letters, people, dogs or whatever, as long as the set is ordered.

Algebra of inequalities

Inequalities involving real numbers are particularly important. There are four types of inequalities, or ordering relations, that are important when dealing with real numbers. They are (together with their symbols) "less than" (<), "less than or equal to" (≤), "greater than" (>), and "greater than or equal to" (≥). In each case the symbol points to the lesser of the two expressions being compared. Since, by convention, mathematical expressions and statements are read from left to right, the statement $x + 2 < 6$ is read "x plus two is less than six," while $6 > x + 2$ is read "six is greater than x plus two." Algebraically, inequalities are manipulated in the same way

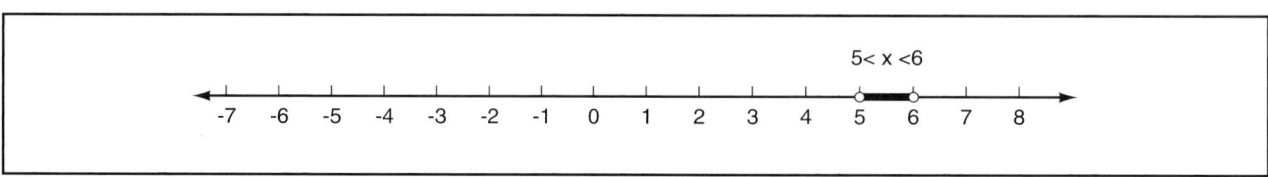

Figure 1. *Illustration by Hans & Cassidy. Courtesy of Gale Group.*

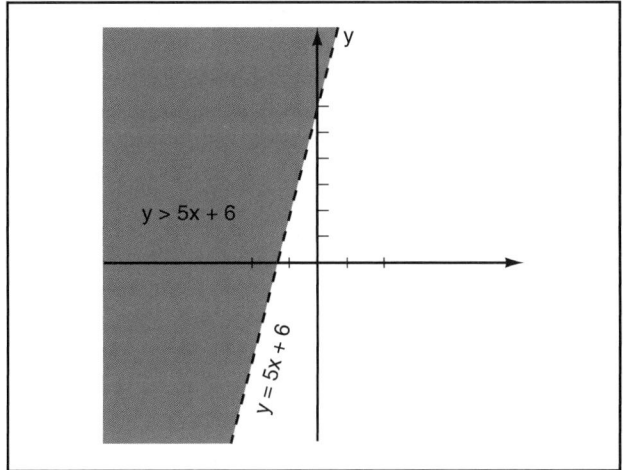

y > 5x + 6

y = 5x + 6

Figure 2. *Illustration by Hans & Cassidy. Courtesy of Gale Group.*

that equalities (equations) are manipulated, although most rules are slightly different.

The rule for **addition** is the same for inequalities as it is for equations:

for any three mathematical expressions, call them A, B, and C, if A > B then, A + C > B + C.

That is, the truth of an inequality does not change when the same quantity is added to both sides of the inequality. This rule also holds for **subtraction** because subtraction is defined as being addition of the opposite or **negative** of a quantity.

The **multiplication** rule for inequalities, however, is different from the rule for equations. It is: for any three mathematical expressions, call them A, B, and C, if A < B, and C is positive, then AC < BC, but if A < B, and C is negative, then AC > BC.

This rule also holds for **division**, since division is defined in terms of multiplication by the inverse.

Examples

As stated previously, an inequality can be a statement about the general location of a member within an ordered set, or it can be interpreted as defining a solution set or relation. For example, consider the compound expression 5 < x < 6 (read "5 is less than x, and x is less than 6") where x is a real number. This expression is a statement about the general location of x within the set of real numbers. Associating each of the real numbers with a **point** on a line (called the real number line) provides a way of picturing this location relative to all the other real numbers.

In addition, this same expression defines a solution set, or subset of the set of real numbers, namely all val-

ues of x for which the expression is true. More generally, an expression in two variables, such as y > 5x + 6, defines a solution set (or relation) whose members are ordered pairs of real numbers. Associating each ordered pair of real numbers with points in a **plane** (called the Cartesian coordinate system) it is possible to picture the solution set as being that portion of the plane that makes the expression true.

See also Cartesian coordinate plane.

Resources

Books

Bittinger, Marvin L., and Davic Ellenbogen. *Intermediate Algebra: Concepts and Applications.* 6th ed. Reading, MA: Addison-Wesley Publishing, 2001.

Davison, David M., Marsha Landau, Leah McCracken, and Linda Thompson. *Prentice Hall Pre-Algebra.* Needham, MA: Prentice Hall, 1992.

McKeague, Charles P. *Intermediate Algebra.* Fort Worth, TX: Saunders College Publishing, 1995.

J.R. Maddocks

Inertial guidance

Inertial guidance is a navigation technology that monitors changes in location by measuring cumulative **acceleration**. In inertial guidance, the **motion** of the object in three-dimensional **space** is measured continuously.

This enables a special computer to provide related real-time information about **velocity** (speed) and location.

An inertial-navigation system (INS) does not use information from an external reference once it has been placed in operation, in contrast to less-sophisticated navigation techniques. Gyrocompasses, older navigation aids that are dependent on the position of the stars or **sun** for guidance, are internally self sufficient, relying on precision gyroscopes for direction reference. However, gyrocompasses will drift with **time** as a result of slow, friction-induced gyrations and must be readjusted occasionally. Radiolocation navigation systems use precisely timed **radio** signals from distant transmitters or satellites. **Radar** mapping and optical terrain matching navigation require interaction with the earth's surface.

In contrast to these navigation tools, inertial navigation systems need only sense the inertial force that results from changing velocity. These forces are not dependent upon external references, but can be measured by accelerometers in a sealed, shielded container.

Inertial navigation was first applied for military uses—guiding deeply submerged submarines, **ballistic missiles**, and airplanes. Inertial navigation gave results that were more accurate than could be obtained with conventional navigation. An inertial-navigation system is effectively immune to deliberate interference, an obvious advantage in wartime.

In addition, inertial navigation functions as well near the earth's poles as it does at the equator. This feature is in marked contrast to the limitations imposed by a magnetic compass's unreliable performance in the Arctic or Antarctic regions of the **earth**. Magnetic compasses are also undependable in the earth's polar regions because of day-to-day variations in the **earth's magnetic field** strength and direction. Magnetic storms caused by solar disturbances that affect the earth are particularly troublesome near the magnetic poles.

The theoretical basis for inertial navigation

Inertial navigation obtains its information from the same type of inertial forces one experiences riding in an **automobile** when turning corners at high speed, accelerating away from a stop sign, or braking. An accelerometer measures these forces continually, and this information is processed by a computer.

An inertial navigation system makes independent measurements along each of the three principal geometric axes, which is collected by a computer. The result is real-time information about velocity and **distance** traveled.

Inertial guidance utilizes a family of relationships from kinematics, the description of motion. The connections between the principal formulas describing acceleration, velocity, and displacement are used. Each of these three aspects of motion contains information about the other two. An inertial navigation system continuously measures acceleration along each of the three dimensions, then calculates the corresponding instantaneous velocity. This can be used to determine the total distance traveled. By measuring acceleration as a function of time, an inertial guidance system calculates instantaneous speed and location without the need to for outside reference.

Inertial navigation and flight

Planes flying over the oceans often rely on inertial navigation to stay on their course. Even in the early 1970s some of the first 747 jets were designed to carry several inertial-guidance systems. When more than one inertial-navigation system is in use, each can monitor the plane's position independently for improved reliability.

On long flights, as from the United States to Japan, an inertial-guidance system can control a plane automatically by providing instructions to the autopilot. At the start of the journey the intended flight path is divided into a **succession** of short segments, perhaps a half dozen. The pilot enters the coordinates of the end points of each of these short flights into a computer. The inertial-guidance system flies the plane to each of these waypoints in turn. The overall route is closely approximated by the series of nearly-straight line segments. The inertial-navigation system's computer knows where the plane is located and its velocity because acceleration is measured continuously. These systems are so accurate that a plane can fly non-stop from San Francisco to Japan under the control of an inertial-navigation system, arriving with a location uncertainty of about 10 ft (3m).

For longer journeys over the surface of the earth, interpreting inertial-navigation data is more complicated. The computer must project the measured acceleration onto the spherical surface of the earth to determine instantaneous position relative to the earth's coordinate system of **latitude and longitude**. There is an additional complication resulting from the **rotation** of Earth. It is not enough to know the direction of a destination when a plane takes off. As Earth rotates, the direction of the planned destination may seem to change. Navigation must continually correct for a plane's tendency to drift off course because of Earth's rotational acceleration, a consequence of the so-called Coriolis force. The inertial-navigation system's computer compensates for these challenges accurately and quickly.

With the advent of the newer Global Positioning Satellite system as an alternative to inertial navigation, inertial navigation may be less significant in the future.

For the near future, navigation by INS will continue to make a valuable contribution to transportation safety and a backup to GPS-guided systems.

Resources

Books

Bolemon, Jay. *Physics: A Window On Our World.* 3rd ed. Needham, MA: Prentice-Hall, 1995.

Periodicals

Stix, Gary. "Aging Airways." *Scientific American* (May 1994).

Donald Beaty

Infection

The term infection refers to the state where a host **organism** has been invaded by another organism, typically a microorganism such as a **virus**, bacterium, **protozoa**, **algae**, or fungus. The invader is able to elude the responses of the host that are designed to kill it. Strategies include rapid multiplication, which can overwhelm the host defenses, or escaping from the host's **immune system** by multiplying inside host cells.

The second aspect of infection is the presence of symptoms. Depending on the type of infection, the symptoms produced can range from the inconvenience of a cold to those that are life threatening.

Until the middle of the twentieth century, infections posed a serious problem even in developed countries. Throughout recorded history, infections often killed millions of people in epidemics of diseases like **bubonic plague** and **typhoid fever**. Even today, infections continue to cause more deaths during times of war and famine than does battle and starvation. Infections can

sweep through a population quickly. For example, the acquired immunodeficiency **syndrome (AIDS)** has only been known for a little over three decades. Yet, AIDS is now the leading cause of death among African males.

Three factors are important in the control of an infection. These include identifying and eliminating the source of the infection, preventing the spread of the infection, and increasing the resistance of the host to the infecting microbe.

The hundreds of different infections that can occur in humans are caused by five major groups of microbes. These groups are the **bacteria**, a group made up of Rickettsiae, Coxiella, and Chlamydiae; viruses; **fungi**; protozoa; and worms known as Helminths. Infections from most of these organisms can be cured or made less severe using antibiotic drugs and anti-fungal medication. However, there is no cure for viral infections.

Most of the infections that humans acquire come from other people, animals or **insects**, and from nonliving objects that have infectious microbes adhering to them. Examples include the passage of a cold virus by kissing or sneezing, transfer of infectious viruses by dog or bat bites (i.e., **rabies**), use of contaminated needles to inject drugs (i.e., **hepatitis** B), unprotected sex with a contaminated partner (i.e., AIDS, syphilis). Infections also arise from drinking contaminated **water** or eating contaminated food.

Infections can become established when the immune system is not functioning properly because of **disease**, **malnutrition**, or treatment for another malady (i.e., chemotherapy for **cancer**). In these cases, microbes that would otherwise be easily defeated are able to proliferate, causing opportunistic infections.

Other infections arise because of a genetic condition in the host that predisposes the host to infection. One example is the persistent lung infections caused primarily by the bacterium *Pseudomonas aeruginosa* in some people who have **cystic fibrosis**. The fluid that accumulates in the lungs enables the bacteria to establish colonies that are resistant to treatment.

Still another route of infection is via the air. This route is especially relevant for bacterial spores, which are so small and light that they can float through the air and be inhaled. A prominent example is *Bacillus anthracis*, the cause of **anthrax**.

The concept of resistance to infection also applies to the host. As some bacteria are able resist host defenses and cause infection, so the host has several mechanisms of resistance. The first line of a host's defense is the various surfaces of the body. The skin, mucous membranes in the nose and throat, and tiny hairs in the nose that act to physically block invading organisms. The uppermost cells of the skin secrete chemicals that are lethal to bacteria such as *Staphylococcus aureus*, a bacterium that can cause skin infections. Microbes can also be washed away from body surfaces by tears, bleeding, and sweating. These are nonspecific mechanisms of resistance.

A host also has a specific defense response, namely the immune system. An invading microbe can be recognized as a foreigner and destroyed. This host resistance can be aided by vaccination, which in some cases provides a life long resistance to a particular organism.

The use of **antibiotics** was thought to be as powerful a deterrent to infection as vaccination. Indeed, when antibiotics were discovered in the middle of the twentieth century, many infections were presumed to have been defeated. However, this has proved not to be the case. The cause of the failure of some antibiotics is the ability of the target bacteria to become resistant to the drug. In the 1990s, this problem became especially evident, with the emergence of several types of infectious bacteria that are resistant to almost all antibiotics. Indeed, a strain of *Staphylococcus aureus* is resistant to all currently used antibiotics.

The development of resistance to antimicrobial agents such as antibiotics can have molecular origins. The membrane(s) of the bacteria may become structurally changed so as to make the passage of drugs across the membrane(s) difficult. Secondly, enzymes capable of degrading the antibiotic are produced. The overuse or inappropriate use of antibiotics (i.e., to treat a viral infection, even thought viruses are not affected by antibiotics) has contributed to the development of bacterial resistance, which can be genetically passed on to subsequent generations

The organization of the infecting **microorganisms** can also be a resistant factor. An example is the resistance that develops as a consequence of the surface growth of bacteria. In this **mode** of growth, which is known as a biofilm, the bacteria grow inside a sugary coating that is excreted by the surface adhering bacteria. Inside the coating the bacteria become almost dormant. The slow chemical activities of the bacteria, combined with the presence of the protective coating, makes biofilm bacteria extremely hardy. An example of the resistance of biofilm bacteria is that of *Pseudomonas aeruginosa*. **Biofilms** of this bacterium cause chronic lung infections in people afflicted with cystic fibrosis, and can grow on artificially implanted material (i.e., urinary **catheters** and **heart** pacemakers.)

See also Lymphatic system; Zoonoses.

Resources

Books

Kaper, J.B., and A.D. O'Brien. *Escherichia coli O157:H7 and Other Shiga Toxin-Producing E. coli Strains.* Washington, DC: American Society for Microbiology Press, 1998.

Salyers, A.A., and D.D. Whitt. *Bacterial Pathogenesis: A Molecular Approach.* 2nd ed. Washington, DC: American Society for Microbiology Press, 2001.

Other

Centers for Disease Control. "National Center for Infectious Disease." [cited November 20, 2002] <http://www.cdc.gov/ncidod/>.

Infertility

Infertility is a couple's inability to conceive a child after attempting to do so for at least one full year. Primary infertility refers to a situation in which pregnancy has never been achieved. Secondary infertility refers to a situation in which one or both members of the couple have previously conceived a child, but are unable to conceive again after a full year of trying.

Currently, in the United States, about 20% of couples struggle with infertility at any given time. Infertility has increased as a problem, as demonstrated by a study comparing fertility rates in married women ages 20-24 between the years of 1965 and 1982. In that time period, infertility increased 177%. Some studies attribute this increase on primarily social phenomena, including the tendency for marriage to occur at a later age, and the associated tendency for attempts at first pregnancy to occur at a later age. Fertility in women decreases with increasing age, as illustrated by the following **statistics**:

• infertility in married women ages 16-20: 4.5%
• infertility in married women ages 35-40: 31.8%
• infertility in married women over age 40: 70%

Since the 1960s, there has also been greater social acceptance of sexual intercourse outside of marriage, and individuals often have multiple sexual partners before they marry and attempt conception. This has led to an increase in sexually transmitted infections. Scarring from these infections, especially from pelvic inflammatory **disease** (PID)—a serious **infection** of the female reproductive organs—seems to be partly responsible for the increase. Furthermore, use of the contraceptive device called the intrauterine device (IUD) also has contributed to an increased **rate** of PID, with subsequent scarring.

To understand issues of infertility, it is first necessary to understand the basics of human reproduction. **Fertilization** occurs when a male sperm merges with a female ovum (egg), creating a zygote, which contains genetic material (DNA) from both the father and the mother. If pregnancy is then established, the zygote will develop into an embryo, then a fetus, and ultimately a baby will be born.

Sperm are small cells that carry the father's genetic material. This genetic material is contained within the oval head of the sperm. Sperm are produced within the testicles, and proceed through a number of developmental stages in order to mature. This whole process of sperm production is called spermatogenesis. The sperm are mixed into a fluid called semen, which is discharged from the penis during a process called ejaculation. The whip-like tail of the sperm allows the sperm motility; that is, permits the sperm to essentially swim up the female reproductive tract, in search of the egg it will attempt to fertilize.

The ovum (or egg) is the **cell** that carries the mother's genetic material. These ova develop within the ovaries. Once a month, a single mature ovum is produced and leaves the ovary in a process called ovulation. This ovum enters the fallopian tube (a tube extending from the ovary to the uterus) where fertilization occurs.

If fertilization occurs, a zygote containing genetic material from both the mother and father results. This single cell will divide into multiple cells within the fallopian tube, and the resulting cluster of cells (called a blastocyst) will then move into the uterus. The uterine lining (endometrium) has been preparing itself to receive a pregnancy by growing thicker. If the blastocyst successfully reaches the inside of the uterus and attaches itself to the wall of the uterus, then implantation and pregnancy have been achieved.

Unlike most medical problems, infertility is an issue requiring the careful evaluation of two separate individuals, as well as an evaluation of their interactions with each other. In about 3-4% of couples, no cause for their infertility will be discovered. The main factors involved in causing infertility, listing from the most to the least common, include: (1) Male factors; (2) Peritoneal factors; (3) Uterine/tubal factors; (4) Ovulatory factors; and (5) Cervical factors.

Male factor infertility

Male factor infertility can be caused by a number of different characteristics of the sperm. To check for these characteristics, a semen analysis is carried out, during which a **sample** of semen is obtained and examined under the **microscope**. The four most basic characteristics evaluated are: (1) Sperm count or the number of sperm present in a semen sample. The normal number of sperm present in just one milliliter (ml) of semen is over 20 million. A man with only 5-20 million sperm per ml of semen is considered subfertile, a man with less than five million sperm per ml of semen is considered infertile. (2) Sperm motility. Better swimmers indicate a higher degree of fertility, as does longer duration of survival. Sperm are usually capable of fertilization for up to 48 hours after ejaculation. (3) Sperm morphology or the structure of the sperm. Not all sperm within a specimen of semen will be perfectly normal. Some may be developmentally immature forms of sperm, some may have abnormalities of the head or tail. A normal semen sample will contain no more than 25% abnormal forms of sperm. (4) **Volume** of a representative semen sample. The semen is made up of a number of different substances, and a decreased quantity of one of these substances could affect the ability of the sperm to successfully fertilize an ovum.

The semen sample may also be analyzed chemically to determine that components of semen other than sperm are present in the correct proportions. If all of the above factors do not seem to be the cause for male infertility, then another test is performed to evaluate the ability of the sperm to penetrate the outer coat of the ovum. This is done by observing whether sperm in a semen sample can penetrate the outer coat of a guinea pig ovum; fertilization can not, of course, occur, but this test is useful in predicting the ability of the patient's sperm to penetrate a human ovum.

Any number of issues can affect male fertility as evidenced by the semen analysis. Individuals can be born with testicles that have not descended properly from the abdominal cavity (where testicles develop originally) into the scrotal sac, or they can be born with only one testicle, instead of the normal two. Testicle size can be smaller than normal. Past infection (including mumps) can affect testicular function, as can a past injury. The presence of abnormally large **veins** (varicocele) in the testicles can increase testicular **temperature**, which decreases sperm count. A history of exposure to various toxins, drug use, excessive **alcohol** use, use of anabolic steroids, certain medications, diabetes, thyroid problems, or other endocrine disturbances can have direct effects on spermatogenesis. Problems with the male **anatomy** can cause sperm to be ejaculated not out of the penis, but into the bladder, and scarring from past infections can interfere with ejaculation.

Treatment of male factor infertility includes addressing known reversible factors first, for example discontinuing any medication known to have an effect on spermatogenesis or ejaculation, as well as decreasing alcohol intake and treating thyroid or other endocrine disease. Varicoceles can be treated surgically. Testosterone in low doses can improve sperm motility.

Some recent advances have greatly improved the chances for infertile men to conceive. Azoospermia (lack of sperm in the semen) may be overcome by mechanically removing sperm from the testicles either by surgical biopsy or needle aspiration (using a needle and syringe). The isolated sperm can then be used for *in vitro* fertilization. Another advance involves using a fine needle to inject a single sperm into the ovum. This procedure, called intracytoplasmic sperm injection (ICSI) is useful when sperm have difficulty fertilizing the ovum and when sperm have been obtained through mechanical means.

Other treatments of male factor infertility include collecting semen samples from multiple ejaculations, after which the semen is put through a process which allows the most motile sperm to be sorted out. These motile sperm are pooled together to create a concentrate which can be mechanically deposited directly into the female partner's uterus at a time that will coincide with ovulation. In cases where the male partner's sperm is proven to be absolutely unable to cause pregnancy in the female partner, and with the consent of both partners, donor sperm may be used for this process. These procedures (depositing the male partner's sperm or donor sperm by mechanical means into the female partner) are both forms of artificial insemination.

Female factor infertility

Peritoneal factors refer to any factors (other than those involving specifically the ovaries, fallopian tubes, or uterus) within the abdomen of the female partner that may be interfering with her fertility. Two such problems include pelvic adhesions and endometriosis.

Pelvic adhesions are thick, fibrous scars. These scars can be the result of past infections, particularly **sexually transmitted diseases** such as PID, or infections following abortions or prior births. Previous surgeries can also leave behind scarring. Complications from appendicitis and certain intestinal diseases can also result in adhesions in the pelvic area.

Endometriosis also results in pelvic adhesions. Endometriosis is the abnormal location of uterine **tissue** outside of the uterus. When uterine tissue is planted elsewhere in the pelvis, it still bleeds on a monthly basis with the start of the normal menstrual period. This leads to irritation within the pelvis around the site of this abnormal tissue and bleeding, and ultimately causes scarring.

Pelvic adhesions contribute to infertility primarily by obstructing the fallopian tubes. The ovum may be prevented from traveling down the fallopian tube from the ovary, and the sperm prevented from traveling up the fallopian tube from the uterus; or the blastocyst may be prevented from entering into the uterus where it needs to implant. Scarring can be diagnosed by examining the pelvic area with a scope, which can be inserted into the abdomen through a tiny incision made near the naval. This scoping technique is called laparoscopy.

Obstruction of the fallopian tubes can also be diagnosed by observing through x ray exam whether dye material can travel through the patient's fallopian tubes. Interestingly enough, this procedure has some actual treatment benefits for the patient, as a significant number of patients become pregnant following this x ray exam. It is thought that the dye material in some way helps clean out the tubes, decreasing any existing obstruction.

Pelvic adhesions can be treated using the same laparoscopy technique utilized in the **diagnosis** of the problem. For treatment, use of the laparoscope to visualize adhesions is combined with use of a **laser** to disintegrate those adhesions. Endometriosis can be treated with certain medications, but may also require **surgery** to repair any obstruction caused by adhesions.

Uterine factors contributing to infertility include tumors or abnormal growths within the uterus, chronic infection and **inflammation** of the uterus, abnormal structure of the uterus, and a variety of endocrine problems (problems with the secretion of certain **hormones**), which prevent the uterus from developing the thick lining necessary for implantation by a blastocyst.

Tubal factors are often the result of previous infections that have left scar tissue. This scar tissue blocks the tubes, preventing the ovum from being fertilized by the sperm. Scar tissue may also be present within the fallopian tubes due to the improper implantation of a previous pregnancy within the tube, instead of within the uterus. This is called an ectopic pregnancy. Ectopic pregnancies cause rupture of the tube, which is a medical emergency requiring surgery, and results in scarring within the affected tube.

X-ray studies utilizing dyes can help outline the structure of the uterus, revealing certain abnormalities. Ultrasound examination and hysteroscopy (in which a thin, wand-like camera is inserted through the cervix into the uterus) can further reveal abnormalities within the uterus. Biopsy (removing a tissue sample for microscopic examination) of the lining of the uterus (the endometrium) can help in the evaluation of endocrine problems affecting fertility.

Treatment of these uterine factors involves antibiotic treatment of any infectious cause, surgical removal of certain growths within the uterus, surgical reconstruction of the abnormally formed uterus, and medical treatment of any endocrine disorders discovered. Progesterone, for example, can be taken to improve the hospitality of the endometrium toward the arriving blastocyst. Very severe scarring of the fallopian tubes may require surgical reconstruction of all or part of the scarred tube.

Ovulatory factors are those factors that prevent the maturation and release of the ovum from the ovary with the usual monthly regularity. Ovulatory factors include a host of endocrine abnormalities, in which appropriate levels of the various hormones that influence ovulation are not produced. Numerous hormones produced by multiple **organ** systems interact to bring about normal ovulation. Therefore, ovulation difficulties can stem from problems with the ovaries, the adrenal **glands**, the pituitary gland, the hypothalamus, or the thyroid.

The first step in diagnosing ovulatory factors is to verify whether or not an ovum is being produced. Although the only certain proof of ovulation (short of an achieved pregnancy) is actual visualization of an ovum, certain procedures suggest that ovulation is or is not taking place.

The basal body temperature is the body temperature that occurs after a normal night's **sleep** and before any activity (including rising from bed) has been initiated. This temperature has normal variations over the course of the monthly ovulatory cycle, and when a woman carefully measures and records these temperatures, a chart can be drawn that suggests whether or not ovulation has occurred.

Another method for predicting ovulation involves measurement of a particular chemical that should appear in the urine just prior to ovulation. Endometrial biopsy will reveal different characteristics depending on the ovulatory status of the patient, as will examination of the mucus found in the cervix (the opening to the uterus). Also, pelvic ultrasound can visualize developing follicles (clusters of cells that encase a developing ovum) within the ovaries.

Treatment of ovulatory factors involves treatment of the specific organ system responsible for ovulatory failure (for example, thyroid medication must be given in the case of an underactive thyroid, a pituitary **tumor** may need removal, or the woman may need to cease excessive **exercise**, which can result in improper activity of the hypothalamus). If ovulation is still not occurring after these types of measures have been taken, certain drugs exist that can induce ovulation. These include Clomid, Pergonal, Metrodin, Fertinex, Follistim, and Gonal F. These drugs, however, may cause the ovulation of more than one ovum per cycle, which is responsible for the increase in multiple births (twins, triplets, etc.) noted since these drugs became available to treat infertility.

The cervix is the opening from the vagina into the uterus through which the sperm must pass. Mucus produced by the cervix helps to transport the sperm into the uterus. Injury to the cervix during a prior **birth**, surgery on the cervix due to a pre-cancerous or cancerous condition, or scarring of the cervix after infection, can all result in a smaller than normal cervical opening, making it difficult for the sperm to enter. Furthermore, any of the above conditions can also decrease the number of mucus-producing glands in the cervix, leading to a decrease in the quantity of cervical mucus. In other situations, the mucus produced is the wrong consistency (perhaps too thick) to allow sperm to travel through. Certain infections can also serve to make the cervical mucus environment unfavorable to the transport of sperm, or even directly toxic to the sperm themselves (causing sperm death). Some women produce antibodies (immune cells) that identify sperm as foreign invaders.

The qualities of the cervical mucus can be examined under a microscope to diagnose cervical factors as contributing to infertility. The interaction of a live sperm sample from the male partner and a sample of cervical mucus can also be examined.

Treatment of cervical factors includes **antibiotics** in the case of an infection, steroids to decrease production of anti-sperm antibodies, and artificial insemination techniques to completely bypass the cervical mucus.

Assisted reproduction comprises those techniques that perhaps receive the most publicity as infertility treatments. These include *in vitro* fertilization (IVF), gamete intrafallopian tube transfer (GIFT), and zygote intrafallopian tube transfer (ZIFT). All of these are used after other techniques to treat infertility have failed.

IVF involves the use of a drug to induce multiple ovum production, and retrieval of those ova either surgically or by ultrasound-guided needle aspiration through the vaginal wall. Meanwhile, multiple semen samples are obtained from the male partner, and a sperm concentrate is prepared. The ova and sperm are then cultured together in a laboratory, where hopefully several of the ova are fertilized. **Cell division** is allowed to take place up to either the pre-embryo or blastocyst state. While this takes place, the female may be given medication to prepare her uterus to receive an embryo. When necessary, a small opening is made in the outer shell (zona pellucida) of the pre-embryo or blastocyst by a process known as assisted hatching. Two or more pre-embryos or two blastocysts are transferred into the uterus, and the wait begins to see if any or all of them implant and result in an actual pregnancy.

The national average success rate of IVF is 27%, but some centers have higher pregnancy rates. Transferring blastocysts leads to a pregnancy rate of up to 50% or higher. Interestingly, the rate of **birth defects** resulting from IVF is lower than that resulting from unassisted pregnancies. Of course, because most IVF procedures place more than one embryo into the uterus, the chance for a multiple birth (twins or more) is greatly increased.

GIFT involves retrieval of both multiple ova and semen, and the mechanical placement of both within the

KEY TERMS

Assisted hatching—The process in which a small opening is made in the outer shell of the pre-embryo or blastocyst to increase the implantation rate.

Blastocyst—A cluster of cells representing multiple cell divisions after successful fertilization of an ovum by a sperm. This is the developmental form that must implant itself in the uterus to achieve pregnancy.

Cervix—The front portion, or neck, of the uterus.

Ejaculation—A spasmodic muscular contraction expelling semen from the penis.

Endometrium—The blood-rich interior lining of the uterus.

Fallopian tubes—In a woman's reproductive system, a pair of narrow tubes that carry the egg from the ovary to the uterus.

Ovary—The female organ in which eggs (ova) are stored and mature.

Ovum (plural=ova)—The reproductive cell of the female which contains genetic information and participates in fertilization. Also popularly called the egg.

Semen—The fluid which contains sperm which is ejaculated by the male.

Sperm—Substance secreted by the testes during sexual intercourse. Sperm includes spermatozoon, the mature male cell which is propelled by a tail and has the ability to fertilize the female egg.

Spermatogenesis—The process by which sperm develop to become mature sperm.

Zygote—The cell resulting from the fusion of male sperm and the female egg. Normally the zygote has double the chromosome number of either gamete, and gives rise to a new embryo.

---

fallopian tubes, where fertilization may occur. ZIFT involves the same retrieval of ova and semen, and fertilization and growth in the laboratory up to the zygote stage, at which point the zygotes are placed in the fallopian tubes. Both GIFT and ZIFT seem to have higher success rates than IVF.

Ova can now be frozen for later use, although greater success is obtained with fresh ova. However, storing ova may provide the opportunity for future pregnancy in women with premature ovarian failure or pelvic disease or those undergoing **cancer** treatment.

Any of these methods of assisted reproduction can utilize donor sperm and/or ova. There have even been cases in which the female partner's uterus is unable to support a pregnancy, so the embryo or zygote resulting from fertilization of the female partner's ovum with the male partner's sperm is transferred into another woman, where the pregnancy progresses to birth.

Chances at pregnancy can be improved when the pre-embryos are screened for **chromosomal abnormalities** and only the normal ones are transferred into the uterus. This method is useful for couples who are at an increased risk of producing embryos with chromosomal abnormalities, such as advanced maternal age or when one or both partners carry a fatal genetic disease.

Multiple ethical issues have presented themselves as a result of assisted reproduction. Some of these issues involve the use of donor sperm or ova, and surrogate motherhood. Other issues include what to do with frozen embryos, particularly when the couple has divorced.

A particularly difficult ethical problem has come about by virtue of the technique of transferring multiple embryos or zygotes into the female. When pregnancy occurs in which there are multiple developing fetuses, there is a greatly increased chance for pregnancy complications, preterm delivery, and life-long medical problems. Techniques allowing only one or two of the fetuses to continue developing may be employed.

See also Reproductive system.

Resources

Books

The Merck Manual of Diagnosis and Therapy. 17th ed, edited by Mark H. Beers and Robert Berkow. Whitehouse Station, NJ: Merck Research Laboratories, 1999.

Speroff, Leon. *Clinical Gynecologic Endocrinology and Infertility.* Baltimore: Lippincott Williams & Wilkins, 1999.

Periodicals

Tesarik, Jan, and Carmen Mendoza. "*In Vitro* Fertilization by Intracytoplasmic Sperm Injection." *BioEssays,* 21 (1999): 791-801.

Yoshida, Tracey M. "Infertility Update: Use of Assisted Reproductive Technology." *Journal of the American Pharmaceutical Association* 39 (1999): 65-72.

Rosalyn Carson-DeWitt
Belinda Rowland

Infinity

The term infinity conveys the mathematical concept of large without bound, and is given the symbol ∞. As children, we learn to count, and are pleased when first we count to 10, then 100, and then 1,000. By the time we reach 1,000, we may realize that counting to 2,000, or certainly 100,000, is not worth the effort. This is partly because we have better things to do, and partly because we realize no matter how high we count, it is always possible to count higher. At this point we are introduced to the infinite, and begin to realize what infinity is and is not.

Infinity is not the largest number. It is the term we use to convey the notion that there is no largest number. We say there is an infinite number of numbers.

There are aspects of the infinite that are not altogether intuitive, however. For example, at first glance there would seem to be half as many odd (or even) **integers** as there are integers all together. Yet it is certainly possible to continue counting by twos forever, just as it is possible to count by ones forever. In fact, we can count by tens, hundreds, or thousands, it does not matter. Once the counting has begun, it never ends.

What of fractions? It seems that just between **zero** and one there must be as many fractions as there are positive integers. This is easily seen by listing them, 1/1, 1/2, 1/3, 1/4, 1/5, 1/6, 1/7, 1/8,.... But there are multiples of these fractions as well, for instance, 2/8, 3/8, 4/8, 5/8, 6/8, 7/8, and 8/8. Of course many of these multiples are duplicates, 2/8 is the same as 1/4 and so on. It turns out, after all the duplicates are removed, that there is the same number of fractions as there are integers. Not at all an obvious result.

In addition to fractions, or rational numbers, there are irrational numbers, which cannot be expressed as the **ratio** of whole numbers. Instead, they are recognized by the fact that, when expressed in decimal form, the digits to the right of the decimal point never end, and never form a repeating sequence. Terminating decimals, such as 6.125, and repeating decimals, such as $1.33\overline{3}$ or $6.53\overline{4}$ (the bar over the last digits indicates that sequence is to be repeated indefinitely), are rational. Irrational numbers are interesting because they can never be written down. The instant one stops writing down digits to the right of the decimal point, the number becomes rational, though perhaps a good **approximation** to an **irrational number**.

It can be proved that there are infinitely more irrational numbers than there are rational numbers, in spite of the fact that every irrational number can be approximated by a **rational number**. Taken together, the rational and irrational numbers form the set of **real numbers**.

The word infinite is also used in reference to the very small, or infinitesimal. Consider dividing a line segment in half, then dividing each half, and so on, infinitely many times. This procedure would results in an infinite number of infinitely short line segments. Of course it is not physically possible to carry out such a process; but it is possible to imagine reaching a point beyond which it is not worth the effort to proceed. We understand that the line segments will never have exactly zero length, but after a while no one fully understands what it means to be any shorter. In the language of **mathematics**, we have approached the limit.

Beginning with the ancient Greeks, and continuing to the turn of the twentieth century, mathematicians either avoided the infinite, or made use of the intuitive concepts of infinitely large or infinitely small. Not until the German mathematician, Georg Cantor (1845-1918), rigorously defined the transfinite numbers did the notion of infinity finally seem fully understood. Cantor defined the transfinite numbers in terms of the number of elements in an infinite set. The **natural numbers** have $\aleph_0$ elements (the first transfinite number). The real numbers have $\aleph_1$ elements (the second transfinite number). Then, any two sets whose elements can be placed in 1-1 correspondence, have the same number of elements. Following this procedure, Cantor showed that the set of integers, the set of odd (or even) integers, and the set of rational numbers all have $\aleph_0$ elements; and the set of irrational numbers has $\aleph_1$ elements. He was never able, however, to show that no set of an intermediate size between $\aleph_0$ and $\aleph_1$ exists, and this remains unproved today.

Resources

Books

Buxton, Laurie. *Mathematics for Everyone.* New York: Schocken Books, 1985.

Dauben, Joseph Warren. *Georg Cantor, His Mathematics and Philosophy of the Infinite.* Cambridge: Harvard University Press, 1979.

KEY TERMS

. . . .

Counting numbers—As the name suggests, the counting numbers are 1,2,3..., also called the natural numbers. The whole numbers are the counting numbers plus zero.

Transfinite numbers—Transfinite numbers were invented by Georg Cantor as a means of expressing the relative size of infinite sets.

Paulos, John Allen. *Beyond Numeracy, Ruminations of a Numbers Man* New York: Knopf, 1991.

Periodicals

Moore, A. W. "A Brief History of Infinity." *Scientific American* 272, no. 4 (1995): 112-16.

J. R. Maddocks

Inflammation

Inflammation is a localized, defensive response of the body to injury, usually characterized by **pain**, redness, **heat**, swelling, and, depending on the extent of trauma, loss of function. The process of inflammation, called the inflammatory response, is a series of events, or stages, that the body performs to attain **homeostasis** (the body's effort to maintain stability). The body's inflammatory response mechanism serves to confine, weaken, destroy, and remove **bacteria**, toxins, and foreign material at the site of trauma or injury. As a result, the spread of invading substances is halted, and the injured area is prepared for regeneration or repair. Inflammation is a nonspecific defense mechanism; the body's physiological response to a superficial cut is much the same as with a **burn** or a bacterial **infection**. The inflammatory response protects the body against a variety of invading **pathogens** and foreign **matter**, and should not be confused with an immune response, which reacts to specific invading agents. Inflammation is described as acute or chronic, depending on how long it lasts.

Within minutes after the body's physical barriers, the skin and mucous membranes, are injured or traumatized (for example, by bacteria and other **microorganisms**, extreme heat or cold, and chemicals), the arterioles and **capillaries** dilate, allowing more **blood** to flow to the injured area. When the blood vessels dilate, they become more permeable, allowing **plasma** and circulating defensive substances such as antibodies, phagocytes (cells that ingest microbes and foreign substances), and fibrinogen (blood-clotting chemical) to pass through the vessel wall to the site of the injury. The blood flow to the area decreases and the circulating phagocytes attach to and digest the invading pathogens. Unless the body's defense system is compromised by a preexisting **disease** or a weakened condition, healing takes place. Treatment of inflammation depends on the cause. Anti-inflammatory drugs such as aspirin, acetaminophen, ibuprofen, or a group of drugs known as NSAIDS (non-steroidal anti-inflammatory drugs) are sometimes taken to counteract some of the symptoms of inflammation.

See also Anti-inflammatory agents.

Inflection point

In **mathematics**, an inflection point is a point on a **curve** at which the curve changes from being concave upward to being concave downward, or vice versa. A concave upward curve can be thought of as one that would hold **water**, while a concave downward curve is one that would not. An important qualification is that the curve must have a unique tangent line at the point of inflection. This means that the curve must change smoothly from concave upward to concave downward, not abruptly. As a practical example of an inflection point consider an "s-curve" on the highway. Precisely at the inflection point the driver changes from steering left to steering right, or vice versa as the case may be.

In **calculus**, an inflection point is characterized by a change in the sign of the second **derivative**. Such a sign change occurs when the second derivative passes through **zero** or becomes infinite.

Influenza

Influenza ("the flu") is a **disease** caused by the influenza **virus**. The disease is easily spread from person to person, typically by inhaling virus that has been expelled into the air by coughing or sneezing. The virus can also be spread by **touch**. For example, if someone touches a doorknob that has influenza viruses clinging to it and then touches their mouth, the virus can pass into their body and cause influenza.

The influenza virus infects the nose, throat, and lungs of people. In contrast to the **common cold**, which is caused by a different virus, the symptoms of the flu develop suddenly. These symptoms include fever, headache and body aches, tiredness, cough, sore throat, and stuffy nose.

Most people who contract the flu recover completely in a few weeks. However, in some people influenza can progress to **pneumonia**, which can be life threatening. Recovery from influenza does not protect someone from future bouts of the disease. This is because the influenza virus readily changes the expressions of its genetic material (i.e., it mutates readily). Thus, the influenza virus that the body's **immune system** responds to one season may be different from the virus that infects the body some months later.

Influenza is a common illness. For example, every year approximately 25 to 50 million Americans (about 10–20% of the population of the United States) contract influenza. Of these, about 20,000 people die of the **infection**, and 114,000 require hospitalization.

There are three types of influenza virus. All three are in the viral group called Orthomyxovirus. The three viral types are called influenza A, B, and C. Influenza A and B cause large numbers of cases of the flu almost every winter, when people are confined indoors and spread of the virus is easier.

Influenza A is further divided into two subtypes called hemagglutinin (H) and neuraminidase (N). H and N are two **proteins** that are found on the surface of the filament-like virus particles. They both protrude from the surface and appear as spikes when viruses are examined under high magnification. The protein spikes function to help the virus invade host cells.

Influenza viruses are resident in animals and **birds** including **pigs**, **horses**, **seals**, whales, a variety of wild birds, and **ducks**. The virus can spread from this reservoir to humans.

Influenza has been part of mankind for millennia. In the twentieth century, there were a number of large outbreaks. For example, in 1918–1919 the "Spanish flu" killed more than 500,000 people in the United States and up to 50 million people around the world. The influenza virus that caused this outbreak was very deadly. Concerns have been raised that the same virus could establish another **epidemic**.

In 1957–1958 the "Asian flu" caused 70,000 deaths in the United States. The same virus remains in circulation today. In 1968–1969, an outbreak of what was dubbed the "Hong Kong flu" killed approximately 34,000 Americans. In 1976, a small outbreak affected soldiers at a military base in Fort Dix, New Jersey. Experts predicted that the influenza, which was known as the "Swine flu," could spread throughout the United States. The subsequent public concern bordered on hysteria, and prompted a vaccination campaign in which 40 million Americans were vaccinated. The outbreak did not materialize.

Vaccination is not a guarantee that all types of influenza will be prevented. Rather, influenza is typically dealt with after it appears. Flu is treated with rest and fluids. Maintaining a high fluid intake is important, because fluids increase the flow of respiratory secretions that may prevent pneumonia. Antiviral medications such as amantadine and rimantadine may be prescribed for people who have initial symptoms of the flu and who are at high risk for complications. This medication does not prevent the illness, but reduces its duration and severity.

A flu **vaccine** is available that is formulated each year against the current type and strain of flu virus. The

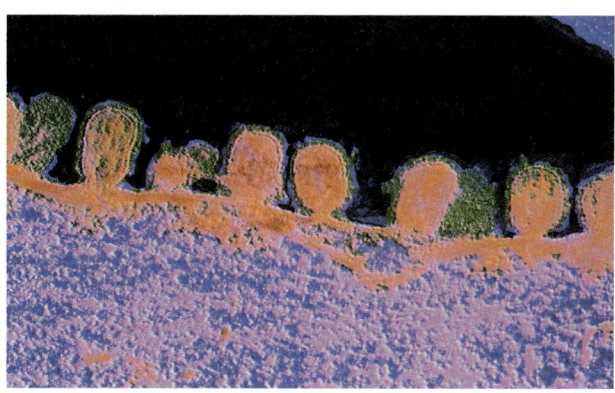

A transmission electron micrograph (TEM) of influenza viruses budding from the surface of an infected cell. *CNRI/Science Photo Library, National Audubon Society Collection/Photo Researchers, Inc. Reproduced by permission.*

virus is grown in chicken eggs, extracted, and then rendered noninfective by chemicals. The vaccine is also "updated" to the current viral strain by the addition of proteins that match the protein composition of the influenza virus type that is currently circulating in a population. The vaccine would be most effective in reducing attack rates if it was effective in preventing influenza in schoolchildren; however, in vaccine trials the vaccine has not been shown to be effective in flu prevention in this age group. In certain populations, particularly the elderly, the vaccine is effective in preventing serious complications of influenza and thus, lowers mortality.

Vaccine research is ongoing. One of the more exciting advances in flu vaccines involves research studies examining an influenza vaccine mist, which is sprayed into the nose. This is predicted to be an excellent route of administration, which will confer even stronger immunity against influenza. Because it uses a live virus, it encourages a strong immune response. Furthermore, it is thought to be a more acceptable immunization route for schoolchildren, who are an important reservoir of the influenza virus.

See also Aerosols; Cold, common.

Resources

Books

Kolata, G.B. *Flu: The Story of the Great Influenza Pandemic of 1918 and the Search for the Virus That Caused It.* New York: Farrar Straus & Giroux, 1999.

Potter, C.W. *Influenza* London: Elsevier Health Sciences, 2002.

Organizations

Centers for Disease Control and Prevention, 1600 Clinton Road, Atlanta, GA 30333 (404) 639–3311. July 29, 2002 [cited November 12, 2002] <http://www.cdc.gov/ncidod/diseases/flu/fluinfo.htm.

Brian Hoyle

Infrared astronomy

Throughout most of history astronomers were confined to using optical **light**, the light we can detect with our eyes. The advent of electronic detectors has, in the past few decades, opened up new vistas to astronomers, allowing them to utilize the entire **electromagnetic spectrum**. Infrared astronomers use traditional optical telescopes equipped with special detectors that can detect infrared light. Earth's atmosphere is, for the most part, only mildly transparent to infrared light, so infrared astronomers work from high, dry mountain tops, airplanes, high altitude balloons, or **space**. The infrared spectral window allows astronomers to probe dusty regions of the universe that obscure optical light.

Electromagnetic spectrum

Light is a form of electromagnetic **radiation**. The electromagnetic waves that comprise electromagnetic radiation consist of **oscillations** in electric and magnetic fields, just as **water** waves consist of oscillations of the water in the **ocean**.

Certain properties describe all types of waves. One is the wavelength, which is the distance between two adjacent peaks in the wave. The **frequency** is the number of peaks that move past a stationary observer in one second. In the case of water waves at the beach, the frequency would be the number of incoming waves that hit a person in one second, and the wavelength would be the distance between two waves. A higher frequency corresponds to a shorter wavelength and vice versa.

The different colors of light that our eyes can detect correspond to different wavelengths—or frequencies—of light. Red light has a longer wavelength than violet light. Orange, yellow, green, and blue are in between. Infrared light, ultraviolet light, **radio waves**, microwaves, and gamma rays are all forms of electromagnetic radiation, but they differ in wavelength and frequency.

Infrared light has slightly longer wavelengths than red light. Our eyes can not detect infrared light, but we can feel it as **heat**. Infrared **astronomy** uses the wavelength range from about 1 micrometer to a few hundred micrometers. Wavelengths near 1,000 micrometers (1 millimeter) are considered **radio** waves and studied by radio astronomers using different techniques than infrared astronomers.

Infrared astronomers divide the infrared **spectrum** into near-, mid-, and far-infrared. The exact boundaries between these regions are indistinguishable, but near-infrared is generally considered to be from one to five micrometers. Wavelengths of 5-20 micrometers are considered mid-infrared. Wavelengths longer than about 20 mircrometers are far-infrared.

Utilizing infrared astronomy

Special infrared detectors must be used to see the infrared universe. These detectors can be mounted on traditional optical telescopes either on the ground or above the atmosphere. The first infrared detector was a **thermometer** used by William Herschel in 1800. He passed sunlight through a **prism** and placed the thermometer just beyond the red light to detect the heat from the infrared light. To detect the heat from distant stars and galaxies, modern infrared detectors must be considerably more sensitive. The infancy of infrared astronomy began with the advent of these detectors in the 1960s.

Modern infrared detectors use exotic combinations of semiconductors that are cooled to either liquid **nitrogen** or liquid helium temperatures. Photovoltaic detectors utilize the **photoelectric effect**, the same principle as the solar cell in a solar powered **calculator**. Light strikes certain materials and kicks the electrons away from the **atoms** to produce an **electric current** as the electrons move. Because infrared light has less **energy** than ordinary optical light, photovoltaic infrared detectors must be made from materials that require little energy to **force** the **electron** from the atom.

Photoresistive thermal detectors work by measuring minute changes in the **electrical resistance** of the detector. The electrical resistance of a wire generally depends on its **temperature**. Infrared radiation striking a photoresistive detector will raise its temperature and therefore change its electrical resistance by a minute amount. A mixture of gallium and germanium is often used. These detectors must be cooled with liquid helium to get the extreme sensitivity required by infrared astronomers.

Early infrared detectors featured a single channel. Accordingly, they could measure the brightness of a single region of the sky seen by the detector, but could not produce pictures. Early infrared images or maps were quite tedious to make. Images were created by measuring the brightness of a single region of the sky, moving the **telescope** a bit, measuring the brightest of a second region, and so on.

In the 1980s infrared arrays revolutionized infrared imaging. Arrays are essentially two dimensional grids of very small, closely spaced individual detectors, or pixels. Infrared arrays as large as 256×256 pixels are now available, allowing astronomers to create infrared images in a reasonable amount of time.

In addition to images, astronomers can measure the brightness of an infrared source at various infrared wavelengths. Detectors record a range of wavelengths, so a filter must be used to select a specific wavelength. This measurement of brightness is called photometry. Both optical and infrared astronomers break light up into its component colors, its spectrum. This can be done on a smaller scale by passing light through a prism. This process, **spectroscopy**, is useful for finding the compositions, motions, physical conditions, and many other properties of stars and other celestial objects. When light is polarized, the electromagnetic oscillations line up. Infrared polarimetry, measuring the amount of polarization, is useful in deducing optical properties of the dust grains in dusty infrared sources.

Ground-based infrared astronomy

Infrared light is heavily absorbed by both **carbon dioxide** and water vapor, major components of Earth's atmosphere. Accordingly, the atmosphere is opaque to many infrared wavelengths. There are a few specific wavelength bands between one and five micrometers, around 10 micrometers, and sometimes near 20 micrometers at which the atmosphere is partially transparent. These bands make up the standard ground based infrared bands. Still, astronomers must build infrared observatories at very dry, high-altitude sites to get above as much atmosphere as possible. One of the best infrared sites in the world is the 14,000-ft (4,200-m) summit of Mauna Kea in Hawaii. On a clear night half a dozen large telescopes may probe the infrared sky, although some of the telescopes are used for optical astronomy. The high altitude at Mauna Kea makes observation at its summit very rigorous.

There are special difficulties to infrared astronomy, especially from the ground. The heat radiation from the telescope, telescope building, and atmosphere are all very bright in the infrared. They combine into an infrared background that is at least a million times brighter than strong astronomical infrared sources. To account for this strong background astronomers rapidly oscillate the telescope field of view from the **star** to a region of sky nearby. Taking the difference of the two intensities allows astronomers to subtract the background.

Airborne and space infrared astronomy

To conduct experiments in infrared astronomy at wavelengths other than those observable from the ground, astronomers must place their telescopes above the atmosphere. Options include mounting telescopes on high-altitude balloons, airplanes, rockets, or satellites. High-altitude balloons are less expensive than the other options, but astronomers cannot ride with the telescope and have little control over the flight path of the **balloon**. Today **aircraft** are more frequently used. Since 1974,

NASA has operated the Kuiper Airborne Observatory (KAO), which is a 36 in (91 cm) infrared telescope in a military cargo plane. It flies at high altitudes in a controlled path with the astronomers along to operate the telescope. Astronomers can make observations at far-infrared wavelengths with more control than from a balloon. Beginning in 2001, NASA is replacing the KAO with the Stratospheric Observatory for Infrared Astronomy (SOFIA), a 100 in (254 cm) telescope that will be flown on a 747.

To record long-term images from space, astronomers must place infrared telescopes on orbiting satellites. Such experiments are quite expensive, but allow astronomers to record a large number of observations. Infrared observatories in space have a more limited lifetime than other space observatories because they run out of liquid helium. Space is cold, but not cold enough for infrared detectors, so they must still be cooled with liquid helium, which evaporates after a year or two. Astronomers must carefully plan their observations to get the most out of the limited lifetime.

In the early 1980s the Infrared Astronomical Satellite (IRAS) surveyed the entire sky at four infrared wavelengths not accessible from the ground (12, 25, 60, and 100 micrometers). The helium ran out in 1983 after a successful mission. Astronomers are still **mining** the vast amounts of data accumulated from that experiment. The **satellite** charted the positions of 15,000 galaxies, allowing a sky survey team to produce a three-dimensional **map** that covers a **sphere** with a radius of 700 million lightyears. Of particular interest to astronomers is the presence of massive **superclusters**, consisting of formed of galactic clusters containing dozens to thousands of galaxies like our own. Between these superclusters lie vast voids that are nearly galaxy-free, provoking great interest from scientists.

In 1995, the European Space Agency launched the Infrared Space Observatory (ISO), an astronomical satellite that operated at wavelengths from 2.5 to 240 micrometers. ISO allowed astronomers to study **comet Hale-Bopp** in detail. The satellite discovered protostars, planet-forming nebula around dying stars, and water throughout the universe, including in star-forming regions and in the atmospheres of the gas giants like **Saturn** and **Uranus**. The telescope was live until 1998, when it ran out of liquid helium.

Future infrared satellites planned include the NASA's Space Infrared Telescope Facility (SIRTF), slated for launch in late 2001.

Infrared view

Infrared light penetrates dust much more easily than

optical light. For this reason infrared astronomy is most useful for **learning** about dusty regions of the universe.

One example is star-forming regions. A star forms from a collapsing cloud of gas and dust. Forming and newly formed stars are still enshrouded by a cocoon of dust that blocks optical light. Infrared astronomers can more easily probe these stellar nurseries than optical astronomers can. The view of the center of our **galaxy** is also blocked by large amounts of interstellar dust. The galactic center is more easily seen by infrared than by optical astronomers.

Many molecules emit primarily in the infrared and radio regions of the spectrum. One example is the **hydrogen molecule** (H_2) which emits in the infrared. Infrared astronomers can study the distribution of these different kinds of molecules to learn about the processes forming molecules in interstellar space and the **clouds** in which these molecules form.

In 1998, using data from the Cosmic Background Explorer (COBE), astronomers discovered a background infrared glow across the sky. Radiated by dust that absorbed heat from all the stars that have ever existed, the background glow puts a limit on the total amount of energy released by all the stars in the universe.

Astronomers began with data acquired by COBE, then modeled and subtracted the infrared glow from foreground objects in our **solar system**, our galaxy's stars, and vast clouds of cold dust between the stars of our **Milky Way**. What remained was a smooth background of residual infrared light in the 240 and 140 micrometer wavelength bands in "windows" near the north and south poles of the Milky Way, which provide a relatively clear view across billions of light years.

The above examples are just a few of the observations made by infrared astronomers. In the past few decades, the new vistas opened in the infrared and other spectral regions have revolutionized astronomy.

See also Stellar evolution.

Resources

Books

Bacon, Dennis Henry, and Percy Seymour. *A Mechanical History of the Universe.* London: Philip Wilson Publishing, Ltd., 2003.

Smolin, Lee. *The Life of the Cosmos.* Oxford: Oxford University Press, 1999.

Periodicals

Gatley, Ian. "An Infrared View of our Universe." *Astronomy* (April 1994): 40-43.

Stephens, Sally. "Telescopes That Fly." *Astronomy* (November 1994): 46-53.

Paul A. Heckert

Infrared radiation *see* **Electromagnetic spectrum**

Inherited disorders

Heredity plays a part in almost all diseases. Recent advances in **gene** research have allowed a steadily increasing number of specific genes and genetic factors to be linked to a wide variety of medical complaints. There are currently approximately 6,000 known genetic diseases. Those that result from simple mutations of single genes are often referred to as hereditary diseases, and they exhibit distinctive patterns of inheritance in families.

Inherited diseases result primarily or exclusively from genetic mutations or genetic imbalance passed on from parent to child at conception. These include Mendelian genetic conditions as well as **chromosomal abnormalities**. A third group of disorders exists wherein both the environment and genetic factors interact to produce—or influence the course of—a **disease**. These conditions are often referred to as having multifactorial or complex inheritance patterns.

Autosomal dominant diseases

Normally there are two working copies of every gene in each **individual**. In the case of a dominant genetic disease, one copy of the gene is altered by **mutation** and causes the disease even though the other gene copy is normal. In autosomal dominant genetic diseases, a parent who shows the trait will pass the mutation on to half of his/her children with an equal chance for sons and daughters to be affected. Children who do not have the trait will generally not pass the disease on to their children. This is sometimes referred to as vertical transmission because it can be observed in each generation, usually without skipping a generation. Examples of autosomal dominant diseases include achondroplasia (a form of dwarfism), neurofibromatosis, and **Huntington disease**.

Autosomal recessive diseases

In autosomal recessive diseases, both parents must be carriers (i.e., they are clinically normal but have one mutation of a particular gene), and both must pass the mutation to a child in order for that child to be affected. This inheritance pattern is distinctive in that the parents and other relatives of the person with the disease appear to be completely normal, while 25% of their brothers and sisters will share the same disease. This is sometimes called horizontal transmission because there is no expression seen in previous generations by the ancestors and relatives who carry the mutation. Rather, the mutation travels unobserved (silently) within the family and is expressed by siblings in a single generation. Examples of autosomal recessive disease include **sickle cell anemia**, **cystic fibrosis**, **Tay-Sachs disease**, and **phenylketonuria**.

Sex-linked diseases

When diseases can be attributed to genes on a sex **chromosome**, either the X or the Y, they are characterized as sex-linked diseases. Human males carry one X and one Y chromosome, and human females carry two X chromosomes.

For example, X-linked recessive diseases are caused by genes on the X chromosome. Because males have only one X chromosome, they tend to express all mutations on the X chromosome they inherit from their mother. Daughters receive an X chromosome from each parent, and they, therefore, have a second copy of each gene that usually compensates for any recessive mutations they might inherit. For this reason, the great majority of patients with X-linked recessive diseases (XLR diseases) are male. The inheritance pattern is characterized by clusters of affected males who are related through apparently healthy female relatives. Examples of XLR diseases include **hemophilia** (types A and B) and Duchenne muscular dystrophy.

Polygenic disorders

When there is interaction between genetic and nongenetic factors, resulting diseases are termed multifactorial, or polygenic disorders. The inheritance patterns can be quite complex. Most chronic illnesses in humans are multifactorial hereditary disorders. Examples include **heart** disease, diabetes, **stroke**, **hypertension**, **cancer**, and most forms of mental illness.

KEY TERMS

· ·

Autosome or autosomal chromosome—Chromosomes other than sex chromosomes. In humans, all chromosomes except the X and Y sex chromosomes.

Chromosomes—Long strands of DNA complexed with proteins, which contain the genetic information. At the time of conception, an extra, missing, or damaged copy of a chromosome or even a part of a chromosome disrupts normal development.

Sex chromosomes—In humans the X and Y chromosomes are termed sex chromosomes. Normal males carry one X and one Y chromosome, normal females carry two X chromosomes in their somatic cells. Somatic cells are all cells other than sex or germ cells (e.g., spermatozoa or ova).

Chromosome abnormalities and disease

Some inherited diseases are attributed to damaged or improperly distributed chromosomes and are termed chromosomal diseases. Chromosomes are long strands of DNA complexed with **proteins** and RNA that condense and allow for equal distribution of the genes when cells divide. Each chromosome contains hundreds or thousands of genes, and every **cell** needs to have two copies of each chromosome in order to maintain genetic balance. At the time of conception, an extra copy or missing copy of a chromosome or even a part of a chromosome disrupts normal development. Most chromosomal abnormalities result from simple accidents of chromosome segregation and, as such, they tend not to recur in families. One example of genetic disorder that results from chromosomal imbalance is **Down syndrome**. This condition is caused by the presence of an extra copy of chromosome 21.

See also Birth defects; Embryo and embryonic development; Genetic engineering; Genetics.

Resources

Books

Jorde, L.B., J.C. Carey, M.J. Bamshad, and R.L. White. *Medical Genetics*. 2nd ed. St. Louis: Mosby-Year Book, Inc., 2000.

Thompson, M.W., R.R. McInnes, and H.F. Willard. *Thompson & Thompson Genetics in Medicine*. Philadelphia: W. B. Saunders Company, 1996.

Robert G. Best

Inoculation *see* **Vaccine**

Inorganic compound *see*
Compound, chemical

Insecticides

Introduction

An insecticide is a substance used by humans to gain some advantage in the struggle with various **insects** that are considered "pests." In the sense used here, a pest insect is considered undesirable, from the human perspective, because: (a) it is a vector that transmits disease-causing **pathogens** to humans (such as those causing **malaria** or **yellow fever**), or other diseases to **livestock** or crop plants; or (b) it causes a loss of the productivity or economic value of crop plants, domestic animals, or stored foodstuffs. The abundance and effects of almost all insect **pests** can be managed through the judicious use of insecticides.

However, the benefits of insecticide use are partly offset by important damages that may result. There are numerous cases of people being poisoned by accidental exposures to toxic insecticides. More commonly, ecological damage may be caused by the use of insecticides, sometimes resulting in the deaths of large numbers of **wildlife**.

Humans have been using insecticides for thousands of years. The Egyptians used unspecified chemicals to combat **fleas** in their homes about 3,500 years ago, and arsenic has been used as an insecticide in China for at least 2,900 years. Today of course, insecticide use is much more prevalent. During the 1990s, more than 300 insecticides were available, in hundreds of different formulations and commercial products (which may involve similar formulations manufactured by different companies).

Almost all insecticides are chemicals. Some are natural biochemicals extracted from plants, while others are inorganic chemicals based on toxic metals or compounds of arsenic. However, most modern insecticides are organic chemicals that have been synthesized by chemists. The costs of developing a new insecticide and testing it for its usefulness, **toxicology**, and environmental effects are huge, equivalent to at least $20-30 million. However, if an insecticide effective against an important pest is discovered, the profits are also potentially huge.

Kinds of insecticides

Insecticides are an extremely diverse group of chemicals, plus additional formulations based on living

microorganisms. The most important groups of insecticides are described below.

- Inorganic insecticides are compounds containing arsenic, **copper**, **lead**, or mercury. They are highly persistent in terrestrial environments, being slowly dispersed by **leaching** and **erosion** by **wind** and **water**. Inorganic insecticides are used much less than in the past, having been widely replaced by synthetic organics. Examples of insecticides include Paris green (a mixture of copper compounds), lead arsenate, and **calcium** arsenate.

- Natural organic insecticides are extracted from plants. They include **nicotine** extracted from tobacco (usually applied as nicotine sulphate), pyrethrum extracted from daisy-like plants, and rotenone from several tropical shrubs.

- **Chlorinated hydrocarbons** (or organochlorines) are synthetic insecticides, including DDT and its relatives DDD and methoxychlor, lindane, and cyclodienes such as **chlordane**, heptachlor, aldrin, and dieldrin. Residues of organochlorines are quite persistent in the environment, having a **half-life** of about 10 years in **soil**. They are virtually insoluble in water, but are highly soluble in fats and lipids. Their persistence and strongly lipophilic nature causes organochlorines to bio-concentrate and to further food-web magnify in high concentrations in **species** at the top of food webs.

- Organophosphate insecticides include fenitrothion, malathion, parathion, and phosphamidon. These are not very persistent in the environment, but most are extremely toxic to **arthropods** and also to non-target **fish**, **birds**, and **mammals**.

- Carbamate insecticides include aldicarb, aminocarb, carbaryl, and carbofuran. They have a moderate persistence in the environment, but are highly toxic to arthropods, and in some cases to **vertebrates**.

- Synthetic pyrethroids are analogues of natural pyrethrum, and include cypermethrin, deltamethrin, permethrin, synthetic pyrethrum and pyrethrins, and tetramethrin. They are highly toxic to **invertebrates** and fish, but are of **variable** toxicity to mammals and of low toxicity to birds.

- More minor groups of synthetic organic insecticides include the formamidines (e.g., amitraz, formetanate) and dinitrophenols (e.g., binapacryl, dinocap).

- Biological insecticides are formulations of microbes that are pathogenic to specific pests, and consequently have a relatively narrow **spectrum** of activity in ecosystems. An example is insecticides based on the bacterium *Bacillus thuringiensis* (or *B.t.*). There are also insecticides based on nuclear polyhedrosis **virus** (NPV) and insect hormones.

Benefits of insecticide use

Humans have attained important benefits from many uses of insecticides, including: (1) increased yields of **crops** because of protection from defoliation and diseases; (2) prevention of much spoilage of stored foods; and (3) prevention of certain diseases, which conserves health and has saved the lives of millions of people and domestic animals. Pests destroy an estimated 37% of the potential yield of **plant** crops in **North America**. Some of this damage can be reduced by the use of insecticides. In addition, insecticide spraying is one of the crucial tools used to reduce the abundance of **mosquitoes** and other insects that carry certain diseases (such as malaria) to humans. The use of insecticides to reduce the populations of these vectors has resulted in hundreds of millions of people being spared the deadly or debilitating effects of various diseases.

This is not to say that more insecticide use would yield even greater benefits. In fact, it has been argued that pesticide use in North America could be decreased by one-half without causing much of a decrease in crop yields, while achieving important environmental benefits through fewer ecological damages. In fact, three European countries (Sweden, Denmark, and the Netherlands) passed legislation in the 1990s requiring at least a 50% reduction in agricultural pesticide use by the year 2000, and similar actions may eventually be adopted in North America.

Because of the substantial benefits of many uses of **pesticides**, their use has increased enormously since the 1950s. For example, pesticide usage increased by 10-fold in North America between 1945 and 1989, although it leveled off during the 1990s. Pesticide usage (including insecticides) is now a firmly integrated component of the technological systems used in modern agriculture, **forestry**, **horticulture**, and public-health management in most parts of the world.

Damages caused by insecticide use

The considerable benefits of many uses of insecticides are partially offset by damages caused to ecosystems and sometimes to human health. Each year about one million people are poisoned by pesticides (mostly by insecticides), including 20,000 fatalities. Although developing countries only account for about 20% of global pesticide use, they sustain about half of the poisonings. This is because highly toxic insecticides are used in many developing countries, but with poor enforcement of regulations, illiteracy, and inadequate use of protective equipment and clothing. The most spectacular case of pesticide-related poisoning occurred in 1984 at Bhopal, India. About 2,800 people were killed and 20,000 seriously poisoned when a factory accidentally released 44 tons (40 tonnes) of vapors of methyl isocyanate to the atmosphere. (Methyl isocyanate is a precursor chemical used to manufacture carbamate insecticides.)

In addition, many insecticide applications cause ecological damage by killing non-target organisms (that is, organisms that are not pests). These damages are particularly important when broad-spectrum insecticides (i.e., that are not toxic only to the pest) are sprayed over a large area, such as an agricultural field or a stand of forest. Broadcast sprays of this sort expose many non-target organisms to the insecticide and cause unintended but unavoidable mortality. For instance, broadcast insecticide spraying causes non-target mortality to numerous arthropods other than the pest species, and birds, mammals, and other creatures may also be poisoned. The non-target mortality may include predators and competitors of the pest species, which may cause secondary damage by releasing the pest from some of its ecological controls.

Some of the best-known damage caused by insecticides involves DDT and related organochlorines, such as DDD, dieldrin, aldrin, and others. These chemicals were once widely used in North America and other industrialized countries, but their use was banned in the early 1970s. DDT was first synthesized in 1874, and its insecticidal properties were discovered in 1939. The first successful uses of DDT were during the Second World War, in programs to control body **lice**, mosquitoes, and other vectors of human diseases. DDT was quickly recognized as an extremely effective insecticide, and immediately after the war it was widely used in agriculture, forestry, and spray programs against malaria. The manufacturing and use of DDT peaked in 1970, when 385 million lb (175 million kg) were produced globally. At about that time, however, developed countries began to ban most uses of DDT. This action was taken because of ecological damages that were being caused by its use, including the **contamination** of humans and their agricultural food web, and the possibility that this was causing human diseases. However, the use of DDT has continued in less-developed countries, especially in the tropics, and mostly in programs against mosquito vectors of diseases.

Two physical-chemical properties of DDT and other organochlorines have an important influence on their ecological damages: their persistence and high **solubility** in fats. Chlorinated hydrocarbons are highly persistent in the environment because they are not easily degraded by microorganisms or physical agents such as sunlight or **heat**. DDT has a typical half-life in soil of about three years. In addition, DDT and related organochlorines are extremely insoluble in water, so they cannot be "diluted" into this abundant solvent. However, these chemicals are highly soluble in fats or lipids (i.e., they are lipophilic), which mostly occur in organisms. Consequently, DDT and related organochlorines have a powerful affinity for organisms, and therefore bio-concentrate into organisms in strong preference to the non-living environment. Moreover, organisms are efficient at assimilating any organochlorines present in their food. As a result, predators at the top of the food web develop the highest residues of organochlorines, particularly in their fatty tissues (this is known as food-web magnification). Both bio-concentration and food-web magnification tend to be progressive with age, that is, the oldest individuals in a population are most contaminated. Although organochlorine residues are ubiquitous in the **biosphere**, much higher concentrations typically occur in animals that live close to areas where these chemicals have been used, such as North America.

Intense exposures to DDT and other organochlorines cause important ecological damages, including poisonings of birds. In some cases, bird kills were caused directly by the spraying of DDT in urban areas during the 1950s and 1960s to kill the beetle vectors of Dutch **elm** disease. So much bird mortality occurred in sprayed neighborhoods that there was a marked reduction in bird song—hence the title of Rachael Carson's (1962) book: *Silent Spring,* which is often considered a harbinger of the modern environmental movement in North America.

In addition to the direct toxicity of chlorinated hydrocarbons, more insidious damage was caused to birds and other wildlife over large regions. Mortality to many species was caused by longer-term, chronic toxicity, often occurring well away from sprayed areas. It took years of population monitoring and ecotoxicological research before organochlorines were identified as the causes of these damages. In fact, the chronic poisoning of birds and other wildlife can be considered an unanticipated "surprise" that occurred because scientists (and society) had not had experience with the longer-term effects of persistent, bio-accumulating organochlorines.

Species of raptorial birds were among the most prominent victims of organochlorine insecticides. These birds are vulnerable because they feed at the top of their food web, and therefore accumulate organochlorines to high concentrations. Breeding populations of various **raptors** suffered large declines. In North America these included the **peregrine falcon** (*Falco peregrinus),* osprey (*Pandion haliaetus),* bald eagle (*Haliaeetus leucocephalus),* and golden eagle (*Aquila chrysaetos).* In all cases, these birds were exposed to a "cocktail" of organochlorines that included the insecticides DDT, DDD (both of which are metabolized to DDE in organisms), aldrin, dieldrin, and heptachlor, as well as PCBs, a non-insecticide with many industrial uses. Research has suggested that DDT was the more important toxin to birds in North America, while cyclodienes (particularly dieldrin) were more important in Britain.

Damage caused to predatory birds was largely associated with chronic effects on reproduction, rather than toxicity to adults. Reproductive damages included the production of thin eggshells that could break under the weight of an incubating parent, high death rates of embryos and nestlings, and abnormal adult **behavior**. These effects all contributed to decreases in the numbers of chicks raised, which resulted in rapid declines in the sizes of populations of the affected birds.

Since the banning of most uses of DDT and other organochlorines in North America, their residues in wildlife have been declining. This has allowed previously affected species to increase in abundance. In 1999, for example, the U.S. Fish and Wildlife Service removed the peregrine falcon from the list of species considered endangered. Although the population recovery of the peregrine falcon was aided by a program of captive-breeding and release, its recovery would not have been possible if their exposure to organochlorines in wild habitats had not been first dealt with.

DDT and related organochlorine insecticides have largely been replaced by organophosphate and carbamate chemicals. These chemicals poison insects and other arthropods by inhibiting a specific **enzyme, acetylcholine** esterase (AChE), which is critical in the transmission of neural impulses. Vertebrates such as **amphibians**, fish, birds, and mammals are also highly sensitive to poisoning of their cholinesterase enzyme system. In all of these animals, acute poisoning of the AChE function by organophosphate and carbamate insecticides can cause tremors, convulsions, and ultimately death to occur.

Carbofuran is a carbamate insecticide that caused much bird mortality during its routine agricultural usage. For this reason, the further use of this chemical was banned in North America during the late 1990s. In 1996, it was discovered that agricultural use of the organophosphate monocrotophos against **grasshoppers** in Argentina was killing large numbers of Swainson's **hawks** (*Buteo swainsoni).* This raptor breeds in the western United States and Canada and winters on the pampas of **South America**. Populations of Swainson's hawks had been declining for about 10 years, and it appears the cause was poorly regulated use of monocrotophos on their wintering grounds. Because of risks of ecological damages caused by its use, monocrotophos has been banned in the United States and was never registered for use in Canada, but it could be legally used in Argentina. These are two examples of non-organochlorine insecticides that cause important ecological damages.

Of course, not all insecticides cause these kinds of serious ecological damages. For example, the toxicity of the

KEY TERMS

Bioconcentration—The occurrence of chemicals in much higher concentrations in organisms than in the ambient environment.

Broad-spectrum pesticide—A pesticide that is not toxic only to the pest but other plant and animal species as well.

Ecotoxicology—The study of the effects of toxic chemicals on organisms and ecosystems. Ecotoxicology considers both direct effects of toxic substances and also the indirect effects caused, for example, by changes in habitat structure or the abundance of food.

Food-web magnification—The tendency for top predators in a food web to have the highest residues of certain chemicals, especially organochlorines.

Non-target organism—Organisms that are not pests, but which may be affected by a pesticide treatment.

Pest—Any organism judged to be significantly interfering with some human purpose.

bacterial insecticide *B.t.* is largely limited to **moths**, **butterflies**, **beetles**, and flies—its is essentially non-toxic to most other invertebrates or vertebrate animals. Other relatively pest-specific insecticides are being developed and are increasing rapidly in use, often in conjunction with a so-called "integrated pest management" (or IPM) system. In IPM, insecticides may be used as a method of last resort, but heavy reliance is also placed on other methods of pest management. These include the cultivation of pest-resistant crop varieties, growing crops in **rotation**, modifying the **habitat** to make it less vulnerable to infestation, and other practices that reduce the overall impacts of pest insects.

The continued development of pest-specific insecticides and IPM systems will further reduce society's reliance on broad-spectrum insecticides and other damaging pesticides. Until this happens, however, the use of relatively damaging, broad-spectrum insecticides will continue in North America. In fact, the use of these chemicals is rapidly increasing globally, because they are becoming more prevalent in less-developed countries of tropical regions.

Resources

Books

Freedman, B. *Environmental Ecology.* 2nd Ed. San Diego, CA: Academic Press, 1995.

Thomson, W.T. *Agricultural Chemicals, Book I, Insecticides.* Fresno, CA: Thomson Publications, 1992.

Ware, G.W. *The Pesticide Book.* 5th ed. Fresno, CA: Thomson Publications, 2000.

Periodicals

Pimentel, D., H. Acquay, M. Biltonen, P. Rice, M. Silva, J. Nelson, V. Lipner, S. Giordano, A. Horowitz, and M. D'Amare. "Environmental End Economic Costs of Pesticide Use." *Bioscience* 42 (1992): 750- 760.

Other

Ware, G.W. *An Introduction to Insecticides.* 3rd ed. University of Arizona. 2000. <http://ipmworld.umn.edu/chapters/ware.htm>.

Bill Freedman

Insectivore

Strictly speaking, insectivores are any predators that catch and eat **insects**. Often, however, insectivorous predators also eat other small **invertebrates**, such as spiders, **millipedes**, **centipedes**, and earthworms.

Some insectivores specialize in catching and feeding upon flying insects, sometimes called aeroplankton. Some prominent examples of this insectivorous feeding strategy include **dragonflies**, smaller **species** of **bats**, flycatchers, swallows, and **swifts**. Insectivores that feed on flying insects must be quick and maneuverable fliers, and they must have acute means of detecting their **prey**. Most species are visual predators, meaning they detect flying insects by sight. Bats, however, feed in darkness at night or dusk, and they locate their prey using **echolocation**, a type of biological sonar.

Other insectivores are gleaners, and they carefully search surfaces for insects to eat. Most gleaners visually examine the surfaces of **plant** leaves and the branches and trunks of trees. Many **birds** that exploit the forest canopy hunt insects in this way, for example, **warblers** and **vireos**; as does the **praying mantis**.

A few species of insectivores specialize by finding their prey inside of **wood**. These insectivores may excavate substantial cavities as they search for food, as is the case of many species of **woodpeckers**, and sometimes **bears** searching for beetle grubs or carpenter **ants**.

Large numbers of insects live in **soil** and in the organic **matter** that sits atop the soil. Many species of burrowing and digging small **mammals** feed on insects and other invertebrates in this substrate, including **shrews**, **moles**, and **hedgehogs** (in fact, the order of these small mammals is called Insectivora). Some birds also hunt in-

The endangered Haitian solenodon (*Solenodon paradoxus*) secretes toxic saliva, which it uses to paralyze its insect prey. *Photograph by N. Smythe. National Audubon Society Collection/ Photo Researchers, Inc. Reproduced by permission.*

sects located in surface litter, for example, **thrushes** and **grouse**. There are also many species of burrowing, predacious insects and **mites** that hunt insects within this zone.

Freshwater lakes, ponds, and **wetlands** can harbor enormous numbers of insects, and these are eaten by a wide range of insectivores. Trout, for example, feed voraciously on aquatic insects whenever they are available in abundance. A few species of birds, known as dippers, actually submerge themselves and walk underwater in mountain streams, deliberately searching on and under stones and debris for their prey of bottom-dwelling insects.

Virtually all insectivores are animals. However, a few plants have also evolved specialized morphologies and behaviors for trapping, killing, and digesting insects and other small invertebrates, and then absorbing some of their **nutrients**. Usually, these plants grow in nutrient-deficient habitats, such as bogs and dilute lakes. Examples of so-called insectivorous plants include the Venus' flytrap, sundews, and pitcher plants.

Insects

Insects are **invertebrates** in the class Insecta, which contains 28 living orders. This class of the phylum Arthropoda is distinguished by a number of anatomical features, including an adult body that is typically divided into three parts (head, thorax, and abdomen), three pairs of segmented legs attached to the thorax, one pair of antennae, and ventilation of respiratory gases through pores called spiracles and along tubes called tracheae. Insect orders in the subclass Pterygota have two pair of wings as adults, but some relatively primitive orders in the subclass Apterygota are wingless.

Insects have a complex life cycle, with a series of intricate transformations (called **metamorphosis**) occurring between the stages, each of which is radically different in morphology, **physiology**, and **behavior**. The most complicated life cycles have four stages: egg, larva, pupa, and adult. Examples of insect orders with this life cycle include **butterflies** and **moths** (Lepidoptera) and the **true flies** (Diptera). Other orders of insects have a less complex, more direct development, involving egg, nymph, and adult. Insect orders with this life cycle include the relatively primitive **springtails** (Collembola) and the **true bugs** (Hemiptera).

Most insects are nonsocial. However, some **species** have developed remarkably complex social behaviors, with large groups of closely related individuals living together and caring for the eggs and young of the group, which are usually the progeny of a single female, known

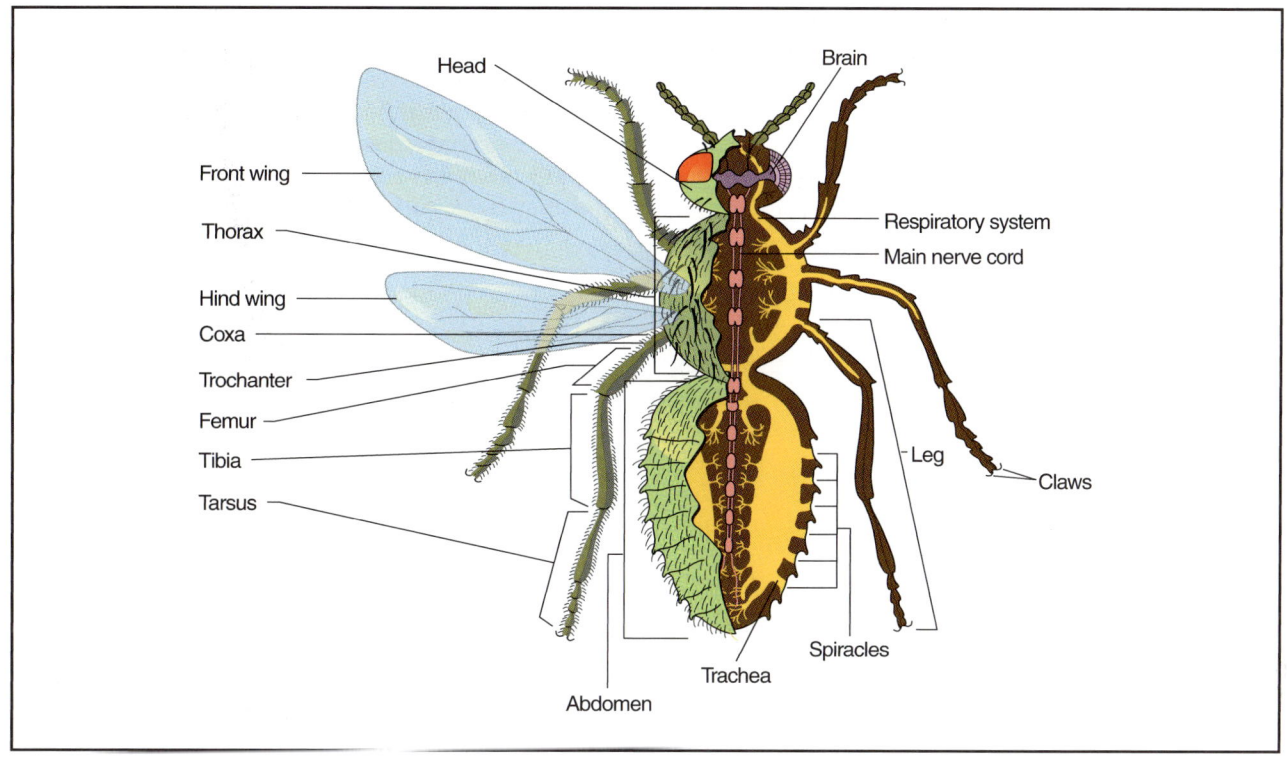

External and internal features of a generalized insect. *Illustration by Hans & Cassidy. Courtesy of Gale Group.*

as the queen. This social system is most common in the **bees**, **wasps**, and **ants** (Hymenoptera), and in the unrelated **termites** (Isoptera).

A few species of insects are useful to humans. Some insects, however, are important factors in the transmission of human diseases. For example, **malaria**, **yellow fever**, **sleeping sickness**, and certain types of **encephalitis** are caused by **microorganisms**, but are transmitted by particular species of biting **flies**, especially **mosquitoes**. Other insects are important defoliators of trees, and can thereby cause substantial damage to commercial timber stands and to shade trees. Insects may also defoliate agricultural plants, or may feed on unharvested or stored grains, thus causing great economic losses. Some insects, particularly termites, cause enormous damage to **wood**, literally eating buildings constructed of that material. Pesticides—chemicals that are toxic to insects—are sometimes used to control the populations of insects regarded as major **pests**.

Taxonomists have recognized and named more than one million species of insects—more than have been recognized in any other group of organisms. Of these, approximately three-quarter of a million have been described in some detail. To lend perspective of the vast number of insect species, there are a mere 6,200 bird and 5,800 reptile species described. There are over 10,000

known species of ants alone. In addition, biologists believe that tens of millions of species of insects remain undiscovered. One estimate is that as many as 30 million species of insects inhabit **Earth**. Most of these are thought to be **beetles** (Coleoptera). In fact, all of the insect orders are poorly known. Most of these undiscovered species of insects occur in tropical rainforests, especially in the canopy.

Globally, there is an enormous abundance and productivity of insects, and an extraordinary richness of species. These exploit a remarkable diversity of habitats, and are ecologically important as herbivores, predators, **parasites**, and scavengers. As a result of these attributes, insects are considered to be one of the most successful group of organisms on Earth, if not the most successful.

See also Pesticides.

Insomnia

The term insomnia applies to a variety of conditions involving lack of **sleep**, difficulty falling asleep, and disrupted or insufficiently restful sleeping patterns. Not only the quantity but the quality of sleep is at issue for those who research insomnia and its effects on **behavior** and productivity during waking hours. Primary insomnias include chronic and temporary sleeplessness. Secondary insomnias are defined as unusual sleeping patterns like sleepwalking or nocturnal eating, "night terrors" or nightmares, and respiratory movement or nervous disorders such as the "restless leg" **syndrome**.

Temporary insomnia

Anyone will, at some point during his or her lifetime, experience a temporary inability to fall asleep. This is officially known as psychophysiological insomnia (**PI**), because the body and mind can react to different kinds of **stress** by developing insomnia. A change in work schedule, jet lag, a recent death in the family, or the use of certain prescription medicines or drugs like **caffeine** can disrupt a person's circadian rhythm. This rhythm is a roughly 24-hour cycle of sleeping and waking, but it can be set off-balance by an all-night study session, during a hospital stay, or by traveling from one **time** zone to another.

Other forms of temporary insomnia accompany stages of the life cycle. Children, pregnant women, and the elderly exhibit sleeplessness in reaction to changes in their body chemistry or of their surroundings. Preschool children commonly find it hard to go to bed on schedule every day. Physicians are wary of prescribing unnecessary drugs to pregnant women, so often they are prevented from relying on their usual sleep aids.

Certain acute medical conditions count insomnia among their symptoms. The endocrine disorder called hyperthyroidism can interfere with **brain** wave patterns, and also obstruct the throat to cause sleep apnea or intermittent breathlessness during sleep. Any medical condition which causes chronic **pain** will also keep people awake, from **ulcers** or angina for instance. Psychiatric causes of insomnia range from **depression** to anorexia-nervosa to psychotic breakdowns. Drug addicts such as alcoholics will encounter insomnia as a withdrawal effect.

Chronic insomnia

Idiopathic, which means "primary," insomnia develops in childhood and persists throughout a person's lifetime. Its true cause is a mystery, though people who exhibit this chronic sleeplessness often suffer from particular brain dysfunctions as well; **dyslexia** for example. These disorders may share a root cause with idiopathic insomnia, but more studies will have to be commissioned before any theories may be formed. Those who suffer from childhood tend to cope more easily than those who develop temporary insomnias. Idiopathic insomniacs are thereby less prone to sleep **phobias** and other psychological reactions which often accompany transient insomnias.

KEY TERMS

Circadian rhythm—The rhythmical biological cycle of sleep and waking which, in humans, usually occurs every 24 hours.

Idiopathic insomnia—Chronic insomnia that begins in childhood and continues into adulthood.

PI—An acronym for psychophysiological insomnia. This term applies to insomnia which may begin in response to emotional distress, illness or disruptions of the daily schedule.

Pseudoinsomnia—A complaint of insomnia or insufficient sleep not supported by "sleep log" reports or medical exams.

Sleep apnea—A disorder which contributes to insomnia, during which a sleeper stops breathing for seconds at a time throughout the night.

Sleep hygiene—A form of therapy which attempts to treat insomnia without using drugs, by instead changing disruptive behavioral patterns of the patient.

Evaluation and treatment

Pseudoinsomnia remains a puzzle for sleep researchers. While the prefix "pseudo" indicates a false impression of sleeplessness, there is a possibility that current monitoring technologies like the EEG may not differentiate clearly enough between sleeping and waking states. Further complications involve variations in sleep needs from person to person. Everyone has heard that Einstein was fond of naps but Thomas Edison hardly slept at all.

Is one person's good night's rest another person's waste of time? Self-described "night owls" function better by working evening shifts and sleeping daytime, but this can be debilitating for "larks" or morning people. Some individuals operate on a 25-hour circadian rhythm, which does not synch with 24-hour clock time.

Self-administered sleep aids can sometimes backfire on an insomniac. **Alcohol** can at first lull you to sleep, but habitual use of this depressant can in turn further disrupt your sleeping patterns. A vicious cycle may soon surface, in which increased use of sleep aids like over-the-counter pills and/or alcohol contributes only to worsening the original condition.

An alternative which does not resort to prescription drugs is known as sleep hygiene. This regimen of behavior modification is designed to lessen exposure to stress and improve the patient's attitude towards sleeping and waking. A "sleep log" is kept to help a therapist pinpoint the probable causes of the patient's insomnia. Then self-monitoring is encouraged, so the patient learns to avoid excitement or heavy meals before bed, curtail the use of stimulants and depressants, and avoid naps. More experimental modification techniques like **biofeedback** may also be prescribed. Sleep hygiene programs are thereby tailored to individual needs.

See also Sleep disorders.

Resources

Books

Kales, Anthony, and Joyce D. Kales, MD. *Evaluation and Treatment of Insomnia.* New York: Oxford University Press, 1984.

Williams, Robert L., MD, Ismet Karacan, MD, DSc, and Constance A. Moore, MD, eds. *Sleep Disorders: Diagnosis and Treatment.* 2nd. ed. New York: John Wiley & Sons, 1988.

Periodicals

"Insomnia: How You Can Conquer It." *Muscle & Fitness* (January 1994).

Jennifer Kramer

Instinct

An instinct is a stereotyped, species-typical **behavior** that appears fully functional the first time it is performed, without the need for **learning**. Such behaviors are usually triggered by a particular **stimulus** or cue, and are not readily modified by subsequent experience. For instance, a kangaroo rat instantly performs an automatic escape jump maneuver when it hears the sound of a striking rattlesnake, even if it has never encountered a snake before. Clearly, instinctive behaviors play an important role in survival, but our understanding of the forces that promote and guide their development in living animals is in fact quite limited.

Classic examples of animal instinct

Researchers of **animal** behavior, ethologists, first named the stereotyped, species-typical behaviors exhibited in particular circumstances fixed action patterns, which were later called instincts. A cocoon-spinning spider ready to lay its eggs builds a silk cocoon in a particular way, first spinning a base plate, then the walls, laying its eggs within, and finally adding a lid to seal the top. The spider performs all these actions in a specific se-

quence, and, indeed, cannot spin its cocoon in any other way. If the spider is relocated after having spun the base plate, she will still make the walls, deposit the eggs (which promptly fall out the bottom), and spin the lid for the top. When ready to begin the next cocoon, if the spider is returned to her original base plate, she will nonetheless begin by spinning a new base plate over the first, as if it were not there.

Many fixed action patterns occur in association with a triggering stimulus, sometimes called a releaser. Baby **gulls** respond to the sight of their parent's bill by pecking it to obtain a tasty morsel of food. The releaser here is a bright red spot on the parent's bill; neither the shape nor the **color** of the adult's head have a significant influence on the response. When a female rat is sexually receptive, rubbing of her hindquarters (the releaser) results in a stereotypical posture known as lordosis, in which the front legs are flexed, lowering the torso, while the rump is raised and the tail is moved to one side (a fixed action pattern). A male rat who encounters a female in lordosis experiences another releaser and initiates copulation. Neither sequence requires any prior experience on the part of the animal.

The role of instinct in learning

Imprinting

In another classic study of instinctive behavior, ethologist Konrad Lorenz showed that baby **ducks** and **geese**, which are observed to closely follow their mother on their early forays away from the nest, could also be induced to follow a substitute. The baby **birds** would form an attachment to whatever individual was present as they opened their eyes and moved about after hatching, regardless of that individual's **species** identity. Young birds that had thus imprinted on Lorenz followed him everywhere as they matured, and as adults, these birds were observed to court humans, in preference to members of their own species.

Lorenz concluded that **imprinting** represented a kind of preprogrammed learning, guided by a mechanism that under normal circumstances would not be corrupted by individuals of the wrong species. In the natural situation, imprinting would facilitate the babies' social attachment to their mother, which later allows them to recognize appropriate mating partners.

Critical periods

Bird song is a largely species-specific behavior performed by males in their efforts to establish and maintain their territories and to attract females. Many songbirds develop their mature songs through a process involving a critical period when, as a nestling, the bird hears the song of its father. The juvenile bird does not sing until the following spring, when it begins to match its immature song to the one it heard from its father during its critical period. If the nestling is prevented from **hearing** adult song during the critical period, it will never develop a species-typical song. Evidently, there is also a strongly instinctive aspect to what may be learned during the critical period; most birds cannot produce every song heard during that time, but appear to be selective toward songs that are produced by other members of their species.

Instincts can be exploited

Some animals have evolved the capacity to take advantage of the reliable, instinctive behavior of others. Avian brood **parasites**, including the North American cowbird and the European cuckoo, exploit the parental behavior of other birds and lay their eggs in the host's nest. The unwitting host feeds the interloper's hatchlings, which are often bigger than its own, and thus may represent a greater releaser of the powerfully instinctive feeding behavior of the parents. The adult brood parasite is literally parasitizing the parental behavior of the host bird, for it exerts no further parental investment in its offspring, leaving them instead in the care of the host.

Instinct and learning: a continuum

We use the term instinct to describe species-typical behavior that is seemingly performed without aid of prior experience, but what we seem to mean is that the animal moves and behaves as if mysterious and unknown forces were guiding it. Many people who study animal behavior argue that the term instinct is not ultimately helpful because it tells us little about the real mechanisms underlying behavior. The use of the term indicates only that the behavior is relatively closed to modification by experience—nothing more. Since **nervous system** tissues are soft, delicate, and often very complex, understanding the operation of these structures in producing behavior presents a great challenge. This, combined with the role of experience in producing many superficially "instinctive" behaviors, makes things even more difficult.

Many behaviors held up as examples of instinct are shown to have an experiential component: for instance, as new gull chicks continue to peck at bill-like objects, the **accuracy** of their pecking improves and the kinds of bill-like objects they will peck at are increasingly restricted. Thus, the wide variety of behavioral patterns observed in living organisms surely represents a continuum, from those not much influenced by learning to those that are greatly influenced by it; a strict "nature versus

KEY TERMS

. . . .

Brood parasite—An animal that deposits its eggs or offspring into the nest of another individual (often of a different species) to be cared for by that individual.

Critical period—A developmental phase in the life of a young animal, usually with a measurable beginning and end, during which some crucial experience must occur if the animal is to develop normally.

Ethologist—A scientist of animal behavior, with particular focus on instinctive behaviors.

Fixed action pattern—Triggered by a particular cue or stimulus, fixed action patterns appear as a sequence of programmed behaviors which are performed to completion once they have been activated.

Releaser—The cue or stimulus that acts as a signal to induce a behavior in an animal.

nurture" dichotomy is probably too simplistic to describe any animal behavior.

The answer to the question "Under what conditions should a behavior be genetically closed, and when should a provision be made for learning?" seems to be related to the situation's predictability in nature. When it is crucial that the correct response to some occurrence be carried out the first time (like a kangaroo rat faced with a striking rattlesnake), natural **selection** should favor a fairly rigid, infallible program to underlie an appropriate response. The existence of a reliable relationship between some environmental cue and a biologically appropriate response permits the development of a releaser for triggering the "right" reaction the first time, whether to a **predator**, potential mates, or one's own offspring.

Resources

Books

Alcock, John. *Animal Behavior: An Evolutionary Approach.* 4th ed. Sunderland, MA: Sinauer, 1989.

Campbell, N., J. Reece, and L. Mitchell. *Biology.* 5th ed. Menlo Park: Benjamin Cummings, Inc. 2000.

Periodicals

West, Meredith J., Andrew P. King, and Michele A. Duff. "Communicating about Communicating: When Innate is Not Enough." *Developmental Psychobiology* 23 (1990): 585-98.

Susan Andrew

Insulin

Insulin is a hormone secreted by the pancreas gland, one of the **glands** in the **endocrine system**. Insulin, working in harmony with other **hormones**, regulates the level of **blood** sugar (glucose). An insufficient level of insulin secretion leads to high blood sugar, a **disease** called **diabetes mellitus**.

Endocrine glands are ductless glands; that is, they pour their products (hormones) directly into the bloodstream. The pancreas, a gland in the upper abdomen, has cells within it that secrete insulin directly into the bloodstream.

History

Prior to the twentieth century, diabetes was a fatal disease. Its cause was unknown, and the method of treating it had yet to be discovered. Not until 1921 did the research of Sir Frederick Banting, a Canadian surgeon, and Charles Best, a Canadian physiologist, reveal that insulin is crucial in blood glucose regulation. The two scientists then isolated insulin, an achievement for which they were awarded the Nobel Prize. Their work was built on previous research by Paul Langerhans, a German pathologist who described the irregular, microscopic collections of cells scattered throughout the pancreas. These later were named the islets of Langerhans and were found to be the source of insulin secretion.

In 1952, a British biochemist, Frederick Sanger, analyzed insulin and discovered it was made up of two chains of amino acids. One chain, called the alpha chain, has 21 amino acids and the second chain, the beta chain, has 30 amino acids. The chains run **parallel** to each other and are connected by disulfide bonds (made up of two **sulfur atoms**). With the discovery of the chemical structure of insulin, efforts to synthesize it began.

Types of diabetes

Diabetes manifests in two types. Type I diabetes is also called childhood onset because it begins in early childhood. Adult onset diabetes or Type II affects adults.

Glucose is a source of **energy** to the muscles of the body. Normally, the glucose content of blood is determined by the demands made by the muscles; the secretion of insulin, which lowers blood glucose; the secretion of glucagon, also manufactured and secreted in the pancreas, which raises blood glucose; extraction by the liver of glucose from the blood to be converted to glycogen and stored; and other hormones secreted by the adrenal and pituitary glands. The secretion of glucagon in the pancreas is a function of the alpha cells of the islets of Langerhans, and the secretion of insulin is the function of the beta cells of the same islets.

If the beta cells fail to function properly or are genetically insufficient in number to provide the needed hormone, diabetes can result. Also, in adults, the damage to, or failure of the cells resulting in lowered insulin secretion can lead to diabetes mellitus.

No cure has been found for diabetes; that is, no way has been found to restore full function to the beta cells in the islets of Langerhans. Individuals who have diabetes must take insulin by injection or take pills that control **carbohydrate metabolism**.

The insulin used by diabetics used to be extracted from porcine (pig) and bovine (cattle) pancreas. However, with the genetic revolution, **genetic engineering** has allowed **bacteria** to be transformed to produce human insulin protein, which is purified from large industrial cultures. Thus, human insulin is obtained from bacteria that have human genes inserted into them. Human insulin technology has advanced to produce new, more effective forms of insulin. One of the problems with insulin therapy is that diabetic patients must inject themselves with insulin regularly—an unpleasant process for many people. The development of Humulin-L, Humulin-N, Humulin-R, and Humulin-U have allowed some patients to reduce the **frequency** of shots required. These new forms of human insulin (from which the name Humulin comes) have effects that last longer than ordinary insulin, making fewer injections possible. As a result, more diabetes patients are likely to stick to their insulin therapy.

Integers

The integers are the positive and **negative** whole numbers... -4, -3, -2, -1, 0, 1, 2,.... The name "integer" comes directly from the Latin word for "whole." The set of integers can be generated from the set of **natural numbers** by adding **zero** and the negatives of the natural numbers. To do this, one defines zero to be a number which, added to any number, equals the same number. One defines a negative of a given number to be a number which, plus the given number, equals zero. Symbolically, for any number n: $0 + n = n$ (additive identity law) and $-n + n = 0$ (additive inverse law). Because **arithmetic** is done with natural numbers, one needs rules which will convert integer arithmetic into natural-number arithmetic. This is true even with a **calculator**. Most simple four-function calculators have no easy way of entering negative numbers, and the user has to apply the rules for

Multiplication	Addition	Law
ab is a unique integer.	a + b is a unique integer.	Closure law
ab = ba	a + b = b + a	Commutative law
a(bc) = (ab)c	a +(b + c)= (a + b) + c	Associative law
(1)(a) = a	0 + a = a	Identity law
	-a + a = 0	Inverse law
If ac = bc (c = 0), then a = b.	If a + c = b + c, then a = b	Cancellation law
	a(b + c) = ab + ac	Distributive law

himself. Rules are often stated using the concept of absolute value. The absolute value of a number is the number itself if it is positive and its opposite if it is negative. For example, the absolute value of +5 is +5, or 5, while the absolute value of -3 is +3, or 3. Absolute values are always positive or zero.

There are two basic rules for **addition**: 1) To add two numbers with like signs, add their absolute values and give the answer the common sign. 2) To add two numbers with opposite signs, subtract the smaller absolute value from the larger and give the answer the sign of the larger.

For example: -4 + (-7) is -11, and -8 + 3 is -5.

There is a single rule for **subtraction**. It does not give the result directly but converts a difference into a sum: To subtract a number, add its opposite. For example, -8 - 9 becomes -8 + (-9), and 4 - (-2) becomes 4 + 2. This latter example uses the fact that the negative (or opposite) of a negative number is positive.

Division and **multiplication** have two simple rules: 1) The product or quotient of two numbers with like signs is positive. 2) The product or quotient of two numbers with unlike signs is negative. For example. (-30)(18) is -540; (-6)/(-3) is 2; and 20/(-4) is -5.

Because the integers include negative numbers, it is possible for every subtraction, as well as every addition and multiplication, to be completed using only integers. The set of integers is therefore "closed" with respect to subtraction, addition, and multiplication. It is not closed with respect to division, however. Three divided by seven is not an integer.

The set of integers form an "integral domain." This is a mathematical system governed by these laws for all integers a, b, and c. Notice that there is no inverse law for multiplication. **Integral** domains do not necessarily have multiplicative inverses, and, consequently, division is not always possible.

Integers are useful in business, where an amount of money can be a loss as well as a gain. They are useful in science when a quantity can be negative or positive, as in the charge borne by electrons, protons and other elementary particles, or in temperatures above and below zero. They show up in games, even, where one can be a number of points ahead or "in the hole." And they are absolutely necessary in **mathematics**, which would otherwise be incomplete and of little interest.

See also Irrational number; Rational number.

Resources

Books

Gelfond, A.O. *Transcendental and Algebraic Numbers.* Dover Publications, 2003.
Klein, Felix. "Arithmetic." In *Elementary Mathematics from an Advanced Standpoint.* New York: Dover, 1948.
Rosen, Kenneth. *Elementary Number Theory and Its Applications.* 4th ed. Boston: Addison-Wesley, 2000.
Stopple, Jeffrey. *A Primer of Analytic Number Theory: From Pythagoras to Riemann.* Cambridge: Cambridge University Press, 2003.
Van Niven, I. *Numbers: Rational and Irrational.* New Mathematical Library, Washington, DC: The Mathematical Association of America, 1975.
Weisstein, Eric W. *The CRC Concise Encyclopedia of Mathematics.* New York: CRC Press, 1998.

Integral

The integral is one of two main concepts embodied in the branch of **mathematics** known as **calculus**, and it corresponds to the area under the graph of a **function**. The area under a **curve** is approximated by a series of rectangles. As the number of these rectangles approaches **infinity**, the **approximation** approaches a limiting value, called the value of the integral. In this sense, the integral gives meaning to the concept of area, since it provides a means of determining the areas of those irregular figures whose areas cannot be calculated in any other way (such as by multiple applications of simple geometric formulas). When an integral represents an area, it is called a definite integral, because it has a definite numerical value.

The integral is also the inverse of the other main concept of calculus, the **derivative**, and thus provides a way of identifying functional relationships when only a **rate** of change is known. When an integral represents a function whose derivative is known, it is called an indefinite integral and is a function, not a number. Fermat, the great French mathematician, was probably the first to calculate areas by using the method of integration.

Definite integrals

A definite integral represents the area under a curve, but as such, it is much more useful than merely a means of calculating irregular areas. To illustrate the importance of this concept to the sciences consider the following example. The work done on a piston, during the power stroke of an **internal combustion engine**, is equal to the product of the **force** acting on the piston times the displacement of the piston (the distance the piston travels after ignition). Engineers can easily measure the force on a piston by measuring the **pressure** in the cylinder (the force is the pressure times the cross sectional area of the piston). At the same time, they measure the displacement of the piston. The work done decreases as the displacement increases, until the piston reaches the bottom of its stroke. Because area is the product of width times height, the area under the curve is equal to the product of force times displacement, or the work done on the piston between the top of the stroke and the bottom.

The area under this curve can be approximated by drawing a number of rectangles, each of them h units wide. The height of each **rectangle** is equal to the value of the function at the leading edge of each rectangle. Suppose we are interested in finding the work done between two values of the displacement, a and b. Then the area is approximated by Area = hf(a) + hf(a+h) + hf(a+2h) +... + hf(a+(n-1)h) + hf(b-h). In this approximation n corresponds to the number of rectangles. If n is allowed to become very large, then h becomes very small. Applying the theory of limits to this problem shows that in most ordinary cases this results in the sum approaching a limiting value. When this is the case the limiting value is called the value of the integral from a to b and is written:

$$A = \int_a^b f(x)\,dx$$

Where the integral sign (an elongated s) is intended to indicate that it is a sum of areas between x = a and x = b. The notation f(x)dx is intended to convey the fact that these areas have a height given by f(x), and an infinitely small width, denoted by dx.

Indefinite integrals

An indefinite integral is the inverse of a derivative. According to the fundamental theorem of calculus, if the integral of a function f(x) equals F(x) + K, then the derivative of F(x) equals f(x). This is true for any numerical value of the constant K, and so the integral is called indefinite.

The inverse relationship between derivative and integral has two very important consequences. First, in many practical applications, the functional relationship between two quantities is unknown, and not easily measured. However, the rate at which one of these quantities changes with respect to the other is known, or easily measured (for instance the previous example of the work done on a piston). Knowing the rate at which one quantity changes with respect to another means that the derivative of the one with respect to the other is known (since that is just the definition of derivative). Thus, the underlying functional relationship between two quantities can be found by taking the integral of the derivative. The second important consequence arises in evaluating definite integrals. Many times it is exceedingly difficult, if not impossible, to find the value of the integral. However, a relatively easy method, by comparison, is to find a function whose derivative is the function to be integrated, which is then the integral.

Applications

There are many applications in business, economics and the sciences, including all aspects of **engineering**, where the integral is of great practical importance. Finding the areas of irregular shapes, the volumes of solids of revolution, and the lengths of irregular shaped curves are important applications. In addition, integrals find application in the calculation of **energy** consumption, power usage, refrigeration requirements and innumerable other applications.

Resources

Books

Abbot, P., and M.E. Wardle. *Teach Yourself Calculus.* Lincolnwood, IL: NTC Publishing, 1992.
Larson, Ron. *Calculus With Analytic Geometry.* Boston: Houghton Mifflin College, 2002.

KEY TERMS

. . . .

Fundamental Theorem of Calculus—The Fundamental Theorem of Calculus states that the derivative and integral are related to each other in inverse fashion. That is, the derivative of the integral of a function returns the original function, and vice versa.

Limit—A limit is a value that a sequence or function tends toward. When the sum of an infinite number of terms has a limit, it means that it has a finite value.

Rate—A rate is a comparison of the change in one quantity with the simultaneous change in another, where the comparison is made in the form of a ratio.

Weisstein, Eric W. *The CRC Concise Encyclopedia of Mathematics.* New York: CRC Press, 1998.

J.R. Maddocks

Integrated circuit

An integrated circuit (IC) is a single semiconducting chip that contains transistors and sometimes, capacitors, resistors, and diodes. These components are connected to create an electrical circuit. Integrated circuits can be found in almost all electronic devices today, including those in automobiles, microwave ovens, traffic lights, and watches.

Just a few years ago, the circuits required to operate a hand-held **calculator** would have taken up an entire room. But today, millions of microscopic parts can fit onto a small piece of silicon capable of fitting into the palm of your hand.

With the invention of the **transistor** in 1948, the need for bulky **vacuum** tubes in computers and other electronic devices was eliminated. As other components were also reduced in size, engineers were able to design smaller and increasingly complex electronic circuits. However, the transistors and other parts of the circuit were made separately and then had to be wired together—a difficult task that became even more difficult as circuit components became smaller and more numerous. Circuit failures often occurred when the wire connections broke. The idea of manufacturing an electronic cir-

cuit with multiple transistors as a single, solid unit arose as a way to solve this problem.

The integrated circuit concept was first suggested by British **radar** engineer G. W. A. Dummer in 1952. He imagined implanting electronic components in a solid layered block of semiconducting material, with connections made by cutting out areas of the layers instead of by wires. In the United States, where the Department of Defense was distributing millions of dollars attempting to miniaturize electronic components, Dummer's idea was made a reality in the late 1950s by two inventors.

In Dallas, Texas, Jack Kilby of Texas Instruments began wrestling with the circuit problem in 1958 and came up with an idea similar to Dummer's. By September 1958, Kilby had succeeded in making the first working integrated circuit—tiny transistors, resistors, and capacitors connected by gold wires on a single chip. Kilby's 1959 patent application added an important feature: the connections were made directly on the insulating layer of the semiconductor chip, eliminating the need for wires.

Meanwhile, Robert Noyce of Fairchild Semiconductor in Mountain View, California, was also pursuing a solution to the miniaturization problem. Working independently of Kilby, Noyce, too, had considered housing an electronic circuit and its connections on a single piece of silicon. Noyce's integrated circuit used a planar technique of laying down alternating layers of semiconductor and insulating material, with photoetching to establish the circuit. Noyce applied for a patent for this technology in 1959.

Despite an ensuing patent dispute, Noyce and Kilby became recognized as co-inventors of the integrated circuit, which completely revolutionized the **electronics** industry. The individual transistor, like the **vacuum tube** before it, became obsolete. The integrated circuit was much smaller, more reliable, less expensive, and far more powerful. It made possible the development of the microprocessor and therefore, the personal computer, along with an array of devices such as the pocket calculator, microwave ovens, and computer-guided **aircraft**.

The early integrated circuits contained only a few transistors. In the era of Small Scale Integration (SSI), IC's typically contained tens of transistors. With the advent of Medium Scale Integration (MSI), the circuits contained hundreds of transistors. With Large Scale Integration (LSI), the number of transistors increased to thousands. By 1970, LSI circuits were in **mass production**, being used for computer memories and hand-held calculators. With the advent of Very Large Scale Integration (VLSI), hundreds of thousands or more transistors could be accommodated in an IC. The year 1986 saw the introduction of the first one megabyte **random** access memory (RAM), containing more than one million transistors.

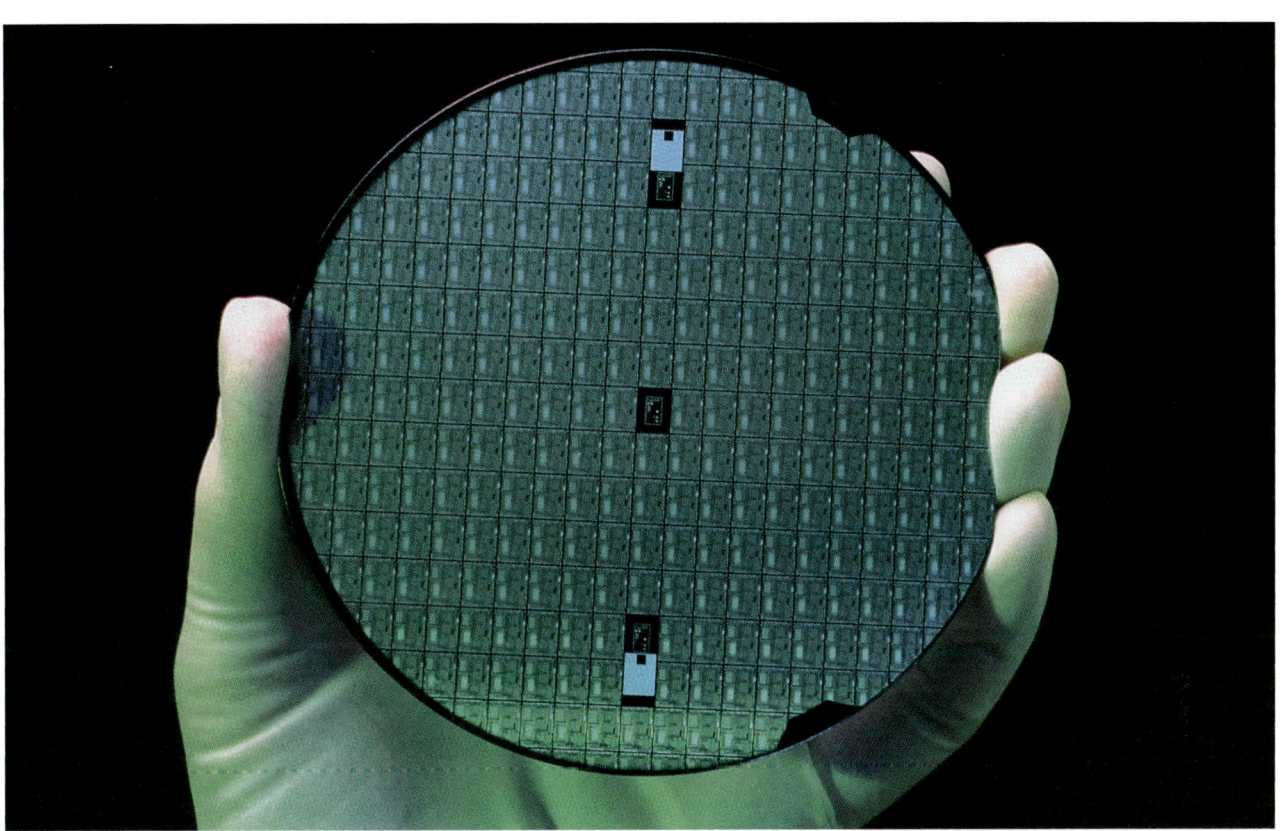

A circular wafer of silicon carrying numerous individual integrated circuits. Multiple circuits are fabricated on one silicon base and later cut from it. *Photograph by Adam Hart-Davis. National Audubon Society Collection/Photo Reserchers, Inc. Reproduced by permission.*

To construct an integrated circuit, a small **rectangle** is first cut from a silicon (or for special applications, sapphire) wafer. This wafer is known as the substrate. Separate areas of the substrate are deposited (doped) with other elements to make them generators of either positive ("p-type") or **negative** ("n-type") carriers. Tracks of polycrystalline silicon or **aluminum** are next etched into layers above the substrate surface. The wafer is then cut up into pieces called dies, and each die is then connected to input and output ports, usually located at the edge of the die using gold wires, to form the "chip."

There are three classes of integrated circuits: digital, analog and **hybrid** (both digital and analog on the same chip). Digital integrated circuits, which are characterized by the presence of logic gates, process information discretely (i.e., in Boolean 1's and 0's). Their small size permits digital IC's to operate at high speed, and with low power dissipation. Digital IC's have the distinct marketing advantage that they are relatively inexpensive to manufacture. In contrast to digital IC's, analog integrated circuits process information continuously, as would be required in a **thermostat** or **light** dimmer switch.

Logic gates are used in devises whose electronic output signals depend only on their input. The input and output values for logic gates are either 0 (False) or 1 (True). Logic gates are used to implement a variety of Boolean functions including AND (e.g., output is 1 when every input signal is 1), OR (e.g., output is 1 when one or more input signals is 1), NAND (e.g., output is 1 when any input is 0, and 0 when all inputs are 1), and NOR (e.g., output is 1 when all input signals are 0, and 0 when at least one input signal is 1). Other examples of logic gates are inverters, flip-flops, and multiplexors.

See also Capacitor; Diode; Electric circuit.

Randall Frost

Integrated pest management

Integrated pest management (IPM) is a system that incorporates many methods of dealing with pest problems. IPM systems may include the use of pest-resistant crop varieties; the modification of **habitat** to make it less suitable for the pest; the use of pest-specific predators, **parasites**, herbivores, or diseases; and **pesticides** when necessary. However, because IMP systems do not have an exclusive reliance on the use of pesticides, they are a key element of any strategy to reduce the overall use of these chemicals, and thereby avoid the toxicological and ecological damages they cause.

Conventional pest control and its problems

Pests can be defined as any animals, plants, or **microorganisms** that interfere with some human purpose. For example, **insects** may be considered pests if they eat crop plants or stored foods, or if they are important vectors in the transmission of diseases of humans or domestic animals. Plants are considered to be pests, or weeds, if they excessively compete with crop plants in agriculture or **forestry**, or if they have an unwanted aesthetic, as is the case of weeds in grassy lawns. Microorganisms are regarded as pests if they cause diseases of humans, domestic animals, or agricultural plants. In all of these cases, humans may attempt to manage their pest problems through the use of pesticides, that is, chemicals that are toxic to the pest.

Very important benefits can be gained through the judicious use of pesticides. For example, agricultural yields can be increased, and stored foods can be protected. Human lives can also be saved by decreasing the **frequency** of diseases spread by **arthropods**; **malaria**, for example, is spread through bites of a few **species** of **mosquitoes**. However, there are also some important **negative** consequences of the use of pesticides to achieve these benefits.

Pesticides are toxic chemicals, and they are rarely specifically poisonous only to the pests against which they may be used. The **spectrum** of pesticide toxicity is usually quite wide, and includes a diverse range of nonpest (or nontarget) species, in addition to the pest. Most **insecticides**, for example, are poisonous to a wide range of insect species, to other arthropods such as spiders and crustaceans, and often to **fish**, **amphibians**, **birds**, and **mammals**, including humans.

Moreover, the operational use of pesticides does not usually achieve a specific exposure of only the pest target—a large number of nonpest species is also exposed. This includes nonpest species occurring on the actual spray site, as well as species elsewhere that are exposed through offsite drift or other movements of the sprayed pesticide. Nontarget exposures are especially important when pesticides are applied as a broadcast spray over a large treatment area, for example, by an **aircraft** or tractor-drawn apparatus.

Some important ecological effects are caused by the typically broad spectrum of toxicity of pesticides, and the extensive exposures to non-pest species whenever broadcast sprays are used. For example, the extensive spraying of synthetic insecticides to manage **epidemic** populations of **spruce** budworm (*Choristoneura fumiferana*), an important defoliator of **conifer forests** in northeastern **North America**, results in huge nontarget kills of diverse arthropod species, and deaths of birds and other vertebrate animals. Similarly, the use of **herbicides** kills large numbers of plants, in addition to the few species that are sufficiently abundant to be considered weeds.

In addition, some pesticides are toxic to humans, and people may be poisoned as a result of exposures occurring through the normal use of these chemicals. The most intense exposures involve accidents, and in rare cases people may be killed by pesticide poisoning. Usually, however, the exposure is much smaller, and the toxic response is milder, and often not easily measurable. Generally, people who are employed in the manufacturing or use of pesticides are subject to relatively intense exposures to these chemicals. However, all people are exposed to some degree, through the food, **water**, and air in their environments. In fact, there is now a universal **contamination** of animals, including humans, with certain types of pesticides, most notably the persistent **chlorinated hydrocarbons**, such as DDT.

Other ecological effects of pesticide use occur as a result of habitat changes. These effects are indirect, and they can negatively influence populations of **wildlife** even if they are not susceptible to direct toxicity from the pesticide. For example, the use of herbicides in forestry causes large changes in the abundance and species composition of the **plant** community. These changes are highly influential on the wildlife community, even if the herbicide is not very toxic to animals.

Obviously, it is highly desirable that alternative methods of pest management be discovered, so that our reliance on the extensive use of pesticides can be diminished.

Integrated pest management

Compared with reliance on the broadcast use of pesticides, integrated pest management is a preferable system of pest control. Through IPM an acceptable degree of pest control can be achieved by using a variety of complementary approaches. These include the following components: (1) Development and use of varieties of crop species that are resistant to the pest or **disease**. If there is genetically based variation for susceptibility to the pest or disease, resistant crop varieties can be developed using standard breeding practices. (2) Attacking the pest biologically, by introducing or enhancing the popu-

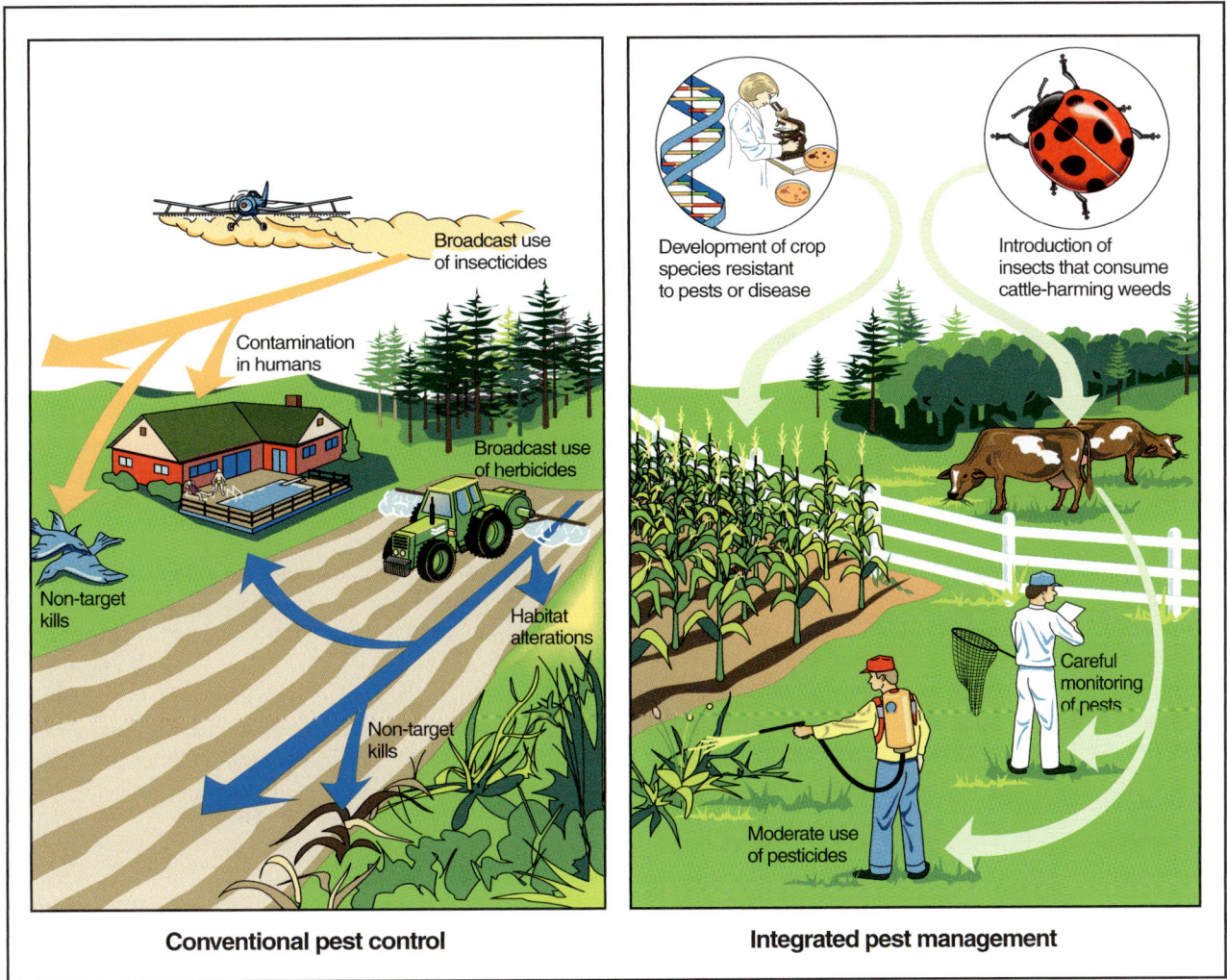

Conventional pest control compared to integrated pest management. *Illustration by Hans & Cassidy. Courtesy of Gale Research.*

lations of its natural predators, parasites, or diseases. (3) Changing other ecological conditions to make the habitat less suitable for the pest. (4) Undertaking careful monitoring of the abundance of pests, so that specific control strategies can be used efficiently, and only when required. (5) Using pesticides as a last resort, and only when they are a necessary component of an integrated, pest-management system.

If a system of integrated management can be successfully designed and implemented to deal with a pest problem, the reliance on pesticides can be greatly reduced, although the use of these chemicals is not necessarily eliminated. For example, a system of integrated pest management has been developed for the control of boll weevil (*Anthonomus grandis*) in cottonfields in Texas. The widespread use of this system has allowed large reductions in the use of insecticides for this purpose. About 19 million lb (8.8 million kg) of insecticides

were used against boll weevil in 1964, but only 2.4 million lb (1.1 million kg) in 1976 after an IPM system became widely used.

Biological control of pests

A very beneficial aspect of integrated pest management is the use of control methods that are highly specific to the pest, whenever this is biologically or ecologically possible. This is important because it allows nontarget damages to be avoided or greatly reduced.

Often, the most useful pest-specific control methods involve the utilization of some sort of biological-control agents, such as a disease, **predator**, or **herbivore** that specifically attacks the pest species. The use of biological agents has been most successful in the case of invasive pests that have been introduced from another **continent**, and that are thriving in the absence of their natural control agents. The utility of biological control is best appreciated by considering some examples.

In the late nineteenth century the cottony-cushion scale insect (*Icyera purchasi*) was accidentally introduced from **Australia** to the United States, where it became a serious threat to the developing citrus industry of California. In one of the first triumphs of biological control, this pest was successfully managed by the introductions of an Australian lady beetle (*Vedalia cardinalis*) and parasitic fly (*Cryptochetum iceryae*).

Because it is toxic to cattle, the klamath weed (*Hypericum perforatum*) became a serious problem in pastures in southwestern North America after it was introduced from **Europe**. However, this weed was controlled by the introduction of two European **beetles** that eat its foliage. A similar success is the control of European ragwort (*Senecio jacobea*) in pastures in western North America through the introduction of three of its insect herbivores.

The common vampire bat (*Desmodus rotundus*) of subtropical parts of the Americas bites cattle and other animals in order to obtain a meal of **blood**, which may weaken the victims or cause them to develop diseases. This serious pest of **livestock** can now be controlled by capturing individual **bats**, treating them with **petroleum** jelly containing a pesticide, and then setting the animals free to return to their communal roosts in caves, where the poison is transferred to other bats during social grooming. This treatment is specific, and other bat species are not affected.

Another serious pest of cattle is the screw-worm fly (*Callitroga ominivorax*), whose larvae feed on open wounds and can prevent them from healing. This pest has been controlled in some areas by releasing large numbers of male **flies** that have been sterilized by exposure to gamma **radiation**. Because females of this species will only mate once, any copulation with a sterile male prevents them from reproducing. The sterile-male technique works by overwhelming wild populations with infertile males, resulting in few successful matings, and a decline of the pest to an economically acceptable abundance.

The future of pest management

Clearly, it is highly desirable to use integrated pest management systems, especially in comparison with broadcast sprays of conventional, synthetic pesticides. This is particularly true of those relatively few pests for which effective biological controls have been discovered, because these methods have few nontarget effects.

It is unfortunate that in spite of ongoing research into their development, effective integrated systems have not yet been discovered for most pest management problems. Because it is important to manage pests, their con-

KEY TERMS

Agroecosystem—A agricultural ecosystem, comprised of crop species, noncrop plants and animals, and their environment.

Drift—Movement of sprayed pesticide by wind beyond the intended place of treatment.

Nontarget effects—Effects on organisms other than the intended pest target of pesticide spraying.

trol must therefore continue to rely heavily on synthetic pesticides. Regrettably, the toxicological and ecological damages associated with a heavy reliance on pesticides will continue until a broader range of integrated tools is available to pest managers.

Integrated pest management is the key for reducing the reliance on pesticides in intensively managed systems in agriculture, forestry, **horticulture**, and public health.

Resources

Books

Briggs, S.A. *Basic Guide to Pesticides. Their Characteristics and Hazards.* Washington, DC: Taylor & Francis, 1992.

Freedman, B. *Environmental Ecology.* 2nd ed. San Diego: Academic Press, 1994.

Bill Freedman

Integumentary system

The integumentary system includes the skin and the related structures that cover and protect the bodies of plants and animals. The integumentary system of plants includes the epidermis, cuticle, **plant** hairs, and **glands**. The integumentary system of **invertebrates** includes shells and exoskeletons as body covering. The integumentary system of **vertebrates** comprises skin, scales, feathers, hair and glands. The human integumentary system is made up of the skin which includes glands, hair, and nails. In humans, the skin protects the body, prevents **water** loss, regulates body **temperature**, and senses the external environment.

Plant integumentary system

The epidermis is the main surface **tissue** of young plants and the covering material of all leaves. Usually the epidermis is one **cell** deep; its cells have thick outer and side walls. On the parts of the plant that are above

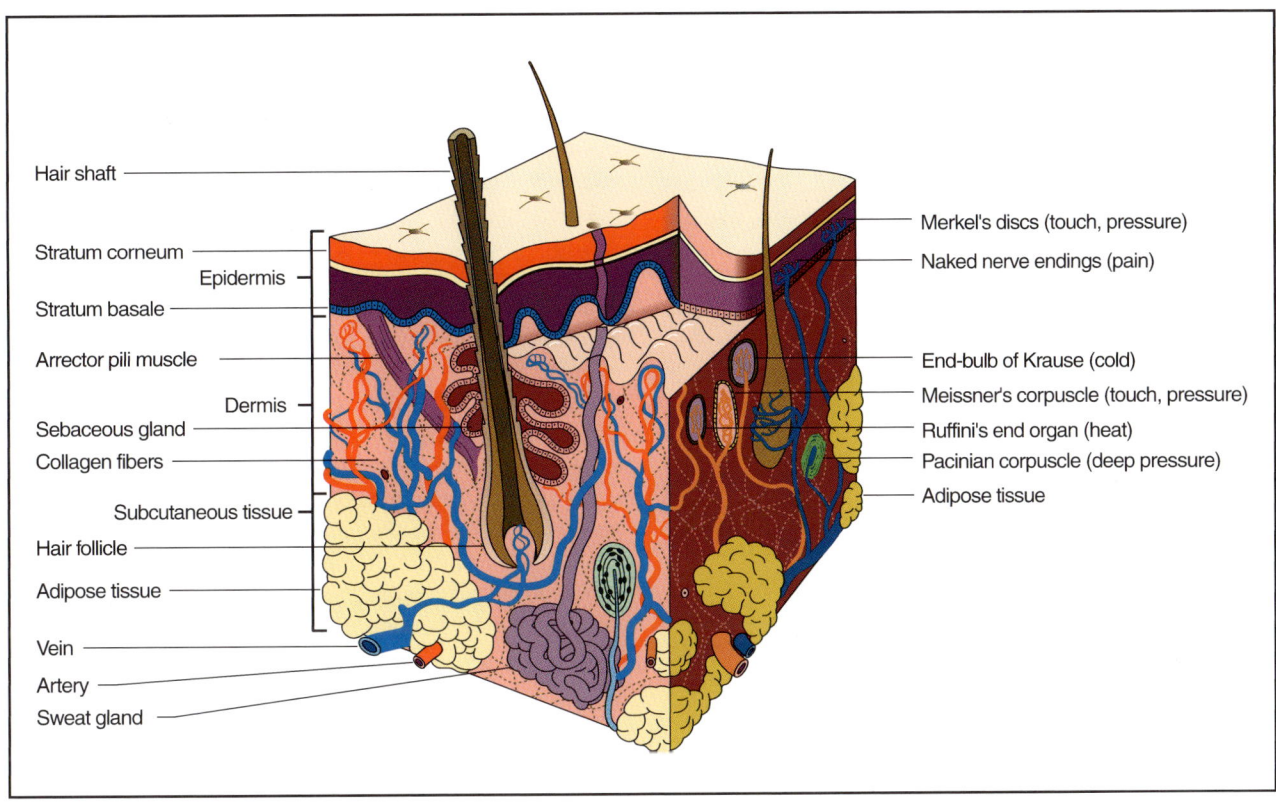

A cross section of the skin. Sensory structures are labeled on the right. *Illustration by Hans & Cassidy. Courtesy of Gale Group.*

ground, the epidermal cells secrete an outer waxy cuticle that is water resistant. The thickened cell walls, together with the cuticle, prevent drying out, injury, and fungus **infection**. The epidermis in aerial parts of the plant gives rise to plant hairs, spines, and glands. In leaves, the epidermis develops guard cells that regulate the size of pores or stomata, which allows the exchange of gases with the atmosphere. The epidermis of roots lacks the waxy cuticle found in the parts of the plant above ground, allowing the root epidermis to absorb water from the **soil**. Root hairs that increase the absorbing surface of the root arise from epidermal tissue. When a plant grows in diameter, the epidermis is replaced by the periderm, in the stem and the roots. The periderm contains **cork** cells whose walls after **cell death** provide a protective waterproof outer covering for plants making up the **bark** of older trees.

Invertebrate integuments

Snails, **slugs**, oysters, and clams are protected by a hard shell made of **calcium carbonate** secreted by the mantle, a heavy fold of tissue that surrounds the mollusc's internal organs. Spiders, **insects**, **lobsters**, and **shrimp**, have bodies covered by an external skeleton, the exoskeleton, which is strong, impermeable, and allows some **arthropods** to live on land. The exoskeleton is

composed of layers of protein and a tough polysaccharide called chitin, and can be a thick hard armor or a flexible paper-thin covering. Arthropods grow by shedding their exoskeletons and secreting a larger one in a process called molting.

Vertebrate integumentary systems

Keratin, an insoluble protein in the outer layer of the skin of vertebrates, helps prevent water loss and dehydration and has contributed to the successful **adaptation** to life on land. Keratin is also the major protein found in nails, hooves, horns, hair, and wool. Feathers, scales, claws and beaks of **birds** and **reptiles** are also composed of keratin.

Human integumentary system

The human integumentary system is made up of the skin, hair, nails, and glands, and serves many protective functions for the body. It prevents excessive water loss, keeps out **microorganisms** that could cause illness, and protects the underlying tissues from mechanical damage. Pigments in the skin called melanin absorb and reflect the sun's harmful ultraviolet **radiation**. The skin helps to regulate the body temperature. If **heat** builds up in the body, sweat glands in the skin produce more sweat which evaporates and cools the skin. In addition, when the body overheats, **blood** vessels in the skin expand and bring more blood to the surface, which allows body heat to be lost. If the body is too cold, on the other hand, the blood vessels in the skin contract, resulting in less blood is at the body surface, and heat is conserved. In addition to **temperature regulation**, the skin serves as a minor excretory **organ**, since sweat removes small amounts of nitrogenous wastes produced by the body. The skin also functions as a sense organ since it contains millions of nerve endings that detect **touch**, heat, cold, **pain,** and **pressure**. Finally, the skin produces **vitamin** D in the presence of sunlight, and renews and repairs damage to itself.

In an adult, the skin covers about 21.5 sq ft (2 sq m), and weighs about 11 lb (5 kg). Depending on location, the skin ranges from 0.02-0.16 in (0.5-4.0 mm) thick. Its two principal parts are the outer layer, or epidermis, and a thicker inner layer, the dermis. A subcutaneous layer of fatty or adipose tissue is found below the dermis. Fibers from the dermis attach the skin to the subcutaneous layer, and the underlying tissues and organs also connect to the subcutaneous layer.

The epidermis

Ninety **percent** of the epidermis, including the outer layers, contains keratinocytes cells that produce keratin, a protein that helps waterproof and protect the skin. Melanocytes are pigment cells that produce melanin, a brown-black pigment that adds to skin **color** and absorbs ultraviolet light thereby shielding the genetic material in skin cells from damage. Merkel's cells disks are touch-sensitive cells found in the deepest layer of the epidermis of hairless skin.

In most areas of the body, the epidermis consists of four layers. On the soles of the feet and **palms** of the hands where there is a lot of **friction**, the epidermis has five layers. In addition, calluses, abnormal thickenings of the epidermis, occur on skin subject to constant friction. At the skin surface, the outer layer of the epidermis constantly sheds the dead cells containing keratin. The uppermost layer consists of about 25 rows of flat dead cells that contain keratin.

The dermis

The dermis is made up of **connective tissue** that contains protein, **collagen**, and elastic fibers. It also contains blood and lymph vessels, sensory receptors, related nerves, and glands. The outer part of the dermis has fingerlike projections, called dermal papillae, that indent the lower layer of the epidermis. Dermal papillae cause ridges in the epidermis above it, which in the digits give rise to fingerprints. The ridge pattern of fingerprints is inherited,

and is unique to each **individual**. The dermis is thick in the palms and soles, but very thin in other places, such as the eyelids. The blood vessels in the dermis contain a **volume** of blood. If a part of the body, such as a working muscle, needs more blood, blood vessels in the dermis constrict, causing blood to leave the skin and enter the circulation that leads to muscles and other body parts. Sweat glands whose ducts pass through the epidermis to the outside and open on the skin surface through pores are embedded in the deep layers of the dermis. Hair follicles and hair roots also originate in the dermis and the hair shafts extend from the hair root through the skin layers to the surface. Also in the dermis are sebaceous glands associated with hair follicles which produce an oily substance called sebum. Sebum softens the hair and prevents it from drying, but if sebum blocks up a sebaceous gland, a whitehead appears on the skin. A blackhead results if the material oxidizes and dries. **Acne** is caused by infections of the sebaceous glands. When this occurs, the skin breaks out in pimples and can become scarred.

The skin is an important sense organ and as such includes a number of nerves which are mainly in the dermis, with a few reaching the epidermis. Nerves carry impulses to and from hair muscles, sweat glands, and blood vessels, and receive messages from touch, temperature, and pain receptors. Some nerve endings are specialized such as sensory receptors that detect external stimuli. The nerve endings in the dermal papillae are known as Meissner's corpuscles, which detect light touch, such as a pat, or the feel of clothing on the skin. Pacinian corpuscles, located in the deeper dermis, are stimulated by stronger pressure on the skin. Receptors near hair roots detect displacement of the skin hairs by stimuli such as touch or **wind**. Bare nerve endings throughout the skin report information to the **brain** about temperature change (both heat and cold), texture, pressure, and trauma.

Skin disorders

Some skin disorders result from overexposure to the ultraviolet (UV) rays in sunlight. At first, overexposure to sunlight results in injury known as sunburn. UV rays damage skin cells, blood vessels, and other dermal structures. Continual overexposure leads to leathery skin, wrinkles, and discoloration and can also lead to skin **cancer**. Anyone excessively exposed to UV rays runs a risk of skin cancer, regardless of the amount of pigmentation normally in the skin. Seventy-five percent of all skin cancers are basal cell carcinomas that arise in the epidermis and rarely spread (metastasize) to other parts of the body. Physicians can surgically remove basal cell cancers. Squamous cell carcinomas also occur in the epidermis, but these tend to metastasize. Malignant melanomas are life-threatening skin cancers that metastasize rapidly.

KEY TERMS

Chitin—Polysaccharide that forms the exoskeleton of insects, crustaceans, and other invertebrates.

Dermis—The internal layer of skin lying below the epidermis. It contains the sweat and oil glands, hair follicles, and provides replacement cells for those that are shed from the outer layer.

Epidermis—The thinner, outermost layer of the skin. Also the thin outermost covering in plants.

Keratin—Insoluble protein found in hair, nails, and skin.

Melanin—Brown-black pigment found in skin and hair.

There can be a 10-20 year delay between exposure to sunlight and the development of skin cancers.

Resources

Periodicals

Czarnecki, D. "10-Year Prospective Study Of Patients With Skin Cancer." *Journal of Cutaneous Medicine and Surgery* 6, no. 5 (2002): 427-429.

Fackelmann, Kathy A. "Skin Cancer's Return: How Big a Threat?" *Science News* (June 27, 1992).

Willis, Judith Levine. "Acne Agony." *FDA Consumer* (July-August 1992).

Other

Skin Deep. Video and videodisc. Princeton, NJ: Films for the Humanities and Sciences, 1995.

Bernice Essenfeld

Interference

Interference is the interaction of two or more waves. Waves move along their direction of propagation characterized by crests and troughs. Wherever two or more waves, either from one source by different paths or from different sources, reach the same point in space at the same time, interference occurs.

When the waves arrive in-phase (the crests arrive together), constructive interference occurs. The combined crest is an enhanced version of the one from the individual wave. When they arrive out-of-phase (the crest from one wave and a trough from another), destructive interference cancels the **wave motion**. The **energy** of the wave is not lost; it moves to areas of constructive interference.

Interference occurs in **sound waves**, **light** waves, shock waves, **radio**, and **x rays**. Waves display crests and troughs like the wiggles along the length of a vibrating jump rope. We see interference when ripples from one part of the pond reach ripples from another part. In some places the combination makes a large wave; in other places the waves **cancel** and the **water** appears calm. Radio, visible light, x rays, and gamma rays are waves with crests and troughs in the alternating **electromagnetic field**. Interference occurs in all of these waves. Interference of sound waves causes some regions of a concert hall to have special behavior. Where the multiple **reflections** of the concert sound interfere destructively, the sound is muffled and appears "dead." Where the reflections are enhanced by adding constructively, the sound appears brighter, or "live." Switching the polarity of the wires on a stereo speaker also can result in the sound appearing flat because of interference effects.

The most striking examples of interference occur in visible light. Interference of two or more light waves appears as bright and dark bands called "fringes." Interference of light waves was first described in 1801 by Thomas Young (1773-1829) when he presented information supporting the wave theory of light.

White light is a mixture of colors, each with a unique wavelength. When white light from the **sun** reflects off a surface covered with an oil film, such as that found in a parking lot, the thickness of the film causes a delay in the reflected beam. Light of some colors will travel a path through the film where it is delayed enough to get exactly out of phase with the light reflected off the surface of the film. These colors destructively interfere and disappear. Other colors reflecting off the surface exactly catch up to the light traveling through the film. They constructively interfere, appearing as attractive

Interference patterns created by the waves from several fallen drops of water. *Photograph by Martin Dohrn. National Audubon Society Collection/Photo Researchers, Inc. Reproduced by permission.*

color swirls on the film. The various colors on **soap** bubbles as they float through the air are another example of thin film interference.

Modern technology makes use of interference in many ways. Active **automobile** mufflers electronically sense the sound wave in the exhaust system and artificially produce another wave out-of-phase that destructively interferes with the exhaust sound and cancels the noise. The oil film phenomenon is used for filtering light. Precise coatings on optical lenses in binoculars or cameras, astronaut's visors, or even sunglasses cause destructive interference and the elimination of certain unwanted colors or stray reflections.

Interferometry

Interferometry uses the principles of **interference** to determine properties about waves, their sources, or the wave propagation medium. Acoustic interferometry has been applied to study the **velocity** of sound in a fluid. **Radio** astronomers use interferometry to get accurate measurements of the position and properties of stellar radio sources. Optical interferometry is widely used to observe things without touching or otherwise disturbing them. **Light** beams are sent through various paths, and they combine at one observation region where interference fringes occur. Interpreting the fringes reveals information about optical surfaces, the precise distance between the source and the observer, spectral properties of light, or the visualization of processes such as **crystal** growth, **combustion**, **diffusion**, and shock **wave motion**.

The Michelson-Morley experiment

The observation and explanation of interference fringes dates back to Robert Hooke (1635) and Isaac Newton (1642-1727), but the invention of interferometry is generally attributed to the American physicist Albert Michelson (1852-1931). The Michelson interferometer consists of two **perpendicular** arms with a half-silvered mirror (a beamsplitter) at the intersection (see Figure 1).

Each arm has a mirror at one end. Light from the source enters the interferometer along one arm, and is equally split at the beamsplitter. Half the light travels to mirror #1 and reflects back toward the beamsplitter. It passes through the beamsplitter and continues to the detec-

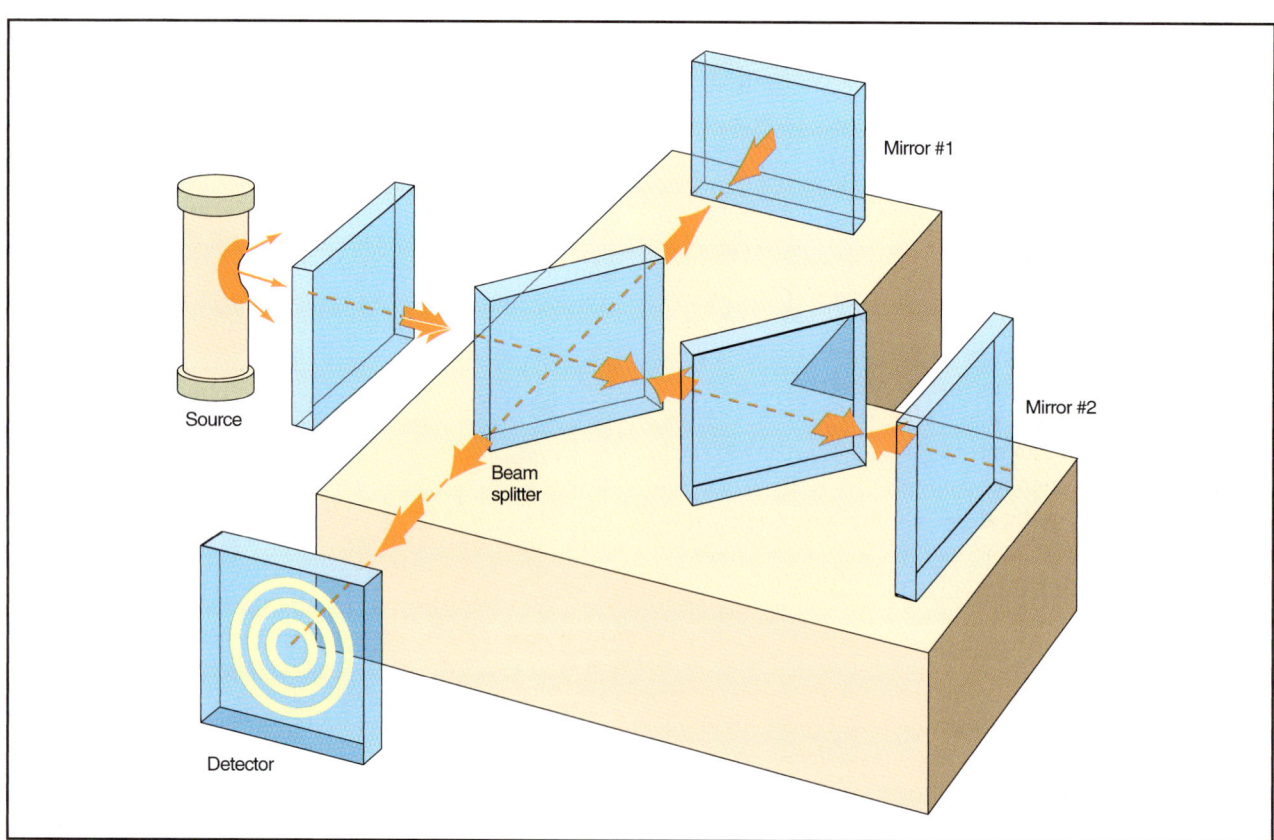

Figure 1. The Michelson interferometer. *Illustration by Hans & Cassidy. Courtesy of Gale Group.*

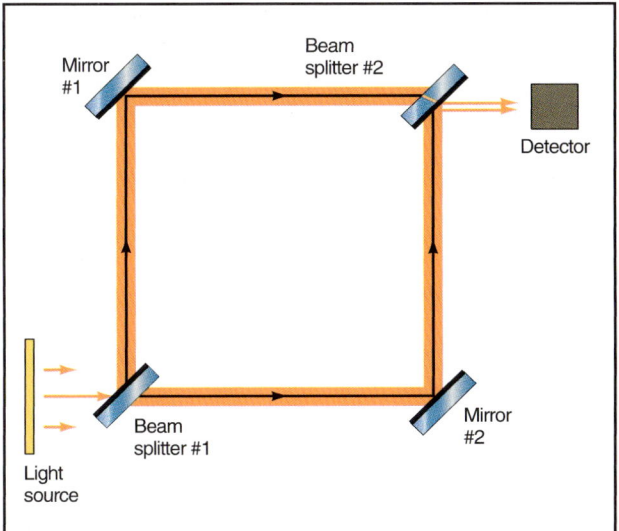

Figure 2. Top view of the interferometer introduced by Mach and Zehnder. *Illustration by Hans & Cassidy. Courtesy of Gale Group.*

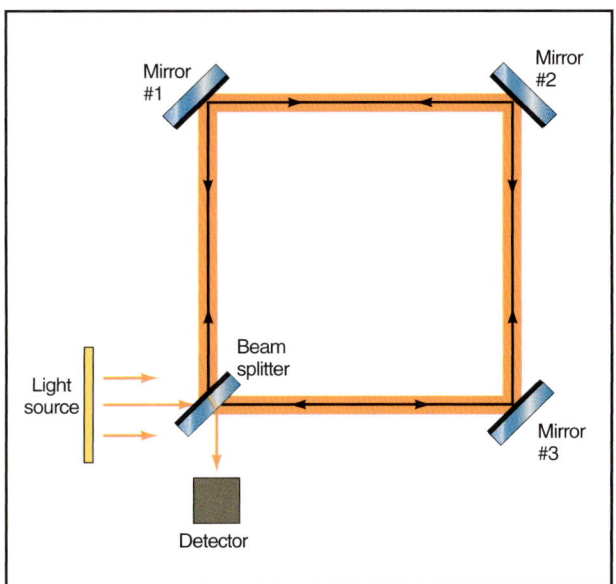

Figure 3. Top view of the Sagnac cyclic interferometer. *Illustration by Hans & Cassidy. Courtesy of Gale Group.*

tor, which can be like a **motion** picture screen. The other half of the light first travels straight through the beamsplitter, reflects off mirror #2, and then returns to the beamsplitter, where it is reflected and sent to the detector. When the two beams combine at the detector, they interfere and produce a pattern of fringes that depends upon the path they traveled and the **time** it took them to travel this path.

In 1887, Michelson, along with the physicist Edward W. Morley (1838-1923), set out to determine if the speed of light was dependent on the speed of the observer. According to the accepted theory of the time, light had to propagate in a medium called the **ether**. The motion of the **earth** traveling through the ether would affect the fringe pattern on Michelson's interferometer, because it would take the light longer to travel over one arm than the other arm. When the interferometer was rotated through 90°, the fringes would shift if the speed of light was not constant. In this most celebrated null experiment in the history of science, Michelson and Morley observed no changes in the fringes after many repetitions. The speed of light appears constant, regardless of the speed of its source, which was later explained by Einstein's theory of relativity.

Interferometers

Interferometers such as Michelson's can be used for many types of measurements. As a spectrometer, they can be used to accurately measure the mirror #1 mounted on a motor, so it can be moved along the direction of the beam at a constant speed. As the mirror moves, a fringe at a point in the interference **plane** will appear as an al-

ternating bright and dark band. The **frequency** of the changing fringes is related to the wavelength, and can be calculated to better than several parts in 100 billion.

The Twyman-Green interferometer

Modifications to the Michelson interferometer were introduced in 1916 by the electrical engineer Frank Twyman (1876-1959) and A. Green for the purpose of testing optical instruments. If the element transmits light, like a **prism** or a **lens**, it is inserted between the beamsplitter and mirror #1 in such a manner that the fringes that are observed become a measure of the element's optical quality. If the element to be tested is a mirror, and reflects light, it is substituted for mirror #1 altogether. Once again, the Twyman-Green interferometer fringes act as a **map** of irregularities of the optical element.

The Mach-Zehnder interferometer.

Another type of interferometer was introduced by L. Mach and L. Zehnder in 1891 (see Figure 2). Light leaves the sources, and is divided by beamsplitter #1. One half travels toward mirror #1, and is reflected. The other half is reflected from mirror #2. The beams are combined by beamsplitter #2, and propagate to the detector, where interference is observed. By virtue of the fact that the beams are separated, the objects to be tested can be quite large. The Mach-Zehnder two-beam interferometer is used for observing gas flows and shock waves, and for optical testing. It has also been used to obtain interference fringes of electrons that exhibit wavelike behavior.

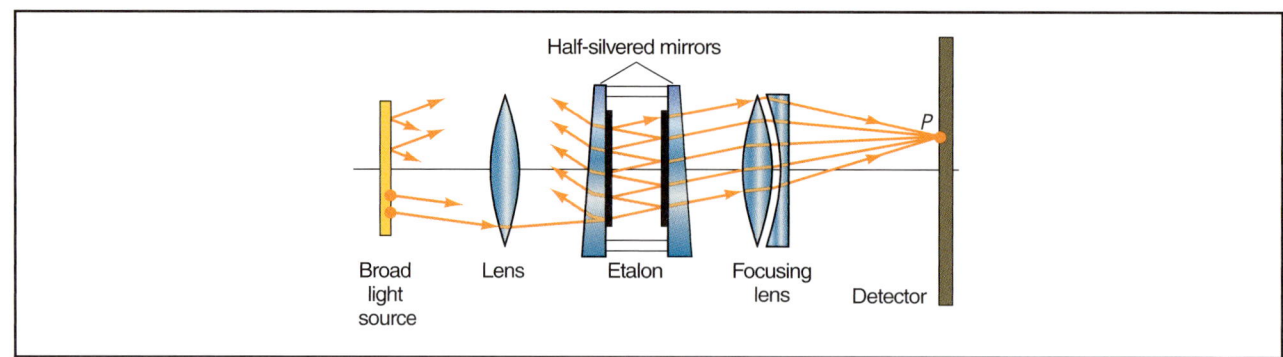

Half-silvered mirrors

Broad light source   Lens   Etalon   Focusing lens   Detector

P

Figure 4. Fabry-Perot interferometer. *Illustration by Hans & Cassidy. Courtesy of Gale Group.*

Cyclic interferometers

Interference can occur if a beam is split, and one half travels in a clockwise path around the interferometer, while the other travels counterclockwise. Figure 3 shows the top view of a cyclic interferometer, named a Sagnac interferometer for its inventor, physicist Georges Sagnac (1869-1928). One half of the beam reflects off the beamsplitter and travels from mirror #1, to mirror #2, to mirror #3, and again reflects off the beamsplitter. The two halves interfere at the detector. The Sagnac interferometer is the basis for **laser** gyroscopes that were first demonstrated in 1963. They sense the interference pattern to determine the direction and the speed of **rotation** in a moving vehicle like an airplane or a spacecraft.

The Fabry-Perot interferometer

In 1899, physicists Charles Fabry (1867-1945) and Alfred Perot (1863-1925) introduced an interferometer designed to produce circular interference fringes when light passes through a pair of **parallel** half-silvered **mirrors**. Figure 4 shows the Fabry-Perot interferometer used with a broad light source. When the plates are separated by a fixed spacer, the interferometer is called a Fabry-Perot etalon. The diameter of the fringes from the etalon is related to the wavelength, so it can be used as a spectrometer. If two beams of slightly different wavelength enter the etalon, the position of the overlap in fringes can be used to determine the wavelength to better than one part in 100,000. If the two plates of the Fabry-Perot interferometer are aligned parallel to each other, the device becomes the device becomes the basic laser resonant cavity, since only certain wavelengths will add constructively as they propagate between the mirrors.

Wavefront splitting interferometry

Rather than splitting the amplitude of the beam by beamsplitters, one part of the beam can be made to interfere with another in the manner of Lloyd's mirror (see Figure 5). One half of the beam from the source propa-

gates directly to the detector. The other half reflects off the mirror and interferes with the direct beam at the detector. Information about the surface of the mirror is contained in the fringes. The surface of crystals can be studied with fringes from **x rays**. The surface of a **lake** or the earth's ionosphere can be studied using interference from **radio waves**.

Wavefront shearing interferometry

A variation of the Mach-Zehnder interferometer, introduced by W. J. Bates in 1947, made it possible to measure the wavefront (phase) of a beam without an error-free reference wave. By rotating one beamsplitter in the Mach-Zehnder configuration, an incoming beam is split into two, and one half is shifted (sheared). Overlapping these beams results in an interference pattern that is a measure of the slope, or tilt, of the wavefront. Shearing interferometers are used in optical testing and in **astronomy** for measuring the distortions of the atmosphere.

Applications

The basic interferometer improves over the years as new technology appears. High-speed cameras and **electronics**, precise **optics**, and computers are brought together to make possible accurate interpretation of the fringes, as well as the extraction of new and exciting information.

Stellar interferometry

Even though we cannot directly photograph and resolve the image of two stars close together, we can use interferometry to measure their separation. First proposed by the physicist Armand Fizeau (1819-1896) in 1868, the method was first applied by Michelson and American astronomer Francis Pease (1881-1938) in 1920, and is commonly called Michelson stellar interferometry.

Light from two sources is collected by two telescopes that are a known distance apart. The light is filtered to restrict the wavelength, and then brought together. Each **star**

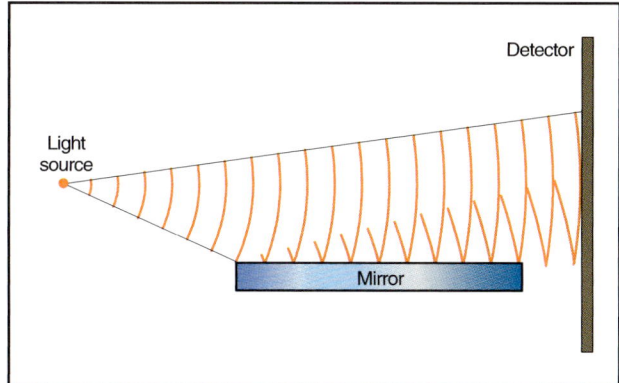

Figure 5. Wavefront splitting interferometer. *Illustration by Hans & Cassidy. Courtesy of Gale Group.*

exhibits a fringe pattern. The fringes will line up if the patterns of the two stars overlap. When the separation of the two telescopes is small, the fringes are visible. When the separation of the telescopes is exactly equal to the wavelength divided by twice the **angle** between the two sources, the fringes will disappear. By varying the separation of the telescopes and observing when the fringes disappear, the separation of the sources is calculated. In a similar way, a single remote star can be thought of as two halves that appear as point sources close together. By using stellar interferometry, the size of a star can be measured.

By placing an opaque screen with holes over the aperture of a **telescope**, each pair of holes will cause interference fringes. Stellar interferometry over a number of simultaneous separations is called aperture plane interferometry.

Radio astronomers R. Hanbury Brown and R. Q. Twiss, the first to use stellar interferometry in the radio region, measured the size of the star Sirius. Today, "Very Long Baseline Interferometry" links radio telescopes around the world to create interference fringes that can be used to measure stellar sizes in fractions of an arcsecond.

Speckle interferometry

Invented by Antoine Labeyrie in 1970, speckle interferometry provides a method for large telescopes to see objects without being limited by the **turbulence** of the atmosphere. A star exposure for less than one hundredth of a second appears speckled, because light from all points of the telescope interferes with each other. The speckle is similar to the speckled pattern of red light that reflects off the **glass** of a supermarket price scanner.

Averaging many short exposures (or taking a long exposure) smears out the speckles because the atmosphere is constantly moving around. Information in the image, smeared out by the blur, is lost. Because individual speckles themselves contain information about the object, the speckle interferometer gathers many short exposures, and a computer processes them to extract the information. Many measurements have been made in the last quarter century. The size and surface features of asteroids and the **planet Pluto** have been determined by speckle interferometry. The size of the nearby star Betelguse has also been measured. The angular separation of binary stars, measured by this technique, leads astronomers to calculate the star masses and develop theories about the **evolution** of the universe.

Holographic interferometry

Holography was invented by Dennis Gabor (1900-1979) in 1948. A hologram is recorded by splitting a light beam and letting half the beam scatter from an object while the other half travels undisturbed. The two beams combine on photographic film, where a complicated fringe pattern is formed. When light shines on the hologram, some of it will pass through the bright fringes, and some will be absorbed by the dark fringes. By observing the light from the hologram one reveals a three-dimensional replica of the original object.

Holographic interferometry is used to view small changes in an object. When two holograms, taken at different times, of the same object are superimposed, fringes will reveal the difference between the two objects. It is possible to see slowly varying changes of a growing **plant**, or rapidly varying changes of a vibrating object such as the face of a violin.

See also Hologram and holography.

Resources

Books

Ditchburn, R.W. *Light.* New York: Dover, 1991.

Hariharan, P. *Basics of Interferometry*. San Diego: Academic Press, 1992.

Newton, Isaac. *Opticks*. First printed, 1704. Reprint, New York: Dover, 1979.

Smith, F.G., and J.H. Thomson. *Opticks*. New York: Wiley, 1988.

Steel, W.H. *Interferometry*. New York: Cambridge University Press, 1967.

Robert K. Tyson

Interferons

Interferons are species-specific **proteins** that induce antiviral and antiproliferative responses in **animal** cells. They are a major defense against viral infections and abnormal growths (neoplasms). Interferons are produced in response to penetration of animal cells by viral (or synthetic) **nucleic acid** and then leave the infected **cell** to confer resistance on other cells of the **organism**.

Interferons are a group of proteins known primarily for their role in inhibiting viral infections and in stimulating the entire **immune system** to fight **disease**. Research has also shown that these proteins play numerous roles in regulating many kinds of cell functions. They can promote or hinder the ability of some cells to differentiate, that is, to become specialized in their function. They can inhibit **cell division**, which is one reason why they hold promise for stopping **cancer** growth. Recent studies have also found that one interferon may play an important role in the early biological processes of pregnancy. Although once thought to be a potential cure-all for a number of viral diseases and cancers, subsequent research has shown that interferons are much more limited in their potential. Still, several interferon proteins have been approved as therapies for diseases like chronic **hepatitis**, genital warts, multiple sclerosis, and several cancers.

The first interferon was discovered in 1957 by Alick Isaacs and Jean Lindenmann. During their investigation, the two scientists found that virus-infected cells secrete a special protein that causes both infected and noninfected cells to produce other proteins that prevent viruses from replicating. They named the protein interferon because it "interferes" with **infection**. Initially, scientists thought there was only one interferon protein, but subsequent research showed that there are many different interferon proteins.

Types of interferons and how they work

Interferons are members of a larger class of proteins called cytokines (proteins that carry signals between cells). Most interferons are classified as alpha, beta, or gamma interferons, depending on their molecular structure. Two other classes of interferons—omega and tau—have also been discovered. So far, more than 20 different kinds of interferon-alpha have been discovered but only beta and one gamma interferon have been identified.

Interferons are differentiated primarily through their **amino acid** sequence. (Amino acids are molecular chains that make up proteins.) Interferon-alpha, -beta, -tau, and -omega, which have relatively similar amino acid sequences, are classified as type I interferons. Type I interferons are known primarily for their ability to make cells resistant to viral infections. Interferon-gamma is the only type II interferon, classified as such because of its unique amino acid sequence. This interferon is known for its ability to regulate overall immune system functioning.

In addition to their structural makeup, type I and type II interferons have other differences. Type I interferons are produced by almost every cell in the body while the type II interferon-gamma is produced only by specialized cells in the immune system known as T lymphocytes and natural killer cells. The two classes also bind to different kinds of receptors, which lie on the surface of cells and attract and combine with specific molecules of various substances.

Interferons work to stop a disease when they are released into the **blood** stream and then bind to cell receptors. After binding, they are drawn inside the cell's cytoplasm, where they cause a series of reactions that produce other proteins that fight off disease. Scientists have identified over 30 disease fighting proteins produced by interferons.

In contrast to antibodies, interferons are not **virus** specific but host specific. Thus, viral infections of human cells are inhibited only by human interferon. The human **genome** contains 14 nonallelic and 9 allelic genes of alpha-interferon (macrophage interferon), as well as a single **gene** for beta-interferon (fibroblast interferon). Genes for any two or more variants of interferon, which have originated from the same wild-type gene are called allelic genes and will occupy the same chromosomal location (**locus**). Variants originating from different standard genes are termed non allelic. Alpha- and beta-interferons are structurally related glycoproteins of 166 and 169 amino acid residues. In contrast, gamma-interferon (also known as immune interferon) is not closely related to the other two and is not induced by virus infection. It is produced by **T cells** after stimulation with the cytokine interleukin-2. It enhances the cytotoxic activity of T cells, macrophages and natural killer cells and thus has antiproliferative effects. It also increases the production of antibodies in response to antigens administered simultaneously with alpha-interferon, possible by enhancing the antigen-presenting function of macrophages.

Interferons bind to specific receptors on the cell surface, and induce a signal in the cell interior. Two induction mechanisms have been elucidated. One mechanism involves the induction of protein kinase by interferon, which, in the presence of double-stranded RNA, phosphorylates one subunit of an initiation factor of protein synthesis (eIF-2B), causing the factor to be inactivated by sequestration in a complex. The second mechanism involves the induction of the **enzyme** 2',5'-oligoadenylate synthetase (2',5'-oligo A synthestase). In the presence of double-stranded RNA, this enzyme catalyses the polymerisation of ATP into oligomers of 2 to 15 adenosine monophosphate residues which are linked by phosphodiester bonds between the position 2' of one ribose and 5' of the next. These 2',5'-oligoadenylates activate an interferon specific RNAase, a latent endonuclease known as RNAase L which is always present but not normally active. RNAase cleaves both viral and cellular single stranded mRNA. Interferons therefore do not directly protect cells against viral infection, but rather render cells less suitable as an environment for viral replication. This condition is known as the antiviral state.

Interferons and the immune system

In addition to altering a cell's ability to fight off viruses, interferons also control the activities of a number of specialized cells within the immune system. For example, type I interferons can either inhibit or induce the production of B lymphocytes (white blood cells that make antibodies for fighting disease). Interferon-gamma can also stimulate the production of a class of T lymphocytes known as suppressor CD8 cells, which can inhibit B cells from making antibodies.

Another role of interferon-gamma is to increase immune system functioning by helping macrophages, still another kind of white blood cell, to function. These **scavenger** cells attack infected cells and also stimulate other cells within the immune system. Interferon-gamma is especially effective in switching on macrophages to kill **tumor** cells and cells that have been infected by viruses, **bacteria**, and **parasites**.

Interferon-tau, first discovered for its role in helping pregnancy to progress in cows, **sheep**, and **goats**, also has antiviral qualities. It has been shown to block tumor cell division and may interfere with the replication of the acquired immune deficiency, or **AIDS**, virus. Since it has fewer unwanted side-effects (flu-like symptoms and decreased blood cell production) than the other interferons, interferon-tau is becoming a new focal point for research.

Commercial fermentation units like these are used to grow cultures of microorganisms for biological products like interferon. *Photograph by S. Stammens. National Audubon Society Collection/Photo Researchers, Inc. Reproduced by permission.*

Interferon's medical applications

In 1986, interferon-alpha became the first interferon to be approved by the Food and Drug Administration (FDA) as a viable therapy, in this case, for hairy-cell **leukemia**. (Interferons are used therapeutically by injecting them into the blood stream.) In 1988, this class of interferons was also approved for the treatment of genital warts, proving effective in nearly 70% of patients who do not respond to standard therapies. In that same year, it was approved for treatment of Kaposi's sarcoma, a form of cancer that appears frequently in patients suffering from AIDS. In 1991, interferon-alpha was approved for use in chronic hepatitis C, a contagious disease for which there was no reliable therapy. Interferon has been shown to eliminate the disease's symptoms and, perhaps, prevent relapse. Interferon-alpha is also used to treat Hodgkin's lymphoma and malignant melanoma, or skin cancer.

In 1993, another class of interferon, interferon-gamma, received FDA approval for the treatment of a form of multiple sclerosis characterized by the intermittent appearance and disappearance of symptoms. It has also been used to treat chronic granulomatous diseases, an inherited immune disorder in which white blood cells fail to kill bacterial infections, thus causing severe infections in the skin, liver, lungs, and bone. Interferon-gamma may also have therapeutic value in the treatment of leishmaniasis, a parasitic infection that is prevalent in parts of **Africa**, America, **Europe**, and **Asia**.

Although all of the disease fighting attributes of interferon demonstrated in the laboratory have not been attained in practice, continued research into interferons will continue to expand their medical applications. For example, all three major classes of interferons are under investigation for treating a variety of cancers. Also,

KEY TERMS

Immune system—That network of tissues and cells throughout the body which is responsible for ridding the body of invaders such as viruses, bacteria, protozoa, etc.

Proteins—Macromolecules made up of long sequences of amino acids. They make up the dry weight of most cells and are involved in structures, hormones, and enzymes in muscle contraction, immunological response, and many other essential life functions.

Type I interferons—A group of interferons that have similar amino acid sequences. They include the alpha, beta, tau, and omega interferons.

Type II interferons—A type of interferon that has a unique amino acid sequence. Interferon gamma is the only interferon in this group.

biotechnological advances making **genetic engineering** easier and faster are making protein drugs like interferons more available for study and use. Using **recombinant DNA** technology, or **gene splicing**, genes that code for ineterferons are identified, cloned, and used for experimental studies and in making therapeutic quantities of protein. These modern DNA manipulation techniques have made possible the use of cell-signaling molecules like interferons as medicines. Earlier, available quantities of these molecules were too minute for practical use.

Another particular area of interest is the use of interferons to enhance other therapies. For example, studies have shown that a combination of interferon-alpha and tamoxifen may be a more effective therapy for breast cancer than either used alone. Future studies will focus more on combining interferons with other drug therapies.

See also Antibody and antigen; Immunology.

Resources

Books

Janeway, Charles A., et al. *Immunobiology.* 5th ed. New York and London: Garland Publishing, 2001.

Periodicals

Johnson, Howard M., Fuller W. Bazer, Brian E. Szente, and Michael A. Jarpe. "How Interferons Fight Disease." *Scientific American* (May 1994): 68–76.
Meulen, Volkerter, and Stefan Niewiesk. "Inhibition of Major Histocompatibility Complex Class II-Dependant Antigen Presentation by Nutralization of Gamma Interferon Leads to Breakdown of Resistance against Measles Virus-Induced Encephalitis." *Journal of Virology* 75 (2000):1–13.
Seppa, Nathan. "Interferon Delays Multiple Sclerosis." *Science News* 158 (November 2000): 280–281.

Other

Multiple Sclerosis Society. Interferons. [cited April 3, 2003] <http://www.interferons.com>.
Worman, M.D., J. Howard. "Interferon Treatment of Viral Hepatitis". Dept. of Gastrointestinal Diseases, Columbia University. April 6, 2002 [January 2003]. <http://cpmcnet.coumbia.edu/dept/gi/intron.html>.

David Petechuk
Judyth Sassoon

Internal combustion engine

The invention and development of the internal **combustion** engine in the nineteenth century has had a profound impact on human life. The internal combustion engine offers a relatively small, lightweight source for the amount of power it produces. Harnessing that power has made possible practical machines ranging from the smallest model airplane to the largest truck. **Electricity** is often generated by internal combustion engines. Lawnmowers, chainsaws, and generators also may use internal combustion engines. An important device based on the internal combustion engine is the **automobile**.

In all internal combustion engines, however, the basic principles remain the same. Fuel is burned inside a chamber, usually a cylinder. The **energy** created by the combustion, or burning, of the fuel is used to propel a device, usually a piston, through the chamber. By attaching the piston to a shaft outside of the chamber, the movement and **force** of the piston can be converted to other movements.

Principles

Combustion is the burning of fuel. When fuel is burned it gives off energy, in the form of **heat**, which creates the expansion of gas. This expansion can be rapid and powerful. The force and movement of the expansion of gas can be used to push an object. Shaking a can of soda is a way to see what happens when gas expands. The shaking **motion** causes a reaction of carbon dioxide—the soda's fizz—which, when the can is opened, pushes the soda's liquid from the can.

Simply burning fuel, however, is not very useful for creating motion. Lighting a match, for example, burns the **oxygen** in the air around it, but the heat raised is lost in all directions, and therefore gives a very weak push. In order for the expansion of gas caused by combustion to be made useful, it must occur in a confined space. This space can channel, or direct the movement of the expansion; it can also increase its force.

A cylinder is a useful space for channeling the force of combustion. The round inside of the cylinder allows gases to flow easily, and also acts to increase the strength of the movement of the gases. The circular movement of the gases can also assist in pulling air and vapors into the cylinder, or force them out again. A rocket is a simple example of the use of internal combustion within a cylinder. In a rocket, the bottom end of the cylinder is open. When the fuel inside the cylinder explodes, gases expand rapidly toward the opening, giving the push needed to force the rocket from the ground.

This force can be even more useful. It can be made to push against an object inside the cylinder, causing it to move through the cylinder. A bullet in a pistol is an example of such an object. When the fuel, in this case gunpowder, is exploded, the resulting force propels the bullet through the cylinder, or barrel, of the pistol. This movement is useful for certain things; however, it can be made still more useful. By closing the ends of the cylinder, it is possible to control the movement of the object, making it move up and down inside the cylinder. This movement, called reciprocating motion, can then be made to perform other tasks.

Structure of the internal combustion engine

Internal combustion engines generally employ reciprocating motion, although gas **turbine**, rocket, and rotary engines are examples of other types of internal combustion engines. Reciprocating internal combustion engines are the most common, however, and are found in most cars, trucks, motorcycles, and other engine-driven machines.

The most basic components of the internal combustion engine are the cylinder, the piston, and the crankshaft. To these are attached other components that increase the efficiency of the reciprocating motion and convert that motion to the rotary motion of the crankshaft. Fuel must be provided into the cylinder, and the exhaust, formed by the explosion of the fuel, must be provided a way out of the cylinder. The ignition, or lighting, of the fuel must also be produced. In the reciprocating internal combustion engine, this is done in one of two ways.

Diesel engines are also called compression engines because they use compression to cause the fuel to self-ignite. Air is compressed, that is, pushed into a small space, in the cylinder. Compression causes the air to heat up; when fuel is introduced to the hot, compressed air, the fuel explodes. The **pressure** created by compression requires diesel engines to be more strongly constructed,

and thus, heavier than gasoline engines, but they are more powerful, and require a less costly fuel. Diesel engines are generally found in large vehicles, such as trucks and heavy construction equipment, or in stationary machines.

Gasoline engines are also called spark ignition engines because they depend on a spark of electricity to cause the explosion of fuel within the cylinder. Lighter than a **diesel engine**, the gas engine requires a more highly refined fuel.

In an engine, the cylinder is housed inside a engine block strong enough to contain the explosions of fuel. Inside the cylinder is a piston which fits the cylinder precisely. Pistons generally are dome-shaped on top, and hollow at the bottom. The piston is attached, via a connecting rod fitted in the hollow bottom, to a crankshaft, which converts the up and down movement of the piston to a circular motion. This is possible because the crankshaft is not straight, but has a bent section (one for every cylinder) called a crank.

A similar structure propels a bicycle. When bicycling, the upper part of a person's leg is akin to the piston. From the knee to the foot, the leg acts as a connecting rod, which is attached to the crankshaft by the crank, or the bicycle's pedal assembly. When power is applied with the upper leg, these parts are made to move. The reciprocating motion of the lower leg is converted to the rotary, or spinning, motion of the crankshaft.

Notice that when bicycling, the leg makes two movements, one down and one up, to complete the pedaling cycle. These are called strokes. Because an engine also needs to draw fuel in and expel the fuel out again, most engines employ four strokes for each cycle the piston makes. The first stroke begins when the piston is at the top of the cylinder, called the cylinder head. As it is drawn down, it creates a **vacuum** in the cylinder. This is because the piston and the cylinder form an airtight space. When the piston is pulled down, it causes the space between it and the cylinder head to become larger, while the amount of air remains the same. This vacuum helps to take the fuel into the cylinder, much like the action of the lungs. This stroke is therefore called the intake stroke.

The next stroke, called the compression stroke, occurs when the piston is pushed up again inside the cylinder, squeezing, or compressing the fuel into a tighter and tighter space. The compression of the fuel against the top of the cylinder causes the air to heat up, which also heats the fuel. Compressing the fuel also makes it easier to ignite, and makes the resulting explosion more powerful. There is less space for the expanding gases of the explosion to flow, which means they will push harder against the piston in order to escape.

At the top of the compression stroke, the fuel is ignited, causing an explosion that pushes the piston down. This stroke is called the power stroke, and this is the stroke that turns the crankshaft. The final stroke, the exhaust stroke, takes the piston upward again, which expels the exhaust gases created by the explosion from the cylinder through an exhaust valve. These four strokes are also commonly called "suck, squeeze, bang, and blow." Two-stroke engines eliminate the intake and exhaust strokes, combining them with the compression and power strokes. This allows for a lighter, more powerful engine—relative to the engine's size—requiring a less complex design. But the two-stroke cycle is a less efficient method of burning fuel. A residue of unburned fuel remains inside the cylinder, which impedes combustion. The two-stroke engine also ignites its fuel twice as often as a four-stroke engine, which increases the wear on the engine's parts. Two-stroke engines are therefore used mostly where a smaller engine is required, such as on some motorcycles, and with small tools.

Combustion requires the presence of oxygen, so fuel must be mixed with air in order for it to ignite. Diesel engines introduce the fuel directly to react with the hot air inside the cylinder. Spark-ignition engines, however, first mix the fuel with air outside the cylinder. This is done either through a carburetor or through a fuel-injection system. Both devices vaporize the gasoline and mix it with air at a **ratio** of around 14 parts of air to every one part of gasoline. A choke valve in the carburetor controls the amount of air to be mixed with the fuel; at the other end, a throttle valve controls how much of the fuel mixture will be sent to the cylinder.

The vacuum created as the piston moves down through the cylinder pulls the fuel into the cylinder. The piston must fit precisely inside the cylinder in order to create this vacuum. Rubber compression rings fitted into grooves in the piston make certain of an airtight fit. The gasoline enters the cylinder through an intake valve. The gasoline is then compressed up into the cylinder by the next movement of the piston, awaiting ignition.

An internal combustion engine can have anywhere from one to twelve or more cylinders, all acting together in a precisely timed sequence to drive the crankshaft. The bicyclist on a bicycle can be described as a two-cylinder engine, each leg assisting the other in creating the power to drive the bicycle, and in pulling each other through the cycle of strokes. Automobiles generally have four-, six-, or eight-cylinder engines, although two-cylinder and twelve-cylinder engines are also available. The number of cylinders affects the engine's displacement, that is, the total **volume** of fuel passed through the cylinders. A larger displacement allows more fuel to be burned, creating more energy to drive the crankshaft.

Spark is introduced through a spark plug placed in the cylinder head. The spark causes the gasoline to explode. Spark plugs contain two **metal** ends, called electrodes, which extend down into the cylinder. Each cylinder has its own spark plug. When **electric current** is passed through the spark plug, the current jumps from one electrode to the other, creating the spark.

This electric current originates in a **battery**. The battery's current is not, however, strong enough to create the spark needed to ignite the fuel. It is therefore passed through a **transformer**, which greatly amplifies its voltage, or strength. The current can then be sent to the spark plug.

In the case of an engine with two or more cylinders, however, the spark must be directed to each cylinder in turn. The sequence of firing the cylinders must be timed so that while one piston is in its power stroke, another piston is in its compression stroke. In this way, the force exerted on the crankshaft can be kept constant, allowing the engine to run smoothly. The number of cylinders affects the smoothness of the engine's operation; the more cylinders, the more constant the force on the crankshaft, and the more smoothly the engine will run.

The timing of the firing of the cylinders is controlled by the distributor. When the current enters the distributor, it is sent through to the spark plugs through leads, one for each spark plug. Mechanical distributors are essentially spinning rotors that send current into each lead in turn. Electronic ignition systems utilize computer components to perform this task.

The smallest engines use a battery, which, when drained, is simply replaced. Most engines, however, have provisions for recharging the battery, utilizing the motion of the spinning crankshaft to generate current back to the battery.

The piston or pistons push down and pull up on the crankshaft, causing it to spin. This conversion from the reciprocating motion of the piston to the rotary motion of the crankshaft is possible because for each piston the crankshaft has a crank, that is, a section set at an **angle** to the up-and-down movement of the position. On a crankshaft with two or more cylinders, these cranks are set at angles to each other as well, allowing them to act in concert. When one piston is pushing its crank down, a second crank is pushing its piston up.

A large metal wheel-like device called a flywheel is attached to one end of the crankshaft. It functions to keep the movement of the crankshaft constant. This is necessary on a four-stroke engine because the pistons perform a power stroke only once for every four strokes. A flywheel provides the **momentum** to carry the crankshaft through its movement until it receives the next power stroke. It does this by using inertia, that is, the principle that an object in motion will tend to stay in motion. Once the flywheel is set in motion by the turning of the crankshaft, it will continue to move, and turn the crankshaft. The more cylinders an engine has, however, the less it will need to rely on the movement of a flywheel, because the greater number of pistons will keep the crankshaft spinning.

Once the crankshaft is spinning, its movement can be adapted to a great variety of uses, by attaching **gears**, belts, or other devices. Wheels can be made to turn, propellers can be made to spin, or the engine can be used simply to generate electricity. Also geared to the crankshaft is an additional shaft, called the camshaft, which operates to open and close the intake and exhaust valves of each cylinder in sequence with the four-stroke cycle of the pistons. A cam is a wheel that is more or less shaped like an egg, with a long end and a short end. Several cams are fastened to the camshaft, depending on the number of cylinders the engine has. Set on top of the cams are push-rods, two for each cylinder, which open and close the valves. As the camshaft spins, the short ends allow the push-rods to draw back from the valve, causing the valve to open; the long ends of the cams push the rods back toward the valve, closing it again. In some engines, called overhead cam engines, the camshaft rests directly on the valves, eliminating the need for the push-rod assembly. Two-stroke engines, because the intake and exhaust is achieved by the movement of the piston over ports, or holes, in the cylinder wall, do not require the camshaft.

Two more components may be operated by the crankshaft: the cooling and lubrication systems. The explosion of fuel creates intense heat that would quickly cause the engine to overheat and even melt if not properly dissipated, or drawn away. Cooling is achieved in two ways, through a cooling system and, to a lesser extent, through the lubrication system.

There are two types of cooling systems. A liquid-cooling system uses **water**, which is often mixed with an antifreeze to prevent freezing. Antifreeze lowers the freezing point and also raises the **boiling point** of water. The water, which is very good at gathering heat, is pumped around the engine through a series of passageways contained in a jacket. The water then circulates into a radiator, which contains many tubes and thin metal plates that increase the water's surface area. A fan attached the radiator passes air over the tubing, further reducing the water **temperature**. Both the pump and the fan are operated by the crankshaft's movement.

Air-cooled systems use air, rather than water, to draw heat from the engine. Most motorcycles, many small airplanes, and other machines where a great deal

of **wind** is produced by their movement, use air-cooled systems. In these, metal fins are attached to the outside of the cylinders, creating a large surface area; as air passes over the fins, the heat conducted to the metal fins from the cylinder is swept away by the air.

The lubrication of an engine is vital to its operation. The movement of parts against each other cause a great deal of **friction**, which raises heat and causes the parts to wear. Lubricants, such as oil, provide a thin layer between the moving parts. The passage of oil through the engine also helps to carry away some of the heat produced.

The crankshaft at the bottom of the engine rests in a crankcase. This may be filled with oil, or a separate oil pan beneath the crankcase serves as a reservoir for the oil. A pump carries the oil through passageways and holes to the different parts of the engine. The piston is also fitted with rubber oil rings, in addition to the compression rings, to carry oil up and down the inside of the cylinder. Two-stroke engines use oil as part of their fuel mixture, providing the lubrication for the engine and eliminating the need for a separate system.

Resources

Books

Schuster, William A. *Small Engine Technology.* Delmar Publishers, Inc., 1993.

Stone, Richard. *Introduction to Internal Combustion Engines.* Society of Automotive Engineers, Inc., 1994.

M. L. Cohen

International Space Station

The International Space Station (ISS), formally designated International Space Station Alpha, is a habitable orbital facility that has been under construction since 1998 and is scheduled for completion in 2006. When finished, it will contain about four times as much working space as the Russian space station *Mir* (1986–2001), the former record holder, and will weigh about one million pounds (453,000 kg). The ISS orbits at an average altitude of 240 mi (387 km). A number of science experiments are to be conducted aboard the ISS in such fields as health effects of **radiation**, molecular and **cell biology**, **earth science**, **fluid dynamics**, **astronomy**, **combustion physics**, and **crystal** growth.

History and structure

The ISS was originally proposed by U.S. President Ronald Reagan in 1984, and was slated to cost $8 billion. Thirty-six U.S. shuttle flights plus nine Russian rocket launches will be required for ISS construction. Today there are 15 major partners in the ISS effort, including the United States, Russia, Japan, Canada, and 11 of the member states of the European Space Agency. The United States, through its National Aeronautics and Space Administration (NASA), is the largest single contributor, bearing approximately $25 billion of the total $50–100 billion cost of building, launching, and operating ISS for at least a decade.

Assembly of the ISS commenced in 1998 with launch of the Russian control module Zarya on a **proton** rocket from Kazahkstan. The U.S. module Unity Node, a connecting segment, was carried into space on the shuttle *Endeavor* later in 1998. This unit is primarily a docking hub to which other sections join. In 2000, another Proton rocket lofted the Russian service module Zvezda, the main Russian contribution to the ISS. Zvezda provided living quarters and life support during the early phases of the ISS's growth; it also provides steering rockets to control the ISS's attitude (orientation in space) and to reboost it to higher altitudes as its **orbit** decays due to **friction** with high-altitude traces of the earth's atmosphere.

The ISS is powered by photovoltaic **electricity**. The first of its four large solar arrays (112 by 39 ft [34.2 by 11.9 m]) was added in 2000. When completed, the ISS will receive about 260 kilowatts of power (peak) from an acre of sun-tracking solar panels. An energy-storage subsystem consisting of six large nickel-hydrogen (Ni-H$_2$) batteries supplies electrical power to the ISS during its passage through the earth's shadow, which lasts about 45 out of every 90 minutes.

In 2001, the U.S. laboratory module Destiny, the largest and most elaborate of the ISS's components, was added using the robot arm of the **space shuttle** *Atlantis*. The U.S. lab module contains 13 equipment racks, on which various scientific experiments will be mounted, and a 20-in (0.5-m) window set in the Earthside wall.

Smaller components were added piecemeal in 2002 by several shuttle flights, and in 2001–2002 several Russian flights ferried passengers and supplies. The

ISS's final configuration will contain a European laboratory module, a Japanese laboratory module, three Russian laboratory modules, a Canadian robot arm to assist in assembly and maintenance, exterior racks for experiments requiring direct exposure to space, and an emergency Crew Return Vehicle on standby. The shuttle has been continuously inhabited since November 2, 2000, and presently houses a full crew of seven.

Science

The ISS is intended to serve as a platform for the performance of scientific experiments that can only be carried out in space. The presence of a crew allows more complex experiments to be performed with simpler equipment than would be possible using purely robotic space vehicles; on the other hand, human beings require much complex gear to survive in space. Further, the ISS is not a particularly efficient platform for astronomical experiments, as it is vulnerable to by vibrations, and experiments that merely require a **vacuum** can be performed economically in vacuum chambers on **Earth**. Yet, the ISS does offer something that cannot be obtained for more than a few seconds at a **time** on Earth: weightlessness, or, more precisely (since the components of the ISS itself create a slight gravitational field), *microgravity*.

Unlike traditional science-fiction space stations, the ISS does not rotate in order to provide a centrifugal equivalent of gravity to its inhabitants. Such an arrangement would require a much more expensive structure due to the stresses imposed by **rotation**; observational science experiments that need to point steadily at one part of the sky would be difficult to operate on a rapidly rotating platform; and rotation would destroy the very microgravitational conditions that make the ISS a unique place to conduct science.

Several of the experiments that have exploited (or will exploit) microgravity are the following:

Dendrite formation in solidifying metals. When metals solidify they tend, like snowflakes, to form branching or tree-like crystalline structures termed dendrites (from the Greek "dendrites," meaning "pertaining to a tree"). Observing the growth of metallic dendrites undeformed by the earth's gravity should help improve mathematical models of dendrite formation, which in turn may help in the design of stronger and more durable alloys.

Bone deterioration. As previous experience with long-term habitation of space has shown, persons living in weightlessness lose about 1% of their bone **mass** per month, even when performing bone-stressing exercises. Generations of small animals raised in space will enable biologists to study the effect of microgravity on genetic mechanisms of bone growth and resorbption. Under-

James Newman, an astronaut on the space shuttle *Endeavor* preparing for mission, working to connect wires on the International Space Station. *AP/Wide World Photos. Reproduced by permission.*

standing these mechanisms may someday make long space voyages (e.g., to **Mars**) medically feasible.

Atomic clock. A French experiment will use microgravity to improve the **accuracy** of an **atomic clock** by a factor of 10 by observing **oscillations** of cesium **atoms** in free fall.

Commercial research. Between 30–40% of the U.S. lab module resources are reserved for use by private corporations, who will pay for access to microgravity research conditions. A slightly lower percentage of lab resources are reserved for commercial buyers in the European laboratory module. However, few corporations have yet purchased lab time on the ISS.

Controversy

The ISS is enthusiastically supported by many people who are interested in space travel for its own sake and by those scientists who hope to fly their own experiments on the platform. However, it has long been heavily criticized by a majority of the scientific community for delivering too little science for the dollar and thus, in effect, diverting money from more effective research. Some scientists argue that the bulk of the research planned for the

ISS addresses technical questions that are peripheral, rather than fundamental. For example, *Science*, the journal of the American Association for the Advancement of Science, complained in 1998 that the ISS's "greatest impact will be felt in the small community already studying problems related to spaceflight—a vital research area only if we assume that increasing numbers of people will someday travel, or even live, outside of normal Earth gravity." In other words, the ISS is an ultimately romantic project that puts astronauts in space in order to figure out how to put *more* astronauts in space.

The claim that the ISS has little to offer science was boosted by Russia's conveyance to the ISS of two private space tourists—officially designated Space Flight Participants—in 2001 and 2002, over loud protests from other ISS participant nations. Two wealthy men, one American and the other South African, paid $20 million apiece to the cash-strapped Russian government in exchange for a trip to the ISS.

Even before the loss of the space shuttle Columbia in February 2003, funding for completion of the ISS was in doubt. Both the United States Congress and the governments of the European Union have long been skeptical of the ISS's costs, and NASA was under such political **pressure** that it admitted it could not guarantee that the station will ever be grown beyond the "core complete" stage, with long-term living quarters for only three astronauts. Three astronauts, however, are not enough to tend the scores of experiments for which the ISS's racks have room, so if the ISS is not expanded much of the science potential already constructed will go to waste. Critics of the ISS argued that continued support for the ISS amounted to throwing good money after bad; supporters of the ISS counter-argued that ISS research is essential for make an eventual trip to Mars and that human space-travel projects generate valuable technological spinoffs.

The Columbia disaster of early 2003 has, as of this writing, made the ISS's future murkier. Although Russian rockets can supply many of the ISS's needs and ferry astronauts back and forth to it, only the space shuttle's cargo hold is large enough to carry many of the components planned for the ISS. Another, more urgent factor is that the ISS loses orbital altitude steadily due to friction with the outer fringes of the earth's atmosphere. Small rockets attached to the station regularly restore its altitude. The fuel for these rockets has been delivered via space shuttle, but after the Columbia disaster, a long delay seemed certain before frequent shuttle flights could be resumed, and Russian spacecraft have not been designed to deliver sufficient fuel. Engineers in both Russia and the United States have proposed alternate solutions, but as of March 2003 no firm course of action had been approved.

See also Gravity and gravitation; Spacecraft, manned.

Resources

Periodicals

Lawler, Andrew. "Can Space Station Science Be Fixed?" *Science* 5572 (May 24, 2002): 1387–1389.

Lawler, Andrew. "Space Station Research: Bigger Is Better for Science, Says Report." *Science* 5580 (July 19, 2002, 2002): 316–317.

Revkin, Andrew. "And Now, the Space Station: Grieving, Imperiled." *New York Times,* February 4, 2003.

"Tension and Relaxation in Space-Station Science." *Nature* 391 (February 19, 1998): 721.

Young, Laurence "The International Space Station at Risk." *Science* 5567 (April 19, 2002): 429.

Larry Gilman

International Ultraviolet Explorer

The International Ultraviolet Explorer **satellite** (IUE) was a joint project of the National Aeronautics and Space

Administration (NASA), the European Space Agency (ESA), and the Planetary Plasma and Atmospheric Research Center (PPARC) in the United Kingdom. NASA provided the spacecraft, **telescope**, spectrographs, and one ground observatory facility. ESA created the solar panels for powering the craft in **orbit**, and the second ground observatory site. The PPARC provided the four spectrographic detectors. In addition to controlling the satellite, the ground sites acted as typical astronomical observatories, except that instead of using telescopes at their locations, their direct participation was by a link to a telescope orbiting far out in space. IUE was the longest lasting and most productive orbiting astronomical observatory up to its time. It was also the first orbiting ultraviolet observatory available to general users, and the first orbiting astronomical observatory in high **Earth** orbit. Because ultraviolet **light** from space is largely absorbed by our atmosphere, observations by IUE provided a whole new range of information not readily available from the ground. Only a small number of high-altitude observatories on Earth can be used with limited effectiveness for ultraviolet studies.

IUE was launched into geosynchronous orbit on January 26, 1978 and remained there until 1996. During these nearly 19 years of operation, it sent to Earth 104,470 images of 9,600 astronomical objects, ranging from **comets** in the inner **solar system** to quasars at the edge of the known universe. IUE was the first scientific satellite that allowed "visiting" astronomers to make real-time observations of ultraviolet spectra with a response time of less than one hour. This provided great flexibility in scheduling observation targets for the satellite. In conjunction with the IUE, simultaneous ground-based observations were performed in wavelengths other than ultraviolet in order to provide measurements of the same objects over a wide range of the **electromagnetic spectrum**. This provided astrophysicists with a new "multi-wavelength" method of looking at objects. The end result was a vast archive of new and more complete information than ever before made available to the scientific community worldwide.

IUE greatly surpassed its expected lifetime and the original science goals set for the mission. These included:

• Obtaining high-resolution spectra of stars of all types in order to determine their physical characteristics. The IUE extended the range of observations available from ground-based observatories into the ultraviolet region.

• Studying streams of gas in and around **binary star** systems, which are difficult to observe from the ground or with standard optical telescopes even from space.

• Observing faint stars, galaxies, and quasars at low resolution, and comparing these spectra to high-resolution spectra of the same objects.

- Obtaining ultraviolet spectra of planets and comets, again extending our knowledge by looking at them in new ways. Such spectra help determine the composition of the atmospheres of planets and gas content of comets.
- Making repeated observations of objects with spectra that change over time in order to reveal new information about them. The long duration of IUE allowed several long-term studies to be performed on objects in areas never before possible.
- Studying the changes of observed starlight passing through interstellar dust and gas. This can reveal how much and what type of gas and dust exists between Earth and the objects from which the light originated.

IUE firsts

IUE contributed to a number of studies and made discoveries that might not have been possible without the long-term availability of a successfully working satellite. One was the discovery of short-term variations in the auroras in the atmosphere of **Jupiter** (which were initially discovered by IUE). Since auroras are caused by the interaction between the upper atmosphere of a **planet** and particles radiated from the **sun**, and the **emission** of these particles increases as the sun becomes more active, the long life of IUE allowed unique studies associating Jovian aurora activity with the solar sunspot cycle. IUE was the first instrument to provide a systematic study of the distribution of different **species** of comets in space. The long life of IUE also enabled the monitoring of variations in the occurrence of different types of comets, the discovery of new material within them, and the classification of comets into groups as a function of age. The behavior and distribution of stellar particle **radiation** (stellar winds) is now beginning to become more clear, and there is hope of understanding the underlying mechanisms driving the stellar winds because of the observations performed with IUE. IUE spectra combined with optical observations have allowed distances to the Magellanic Clouds, the closest galaxies to the **Milky Way**, to be determined. Many other studies within and outside our **galaxy** were also conducted adding significant data to the science of **astrophysics**. Volumes have already been published on these and many other topics. With the importance of the IUE observations and the concurrent development of the Internet while the data was being received and analyzed, the IUE data archive has become the most heavily used astronomical archive in existence.

Possible future needs were identified during and after the IUE mission, which brought about the concept of creating a World Space Observatory that could provide flexible access to space-based observatories and observation times for astrophysicists world-wide. A working group was formed to further study the associated problems and opportunities.

With all its success, IUE had a few serious problems during its very long mission. All of these came from the fact that five of the six gyroscopes in its attitude control system failed over the years. After the fourth one failed in 1985, IUE continued operations because of the use of its fine sun sensor as a substitute to controlling the attitude of the spacecraft. Even when another gyro was lost in the final year, IUE could still be stabilized in 3-axes, with only one remaining **gyroscope**, by adding **star** tracker measurements to other guidance parameters. Until October 1995, IUE was in continuous operation, controlled 16 hours a day from the Goddard Space Flight Center in Greenbelt, Maryland, and eight hours from ESA's Villafranca Satellite Tracking Station (VILSPA) west of Madrid, Spain. After that, ESA took on the role of redesigning control schemes to make it feasible to cover the science operations fully controlled from VILSPA. But then, only 16 hours were used for scientific operations, with eight hours used for spacecraft housekeeping. IUE

KEY TERMS

Geosynchronous orbit—When placed in orbit at an altitude of 22,241 mi (35,786 km) above the surface of Earth, a satellite orbits the earth once each day. This means it remains stationary over a specific location on Earth and is said to be synchronized with Earth. Communications satellites can be found in geosynchronous (also called geostationary) orbit above the equator.

Gyroscope—A device similar to a top, which maintains rotation about an axis while maintaining a constant orientation of that axis in space. The child's toy gyroscope is a very simple version of the gyros used to provide a frame of reference for guidance and attitude control systems in spacecraft.

High Earth orbit—The region around Earth above 500 mi (380 km) from the surface. This is where the communications and many other satellites are found. The Space Shuttle orbits Earth in low Earth orbit, about 300 mi (460 km) above the surface of Earth.

Magellanic Clouds—Two small irregular galaxies that are relatively close to our own. They can be seen in the sky from low northern and all southern latitudes as small fuzzy patches of light.

Spectrum—A display of the intensity of radiation versus wavelength.

---

remained operational until its attitude control fuel was deliberately released into space, its batteries drained and its transmitter turned off on September 30, 1996.

Resources

Books

Pasachoff, Jay M. *Contemporary Astronomy*. Saunders College Publishing, 1989.

Other

Starchild Project Team. *IUE Home Page*. [cited 2003]. <http://starchild.gsfc.nasa.gov/docs/StarChild/space_level2/iue.html>

Clint Hatchett

Internet file transfer and tracking

Internet messages (e-mail, instant messages, etc.) and file transfers leave an electronic trail that can be traced. Tracing is a process that follows the Internet activity backwards, from the recipient to the user. As well, a users Internet activity on web sites can also be tracked on the recipient site (i.e., what sites are visited and how often, the activity at a particular site). Sometimes this tracking and tracing ability is used to generate e-mail to the user promoting a product that is related to the sites visited. User information, however, can also be gathered covertly.

Techniques of Internet tracking and tracing can also enable authorities to pursue and identify those responsible for malicious Internet activity. For example, on February 8, 2000, a number of key commercial Internet sites such as Yahoo, Ebay, and Amazon were jammed with incoming information and rendered inoperable. Through tracing and tracking techniques, law enforcement authorities established that the attacks had arisen from the computer of a 15-year-old boy in Montreal, Canada. The youth, whose Internet identity was "Mafiaboy," was arrested within months of the incidents.

Law enforcement use of Internet tracking is extensive. For example, the U.S. Federal Bureau of Investigation has a tracking program designated Carnivore. The program is capable of scanning thousands of e-mails to identify those that meet the search criteria.

Tracking tools

Cookies

Cookies are computer files that are stored on a user's computer during a visit to a web site. When the user electronically enters the web site, the host computer automatically loads the file(s) to the user's computer.

The cookie is a tracking device, which records the electronic movements made by the user at the site, as well as identifiers such as a username and password. Commercial web sites make use of cookies to allow a user to establish an account on the first visit to the site and so to avoid having to enter account information (i.e., address, credit card number, financial activity) on subsequent visits. User information can also be collected unbeknownst to the user, and subsequently used for whatever purpose the host intends.

Cookies are files, and so can be transferred from the host computer to another computer. This can occur legally (i.e., selling of a subscriber mailing list) or illegally (i.e., "hacking in" to a host computer and copying the file). Also, cookies can be acquired as part of a law enforcement investigation.

Stealing a cookie requires knowledge of the file name. Unfortunately, this information is not difficult to obtain. A survey, conducted by a U.S. Internet security company in 2002, on 109,212 web sites that used cookies found that almost 55% of them used the same cookie name. Cookies may be disabled by the user, however, this calls for programming knowledge that many users do not have or do not wish to acquire.

Bugs or beacons

A bug or a beacon is an image that can be installed on a web page or in an e-mail. Unlike cookies, bugs cannot be disabled. They can be prominent or surreptitious. As examples of the latter, graphics that are transparent to the user can be present, as can graphics that are only 1x1 pixels in size (corresponding to a dot on a computer monitor). When a user clicks onto the graphic in an attempt to view, or even to close the image, information is relayed to the host computer.

Information that can be gathered by bugs or beacons includes:

- the user's IP address (the Internet address of the computer)
- the e-mail address of the user
- the user computer's operating system (which can be used to target viruses to specific operating systems)
- the URL (Uniform Record Locator), or address, of the web page that the user was visiting when the bug or beacon was activated
- the browser that was used (i.e., Netscape, Explorer)

When used as a marketing tool or means for an entrepreneur to acquire information about the consumer, bugs or beacons can be merely an annoyance. However, the acquisition of IP addresses and other user information can be used maliciously. For example, information on active e-mail addresses can be used to send "spam" e-mail or virus-laden email to the user. And, like cookies, the information provided by the bug or beacon can be useful to law enforcement officers who are tracking down the source of an Internet intrusion.

Active X, JavaScript

These computer-scripting languages are automatically activated when a site is visited. The mini-programs can operate within the larger program, so as to create the "pop-up" advertiser windows that appear with increasing frequency on web sites. When the pop-up graphic is visited, user information such as described in the above sections can be gathered.

Tracing e-mail

E-mail transmissions have several features that make it possible to trace their passage from the sender to the recipient computers. For example, every e-mail contains a section of information that is dubbed the header. Information concerning the origin **time**, date, and location of the message is present, as is the Internet address (IP) of the sender's computer.

If an alias has been used to send the message, the IP number can be used to trace the true origin of the transmission. When the originating computer is that of a personally owned computer, this tracing can often lead directly to the sender. However, if the sending computer serves a large community—such as a university, and through which malicious transmissions are often routed—then identifying the sender can remain daunting.

Depending on the e-mail program in use, even a communal facility can have information concerning the account of the sender.

The information in the header also details the route that the message took from the sending computer to the recipient computer. This can be useful in unearthing the identity of the sender. For example, in the case of "Mafiaboy," examination of the transmissions led to a computer at the University of California at Santa Barbara that hade been commandeered for the prank. Examination of the log files allowed authorities to trace the transmission path back to the sender's personal computer.

Chat rooms are electronic forums where users can visit and exchange views and opinions about a variety of issues. By piecing together the electronic transcripts of the chat room conversations enforcement officers can track down the source of malicious activity.

Returning to the example of "Mafiaboy," enforcement officers were able to find transmissions at certain chat rooms where the upcoming malicious activity was described. The source of the transmissions was determined to by the youth's personal computer. Matching the times of the chat room transmissions to the malicious events provided strong evidence of the youth's involvement.

While Internet tracking serves a useful purpose in law enforcement, its commercial use is increasingly being examined from the standpoint of personal privacy. The 1984 Cable Act in the United States permits the collection of such information if the information is deemed to aid future commercial developments. User consent is required. However, the information that is capable of being collected can exceed that needed for commerce.

See also Computer languages; Computer memory, physical and virtual memory; Computer software; Computer, analog; Computer, digital; Internet and the World Wide Web.

Resources

Books
Bosworth, Seymour (ed.), and Michel E. Kabay. *Computer Security Handbook.* New York: John Wiley & Sons, 2002.
National Research Council, Computer Science and Telecommunications Board. *Cyber Security Today and Tomorrow: Pay Now or Pay Later.* Washington, DC: The National Academies Press, 2002.
Northcutt, Stephen, Lenny Zeltser, Scott Winters, et al. *Inside Network Perimeter Security: The Definitive Guide to Firewalls, Virtual Private Networks (VPNs) Routers, and Intrusion Detection Systems.* Indianapolis: New Riders Publishing, 2002.

Brian Hoyle

Internet and the World Wide Web

Overview of the Internet

The Internet was born in 1983 as a product of academic and scientific communications. Universities and other academic institutions formed a network to connect their internal networks to a larger system, and these communications were built on standards or protocols for addressing systems and for exchanging data. Called the Transmission Control Protocol/Internet Protocol (TCP/IP), these included the word Internet that came to identify the huge, global network in use today.

By linking their communications, the original users of the Internet were able to exchange electronic mail (now known as e-mail), use file transfer protocol (ftp) to exchange data, obtain access via **telephone** lines to computers at other locations (through telnet), and to converse using newsgroups and bulletin boards. By the 1990s, the Internet was the common bond among millions of computers.

Internet history

The Internet did have a parent in a program called ARPANET, the Advanced Research Projects Agency Network. The United States Department of Defense developed ARPANET in 1969 as a network for organizations involved in defense research and as a secure communications system that would also survive attack. One of the characteristics of ARPANET was that its data were transmitted in so-called packets that were small parts of the longer messages the computers were exchanging. By segmenting the data and sending it by packet-switching, fewer problems in data transfer occurred. The system also had fault tolerance, which meant that communication errors could happen without shutting down the whole system.

When researchers began extending ARPANET into other applications, the National Science Foundation (NSF) adapted ARPANET's TCP/IP protocols to its own NSFNET network with many potential layers and the ability to carry far more communications. In fact, many other education and research organizations formed other networks in the 1980s; the Computer + Science Network (CSNET), the National Aeronautics and Space Administration (NASA), and the Department of Education (DOE) were among these. The need to make these networks a seamless operation was addressed when the National Research and Education Network (NREN) was formed, and it smoothed operations to make the Internet the network of all networks. By 1990, ARPANET ceased to exist because it had been fully replaced by the Internet.

While government and academic entities were developing networks that eventually combined under the infrastructure of the Internet, some businesses created successful networks of their own. Perhaps the most famous of these was Ethernet, a creation of Xerox Corporation that, in 1974, enabled all the machines in a single location to communicate with each other. In 1991, the Commercial Internet Exchange or CIX was formed by businesses with their own large networks. CIX is a high-speed interconnection point that allowed the member networks to exchange information for commercial purposes. CIX was largely independent of the NSFNET. Today, the Internet seems like one massive entity, and these separate networks are not easily distinguished in the global workings of the Internet.

The NSF remains actively involved in the operations and future of the Internet as one of several organizations that administers the Internet. The Internet Network Information Center (InterNIC) and the Internet Architecture Board (IAB) name networks and computers and resolve conflicts. Other organizations develop and administer protocols and engineer the complex interrelationships of networks.

While the Internet was evolving, need arose for methods for independent computer users to access the Internet. Within businesses, educational institutions, and government organizations, the Internet is accessed through a LAN or Local Area Network that provides service to all the employees of a company, for example, and is also a stepping-off point for Internet access. Independent users contract with commercial access providers to obtain Internet access. The commercial access providers are hosts to the Internet. They include America Online, Compuserve, Netcom, AT&T, and many other nationwide and local providers.

Internet communications use a number of other technologies. Services are transmitted by **television** cables, satellites, **fiber optics**, and **radio**. Cable television wires are steadily becoming more popular especially among users who want high-speed Internet services termed "broadband" services. Most consumers use modems (devices that translate electrical signals to sound signals and back) as the means of accessing the Internet through telephone lines. Special cable modems have speeds of 1.5 million bits (units of computer information) per second compared to the 56,000 (56k) bits per second (bps) of standard modems. Telephone companies also provide Digital Subscriber Line services that use a wider range of frequencies over regular telephone lines and can transmit data at 7 million bps (Mbps). Interest in cable net connections is outpacing the introduction of technologies like **color** television or cellular phones.

Evolution of the World Wide Web

In 1990, Tim Berners-Lee and other scientists at the international organization called CERN (European Center for Nuclear Research) in Geneva, Switzerland, developed a computer protocol called the HyperText Transfer Protocol (HTTP) that became the standard communications language between Internet clients and servers. Exchanges of information on the Internet take place between a server (a computer program that both stores information and transmits it from one computer to another) and a client (also a computer program but one that requests those transmittals of documents from the server). The client is not a person; the person giving instructions to the client is called a user. The first Web server in the United States was the Stanford Linear Accelerator Center (SLAC) in Palo Alto, California. To be able to look at retrieved documents, the user's computer is equipped with browser software. The programmers at CERN also developed a text-based Web browser that was made public in 1992; they also proposed the name World Wide Web for their system.

Documents that comply with the HTTP protocol are called hypertext documents and are written in HyperText Markup Language (HTML) which includes both text and links. Links are formally called hypermedia links or hyperlinks that connect related pieces of information through electronic connections. Through links, users can access arrays of documents identified by these shared links. Documents consisting of text are identified through hypertext; and other kinds of information like photos and other images, sounds, and video are identified as hypermedia. Users find and access hypertext or hypermedia through addresses called Uniform Resource Locators or URLs. URLs often contain the letters "http," "www," and "html" showing that, within the HTTP rules, they want to access the World Wide Web by speaking in HTML.

Web browsers

The World Wide Web helped new users to explore the Internet and became known as the Web or www. The World Wide Web is a graphical **map** for the Internet that is simple to understand and helps the user navigate around Internet sites; without the Web, the Internet would have remained a mystery to those without computer training. Web browsers have made the huge blossoming of use of the Web possible. Following CERN's pioneering work, the National Center for Supercomputing Applications at the University of Illinois developed Mosaic, a web browser that adapted the graphics, familiar icons (picture symbols), and point-and-click methods which were available on personal computers in 1993 to the Web. In 1994, Marc Andreesen, one of Mosaic's creators, helped form the Netscape Communications Corporation and devised Netscape Navigator, a highly successful Web browser that gave users comfortable access to the Web by using a mouse to click on familiar picture icons and search for information through links. These easy steps eliminate the need for the average user to understand **computer languages** and programming.

The cyberspace explosion

In two years (from 1993-1995), the World Wide Web exploded from an unknown entity to one which pervades every aspect of life: access to libraries around the world, recipes and coupons for tonight's dinner, medical advice, details on how to build your own **space shuttle**,

---

KEY TERMS

Cyberspace—the computer universe including software and data.

Hypermedia, hypermedia links, or hyperlinks—Computer sound, video and images that comply with HyperText Transfer Protocol (HTTP).

Hypertext—Computer text documents complying with HyperText Transfer Protocol (HTTP).

Internet—The huge network connecting all other networks.

Links—The electronic connections between pieces of information.

Local Area Network (LAN)—The private network used within a company or other organization.

Modem—A device that modulates electrical computer signals from the sender into telephone tones and demodulates them back to computer signals at the receiver's end.

Network—A system made up of lines or paths for data flow and nodes where the lines intersect and the data flows into different lines.

Packets—Small batches of data that computers exchange.

Protocols—Rules or standards for operations and behavior.

Web browser—Software that allows the user to access the World Wide Web and the Internet and to read and search for information.

---

and shopping for everything from music to mortgages. By 1997, 47 million Americans had attempted to access the Internet, prompting high-tech executives to classify the Internet as "mass media." Colleges are using the Internet to market their facilities, recruit students, and solicit funds from alumni. In 2003, the Internet search engine "Yahoo!" reviewed 4,000 campuses and identified the top 100 schools as the "most wired" with access to library catalogs, access to the Web for students, computer connections available to every dormitory resident, and a range of other services. Programs for younger students sponsored by the NSF and NASA let grade schoolers go on "electronic field trips" through closed-circuit television broadcasts from **Mars**, the South Pole, and other places far beyond the classroom.

A survey conducted in 2003 showed the average Internet user spent 11.2 hours per week using the Internet and that 25% of Internet users in the U.S. used broadband connections.

Resources

Books

Dern, Daniel P. *The Internet Guide for New Users.* New York: McGraw- Hill, Inc., 1994.

Falk, Bennett. *The Internet Roadmap.* San Francisco: SYBEX, 1994.

Ross, John. *Discover the World Wide Web.* Foster City, CA: IDG Books Worldwide, Inc., 1997.

Periodicals

Marklein, Mary Beth. "High-tech: Best-wired Schools Get Nod from 'Yahoo!'" *USA Today,* April 9, 1997.

Snider, Mike. "High-tech: Growing online Population Making Internet 'Mass Media.'" *USA Today* <http://www.usatoday.com> February 19, 1997.

Toon, Rhonda. "Technology & You: A Class Act on the Net." *Business Week,* July 28, 1997: 18.

Other

History of the Internet and World Wide Web. NetValley (2003). http://www.netvalley.com/intval.html

"Internet Speed Goes Up: Most Broadband Subscribers Use Cable Net Connections." [cited January 28, 1999] <http://www.abc.com >.

Gillian S. Holmes

Interstellar matter

On a clear winter night go outside to a dark location and look for the **constellation** Orion, the hunter. A row of three stars makes up his belt. Hanging from his belt is his sword, a smaller row of three fainter stars. If you look at the center **star** in the sword with a pair of binoculars or a small **telescope**, you will see a small fuzzy patch of interstellar gas and dust, called the Orion Nebula. **Space** is not empty. The **matter** in the space between the stars is called interstellar matter or the interstellar medium. The interstellar medium consists of **atoms**, ions, molecules, and dust grains. It is both concentrated into **clouds** and spread out between stars and the clouds. The interstellar medium is tenuous enough to qualify as a **vacuum** on the **earth**, but it plays a crucial role in the **evolution** of the **galaxy**. Stars are born out of the interstellar medium, and when stars die they recycle some of their material back into the interstellar medium.

Components of the interstellar medium

The interstellar medium can be broadly classified into gas and dust components. The average **density** of the interstellar gas is roughly one **hydrogen** atom per cubic centimeter. This density can however vary considerably for different components of the interstellar gas.

The components of the interstellar gas include: cold atomic gas clouds, warm atomic gas, the coronal gas, HII regions, and molecular clouds.

Gas

The cold atomic gas clouds consist primarily of neutral hydrogen atoms. Astronomers refer to neutral hydrogen atoms as HI, so these clouds are also called HI regions. These gas clouds have densities from 10–50 atoms per cubic centimeter and temperatures about 50–100K (-369.4– -279.4°F [-223– -173°C]). They can be as large as 30 **light** years and contain roughly 1,000 times the **mass** of the **sun**.

The warm atomic gas is much more diffuse than the cold atomic gas. Its density only averages one atom per ten or more cubic centimeters. The **temperature** is much warmer and can range from 3,000–6,000K (4,940.6– 10,340.6°F [2,727–5,727°C]). Like the cold atomic clouds, the warm atomic gas is primarily neutral hydrogen. For both the warm and cold atomic gas 90% of the atoms are hydrogen, but other types of atoms are mixed in at their normal cosmic abundances. The atomic gas accounts for roughly half the mass and **volume** of the interstellar medium. The warm diffuse gas is spread out between the clumps of the cold gas clouds.

The coronal gas is named for its similarity to the sun's corona, which is the outermost layer of the sun. The coronal gas like the sun's corona is both very hot and very diffuse. The average temperature and density of the coronal gas are roughly 1,799,541°F (99,727°C) and one atom per 1,000 cubic centimeter. The coronal gas is most likely heated by **supernova** explosions in the galaxy. Because the temperature is so high, the hydrogen atoms are ionized, meaning that the electrons have escaped from the nuclei.

Astronomers often call ionized hydrogen HII, so HII regions are clouds of ionized hydrogen. HII regions have temperatures of roughly 17,541°F (9,727°C) and densities of a few thousand atoms per cubic centimeter. What causes these HII regions? They are generally associated with regions of **star formation**. Newly formed stars are still surrounded by the clouds of gas and dust out of which they were formed. The hottest and most massive stars emit significant amounts of ultraviolet light that has enough **energy** to eject the electrons from the hydrogen atoms. An ionized HII region forms around these stars. Like the other atomic clouds, 90% of the atoms in HII regions are hydrogen, but other types of atoms are also present. These other types of atoms also become ionized to varying degrees.

The ionized atoms emit visible light so many HII regions can be seen in small telescopes and are quite beautiful. The Orion Nebula mentioned in the opening paragraph of this article is the closest example of a glowing HII region that is heated by newly formed stars. These HII regions are also called **emission** nebulae. Molecular clouds are also associated with star formation. Giant molecular clouds have temperatures below -369.4°F (-223°C), but can contain several thousand molecules per cubic centimeter. They can also be quite large. They range up to 100 light years in size and typically contain 100,000 times the mass of the sun. These clouds appear dark because they block the light from stars behind them. The most massive contain as much as 10 million times the mass of the sun. Roughly half the mass of the interstellar medium is found in molecular clouds. Like the atomic gas, most of the molecules are hydrogen molecules, but hydrogen molecules are difficult to detect. Molecular clouds are therefore most commonly mapped out as **carbon monoxide** (CO) clouds because the CO **molecule** is easy to detect using a **radio** telescope.

So far more than 80 different types of molecules have been found in molecular clouds, including some moderately complex organic molecules. The most common molecules are the simplest ones, containing only two atoms. These include molecular hydrogen (H_2), some carbon monoxide (CO), the hydroxyl radical (OH), and carbon sulfide (CS), followed by the most common three-atom molecule, **water** (H_2O). More complex **species** are relatively rare. However, molecules having as many as 13 atoms have been identified, and even larger species are suspected.

How can all these molecules form in interstellar space? For molecules to form atoms have to get close together. In even the densest interstellar clouds the atoms are too spread out. How can they get close? The details are poorly understood, but astronomers think that dust grains play a crucial role in interstellar **chemistry**, particularly for such important species as molecular hydrogen. The atoms on the surface of the dust grains can get close enough to form molecules. Once the molecules form, they do not stick to the dust grains as well as atoms so they escape the surface of the dust grain.

Dust

In addition to gas, the other major component of interstellar matter is dust. Dust grains permeate the entire interstellar medium, in clouds and between them. Interstellar dust grains are usually less than a millionth of a meter in radius. Their compositions are not well known, but likely compositions include silicates, ices, carbon, and **iron**. The silicates are similar in composition to the silicate **rocks** found on the **Moon** and in the earth's mantle. The ices can include **carbon dioxide**, methane, and

KEY TERMS

Dark cloud—A cloud of dust that block light from stars behind it.

HI region—A cloud of neutral hydrogen.

HII region—A cloud of ionized hydrogen.

Interstellar medium—The matter between the stars.

Ion—An atom that has lost or gained one or more electrons. In astronomy it will virtually always have lost electrons.

Molecular cloud—An interstellar cloud of molecules.

Nebula—An interstellar cloud of gas and/or dust.

Reflection nebula—A cloud of dust that glows from reflected starlight.

ammonia ice as well as water ice. Astronomers think that a typical grain composition is a silicate core with an icy mantle, but pure carbon grains may be present as well.

Dust exists in diffuse form throughout the interstellar medium. In this diffuse form each dust grain typically occupies the volume of a cube the length of a football field on each side (one million cubic meters). We detect this diffuse interstellar dust by the **extinction** and reddening of starlight. The dust grains block starlight, creating extinction, and they also preferentially block blue light over red light, causing reddening. Stars therefore appear redder in **color** than they otherwise would. This extinction and reddening is similar to the effect that makes sunsets red, especially over a smoggy city.

We can see dust grains more directly in dense regions, that is, in interstellar clouds. Two types of clouds showing the effects of dust are dark clouds and reflection nebulae. We see dark clouds by their effect on background stars. They block the light from stars behind the cloud, so we see a region of the sky with very few stars. Reflection nebulae are dust clouds located near a star or stars. They shine with reflected light from the nearby stars, and are blue in color because the grains selectively reflect blue light.

Significance of the interstellar medium

Neutral hydrogen atoms in the interstellar medium emit **radio waves** at a wavelength of 8 in (21 cm). Studies of this 8 in (21 cm) emission are not just important for studying the interstellar medium. Mapping the distribution of this interstellar hydrogen has revealed to us the **spiral** structure of the **Milky Way** galaxy.

The interstellar medium is intimately intertwined with the stars. Stars are formed from the collapse of gas and dust in molecular clouds. The leftover gas around newly formed massive stars forms the HII regions. At various times stars return material to the interstellar medium. This **recycling** can be gentle in the form of stellar winds, or it can be as violent as a supernova explosion. The supernovas are a particularly important form of recycling in the interstellar medium. The material recycled by supernovas is enriched in heavy elements produced by **nuclear fusion** in the star and in the supernova itself. With **time** the amount of heavy elements in the composition of the interstellar medium and of stars formed from the interstellar medium slowly increases. The interstellar medium therefore plays an important role in the **chemical evolution** of the galaxy.

See also Stellar evolution.

Resources

Books

Bacon, Dennis Henry, and Percy Seymour. *A Mechanical History of the Universe.* London: Philip Wilson Publishing, Ltd., 2003.

Morrison, David, Sidney Wolff, and Andrew Fraknoi. *Abell's Exploration of the Universe.* 7th ed. Philadelphia: Saunders College Publishing, 1995.

Verschuur, Gerrit L. *Interstellar Matters.* New York: Springer-Verlag, 1989.

Periodicals

Knapp, Gillian. "The Stuff Between The Stars." *Sky & Telescope* 89 (May 1995): 20-26.

Paul A. Heckert

Interval

An interval is a set containing all the **real numbers** located between any two specific real numbers on the number line. It is a property of the set of real numbers that between any two real numbers, there are infinitely many more. Thus, an interval is an infinite set. An interval may contain its endpoints, in which case it is called a closed interval. If it does not contain its endpoints, it is an open interval. Intervals that include one or the other of, but not both, endpoints are referred to as half-open or half-closed.

Notation

An interval can be shown using set notation. For instance, the interval that includes all the numbers between 0 and 1, including both endpoints, is written $0 \leq x \leq 1$, and read "the set of all x such that 0 is less than or equal

to x and x is less than or equal to 1." The same interval with the endpoints excluded is written $0 < x < 1$, where the less than symbol $(<)$ has replaced the less than or equal to symbol $(\leq)$. Replacing only one or the other of the less than or equal to signs designates a half-open interval, such as $0 \leq x < 1$, which includes the endpoint 0 but not 1. A shorthand notation, specifying only the endpoints, is also used to designate intervals. In this notation, a square bracket is used to denote an included endpoint and a parenthesis is used to denote an excluded endpoint. For example, the closed interval $0 \leq x \leq 1$ is written [0,1], while the open interval $0 <: x < 1$ is written (0,1). Appropriate combinations indicate half-open intervals such as [0,1) corresponding to $0 \leq x < 1$.

An interval may be extremely large, in that one of its endpoints may be designated as being infinitely large. For instance, the set of numbers greater than 1 may be referred to as the interval $1 < x < \infty$, or simply $(1,\infty)$. Notice that when an endpoint is infinite, the interval is assumed to be open on that end. For example the half-open interval corresponding to the nonnegative real numbers is $[0,\infty)$, and the half-open interval corresponding to the nonpositive real numbers is $(-\infty,0]$.

Applications

There are a number of places where the concept of interval is useful. The solution to an **inequality** in one **variable** is usually one or more intervals. For example, the solution to $3x + 4 \leq 10$ is the interval $(-\infty,2]$.

The interval concept is also useful in **calculus**. For instance, when a **function** is said to be continuous on an interval [a,b], it means that the graph of the function is unbroken, no points are missing, and no sudden jumps occur anywhere between $x = a$ and $x = b$. The concept of interval is also useful in understanding and evaluating integrals. An **integral** is the area under a **curve** or graph of a function. An area must be bounded on all sides to be finite, so the area under a curve is taken to be bounded by the function on one side, the x-axis on one side and vertical lines corresponding to the endpoints of an interval on the other two sides.

See also Domain; Set theory.

Resources

Books

Bittinger, Marvin L, and David Ellenbogen. *Intermediate Algebra: Concepts and Applications.* 6th ed. Reading, MA: Addison-Wesley Publishing, 2001.

Gowar, Norman. *An Invitation to Mathematics.* New York: Oxford University Press, 1979.

Larson, Ron. *Calculus With Analytic Geometry.* Boston: Houghton Mifflin College, 2002.

J. R. Maddocks

Introduced species

Some **species** of plants, animals, and **microorganisms** have been spread by humans over much wider ranges than they occupied naturally. Some of these introductions have been deliberate and were intended to improve conditions for some human activity, for example, in agriculture, or to achieve aesthetics that were not naturally available in some place. Other introductions have been accidental, as when plants were introduced with **soil** transported as ballast in ships or **insects** were transported with timber or food. Most deliberate or accidental introductions have not proven to be successful, because the immigrant species were unable to sustain themselves without the active intervention of humans. (In other words, the introduced species did not become naturalized.) However, some introduced species have become extremely troublesome **pests**, causing great economic damage or severe loss of natural values. One study estimated that there were more than 30,000 introduced species in the United States, and that the damaging ones caused $123 billion in economic losses.

Deliberate introductions

The most common reason for deliberate introductions of species beyond their natural range has been to improve the prospects for agricultural productivity. Usually this is done by introducing agricultural plants or animals for cultivation. In fact, all of the most important species of agricultural plants and animals are much more widespread today than they were prior to their domestication and extensive cultivation by humans. **Wheat** (*Triticum aestivum*), for example, was originally native only to a small region of the Middle East, but it now occurs virtually anywhere that conditions are suitable for its cultivation. Corn or maize (*Zea mays*) originated in a small area in Central America, but it is now cultivated on all of the habitable continents. **Rice** (*Oryza sativa*) is native to Southeast **Asia**, but is now very widespread under

cultivation. The domestic cow (*Bos taurus*) was native to Eurasia, but it now occurs worldwide. The turkey (*Meleagris gallopavo*) is native to **North America**, but it now occurs much more extensively. There are many other examples of **plant** and **animal** species that have been widely introduced beyond their natural range because they are useful as agricultural **crops**.

Other species have been widely introduced because they are useful in improving soil fertility for agriculture or sometimes for **forestry**. For example, various species of nitrogen-fixing **legumes** such as clovers (*Trifolium* spp.) and alfalfa (*Medicago sativa*) have been extensively introduced from their native Eurasia to improve the fertility of agricultural soils in far-flung places. In other cases, species of earthworms (such as the European nightcrawler, *Lumbricus terrestris*) have been widely introduced because these animals help to humify organic **matter** and are useful in aerating soil and improving its structural quality. There have also been introductions of beneficial microorganisms for similar reasons, as when mycorrhizal **fungi** are inoculated into soil or directly onto **tree** roots. When their roots are infected with a suitable root **mycorrhiza**, plants gain significant advantages in obtaining **nutrients**, especially **phosphorus**, from the soil in which they are growing.

In some cases, species of animals have been introduced to improve the prospects for hunting or fishing. For example, Eurasian gamebirds such as the ring-necked pheasant (*Phasianus colchicus*) and gray or Hungarian partridge (*Perdix perdix*) have been widely introduced in North America, as have various species of **deer** in New Zealand, especially red deer (*Cervus elaphus*). Species of sportfish have also been widely introduced. For example, various species of Pacific **salmon** (*Oncorhynchus* spp.) and common **carp** (*Cyprinus carpio*) have been introduced to the Great Lakes to establish fisheries.

Species of plants and animals have also been widely introduced in order to gain aesthetic benefits. For example, whenever people of European cultures discovered and colonized new lands, they introduced many species with which they were familiar in their home countries but were initially absent in their new places of residence. Mostly, this was done to make the colonists feel more comfortable in their new homes. For example, parts of eastern North America, especially cities, have been widely planted with such European trees as Norway maple (*Acer platanoides*), linden (*Tilia cordata*), **horse chestnut** (*Aesculus hippocastanum*), Scots pine (*Pinus sylvestris*), Norway **spruce** (*Picea abies*), as well as with many exotic species of shrubs and herbaceous plants. The European settlers also introduced some species of **birds** and other animals with which they were familiar, such as

Growth of introduced kudzu in Tennessee. *JLM Visuals. Reproduced by permission.*

the starling (*Sturnus vulgaris*), house sparrow (*Passer domesticus*), and pigeon or rock dove (*Columba livia*).

Accidental introductions

Humans have also accidentally introduced many species to novel locations, and where the **habitat** was suitable these species became naturalized. For example, when cargo ships do not have a full load of goods they must carry some other heavy material as ballast, which is important in maintaining stability of the vessel in rough seas. The early sailing ships often used soil as ballast, and after a trans-oceanic passage this soil was usually dumped near the port and replaced with goods to be transported elsewhere. In North America, many of the familiar European weeds and soil **invertebrates** probably arrived in ballast, as is the case for **water** horehound (*Lycopus europaeus*), an early introduction to North America at the port of New York. In addition, ships have used water as ballast since the late nineteenth century, and many aquatic species have become widely distributed by this practice. This is how two major pests, the zebra mussel (*Dreissena polymorpha*) and spiny water flea (*Bythothrepes cederstroemii*), were introduced to the Great Lakes from European waters.

An important means by which many agricultural weeds became widely introduced is through the **contam-ination** of agricultural seed-grain with their **seeds**. This was especially important prior to the twentieth century when the technologies available for cleaning seeds intended for planting were not very efficient.

Introduced species as an environmental problem

In most places of the world, introduced species have caused important ecological damage. There are so many examples of this phenomenon that in total they represent a critical component of the global environmental crisis. A few selected examples can be used to illustrate problems associated with introduced species.

Several European weeds are toxic to cattle if eaten in large quantities, and when these plants become abundant in pastures they represent a significant management problem and economic loss. Some examples of toxic introduced weeds of pastures in North America are common St. John's wort (*Hypericum perforatum*), ragwort (*Senecio jacobaea*), and common milkweed (*Asclepias syriaca*).

Some introduced species become extremely invasive, penetrating natural habitats and dominating them to the exclusion of native species. Purple loosestrife (*Lythrum salicaria*), originally introduced in North

America as a garden ornamental, is becoming extensively dominant in **wetlands**, causing major degradation of their value as habitat for other species of plants and animals. In Florida, several introduced species of shrubs and trees are similarly degrading habitats, as is the case of the bottlebrush tree (*Melaleuca quinquinerva*) and Australian oak (*Casuarina equisetifolia*). In **Australia**, the prickly pear **cactus** (*Opuntia* spp.) was imported from North America for use as an ornamental plant and as a living fence, but it became a serious weed of rangelands and other open habitats. The cactus has since been controlled by the deliberate introduction of a moth (*Cactoblastis cactorum*) whose larvae feed on its tissues.

Some introduced insects have become troublesome pests in **forests**, as is the case of the gypsy moth (*Lymantria dispar*), a defoliator of many tree species introduced to North America in 1869 from **Europe**. Similarly, the introduced **elm bark** beetle (*Scolytus multistriatus*) has been a key factor in the spread of Dutch elm **disease**, caused by an introduced fungus (*Ceratocystis ulmi*) that is deadly to North American species of elm trees (especially *Ulmus americana*). Another introduced fungus (*Endothia parasitica*) causes **chestnut** blight, a disease that has eliminated the once abundant American chestnut (*Castanea dentata*) as a canopy tree in deciduous forests of eastern North America.

Other introduced species have caused problems because they are wide-feeding predators or herbivores. Vulnerable animals in many places, especially isolated oceanic islands, have been decimated by introduced predators such as **mongooses** (family Viverridae), domestic **cats** (*Felis catus*), and domestic dogs (*Canis familiaris*), by omnivores such as **pigs** (*Sus scrofa*) and **rats** (*Rattus* spp.), and by herbivores such as **sheep** (*Ovis aries*) and **goats** (*Capra hircus*). The recent deliberate introduction of the predatory Nile **perch** (*Lates niloticus*) to Lake Victoria, Africa's largest and the world's second largest **lake**, has recently caused a tragic **mass extinction** of native fishes. Until recently, Lake Victoria supported an extremely diverse community of more than 400 species of **fish**, mostly cichlids (family Cichlidae), with 90% of those species occurring nowhere else. About one-half of the **endemic** cichlid species are now extinct in Lake Victoria because of predation by the Nile perch, although some species survive in captivity, and a few are still in the lake.

Ecologically, it is reasonable to consider humans and their symbiotic associates (that is, the many species of plants, animals, and microorganisms with which humans have intimate, mutually beneficial relationships) as the ultimate in **invasive species**. Humans are, in fact, widely self-introduced.

Resources

Books

Devine, R.S. *Alien Invasions: America's Battle with Non-native Animals and Plants.* Times Books, 1998.

Freedman, B. *Environmental Ecology.* 2nd ed. San Diego: Academic Press, 1994.

Goudie, A. *The Human Impact on the Natural Environment.* 3rd ed. Cambridge, MA: The MIT Press, 1990.

Luken, J. O., and J.W. Thieret, eds. *Assessment and Management of Plant Invasions.* Springer-Verlag, 1997.

Randall, J. M., and J. Marinelli, eds. *Invasive Plants: Weeds of the Global Garden.* Brooklyn Botanic Garden, 1997.

Bill Freedman

Invariant

In **mathematics** a quantity is said to be invariant if its value does not change following a given operation. For instance, **multiplication** of any real number by the **identity element** (1) leaves it unchanged. Thus, all **real numbers** are invariant under the operation of "multiplication by the identity element (1)." In some cases, mathematical operations leave certain properties unchanged. When this occurs, those properties that are unchanged are referred to as invariants under the operation. Translation of coordinate axes (shifting of the origin from the point (0,0) to any other point in the **plane**) and **rotation** of coordinate axes are also operations. Vectors, which are quantities possessing both magnitude (size) and direction, are unchanged in magnitude and direction under a translation of axes, but only unchanged in magnitude under rotation of the axes. Thus, magnitude is an invariant property of vectors under the operation of rotation, while both magnitude and direction are invariant properties of a vector under a translation of axes.

An important objective in any branch of mathematics is to identify the invariants of a given operation, as they often lead to a deeper understanding of the mathematics involved, or to simplified analytical procedures.

Geometric invariance

In **geometry**, the invariant properties of points, lines, angles, and various planar and solid objects are all understood in terms of the invariant properties of these objects under such operations as translation, rotation, reflection, and magnification. For example, the area of a triangle is invariant under translation, rotation and reflection, but not under magnification. On the other hand, the interior angles of a triangle are invariant under magnification, and so are the proportionalities of the lengths of its sides.

The **Pythagorean theorem** states that the **square** of the hypotenuse of any right triangle is equal to the sum of the squares of its legs. In other words, the relationship expressing the length of the hypotenuse in terms of the lengths of the other two sides is an invariant property of right triangles, under magnification, or any other operation that results in another right triangle.

Very recently, geometric figures called fractals have gained popularity in the scientific community. Fractals are geometric figures that are invariant under magnification. That is, their fragmented shape appears the same at all magnifications.

Algebraic invariance

Algebraic invariance refers to combinations of coefficients from certain functions that remain constant when the coordinate system in which they are expressed is translated, or rotated. An example of this kind of invariance is seen in the behavior of the **conic sections** (cross sections of a right circular cone resulting from its intersection with a plane). The general equation of a conic section is $ax^2 + bxy + cy^2 + dx + ey + f = 0$. Each of the equations of a **circle**, or an **ellipse**, a **parabola**, or **hyperbola** represents a special case of this equation. One combination of coefficients, $(b^2\text{-}4ac)$, from this equation is called the discriminant. For a parabola, the value of the discriminant is **zero**, for an ellipse it is less than zero, and for a hyperbola is greater than zero. However, regardless of its value, when the axes of the coordinate system in which the figure is being graphed are rotated through an arbitrary **angle**, the value of the discriminant $(b^2\text{-}4ac)$ is unchanged. Thus, the discriminant is said to be invariant under a rotation of axes. In other words, knowing the value of the discriminant reveals the identity of a particular conic section regardless of its orientation in the coordinate system. Still another invariant of the general equation of the conic sections, under a rotation of axes, is the sum of the coefficients of the squared terms $(a+c)$.

Resources

Books

Larson, Ron. *Calculus With Analytic Geometry.* Boston: Houghton Mifflin College, 2002.

J.R. Maddocks

Invasive species

An exotic **species** is one that has been introduced into a **habitat** it would not normally populate. This introduction can be intentional or unintentional. Exotic species have also been called introduced, nonnative, nonindigenous, or alien species. An invasive species is an exotic species that thrives in its new environment, disrupting the natural **ecosystem**. The majority of exotic species have been introduced unintentionally. "Hitchhiker" organisms such as **seeds** or **insects** attach to people's shoes, clothes, or luggage when they travel. When the humans return to their native land, they arrive bearing these nonnative species. Sometimes people bring beautiful plants and flowers home with them for ornamental purposes. These intentional introductions occur less frequently, but can have the same disastrous effects. Most of the time, the exotic species cannot survive in its new environment. Changes in climate, resources, and **competition** simply do not favor survival, and the **organism** eventually dies out. Occasionally, the **introduced species** ends up being invasive, out-competing the natural habitat for resources, displacing native **flora** and **fauna**, and wreaking economic havoc on a community.

Survival of exotic species

When a new species is introduced into an ecosystem, there are four different interspecific interactions that can occur. These are interactions that occur between two different species living in the same community. These interactions could have positive, **negative**, or neutral effects on the involved organisms. One such interaction is predation. This is when one species, the **predator**, uses the other, the **prey**, for food. This interaction obviously benefits the predator, but not the prey. When an introduced species is a predator, it may become invasive if it can outcompete the native predators. This competition is the second interspecific interaction. This interaction hurts both species involved. When two species are both competing for the same resources, neither will be as successful as they would be alone. The last two possible interactions, **commensalism** and **mutualism**, would not make an introduced species invasive. Commensalism is when one species benefits and the other is unaffected, and mutualism is when both species benefit from the interaction. Most of the time, an introduced species cannot compete with native populations and does not survive. Of the species that do become successful in their new environment, the majority of these organisms have no effect on the ecosystem. For example, the pheasant is a bird that was introduced to **North America** from **Asia**. These **birds** have had no impact on native species. The species is considered invasive when it can out-compete and displace other species already present in the ecosystem.

An introduced species must exhibit certain characteristics in order to become invasive. For one, the organism must be able to reproduce in their new environment. They must also be able to out-compete the native populations. They must not be susceptible to herbivores or diseases, especially if these types of organisms characterize the ecosystem to which they have been introduced. Lastly, they must be able to survive in their new climate with the available resources.

Effects of invasive species

When an introduced species becomes invasive, the effects can be terrible. In the United States alone, invasive species cause more than $123 billion in damages per year. It has been estimated that over 68% of the organisms listed as threatened or **endangered species** by the International Union for the Conservation of Nature and Natural Resources have been classified as a result of invasive species. Over 50,000 of the 750,000 species in the United States are exotic. Approximately 5,000 of these are considered invasive. Because it is difficult to gather these data, these are probably underestimations of the actual effects of invasive species. Invasive species not only displace native flora and fauna, they homogenize existing ecosystems, greatly reducing the number of available biological resources.

Examples of invasions

There are countless examples of invasive species and the problems they cause. The **water** hyacinth was introduced into the United States from South **Africa** in 1884. This was an intentional introduction; travelers brought back the **flower** for its ornamental beauty. These flowers grow quite rapidly, and without any natural predators in their new environments, they quickly overpopulated their new environments. As a result, they clogged waterways, out-shaded natural vegetation, and displaced several native species. A well-known example in the southern United States is kudzu. This legume was introduced from Asia, where it is considered an ornamental vine. In the United States, it has taken over the land. It grows over anything in its path, including trees, shrubs, and even houses.

Many disease-causing organisms are invasive species. For example, the fungus *Ophiostoma ulmi*, the pathogen that causes Dutch **elm disease**, and the **bark** beetle, which carries the pathogen, were both introduced to the United States from **Europe**. They were both imported on infected **wood**, first the beetle in 1909, and then the fungus in 1930. The combination of these two organisms has caused the destruction of millions of elm trees. The **chestnut** blight fungus, *Cryphonectria parasitica*, was introduced into the United States from Asia on nursery plants in 1900. This fungus has caused the destruction of almost all of the eastern American chestnut trees. Both of these **pathogens** have caused great disruptions in forest ecosystems.

Starlings and English sparrows were both introduced intentionally to the northeastern United States from Europe in the 1800s. They can now be found just about anywhere in North America, and have displaced the native birds in many communities, caused significant crop damage, and contributed to the spread of certain swine diseases. **Deer** were introduced to Angel Island in the San Francisco Bay from the mainland in the early 1900s. This island was a game reserve with no natural deer predators. The deer population exploded and soon outgrew the meager food supply on the small island. People who visited the island felt sorry for the starving deer and fed them bits of their picnic lunches. Therefore, the deer survived and continued to multiply, despite their limited resources. The deer were eating the native **grasses**, tearing the leaves off of seedlings, and killing trees by eating bark. The deer had to be removed, a project that cost the State of California over $60,000. Sea lampreys from the North Atlantic Ocean were introduced to the United States through the Erie Canal in the 1860s, and again through the Welland Canal in 1921. These organisms have displaced the lake trout and whitefish from the Great Lakes, and have cost the United States and Canada over $10 million a year.

Argentine fire **ants** were introduced to the southern United States from a coffee shipment from Brazil in 1891. These organisms damage **crops** and disrupt ecosystems. They have been spreading steadily northward since their introduction. Examples of their destruction include the reduction of native ant species in one part of Texas from 15 to 5 species, and their killing of brown pelican hatchlings (a threatened species) in **wildlife** refuges. The Japanese beetle was introduced to the United States from Japan on a shipment of iris or azalea flowers in 1911. These **beetles** have caused the destruction of over 250 native **plant** species. Gypsy **moths** were once contained in a research facility on the east coast of the United States until they escaped in 1869. These moths have caused the destruction of entire **forests** by eating the leaves off of the trees, with damages estimated near $5 million.

A recent invasive species that has received much media attention is the zebra mussel. This organism was imported from the Caspian Sea to the United States via a cargo ship that emptied its ballast water into the St. Lawrence Seaway in the mid 1980s. By 1993, zebra mussels could be found as far south as New Orleans. Zebra mussels compete with native shellfish and **fish** for food and shelter. They also clog waterways. It is not known what effects these organisms will eventually have on the ecosystems they have invaded, but it is likely that the shellfish and native fish will suffer. It has been estimated that if the zebra mussel population is not controlled, damages will reach $5 billion by the year 2002.

The United States is not the only nation to suffer the effects of invasive species. Well-intentioned Europeans, for the purpose of providing food and income to natives, introduced the Nile **perch** into Lake Victoria in East Africa. Lake Victoria was the home of many native fish, including cichlids. These fish feed on detritus and plants at the bottom of the lake. The addition of the Nile perch introduced a new predator, which fed on the cichlids. Eventually, all of the cichlids disappeared, and once this happened, the perch had no food. The perch ended up dying off as well, leaving the native people in an even worse situation, with nothing but a **lake** overgrown with detritus and plants. In 1859, an Australian released two dozen English rabbits for hunting. Without any natural predators, the rabbit population grew to over 40,000 in only six years. The rabbits displaced many natural animals, including kangaroos. The Australians tried building a 2,000-mi (322-km) long fence to contain the rabbits, but some had traveled past the fence before its completion. The myxoma **virus** was introduced to the rabbit population in 1951 in hopes of controlling its growth. Recently, a new population of rabbits that are resistant to this virus has begun growing, and the problem is far from being solved.

As the use of **genetic engineering** technologies increases, the threat of a new type of invasive species emerges. Genetically engineered organisms, if introduced into the wild, could also alter ecosystems in many ways. Genetically engineered plants have acquired such traits as herbicide resistance, pest resistance, faster growth, and tolerance of extreme climatic changes. If these engineered organisms were accidentally released, they would have a competitive advantage over native species, and could become invasive.

Management

There are many more examples of the destruction invasive species can cause. The problem facing environmentalists and naturalists is management of these organisms. Management is important because invasive species can disrupt entire ecosystems, reduce **biodiversity**, endanger plants and animals, destroy landscapes and habitats, and transmit diseases. The United States National Park Service has proposed over 500 projects to eradicate invasive species in over 150 different parks. The cost of these projects would be over $80 million. The Park Service has also established an Integrated Pest Management Program with the agenda of controlling the introduction of new invasive species. Some National Park Service sites are being used as insect nurseries, where insects that could be used as biocontrol agents are harvested and distributed. In 1993, the Bureau of Land Management developed the Federal Interagency Committee for Management of Noxious and Exotic Weeds to eradicate invasive plants on federal lands and to provide help to similar projects on public lands. In 1998, the Fish and Wildlife Service started a program in North and South Dakota, Nebraska, Kansas, Oklahoma, and Texas to halt the spread of zebra mussels into these states.

In 1999, President Clinton signed an executive order to address the growing problem of invasive species in the United States. This order created an Invasive Species Council that will develop a proposal to minimize the detrimental effects caused by, as well as to prevent, the introduction of new invasive species. This council will work with groups at the state, university, and local levels to solve the problems these organisms can cause. The Council's budget for the year 2000 was approximately $30 million, to be used for program implementation as well as research. The agenda also includes the reintroduction of native species into their original habitats. Federal legislation has already been created to begin the process of restoration. The USDA now has over 1,300 inspectors at 90 ports of entry, assisted by the "Beagle Brigade," beagles trained to **smell** agricultural products being transported into the country. The USDA has also prohibited importing untreated wood packing material from China, which can carry the Asian long-horned beetle. A proposal has been made to

KEY TERMS

Ballast—An area of a ship filled with water to help stabilize the ship.

Biocontrol agent—An organism that can itself be used to control unwanted organisms, usually by feeding on the unwanted species.

Community—In ecology, a community is an assemblage of populations of different species that occur together in the same place and at the same time.

Detritus—Dead organic matter.

Ecosystem—All of the organisms in a biological community interacting with the physical environment.

Fauna—Animals or animal life.

Flora—Plants or plant life.

Homogenize—To create an area made entirely of same or similar things.

Myxoma virus—A fatal virus that infects rabbits.

Pathogen—A disease-causing agent.

other countries to enact this ban as well. A barrier is being built in the Chicago Ship Canal to stop the spread of invasive species from Great Lakes to the Mississippi River. The National Oceanic and Atmospheric Administration is funding research into possible ballast water treatments that could eliminate the introduction of invasive species from cargo ships. These measures will help eliminate some invasive species, but not all. More research is needed, and international programs need to be implemented in order to completely eradicate the problem of invasive species.

Resources

Books

Campell, Neil A. *Biology*. Menlo Park, CA: The Benjamin/ Cummings Publishing Company, 1996.

Starr, Cecie. *Biology: Concepts and Applications*. Belmont, CA: Wadsworth Publishing Company, 1997.

Periodicals

"Costly Interlopers." *Scientific American* (February 15, 1999).

Gordon, Doria R. "Effects of Invasive, Non-indigenous Plant Species on Ecosystem Processes: Lessons from Florida." *Ecological Concepts in Conservation Ecology* no. 84, (1998).

Jennifer McGrath

Invertebrate paleontology *see* **Paleontology**

Invertebrates

Invertebrates are animals without backbones. This simple definition hides the tremendous diversity found within this group which includes **protozoa** (single-celled animals), corals, **sponges, sea urchins, starfish, sand dollars**, worms, **snails**, clams, spiders, **crabs,** and **insects**. In fact, more than 98% of the nearly two million described **species** are invertebrates, ranging in size from less than a millimeter to several meters long. Invertebrates display a fascinating diversity of body forms, means of locomotion, and feeding habits.

Invertebrates are an essential part of every **ecosystem** on this **planet**. We could not function without them. They are responsible for the **decomposition** of organic waste, which allows the **recycling** of the chemicals in the ecosystem. Invertebrates also are involved with the **pollination** of plants, and are crucial as links in food chains where herbivores convert the **energy** in plants into energy that is available to animals higher up the food web.

Most invertebrates live in **water** or have some stage of their life in water. The external layers of aquatic invertebrates are generally thin and are permeable to water, allowing the exchange of gas, although some have specialized respiratory structures on their body surface. Aquatic invertebrates feed by ingesting directly, by filter feeding, or actively capturing **prey**.

Some groups of invertebrates such as earthworms, insects, and spiders live on land. These invertebrates need to have special structures to deal with life on land. For example, because air is less buoyant than water, earthworms have strong muscles for crawling and burrowing while insects and spiders move by means of several pairs of legs. Drying out on land is a problem so earthworms must secrete mucous to keep their bodies moist, while insects and spiders are waterproof and are physiologically adapted to conserve water.

Iodine *see* **Halogens**

Iodine-131 *see* **Radioactive pollution**

Ion exchange

Ions are electrically charged **atoms** or groups of atoms. Ion exchange is the phenomenon of replacing one ion with another of similar charge. Ion exchange **resins** are solids containing strongly bonded charged atom groups. (Positively charged atoms are called cations; negatively charged, anions.) They occur as natural materials and can

Invertebrates make up 97% of the animal kingdom, the largest of the five kingdoms of living organisms. *Illustration by Hans & Cassidy. Courtesy of Gale Group.*

be synthetically made. An ion of opposite charge is loosely bound to a charged group. When placed in contact with an ionic **solution** (a solution of an ionic substance), the loosely bound ions are replaced by those in solution and are retained on the solid. Ion exchange membranes are sheet-like films which allow the passage of ions while restricting the crossing of larger, uncharged molecules.

Ion exchange resins

If an ionic solution is brought into contact with a solid having ions that are only weakly joined in its crys-

talline structure, it is possible for ions from the solution to interchange with those of the same charge in the solid. Electrical neutrality is maintained throughout this exchange; that is, the total number of positive charges equals the total number of **negative** charges in the solid and the solution at all times. What changes is the type of ion that then resides with the solid and in the solution. A solid that has loosely bound **sodium** ions, when placed in a solution of potassium chloride, will interchange some of its sodium ions for potassium ions. The result is that the solid and the solution each have sodium and potassium ions in some **ratio** determined by the inherent capacity of the solid to undergo the exchange process. Equation 1 illustrates this interchange between cations initially attached to a solid interacting with cations initially in solution.

$$Na^+(solid) + K^+(aq) \rightleftharpoons Na^+(aq) + K^+(solid) \quad (Eq. 1)$$

Water solution is indicated by (aq) for aqueous.

The exchange continues until the ratio of each **cation** in the solid and the solution remains constant. For this example:

$$\frac{K^+ (solution)}{K^+ (solid)} = \text{a constant} \quad (Eq. 2)$$

Because the solid in equation 1 exchanges cations with the solution, it is termed a cation exchange solid. Other solids with exchangeable negatively charged ions are called **anion** exchangers.

There are several naturally occurring materials that function as ion exchangers. Many synthetic ion exchange materials are also available. Many of these synthetic materials are tailor-made to serve a specific purpose and be selective in the type of ions with which they exchange. Zeolites are a naturally occurring class of **minerals** containing **aluminum**, silicon, **oxygen** and a loosely held cation from group 1 or group 2 of the **periodic table** (e.g., sodium or **magnesium**). When placed in a solution of an ionic compound, exchange occurs between the loosely held zeolite cation and the dissolved cation in water. Various clay and **soil** materials also possess ion exchange capabilities. Most often an ion exchange reaction uses a synthetic ion exchange material specifically designed to achieve the desired separation.

Synthetic ion exchangers are composed of a charged group attached to a rigid structural framework. One end of the charged group is permanently fixed to the frame while a positive or negative charged portion loosely held at the other end attracts other ions in solution. Common materials for these ion exchange resins are styrene and divinylbenzene. Molecules of these organic substances can join together forming a divinylbenzene **polymer** consisting of long rows of styrene crosslinked, that is attached, by divinylbenzene.

The extent to which divinylbenzene is crosslinked affects the ability of the resin to undergo ion exchange with an ion in solution. Resins that are only slightly crosslinked may have sufficient open space to allow solution ions to pass through and avoid contact with the fixed, exchangeable groups. Resins that are too highly crosslinked may not have openings big enough for solution ions to penetrate. This prevents them from contact with the fixed exchangeable groups. Table 1 lists various chemical groups that can be joined to the resin framework for attracting ions in solution.

Cation resins often are prepared in their **hydrogen** ion form. In this state exchange occurs when the resulting product in solution is the acid corresponding to the dissolved solid. An example of this type of exchange is shown in equation 3 where a strong cation resin in the hydrogen form interacts with a **sodium chloride salt** solution to yield the sodium form of the resin and hydrochloric acid.

$$ROSO_3^- H^+(resin) + Na^+ Cl^-(aq) \rightleftharpoons ROSO_3^- Na^+(resin) + HCl(aq)$$
$$(Eq. 3)$$

A similar exchange between dissolved sodium chloride and a strong anion resin in the hydroxide (basic) form yields dissolved **sodium hydroxide**, a strong base.

TABLE 1 SELECTED CHARGED ION EXCHANGE GROUPS	
Cation exchange groups joined to an ion exchange framework[1]	
R-OSO$_3^-$ M$^+$	sulfonic acid group
R-COO- M$^+$	carboxylic acid group
Anion exchange groups joined to an ion exchange framework	
R-CH$_2$NCH$_3$)$_3^+$ X$^-$	quaternary ammonium group
R-NH(CH$_3$)$_2^+$ X$^-$	ternary ammonium group

[1]R represents the divinylbenzene polymer framework
M$^+$ represents an exchangeable cation.
X$^-$ represents an exchangeable anion.

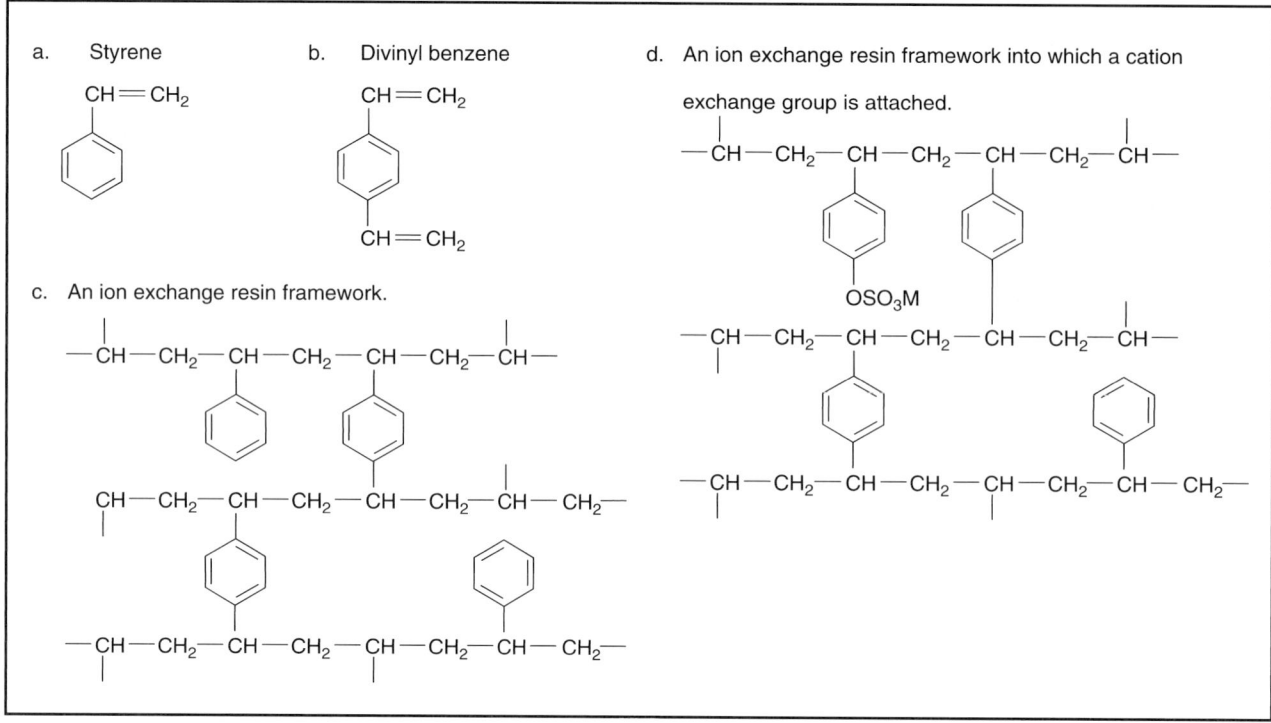

Figure 1. Structure of a synthetic ion exchange resin. *Illustration by Hans & Cassidy. Courtesy of Gale Group.*

$$RCH_2N(CH_3)_3{+}OH^-(resin) + NaCl(aq) \rightleftharpoons$$
$$RCH_2N(CH_3)_3{+}Cl^-(resin) + NaOH(aq)$$
$$(Eq.\ 4)$$

Complete exchange of solution ions (cation or anion)—that is, complete absorption on the resin—can occur if the **sample** solution is poured slowly through a packed column of resin material. This allows the sample to come into contact continually with fresh resin; the exchange occurs until none of the original exchangeable ions remains. These ions then can be collected by running another solution through the column, a solution that removes, or elutes, the absorbed ions from the resin.

Applications

Ion exchange and exchange resins have numerous applications. In scientific studies, exchange resins are used to isolate and collect various ionic **species**, cations on a cation resin, and anions on an anion resin. In industry resins are used to purify water by removing all ions from it. Upon passage of a water sample through both a cation resin (H^+ form) and an anion resin (OH^- form) the cations and anions in the water are retained. The H^+ and OH^- ions released from the resin then combine to form additional water. Deionized water also is a source of pure water containing no ionic chemical compounds.

$$H^+(aq) + OH^-(aq) \rightarrow H_2O\ (liquid)$$
$$(Eq.\ 5)$$

Ion exchange also is used to remove ionic compounds from boiler water used in the steam generation of electric power. Ion exchange resins also are used in the separation and purification of various chemicals. Rare **earth** elements are separated from their ores and purified in this manner.

Ion exchange membranes

Various **membrane** materials, both natural and synthetic, have the ability to selectively allow or retard passage of charged and uncharged molecules through their surface. These semipermeable membranes are extremely important in ion transport within living systems and have many industrial applications.

The balance between sodium ion, Na^+, and potassium ion, K^+, within the cells of living organisms is essential for life. The transport of these ions across the **cell** membrane allows this proper balance to be maintained.

Semipermeable membranes are used in the purification of large organic molecules. They allow small ionic compounds to pass through, separating them from the larger molecules. This procedure, known as **dialysis**, is the principle upon which patients with nonfunctioning kidneys can remove harmful waste products artificially.

Commercially, the forced separation of ions from seawater by passing them through a semipermeable

KEY TERMS

· ·

Anion resin—A solid material with tightly bonded positively charged ions and loosely bonded negative counter ions that will exchange the negative ions for dissimilar negative ions in solution.

Cation resin—A solid material with tightly bonded negatively charged ions and loosely bonded positive counter ions that will exchange the positive ions for dissimilar positive ions in solution.

Ion exchange membrane—A flat sheet-like semipermeable material that allows ions to pass unrestrictedly while serving as a barrier to larger, uncharged molecules.

membrane is an economical means of transforming seawater into potable water (water safe for drinking). This technique is known as reverse **osmosis** and is in use by countries bordering oceans or seas to obtain fresh drinking water.

Resources

Books

Gross, M.L., R. Caprioli, and P.B. Armentrout. *The Encyclopedia of Mass Spectrometry: Ion Chemistry and Theory.* Oxford: Pergamon Press, 2001.

Joesten, Melvin D., David O. Johnston, John T. Netterville, and James L. Wood. *World of Chemistry.* Philadelphia: Saunders College Publishing, 1991.

Simon, George P. *Ion Exchange Training Manual.* New York: Van Nostrand Reinhold, 1991.

Gordon A. Parker

Ion and ionization

Ionization is the process in which one or more electrons are removed from an atom or **molecule**, thereby creating an ion. The word ionization is also used for the process in which an ionic solid, such as a **salt**, dissociates into its component ions upon **solution**. In order to remove an **electron** from an atom, enough **energy** must be supplied to break the bond between the negatively charged electron and the positively charged nucleus; this is the ionization energy. Ionization can be induced by high energy **radiation** such as **x rays** and ultraviolet **light** (photoionization), bombardment by high energy electrons (electron impact ionization) or small molecular ions (chemical ionization) and by exposure to high electric fields (field ionization). Ionization is employed in many important analytical techniques used to study the character of **atoms** and molecules including **mass spectrometry**, photoelectron and Auger electron **spectroscopy**, and multiphoton ionization spectroscopy.

Ionization energy

In the **Bohr model** of atomic structure, electrons **orbit** the nucleus at fixed distances, similar to the orbits of the planets around the **sun**. For every element, the distances of the electron orbitals are fixed and unique to that element. Normally, the electrons occupy the orbits closest to the nucleus. This is the most stable configuration of the atom and is known as the ground state. To move an electron to an orbital further from the nucleus requires the input of energy. Atoms which have an electron in a higher orbit are said to be in an excited state.

The strength of attraction between a negatively charged electron and the positively charged nucleus is greater the closer together they are. The energy needed to move an electron from one orbit to a higher energy one is equal to the difference in the attraction between the two configurations; it takes increasing amounts of energy to move an electron to orbits further and further from the nucleus. The energy needed to move electrons from one orbit to another can be thought of like the energy needed to move between rungs on a ladder; to move from a lower rung to a higher rung requires the input of energy, and the more rungs you move up, the more energy it takes. However, if the electron is moved too far from the nucleus, the attraction between the electron and the nucleus is too small to hold the electron in its orbit, and, analogous to stepping of the top rung of the ladder, the electron is separated from the atom leaving behind a positively charged atom; the atom has been ionized.

The ionization of an atom can be represented by:

$$X + energy \rightarrow X^+ + e^-$$

where X is a single atom of any element and e< is the ejected electron. The amount of energy required for this process is called the ionization energy. The ionization energy is a measure of how difficult it is to remove the electron from the atom—the more strongly the electron is attracted to the nucleus, the higher the ionization energy. Although in theory it is possible to remove any of the electrons from an atom, in practice, the electron in the outermost orbit is typically the first to be removed. The energy required to remove the first electron is called the first ionization energy.

For many electron atoms it is possible to remove more than one electron. A second electron can be re-

TABLE I. IONIZATION ENERGIES (EV) OF THE ELEMENTS IN THE FIRST THREE ROWS OF THE PERIODIC TABLE.			
Atomic Number, Z	Element	First Ionization Energy $X + energy \rightarrow X^+ + e^-$	Second Ionization Energy $X+ + energy \rightarrow X^{2+} + e^-$
1	H	13.595	
2	He	24.481	54.403
3	Li	5.39	75.619
4	Be	9.32	18.206
5	B	8.296	25.149
6	C	11.256	24.376
7	N	14.53	29.593
8	O	13.614	35.108
9	F	17.418	34.98
10	Ne	21.559	41.07
11	Na	5.138	47.29
12	Mg	7.644	15.031
13	Al	5.984	18.823
14	Si	8.149	16.34
15	P	10.484	19.72
16	S	10.357	23.4
17	Cl	13.01	23.8
18	Ar	15.775	27.62

moved from a singly charged ion X^+ to yield a doubly charged ion, X^{2+}. This process can be written as:

$$X^+ + energy \rightarrow X^{2+} + e^-$$

The energy required to remove the second electron is called the second ionization energy. Following the removal of the first electron, the atom has one more positively charged **proton** in the nucleus than it has negatively charged orbiting electrons. This charge imbalance causes the remaining electrons to be held even more tightly to the nucleus. Consequently, more energy is required to remove the second electron than was required to remove the first. The removal of subsequent electrons, creating X^{3+}, X^{4+}, and so on, requires ever increasing amounts of energy. This effect is rather like a small child with a collection of toys. The child might be easily persuaded to give the first toy away, but will hold on to each remaining toy with increasing vigor, thereby requiring increasing amounts of persuasion to give away each subsequent toy.

The first and second ionization energies of the elements in the first three rows of the **periodic table** are listed in Table I.

Note that the second ionization energy in all cases is larger than the first ionization energy. The **hydrogen** atom, however, having only one electron, only has a first ionization energy. Note that the ionization energy, in general, increases with increasing **atomic number** for elements within the same row of the periodic table. The ionization energy is smallest for the alkaline **earth** elements, Li, Na, K, etc., increasing with atomic number and reaching a maximum at the end of each row, corresponding to the noble gases, Ne, Ar, Kr, etc. This effect is related to the way in which atomic orbitals are filled. The noble gases have filled electronic orbitals, which are very stable.

Molecules can be ionized in a manner analogous to atoms. However, because electrons form the bonds that hold molecules together, their removal may result in the bond being weakened, or even broken. The ionization energies of some simple molecules are listed in Table II.

TABLE II. IONIZATION ENERGIES (EV) OF SELECTED MOLECULES

Molecule	Ionization Energy (eV)	Molecule	Ionization Energy (eV)
N_2	15.576	CH_4 (methane)	12.6
O_2	12.063	C_2H_6 (ethane)	11.5
CO_2	13.769	C_3H_8 (n-propane)	11.1
CH_3F	12.85	C_4H_{10} (n-butane) $CH_3CH_2CH_2CH_3$	10.63
CH_3Cl	11.3	C_4H_8 $CH_2{=}CHCH_2CH_3$	9.6
CH_3I	9.54	C_4H_6 $CH_2{=}CHCH{=}CH_2$	9.07

Note that in general, the ionization energies of molecules have values the same order of magnitude as the first ionization energies of isolated atoms. Molecules with only a few atoms, such as N_2, CO_2 and H_2O, tend to have the highest ionization energies. Within a group of similar molecules, such as the alkanes listed in the table, the ionization energy decreases with increasing size. This effect is due to the fact that in larger molecules, there are more electrons available for ionization without disrupting the bonding stability of the molecule. Again, this is analogous to persuading a child to give up its toys; the more toys the child has, the easier it will be to persuade it to give one up.

Ionization methods

Ions, being electrically charged, are much easier to manipulate and detect than electrically neutral atoms or molecules. The direction or speed of ions can be changed by application of electric and magnetic fields, similar to the way a magnet can move small pieces of a magnetic material. Ions can be detected simply by measuring the **electric current** produced by their movement. Consequently, ionization is frequently employed in scientific apparatus to transform neutral **species** to charged species so that they may be more easily studied.

Mass spectrometry is a powerful analytical technique based on the transformation of the neutral components of a **sample** to ions which are then separated according to their mass-to-charge **ratio**. The structure and composition of molecular species can be deduced by studying the masses of the molecular ion along with the smaller, fragment ions which are sometimes formed.

There are several different types of ion sources which are used in mass spectrometers: electron impact, field ionization and chemical ionization are the most common, and are described below. Other instrumental analytical techniques which are based on ionization include spectroscopies which study the energy of the ejected electron as well as the positively charged ion. These methods typically employ electromagnetic radiation to supply the ionization energy (photoionization). The basic requirement for all ionization sources is that sufficient energy must be supplied to remove at least one electron.

Electron impact source

The most common method of producing ions for mass spectrometry is by bombarding a gaseous sample with a stream of fast moving electrons. The stream of electrons, produced by an electron gun (a heated tungsten wire from which electrons are emitted—thermionic **emission**), bombard the sample and "kick" out additional electrons. The process of electron impact ionization is not very efficient. Because of the very small size of electrons and the relatively low **density** of electrons around molecules, electron-electron impacts are rare. Nonetheless, electron impact is the most widely used ion source in commercial mass spectrometers. Electron guns can produce vast quantities of electrons, so even if one in a million is successful, enough ions can be generated. When the impact is effective in producing ionization, typically there is more energy supplied by the impact than is needed to remove the electron. The excess energy may result in the ion breaking up into smaller fragment ions. The intact molecular ion is referred to as the parent ion, and the fragment ions are called daughter ions.

Field ionization

Ionization may also be produced by subjecting a molecule to a very intense electric field. This process is called field ionization. A familiar example of field ionization is the small blue spark that jumps from the tip of your finger to any grounded surface on a dry day when static **electricity** can build up. The strong electrostatic field actually pulls electrons out of your finger. Electric fields are strongest at the tips of pointed conductors. To produce electric fields of sufficient magnitude to ionize molecules, very fine, sharpened wires are used. Field ionization sources are relatively gentle compared to electron impact sources in that they do not deposit as much excess energy into the parent ion. Therefore, field ionization sources are usually employed when we do not want to damage the ionized specimen too much.

Chemical ionization

Chemical ionization is similar to electron impact ionization except that a beam of positively charged molecular ions, rather than electrons, is used to bombard and ionize the sample. The bombarding ions are usually small molecules such as methane, propane, or **ammonia**. Because of the much larger size of a molecular ion compared to an electron, these collisions are highly reactive and generally produce less fragmentation than electron impact ionization with comparable efficiency. Chemical ionization is widely used in commercial mass spectrometers, and many instruments are equipped with a source which is capable of both electron impact and chemical ionization.

Photoionization

If the ionization energy is supplied by electromagnetic radiation, the ionization is called photoionization, referring to the fact that a **photon** of radiation produces the ionization. However, not all electromagnetic radiation has sufficient energy to cause ionization. Generally, only radiation with wavelengths shorter than visible light, that is, radiation in the ultraviolet, x ray, and gamma ray regions of the **electromagnetic spectrum** can produce ionization.

Ultraviolet radiation can cause ionization of many small molecules, including **oxygen**, O_2. In fact, short wavelength solar radiation causes ionization of molecular oxygen and molecular **nitrogen** found in the upper atmosphere; these processes are important to the **chemistry** of the earth's atmosphere. In the laboratory, ultraviolet light from special lamps or lasers is used to ionize molecules in order to study them. Ultraviolet photoelectron spectroscopy (UPS) measures the energy of the departing electron.

The high energy carried by x rays can easily cause ionization of isolated atoms. X rays are therefore frequently referred to as **ionizing radiation**. X ray photoelectron spectroscopy and Auger spectroscopy are two techniques which, like ultraviolet photoelectron spectroscopy, study the ejected electron to gain information about the atom from which it came.

Flame ionization

Probably the simplest way of supplying energy for ionization is by subjecting the atoms or molecule to a flame. However, temperatures of several thousand degrees are usually required to achieve an appreciable degree of ionization. Specialized flames, such as an electrical arc, spark or **plasma**, can produce the necessary temperatures in a controlled manner.

See also Dissociation.

Resources

Books

Gross, M L., R. Caprioli, and P.B. Armentrout. *The Encyclopedia of Mass Spectrometry: Ion Chemistry and Theory.* Oxford: Pergamon Press, 2001.

Oxtoby, David W., et al. *The Principles of Modern Chemistry.* 5th ed. Pacific Grove, CA: Brooks/Cole, 2002.

Periodicals

Letokhov, V. S. "Detecting individual atoms and molecules with lasers (Resonance-ionization Spectroscopy." *Scientific American* 259 (September 1988): 54-59,

Karen Trentelman

Ionizing radiation

Ionizing radiation is any **energy** that causes the ionization of the substance through which it passes. As the **radiation** is emitted from a source, it detaches a charged particle from an atom or **molecule**, leaving the atom or molecule with an excess charge. This charged particle is called an ion.

To remove an **electron** from an atom or molecule, the ionizing particles must have a kinetic energy exceeding the binding energy of the target **species**, typically a few electron volts. (An electron volt is a unit of energy defined as the work it takes one electron to move across a voltage difference of one volt.) Radiation of sufficient energy for this process to occur is commonly produced in nature.

Some common ionizing charged species are electrons, positrons, protons, and β particles (Helium nuclei). Electrons, positrons, and β particles are emitted by radioactive elements. Photons, the uncharged particles of **light**, can also be emitted by radioactive nuclides, but can also be generated by x-ray devices. All the charged species, as well as neutrons, are currently produced at man-made particle **accelerators**, and lasers now generate photons of sufficient energy to exceed the binding energy of many **atoms** and molecules.

Most elements formed during the very early expansion of the universe were radioactive in the past, but over time became stable. Some, such as **uranium**, thorium, radium, and **radon** are still unstable, and spontaneously emit ionizing radiation. Here on **Earth** many **rocks** and **minerals** emit radon gas, a radioactive gas formed by the decay of radium. Other radioactive elements (^{3}H [Tritium] and ^{14}C) can be produced by atmospheric interactions with cosmic rays (energetic particles continuously entering the earth's atmosphere from outer **space**).

Ionizing radiation is more damaging to human **tissue** than non-ionizing thermal-type radiation, as it is more likely to be localized and have a higher intensity (energy deposited per area per second). The damage is initialized by the ionizing particle when it knocks an electron off an atom or molecule in a living system, leaving an unpaired electron behind. The target atom or molecule is then a free radical, a highly reactive type that can spawn many more free radicals in the body. The in-duced chemical changes have been shown to cause **cancer** and genetic damage.

A unit called the rem (roentgen equivalent man) is used to measure the absorbed dose of ionizing radiation in living systems. An absorption of 0.5 rem annually is considered safe for a human being. By comparison, about 0.1 to 0.2 rem per year is contributed by natural sources, about 0.002 rem comes from dental **x rays**, and about 0.05 from a chest x ray.

Ionosphere *see* **Atmosphere, composition and structure**

Iridium *see* **Element, chemical**

Iris family

Irises are plants in the family Iridaceae which contains 1,500-1,800 **species** and 70-80 genera. The center of diversity of this family is in southern **Africa**, but species are found on all of the habitable continents. The largest groups in the family are the true irises (*Iris* spp.) with 200 species and gladiolus (*Gladiolus* spp.) with 150 species.

Many species in the iris family have large, attractive flowers. The major economic importance of this family involves the cultivation of many species in **horticulture**. In France and Quebec the iris is generically known as the *fleur-de-lis*, and it is an important cultural symbol.

Biology of irises

Most species in the iris family are perennial herbs. These plants die back to the ground surface at the end of the growing season and then redevelop new shoots from underground rhizomes, bulbs, or corms at the beginning of the next growing season. A few species are shrubs.

The leaves of species in the iris family are typically long, narrow, and pointed at the tip with **parallel veins** and sheathing at the base of the **plant** or shoot. The flowers are erect on a shoot and are large, colorful, and showy, and they contain both female (pistillate) and male (staminate) organs. The floral parts are in threes: three petals, three sepals, three stamens, and a pistil composed of three fused units. The sepals are large and petal-like, and they enclose the petals which are erect and are fused into a tube-like structure in some species. The flowers may occur singly or in a few-flowered inflorescence, or cluster. The flowers produce **nectar**, are pleasantly scented, and are pollinated by flying **insects** or **birds**, although some species are wind-pollinated. The **fruits** make up a three-compartmented capsule containing numerous

A Rocky Mountain iris (Iris missouriensis). *JLM Visuals. Reproduced by permission.*

seeds. The leaves and the rhizomes of *Iris* species contain an irritating chemical which is poisonous if eaten.

Native species of North America

Various species in the iris family are native to wild places in **North America**. Wild irises are most commonly found in moist habitats beside lakes, ponds, **rivers**, and seashores. Some of the more widespread species of iris include the blue-flag (*Iris versicolor*), violet iris (*I. verna*), water-flag or western blue-flag (*I. missouriensis*), western iris (*I. tenax*), and the beachhead-iris (*Iris setosa*). Another widespread group in the iris family is the blue-eyed **grasses**, for example, *Sisyrinchium montanum*. The blue-eyed grasses are found in a wide range of moist habitats and sometimes beside roads and other disturbed places.

Horticultural irises

Many species and cultivars in the iris family are grown in gardens and greenhouses for their beautiful flowers. These plants are typically propagated by splitting their rhizomes, bulbs, or corms, and sometimes by seed.

Various species of iris are cultivated in gardens. These include the yellow-flowered water-flag (*Iris pseudacorus*) and blue-flowered species such as the true fleur-de-lis (*I. germanica*), the Siberian iris (*I. sibirica*), the stinking iris (*I. foetidissima*), and the butterfly iris (*I. ochroleuca*). Some cultivated species of iris have become naturalized in parts of North America and can be found in wild habitats and in old gardens near abandoned houses.

Many of the approximately 80 species of crocuses are grown in gardens. In places where there is a snowy winter, crocuses are often planted in lawns where they bloom very soon after the snow melts and air temperatures become mild. The most commonly cultivated species is the European spring crocus (*Crocus verna*).

Another commonly cultivated group is the gladiolus, including *Gladiolus byzantinus* from southwestern **Asia** and many horticultural hybrids. The tiger **flower**

(*Tigridia pavonia*) is native to Mexico and is sometimes cultivated in temperate gardens.

Other economic products

The world's most expensive spice is said to be saffron, a yellow substance made from the blue-flowered saffron crocus (*Crocus sativa*) of the eastern Mediterranean region. The major expense of saffron is in labor costs because it takes the floral parts 600-800 crocus flowers to make 0.035 oz (1 dry gram) of the spice. Saffron is mainly used to flavor foods and also as a yellow colorant of certain cooked foodstuffs, as in saffron **rice**.

The rhizomes of the orris (*Iris florentina*) are used to manufacture perfumes and cosmetics. The rhizomes must be peeled and dried before their odor, much like that of violets (*Viola* spp.), will develop.

Resources

Books

Judd, Walter S., Christopher Campbell, Elizabeth A. Kellogg, Michael J. Donoghue, and Peter Stevens. *Plant Systematics: A Phylogenetic Approach.* 2nd ed. with CD-ROM. Suderland, MD: Sinauer, 2002.

Klein, R. M. *The Green World: An Introduction to Plants and People.* New York: Harper and Row, 1987.

Raven, Peter, R. F. Evert, and Susan Eichhorn. *Biology of Plants.* 6th ed. New York: Worth Publishers Inc., 1998.

Bill Freedman

Iron

Iron is a metallic chemical element of **atomic number** 26. Its symbol is Fe, **atomic weight** is 55.847, specific gravity is 7.874, melting point is 2,795°F (1,535°C), and **boiling point** is 4,982°F (2,750°C).

Iron is one of the transition metals, occurring in group 8 of the **periodic table**. Four naturally occurring isotopes exist with atomic weights of 54 (5.8%), 56 (91.7%), 57 (2.2%), and 58 (0.3%). In addition, six radioactive isotopes have been prepared, with atomic weights of 52, 53, 55, 59, 60, and 61. The element was originally known by its Latin name *ferrum*, from which its chemical symbol is derived.

General properties

Iron is a silver-white or gray **metal** that is malleable and ductile. In a pure form, it is relatively soft and slightly magnetic. When hardened, it becomes much more magnetic. Iron is the most widely used of all metals.

Prior to its use, however, it must be treated in some way to improve its properties or it must be combined with one or more other elements to form an **alloy**. By far the most common alloy of iron is **steel**.

One of the most common forms of iron is pig iron, produced by smelting iron **ore** with coke and limestone in a blast furnace. Pig iron is approximately 90% pure iron and is used primarily in the production of cast iron and steel.

Cast iron is a term used to describe various forms of iron that also contain **carbon** and silicon ranging in concentrations from 0.5-4.2% of the former and 0.2-3.5% of the latter. Cast iron has a vast array of uses ranging from thin rings to massive **turbine** bodies. Wrought iron contains small amounts of a number of other elements including carbon, silicon, **phosphorus**, **sulfur**, chromium, nickel, cobalt, **copper**, and molybdenum. Wrought iron can be fabricated into a number of forms and is widely used because of its resistance to **corrosion**.

Sources of iron

Iron is the fourth most abundant element in the earth's crust and the second most abundant metal, after **aluminum**. It makes up about 6.2% of the crust by weight. In addition, iron is thought to be the primary constituent of the earth's core as well as of siderite meteorites. **Soil** samples taken from the **Moon** indicate that about 0.5% of lunar soil consists of iron.

The primary ores of iron are hematite (Fe_2O_3), magnetite (Fe_3O_4), limonite ($FeO(OH) \cdot nH_2O$), and siderite ($FeCO_3$). The element also occurs as the sulfide, iron pyrite (FeS), but this compound is not used commercially as a source of iron because of the difficulty in reducing the sulfide to the pure element. Iron pyrite has a beautiful golden appearance and is sometimes mistaken for elemental gold. This appearance explains its common name of fool's gold. Taconite is a low-grade ore of iron that contains no more than about 30% of the metal.

In nature, oxides, sulfides, and silicates of iron are often converted to other forms by the action of **water**. Iron(II) sulfate ($FeSO_4$) and iron(II) bicarbonate ($Fe(HCO_3)_2$) are the most commonly found of these.

How iron is obtained

Iron is one of the handful of elements that was known to ancient civilizations. Originally it was prepared by heating a naturally occurring ore of iron with charcoal in a very hot flame. The charcoal was obtained by heating **wood** in the absence of air. There is some evidence that this method of preparation was known as early as 3,000 B.C., but the secret of ore smelting was carefully guarded within the Hittite civilization of the Near East for almost two more millennia.

When the Hittite civilization fell in about 1200 B.C., the process of iron ore smelting spread throughout eastern and southern **Europe**. Ironsmiths were soon making ornamental objects, simple tools, and weapons from iron. So dramatic was the impact of this new technology on human societies that the period following 1200 B.C. is generally known as the Iron Age.

A major change in the technique for producing iron from its ores occurred in about 1773. As trees (and therefore the charcoal made from them) grew increasingly scarce in Great Britain, the English inventor Abraham Darby (1678?-1717) discovered a method for making coke from soft **coal**. Since coal was abundant in the British Isles, Darby's technique insured a constant supply of coal for the conversion of iron ores to the pure metal. The modern production of iron involves heating iron ore with coke and limestone in a blast furnace, where temperatures range from 392°F (200°C) at the top of the furnace to 3,632°F (2,000°C) at the bottom. Some blast furnaces are as tall as 15-story buildings and can produce 2,400 tons of iron per day.

Inside a blast furnace, a number of **chemical reactions** occur. One of these involves the reaction between coke (nearly pure carbon) with **oxygen** to form **carbon monoxide**. This carbon monoxide then reacts with iron ore to form pure iron and **carbon dioxide**. Limestone is added to the reaction mixture to remove impurities in the iron ore. The product of this reaction, known as slag, consists primarily of calcium silicate. The iron formed in a blast furnace exists in a molten form known as pig iron that can be drawn off at the bottom of the furnace. The slag is also molten but less dense than the iron. It is drawn off from taps just above the outlet from which the molten iron is removed.

Efforts to use pig iron for commercial and industrial applications were not very successful. The material was quite brittle and objects of which it was made tended to break easily. Cannons made of pig iron, for example, were likely to blow apart when they fired a shell. By 1760, inventors had begun to find ways of toughening pig iron. These methods involved remelting the pig iron and then burning off the carbon that remained mixed with the product. The most successful early device for accomplishing this step was the Bessemer converter, named after its English inventor Henry Bessemer (1813-1898). In the Bessemer converter, a blast of hot air is blown through molten pig iron. The process results in the formation of stronger forms of iron, cast and wrought iron. More importantly, when additional elements, such as manganese and chromi-

um, are added to the converter, a new product—steel—is formed.

Later inventions improved on the production of steel by the Bessemer converter. In the open hearth process, for example, a charge of molten pig iron, hematite, scrap iron, and limestone is placed into a large **brick** container. A blast of hot air or oxygen is then blown across the surface of the molten mixture. Chemical reactions within the molten mixture result in the formation of either pure iron or, with the addition of alloying metals such as manganese or chromium, a high grade of steel.

An even more recent variation on the Bessemer converter concept is the basic oxygen process (BOP). In the BOP, a mixture of pig iron, scrap iron, and scrap steel is melted in a large steel container and a blast of pure oxygen is blown through the container. The introduction of alloying metals makes possible the production of various types of steel with many different properties.

How we use iron

Alloyed with other metals, iron is the most widely used of all metallic elements. The way in which it is alloyed determines the uses to which the final product is put. Steel, for example, is a general term used to describe iron alloyed with carbon and, in some cases, with other elements. The American Iron and Steel Institute recognizes 27 standard types of steel. Three of these are designated as carbon steels that may contain, in addition to carbon, small amounts of phosphorus and/or sulfur. Another 20 types of steel are made of iron alloyed with one or more of the following elements: chromium, manganese, molybdenum, nickel, silicon, and vanadium. Finally, four types of stainless and heat-resisting steels contain some combination of chromium, nickel, and manganese alloyed with iron.

Steel is widely used in many types of construction. It has at least six times the strength of **concrete**, another traditional building material, and about three times the strength of special forms of high-strength concrete. A combination of these two materials, reinforced concrete, is one of the strongest of all building materials available to architects. The strength of steel has made possible some remarkable feats of construction, including very tall buildings (skyscrapers) and **bridges** with very wide spans. It has also been used in the manufacture of **automobile** bodies, ship hulls, and heavy machinery and machine parts.

Metallurgists have also invented special iron alloys to meet very specific needs. Alloys of cobalt and iron (both magnetic materials themselves) can be used in the manufacture of very powerful permanent magnets. Steels that contain the element niobium (originally called columbium) have unusually great strength and have been used, among other places, in the construction of nuclear reactors. Tungsten steels are also very strong and have been used in the production of high-speed metal cutting tools and drills. The alloying of aluminum with iron produces a material that can be used in AC magnetic circuits since it can gain and lose **magnetism** very quickly.

Metallic iron also has other applications. Its natural magnetic properties make it suitable for both permanent magnets and electromagnets. It is also used in the production of various types of dyes, including blueprint **paper** and a variety of inks, and in the manufacture of **abrasives**.

Biochemical applications

Iron is essential to the survival of all **vertebrates**. Hemoglobin, the **molecule** in **blood** that transports oxygen from the lungs to an organism's cells, contains a single iron atom buried deep within its complex structure. When humans do not take in sufficient amounts of iron in their daily diets, they may develop a disorder known as **anemia**. Anemia is characterized by a loss of skin **color**, a weakness and tendency to faint, palpitation of the **heart**, and a general sense of exhaustion.

Iron is also important to the good health of plants. It is found in a group of compounds known as porphyrins that play an important role in the growth and development of **plant** cells. Plants that lack iron have a tendency to loose their color, become weak, and die.

Chemistry and compounds

Iron typically displays one of two valences in forming compounds, 2^+ and 3^+. According to the older system of chemical nomenclature, these classes of compounds are known as the ferrous and ferric salts, of iron respectively. Because of the abundance of oxygen in the atmosphere, most naturally occurring iron compounds tend to be in the higher (3^+) **oxidation state**.

One of the most widely used of iron compounds is iron(III) (or ferric) chloride, $FeCl_3$. When added to water, it reacts with water molecules forming a thick, gelatinous precipitate of iron(III) hydroxide. The compound is used in the early steps of water purification since, as the precipitate settles out of **solution**, it traps and carries with it organic and inorganic particles suspended in the water. Iron(III) chloride is also used as a mordant, a substance used in dyeing that binds a dye to a textile. In gaseous form the compound has still another use. It attacks and dissolves metal and can be used, therefore, for etching. Printed circuits, for example, are often first etched with iron(III) chloride.

> **KEY TERMS**
> ....
>
> **Blast furnace**—A structure in which a metallic ore (often, iron ore) is reduced to the elemental state.
>
> **Ductile**—Capable of being drawn or stretched into a thin wire.
>
> **Isotopes**—Two molecules in which the number of atoms and the types of atoms are identical, but their arrangement in space is different, resulting in different chemical and physical properties.
>
> **Malleable**—Capable of being rolled or hammered into thin sheets.
>
> **Transition metal**—An element found between groups IIA and IIIA in the periodic table.

Iron(II) (ferrous) compounds tend to oxidize rather easily and are, therefore, less widely used than their 3^+ cousins. Iron(II) (ferrous) sulfate is an important exception. In solid form, the compound tends not to oxidize as readily as other Fe^{2+} compounds and is used as an additive for **animal** feeds, in water purification, in the manufacture of inks and pigments, and in water and **sewage treatment** operations.

From a commercial standpoint, probably the most important chemical reaction of iron is its tendency to oxidize. When alloys of iron (such as the steels) are used in construction, a major concern is that they tend to react with oxygen in the air, forming a coating or iron oxide, or rust. The rusting process is actually a somewhat complex process in which both oxygen and water are involved. If one or the other of these materials can be prevented from coming into contact with iron, oxidation will not occur. But if both are present, an electrochemical reaction is initiated, and iron is converted to iron oxide.

Each year, billions of dollars are lost when iron-containing structural elements degrade or disintegrate as a result of oxidation (rusting). It is hardly surprising, therefore, that a number of techniques have been developed for reducing or preventing rusting. These techniques include painting, varnishing, galvanizing, tinning, and enameling.

Resources

Books

Greenwood, N. N., and A. Earnshaw. *Chemistry of the Elements.* 2nd ed. Oxford: Butterworth-Heinneman Press, 1997.

Hawley, Gessner G., ed. *The Condensed Chemical Dictionary.* 9th ed. New York: Van Nostrand Reinhold, 1977.

Joesten, Melvin D., et al. *World of Chemistry.* Philadelphia: Saunders, 1991.

Knepper, W. A. "Iron." *Kirk-Othmer Encyclopedia of Chemical Technology.* 4th ed. Suppl. New York: John Wiley & Sons, 1998.

Seely, Bruce Edsall, ed. *Iron and Steel in the Twentieth Century.* New York: Facts on File, 1994.

David E. Newton

Irrational number

An irrational number is a number that cannot be expressed as a fraction, that is, it cannot be written as the quotient of two whole numbers. As a decimal, an irrational number is shown by an infinitely long nonrepeating sequence of numbers. Examples of irrational numbers are π (**pi**, the **ratio** of circumference to diameter of a **circle**), e (base of the natural **logarithms**), and $\sqrt{2}$ (that number which multiplied by itself equals 2).

See also e (number); Rational number.

Irrigation

The practice of diverting **water** from natural resources to **crops** has been practiced for at least 7,000 years. The earliest methods, as practiced in places like the areas surrounding the Nile river **basin**, included digging channels to allow water from the river during flood periods to reach cultivated fields along the river's banks. Ancient farmers also built dikes to help retain the water on the flooded land. Other early irrigation techniques included the construction of diversion **dams** and the use of machinery to lift the water and irrigate land that was higher than the flood plains. Evidence of early irrigation systems has been found in North and **South America**, the Middle East, and in China and India.

Surface irrigation system techniques include surface **flooding**, furrow flooding, and dead-level surface flooding. In surface flooding the whole land area to be irrigated is flooded with water. This technique is good, for instance, for growing **rice**. Furrow flooding involves planting trees or crops between shallow trench-like channels and flooding the area. In arid regions, dead-level surface irrigation, where fields are leveled to a **zero** slope, is practiced.

Closed-conduit irrigation includes sprinkler systems, bubbler irrigation, and drip or trickle irrigation. Gardeners, as well as farmers, commonly use these techniques. Sprinkler systems pump water through pipes or hoses to the sprinkler, which can be fixed or mobile. Bubbler and drip systems periodically supply water to the roots of one or more plants. These systems are constructed of tubing or pipes. Drip systems deliver water slowly and are the most conservative users of water resources. They are particularly favored in arid regions, such as the southwest area of the United States, **Australia**, and the Middle East.

There are more than 600 million acres (243 million ha) worldwide, about 17% of agriculturally productive land, that are routinely irrigated. More than 60% of the irrigated land is contained within a few countries—China, India, Pakistan, the United States, and parts of the former USSR. Since becoming independent in 1947, India has developed over 700 irrigation projects, more than doubling the amount of land they irrigate, which exceeded 100 million acres (41 million ha) by the late 1980s.

In China, where irrigation has been used since the third century B.C., irrigated land doubled and tripled in some areas after the completion of dam projects undertaken since World War II. The primary irrigation crop in China is rice, but they also irrigate their **wheat** and **cotton** fields. One dam, the Tujiang on the Min River, was built around 300 B.C. and is still in use. Tujiang Dam is the source of water for 500,000 acres (202,000 ha) of land.

While the purpose of irrigation is to produce a better crop yield, the need for irrigation varies depending upon seasonal and climatic conditions. Some regions need crop irrigation all year, every year; some only part of the year and only in some years; and others need to irrigate only during **seasons** of water shortage from rainfall. In Iraq and India, for instance, irrigation is absolutely necessary in order to grow crops, since rain cannot be depended upon in those regions. In other areas, irrigation may be used only as a backup in case there is not sufficient rainfall during a crop's growing season. This is termed supplemental irrigation.

The problem of salinization

Salinization is a major problem associated with irrigation, because deposits of salts build up in the **soil** and can reach levels that are harmful to crops. In addition, the salts can make ground water, which may be in use for drinking, saltier and unsuitable for drinking. It is mostly in arid and semiarid regions where the problem of high **salt** content deposited from irrigation threatens crops.

Drip irrigation is a technique that can be used in areas where the ground water level is high and in danger of suffering from a high salt content. Where salinization is a problem to plants, enough water can be added to the irrigation process to leach salts away from **plant** roots. When the danger of salinization is to the water table, it is necessary to add drainage to the irrigation system away from the water table.

Crops have different salt tolerances and their **selection** in relation to the salinity of the soil is an advisable practice. Among the common crops that have a high salt tolerance are red beets, **spinach**, kale, asparagus, sugar beets, **barley**, cotton, date **palms**, and some **grasses** used for **animal** feed, such as wild rye and wheat grass. Crops that have a low tolerance for salinity include radishes, celery, green beans, **fruits** such as pears, apples, oranges, grapefruit, plums, apricots, peaches, strawberries, lemons, and avocados, and a number of clovers that are used for animal grazing.

Areas in the world where farming is threatened by high salinity include the Indus Basin in Pakistan where they also face the problem of a rising water table. The Imperial Valley in California, formerly productive agricultural lands in South America, China, India, Iraq, and many other regions throughout the world are all facing the threat of losing fertile land because of salinization. After the building of the Aswan Dam in Egypt, the Nile River and the surrounding fields that had been irrigated successfully for over 5,000 years became threatened by high salinity in the water.

The main technique used to reclaim land that has developed a high salt content from irrigation is a **leaching** process. This is based on a careful analysis of the soil and the amount of water that must be applied to reach a level of acceptable salt content. One problem of leaching is that other **nutrients** needed by the crops, besides the undesirable salts, may also be leached from the soil. Consequently, nutrients often need to be replaced after an area is reclaimed from high salinization.

Irrigation systems

The planning of irrigation systems is highly specialized and requires the help of agricultural engineers who understand not only the design and construction of irrigation systems, but also farm management and mechanization, soil science, crop husbandry, and the economics of farming. The engineer's education in these related fields is important so that he or she is able to design an irrigation system that is appropriate to the type of farming in the area that is to be irrigated.

Before an irrigation system can be built, a number of important studies must be made. Among them would be a survey of land and water resources, a study of the current uses of the area, a proposal for an irrigation system, cost estimates of the project, and a projection of its economic benefits. A large regional or national project might also include the economic and material resources for the project that are available by the particular entity, the cost of construction and administration of the project, the financing and marketing of the project to individual farmers, and the training of personnel to carry out the project.

Among the specific surveys that must be made before an irrigation system is constructed are soil, water, and topographic surveys. Sometimes critical decisions have to be made about the destruction of monuments from antiquity. For instance, when the Aswan Dam was built, some important statues from ancient Egyptian culture were lost because they were covered by water. Another important consideration for building a new irrigation project is whether it will change the current farm practices, and if so, how to educate farmers to new methods. Foremost at issue is the consideration of how an irrigation system will impact the farmers and farming in the area.

Surface irrigation

In surface irrigation systems, the area to be covered with water is sloped away from the supply channel so that the water will flow over the entire area with the water moving both across the surface to be irrigated and filtering down to the root bases of the plants in the field. Among the variations of surface irrigation are the techniques of furrow irrigation, border strips, basins, and wild flooding.

Furrow irrigation has the advantage of allowing the crops to be tended shortly after watering periods. The system is useful for crops that are grown in rows that can be separated by furrows (shallow ditches) along the rows. The furrows are usually dug along the line of the slope, but sometimes they run **perpendicular** to it. The problem with cross cutting the furrows on a slope is that it may collapse during irrigation periods from the force of the water.

The preferred method of supplying water to the furrows is to siphon water from a main source and carry it through plastic or **aluminum** pipes set in a main ditch at the head of the field. Another ditch at the end of the furrows collects excess water and runs it along to lower lying fields. The best incline for furrow irrigation is one from 0-5% slope. Crops are usually planted on the rise between the furrows, but sometimes trees are planted at the bottom of the furrow. Since there is less water surface in this method, **evaporation** of water is less than in surface flooding.

In pastures where there are crops that grow closely together border strips may be used. In this system a main ditch is constructed along the highest end of the slope and banks, called checks, which can be built as much as 70 ft (21 m) apart. Water is then siphoned from the main ditch onto the strips where the crops are grown. Sometimes the banks are replaced with border supply ditches, which allow more control over the release of water. This system is often used in research studies. In hilly areas contour ditches are built that follow the contour of a hill. They are carefully graded to control the flow of water.

For landscapes, gardens, and the watering of individual trees, the use of basins may be a suitable method of irrigation. The area to be irrigated is surrounded by banks (checks) and then watered from a main source along a high point in the basin. A drain is also placed along the major depression of a basin to allow water to run off. This system is easy to build since it requires very little movement of Earth. It is usually built around the natural contours of the area. Where the land is extremely steep, an **adaptation** of basin irrigation called **terracing** can be used. Here basins are created in a step fashion along the slope of the hill. At the end of each basin step a check is built. Basin watering is not generally recommended for flat ground.

While wild flooding is still practiced, it is not recommended by agricultural engineers because the water distribution is uneven and can lead to high saline contents in the soil and to waterlogging. The crop yields are consequently unpredictable.

Sub-irrigation

In areas where the topsoil is of high quality and porous and there is an underlay of clay soil that absorbs water slowly, conditions exist for natural sub-irrigation, provided that the water table is high. Ditches dug along the fields can be used to monitor the water level and to also replenish the water supply when it is low. Where there is little or no rainfall and the salts in the water build up on the surface of the soil, leaching is carried out. To overcome excess rain in areas where sub-irrigation systems are in use, water can be removed by pumping or using natural gravity features available in the terrain, that is, slopes and depressions in the ground.

When sub-irrigation is desired but the conditions are not available naturally, pipes with evenly distributed punctures can be buried underground. A difficulty involved with these systems is that they can be damaged when the soil is being cultivated. These systems also work by the use of natural sloping features in the terrain or by pumping water through the pipes.

Drip irrigation, which is not actually sub-irrigation, but uses some of the same principles as in sub-irrigation, delivers water slowly to the root areas of plants. Here, too, pipes are used as the channels for transporting the water and emitters are placed to water plants directly. While it is economical to use because there is little waste of water and evaporation is at a minimum, initial costs of installing drip irrigation systems are higher than other methods. There is also a tendency for emitters to become clogged

by the salts in the water. Salts, however, do not build up around the roots of plants in drip irrigation systems.

Overhead irrigation

These systems use a pumping unit, conveyor pipes, and some form of sprinkler mechanism. Of all the irrigation systems, they most resemble natural rainfall. Some systems are fixed and use pipes laid on the ground with risers that have a sprinkling nozzle at the top that rotates 360 degrees. The size of the water droplets, the speed of **rotation**, and the evaporation **rate** are considerations in selecting sprinkler systems, since these all have an effect on the soil. An added use of sprinkler systems is that they can in some situations be used for frost protection.

Besides fixed systems, mobile sprinkling systems are in use in the United States and Great Britain. Portable systems use a pump at the water source to pump the water into a main line that is laid throughout the field. The sprinkler units are moved from field to field for irrigation of crops. Other mobile sprinkling systems use a device called a rain gun, which has a nozzle with a large diameter.

Resources

Books

Crouch, Dora P. *Water Management in Ancient Greek Cities.* New York: Oxford University Press, 1993.

Doolittle, William E. *Canal Irrigation in Prehistoric Mexico.* Austin: University of Texas Press, 1989.

Guillet, David. *Covering Ground: Communal Water Management and the State in the Peruvian Highlands.* Ann Arbor: University of Michigan Press, 1992.

Kluger, James R. *Turning on Water with a Shovel: The Career of Elwood Mead.* Albuquerque: University of New Mexico Press, 1992.

Shortle, J. S., and Ronald C. Griffin, eds. *Irrigated Agriculture and the Environment.* Northampton, MA: Edward Elgar, 2001.

Wallace, Henry A. *Henry A. Wallace's Irrigation Frontier: On the Trail of the Corn Belt Farmer.* Norman: University of Oklahoma, 1991.

Other

American Society of Agricultural Engineers. *National Irrigation Symposium: Proceedings of the Fouth Decennial Symposium.* November 14-16, 2000.

Vita Richman

Island

An island is an area of land, smaller than a **continent**, that is entirely surrounded by **water**. That distinction, although somewhat artificial, suggests different geologic forces acting to create and maintain islands versus continents. Islands further differ from continents in their natural environments—in the biological systems they support, in their **rate** of response to change, in their ability to recover from ecological disaster.

The plants and animals found on islands often seem an odd assemblage. Some in fact are odd, in the sense that they live nowhere else. The seemingly skewed distribution of populations on islands, compared with that on a mainland, results in part from the small size of islands; they cannot carry a representative zoo, and, without **migration** in from outside, the **extinction** of one life form might leave a gaping hole in the biota. Also, the water surrounding islands acts as a barrier to the passage of some life forms, particularly large **mammals**, but encourages the migration of others, such as **birds** and **insects**. Because of their relative isolation and the potentially unique biota that may be established on them, islands have been known, at least since the time of Charles Darwin, as natural laboratories of **evolution**.

Island types

The islands discussed here are of three kinds: continental islands, oceanic islands, and coral islands. Not discussed are inland islands, such as islands found in the middle of a **lake**.

Continental islands

Continental islands are parts of the **continental shelf** that rise above the surrounding water. That is, they are situated on the shallow water margin of a continent, usually in water less than 600 ft (200 m) deep. Greenland, the largest island in the world, and Newfoundland are examples of continental islands. A drop in **sea level** would be sufficient to connect these islands to the North American continent.

Another, rarer kind of continental island consists of small pieces of continental material that broke away from a land mass. These islands are now part of a separate crustal plate that is following an independent path. The Seychelles in the Indian Ocean were once associated with the Madagascar-India portion of the supercontinent Pangaea. With the breakup of Pangaea about 200 million years ago, the Seychelles began their independent existence. Their continental basement structure, however, clearly associates them with the continents rather than with oceanic islands of volcanic origin.

Oceanic islands

Oceanic islands arise from volcanic action related to the movement of the lithospheric, continent-bearing crustal plates. Unlike continental islands, oceanic islands grow from oceanic crust. Oceanic islands are not scattered haphazardly about the deep ocean waters but are aligned along converging oceanic plate boundaries or along the mid-ocean ridges, or diverging oceanic plate boundaries, associated with sea-floor spreading. In addition, some arose as oceanic plates moved over fixed hot spots in the deeper mantle.

Coral islands

Coral islands are distinct from both continental islands and oceanic islands in that they are formed of once living creatures, the corals, which colonize in place to form coral reefs.

Barrier islands

Unrelated to this three-part classification of islands are islands of a fourth kind, **barrier islands**. Barrier islands occur in shallow-water coastal areas and are composed of unconsolidated sediment, usually **sand**. Barrier islands form 15% of all the coastline in the world, including most of the coastline of the continental United States and Alaska, and also occur off the shores of bays and the Great Lakes. Some barrier islands are stable enough to support houses or an airport runway; others are short-lived, ripped up annually by winter storms and reestablished by wave and tidal action. As their name suggests, they afford some protection to the mainland from **erosion**.

How many islands?

Islands are intrinsically impermanent. The more stable oceanic islands last a relatively brief time of 5-10 million years. Some islands drown, a result of erosion, **subsidence** of the ocean crust, or rising sea level. Sea levels are related in part to the amount of water bound up in the **polar ice caps** or released into the oceans; and the size of the polar **ice** caps is related to a variety of factors including variations in the positions of continents, the orientation of the earth's axis, and the amount of cloud cover.

Sea level is fairly high now; it was lower during the Little Ice Age, circa fourteenth to nineteenth century, and even lower about 18,000 years ago. A lowering of sea level brings back into view drowned islands.

Ongoing volcanism continues to add to existing islands and create new ones; an example is Surtsey, off the southern coast of Iceland, which came into existence with a **submarine** volcanic explosion on November 14, 1963, and has continued to accrete surface area as the ongoing lava flows cool. There are also islands that appear intermittently.

Because islands come and go, the number of islands in existence cannot be established except in relation to a proscribed time period—a human generation, or a century or two. With the discovery of some islands in the Russian Arctic in the mid-twentieth century, however, it is thought that no islands remain to be discovered in our time. **Satellite** and ship-based scanning equipment is now being used to search for islands whose positions appear on nautical maps but which have themselves disappeared, and to identify underwater sites of new island formation.

Island formation

The classification of islands based on their foundation—continental crust versus oceanic crust versus corals—as given above has been around in some form since it was first addressed by Darwin in 1840. A study of island formation, however, shows different geologic events contributing to the genesis of different kinds of islands. The following discussion of island origins is limited to islands that do not have the geological characteristics of continents, namely, oceanic islands and coral islands.

Oceanic islands

The development of **plate tectonics** theory in the 1960s greatly aided scientists' understanding of the genesis of islands. Oceanic islands originate in volcanic action typically associated with the movement of the lithospheric plates.

The **lithosphere** is the major outer layer of the **earth**. It consists of the crust—both continental and oceanic-and upper mantle, and ranges from the surface to 60 m (100 km) deep, although subducted crust has been remotely detected at depths of 620 mi (1,000 km). (For comparison, the average radius of the earth is 4,173 mi [6,731 km].) The lithosphere is divided into rigid, interlocking plates that move with respect to one another.

There are 11 major plates (two of which seem to be fracturing) and many smaller ones. The plates move over the next lower layer, the **asthenosphere** (a sort of crystalline sludge), perhaps by thermal **convection**, at an average rate of 4-4.5 in (10-11 cm) per year. The plate boundaries tend to move away from each other at mid-ocean ridges and to approach each other at the edges of continents.

At the mid-ocean ridges, **magma** wells up and cools, forming **mountains**. At the same time, existing sea floor spreads apart, and new sea floor is created. As sea-floor spreading continues, the mountains, which sit on ocean crust, are carried away from the mid-ocean ridge and therefore away from the source of new lava deposits. Many of these underseas mountains, as a result, never grow tall enough that their tops could emerge as islands. These submarine mountains are known as **seamounts**. It is possible to date with some **accuracy** the age of seamounts by measuring their **distance** from the ridge where they were born.

A major exception to the nonemergence of mountains formed at mid-ocean ridges is the volcanic island of Iceland, which has had a more complicated history; it was formed both by **upwelling** magma from the mid-Atlantic ridge and by volcanism over a **hot spot** deeper in the mantle, which also contributed to upwelling magma in the same area. The hot spot, active for about 55 million years, has since cooled, and the mid-Atlantic ridge has shifted abruptly from its previous position to a position somewhat eastward. It is still active, producing lava from volcanoes, the flows of which sometimes close harbors. Iceland is one of the few places on Earth where a mid-ocean ridge has risen to the land's surface and become visible.

Volcanoes producing lava flows and occasional seamounts, and more rarely emerged islands, characterize divergent plate boundaries at the mid-ocean ridges. However, converging plate boundaries—two plates coming together—are characterized by volcanoes that often produce emerged islands, as well as by forceful earthquakes and deep oceanic trenches, such as the Marianas Trench. Most plates that converge do so at the edges of continents.

When two plates meet, the plate carrying the heavier oceanic crust dips under, or subducts, and the plate carrying the lighter continental crust rides over it. At the point of subduction a deep trench develops; and **parallel** to it, on the lighter plate, volcanic action produces a row of islands. (The magma involved in the volcanism comes from melting of the oceanic crust as it is subducted.) These island groups are called island arcs, after their curved pattern. An island arc with active volcanism is called a back arc. Between the back arc of the system and its associated trench there may be a second, nonvolcanic arc of islands, called the front arc, that is thought to be caused by upthrust of crust from the lighter plate. The front arc may lie below the surface of the water and not be readily visible.

Island formation occurs at intraplate locations (anywhere between the boundaries) as well as at plate boundaries. It is reasoned, with strong scientific support, that some of these mid-plate volcanic islands resulted from passage of a lithospheric plate over a thermal plume rising from a fixed hot spot deeper in the mantle layer.

The formation of the Hawaiian-Emperor island chain in the north Pacific is attributed to this mechanism. The hot spot is thought to be located near the Loiki Seamount and to be causing the currently active volcanoes of Mauna Loa and Kilauea on the island of Hawaii. The chain trends northwestward, with the oldest islands at the northwestern end. As the Pacific plate drifts, at a rate of 4 in (10 cm) per year, areas that had been positioned over the hot spot and so subject to volcanism, lava accumulation, and island formation, move off the hot spot and cool down. Their position is then taken by new areas of lithosphere that drift over the thermal plume and a new island begins to form on the sea floor.

This scenario would explain why some of the animals populating the Hawaiian islands are older than the islands themselves. Because of the relative closeness of the islands to each other and the leisurely pace of the Pacific plate's drift, animals theoretically could have had time to raft, swim, or fly from the older, now submerged islands at the northern end of the chain toward the younger, more southerly islands, which are not submerged. By island hopping over geological time, some of the original **species** may eventually have made their way to islands that had not yet come into existence at the time the animals established a presence on the islands.

The hot spot model works well for the Hawaiian-Emperor island chain but does not explain all intraplate clusters or chains of islands. Indeed, one of the problems

An island off the east coast of the Malay Peninsula. *Photograph by R. Ian Lloyd. Stock Market. Reproduced by permission.*

facing oceanographers is the association of islands whose formation can easily be described by global and local conditions with islands that appear not to have been produced by the same processes.

Coral islands

Coral islands are (usually) low-lying islands formed by hermatypic, or reef-building, corals, chiefly scleractinian corals and hydrocorallians. Reef-building corals occur in a broad band stretching around the globe from 25 degrees north of the equator to 25 degrees south of the equator and require an average water **temperature** of about 68–77°F (20–25°C). They do not grow below 165 ft (50 m) in depth. They also have specific needs for water salinity, clarity, calmness, and sunlight. Sunlight aids in formation of the living corals' exoskeleton, and so aids in reef-building. Corals anchor on something—seamounts, submarine slopes of islands, or debris such as abandoned army vehicles and bedsprings—and therefore are generally found at the edges of continents or existing islands. If the surface of a reef emerges into the air—through, for example, a slight drop in sea level—the creatures dry up and die. The exposed, dead surface of the reef then serves as a platform for the accumulation of sediment, which may in turn become sufficient to support **plant** and **animal** life. Thus, offshore islands in tropical and semitropical zones around the world often have a core of emerged, dead coral reef. For example, a reef that emerged in about 3450 B.C. provided the base on which all of the islands in the Maupihaa Atoll, in the Society Islands, are founded. Indeed, study of the rate of **uplift** of emerged coral reefs has helped scientists determine local sea levels in past eras.

Island biogeography

Islands may be regarded as closed ecosystems. Although this is not true in every case—witness the island-hopping of species on the Hawaiian-Emperor island chain—or at all times for any given island, the relative isolation of islands has made them an ideal setting in which to explore theories of evolution and **adaptation**.

Two words frequently used in relation to island environments are equilibrium and change. Ecosystems in equilibrium are assumed to have reached steady state, with very slow rates of change. No more is taken out of the **ecosystem** than is replenished; predator-prey relationships remain constant, and die-offs are balanced by new colonization. Whereas the **individual** species involved in these interactions may change, the overriding patterns do not. The equilibrium model of insular biogeography was formally stated in the 1960s by R. H. MacArthur and E. O. Wilson and has been used to, among other things, establish and manage natural preserves on both islands and mainlands.

Environments in equilibrium are, of course, subject to change. A catastrophic **storm** can destroy a large sector of the biota; land **bridges** come and go. The changes introduced by the entry of other life forms into the closed system of an island, however, can have dramatic immediate effects. Island environments may have permissive or controlling effects on organisms that attempt to establish a presence. Among the permissive effects, or opportunities for colonization, is the availability of unoccupied biological niches. Unfilled niches appear to hasten organismal **radiation**, the evolutionary branching of species. On the Hawaiian islands, for example, the native, or **endemic**, family of birds known as **honeycreepers** has branched into 23 species, many adapted to different feeding niches—seeds, insects, and so on. Further, the beaks of the birds have adapted to extracting the different diets from different **tree** species—a remarkable series of adaptations.

Among the controlling effects that islands have on would-be immigrants is simple inhospitality. Volcanic or coral islands lacking sufficient layers of sediment to grow plants, for example, would not be attractive or even feasible as a home for many kinds of animals, including agricultural humans.

Once a breeding pair of immigrants has successfully penetrated the isolation of an island and taken up residence there, it can profoundly affect the existing dynamics of the island's ecosystem. Human-induced change is particularly devastating to islands. The domestic **goats** and rabbits introduced by human colonizers can denude a small island of succulent vegetation in less than a year, and dogs can turn every small mammal into **prey**. If plants to a goat's liking are not available or if steep ravines effectively corral dogs' activity, the ecosystem effects may be finite—if one does not consider the ticks and other **disease** carriers that may be introduced with the immigrants. Thus, the interactions between islands and migrant species are a two-way street; migrant species propose entering (and potentially changing) the closed system of an island's environment, and the environment permits or controls the success of such entry.

The question of closed versus open systems becomes highly interesting in the case of endemic island species—species native to an island, and perhaps found nowhere else. Did they evolve in place from an extinct ancestor? Did they island-hop from now drowned islands? If a land bridge was ever available, were the species around to use it? Local conditions rather than grand theories are usually called on to answer such questions, although the answers may in turn support grand theories. In many cases the answers remain perplexing. For example, it is estimated that more than 40% of the species of marine molluscs on the shores of Easter Island and a neighboring island are endemic. This is a startlingly high figure, for the islands are considered too young for evolution alone to have resulted in such prolific and successful branching. The molluscs may have originated from the shores of drowned islands in the region. The finding of old endemic species on young islands has led to some changes in the temporal boundaries of the **geologic time** scale, which are linked to index species.

The closed-system model of islands is useful for making inferences about evolution and adaptation but does not necessarily agree with reality. Immigrant species do colonize islands, with the colonization rate correlated with closeness to the mainland and size of the animal. Animals reach islands by swimming, rafting (on floating logs or matted leaves), flying, transport by carriers (a tick on a dog), or walking on frozen ice. The success rate need not be very high to develop thriving animal populations on islands. It is estimated that the arrival of one breeding pair every few hundred thousand years would have been enough to build the rich species diversity of the Philippine islands. Over geological time, that amounts to a large number of accidental tourists.

A second metaphor has therefore arisen, that of islands (more specifically, interisland distance) as a filtering mechanism. The filter has, again, permissive and controlling effects. In a hypothetical series of five islands extending outward in a line from a species-rich mainland, large mammals may never get beyond the first island, small mammals and tortoises may be filtered out by the third island, and the fifth island may be colonized only by fliers— birds, **bats**, and insects. Such a filtering effect has been recorded in islands extending eastward from New Guinea; the last wallabies and **marsupials** occur on the close islands New Britain and New Ireland, the last **frogs** on the Solomon islands, the last **snakes** on Fiji, and the last lizards on the island of Tonga. Biologists studying the biota on filtering islands consider the **energy** expenditure needed for animals to reach a distant goal and the adaptations species may have had to make to consume local food.

Island economics

Islands provide a variety of economic features. In addition to **fish** (and animals that feed on fish) as a food source, shells have been used as money and exported in jewelry. Coral has many uses, including manufacture into road-building material, jewelry, and small implements. Harbors promote ocean trade. Snorkeling draws tourists, and some tropical woods are in high demand.

Island ecosystems, however, are coming under intense **pressure** from human use as industrialization continues. Management of island resources by legislation that prevents or limits certain activities has not worked well in developing countries, where individuals increasingly rely on harvest of local resources for subsistence or to improve their standard of living. Islands with developing economies may also lack scientists and government ministers trained in the long-term care of island ecosystems. Such situations are being addressed on several fronts. International attention has been directed toward the renewable use of resources and the training of island biologists. Island and marine parks have been proposed. As some island species are approaching extinction before their origins are known, scientists are increasingly concerned about raising awareness of the special features of islands and their contributions to geological and evolutionary knowledge.

Resources

Books

Bakus, Gerald J., et al. *Coral Reef Ecosystems*. Rotterdam: A. A. Balkema, 1994.

Davis, Richard A., ed. *Geology of Holocene Barrier Island Systems.* New York: Springer-Verlag, 1994.

Dubinsky, Z., ed. *Coral Reefs. Ecosystems of the World 25.* (series ed., David D. Goodall). Amsterdam: Elsevier, 1990.

Nunn, Patrick D. *Oceanic Islands.* Oxford, England: Blackwell Publishers, 1994.

Marjorie Pannell

Isobars

Isobars are lines that connect points of equal **atmospheric pressure** on **weather** maps. Isobars are similar to height lines on a geographical **map**, and they are drawn so that they can never cross each other. Meteorologists use isobars on weather maps to depict atmospheric **pressure** changes over an area and to make predictions concerning **wind** flow.

The term "isobar" originates from the Greek, *isos* (equal) and *baros* (weight).

The lines are drawn using data from mean sea-level pressure reports. Because most of the weather stations are not located at **sea level**, but at a certain elevation, the pressure measured at every location has to be converted into sea level pressure before the isobars are drawn. The normal atmospheric pressure at sea level is defined as 1 atm of pressure or 29.92 inHg (760 mmHg or 760 torr). This normalization process is necessary because atmospheric pressure lapses (decreases) with increasing altitude, and the pressure difference on the maps has to be due to the weather conditions, not due to the elevation differences of the locations.

Wind is a direct consequence of air pressure differences. The greater the pressure contrast over an area, the shorter the distance between isobars on a weather map depicting the area. Wind blows from areas of high to low pressure. The greater the contrast in pressure difference between two areas, the faster the wind will blow, so closer isobars on a weather map predict higher **velocity** winds.

Although the wind initially is controlled by the pressure differences, it is also modified by the influence of the **Coriolis effect** and **friction** close to Earth's surface. This is why isobars can only give a general idea about the wind direction and wind strength.

A rule observed first in 1857 by Dutch meteorologist Christoph Buys-Ballott (1817–1890) described the link between isobars and wind: In the Northern Hemisphere, if you stand with your back to the wind, the low pressure area is located on the left. In the Southern Hemisphere, standing with your back to the wind means that the low-pressure area is on the right. This is called Buys-Ballott's law.

Isobars can form certain patterns, making it useful for weather analysis or forecast. A cyclone or depression is an area of curved isobars surrounding a low-pressure region with winds blowing counterclockwise in its center in the Northern Hemisphere. An anticyclone is an area of curved isobars surrounding a high-pressure area, and the wind blows clockwise in the center of an anticyclone in the Northern Hemisphere. Open isobars forming a V-shape define a through of low pressure while high-pressured, N-shaped, open isobars define a ridge of high pressure. These features are usually predictable, and associated with certain kinds of weather, making it easier to forecast weather for a particular area.

See also Aerodynamics; Atmosphere, composition and structure; Atmospheric circulation; Barometer; Weather forecasting; Weather mapping.

Isomer

Isomer is the term used to describe two or more chemical compounds which can be represented by the same **chemical formula**. There are two main types of isomers: structural isomers which differ from one another by the attachment of **atoms** on the **molecule**; and stereoisomers which differ from on another by the location of the atoms in **space**.

Chemical compounds can be represented by a formula which qualitatively and quantitatively describes its component elements. For example, the formula for **water** is H_2O, which indicates that the compound contains two **hydrogen** atoms attached to one **oxygen** atom. In the early 1800s two chemists, Friedreich Wohler and Justus Liebig, realized that two chemical compounds might have

the same elemental composition yet differ in the order in which the atoms were linked together. Therefore, it is possible that a given chemical formula may describe more than one compound. For example, propyl alcohol and isopropyl alcohol both are represented by the same formula ($[CH_3]_2CHOH$), but they are different compounds with different properties depending on whether the **alcohol** group (also known as the hydroxyl group) is located on a terminal (end) **carbon** atom or on the middle carbon atom. This form of isomerism is known as positional, or structural, isomerism. Positional isomerism occurs because the various sites where groups are attached are not equivalent. This principle is demonstrated by the molecule known as **benzene** which consists six carbon atoms arranged in a ring. These carbon atoms provide benzene with six different positions where other chemical groups can be substituted for hydrogen atoms. A substituted benzene ring, such as toluene, can accept another substituent on any of the other five carbon atoms; but because two pairs are equivalent, there are only three possible isomers. These are designated as ortho, meta, and para.

Chain isomerism, another type of structural isomerism, occurs among chemical compounds known as alkanes, which consist of chains of carbon atoms. These carbon atom chains can be configured as either a straight or branched chain with exactly the same overall chemical formula. These different structural configurations are isomers of each other. Although the properties of isomers of a given formula are similar, the compounds are nonetheless distinct. Similarly, the location of the double bond in alkenes and the triple bond in alkynes determines another form of positional isomerism.

Isomers may also be stereoisomers which differ from one another by spacial position of their atoms. There are two subcategories of stereoisomers, geometric isomers and optical isomers. Both geometric and optical isomers occur in molecules in which the atoms are attached in the same order but have different spatial relationships. For example, picture a chain of carbon atoms which has two hydroxyl groups attached to the first two positions. Furthermore, assume there is a double bond present between the 1 and 2 carbons. The hydroxyl groups can be attached to the same side of this double bond, (e.g., both on the "top" or both on the "bottom"). Or they may be oriented on opposite sides such that one resides on top and the other on the bottom. If the groups are aligned on the same side of the double bond the compound is said to be a cis isomer (from the Latin "on this side"); if they are on opposite sides it is trans (from Latin "across.") In certain compounds, the atoms are free to rotate about this double bond, which gives rise to multiple isomeric configurations. If the mirror images of these different configurations are superimposable, the isomers

KEY TERMS

. . . .

Geometric isomers—Stereoisomers whose mirror images are superimposable.

Optical isomers—Stereoisomers whose mirror images are nonsuperimposable.

Stereochemistry—The study of stereoisomers.

Stereoisomers—Isomers which differ from on another by the location of the atoms in space.

Structural isomers—Isomers which differ from one another by the attachment of atoms on the molecule.

are said to be geometric. If the mirror images are non-superimposable they are said to be optical. Optical isomers are distinguishable by the way they interact with a beam of polarized **light**. Such isomers are the subject of the branch of **chemistry** known as **stereochemistry**.

Stereochemistry is the study of the spatial arrangement of atoms in molecules and the effect of their orientation on the physical and chemical properties of those compounds. However, the three dimensional nature of this spacial orientation was not really understood until 1871 when two independent chemists, Hendricus van't Hoff and Joseph Achille le Bel, proposed their theories of stereoisomerism, or how isomers are structured.

The study of isomers provides information which is useful in improving the efficiency of the reactions or in the search for new types of reactions or chemical **species**. Through evaluation of isomeric compounds, chemists gain useful information about **chemical reactions** and learn how certain bonds are broken and formed or what kinds of intermediates are involved.

See also Formula, chemical; Formula, structural.

Perry Romanowski

Isopropyl alcohol *see* **Alcohol**

Isostasy

Isostasy is the term describing the naturally occurring balance of masses of Earth's crust that keeps the planet's gravity in equilibrium. Isostasy is not a **force** or a process; it is only the term for the phenomenon of adjustments **Earth** makes to stay balanced in **mass** and gravity.

Isostasy as a description of the earth's balance

Nature is a perfect system of balances. **Matter** and **energy** exist in finite (specific) amounts and cannot be created or destroyed. The earth is a perfect example of nature's balance system. Rock particles are eroded from the mountain top, deposited in valleys or stream channels, compacted under their own weight into rock, and uplifted by mountain-building processes until they again rise to the top of the mountain.

Deeper within the earth, balancing processes also take place as major shifts in the upper part of the earth's crust change the planet's gravitational balance. Under mountain ranges, the thin crust slumps or bows deeper into the upper mantle than where the land mass is thinner across continental plains. The land masses float on the crust and mantle-like **icebergs** float in seawater, with more of the mass of larger icebergs below the **water** than smaller ones. This balance of masses of the earth's crust to maintain gravitational balance is called "isostasy."

Isostasy is not a process or a force. It is simply a natural adjustment or balance maintained by blocks of crust of different thicknesses to also maintain gravity. Isostasy uses energy to balance mass. The energy comes from the **hydrologic cycle**, which is the path of a drop of water that originates in the **ocean**, evaporates to form a cloud, falls on the mountain as a raindrop, and flows back to the sea carrying particles of rock and **soil**. The hydrologic cycle derives its energy from gravity and solar **radiation**. As water flows or a glacier slowly grinds over land, energy is lost in that now-isolated system.

Within Earth, energy comes from radioactive energy that causes **convection** currents in the core and mantle. Opposing convection currents pull the crust down into geosynclines (huge structural depressions). The sediments that have collected (by the processes of deposition that are part of the hydrologic cycle) are squeezed in the downfolds and fused into **magma**. The magma rises to the surface through volcanic activity or intrusions of masses of magma as batholiths (massive rock bodies). When the convection currents die out, the crust uplifts and these thickened deposits rise and become subject to **erosion** again. The crust is moved from one part of the surface to another through a set of very slow processes, including those within Earth (like convection currents) and those on the surface (like **plate tectonics** and erosion).

In isostasy, there is a line of equality at which the mass of land above **sea level** is supported below sea level. So, within the crust, there is a depth where the total weight per unit area is the same all around the earth. This imaginary, mathematical line is called the "depth of

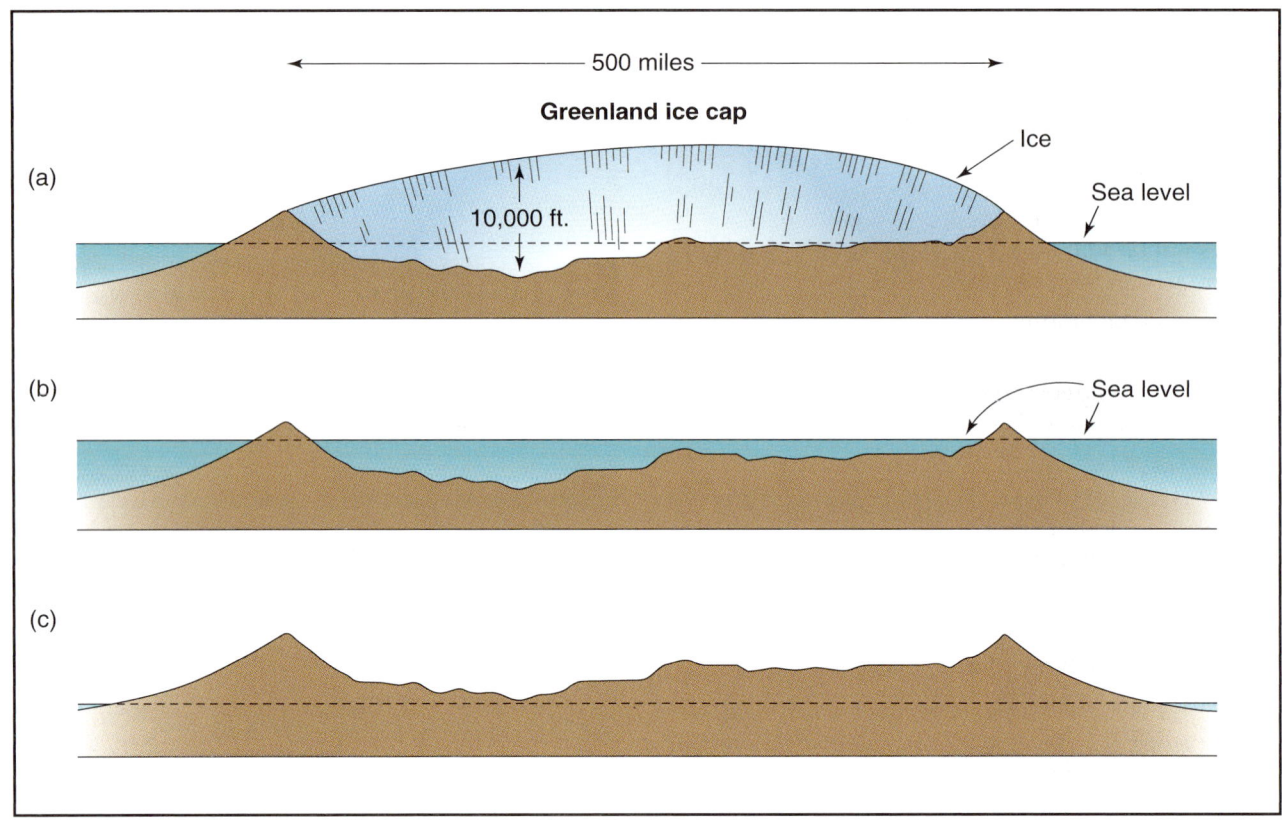

Establishing isostatic balance. (a) The weight of the ice covering Greenland pushes the land below sea level. (b) As the ice melts, the land mass rises as the pressure is removed but remains below the recently elevated sea level. (c) Normal isostatic balance is restored when the land mass is raised above sea level. *Photo Researchers, Inc. Reproduced by permission.*

compensation" and lies about 70 mi (112.7 km) below the earth's surface.

Isostasy describes vertical movement of land to maintain a balanced crust. It does not explain or include horizontal movements like the compression or folding of rock into mountain ranges.

Greenland is an example of isostasy in action. The Greenland land mass is mostly below sea level because of the weight of the **ice** cap that covers the **island**. If the ice cap melted, the water would run off and raise sea level. The land mass would also begin to rise, with its load removed, but it would rise more slowly than the sea level. Long after the ice melted, the land would eventually rise to a level where its surface is well above sea level; the isostatic balance would be reached again, but in a far different environment than the balance that exists with the ice cap weighing down the land.

The theory of isostasy

Scientists and mathematicians began to speculate on the thickness of the earth's crust and distribution of land masses in the mid 1800s. Sir George Biddell Airy (1801-

1892) assumed that the **density** of the crust is the same throughout. Because the crust is not uniformly thick, however, the Airy hypothesis suggests that the thicker parts of the crust sink down into the mantle while the thinner parts float on it. The Airy hypothesis also describes the earth's crust as a rigid shell that floats on the mantle, which, although it is liquid, is more dense than the crust.

John Henry Pratt (1809-1871) also proposed his own hypothesis stating that the mountain ranges (low density masses) extend higher above sea level than other masses of greater density. Pratt's hypothesis rests on his explanation that the low density of mountain ranges resulted from expansion of crust that was heated and kept its **volume** but at a loss in density.

Clarence Edward Dutton (1841-1912), an American seismologist and geologist, also studied the tendency of the earth's crustal layers to seek equilibrium. He is credited with naming this phenomenon "isostasy."

A third hypothesis developed by Finnish scientist Weikko Aleksanteri Heiskanen (1895-1971) is a compromise between the Airy and Pratt models. But it is the Hayford-Bowie concept that has been most widely

KEY TERMS

. . . .

Batholith—A huge mass of igneous rock that is intruded (forced by pressure) into the earth's crust but may not reach the surface.

Convection current—Massive currents within the semi-molten mantle of the earth that move due to differences in temperature.

Density—The amount of mass of a substance per unit volume.

Depth of compensation—The line at which the earth's land masses above the line are balanced by those below.

Geodesy—The mathematics of measurements of the earth including its size, shape, and location of points on its surface.

Geosyncline—A massive downward bend in the earth's crust; the opposite of an anticline, which is a huge upward flex in the earth's surface.

Gravity—The force of attraction of Earth's mass for objects near it.

Hydrologic cycle—The continuous, interlinked circulation of water among its various compartments in the environment.

Magma—Molten rock within the earth. When magma reaches the surface, it cools and forms igneous rock.

accepted. John Fillmore Hayford (1868-1925) and John William Bowie (1872-1940) were American geodesists who studied gravitational anomalies (irregularities) and first began surveying gravity in the oceans. Geodesists, or specialists in geodesy, are mathematicians who study the size, shape, and measurement of the earth and of Earth forces, like gravity. Hayford and Bowie were able to prove that the anomalies in gravity relate directly to topographic features. This essentially validated the idea of isostasy, and Hayford and Bowie further established the concept of the depth of isostatic compensation. Both gentlemen published books on isostasy and geodesy. Hayford was the first to estimate the depth of isostatic compensation and to establish that Earth is an oblate sphere (a bowed **sphere**) rather than a true sphere.

Resources

Books

Marshall, Clare P., and Rhodes W. Fairbridge. *Encyclopedia of Geochemistry*. Boston: Kluwer Academic Publishers, 1999.

Skinner, Brian J., and Stephen C. Porter. *The Dynamic Earth: An Introduction to Physical Geology*. 4th ed. John Wiley & Sons, 2000.
Woodhead, James A. *Geology*. Boston: Salem Press, 1999.

Gillian S. Holmes

Isotonic *see* **Osmosis**

Isotope

An isotope is one of several kinds of **atoms** of the same element that have different masses. These atoms have the same number of protons in their nuclei, but different numbers of neutrons, and therefore different **mass** numbers. The term isotope comes from the Greek *isos topos*, which means same place, because isotopes of the same element have the same **atomic number** and therefore occupy the same place in the **periodic table**.

The element **carbon**, for example, has two stable isotopes, carbon-12 and carbon-13, symbolized as ^{12}C and ^{13}C. The numbers 12 and 13 are the mass numbers of the isotopes—the total numbers of the protons plus neutrons in their nuclei. Because all carbon atoms have six protons in their nuclei, ^{13}C must have seven neutrons (13-6) in its nucleus, and ^{12}C has six (12-6). The element carbon as we find it in nature consists of 98.89% ^{12}C atoms and 1.11% ^{11}C atoms. Carbon in living organisms contains also a very small amount of ^{14}C, a radioactive isotope that is used in radiocarbon dating.

Most elements have between two and six stable isotopes (as opposed to unstable, or radioactive ones). Twenty elements, including fluorine, **sodium**, **aluminum**, **phosphorus**, and gold consist of only one stable isotope each. Tin, however, has ten—more than any other element. The number of stable isotopes an element has is determined by the relative stabilities of various numbers of neutrons and protons in their nuclei.

Only two isotopes have been given distinctive names, both isotopes of **hydrogen**. The stable isotope ^{2}H is known as **deuterium**, or heavy hydrogen, and the radioactive isotope ^{3}H is called **tritium**.

Because isotopes of the same element have identical chemical properties, they cannot be separated by chemical methods, but only by methods that are based on their mass differences, such as **mass spectrometry**. One of the extraordinary accomplishments of the Manhattan Project, which created the atomic bomb during World War II, was the successful separation of large amounts of

^{235}U, the highly fissionable isotope of **uranium**, from the much more abundant ^{238}U by allowing a gaseous uranium compound to diffuse through porous barriers. Being heavier, the ^{238}U-containing molecules move more slowly through the barriers.

Over 1,000 radioisotopes—radioactive isotopes—either exist in nature or have been made artificially by bombarding stable isotopes in particle **accelerators**. They are useful in so many applications that the word isotope is commonly used to mean radioisotope, as if stable isotopes did not exist.

See also Dating techniques; Mass number; Nuclear fission; Nuclear medicine; Proton; Radioactive tracers.

Isthmus

A narrow strip of land, an isthmus, connects two wider sections of land. The isthmus of Panama, which connects **South America** to Central/North America, and the Sinai **peninsula**, which connects **Africa** to **Asia**, illustrate better than any other examples.

Moving plates create many isthmi (isthmuses). The earth's outer shell, the crust, breaks into sections, plates, that slowly slide around the **earth**. When two continents collide, they can build an isthmus connecting the two continents (Sinai), or when a **continent** collides with a sea floor plate, enough volcanism can result to build land (Panama region).

During **ice ages**, **glaciers** hold much **water** in the form of **ice** and cause sea level to fall as much as 427 ft (130 m). With **sea level** down, more isthmi appear connecting, for example, the British isles to each other and to continental **Europe**. Rising oceans later flood these temporary **bridges**.

Remnants of a former isthmus linking Africa to Europe lie at Gibraltar. The collision of Africa with Europe around eight million years ago closed the Strait of Gibraltar creating a dam which cut off the Mediterranean Sea's main source of water, the Atlantic Ocean. While the dam remained intact, the Mediterranean evaporated in the **desert** climate leaving a large, arid **basin**. Eventually, the Atlantic eroded the isthmus, and the Mediterranean refilled.

In the early 1900s, when Alfred Wegener proposed his **continental drift** theory—that continents move around Earth—geologists of the time rejected his fossil evidence based on their belief in submerged isthmi. Wegener stated that the distribution of fossil animals in South America and Africa could only be explained by the two continents being formerly one bigger continent. Opponents countered: those organisms used now-submerged land bridges (isthmi) to cross the Atlantic from South America to Africa, explaining why geologists discover fossils of animals that could not swim that far on both continents. After World War II, when improved technology mapped bridgeless sea floors, the submerged isthmus belief perished, vindicating Wegener.

While an isthmus exists, organisms can travel across it freely—mixing, breeding, evolving, preying, and hiding. Eliminating the connection lets organisms develop, evolve, and die out separately. The isolation of **Australia** from other continents millions of years ago, for example, produced animals and plants found nowhere else in the world.

Iteration

Iteration consists of repeating an operation of a value obtained by the same operation. It is often used in making successive approximations, each one more accurate than the one that preceded it. One begins with an approximate **solution** and substitutes it into an appropriate formula to obtain a better **approximation**. This approximation is subsequently substituted into the same formula to arrive at a still better approximation, and so on, until an exact solution or one that is arbitrarily close to an exact solution is obtained.

An example of using iteration for approximation is finding the **square root**. If s is the exact square root of A, then A |6-8| s = s. For example, since 8 is the square root of 64, it is true that 64 |6-8| 8 = 8. If you did not know the value of |5-14| 64, you might guess 7 as the value. By dividing 64 by 7, you get 9.1. The average of 7 and 9.1 would be closer. It is 8.05.

Now you make a second iteration by repeating all the steps but beginning with 8.05. Carry out the **division** to the hundredths place; 64 |6-8| 8.05 = 7.95. The average of 8.05 and 7.95 is 8. A third iteration shows that 8 is the exact square root of 64.

Finding the roots of an equation

Various methods and formulas exist for finding the roots of equations by iteration. One of the most general methods is called the method of successive bisection. This method can be used to find solutions to many equations. It involves finding solutions by beginning with two approximate solutions, one that is known to be too large and one that is known to be too small, then using their average as a third approximate solution. To arrive at a fourth approximation, it is first determined whether the third approximation is too large or too small. If the third approximation is too large it is averaged with the most recent previous approximation that was too small, or the other way around; if approximation number three is too small it is averaged with the most recent previous approximation that was too large. In this way, each successive approximation gets closer to the correct solution.

Testing each successive approximation is done by substituting it into the original equation and comparing the result to **zero**. If the result is greater than zero then the approximation is too large, and if the result is less than zero, then the approximation is too small.

Iteration has many other applications. In **proof**, for example, mathematical induction is a form of iteration. Many computer programs use iteration for looping.

Resources

Books

Larson, Ron. *Calculus With Analytic Geometry*. Boston: Houghton Mifflin College, 2002.

J.R. Maddocks

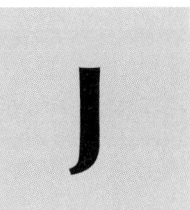

J

Jacanas

Jacanas are eight **species** of distinctive **birds** that inhabit the marshy edges of ponds, lakes, **rivers**, and swamps, and that make up the family Jacanidae. Jacanas are tropical birds, breeding in central and northern **South America**, sub-Saharan **Africa**, Madagascar, South and Southeast **Asia**, and Australasia. Jacanas do not migrate, but they may wander widely if their local aquatic habitats dry out.

Jacanas range in body length from 6-20 in (15-51 cm). They have rather short, rounded wings, with an unusual, spiny projection from the wrist, and a stubby tail. Their legs are large, and their unwebbed toes and claws are extraordinarily long, and useful for supporting the weight of these birds as they walk gingerly over the surface-floating foliage of aquatic plants. Jacanas usually have a brightly colored patch of bare skin, known as a frontal shield, in front of their eyes and above the beak. The body of these attractive birds is generally colored cinnamon-brown, with bold patterns of black, white, or yellow. The sexes are similarly colored, but female jacanas are considerably larger than the males.

Jacanas fly weakly, but they run, swim, and dive well. They are commonly observed walking on aquatic plants, such as the floating leaves of **water lilies** and lotus. Appropriately, alternative common names for these birds include "lily-trotters" and "lotus-birds." Jacanas carefully glean this **habitat** for their food of aquatic **insects**, crustaceans, molluscs, and **plant** seeds. Jacanas often forage in loose groups.

Jacanas lay four eggs in a nest built of aquatic plants, which is often partially afloat. In most species, the male incubates the eggs and cares for the young birds. In fact, some species of jacanas are polyandrous, meaning that a single female will mate with several males and then lay eggs in each of their nests. In most species of jacanas, a female consorts with two to four males, although the number can be as many as 10 in the case of the pheasant-tailed jacana. Polyandry is a rare breeding system in birds.

Jacanas do not occur in **North America**, although the northern jacana (*Jacana spinosa*) of Central and northern South America is an occasional visitor to south Texas and Florida. The closely related wattled jacana (*Jacana jacana*) is a widespread species in the tropics of South America. Some taxonimists consider these two to be variants of the same species.

The African jacana (*Actophilornis africanus*) is a widespread bird of ponds and other **wetlands** in sub-Saharan Africa. The closely related Madagascar jacana (*Actophilornis albinucha*) only occurs on the **island** of Madagascar. The lesser jacana (*Microparra capensis*) is a rarer African species.

The bronze-winged jacana (*Metopidius indicus*) is widespread from India to Indonesia, as is the pheasant-tailed jacana (*Hydrophasianus chirurgus*), with its very long and distinctive tail feathers. The lotus-bird or comb-crested jacana (*Irediparra gallinacea*) ranges from Borneo and the Philippines through New Guinea and eastern **Australia**.

Bill Freedman

Jack-in-the-pulpit *see* **Arum family (Araceae)**

Jackals *see* **Canines**

Jacks

Jacks, also called scads, trevallys, and crevalles, are marine bony fishes living in open waters. Amberjacks, runners, and pompanos also belong to the same family Carangidae, order Perciformes. Jacks are swift, predatory fishes, found widely in warm and tropical seas. The younger **fish** tend to travel in vast schools, but the older ones may be solitary.

Many jacks are valued for commercial and sport fishing, and some **species** are successfully exhibited in public aquariums. The Florida pompano (*Trachinotus carolinus*), which grows to 1 ft (30.5 cm) in length, is considered a seafood delicacy. The crevalle jack (*Caranx ruber*) may reach the size of 2 ft (61 cm) and weigh over 20 lb (9 kg). It is the most common jack of the West Indian waters, and is often seen near coral reefs. In the summer, large schools of this species cruise by the Bahamas, where it is known as the "passing jack." The more than 200 species in the family Carangidae vary greatly in form, from long and streamlined to deep-bodied and very thin from side to side. Generally they share the following features in common: two dorsal fins (the first may be greatly reduced in size); anal and second dorsal fins usually high in the front; slim, often sickle-shaped pectoral fins; a strong, fork- or crescent-shaped tail with a slim base; small scales. Many species in this family are quite small, but some can reach very large sizes. For example, the amberjack (*Seriola dumerili*) can grow to 6 ft (1.8 m) in length and can weigh as much as 150 lb (70 kg). Some jacks have a series of scutes (comb-like scales) along the caudal peduncle (the fleshy part of the tail), which reinforce it for fast swimming.

Most carangids are silvery in **color**, but some exhibit lovely colors or markings. The rainbow runner *(Elagatis bipinnulatus)* of the tropical Atlantic and Indo-Pacific has beautiful blue bands on the sides. It is a hard-fighting sports fish, and supposedly very tasty. Color changes may occur in some species as the fish mature. The Indo-Pacific blue-banded golden jack *(Gnathanodon speciousus)* is solid yellow when young. The African pompano and other threadfins of the genus *Alectis* have streaming fins which trail behind them, resembling the long tentacles of **jellyfish**. When adult fish reach the size of about 3 ft (91 cm), the fins appear shorter. Palometas (*Trachinotus goodei*) are silvery jacks that tend to form schools in shallow **water**, and often approach wading people. In public aquarium exhibits they form attractive schools.

The permit (*Trachinotus falcatus*) lives in shallow Atlantic waters near reefs and sandy flats. It may grow to the size of 3.5 ft (1.1 m). When young, it lives in sheltered waters and feeds on small crustaceans; later, it includes molluscs and **sea urchins** in the diet. The greater amberjack is the most common species of the genus *Seriola* in the tropical and subtropical waters of western Atlantic. It has a lengthwise brassy stripe on the side of the body at the level of the **eye**. The back above the stripe is olive to blue, and the body below the stripe is silvery white. There is also a diagonal dark band running from the snout, through the eye, to the nape. The greater

amberjack and several other jack species may at times carry a toxic substance in their flesh that causes ciguatera poisoning when these fishes are eaten by humans. This toxin comes from **algae** that the fish ingests either directly or in the smaller fish it consumes.

Jaguar *see* **Cats**

Jaguarundi *see* **Cats**

Jaundice

Jaundice is not a **disease** but a symptom of an underlying disease or condition. It is caused by too much bilirubin in the **blood** stream, and is characterized by yellowness of skin, sclera (white of eyes), mucous membranes, and of body fluids such urine and blood **plasma**. The resulting yellow **color** (*jaune* means yellow in French) is also described by the Latin term *icterus*.

Most bilirubin, which is a reddish pigment, is a byproduct of red blood cells. When a red blood **cell**, which has a lifespan of about 120 days, is no longer functional. It is "recycled" by organs such as the spleen and liver. Hemoglobin, the red pigment in a red blood cell, is broken down, or catabolized, into several substances that are either used to make new blood cells, **proteins**, or other pigments. After the bilirubin is liberated from the red blood cell, it is bound to albumin and transported via blood plasma to the liver, where it is conjugated or joined to glucuronic acid, and excreted into the bile ducts. Bilirubin is also the prominent pigment of bile, a digestive substance secreted by the liver into the gallbladder and small intestine (jejunum), and it is also responsible for the brown color of feces.

There are several ways in which a person can have too much bilirubin, or become jaundiced. Hemolytic jaundice occurs when the liver is overloaded with bilirubin. In conditions and diseases causing hemolysis (the separation of hemoglobin from the red blood cells), including hemolytic anemias, incompatible blood transfusion, and extreme **heat** or cold, the liver is unable to remove enough bilirubin from the blood stream. Liver cells which are damaged by hepatocellular disease (**hepatitis, cirrhosis**), toxins, tumors, or inflammatory conditions, are unable to conjugate bilirubin, thus preventing normal excretion. Neonatal jaundice, especially in premature infants, is fairly common. This type of jaundice is caused by the fact that the liver is immature and lacks specific enzymes to conjugate bilirubin, and a large amount of bilirubin is consequently excreted into the blood stream instead of being incorporated into bile.

The most common cause of jaundice is obstruction of bile flow through the biliary system. For example, a gallstone, a liver **tumor**, or a pancreatic tumor can block a biliary duct. The treatment of jaundice depends on the cause.

Jays *see* **Crows and jays**

Jellyfish

Jellyfish, also called medusae, are free-swimming, marine **invertebrates** in the class Scyphozoa (phylum Coelenterata). They have a gelatinous, translucent, dome-shaped body and occur most commonly in warm, tropical seas, although they are found in all the world's oceans. Jellyfish feed on small planktonic animals or **fish** which they sting and paralyze with special cells called nematocysts located on the tentacles that hang from the edge of their dome-shaped bodies. The body of a jellyfish is 99% **water**; when washed onto dry land, these animals die and rapidly disappear as the water in their body evaporates.

About 200 **species** of true jellyfish are known, ranging in size from 0.06 in (1.5 mm) to 6.5 ft (2 m). All jellyfish have a prominent dome; the shape of the dome varies from a shallow saucer to a deep bell. In the subclass Cubomedusae, the dome is cube-shaped. Hanging from the edge of the dome are nematocyst-bearing tentacles; the number and length of these tentacles varies greatly from species to species. On the underside of the dome is a feeding tube (the manubrium or proboscis) with the animal's mouth at its free end. Radial canals (usually four in number or some multiple of four) extend from the jellyfish's four-chambered stomach to the dome's margin where they connect with the ring canal. This system of canals serves to distribute food to the outer parts of the jellyfish's body. Light-sensing organs (eyespots), balance organs (statocysts), and other sensory organs are located at the base of the tentacles. Jellyfish move through the water by pulsating contractions of the muscles on the lower edge of the dome.

Jellyfish release eggs and sperm through the mouth into the water, where **fertilization** occurs. Fertilization results in free-swimming, ciliated (with tiny, hair-like structures) larvae called planulae; these larvae settle on a surface, such as a rock, and turn into a polyp (a hollow cylinder with tentacles and a mouth at one end) or a strobila (a hollow structure that looks like a stack of upside-down saucers). Strobila develop into adult jellyfish after passing through another free-swimming phase during which they are called ephyrula. Polyps produce adult jellyfish by bud-

A jellyfish. *Photograph by Mark A. Johnson. Stock Market. Reproduced by permission.*

ding. The average life span of a jellyfish is one to three months; the largest may live for about one year.

The most common jellyfish on the coasts of **North America** and **Europe** is the moon jellyfish (*Aurelia aurita*). This 6–8-in (15.2-20.2 cm) species is found at depths of 0-20 ft (0-6 m). It is whitish, often shaded with pink or blue, and has a saucer-shaped dome with a fringe of numerous, short tentacles around the margin. Its sting is mildly toxic to humans, occasionally producing an itchy rash. In the same order (Semaeostomeae) as the moon jellyfish is the giant pink jellyfish (*Cyanea capillata*); this species is common in the waters of the Northern Hemisphere where it reaches about 6.5 ft (2 m) across.

The 16 species in the subclass Cubomedusae are commonly called box jellyfish; they live in the warm waters of the continental shelves. The largest species in this subclass, the sea wasp (*Carybdea alata*), is found in tropical harbors and river mouths. It reaches a diameter of 9.75 in (25 cm) and sometimes eats fish much larger than itself. The sting of this and many other box jellyfish can be highly toxic producing a reaction in humans that may include skin welts, muscle cramps, and breathing difficulty. Two genera, *Chiropsalmus* and *Chironex*, found in the Indian Ocean produce a toxin so potent that contact with their nematocysts can kill a person within several minutes.

The 31 species of deep-sea jellyfish (order Coronatae) are heavily pigmented in colors ranging from red and violet to brown and blackish. They are found at extreme depths; for example, *Nausithoe* has been found at a depth of 23,000 ft (7,100 m). The order Rhizostomeae includes about 80 species commonly known as many-mouthed jellyfish. In these species, the feeding tube has many small pores rather than one large opening. The genus *Stromolophus* of this order is common on the southeastern coast of the United States where it reaches a diameter of 7 in (18 cm).

Other marine animals in the class **Hydrozoa**, order Siphonophora, superficially resemble true jellyfish and are often confused with them. The most well-known member of this order is the Portuguese man-of-war (*Physalia physalis*). Each Portuguese man-of-war (so named because it resembles an eighteenth century warship) is a colony composed of four kinds of polyp. The main polyp is a gas-filled float that measures up to 12 in (30 cm) long. This float has a high crest which serves as a sail to catch the **wind** and its **color** varies from blue to purple to red. Hanging below the float in the water are other polyps; some of them are concerned with feeding while other are concerned with reproduction. Also below the float are trailing tentacles (up to 40 ft/12 m long) armed with nematocysts. The **animal** uses these tentacles to catch the fish and other sea creatures that it eats. The toxin produced by this species is also very potent; humans have been known to die from a Portuguese man-of-war sting. Usually when humans are stung, redness, skin welts, and blisters result. When washed ashore, this animal remains toxic for some **time**.

Resources

Books

Cousteau, J. *The Ocean World of Jacques Cousteau.* Danbury, CT: Grolier, 1975.
Whiteman, Kate. *World Encyclopedia of Fish & Shellfish.* New York: Lorenz Books, 2000.

Christine B. Jeryan

Jerboas

Jerboas are small kangaroo-like **rodents** with large hind legs that make up the family Dipodidae. Three **species** of jerboas occur in North **Africa**, and a number of other species occur in **Asia**.

Jerboas are typically pale-colored, with large eyes, immense ears, a long tail, small front legs and paws, and distinctively large hind legs and feet, which are used for jumping. Although the body length of a typical jerboa is only about 2-4 in (5-10 cm), these animals can cover as far as 6.5-10 ft (2-3 m) in a single leap, using their long tail for balance. The remarkable jumping ability of jerboas is likely adaptive for avoiding their predators. When they are in less of a rush, jerboas move about using short hops, or even with alternate strides of the hind legs.

Jerboas typically live in arid and semi-arid habitats. They generally avoid the **heat** of day and conduct their foraging activities at night when it is relatively cool. During the day, jerboas retreat to their underground tunnels, which they tightly plug to keep the hot air out and the moisture in.

Jerboas mostly feed on succulent **plant** tissues and **seeds** as well as **insects** when they are available. Jerboas sometimes cause significant damages to **crops** in fields and gardens. Jerboas can satisfy all of their requirements for moisture through **water** that is produced metabolically when foods are oxidized during **respiration**. However, these animals will drink readily when water is available.

The desert jerboas (*Jaculus* spp.) are four species occurring in North Africa and southwestern **Europe**. The greater Egyptian jerboa (*Jaculus jaculus*) is a widespread species and was dubbed the "desert rat" by soldiers during the Second World War.

The hairy-footed jerboa (*Dipus sagitta*) is a widespread and relatively abundant Asian species. The earth hares (*Allactaga* spp.) are nine species of the steppes and deserts of Asia and Egypt, which sometimes cause minor agricultural damages. The three-toed dwarf jerboas (*Salpingotus* spp.) are four little-known species of the deserts of central Asia.

Bill Freedman

Jet engine

A jet engine is a **heat** engine that is propelled in a forward direction as the result of the escape of hot gases from the rear of the engine. Two general types of jet engines exist: the air-breathing jet engine and the rocket. In an air-breathing jet engine, air entering the front of the engine is used to burn a fuel within the engine, producing the hot gases needed for propulsion. In a rocket, air is not needed for propulsion. Instead, some type of chemical, nuclear, or electrical reaction takes place within the rocket engine. Hot gases formed as a result of that reaction exit the engine from the rear, providing the engine with its thrust, or forward **momentum**. Some authorities reserve the term jet engine for the first of these two types, the air-breathing jet engine. Air-breathing jet engines are used for the fastest commercial and military **aircraft** now available.

Scientific principle

The scientific principle on which the jet engine operates was first stated in scientific terms by Sir Isaac Newton in 1687. According to Newton's third law, for every action, there is an equal and opposite reaction. That principle can be illustrated by the **behavior** of a **balloon** filled with air. As long as the neck of the balloon is tied, gases (air) within the balloon push against all sides of the balloon equally, and no **motion** occurs. If the neck of the balloon is untied, however, gases begin to escape from the balloon. The escape of gases from the balloon is, in Newton's terms, an "action." The equal and opposite reaction resulting from the escape of gases is the movement of the balloon in a direction opposite to that of the movement of the gases. That is, as the air moves outward in one direction, the balloon moves outward in the opposite direction.

Rockets

Rockets can be broadly classified into one of two categories: those that use a chemical reaction as their **energy** source, and those that use some other kind of energy source. An example of the former are rockets that are powered by the chemical reaction between liquid **oxygen** and liquid **hydrogen**. When these two chemicals react with each other, they produce very hot steam (**water** vapor). The escape of steam from the back of the rocket provides the propulsive **force** that drives the jet engine forward.

Chemical rockets make use of either liquid fuels, such as the rocket described above, or of solid fuels. An example of the latter are the solid rockets used to lift a **space shuttle** into **orbit**. These rockets contain a fuel that consists of a mixture of **aluminum metal** (the fuel), ammonium perchlorate (the oxidizer), and a plastic resin (the binder).

Nuclear and electric rockets are examples of jet engines that make use of a nonchemical source of propulsion. In a nuclear rocket, for example, a source of nuclear energy, such as a fission or fusion reactor, is used to heat a working fluid, such as liquid hydrogen. The hot gases formed in this process are then released from the rear of the rocket, providing its forward thrust.

Various kinds of electrical rockets have been designed. In one type, a fluid within the engine is first ionized. The ions thus formed are then attracted and/or re-

A jet aircraft with the engine cover open. The engine works by sucking air into one end, compressing it, mixing it with fuel and burning it in the combustion chamber, and then expelling it with great force out the exhaust system. *Photograph by George Haling. Photo Researchers, Inc. Reproduced by permission.*

pelled by strong electrostatic fields created within the engine. The escape of the ionized fluid provides the rocket with its forward thrust.

Ramjets

The simplest of all jet engines is the ramjet. The ramjet consists of a long cylindrical tube made of metal, open at both ends. The tube bulges in the middle and tapers off at both ends. This shape causes air entering the front of the engine to expand and develop a higher **pressure** in the center of the engine. Within the engine, the compressed air is used to burn a fuel, usually a kerosene-like material. The hot gases produced during **combustion** within the engine are then expelled out the back of the engine. These exiting gases can be compared to the air escaping from a rubber balloon. As the gases leave the back of the jet engine (the nozzle exit), they propel the engine itself in a forward direction.

When the ramjet engine is at rest, no air enters the front of the engine, and the engine provides no thrust. Once the engine is moving through the air, however, it begins to operate more efficiently. For this reason, the use of ramjet engines is usually reserved for aircraft that travel at very high speeds.

A typical ramjet engine today has a length of about 13 ft (4 m), a diameter of about 39 in (1 m), and a weight of about 1,000 lb (450 kg). A ramjet engine of this design is capable of producing a thrust of 9,000 Newtons (N) (about 2,000 lb), giving a maximum **velocity** of about Mach 4 at higher altitudes.

Turbojets

One might guess that one way to improve the efficiency of a jet engine would be to increase the speed at which exhaust gases are expelled from the engine. In fact, that turns out not to be the case. Aeronautical engineers have discovered that a larger **mass** of gas moving at a lower velocity produces greater thrust in the engine. The modification that was developed to produce this effect is called a turbojet. The major difference between a turbojet and a ramjet is that the former contains a compressor attached to a **turbine**. The compressor consists of several rows of metal blades attached to a central shaft. The shaft, in turn, is attached to a turbine at the rear of the compressor. When air enters the inlet of a turbojet engine, some of it is directed to the core of the engine where the compressor is located. The compressor reduces the **volume** of the air and sends it into the combustion chamber under high pressure.

The exhaust gases formed in the combustion chamber have two functions. In the first place, they exit the rear of the chamber, as in a ramjet, providing the engine with a forward thrust. At the same time, the gases pass over the blades of the turbine, causing it to spin on its axis. The spinning turbine operates the compressor at the front of the engine, making possible the continued compression of new incoming air. Unlike a piston-powered engine, which has power strokes and exhaust strokes, the turbojet engine described here operates continuously. It is not subject, therefore, to the kind of vibrations experienced with a piston-powered propeller airplane.

Turbofan jets

A jet engine can be made more efficient by the addition of a large fan surrounded by a metal cowling at the front of the engine. The fan is somewhat similar to a propeller except that it has many more blades than a simple propeller. The fan is attached to a shaft that is also powered by the turbine at the rear of the engine. When exhaust gases from the compression chamber cause the turbine to spin, rotational energy is transmitted not only to the compressor, as described above, but also to the fan at the front of the engine.

The spinning fan draws more air into the engine, where some of it follows the pathway described above. Some of the air, however, bypasses the compressor and flows directly to the back of the engine. There, it joins with the exhaust gases from the combustion chamber to add to the engine's total thrust.

The turbofan jet engine has the advantage of operating more efficiently and more quietly than turbojet engines. However, they are heavier and more expensive than are turbojets. As a consequence, turbofan engines are usually found only on larger commercial and military aircraft (such as bombers), while turbojets are the preferred engine of choice on smaller planes, such as smaller commercial aircraft and military fighters.

Afterburners

Combustion within any type of air-breathing jet engine is quite inefficient. Of all the oxygen entering the front of the engine, no more than about a quarter is actually used to burn fuel within the engine. To make the process more efficient, then, some jet engines are also equipped with an afterburner. The afterburner is located directly behind the turbine in the jet engine. It consists of tubes out of which fuel is sprayed into the hot exhaust gases exiting the tubing. Combustion takes place in the afterburner, as it does in the combustion chamber, providing the engine with additional thrust. In a typical jet engine of moderate size, an afterburner can increase the takeoff thrust from about 50,000 N (11,000 lb) to about 70,000 N (15,500 lb).

Turboprop engines

When jet engines were first introduced in the 1940s, they were not very efficient. In fact, the cost of operating a jet airplane was so great that only military uses could be justified. At the time, commercial airline companies decided to compromise between the well-tested piston engines they were then using and the more powerful, but more expensive, jet engines. The result was the turboprop engine. In a turboprop engine, a conventional propeller is attached to the turbine in a turbojet engine. As the turbine is turned by the series of reactions described above, it turns the airplane's propeller. Much greater propeller speeds can be attained by this combination that are possible with simple piston-driven propeller planes. The problem is that at high rotational speeds, propellers begin to develop such serious eddying problems that they actually begin to slow the plane down. Thus, the maximum efficient speed at which turboprop airplanes can operate is less than 450 MPH (724 km/h).

KEY TERMS

. . . .

Afterburner—A device added at the rear of a jet engine that adds additional fuel to the exhaust gases, increasing the efficiency of the engine's combustion.

Ramjet—A simple type of air-breathing jet engine in which incoming air is compressed and used to burn a jet fuel such as kerosene.

Turbofan engine—A type of air-breathing jet that contains a large fan at the front of the engine operated by the turbine at the rear of the engine.

Turbojet—A type of air-breathing jet engine in which some of the exhaust gases produced in the engine are used to operate a compressor by which incoming air is reduced in volume and increased in pressure.

Turboprop—An engine in which an air-breathing jet engine is used to power a conventional propeller-driven aircraft.

Resources

Books

Boyne, Walter, Terry Gwynn-Jones, and Valerie Moolman. *How Things Work: Flight.* Alexandria, VA: Time-Life Books, 1990.

Cumpsty, Nicholas A. *Jet Propulsion: A Simple Guide to the Aerodynamic and Thermodynamic Design and Performance of Jet Engines.* Cambridge: Cambridge University Press, 1998.

Gunston, Bill. *The Development of Jet and Turbine Aero Engines.* 2nd ed. New York: Haynes Publishing, 1998.

"Jet Engine." In *The Rand McNally Encyclopedia of Transportation.* Chicago: Rand McNally, 1976.

"Jet Engines." In *How It Works.* New York: Simon and Schuster, 1971.

Karagozian, A. R. "Jet Propulsion." In *Encyclopedia of Physical Science and Technology,* edited by Robert A. Meyers. Orlando Academic Press, 1987.

Shaw, John M. "Jet Engines." In *Magill's Survey of Science: Applied Science Series,* edited by Frank N. McGill. Pasadena, CA: Salem Press, 1993.

David E. Newton

Jet stream

The jet stream is a narrow, fast, upper atmospheric **wind** current, flowing at high altitudes around **Earth**. Although often erroneously applied to all upper-level winds, by definition jet stream wind speeds are in excess of 57 MPH (92 km/h). The jet stream may extend for thousands of miles around the world, but it is only a few hundred miles wide, and usually less than a mile thick.

Undulating jet stream movements often greatly influence **storm** formation and **weather** changes. Research sponsored by the National Aeronautics and Space Administration (NASA) culminated in a 2001 report that also correlated solar activity, jet stream **migration**, and **precipitation** patterns over **North America**.

The wind speeds in the core of the stream sometimes can reach 200–300 MPH (322-483 km/h). These wind speeds within the jet stream that are faster than the surrounding regions are called jet streaks. On average, the jet stream flows from east to west, but it often meanders into northern or southern moving loops. Jet streams occur in both hemispheres, but the Southern Hemisphere jet streams show less daily variability. Jet streams can be detected by drawing isothachs (the lines connecting points of equal wind speed) on a weather **map**.

Jet streams form in the upper troposphere, between 6 and 9 mi (10 and 14 km) high, at breaks in the tropopause, where the tropopause changes height dramatically. Jet streams are located at the boundaries of warm and cold air, above areas with strong **temperature** gradients. For example, the polar front, which separates cold polar air from warmer subtropical air, has a great temperature contrast along the frontal zone, leading to a steep **pressure** gradient. The resulting wind is the polar jet stream at about 6 mi (10 km) high, reaching maximum wind speed in winter. Sometimes the polar jet can split into two jets, or merge with the subtropical jet, which is located at about 8 mi (13 km) high, around 30 degrees latitude. A low-level jet stream also exists above the Central Plains of the United States, causing nighttime **thunderstorm** formation in the summertime. Over the subtropics, there is the tropical easterly jet, at the base of the tropopause in summertime, about 15 degrees latitude over continental regions. Near the top of the stratosphere exists the stratospheric polar jet during the polar winter.

Detailed knowledge about the jet stream's location, altitude, and strength is essential not only for safe and efficient routing of aircrafts, but also for **weather forecasting**.

See also Atmosphere observation; Atmosphere, composition and structure; Atmospheric circulation; Global climate; Solar activity cycle; Solar illumination: Seasonal and diurnal patterns.

Jimson weed *see* **Nightshade**
Josephson effect *see* **Superconductor**

View of the jetstream over the Sahara Desert in Egypt. *NASA/Science Photo Library. Reproduced by permission.*

Josephson junctions *see* **Superconductor**

Joshua tree *see* **Amaryllis family (Amaryllidaceae)**

Juniper

Juniper is the common name for a large group of evergreen shrubs and trees belonging to genus *Juniperus*, in the family Cupressaceae (Cypress), order Pinales (pine). There are more than 50 **species** of *Juniperus*. They can be low creeping ground cover, broad spreading shrubs, or tall narrow trees. Both low growing and tall varieties are cultivated for ornamental purposes.

Junipers have thick, dense foliage and some species can be trimmed or sculpted to unusual shapes. Tall varieties, with their thick foliage, are quite **wind** resistant and are often planted in rows, as "windbreaks." Junipers grow throughout the world in many climates, from arctic regions, northern temperate areas, to the subtropics. Junipers are conifers, but they differ from typical cone bearing trees, which produce both male and female cones—junipers are either male or female. The female cones turn into fleshy, aromatic berries that are used for a variety of medicinal and culinary purposes. Junipers have two different types of leaves which, depending on the species, range from dark-green to a light shade of blue-gray. Some leaves are needle-like, similar to other conifers. The other type are scales that are pressed close to the twigs. Most species have a combination of the two types of leaves; young branches typically have needles, while the more mature branches have scales. *Juniperus communis*, or common juniper, is the one species that has only the needle type leaves.

The common juniper is a variable species, as it can occur as a shrub (3-4 ft/1-1.3 m) or tall **tree** (30-40 ft/10-

13 m). Native to **Europe**, it is now widely distributed in the northern temperate zones. The **color**, size, and shape depend on the variety, climate, and **soil**. The sharp leaves, 0.7-1 in (5-15 mm) long, grow in whorls of three. Small yellow (male) or blue-green (female) cones grow at the base of the leaves. The scales on the female cones grow together and develop into fleshy, aromatic, pea-sized berries that contain two to three **seeds**. The berries take about one or three years to mature and turn a dark blue-black color when ripe.

The best known use of the oil obtained from the berries is the flavoring for gin, an alcoholic beverage invented by the Dutch. The name gin is derived from *jenever*, the Dutch word for juniper. The berries have a strong flavor and are thought to stimulate the appetite. They are also used to flavor soft drinks, meat dishes, and condiments (they show up in jars of dill pickles). The principal medicinal use of juniper berries has been as a diuretic (an agent that promotes urination). Juniper berries can be toxic; pregnant women and people with kidney ailments should not ingest them. Poultices made of leaves and berries have been used for bruises, **arthritis**, and rheumatism. The berries have also been used as a substitute for **pepper**, and when roasted, a substitute for coffee. Fabric dyes are also obtained from the berries.

Some species such as, the Mediterranean *J. thuridera* and *J. excelsa* grow into large trees and are an important source of timber. *Juniperus virginiana*, also called eastern red cedar or Virginia juniper, found in the eastern United States, was used extensively for building houses in the early nineteenth century—its aromatic **wood** was an excellent bug repellant, particularly the bedbug. The wood for *J. virginiana* is also used to make high quality pencils.

Jupiter

Jupiter, the fifth **planet** from the **Sun**, is the largest and most massive planet in our **solar system**. One of the gas giants, it is composed of mostly **hydrogen** and helium. The Jovian atmosphere provides a rich laboratory for the study of **planetary atmospheres**. Its most famous feature, the Great Red Spot, has been visible for hundreds of years, and many smaller features are visible in its atmosphere. Thirty-nine satellites (or moons) of Jupiter have been discovered (the most of any known planet), ranging in size from larger than Mercury to tiny bodies with radii of less than 6 mi (10 km); in addition, the planet has thin rings composed of small particles.

Planetary probes, such as the two Pioneer and Voyager spacecrafts, have flown through the Jovian system

Jupiter's Great Red Spot, a tremendous atmospheric storm twice the size of Earth, has been visible since the earliest telescopic observations of the planet. The spot rotates counterclockwise, completing a full rotation once every six days. *U.S. National Aeronautics and Space Administration (NASA).*

and provided enormous detail on its physical and chemical properties, while ground-based and space-based observations have been used to monitor the planet for centuries. Launched in 1989, the Galileo spacecraft arrived at Jupiter in 1995, and began sending back pictures and data from the Jovian system for the next seven years. In 1994, fragments of Comet Shoemaker-Levy 9 crashed into Jupiter, giving a once-in-a-lifetime opportunity to study the atmosphere of the planet, and to learn about the effects of impacts on planets in general.

The Jovian system

The planet Jupiter, due to its size and brightness, was known to the ancients. It is named after Jupiter, the chief god in Roman mythology. This name is appropriate, since Jupiter is the largest and most massive planet in our solar system. It has a **mass** of (1.8988×10^{27} kg), more than three times that of **Saturn**, and more than 300 times that of **Earth**. Its equatorial radius 44.4 mi (71,492 km) is more than 11 times that of Earth.

Jupiter is at a distance of 7.783×10^8 km from the Sun, or about 5.2 times the distance of Earth. Because the planet is so far from the Sun, it receives much less solar **radiation** than Earth, and is consequently much colder, with a **temperature** of only -184°F (-120°C) at the top of its **clouds**. This temperature is actually higher than would be expected from the input of solar **energy** alone, since the planet generates some of its own **heat** internally, most likely due to the energy from its gravitational collapse.

Jupiter is one of four gas giants in the outer solar system. These planets differ substantially from the rocky bodies found closer to the Sun. Jupiter does not have a solid surface at all, and is hypothesized to have a lower atmosphere of molecular (gaseous) hydrogen that is 8.699–12,427 mi (14,000–20,000) km thick, which is underlain by a mantle of metallic (liquid) hydrogen that is 18,641 mi–24,855 mi (30,000–40,000 km) thick. This mantle surrounds an inner core of rock and **ice** thought to be 6,214 mi (10,000 km) thick. When we look at the planet, we see only the **ammonia** ice clouds in a hydrogen-helium mixture at the top of the atmosphere. These clouds rotate with different periods. The Jovian cloud structure consists of bands (divided into zones [light **color**] and belts [dark color]) and inter-band shear zones (characterized by streaks, loops, plumes, and spots [storms]). Because of the atmospheric **motion**, there is no single **rotation** period that can be associated with visible features on this gaseous planet like there would be with a terrestrial planet.

The most commonly used rotation period, referred to as System III, corresponds to the period of the planet's periodic **radio** emissions, which is the **rate** of rotation of the interior of the planet. This period is 9 hours, 55 minutes, 30 seconds. The Jovian year, which is the time for the planet to complete an **orbit** about the Sun, is 11.86 Earth years long.

Observations from Earth and space

Although Jupiter is visible to the naked **eye**, and has thus been known for thousands of years, much more detail is visible through even a small **telescope**. Galileo Galilei constructed one of the first telescopes of sufficient quality to make astronomical observations, and turned it toward the Jupiter in 1610. By doing so, he was the first to see the band structure in the planet's atmosphere, and was the first to see its satellites, or moons. In even a low-power telescope, these bands are visible. In a typical pair of binoculars, the four Galilean satellites (described below) are visible.

Today, astronomers observe the Jovian system regularly using both ground- and space-based telescopes. These observations are made in many wavelength regions, since each reveals distinct details. Ultraviolet wavelengths, for instance, are particularly useful for observing phenomena such as the aurora, while infrared observations are used to monitor the temperature of the planet.

In situ measurements

The first man-made objects to travel to Jupiter were Pioneer 10 in December 1973 and *Pioneer 11* in December 1974. These were very simple spacecraft, which provided preliminary information about the Jovian system. They made measurements of the radiation belts, the magnetic field, and obtained rather crude images of the planet and its satellites.

The more detailed measurements were made by the two **Voyager spacecraft**, which passed by Jupiter in March and July, 1979. These very capable spacecraft included a large array of scientific instruments to measure properties of the planet and its environment. Although it is the photographs taken by the Voyagers which are most widely known, much more information was acquired. Ultraviolet and infrared spectra, charged particle counts, and magnetic field measurements were also obtained, in order to provide a more complete view of the entire Jovian system. Since the encounters were separated by several months, information obtained from *Voyager 1* was used in the planning of the *Voyager 2* observations, in order to maximize the scientific return. Similarly, the results of Pioneer and ground-based measurements were used to plan the Voyager observations. Thus the observations of the future build on the measurements of the past.

In 1989, the Galileo spacecraft was launched, with objective of orbiting and studying Jupiter. The mission suffered many setbacks, however, including a malfunction that prevented the main **antenna** from deploying. Nonetheless, NASA scientists were able to reprogram Galileo's software remotely, allowing the probe to meet approximately 70% of mission objectives. In late 1995, the spacecraft reached Jupiter, settling into orbit to study the planet and its satellites at close range. Galileo also released a probe carrying six instruments into the Jovian atmosphere to make direct measurements of the planet's composition. Galileo has had a remarkable record of success while functioning in the Jovian system from 1995-2002. Many thousands of photographs of Jupiter and its moons have been returned along with much additional data. Galileo has been instrumental in helping find over 20 new moons of Jupiter.

The planet

Formation and composition

Jupiter is believed to have formed in a manner similar to the other gas giants in the outer solar system. Rocky and/or icy planetesimals condensed from the solar nebula when the solar system was formed more than 4.5 billion years ago. The condensation process continued until the planetesimals were roughly Earth-sized. Although the temperatures close to the Sun in the inner solar system drove off the volatile (easily-vaporized) gases from bodies formed there, at the much greater distance of Jupiter this did not occur, and the planetesimals continued to collect gas. As the **force** of gravity slowly crushed the material in the center of the planet, energy was released, which is still seen today as extra heat released from the planet as it cools.

The current composition of the whole of Jupiter is believed to be 24% helium and otherwise primarily molecular hydrogen, with less than 1% of other constituents, including methane (CH_4), ammonia (NH_3) and **water** (H_2O). This composition is similar to the primordial composition of the solar system, and similar to the makeup of the Sun today. A rocky and/or icy core is thought to exist at the center of the planet, with a size similar to that of Earth. Hydrogen and helium in metallic and molecular form make up the majority of the atmosphere further from the center.

Atmosphere

The most striking feature in Jupiter's atmosphere is the Great Red Spot, a huge **storm** several times larger than Earth's diameter, which has been observed for more than 150 years (or perhaps much longer—because the observations were not continuous, it is unclear if spots observed before then were of the present-day spot). Jupiter's atmosphere has many storm systems, but the Red Spot is the most prominent. Early theories suggested that it was due to clouds colliding with a feature on the surface of the planet, but since there are no such features (and no solid surface, except perhaps at great depths in the core of the planet), this has since been abandoned. The Spot is also not static, but varies over both short and long timescales. Winds on both the north and south sides of the spot prevent it from varying in latitude, but variations in longitude are seen.

Jupiter's storm activity is quite violent. Although data from Galileo showed **lightning** activity is only about 10% as frequent per given area as on Earth, the intensity is 10 times as high. It is the strong winds that dominate the atmosphere of Jupiter, however. These winds are found in bands, with the speed and direction varying greatly with latitude. Both the northern and southern hemisphere have at least 10 bands of alternating **wind** direction. The strongest of these winds is more than about 400 miles/hour (600 km/hr). The winds are very stable, although the fine details they cause, such as small white spots, can come and go in just days. Other features, such as the Red Spot, and three nearby white spots which have been visible since 1938, can last months or years. The long-lived spots are seen to vary with longitude, but not latitude. Astronomers make computer models of the atmosphere in order to understand the processes occurring there, but they are not yet able to explain the persistence of these features.

The other primary feature visible in the atmosphere are the clouds. Three main cloud layers are seen, composed of ices of water, ammonium hydrosulfide, and ammonia. Since the different cloud layers are at different

Jupiter, as seen by *Voyager 1.* The planet is the most massive object in the solar system after the Sun; its mass is greater than that of all the other planets combined. The latitudinal bands in the Jovian atmosphere may be partially the result of the rapid rotation of the planet which, despite its enormous size, rotates once every 9 hours 55 minutes. *U.S. National Aeronautics and Space Administration (NASA).*

heights in the atmosphere, each represents a different temperature region. The variation in the colors of the clouds are also thought to be due to different **chemistry**. Theories suggest that the colors are due to **sulfur** or **phosphorus** in the atmosphere, but this has not been verified.

The Jovian system

Satellites

Satellites of Jupiter have an organizational structure that puts them into one of several classes. The inner satellites have circular orbits, they move in a prograde direction, and they lie in Jupiter's equatorial plane (e.g., Adrastea and Metis (which share one orbit), Amalthea, and Thebe. The Galilean satellites are slightly more distant than the inner satellites and occupy nearly circular orbits, which are prograde. All Galilean satellites (Io, Europa, Ganymede, and Callisto) are locked in a 1:1 spin orbit couple with Jupiter (the same face toward Jupiter all the time, like Earth's **Moon** to Earth). An inner group of outer satellites are farther out and are all prograde, but have inclined orbits with respect to Jupiter's equatorial plane (e.g., Leda, Himalia, Lysithea, Elara). An outer group of outer satellites have a mixture of prograde and retrograde orbital motions and also have highly inclined orbits.

When Galileo turned his telescope toward Jupiter in 1610, he discovered what he called the four new "Medician stars," after Cosimo of Medici, a former pupil. These four satellites of Jupiter, now known as the Galilean satellites in his honor, today carry the names Io, Europa, Ganymede, and Callisto, after Jupiter's (the Greek god's) lovers. These bodies are planets in their own right, with sizes that rival or exceed those of the inner planets.

Galileo's discovery of a solar system in miniature lent strong support to the idea of the Copernican system, which postulated that the planets were in orbit about the Sun, rather than Earth. Until Galileo's observations, there was no physical evidence for such a system. By finding a planet with its own satellites, the possibility that there can be centers of motion other than Earth appeared much more likely.

Since Galileo's time, astronomers have discovered 35 additional satellites of Jupiter. All of these are much smaller than the Moon-sized Galilean satellites, and hence are much more difficult to observe. Most were not discovered until the Voyager spacecraft reached Jupiter. These smaller objects are commonly irregularly shaped and in inclined (tilted), eccentric orbits. It is likely that many of them have been captured by the planet's gravitational field, rather than having formed in the vicinity of Jupiter like the Galilean satellites.

Io

Io is the most striking of the Jovian satellites and it is the most volcanically active body in the solar system. The Voyager spacecraft discovered active volcanoes on the surface, with plumes as high as 186 mi (300 km); eight were recorded by *Voyager 1*, and six of these were seen to be still active when *Voyager 2* arrived several months afterward. Based on the colors of the surface of the moon (yellow, orange, red, and black), scientists believe that the surface consists mostly of sulfur and sulfur compounds.

Most of the surface of the **satellite** had been transformed by the time Galileo arrived, 17 years later. The color and contours of the surface in the southern hemisphere had changed significantly, giving evidence of nearly continuous volcanic activity. Io has the most rapid planetary resurfacing process in the solar system, and hence the youngest overall surface of any known planetary body. Volcanic features abound on Io, including lava flows, ash falls, and volcanic vents and caldera.

Only in a few, isolated spots is there a hint of older (perhaps not volcanic) crust on Io. These spots are called massifs and plateaus, and consist of highly fractured rocky crust. Impact craters are absent on Io, suggesting—again—that the crust is very young.

Current theories suggest that the volcanism on Io is caused by its proximity to Jupiter, and hence the strong gravitational forces, which continually squeeze and stretch the satellite. Because of this continual reprocessing of the surface, impact craters are not seen as on the other satellites. Io also has an important interaction with the Jovian **magnetosphere**.

It is hypothesized that Io has an internal structure consisting of a liquid iron-sulfur core, a mantle of silicate rock that may be partially molten, a **lithosphere** of brittle silicate rock, a "thiosphere" (sulfur layer) of liquid or plastic sulfur, and a crust of solid sulfur and sulfur compounds.

Europa, Ganymede, and Callisto

Europa is an ivory, gray, and brown world with a water ice surface. Few craters are visible, and the surface is relatively smooth, but dissected by global networks of fractures. Some areas of Europa (bright plains) have a high **density** of fractures, whereas other areas (mottled terrains) have less obvious fractures. Icy "lava" from below Europa's ice crust may have flowed out and resurfaced parts of the mottled terrains. This suggests that processes have been (and may be still) at work to renew and re-coat the surface of Europa.

Even more exciting is evidence suggesting that Europa may possess the right combination of conditions to sustain life. Based on data from Galileo, paleontologists now believe that the satellite features liquid (or at least slushy) oceans beneath its icy crust, warmed by volcanic activity, geysers, and other thermal outflow generated by tidal stretching and squeezing (like Io, on a smaller scale). In particular, the surface in many areas appears fractured and segmented, indicating ice floes moving over liquid water. In conjunction with organic compounds, the water and heat may have created a biologically viable environment, similar to the **hydrothermal vents** that on Earth have been shown to support organic activity.

Europa has but five clearly defined impact craters, but there are many other impact-crater-like features called crater palimpsests, which look like flat, circular, brown spots that may be craters filled by water lavas. The largest of these features is called Tyre Macula, and is 62 mi (100 km) in diameter.

It is hypothesized that Europa has a solid iron-sulfur core, a solid silicate mantle, an **asthenosphere** of silicates, ice, and water, and a cryosphere (ice crust) of water ice that is approximately 62 mi (100 km) thick.

Ganymede, the largest of the moons, is an ivory and grey water-ice world. Impact craters are visible on the surface, but there is a mixture of both dark heavily cratered (older terrains) and sparsely cratered but highly fractured regions (bright and grooved terrains). This suggests that part of the surface is very old (heavily cratered regions) and part has experienced resurfacing since its formation (bright and grooved terrains). Ganymede has a density that suggests a silicate rock core surrounded by a mantle and cryosphere somewhat like that of Europa (but perhaps not as warm). Ganymede has a large, very old **impact crater basin** that has been partially covered by ice lavas, known as Memphis Facula (about 311 mi [500 km] in diameter).

Callisto is the outermost of the Galilean satellites. Callisto is another gray water ice world, but it is darker than the Europa and Ganymede. Impact craters are visible over the entire surface, suggesting that little resurfacing has occurred. The density of craters on Callisto is higher than on Earth's moon, suggesting most of the surface is very old. There is a huge, multi-ring impact crater basin on Callisto called Valhalla, which has 25 concentric rings extending out to a diameter of 621 mi (1,000 km). Valhalla may be the largest impact crater in the solar system. Callisto has an internal core of silicate rock and a mantle and cryosphere like the other Galilean satellites.

Several other small satellites of Jupiter have been imaged (e.g., Sinope, Amalthea, and Thebe) and these appear to be captured chondritic asteroids. This lends credence to the theory that all irregularly shaped (non-Galilean) satellites of Jupiter are captured asteroids of one kind or another.

Rings

A remarkable result of the Voyager spacecraft's encounter with Jupiter was the discovery of a ring of particles orbiting Jupiter. This ring is much smaller and simpler than the familiar rings of Saturn, and they are not visible directly from Earth. Since their discovery, however, astronomers have made measurements of them by watching the light from stars as they pass behind the rings. In this way they can measure their extent and density.

The rings extends to approximately 1.7 Jovian radii from the center of the planet. Small "shepherd" moons

The four largest satellites of Jupiter are the Galilean satellites, named after seventeenth century Italian astronomer Galileo who was the first to observe them with a telescope and describe them as moons. This composite image shows the four satellites to scale: Europa (upper left), Callisto (upper right), Io (lower left), and Ganymede (lower right). All but Europa are larger than our moon. Ganymede is larger than Mercury. *U.S. National Aeronautics and Space Administration (NASA).*

(Metis and Adrastea) may act to help gravitationally confine the ring material, creating a sharp outer boundary. Indeed, debris lost from these shepard satellites is thought to be the ring's origin. The rings are about 3,728 mi (6,000 km) wide and about 19 mi (30 km) thick. They are dark and there is some internal band structure to the rings. The size of material is very small, a few microns in size (like fine dust).

Magnetosphere

Jupiter has a magnetic field more than ten times stronger than that of Earth. The magnetosphere, which is created due to this field and its interaction with the **solar wind**, has major effects on the Jovian system. The size

KEY TERMS

Comet Shoemaker-Levy 9—A comet which crashed into Jupiter in 1994.

Copernican system—The description of the solar system which has the planets orbiting the Sun, proposed by Nicholas Copenicus.

Galilean satellites—The four largest satellites of Jupiter, discovered by Galileo in 1610.

Gas giant—One of the large outer planets, including Jupiter, composed primarily of hydrogen and helium.

Great Red Spot—A large, storm in Jupiter's atmosphere, which has been visible for more than 150 years.

Io plasma torus—A region of charged particles which are trapped in Jupiter's magnetic field.

Planetesimal—Small bodies from which planets formed.

Voyager—Two unmanned planetary probes which flew by Jupiter and its satellites in 1979.

of the magnetosphere is larger than the Sun; since the innermost Jovian satellites are embedded deep in the magnetosphere, they are particularly affected by the magnetic field. One of the most complex results is the interaction of Io with this field.

Atoms such as sodium and sulfur have been discovered in a cloud around Io as it orbits Jupiter. This material is believed to originate on the surface of the satellite, and then reach space after being driven from the surface due to collisions with high energy particles. Through a not yet understood process, these atoms can become charged, and thus get trapped by the magnetic field of Jupiter. The result is a "plasma torus," a doughnut-shaped region of charged atoms, which rotates with the planet's magnetic field. Observations of these emissions in the ultraviolet since the 1970s have shown both a time variability and a spatial asymmetry in brightness from the torus.

Particles from the region around Io are also thought to be responsible for the aurora (similar to Earth's northern lights) seen at Jupiter's poles, but once again, the process is not clearly understood. The auroras appear to be caused by particles from the torus region, which rain down on the atmosphere at the poles of the planet, creating emissions.

Radio telescopes first detected emissions from Jupiter in the 1955. These radio waves are created when electrons travel through the planet's magnetosphere. Measurements show both short- and long-term variability of the radio emissions.

Comet Shoemaker-Levy 9 collision

In early 1993, Eugene and Carolyn Shoemaker and David Levy discovered a comet moving across the night sky. They were surprised at its appearance, since it seemed elongated compared to other comets they had seen. Further observations showed that the comet consisted of a large number of fragments, apparently torn apart during a close encounter with Jupiter during a previous orbit. Calculations showed that this "string of pearls" would collide with Jupiter in July, 1994.

A worldwide effort was mounted to observe the impacts with nearly all ground-based and space-based telescopes available. Although astronomers could not predict what effect the collisions would have on Jupiter, or even whether they would be visible, the results turned out to be spectacular. Observatories around the world, and satellite telescopes such as the Hubble Space Telescope observed the impacts and their effects. Galileo, en route to Jupiter at the time, provided astronomers with a front-row seat at proceedings. Even relatively small amateur telescopes were able to see some of the larger impact sites. Dark regions were visible in the atmosphere for months.

See also Space probe.

Resources

Books

Christiansen, E.H., and W.K. Hamblin. *Exploring the Planets.* 2nd ed. Englewood Cliffs, NJ: Prentice-Hall, 1995.
de Pater, Imke, and Jack J. Lissauer. *Planetary Sciences.* Cambridge, UK: Cambridge University Press, 2001.
Morrison, D., and Tobias Owen. *The Planetary System.* 3rd ed. Addison-Wesley Publishing, 2002.
Taylor, F.W. *The Cambridge Photographic Guide to the Planets.* Cambridge University Press, 2002.

Periodicals

Beebe, Reta F. "Queen of the Giant Storms." *Sky & Telescope* (1990): 359-364.
O'Meara, Stephen James. "The Great Dark Spots of Jupiter." *Sky & Telescope* (1994): 30–35.

Other

Arnett, B. SEDS, University of Arizona. "The Nine Planets, a Multimedia Tour of the Solar System." November 6, 2002 [cited February 8, 2003]. <http://seds.lpl.arizona.edu/nineplanets/nineplanets/nineplanets.html>.

David Sahnow
David T. King, Jr.

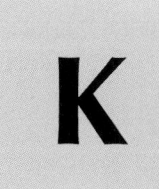

K

K-T event (Cretaceous-Tertiary event)

The K-T event (Cretaceous-Tertiary event) refers to the **mass extinction** of the dinosaurs that took place approximately 65 million years ago (mya). In addition to the dinosaurs, most large land animals perished and an estimated 70% of **species** became extinct.

In the early 1980s, a team of physicists and geologists documented a band **sedimentary rock** in Italy that contained an unusually high level of the rare **metal** iridium (usually found on Earth's surface only as a result of meteor impacts). The scientists eventually argued that that the iridium layer was evidence of a large asteroid impact that spewed iridium contaminated dust into the atmosphere. Blown by global **wind** currents, the iridium eventually settled into the present thin sedimentary layer found at multiple sites around the world. Given the generalized dispersion of iridium the researchers argued that the impact was large and violent enough to cause dust and debris particles to reach high enough levels that they seriously occluded **light** from the **Sun** for a large expanse **Earth**.

The subsequent reduction in **photosynthesis** was sufficient to drastically reduce land **plant** population levels and eventually drive many plant species to **extinction**. The reduction in plant population levels also provided evolutionary **pressure** on species nutritionally dependent upon plant life. Large life forms with especially high-energy demands (e.g., dinosaurs) were especially sensitive to the depleted dietary base. The adverse consequences of population reductions and extinctions of plant-eating life forms then rippled through the ecological web and food chain—ultimately resulting in mass extinctions.

Calculations of the amount of iridium required to produce the observable layer (on average about a centimeter thick) yield estimates indicating that the asteroid measured at least 6 mi (10 km) in diameter. The **impact crater** from such an asteroid could be 100 mi (161 km) or more in diameter. Such an impact would result in widespread firestorms, earthquakes, and tidal waves. Post-impact damage to Earth's **ecosystem** occurred as dust, soot, and debris from the collision occluded the atmosphere to sunlight.

Based on **petroleum** exploration data, Canadian geologist Alan Hildebrand identified a major impact crater in the oceanic **basin** near what is now the Yucatan Peninsula of Mexico. The remains of the impact crater, termed the Chicxulub crater, measures more than 105 mi (170 km) in diameter. Argon dating places the Chicxulub impact at the expected Cretaceous-Tertiary **geologic time** boundary, approximately 65 mya.

Other geological markers are also indicative of a major asteroid impact approximately 65 million years age (e.g., the existence of shock quartz, ash, and soot in sedimentary layers dated to the K-T event). Tidal waves evidence surrounding the Gulf of Mexico basin also dates to 65 mya.

Other scientists have argued that it was not a solitary impact—that alone caused the mass extinction evidenced by the fossil record. At end of the prior Cretaceous period and during the first half of the Tertiary period, Earth suffered a series of intense and large impacts. Geologists have documented more that 20 impact craters greater than 6.2 mi (10 km) in diameter that date to the late Cretaceous period. Large diameter impact craters were especially frequent during the last 25 million years of the Cretaceous period. (i.e., the Senonian epoch).

The extinction of the dinosaurs and many other large species allowed the rise of **mammals** as the dominant land species during the Cenozoic era.

See also Astronomy; Catastrophism; Evolution, evidence of; Historical geology.

Kangaroo rats

Kangaroos rats are small burrowing **mammals** with fur-lined cheek pouches, making up the rodent family

Ord's kangaroo rat. *Photograph by Larry L. Miller. The National Audubon Society Collection/Photo Researchers, Inc. Reproduced by permission.*

Heteromyidae, found principally in North and Central America. There are five genera of **rodents** with external pouches in this family but only two of them are given the popular name of kangaroo. It derives from the fact that their front feet are very small and the back legs are quite large and strong, adapted for two-legged leaping. Its thick tail helps in balance and serves as a third "leg" when standing still.

Kangaroo rats belong to the genus *Dipodomys,* of which about 20 **species** are found from southern Alberta to central Mexico. Their heads and bodies may be 4-8 in (10-20 cm) long, plus a tail that is even longer. They are brown to yellow in **color**, with a lighter belly.

Unlike other rodents with cheek pouches, such as pocket **gophers**, kangaroo rats and **mice** live in very open country. They require unobstructed space in which to make their leaps, which may be as much as 6 ft (2 m) or more. They generally come out only at night to forage for **seeds**, fruit, and even small **insects**. The food is carried back to their burrows where it may be stored. Each adult has a separate burrow system.

Kangaroo rats do not consume **water**. Rather, they receive fluids from food. They have very special kidneys that function efficiently at ridding the body of poisons. An important part of their lives is dust bathing. They have an oil-secreting gland located on the back between the shoulders, and if they are not allowed to bathe in dust to remove the excess oil, their fur and skin becomes matted and irritated.

During the mating process, a male kangaroo rat seeks the attention of a female by thumping on the ground with his back feet. The female will produce one to six offspring in one of her two or three litters each year.

Several species of kangaroo rats are regarded as threatened or even endangered due to the destruction of their **desert** habitat. The growth of agriculture through **irrigation** and the development of new residential areas can decimate a subspecies or even a species almost overnight.

The two species of smaller kangaroo mice belong to the genus *Microdipodops,* which live in the desert sands of Nevada. They look much like small furry balls with eyes and a single pair of long legs. Their fur is longer and silkier than the kangaroo rats'. Even their hind feet have special fur, called fringe, that broadens the base from which they spring. Perhaps because they burrow into sandy **soil**, their burrows are not as elaborate as those of other burrowing rodents. They are most active at night.

There are several other genera of pocket mice but they are not the leapers that kangaroo rats and mice are. The pocket mice of Mexico and the southwestern United States make up the genus *Perognathus.* Two genera of spiny pocket mice, *Liomys* and *Heteromys,* can be found in the arid regions and forest areas of central America. These animals have much stiffer fur than the other members of the family.

Resources

Books

Caras, Roger A. *North American Mammals: Fur-Bearing Animals of the United States and Canada.* New York: Meredith Press, 1967.

Hanney, Peter W. *Rodents: Their Lives and Habits.* New York: Taplinger Publishing Co., 1975.

Knight, Linsay. *The Sierra Club Book of Small Mammals.* San Francisco: Sierra Club Books for Children, 1993.

Jean F. Blashfield

Kangaroos and wallabies

Kangaroos and wallabies are pouched **mammals**, or **marsupials**, of **Australia** and nearby islands. Kangaroos and wallabies have hind legs enlarged for leaping. Most **species** live on the ground, and some in trees. The name kangaroo is usually used for large species, and wallaby for smaller ones. They all belong to the family Macropodidae, meaning "big footed," and they are herbivorous, or plant-eating animals.

Kangaroos and wallabies hold the same place in the **ecosystem** as ruminants, such as **deer**. They graze and have similar mechanisms for chewing and digesting plants. Most members of the family are nocturnal, feeding at night.

The kangaroo's hand has five clawed fingers, all approximately the same length. It can be used for grasping.

Red kangaroos in Australia. © Len Rue, Jr./The National Audubon Society Collection/Photo Researchers, Inc. Reproduced by permission.

The hind feet are quite different, being extremely large and having only four toes. The first two are tiny and are joined together at the bone but not at the claw. These claws are useful for grooming. The third toe is huge, with a strong, sharp claw. When fighting, the **animal** may use this claw as a weapon. The fourth toe is again small, but not as small as the grooming toes.

Kangaroos are famous for their prodigious leaps—sometimes up to 30 ft (9.2 m) long and 6 ft (1.8 m) high by the gray kangaroo. Because the spring-like tendons in their hind legs store **energy** for leaps, they are sometimes called "living pogo sticks." It has been calculated that kangaroos actually use less energy hopping than a horse uses in running. When they are grazing, kangaroos tend to move in a leisurely fashion using all four feet plus the hefty tail for balance. They move the hind legs while balancing on the front legs and tail, then move the front legs while balanced on the hind legs, rather like a

person walking on crutches. They often rest by reclining on their side, leaning on an elbow.

Most kangaroos are unable to walk in normal fashion, moving the hind legs at separate times. However, tree-dwelling kangaroos have the ability to move their hind legs at different times as they move among the branches.

Like all marsupials, female kangaroos have a protective flap of fur-covered skin that shields the offspring as they suckle on teats. The kangaroo's marsupium, as this pouch is called, opens toward the head. The pouch is supported by two bones, called marsupial bones, attached to the pelvis. No other mammals have these bones, but even male kangaroos do, despite the fact that they do not have pouches.

Some kangaroos live in social groups and others are solitary. In general, the larger animals and the ones that live in open **grasslands** are more social. Within a group, called a mob, the individuals are more safe. In a mob, the dominant male competes with the others to become the father of most of the offspring, called joeys. Because the dominant male is larger than the other males (called boomers), over many generations, through sexual **selection** the males have evolved to become considerably larger than the females (called does).

The difficult life of a newborn kangaroo

For such a large animal, the gestation period of kangaroos is incredibly short. The longest among the kangaroos is that of the eastern gray (*Macropus giganteus*), in which the baby is born after only 38 days. However, it is less than an inch long, blind, and hairless like the newborns of all marsupials. It may weigh as little as 0.01 oz (0.3 g).

The kangaroo has virtually no hind legs when born. In fact, the front legs, which are clawed, look as if they are going to be mammoth. These relatively large front paws serve the purpose of pulling the tiny, little-formed creature through its mother's fur and into her pouch. **Instinct** guides the tiny infant, who moves with no help from its mother. If it moves in the wrong direction, the mother ignores it. If it moves slowly, it may die from exposure. These tiny creatures are born with disproportionately large nostrils, so **smell** apparently plays a major role in guiding the path to the mother's pouch.

A newborn kangaroo has a longer distance to travel than most marsupials. Most others have a pouch that faces backward, or is an open flap of skin where it is easier for the baby to find the teats. In the kangaroos, the baby must climb up to the top of the pouch, crawl over the edge, and then back down inside to reach a teat.

If successful in reaching the pouch, the baby's tiny mouth clamps onto a teat, which swells into the mouth so that the infant cannot release it. The infant's esophagus expands so it can receive both nourishment and **oxygen**. The baby, now called a pouch embryo, cannot let go even if it wants to. It will be a month or more before its jaw develops enough to open. Only the teat that the baby is attached to actually produces milk.

After the infant is born and moves into the pouch, a female red kangaroo (*Macropus rufus*) may mate again. This time, however, the fertilized egg goes into a resting state and does not develop until the female stops nursing the first young. Her body signals that change to the zygote, which then starts developing again. This time lag, called diapause, has great advantages to the species in that if one young dies, another embryo can quickly take its place. Diapause does not occur in the eastern gray kangaroo. The pouch embryo will continue to develop as it would if inside a uterus. In the big kangaroos, it takes

10 months or more before the joey emerges for the first time (often falling out by accident). It gradually stays out for longer and longer periods, remaining by its mother's side until about 18 months old. A male great kangaroo reaches sexual maturity at about two or three years, a female not for several years more.

The great kangaroos

One fossil kangaroo, *Procoptodon goliah*, was at least 10 ft (3.1 m) tall and weighed about 500 lb (227 kg). Today, the largest of the species is the male red kangaroo, which may have a head-and-body length of almost 6 ft (1.8 m), with a tail about 3.5 ft (107 cm) long. It may weigh 200 lb (90 kg).

Fourteen species of living kangaroos belong to the genus *Macropus*. They include the largest living marsupials. In varying contexts and times they have been regarded as **pests**, or as among the treasures of Australia. Farmers have argued that kangaroos take food from **sheep** and cattle, but actually kangaroos tend to select different plants from domestic **livestock**. Today, only a few are seen near urban areas, but they are widespread in the countryside, where they are still a favorite target of hunters, who sell their meat and skins.

The eastern gray kangaroo and its western relative (*M. fuliginosus*), which is actually brown in **color**, occupy forest areas throughout the eastern half and the southwest region of Australia. The forest-living species mostly eat **grasses**. Their young are born at more predictable times than those of the red kangaroos and they take longer to develop. They spend about 40 weeks in the pouch, and the mother does not mate again until the joey becomes independent and mobile.

The red kangaroo shares the western gray kangaroo's **habitat**. As European settlers developed the land, **forests** were cut, reducing habitat for the gray kangaroos, but increasing the red kangaroo populations. Only the male red is actually **brick** red; the female is bluish gray, giving it the nickname "blue flier." It has the ability to care for three young at different stages of their lives at once. It can have an egg in diapause in its uterus, while a tiny pouch embryo can be attached to one teat. Then, another teat elongates so that it is available outside the pouch, where an older, mobile offspring can take it for nourishment. This situation probably evolved in response to the dryness of the red kangaroo's **desert** environment, which can easily kill young animals.

In the continental interior, the red kangaroo lives in open dry land, while wallaroos, also called euros (*M. robustus*), live around rock outcroppings. The wallaroos, which have longer and shaggier hair than the larger kangaroos, are adapted for surviving with minimal **water** for nourishment. When water is not available, the animal reduces the body's need for it by hiding in cool rock shelters, and their urinary system concentrate the urine so that little liquid is lost.

The smaller wallabies

Smaller kangaroos are usually called wallabies. The name is especially used for any kangaroo with a hind foot less than 10 in (25 cm) long. Several species of *Macropus* are regarded as wallabies as well as other genera. The two smallest are the tammar wallaby (*M. eugenii*) of southwestern Australia and adjacent islands and the parma wallaby (*M. parma*) of New South Wales. Their head and body are about 20 in (50 cm) long with tail slightly longer. The tammar wallaby has been known to drink **saltwater**. The whiptail wallaby (*M. parryi*) is the most social of all marsupials. It lives in mobs of up to 50 individuals, and several mobs may occupy the same territory, making up an even bigger population.

Rock wallabies (*Petrogale* species) have soft fur that is usually colored to blend in with the dry, rocky surroundings in which they live. However, the yellow-footed rock wallaby (*P. xanthopus*) is a colorful gray with a white strip on its face, yellow on its ears, dark down its back, yellow legs, and a ringed yellow-and-brown tail. Rock wallabies have thinner tails than other wallabies and use them only for balance, not for propping themselves up. They are very agile when moving among the **rocks**. Some have been known to leap straight up a rock face 13 ft (4 m) or more. Rock wallabies have sometimes been kept in zoos, where they live and breed in communal groups.

The smaller hare wallabies are herbivores that feed mostly on grasses and can run fast and make agile jumps. Close study of the hare wallaby called the quokka (*Setonix brachyurus*) provided naturalists with their first solid information about marsupials. With a head-body length of about 20 in (50 cm), plus a 10-in (25-cm) tail, this rodent-like creature lives in swampy areas in southwestern Australia. Today it lives mainly on neighboring islands. Several other species are rapidly disappearing and one, the central hare wallaby (*Lagorchestes asomatus*), is known from only one specimen. However, its range of the Northern Territory has not been adequately explored, and it may not be extinct.

Several wallabies that were widespread in the past are probably extinct. Nail-tailed wallabies (*Onychogalea* species) had tough, horny tips to their tails. These 2 ft (61 cm) tall marsupials lost their habitat to grazing livestock and agricultural pursuits, and were also hunted. Nail-tails were also called organ grinders, because their forearms rotated while they were hopping.

Five species of wallabies (*Dorcopsis* species) live only in the **rainforest** of New Guinea. They do not hop as well as other kangaroos because their hind feet are not much larger than their front. Somewhat smaller kangaroos called pademelons (*Thylogale* species) live in New Guinea as well as on the Australian **continent**.

Tree kangaroos

The tree kangaroos (*Dendrolagus* species) are herbivores that live in trees in mountainous forest of New Guinea and Australia. They have fairly long fur and live in small groups. Some of them have the ability to leap between strong branches of trees as much as 30 ft (9.2 m) apart.

Tree kangaroos have longer forearms and longer tails. Although their tail is not truly prehensile, or grasping, they may wrap it around a branch for support. Unlike other kangaroos, their tail is about the same thickness from base to tip. Tree kangaroos are hunted as food and so they are decreasing in numbers. The single young stays in the pouch for almost a year and suckles even longer, so the **rate** of reproduction is rather slow.

Rat-kangaroos

A subfamily of smaller, more ancient marsupials is called rat-kangaroos. Some scientists classify the rat-kangaroos in a separate family, the Potoroidae. These animals are omnivorous, eating a variety of foods.

The musky rat-kangaroo (*Hypsiprymnodon moschatus*) is the smallest member of the kangaroo family, with a head-body length of only about 10 in (25 cm) plus a furless tail (the only one in the family) of about 5 in (12.7 cm). This species also has front and hind feet closer to the same size than any other member of the family. It eats some **insects** along with grasses and other plants. The potoroos (*Potorous* species) are about twice as large as the musky rat-kangaroos and display a more advanced leaping ability.

The desert rat-kangaroo (*Caloprymnus campestris*) was first seen in southern Australia in 1843, but not again until 1931. Consequently, little is known about it. Similarly, the northern rat-kangaroo (*Bettongia tropica*) of Queensland was observed in the 1930s, but not again until 1971. It has huge hind feet, which cover half the length of its body. Both are **endangered species**.

The bettong, also called the woylie or brush-tailed rat-kangaroo (*Bettongia penicillata*), has a prehensile tail, which it uses to carry the dry grasses used in building a nest. Woylies were quite common over southern Australia, but as human populations have increased, it has become extirpated over most of its original range.

KEY TERMS

Diapause—A period during which a fertilized egg does not implant and start to develop. A change in the mother's hormone system, perhaps triggered by favorable weather conditions, signals the egg to start developing.

Embryo—A stage in development after fertilization.

Herbivorous—An animal that only eats plant foods.

Marsupium—The pouch or skin flap that protects the growing embryo of a marsupial.

Prehensile—Of a tail, able to be used for grasping.

Ruminant—A cud-chewing animal with a four-chambered stomach and even-toed hooves.

Uterus—Organ in female mammals in which embryo and fetus grow to maturity.

Similarly threatened is the boodie or short-nosed rat-kangaroo (*B. lesueur*). The only kangaroo that digs burrows, where it gathers in a family group, it is now restricted to several islands in western Australia's Shark Bay. Unlike the other members of the kangaroo family, the boodie never uses its front feet while walking.

Clearly, many of the smaller members of the kangaroo are endangered and even nearing **extinction**. Apparently, they are vulnerable to even small changes in their habitat. The great kangaroos, on the other hand, appear to be thriving as long as their habitats remain protected and hunting for their skin and meat is conducted on a sustainable basis.

Resources

Books

Arnold, Caroline. *Kangaroo.* New York: William Morrow & Co., 1987.

Knight, Linsay. *The Sierra Club Book of Great Mammals.* San Francisco: Sierra Club Books for Children, 1992.

Lavine, Sigmund A. *Wonders of Marsupials.* New York: Dodd, Mead & Co., 1978.

Lyne, Gordon. *Marsupials and Monotremes of Australia.* New York: Taplinger Pub. Co., 1967.

Stidworthy, John. *Mammals: The Large Plant-Eaters.* Encyclopedia of the Animal World. New York: Facts On File, 1988.

Tyndale-Biscoe, Hugh. *Life of Marsupials.* New York: American Elsevier Publishing Co., 1973.

Jean F. Blashfield

Karst topography

Karst is a German name for an unusual and distinct limestone terrain in Slovenia, called Kras. The karst region in Slovenia, located just north of the Adriatic Sea, is an area of barren, white, fretted rock. The main feature of a karst region is the absence of surface **water** flow. Rainfall and surface waters (streams, for example) disappear into a drainage system produced in karst areas. Another feature is the lack of topsoil or vegetation. In **geology**, the term karst topography is used to describe areas similar to that found in Kras. The most remarkable feature of karst regions is the formation of caves.

Karst landscapes develop where the **bedrock** is comprised of an extremely soluble **calcium carbonate** rock, for example: limestone, gypsum, or dolomite. Limestone is the most soluble calcium carbonate rock. Consequently, most karst regions develop in areas where the bedrock is limestone. Karst regions occur mainly in the great sedimentary basins. The United States contains the most extensive karst region of the world. The Mammoth **cave** system is located in this area. Other extensive karst regions can be found in southern France, southern China, Central America, Turkey, Ireland, and England.

Karst regions are formed when there is a chemical reaction between the **groundwater** and the bedrock. As rain, streams, and **rivers** flow over the earth's surface, the water mixes with the **carbon dioxide** that naturally exists in air, and the **soil** becomes acidic and corrodes the calcium carbonate rock. The carbonate **solution** seeps into fissures, fractures, crevices, and other depressions in the rock. **Sinkholes** develop and the fissures and crevices widen and lengthen. As the openings get larger, the amount of water that can enter increases. The **surface tension** decreases, allowing the water to enter faster and more easily. Eventually, an underground drainage system develops. The bedrock is often hundreds of feet thick, extending from near the earth's surface to below the water table. Solution caves often develop in karst regions. Caves develop by an extensive enlargement and **erosion** of the underground drainage structure into a system of connecting passageways.

There are many variations of karst landscape, often described in terms of a particular **landform**. The predominant landforms are called fluviokarst, doline karst, cone and tower karst, and pavement karst. Some karst regions were etched during the Ice Age and may appear barren and very weathered (pavement karst). Other karst areas appear as dry valleys for part of the year and after seasonal floods, as a **lake** (one example of fluviokarst). In tropical areas, karst regions can be covered with **forests** or other thick vegetation. Sometimes, the under-

Tower karst topography like this along the Li River in southern China is formed when karst sinkholes deepen faster than they widen. *JLM Visuals. Reproduced by permission.*

ground drainage structure collapses, leaving odd formations such as natural **bridges** and sinkholes (doline karst). Tall, jagged limestone peaks are another variation (cone or tower karst).

Karyotype and karyotype analysis

A karyotype is a technique that allows geneticists to visualize chromosomes under a **microscope**. The chromosomes can be seen using proper extraction and staining techniques when the chromosomes are in the metaphase portion of the **cell** cycle. Detecting **chromosomal abnormalities** is important for prenatal **diagnosis**, detection of carrier status for certain genetic diseases or traits, and for general diagnostic purposes.

Karyotype analysis can be performed on virtually any population of rapidly dividing cells either grown in **tissue** culture or extracted from tumors. Chromosomes derived from peripheral **blood** lymphocytes are ideal be-

cause they can be analyzed three days after they are cultured. Lymphocytes can be induced to proliferate using a mitogen (a drug that induces **mitosis**) like phytohemagglutinin. Skin fibroblasts, bone marrow cells, chorionic villus cells, **tumor** cells, or amniocytes also can be used but require up to two weeks to obtain a sufficient amount of cells for analysis. The cultured cells are treated with colcemid, a drug that disrupts the mitotic spindle apparatus to prevent the completion of mitosis and arrests the cells in metaphase. The harvested cells are treated briefly with a hypotonic **solution**. This causes the nuclei to swell making it easier for technicians to identify each **chromosome**. The cells are fixed, dropped on a microscope slide, dried, and stained. The most common stain used is the Giemsa stain. Other dyes, such as fluorescent dyes, can also be used to produce banding patterns.

Chromosome spreads can be photographed, cut out, and assigned into the appropriate chromosome number or they can be digitally imaged using a computer. There are seven groups (A-G) that autosomal chromosomes are divided into based on size and position of the centromere. The standard nomenclature for describing a karyotype is based on the International System for

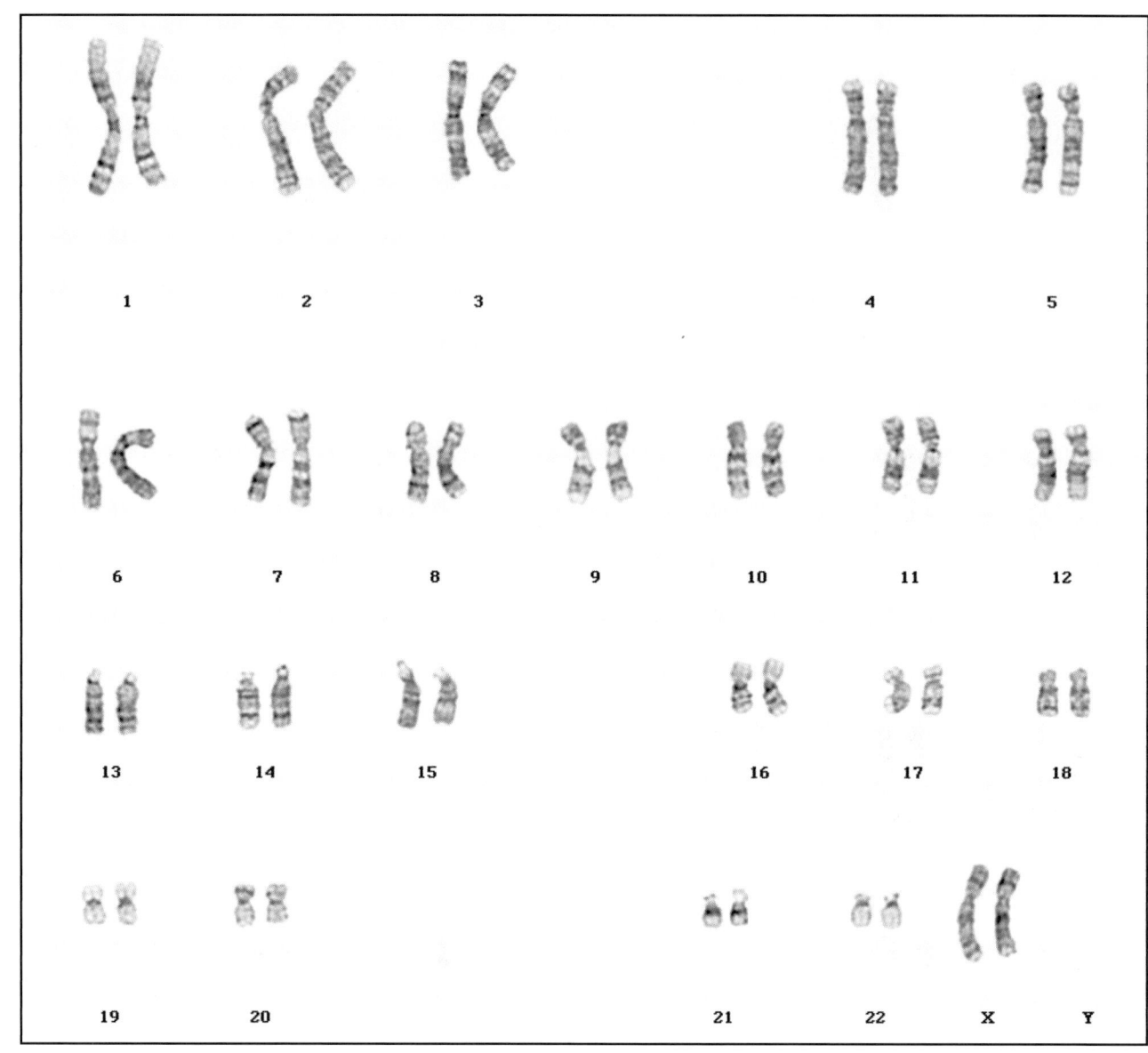

Karyotype of a normal human female. *Courtesy of Dr. Constance Stein.*

Human Cytogenetic Nomenclature (ISCN). First, the total number of chromosomes are written followed by a comma, then the sex chromosome constitution and any abnormality written in parentheses. Many genetic abnormalities cannot be detected by karyotype analysis. These include small, esoteric aberrations such as point mutations, frameshift mutations, nonsense mutations, or single nucleotide polymorphism's.

Genetic counselors rely on karyotypes to diagnose abnormal pregnancies. **Amniocentesis** is a routine procedure used in prenatal screening that involves removing amniotic fluid for karyotype analysis. A karyotype can pick up aneuploidy (i.e., Trisomy 21 or **Down syn-**

drome) and rearrangements such as deletions, duplications, and inversions that might be helpful in prenatal diagnosis. It also can be helpful in certain cases to obtain karyotypes from parents to determine carrier status, which can be relevant to recurrence risks in future pregnancies. Karyotypes also may help determine the cause of **infertility** in patients having reproductive difficulties.

Many sports organizations, including the Olympics, use karyotype analysis for "gender verification" purposes in order to prevent male athletes from competing in female sports events. To prevent an unfair competitive advantage by male imposters, the International Olympic Committee (IOC) in 1968 required that all female athletes

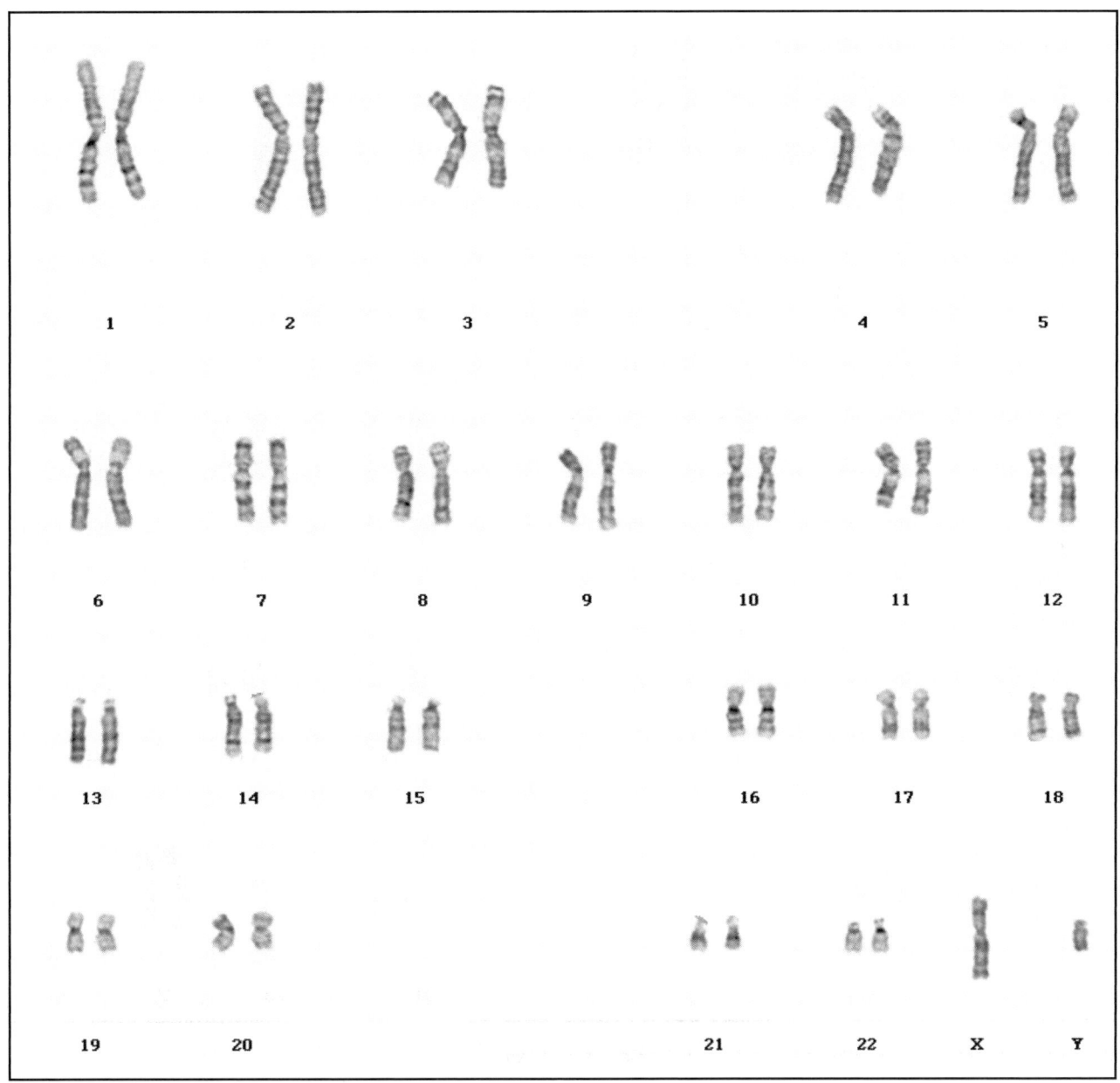

Karyotype of a normal human male. *Courtesy of Dr. Constance Stein.*

undergo a controversial gender verification testing using buccal smears (cheek cells) to karyotype individuals. Unexpectedly, athletes that had genetic abnormalities were detected. Some of these individuals had part or all of a Y chromosome and appear phenotypically to be female due to different genetic conditions that results in ambiguous external secondary sex characteristics or degenerate internal genitalia. As a result of sex testing, many of these individuals suffered from public disgrace and humiliation, loss of titles and scholarships, and were banned from future competitive events. It was not until June 1999, over 30 years later, that the IOC Athletes' Commission discon-

tinued gender verification on a trial basis. The proposal, similar to the International Amateur Athletic Federations plan adopted in 1992, allows for such testing by the appropriate medical personnel only if there is a question of gender identity. Most competitive sports organizations now require only individuals suspected of being male imposters to undergo sex testing. The IOC allows genetically abnormal individuals to compete only after confirmation of testing by medical professionals and the appropriate counseling has been completed. In 2002, the IOC suspended gender verification procedures for the Olympics in Sydney, **Australia** citing potential harm to "women athletes born with relatively rare genetic abnormalities that affect development of the gonads or the expression of secondary sexual characteristics."

See also Chromosome mapping; Genetic disorders; Genetic testing; Genetics; Genotype and phenotype; Hermaphrodite.

Kelp forests

Kelp

Brown **algae**, also known as kelps, are a group of seaweeds in the order Phaeophyta. They attach to **rocks** on the sea bottom by a **tissue** known as their holdfast, from which their flexible stems (known as a stipe) and leaf-like tissue (or fronds) grow into the **water** column. In some **species** the fronds are kept buoyant by gas-filled bladders. Kelp tissues are extremely tough only the strongest storms are capable of tearing their fronds or ripping their holdfasts from the rocky bottom. When this happens, however, large masses of kelp **biomass** can float around as debris known as "paddies," or wash onto the shore as "wrack."

Kelp forests

In some temperate marine habitats large species of brown algae can be extremely abundant. These ecosystems are known as kelp **forests**. Because they are extremely productive ecosystems, and have a great deal of physical structure associated with their seaweed biomass, kelp forests provide **habitat** for a wide range of marine organisms. These include a diversity of species of smaller algae, **invertebrates**, **fish**, marine **mammals**, and **birds**. The kelp forests of the Pacific coast of **North America** are estimated to support more than 1,000 species of marine plants and animals.

Kelp forests occur in many parts of the world, including the Atlantic and Pacific coasts of North America. However, the tallest, best-developed kelp forests are in waters 20-210 ft (6-70 m) deep over rocky bottoms off the coast of California. This **ecosystem** is dominated by the giant kelp (*Macrocystis pyrifera*), which ranges from central California to Baja California (the genus also occurs on the west coast of **South America**, and off South **Africa**, southern **Australia**, and New Zealand). This enormous seaweed is also known as the giant bladder kelp because of the flotation structures attached to its fronds. The giant kelp begins its life as a microscopic **spore**, but can grow as immensely long as 200 ft (60 m)

and live for 4-7 years. Most of its photosynthetic activity occurs in the upper part of its tall canopy, because the lower areas are intensely shaded and do not receive much sunlight.

Other, somewhat smaller species of *Macrocystis* occur more widely along the Pacific coast, as far north as southern Alaska. Other giant seaweeds of kelp forests of the Pacific coast include the bull kelp (*Nereocystis leutkeana*), the elk horn kelp (*Pelagophycus porra*), the feather boa kelp (*Egregia menziesii*), and the Fucalean alga (*Cystoseira osmundacea*).

Sea urchins, sea otters, and kelp

Sea urchins are marine invertebrates that feed voraciously on kelp biomass (they are herbivores meaning that plants are their primary source of food). Periodically, sea urchins of the genus *Strongylocentrotus* can become extremely abundant and cause an intense disturbance to the kelp-forest ecosystem. They do this by feeding on the holdfasts and causing the kelp to detach from their rocky anchors, resulting in an ecosystem known as an "urchin barren" because it sustains so little biomass of seaweeds or other species. This sort of natural ecological damage has been observed numerous times, in various parts of the world.

Off the coast of western North America, however, sea **otters** (*Enhydra lutris*) feed on the urchins and can prevent them from becoming too abundant, thereby keeping the kelp forests intact. This ecological balance among sea urchins, sea otters, and kelps became upset during the nineteenth century, when the populations of the otters were virtually wiped out by excessive hunting for the fur trade. Because of the collapse of otter populations, the urchins became more abundant. Their excessive feeding on kelps greatly reduced the extent and luxuriance of the kelp forests. Fortunately, this balance has since been restored by the cessation of the hunting of sea otters, allowing them to again control the abundance of the urchins. In turn, the productive kelp forests have been able to redevelop.

Kelp forests as an economic resource

Seaweed biomass contains a number of useful chemicals, such as alginates used as thickeners and gelling agents in a wide variety of manufactured products. A more minor use is as a food supplement. In some regions kelps are being harvested as an economic resource to supply these industrial chemicals. Off the coast of California, for example, kelp harvesting amounts to as much as 176,000 tons (160,000 metric tons) per year. If the harvesting method takes care to not damage the holdfasts and other deeper tissues of the kelps, then the forest can regenerate quite well from the disturbance. In California, for example, kelp harvesters are only allowed to cut in the top 4 ft (1.4 m) of the water column, leaving the deeper parts of the forest intact. The kelp harvesting is done using a large barge-like apparatus, which can collect as much as 605 tons (550 metric tons) of kelp per day.

Kelp forests also have an extremely large indirect value to the economy, by serving as the **critical habitat** for many species of fish and shellfish that are harvested in the coastal fishery. The forests are also critical habitat for many species of indigenous **biodiversity**. This has an indirect benefit to the coastal economy, through recreational activities associated with **ecotourism**.

Resources

Books

Connor, J., and C. Baxter. *Kelp Forests.* Monterey Bay Aquarium Foundation, 1980.
Foster, M.S., and D.R. Scheil. *The Ecology of Giant Kelp Forests in California: A Community Profile.* U.S. Fish and Wildlife Service, Report 85 (7.2), 1985.
Lobban C.S., and P.J. Harrison. *Seaweed Ecology and Physiology.* Cambridge University Press, 1996.
McPeak, R. *Amber Forest: The Beauty and Biology of California's Submarine Forest.* Watersport Publishers, 1988.

Bill Freedman

Kelvin *see* **Temperature**

Kepler's laws

Johannes Kepler made it his life's work to create a heliocentric (sun-centered) model of the **solar system** which would accurately represent the observed **motion** in the sky of the **Moon** and planets over many centuries. Models using many geometric curves and surfaces to define planetary orbits, including one with the orbits of the six known planets fitted inside the five perfect solids of Pythagoras, failed.

Kepler was able to construct a successful model with the **earth** the third **planet** out from the **Sun** after more than a decade of this trial and **error**. His model is defined by the three laws named for him. He published the first two laws in 1609 and the last in 1619. They are:

1. The orbits of the planets are ellipses with the Sun at one focus (F_1) of the **ellipse**.

2. The line joining the Sun and a planet sweeps out equal areas in the planet's **orbit** in equal intervals of time.

3. The squares of the periods of revolution "P" (the periods of time needed to move 360°) around the Sun for the planets are proportional to the cubes of their **mean** distances from the Sun. This law is sometimes called Kepler's Harmonic Law. For two planets, planet A and planet B, this law can be written in the form:

$$\frac{P_A^2}{P_B^2} = \frac{a_A^3}{a_B^3}$$

A planet's mean distance from the Sun (a) equals the length of the semi-major axis of its orbit around the Sun.

Kepler's three laws of planetary motion enabled him and other astronomers to successfully match centuries-old observations of planetary positions to his heliocentric solar system model and to accurately predict future planetary positions. Heliocentric and geocentric (Earth-centered) solar system models which used combinations of off-center circles and epicycles to model planetary orbits could not do this for time intervals longer than a few years; discrepancies always arose between predicted and observed planetary positions.

Newton's generalization of Kepler's laws

The fact remained, however, that, in spite of Kepler's successful modeling of the solar system with his three laws of planetary motion, he had discovered them by trial and error without any basis in physical law. More than 60 years after Kepler published his third law, Isaac Newton published his *Principia*, in which he developed his three **laws of motion** and his theory and law of universal gravitation. By using these laws, Newton was able to derive each of Kepler's laws in a more general form than Kepler had stated them, and, moreover, they were now based on physical theory. Kepler's laws were derived by Newton from the basis of the two-body problem of **celestial mechanics**. They are:

1. The orbits of two bodies around their center of mass (barycenter) are **conic sections** (circles, ellipses, parabolas, or hyperbolae) with the center of mass at one focus of each conic section orbit. Parabolas and hyperbolas are open-ended orbits, and the period of revolution (P) is undefined for them. One may consider a circular orbit to be a special case of the ellipse where the two foci of the ellipse, F_1 and F_2, coincide with the ellipse's center (C), and the ellipse becomes a **circle** of radius (a).

2. The line joining the bodies sweeps out equal areas in their orbits in equal intervals of time. Newton showed that this generalized law is a consequence of the **conservation** of angular **momentum** (from Newton's third law of motion) of an isolated system of two bodies unperturbed by other forces.

3. From his law of universal gravitation, which states that two bodies of masses, M_1 and M_2, whose centers are separated by the distance "r" experience equal and opposite attractive gravitational forces (F_g) with the magnitude

$$F_g = \frac{GM_1M_2}{r^2}$$

Where G is the Newtonian gravitational factor, and from his Second Law of Motion, Newton derived the following generalized form of Kepler's third law for two bodies moving in elliptical orbits around their center of mass where π is the **ratio** of the circumference of a circle to its diameter, "a" is the semi-major axis of the *relative* orbit of the body of smaller mass, M_2, around the center of the more massive body of M_1.

$$P^2 = \left[\frac{4\pi^2}{G(M_1 + M_2)} \right] a^3$$

Some of the applications of these generalized Kepler's laws are briefly discussed below.

Applications of the generalized forms of Kepler's laws

Let us first consider applications of Kepler's third law in the solar system. If we let M_1 represent the Sun's mass and M_2 represent the mass of a planet or another object orbited the Sun, then if we adopt the Sun's mass ($M_1 = 1.985 \times 10^{30}$ kg) as our unit of mass, the **astronomical unit** (a.u.; 1 a.u. = 149,597,871 km) as our unit of length, and the sidereal year (365.25636 mean solar days) as our unit of time, then $(4\pi 2/G) = 1$, $(M_1 + M_2) = 1$ (we can neglect planet masses M_2 except those of the Jovian planets in the most precise calculations), and the formidable equation above is reduced to the simple algebraic equation $P^2 = a^3$ where "P" is in sidereal years and "a" is in astronomical units for a planet, asteroid, or comet orbiting the Sun. Approximately the same equation can be found from the first equation if we let Earth be Planet B, since $F_B = 1$ sidereal year and a_B is always close to 1 a.u. for Earth.

Let us return to the generalized form of Kepler's third law and apply it to planetary satellites; except for Earth-Moon and Pluto-Charon systems (these are considered "double planets"), one may neglect the satellite's mass ($M_2 = 0$). Then, solving the equation for M_1,

$$M_1 = \left(\frac{4\pi^2}{G} \right) \left(\frac{a^3}{P^2} \right)$$

Measurements of a satellite's period of revolution (P) around a planet and of its mean distance "a" from the planet's center enable one to determine the planet's mass (M_1). This allowed accurate masses and mean densities to be found for **Mars**, **Jupiter**, **Saturn**, **Uranus**, and **Neptune**. The recent achievement of artificial satellites of **Venus** have enabled the mass and mean **density** of Venus to be accurately found. Also the total mass of the Pluto-Charon system has been determined.

Now we consider the use of Kepler's laws in stellar and galactic **astronomy**. The equation for Kepler's third law has allowed masses to be determined for double stars for which "P" and "a" have been determined. These are two of the orbital elements of a visual doublestar; they are determined from the doublestar's true orbit. Kepler's second law is used to select the true orbit from among the possible orbits that result from solutions for the true orbit using the doublestar's apparent orbit in the sky. The line joining the two stars must sweep out equal areas in the true and apparent orbits in equal time intervals (the time **rate** of the line's sweeping out area in the orbits must be constant). If the orbits of each **star** around their center of mass can be determined, then the masses of the individual stars can be determined from the sizes of these orbits. Such doublestars give us our only accurate information about the masses of stars other than the Sun, which is very important for our understanding of star structure and **evolution**.

In combination with data on the motions of the Sun, other stars, and interstellar gases, the equation for Kepler's third law gives estimates of the total mass in our **Milky Way galaxy** situated closer to its center than the stars and gas studied. If total mass ($M_1 + M_s$) is constant, the equation predicts that the orbital speeds of bodies decrease with increasing distance from the central mass; this is observed for planets in the solar system and planetary satellites. The recently discovered fact that the orbital speeds of stars and gas further from the center of the Milky Way than the Sun are about the same as the Sun's orbital speed and do not decrease with distance from the center indicates much of the Milky Way's mass is situated further from the center than the Sun and has led to a large upward revision of the Milky Way's total estimated mass. Similar estimates of the mass distributions and total masses of other galaxies can be made. The results allow estimates of the masses of clusters of galaxies; from this, estimates are made of the total mass and mean density of detectable

matter in the observable part of our universe, which is important for cosmological studies.

When two bodies approach on a parabolic or hyperbolic orbit, if they do not collide at their closest distance (pericenter), they will then recede from each other indefinitely. For parabolic orbit, the relative **velocity** of the two bodies at an infinite distance apart (**infinity**) will be **zero**, and for a hyperbolic orbit their relative velocity will be positive at infinity (they will recede from each other forever).

The parabolic orbit is important in that a body of mass M_2 that is insignificant compared to the primary mass, M_1 ($M_2=0$) that moves along a parabolic orbit has just enough velocity to reach infinity; there it would have zero velocity relative to M_1. This velocity of a body on a parabolic orbit is sometimes called the parabolic velocity; more often it is called the "escape velocity." A body with less than escape velocity will move in an elliptical orbit around M_1; in the solar system a spacecraft has to reach velocity to orbit the Sun in interplanetary **space**. Some escape velocities from the surfaces of solar system bodies (ignoring atmospheric drag) are 2.4 km/sec for the Moon, 5 km/sec for Mars, 11.2 km/sec for Earth, 60 km/sec for the cloud layer of Jupiter. The escape velocity from Earth's orbit into interstellar space is 42 km/sec. The escape velocity from the Sun's photosphere is 617 km/sec, and the escape velocity from the photosphere of a **white dwarf** star with the same mass as the Sun and a photospheric radius equal to Earth's radius is 6,450 km/sec.

The last escape velocity is 0.0215 the **vacuum** velocity of **light**, 299,792.5 km/second, which is one of the most important physical constants and, according to the Theory of Relativity, is an upper **limit** to velocities in our part of the universe. This leads to the concept of a **black hole**, which may be defined as a **volume** of space where the escape velocity exceeds the vacuum velocity of light. A black hole is bounded by its Schwartzchild radius, inside which the extremely strong **force** of gravity prevents everything, including light, from escaping to the universe outside. Light and material bodies can fall into a black hole, but nothing can escape from it, and theory indicates that all we can learn about a black hole inside its Schwarzschild radius is its mass, net electrical charge, and its angular momentum. The Schwartzschild radii for the masses of the Sun and Earth are 2.95 km and 0.89, respectively. Black holes and observational searches for them have recently become very important in **astrophysics** and **cosmology**.

Hyperbolic orbits have become more important since 1959, when space technology had developed enough so that spacecraft could be flown past the Moon.

KEY TERMS

Conic section—A conic section is a figure that results from the intersection of a right circular cone with a plane. The conic sections are the circle, ellipse, parabola, and hyperbola.

Double star—A gravitationally bound system of two stars which revolve around their center of mass in elliptical orbits.

Jovian planets—Jupiter, Saturn, Uranus, and Neptune. They are the largest and most massive planets.

Mass—The total amount of matter (sum of atoms) in a material body.

Mean density—The mass of a body divided by its volume.

Volume—The amount of space that a material body occupies.

White dwarf—A star that has used up all of its thermonuclear energy sources and has collapsed gravitationally to the equilibrium against further collapse that is maintained by a degenerate electron gas.

Spacecraft follow hyperbolic orbits during flybys of the Moon, the planets, and of their satellites.

See also Satellite.

Resources

Books

Beer, A., ed. *Vistas in Astronomy: Kepler.* Vol. 18. London: Pergamon Press, 1975.

Morrison, David, and Sidney C. Wolff *Frontiers of Astronomy.* Philadelphia: Saunders College Publishing, 1990.

Motz, Lloyd, and Anneta Duveen. *Essentials of Astronomy.* Belmont, CA: Wadsworth, 1975.

Frederick R. West

Ketone *see* **Acetone**

Keystone species

A keystone species is a particular species that has a great influence on the structure or functioning of its ecological community. This influence is far out of proportion to the relative **biomass** or productivity of the keystone species within its community. Most keystone species are top predators, although a few are influential because they play a critical role as herbivores or in nutrient cycling.

In **engineering**, the keystone is a wedge-shaped stone that occurs at the top of a self-supporting stone arch or dome. The keystone is extremely important in the physical integrity of the structure, and if it is removed, the arch or dome will collapse. The keystone metaphor in **ecology** is used to refer to **species** that have a similarly critical influence on the functioning or structure of their community.

The importance of particular keystone species can often be deduced by careful examination of their interactions with other species or by measuring their functional attributes, especially those important in nutrient cycles. The role of keystone species can also be demonstrated by conducting experiments in which these organisms are removed from their community, and the resulting ecological changes are monitored.

Keystone predators and herbivores

The first use of the keystone-species metaphor in ecological literature was in reference to certain temperate intertidal communities on the west coast of **North America**. In this **ecosystem**, experimental removal of a predacious **starfish** (*Pisaster ochraceous*) was found to result in a rapid increase in the growth and biomass of a filter-feeding mussel (*Mytilus californianus*), which then managed to crowd out other species and strongly dominate the community. In this case, the starfish was described as a keystone **predator** that prevented the mussel from achieving the full degree of community dominance that it was capable of developing on the basis of its competitive superiority over other species. Interestingly, the starfish could not eliminate the mussel from the community because it was not able to predate upon the largest mussels. Therefore, predation on mussels by the starfish allowed other species to occur in the intertidal zone, so the community could maintain a greater richness of species and was more complex in structure because the development of a **monoculture** of mussels was prevented.

Another case of a predator having a crucial influence on the structure of its community involves the sea otter (*Enhydra lutris*) of western North America. These marine **mammals** mostly feed on **sea urchins**, which are herbivores of the large **algae** known as kelps. By keeping urchin densities relatively small, the seaotters allow the kelps to maintain a large biomass, and the community develops into a so-called "kelp forest." In the absence of the seaotters, the urchins are capable of developing populations large enough to overgraze the kelps. The ecosystem would then maintain a much smaller biomass and productivity of these seaweeds and would become much

more open in structure. This capability of the urchins has been demonstrated in experiments in which these herbivores were removed by ecologists with the result that kelps flourished. The role of the **otters** was demonstrated indirectly through the ecological changes associated with the widespread extirpation of these animals from almost all of their range as a result of overharvesting for their rich, lustrous fur during the eighteenth and nineteenth centuries. With the otters gone, the **kelp forests** declined badly in many places. Fortunately, seaotters have been colonizing many of their former habitats since about the 1930s, and this recovery has led to a return of the kelp forests in many of those places.

Another example of a keystone species is the African **elephant** (*Loxodonta africana*), an herbivorous species that eats a wide range of herbaceous and woody plants. During its feeding on the foliage of shrubs and trees, elephants commonly knock these woody plants over, which often kills the plants. By feeding in a manner that is destructive to shrubs and trees, elephants shift the balance of the **savanna** ecosystem toward a greater dominance of herbaceous species. This keeps the **habitat** in a relatively open condition. This ecological change is not, however, necessarily to the benefit of the elephants because they require a mixture of herbaceous and woody plants for a balanced **nutrition**.

The beaver (*Castor canadensis*) also has an enormous influence on the structure of its habitat. **Beavers** create extensive **wetlands** by damming streams, causing them to flood low-lying areas. By doing so, beavers create fertile open-water wetlands for their own use as well as for many other species that otherwise might not be able to utilize the local habitats.

Keystone species in nutrient cycling

Some keystone species are important because they play a crucial role in nutrient cycling, particularly if that function cannot be carried out by other species. Good examples of these sorts of keystone species are those that play unique roles in the **nitrogen cycle**, in particular in the ecologically important process known as **nitrification**.

Nitrification is a process during which highly specialized **bacteria** oxidize the positively charged ion ammonium (NH_{4+}) to the negatively charged ion nitrite (NO_2-) and then to nitrate (NO_3-). Nitrification is a very important component of the larger **nitrogen** cycle because most plants prefer to utilize nitrate as the chemical form by which nitrogen, an essential nutrient, is absorbed from **soil** or **water**. Because this preference for nitrate is true of most agricultural species of plants, nitrification is also an ecological function that is important for human welfare.

KEY TERMS

. . . .

Community—In ecology, a community is an assemblage of populations of different species that occur together in the same place and at the same time.

Competition—An interaction between organisms of the same or different species associated with their need for a shared resource that is present in a supply that is smaller than the potential, biological demand.

Keystone species—A species that plays a crucial role in the functioning of its ecosystem, or that has a disproportionate influence on the structure of its ecosystem.

Nitrification occurs in two discrete steps. The first stage is the oxidation of ammonium to nitrite, a process that is only carried out by specialized bacteria in the genus *Nitrosomonas*. The nitrite formed is then quickly oxidized to nitrate by other specialized bacteria in the genus *Nitrobacter*. Neither *Nitrosomonas* or *Nitrobacter* are abundant in soils or water. These **microorganisms** can, however, be viewed as keystone organisms because nitrification is such an important aspect of the nitrogen cycle in ecosystems, and it is only carried out by these bacteria.

Resources

Books

Krebs, C.J. *Ecology. The Experimental Analysis of Distribution and Abundance.* Harper and Row, New York: 1985.

Bill Freedman

Killifish

Killifish (*Fundulus* spp.) are small **fish** tolerant of a wide range of temperatures and salinity, found throughout temperate and tropical waters on every **continent** except **Australia** and **Antarctica**. Not to be confused with the other large group of small fish in the same order (Cyprinodontiformes) known as **minnows**, killifish differ in having an incomplete lateral line, often not extending past the head, and a protruding lower jaw which allows them to feed from the surface. This characteristic gave rise to their other common name "top-minnow." Mummichogs (*Fundulus heteroclitus*) are the most well known **species** of killifish. Millions of these common in-

tertidal fish have been removed from the East coast of **North America** since the 1800s for use in a wide variety of experiments, including a recent trip into outer **space** to determine the effect of weightlessness on fish development and locomotion.

Evolution and taxonomy

The earliest fossils of the order Cyprinodontiformes (meaning toothed **carp**) were found in **Europe** and date from the early Oligocene (26-37 mya). The most likely ancestor of this order is traced to fossils from the Tethys Sea in the Early Cretaceous (65-136 mya) time period. More recent Miocene (12-26 mya) fossils have been located in Kenya. Over time, the number of species diversified as the continents drifted apart, giving rise to some of the most resilient and tolerant species of fish known on **Earth**.

Because of this incredible diversification, the class Osteichthyes (bony fishes) has been broken down into several sub-classes, including Teleostei, which includes the orders for minnows, killifish, silversides, and others. The **taxonomy** of the killifish is quite complex due to the 900 or more different species in the order Cyprinodontiformes and the high degree of endemism resulting from isolated populations. Like other Cyprinodontes, killifish have one dorsal fin lacking spines. The killifish belongs to the sub-order Cyprinodontoidei, which includes such interesting families as the cavefishes (Amblyopsidae) of North America; the live-bearing top minnows, including familiar aquarium fish such as guppies, swordtails, and mollies (Poeciliidae); and the four-eye fishes that can see simultaneously above and below **water** (Anablepidae).

The killifish family is known as either Cyprinodontidae or Fundulidae, depending on the taxonomic source. Including over 200 species, the genus Fundulus has been further divided into five sub-genera classified according to anatomical or geographical characteristics. Among the 34 species in North America, there is a distinct separation of species from East to West, and genetically within species from North to South along the Atlantic coast, as seen in the mummichog (*F. heteroclitus*). This is probably a result of the combined influences of glaciation and disjunct **habitat** distribution.

The East Coast species are most numerous in Florida and the Gulf of Mexico, but do include some species from more northern **freshwater** and **brackish** regions extending as far north as Newfoundland and the Gulf of St. Lawrence. There are several species found exclusively in the Midwest and West. California has six native species found primarily in the **desert**, and one **introduced species**, the rainwater killifish (*Lucania parva*), native to the Atlantic and Gulf of Mexico. Due to the geographic limitations of their habitat and its further disappearance in the wake of development and pumping of **groundwater** reserves, some of these California species known as pupfish have the most restricted ranges of any known vertebrate. The Devil's Hole pupfish (*Cyprinodon diabolis*) is limited to a 215 sq ft (20 sq m) area of an underwater limestone shelf in a Nevada spring. For at least 10,000 years, a small population (200-700 individuals) has perpetrated itself in this tiny **niche**.

Ecology

Characteristic of euryhaline and freshwater habitats, extremes of temperatures, variations in dissolved **oxygen** and high degree of habitat disturbance are the main parameters shaping the lives of killifishes. The ability to move from fresh to **saltwater** requires tremendous osmoregulation adjustments. The sheepshead minnow (*Cyprinodon variegatus*) is found in marsh ponds from Maine to the West Indies, and can tolerate a wider range of salinity than any other fish. Killifish kidneys have specially adapted renal tubules to process **sodium chloride** salts as **concentration** increases from freshwater to saltwater. Special chloride cells located in the opercular epithelium of the gills help facilitate this effort for the mummichog (*F. heteroclitus*). Metabolic regulation of intercellular activity provides for greater hemoglobin-oxygen binding efficiency in mummichogs, allowing them to respond quickly to changes in dissolved oxygen levels, regulate **pH**, and thus effectively deliver oxygen to muscle tissues. This model may be the same for all killifishes, although a genetic component is probably involved, explaining the variety of tolerance levels limiting distribution of certain species.

The pupfishes of the California deserts show the most amazing **temperature** tolerance, commonly withstanding months of summer water temperatures between 35–40°C (95–104°F), with extremes of 47°C (116.6°F) not uncommon. Very few **vertebrates** can survive at this thermal level. Their small size (60-75 mm, 2-3 in), while providing a distinct advantage for survival in shallow, densely populated areas with limited food resources, would seem to present a major thermoregulatory challenge. Desert nights can experience temperatures below freezing as well, at which times the pupfish burrows into detritus on the bottom and remains dormant until temperatures rise again. Since all pupfish species are considered rare or endangered, further study of this remarkable ability awaits better captive rearing techniques.

Another important **adaptation** necessary for survival in **salt** marsh habitats is the ability to tolerate high levels of dissolved **hydrogen** sulfide. This gas is released as a by-product of **anaerobic** marsh **decomposition** and can achieve potentially toxic levels. Experiments with the California killifish (*F. parvipinnis*) indicate that metabolic tolerance is achieved by response of cellular mitochondrial oxidation, allowing a tolerance two to three times greater than that of most fishes.

Biology

In general, female killifish are larger than the males (ranging from 50-155 mm, 2-6 in), although less brightly colored. As the days lengthen and temperatures rise during the spring/summer reproductive season, **hormones** cause bright colors to appear on males. These aggressive males fight with each other for the privilege of fertilizing a female, preferably a large one capable of laying up to 50,000 eggs per season. Deposition of eggs is closely tied to the tidal cycle for intertidal fishes like the mummichog. Preference is for placing the sticky mass of eggs above the strand line on stalks of marsh grass (*Spartina alterniflora*) or in empty mussel shells (*Geulensia demissa*) during a spring tide, allowing them to develop in air until the next spring tide washes them back into the water, stimulating them to hatch. Thus the eggs themselves are tolerant of a wide salinity range.

Young fish fry emerge from the eggs nine to 18 days after laying and are considered more developmentally advanced than most other fish species. Mortality among the larva and fry is quite high, despite their tendency to take refuge in the shallow intertidal areas beyond reach of the larger fishes and **crabs**. Given warm temperatures and adequate food resources, size increases throughout the season. While most species achieve reproductive maturity after the first season, few seem to live for more than two to three years. During the winter, many species take refuge in salt marsh pools which provide slightly warmer temperatures than intertidal creeks.

Killifish are generally carnivorous, but a few species also consume **algae** and other marine plants. Small **invertebrates**, especially brine **shrimp**, insect larvae, worms, and various **zooplankton** form the bulk of their diet. The killifishes in turn become food to many larger fish (such as flounders, **perch**, eels, and **bass**), and along the upper edges of the water, feed foraging **birds** and land crabs. Protection from predation varies with species, but includes cryptic coloring, small, elongated bodies to take advantage of hiding places and identification of potential danger by recognizing **color** contrasts from either above or below. The upturned placement of the mouth indicates their preference for swimming just under the surface and skimming for **prey**. **Competition** for a limited food resource between species of killifish may explain the distribution pattern that limits their co-occurrence in any given habitat.

Unlike the majority of killifishes, the small (30-75 mm, 1.17-3 in) pupfish have a more rounded shape and have omnivorous feeding habits. Its mouth is located on the terminus of the body. They also differ in their breeding strategy. Same age class schools forage together until the spring breeding season (temperatures above 20°C [68°F] from April-October), at which time the males establish territories, leaving females and juveniles in the group. When a female is ready to spawn, she leaves the school and ventures into a territory. Here she indicates her intention by biting a piece of the bottom and spitting it out. The male joins her and she begins laying a single egg at a time. Over a season, she may deposit from 50-300 eggs, depending on her size. High temperatures can speed up the life cycle, making it possible for up to 10 generations a year to occur in hot springs.

Economic importance

Many species of *Fundulus* have been used extensively as bait fish, but perhaps their best known use is in the laboratory. Beginning in the 1800s, the mummichog (*F. heteroclitus*) was used to study fish **embryology**. The transparent eggs were stripped of their protective covering, opening a window on the developmental process and allowing easy manipulation with a variety of chemicals to assess endocrinological and biological responses. Capitalizing on the ability of these fish to withstand a wide range of temperatures and salinities, and its natural distribution near **pollution** sources, the mummichog also became the main study **animal** for intertidal and nearshore pollution tolerance.

Other experiments have used killifish and mummichogs to test **learning** in fish (they are able to navigate a simple maze), examine pigmentation and the function of chromatophores in response to different physical and chemical stimuli, chemical responses allowing survival at sub-zero temperatures, and most recently in determining population stability in response to environmental stresses.

Other recent efforts have focused on two different extremes: 1) use of the hermaphroditic killifish (*Rivulus marmoratus*) from Florida as well as other killifish species as a biocontrol agent for reducing populations of mosquito larvae and eggs; and 2) as part of the reestablishment of endangered populations of the Desert pupfish (*Cyprinodon sp.*) species in the California deserts. Both sets of experiments have important impacts on **land use** planning and local environmental stability.

While few of the many species of killifish are considered rare or endangered, it was not until the 1960s that a technique for captive rearing was refined and the removal of thousands of fish from the wild discontinued.

KEY TERMS

Anaerobic—Describes biological processes that take place in the absence of oxygen.

Brackish—Water containing some salt, but not as salty as sea water.

Chromatophores—Components of cells responsible for allowing color changes.

Endemic—Refers to species with a relatively local distribution, sometimes occurring as small populations confined to a single place, such as an oceanic island. Endemic species are more vulnerable to extinction than are more widespread species.

Endocrinological—Refers to the function of hormones in regulating body processes.

Euryhaline—Waters that change salinity in response to tides or fresh water influx.

Lateral line—A row of sensors found along the sides of fish.

Mitochondria—An organ inside cells that releases energy.

Operculum—The protective covering over the gills.

Osmoregulation—The process regulating the water content of cells in relation to that of the surrounding environment.

Salinity—The amount of dissolved salts in water.

Even today, sampling of wild populations is still conducted fairly regularly in order to further understanding of their remarkable tolerance ranges.

Ecological importance

Maintaining species diversity in the nearshore, intertidal, and freshwater systems inhabited by the many kinds of killifish has become increasingly important as the twentieth century draws to a close. Because of their incredible ability to withstand a wide range of temperatures, salinities, and pollutants, especially organochloride **pesticides** and **fertilizers**, killifish seem to be among the most persistent species. In 1990, five species of killifish were the only fish remaining in Mullet Pond, North Carolina, which had 27 fish species present in 1903 and 1914. In the 1960s, the entire remaining population of the Owen's Valley pupfish (*Cyprinodon radiosus*) was transferred in a few buckets by Fish and Wildlife agents, as their native pool drained away. Perhaps because of their widespread distribution and ability to adapt to whatever conditions exist, the killifish are among the most important fish species for monitoring the future of our environment.

Rosi Dagit

Kilogram *see* **Metric system**

Kinetic energy *see* **Energy**

Kingfishers

There are 87 **species** of kingfishers (family *Alcedinidae*) which are brightly-colored **birds** ranging in size from the 4 in (11 cm) long malachite-crested kingfisher, to the laughing kookaburra of **Australia**, which is 18 in (46 cm) long, weighing 2 lb (0.5 kg).

Kingfishers have a stocky body, with a large head equipped with a large, stout, dagger-like bill for grasping their food of **fish** or other small animals. The three front toes of kingfishers are fused for at least half of their length, but the adaptive significance of this trait is not known.

All kingfishers nest in cavities, usually digging these in earthen banks, or in rotten trees. Kingfishers are monogamous and pair for life. Kingfishers generally hunt from perches, although many species will also hover briefly to find their **prey**. The aquatic kingfishers plunge head-first into the **water** in pursuit of their prey.

Most kingfishers occur in the vicinity of a wide range of aquatic habitats, both fresh and estuarine, where they typically feed on fish and **amphibians**. Other species live in essentially terrestrial habitats, including mangrove **forests**, upland tropical forests, and **savanna**. The relatively terrestrial species of kingfishers eat a wide variety of foods, ranging from **arthropods**, to amphibians, **reptiles**, and small **mammals**. The prey is usually killed by repeatedly battering it against a branch or other hard substrate, and it is then eaten whole. One species, the shoe-billed kingfisher (*Clytoceyx rex*) of tropical forests of New Guinea, is a terrestrial bird that is specialized for digging earthworms, and has evolved a flat, stout, shovel-like bill.

Kingfishers typically occur in tropical and sub-tropical habitats, with only a few species nesting in the temperate zone. The greatest richness of species of kingfishers occurs in southeast **Asia**.

The most widespread species in **North America** is the belted kingfisher (*Megaceryle alcyon*), occurring over the entire **continent** south of the boreal forest. The belted kingfisher utilizes a wide range of aquatic habitats, ranging from estuaries to **freshwater** lakes, wet-

A common or small blue kingfisher (*Alcedo atthis*). *Photograph by E. Hanumantha Rao. Photo Researchers, Inc. Reproduced by permission.*

lands, and even large ditches. This species has a crest, a blue back, and a white breast with a blue horizontal stripe, and a familiar, rattling call that is often heard before the bird is seen. The female of this species is more brightly colored than the male, having a cinnamon stripe across her breast, a coloration that the male lacks. The belted kingfisher nests in chambers at the end of a 3-6.5 ft (1-2 m) long tunnel dug into an exposed, earthen bank, usually beside water. This species is frequently seen perching on overhead wires, posts, and **tree** branches in the vicinity of aquatic habitats. The belted kingfisher is a migratory species, wintering in the southern parts of its breeding range, or in Central America and the Caribbean. The green kingfisher (*Chloroceryle americana*) occurs only in south Texas and Arizona, and more widely in Mexico.

Most of the 10 Australian species of kingfishers are terrestrial, the laughing kookaburra (*Dacelo gigas*) being the best known species to most people. This is a large bird that makes its presence noisily known, and has garnered at least 25 common names in various parts of that country, most of which describe its raucous cries. The laughing kookaburra feeds largely on **snakes** and lizards, and some people feel that the species is beneficial for this reason. However, the kookaburra sometimes raids farmyards for young chickens and ducklings, and is then regarded as a minor pest.

Sometimes, particular kingfishers learn to feed at commercial trout farms or other sorts of aquaculture facilities, where these birds can become significant **pests**. However, the damage caused by kingfishers and other fish-eating birds can be easily dealt with by suspending netting over the aquaculture ponds.

Resources

Books

The Cambridge Encyclopedia of Ornithology. M. Brooke and T. Birkhead, eds. Cambridge University Press, 1991.

Forshaw, Joseph. *Encyclopedia of Birds.* New York: Academic Press, 1998.

Fry, C.H., K. Fry, and A. Harris. *Kingfishers, Bee-eaters, and Rollers: A Handbook.* London: Helm, 1992.

Bill Freedman

Kinglets

Kinglets are small forest **birds** in the subfamily Silviinae, family Muscicapidae, within the largest of the avian orders, Passeriformes, the perching birds. Kinglets are the most common North American representatives of their rather large subfamily, which is much more diverse in **Europe**, **Asia**, and **Australia**, and includes some 279 **species**.

There are two species of kinglets in **North America**, both of which are very active insectivorous feeders that breed in northern and montane conifer-dominated **forests**. The golden-crowned kinglet (*Regulus satrapa*) is a small bird, only about 3-4 in (9-10 cm) long, with an olive-green body and a white eye-stripe. The bird is a very active hunter of **insects**, spiders, and other small **arthropods**. Its song is a high-pitched series of notes followed by a chatter. The kinglet's nest is generally located within dense foliage high in the outer part of the crown of a **conifer** tree, and can contain as many as 10 eggs, which together may weigh more than the female bird. Both sexes have a bright yellow head-cap, but in male birds this feature is crowned by an orange-red cap, which is apparent only during a **courtship** display. Although insectivorous feeders, the golden-crowned kinglets can winter in boreal conifer forests as well as in temperate broadleaf forests, where, despite the cold, the kinglet successfully removes **bark**, twigs, and foliage in search of hibernating arthropods and their eggs. This species also winters as far south as parts of Central America. Wintering birds can often be seen in mixed flocks with other small species of birds, such as chickadees, **nuthatches**, creepers, and smaller **woodpeckers**.

The ruby-crowned kinglet (*Regulus calendula*) is also a common breeding bird in many coniferous forests of North America. This small, 3-4.3 in (9-11 cm) long bird also has an olive-green body and a distinctive white eye-ring, and is a very energetic gleaner of small arthropods, sometimes hunting at branch tips using brief fluttering flights. The song is somewhat of an ascending chatter, with elements often repeated in threes, and the final parts

are amazingly loud for such a tiny bird. The male ruby-crowned kinglet has a vermillion patch on the top of its head which is only apparent when the bird raises the crest during courtship display. This species winters farther to the south than the golden-crowned kinglet, mostly in southern North America and northern Central America.

Two very closely related species of Eurasia are the goldcrest (*Regulus regulus*) and the firecrest (*Regulus ingicapillus*), which also breed in northern or temperate conifer-dominated or mixed-wood forests.

Kiwis *see* **Flightless birds**

Koalas

The koala (*Phascolarctos cinereus*) is a tree-living Australian marsupial, or pouched mammal, which early English settlers in **Australia** called the native bear. The koala is not a bear, but is the only living **species** in the family Phascolarctidae, though fossils indicate that there were once were a number of species of koala. The name is derived from an Aboriginal word meaning "animal that does not drink," for koalas get their **water** from the leaves they chew. Koalas are found primarily in dry forest of eastern Australia, and their closest relatives are **wombats**. Koalas were once hunted for their fur, and millions were killed, rendering it an **endangered species**.

Koalas weigh up to 30 lb (13.6 kg), with males being considerably larger than females, and measuring from 23-33 in (60-85 cm) long. The thick fur on their round, compact body is primarily dark gray with white markings. The furry ears have a large white fringe and there is a white "bib" on the chest and white on the bottoms of the arms and legs. The nose is black and leathery and the eyes are black, giving the koala a button-eyed look typical of toy plush animals.

Unlike most climbing **marsupials**, the koala has only a small tail almost hidden in its fur. Koalas climb by means of their large hands and feet, which are equipped with long, strong, curved claws. The hand has three fingers and two thumbs, while on the hind foot, a big clawless thumb works separately from the fingers, and the first two toes are fused and longer than the others. Koalas use these two fused toes (sometimes called toilet claws) for combing their thick, woolly fur. Koalas do not have a sweet disposition and will use their claws against animals or people who molest them.

Koalas usually sit upright in a **tree**, as if they were perched on a chair. They leave their tree awkwardly, in order to go to a different one. Koalas depend on the trees

A koala (*Phascolarctos cinereus*). The animal has a highly specialized diet that consists of around 2.5 lb (1.1 kg) daily of about 12 species of eucalyptus leaves. *Photograph by Robert J. Huffman. Field Mark Publications. Reproduced by permission.*

in which they live because they have one of the most specialized diets of any mammal. They eat only the leaves of several dozen species of eucalyptus (or gum) trees, and then only at certain times of the year. They sniff each **leaf** carefully before consuming it. Scientists are not yet certain what chemicals in the leaves cue the koalas to accept them only some of the time and reject them otherwise. Apparently, certain oils become poisonous as the leaves mature. If their tree becomes unacceptable, a koala climbs down and goes in search of another tree with more appealing leaves. Koalas are found in the largest numbers in forest dominated by manna gum trees *(Eucalyptus viminalis).*

Most animals cannot digest tough eucalyptus leaves, but koalas have a long sac (called the caecum) at the point where the small intestine meets the large intestine. The caecum contains **bacteria** that help break down the

cellulose in the leaves, releasing organic acids and other useful chemicals.

Their specialized diet is the reason that koalas are so difficult to keep in captivity. The correct species of eucalyptus tree has been planted in southern California, and the San Diego Zoo is the only foreign place to which the Australian government will allow koalas to be exported.

Koalas spend at least two-thirds of the day resting or asleep in their tree. At night, when active, koalas do not move hurriedly, nor do they travel far, perhaps only to the next branch or so, feeding primarily after dusk and before dawn. When a male challenges another for his tree or mate, koalas can move quickly, grabbing the other's arm or biting the elbow. Because of their unwillingness to move quickly, koalas have long served as easily-caught food for aboriginal Australians.

Mating among koalas is timed to insure that food will be most abundant when the young emerge from the pouch. Usually, adult koalas live a solitary life, but during the mating season males will issue loud bellows, which draw females from the nearby area. About 35 days after mating, the female gives **birth** to one (rarely two) young in late spring or early summer. The young are less than an inch (about 19 mm) long when born, and remain in the backward-facing pouch for six months, by which time it has fur, teeth, and open eyes. Once it is out of her pouch, the mother will carry her young koala on her back for another six months as it learns to eat leaves, nestling on her belly when sleeping. Adolescent females stay near their mother, often producing their own first young at about two years of age. Young males gradually disperse through the forest, not mating until they are older and stronger. Koalas probably live 12-14 years in the wild, though they have reached 16 years in captivity.

The population of koalas dropped drastically during the early part of the twentieth century because they were over-hunted for their fur. In addition, diseases of their reproductive tract limited their fertility, and the destruction of their forest **habitat** has also played a big role in their population decline. Koalas are now protected and their numbers are increasing again. When the number of koalas in a particular area becomes too large for the local food supply, government naturalists move some animals to habitat elsewhere. This prevents excessive defoliation of the food-trees, and helps to returns the species to its former habitat.

Resources

Books

Arnold, Caroline. *Koala.* New York: William Morrow & Co., 1987.

Koalas. Zoobooks series. San Diego: Wildlife Education, 1988.

KEY TERMS

Caecum—A sac, open at only one end, in the digestive system of the koala. It is apparently used to help digest the tough fibers of eucalyptus leaves.

Defoliation—Removal of leaves from a tree.

Fertility—The ability to reproduce.

Sternal gland—A gland, located on the chest (or sternum) of a male koala, which secretes a smelly fluid used in marking his territory.

Lavine, Sigmund A. *Wonders of Marsupials*. New York: Dodd, Mead & Co., 1978.

Lee, Anthony, and Roger Martin. *The Koala: A Natural History*. Australian Natural History series. Kensington, NSW: New South Wales University Press, 1988.

Jean F. Blashfield

Kola

Kola is a member of the tropical family Sterculiaceae, and it grows as a **tree** form. Kola nuts from two **species**, *Cola nitida* and *C. acuminata*, have been important objects of trade for at least 1,000 years. These nuts are perhaps most well known now as a constituent of soft drinks.

There are over 50 species of kola. Of these seven have edible nuts, but only two have been commercially exploited (*C. nitida* and *C. acuminata*). The most important is *C. nitida*. The main centers of production are in **Africa**, particularly in Nigeria, Ghana, and the Ivory Coast. Annual production from these countries alone is in excess of 250,000 tons.

Generally, kola trees grow up to 40 ft (12 m) tall, although specimens in excess of 75 ft (25 m) are known. They produce small buttress roots and have a very dense foliage. The flowers are white or cream usually with red markings at the base. Two types of **flower** are produced—a **hermaphrodite** flower with both male and female reproductive structures and a smaller male-only flower. They are quite similar in coloring, but are easily identifiable from a distance by their difference in size. The hermaphrodite flower is up to 3 in (7.5 cm) and the male flower is rarely above 1 in (2.5 cm). **Insects**, attracted by a particularly penetrating aroma, fertilize the flowers.

The **seeds** are produced as quite hard nuts. These can be of various colors but are all about 2-3 in (5-7.5 cm) long. Nuts are not produced by the tree until it is six or seven years old. Peak production does not start until the tree reaches 15 years of age. Estimates for the number of nuts produced annually per tree vary due to the age and location of the trees. However, a top figure of 120,000 nuts is often given. The nuts are generally produced between November and December for *C. nitida* and from April to July for *C. acuminata*.

It is believed that kola trees are native to Ghana and the Ivory Coast and that their spread has been brought about by humans. Kola trees were introduced to **South America** in the sixteenth century. This spread was brought about by the stimulating and sustaining properties of the kola **nut**. They grow best in tropical lowlands below about 600 ft (200 m). Kola trees are all evergreen, but they will start to shed their leaves at times of **water** shortage quite readily. The seeds will die if they are allowed to dry out, and they generally remain at the foot of the parent tree. In the wild this produces isolated groves of kola trees.

Even though everyone knows kola nuts from their use in soft drinks, they are present in these beverages only in minute quantities. In 8 gal (30 l) it is not uncommon to have less than 0.01 oz (0.4 g) of kola nut. The kola nut contains **caffeine** and theobromine. Caffeine is a mild stimulant and is widely used to wake people up, particularly when engaged in boring or repetitive tasks. Theobromine (which means food of the gods) is used in the treatment of coronary **disease** and headaches. The name kola comes from the eighteenth century and is probably a derivative of the West African *kolo*, the native name for the trees.

The commercial production of kola nuts is frequently carried out using clonal propagation. The plants thrive in half-shaded environments. Seed collection is still generally carried out manually in Africa, using hook-ended poles to cut the nuts down. Kola trees can be susceptible to attack by a number of species of **fungi**, and this becomes a major problem with the large scale farming that is now carried out. Insects can also cause great damage to kola trees. If the nuts are poorly stored, they may become infested with fungus or kola **weevils**.

Resources

Books

The American Horticultural Society. *The American Horticultural Society Encyclopedia of Plants and Flowers*. New York: DK Publishing, 2002.

Heywood, Vernon H., ed. *Flowering Plants of the World*. New York: Oxford University Press, 1993.

Gordon Rutter

Komodo dragon *see* **Monitor lizards**

Korsakoff's syndrome

Korsakoff's syndrome is a **memory** disorder which is caused by a deficiency of **vitamin** B1, also called thiamine. In the United States, the most common cause of such a deficiency is **alcoholism**. Other conditions which cause thiamine deficiency occur quite rarely, but can be seen in patients undergoing **dialysis** (a procedure during which the individual's **blood** circulates outside of the body, is mechanically cleansed, and then circulated back into the body), pregnant women with a condition called hyperemesis gravidarum (a condition of extreme morning sickness, during which the woman vomits up nearly all fluid and food intake), and patients after **surgery** who are given vitamin-free fluids for a prolonged period of time. In developing countries, people whose main source of food is polished **rice** (rice with the more nutritious outer husk removed) may suffer from thiamine deficiency.

An associated disorder, Wernicke's syndrome, often precedes Korsakoff's syndrome. In fact, they so often occur together that the **spectrum** of symptoms produced during the course of the two diseases is frequently referred to as Wernicke-Korsakoff syndrome. The main symptoms of Wernicke's syndrome include ataxia (difficulty in walking and maintaining balance), paralysis of some of the muscles responsible for movement of the eyes, and confusion. Untreated Wernicke's Syndrome will lead to **coma** and then death.

Symptoms of Korsakoff's syndrome

An individual with Korsakoff's syndrome displays difficulty with memory. The main area of memory affected is the ability to learn new information. Usually, intelligence and memory for past events is relatively unaffected, so that an individual may remember what occurred 20 years previously, but be unable to remember what occurred 20 minutes previously. This memory defect is referred to as anterograde **amnesia**, and leads to a peculiar symptom called "confabulation," in which an individual suffering from Korsakoff's fills in the gaps in his/her memory with fabricated or imagined information. An individual may insist that a doctor to whom he/she has just been introduced is actually an old high school classmate, and may have a lengthy story to back this up. When asked, as part of a memory test, to remember the name of three objects which the examiner listed 10 minutes earlier, an individual with Korsakoff's may list three entirely different objects and be completely convincing in his/her certainty. In fact, one of the hallmarks of Korsakoff's is the individual's complete unawareness of his/her memory defect, and complete lack of worry or concern when it is pointed out.

Why alcoholism can lead to Korsakoff's

One of the main reasons that alcoholism leads to thiamine deficiency occurs because of the high-calorie nature of **alcohol**. A person with a large alcohol intake often, in essence, substitutes alcohol for other, more nutritive **calorie** sources. Food intake drops off considerably, and multiple vitamin deficiencies develop. Furthermore, it is believed that alcohol increases the body's requirements for B vitamins, at the same time interfering with the absorption of thiamine from the intestine, and impairing the body's ability to store and use thiamine.

Thiamine is involved in a variety of reactions which provide **energy** to the neurons (nerve cells) of the **brain**. When thiamine is unavailable, these reactions cannot be carried out, and the important end products of the reactions are not produced. Furthermore, certain other substances begin to accumulate, and are thought to cause damage to the vulnerable neurons. The area of the brain believed to be responsible for the symptoms of Korsakoff's syndrome is called the diencephalon, specifically, structures called the mammillary bodies and the thalamus.

Diagnosis

Whenever an individual has a possible **diagnosis** of alcoholism, and then has the sudden onset of memory difficulties, it is important to seriously consider the diagnosis of Korsakoff's syndrome. There is no specific laboratory test to diagnose Korsakoff's syndrome in a patient, but a careful exam of the individual's mental state can be revealing. Although the patient's ability to confabulate answers may be convincing, checking the patient's retention of factual information (asking, for example, for the name of the current president of the United States), along with his/her ability to learn new information (repeating a series of numbers, or recalling the names of three objects 10 minutes after having been asked to memorize them) should point to the diagnosis. Certainly a patient known to have just begun recovery from Wernicke's syndrome, who then begins displaying memory difficulties, would be very likely to have developed Korsakoff's syndrome.

Treatment

Treatment of both Korsakoff's and Wernicke's syndromes involves the immediate administration of thiamine. In fact, any individual who is hospitalized for any reason and who is suspected of being an alcoholic, must receive thiamine. The combined Wernicke-Korsakoff syndrome has actually been precipitated in alcoholic patients hospitalized for other medical illnesses, by the administration of thiamine-free intravenous fluids (intravenous fluids are those fluids containing vital sugars and salts which are given to the patient through a needle inserted in a vein).

KEY TERMS

Amnesia—Inability to remember events or experiences. Memory loss.

Anterograde amnesia—Inability to retain the memory of events occurring after the time of the injury or disease which brought about the amnesic state.

Confabulation—An attempt to fill in memory gaps by fabricating information or details.

Retrograde amnesia—Inability to recall the memory of events which occurred prior to the time of the injury or disease which brought about the amnesic state.

Fifteen to 20% of all patients hospitalized for Wernicke's syndrome will die. Although the degree of ataxia nearly always improves with treatment, half of those who survive will continue to have some permanent difficulty walking. The paralysis of the **eye** muscles almost always resolves completely with thiamine treatment. Recovery from Wernicke's begins to occur rapidly after thiamine is given. Improvement in the symptoms of Korsakoff's syndrome, however, can take months and months of thiamine replacement. Furthermore, patients who develop Korsakoff's syndrome are almost universally memory-impaired for the rest of their lives. Even with thiamine treatment, the memory deficits tend to be irreversible, with less than 20% of patients even approaching recovery. The development of Korsakoff's syndrome often results in an individual requiring a supervised living situation.

Resources

Books

Andreoli, Thomas E., et al. *Cecil Essentials of Medicine.* Philadelphia: W.B. Saunders Company, 1993.

Berkow, Robert, and Andrew J. Fletcher. *The Merck Manual of Diagnosis and Therapy.* Rahway, NJ: Merck Research Laboratories, 1992.

Isselbacher, Kurt J., et al. *Harrison's Principles of Internal Medicine.* New York: McGraw Hill, 1994.

Rosalyn Carson-DeWitt

Krebs cycle

The **citric acid** cycle (also called the tricarboxylic acid cycle) is the common pathway by which organic fuel molecules of the **cell** are oxidized during cellular respiration. These fuel molecules, glucose, **fatty acids**, and amino acids, are broken down and fed into the Krebs cycle, becoming oxidized to acetyl coenzyme A (acetyl CoA) before entering the cycle. The Krebs cycle is part of the **aerobic** degradative process in eukaryotes known as cellular respiration, which is a process that generates **adenosine triphosphate** (ATP) by oxidizing energy-rich fuel molecules.

The Krebs cycle was first postulated in 1937 by Hans Krebs, and represents an efficient way for cells to produce **energy** during the degradation of energy-rich molecules. The electrons removed from intermediate metabolic products during the Krebs cycle are used to reduce coenzyme molecules nicotinamide adenine dinucleotide [NAD$^+$] and flavin mononucleotide [FAD]) to NADH and FADH$_2$, respectively. These coenzymes are subsequently oxidized in the **electron** transport chain, where a series of enzymes transfers the electrons of NADH and FADH$_2$ to **oxygen**, which is the final electron acceptor of cellular **respiration** in all eukaryotes.

The importance of the Krebs cycle lies in both the efficiency with which it captures energy released from nutrient molecules and stores it in a usable form, and in the raw materials it provides for the biosynthesis of certain amino acids and of purines and pyrimidines. Pyrimidines are the nucleotide bases of **deoxyribonucleic acid (DNA)**.

In the absence of oxygen, when **anaerobic** respiration occurs, such as in **fermentation**, glucose is degraded to lactate and **lactic acid**, and only a small fraction of the available energy of the original glucose **molecule** is released. Much more energy is released if glucose is fully degraded by the Krebs cycle, where it is completely oxidized to CO_2 and H_2O.

Before glucose, fatty acids, and most amino acids can be oxidized to CO_2 and H_2O in the Krebs cycle, they must first be broken down to acetyl CoA. In **glycolysis**, the 6-carbon glucose is connected to two 3-carbon pyruvate molecules, and then to the 2-carbon acetyl-CoA. In eukaryotic cells, the enzymes that are reponsible for this breakdown are located in the mitochondria, while in procaryotes they are in the cytoplasm.

The two **hydrogen atoms** removed from the pyruvate molecule yield NADH which subsequently gives up its electrons to the electron transport chain to form ATP and **water**.

The breakdown of pyruvate irreversibly funnels the products of glycolysis into the Krebs cycle. Thus, the transformation of pyruvate to acetyl-CoA is the link between the metabolic reactions of glycolysis and the Krebs cycle.

The enzymatic steps of glycolysis and the subsequent synthesis of acetyl-CoA involve a linear sequence, whereas the oxidation of acetyl-CoA in the Krebs cycle is a cyclic sequence of reactions in which the starting substrate is subsequently regenerated with each turn of the cycle.

The **carbon** atom of the **methyl group** of acetyl-CoA is very resistant to chemical oxidation, and under ordinary circumstances, the reaction would require very harsh conditions, incompatible with the cellular environment, to oxidize the carbon atoms of acetyl-CoA to CO_2. However, this problem is overcome in the first step of the Krebs cycle when the **acetic acid** of acetyl-CoA is combined with oxaloacetate to yield citrate, which is much more susceptible than the acetyl group to the dehydrogenation and decarboxylation reactions needed to remove electrons for reduction of NAD^+ and FAD^+.

Each turn of the Krebs cycle therefore begins when one of the two acetyl-CoA molecules derived from the original 6-carbon glucose molecule yields its acetyl group to the 4-carbon compound oxaloacetate to form the 6-carbon tricarboxylic acid (citrate) molecule. This reaction is catalyzed by the **enzyme** citrate synthetase. In step two of the Krebs cycle, citrate is isomerized to isocitrate by means of a dehydration reaction that yields *cis*-aconitate, followed by a hydration reaction that replaces the H^+ and OH^- to form isocitrate. The enzyme aconitase catalyzes both steps, since the intermediate is *cis*-aconitate.

Following the formation of isocitrate there are four oxidation-reduction reactions, the first of which, the oxidative decarboxylation of isocitrate, is catalyzed by isocitrate dehydrogenase.

The oxidation of isocitrate is coupled with the reduction of NAD^+ to NADH and the production of CO_2. The intermediate product in this oxidative decarboxylation reaction is oxalosuccinate, whose formation is coupled with the production of $NADH + H^+$. While still bound to the enzyme, oxalosuccinate loses CO_2 to produce alpha-ketoglutarate.

The next step is the oxidative decarboxylation of succinyl CoA from alpha-ketoglutarate. This reaction is catalyzed by the alpha-ketoglutarate dehydrogenase **complex** of three enzymes, and is similar to the conversion of pyruvate to acetyl CoA, and, like that reaction, includes the cofactors NAD^+ and CoA. Likewise, NAD^+ is reduced to NADH and CO_2 is formed.

Succinyl CoA carries an energy-rich bond in the form of the thioester CoA. The enzyme succinyl CoA synthetase catalyzes the cleavage of this bond, a reaction that is coupled to the phosphorylation of guanosine diphosphate (GDP) to produce guanosine triphosphate (GTP). The phosphoryl group in GTP is then transferred to **adenosine diphosphate** (ADP) to form ATP, in a reaction catalyzed by the enzyme nucleoside diphosphokinase.

This reaction, which is the only one in the Krebs cycle that directly yields a high-energy phosphate bond, is an example of substrate-level phosphorylation. In contrast, oxidative phosphorylation forms ATP in a reaction that is coupled to oxidation of NADH and $FADH_2$ by O_2 on the electron transport chain.

The final stages of the Krebs cycle include reactions of 4-carbon compounds. Succinate is first oxidized to fumarate by succinate dehydrogenase, a reaction coupled to the reduction of FAD to $FADH_2$. The enzyme fumarate hydratase (fumarase) catalyzes the subsequent hydration of fumarate to L-malate. Finally, L-malate is dehydrogenated to oxaloacetate, which is catalyzed by the NAD-linked enzyme L-malate dehydrogenase. The reaction also yields NADH and H^+.

Oxaloacetate made from this reaction is then removed by the citrate synthase reaction to produce citrate, which begins the Krebs cycle anew. This continuous removal of oxaloacetate keeps the **concentration** of this metabolite very low in the cell. The equilibrium of this reversible reaction is thus driven to the right, ensuring that citrate will continue to be made and the Krebs cycle will continue to turn.

Each turn of the Krebs cycle represents the degradation of two 3-carbon pyruvate molecules derived either from the 6-carbon glucose molecule or from the degradation of amino acids or fatty acids. During each turn, a 2-carbon acetyl group combines with oxaloacetate and two carbon atoms are removed during the cycle as CO_2. Oxaloacetate is regenerated at the end of the cycle, while four pairs of hydrogen atoms are removed from four of the Krebs cycle intermediate metabolites by enzymatic dehydrogenation. Three pairs are used to reduce three molecules of NAD^+ to NADH and one pair to reduce the FAD of succinate dehydrogenase to $FADH_2$.

The four pairs of electrons captured by the coenzymes are released during the oxidation of these molecules in the electron transport chain. These electrons pass down the chain and are used to reduce two molecules of O_2 to form four molecules of H_2O. The byproduct of this sequential oxidation-reduction of electron carriers in the chain is the production of a large number of ATP molecules. In addition, one molecule of ATP is formed by the Krebs cycle from ADP and phosphate by means of the GTP yielded by substrate level phosphorylation during the succinyl-CoA synthetase reaction.

The Krebs cycle is regulated by several different metabolic steps. When there is an ample supply of ATP, acetyl-CoA, and the Krebs cycle intermediates to meet the cell's energy needs, the ATP activates. This enzyme uses the ATP to phosphorylate the pyruvate dehydrogenase into an inactive form, pyruvate dehydrogenase phosphate. When the level of ATP declines, the enzyme loses its phosphate group and is reactivated. This reactivation also occurs when there is an increase in the concentration of Ca^{2+}.

The pyruvate dehydrogenase complex is also directly inhibited by ATP, acetyl-CoA, and NADH, the products of the pyruvate dehydrogenase reaction.

In the Krebs cycle itself the initial reaction, where acetyl-CoA is combined with oxaloacetate to yield citrate and CoA, is catalyzed by citrate synthase, and is controlled by the concentration of acetyl-CoA, which in turn is controlled by the pyruvate dehydrogenase complex. This initial reaction is also controlled by the concentrations of oxaloacetate and of succinyl-CoA.

Another rate-cautioning step in the Krebs cycle is the oxidation of isocitrate to alpha-ketoglutarate and CO_2. This step is regulated by the stimulation of the NAD-linked enzyme isocitrate dehydrogenase by ADP, and by the inhibition of this enzyme by NADH and NADPH.

The rates of glycolysis and of the Krebs cycle are integrated so that the amount of glucose degraded produces the quantity of pyruvate needed to supply the Krebs cycle. Moreover, citrate, the product of the first step in the Krebs cycle, is an important inhibitor of an early step of glycolysis, which slows glycolysis and reduces the **rate** at which pyruvate is made into acetyl-CoA for use by the Krebs cycle.

In addition to its energy-generating function, the Krebs cycle serves as the first stage in a number of biosynthetic pathways for which it supplies the precursors. For example, certain intermediates of the Krebs cycle, especially alpha-ketoglutarate, succinate, and oxaloacetate can be removed from the cycle and used as precursors of amino acids.

Resources

Books

Alberts, Bruce, Dennis Bray, and Julian Lewis, et al. *Molecular Biology of The Cell.* 2nd ed. New York: Garland Publishers, 1989.

Lehninger, Albert L. *Principles of Biochemistry.* New York: Worth Publishers, 1982.

Krypton *see* **Rare gases**

Kuiper belt objects

Kuiper belt objects (KBOs) are chunks of rock, dust and **ice** found in the area of the **solar system** just beyond the **orbit** of **Neptune**, starting at about 30 astronomical units (AU) to about 50 AU. In 1992, astronomers proposed that there must be at least 70,000 of these objects with diameters larger than 60 mi (100 km). It is estimated that there are many more such bodies beyond 50 AU, but these are so small and faint that they are outside the limits of detection by present-day instruments. Observations do show that the majority of the known KBOs are found within a few degrees of the ecliptic, or the **plane** of the solar system, just like all the planets except **Pluto**. This band of objects has been named the Kuiper belt after Gerard Kuiper, the astronomer who, in the 1950s, first hypothesized the existence of such a "ring" around the solar system. Pluto and its **moon** Charon are composed of much more ice than the other planets and orbit the **Sun** in a much less circular orbit at a high inclination, or tilt (about 17°). Because of this and fact that their composition seems to resemble that of the KBOs, they have been called, by some, the largest known members of this class of objects. In the late 1990s there was even a heated debate among astronomers as to whether Pluto should be reclassified and called a KBO rather than a **planet**. One argument against this idea, that a minor solar system body (a KBO, an asteroid, or a comet) cannot have a moon had been disproved years before when small asteroids orbiting larger ones had been found. However, the official decision, as of the year 2003 at least—though it continued to meet with disagreement—was that the solar system's complement of planets would remain at nine with Pluto as the mysterious oddball most distant from the sun.

Years before Kuiper proposed the existence of these objects, Dutch astronomer Jan Oort had noticed that orbital calculations revealed no **comets** arrived within detection range of **Earth** from outside the solar system. He also determined that many of them originally came from distances as close as just beyond the orbit of Pluto and as far away as a light year from the sun.

Oort proposed a huge **sphere** of icy, rocky objects surrounded the solar system that became known as the Oort cloud. Occasionally, these chunks are pulled by the gravity of one of the planets into a new orbit, which brings it close enough to be observed as a comet from Earth as its ice is warmed and evaporates, releasing gases and dust to form the well-known cometary tails. Most of the objects from the Oort cloud never come into the solar system at all, and still others probably leave the solar system due to the gravitational pull of nearby stars. The material in the Oort cloud appears to account for most long-period comets—those with that take more than 200 years to orbit the sun. The Kuiper belt seems to be responsible for the shorter-period comets. Both groups of objects taken together make up the main body of leftover debris from the formation of the solar system. As such, especially when observed at the distances of Neptune and Pluto where they remain in their original frozen state, the study of these objects is very important to understanding the early phases of solar system formation.

Most of the observed KBOs remain far from Neptune, even at perihelion, their closest approach to the sun. These are called "classical" (CKBOs) because they follow nearly circular orbits, as do most of the planets. This is what would be expected if they formed with the rest of the solar system. Some KBOs have much larger and more elliptical, tilted orbits that have perihelion distances near 35 AU. These are called "scattered" Kuiper Belt Objects (SKBOs), the first of which was discovered in 1996. Three more were discovered in wide field scans of the solar system in 1999 and it is expected that many more will be discovered as improved technology allows astronomers to probe larger areas of the sky in ways that allow fainter objects to be seen.

Since the SKBOs reach perihelion distances that are smaller than those of CKBOs, the gravitational pull of Neptune can deflect them into new orbits. This can have the same effect as the outer planets do on Oort cloud objects. It can send them into the inner solar system where they are eventually classified as comets. Other possibilities are that their orbits can change in a way that they remain in the distant reaches near Neptune but in the elliptical, tilted orbits that define them as "scattered," or they are ejected into the Oort cloud or out of the solar system into interstellar **space**. The SKBOs seem to form a fat doughnut that surrounds the classical KBOs in their flatter ring-like region, extending a little closer and also to much larger distances from the sun.

Both the **Hubble Space Telescope** (HST) and ground-based observatories have detected these populations of comet-like material at the cold fringe of the solar system. We now know, conclusively, that our solar system does not end at Neptune and Pluto. Obviously, the larger objects are easier to find and just as there are more pebbles than boulders on a beach, it is expected that many millions, billions, or even trillions of much smaller objects exist than may ever be found in the Kuiper belt and the Oort cloud. Detecting even the larger bodies in their distant icy state, at the dim edge of the solar system, pushed Hubble Space Telescope to its limits. One astronomer compared it to trying to see something the size of a mountain, draped in black velvet, located four billion miles (6.4 billion km) away.

The recent discovery of the Kuiper belt and the even newer information about the number and distribution of the objects in it fueled an interest in the possibility of using the Pluto Express spacecraft, already scheduled to fly past Pluto in 2012, to also explore this region of the solar system. The main scientific reason for attempting KBO flybys is the opportunity to explore a whole new region of the planetary system. The mounting evidence that the Kuiper belt is a region where planet-building processes ended is also an intriguing aspect of solar system **evolution** to study. The opportunity to study comet nuclei that have been undisturbed by the warming influence of the sun is an additional important goal for the mission. Such study may reveal many secrets about the formation of the sun and planets. With these compelling motivations in mind, the Pluto Express project has now been renamed the Pluto-Kuiper Express and plans are in place to continue observations beyond the most distant known planet.

The Pluto-Kuiper Express twin spacecraft are well suited for possible flybys of KBOs because, since the composition of Pluto is similar and the planet is also out on the dim, cold edge of the solar system, the scientific instrument packages already installed can adequately observe them. The high-resolution imager, infra-red spectral mapper, and the ultraviolet spectrometer, should be able to provide detailed information about many aspects of these as yet mysterious objects. Maps obtained even from many tens of thousands of miles away would have a feature resolution of a few miles across. This would provide geological and **color** information about of surface features. A Kuiper belt extension to the already ground breaking Pluto Express mission would be scientifically valuable and unquestionably historic. Not only will they be the first spacecraft from earth to observe Pluto at close range, but the only ones to travel to this distant region of the solar system since *Voyagers 1* and *2* crossed the distance of Pluto's orbit in the early 1990s. It is also the only mission planned to reach this area in the first two decades of the twenty-first century.

While there currently are no KBOs identified for the flyby phase of the extended mission, there is at least a decade during which to find more and identify suitable targets for the Pluto-Kuiper Express. It is very likely that the course of one or both spacecraft can be changed after the Pluto encounters to allow close encounters of Kuiper belt objects. Even one of the 60 mi (96 km) diameter objects already detected by ground-based telescopes may be within reach if enough fuel remains in the spacecraft for a sufficient change in course. In theory, reaching one of the much smaller comet-sized objects in the Kuiper belt will be even easier because they are so much numerous than the intermediate- and large-sized objects. However, it will be more difficult to identify and to determine the orbit of smaller bodies in advance. But since the actual selection of specific objects to visit may not need to be made until well into the mission, the probability of reaching this goal is quite high. Success in this mission will mark the greatest milestone for solar system science

KEY TERMS

Astronomical unit—The average distance between the Sun and Earth. One astronomical unit, symbol AU, is equivalent to 92.9 million mi (149.6 million km).

Comet—An object usually seen in the inner solar system that results when a dusty, rocky chunk of ice left over from the formation of the solar system moves close enough to the sun that its ices evaporate. The resulting release of gases and dust surrounds the original object with a cloud called the coma and a tail that can extend for 100 million miles (161 million km) across space. This creates the sometimes spectacular objects observed from Earth known as comets.

Ecliptic—In the sky, the ecliptic is the apparent path of the sun against the star background, due to the earth orbiting the sun. The term ecliptic plane is used to describe the average location in space of the orbits of the planets of our solar system, except Pluto which orbits the sun at a 17° angle to the others.

Inclination—The orbital "tilt" of a planet or other object in the solar system. The first eight planets with very low inclinations to the ecliptic plane. Pluto's orbit has a 17° tilt or inclination.

Infrared Spectral Mapper—A device used to detect heat (frequencies lower than visible light) and

"map" the intensity of the radiation received in order to determine chemical processes at work in the object being observed.

Light year—A unit of measure used between stars and galaxies. A light year is the distance that light travels (at about 186,272 mi or 300,000 km per second) in one year. One light year is equal to about six trillion miles (9.6 trillion km).

Perihelion—The closest approach of an object to the sun in its orbit.

Ultraviolet spectrometer—A device that receives, and breaks into its component frequencies, electromagnetic radiation in the region "above" (of higher frequency) than visible light. The spectrometer splits up the received energy allowing analysis of chemical elements and processes that caused the radiation.

Voyager spacecraft—A pair of unmanned robot spacecraft that left earth in 1977 to fly by all the gas giant planets (Jupiter, Saturn, Uranus, and Neptune). The original mission called for them to also fly past Pluto in what was to be called the "Grand Tour" of the solar system, but mission delays made that impossible, necessitating the Pluto Express mission. These craft were the second set to reach Jupiter and Saturn and the only, so far, to reach Uranus and Neptune.

that has been reached since the late 1970s when the Voyagers first visited all the gas giant planets.

Resources

Books

Sagan, Carl, and Ann Druyan. *Comet*. Random House, Inc., 1985.

Zeilik, Michael. *Astronomy*. 7th ed. Wiley and Sons, Inc., 1994.

Other

Jewitt, David. *Kuiper Belt Page*. University of Hawaii. <http://www.ifa.hawaii.edu/faculty/jewitt/kb.html>.

Clint Hatchett

▍Kuru

Kuru, a **disease** once **endemic** to Papua New Guinea and now virtually extinct, is one of several types of dis-

eases called spongiform encephalopathies, all thought to be caused by abnormal **proteins** called **prions**, which riddle the **brain** with holes. According to proponents of the prion hypothesis, these diseases can arise by direct **infection** with prions, by inheriting genes that produce faulty proteins, or by accidental genetic **mutation**. While prion diseases are more frequently seen in animals in the form of scrapie (**sheep** and **goats**), transmissible **mink** encephalopathy, and bovine spongiform **encephalitis** ("mad cow disease"), human prion diseases are relatively rare.

Kuru occurred among the Fore highlanders of Papua New Guinea, who called it the "laughing death". It was first noted by Vincent Zigas of the Australian Public Health Service and D. Carleton Gajdusek of the U.S. National Institute of Health in 1957. The disease caused its victims to lose coordination and often to develop **dementia**. This disease affects the brain, and it was probably spread by the Fore practice of honoring the dead by eating their brains. When the Fore highlanders were persuaded to cease consuming human brains, kuru disappeared from Papua New Guinea. About 2,600 cases were identified before the Fore highlanders ended this custom.

The course of the illness runs from three months to one year. Among the major signs of kuru are cerebellar abnormalities such as rigidity of the limbs and clonus (rapid contractions and relaxations of muscles). Often, the victim bursts out in wild laughter for no obvious reason. Toward the end of the disease, the person with kuru is very calm and quiet and unresponsive to stimulation. Finally, the victim succumbs to severe skin **ulcers** caused by lying in one position for extended periods of time; or to **pneumonia** caused by stagnation of the **blood** in the lungs.

Among the other human spongiform encephalopathies caused by prions are **Creutzfeldt-Jakob disease** (millions of cases worldwide characterized by dementia and loss of coordination); Gerstmann-Straussler-Scheinker disease (found in 50 extended families by 1995); and fatal familial **insomnia** (trouble sleeping, followed by dementia; found in nine extended families by 1995).

The concept of prions, a term coined by Stanley B. Prusiner as an acronym for "proteinaceous infectious particles," was originally met with great skepticism by most scientists when Prusiner and his co-workers proposed the existence of these proteins 15 years ago. The controversy continues today, although additional evidence has accumulated to support the hypothesis that spongiform encephalopathies are caused by prions rather than viruses.

According to current theory, prion proteins multiply by inducing benign protein molecules to convert themselves into the dangerous form of the **molecule** simply by changing their shape. In addition, prions underlie both inherited (i.e., familial forms) and communicable forms of diseases. This dual nature of prions-inducing other proteins to become prions, while also being the basis of inherited disease-is otherwise unknown to medical science. Prions can also cause sporadic (i.e., non-communicable, non-inherited) neurodegenerative diseases.

Creutzfeldt-Jakob disease and kuru had been known for many years to be experimentally caused by injecting extracts of diseased brains into the brains of healthy animals. Although these infections were at first thought to be caused by a slow-acting **virus**, such an agent was never found. Moreover, ultraviolet and **ionizing radiation**, which destroys genetic material, did not eliminate the ability of brain extracts to cause disease. Prusiner's group eventually determined that scrapie prions contained a single protein that they called PrP ("prion protein"). Further studies showed that PrP is harmless in its so-called benign state, i.e., when the backbone of the protein is twisted into many helices (spirals). PrP converts into its prion form when the backbone stretches out, flattening the overall shape of the protein.

The prion protein multiplies in the brain by a process Prusiner describes as a "domino effect." In one particularly favored hypothesis of prion propagation, a molecule of PrP contacts a normal PrP molecule and induces it to refold into the abnormal, flattened form. The newly transformed proteins then force other proteins to refold into the abnormal form until the prion protein form accumulates to destructive levels.

The discovery that kuru was caused by consumption of infected brains has serious implications in developed countries. For example, **cats** in England have been infected by eating pet food made from contaminated beef. This has prompted concern that humans might get a prion disease by eating meat from infected cows.

Resources

Periodicals

Kolata, Gina. "Viruses or Prions: An Old Medical Debate Still Rages." *The New York Times* October 4, 1994.

Prusiner, S. B. "The Prion Diseases." *Scientific American* 272 (January 1995): 48-57.

Marc Kusinitz